THE HISTORY OF
SCIENCE AND
TECHNOLOGY

A Narrative Chronology

Volume 1
Prehistory—1900

THE HISTORY OF
SCIENCE AND
TECHNOLOGY

A Narrative Chronology

Volume 1
Prehistory—1900

Facts On File
New York • Oxford

Edgardo Marcorini, editorial director

The History of Science and Technology:
A Narrative Chronology

Copyright © 1988 by Facts On File for English Language

Copyright © 1975 Arnoldo Mondadori Editore, S.p.A., Milan

Facts On File, Inc.
460 Park Avenue South
New York, New York 10016

Library of Congress Cataloging-in-Publication Data

Scienza e tecnica. English.
The History of science and technology.
p.   cm.
Translation of: Scienza e tecnica.
Includes index.
ISBN 0-87196-477-5
1. Science—History. 2. Technology—History. I. Title.
Q125.S43713 1988
509—dc19   88-26052
                                                          CIP

British Library Cataloguing in Publication Data

The History of science and technology.
1. Science—History
I. Macorini, Edgardo II. Scienza e
technica. *English*
509 Q125

ISBN 0-87196-477-5 Set
ISBN 0-87196-475-9 Vol. 1
ISBN 0-87196-476-7 Vol. 2

Printed in the United States of America

10 9 8 7 6 5 4 3 2 1

CONTENTS

CONTRIBUTORS

Evandro Agazzi

Lilia Alberghina

Massimiliano Aloisi

F. Tito Arcchi

Marcello Baldo

Enrico Bellone

Paolo Beltrame

Luigi Bulferetti

Francesca Calabi

Piero Caldirola

Federico Canobbio-Codelli

Vittorio Cantù

Vincenzo Cappelletti

Giuseppe Cena

Tullio Chersi

Mario Ciampolini

Jean Dieudonné

Gianfranco Faina

Gian Arturo Ferrari

Ettore Fiorini

Marco Fontana

Aldo Gaudiano

Ludovico Geymonat

Gaetano Giaquinta

Giulio Giorello

Alberto Girelli

Francesco Grasso

Carlo Gregolin

Mirko D. Grmek

Giulio Lanzavecchia

Carlo Maccagni

Paolo Maffei

Nunzio Antonio Mancini

Erasmo Marré

Giorgio Paolo Morpurgo

Ferruccio Mosetti

Walter Nicodemi

Pietro Omodeo

Federico Parisi

Giuliano Piccoli

Adolfo Quilico

Luciano Reatto

Erasmo Recami

Emanuele Rimini

Enrico Sermonti

Giuseppe Sermonti

Noris Siliprandi

Mario Silvestri

Elio Sindoni

Sergio Tonzig

Prehistory and Protohistory

History and science, two of the main products of written communication, have only appeared very recently if we consider the entire time that the human race has existed. But we have evidence of manufactured products—materials that were intentionally modified for an increasingly widening and more specialized range of uses—from hundreds of thousands of years ago. The seeds of cultivated plants and the bones of domesticated animals appear much later, signs of a revolution in techniques and production based on the ability to predict and control biological events. Still later, the remains of town walls trace the boundaries of cities, in which a hostile outer world was separated from an inner world dominated by the temple.

The precarious world of a predatory economy

The transition from a vegetarian to a carnivorous or predominantly carnivorous diet considerably modified man's energy potential, by reducing the amount of food needed by each individual and freeing him from the need to eat almost continuously. His attention turned toward procuring prey. This was the basic element of Paleolithic economy, along with the gathering of vegetables as a supplement to the carnivorous diet or as a substitute for it in times of shortage. Paleolithic man had no means of preserving the food that he hunted or gathered; he was unable to accumulate collective, much less individual, surpluses. Only the introduction of meat smoking at the end of the Upper Paleolithic made small reserves possible. Before that time, the supply had to be replenished almost daily, and this accounted for virtually all available time and energy. On the other hand, a tendency to procure medium or large-sized prey developed, requiring hunting on a group basis. The hunting team, which probably had a precise hierarchical structure, was the first organized form of work. Later, there developed a division of economic roles between men and women, with men assigned to hunting and women to gathering. Such tasks as making implements, finding materials for tools, and preparing sites and shelters were shared by all members of the Paleolithic community. These jobs became more specialized toward the end of the Upper Paleolithic with the increase of trade, which often involved traveling long distances for materials such as obsidian, which had exceptional technological characteristics.

Paleolithic communities were small: the inhabitants of what is now England numbered only a few hundred. The distances between settlements rendered communication impossible or extremely slow, but also prevented the spread of epidemics.

Survival and the source of raw materials

A long tradition has it that prehistory developed with the evolution of stone technology. But in a strict sense, the basic raw material of the Paleolithic period was not stone; it was meat, in the form of prey. During the course of this period, other animal resources began to be exploited. Bones, horns, fats, and skins were used in increasingly varied ways that took advantage of their technological features. Animal bones, for instance, were used in a number of ways during the Lower Paleolithic. Bone splinters became primitive cutting implements or fuel for the hearth. Particularly large bones were used for construction. In the Upper Paleolithic, the specific characteristics of bones and horns—their compactness and the ease with which they could be worked—were utilized. Bones were reshaped with stone cutting tools into particularly delicate pointed implements such as needles, awls, and arrowheads. The Magdalenian period was marked by the widespread use of bone implements. Saws were used to manufacture bone harpoons, bows, and staffs. Horns were also drilled during this period, probably by means of the technologically advanced bow-drill. The first containers, used to hold dyes or as lamps, were made from bones, especially from skulls.

Beginning in the Upper Paleolithic, skins were tanned and used for clothing and as coverings for temporary and permanent huts. The hunting team depicted on the walls of the Eli Secans cave at Lerida, Spain, wears leather "skirts" and "trousers." A Siberian figurine of the same era wears close-fitting, stitched clothing. Convex bone scrapers, from the earliest phases of the Lower Paleolithic, suggest that animal skins were being treated that far back. Tanning techniques, which transformed skins from being rigid and decomposable to being soft and nondecomposable were first based on smoking and then on the use of salts. In addition to bones and skins, animal fats also had various uses. In a culture that did not know how to produce vegetable oils, animal fats were the only fuels used for illumination, and as such were burned in lamps made of stone or bone. They also served as a base for dyes, whenever water was not used.

Stone, the material of technological innovation

Paleolithic man had access to the rich and varied source of raw materials provided by prey because he had special implements that could concentrate his limited strength onto a very small cutting surface used to attack the animal. The unifying principle in the extremely slow progress of technological innovation during the more than 490,000 years of the Paleolithic may be found in the successive solutions brought to bear on the problem of how to increase the efficiency of the hunt. At least some hunting was done without implements of the type described above, since traps and pits were certainly used. There was, however, a constant tendency to develop better implements and devices for attacking animals and to investigate the many possibilities of stone, the only material that was hard and durable enough for the task.

Stoneworking techniques

Paleolithic communities did not confine themselves to any one type of stone for making their implements and weapons. They utilized quite a wide range, including flints, large-grain granites, quartzites, and obsidians. The choice was mainly determined by the proximity of deposits to the settlement. During the Upper Paleolithic, increased trade facilitated the procurement of materials such as Aeolian obsidian and encouraged the establishment of permanent quarries. Flint quarries at Grimes Graves in England, as well as at various places in France, Belgium, Sweden, and Portugal, were equipped with wells and short drainage tunnels, making intensive exploitation possible. Stone was worked by two basic methods, spalling and cleavage, which remained unchanged throughout the Paleolithic and Mesolithic eras. On the other hand, there was a gradual evolution in the techniques used for preparing the original block, in the tools used for cutting it, and in the determination of the angle of incidence. The original technique, developed at the dawn of the human race, was to hit a stone block with another piece of stone. Paleolithic man was increasingly able to control spalling and obtain the implements he desired as he came closer to discovering the almost constant angle (about 120 degrees) between the plane of incidence and that of cleavage. The Chellean culture (from the site at Chelles, France) developed an anvil technique, using a mass fixed to the ground, for the same purpose. The earliest stoneworking tool was a cylindrical hammer (a branch or a bone) which was used to chip pieces off the edge of a larger chip in order to get a straight cutting edge. A universal implement used for attacking, cutting, and digging, the hand-axe (or amygdala) was the basic tool of the Acheulean culture, which takes its name from the site at St. Acheul near Amiens. The so-called Levalloisian cultures (from Levallois–Perret) made a significant advance in cutting down labor time when they were able to predetermine all the cleavage planes on the original stone block. The entire axe would come out nearly finished with one final blow. Prepared core techniques evolved further in the cultures of the Upper Paleolithic. From a roundish block of flint with a lateral protuberance, two halves, neatly separated by a uniform plane normal to the main axis of the block, could be obtained by hitting the protuberance at a plane of incidence or cleavage angle of about 130 degrees. Incisions were then made in rings on the circular cross-section of each half, each ring being a sort of chord of the circle of cross-section. A final blow was then dealt each of the incisions to break off a section that was as tall as the entire block. The result was a long, narrow piece of stone with cutting edges. The blades obtained by this method were either finished tools for cutting and incision, or partially worked products that could become burins, scrapers, or tools for working horn and bone.

The hunter's tools

The Magdalenians, hunters of large animals, were skilled in making blades and had a vast array of specialized implements. They could get almost a meter of cutting surface from a kilogram of material, compared with their ancestors' few centimeters. Each implement had a particular use. While the early hand-axe was a specialized tool, and stone anvils, stone chips for working skins, and slicing axes had specific uses as early as the Lower Paleolithic, Magdalenian implements were specialized in the much more precise sense that each was used for a specific operation. There were tools for engraving, cutting horn, or making a notch, as well as harpoons for fishing. At the threshold of the Neolithic, when levigation, another important process, was overshadowed by new and more radical changes in other

sectors, stone technology was quite complex and varied. Many implements had no direct use, but were used to make other implements. Paleolithic man used mechanical principles to apply muscle power from a distance, through systems of levers and springs, as well as from close-up. Bows, man's first strategic weapons, were used extensively by the end of the Paleolithic. The bow-drill, a mechanical device for transforming translational motion into rotary motion, was a closely related implement. Two types of bow-drill were common, one for boring holes and the other for making fires.

Combustion

Fire was known throughout the hundreds of thousands of years of the Stone Age. There is evidence of it in the Zhou kou dian caves and in the camps of the mammoth hunters; there are the lamps of the painters of sacred symbols and the ovens of northern Scandinavia, where the first sedentary populations of Europe learned to bake clay. Fire was not only used for domestic purposes, such as heating, cooking, illumination, and defense, but also for hardening the points of wooden implements and for working stone. Later, during the Mesolithic and the early Neolithic, the first clay pots were fired and used to cook seafood by the so-called leftover culture—the name deriving from the large number of shells they left behind. At a certain point, Paleolithic man learned how to kindle fires, as opposed to merely keeping them lit. Nodules of iron pyrites were struck together to make sparks to light a tinder of dried moss. Later, linen was used as tinder. During the Neolithic period, bow- or belt-type fire-making drills were commonly used. Animals' bones, rather than wood, were the original fuel; wood charcoal came later. Animal fats were burned in stone lamps with wicks in order to illuminate caves. Mammoth hunters' huts sometimes had rudimentary chimneys to help draw the fire and let out smoke. Smoking as a technique for preserving meat and skins had been known for some time. The location and arrangement of fireplaces varied; they were usually just small depressions in the ground lined with stones. But if the earth was claylike, it would harden with the heat. During the last glacial period, the mammoth hunters of Dolní Věstonice in Moravia built fireplaces in which they baked animal figures made of clay.

Wood, the lost raw material

Prehistoric technological innovation concerned itself almost exclusively with the working of stone. Other materials, such as bone, horn, and wood, were certainly widely used but were never more important than stone. Wood, however, used in various ways, frequently appears in cave paintings. Bows, clubs, and spears, which must have been made of wood, are often depicted; and a large spearhead made of yew was uncovered at the very ancient site of Clacton-on-Sea in England. The late Paleolithic even developed specific woodworking tools, such as the concave scraper, which is very similar in form and use to a two-handled plane.

Wooden objects from these early times, however, are rarely found. Of course many of them must have decayed fairly quickly, but this cannot completely account for the scarcity of wooden objects from early cultures. Well-defined technological limits prevented the widespread use of wood among Paleolithic hunters. Dull cutting edges meant that felling tall trees and rough-hewing large branches were impossible or very lengthy tasks. Only during the Mesolithic and Neolithic were smoothing and grinding techniques developed, making it possible to produce tools capable of cutting down trees about 15 centimeters in diameter fairly quickly.

Shelters, caves, and huts

The outermost chambers of natural caves and shelters under rocks provided Paleolithic man with his first home. The entrance to the dwelling was often reinforced by a dry-stone wall with trunks and branches placed over it. These were temporary dwellings, as precarious as the rest of the world of the hunters. Besides these cave dwellings, there were a few summer huts, temporary shelters that were rebuilt every year. More complicated winter dwellings were developed by western European hunters during the Paleolithic. Pits were dug in the ground, and a floor of stone and wide mammoth bones was laid. The dwelling was covered with logs and earth, or with skins held up by long mammoth bones. Living in the ground, the hunters of Gagarino on the Don, Kostienki, Timonovka, and Siberia heated their huts with numerous fireplaces. The first complex dwelling was built at Timonovka. Several rooms were used for storage and others for working flint. A few fireplaces were built outside the shelters, and those inside had chimneys.

Mesolithic technology made it possible to use wood regularly and fashion it for particular needs. As a result, the quality of shelters soon improved. The huts of the settlement on the Federsee in Württemberg had frames made of branches; in England, wooden floors were made by hewing birch trees thirty-five centimeters in diameter. Later, in Germany and England, huts were built entirely above ground. They were supported by wooden poles, covered with leafy branches and pieces of turf, and had gravel floors. People lived for long periods of time inside these huts, as attested by the left-over fragments of flint that collected over the years on the ground nearby.

Images and symbols. The beginnings of abstract communication

The patriarchal totemic clan that formed hunting teams had a rich array of images around which collective communication revolved. Images of the hunt, men, weapons, and, above all, large animals were the first significant large iconic productions. They were painted on the walls of the innermost rooms of the caves, engraved on slabs of polished stone and sometimes sculpted in clay. As opposed to signs casually traced by hunters on the clay roofs of grottoes, this system of social communication required complex organization. Coloring materials had to be found and prepared, suitable painting or etching instruments had to be made, and several members of the community had to be chosen to depict the image or series of decorations. Thus the totemic clan invested a considerable amount of time and energy in ritualized social communication. Iron and magnesium oxides (for red ochres and blacks), yellow iron carbonates, and lampblacks were ground in stone mortars with ivory pestles. The powders were diluted with water or mixed with animal fats, and applied with sharpened sticks, plant brushes, and fur pads. Sometimes the paints were blown through tubes of bone

to produce the negative image of a hand placed on the wall. The epic of the bow, the new weapon that transformed the hunt, was often painted and engraved, but there was also an increasing tendency toward stylization. Detailed, realistic images were transformed into simplified symbols that evoked the entire subject, although it was no longer reproduced as a whole. The representations for male and female displayed the most pronounced symbolic evolution. Although there is an immense difference between a stylized symbol and an ideogram, in which the association between signifier and signified is only a convention, the hunters' abstract symbols were nevertheless a revolutionary step in the development of visual communication.

Magic and ritual: the renewal of presence

When the Neanderthals of Mugharet es-Skhul in Israel placed their dead at the backs of caves, they made it a point to leave certain things with them: food for survival and flint weapons for continuing the hunt. The funeral ceremony with the infinite variations and additions that served to articulate the microcosm of the hunters. It sought by means of ritual to arrange things in such a way that hunting—that is, survival—would still be possible. If the dead had around them everything (and only those things) necessary for the hunt, then they would surely return to hunting. And if the hunting scene was convincingly depicted on the walls of the cave, it would actually come true. The huge animals, so well represented, would appear to the hunters, especially if the artist could re-create a dynamic situation with movement and postures. Thus the great shaman dressed as an elk in the cave of Les Trois Frères at Ariège is calling to the herd with unmistakable imperiousness. Magic, then, consisted mainly in the ability to re-create a certain situation, to induce an event to repeat itself. It contrasted with the incurable tendency of the precarious world to dissolve, escape, and slip out of reach, condemning the hunting community to a lack of identity that also meant biological annihilation. Magic did not do violence to reality; it was reality. The strict coercion of ritual brought the world closer, gave it an ampler presence, made it truly workable and, as a result, understandable. This understanding, however, by its very nature could not be codified in the form of knowledge, in an autonomous memory apart from the here-and-now. In fact, it had to be continuously reactivated in order to prove its permanent effectiveness and confirm belief in it. In a world where the only permanent thing was precariousness, and where cultural and technological progress was imperceptible, magic established and renewed an order that was always on the point of disintegrating. It did not create time and space, but produced repeated temporal discontinuities and qualitative localizations. It did not proclaim a law of nature, but connected events in fixed associations which, repeated for millennia, constituted the only form of collective memory. Under the sign of magic and within the system of its beliefs, an ambiguous balance between power and being was established, a balance that permitted social structures to endure and made possible those events that we now call technological innovation.

Agriculture, the new mode of production

Food gathering had a firm, though secondary, position in the hunter's economy; but a revolution in this basic mode of production took place in the course of just a few millennia. This revolution was an accomplished fact by around 5000 B.C. The farmers of Jarmo on the hillsides of northeastern Iraq, who grew emmer (*Triticum dicoccum*) and barley (*Hordeum spontaneum*), followed the basic rule of this new economy, which was not to use the entire harvest for food, but to set some of it apart and replant it. Behind this new procedure, which may seem obvious, lay a series of complex thought processes and a courageous decision. First of all, the gatherer-farmers had learned that the growing cycle was constant, that it repeated itself at regular intervals. Then they learned which of the edible plants had particularly high yields. Finally, they made the seemingly senseless choice of literally taking the bread from their mouths and burying it, hoping to multiply their investment in a few months. The farmers' "Copernican revolution" was so great and impressive that for centuries myths recorded their infinite surprise at the seed which, when buried, did not die, but was reborn and gave fruit. The agricultural revolution provided Neolithic farmers with harvests that had previously been absolutely unthinkable. At the same time, it forced them (though it would be better to say "made it possible for them") to set up permanent residences and vary their activities according to the season. Most of the harvest was used as food, and part of this went into a communal surplus which greatly exceeded the subsistence levels of earlier hunting communities. The surplus was used for more than simply feeding those who produced it. One of the first forms of investment was in demographic growth: the farmers could feed an increasing population, which in turn could cultivate new land. In addition, with a constant supply of agricultural products on hand, domestication—which had already been practiced in a marginal way by some Mesolithic hunters—could become permanent and could extend to new species. Domestic animals took on their dual role as suppliers of physical power and as living food reserves. Thirdly, the existence of a constant communal surplus allowed some members of the community to be freed from primary production and used for specialized jobs, such as making implements. Finally, the surplus could be traded for materials or merchandise that the community itself did not possess or used to enlarge the sphere of individual needs to include not only survival but clothing, tools, and household goods. In the case of clothing, for example, an increased demand resulted in the cultivation of fiber plants, such as linen, cotton, and hemp. And the massive investments necessary for locating and working metals could only be made when large food surpluses and a lively demand for implements existed. The demand for household goods, as well as the need for adequate containers in which to store provisions, brought about a spectacular development in ceramics and various basket-weaving techniques. Neolithic farmers had triggered a mechanism that automatically increased the total amount of goods available. It was also a mechanism that was difficult to stop, and it eventually created an imbalance between the small size of Neolithic communities and their need to accumulate increasingly larger surpluses fin order to satisfy growing demand. It was not by accident that, in the zones where cultivation was first practiced—in the Nile and Indus valleys, in Mesopotamia, and in China—the rural villages of Neolithic shepherds gave way, after little more than a thousand years, to cities.

The new stone

The two basic technical innovations that made it possible to transform Paleolithic hunting equipment into tools for the Neo-

lithic farmer were levigation and the placing of handles at an angle. Levigated stone blades reached a high degree of cutting efficiency, though copper ore was soon used to make better ones. Levigated points, lunettes, and blades mounted on wooden handles provided manageable implements. Handles were placed at an angle on tools such as hatchets, increasing their effectiveness. The universally used bow-drill became even further specialized when abrasives began to be used with it. Holes could then be bored quite deeply into levigated stone, and could even be made lengthwise in axe-blades so that wooden handles could be inserted. Neolithic axes thus became quite similar to the ones we use today. Much later, during the Iron Age, when metal tools were widely used, they were still some artisans so skilled at making stone axes and hammers that they were able to imitate the metal tool in stone.

But above all the Neolithic stone industry developed agricultural tools. The basic implement was the hoe, which evolved from the ancient gatherer's stick. With a hoe, the surface of the earth could be worked enough for the first seeds to be sown. But the hoe alone could not sustain a continuous agriculture, since the soil would soon be exhausted, so it was pulled by a rope and became a rudimentary plow that made shallow furrows in the ground. It finally became a real, if simple, plow when animals were used to pull it. Another basic tool was the scythe. Around 8000 B.C., the Natufians of Mount Carmel in Palestine and other nearby cultures used small, toothed flint scythes mounted on bone handles. These scythes were not yet used for regular harvests, but for gathering edible plants at irregular intervals. However, the small stone scythes of Merimda and the sites around Lake Fayum in Egypt were regularly used to harvest emmer and barley. These first modest crops, still halfway between being gathered and being cultivated, were ground in mortars with pestles. Later, hand-turned grindstones were used.

When trades became more specialized during Neolithic times, productivity increased in the already existing flint quarries. The miners at Grimes Graves, who still cultivated their own fields, used an array of quite specialized tools: hammers, sledge-hammers, wedges, shovels, rakes, and picks made from deer horns, in which one of the antlers was used and the others cut off.

The decline of stone and the superiority of metal

Small objects made of copper ore, pins and clasps of meteoric iron, and a few gold ornaments were the first signs of the end of stone technology. Paleolithic farmers had not yet learned the complicated technology of extracting metal from ore, so they satisfied themselves with what they could find in its natural state: a few nodules of copper in auriferous waters, a few dendritic or lamellar formations of copper, irregular lumps of gold in veins of quartz, and meteoric iron. The farmers immediately began to work the small amounts they found. They soon discovered the basics of cold-working, finding that the metals would harden when hammered and return to their original state when heated. With this small store of knowledge, Neolithic farmers, who still had nothing in common with the highly specialized artisans of a thousand years later who were able to find metals and melt them, were nonetheless able to make several metal implements. Their hatchets made of natural copper turned out to be better and more in demand than those made of levigated stone.

From organized gathering to cultivation

Grain, sign of a new era, grew wild in a wide area of the Middle East. The Natufians of Mount Carmel, the inhabitants of Mallaha in the Jordan valley, Tell Muzeybai in Syria, and Zawi Chemi Shanidar and Karim Shahir in Iraq, searched it out and gathered it with special care. They also built permanent settlements, so that they would not have to wander far from where the grain grew, and clay storage wells for preserving the precious harvests that fed them. Wild strains of wheat and barley were harvested with small levigated stone scythes during only a very short period of the year, but the crop did not spoil and could be stored for long stretches of time. In the area extending from the Jordan valley north to what is now the border of Turkey and then south along the left bank of the Tigris and Euphrates, the inhabitants of permanent pre-agricultural villages had an economy based entirely on grains. The legumes, another large family of plants with very big seeds and, most important, a high protein content, soon appeared alongside the grains. The Fertile Crescent was the obvious locale for the transition from organized gathering to planting and cultivation. The relative security that stored reserves guaranteed to these early farmers enabled them to set aside part of their provisions for use in replanting, instead of for immediate consumption. New seeds from cultivated plants were stored in the granaries of Jarmo, Zawi Chemi Shanidar, and Karim Shahir around 6500 B.C. But the wheat at Nathal Oreu in Israel had alrady undergone the change from wild to cultivated as early as 9000 B.C.

Cultivation did not just mean the ability to understand and exploit the biological cycle. It also inevitably implied choosing to grow those plants that promised the best yields or lasted longest in storage. Plants thus began to change, to grow large and hardier, and new varieties were created by the selection the farmers made.

In late preagricultural and agricultural times, the Fertile Crescent was populated by small permanent settlements where wheat and barley were gathered or cultivated. From there, the agricultural revolution spread, first to the Mediterranean basin, then to continental Europe, reaching Greece in 6500 B.C. and the plains of the Danube a thousand years later. Around 4000 B.C., farms appeared in Spain, on the shores of the North Sea, and in the southern part of the British Isles. In the third millennium B.C., emmer and barley were the grains most frequently cultivated in Mesopotamia, in lower Egypt at Merimde and Maadi, in upper Egypt at El-Fayum, in the area of the Danube, in the British Isles, and even in Scandinavia. Cultivated plants displayed the decisive feature of adaptability to habitats very different from their original ones.

The agricultural revolution spread southeastward from the Fertile Crescent more slowly. It reached Afghanistan and Pakistan around 3500 B.C.; in 3000-2500 B.C. it was established in the Indus valley. The huge granaries at Mohenjo Daro and Harappa bear witness to the gigantic harvests produced under the favorable hydric and climatic conditions of that area. China did not wait for wheat and barley to arrive from Jarmo. Millet was grown, and pigs were raised at Ban po, a permanent settlement on the semiarid loess hills bordering the upper course of the Yellow River as early as 4000 B.C. The Chinese agricultural revolution occurred on its own. Millet was grown in Greece as early as 5000 B.C., but did not have time to reach China. And the particular combination of millet and pigs, with-

out sheep or cattle, the animals associated with the agricultural revolution in the Middle East, attests to the independence of the farming revolution in China.

Both the Fertile Crescent and the loess regions of the upper Yellow River valley were semiarid, temperate, hilly zones. Conditions were different, however, in southeast Asia, the third site of the agricultural revolution. In 9500 B.C., the inhabitants of Spirit Cave in northwestern Thailand gathered and stored almonds, betel nuts, peppercorns, squashes, beans, peas, and cucumbers. A few millennia later, yams (genus *Dioscorea*) and taro (*Colocasia esculenta*) began to be systematically cultivated throughout the entire humid, semitropical area of southeast Asia. By 4500 B.C., the village of Non Nok Tha was using advanced agricultural technology, raising cattle, and producing pottery. Its flourishing economy was based on the dry cultivation of rice. Around a thousand years later, the rice fields at Ban Chiang were regularly being submerged, and a complex irrigation system was in operation. Rice and soybeans began to advance northward, arriving in China around 2000-1000 B.C., at the same time that barley and wheat were coming in from the opposite direction, the Middle East.

Vegetarian diet

One of the most striking consequences of the agricultural revolution was that humans settled in areas where cattle, their main source of fats and protein, could not survive. Neolithic farmers had learned to grow two large groups of plants other than grains—the legumes and the oil-bearing plants—that provided effective substitutes for animal foods. In general, these plants supplemented a varied diet. But in some cases, for example in Asia, the survival of entire populations depended wholly on the cultivation of soybeans (*Glycine max*) or urds (*Phaseolus mungo*) and mung beans (*Phaseolus aureus*). Neolithic farmers cultivated many legumes that are unknown to us today. The size of the seeds and the ease of storage and transportation had already caught the attention of gatherers. Farmers tried to develop species with increasingly larger seeds, and found uses for the entire plant, including the pod. Peas, cultivated in Merimde around 4400 B.C., spread to Europe together with grains and plant cultivation; during the Iron Age they reached Scandinavia. Along with peas, fava beans kept pace with the agricultural revolution, and became part of the regular diet of primitive farmers everywhere. Lentils (*Lens esculenta*) appeared at Khafaje in southern Iran around 3000 B.C. At Ashur, at Troy II, and in Egypt, they were the main source of sustenance for the poor until the Classical Age.

Even more necessary than legume proteins were the oils from oil-bearing plants, which rendered farmers independent of animal fats. The oils were used for many other purposes besides that of food.

The basic oil of primitive times in Mesopotamia and Egypt was sesame, rather than olive oil. Harappa, in the Indus valley, was the main center for the production of *Sesamum indicum*. From there it followed the path of grains, but in the opposite direction. Sesame and soybean oils were used mainly in Asia, while flax seeds (*Linum usitatissimum*) were widely used in the Mediterranean basin. Flax was soon used as a textile plant as well as for its oil. Many crociferae, whose leaves and roots were edible, were also cultivated for oil production. Rape

(*Brassica Napus*) and radishes were grown in Egypt, and turnips (*Brassica rapa*) in Europe. *Brassica* seeds were soon replaced by other sources of oil, such as the castor-oil palm and the sesame plant in Mesopotamia, flax and the castor-oil palm in Egypt, and olives in the Mediterranean basin. However, *Brassica* remained a sort of ever-present reserve in case of shortages and a source of inexpensive oil. Wild olives from Afghanistan and Baluchistan soon reached Syria and Palestine, where they developed into the cultivated species (*Olea europaea*). Olives spread throughout the Mediterranean basin from the Asiatic end. Egypt was still importing olive oil from Palestine and Syria during the Fourth Dynasty.

The decline of the beast of prey and the rise of the domesticated animal

Cultivation, the new method that Neolithic farmers used to acquire food, was not an isolated or limited phenomenon, but one which caused a change in the entire ecosystem. Wild plants, among them the ancestors of wheat and barley, were the main source of food for sheep and goats. When the farmers' permanent villages were encircled by fields of grain, animals were forced into closer contact with people. Hunting was no longer a suitable activity, since all the animals near the settlement might be killed off and the farmers would thus be deprived of a precious source of energy. The spread of grain cultivation made it possible, however, to reestablish an equilibrium without disadvantages to either humans or animals. The villagers did not consume the entire cultivated plant, and animals could eat the parts that were left over. In this way animals could survive near settlements, and humans could use them in a systematic way, without the risk of exterminating them entirely. The choice of domesticable species, as well as the selection process used to improve them, was much easier in the case of animals than it was for plants. Through cross-breeding, the individual species soon manifested infinite variety. Animal domestication thus developed along with plant cultivation. At Zawi Chemi Shanidar domestic sheep appeared shortly after 9000 B.C., and there were domesticated goats at Ali Kosh in Iran around 7500 B.C., along with barley and emmer. Dogs, being basically carnivorous, probably did not have to wait for the agricultural revolution in order to establish a close relationship with people. There were pigs in the agricultural village of Cayönü, in Turkey, in 7000 B.C. Cattle, the typical animal of the agricultural revolution, entered into symbiosis with man much later. Around 5500 B.C. the first domestic herds grazed in Thessaly and Anatolia. The fundamental cycle of domestication ended with the domestication of cattle, which because of their size, constituted a considerable food reserve. Neolithic farmers rarely, if ever, used them as sources of power, and, for the time being, cattle provided nothing but food. They were butchered for meat, and their milk like that of sheep was used to make a wide variety of dairy products.

Villages and houses, the buildings of a permanent agriculture

The non-temporary settlements necessary for cultivation created a new technological problem for Neolithic farmers. Permanent shelters were needed not only for people but also for storing and working on the crops that were harvested. The problem existed wherever the agricultural revolution had arrived, but the nature of the problem made it impossible for the same solutions to be adopted everywhere. Climatic conditions varied, and permanent dwellings, much more so than temporary ones, had to be built with this factor in mind. The building materials available varied even more from place to place. With the agricultural revolution, houses first became differentiated, as they have continued to do throughout history. Although they all served the same purpose, they took on different forms and were structured differently depending on the materials the farmers could find near their fields. In the Fertile Crescent, where wood and stone were unobtainable, the basic building material was clay. The agricultural or immediately pre-agricultural communities at Tell Mureybat, Zawi Chemi Shanidar, and Jarmo had houses with straight, compact clay walls and cane floors. Their granaries were pits dug into the ground and lined with clay, which was hardened by lighting fires in the pits. Along the banks of the great rivers, at Badari in upper Egypt and El-Ubaid in southern Iraq, canes stuck together with mud were used instead. Cane structures, however, were soon replaced by more solid pressed clay walls. In the forested areas of Europe that were being cleared by fire in order to make room for fields the most convenient and plentiful building material for domestic architecture was wood. Polished stone axes were the farmers' basic tools, used mainly to build the frames of huts. The first Neolithic farmers of the lower Danube made the walls of their sturdy rectangular houses out of tree trunks, filling in the chinks with woven materials and plaster. The inhabitants of the fertile loess zones of central Europe used the same technique to build their collective huts, which were sometimes as long as thirty-two meters. Carpenter-farmers often used pitch for waterproofing and glue for building fences; both substances were produced from tree bark.

The savage and the civilized

At the same time that the agricultural revolution divided plants and animals into two large categories, almost two cultural types, it also created an analogous and permanent division among people. Inequality, which up to then had existed solely within the social system, began to apply to systems and modes of production and affected all aspects of culture and technology. As farmers advanced, they modified their surroundings, burning forests, terracing mountains, changing the course of rivers, and killing off animals—mainly carnivores—that hindered the productive process. An unbridgeable gap was created between these people and those who progressively withdrew or did not adapt themselves to the new technology. It was thus that the savage was born. Along with this large-scale distinction, the agricultural revolution created other inequalities in the zones it progressively conquered. In predatory cultures, the rate of diffusion was always greater than the rate of innovation. In other words, a new technological discovery could spread completely before it was replaced by a successive development. In agricultural societies this relationship was inverted: the new mode of production offered such varied possibilities that methods, systems, and tools were replaced by others before they could reach all the potentially receptive areas. The distance between the centers and outlying areas, between the cities and the countryside, widened until it became an abyss. Diffusion was also slowed down by the fact that the spread of an agricultural way of life depended on plants and animals becoming acclimatized to regions with very different characteristics. The introduction of agriculture into Europe implied substantial modifications in the original Neolithic system. The process of urbanization was also quite different and much slower there than it had been in the Middle East. The valleys of the great rivers near the Fertile Crescent, where the agricultural revolution had originated—the Tigris and the Euphrates, the Nile and the Indus—remained the poles of technological innovation.

The farmer's view of things: the cycle and the waiting period

Those who sowed the earth, dominated animals and held them in captivity, and found their own place in the world—a permanent residence—developed not only a new mode of production, but a new view of things. Unlike hunters, they no longer asked that their magic evoke favorable conditions in the immediate present, but that it guarantee them more security during the long waiting periods. The farmers were, at the same time, witnesses and protagonists of their magic—the rebirth of the seed. This magic was stronger than its predecessor, and radically different. It lived in anticipation and projected its success into a dimension other than the present. Time extended toward a future which, as it became the present, reconstituted the original situation. The farmers' basic request was that the cyclical process be repeated. The elements that guaranteed this—the stars in the sky, the alternating seasons, and the earth that preserved its fertility—became detached from the uniformity of the world and took on a value of their own. They guaranteed not so much the existence of the world as its precise order, to which the farmers confidently entrusted themselves. It was essential to invoke these masters with great care, to divine their intentions, and to conform to the order they imposed. They soon took up residence alongside the people. The stones of their abodes, such as those in the megalithic temples of Malta, were engraved with the interminable spirals of their continuity, with the abstract signs of their presence. They could thus be supplicated (the only suitable way to address them) in the circular enclosures where they actually lived.

The founding of cities

From about two hundred and fifty meters above sea level, near the perennial springs of an oasis, Jericho dominated the trade in salt extracted from the Dead Sea. The walls, interspersed with towers, enclosed an area of about five hectares; two thousand people lived inside. The walls of this, the first city in the history of mankind were built around 8000 B.C. One thousand two hundred and fifty years later, the city of Catal Hüyük in Turkey was built with sun-baked bricks; it covered about fifteen hectares. During the next few millennia, domestic cattle spread throughout the Fertile Crescent. Beginning around 4000 B.C., they were yoked to plows, making it possible to cultivate extensive areas of virgin land. Demographic pressure increased within the fertile areas of the southern Tigris and Euphrates valley. The land was fertile, but rain was scarce, and the survival of the barley and wheat fields depended on artificial irrigation. Villages grew and became cities. In 4350 B.C., monumental temples were built in the middle of large cities such as Eridu and Al Ubaid, displaying a variety of techniques: dry walls, baked and plastered bricks, and stone structures and

facings. Around 3800 B.C., Uruk covered an area of several hundred hectares with its streets, houses, temples, and palaces. Thousands of well-watered, cultivated hectares were needed to keep it alive. Cities, those great machines for survival in periods of strong demographic pressure, soon became the places where metals were worked, fibers were woven, and pottery was made on a large scale. The dual function of ancient cities was simply this: to guarantee survival and provide a site for the development of technology and a diversified labor force.

The need for centralized control over water distribution soon merged with a potential for accumulating food surpluses that would have been inconceivable in earlier village economies. The giant granaries of Mohenjo Daro and Harappa required a large group of specialists, mostly priests, to supervise the accumulation, administration, and investment of the surpluses. Accumulation also had a religious motive, almost as though it represented a tribute to the divinity. Thus a religious-aristocratic class was soon added to that of the farmers. This class administered the community's surplus, and naturally tended to enlarge the city and its sphere of influence so that more crops could be harvested. Social wealth, which the farmers themselves would never see, served mainly to support individuals and classes that did not participate directly in basic production. Besides the priests and nobles, there were new members of society who did necessary non-agricultural work. Artisans were the direct offspring of the city. They worked metals, the most expensive, and the most difficult to find, of all raw materials. In addition, metal technology required specific preparation and training for those who practiced it. Metal craftsmen were thus the first specialists to detach themselves completely from working the land and to be supported by the communal surplus. This was the only way they could practice their trade profitably.

Metals, the raw materials of the city

The urban revolution necessitated a much wider and more varied range of raw materials than had been required by Neolithic farmers. There was also a massive conversion of superfluous items into necessary ones, of luxuries into basic commodities. Building and bonding materials, ceramics, glass, coloring pigments, and textile fibers all became part of the city's raw materials. From a quantitative point of view, metals did not have a particularly important place within this new transition. If there were some method of weighing them, the dyes of ancient civilizations would certainly far outweigh all the metals. And although the technology of metalworking achieved a high degree of development and extremely complex results, it was never of primary importance to the basic economy—that is, it never induced any changes in the modes of production. The ancient world remained a great sea of agriculture with a few islands of metal. Metals acquired specific uses in war and the arts, and only very slowly, especially after iron was introduced, became the raw materials for producing agricultural implements. Metallurgy, however, soon became the area in which the most important and rapid technological innovations took place. In metallurgy, ancient technology carried out the first procedures that required precise, simultaneous control over extremely different variables and followed steps which seemed to have no connection with the desired results. The process for producing levigated stone implements was a linear one: the artisan produced the desired object by hewing his raw material on the site where it was extracted. The process for making metal implements, on the other hand, was fragmented in time and space and involved several agents. Prospectors discovered the ore that contained the right metal, miners dug it out, and founders reduced the metal at the site in ovens or crucibles. This raw material, in transportable form, then passed into the hands of merchants who carried it on long journeys to its destination, where a second cycle of work began with the smith. This artisan refined, melted, and finally cast the metal in terracotta molds. The product was not always finished even at this stage; often another artisan entered upon the scene to hammer it, weld it, make it into metal plate which might be embossed or engraved, or transform it into wires and granules for filigree. But in spite of this long labor process, metals offered what no stone ever could: inexhaustible plasticity combined with a hardness and resistance that no other material possessed. Not only could every existing implement be made even better in metal, but new ones could be invented in an infinite variety of forms and for an infinite variety of uses. King Alcinous's metal palace, regarded with wonder by the Greek merchant-hero, was an epic hyperbole for the awe with which the entire world viewed the new material.

The earliest metallurgy

Metallurgy in the strict sense began when people learned that metals could be given new and controllable forms by heating, melting, and casting them and that there was a far from casual connection between certain ores and the metals derived from them. Early metallurgists worked simultaneously, or almost so, with gold, silver, lead, antimony, tin, and bronze alloys—a group of metals with quite similar characteristics. For each of them, the final result depended largely on the properties of the alloy or the ore used, rather than on the treatment it underwent, while the opposite was true for iron. Towards the end of the fourth millennium B.C., a basic technology had already been developed. In the most favorable instances, gold and copper were found in their natural states, though the small deposits of natural copper were soon exhausted. Gold needed only to be washed; copper had to be annealed in order to strengthen it after it was cold-worked. This procedure was followed in Egypt, Asia Minor, Mesopotamia, and Palestine until around 3000 B.C., when natural copper was exhausted and copper ores had to be found. Malachite, cuprite and azurite were the first ores to be used. They were reduced in crucibles and ovens to obtain molten metal, which was then cast in terracotta molds. The first molds were open, and later ones were closed. The metal age began with these first metallurgical ovens, which were about thirty centimeters in diameter.

The second decisive step occurred when sulfurous ores, chalcocite, yellow pyrite, bornite, and covellite, began to replace copper carbonates and oxides. These ores were fairly common in the Sinai peninsula, Syria, Baluchistan, the area of the Caucasus, on Cyprus, and in Macedonia and central Europe. Sulfurous ores could not be reduced by a simple process, so a new and quite complex technology had to be developed. A series of roastings and meltings brought the metal to ninety-five percent purity. The process for working sulfurs, however, was already known from the time when silver and lead were extracted from galena in the land of the Chalybes, the mining zone of the Hittites. Asia Minor, as a matter of fact, held a sort of monopoly

over silver. Sargon of Akkad and Gudea of Lagash, the great Mesopotamian lords, sent out expeditions to buy raw and refined silver and pure lead ingots. Galena was worked in simple open-hearth ovens. Some sulfur evaporated in the form of sulfur dioxide, while the rest remained as untransformed galena or lead sulfate. The final product was a lead-silver alloy, harder than hard lead and containing a large number of impurities. Silver was then extracted from this argentiferous lead by cupellation. The alloy was melted in a cupel, a porous clay crucible in which the lead and other metals were oxidized by a current of air; the walls of the cupel then absorbed the oxides. The same method that proved effective for silver was also used in gold refining.

Copper metallurgy developed copper alloys, tin bronze in particular. The first workers in metallurgy used lead and antimony to make copper pour more easily. Tin, reduced from nodules of native cassiterite by the usual procedure, was considered a variety of lead and therefore fused with copper. This produced bronze, a new "copper" that was harder and mechanically more resistant, and consequently quite a bit more useful than unalloyed copper. From then on, the tin deposits of the Near East were so intensively exploited that they were soon completely exhausted. Deposits in Cantabria, Brittany, and Cornwall were then used, creating a flourishing maritime trade centered on the island of Cyprus. Tin was extracted from cassiterite in ovens filled with alternating layers of ore dust and wood charcoal. Sumerian metal culture reached the Danube regions and then Bohemia and Saxony in the wake of pioneers who ventured throughout Europe in search of tin deposits. Various alloys of tin, however, only began to be produced for the manufacture of weapons, mirrors, and statues around 1500 B.C.

Knowledge of metallurgical procedures spread rapidly. The inhabitants of the Iranian plateau east of the Caspian Sea who discovered and introduced the technology soon carried it via their mining prospectors into the Indus valley and the copper-rich Caucasus region. From there it was a short step to the Mediterranean basin, and maritime transport quickly took it beyond the coasts of Asia Minor. The basic steps for extracting raw or partially refined metals from ores required large amounts of fuel and were therefore carried out in wooded areas near the mining sites. After the initial process, however, the bars and ingots of raw metal could be worked wherever the finished product was to be used. Metal workers were thus divided into two distinct categories. For a long time they worked in separate places and were not in communication with each other. On the one hand there were the mining prospectors, the miners, and above all the metallurgists who knew how to transform the colored stones into molten metal; on the other there were the artisans in the cities, the smiths who refined and worked the metals by means of ancient, unchanging methods. The former of these—mysterious soot-covered inhabitants of the forests who knew the magic secrets for changing the state of stones and transforming them into metal—were looked upon with suspicion and fear by city culture. However, it was from them that decisive innovations in the development of metallurgy as well as the apparently senseless procedures for producing steel, came. On the other hand, the artisans who adapted metal to the demands and needs of the people with whom they were in contact, soon became a permanent part of urban life; but their story goes nowhere. Limited to repeating a few unvarying procedures handed down only orally, they soon lost all innovative ability. They were the first true laborers in history.

Iron, the anomalous metal

Copper and bronze foundrymen had fixed in their minds a precise series of procedures which worked for all the metals they knew: several varieties of colored stone could be melted in an oven fueled with wood charcoal to produce a flow of metal. However, if the stones were magnetite, hematite, limonite or siderite, the process did not work. Instead of a molten flow, only a spongy, pock-marked mass was produced. The small pasty globules of soft iron inside the mass of slag solidified immediately after reduction due to the high melting point of iron (1535 degrees centigrade) and remained hidden. There seemed to be nowhere to go from there; the spongy bloom looked like an unusable by-product. Besides, the small ovens could never reach the melting point of the ore. Ancient metallurgists never saw cast iron. The discovery that the bloom could be heated to red-hot and hammered to eliminate the slag and get a lump of iron was only made around 2000 B.C., on the Armenian and Iranian plateaus. But even then the new metal had no technological features that made it superior to copper or, especially, bronze. It was harder to work, required more fuel, and blades hammered out of it easily lost their edge. It was only later that the Chalybes, subjects of the Hittite kings, made the decisive discovery that hammering and heating the soft iron and then immersing it in cold water gave a surface much harder than bronze. Contact with the fuel during heating carburized the surface, and the diffusion of carbon transformed it into steel. The steeling of welded iron was the exclusive property of the Chalybes, and therefore under the monopoly of the Hittite kings, from 1400 to 1200 B.C.

The Thracian and Phrygian invasions of Asia Minor and the ensuing fall of the Hittite empire were the main factors in the spread of steel technology. Tribes of foundrymen were pushed eastward and southward. Philistines from Armenia settled in the coastal zone of Canaan, and the Midianites near the iron deposits of Edom and Midiam. As soon as the new techniques reached the shores of the Mediterranean, they began to spread much more rapidly than the preceding copper and bronze metallurgies, finding roads already opened by earlier attempts at melting iron. After only a few centuries, a complete iron technology reached Europe, by way of southern Italy and the Umbrians located near Bologna. But only after contacts were established between the invaders of Asia Minor and the ironworkers were the new techniques transmitted through Greece and the Balkans to Noricum, in the eastern Alps, which became the major metallurgical center of ancient times.

A totally different technology was developed in the Indus valley, where the melting of iron in a crucible provided the ancient world with fine cast steel, the most sought-after type. The crucible technique soon spread from the Indus valley to Persia. A third phase, tempering, was eventually added to carburization and hardening, the two already existing techniques of the steel-producing process. Heating the steel to a moderate temperature and cooling it slowly canceled some of the effects of the hardening process, but it made it more pliant and less fragile. The quality of the end product depended entirely on the artisan's ability to control the length of each phase in order to obtain the desired results. Uniform melting gave way to a series of techniques by which the amount of fuel used for carburization, the system used for hardening and its speed, and the tempering time and temperature were all regulated. It was the age of smiths, hammering on anvils and working at forges.

Water, the raw material of the countryside

Water regulation and the ability to get enough water were necessary conditions for well-irrigated crops and consequently for the cities whose survival and accumulation of surplus depended on them. Hydric technology thus became the major area of collective investment in the great river valleys as well as in semiarid lands, resulting in the most grandiose projects of those regions. The water supply system for irrigating small areas, especially gardens or specialized cultivations such as date palms or vineyards, consisted in permanent wells, lined with stones or bricks or, more simply, with clay plaster. A lifting device, which was, in effect, one of the first machines of antiquity, was used with the wells. This device, the *shaduf*, consisted of a balancing system that lifted the water to a height of about two meters. At the end of its course, the container was emptied into the irrigation canal. Around 2000 B.C., one person could draw about twenty-seven hundred liters of water in a day. For larger areas of land, huge reservoirs enclosed by dams were used instead of simple wells and cisterns. The Orontes valley in Syria was closed off around 1300 B.C. by a dam two kilometers long; the Assyrian dams above Nineveh were about three meters high. In the Nile, Tigris, and Euphrates valleys, the irrigation problem was connected with that of the annual floods. The arid land had to get enough water for the rest of the year, but the floods had to be kept from destroying the crops. Their ability to forecast the date and the height of the Nile flood accounted for the pride and prestige of the Egyptian priestly caste. A series of flood gauges, called Nilometers, were placed along the entire course of the river. These were among the earliest scientific measuring instruments. The flood itself was contained by a complex system of stone and mud dams. The water was directed through flood gates and sluices into rectangular reservoirs, which varied in size from 400 to 16,000 hectares and had a water level of 0.80 to 1.80 meters. After about a month, when the soil was saturated, the water was allowed to flow off into a lower reservoir, and then back to the Nile. A network of canals brought water to the land that the flood did not cover. A rather different situation prevailed on the Mesopotamian plains. The Tigris and Euphrates floods were irregular and dangerous; they occurred at the beginning of summer, too early to water the land sufficiently. Caught between the terror of the huge, damaging floods and the need to save their crops, the inhabitants of Mesopotamia carried out a vast project—that of continuously irrigating their lands. A series of canals cut across the plain; the largest of these, the Nahrwan, was 120 meters wide and about 300 kilometers long. A system of smaller canals and ditches carried the water to each individual plot of land. The system was based on various insights into hydraulic engineering. The level of the main canal always had to be higher than ground level, and its flow had to be regulated so that it was neither too fast (which would destroy the embankments) nor too slow (the water would pool and stagnate). Besides a centralized water administration, the system required a large labor force dedicated exclusively to its maintenance.

Construction materials and techniques

Building technology in preurban agricultural Europe had no real opportunities for making decisive progress. Wooden houses remained the basic dwelling units, and the only other structures were places of worship; but all these buildings remained outside of an urban context. Nevertheless, Copper- and Bronze-Age farmers devoted particular care to their places of worship. Around the second millennium B.C., gigantic houses of worship (or, more probably, astronomical observatories) constructed of enormous stone blocks, called megaliths, arose from Malta to the Atlantic shores, and from the Iberian peninsula to France and the British Isles. The stones for these sacred enclosures were sometimes excavated several dozen kilometers away from the construction sites. Teams of laborers transported them on rollers and dragged them with ropes to the site, where a forked pole was used to stand them upright. The architraves were dragged up earthen ramps or lifted on wooden scaffolds. Megalithic technology was practiced on an exorbitant scale, but its innovations in construction were modest. The builders of Stonehenge merely reproduced wood technology in gigantic dimensions. Architraves were fastened to pillars with tenon and mortise joints, almost as if they had been wooden models. In this sphere the Orient was once again the cradle of technological innovation. Clay was the basic building material of the Mesopotamian floodplains for over six thousand years. Air-dried bricks were still used at Jericho, and holes for inserting support beams were drilled in some of them. Soon, however, sun-dried bricks and, later, oven-baked bricks replaced pressed clay. The clay for making bricks was usually mixed with chopped straw or manure to keep it from splitting; then the bricks were molded two at a time in rectangular wooden molds and left to dry in the sun. The same clay mixture was used as plaster. Oven-baked bricks were much more expensive to produce, so for a long time they were used only for waterproof flooring, facings, and other special uses. Mesopotamian walls were made with slightly curved bricks, laid with their convex side up at an angle of a few degrees to form a sort of herringbone pattern. Bitumen was the preferred bonding material at Ur, and large quantities of it were extracted from the asphalt blocks of the upper Euphrates. Mesopotamian builders were already familiar with both the false arch, or corbel, and the true arch, which they used over doorways. The Assyrians, arriving from the north, introduced stone into Mesopotamia, and thus the stonemason joined the bricklayer. An analogous transition had taken place in Egypt at the beginning of the third millennium B.C. There, however, the distinction between stone and clay was sanctioned beyond any doubt: stone was the noble material, a monopoly of the state and used for its buildings, whereas clay was used for the private dwellings of the people. The main phase of labor in the great Egyptian construction works took place in the stone quarries. There, in tunnels hundreds of meters long, miners determined the size of the blocks needed and inserted wooden wedges into chiseled cracks. They then wetted the wedges, which proceeded to swell until the block broke off. The builders of Egyptian sacred buildings were not familiar with arches, and, since they used limestone for the architraves, the distance between pillars had to be less than three meters. Later, Silsila sandstone made it possible to extend this to about nine meters. Egyptian stoneworking techniques were extremely precise: an array of instruments, such as sighting poles and bricklayers' squares, were used, rendering the process almost a scientific one. The builders were able to obtain large monoliths, called obelisks, and they even made columns consisting of a series of drum-like sections. At the beginning of the first millennium B.C., Mesopotamian builders introduced a new bonding material, chalk plaster, and a new building material, limestone. The latter was used to build Sennacherib's aqueduct, which transported water to Nineveh, eighty kilometers away. This aqueduct was waterproofed with bitumen. Brick arches gradually developed into vaults, in which each semicircle of bricks was slightly inclined backwards so that keying was unnecessary. In Crete, building techniques were heterogeneous. Stone debris was used for the foundations of walls and sun-dried bricks for the upper part. The upper floors of dwellings were supported by stone pillars, but the structures themselves were still made of wood. A certain style of house that originated in Crete later became very popular in the Mediterranean basin and all over Europe. The Minoans developed stairs, situated their bedrooms on the upper floors, and made flat terrace roofs. Bathrooms, facilities with running water, terracotta household water pipes, and sloping gutters were also features of the Minoan house.

The growth of great empires, with their vital needs for transport, gave rise to the technology of bridges and roads. Nebuchadnezzar had a wooden structure that was sustained by more than a hundred brick pillars built over the Euphrates. Sennacherib's great aqueduct had an underground pipe nineteen kilometers long and 2.75 meters wide. Great paved roads were often linked to some other building enterprise: the materials for the Great Pyramid were carried over a wide stone road. Egyptian and Mesopotamian cities, as well as some stretches of the streets of Crete and of cities under Minoan influence, were paved with stone slabs. Bogazköy, the Hittite capital, built the first Holy Road in 1200 B.C., and Mesopotamian cities soon followed the example. The Holy Road of Babylonia, built high upon the plain, was entirely paved with ninety-centimeter square slabs of red and white stone.

Potter's clay, the raw material for a new specialist

Hunters and the earliest farmers were already familiar with clay and the elementary technique for working it, hardening by fire. But the procedures were in a certain sense casual; they were not fixed, and the technology was irregular. Various factors made clay a material of considerable importance in the economy of ancient cities. First of all, the establishment of permanent settlements and the need to preserve provisions created a demand for vessels and containers that natural containers could not fulfill. In addition, the permanent houses of the country and the city had new needs for specific household equipment and implements. At first, metal appeared to be the best material for satisfying these needs. But finished products in metal were very costly, and although Mycenaean nobles may have been able to drink from embossed metal chalices, their subjects had to settle for substitutes. Clay was the best choice in many respects. It was easy to obtain and extremely malleable, so that it could be modeled into any shape at all.

Baking ovens, at least for the coarser types of crockery, did not require excessive temperatures. And the end product had remarkable characteristics: it was hard, waterproof and heat refractory. Its main defect was its fragility, although the reduced risks of sedentary life and the low cost of eventual replacement alleviated the problem. Clay working, like basket weaving, was a domestic, family-based technology. But basket weaving remained unchanged for millennia, whereas clay technology developed into an industry, as first a city-wide, then an international, market demonstrated the advantages of large-scale production, in terms of reduced labor time and cost. Armed with their wheels—the first industrial machines—potters soon became specialists who dedicated all their time to ceramics. Later—first at Corinth then at Athens—ceramics were manufactured for a vast foreign market. Levels of industrial concentration and technology were attained that would not be equaled until the fourteenth century.

Clay technology

At Mohenjo Daro, just after 3000 B.C., dry clay was still being pounded and sifted in order to obtain the raw material for making pots. But at the same time, the technique of washing and decanting clay in a series of basins dug into a hillside was being practiced at Tel Beit Mirsim in Palestine. The clay from the middle basins was used to make the body of the vases, whereas the finer clay from the bottom basins was applied to the surface to serve as a thin waterproof coating.

Ancient potters often added non-plastic materials, such as sand, flint, powdered quartz, and ground-up shards, to natural clays in order to reduce their extreme plasticity. They also worked the wet clay with their feet in order to distribute the water evenly throughout the mixture and to eliminate air bubbles. After the vases were modeled on the wheel and dried until the water content was eight to fifteen percent, they were burnished to eliminate pores in the surface.

The biggest technological problem lay in baking. Primitive potters' ovens reached temperatures of only 450 to 700 degrees centigrade which was only sufficient to eliminate the dampness of the clay. The chemical reactions that made clay more resistant and less porous could occur only later, when temperatures from 750 to 800 degrees were reached. The ovens of late pottery industries reached temperatures of around 1000 degrees centigrade. Mycenaean potters baked their stirrup vases, made for export, at 1030 degrees centigrade. To obtain the temperatures of around 800 degrees required for common crockery, primitive potters used great open-air fires that burned for several days under a sealed layer of either earth or branches, depending on the coal-dealer's method.

The first ovens made especially for ceramics appeared during the fourth millennium B.C. They were able to quadruple the 200 degree centigrade heat of normal domestic ovens. Vertical ovens spread to those countries where bricks were used for building. Only bricks were sufficiently refractory to allow the construction of vaults, in which the hot gases generated in the furnace escaped through chimneys. The vases were placed inside the ovens on grills of perforated clay. The grills at Susa, in the fourth millennium B.C., were 1.80 meters in diameter and were held up by a structure of arches 0.90 meters high. The fire, fueled from a lateral opening, was located under the arches. Because of indirect heating, the vases were neither contaminated by smoke nor stained by the fuel.

Plants and their industrial uses. Basket and textile weaving

Techniques utilizing vegetable fibers satisfied needs similar to those that revolved around pottery, and, in the same way, made use of the new conditions created by the agricultural revolution. But the technology soon split into two branches, which tended to diverge increasingly from each other as time went on. Basket weaving and its products, baskets and mats, developed styles that did not undergo any further changes. Relegated to auxiliary functions and confined to the area of family industry, basket weaving remained a static technology, lacking a market as well as innovative force. Fabric weaving, on the other hand, soon developed mechanical devices which, despite their relative fixedness, increased production enough to justify the cultivation of plants used in making textiles, the first agricultural products not destined for use as food. The textile market eventually grew to international dimensions, and the related technologies of dyes and mordants widened the range of raw materials and created large new markets.

The first breadbaskets in the pre-agricultural and agricultural villages of the Fertile Crescent were produced by the already well-developed technique of spiral weaving, by means of which bundles were sewn into shape working from bottom to top. Mats were made of rushes, and larger baskets were woven on pole frames. Outside the domestic environment these techniques found an important use in the manufacture of ropes. Both maritime transport and large building projects, especially those in stone, depended on the effectiveness of ropes, which were made of reeds, palm fibers, or esparto grass. Fibers were twisted into threads, which were then twisted into strands. The strands were then wound together, but in the direction opposite to that of the internal twist, so that the rope would not come apart.

The first plant cultivated for textile use was flax. Cotton was cultivated at Mohenjo Daro around 3000 B.C. It appeared slightly later in Peru, along with flexible leaf fibers. Linden fibers were soon used too, whereas wool, considered impure in Egypt, was originally used in Mesopotamia from 1000 B.C. on. Large amounts of wool were also used in Scandinavia. Hemp, the first Chinese fiber plant, reached eastern Europe only around 500 B.C.

The preliminary technological problem in weaving was transforming vegetable fibers, usually fairly short, into a continuous thread. The fibers had to be laid lengthwise and made to adhere to one another so that the thread would not come apart. The problems of stretching and twisting were both effectively solved with the invention of the spindle, which could be used in different ways depending on the desired result. It could be suspended, which resulted in a fine, uniform thread, used for the most precious linens, or held and rotated by hand to produce a rough thread.

Looms, the basic machines for weaving, appeared in Egypt around 3000 B.C. The earliest of these were horizontal floor looms, with the warp beams fixed to poles stuck in the ground. The two essential elements, the shaft rod and the heddle, were already present. Vertical looms, with the warp attached to an upper beam held taut by a series of weights, were being used in Troy by 2500 B.C. Vertical looms with two beams, enclosed in a rectangular wooden frame, became common in Egypt around 1400 B.C. The weaver, who could work sitting down, wove the woof starting at the bottom. Penelope's web was woven and unwoven on a loom like this, and her suitors never understood its technology.

An incredible variety of fabrics could be woven on ancient looms, depending on whether the warp or the woof threads were made to predominate, whether the colors of the threads were changed or not, and whether a shaggy or a diagonal-weave fabric was desired. Developments in this field began around the middle of the third millennium B.C., and proceeded so rapidly that by the middle of the second millennium B.C. the body of Pharaoh Tutankhamen was wrapped in linen with a count of 12 × 32 threads per square centimeter. Two centuries later, the sash of Ramses III was made with two different weaves with a count of 136 × 25 threads per square centimeter. Scandinavian wool weaving shows the obvious signs of technological backwardness with a count of 5 × 4 threads per square centimeter. But Chinese pure silk fabrics of the Yin period were damask-woven with a diagonal pattern, the result of superior weaving skills, as early as 1000 B.C. Chinese silk only reached the Mediterranean via the pathways of the silk market during the Han dynasty, in the first two centuries A.D. Still more centuries passed before the jealously-kept secrets of its manufacture followed the same pathways and spread to Mediterranean countries.

Glass, a conservative technology

Crucible melting, the basic procedure for manufacturing glass, related this technology to metallurgy, whereas the materials used and the type of implements produced rendered it similar to pottery. Glassmaking also depended on high oven temperatures, and for this reason alone it soon became the patrimony of specialists. But in comparison to their colleagues, the potters and metalworkers, glassmakers could boast of a more difficult, complex technology, in which the end product resulted from simultaneous control over a much larger number of variables. First of all, they did not begin with any one material, but rather with various materials that could be combined in a number of ways to obtain different types of glass (soda or potash). And the same type of glass could be made in different colors, depending on what traces of metal oxides were added. Secondly, once the melting point was reached, the process was far from over, and the real difficulties began. The glass had to be cooled very slowly in a special tempering oven, to keep the silicates from crystallizing and making the mixture opaque and fragile. Also, the mixture of soda, lime, and sand did not have a definite melting point, so the entire process required constant vigilance and a series of shrewd calculations. Glassmakers thus developed their own recipes. In Mesopotamia, between the second and first millennia B.C., recipes were written from memory on tablets in cuneiform; but in Egypt, the real center of the glass industry, the secrets of the art were only passed on orally, so that others could not get hold of them. Glazing was originally the casual result of heating ceramic products in ovens, but around 4000 B.C. glazing steatite beads was already a well-known, widespread practice. Later, glazes made with the same ingredients as glass, but with more sand and less lime, were applied to the surface of other materials in the form of wet

powders. Egyptian faience was the result of quartz powder glazes on clay surfaces. Glass began to be produced around 2500 B.C., and for about a millennium it remained as precious as the stones it imitated. Around 1350 B.C., glass vessels and vitreous veneering became more common. Balsam and ointment holders, bottles, and cold-worked objects, came from Egyptian workshops at El-Amarna, Lishr, and Gurob. Small glass objects were made by pouring and pressing molten glass into clay molds, whereas blocks of coarse glass were carved and polished like precious stones. A different technique was used to make small-mouthed containers: a bag of sand was dipped in a crucible full of melted glass, then rotated on a stone bench to give it shape. Once the container was cool, the sand could be emptied out. Around 1000 B.C., glass disappeared for about four hundred years. It became common again in the eighth and seventh centuries B.C., when Phoenician merchants carried it as far as the Atlantic coast, while the production of beads was spreading throughout Europe.

Collateral chemical transformations and the range of raw materials

After the cultivation of plants, the greatest technological innovations occurred in the area of transforming the chemical states of substances. Agricultural surpluses were invested in chemical transformations and technologies; metallurgy, ceramics, and glassmaking were the most important of these, though not the only ones. Once the problem of survival had been overcome, the range of usable materials was unlimited and rapidly spread to include an array of items that, with a few exceptions, remained unchanged until the industrial revolution.

Population growth, increased food needs, and the practice of boiling food all increased the demand for salt, which became a basic commodity. Sodium chloride was first obtained from salt-marshes, but was soon being extracted from deposits of rock salt as well. It was sometimes carried over long distances on the Salt Roads, and bars of salt were used as money. Around 1000 B.C., there were fully active salt mines in Spain, as well as in Hallstatt in Austria. Sometimes the ashes of plants rich in salt, such as *Salicornia*, were also used. In Egypt, natron was used much more than common salt. An impure sodium carbonate, natron was the residue of Nile-water infiltrations and came from three centers, the largest of which was Wadi Natrum. Besides its use in food preparation, natron was used industrially in the manufacture of glass, enamels, and green and blue frits used as pigments. In Egypt and Palestine it was also used in worship as a holy sign of purification: incense burned to the gods was mixed with natron; celebrants used it to purify themselves before the rites, chewing it or washing out their mouths with it; and in mummification techniques it preserved the body from corruption and rendered it a fit dwelling for the soul that wanted to return to the world of the living.

Controlled fermentation soon made the transition from domestic practice to organized technology, and added the pleasant complement of a series of alcoholic beverages to the city dwellers' diets. The Egyptians called them "bringers of joy," "additions to meals," "heavenly," and "good and beautiful." The alcoholic fermentation of sugary fruit juices and honey had already been noticed by prehistoric gatherers. Palm and date wines were the first produced, along with hydromel, the traditional nectar of the gods, which had a low alcoholic content. But alcoholic beverages spread rapidly as the result of intensive grain cultivation. Malting first made oven-baked bread tastier, but cakes of malted grain were soon used to make beer. In the third millenium B.C., forty percent of the grain production of Mesopotamia was used for making beer, and leavened bread became common everywhere. Beer was the popular alcoholic drink of the ancient empires, whereas wine was reserved for nobles and holy libations. Vineyards had existed in the Fertile Crescent from the beginning of the agricultural revolution, but they began to be intensively cultivated around 2100 B.C. Gudea of Lagash built well-watered terraces protected by trees for his vineyards, and Assyrian and Egyptian nobles soon followed his example. They all appointed special functionaries to supervise the bottling procedures, and required that the date, place of vintage, and quality of the wine be stamped on the clay corks. Hellenic civilization takes historical credit for making wine a popular drink, replacing beer. Vinegar, the strongest acid known in ancient times, was also derived from the fermentation of grapes. It was used as a solvent for herbal medicines and as a very effective preservative and had various industrial uses, including the manufacture of white lead.

Painters and decorators, unlike other artisans, were accustomed to working with a variety of materials, which they needed to make the pigments they used on vases, walls, tiles and metal. The preparation of pigments was the most specialized chemical practice of the times. Exact procedures had to be followed, different ingredients had to be mixed in varying amounts, and the changes in color caused by oxidation and the high temperatures of the ovens had to be foreseen. Specialized instruments were developed, some of which would pass unchanged into alchemy and even modern chemistry laboratories. Pigments were ground in special mortars with pestles. By 2200 B.C., filtering finely ground solids through cloth to separate them was already a specialized practice, and the sieve was a common instrument. Kilns, industrial furnaces, and small crucibles were specially made for the delicate process of preparing colors. Not all colors, however, required procedures more complicated than grinding, mixing, baking, and adding an adhesive at the last moment. Casein was the adhesive used in cabinetmaking and whitewashing; for other purposes, gum arabic, egg whites, or gum tragacanth were used. Reds, yellows, and browns were made from natural ochres, sometimes mixed with chalk (calcium sulfate). Powdered malachite (basic copper carbonate) or chrysocolla (copper silicate) could be used for greens. Blue could be made as well by grinding lapis lazuli, but the resulting pigment was quite inferior. There was plenty of chalk and limestone for whites. Some pigments, however, required great ability and chemical precision. Red lead was made in Mesopotamia by heating lead with basic lead carbonate, which was obtained by adding natron to a lead salt solution. Artificial greens were sometimes made in Egypt and Mesopotamia by melting sand, alkalis, and copper ores together. Verdigris, unknown in Egypt, was prepared and used in Mesopotamia. Around 2500 B.C. various attempts were made to find a suitable blue pigment. Egyptian painters developed a lapis lazuli substitute, a mixture of silica, malachite, calcium carbonate and natron, which formed a blue compound when melted at 830 degrees centigrade. Its success was immediate, and even the Mesopotamian market requested the artificial lapis lazuli. A special factory was set up

at El Amarna, and the secret for preparing the blue pigment spread from there to Mesopotamia and the Mediterranean, where it was lost around 500 B.C. This blue was the Greek *kyanos* and the Latin *caeruleum*.

The textile industry, on the other hand, used organic dyes, and the techniques for preparing them were usually simpler. The scattered geographic location of dye-producing plants created a flourishing trade. Blue was derived from *Indigofera tinctoria*, the indigo of Indian origin, or from *Isatis tinctoria*, or dyer's woad. The method of preparation was similar: the plants were ground and left to soak and ferment until the blue color appeared. After a certain period of time, the indigo precipitated in the form of insoluble indigotin that was collected and dried. Reds were generally derived from madder (*Rubia tinctorum*), which was used to dye the cotton cloth of Mohenjo Daro. But the most stable reds were obtained by drying and crushing kermes, cochineals that lived on the so-called kermes oak. Tiglath Pileser I introduced this dyestuff into Assyria around 1100 B.C.. The best yellows came from safflower (*Carthamus tinctorius*) and dyer's rocket or weld (*Resedla luteola*), whereas yellow–orange came from saffron (*Crocus sativus*), which was produced in Syria, Crete, Egypt, and Cilicia.

The most prized and fabulously expensive dye was extracted from the *Purpura* and *Murex* molluscs by the Phoenicians of Tyre. Only a very small part of the animal was used. It was left to soak and then boiled in a one-percent salt solution for three days until the liquid was reduced to one-sixth of its original volume. Methods and procedures were tightly kept secrets. Although the manufacturers of Tyre sometimes procured their molluscs from faraway countries, all production of the dyestuff was centralized. The color, a dark, brownish violet, was used from 1500 B.C. on to dye small quantities of wool, which sold at fabulous prices. Dyers were recognized specialists, and in 2000 B.C., in Egypt, they became temple-adepts. In the sacred workshop, dyers colored not only the vestments of priests, but those of the gods as well. In Palestine and Syria, dyeworks were established in several cities later known for only one industry: ''Magdala of the dyers''; Beth-Asbea, ''the city of byssus''; and Luz, the blue cloth producing center. Dyers had secret recipes not only for applying the dyes but for washing and softening the cloth as well. The principal dye detergent was *Saponaria officinalis*, whereas alums, copper and iron salts, soil, and organic substances were used as softeners. Indigo, which did not require a softener, was often used as a softener for other colors.

The building industry also contributed to expanding the range of raw materials. In Egypt, conventional clay plaster was substituted by chalk plaster when walls were to be decorated. The plaster was prepared by heating raw chalk from the El Fayum mines to about one hundred and thirty degrees centigrade, which reduced it to a powder, and then quenching the powder with water. In Mesopotamia, limestone plaster was preferred. The first furnace for roasting limestone and clay to eliminate carbon dioxide was operating at Khafaje in 2500 B.C. White lime mortar was mixed with ashes and applied in layers about four millimeters thick. Fresco painting was not done in Egypt and Mesopotamia, but as early as 2500 B.C., Cretan paintings were executed on wet plaster made caustic with lime.

Mesopotamian technology was familiar with petroleum products, thanks to the geological conditions of the soil in that area. Although crude petroleum and natural gas were only objects of curiosity or divinatory practices, a specialized technology, which had a great influence on Babylonian construction and architecture, developed around heavy petroleum products. Some bitumen was extracted from asphalt rocks by means of a refining process that the alchemists called *distillatio per descensorium*, a procedure that consisted of separation by melting and draining. However, most Mesopotamian bitumen came from natural deposits near the city of Hit, where, during the floodings of the Euphrates, great masses of bitumen floated on the current. Bitumen was traded in its crude state. It was melted and mixed with sand and a feed stock of marl, lime, or limestone and used as mastic. Vegetable fibers were often used instead of a feed stock. The amount of bitumen in these compounds varied from twenty five percent to thirty five percent.

In the hands of Babylonian bricklayers, mastic became the best bonding material of ancient times. It gave their brickwork a solidity and strength that could not have been attained otherwise. When heated, it liquefied and penetrated the surface pores of bricks that had been baked at low temperatures (550–600 degrees centigrade). Once it cooled, it was virtually impossible to separate from the basic building material. The results were excellent, but the procedure was quite costly, so mastic was used only for palaces and temples, the principal public buildings, or for great walled works such as dams, embankments, and breakwaters. These latter were coated with mastic to make them waterproof, a precious feature in countries where the water supply was of special concern.

Power and human labor

Ancient technology did not deal with the problem of energy sources. Only biological engines—animals or more frequently humans—were utilized. Thus, behind the increase in productive activity, there lurked a sentence to hard labor. One could free oneself from this condition only by reducing other human beings to the state of pure sources of energy: teams of slaves, usually prisoners of war, rolled the gigantic monoliths, dug the metal ores, and later rowed the ships. Slavery became the great fear of the ancient world, whereas freedom from physical labor became a standard for happiness. All productive activity became charged with negative connotations because of the energy required. The blessed idleness of the gods and nobles consisted not so much in their escape from work as in their freedom from the inhuman degradation of the physical labor that went along with it.

Domesticating animals brought about only partial changes. Farmers benefited from learning to yoke oxen and use them for plowing and for carrying heavy loads. But beyond the farm, oxen were too slow for efficient overland transport or use in the city workshops. The equines—donkeys, mules, and especially horses—were more promising. Horses could drag sixteen times more weight than humans, and were quite fast. However, ox-harnesses were also used on equines, and wagons had beams and yokes rather than shafts. The animals were connected to the yoke by a breastband that, as they pulled, worked its way up and constricted their windpipes, reducing their power to four times that of humans. The indiscriminate use of this breastband kept draught-horses from being used regularly, since a horse and a team of slaves ended up being equally economically feasible. Yoked with the ancient harness and unshod, a horse

could pull a load of around sixty-two kilograms. Long-distance overland transport was thus quite costly, and was used only for valuable merchandise. Hydraulic energy was completely ignored. The wind's energy was trapped by ships' sails, but this did not suggest a more generalized device. Only later did Alexandrian engineers regard pneumatic phenomena with scientific curiosity.

The potter's wheel, industry's rotary machine

The thick, bitumen-coated disc with its spindle-hole that the potters of Uruk used to model their clay from 3250 to 3000 B.C. was a prototype for rotary industrial machines and had evolved from various rotary devices and instruments for drawing perfect circles. The bow-drill had been perfected for some time. It had become a reamer, a hollow tubular drill that made its perforations with the edge of a small stone cylinder rather than with the entire drill. It also had a device for centering the hole. Continuous rotary motion had been used in industry since the agricultural revolution. A perforated stone spinning wheel, which acted like a fly-wheel to accumulate kinetic energy, had been added to spindles. The fixed-point compass was used long before 3000 B.C.; shortly after 2500 B.C. it was used to make decorative circles on the bricks and vases of Harappa, in the Indus valley. The potter's wheel combined all these discoveries into a single device. The wheel was heavy enough to preserve kinetic energy once it was centered on a spindle and started rotating. When the wheel reached a speed of at least a hundred revolutions per minute, the potter threw a piece of clay onto the middle. Centrifugal force made the clay rise at a light touch of the fingers, and any form with a circular cross-section could be produced. In addition to saving energy, a potter could make, in ten minutes, a container that would take at least ten hours to make by hand. By around 2000 B.C. the potter's wheel was perfected: a second wheel beneath the first was turned by foot or with a stick, or sometimes a child would pull a belt to keep it moving.

Overland transport

Domesticated animals freed women from the mainly female job of dragging loads or carrying them on their backs. Donkey caravans became the symbol of a transport network that finally went beyond the narrow domestic environment, extending over thousands of kilometers. Sledges or drags attached to yokes were among the first vehicles pulled by people and later by animals. Huge weights could be moved and transported when friction was reduced by smoothing the skids of the sledge and decreasing the drag's points of contact to two. Even after carts with wheels became common, drags continued to be used for heavier loads and around farms. The wheel, the biggest innovation in overland transport, was developed at the same time as the potter's wheel. Around 3500 B.C. the Sumerians of Erech had already attached axles with four massive wheels on the ends to their sledges. Full wheels were made of three wooden boards cut with a metal saw to form a circle. This method was used to make wheels where trees with large trunks were not available. The pulling power of the ox-team was transferred from the plowbeam to the sledge mounted on wheels. At first, the outside rims of the wheels were reinforced with copper nails to keep them from wearing out too quickly. Copper hoops were already being used on the wheels of the war-chariots of Susa around 2000 B.C. The chariot was a sign of supreme prestige, royalty, or divinity; the nobles of Ur, Kish, and Susa took their chariots with them to their graves. By 1800 B.C. the basic innovation had already taken place; the full wheel, strong and perfectly efficient for transporting merchandise, was replaced by the swift, maneuverable spoked wheel. The team was no longer made up of oxen, but of horses; and the new vehicle was the light, two-wheeled chariot, more of a deadly war weapon than a means of transport. The Sumerian onager-cart, with its small, heavy, full wheels, gave way to the new weapon. But chariots soon had another important function: they made possible the central administration of vast expanses of territory. Egyptian, Hittite, and Assyrian nobles could organize their empires and render them more stable and enduring than the dominions of the kings of Akkad and Ur by using the swift, horse-drawn chariots to insure rapid communication. Ordinary wagons continued to be used in the countryside and for transport. This latter use, however, was limited whenever possible to carrying merchandise from the site of production or harvest to the nearest waterway, since water transport was the only means that was sufficiently rapid and economical.

River transport and the Aegean international market

The world of the agricultural revolution and the ancient empires was a world of journeys but not of roads. Lines of communication, from village paths to the Salt and Amber Roads that crossed continents, were tenaciously kept open, but nothing was done to modify them or adapt them to their new functions. Transport was always tedious and uncertain, a precarious compromise between the weight, size, and characteristics of the merchandise being transported and the harshness of nature. River currents, however, solved many of these problems. The right float could hold any load, and the river itself would do the work of transporting it. Paleolithic and early Neolithic peoples had already taken advantage of river transport, using float-boats made of the most readily available materials—tree bark, cane, or animal skins blown up and spread over simple wooden structures. The first boats were steered and sometimes propelled with paddles, making it possible to travel short distances along the riverways and to fish offshore.

The beginnings of the agricultural revolution and the rise of cities and states caused an extraordinary increase in the production of goods and, above all, made it necessary to transport them for purposes of trade. The boats had to fulfill collective rather than individual needs, and room had to be provided on them for ores and metal implements, vases, and woven and dyed cloth. In Egypt, boats made of papyrus bundles grew longer and acquired a new revolutionary device, the sail, which enabled them to travel up the Nile by exploiting the steady winds that blew from the north. A square sail was stretched on a short yard and suspended at the top of a double mast, which made

shrouds unnecessary and distributed the tension generated by the sail over the entire hull. Cane boats were steered by two long oars fastened to the sides—not a very efficient system, but good enough for the calm waters of the Nile. Egyptian shipbuilders remained faithful to this basic model even when, around 2600 B.C., they undertook the construction of hulls made entirely of wood. Egyptian hewing-masters reproduced the structure of cane boats in wood, and ships reached a length of thirty-five meters around 2500 B.C. Five hundred years later they were fifty-four meters long, eighteen meters wide, and carried a crew of 120, but, like primitive boats, they had no keel and were not very strong longitudinally. To improve them, a stay was stretched between the stern and the bow and subjected to tension by a rod inserted in the middle. The main advantage of wooden boats, besides increased size, was that the double mast could be replaced by a one-piece mainmast, reinforced at the base to distribute stress. The ships that Queen Hatshepsut ordered for the expedition to Punt, around 1500 B.C., showed several new improvements. The mainmast, located in a forward position before, was now exactly in the center of the boat, almost at the center of lateral resistance, making it easier to steer. The sail was wider and pulled taut by two very long yards at the top and bottom. Numerous lifts were used to fasten it to the mast. Rudders were installed on the gunwales and protruded from each side of the stern, and a tiller was attached to them. But even with these improvements the great Egyptian ships were too fragile and expensive to use for trade in the Mediterranean basin. The Aegean merchants who transported tin from the West to the East, took on obsidian at the Lipari Islands, gold at Siphnos, marble at Naxos, and copper at Cyprus, needed smaller, stronger, more economical ships. Their ships were descendants of the pirogues, the primitive boats of areas where long-stemmed plants grew, and their distinctive feature, the keel, was derived from these forebears. The longitudinal strength of the new vessels was increased by a smooth sheathing of tight-fitting boards fastened by wooden pins.

Mercantile ships soon assumed the form that would distinguish them for many centuries. They had tall sides, a rather stocky build, and wide bottom; they were partially covered and had a square sail that was their main means of propulsion, since the use of a crew of rowers would have rendered transport uneconomical. The Phoenician war ships that roamed the Mediterranean around 1000 B.C. had the one-piece keel, but little else, in common with the transport ships. Their sides hung low over the water line, and they had a pointed ram on the prow, behind which was a row of oars, which freed them from dependence on the wind.

Control over maritime trade thus fell into the hands of those who had access to vast quantities of wood and, even more important, had city and state organizations behind them that guaranteed the accumulation of capital needed to arm and furnish their war galleys with large crews of rowers.

From Material Culture to Scientific Tradition

The technology of the intellect

The great achievements in domination and technology that give protohistory its particular character—a complex, organized world within a precise framework—would never have been possible without the written word, a technology that introduced a completely new form of communication. Through writing, language acquired a totally autonomous dimension, independent of the presence of a subject to speak it. A precise material correlative, a set of visible signs, was evidence of its concreteness. In more advanced alphabetic forms of writing, these signs lost their residual connotative capacity, and, along with it, that sort of halo effect that ideographic writing still preserved. The signs were completely transparent and neutral for those who knew them—simply supports for sounds—but totally incomprehensible and mute for those who were not trained to recognize them. The gap between literate and illiterate became incomparably wider than that which separated the shaman from the rest of the totemic clan. Under the new rules of communication, speech acquired completely new characteristics. Via its material correlative, it could be transmitted in space and preserved unchanged in time. It had an elasticity and a permanence that rescued it from the transitory nature of oral communication, but at the same time placed it in direct and ruthless competition with the latter. Writing won out, not by doing away with oral communication, but by establishing a precise hierarchy between the written and the spoken word, to the total advantage of the former. The potentials of this new instrument of communication simultaneously affected the entire range of human activities, from politics to economics, from administration to religion. Communication between the capitals and the provinces of empires was possible only because of writing as was the establishment of centralized bureaucracies, especially in financial matters. And only writing could endow the will of a god with the permanence that transformed it into law. The mighty of the earth proclaimed themselves the authors of the writings on stone that laid down its norms. Differences soon arose between written law and unwritten law, and only an optimist could believe it would be resolved in favor of the latter.

Domination and the book

If writing could survive time, then the words that it transmitted were pronounced for eternity. When temple priests put into writing the traditional contents of belief in their gods, oral epics, national sagas, proverbial wisdom, prescriptions for rituals, cosmogonies, and hygienic rules, they created both a new institution and a new mode of belief. Religions of the book were based on obeying the written word, and therefore on conversion and exclusion. Literate religions had fixed points of reference, special means of supernatural communication, and a secret literature of ceremonial texts that in Egypt and Mesopotamia were not meant to be read by human eyes, since they were essentially communications between god and man and not between one man and another. And the book religions often exploited the widespread use of the written word to give their message a pretense of universality.

The knowledge of writing

From the very beginning, the close relationship between writing and the ruling system was obvious, and writing soon proved its effectiveness as a tool for social division and power. What was perhaps its most explosive potential, however, remained hidden much longer and was fully revealed only when the alphabetic form of writing had been consolidated and brought into contact with realities that were peripheral and marginal to the great empires, as happened in archaic Greece. This was the possibility that writing provided of creating a sort of objective, cumulative memory, an archive within which one could move about without having to stay within a single traditional vein. Likewise direct channels of communication were opened, in which the mediation of speech was not necessary and which were available for individual expression. A new world was created, whose only inhabitants were authors, and one could carry on a direct relationship with them thanks to the permanence of writing. Perhaps writing's greatest contribution to setting up this independent new world of knowledge was the magnificent metaphor it created: to the eyes of the reader, reality itself became increasingly similar to an immense writing, a complex system of signs that the reader had to decipher. The skies were scrutinized by those who knew how to "read" the signs, and they appeared to be full of precise symbols with hidden meanings that only the "illiterate" could not understand. In divinatory practices, a large number of natural phenomena began to be "read." But Babylonian astronomers already knew that numbers were the key to reading the skies and to predicting the movements of the stars, their conjunctions and eclipses. And when geometric reading had already transformed the universe into a sphere with the earth as its midpoint, Plato would insist that the heavens were given to mankind so that we might learn to read there our numbers.

Knowing how to read and write

In *The Instruction of Khety, Son of Duauf*, a poem which goes back to the Eleventh Dynasty, a father explains the characteristics of the various professions to his son: blacksmiths live at the mouths of furnaces and their fingers become like those of crocodiles; masons end up with their backs and knees broken by labor; and farmers are prisoners of the fields and have only rags on their backs all their lives. None of them has ever become an ambassador. At the end of his story, the father accompanies his son to the school of the scribes, begging him to study diligently, since "a day at school is a gain for all eternity." In Egypt and Sumer, learning the art of writing, which could change a person's destiny, required lengthy training that began during childhood. In Egypt the scribal schools were invariably connected with the temple and very restricted, but in Mesopotamia the number of literates was much higher because of the divisions between higher and lower scribes, temple and palace scribes, administrative and specialized scribes, and teachers and notaries. Mesopotamian scribes numbered in the thousands, thanks also to numerous flourishing private schools. An average school at Ur in about 1780 B.C. had twenty-four students and a teaching staff of two scribes (one of whom was the mathematics and land-surveying scribe) and three assistants (teachers of drawing and Sumerian). On the backs of their tablets, the students reproduced the lines of the classics that the teacher had written on the fronts, extracted square and cube roots, and solved practical geometry problems, such as the amount of earth that had to be removed, given the measurements of a pit. Almost all the students came from the upper administrative class and were the children of ambassadors, temple administrators, officials, rulers, priests, and scribes. The privilege of writing was rigidly determined, and the mere fact that it was imparted tended to perpetuate it as a tradition.

Writing: from the symbols of memory to the signs of communication

The priest-functionaries of Sumerian cities developed a system of graphically and synthetically representing things and their corresponding words in the spoken language mainly in order to remember data for economic purposes. Around 3500 B.C. at Uruk, Jemdet-Nasr, and other cities, the priests who supervised the collecting of foodstuffs for their god wrote down the revenues, gave out receipts for payments, and briefly indicated where the tributes were going. The growth and diversification of cities made this combination of economic information necessary. It had to be kept for a considerable length of time—from one harvest to the next—longer, in any case, than simple mnemonic systems would work. So the priests devised a new system. They drew the highly stylized, schematic symbol of the head of a cow, a spike of grain, or a fish on a clay tablet, using a small reed. Beside it they drew one or more circular signs which stood for the number of times, that is, the amount, the number. This accounting system was efficient and, above all, alluring. It encouraged people to record more complex information, data that could not simply be reduced to logograms, word-signs, or numbers. For this purpose, the number of signs had to be greatly increased, and relations and functions (verbs, for example) that existed in the spoken language, but could not be reduced to logograms, also had to be rendered in visible form. The solutions to this problem all tended toward simplification, since otherwise the symbols would continue to increase in number until communication broke down. The most effective solution was in use at Jemdet-Nasr by around 3300 B.C. Some symbols were given an exclusively phonetic value, independent of their original meaning. To write "en-lil-ti" ("god lets you live"), for example, the scribe used the sign of the arrow for TI, since "arrow" and "life" had the same phonetic value in Sumerian, namely TI. Analogously, the name of the pharaoh Narmer was written in 3100 B.C. by combining the signs for "fish" (phonetic value "nar") and "chisel" (phonetic value "mer"). Thus, at least in part, signs stopped signifying things, and consequently words, and began to indicate syllabic sounds instead. There was a dramatic decrease, from the 2000 signs used at Uruk around 3600 B.C., to the 800 in use three hundred years later; and this decreased to 200 in 2900 B.C. The Sumerian syllabary included both true syllabic signs, which had only a

phonetic value, and determiners, that is, signs that denoted classes of people and things, as well as archaic word-signs. With these developments, writing was no longer merely a precise fixating, but became an articulated entity with a continuous quality resembling speech. The signs were no longer drawn at random, but followed a precise order on successive lines of tablet sections. Around 3000 B.C., written words followed the same order as spoken words. The increased expressive possibilities of writing brought about its quantitative increase, and quicker ways of writing things down were needed. The professional writer was born. In Mesopotamia, where the only common material on which to write was clay, scribes had to contend with the difficulties of drawing the curved lines of pictograms on this material. The original designs were thus abandoned and replaced with schematic ones that were a combination of several basically similar components. Cuneiform styluses replaced sharpened reeds. The stylus was pressed into the clay several times until a group of wedge shapes of differing thickness and size formed the proper design. This combination of shapes into symbols made cuneiform writing elastic, and freed it from a strict tie to the language that developed it. As a matter of fact, Semitic invaders adopted it to write a language with completely different sounds and words. The diplomatic correspondence between pharaohs and Hittite kings, Hurrite princes and Canaanite leaders around the middle of the second millennium B.C. was all carried on in cuneiform.

Egyptian hieroglyphics, on the other hand, the great monumental writing that recorded the names and titles of the gods, the pharaohs, their priests and functionaries, and the illustrious dead, never gave up its peculiar figurative appearance, even though it had syllabic and determinant signs analogous to those of Sumerian and Semitic cuneiform. Egypt needed administrative and documental writing, suitable for the administrative needs of a great empire, as early as the First Dynasty, and a cursive form of hieroglyphic writing, called hieratic, was created. The faster demotic writing style was developed in the eighth century B.C. Egyptian scribes did not use clay and did not have to worry about simplification. Their writing material came from the pulp of the papyrus sedge (*Cyperus papyrus*). Sliced, compressed, and glued or pasted together into long strips, papyrus was an excellent recipient of the carbon black and gum ink that were used with reed pens and brushes.

Ancient writings were all sacred. Scribes worked in the temples or in the courts of the deified lords, or, as in Egypt, in the tombs where the written word perpetuated the spoken word and made it eternally effective. The god of the Jews was the first to declare his will in writing.

In spite of the high degree of codification and the smaller number of signs, syllabic writing remained extremely complicated, accessible to only a few scribes and priests. The decisive step that transformed the mystery of writing into clarity, and deciphering into reading, occurred around 1800 B.C., when the inhabitants of Ugarit, who spoke a dialect that was a mixture of Phoenician and Hebrew, finally reduced the necessary signs from around three hundred to thirty. A novel principle was used to effect this drastic change: the signs were no longer associated with entire words or syllables, but with single sounds, or phonemes. Signs became letters and syllabaries became alphabets. The Ugaritic alphabet was similar to the Arabic and was related to both cuneiform and Egyptian writing. But its success was short-lived. It was still written on clay, a writing material not native to Palestine, and was succeeded by the Phoenician alphabet after the destruction of the Ugaritic culture in 1200 B.C. Between 1700 and 1500 B.C., King Shaphatbaal of Byblos had inscriptions written in the new Phoenician alphabet, which consisted of twenty-two letters (all consonants), and was written from right to left. This alphabet reached Syria, Palestine, and Arabia at the beginning of the first millennium B.C. One of its variants caught on slightly further south, eventually resulting in the Arabic alphabet. This linear alphabet of twenty-two letters spread to the coasts of the Mediterranean along Phoenician trade routes. Around the middle of the ninth century B.C. it spread to Greece and from there passed, in modified form, into Etruria. The fabulous Cadmean letters, which reached Greece from the Aegean, carried the mark of their origin in their name: Cadmus was the Hellenized form of the Semitic root *qdm*, which means east.

The introduction of alphabetic writing was the last step in a rapid evolutionary process. In the span of two thousand years, figurative pictograms had developed into signs for phonemes. The Phoenician alphabet had the profoundly unnatural feature of breaking the semantic unity of words into their phonetic components. But it accomplished something no other form of writing had ever done—a spoken language, with all its meanings and uses, as well as with all its variants, could now be recorded, with nothing left out, on a material surface. Relatively easy to learn, it took writing out of the enclosure of the temple. Those who learned this little abstract mechanism wrote the words that shaped our culture.

Weights and measures: the relationships between goods

When Gudea, lord of Lagash, had himself portrayed in a regal pose holding a measuring-rule across his knees, the problems of regulating the complex economic and commercial relationships within ancient cities had already found a satisfactory solution. The possibilities of exchange were based on both parties acknowledging certain basic characteristics of merchandise, such as weight, length and volume. The first systems of measurement, especially those for length, were largely anthropomorphic. Measurements were made with arms, or rather forearms, (cubits), feet, hands (the span), or their multiples (one toise = four cubits; one foot = four spans; one cubit = seven spans). Around 3000 B.C., the need arose to fix the values of measurements of length and weight more precisely. Suitable exemplars were made, on which the nobles set their names. They were then deposited in the main temple as national units of reference. The systems varied from state to state and from city to city, not so much in the choice of basic units, which for the most part were the same, but in their values. The political relationships arising from conquest and subjugation were to unify the values of the basic units over a large part of Europe and the Mediterranean basin, even though some individual multiples and appellations survived for five thousand years to the enlightened calibration of the standard meter. Besides differences associated with topography and geography, others were created by the fact that the crafts were tightly closed, and each one had its own nomenclature. Nor did a world dominated by division feel any need for universal units. The basic unit for measuring length was the cubit. The royal Egyptian

cubit was about fifty-one centimeters long, the Sumerian cubit was fifty centimeters in length, and the later Greek Olympic cubit measured forty-six centimeters. Geographical distances were only measured in days of travel, since this was more important than abstract linear measurements, given the variety of routes and the problems of transport. Measurements of area, more closely associated with agricultural work than with commerce, remained less clearly defined. They were mainly based on the amount of land that a pair of oxen could plow in a day.

The basic crops, such as grains, and all liquids were measured by volume rather than by weight. The Egyptian unit was the *hon*, slightly less than twenty ounces. The Syrian *kotyle* was widely used, and corresponded to 12.5 ounces. Measurements of weight spread more slowly, since they depended on scales, which required accurate manufacture and perfect calibration. Small scales were used as early as 5000 B.C. for weighing gold dust; in 2500 B.C. scales were being used in commercial transactions at Mohenjo Daro and Harappa, but the practice reached Egypt only around 1350 B.C. Units of weight were related to their original use in measuring precious metals. The shekel, the basic unit in Mesopotamia and the Middle East, weighed from 7.78 to 14.13 grams. Its first multiple was the mina (twenty-five to one hundred shekels) and then the talent (fifty minas). A Sumerian talent corresponded to about 50.5 kilograms.

Measuring time, a sacred practice

The agricultural revolution involved Neolithic farmers in a cycle of annual forecasts. They learned to observe, and maybe even count, the signs of the approach of the harvest, and those that signaled the times to shift agricultural operations. They also soon learned to distinguish a series of minor cycles within the great annual cycle; these were linked to the lunar phases and were also important in their work. When cities were built, and the main jobs of administering and governing the community were assigned to priest-functionaries, the farmers' observations were codified, and a unified, collective system of measuring time, the calendar, was developed. In Egypt, astronomical observation and the related activity of measuring time took on particular importance because of the need to forecast when the Nile flood would begin. Egyptian priests thus developed means, including their Nilometers, that would enable them to make a precise forecast. In both Egypt and Mesopotamia the year was made up of twelve months of twenty-nine or thirty days each, which corresponded to the lunar cycles. Only in 300 B.C. were Babylonian astronomers able to predict accurately the length of the lunar month. In Egypt, the beginning of the year was determined by "the clear rising of Sirius," which announced the imminence of the flood. However, the discrepancy between the astronomical year of $365\frac{1}{4}$ days and the year of twelve lunar months, which was about eleven days shorter, remained a basic problem for the astronomer-priests. At the beginning of the third millennium B.C. the Egyptians introduced a special civil calendar made up of twelve thirty-day months, with five days added as a special period at the beginning of the year. But even the "Nile year" was imprecise, due to a six-hour difference each year. In 2500 B.C. a special lunar calendar of thirteen months was proposed.

The division of the day into twelve hours came from Egypt, by analogy with the twelve hours of the night, which were determined by the successive rising of certain stars called decans. Minutes, on the other hand, arose during the Hellenistic era, when the twenty-four Egyptian hours were combined with the number system based on sixty of the Bablyonian astronomers. Sundials were used to measure the time of day from 1450 B.C. on. Later models had different scales for different times of the year. Clepsydras (water clocks) and hourglasses were also used. None of these instruments was very accurate, however, and astronomical observations based on time reflected this shortcoming. Although Babylonian astronomers were able to solve this problem around 300 B.C., the Egyptians never attained a high level of accuracy. Their most advanced proposal was a cycle that alternatingly distributed short and long lunar years, with respect to the years of the civil calendar, over a period of 309 months.

Writing numbers

The first step towards establishing sciences such as mathematics and astronomy was the formulation of a special system of signs (notation) to indicate numbers. This explains why, in spite of the consolidation of syllabic writing around 3000 B.C., the first mathematical texts only date back to around 1800-1600 B.C., when numerous ancient Babylonian cuneiform tablets and the most important mathematical papyruses of ancient Egypt were written.

In a very general sense, special notations for designating numbers originated in different ways, depending on whether civilizations had alphabetic writing or not. If they did, several transitions occurred fairly naturally. After a period in which numbers were written out in letters, they began to be indicated with just the initials of their names. This is called acrophonic notation and is found in several Greek inscriptions. Finally, a series of basic numbers began to be indicated by means of their corresponding letters; this was the basic alphabetical notation which was used, for example, in Greek mathematics. The situation was different where there was no alphabet—for example, in the Babylonian and Egyptian civilizations. Special signs were introduced to indicate certain basic numbers, whereas the rest could be written by following special rules. This was symbolic notation, which was later replaced by the Greco-Latin alphabetic system, but which reappeared in the West when Indo-Arabic ciphers were introduced in the second century A.D.

All the notational systems used, however, had a finite number of signs (either alphabetic letters or special symbols) to denote certain basic numbers, which could be used to write the rest of the numbers, following some procedure of interaction. Thus the problem of the choice of a base for number systems existed in the simple writing of numbers, even before computation had any influence on it. In most civilizations, the number ten was chosen. Aristotle pointed out this meaningful coincidence and explained it by the observation that the fingers of both hands, everyone's primordial aid for counting, totaled, in fact, ten. But the Babylonians soon introduced a numeration with base sixty, which on a practical level was combined with the base-ten system by writing numbers up to sixty in decimal notation, and those over sixty in sexagesimal notation. The sexagesimal system was used for cuneiform inscriptions of a scientific nature, whereas the more elementary decimal system continued to be used (mixed with other bases as well) on an everyday level.

The choice of a base, however, was only a necessary rather than a sufficient condition for writing numbers effectively. Successive numbers could be generated by starting with the basic signs and using a subtractive or additive notation—a well-known example, which used both, is the Roman system, which wrote, for example, IV and VI for four and six respectively.

There was, however, a method that proved to be much more powerful than the rest in the history of civilization, namely positional notation. The value of a digit depended on the position it occupied in the number. For example, we say that 23 and 32 are different numbers even though they are written using the same digits arranged differently. In positional notation, writing 125 expresses the number obtained by adding one unit of the third order and two units of the second order to five units of the first order. If the base is ten, we will obtain the number 125, whereas if the base is sixty, we will have the number that, in our decimal notation, is determined by $5 + 2 \times 60 + 1.60^2 = 5 + 120 + 3600 = 3725$. One of the major scientific merits of ancient Babylonian mathematics was its adoption of positional notation for writing numbers.

The different solutions found for the problem of writing numbers can be useful for understanding the history of mathematics in later epochs. When Babylonian tradition met Greek tradition in the melting pot of Hellenistic civilization, the systems described above were synthesized. In the *Almagest*, Ptolemy used Greek alphabetic notation except for computing astronomical tables, in which he used the Babylonian sexagesimal base for fractions of degrees of arcs (or angles), combined with base ten for whole degrees of arcs. The same system is still used today. Alphabetic notation did not disappear even after the so-called Indo-Arabic digits were introduced during the Middle Ages. Moslem mathematicians and astronomers continued to use alphabetic notation in astronomical tables for a long time, using Indo-Arabic digits only in specifically and strictly mathematical contexts. The arrival of Indo-Arabic digits also reintroduced positional notation in the West, where it had not been used since the time of the Bablyonians.

Ancient Babylonian mathematics and astronomy

There are no signs of any preparation for the remarkable patrimony of mathematical knowledge produced during the ancient Babylonian period in any of the available documents. However, examples of rapid and prolific developments in the course of a few decades, preceded and followed by long periods of inactivity, are frequently found in the history of mathematics. The sudden rapid development in Mesopotamian mathematics may be explained by the exceptional importance of positional notation, which, once introduced, led its inventors to achieve conspicuous progress in a short period of time.

The cuneiform texts that have come down to us can be divided into two classes, the tables and the problems. The first are typical computational tables, often accompanied by tables of weights and measures, attesting to the practical uses of Mesopotamian mathematics, characteristically linked to economic activity. They are mostly multiplication and division tables, and were used along with tables of reciprocals, which made them extremely versatile. Even calculating fractions, generally very complex in ancient mathematics, became surprisingly easy and generalized with these tables. Between A.D. 350 and 400, Theon of Alexandria dedicated entire pages of his commentary on the *Almagest* to explaining Ptolemy's sexagesimal computations, whereas two thousand years earlier a simple Babylonian scribe would have considered it all a matter of elementary problems. The skill in computation that these scribes developed is further attested to by the numerous tables of squares and cubes (and their respective roots), of sums of squares and cubes, and of exponential functions that have come down to us. These could be used to solve several particular types of third- and fourth-degree equations, as well as to calculate compound interests. The tables thus seem to be geared toward solving the various types of problems that are frequently found in cuneiform mathematical documents.

The levels that Mesopotamian mathematics reached in solving problems of approximation are also of great interest. The value of the square root of two was calculated up to the fifth decimal place. Almost two thousand years later, Ptolemy himself used this Babylonian approximate value in computing his tables of chords. This mastery of the value of the square root of two, together with the existence of tables containing copious lists of the so-called Pythagorean numbers (numbers related by the formula $a^2 + b^2 = c^2$), demonstrates that the mathematicians of that period were already familiar with the so-called Pythagorean theorem, at least in practice, even though they probably did not have a demonstration of it for the general case. This is not surprising, if we remember that these mathematicians did not show any interest in pure geometry. Even their solutions to the problem of finding trios of Pythagorean numbers appear to be the result not of geometric processes but of the great skill they had acquired in dealing with quadratic equations, attested to by numerous and surprising examples. In fact, we can safely say that an algebraic mentality characterized by an interest in pure computation, independent of the actual results and even of the possibility, in many cases, that these calculations could even have any concrete application, dominated among these scribes, who failed only in not taking the final (though decisive) step of using a consciously algebraic notation. It was this emphasis on the arithmetical aspect of inquiries that made geometry appear simply as one of the many sources of concrete problems, without lifting it to the rank of a demonstrative science. Geometry remained a field like any other, within which the numerical relations between objects of everyday life could be discovered.

This explains fairly well how these mathematicians could know, for example, certain characteristic relationships between lines, areas, and volumes with a fair degree of accuracy, and why even when (however rarely) their approach began to resemble the geometric methods which later became common in Western civilization, their interest always lay in calculating geometric constants. For example, even though the relationship between circumference and diameter was almost always taken to be equal to 3, one cuneiform text, more scientific than the others, offers the remarkable approximation of $\pi = 3\frac{1}{8}$.

In spite of these considerable developments, it must be said that ancient Babylonian mathematics remained at a profoundly elementary state, not only because of the types of problems it attacked, but above all because of its lack of conceptualization, of a capacity to abstract, and of logical argumentation. Only neo-Babylonian mathematical astronomy operated with authentic scientific rigor equal to that of contemporary Greek mathematics. From a strictly technical point of view, however, it

must be said that nothing of substantial importance was to be added, during ancient times, to mathematical knowledge. The most important (and certainly most significant) later innovation was the introduction of zero, which rendered positional numeric notation more effective.

Ancient Babylonian astronomy, on the other hand, attained high levels of development. With the rise of ancient rural civilizations and the resulting need to regulate agricultural work, astronomical observation became systematic. The observations were conducted by priests, and they involved fairly simple instruments (gnomons, rods, solar clocks, water clocks, etc.). The cuneiform archives of ancient Babylonia contained many astronomical observations and descriptions of constellations. Systematic observation, over the course of many years, made it possible, even in ancient times, to calculate the periods of the apparent motion of the moon and sun, such as the length of the synodic lunar month (29.5 days) and the lunar and tropical years (354 and 365.25 days), as well as the lunar and solar cycles. These data were based on different calendar systems. Long before Hipparchus (see ASTRONOMY, 200–100 B.C.), Babylonian astronomers discovered the precession of the equinoxes and determined the cycles of the apparent motion of Venus (1650 B.C.), as well as those of the other planets (eighth to seventh centuries B.C.).

By noting the recurrence of eclipses, the ancient Babylonians were able to predict them as early as 1650 B.C. They started from the hypothesis that the sun's apparent motion was regular, and by the eighth to seventh centuries were taking its irregularity into account as well. The rise of religious ideas connected with various astral cults gave birth to astrology, a field of considerable development in ancient Babylonia.

Egyptian mathematics and astronomy

While such conspicuous activity in the field of mathematics was taking place in the area between the Tigris and Euphrates, an exact science of mathematics and astronomy was being developed, though to a much lesser degree, in the Nile valley. Neither of these disciplines, however, appear to have exercised much influence on the life of Egypt.

Mathematics was directed towards practical questions, but the elementary needs of the Egyptian economic system stimulated it very little, and it thus remained at a very basic level. Astronomy also arose from practical needs, such as those of measuring time and finding one's bearings on dry land, and became highly applicative in nature. In Egypt, where agricultural life depended entirely on the Nile floods, the priest-astronomer's main job was to forecast this phenomenon and regulate agricultural labor. On the basis of systematic observations, over the course of many years, with sundials, solar clocks, clepsydras, and sighting instruments attached to plumb lines, Egyptian astronomers developed the solar calendar and the decanate system (the thirty-six ten-degree sections of the zodiac), worked out methods for forecasting the Nile floods, discovered the precession of the equinoxes, and established that the planets nearest the sun revolved around it, all as early as the third millennium B.C. The development of astronomic sciences later declined in Egypt. However, astronomy greatly influenced cultural and religious life, and quite a while later also became a mainstay of astrology.

Although Egyptian mathematics did not attain a high level of development, it did have an influence on later epochs. Its use of unit fractions (typical of this mathematics, as we shall see) even influenced the *Almagest*, in which computations were made using the sexagesimal method, but the final results were often expressed in unit fractions, sometimes even at the expense of accuracy. Although its mathematics did not have a very positive and stimulating influence, Egyptian astronomy endowed posterity with at least one important product, the Egyptian calendar, which became quite a useful tool in ancient and medieval astronomy.

Egyptian mathematics had two distinctive features, "summation" and the development of fractional computation. In summation, multiplications and divisions were done by finding a suitable number of multiples and then adding them together, sometimes with the aid of unit fractions. These mathematicians used special signs for some of the basic unit fractions, such as $\frac{1}{2}$, $\frac{1}{3}$, and $\frac{1}{4}$, putting them on the same level as the basic integers. Beginning with these unit fractions, they were able to calculate others with remarkable skill. Using successive divisions by two and then trying out various computations, they arrived not only at more complex unit fractions, but at other types of fractions as well. They could express almost any fraction as a sum of unit fractions.

Geometry was also at a very elementary level and directed only toward solving practical problems, much like Mesopotamian geometry.

The beginnings of medicine

The earliest evidence of therapeutic activity comes from the very bones that signal the evolution of the human race. Bone injuries were without a doubt very common, and the arrangement of the fragments and the calluses of several well-healed fractures show that prehistoric therapists had an excellent knowledge of various treatment procedures, ranging from immobilization to, perhaps, traction of the fractured limb. As early as the Neolithic, surgical procedures of considerable difficulty were performed: skull trepanations, amputations, castration, incisions of abscesses, and probably Caesarean sections, at least on pregnant women who had died.

Skull trepanation was probably connected with magical practices that were not exclusively therapeutic, but it was most likely used mainly for curative purposes. The patient often survived these operations, and the bone edges thus began to heal. The percentage of successful operations during the Neolithic was surprisingly high, better for example than that of European surgeons in the eighteenth century. Neolithic surgeons operated using good systems of disinfection and perhaps even anesthesia. Their instruments, flint knives, were probably largely responsible for the asepsis of their surgical interventions.

Besides being surgeons, the physicians of prehistoric and protohistoric times were primarily herbalists. The pharmacological properties of herbs were well-known and regularly used. Health-giving herbs were, for that reason, sacred; they were administered along with complex magical rituals aimed at driving out the disease—that foreign body that had crept into the body of the sick person and exercised control over it. The disease was thus a separate being in itself, violent and destructive, and naturally opposed to air and blood, the vital principles of pre-

historic physiology. Homeric epic and the Bible still show traces of the privileged status of blood in the most ancient therapeutic notions, which were obviously derived from the everyday experience of hunters and warriors. Later, the human ecosystem and all the diseases to which it gave rise was radically altered by the domestication of animals and the close contact with them this involved. It was also changed by the demographic expansion caused by the agricultural revolution and by the appearance of the first urban centers. The disease of the city was no longer the wound, the loss of vitality that flowed out with the blood, but something more immediate and treacherous, something more related to the new way of living in close contact with other people and animals. Epidemics meant the death of an entire community, not just that of an individual. The Homeric poems record in a significant way the transition between these two worlds of disease and, at the same time, between the two worlds of patients and physicians.

The warriors and great lords suffered various types of injuries, but the physicians were able to describe them with admirable anatomical accuracy. They were also able to formulate surprisingly exact prognoses. In the face of epidemics, however, which did not strike individuals but "the people," condemning them without distinction to the same, unexplainable death, the physicians were disarmed. Only the priests, with their generalized prescriptions of hygienic practices could effect a cure. They were, in effect, the initiators of a social medicine that tried to avert the consequences of a drastic drop in population: loss of power, defeat, and, ultimately, slavery for the survivors.

A disproportionate interest in surgery also characterized medicine in Mesopotamia and Egypt. The code of Hammurabi, around 1780 B.C., regulated the activities of surgeons. It set their hours, forbade them to intervene in cases that were judged incurable or in which the successful outcome of an operation was doubtful, and prescribed the punishment of having a hand cut off for breaking the law. Mesopotamian surgical interventions were superficial and were not based on solid anatomical knowledge. In Egypt, on the other hand, the internal image of the human body was well known and followed a precise hierarchy: the central organ was the heart, and "the first thing in medical science," as the Ebers papyrus declares, "is to know the heart's movement and the heart itself." Even though Egyptian surgeons classed nerves, muscles, arteries, and veins all in the same category, they took advantage of the art of writing by beginning to accumulate a technical literature that was the new form of knowledge. The surgical papyruses evince an abundant casuistry determined by consistent criteria. There is the title, the observed symptom, the diagnosis, an evaluation of the appropriateness of the treatment and, if the latter is judged affirmatively, the therapy.

The field of internal medicine was less consolidated. Medical practice in both Egypt and Mesopotamia consisted mainly of incantations and ritualistic touching, with some administering of medicines. In times of established religions, magical practices in the strict sense—those aimed at reactivating a presence, in this case, that of health—survived in medical ceremonies. Sumerian pharmacology, on the other hand, developed along the lines of the written archive. The extant recipes list 550 simple ingredients, 120 of which are minerals. Plants were the main ingredients of these recipes. Among the narcotics were opium, belladonna, and mandrake; camomile was prescribed for stomach problems, mustard as an emetic, and hellebore for internal and external uses and as a fumigant.

The general situation around 1000 B.C.

Around 1000 B.C., there came to an end the great innovative period that during the preceding 7000 years had transformed gatherers into farmers, brought them together into villages and then into cities, given them metal weapons and implements, driven them to crossing seas, and enabled them to record events and transmit them to posterity as well as to forecast the motion of the stars and to measure and calculate more immediate realities. The innovative cycle of prehistory and protohistory, as we have already mentioned more than once, produced a people and a world that were to remain unchanged for two thousand years.

The material cultures of classical antiquity and the European Middle Ages continued for centuries to imitate very ancient models, practices, and procedures, even though contact with their origins had been lost.

This was the general setting for the history of technological innovation after the year 1000 B.C. The processes of change, of the perfecting of procedures, and of growth took place gradually over long periods of time until the radical acceleration of the modern age. But the multiplicity of diverse cultures and specific situations makes this picture more like a heterogeneous mosaic than a uniform design. The main factor behind this diversity was the way in which technological discoveries spread. The history of each discovery was partly determined by its own characteristics and partly by change. The result, around the year 1000 B.C., at the beginning of recorded history, was an obvious lack of homogeneity, even over extremely small areas. This phenomenon was particularly evident in areas that were marginal to the great middle eastern center where the agricultural revolution and its consequent effects had originated. The periphery of the urban world was explored and crossed in the search for metals and ores, the most precious raw materials, while its boundaries were being affected by the gradual advance of an agricultural technology that sought to increase the production of food. The geographical distance that separated the coastal regions from the hinterland was often much less than the technological and cultural gap. The era of technological innovation never fully reached the outlying regions; changes there came in spurts, discovery by discovery, and those that did reach that far did so only because they could be adopted rapidly and effectively. In these areas, unforeseen developments would often change the scope of certain techniques, or magnify and emphasize aspects that had remained secondary before.

A similar slow distortion was at the source of larger changes, such as the progressive decline in the innovative function of the old urban centers and the rise of a new form of culture, which no longer identified itself, as it had in the past, with the sum of the techniques it possessed, but was instead founded on the development of a particular skill, that of recording events and representing reality by writing. The birth of written culture was to be viewed by much of posterity as such an important event that it was often considered to signal the full-fledged birth of *Homo sapiens*. More often, this impressive event has seemed so radical that its origins, its ties with the specific techniques and technological innovations that made it possible, have been forgotten. Even when the ties have been recognized, the break created by the development of writing has been so deeply felt that it has been impossible to see what came before it as anything more than a certain "something before," a prehistory to our history.

The Hallstatt period in Europe

Around 1000 B.C., the inhabitants of Val Camonica, armed with bronze swords, spears, and arrowheads, and protected by wooden and leather shields, sought to defend their territory from outside attack. They depicted themselves in martial poses in a huge series of wall drawings and designs. They are an example of the "barbarian" European populations, who inhabited an outlying area where only some of the decisively important technological innovations had been accepted and absorbed. European smiths were able to make not only bronze weapons, but also axes, knives, scythes, and carpentry tools. During about the same period, on the plains of southern Russia, Scythian goldsmiths embossed large sheets of gold with totemic and votive animals as decorations for the centers of their shields.

Above all, the adoption of metal technology brought about improvements in agricultural techniques, and resulted in the possibility of reclaiming more land from the forests for purposes of cultivation. In southern Britain, settlements on the calcareous hills consisted of small fields that could be plowed in a single day, enclosed within stake fences and peat walls, and connected to the farmhouses by a small but efficient system of paths. In more mountainous zones, the farmers built their huts with stone foundations, constructed permanent enclosures for their cattle, and marked boundaries with dry walls or, sometimes, with long trenches. European civilization was a farming culture dependant on bronze metallurgy, but it was already able to set aside a part of its surplus toward the production of religious ornaments or the improvement of social status. An extraordinary acceleration of this process was caused by the spread of iron technology. After the fall of the Hittite empire in 1200 B.C., iron technology reached the Aegean Islands, moved up through Greece, and entered Europe, going as far as Noricum, where there were rich deposits of siderite. Hallstatt became the principal mining and metallurgical center of ancient times. Around 900 B.C., about a century after the introduction of early iron technology, vast deposits of rock salt began to be intensively utilized. The salt extracted at Hallstatt was destined not only for food and domestic uses; much of it went to metallurgy, where it was used to treat metallic ores. Mines soon became larger than ever before. Excavation extended as far as 390 meters horizontally from the shaft entrance, and tunnels penetrated to a depth of 100 meters. Hallstatt was a great mining center, but above all it was the city of iron. The first products were new, slender swords that replaced the large bronze broadswords. Daggers, spearheads, and battle-axes were to follow. But above all, a new series of implements were developed at Hallstatt. Some of them were necessary to work the iron iself: hinged pincers, bucksaws, anvils for making nails, and wire-drawing blocks. The array of manual tools produced at Hallstatt around 500 B.C. included almost all those used today. Hatchets, chisels, and gouges were improved by the introduction of hardening and tempering, and drills, awls, borers and hammers were produced in a wide assortment of forms, depending on their specific uses. The barbarian outskirts had surpassed the old urban centers in the most sensitive and important field of technology.

Iron, the alphabet, money: the birth of Greece

If Europe could stilll be considered a marginal region at the turn of the millennium, Greece was a totally negligible entity. The great Aegean civilization with its capital at Crete was no more than a legendary memory, half a millennium old. This civilization had ended around 1480 B.C., when the volcano of Thera (now Santorini) erupted, causing a tidal wave that swept the Cretan cities right off the coast. The Mycenaean nobles, enclosed within their fortresses, intent on carving out small monarchies after a basically oriental model, could not resist the successive waves of invasions that came down from the North. The last of these, the Dorian invasion, caused a migration from Greece toward the coasts of Asia Minor, where the Ionians, as they were later called, ended up settling. Knowledge of the sacred writing that was used to record taxes and the contents of the royal warehouses was also lost in the fall of the Mycenaean citadels. Until 700 B.C., Greece was merely a region of poor farms, agricultural villages, and infrequent communication. There too, as in central Europe, the decisive event was the introduction of iron technology. Between the eighth and the seventh centuries B.C., diggings in the small deposits of iron ores had already produced enough of the metal to free agricultural communities from their dependence on those who controlled the vast trade routes of copper and tin. From the smiths' forges came highly resistant, and thus less costly, agricultural tools. When Hesiod addressed the farmers of Boeotia, he was already able to call them "the iron race," certainly not referring to their non-existent bellicosity, but to the new era opened by the use of the revolutionary metal , even in Greek villages.

Another crucial technology, Phoenician alphabets writing, arrived in Greece almost at the same time as iron technology. Greece's marginal status was an unexpectedly favorable factor in the adoption and development of this new technology. The fall of Mycenae had buried ideographic writing, and aristocratic culture was exclusively oral in nature. There was no social structure uniting the entire territory, and it was thus impossible to centralize the knowledge and use of the new technique. It was the Greek merchants, strange to say, who brought Phoenician alphabets writing into Greece. Th intrinsic value of Phoenician writing lay in the ease with which it could be learned. This favored an incomparably more rapid diffusion than was possible for any other type of writing. Even though elementary instruction was made available to all citizens only in the fifth century, literacy became a widespread and typically urban phenomenon much earlier. In the Ionian trade centers, entire classes for the first time had access to the instrument of writing and could use it in any way they considered useful or interesting, since there was no centralized control. The Greek alphabet contributed a significant feature to that adopted from the Phoen-

icians, a feature which made writing even easier. Signs for vowels were added to those for consonants by the use of letters that already existed in the Phoenician alphabet but that represented consonants that the Greek language did not have. Originating in sacred ideograms that were indecipherable to the uninitiated, writing had become a complete transcription of the phonetic structure of words, as easy to use as the words themselves.

Between 750 and 550 B.C., the Greeks contributed a controlled migratory flow to the original Ionian migration that had formed the main trade centers on the coast of Asia Minor. At first this was aimed at increasing the food resources of the mother city; later its goal was to build trade centers in the most favorable locations. This system of colonization created a trade network throughout the Mediterranean and nearby seas, which was based on the conviction that an increase in trade meant an increase in wealth and, consequently, in freedom from scanty harvests and aristocratic proprietors, the twofold slavery of farming. Around the seventh century B.C. the Greek merchants of Asia Minor, pressured by the rapidity and frequency of exchanges, began the custom of doing business, using precious metals cast into small ingots with the merchants' own characteristic marks stamped on them. Electrum—a natural alloy of gold and silver—was the metal most frequently used for this purpose. The next step towards a monetary economy was taken by the kings of Lydia, who owned large gold mines. They replaced the ingots of electrum with gold ingots of a fixed weight, guaranteed by the royal authority whose lion's-head seal was stamped directly into the ingot. This new discovery, money, had exceptional features. First of all, money drastically reduced the material volume of transactions, rendering them easier and smoother. Secondly, the base unit could be divided into submultiples and combined into multiples, so that any value would still have a small material volume. Finally, since money was itself a form of wealth, it soon became *the* form of wealth. A new type of financial power arose, one that could hide itself and become invisible. A new type of rich person was also created, one who could appear humble but possess a fatal heap of the new symbol of true wealth, a hoard of money, hidden in the house or buried in the garden, as classical drama never tired of repeating. The invention of the Ionian merchants and the kings of Lydia soon spread to all the Greek cities and replaced the cumbersome iron ingots previously in circulation. The birth of a monetary economy was the major factor in the development of Ionian communities. On the hinge of the great trade routes between East and West, and with their sheltered, well-equipped harbors, they were able to derive vast profits from the soaring increase in international trade. Small and medium farmers, still tied to bartering, grew weaker, while new classes less directly tied to the land, arose. These traders, money-changers, money-lenders and technicians no longer exchanged their services for corresponding amounts in kind, but for money, an instrument of much greater efficacy in society.

The Classical World (from the sixth century B.C. to the fifth century A.D.)

600–500 B.C.

From the Mycenean civilization to the polis

The Mycenean world, which is essentially the one reflected by Homer's poems, collapsed towards the 11th century. Until the end of the eighth century no new unitarian social organization replaced it; the "Greek Middle Ages" was characterized by isolated villages gathered around the palace of the local lord (basileus), closed economies, and lack of artistic development. Between the eight and the seventh century a sudden series of technical and economic innovations were disseminated to continental Greece across the Aegean zone of Asia Minor; an echo of which was already perceivable in Hesiod's poems at the end of the eighth century. Later in the Greek culture it was an almost universal tendency to attribute an "antique knowledge" to the Orient, herald of welfare and civilization, of rational inquiry.

Modern scholars have offered various theses about the "oriental" origin of science and philosophy. In the complex and multifaceted interaction between the Orient and the Occident in the seventh century and the following century, a process of which we only have fragmented testimonies, the Ionic cities played a central role as a refuge in the epoch of great migrations. Due to its geographic position, Ionia on the one hand benefited from the possibilities which the Aegean sea offered; on the other hand it benefited from the contacts with the Asian people. Thus, it very soon acquired various techniques ranging from the processing of iron to the minting of metal coins. At the same time it could compare the tradition that was handed down from the Mycenean splendors to the political and cultural experience of the oriental countries and Egypt. The most important cities in the Aegean basin were cities like Miletus, Ephesus, Colophon, Clazomene, and Focea. The first philosophers and scientists were Ionians; figures like Pythagoras and Xenophanes were also of Ionian origin and gave birth to relevant political and intellectual experience in southern Italy.

The creation and structure of the poleis

The transformation from primitive agricultural communities to cities (poleis) with mixed economies first occured in Ionia. The Ionic polis was bipolar: it was governed by an aristocracy and its institutional structure was concentrated in the acropolis (from a religious viewpoint it contained the biggest temples; from a political viewpoint it housed the city Senate; from a military viewpoint the acropolis was above all a fortress). Its main economic reference was the earth as the principal source of wealth. Opposed to the acropolis was the marketplace (agorà). The rise of the urban demos was not only an element of social tension but also of cultural revolution. One essential element of the cultural revolution was the passage from the traditional oral culture to a written one.

The Mycenean writing that derived from the ideographic models of the great oriental monarchies was the heritage of priests and writers of the palace. With the crisis of the power system and the Mycenean culture, Greek culture was reduced to oral transmission, works of priests and poets who sang their epos in the aristocratic palaces. The alphabetic writing imported from the Phoenicians, however, was easy to learn and, thus, rendered the techniques of memory and oral transmission less important. Even though the teaching of the alphabetic writing was not introduced until the second half of the fifth century, its diffusion constituted an effective instrument for the knowledge about traditional culture and for the codification of new forms of intellectual life: the drafting of valid laws for the polis and rules for the agorà, the composition of professional manuals, tales and reports of journeys, etc.

Culture of the temple and the square. The texts.

Naturally, this passage from oral tradition to written culture was not a matter of a sudden change. Socrates in the fifth and fourth century still preferred oral teaching and Plato wrote in dialogue form.

One should not talk about written works in terms of modern treatises. In most cases the philosophic text still represented a transcription from an oral communication. Later on, manuals were introduced, particularly in disciplines that gained autonomy and specificity such as mathematics and medicine. Poetic works, generally linked to the aristocracy, also appeared in written form.

The new naturalistic knowledge: Thales, Anaximander, Anaximenes

Already in previous centuries elements of reflection on nature are present in myths, the epic poems, the production of lyrics and in the widespread genre of the judgments. "This was the form of philosophy of the ancients," Plato wrote. But this form of poetic and gnomic reflection occurred again in subsequent ages and was taken up in an asystematic way by the so-called physiologists, scholars interested in nature (physis). These scholars tended to take the place of priests and theologists, "forcefully submitting nature to examination so that it reveals its secrets" as a fifth-century physiologist said. Hence, Xenophanes of Colophon rummaged the earth for fossil traces and the first students of medicine examined the internal organs of animals to understand the functioning of their organism. In the end scholars tried to reduce the multiplicity of observed phenomena to one principle (arché): according to Thales of Miletus (approximately 624–546 B.C.) this principle was water; according to Anaximander, however, the origin of the world was to be found in a primary, undetermined substance (apeiron); according to Anaximenes, also from Miletus (586–528 B.C.), the air was the origin of the world.

We only have fragmented information about the Greek philosophers of nature, at least until Plato, and our analysis must be drawn from the few fragments of their original works and from traditional assessments made after the figure assumed legendary proportions. Thus, for instance, Thales is traditionally considered one of the seven sages whose astronomical knowledge had to be set against the background of cosmological themes. In confronting the problem of the first principle (*arché*), Thales seems to have accepted the most ancient theories of the origin of the world from water, theories already expounded by Homer and Hesiod.

Anaximander's concept appears more articulate. He placed the earth in the center of the world in an indifferent equilibrium. The stars were in the air surrounding the earth, set in concentric rings or rather flat cylinders similar to "cart-wheels." Through a flaming opening they diffused light and warmth. According to Anaximander the universe developed in cycles that repeated themselves endlessly.

All three authors cited above are important in other areas. This is not by chance. Linked to a city in which the experience of navigation and commerce was bringing to light a world whose outlines faded away in the unknown, the thinkers of Miletus sensed the necessity of supplying their fellow-citizens with secure terms of references in the same way that the variety of phenomena was explained with the help of a unitarian principle.

Pythagoras and the Pythagoreans

The Miletus philosophers' rational inquiry of nature did not conflict with the religious tradition of Ionia, an olympic religion of myths with prevailing naturalistic characters, averse to mystic and magic inclinations. However, Orphism, an element of religious renewal common to many Greek cities during this period introduced new elements into scientific thought. Orphic motifs can be found particularly in Pythagoras' (approximately 570–497 B.C.) doctrines. They are, for instance, division between divine and bestial elements in man; longing for the purification which is supposed to free man from the "wheel of destiny and generation," from the "cycle of births and misery"; and the doctrine of the transmigration of souls.

The subsequent Neopythagorean tradition, which asserted itself from the first century after Christ, portrayed him as a *thaumaturge*, a prophet. But little is known about him, and Porphirius (see Thought and Scientific Life 200–300 after Christ) admits that "what Pythagoras said to his companions cannot be reported with certainty." What the subsequent doxographers called the "Pythagorean School," was in reality a religious-sacerdotal institution. It had its roots in the temple, and its doctrines were transmitted only orally, at least until Philolaos of Thebes (approximately 480–400 B.C.) first wrote an exposition of the system. This lack of knowledge makes it even more difficult to distinguish the work of Pythagoras, or his first disciples, from the contribution of subsequent generations.

The fundamental thought that numbers are the cause to all things and the consequent development of a theory of whole numbers understood as a collection of units represented by points as a means of understanding how all beings are generated by the collection of a number of units, seems to be part of the original nucleus of Pythagoreanism. The laws of the formation of numbers were considered the laws of the formation of things: the principal task of philosophy (the term seems to appear for the first time with the Pythagoreans) was to investigate these laws. Philosophy comprised arithmetic and geometry; astronomy was understood as geometry applied to the universe, acoustics as a relation of numbers, that is, applied arithmetic. The way of purification is also based on philosophy. The mystic character of Pythagorean mathematical research, be it pure or applied, impelled mathematical development on the one hand but impeded it on the other because mathematics was not considered a purely scientific discipline. Despite this impediment, the Pythagorean school contributed significantly to mathematic research.

The determination of the sum of the inner angles of a triangle and the fundamental properties of the rectangular triangle go back to the Pythagorean school. This was already known for specific cases but it had never been demonstrated generally. The most important achievements of the school were the discovery of irrational numbers and the development of geometric algebra.

The Pythagoreans also made important contributions to the theory of proportions: they studied arithmetic, geometric and harmonic proportion related to whole numbers.

Xenophanes and Alcmeone

The picture of the cultural life of the VI century is completed by Xenophanes (565–480 B.C.) and Alcmaeon of Croton (approximately 500 B.C.). Xenophanes is, according to the scholastic and dogmatic tradition, the founder of the so-called Eleatic school. However he seems to connect to the Ionic thought of the first half of the century; "all things" says one of his fragments "come from the earth and will end in the earth." And another: "the things which are born are of earth and water." But Xenophanes' importance does not lie as much in his modifications of the Ionic doctrines as in his perception of the incompatibility of these doctrines with the tradition of myth and poem. The polemic against Homer and Hesiod is a polemic against religious anthropomorphism. Scientific research takes the place of divine revelation as a guarantee of growth of knowledge.

Alcmaeon, who, according to some sources was a disciple of Pythagoras, seems to represent a school of scientific, especially medical, observation in many ways autonomous from the Ionians. The discovery of the connection between the sense organs and the brain, the distinction between knowledge and sensation, the dynamic conception of health as an equilibrium of opposed qualities and of disease as a supremacy of one of the polar qualities of the organism, and development of theories through empirical research make Alcmeone the precursor of Hippocrates. With Xenophanes he shares the idea of knowledge as a process. Xenophanes characterizes scientific research "as a proceeding for evidence" in contrast to divine knowledge that is of little use because man cannot draw on it.

500–400 B.C.

The Persian Wars

Greek history in the first half of the fifth century was dominated by the encounter with the Persians who, after conquering Ionia, aimed at controlling continental Greece. But the Persian Wars, which provided Greek culture with its model for the united patriotic resistance of a "free" people against "barbarians" accustomed to "slavery" (Aeschylus' attitude in *The Persians* is a typical example), ended with a Greek victory in 479. The control of the Aegean was a decisive factor in the outcome of the war, and the Athenian fleet's dominance sanctioned the moral and political superiority of Athens. During the first half of the century, the city was still ruled by the aristocracy; but soon the various classes that made up the *demos*, or populace, who had become familiar with the basic workings of trade and fleets, ascended to power. The *demos* had considerable weight as well in the Ionian cities, where Athenian influence was being felt more and more. The landed and military aristocracy, however, continued to exercise their hegemony over most of the territories of the Greek motherland and in Magna Graecia. This geographic distribution was also reflected on a cultural level. Aristocratic thinkers often left their Ionian cities of origin for Magna Graecia, as Pythagoras did in the sixth century B.C., whereas physicists and physicians, thinkers more closely tied to the knowledge of the *agora* (market place) than to the knowledge of the temple, migrated in the opposite direction. The Pythagoreans in particular, justified and inspired by the knowledge of which they felt themselves to be trustees, held power in Magna Graecia, though with varying success, for about a century after the foundation of their sect. Their first great political and scientific crisis was linked to the legendary figure of Hippasos. In spite of his royal descent, Heraclitos (flourished c. 500 B.C.) was unable to do what the Pythagoreans were at least partially able to do. His political and cultural attempt to restore the aristocracy in the democratic Ionian city of Ephesos ended in failure.

The renewal of aristocratic-sacerdotal knowledge: Heraclitos and the Eleatic school

Heraclitos of Ephesos, called "the Weeping Philosopher" by later doxographers, actually belonged to the oracular tradition of priestly knowledge and the great mystic initiations. But within the context of the above-mentioned political and cultural restoration he devised a complex and highly refined theory of the constant becoming of all things. The heavenly bodies were thus produced by the continuous exchange of matter between the earth and the heavens; they were fires gathered in the hollows of the sky, from whence their light and warmth reached the earth. All phenomena recurred cyclically during the course of a Great Year. Becoming and contradiction were essential features of both the natural and the human worlds, because strife, conflict, and war dominated everything. "Strife," according to a Heraclitean fragment, "is the father of the world." Strife among human beings, "which makes the slaves free and the free slaves," corresponded to the strife among the elements, which alternately suppressed each other. The only guide for the sage, who rejected the apparent knowledge of the mathematical and empirical disciplines (Heraclitos condemned both the physicians and the Pythagoreans), was the *logos*, the law of the world, the reason that understood this law and the language that formulated it.

The common theme that linked Heraclitos and his contemporary, Parmenides, was that of truth as revelation and prophecy. Parmenides was the real founder of the so-called Eleatic school of which Zeno and Melissos were also representatives. Parmenides lived on the western coast of Magna Graecia, at the opposite end of the Greek world from Heraclitos' Ephesos. Despite the radical differences between their doctrines, Parmenides—a legislator in his city and the founder of a dynasty of priest-physicians connected with the cult of Apollo—also gave rise to an aristocratic cultural policy. Melissos of Samos was also a part of this same aristocratic trend; in 442 B.C. he led the Samian fleet against Athens, which was then under democratic control. The Eleatic experiment and the attempts to transport it to Ionia were doomed—as was early Pythagoreanism—to a catastrophic outcome. But Pythagoreanism experienced a rebirth with Archytas and Philolaos (see below), whereas Eleatism disappeared, even in Sicily, after the unsuccessful attempt by Zeno (flourished c. 460) to oppose a "tyrannic" government, which was most probably a democratic one that tended to reduce drastically the traditional "freedoms" of the aristocracy. But apart from its political goals, Eleatism exercised a massive influence on later Hellenic thought. It rejected all the various ways of fragmenting the world that characterized the philosophical tradition, denouncing both the Pythagorean mathematical interpretation of the cosmos as well as Heraclitos' dialectical representation of becoming. It proposed the identity of all being as the only authentic reality, in contrast to the mutable world of appearance.

Neither Heraclitos' doctrine nor Eleatic thought constituted restatements of the old Delphic sacerdotal wisdom. They were, rather philosophical doctrines built on the essentials of aristocratic tradition. But these doctrines consisted of refined arguments against the knowledge that was gradually emerging from naturalistic research and technology. They were able to take up their adversaries' themes and assimilate them, and to replace mythological wisdom with rational thought. Thus, it is not surprising that, by means of their polemics, these schools of thought developed a whole series of argumentative techniques that were able to transform apparent confutations of their doctrines into support for them. Combined with this was a systematic critique of rival doctrines. The famous arguments of Zeno concerning the impossibility of understanding plurality and motion are an example of this method. These arguments, which were perhaps directed against the Pythagoreans and the atomists, constituted the first rigorous inquiry into the concept of infinity.

The rational science of nature: Empedocles, Anaxagoras, and Leucippos

Rational reflection on nature could not renounce the scientific ideal pursued by the Ionians, but it also had to contend with Eleatic critiques. Arguments such as Zeno's, for example, pointed out the logical absurdity inherent in any attempt to speak of multiplicity, change, or time. Natural phenomena had to be explained with schemes that were more consistent than those of the Ionians, and attempts were often made to reconcile democratic cultural aspirations with religious tradition.

Empedocles (c. 490-430 B.C.), born in the democratic city of Agrigentum, was physicist and politician, physician and technologist, prophet, and magician. According to Empedocles, prophecy announced the road that the soul had to follow in order to attain the universal harmony that was a law of physics and that political (democratic) action attempted to establish in the *polis*. Magic expressed, in a supernatural form, the control over nature that physics achieved on the technological level. These various ideas coalesced in an articulated theory based on the primary elements of the cosmos—water, air, earth, and fire. The elements were sometimes deified, but more often they were rationalized within a framework that attempted, through the idea of love and hate as universal forces governing the combination and separation of the elements, to explain multiplicity and becoming by means of entities (the elements) that in themselves were uniform and eternal.

Anaxagoras of Clazomenae (c. 500-428 B.C.), born in an Ionian city but a resident of Athens for many years (probably 465 to 432 B.C.), totally renovated Ionian cosmology and cosmogony. According to Anaxagoras, in the beginning, all things were together, but the seeds for generating all things were present in this original indistinct mass. These seeds were already qualitatively differentiated, but they were not yet combined in the proportions and relationships that define each individual thing. The seeds of each thing were unlimited and identical in quantity and quality and had been so since the beginning of the world's history. However, during that history, they were combined in different ways and different proportions. A complex phase of transition from original indistinctness to organization characterized the present state of the world, which, according to Anaxagoras, came about through the intervention of an ordering principle, the *Nous*. Later this principle was understood as "thought," "mind," "intelligence," which did not so much generate the world as shape it according to a rational plan. Anaxagoras' *Nous*, however, was definitely a material substance, even though it was "the most subtle and rarefied" of all.

Leucippos, who lived around the middle of the century, was a native of Miletos and was shaken by the aristocratic reaction of 450 B.C. His thinking critically re-examined certain Eleatic theories, with which he came into contact during a period of exile at Elea. Certain characteristics that the Eleatics attributed to being, he attributed to the atom. Leucippos pursued a doctrine "in harmony with sense-perception, by which neither the birth, nor the destruction, nor the motion, nor the multiplicity of things was abolished" (Aristotle). The theory of atoms and the void, which Democritos systematically developed in the following century, arose.

Changing techniques in astronomy and medicine

In scientific life, the rebirth of a more critical reflection on nature was accompanied by a change in the organization of research. According to Plato, Anaxagoras' works were probably the result of coordinated, specialized research, conducted by those who studied with Anaxagoras. They constituted a unified, articulated plan for a rationalistic and democratic culture. While in Athens as political adviser to Pericles, Anaxagoras was tried for impiety because of the novelty of his astronomical doctrines. The trial was an indication of the political interest in Anaxagoras' thought, of his prestige among the elite of *sophoi* that surrounded Pericles, and of the complex institutional, political, economic, and intellectual dialectic between conservative and innovative forces in Athens at that time. Anaxagoras' theories as well as those of the other philosophers of nature, greatly influenced the development of the individual sciences and their techniques.

Anaxagoras' cosmogony, in which his well-known theory of "everything in everything" can already be glimpsed, is quite interesting as far as astronomy is concerned. His cosmogony is permeated with a lively rationalistic argumentation, which he applied even to extremely complex questions, such as the plurality of worlds (a notion that some attribute to Anaximander). The philosophy of the Milesian school had already implicitly contained a subversive element, since it attempted to provide a rational explanation for cosmic phenomena that was independent of divine agency, at least as this agency was anthropomorphically interpreted by the official Greek religion.

Parmenides, as a consequence of his negation of the void, seemed to believe that the sphere of the universe was infinite and immobile; its rotation was pure illusion. The earth was at the center of this universe, surrounded by layers of ether and fire, in which were located, in the order of their distance from the earth: the moon, which shone only by reflected light, the sun, the planets, and the fixed stars. Parmenides thus shared with his contemporaries, the Pythagoreans, the belief that the heavenly bodies were arranged within a "cosmos," that is, a universe in which rationality and order had replaced chaos.

Anaxagoras coherently developed these naturalistic ideas. "The sun, the moon, and all the stars are fiery rocks, moved by the rotation of the heavenly vault," he wrote. If we do not feel the heat of the stars, it is because they are very far from the earth. These great distances deceive us as to the true size of the stars. Anaxagoras' theory of eclipses was the first correct explanation of this phenomenon. He also made accurate studies of the moon's phases. In addition, Anaxagoras proposed an interesting theory of the Milky Way, which he conceived of as a projection of the earth's shadow into the sky when the sun passed behind the earth at night. When he moved from Ionia to Athens, Anaxagoras found himself enmeshed in the complex political situation of Pericles' city. Diophebes, an interpreter of the oracles, accused him of impiety, stigmatizing him as Plutarch wrote, as one of those "who do not practice the religion of the city or who give lessons on heavenly phenomena."

Empedocles, who was only a few years younger than Anaxagoras, considered biological as well as cosmological problems. His theory of the four elements greatly influenced medicine. According to Empedocles, organs and organisms are the result of adaptation to environment. Living beings must therefore be considered as functions of their environments.

The schools of medicine

The beginnings of the so-called Sicilian school of medicine were linked to the thought of Empedocles, who, unlike Alcmaeon, assigned pre-eminence to the heart, source of the flame of life. Sensation and thought took place in the blood, particularly in the areas near the heart, because only the heart was able to achieve the perfect fusion of the elements.

Anaxagoras applied himself very specifically to medicine and the dissection of animals. His theory of knowledge influenced "technicians," such as Hippocrates (c. 460–377 B.C.), who wanted to move beyond the great ancient methods by means of concrete research that could be verified in everyday experience.

Anaxagoras had a rationalizing influence on all forms of knowledge he studied. In his philosophy, knowledge developed through experience, memory, learning, and method; humans were the most intelligent animals because they had hands. His influence was felt in the field of medicine, where, as early as the end of the sixth century B.C., especially in the Ionian poleis, lay physicians had already set themselves up in opposition to the priest-healers. The teachings of these physicians provided homogeneous doctrinal guides to the etiology and treatment of diseases. These men were the first nuclei of the later schools of medicine (for example, those of Cnidos and of Cos), which began to be organized when permanent centers and libraries were set up around the end of the century. As early as the middle of the fifth century B.C. lay physicians took the offensive against religious medicine, which they accused of charlatanism because of its inability to provide explanations for its methods of treatment. This was a typical theme of Anaxagoras, and it appears, for example, in Περὶ ἀέω, ὑδάτω, τνπω' (De aëre, aquis et locis) and Περὶ ἱερῆσ νοϭον (De morbo sacro), both of which were part of the so-called Corpus hippocraticum, or Hippocratic Collection where religious and mystic explanations of epilepsy were replaced by a rational one that encouraged the first attempts at therapy.

After the great crisis of the Athenian plague in 429 B.C., the group that had gathered around Hippocrates, considered the founder of the School of Cos, began to develop scientific medicine. The new feature of Hippocratic research was the attention it gave to problems of method, which was described as the path to follow in medicine and which was located between observation and reasoning. This concern was valid not only for medicine, since it emphasized the need to mediate between the two poles of research and to review critically all acquisitions from the past in a new understanding of the historical nature of scientific knowledge itself.

Approximately sixty medical treatises have been attributed to Hippocrates. With the Hippocratic Collection, medical tradition, almost exclusively oral before, began using written documents. Hippocrates showed little concern for doctrinal systemization, but rather emphasized direct clinical observation. He analyzed physiological and pathological processes in great detail, interpreting them by means of accurate, rational explanations. He studied the organism in close relation to its environment, attributing character differences between Europeans and Asians to climatic factors.

The Hippocratic Collection mingles together the works of many authors, and it reflects the ideas of at least three different schools of medicine, the Italian school and those of Cnidos and Cos. The Hippocratic school of Cos has already been mentioned. The school of Cnidos, in Asia Minor, attempted to identify the different diseases and treat them using various complex methods.

Euryphon, one of the authors of the Cnidian Sentences, exemplifies the thought of this school quite well. He established an analogy between the growth of plants and that of human embryos. He explained impregnation, embryonic development, and heredity by means of acute observations and a number of technical considerations, though from a viewpoint saturated with prejudices.

The fortunes of the Pythagorean school are complex and difficult to reconstruct. It was torn by internal crises and was the object of Eleatic polemics. The most famous crisis, which was probably both political and intellectual, was that connected to the discovery of the incommensurability between the sides and the diagonal of a square, that is, the discovery of irrational numbers. This new discovery was spread outside the sect, as was already mentioned, by Hippasos of Metapontum. The crisis was further intensified by Zeno's paradoxes concerning motion and divisibility. On a cultural level, it caused a radical division between geometry and arithmetic, with the former viewed as the study of continuity and the latter as that of discontinuity.

The most important Pythagorean of this century was Philolaos of Crotona, who lived in the last half of the century, whereas Archytas of Tarentum (see APPLIED SCIENCES, 400–300 B.C.) flourished in the fourth century. Archytas, an influential political figure in his native city, was especially interested in problems of political philosophy, but was also a student of mathematics, mechanics, and harmony. As for Philolaos, he was exiled to Thebes after the democratic revolt of 454 B.C.; there he reestablished a Pythagorean thiasos. He developed an original revision of the Pythagorean theory of oppositions, and elaborated a cosmology that in some ways prefigured the discarding of the geocentric system.

Philolaos' cosmology completely renovated Pythagorean celestial mechanics. He was motivated by both the astronomical need to explain eclipses and the mystic need to assign the central place in the universe to fire. In this context, Philolaos considered the earth an eccentric mobile planet like all the others, anticipating by centuries the theories of Aristarchos.

Athenian culture in the second half of the fifth century. The Sophists and Socrates

Athens, under the government of the sophoi who surrounded Pericles, was the liveliest center in Greece during the second half of the century. Anaxagoras and his influence on medicine have already been mentioned. But physicians were not the only "technicians" of cultural importance. There was also, for example, Hippodamos of Miletos, the city planner who directed the construction of the port of Piraeus and who was the first to develop systematic ideas on the relationship between urban planning and the social system. In the broader terms of the economic and social maturity of a democracy that had laboriously achieved hegemony, the rapid development of scientific-technical knowledge was reflected in a series of speculations on the progress of humanity, on the progressive achievement of a higher level of intellectual life and social harmony, and on the possibilities of exploiting nature for humanity's ends and of organizing a better life on this basis. This theme had already been sounded at the beginning of the century in works such as Aeschylus' Prometheus Bound and Sophocles' Antigone, and was later taken up by authors like Archelaos (a disciple of Anaxagoras), Prodicos, Protagoras, and Democritos, the great master of atomism.

These writers developed a relatively new theme, that of the profound unity between scientific and technical progress, on the one hand, and the newly discovered technology of life in the aggregate, or politics on the other. From this perspective (already prefigured in Aeschylus' The Eumenides), the democratic polis, closely connected with technical knowledge and its creators, was viewed as the apex of human history. This was the age of the great historians, such as the Ionian Herodotos (c. 484-425 B.C.) and the Athenian Thucydides (c. 471-400 B.C.) of the flourishing of tragical and comical drama; and of the first systematic studies in pedagogy, intended mainly as a preparation for political life and for putting into action the cultural ideals of the *polis*.

The need for the formation of a new ruling class was also being felt, in a situation in which the upper echelons of the *polis* always came from the great Athenian families. The pressure of the *demos* was imposing a new structure based on consent, different from that of the aristocratic *polis* of half a century before. From this situation arose the importance of using language in a convincing way. The art of persuasion was needed in every manifestation of civil life, from the assembly to the council, from court sessions to meetings; and a group of intellectuals emerged to fulfill this need. They were of various philosophical extraction, and they dedicated themselves professionally to preparing the ruling class and teaching rhetoric. These were the so-called Sophists, professional scholars who were all foreigners, since no Athenian citizen of good social standing would ever sell his professional services. A new form of culture arose. In fact, the concept of culture as a body of skills and knowledge distinct from priestly knowledge, scientific endeavors, and the skills of specialized technicians was born. This is not to say that the above-mentioned forms of intellectual activity were unfamiliar to the Sophists. Many of them demonstrated a profound knowledge of medicine and astronomy, effectively studied mathematics, were expert technicians, and sometimes even drew on the repertory of myths, often interpreting them in a critical light, as was done in the tragedies of Euripides. However, the aim of this new culture was to form politicians, who were alternatively the objects and subjects of political decisions and practices in the *polis*. Several figures may be viewed in this light. Protagoras of Abdera (c. 485-410 B.C.) insisted that "man is the measure of all things, those that are, inasmuch as they are, and those that are not, inasmuch as they are not." Gorgias of Leontini (c. 485-380 B.C.) was a critic of the absolutism of Parmenides and Melissos, and, in general, of any notion of an absolute reality. Prodicus of Ceos (born c. 465 B.C.) was the theoretician of *orthoepeia*, that is, of the rigor and propriety of discourse. Critias (c. 480-403 B.C.), one of the aristocrats named one of the Thirty Tyrants after the Peloponnesian War (404 B.C.), carried further the relativism and religious agnosticism of Protagoras, clearly following the early Sophists (even though, as a member of the government, he saw fit to prohibit their activity). Thrasymachus of Chalcedon (born c. 460 B.C.) defined justice as that which is useful to those in power. Hippias of Elis, a brilliant mathematician discussed below, and Antiphon of Athens enunciated the opposition between human laws, which could be traced back to convention, and the laws of nature, or "necessary precepts." This opposition between *nomos* and *physis* was not to be reformulated until the end of the eighteenth century.

Finally, as shall be seen below, the ideas of Socrates (born in Athens in 470 B.C.), as well as his teaching practices (including his famous maieut method), were very much a part of the political and social framework in which Sophism developed. Socrates held debates with the major Sophists, and shared with them the same interlocutors, the young aristocrats of Athens. He also focused his interest on the same questions practical ethics connected with the political life of the city. Like many of the Sophists, Socrates helped shift the focus of research from "things" (that is, natural phenomena, the domain Ionian scientific reflection) to "discourse," that is, to the logico-linguistic interests that were closer to the political and rhetorical debates of the city.

Mathematics: Hippocrates of Chios

The work of Hippocrates of Chios was particularly important during this period, when the trend was toward intensive research in specific fields of knowledge. Hippocrates, who lived in the second half of the century, compiled the first *Elements of Geometry*. Although the work has been lost, it is important to note that the systematic, theorizing mentality of Greek mathematics, which first appeared in the work of the Ionian and Pythagorean schools of the preceding century, had developed little more than a century later, a body of knowledge that could be organized within a treatise. Indeed, the very act of presenting a discipline in the form of a treatise was a totally new and highly significant occurrence in the history of science. Hippocrates introduced two famous mathematical problems: the doubling of the cube and the squaring of the circle. The first is also known as the Delian problem because, according to legend, the oracle of Apollo at Delos had pronounced that, in order to put an end to the plague, a cubic altar with a volume twice that of the cubic altar already in the sanctuary had to be built. Hippocrates found that the solution could be reduced to the problem of inscribing two mean proportionals between two straight lines, (Archytas of Tarentum was the first to offer a solution of this second problem—see APPLIED SCIENCES, 400–300 B.C.).

As for squaring the circle, Hippocrates approximated a solution by studying the possibilities of squaring various lunes, formed by circular arcs. His work shows a great familiarity with quite advanced geometric methods. The problem of squaring the circle thus came to the attention of mathematicians. Two Sophists, Antiphon and Bryson (who succeeded each other closely in this century), thought they had solved it. Antiphon considered the class of polygons inscribed in a circle, and believed that by increasing the number of sides, a polygon with sides so small that they would coincide with the circumference could be attained. Bryson considered both inscribed and circumscribed polygons, in which the number of sides would increase indefinitely. This in some ways anticipated the central idea of the method of exhaustion, which would be rigorously developed and fruitfully applied by Eudoxos of Cnidos (see THOUGHT AND SCIENTIFIC LIFE, 400-300 B.C.) and by Archimedes (see MATHEMATICAL SCIENCES, 300-200 B.C.). Finally, another Sophist, Hippias of Elis, attacked the same problem by means of a clever geometric stratagem, the invention of a new curve called a quadratrix.

One of the most important mathematical developments of the fifth century was the introduction of an alphabetic notation for writing numbers to replace the earlier acrophonic notation. The decimal base was used; and the twenty-four letters of the Greek alphabet, as well as three ancient letters no longer in use (stigma, koppa, and sampi), provided the 27 symbols needed to write numbers up to 1,000. From 1,000 on, the first series was repeated with bars added. Numbers that are not powers of ten were written by using additive notation (for example: $\sigma\chi\beta = 200 + 20 + 2 = 222$).

400–300 B.C.

The trial of Socrates and the crisis of Athens

Socrates' philosophic activity, which took place almost entirely within the political framework of democratic Athens, ended in 399 B.C. with his trial and death sentence. The philosopher was accused of impiety, of corrupting youth, and of having ties with the enemies of democracy (which had just been restored a few years earlier, after Athens' defeat in the Peloponnesian War and the dictatorship of the Thirty Tyrants appointed by the Spartans). The political reason for his condemnation can be found in the general situation in Greece after the debilitating Peloponnesian War, as well as in the weakness of the new Athenian democracy, distrustful of internal and external enemies and unable, with its reduced military force, to seek a solution to its problems by imperialistic expansion, as it had done in the fifth century B.C. Symptoms of malaise had already appeared with the Sophists: in Critias, and in Hippias and Antiphon's opposition between *physis* and *nomos*, which attacked the juridical order of the *polis*. Within this context, Socratic teaching, which interpreted this malaise from the viewpoint of the great aristocratic priestly philosophies, could have been seen as an intellectual justification of aristocratic pretensions. It was not an accident that Socrates' principal accuser, Anytos, was a prominent figure in the re-established democracy—a form of government whose grave limitations Socrates had clearly outined in his public teachings.

This argument, however, is not unanimously accepted by historians. According to some, Socrates' attitudes may have been dangerous to the democracy during its period of adjustment and instability after the fall of the Thirty Tyrants, when it was unable to understand and absorb his innovations.

In any case, the philosopher's death dramatically underlines the narrowness of the new Athenian democrats, who were unable even on a cultural level to achieve the spirit of enlightened mediation that had been characteristic of political leaders at the time of Pericles. The government of the *sophoi* had been able to promote developments that were mostly favorable to the *demos*, without ever identifying with it. But those days were over. The new democratic rulers were faced with an unresolvable social crisis, one in which the opposition between the landed aristocracy and the *demos* of the city was aggravated by the objective limits of ancient democracy, which was unable to integrate metics, small merchants, and artisans into a stable political format. And then there was the problem of slavery. The crisis was, of course, reflected on the cultural level, where authoritarian and coercive solutions often replaced subtler techniques of achieving a consensus.

The legacy of Socratic teaching: Plato's Academy

The Athenian crisis did not, however, mean the end of politico-cultural debate or the decline of philosophical and scientific thought. In fact, several of the weightiest theoretical systems of Greek thought appeared during the fourth century B.C. After the death of Socrates, there was no shortage of thinkers to take up his legacy, though from different perspectives. Socrates became a symbol of the teacher, a symbol which later took on legendary and mythic proportions. His refusal to write enabled others to use Socratic theme in contexts that became progressively more autonomous of his actual thought. The idea of the rights of the individual soul, as opposed to the claims of scientific knowledge and political life, was one of Socrates' teachings that survived. The founder of the Cynic school, Antisthenes of Athens (c. 444-370 B.C.), may be seen in this light. Other themes dear to the Sophist thought of Hippias and Antiphon also survived; these later emerged in the stoicism and skepticism of the third century B.C. Among the various minor Socratics, as they were traditionally called, were Eucleides of Megara, founder of the Megarian school, and Aristippos (c. 435–356 B.C.), founder of the school of Cyrene. These thinkers often harked back to the themes of Sophism, particularly to Protagoras' relativism and Gorgias' critique of an absolute reality, and they distrusted the great philosophical systems of the time, beginning with Plato's.

The concept of philosophy as a methodical search for the good and the true—which was secondary or sometimes even non-existent in the thought of the minor Socratics—was central to the aristocratic and idealistic philosophy of Plato (c. 427–347 B.C.), the most important philosophical system that looked back to Socrates. Plato developed the logical aspects of the Socratic method, as well as its connection between goodness and truth, but his great system was far different from Socrates' "knowledge of not knowing." On a political and social level, Plato, who came from an aristocratic family that claimed the great legislator Solon among its ancestors, was closer to Critias than to Socrates. This is not the place to discuss Plato's political aspirations or his unsuccessful attempt to set up an ideal state in Magna Graecia. More relevant to the development of scientific life was the founding of the Academy (named after the gymnasium and gardens dedicated to the hero Academos) in Athens in 387. Whereas the schools of the Cynics, the Cyrenaics, and the Megarians were primarily associations of those of Socrates' disciples who sought to perpetuate his teachings, Plato's Academy was a complex institution from the very beginning. From a legal point of view, it was a religious association dedicated to the cult of the Muses. In reality it was a political organization that attracted young aristocrats from Athens and elsewhere and a center for scientific research that served as a point of reference not only for philosophers, but for mathematicians, astronomers, and students of other scientific disciplines as well. A characteristic feature of the Academy was the communal life that existed among teachers and disciples—as in the Pythagorean *thiasos*—all of whom were convinced of the need to reform both the *polis* and the individual by the acquisition of philosophical and scientific knowledge. The Academy was also the first great institute of higher education.

Plato's teachings

We shall not discuss the problem of classifying the dialogues or of establishing the authenticity of Plato's *Epistles*, nor will we deal with the early Socratic dialogues. We are more interested in showing how Plato's great later works, such as the *Republic*, the *Parmenides*, the *Philebos*, and the *Laws*, may be interpreted as programmatic documents and as contributions to the internal debates of the school. It is in works such as these that Plato developed his theory of *forms*, which, although it approached the status of a complete philosophical system in his final works, was at this stage characterized by complex developments, reworkings, and revisions. This process was consistent with the Platonic concept of the need for philosophical mediation between the pure knowledge of the inspired sage and the false opinions of the majority, since the mediation would be accomplished by those who were capable of persuading others to establish their opinions on the basis of truth, by means of a comprehensible language that even "the many" could—though with difficulty—understand. This is why Plato chose the dialogue, an intermediate form between oral and written culture. The Platonic dialogues purport to be transcriptions of philosophical discussions, with one or more speakers representing the popular knowledge associated with public opinion and opposed to *theoria*. But another character, the philosopher, mediates between the two poles, guiding opinion towards truth by means of a dialectical process ("dialectical" comes from *dialeghesthai*, which means "to have a dialogue, to discuss"). We shall not examine the Platonic dialectic in detail here, nor enter into the details of Platonic doctrine, in which objects of the natural world are the reflections details of Platonic doctrine, in which objects of the natural world are the reflections of ideas or forms, immutable and eternal essences. However, we should mention that in addition to his written works (in particular the *Dialogues*), Plato developed in the latter part of his life an oral doctrine concerning several supreme principles (especially those of "the one" and "the indeterminate") that were the constituents of the forms. But little is known of these oral teachings of Plato (often known as the doctrine of "number-forms"), and it is difficult to distinguish genuine Platonic thought from the teachings of Plato's successors, in particular Speusippos, who directed the Academy until 339 B.C., and Xenocrates, who succeeded him and headed the school from 339 to 314 B.C.. We do know that themes of an initiatory nature, derived from Pythagoreanism, were emphasized during this period. Typical was the reduction of the theory of forms to a few very general principles, which could not be fully understood through discursive procedures but only through an intuitive and direct vision.

Although in many ways the Academy under Plato (during the period of his later *Dialogues*) and under his successors followed the model of the Pythagorean *thiasos*, it distinguished itself on both the political and pedagogical level by one extremely important feature. Confronted with the opposition of "the many" who resisted the true doctrine, the Pythagoreans were only able to impose their authority forcibly. The Platonists of the Academy, on the other hand, attempted to convince. It was this flexible attitude that attracted politicians, orators, and scientists to the teachings of Plato and his immediate successors. Science, hierarchically arranged, with mathematics at the top and the various technologies with their empirical focus at the bottom, was the essential element of this mediation. Hence the importance of stars, numbers, and geometric figures in the Academy even after Plato. The theory of the divinity of the stars (dear to Philipos of Opos, another of Plato's disciples, the editor of the master's last great work, the *Laws*, and the probable author of the *Epinomis*) also had its roots in this concept. The same notion also inspired the astronomy of Heraclides of Pontos.

Plato made several fairly unimportant technical contributions to mathematics, such as a generalization of the analytic method, a definition of the concept of geometric flocus, and the construction of an instrument for mechanically inserting two proportional means between two given straight lines. His greatest

contributions to the field, however, lay in the emphasis he placed on the abstract character of mathematical research, on its hypothetical-deductive procedures and the elevation of mathematics to a paradigm of rigorous knowledge. Mathematics thus conceived opened the way to philosophy, which Plato considered the most perfect form of knowledge. For mathematics to fulfill this function, however, it had to be approached with purely speculative aims in mind. Practical needs, such as the computations of commerce, the applications of geometry to land surveying, or other technical uses, had to be rigorously excluded. Plato's attitude had a great influence on the development of Greek mathematics (especially geometry) into a pure science, but it also stood in the way of discovering applications for this science. In addition, the Pythagorean influence on Plato often led him to search for the mystic-symbolic meanings of mathematical concepts. This had positive effects during certain periods that were little inclined toward science, when admiration for Plato created a resurgence of interest in mathematics, but it also distracted the study of mathematics from its rigorously scientific foundations.

The profound influence of Platonism on mathematics can also be seen indirectly by considering the number of important mathematicians among the students of Plato: Laodamas, Leon (who wrote a text of *Elements*), and Theaetetos (credited with important research on irrational numbers and regular solids). Nothing, however, has survived of their writings.

Platonic astronomy

Although astronomy was not a fundamental part of Plato's thought, he evinced a lively interest in it. In the Timaeus he adopted the Pythagorean geocentric system, the most antiquated component of Pythagorean cosmology. The universe was comprised of eight circles or bands, one for each of the moving heavenly bodies. The earth was not counted among these, since it was immobile at the center. The motion of the most distant band, that of the fixed stars, imparted movement to the rest, though all the bands had autonomous retrograde motions of varying speeds. (This furnished an explanation for the characteristic motion of the planets.) Speed and distance were expressed by harmonic chords. The four elements of the Presocratics—earth, water, air, and fire—corresponded to the Pythagoreans' four regular solids—tetrahedron, icosahedron, cube, and octahedron—according to the theory proposed in the *Epinomis*. Since there was no fifth element, the fifth regular solid, the dodecahedron (also studied by the Pythagoreans but later than the others) corresponded to the model that the deity used for the universe. The theory of Platonic corpuscles exercised a great influence on Pythagorean and Neoplatonic thought, and continued to fascinate mathematicians and astronomers—Kepler, for example—for some time.

In astronomy, as in mathematics, Plato made a distinction between practical study, which concerned the motion of the heavenly bodies as it appeared to the senses, and theoretical study, dedicated to the purely mathematical structure of which observable astronomical phenomena were only vulgar manifestations. Thus the door was opened to the search for a mathematical-mechanical model of the universe, which did not necessarily have to correspond to the reality of the heavens as long as it could account for observed phenomena. Simplicios tells how one day Plato gave his disciples the problem of finding "those uniform and orderly movements that would make it possible to explain the apparent motions of the planets," and he concludes that the earliest solutions were proposed by Eudoxos and Heraclides of Pontos.

Democritos and atomism

We should, however, not let the great prestige of Platonism and the Academy make us forget, the extreme importance of atomism in this century. This was mainly due to the teachings of Democritos of Abdera (c. 465-370 B.C.). Developing a theme probably already present in the thought of Leucippos, Democritos proposed a theory of reality made up of eternal, immutable entities, devoid of properties other than those that could be described in geometric terms. The entities were indivisible (as indicated by the word *atomos*), the ultimate limits to all possible divisions of material bodies. (Democritos thus avoided Zeno's *apeiron*.) But in order to explain natural processes, it was necessary to ascribe motion to the atoms, and this necessitated the hypothesis of an empty space through which the atoms could move. More precisely, each atom continued to move in one direction until it collided with other atoms. The collision, random in itself, determined a series of mechanically necessary consequences. This was an extremely sophisticated doctrine which, on the one hand, considerably influenced the development of physics, and, on the other, reinforced the role of mathematics in scientific explanations. Democritos' universe was ruled entirely by mechanistic necessity, which depended on the purely quantitative characteristics of matter. The movement of atoms and their groupings into bodies explained the origins of the world, or rather, of the infinite worlds that populated empty space and that arose from the vortices caused by the encounter of the bodies themselves. The earth was flat and at the center of our system, surrounded by the heavenly bodies: the moon, Venus, the sun, Mercury, Mars, Jupiter, Saturn, and the fixed stars. There were inner and outer planets (with respect to the sun), and the Milky Way was a conglomeration of stars.

Democritos' thought was thus characterized by a secular materialism that excluded the intervention of providence or divinity in the universe. There was also a tendency towards reductionism implicit in Democritos' atomism, a tendency to reduce all the great scientific systems, from astronomy to biology—and perhaps even ethics—to the atomic theory. This remained a characteristic feature of atomism for centuries and was at the same time its strongest and weakest point. It was opposed by the adherents of rival philosophical schools (beginning with the Platonists, who, according to tradition, had Democritos' works burned), as well as by scholars of the various scientific disciplines, who sought to preserve the specificity of their own research. But this was also the source of its fascination; like Platonic theory, it assigned mathematics the task of rationally structuring the explanations of phenomena, but in a materialistic key on the level of ontology, and in a democratic one on the political level. After all, the basic idea of atomism, borrowed from Empedocles, was that scientific explanation consisted in reducing the complexity of phenomena to simple principles. This was an embryonic form of the hypothetical-deductive concept of scientific theory, which also reappeared in Plato's writings. In the Timaeus, when Plato attempted to draw an all-inclusive picture of the relationships among forms, numbers, and physical reality, he used the models of Empedocles and Democritos, combining them with Philolaos' Pythagoreanism. Even mathematics, as systematized by Eudoxos and, later, by Euclid (see MATHEMATICAL SCIENCES 300-200 B.C.), adopted this type of simplification. By means of suitable procedures of demonstration, all the theorems of mathematical theory could be derived from a few principles (axioms and postulates).

On a level other than that of physics and of mathematics applied to the knowledge of reality, Democritos elaborated an antithetical concept, according to which space was made up of infinitely small parts. By applying this idea to solid figures, it seems that he was able to show that the volume of a cone was equal to one-third the volume of a cylinder with the same base and height.

Eudoxos of Cnidos (408-355 B.C.) was, however, a much more accomplished student of geometrical volume, and is credited with writing the fifth book of Euclid's *Elements*. He established a rigorous theory of proportions, and also, it seems, arrived at the famous axiom (called the Eudoxos-Archimedes axiom) that was one of the most successful attempts at formulating a correct concept of continuity. With the aid of this principle, it seems that he was able to prove several theorems concerning area and volume, thereby inaugurating the method of exhaustion, which Archimedes later carried to its greatest heights, and which remained the basis of the geometry of measurements until the invention of integral calculus.

Another noteworthy geometrician, Menaechmos, also belongs to this period. He was the first writer to deal with conic sections; his work however, has been lost. Archytas of Tarentum (c. 430-365 B.C.) has already been mentioned in the discussion of Philolaus; he is credited with solving the problem of the doubling of the cube. This would seem to place him in the tradition of pure geometry; however, he arrived at the solution by attempting to insert two mean proportionals between two given straight lines, thereby using purely algebraic calculations. He also stated that arithmetic, not geometry, could provide satisfactory proofs—an assertion that helps us understand that the nature of early Greek mathematics was more arithmetical than geometrical.

The astronomy of Eudoxos

Eudoxos' work in astronomy, which in some ways seems very similar to Plato's, is perhaps even more famous than his work in mathematics. In any event, all the sources that testify to the similarity between the views of Plato and Eudoxos (Simplicios and Proclos, for example) are biased. Without going into the complex question of Eudoxos' relationship to the Academy, we shall simply point out that the atomists and many exponents of Sophism, who were also excellent scholars of mathematics, considered mathematical models important in explaining the world of sensory appearances in purely rationalistic terms. This did not necessarily imply, however, the acceptance of Platonic mathematical ontology. Eudoxos' theory of astronomy consisted in the geometric explanation of the motion of the heavenly bodies, apart from any considerations of a physical nature (for example, those concerning the transmission of motion of the nature of the heavenly spheres).

Developments in medicine

Specific scientific developments also occurred that were not part of any all-embracing philosophical scheme. Among these, the later thought of Hippocrates is particularly important, since his teachings progressively acquired the form of a discipline complete in itself. For Hippocrates, speculation was more important than observation; this tendency appeared later in some of his disciples, especially in his son-in-law, Polybios. Pathology was dominated by the theory of the four humors—blood, phlegm, yellow bile, and black bile—which was discussed in Περὶ ψύσιοσ ὑνθρὸπου (De natura hominis).

As opposed to the individualistic focus on the localization of disease of the Cnidian school, the physicians of Cos conceived of the organism only as a whole, and did not subscribe to an ontological notion of disease. They adopted a conservative approach toward therapy, which was based more on diet than on medicines. Health was seen as the proper balance of the bodily humors, whereas disease consisted in an imbalance (dyscrasia). The strict relationship between the organism and the environment was also emphasized. The ideas of the Coan master were gradually made part of a rigid system, and the so-called dogmatic school was formed.

In the Sicilian school of medicine, Philistion of Locri (fourth century B.C.) continued to develop the thought of Empedocles. It was this medical school that influenced Plato in the vague references to the life sciences that appear in the Timaeus.

Aristotle and the founding of the Lyceum

The Athenian crisis grew worse towards the middle of the century. Tension increased between the aristocrats and the demos. Financial resources had diminished with the decreased exploitation of the allies and the currents of international trade, and the rulers were trying to raise money by means of public and private borrowing, taxes on the metics (foreigners without political rights), and commercial speculation. But the entire system of the poleis showed serious signs of weakening. Spartan hegemony had disappeared with the battle of Leuctra (371 B.C.), and Thebes, Athens, and Sparta were all unable to launch a consistent campaign to consolidate the various poleis, at a time when Macedonian power was growing on the northern borders of Greece. Philip, and his son Alexander after him, were the dynamic elements of the Greek political situation. They first subjugated the various poleis, then built a vast empire that absorbed the Persian Empire and pushed as far as the Persian Gulf and India. This development in political history rendered meaningless the political aspects of the Academy little more than a decade after Plato's death (347 B.C.). Macedonian power gradually put an end to the social conflicts within Greek cities and guaranteed to the existing order a stability that it had never before known. But it also replaced the ideal of the wise ruler with that of an efficient bureaucratic and military administration. The philosopher's new role—that of independent thinker or of counselor to the monarchy—was much different from what it had been in the polis, as Plato had envisioned it.

It is not surprising, then, that the Academy's importance was considerably diminished, at least on a political level. But this did not impede the development of new ways of organizing knowledge, with different views of the relationship between culture and political life, which in certain respects prefigured the Alexandrian age. The Lyceum, founded by Aristotle (c. 384-322 B.C.) in 355 B.C. at Athens after he had left the Academy, was one of the places where these ideas were formed. The gymnasium in which Aristotle founded his school of philosophy was dedicated to Apollo Lyceios, but the school was also called Peripatetic, from the custom of holding discussions while walking in the gardens surrounding the Lyceum. Its organization was quite different from that of Plato's Academy. The students no longer had any shared religious ties, common rules of life, or political projects that they could work on together. Many of them, like their master, were metics, and therefore excluded from active participation in the life of the polis. The Lyceum was also less open to dialectical discussion than the Academy had been, since the doctrine and teachings of Aristotle quickly assumed an official character, but it thus left more room for the specialized research of scholars under the guidance of the master.

Aristotle's encyclopedia

The impressive series of works by Aristotle, which cover almost all the fields of knowledge, is closely connected with his teaching in the school. Some of his earlier works were dialogues similar to Plato's, written for publication but no longer extant. But the core of his knowledge is contained in treatises, consisting of transcriptions of the courses he taught, first at the Academy and then at the Lyceum. Thus there arose a new literary form for philosophical and scientific writing, since the treatises were not intended for publication, but for the students of the school. They contained no secret knowledge, as a certain later tradition would have it, but actually formed an essential encyclopedia for the students of the Lyceum. Aristotle's purpose was to compile a body of universal knowledge that brought together and arranged all the knowledge then available to mankind, without necessarily subscribing to the deductive-reductionistic model—such as that of Democritos or of Plato in the Timaeus—that attempted to derive explanations of phenomena from a theory of the basic elements of reality. Aristotle was critical of atomism and Platonism, preferring instead an investigation of the empirical world, both natural and human, that would preserve the autonomy of the individual disciplines and at the same time assign a greater role to experimental research. This is why Aristotle and his pupils dedicated themselves to an encyclopedic project. They collected a vast quantity of scientific material, which they organized and presented, in strictly logical form, in the courses and treatises of the school. Hence the interest in logic, which made of Aristotle's Organon the first great systematic treatise on the subject; hence the works on biology, physics, astronomy, and psychology, as well as the profound interest in ethics, politics, and rhetoric. Hence too the critique of Platonism and the elaboration of a new ontology in the Metaphysics based on the notions of matter and form, potentiality and actuality. Aristotle's encyclopedic method, which he opposed to the deductive methods of both Democritos and Plato, did not, however, signal the end of the search for a unified form of knowledge. Far from being merely an accumulation of disparate notions, the encyclopedia was a project that found its unity in philosophy, the supreme theoretical activity, which was capable of finding the thread connecting the various autonomous sciences and, through them, of recognizing the structure and order of the world. Philosophy was therefore a reflection on language (and thus a logic, an epistemology, an ontology), which finally became a theology, a unified philosophical discourse that ended with a discussion of God. It is impossible to provide here a brief summary of Aristotle's major concerns: they extend from his discoveries in logic to a theory of science, from an impartial analysis of nature, coming-into-being, and life to the fusion of cosmology and theology in considering the cosmos, from a conception of God as the "thought of thought" to a rich anthropological research.

His reflections on infinity arrived at the famous distinction between actual infinity and potential infinity. Aristotle recognized only the latter as a concern of science, a thesis that remained influential until the nineteenth century.

Aristotle's inquiries into continuity were also tied to this distinction, where continuity was seen as a potential, not an actual, factor. On the basis of this conception, Aristotle also ventured upon a computation of Zeno's paradoxes.

The development of the Lyceum

Even after the death of Aristotle, the Lyceum remained the main center of cultural life in Athens which had come under Macedonian rule. Aristotle's successor as head of the school was Theophrastos of Eresos (c. 372-287 B.C.), assisted by Straton, Eudemos, and Dicaearchos. Theophrastos had already been officially responsible for the school while the master was still alive, since Aristotle was a metic, that is, a non-Athenian, and could not legally direct the school. After Aristotle's death, however, Theophrastos was not able to maintain a balance between specialized scientific research and the unifying activity of comprehensive philosophical thought—a balance that had united the various pursuits of the Lyceum.

Theophrastos, who wrote basic treatises on botany and who can in some ways be considered the initiator of doxography (the history of doctrines), was sometimes citrical of the "first philosophy" (metaphysics), which had been a major unifying factor in Aristotle's thought. The same passion for highly specific work, for the vast and systematic collection of documents, reports, and information, and the same caution with respect to premature speculative syntheses, also animated Aristotle's other early disciples, such as Eudemos of Rhodes, who wrote *histories of geometry*, *arithmetic*, and *astronomy*; Aristoxenos of Tarentum, author of numerous books on music; and Dicaearchos of Messene, who wrote, among other works, *Life in Greece*. It is extremely interesting that Dicaearchos posited the superiority of politics to the contemplative life, while a later Aristotelian, Straton of Lampsacos, considered scientific research the theoretician's first concern. But whereas Straton, the philosopher-scientist, had a long and prosperous career in the Museum of Alexandria, Dicaearchos, the philosopher-politician, became an idol of the schools of rhetoric rather than a major force in the new Hellenistic world. As a matter of fact, the Lyceum itself became a school of rhetoric and grammar under the direction of Lyco.

Subsequent developments in astronomy and medicine

Whereas there do not seem to have been any innovations worth mentioning in the field of mathematics between Eudoxos and Euclid, the same does not hold true for astronomy and medicine, which were closely connected to earlier research.

The followers of Eudoxos were faced with the task of trying to improve upon the master's explanations without giving up his basic astronomical concepts.

Callippos of Cyzicos (c. 370-325 B.C.), who also lived in Athens, where he worked with Aristotle for a while, increased the number of the spheres of Mars, Venus, and Mercury, thus eliminating the inconsistencies that were becoming progressively more evident through more precise observation. The works of Eudoxos and Callippos put an end to the concept of astronomy as merely a cosmology or a cosmogony grafted onto a philosophy of nature. It became instead the development of mathematical theories to explain phenomena, shaped by lengthy, accurate, systematic observation, which in turn was used to verify the theories themselves.

Whereas Eudoxos and Callippos presented a mathematical system of the world, Aristotle provided an exclusively physical interpretation of Eudoxos' theory of the spheres (which had increased to fifty-five from Callippos' thirty-three). The spheres were considered solids, to which their respective heavenly bodies were rigidly joined. Motion was transmitted mechanically from one sphere to another within each system, but this did not affect the motion of neighboring systems, since for each sphere there was a corresponding one moving in the opposite direction, which neutralized these potential effects. The earth was round, according to Aristotle, as evinced by the shape of its shadow projected onto the moon during eclipses. It was immobile and at the center of the universe, in a space delimited by the sphere of the moon. Meteorological and astronomical phenomena such as shooting stars and comets occurred in this sublunary region, since the only possible form of change beyond it was uniform circular motion. Aristotle's cosmos obviously corresponded more to the needs of his general philosophical system than to a scientific theory for organizing and explaining a particular set of phenomena.

Oenopides of Chios, roughly a contemporary of Democritos, is sometimes credited with the discovery of the obliquity of the ecliptic. He defined the length of the Great Year as fifty-nine solar years, each of which he calculated to consist of 365 days.

Our picture of fourth century astronomy would not be complete without mentioning Heraclides of Pontos (c. 388-315 B.C.), a Platonist who was a contemporary and adversary of Eudoxos. Heraclides' system was an important attempt, though limited to the inner planets, to resolve the problem of the varying distance of the planets from earth (obvious from the periodic variations of their brightness). He proposed that the planets Venus and Mercury revolved around the sun, rather than the earth. The sun itself revolved around the earth, and the two planets revolved in the same direction around the sun, in spheres with smaller radii. This was the first formulation of a partially heliocentric system. Under the name of the Egyptian system, it was proposed as a compromise solution during the great debate of the sixteenth and seventeenth centuries on the true nature of the cosmos by such astronomers as Tycho Brahe.

In medicine, Diocles of Carystos, a contemporary of Aristotle, was considered in antiquity to be the most famous physician between Hippocrates and Galen. He was instrumental in bridging the gap between the Hippocratic system and Peripatetic philosophy. Toward the end of the century, Praxagoras of Cos was the most renowned representative of the dogmatic school. He did research on the pulse and distinguished arteries from veins.

As mentioned above, Theophrastos of Eresos, Aristotle's successor as director of the Lyceum, compiled works on botany and mineralogy, thereby providing important information on the nature, classification, and properties of various plants, as well as on phenomena dealing with vegetation and cultivation.

Other developments in philosophical thought

During the last decades of the century, in addition to the schools of Plato and Aristotle, the minor Socratic schools were also actively developing their doctrines. The teachings of Aristippos the Younger, grandson of the founder of the Cyrenaic school, constituted a truly hedonistic system. Diogenes of Sinope (c. 413-323 B.C.), nicknamed "the mad Socrates," enjoyed considerable popularity in the Cynic school. He emphasized individualism in ethics and inveighed against social institutions, as Antisthenes had already done. A distinct interest in logico-linguistic discourse persisted in the school of Megara. There Eubulides of Miletos, a disciple of Euclid, developed several arguments that emphasized the difficulties of demonstrative discourse, such as the famous antinomy of the liar and the argument by the "sorites" (or "heap"). Diodoros Cronos developed a critique of Aristotle's theory of contingent futures, and Stilpo (flourished c. 380-300 B.C.) criticized Plato' theory of forms as well as Aristotle's theory of predication.

Despite the considerable differences among them, the Platonic, Aristotelian, and minor Socratic schools were all part of a purely Greek tradition. The school founded by Pyrrhon of Elis (c. 360–270 B.C.) in 324 B.C., on the other hand, was perhaps more directly influenced by the new political situation created by the Asian conquests of Alexander the Great. Although Pyrrhon was in contact with disciples of Democritos and with the Megarians, he seems to have been influenced by the Indian gymnosophists. His "skepsis"—an investigation aimed at discrediting all philosophical doctrines that tend to fix the nature of things by means of rigid formulas—seems to have been inspired by Indian teachings. With the disintegration of the old ethico-political tradition, and in light of the new situation, Pyrrhon advocated individual happiness through self-control and mastery over the passions, as well as through the renunciation of all sciences with any pretensions to objective knowledge.

Epicureanism and Stoicism arose shortly after the Skeptic school, with which they shared several characteristics.

300–200 B.C.

UNITY OF KNOWLEDGE AND THE AUTONOMY OF THE SCIENCES

Fundamental to Aristotle's concept of science was his theory of the autonomy of the single sciences. Far from being hierarchically arranged in a pyramid, the sciences maintained their reciprocal independence without any one of them dominating the others. Each science had its object of investigation as well as its own principles that were not reducible and could not be derived from other fields of knowledge. Sciences could not be derived from each other; at the most, placed side by side they constituted an entire range of fields within which Aristotle attempted to compile an encyclopedia that included all knowledge arranged in an orderly and analytical manner.

The work done at the Lyceum to define the various fields of knowledge and the principles of each science was carried out by a number of researchers doing specialized research, analyzing and re-elaborating the material. The method used at the Lyceum consisted of collecting information, classifying it, analyzing its historical development, and carrying investigations increasingly further by using new material and specializing even more. The search for specialized knowledge and the process of organizing Peripatetic knowledge into researches devoted to particular studies professionalized scientific research in a way that was directly antithetical to the Socratic concept, according to which education should be communal, tied to the *polis*, non-institutionalized, and directly functional to the policies of the city.

Aristotle's concept of the autonomy of the sciences—radically different from previous structures, especially Plato's essentially deductive and hierarchical idea of science—did not, however, lead to a negation of the unity of knowledge. There were elements in encyclopedic ordering that formed a framework into which specific events could be fit. The entire order of reality, in all its stability and rationality, was guaranteed by the unifying and organizing nature of philosophy, which overcame the disconnectedness of the specific concerns of each separate science.

In his theorizing, as well as in his organization of the research conducted at the Lyceum, Aristotle was able to maintain a balance between the development of specialized research in each field of science and the unifying moment of philosophical reflection. Theophrastos, however, and Straton even more, abandoned all attempts at unification in favor of specific dedication to the individual sciences.

Thus within the Lyceum under Aristotle's immediate successors, unity of knowledge gave way to the development of the various sciences in their specificity. This tendency was even more marked at the Museum at Alexandria, where sciences such as astronomy and biology became very highly developed (this trend will be discussed in THOUGHT AND SCIENTIFIC LIFE 300–200 B.C.). The thrust that specialized studies gave to the individual disciplines was very much a part of the more general separation of scientific research from political and social life, and of "pure" research from the technical applications of scientific knowledge. Research was divided into increasingly more specific fields that no longer claimed to be tools for understanding all forms of reality. In our analysis of pre-Aristotelian knowledge, it was difficult to distinguish the different fields and follow the development of various disciplines and sciences. In post-Aristotelian theory, on the other hand, we can single out various sciences that progressively gained autonomy at the cost of a unitary discourse on reality as a whole. So while our discussion of pre-Aristotelian theory could only give a total unitary picture of knowledge in its various forms, from the third century on the boundaries that distinguish the various fields of knowledge can be much more clearly seen.

This is why we have treated science as a whole up to this point, while from here on we must divide it into disciplines. Our format will now consist of various sections, each dealing with a particular science.

THOUGHT AND SCIENTIFIC LIFE

The transformation of the ancient world: Hellenism

Alexander the Great's ventures brought about a radical transformation in the eastern Mediterranean. Political, economic, and cultural horizons were considerably widened. Huge masses of people moved from Greece into Egypt and Asia Minor: first the soldiers, who remained in the newly-conquered lands, then the politicians, engineers, artists, merchants, contractors, and artisans who were drawn by the possibilities of greater wealth or solicited by the Hellenistic government, which wanted more Greeks in those areas. But while the *polis* were declining in Greece, unable in the long run to compete with the great Hellenistic states, and the main roads of communication became those running toward the south and the east, the cities of Asia Minor, such as Pergamos, Antioch, and Rhodes, began to flourish anew. Alexander's new urban centers, such as the Egyptian city that was named after him, also became increasingly important. The institutions which developed in these cities were very different from those in the Greek *polis*. A broader political struggle replaced the dialectical exchange between the aristocracy and the *demos*. The situation was mediated by creating an impressive state apparatus, with an absolute monarchy at the top and a capillary bureaucratic administration as its base, following an oriental, rather than a Greek, model. This restructured state excluded any active political intervention on the part of the population and did not even consider taking advice from the philosophers. As a result, political eloquence and politico-social comedy declined, while historiography was reduced to rhetoric or scholarly works.

Literature, like theater, was no longer written for an assembly of the *demos*, but for small circles of scholars. Their tie with the *polis* destroyed, philosophers tended to transform themselves into wise citizens of the world, but they were isolated with respect to the centers of power. Scientific research, which had developed considerably during the fourth century B.C. even in the midst of grave political and economic crises (a large part of the *Hippocratic Collection* had been edited, Eudoxos had made extremely important contributions to mathematics and astronomy, and Aristotle and Theophrastos had furthered investigations in biology), now leaned increasingly toward specialization.

The Museum and Library at Alexandria

One of the phenomena that characterized the new epoch the most was the development of Alexandria. Under the reign of Ptolemy I Soter, artists, scientists and technicians began to flock to the city from all over the Greek world. Demetrios of Phalera, the main representative of the pro-Macedonian party in Athens, who had been exiled in 301 B.C. during a democratic revolt, was appointed cultural organizer.

Demetrios, a disciple and friend of Theophrastos, invited Straton of Lampascos to Alexandria to teach the heir to the throne. Straton was thus completely separated from Athens. A large part of the Lyceum library was moved to the new capital of Egypt, and the Lyceum was deprived of the best part of its patrimony. At the same time, Demetrios conceived of a plan to concentrate all the scientific and bibliographic material that could be found in the various cultural centers of Greece and Asia Minor into one huge institution, modeled after the Lyceum but on a much larger scale. The material would be made available to regularly salaried teachers. The Museum and Library at Alexandria (the Pythagoreans' refectories were originally called "museums") had their roots in this idea, though they were only officially founded under the reign of Ptolemy II.

The museum soon had comfortable reading rooms, anatomy rooms, an astronomic observatory, and zoological and botanical gardens. The library grew rich with precious volumes, which were well-ordered and well-preserved, even though each one was a large roll of papyrus and took up a considerable amount of space.

Features of Hellenistic science. Specialization

The great flourishing of Hellenistic science in the third century took place in such institutions as the Museum at Alexandria and others like it that were built shortly after in Rhodes, Pergamos, and other cities.

It was the age of Euclid, Eratosthenes, Archimedes, and Appolonios, of bold conjectures on the solar system and great research programs in mathematics and mechanics. The sciences acquired a definite autonomy with respect to philosophy. Specialization also meant thoroughness and rigor, the acquisition of new sectors of knowledge, and the discovery of new and revolutionary facts. But the process was accompanied by a progressive loss of awareness of the close ties between science and philosophy so that scientific findings were considered in their particulars apart from any reference to a philosophy or a unitary global view. What resulted was not only a new concept of the scholar, but an increasingly clear division between philosophic activity and scientific research. Philosophy began to close in on itself, concentrating essentially on moral questions, while science was broken down into increasingly more specialized fields of research, which lead finally to pure erudition.

Philosophy: Epicurians and Stoics

The effects of this division were mostly felt during later centuries. But even within this epoch, traditional philosophy was centered in Athens, while science, the Aristotelian heritage, was centered in Alexandria. This did not mean that the new philosophical currents were not concerned with science, for example physics, logic, and psychology. But the often penetrating theoretical research into logic and the empirical sciences was almost entirely motivated by the belief that people could better face the basic problem of life after becoming aware of what the world really was and freeing themselves from fears and illusory hopes. Thus Epicuros of Samos (342-270 B.C.), who founded a school in Athens in 306 B.C. (the so-called Garden), was mainly interested in morals, and combined Skeptic and Cyrenaic themes. The doctrine of pleasure (*hedone*) was linked to the problem of criteria, i.e. how to distinguish between genuine knowledge on the one hand and pseudo-science and superstition on the other. Epicuros' concept of the world seems to be taken primarily from atomism. He was concerned with a vast and varied set of problems centered around the main problem of human happiness, where human beings were taken as individuals apart from both their acquisition of objective knowledge and any political context.

Many elements of Epicureanism (knowledge through the senses, the theory of anticipation, the essential function of memory) were shared by Stoicism, which also aimed at individual happiness. But as Cicero pointed out in the first century B.C., what the Epicurians called pleasure, the Stoics called virtue. In any case, the Stoics of this period still showed an interest in learning about nature. Recollections of non-Greek concepts often appear, as in the works of Zeno. Around 300 B.C., Zeno of Citium (c. 335-263 B.C.), founded his school in a porch (Stoa) in Athens that was decorated with the paintings of Polygnotos. Zeno was called "the little Phoenician" by his contemporaries, because of his clearly non-Greek countenance. With Zeno and his successors Cleanthes of Assos (c. 330-232 B.C.) and Chrysippos of Soloi (c. 281-208 B.C.), theories on the eternal fire, the cyclic nature of the universe, and the eternal return of human events contributed to a finalistic view of the universe which mixed both Greek and Asian themes. The criteria for human conduct were to be found by recognizing this finality.

Unlike Epicuros and his disciples, the Stoics maintained that correct discourse was an essential factor in the ascent towards wisdom. This is where the interest in logic displayed by Zeno and especially Chrysippos, the last of the so-called old Stoa, came from. They studied Aristotelian logic and revived it in an extremely original way. Their ideas influenced studies in logic at least through the Middle Ages. As the events in the life of Zeno, who was a friend of the Macedonian king Antigonus, exemplify, the Stoics did not preach an Epicurean-type hidden life, far from the affairs of public life, but rather developed a theory of political life that included the new cosmopolitan themes of Alexandrian culture. Although they showed a keen interest in the philosophy of nature, the Stoics too were unable to develop ties with specialized scientific research. Cleanthes' harsh criticism of Aristarchos (see ASTRONOMY 300–200 B.C.) shows the philosophers' lack of understanding of the more daring doctrines of Hellenistic science.

Skeptics and Academicians

The more radical post-Socratic schools—the Cynics, the Cyrenaics and the Megarics—still flourished during this century. Their relation to Socrates, however, became increasingly paradigmatic, as they rejected a school, which was by then fixed in rigid tradition, and at the same time also refused civil cohabitation in a sometimes desperate search for individual freedom. The negative aspects of the doctrine of pleasure and happiness ("the wise will not so much look for good as avoid evil") were particularly accentuated in the Cyrenaic school, with Theodoros the Atheist, Hegesias "the persuader of death," and Anniceris. There were also criticisms of religion, often in relation to Epicureanism, which viewed religion as the source of fears and superstitions that had to be done away with. This seems to have been one of Theodoros' central themes, and Euhemeros' famous Ἱερὰ ἀναγραφή (Sacred History), written around 280 B.C., is also related to Theodoros' ideas. Euhemeros' work explains that myths are created by people's admiration for those among them who particularly excel in power and virtue.

Many radical Socratic themes reappeared among the Skeptics. Their moral commitment to total liberation was based on the belief that an unbiased critique of all science was the most solid foundation for ethics. In this vein, Timon of Phlius (c. 320-230 B.C.) revived Pyrrhon's ideas and in his Silloi, or Satires,

ridiculed dogmatic philosophers, Stoics and Epicureans included. But the most interesting cultural novelty in this current was the doctrine of Arcesilaos of Pitane (c. 315–241 B.C.), a scholarch of the Academy, who brought Skeptic interests and teachings into the school. Most of his ideas were based on Plato's early dialogues; the dialectic use of Socratic irony to answer the adversary's arguments became a prototype for correct philosophic behavior. The philosopher, taking into account the impossibility of distinguishing between true and false representations, moved from a critique of tangible certainty to epoche, or suspension of judgment. With Arcesilaos, who founded the so-called Middle Academy on this Skeptic note, philosophers no longer attempted to define, even in a negative sense, the nature of things, but rather argued that nothing at all could be said with certainty in the realm of philosophy.

MATHEMATICAL SCIENCES

The apogee of Hellenic mathematics

Alexandria's cultural predominance during the centuries between the death of Alexander the Great and the Roman conquest was also felt in the field of mathematics. Not only did regular mathematical research and teaching go on in the two famous cultural institutions of the city, the library and the museum (Eratosthenes and Apollonios were both directors of the library), but above all because the city itself had attracted almost all of the major scientists of the time. Euclid (flourished c. 300 B.C.) taught at Alexandria. His Στοιχεῖα (Elements) is certainly his most famous work, and the one that has had the most long-lasting influence on the entire history of science. The thirteen books of the Elements give an organic synthesis of all previous geometric, as well as arithmetic and algebraic, mathematical knowledge. For this reason it is difficult to establish which of the theories of the treatise are actually Euclid's; in fact, it seems that almost all of its contents were already known. In any case, the work of synthesis and the absolutely rigorous and inflexibly demonstrative arrangement are unequaled and have made the Elements a model for future generations. Euclid applied the axiomatic method almost perfectly; on the basis of the definition given by Aristotle in the Posterior Analytics (see the section 400-300 B.C.), this was the most perfect form of science, and Euclid's work explicitly established the notion of mathematical demonstration, which remained definitive for centuries. The method started with evident propositions (axioms and postulates) and arrived at others through proper formal arguments, without using any information that was not given in the initial propositions or others that had already been demonstrated. In this way, the demonstration was an absolute guarantee of the truth of its results. Euclid's own faithfulness to the method caused him to question the evidence of his postulate on parallel lines. He was not totally convinced of it, and not being able to demonstrate it, did not use it until the twenty-ninth proposition of the Elements. Hosts of mathematicians labored over this problem for centuries after. The result was the creation of non-Euclidian geometry in the twentieth century, and the abandonment of the criterion of proof as the basis for mathematical axioms.

Many mathematicians have attempted to reconstruct the contents of Euclid's Δεδομένα (Data) and his Πορίσματα (Porisms), two lost works. Euclid is also credited with writing an 'Οπτιχὰ (Optics) and a Κατοπτριχὰ (Catoptrics).

The work of Archimedes

Euclid's work was typically Alexandrian in its scholarly, systematic and treatiselike nature, and in the way it codified already-known information rather than presenting the results of original research. Archimedes (287-212 B.C.), on the other hand, worked to solve new problems and old questions that had not yet been answered.

Archimedes' contributions to traditional problems included some concerning circles. He did not try to square the circle using ingenious constructions, but only attempted to measure it with an extremely high degree of accuracy. By considering both inscribed and circumscribed polygons, he reached very close approximations of the value of π.

These findings served to refine Euclid's Elements, as did Archimedes' treatise Περι σφαιρασ χυι χυλινδρον (On the Sphere and Cylinder). Quite advanced considerations on different types of surfaces of rotation are also developed in that work, as well as in Περι χοινοειδειον χαλ αφαιρο̧ειδεοη (On Conoids and Spheroids).

Other famous works by Archimedes are Περὶ ελνχων (On Spirals) and Τετραγωνισμοσ παραβοληρσ (On the Quadrature of the Parabola). The former defines the famous logarithmic spiral, while the latter gives rigorous geometric proofs of findings on the spiral that Archimedes had made previously using statics. These are explained in his 'Εφοδοσ (The Method). Archimedes had actually founded statics with his two works 'Επιπεδοι ισορροπολ (On the Equilibrium of Planes) and Περι ὀχουμεων (On Floating Bodies); the latter also laid the foundations of hydrostatics).

Thus Archimedes can be considered the most complete mathematician of the ancient world. His geometric and deductive rigor was perfect, and he was extremely inventive and surprisingly free in his choice of objects for investigation. In his numerous calculations of areas, volumes and barycenters, he treated the problem of infinity with unequalled logical rigor. In addition to all this, he possessed an engineering mentality, likening him to several illustrious representatives of Alexandrian science (e.g. Heron). Nor did he lack interest in arithmetical questions: in Ψαμμιτησ (The Arenarius) he demonstrated—with a procedural rigor and sureness that exemplify his greatness in this area of mathematics, as well—how the Greek notation in use at that time could, with the introduction of suitable ingenious conventions, express numbers of any size whatsoever.

Appolonios

The other great member of the exceptional mathematical trio of this period was Apollonios of Perga (c. 262-190 B.C.). Besides several lost works on geometry, he also wrote a treatise called Τὰ χωνιχά (Conic Sections), which completely surpassed all previous works in its perfection (it seems that Euclid and Archimedes had also written works on the subject). The work gave nomenclature and demonstrations for all the essential geometric properties of conic sections, which remained basically unchanged until the invention of analytical geometry. The work's purity of style and demonstrative rigor, in addition

to its lofty scientific content, make it one of the finest products of mathematics (not only Greek); without a doubt it represents the apex of Hellenic geometry.

Apollonios' exceptional mathematical competence explains his greatness in the field of astronomy, as well (this was also true of Eudoxos).

Other noteworthy contemporaries have been somewhat overshadowed by these great men. One of these is Eratosthenes (see ASTRONOMY 300-200 B.C.), a friend and correspondent of Archimedes, known for his studies on the calendar. He also wrote a lost work on proportions, and invented the "sieve" for finding successive prime numbers.

Notable contributions to mathematics also came from other disciplines in this century of ardent mathematical research: Aristarchos' almost-trigonometric calculations of ratios (see ASTRONOMY 300-200 B.C.) and Heliodoros of Larissa's mathematical considerations in his treatise on perspective are good examples.

ASTRONOMY

Aristarchos of Samos' heliocentric theory

Astronomy was a particularly flourishing discipline at Ptolemy Soter's splendid cultural center, the Museum at Alexandria. Alexandrian astronomers were by then very different from the Ionic philosophers of nature, as well as from the creators of the great philosophical syntheses, such as Democritos, Plato, and Aristotle. They were real specialists who needed very elaborate structures for their research. The great library and the observatory provided these structures. All the observations used by Claudius Ptolemy (second century A.D.) in his great astronomic synthesis were made and systematically recorded in the Alexandrian observatory.

One of the first teachers at Alexandria was Aristarchos of Samos (c. 320-250 B.C.). Very little is known about him. He was a pupil of Straton of Lampsacos, "the physicist" and third scholarch of the Lyceum after Aristotle and Theophrastos. Ptolemy tells us that Aristarchos observed the summer solstice in 280 B.C., the date that is usually given for his akmé. His only extant work is On the Magnitudes and Distances of the Sun and the Moon, a brief treatise in which the measurements referred to in the title are calculated using the angular distance between the sun and the moon when the moon is half-full, i.e. when the earth, moon and sun form a right-angle triangle. The work combines direct observation and mathematical methods in a style typical of the Alexandrian school of astronomy. But Aristarchos is best known for his heliocentric theory, which started from the idea of the sun as the central fire, proposed by Philolaos and the Pythagoreans and extended the heliocentric theory advanced by Heraclides of Pontos to all the planets. The work in which Aristarchos put forth this theory, a prelude to Copernicus, has been lost; and we have only indirect information about it. Archimedes' testimony is particularly valuable, although he does not seem to have understood the importance of the theory. In the Arenarius, Archimedes summarizes Aristarchos' thought with the following comment: "his theory is that the fixed stars and the sun remain immobile, that the earth revolves around the sun following the circumference of a circle, and that the sun is at the center of this orbit.'' We do not know what arguments Aristarchos advanced in support of his theory, but as we have already said, its importance was not at all recognized. Aristarchos was essentially isolated, and was often the brunt of harsh attacks. For example, the Stoic philosopher Cleanthes (head of the Stoa from 264 to 232 B.C.) attacked the heliocentric theory as contrary to a finalistic view of the universe. In his *Against Aristarchos*, he argued that the Greeks ought to charge Aristarchos of Samos with blasphemy, for trying to change the center of the universe.

Eratosthenes, Apollonios, and Aratos

Eratosthenes (c. 275–195 B.C.), a librarian under Ptolemy III Evergetes, attempted in his *Περὶ τῆς ἀναμετρήσεως τῆς γῆς* (*On the Measurement of the Earth*) to measure the earth using methodological criteria analogous to Aristarchos'. He also set the value of the obliquity of the ecliptic at $23°51'$.

Ptolemy mentions Apollonios of Perga (see MATHEMATICAL SCIENCES 300–200 B.C.) in connection with a new system for explaining the motion of the planets. This was the theory of epicycles, which we shall return to when we talk about Hipparchos; it later replaced the theory of eccentrics that Pythagorean astronomers used at the time of Heraclides of Pontos and immediately thereafter.

The poem *Φαινόμενα* (*Phenomena*) by Aratos of Soli (c. 315–245 B.C.) was a popularizing work about astronomy and meteorology, written in the Stoic philosophical tradition (Aratos was a disciple of Zeno the Stoic). In later epochs, Aratos' poem was considered more a book for study than for pleasure, and famous astronomers such as Hipparchos wrote scientific commentaries on it. In Rome, it not only influenced scholarly poetry (Virgil's *Georgics*, for example), but also aroused an interest in problems of astronomy among the learned elite. Cicero, Germanicus, and others translated it into Latin, and, through these translations, it remained well-known in the West throughout the Middle Ages.

It was also during this century that the Greeks began dating years by olympiads (the first olympiad was considered to have begun July 1, 776 B.C.).

PHYSICAL SCIENCES

The physical sciences and Alexandrian culture

Before the Hellenistic period, it is impossible to find a set of theoretical ideas that could be called an early form of physical science as the term is commonly used, except perhaps in some of the doctrines of Aristotle's *Corpus*. Unlike mathematics, medicine, and astronomy, a specifically physical sector of research cannot be found prior to Aristotle (see the section 400-300 B.C.), or even perhaps prior to his successors or the first great Alexandrian scientists.

Only during this period was physical research done for the purpose of establishing articulated theoretical systems that could be checked against systematic observations and experiments on phenomena and specific bodies. Any attempt at explaining why no such research took place before this time is pure conjecture, since adequate sources are lacking. However, the fact that a physical science aware of its specificity—and therefore not directly connected to a particular philosophy, cosmology, or metaphysics—emerged at this time is certainly in keeping with the more general context of Alexandrian scientific culture. Knowledge was continually being divided into distinct categories, and rigorous specialized research had replaced the vast syntheses and the great, generalized philosophical systems of earlier times. It is not surprising, therefore, that a taste for experimentation and checking out theories against experiments developed in this type of scientific atmosphere.

Reference points of Alexandrian science: The great systems of Democritos and Plato

The great systems of Democritos and Plato had quite a different and much more important influence on Hellenistic science. According to Democritos (see the section 400-300 B.C.), who was perhaps polemicizing with Heraclitos' concept, all things were made of full atoms separated by space or a void. The atoms and the void were thus primary realities, while the qualities perceived in things were produced by the movement and the different ordering of the atoms, which could vary in shape, position and order. Since everything was made up of eternal and immutable atoms, all birth and death was nothing more than different combinations of atoms. The Platonists' constant struggle against Democritos' doctrine, the later introduction of a teleological view—that of Aristotle, for example—and the hostility of numerous philosophical circles toward materialism created a tradition according to which Democritos attributed the cause of the world's existence to chance. This tradition lasted for the entire Alexandrian period and even longer. What Democritos really believed, however, was that everything concerning the formation and development of the universe was regulated by a mechanical necessity (the movement of matter was original and eternal). This natural necessity could be considered chance only by those who proposed the intervention of an external force or chance as an alternative. Democritos' system did not propose such an alternative; instead it represented an articulated mechanistic system, where a sort of distinction between primary and secondary qualities (to use the classical terms of the seventeenth century scientific revolution) opened the way for applying mathematics to the study of natural phenomena. During the third century B.C. Democritos' ideas on physics were revived in the teachings of Epicuros.

In many ways, Plato's natural philosophy (see section 400-300 B.C.), more than Democritos' atomism, represented an extension of ideas that had already been held by earlier thinkers, especially the Pythagoreans. In particular, the methodological theory that dominated theoretical astronomy for centuries under the name of the Platonic axiom was motivated by the unconditional support of many Pythagorean ideas. The method aimed at discovering the ideal mathematical system of uniform circular motion that was the true reality of the cosmos underlying the confused irregularity of planetary motion. This was the reason for the interest in astronomy and music, which were considered applications of mathematics. But there is, in the *Timaeus*, also a corpuscular theory of matter which is in many ways similar to Democritos' theory. Plato also thought that atoms were defined portions of a primordial material whose only features were those of extension in space and consequently impenetrability. Democritos' and Plato's theories are also similar in the way these elementary bodies are used to explain concrete physical phenomena. But they differ radically in that Plato's bodies seem not to be indivisible. The theories of the Demiurge and harmony are further points of difference.

Points of reference of Alexandrian science: Aristotle's physics

Throughout antiquity and the Middle Ages, the ideas from Plato's *Timaeus* prevailed over those of Democritos. But Aristotelian tradition soon joined the Platonic. For our purposes, the most interesting treatises are those containing Aristotle's physical doctrine, i.e. the *Physics*, *On the Heavens*, *On Generation and Corruption*, and the *Meteorology*, to which we can add the *Problems* and the *Mechanics*, two collections of mechanical problems in the form of questions and answers (both works are of doubtful authenticity). Altogether, they form an ample synthesis—an attempt to arrange all the physical knowledge of the times, referring to and, when necessary, confuting, prior beliefs. These works were an obligatory point of reference for Alexandrian science, which seems natural if we keep in mind how Aristotle's teachings and the structures of the Lyceum were all a part of the cultural policies of Alexander and his successors. The critical and philological attitudes of Alexandrian scholars in the third century and after caused physicists to focus their interests on Aristotle's doctrine in particular.

It is worth noting that the attention Aristotle gave to meteorology precociously raised it to the status of an autonomous science and added it to the set of disciplines considered worthy of being studied and taught during the eras of medieval and Renaissance Aristotelianism. It may also explain the privileged position that meteorology held until near the end of the nineteenth century in all of European culture, from the great naturalists and physicists down to elementary education.

One of the distinctive features of Aristotle's doctrine that remained paradigmatic for later Hellenistic research was the belief (in antithesis to the atomists) that matter was continuous. Furthermore, all earthly things were made up of four elements— earth, air, water, and fire—which in turn contained the four qualities of heat, cold, dryness, and humidity, combined in pairs. Here Aristotle had evidently adopted an Empedoclean concept that would suit Christian as well as Jewish and Moslem thought. The stars and planets moved with a circular motion in concentric crystalline spheres around the earth. The circular movement was perfect because the circle is the perfect figure. It represents the eternal and unchangeable order of the heavens in contrast to the rectilinear movement that prevails on our changeable and imperfect earth (''where imperfection ends the heavens begin''). This doctrine dominated astronomy until the time of Kepler (see ASTRONOMY 1600-1610). Finally, Aristotle considered the universe to be limited in space, in the sense that it was contained within an external sphere. However, it was unlimited in time, not being subject to creation and destruction. This was another concept that dominated during Hellenistic-Roman antiquity and the Middle Ages. Only late-Renaissance cosmologists (Giordano Bruno, for example), would question the limitedness of the universe.

Specific characteristics of Aristotelian dynamics

The Aristotelian science of motion dominated physics for around fifteen hundred years, up until the time of the modern scientific revolution. By motion, Aristotle meant any quantitative or qualitative change that caused a phenomenon to happen. The change in position of a body in time caused local motion, and a distinction was made within local motion between natural motion and violent motion. Aristotle thus broke down the homogeneity of motion and, as a result, the homogeneity of the world. In the sublunary world, things were born, corrupted, and died, while in the etherial world, order was incorruptible and eternal. Natural motion was divided into rectilinear motion and circular motion. Circular motion took place in the heavens, while rectilinear motion was found in the sublunary world (the falling of heavy objects, the rising of light bodies). The explanation was teleological: celestial motion was explained by the complex doctrine of the motionless mover; but for rectilinear motion, Aristotle introduced the theory of natural places (flames naturally move upward, rocks naturally fall downward).

Some aspects of Aristotle's kinematics clearly foreshadowed modern theories, such as his description of the trajectory of missiles. According to Aristotle, the trajectory was made up of three parts: one rectilinear and oblique, one rectilinear and vertical, and one circular, which connected the other two. This description lasted until Niccolo Tartaglia (see PHYSICAL SCIENCES 1550-1600) wrote *Quesiti et inventioni diverse* in 1546.

Aristotle's concepts diverge most radically from modern ones in his explanations of dynamics. According to Aristotle, a missile did not remain in motion once it was launched because of the missile or the mover, but because of the medium. The body that was launched was continuously pushed by the air that rushed into the void left by the missile in motion. A body in motion was always under the influence of a force, and its speed was directly proportional to the force applied and indirectly proportional to the resistance of the medium. In a void, since the resistance of the medium would be null, the speed would be infinite and the body would be omnipresent. From this theory came the negation of the void, the *horror vacui* that was central to the Peripatetics, who were staunch adversaries of the Democritan and Epicurean theories. The *horror vacui* was a mainstay of traditional dynamics until the time of Evangelista Torricelli (see PHYSICAL SCIENCES 1640-1650).

Another idea of Aristotelian dynamics was that the speed of a falling body in a given medium was proportional to the weight of the body. This explanation was contested by late-Scholastic physicists, who, according to some scholars, were the real discoverers of the law of falling bodies (Pierre Duhem, for example, contended that the works of Galileo were based heavily on medieval scientific tradition). And while we are speaking of forerunners, we should not forget that E. Hoppe and other authors actually saw an early formulation of the principle of inertia in a passage of the *Physics* (IV,8,215–219) that openly polemicizes on the question of the void.

Straton of Lampsacos

By the beginning of the third century B.C., the early scholarchs of the Lyceum seemed to have reconciled Aristotelian heritage with Alexandrian science. One of the outstanding figures of the time is Straton of Lampsacos, a student of Theophrastos who was at the head of the Lyceum from 288 to 269 B.C., after which he was at the court of Ptolemy I Soter in Alexandria. We know very little about his astronomical theories (some authors, such as Diels, attribute the heliocentric theory to him, but the idea is highly questionable); but somewhat more is known about his ideas on physics. For example, while Aristotle held that the four elements by nature had opposite motions (upward and downward), Straton denied the existence of a quantitative difference in movement.

All elements obeyed a single law of falling bodies. If air seemed to rise upward, it was because it weighed less than the other elements, just as bodies that were lighter than water rose when immersed in it. Weight was therefore a qualitative characteristic of all things. Straton's theories in the field of hydrostatics and some of his intuitions in the field of statics are similar to those of Archimedes.

Archimedes and statics

Archimedes of Syracuse was definitely linked to the Alexandrian environment, although his themes were in some ways typical of the culture of Sicily and southern Italy, and his life in general was quite different from the image of the Alexandrian scholar closed within the boundaries of one particular field (see MATHEMATICAL SCIENCES 300-200 B.C.).

Mathematical elaboration, inquiries into physics, experimentation, and great technological skill were inseparable elements of Archimedes' research. Tradition also credits Archimedes with the invention of a vast array of technological devices: the cochlea or Archimedian screw, the endless screw, the compound pulley, the plans for a planetarium, and even the famous burning mirrors that were supposedly used to defend Syracuse. Archimedes, like Aristotle, contributed to a trend towards research into mechanics and applied physics.

His theoretical research included his treatise *On the Equilibrium of Planes*, a geometric exposition based on postulates derived from experiments that Archimedes does not describe. But his treatment of the principles of levers and centers of gravity seems to imply that he carried out a vast number of experiments. On the one hand, Archimedes' combination of geometry and applied physics seems to go back at least to the Pythagorean Archytas of Tarentum (see the section 400–300 B.C.). On the other hand, Aristotle's *Mechanics* already contained the condition for balancing levers, as well as an early formulation of the principle of virtual forces (there were also accurate descriptions of scales and pulleys), all mixed with his theories of dynamics. In Archimedes' work, the condition for balancing levers is deduced from postulates that seem to come from experiments actually carried out with levers.

The appearance of the idea of centers of gravity is very important. The fact that Archimedes does not give an explicit definition makes it seem that the concept must have been introduced by an unknown previous scholar or in an earlier work of Archimedes' that has since been lost. In any case, Archimedes can be considered the founder of the rational theory of centers of gravity. The introduction of another basic mechanical concept, the momentum of a force with respect to a line or a plane, is also related to the elaboration of the theory of centers of gravity. From a letter to Eratosthenes about *The Method* (discovered by Heilberg in 1906), it is clear that Archimedes knew that two weights suspended on the arms of a lever would balance when the products of their areas or volumes times the distance of their centers of gravity from the fulcrum were equal. Archimedes was using a basic heuristic method, which he also applied to finding areas and volumes, foreshadowing the infin-

itesimal methods of the sixteenth and seventeenth centuries.

Hydrostatics

Archimedes' contributions to hydrostatics were equally important. Apart from the legend that some authors, such as Vitruvius (see PHYSICAL SCIENCES 100 B. C. – 0), tell about the crown of Hieron, tyrant of Syracuse, Archimedes' theoretical treatment in *On Floating Bodies*, as well as in his works on statics, all of which are done *more geometrica* [in the geometric manner], is quite interesting. From two essential hypotheses— that in all fluids the part that is less compressed gives way to the part that is more compressed, and that solids immersed in a fluid are pushed upward along a vertical line that passes through their center of gravity—Archimedes deduced that the surface of any fluid at rest was part of a spherical surface with its center at the center of the earth, so that sea level was the same everywhere.

He then introduced the concept of relative specific weights, which seems to have been unknown to his predecessors (Aristotle, Straton, etc.). A solid body immersed in a fluid of equal weight and volume would sink until its entire surface was covered and no further. Archimedes then studied bodies specifically lighter and heavier than the fluid in which they were immersed. He finally stated his famous principle: "Solids heavier than the fluid, when thrown into the fluid, will be driven downward as far as they can sink and they will be lighter in the fluid by the weight of the portion of the fluid having the same volume as the solid." This briefly sums up the contents of the first book of *On Floating Bodies*. The second book is dedicated to the conditions of floating, especially to the conditions of equilibrium of a right-angled segment of a paraboloid.

Pneumatics

Archimedes' physics was without a doubt mathematical, but it should not for this reason be considered an *a priori* construction, detached from any practical applications of it. Archimedes' theoretical reflections were closely connected to the work of Alexandrian engineers. One of these was Ctesibius (c. 310-240 B.C.), a contemporary of Archimedes and the founder of the Alexandrian school of mechanics. Vitruvius, in his *De Re Aedificatoria*, credits Ctesibius with inventing many early toys, the first hydraulic and pneumatic machines, as well as the water clock. In particular, according to Vitruvius, Ctesibius (see APPLIED SCIENCES 300–200 B.C.) invented a force pump for pumping water, which was modified into a fire pump. Until the Renaissance this type of pump was called a *ctesibia machina*. A (doubtful) fragment also attributes to Ctesibius a hydraulic organ, in which air compressed by water made reeds of various heights vibrate.

The Alexandrian engineers

From the work of Archimedes on, research into mechanics (which was almost applied mechanics, or even engineering) seems to have been focused on the ways that compressed air could be used. Hence the name pneumatics for both the theoretic study and the applications. *Pneumatics* was also the title of one of the books in a treatise on mechanics written by Philon of Byzantium (c. 290-220 B.C.), a disciple of Ctesibius who lived in Alexandria. Most of the work, which was written around

250 B.C. is extant. The first three books after the introduction are about the applications of levers, harbor engineering, and the engineering of hurling-type war machines. The following books concern compressed air machines, automatons, the defense and siege of cities, and other military subjects. Philon's theoretical knowledge was remarkable; the books show that he knew how to utilize atmospheric and steam pressure to their fullest extent and that he was completely familiar with the laws of siphoning.

The experimental knowledge gathered by the scientists who dedicated their studies to pneumatics led the Alexandrians to take an intermediate position between those who supported and those who critized the theory of the void. According to the Alexandrians, there could not be large areas of void, but only a void distributed among the various particles of matter (the Latins called this *vacuum intermixtum*). This particular hypothesis seems satisfactorily to explain the compressability and elasticity of air: when a volume of air decreases, the particles come closer together, but are in a forced position and tend to return to their original position, which is where the force of compressed air comes from. The Alexandrians also thought that fire behaved analogously, insinuating itself between one particle and another.

Developments in optics

Another merit of Alexandrian science was that it gave momentum to studies in optics. The problem of vision had already been considered before. The Pythagoreans had advanced the theory of a special fluid that left the eyes, came into contact with objects and carried the sensations back to the eyes. The atomists upheld the theory that objects emitted simulacra (*eidola*) which struck the eyes producing visions of forms and colors. Epicuros adopted this theory, and it reappeared in Lucretius' *De rerum natura* as well. Plato, perhaps taking his ideas from Empedocles, elaborated the theory of "the light of the eyes," a "smooth, dense" fluid that flowed out of the pupils and encountered the fluid emitted by objects. In a passage in *De anima* (II,7), of questionable interpretation, Aristotle mentions a theory of light propagation based on a change in the medium between the eye and the object.

Alexandrian research proceeded much differently. The oldest document from the museum is a treatise on optics attributed to Euclid. It is divided into two parts, the *Optics* and the *Catoptrics*. According to some authors, the second book was written after the great geometer's akmē (which was around 300 B.C.) and should thus be attributed to a later author. The *Optics* is written *geometrico more*: from several premises—that the rays emitted by the eye travel in a straight line, and that the figure outlined by the visual rays is a cone with its vertex at the eye and its base the edges of the object—explanations of shadows, images produced through cracks, the apparent sizes of objects and their relationship to the distance from the eye are deduced. The *Catoptrics* deals with phenomena produced by flat and spherical mirrors (see MATHEMATICAL SCIENCES 300–200 B.C.).

There is also an extremely interesting theory on reflection, which seems to foreshadow modern explanations. Although the subject is dealt with geometrically, there are references to experiments as well. Proposition XXXI, for example, says that fires can be lit using concave mirrors directed at the sun. This area of research undoubtedly goes back to Archimedes, who

wrote books on catoptrics that have not survived but that must have been fairly well-known in the ancient world. A story that certainly must be exaggerated tells how Archimedes used burning mirrors to set the Roman ships on fire from a distance during the siege of Syracuse.

LIFE SCIENCES

The Alexandrian school: Herophilos and Erasistratos

A medical school, which was primarily concerned with anatomical research and completely opposed to the opinions of the dogmatics, i.e. to Hippocratic tradition, was set up at Alexandria. The principal exponents of this school were Herophilos (fl. 300 B.C.) and Erasistratos (c. 304-250 B.C.). Herophilos mainly studied the anatomy and physiology of the nervous system, and attributed much importance to the pulse.

With Erasistratos, the mechanistic concept triumphed. A model of a pneumatic machine, made up of three systems of tubes, represented the living organism. The first accurate anatomical observations using dissection were made by the physicians of this school. Although the school of Alexandria developed in Egypt, the language it used, as well as its tendencies and its whole philosophical tradition, were Greek.

Greek medicine was brought to Rome toward the end of the century by several practicing physicians, but its success was mediocre at first.

APPLIED SCIENCES

Mechanics

Greek ideas of the geometrization of space and their quantitative abstract speculation (especially of the Pythagorean and Platonic schools) found concrete expression in their development of mechanical technologies. This in turn furthered the slow process of freeing people from heavy manual labor. In spite of the continued massive use of slaves, machines tended to play an increasingly important role, and mechanization took over various fields of activity, multiplying human and animal power or controlling their transformation. The problem of lifting large weights had been satisfactorily solved with the invention of pulleys. During this period, important improvements on this basic tool were made—for example, the triple-pulley crane that geared down the tension on the cord was perfected by Archimedes. He also developed several applications of screws to hydraulic machines and invented toothed wheels and gears (see PHYSICAL SCIENCES 300–200 B.C.).

The use of the rotary millstone spread throughout Greece. This was the first and most significant application of rotatory motion since the invention of the potter's wheel and the lathe. Hand mills for grains also became fairly widespread during this period.

The most important development in woodworking was the diffusion of the lathe. Lathes had already been known for about a millennium, and were used in furniture making and for intricate work on small pieces of wood, but during this century they became a basic tool. The art of lathe-working flourished especially in Rome, where artisans made elaborate and even extravagant turnery. Some mechanical improvements were also made in ceramic production; the pedal-operated potter's wheel appears to have been invented during this century.

Less important developments took place in the spinning and weaving industries. The Greeks and Romans elevated wool-weaving to a remarkable level, but the improvements were mainly due to the use of better quality fleece, a result of selective breeding. The fact that the Romans were the first to use shears for clipping sheep shows that they focused quite a bit of attention on this industry. The shape of their shears was almost identical to that of modern shears, with two triangular blades attached to a U-shaped spring.

Chemical technology

There is very little information about the dye-stuffs used during the Classical period. Theophrastos describes a white lead pigment that was made by putting sheets of lead and a little vinegar into earthenware containers. Red lead, or minium, could be made by roasting white lead in the presence of air. The Romans were also familiar with malachite for green pigment, orpiment for golden yellow, and yellow ochre. The Romans did not use indigo as a dye-stuff but only as a pigment, probably because they did not know how to dissolve the indigo. For red, the Romans used madder, and any astringent mineral substance could be used as an alum.

Overland communication

The Romans' skills in road construction advanced at the rate of their conquests. After the Appian Way had connected Rome to the south, the via Flaminia, finished under the censor Gaius Flaminius in 220 B.C., was built to connect Rome and Rimini. The techniques used were similar to those that Statius described for the via Domitiana: "The first task is to start the tracks (sulci), to open the way, and then to dig deeply to remove the earth. When this is done, the excavation is filled with other materials and a foundation (gremium) is prepared for laying the road surface, unless the ground gives way or crumbling soil provides an unsafe roadbed. Then the road is laid with stones tightly fit together on both sides (umbones) with many hooks."

The paved surface of the via Flaminia is four to five meters wide near Rome. The division of roads into miles, which Plutarch erroneously attributed to Gaius Gracchus, probably dates back to this century as well.

One of the most interesting features of the Roman roads is how straight they are. Many cuts and tunnels had to be made; the thirty six-meter-deep slash on one side of the Appian Way near Terracina is one of the most spectacular of these.

In view of the remarkable developments in the road network, it is surprising that no important changes were made in the means of transport. The technical efficiency of horses was as bad as it had been in the preceding period, since the harnesses still kept horses from pulling with all their force. In Rome, unlike Greece, four-wheeled vehicles were as common as spoked-wheel chariots. The art of horseback riding made some progress when spurs and the chain bit, which made it possible to stop a horse by forcing its head upward, began to be used.

Military technology

Interest in mechanics was particularly keen in the area of military development. We know how Archimedes used his triple-pulley crane in the defense of Syracuse. In telling of the Roman siege of the city, Plutarch says that several Roman ships were hooked with iron claws and lifted by means of a counterweight, then sunk down to the bottom. One often saw the frightening spectacle of a ship lifted into the air and hurled against the fortifications or dropped into the sea when the claws were unhooked. Ctesibius (c. 310–240 B.C.), one of the promotors of the Alexandrian school of engineering during this century, also designed plans for military machines. The most fantastic of these plans, the aerotonon, was a war machine whose spring power came from air compressed in a pair of cylinders. This project gives credit to the theory that Ctesibius discovered the compressability of air and invented the suction and force pumps (see PHYSICAL SCIENCES 300–200 B.C.) used in the following century. He also invented the clepsydra and the hydraulic organ. Ctesibius supposedly designed a catapult, the chalcotonon, which is described by Philon of Byzantium (c. 290-220 B.C.). Philon's work on war machines is also worthy of mention. Above all, Philon attempted to perfect catapults. He was quite familiar with the difficulty of keeping the skeins of fibers in manageable wooden frames that would resist the effects of torsion. Philon's solution was to eliminate the twisting by winding the cords around the frames and tightening them by inserting wedges. He also designed an automatic cross-bow that had a magazine of arrows among other features.

The Alexandria lighthouse

Among the great civil achievements of the third century B.C. was the Alexandria lighthouse, remembered as one of the Seven Wonders of the World. The work of Sostratos of Cnidos, the lighthouse was always lit to signal the position of the great Mediterranean port.

200–100 B.C.

THOUGHT AND SCIENTIFIC LIFE

Science and philology in Alexandria and other Hellenistic cities

Rome's political and economical advance into the territories of Hellenistic civilization was relentless. The kingdom of Macedonia fell in 168 B.C., and in 146 B.C. the Romans conquered Corinth and subjugated Greece. In the same year, the destruction of Carthage guaranteed the Romans control over northern Africa; and Asia Minor was conquered a decade later. The Hellenistic states, born out of the break-up of Alexander's vast empire, did not seem to be able to resist Roman penetration effectively, even though they often contained it temporarily with skillful political moves. What took place was an extremely complicated process in which two very diverse worlds with different cultures were intermingled for the first time.

Nothing in Rome during those times, except perhaps its already impressive juridicial tradition, could compare with the cultural products of Hellenism, such as the refined philology that flourished in Alexandria, where Aristophanes, who succeeded the great Eratosthenes as director of the library, lived from 257 B.C. to 180 B.C. While Aristophanes helped perfect methods of textual criticism and arrange criteria for punctuation, his successor Aristarchos of Samothrace (c. 220–145 B.C.) wrote numerous commentaries and critical studies on Greek literature, among them important works on Homer. Beginning in the third century B.C. all work in philology had accompanied a phenomenon that reached its splendor in the first half of the second century B.C.: books as we know them today, the precious instruments of intellectual communication, were born. This explains the great flourishing of commentaries on the classics, critical editions, and doxographical works. It also explains the considerable development of science, by then thought of as the cumulative acquisition of specific truths. But the rigid compartmentalization of society and the immobility of Hellenistic structures led learned citizens to keep themselves apart both from the political world and the context of productive forces. This caused a progressive separation between science and technology. What resulted was a cultural environment that for all of this century and the next, could boast of the most interesting schools of applied mechanics, but that made automatons, devices, and machines that were intended to be objects of wonder and surprise for rulers and scholars more than technological instruments in the modern sense.

Alexandrian science received a heavy blow when the museum was sacked in 145 B.C. and the declining scientific activity had to contend with outside difficulties as well. During the first half of the century, other centers such as Pergamos and Rhodes became more important. The astronomer Hipparchos of Nicaea (c. 194-120 B.C.), who mainly taught at Rhodes, flourished in this epoch. Crates directed the new library at Pergamos (where there was no papyrus and a new writing material, *pergamēnē* or parchment, was introduced), and gave a strong push to research in geography. Theoretical development in this field was increasingly aided by the large amount of empirical data that was being gathered during journeys, military and civil expeditions, and great commercial enterprises. Developments in medicine were just as important. At Alexandria, Serapion founded an Empirical school of medicine that argued against all dogmatisms, including that of the *Hippocratic Collection*, and attempted to found a practice based exclusively on experience. Many themes of the empirical school of medicine later influenced Skepticism.

The development of philosophy: Carneades and Panaetios

We must still look to Athens for philosophy. Stoicism was articulately organized by Chrysippos and became extremely popular. The lexicon and categories invented by the Stoics entered into Hellenistic language and culture overwhelmingly, often to the complete disadvantage of rival philosophies, such as Epicureanism. They created new disciplines, such as linguistics, and revived areas of research that were already known, such as logic. They were also the main point of reference for philosophical debates, which were increasingly detached from the sciences. However, the structure of reality that Stoicism proposed was rigid, and its practical applications were not at all flexible. The ethical doctrines required total participation. Motives of rivalry between schools were certainly not extraneous

to the harsh attack that Carneades (c. 215–219 B.C.), head of the Academy half a century after Arcesilaos, launched against Stoicism, which he equated with dogmatism. The polemic was not waged in the name of an alternative gnoseological system (such a system had no reason to exist for the Academicians), but was rather aimed at perfecting a series of counter-examples to match Stoic examples point for point. The importance of the adversary also caused the confutations to be codified, and Skeptic tradition until the Sixth Empire always accurately formulated the different modes ("tropes") used to attach Stoic theories. At the same time that the Stoic system went into crisis as a result of Academic confutation, a profound transformation took place in the Athenian school of philosophy, which had become the main point of reference for the young aristocrats of the Hellenic empires (and of the Roman state itself in the next century) who were to have careers as high government officials, politicians, military officers, and lawyers. It is therefore understandable why Panaetios of Rhodes (c. 180-109 B.C.), a scholarch of the Middle Stoa, put aside logico-linguistic themes and later even physical theory to focus his pronouncements on ethics. But the dogmatic tones were muted in this area as well. In contrast to rigid norms, a series of reflections, arguments, and considerations free from the context of a general system were proposed. As a result, culture became less of an exasperated opposition of rival theories and more of a conciliation. This was the origin of an increasingly vast interaction of Academic, Skeptic, and Stoic ideas that led to the Eclecticism of the first century B.C.

Culture and philosophy in Rome

The Roman conquest of Macedonia mentioned earlier brought the Romans into direct contact with Greece and greatly increased their contact with Hellenic culture. But soon after the penetration of Greek and Hellenistic (the Romans did not always make the distinction) art, philosophy, and science into Rome, conservatives like Cato, who saw these influences as factors that would corrupt the *res publica*, began to oppose them. In 161 B.C. a resolution of the Senate ordered that the rhetoricians and philosophers who had come to Rome as exiles from Macedonia be driven out of the city. Five years later, however, Athens sent to Rome a diplomatic mission of three philosophers: Critolaos, representing the Lyceum, Diogenes the Babylonian, representing the Stoa, and the above-mentioned Carneades. They expounded their doctrines in public and had considerable success among the youth, despite the conservatives' opposition. Among the most prominent families, the custom of sending their children to Athens grew. Panaetios spent quite a bit of time in Rome, where he was closely connected to the pro-Hellenic Scipionic circle, which besides the great Greek historian Polybios (c. 205–125 B.C.), included the major representatives of Roman culture of the times: Terentius, Lucilius, Gaius Laelius, and Quintus Elius Tuberon. Panaetios' ethics, which were far from the rigor and dogmatism of earlier Stoic tradition were based on the criterion of *humanitas*—the civil sense of cohabitation and the development of personal abilities—and were assimilated especially quickly.

It would be a mistake, however, to say that Hellenic or Hellenistic culture, even with the success it enjoyed, ever succeeded in imposing itself fully in Rome. Language is a typical example. While the Greek language spread quickly all through the eastern Mediterranean, where it became a real *koinē*, at least for science and culture, nothing similar happened in the West. The resistance of those like Cato prevailed, at least on a linguistic level, over the pro-Hellenics. Although the Romans devised new words by coping Greek ones, they contined to write in Latin. With the Roman conquests of the following century bilingualism became a determining element of Mediterranean civilization.

MATHEMATICAL SCIENCES

The descendents of Greek geometry

After the great efforts that had been made previously, Greek geometry showed signs of exhaustion. The personalities of this period are known more for their specific contributions than for the sort of fecundity of projects and generality of results that distinguished the earlier great authors. Hypsicles, who lived in Alexandria around 175 B.C., wrote a manuscript which was later printed as Book XIV of the *Elements*. Special plane lines were discovered; devised to improve some of the solutions to the famous "three problems," they foreshadowed the establishment of a general theory of curves. Nicomedes' conchoid, Diocles' cissoid, and Perseus' spirics were the new inventions. During the same period, Zonodoros' perimeters demonstrated that a circle is the figure that has the greatest area of all plane figures with equal perimeters.

At the same time, astronomy continued to make contributions to mathematics. Hipparchos of Nicaea (c. 190-126 B.C.) founded trigonometry by constructing tables of chords and arcs. But by Hipparchos' time the influence of Babylonian mathematics and astronomy had already reached Egypt, thanks to the cultural unity of the Hellenistic world; and Hipparchos himself seems to have been familiar with the findings of Babylonian astronomy. As much as these influences may have directly concerned astronomy, they were also very important to mathematics, since the Babylonian Seleucidian astronomy that came into contact with Greek culture was a highly developed form of mathematical astronomy based on purely arithmetical, not geometrical, methods, the heritage of the great Mesopotamian tradition of computation. In this way, Babylonian influences strengthened the arithmetic component of Greek mathematics, which had been in the shadows for quite some time (especially since Plato and Aristotle had distinguished true arithmetic, i.e. the theory of numbers, from logistics, the practical arithmetic of accountants, denying any scientific dignity to the latter). Arithmetic was revived during this period, although its scholars have remained anonymous. Greek papyruses abundantly attest to the development of computation in the Alexandrian world.

ASTRONOMY

Hipparchos. The precession of the equinoxes and the theory of planetary motion

Hipparchos of Nicaea (c. 190-125 B.C.), considered the greatest astronomer of the second century B.C. and of all antiquity, was only indirectly connected to the Museum at Alexandria. Most of his activity took place in the important cultural center of Rhodes, where he promoted the establishment of an observatory. Hipparchos not only used the results of his Alexandrian predecessors, but those of Babylonian tradition as well. He personally conducted numerous astronomic observations, using optical instruments invented especially for the purpose, such as the diopter, the so-called universal instrument, and others. He systematically used the branch of geometry that later developed into trigonometry to perfect his calculations. When he discovered a new star in the constellation of Scorpio in 143 B.C., he got the idea of cataloging the stars. He recorded the latitude, longitude and brightness of about eight hundred and fifty stars. Furthermore, by comparing his results with observations made 150 years earlier, Hipparchos discovered that during that period of time changes in the distances of the stars from fixed celestial points had taken place. These could only be explained by the rotation of the earth's axis in the direction of the apparent daily motion of the stars, so that the equinoxes fell slightly earlier each year. Hipparchos also determined the value of the precession of the equinoxes, and found that 26,000 years were needed for a complete cycle.

Ptolemy's *Almagest* gives quite a thorough explanation of Hipparchos' astronomic system. Hipparchos refused the Eudoxian theory of the spheres, and at least partly adopted Heraclides of Pontos' hypothesis on eccentricity and his idea of the planets rotating around a body that is itself rotating. But while Heraclides thought of the planets as rotating around the sun, which was also assumed to be rotating, Hipparchos said that the planets, as well as the sun and the moon, moved along circles (of lesser radii) that rotated in a circular orbit (of greater radius) around the earth. These lesser circles were called epicycles, because they were superimposed onto another circle (the circular orbit just mentioned) which was eccentric with respect to the earth (that is, its center was not the earth). This larger circle was called the deferent.

Developments after Hipparchos

With Hipparchos, astronomy reached a level where lunar eclipses could be forecast with an accuracy of within one or two hours (a wider margin of error persisted for solar eclipses). Astronomy had come a long way from the first Babylonian attempts at predicting eclipses, which as Herodotos observed, were made by the year, without accurately specifying where on the earth the eclipses would be seen.

Around 160 B.C., the Stoic philosopher Diogenes the Babylonian arranged the heavenly bodies in an order that was maintained until the Copernican revolution: the moon, Mercury, Venus, the sun, Mars, Jupiter, Saturn and the fixed stars. The so-called *Ars astronomica Eudoxi* was probably written at the end of this century. A papyrus with notes, made perhaps by a student, it is full of valuable astronomic data and theoretical information. It is a significant piece of evidence of the amplitude and diffusion of knowledge in astronomy at that time. This is also attested to by scholastic texts such as those written by Geminos and Cleomedes.

PHYSICAL SCIENCES

Physics and technology in the Alexandrian world

The four great traditions of thought that dominated philosophy—Peripatetic, Platonic, Epicurean, and Stoic—all showed a considerable amount of interest in the physical world, although none of them produced physical doctrines that were comparable to Archimedes' mechanical theories or Euclid's optics. Rather there was a keen interest in questions of music, meteorology, physics, and applied mechanics among the supporters of the various doctrines. They confronted problems such as the diffusion of sound and light, reflection, the weight of air, rainbows, etc. It is often difficult to date and attribute opinions or even physical theories properly and to establish their chronology, since tradition has not always been clear and precise.

Finding practical applications for mechanics, especially fluid mechanics, was definitely a typical feature of Alexandrian scientific culture in the second century B.C. But although the tendency sometimes seemed to be towards remarkably interesting practical applications (war machines, for example), more often it seemed to be aimed at creating devices that would excite astonishment and wonder. The last great representative of the Alexandrian engineering school, Heron, whose akmē was probably in the first half of the first century B.C., belonged to this trend. If we look at these as "characteristic features of the early history of mechanics," as Mach put it, we can see how complicated the relationship between philosophy, physics, and technology was during this period, even in light of the argument that technical devices imitated nature or even made it possible to get to the core of the phenomena. A culture that had known the grandiose theories of someone like Archimedes was not able to achieve a technology on the same level, so that sporadic episodes provided the only link between scientific theorizing and technological applications.

Characteristics of Roman science

With the conquest of Greece and the destruction of Carthage in 146 B.C. Rome extended its power beyond the western Mediterranean. The potential for development lay in moving eastward, at the expense of the Hellenistic kingdoms. There were several particular features in the Romans' attitude toward science. While the practical uses of statics and hydrostatics were of relative importance to the Alexandrian engineers, technological applications were a priority in the Roman world. The Romans soon learned everything they could from the Greek scholars about navigation, building harbors, aqueducts, roads, and war machines.

As for mechanics in particular, we know that the Romans were already using steelyards in the third century B.C. During the second century B.C., they developed devices that operated using the principles of pulleys to lift heavy building materials. Suction and force pumps were also well known.

LIFE SCIENCES

The Empirical School

A school known as the Empirical school was formed at Alexandria in reaction to the crystallized dogmatism of Hippocrates' followers and to certain abstract, categorical theoretical positions of the disciples of the Alexandrian school. According to Philinus (fl. middle of the third century B.C.), a disciple of Herophilos and the founder of the Empirical school, physicians should be formed through their own experience with patients, and not through verbal teachings full of philosophical abstractions. The school reached its highest point of development under Philinus' successor, Serapion (fl. first half of the second century B.C.).

The concept of household medicine still prevailed in Rome. The writings of Marcus Porcios Cato (234–149 B.C.) are good examples of an adherence to ancient practices and hostility toward the contemporary Greek medicine.

APPLIED SCIENCES

Agriculture

The Romans were faced with the problem of working with heavy, compact soil that could not be moved as easily as sandy soil. Cato (234-139 B.C.) tells us that iron plowshares were used throughout the northern Roman provinces as well as in Italy. These evolved into plows that cut the clods of soil vertically and turned them over. This was accomplished by adding a keel to the stock of the plow itself in order to keep it inclined.

Cato also tells us how flour mills became common during this period. He lists as equipment that was necessary for outfitting a farm three mills operated by donkeys and one hand mill.

The Romans were among the first to use animal excrement as a fertilizer, and to establish a three-year crop rotation. The first year, the land was prepared by leaving it fallow; then grain was grown, followed by a year of nourishing rest (meadow or natural grazing). By decreasing the extent of the system and introducing fertilizer, the areas lying fallow progressively became less. Thus the foundations were laid for the principle of continuous rotation, according to Virgil's idea that the soil rests while cultivation continues, as long as the crops are changed.

Hydraulics

New drainage methods introduced by the Romans improved mining methods, making it possible to dig deeper mines, beyond the limits of water tables. Water wheels with an average diameter of 4.5 meters and worm conveyors, an invention of this century, have been found in Spanish and Portuguese mines. The great aqueducts built in Spain furnished enough water to carry ores, making wide-scale production of tin possible. The Greeks were more interested in city water supplies than in mines. The methods proposed by the hydraulic engineers were often very clever, such as the worm conveyor, a water lifter inspired by Archimedes' worm screw, and the saqiyeh, a device based on the principle of the toothed wheel, another of Archimedes' inventions. An interesting and daring solution was found for the problem of transporting water at Pergamos. Water was piped from a cistern located at 375 meters altitude across two valleys at 172 meters and 195 meters altitude, to the city, which was at an altitude of 332 meters. The water pressure at the bottom of the valleys must have reached twenty and seventeen atmospheres respectively, pressures which could only have been contained by bronze or lead tubes.

Raw materials

Parchment (*Pergamēne*) was also invented in Pergamos at the time of King Eumenes II (ruled c. 197–159 B.C.), inspired by the desire to limit imports of costly papyrus from Egypt. Animal skins were subjected to a lengthy process of treatment which made them into useful writing materials. First they were washed, then soaked in pure water for one day; after this they were treated with lime and left to soak for eight to fifteen days in a pasty liquid containing thirty per cent lime. At the end of this procedure, they were laid out to dry.

Paper, such an enormously important substance in the modern world was invented during this period in China, although it was unknown in the Greco-Roman world. It was "recuperated" by the Arabs ten centuries later.

100 B.C.–0

THOUGHT AND SCIENTIFIC LIFE

Academicians, Stoics, and Epicureans

The Stoic philosopher Posidonios of Apamea (c. 135–51 B.C.) worked to bring natural science back to *humanitas*. The fact that this coincided with the decline of Alexandria, the major scientific center of the time (Egypt fell under Roman rule toward the middle of the century), leads us to believe that the Stoic had a project for the complete unification of knowledge to rival, in size if not in rigor, the Peripatetic synthesis (Andronicos of Rhodes' critical edition of the *Corpus aristotelicum* also came out during this century). Posidonios' encyclopedia was compiled following the usual criteria for summarizing: mathematics, geography, and the natural sciences were combined with the esoteric and mystical theories of the flourishing religious sects, which in many cases had been imported from the East. The basic idea was not to let any human manifestations escape consideration, although by so doing, philosophical reflection was opened up not only to debates on mathematics and astronomy, but also to those mystical and magical currents that had continued to flourish in the East but had not yet been able to gain control in all the regions of Hellenistic civilization.

The Academy also underwent important developments with Philon of Larissa as its head from 110 to 85 B.C., and Antiochos of Ascalon from 85 to 69 B.C. During this time, the battle against the Stoics, in which Carneades had been so keenly involved, subsided, and several terms were proposed to reconcile Stoicism with the heritage of Plato, disavowing Arcesilaos and Carneades. We should also note, however, that, faced with this attempt at bringing the Academy back to a renewed form of dogmatism, perhaps based on the reconciliation of various schools of thought, Aenesidemos of Alexandria programatically restored Pyrrhonian tradition toward the end of the century. His Πυορωνειοι λογοι (*Pyrrhonian Discourses*) contained numerous arguments against the Stoics, the Academicians, and even the Epicureans. The latter, as we have seen, were small elite groups, followers of a hidden life, and did not always keep up the rivalry with the Academicians and the Stoics.

One of the most representative figures of Epicureanism was Philodemos of Gadara, Syria (110–35 B.C.), who lived in Rome for a time and defended Epicuros' doctrines against the attacks of the Stoics and the confutations of the Academicians. He contributed to the spread of Epicureanism, often taken up as a literary theme (by Horace, for example) in Rome. The most prominent Epicurean of this century was Titus Lucretius Carus (95–55 B.C.), the author of De rerum natura, a poem that expounds Epicurean themes with remarkable originality. Lucretius also drew from Empedocles, Democritos, Aristotle, and Theophrastos. His Epicureanism was above all a view of the world, in which atomism played an essential role. The doctrine of the clinamen, the ability of atoms to deviate from a straight line, is also attributed to Lucretius. Lucretius' concept of the world was materialistic and rationalistic and opposed to a religious or superstitious world view.

Eclecticism: Cicero

With Lucretius, we enter into a Roman enviroment in which a Latin philosophical culture was developing. The major author of this new current was Marcus Tullius Cicero (106–43 B.C.), a lawyer and politician who ended up being defeated in the *bellum civile* following the death of Julius Caesar in March of the year 44 B.C. Cicero's teachers were Philon and Antiochos, but he also heard Posidonios and Philodemos of Gadara. He sought by means of Eclecticism to reduce all the conflicts in philosophical literature by looking for *loci communes*, i.e., the opinions that the representatives of the various schools agreed upon. According to Cicero, a unitary culture could be derived from a synthesis of Platonic, Aristotelian, and Stoic thought, often within the framework of a moderate skepticism. This culture was fundamental to the civil society that Cicero attempted with his practical policies to set up in Rome.

The growing popularity of encyclopedias: Varro

Cicero's friend, Marcus Terentius Varro (116-27 B.C.), was also an eclectic. Varro was a student of Antiochos of Ascalon, and became, during the time of Caesar the head of the first Roman library. Among other scholarly works, he compiled a vast encyclopedia, written in Latin, of the geographical, historical, and political knowledge found in Hellenistic literature. Among his works that have not survived was *Disciplinarum libri IX*, in nine books on grammar, dialectics, rhetoric, geometry, arithemetic, astrology, music, medicine, and architecture, in that order. The books remained cannons of Hellenistic-Roman culture nutil Martianus Capella reduced the *artes* to seven.

The function and popularity of the encyclopedic genre was very important, and encyclopedias became permanent products of culture from this time onward. The era of Varro and Cicero was also the era of historians such as Sallustius (86–34 B.C.), of geographers such as Strabo (c. 63 B.C.–25 A.D.), of a methodical trend in medicine, and of research in engineering and mechanics. The Romans' interest in putting theoretical science to practical use is also quite obvious in the *De re aedificatoria* by Vitruvius (second half of the first century B.C.), a manual of architecture with remarkable cultural range. Castronomy Posidonios' interest in astronomy

ASTRONOMY

Posidonios' interest in astronomy

Posidonios (c. 135-51 B.C.), a representative of the "Middle Stoa," had more of a sense for scientific experimentation and a taste for observation than the other Stoics, even though he was more open than others to mystic and syncretistic currents. None of his works have survived, but from indirect sources we know of his attempts to measure the real diameter of the sun, by considering physical and geometrical factors.

Astronomy in Rome

Alexandria had experienced its first crises during the second half of the preceding century (the Museum was first sacked in 145 B.C. during an attack). During this century, Alexandrian science, so brilliant in the past, underwent a slow process of disintegration. We must not forget that Alexandria entered into the Roman orbit in the year 50 B.C. when Rome was aspiring to become the center of culture and scientific life, especially with Julius Caesar's ambitious projects.

But Roman science had extremely peculiar characteristics which tended to accentuate practical aspects more than profound theoretical inquiries. On the other hand, Stoicism was the most popular philosophy in Rome, at least in the ranks of the ruling class, and Stoic dogmatism was not always favorable to the development of impartial scientific investigation. Hence no tradition of studies in astromony developed in Rome, although there were intellectuals who were interested in astronomy. Cicero was one of these. He translated Aratos' poem into Latin, and was exposed to problems in physics and astronomy through the teachings of Posidonios at Rhodes. The Epicureanism of Lucretius' De rerum natura also displays a considerable interest in celestial phenomena, following the atomistic tradition. Julius Caesar himself associated with various scholars in Alexandria, from whom he learned especially about the motions of the stars. According to Macrobius, he left important works on this subject, however these works have not survived. Tradition has it that Caesar conceived of the idea of reforming the Roman calendar while he was in Egypt, and entrusted the project to what Plutarch in his life of Caesar calls the most famous philosophers and mathematicians. The Egyptian astronomer Sosigenes (flourished middle of the first century B.C.) coordinated the reform, which went into effect on January 1 of the year 45 B.C.

Although the Romans were backward in theoretical astronomy, they soon learned how to use several basic instruments, such as the sundial (which they had known about since the third century B.C.), employing them to compute the calendar. Around the year 10 B.C., the architect Vitruvius wrote instructions on how to build sundials, and described various kinds of sundials that were in use at the time. Building these instruments required considerable mechanical skills and a certain amount of ability at making various basic astronomical observations. Sundials for travelers were quite common, and Vitruvius also describes an extremely simple and practical water clock.

PHYSICAL SCIENCES

Heron and mechanics

Heron, who flourished around 62 B.C., belonged to the brilliant tradition of Alexandrian engineers. His work on pneumatics (which was widely read by Renaissance scientists) was explicitly based on the works of Ctesibius and Philon. Heron applied the knowledge of air compressability to a number of devices, one of which was the famous eolipile, the first steam machine that actually worked. In his Mechanics, the only complete version of which has reached us through an Arabic translation, Heron goes into great detail in describing simple machines (winches, levers, pulleys, wedges, and screws), gears, and other more complicated machines. His Diopter deals with the construction and uses of instruments for measuring angles, distances, levels, etc. It also contains a description of an odometer, an instrument for measuring distances on a road.

Alexandrian mechanics ended with Heron. Egypt entered the Roman orbit in the year 50 B.C.; and Alexandria's importance diminished, although a talent for organizing local initiatives, together with a certain degree of scientific curiosity and a considerable disposition toward scholarly works of compilation persisted. The work of Heron, even more than that of his predecessors, exemplifies the paradox that existed in mechanics during the Alexandrian age: an ability for both very subtle theorization and bold achievements, but an inability to reach a systematic balance between the two. Alexandrian mechanics was thus confined within a cultural context that for various theoretical, institutional, and economical reasons ended up hindering its development.

Heron's catoptrics

Heron's name is also associated with a brief treatise on catoptrics, of which only a Latin version is extant. The study of mirrors, in line with the tendency of the period, is dedicated to illustrating the amazing effects that they can produce. It also contains some interesting geometrical observations, for example proposition IV, according to which ''of all incident rays coming from the same point and reflected in the same point, those that reflect at equal angles in flat and spherical mirrors are the shortest.'' In modern times, these geometrical observations contributed to the first principles of variations (especially Fermat's principle).

Physics and philosophy in Rome

We have already discussed the relationship between the Greek philosophical tradition and Greek research in physics. Now we shall point out some of the specific instances of that relationship in the Roman world.

During this period, Stoicism expressed the speculative attitude towards nature that prevailed among the Latin ruling class. Several Stoic thinkers had a keen interest in natural studies, physics obviously among them. However, moral interest dominated Stoic ideas, and the Latin tradition of virtus tended to emphasize issues of ethic behavior and duty. The Roman's love of rhetoric created a taste for a type of argumentation that was quite different from what would have resulted from following the geometric rigor of Archimedes or the Alexandrians. There are a number of references to natural phenomena in the works of Latin authors of this period, often accompanied by original observations; and these thinkers were inclined towards experimenting. However, the typical lack of interaction between mathematics and experimentation which had existed to a great extent in the environment of Hellenistic physics persisted.

While Cicero (106–43 B.C.), who had heard Posidonios the Stoic at Rhodes, was introducing Stoicism, mitigated by a modest Skepticism, into Rome as a point of reference for intellectual debate, Epicureanism, a much less popular doctrine, found a representative in Titus Lucretius Carus (95–55 B.C.).

The revival of atomism, the reduction of all phenomena to matter and motion, and the concept of the eternity of atoms and the void (emptiness), are particularly important to our discussion. Lucretius (see THOUGHT AND SCIENTIFIC LIFE 100 B.C.) was not so much interested in giving explanations of individual phenomena (much of his material came from Democritos, as well as from Epicuros) as in reestablishing a materialistic concept, an extremely consistent mechanistic materialism, where rigid determinism was tempered only by the doctrine of *clinamen* which was of uncertain origin.

Although Lucretius' ideas became the target of attacks from other philosophical schools especially the Peripatetics, who upheld the continuity of matter and refused the doctrine of the void and later from Christian authors from whom Epicureanism was synonymous with blasphemy—during the Renaissance they were a powerful stimulus for the revival of atomistic theories. This gives an idea of the importance of the *De rerum natura* in the history of world physics.

Physics and technology in Rome

As we have already mentioned, a typical feature of Roman scientific culture was the emphasis placed on the practical aspects of scientific inquiry. *De re aedificatoria*, by Vitruvius (second half of the first century B.C.), is probably the most typical Roman technical work. Vitruvius had a wide variety of cultural and scientific interests, and most of the problems he confronted required a considerable knowledge of physics. His knowledge, however, was always conditioned by utility; it is not a coincidence that Vitruvius was less accurate than the Alexandrian scientists in many instances (for example, he gives a much less precise value for π than Archimedes did in his time).

LIFE SCIENCES

Asclepiades, medicine and the atomic theory

Greek scientific medicine took root in Rome when Asclepiades of Bithynia, a great master of rhetoric and a very skilful physician settled there in 91 B.C. In the capital, Asclepiades taught a completely new medical doctrine, based on the atomism of Epicuros (341–270 B.C.). He applied an interpretation of pathological phenomena.

The Methodic School

Asclepiades' doctrine was simplified and organized by Themison of Laodicea, the major exponent of the so-called Methodic school.

According to Methodic theory, illnesses were essentially mechanical phenomena. It therefore followed that hygiene and therapeutics should be based on the physical functions and on diet.

Glaucias of Tarentum's empirical tripod

One of the major products of the Empirical school, which flourished during this period, was Glaucias of Tarentum's Empirical tripod, according to which medical practice should be based on observation, tradition (or the passing down of observations made by ancient physicians) and analogisms, an intellectual process for deciding on the appropriate therapy when both personal observation and a precise tradition of a disease are lacking. This school discovered many new medicines and used a wide range of drugs.

The encyclopedic collections

Along with the categorization and progressive enrichment of science came the first encyclopedic collections of medical knowledge. The trend, along with a strong inclination for didactics, mainly developed among Latin authors. Marcus Terentius Varro (116–27 B.C.) compiled his *Disciplinarum libri IX,* in which medicine was given considerable importance.

Varro also wrote a treatise on agriculture, in which he codified plant cultivation as well as the breeding of domestic animals and bee-keeping.

APPLIED SCIENCES

Heron's Automata

The most important figure in the field of technological inventions during the Alexandrian age was Heron, who flourished around 62 B.C. Besides his treatises on pneumatics and mechanics (see PHYSICAL SCIENCES 100 B.C.–0), where numerous examples of machines, implements, and instruments are given along with theory, Heron wrote a successful little work entitled *Automata,* which is worthy of mention as evidence of the era's taste for the strange and artful more than for its intrinsic value. It describes several complicated automatic toy theaters with mythical scenes full of gods, demigods, ships, rowers, etc., which operated on the principle of the force of gravity acting on weights attached to cables, and using pulleys, rollers, and toothed wheels.

The Vitruvian mill

The earliest evidence of the existence of water-mills is found in a poem by Antipatros of Thessalonica, who probably lived in the first century B.C. Strabon is the next to mention them, when he tells of a water-mill that Mithridates had built in 65 B.C. The Greek mill had a horizontal wheel with a vertical axis. Since it was not very efficient, it was mainly a family tool; it could meet the limited needs of a farm, but could not be used for the commerical production of flour. During the first century B.C., this low-yield Greek mill was transformed by Roman engineers into the more efficient Vitruvian mill, which had a vertical wheel with a horizontal axis. The grinding stones made five revolutions for every revolution of the wheel, while the Greek model only made one revolution for every turn of the wheel. Increased efficiency compensated for a more complex construction (gears had to be used to connect the horizontal axis to the vertical axis). A mill of this type, found at Venafro near Naples, was powered by an aqueduct which produced energy equivalent to three horsepower. The stones made forty six revolutions per minute and could grind 150 kilograms of grain in an hour, a considerable improvement over the 7 kilo grams per hour that two slaves or the one-half horsepower produced by a donkey-powered mill could grind. But in spite of all this, the Vitruvian mill was not widely used until the third century A.D., and even then was only used for grinding grain.

Raw materials

In Spain, the Romans used the liquation process for extracting gold; the metals were separated by gradual melting, utilizing the fact that gold and silver are more soluble in lead than in copper. Copper containing silver and gold was mixed with three or four times as much lead; the gold and silver were incorporated into the lead, then recuperated by cupelation. Roman founders were able to unsilver lead up to 0.01 percent. They used pit ovens dug at the site and equipped with insufflator tubes to blow air onto the surface during cupelation so that the iron would oxidate and be freed from the slag.

The production of copper on an industrial scale was more complex. The old melting techniques could be used for small deposits, but most production came from rather complex ores and required a complicated process of purification. Any material that combined easily with the earthy gangue was used as a flux, while refining was usually done in cupola furnaces similar to those used for ceramics. The Romans also began the regular production of brass, the most important alloy of copper in use during the Classical Age. Brass was made by the ancient case-hardening method, where the copper was heated in a mixture of powered zinc ores and wood charcoal. Part of the zinc that was freed in the presence of the copper diffused itself in it, forming a layer of brass.

Ceramics

Ceramics had already reached high levels of sophistication; and the Romans' only contribution was to introduce the systematic use of milling for mass producing huge quantities of earthenware, often decorated with relief work.

In the glassmaking industry, both the Romans and the Greeks used the ancient Egyptian technique of milling around a core of pressed sand until the first century B.C. During the first century B.C., glass-blowing, the most important invention of antiquity, appeared. At first glass was blown into molds, but the method later evolved into free blowing. The invention was of Syriac origin.

Measuring time

During this period, the Romans also became rather skilful in making instruments such as sundials which measured time based on astronomic observations.

This subject is discussed in more detail in the section ASTRONOMY 100–0 B.C.

0–100

THOUGHT AND SCIENTIFIC LIFE

The political-cultural unification of the Mediterranean world

Rome, with its vast agricultural hinterlands that no Greek *polis* could ever have controlled, a basically rural population, and the ability to form a solid military apparatus, expanded rapidly into the entire Mediterranean basin, at the expense of the backward populations of western Europe including Italy itself, the debilitated Hellenistic powers, and the old oligarchies, such as the Etruscan and Carthaginian states. Augustus's empire had political economic and military control over an immense expanse of territory and a wide variety of peoples. Only the Germans to the north, after the battle of Teutoburg, remained outside the Roman sphere of influence; and both sides limited their expansion eastward. In all other directions Roman expansion continued: power was consolidated in northern Africa, and Britain (what is now southeast England became a province in 43-44).

The political and social contradictions that had exploded into civil war in the previous century reemerged on a wider scale. The landed aristocracy, thanks to imperial protection, developed enormous estates; but on a more specifically political level, the aristocracy was an ambiguous ally for the imperial government, since it was always ready to claim its traditional right to rule over public affairs. The conflict between the senatorial oligarchy and the emperor was an important cause of political instability for the entire century, and even later. At the same time, landed estates and fiscal pressures were among the main causes of a stagnation in the production that began to manifest itself at the core of the Empire—in Italy, for example, which had been devastated by the civil war. A declining standard of living among free laborers (small tenants, small merchants, and artisans), who were fast approaching the level of slaves; the weight of the central bureaucracy; and corruption among officials were all symptoms of a crisis that grew throughout the empire during the following centuries. However, we should also remember that the unity and relative *pax* guaranteed by imperial authority allowed ideas, religious creeds, and philosophical systems to circulate in a way that would have been inconceivable before.

Here we are speaking of the most diverse intellectual experiences. The traditional doctrines that had aimed at the formulation of rational criteria to guide people in learning about the world around them and to direct their behavior remained the patrimony of small intellectual groups. Great masses of people, on the other hand, believed they had found firm guidance in the new rites chaotically imported from the East due to increased communication and the transfer of slaves. Thus the cults of Isis, Osiris, Cybele, and Mithras swept into the West, reaching as far as Britain and Germany.

The official culture. The institution of imperial chairs

The central authorities were not unaware of the need to direct this enormous variety of cultural and religious stimuli back into some kind of ideological and cultural unity, in order to create a political unity based on persuasion and consent rather than simply military control. The imperial authority tolerated a wide range of religions and philosophies on the condition that they agree to a few common principles. One of these was to recognize the divine and providential nature of the principality's power, which was a way of giving the entire state apparatus rational and religious security.

Following the first great crisis of the principality, which culminated at Nero's death in the extinction of the Julian-Claudian dynasty and the civil war which followed (68-69), the empire placed at the service of the official culture and the task of universalizing ideology an institutional structure that was absolutely new to the ancient world. Beginning with the reign of Vespasian, there were instituted in Rome and in other centers of the empire chairs paid for by the state to which major intellectuals of the time were called to teach regular lessons. A permanent, higher-level scholastic structure was thus created, which progressively spread to all the territories of the state. Rhetoric, grammar, and philosophy were taught in these centers. These disciplines were the vehicles of cultural unification but were also indispensable to the curriculum of high imperial officials. Quintillian's famous *Institutio oratoria* was written expressly for the purpose of forming this type of official. Quintillian, like Seneca, was convinced by Roman cultural ideals and held the first chair of rhetoric that Vespasian instituted in Rome. In the twelve books of the *Institutio*, Quintillian arranged the cultural contents of Cicero's work into an educational program of encyclopedic formation that included grammar, rhetoric, dialectics, geometry, arithmetic, music, and astronomy. Quintillian agreed with Cicero that universal order depended on the proper functioning of the state machine, although he was no longer referring to the *res publica* of the first century B.C., but rather to the new structures of the principality. The way the imperial bureaucracy organized philosophical teaching is particularly interesting. Four basic chairs were instituted for teaching Platonism, Aristotelianism, Stoicism and Epicureanism respectively—the doctrines of the four officially recognized "sects." A dogmatic rigidity soon arose around this type of organization. In order to hold a chair, one had to prove allegiance to the doctrines of a particular school. Since holding a chair was almost the only guarantee of social and economic security, conformism toward the central authority increased. In the long run, the cultural function of philosophy diminished considerably compared to rhetoric and even law, a discipline of vital importance for the functionaries who were sent to the imperial provinces.

Stoicism from Seneca to Epictetus

Among the official philosophies, the influence of those that did not in principle exclude a mystically-oriented interpretation grew, while currents that could not easily be reinterpreted in a mystical-transcendent sense declined. The moral themes of Stoicism were particularly popular during this century, especially among the intellectuals connected to the traditions of the senatorial aristocracy, which was always suspicious of the centralization of power in the hands of the *princeps* and jealous of its own traditional privileges. These privileges took on the value of authentic *libertas* in the eyes of the principality's opponents. The idealization of Cato the Utican, who committed suicide "for freedom" after his unsuccessful resistence to Caesar, was typical.

Stoicism was also well-received in the environment of the imperial court, especially where a moderate principality was favored. The Stoicism of Lucius Anneus Seneca (36-5), Nero's teacher and counselor, whom Nero forced to commit suicide at the time of the conspiracy of Piso, was focused on pedagogical and moral problems. His famous *Epistulae ad Lucilium* is important from this point of view. Seneca also had a keen interest in physical phenomena (see ASTRONOMY and PHYSICAL SCIENCES

0–100), and wrote a *Naturales quaestiones*, the contents of which are not, however, original. In it he discusses lightning, rainbows, clouds, wind, and earthquakes and takes up the eclectic theories of the Middle Stoa.

The moral theme is also dominant in the writings of Epictetus, who was born in Hierapolis, Phrygia, around the year 50. He was brought to Rome as the slave of one of Nero's freedmen and was later himself freed by Nero. A pupil of the Stoic Musonius Rufus (30-102), Epictetus taught philosophy in Rome beginning around 92. During the rule of the Flavians, the relations between philosophers and the imperial authority were not always easy. The philosophers, except for Musonius Rufus, had all been temporarily sent away from Rome by Vespasian in 71. A later edict issued by Domitian banished "philosophers, astrologers, and mathematicians" from Rome. It was following this edict that Epictetus withdrew to Nicopolis in Epirus, where he founded his own school. His teachings were collected by his disciple Arrian in Δωτοιβαι' επαχττου (*Dissertations of Epictetus* in eight books, only four of which are extant, and in ἐγκειοιπριδιον ἐπικ (*Manual of Epicetus*). The teacher's Stoicism had a cynical tone, in many ways more faithful to Greek tradition than Seneca's.

Platonism, Aristotelianism and encyclopedic culture

While eastern mystical and religious themes tended to prevail in the cultural *koinē* of the imperial period, Platonism and Aristotelianism tried to maintain their autonomy, often focusing on gnoseological themes and the understanding of nature, in opposition to both Stoicism and the mystical-religious sects. Extensive contamination had substantially changed the thoughts of Plato and Aristotle into a new doctrine, which was now being proposed by Platonists and Aristotelians. The work *De mundo* came out of the Aristotelian current; although it has reached us under the name of Aristotle, it was written during this century. It makes several concessions to Stocism and Platonism in its treatment of the cosmic order and the divine.

Plutarch of Chaeronea (c. 46-125) was also linked to Platonism. His Βίοαι παοαλληλοι (*Lives*), forty six paired biographies of Greeks and Romans, plus four individual ones, became an educational paradigm for civil and intellectual life. Plutarch also wrote a wide range of philosophical, moral, and religious works (the so-called *Moralia*).

Important in Plutarch's writings are the polemics against the Stoic concept of a divinity that is immanent in the world, return to the teachings of Plato in the light of Pythagorean tradition, and an attempt to restore Greek religious tradition. Plutarch even used elements of direct religious experience for this purpose, including divination and demonology, often disagreeing with rationalistic pretenses he considered excessive.

A trend toward writing commentaries on authors who were already considered classics, a desire for a global synthesis of knowledge, the usage of summaries, and a need for encyclopedias were all distinctive features not only of the philosophical tradition of the official sects, but also of a large part of scientific culture during this epoch. Specialization and encyclopedism came together in a wide variety of ways depending on the author. Seneca's interest in scientific studies of nature was not an isolated phenomenon. While the various sciences continued to be cultivated in Alexandria and the other centers of Hellenistic tradition, scientific and technical interests were flourishing in Rome and other cities of the empire, although the tendency was more one of popularizing discoveries that had already been made than of making original contributions. Aulus Cornelius Celsus, for example, who flourished during the reign of Tiberius (14-37), wrote a sort of encyclopedia, entitled *Artes*, which dealt with rhetoric, philosophy, law, military science, agriculture, and medicine. Columella wrote his *De re rustica* and Dioscorides, his medical encyclopedia, while arithmetic began to be treated as an autonomous discipline with the work of Nicomachos of Gerasa.

The unofficial sects. Neo-Pythagoreanism and Gnosis

The mystical and religious ferment of the times did not find its full expression in the official culture. Even Stoicism, which was quite open to moral problems, issues of human destiny, man's place in the universe, and a pantheistic view of nature, remained a basically rationalistic doctrine. The more typical themes of mysticism were still extraneous to it. Religious unrest tended to find outlet in those sects that made up a sort of counterculture, for example the Neo-Pythagoreans, who had already existed in Alexandria during the second and first centuries B.C., now spread into the Roman world. In many ways the Neo-Pythagorean sect constituted an actual church, whose belief in the transmigration of souls, incarnation as atonement for sins, and magical practices as a means of freedom and salvation came from ancient Pythagoreanism. The practice of philosophy was seen above all as a tool for purifying the soul of its stains so that it could break out of its cycle of reincarnations and return to the divinity.

In this context, we can understand how referring back to the master (Pythagoras) was transformed into the evocation of his spirit (in dreams and initiation rites), and the revival of his doctrine took on the aspect of transmitting a truth that the master revealed to his disciples in "sacred writings" that contained "the word." It is from this that the myth of Egypt and the other countries of the East came from, where the priests were supposedly the depositaries of knowledge that in ancient times had been revealed to Pythagoras, Democritos, and Plato. This explains the myth of Pythagoras, the wizard and master who revealed himself to his "children," i.e. to his initiated disciples. The Pythagoreanism of Publius Nigidius Figulo (died c. 45 B.C.), a Roman of the previous century, had been of this type, as attested to by some fragments of his writings on dreams and the wisdom of the Magi. The Pythagorenas, hindered by the majority of educated Romans, who considered them dangerous to the order of the state, were notably successful in certain circles; the discovery that dates back to this century of a great underground Pythagorean basilica in Rome, under the Porta Maggiore, is evidence of their popularity. The many prophets and priests who travelled from town to town preaching about how human beings could become transfigured into God often presented themselves as examples. Apollonios of Tyana, glorified for his miracles and for the mysterious revelations with which he had supposedly been entrusted, was one of these. The main source of information on this legendary figure is the *Life of Apollonios*, written in the third century by the rhetorician Philostratos. It is significant that the work emphasizes the theme of the dignity and freedom that the philosopher enjoyed as

opposed to the fear that the despots (i.e. the emperors) and slaves of passion in general felt towards philosophy.

The theme of the presence of evil in the world, which already existed in Neo-Pythagoreanism, was enlarged upon by the Gnostics, members of a sect that was popular sometime before Christianity. The name came from the true knowledge (gnosis) of which the Gnostics claimed themselves depositaries. The world was slavery, oppression, exploitation, and the reign of evil. The Gnostics were thus in complete antithesis to the optimism of the Stoics, who saw the realization of divine providence in the world.

Judaism and Christianity

The Hebrew culture of the diaspora had very special features. The mediation between Jewish tradition and official Greco-Roman culture performed by Philon of Alexandria (25 B.C. A.D. 41), author of an allegorical commentary on the *Bible*, is significant. The Scriptures were the word of God, and had been given to Moses and the other prophets of the Hebrew people. But according to Philon, Plato was also inspired by Moses and by God himself, so that in principle there were no obstacles to reconciling revealed knowledge with the rational truth of philosophy. Philon's writings on philosophical subjects, such as providence and the eternity of the world, influenced by Platonism and Stoicism, are also along these lines. Other themes, such as God's being unknowable through human reason, the transcendency of the divine with respect to nature, and the resulting reduction of philosophy to a preparatory study for an entirely different kind of knowledge, transmitted by prophets who were directly inspired by the divine word, connected Philon to the Jewish sects that were not integrated into the official culture—for example the Essenes and the Therapeutics—and to the Gnostic and Neo-Pythagorean circles of the Alexandrian environment. In the short run, Philon's conciliatory program was destined to fail in the face of the harsh anti-Jewish persecutions of imperial times. This culminated in the destruction of the temple of Jerusalem under Titus. But in later centuries, Philon's solution to the relationship between philosophy and revealed truth influenced the more sensitive Christian thinkers.

The idea of the *logos* as a mediator between the divine creator's being and the being of the creations was not new. It had already appeared in the Old Testament Book of Wisdom and, as we have just mentioned, was a central theme in Philon's thinking. The novelty of Christianity lay in its identification of the *logos* in a concrete historical person, Jesus. Jesus' preachings, unlike Philon's synthesis, were not directed toward intellectual elites, but rather toward the Palestinian masses, whose only cultural reference was Judaism and for whom the presence of the Romans was only the last link in a chain of slavery that the redeemer would break. Thus it was natural that Jesus' preachings, which have come down to us in the three *Synoptic Gospels* (Matthew, Mark, and Luke), continually referred to the themes of Judaism, though with considerable originality and a detachment from all the religious trends within Judaism. Christian preaching was in fact poles apart from the attitude of Sadducees, who were satisfied with a formal observation of Mosaic Law and still accepted Hellenism's purely mundane view of life, and of the Pharisees, who insisted on a strict, literal observation of the law. It also distanced itself from the asceticism of the Essenes. Asceticism was an end in itself for the Essenes, while in Christianity, distancing oneself from the world had meaning only in the prospect of the Kingdom of God. The death sentence and crucifixion soon gave asceticism a divine value and meaning that was echoed in the early developments of the Christian community. The doctrine was developed much more fully during the following decades, with the decisive contribution of Paul, a native of Tarsos in Cilicia, who studied in Jerusalem and Antioch. During his long journeys in Asia Minor, Thrace, and Greece, Paul spread the teachings of Jesus among the gentiles— those who were outside the Jewish community. His *Epistles* to the communities that were founded during these journeys took on particular importance. The orginality that Paul saw in Jesus' preachings with respect to Hellenistic culture as well as to Judaic religious tradition emerges in the more reliably authentic *Epistles* (those written to the Corinthians, the Galatians, and the Romans around 55-58). Typical themes of Pauline Christianity are a strong egalitarianism ("there is neither Jew nor Greek, there is neither slave nor free, there is neither male nor female; for ye are all one in Christ Jesus"—Epistle to the Galatians), faith in the imminence of universal liberation through the coming of the kingdom of God, and the contrast between Christian virtues (hope, simplicity, and love) and the values of pagan intellectualism and Judaic formalism. It was because of these themes that Christianity spread rapidly among very different social classes, who were all linked, however, by a sense of oppression, desperation, and slavery. Paul was also aware of the need for Christians to confront the ruling culture and set up a hierarchically structured organization dedicated to developing its doctrinal orthodoxy.

Pressed by the needs of an apostolate within an imperial society, outside its original environment of Palestinian sects, Christianity took on some of the linguistic and conceptual elements of Hellenism. In the fourth Gospel, which according to some authors John wrote at the end of this century, Christ is definitely identified with the *logos*.

MATHEMATICAL SCIENCES

The influences of astronomy on mathematics

While pure mathematical research had almost completely died out by this century, that part of geometry that was more or less directly tied to astronomy continued to develop. Since the goal was to construct a kinematic model of the cosmos based on the circular motions of the planets, there was a need for abundant information on the geometric properties of spheres. For quite some time, various authors had been gathering and elaborating detailed information under the generic name of spherics. Autolycos of Pitane, a contemporary of Euclid, was the first scholar of this type known to us. Another author of a famous *Spherics* was Theodosius of Tripoli, who flourished around 150 B.C. But the apogee of these studies was reached in the work of Menelaos of Alexandria, who lived between the first and second centuries of the Christian Era. In his *Spherics*, the theory of figures drawn on spheres took on a form that was to become definitive. In particular, he introduced the study of spherical triangles and determined many of their basic properties.

Applied mathematics

Along with mathematical studies of a philosophical nature, applied mathematics also developed, mainly due to the Romans, who were particularly interested in land surveying, which they called gromatica from the "groma," an instrument used in measuring land. Special schools of gromatica were set up to turn out functionaries educated in practical geometry.

Neo-Pythagoreanism and Neoplatonism

The birth of Neo-Pythagoreanism and Neoplatonism during this epoch also influenced mathematics. The sciences had lost some of the autonomy with respect to philosophy that had characterized the golden age of mathematics. Philosopher-scientists whose philosophical interests prevailed over their scientific interests had begun to reappear. However, mathematics held a place of honor in these new philosophies, and this fact indirectly caused a certain amount of development to take place in the discipline, especially in arithmetic. The Neo-Pythagorean Nicomachos of Gerasa (c. 50-110), for example, wrote an 'Αο-ιθμητικη εισαγωγη (*Introduction to Arithmetic*) which, besides its many philosophical speculations, includes several interesting inquiries into various classes of numbers and their relationships. Iamblichos of Chalcis was one of the Neoplatonists who did research in arithmetic during the following century.

ASTRONOMY

Astronomy and measuring time in Rome

The Romans were well aware of the differences in the lengths of the days at various latitudes, therefore, from the fact that the longest day lasted fourteen hours in Alexandria, fifteen hours in Italy, and seventeen hours in Britain, Pliny the Elder (23-79) deduced that countries near the pole had a twenty-four - hour day in summer and a twenty-four - hour night in winter.

Pliny and Seneca

Pliny's *Naturalis historia* contains many astronomical themes, including a debate on the shape of the earth which makes references to previous discussions on the same subject. Seneca (4-65) also displayed enthusiasm for astronomical research in his *Naturales quaestiones*, which contain a considerable amount of information. However, both authors rarely dealt with original research. Within the context of the great encyclopedias, astronomy, like all the other scientific disciplines, was usually treated by comparing the various existing authorities.

Astrological themes

In Pliny's time, the Romans already attributed an influence on human affairs to the moon and fixed stars. "Who does not know," wrote Pliny, "that when the constellation of Canis rises, it exercises an influence on a great part of the earth?" The influence of the constellation of Canis was an idea that went back to time of Hesiod. Also, the well-known relationship between the moon and the tides had been extended to the human body as well. Astrological divination, which Cicero had mocked in his *De divinatione*, was quite a common practice by this time. The idea that all events in human life were subordinate to the influence of the fixed stars began to gain a hold and be accepted by both the educated and the illiterate, according to Pliny. Many of these themes already existed in Greek and Hellenistic thought, but until the first century B.C. the Romans did not know that these ideas, including astrology itself, had arisen in ancient Babylonia. Their influence was intensified during the first century of the Christian Era by the diffuse climate of mysticism that spread throughout the empire and that appeared to have come from the Eastern provinces. This explains the common opinion, which Vitruvius, for example, accepted, that astrology had Chaldean origins.

PHYSICAL SCIENCES

The great scientific encyclopedias

In the area of natural science, works of synthesis and arrangement, great encyclopedias, and compendia, rather than original investigations, were the rule during the time of the Roman Empire. The tendency toward this type of compilation had already existed during the Hellenistic Age (second and first centuries B.C.), and was now taken up by Latin scholars. Thus the *Naturales quaestiones*, by Seneca (4-65), and the Naturalis historia, by Pliny the Elder (23-79), contained considerations on physics. The amplitude of information in works like these often existed at the expense of a critical attitude. Pliny, for example, said about the remora (a Mediterranean fish) that it could dominate the ocean's violence. Despite the blowing of the wind and the hurricane's rage, the fish, he said, would resist their power and even stop ships just by attaching itself to them. His lack of an adequate theoretical foundation in physics is evident.

Optics: Cleomedes

Greek scientific culture produced a considerable body research on problems in optics. Earlier, Heron, in his *Diopter*, had applied the theory of reflection to designing measuring instruments which utilized the fact that the angle of incidence and the angle of reflection are equal. The study of refraction had also been improved in many ways.

During the first century A.D., Cleomedes explained how an object that is immersed in an opaque tank and is just hidden by the edge of the tank can be made visible by pouring water into the tank. To explain this, he applied a principle concerning the behavior of light when it passes from one medium to another medium of a different density (for example, from air to water), likening this phenomenon to the apparent deformation of oars and poles immersed in water, which had been known for centuries. He then applied the same principle to the atmosphere, and formulated the theory that even when the sun was below the horizon, it could be seen under certain circumstances.

Acoustics: Nicomachos of Gerasa

The study of harmony had been an essential part of Greek culture since the time of the Pythagoreans. Acoustics was both and art and a science: an art for musicians, who were basically interested in auditory sensations and only secondarily in the physical aspects of sound, and a science for those mathematicians whose interest lay in the physical study of sounds and in efforts

to build a scale. Aristotle himself had attempted to solve some problems of acoustics in his *Problemata*. Interest in this type of research, which belonged to both Pythagorean and Peripatetic tradition, was again awakened around the year 100 by Ἐγχειρίδιον ἁρμονιχης (*Manual of Harmony*), written by the mathematician Nicomachos of Gerasa (c. 50–110). Important developments took place in acoustical studies during the following century, although they were not up to the level of studies in optics.

LIFE SCIENCES

The Pneumatic School

Influenced by Stoicism, Athenian physicians upheld the existence of a close relationship between substance and form, which were supposedly united in living beings by an aerial principle, the *pneuma*. According to the so-called pneumatic school of medicine, the essence of life did not reside in the humors, nor in the solid structures, but in this gaseous, semimaterial principle.

Aulus Cornelius Celsus

In the area of practical technology, the Athenians discovered water filtration during this period. Encyclopedias of natural history and medicine also reached their highest level of diffusion. In the first half of the century, Aulus Cornelius Celsus compiled a large encyclopedia, of which only the part specifically dedicated to medicine is extant. It is a work of remarkable clarity, accuracy and objectivity of critical judgment.

Pliny the Elder's Historia naturalis

In his thirty-seven-volume *Historia naturalis*, Pliny the Elder (2379) displayed boundless credulity. In any case, his work is a precious document that preserved for posterity much of the knowledge and many of the opinions of natural science during his time.

Practical medical treatises

Pedanius Dioscorides, a military physician of Greek origin who wrote in Greek although he belonged to the Roman world, united all the existing classical knowledge of medicine into one treatise. His work, which is rich in practical experience, is meticulous and scholarly as well as unbiased and contains valuable information on plants, animals, and minerals.

Scribonius Largus compiled a collection of medical prescriptions. Marcus Apicius' work on the culinary arts also belongs to this genre of treatises. Toward the middle of the century, Thassalus of Tralles made a clear distinction between acute and chronic illnesses.

APPLIED SCIENCES

The extraction of ores

Strabo (c. 63 B.C.–c. 21 A.D.) describes the procedures followed in Greek gold mines. He states that at that time more gold was extracted from flood sands than from underground and that the sand was sifted through basket-shaped woven sifters. Red ochre, whose color comes from iron oxides, was extracted from quarries in Cappadocia, according to Strabon and Dioscorides. Other pigments were accurately described by Pliny the Elder (23–79), who also mentions Tyre purple, a dyestuff extracted from the mollusc *Murex brandaris*.

The mining activity of the Romans was always intense, especially in Spain, where mercury ores, especially cinnabar (mercuric sulphide), began to be extracted. Cinnabar was sometimes confused with minium, at least in name. The production of mercury did not present any great problems, since it was so mobile; the ore was roasted and the metal distilled off. One of its main uses was in the treatment of auriferous ores when amalgamation, a new method of refining gold, was introduced. In this process, crushed auriferous ores were treated with mercury, and the resulting amalgam was often separated from the gangue by squeezing it through skins.

Marble as a building material

Marble quarries were being worked throughout the Roman Empire. Up until this period, the austere Romans had rarely used marble, although they built with excellent stone. But with the coming of the empire and its commemorative needs, the use of marble spread quickly, to the extent that porphyry, extracted from the Egyptian quarries opened by the emperor Claudius, became known as the Roman stone. The quarries of Carrara provided Rome with a pure white marble called Luna, after the port of that name from which it was shipped. Marble was cut with hand-operated saws, although water mills may also have been used.

It has been noticed that the building stones from this period have remained newer looking than those used in later times. The secret of their freshness lies in the methods of production that were used and later lost. The main source of information about Roman building techniques is Vitruvius' *De architectura* (see PHYSICAL SCIENCES 100 B.C.–0).

According to Vitruvius, the successful aging of building stones depended on the care with which they were selected and seasoned. He describes the sequence of steps in the procedure as follows: "The stone must be extracted two years before [its use], in the summer and not in the winter, and left exposed. The stones that appear damaged by the weather after this period should be used for foundations. Those that are not defective have been tested by nature and, used in above-ground construction, can last for a long time."

Roman architectonic techniques

Unlike the Greeks, the Romans used trabeated architecture more for decorative than for structural purposes. However, Roman architecture did not develop the forms and techniques that later characterized it until the first century A.D., and these evolved from Etruscan rather than Greek building technology. From the Etruscans, the Romans mainly learned the art of building with arches. During the imperial period, they developed this art in an original way, constructing brick domes and vaults. Vaults are continuous arches; the limits of their use in constructions lie in the difficulties encountered in supporting them, as well as in the enormous thrust they exert outward on supporting walls and pillars. However, the Romans succeeded in building spans of thirty meters and more, by adding cross vaults to reinforce the support walls and pillars of the main

vault, as well as by lightening the weight of the vault itself by using light-weight puzzolan concrete, a natural cement of volcanic origin found at Pozzuoli (near Naples) and in the Albanian mountains, between the brick arches. The framework for this type of structure could be lighter-weight and therefore less costly. Vitruvius tells us that the Romans preferred flat terracotta roof tiles covered with curved semi-circular tiles, a system similar to that used in the Greek world, and that they excelled in the art of plastering.

Frontinus' treatises

Sextus Julius Frontinus (c. 30–104), a writer and soldier, was the author of the *De aquis urbis Romae*, a history and description of the aqueducts that supplied the Empire's capital. By then the Empire had a vast network of roads, with a total length of sixty to seventy thousand kilometers, making it possible to move troops rapidly from one part of the empire to another.

Frontinus also wrote *De agrimensura*, a discussion of land measurement, as well as *De re militari* and *Stratagemata*, treatises on military topics.

100–200

THOUGHT AND SCIENTIFIC LIFE

The cultural policies of the Antonines. The Late Stoa

The establishment of power in the empire by structuring it and providing it with a constitution was just as ambiguous with respect to the Senate and the people of Rome as the founding of the empire had been at the time of Augustus. The grave crises that developed during the period between Nero and Domitian (who died in 96), led their successors, the so-called emperors by adoption (the Antonines) to define the functions of the prince and the state through a whole series of mediations that had not existed in the policies of the Julians or Claudians emperors or Flavians. The centralization of power into the hands of one ruler was not only not supposed to lead to tryanny or an eastern-type monarchy (to which both Caligula and Nero had aspired), but was to be defined as the supreme magistrature of the *res publica*. Along with this redefinition of the functions of imperial power, provincialization was increased, citizenship requirements were broadened, and the Senate was opened up to men of different origins, until the logical outcome was reached with the *Constitutio antoniniana*, written in 212 under Caracalla.

At the same time, for the entire century, from Trajan to Marcus Aurelius (who died in 180), imperial political propaganda aimed in various ways to show how the freedom and autonomy of the ancient Greek city-states, the great centers of Hellenistic culture, and the eastern regions could co-exist with Rome's increasing need for centralization.

The idea of a political order that could harmoniously unite different peoples and cultures into a single state had its correlative the view of a cosmic order ruled by a single principle which was the *raison d'être* of everything. Stoicism's optimistic and providentialistic concept became the typical philosophical point of reference for this political and cultural experiment. The most original features of the Late Stoa appeared during this period, while increasingly less attention was being paid to the gnoseological and logical problems which had been so important in the doctrine of Stoicism's founders. Even the theory of the universal fire was slowly transformed from a physical concept of the world into a pantheistic-type metaphysics that had the function of rationalizing the growing mystical-religious currents. Seneca's work already contained this aspect, and it was even more pronounced in the reflections of Marcus Aurelius, the emperor-philosopher who instituted a chair of philosophy in Rome in 176 and wrote, in Greek, the twelve books Tὰ εἰζ ἑαϯόν (*To Himself*), also known as *Meditations*.

Developments in Judaism and Christianity

While Stoic moral philosophy aimed at the attainment of serenity, the Christian doctrine proposed to save souls. There was an abyss between the ''rationalistic religiosity,'' as Mondolfo calls it, of the Roman Stoics and Christian voluntarism, even though certain Pauline attitudes toward the failure of worldly science compared to the salvation preached by Jesus had softened. The translation of the Hebrew books of the Bible into Greek was an event of considerable cultural importance. This version came to be called Of the Seventy because of the large number of translators who supposedly worked on it. The translation fostered an interest in Jewish culture among many exponents of Greek thought.

The dialectic between Gnostics and Christian thought was extremely complex. During this century, Gnosticism matured in the work of Basileides, who taught at Alexandria between 120 and 140, and of Valentinus, who was quite influential between 130 and 160 in both Italy and Egypt. Without going into descriptions of the various Gnostic doctrines, we shall only mention that the pessimistic interpretation of several Platonic themes and the mystic and ascetic spirit that accompanied them was a point which Gnosticism had in common with several of the currents within the Christian community. As the persecutions worsened, these Christians tended to see the world, civil society, and the State as irreparably evil, and interpreted their own faith as an escape toward the divine. A new *gospel*, written by the Gnostic Marcion (died c. 160 in Rome) who organized a vast religious community inspired by severe rigorism, came out of this environment. It is not surprising, therefore, that Christianity often seemed to be a version of Gnosticism, and that many Gnostics became Christians. The Church, however, soon condemned Marcion's *gospel*.

Justin Martyr and the apologists

Apart from their relations with the Gnostics, the Christians increasingly felt the need to find a productive meeting ground between Jesus' preachings and the cultural philosophical circles, and attempted to find connections uniting Christian truth and the philosophy of the Gentiles. A reconciliation of Plato with Christ, of the *Timaeus* with *Genesis*, had to be possible, according to Justin, who was born at the beginning of the century and lived at Ephesos and Rome (where he died a martyr in about 145). According to Justin, Socrates and Abraham were ''Christians before Jesus,'' inasmuch as they were guided by the *logos*. Once the *logos* was reincarnated in Christ, the continuity between human and divine reason was guaranteed, thanks to the intervention of faith.

Justin's *Apologies* belong to a typical genre of Christian literature called apologetic literature. The current had begun several years before and continued long after Justin. The first apologists were Quadratus and Aristides, who each wrote an apology to the Emperor Hadrian between 120 and 130. Quadratus' work has been lost, but from the extant pages of Aristides' apology, we can see how the author maintained that a comparative study of barbarian, Greek, and Jewish ideas of the divine would show that only the Christians had a correct concept of God. Around 140, Ariston of Pella wrote *Altercatio Jasonis et Papisci* on Christ.

Tatian, a pupil at Justin's school of Christian philosophy in Rome, also belonged to the apologetic current. He wrote *Discourse to the Greeks* sometime between 166 and 171; in the polemic against Greek religion and mythology, he develops the hypothesis that the Greeks derived many of their philosophical theses from the *Bible*. Around 177, Athenagoras of Athens addressed *Apology for the Christians* to Marcus Aurelius and his son, Commodus, in which he rejected the accusation of atheism made against the Christians and tried to demonstrate dialectically God's uniqueness. Shortly after 180, Theophilos of Antioch sent three books *Ad Autolycum* (no longer to an emperor, but to a private adversary), which defended Christian theories of the essence of God and of providence.

The pagan reaction to Christianity. Celsus

The conflict between the imperial authority and the Christians during the previous century was due to the Roman world's hostility toward the Jews and the Christians, who were confused with them. These peoples represented dynamic forces that were not easily tolerated by the *pax romana* in the East (according to Tacitus, for example, the Jews were *deterrima gens* and Christianity was *execrabilis superstitio*). During this century, the new doctrine assumed increasingly marked features, which helped bring it to the forefront with respect to Judaism. In spite of the Christians' decision to recognize the empire's authority over worldly things, the establishment of a church that claimed autonomous power, even though it was only in the realm of the spirit, refused to accept the deification of the emperor, and made a distinction between political obedience and freedom of conscience, caused the central power, which was usually tolerant of various religious creeds, to see Christianity as much more dangerous than Stoic-type Roman traditionalism or Jewish independence. By questioning some of the bastions of the Roman concept of the state, especially at a time of delicate mediation between the *auctoritas* of the *princeps* and the power of the Senate, the Christians, in the eyes of the elite of philosophers, jurists, bureaucrats, and Roman scholars, placed themselves outside human society itself and violated the rules of reason and civil life. Thus the unyielding persecutions were often promoted not by bloodthirsty autocrats (as later Christian tradition, especially Lactantius, would have us believe), but by emperors concerned with the good of the state.

As Christianity continued to organize for its survival (often with methods that other non-official sects had already used, such as the *collegia funeraticia*, underground places of worship, forms of mutual solidarity, etc.), and in fact began to spread, numerous pagan intellectuals—some of them tied to the traditions of official sects, and others sensitive to the mystical trends that came from the East—began to feel a religious and moral duty, even more than a political duty, to combat a doctrine that to them seemed both antirational and antisocial. A typical representative of this trend was Celsus, who probably lived in Egypt, and who wrote Ἀληυῆζ λόγοζ (*True Word*), in about 180. We know of this work through Origen's refutation in Κατὰ Κέλσο'(*Against Celsus*). Celsus' work, in four books, was the first systematic critique of Christian thought written by an author who was extremely well-informed about the new religion and had a considerable knowledge of philosophy.

Middle Platonism

Aetios' *Placita*, which took the place of an older collection based on the Θυιχωῦ δόξαι (*Opinions of the Physicists*) by Theophrastos, is representative of the doxographical literature of this same period. The trend continued in the next century as well, culminating in Diogenes Laërtos' *Lives of the Philosophers*. Plato was studied the most, with commentaries and manuals. A sentence taken from Porphyry's *Life of Plotinus* mentions "the commentaries of Severus, Cronius, Numenius, Gaius, and among the Peripatetics, Aspasios, Alexander and Adrastos."

Among the Platonic commentators, according to Proclos, Severus (the author of a lost *De anima*) wrote a commentary on the *Timaeus*, in which he proposed original theories on the creation of the world. We should also add the names of Atticus, a Greek and a Platonist who may have headed the school of Athens at the time of Marcus Aurelius (Atticus was praised by Proclos as a "very hard worker"), and Albinus, mentioned frequently in antiquity. *Tertullian* discusses Albinus' doctrine of the soul; Galen (129–200) calls him his teacher (at Smyrna, around 151–152); Iamblichos appeals to his authority; and Proclos sees in him one of the "leaders of the Platonists." Only a *Prologue* to Plato's dialogues has remained of Albinus' writings, although he may be the Alcinous who wrote the *Epitomē*, or *Didaskalikos*, that is included in various Platonic manuscripts.

The general tendency of the "middle Platonism" that began with Plutarch, was to integrate the religious needs that came from the East with traditional philosophical doctrines, in a synthesis of the Platonic doctrine with touches from the Stoic and Aristotelian texts. It is not an accident that many commentators of Plato were also commentators of Aristotle, and that all of them were attempting to establish a common concept of the world and the divine. An important factor in the interpretations of Plato made by Albinus and Apuleius of Madadura—author of the famous *Metamorphoses*, or *Asinus aureus*, and of a *De Platone eiusque dogmate* (*Doctrine of Plato*), and the Latin translator of the pseudo-Aristotelian *De mundo*—is the continuous attempt to demonstrate rationally Platonic theses using the Aristotelian method. Both Albinus and Apuleius derive Plato's logic from Aristotle, through investigations that some historians say had a great influence on later developments in the study of logic. But at the same time, these investigations were tied to an attempt to propose Plato as a theologian, reread in the light of Neo-pythagoreanism and the influences of Eastern cults, to which Apuleius gave particular attention.

The spread of the occult sciences

The tendency toward irrational beliefs that had already existed for some time began to spread. It showed up in the attitudes toward problems of explaining nature and human beings, and

can even be found in those authors who contributed to making classical scientific tradition flourish again during this period. Galen, for example, revived and strengthened research in biology, even though he stated his concepts in an extremely unsystematic way. Although it appears certain that the accusations of magic that several contemporary physicians made against him were the product of professional envy and that he was often skeptical of medical superstitions and popular medicine, it is also true that his pharmacopea and therapeutics were much less accurate than his anatomy and physiology. He gave a lot of credibility to commonly accepted remedies, insufficiently justified medical practices and fantastic, unbelievable stories. Ptolemy (c. 100–170), the scholar who arranged and codified astronomy, was also the author of the most famous treatise on astrology of all antiquity, the *Apotelesmatika* (or *Tetrabiblos*). He tried to give the subject a scientific dimension, but on the whole his attempt proved useless and inadequate.

Magic, astrology, and alchemy, which almost replaced scientific research during this period, were the different forms which the irrational attitude of the times took. Their origins are traditionally attributed to the East and the knowledge of ancient priests. According to a legend voiced as early as Pliny the Elder, magic—a set of precepts with the goal of provoking events contrary to the regular course of nature using supernatural forces—went back to Zoroaster, who supposedly entered the West when he accompanied Xerxes in his expedition against Greece. Astrology—a type of divination that attempts to derive knowledge about the future from the positions and motions of the stars—was based on the observations of the heavens that had supposedly been made in the temples of Chaldea since the earliest times. Alchemy—a collection of methods that were believed to transform base metals into gold and silver—came from the collections of artisans' recipes for working metals, real secrets that had supposedly been jealously kept by the priests of Egypt.

The first treatise on alchemy goes back to the second century. It violated the ban against divulging secret practices. The essential core of these doctrines came from Greek and Roman culture, and the emphasis on their Eastern origins arose from the typically Greek tendency of considering the East as the mythical source of all wisdom. The texts on magic, alchemy, and astrology are, with rare exceptions, really religious texts. The theme of getting closer to the divine was actually an element of unification among various cultural traditions.

MATHEMATICAL SCIENCES

The work of Ptolemy

Astronomy's contribution to the development of mathematics reached its acme in the *Almagest* (see ASTRONOMY 100–200) by Claudius Ptolemy (c. 100–700), which also contains the most complete evidence of the classical world's knowledge of spherical and plane trigonometry.

The section on spherical trigonometry (the part that is directly applicable to astronomy) definitely prevails over the plane, further confirming the interest in practical applications (though in a lofty sense) that guided mathematics during this period. Spherical trigonometry thus preceded plane trigonometry which, although it was easier on a theoretical level, did not have any important applications in astronomy.

ASTRONOMY

The great Ptolemaic synthesis

In his Μεγάλη μαθηματιχὴ σὺνταξιζ (Great Mathematical Syntaxis, or Syntaxis), also known as Μεγίοτη σὺνταξιζ (al-Magistī in Arabic, Almagestum in Latin, and our Almagest), Claudius Ptolemy presented a vast astronomical synthesis that took into account all of the planetary motions, observed with an accuracy slightly better than normally possible with the naked eye, or within five arc seconds. Although a great part of Ptolemy's success was due to his intelligent utilization of previous authors (Hipparchos among them), the great achievement of the Alexandrian astronomer and geographer was his grandiose scheme for explaining the entire universe. Another of Ptolemy's specific contributions was to situate the earth within the principal circle of planetary orbits but in an eccentric position, so that the planets rotated around another point, the equant, which was located at the same distance from the earth on the opposite side of the center of the circle.

In the thirty books of the Almagest, Hipparchos' geocentric theory (see ASTRONOMY 200–100 B.C.) on the motions of the sun, the moon and the known planets (Mercury, Venus, Mars, Jupiter, and Saturn), which Ptolemy improved, is considered in detail. The theory is represented by a kinematic mechanism made up of a system of circles (deferents and epicycles), devised by Apollonius of Perga. The catalog included in the Almagest lists the positions of more than a thousand stars. It is based on Hipparchos' star catalogue, to which Ptolemy added 170 stars, after recalculating the apparent sizes (subdivided into six classes) of all the stars visible to the naked eye. The Almagest also contains tables of the apparent motions of the sun, the moon and the planets, tables of chords of angles, and descriptions of instruments for making observations and computations.

Ptolemy's encyclopedic work was very popular in a number of cultures up until the fourteenth century.

Ptolemy's Geography also became very famous; it includes a description of the entire known world of the time, as well as the geographical coordinates of many places.

The progressive spread of astrology

Ptolemy himself did not escape the mystical and religious climate of his time. His Tetrabiblos was dedicated to the study of the influence of the stars on human events (see THOUGHT AND SCIENTIFIC LIFE 100–200) although he proposed his work as a scientific investigation into controversial astrological topics. The vast diffusion throughout the empire of innumerable eastern cults were magical rituals and astrological doctrines played an essential role, ended up making astrology dominant over mathematical astronomy. Medical practices also became increasingly more closely tied to astrology. The phenomenon was widespread and involved other sciences as well; theoretical-mathematical elaboration and systematic observation were often abandoned for unverifiable procedures.

It is important to remember that early Christian writers tended to reject astrology uncompromisingly, since in it they saw the negation of the doctrine of free will. All of the Church fathers, from Tertullian and Lactantius to Augustine, incessantly polemicized against divinatory practices and especially against astral dogmatism.

PHYSICAL SCIENCES

Ptolemy's optics

In the midst of the revival of Alexandria as the center of scientific research, directly favored by the imperial authority, the predominant figure in physics was again the geographer and astronomer Claudius Ptolemy.

His treatise on optics minus the first book, has come down to us in a translation into Latin from the Arabic. The vicissitudes of the text (the extant version is certainly not without errors) make it difficult to judge the mathematical importance of the work. In any case, it is of unquestionable interest to physics. Following Euclidean tradition, there is an extensive theoretical treatment of flat, concave, convex, spherical, and cylindrical mirrors. It also contains a theory on vision and a study of optical illusions. The most original part appears to be the treatment of light refraction in air/water, air/glass, and water/glass media. The approach is both theoretical and experimental. Ptolemy describes several observations made with an instrument for measuring angles, a graduated copper disc with a mirror at the center and an index for reading the values of the angles. The observations were used to check what he had deduced geometrically from principles that we know did not correspond to the modern law of refraction (established in the seventeenth century), but still made it possible to associate variations in the ratio between the angle of incidence and the angle of refraction with variations in the angle of incidence. Ptolemy's successors, however, assumed this ratio to be constant, performing a gross simplification.

Another important contribution that Ptolemy made, in the tradition of Cleomedes, was an accurate study of astronomic refraction. Ptolemy deduced that the stars were elevated in appearance due to astronomic refraction, so that stars that had not yet risen, on one hand, and stars that had already set, on the other, were visible on the horizon.

Acoustics

While the study of optical phenomena was evolving through a series of experimental observations, physical, physiological, and psychological theories, and metaphysical speculations on geometrical optics, experimental observation was also beginning to be used to check acoustical theories more carefully, again thanks to Ptolemy. In his work on acoustics, as well as that on optics, Ptolemy was able to synthesize various traditions, utilizing the work of both musicians and mathematicians. His work on acoustics notably influenced medieval thought.

LIFE SCIENCES

Eclecticism and the flourishing of surgery

The history of medicine in this century was characterized by a refined eclecticism (Rufus and Soranos of Ephesos, Galen of Pergamos [c. 129–200]), and the flourishing of surgery, with Antyllos (second century) as its most illustrious representative. Rufus did an important study on pulses and several studies on anatomy. Soranos (second century) was the leader of the methodists and an especially accomplished practicing physician. His works on obstetrics are among the best in classical Greek medicine.

Aretaeos of Cappodocia

At the end of the century, Aretaeos of Cappadocia (end of the second century) wrote a manual on the etiology, symptoms and treatment of acute and chronic illnesses. An eclectic, Aretaeos excelled in the clinical description of diseases (diabetes, diphtheria, hemiplegia, etc.).

Galen. The organization and unification of Greek medical thought

The predominant figure of Galen overshadows the work of his contemporaries in medicine and biology. Galen considered himself a disciple and continuator of Hippocrates. But, being an eclectic and very versatile, he was able to profit from the anatomical discoveries of the Alexandrian school, his predecessors' clinical descriptions, and the intellectual tools of Platonic and Aristotelian philosophies. He wrote numerous treatises using research methods that remained a standard in anatomy, physiology, and humoral pathology for more than a millennium.

APPLIED SCIENCES

Agriculture

Methods of cultivation did not undergo any important improvements during this period. As a matter of fact, data on the yields of the best pieces of land in the empire indicate that the harvest hardly ever exceeded four times the amount of seed used.

The cultivation of linen developed somewhat, leading to a decrease in the use of wool for clothing.

Bronzes and ceramics

Crafts such as bronze- and ceramic-working continued to become more specialized, in order to fulfill the needs of a constantly growing population. Bronze-working had begun to flourish in the fourth century B.C.; and a remarkable variety of techniques (fusion, lamination, the application of *emblemata*, etc.), motifs, and shapes had developed. During the second century, the more industrialized techniques such as fusion and grinding prevailed, shapes became heavier, applications were applied more hastily, and decorations became more geometric.

The ceramics industry became very important in the regions of Campagna and Umbria in Italy, as attested to by the huge mixing tanks, capable of holding up to 45,000 liters, found in several factories. An analysis of the ceramic materials has revealed that the work was quite specialized and that various phases of the process were entrusted to different artisans.

Glass

By the second century, the glass industry already rivaled the ceramics industry, due to the decisive improvements brought about by the invention of glass blowing. The Roman glass industry, centered around the Volturno River delta, where a suitable type of sand could be found and had the aspect of a regional monopoly from the very start. The Roman artisans were very good at making all kinds of shapes, and produced a white translucent glass, similar to crystal.

Engineering projects

By Trajan's reign, the old Claudian harbor of Ostia had become inadequate, so construction was begun on a new harbor, connected to the Tiber by a canal that would keep the river from silting up. The new hexagonal port covered an area of thirty six hectares, was about five-and-a-half meters deep and had a pier that was two km long. We should also mention the grandiose bridge over the Danube, which had twenty square-shaped pylons that were forty-five meters high and fifty meters apart.

Oriental inventions: the pedal and the stirrup

Two fundamental inventions made their first appearance in the Orient during this period. In China, pedals began to be used to operate looms, and in India a sort of rudimental stirrup, a belly band in which riders put their feet, appeared. A hook-shaped stirrup was used in the northernmost regions, as a cameo from the period shows.

200–300

THOUGHT AND SCIENTIFIC LIFE

Sextos Empiricos and the Skeptic synthesis

There were no significant exponents of the Skeptic school after Sextus Empiricus, who lived between the end of the second and the beginning of the third century. Sextus Empiricus belonged to a generation that had seen Skepticism revived in the first century B.C. by Aenesidemos and continued by Agrippa in the first century A.D. and by Menodotos, Theodas, and Favorinos of Arles in the second century A.D. Sextus' main source seems to have been Aenesidemos, although he makes a number of references to the other authors mentioned, as well as to Timon of Phlius and the Academicians (Clitomachos and Antiochos of Laodicea). Sextus extant works are the Πρρώνειοι ἱοπἱώσειζ (*Outlines of Pyrrhonism*) and Πρὸζ μαἡγατιχοζ (*Adversus mathematicos*, or *Against Scientists*). The *Outlines*, in 3 books, present an elementary summary of Skepticism. The first directly defines and justifies the doctrine, while the other two confute dogmatism. *Adversus mathematicos* is a combination of two works, one against science in general and the other against dogmatic philosophers. The first, in six books, reviews all the sciences of the time (grammar, rhetoric, geometry, arithmetic, astronomy, and music), showing their success to be unfounded and demonstrating that contrary affirmations could just as well be proposed against them. The second part, in five books, also called Πρὸζ δογματιχοζ (*Adversus dogmaticos*), refers to philosophers; the first two books confute the logicians, the next two the physicists, and the last the moralists.

Commentators on Aristotle and doxographers

Alexander of Aphrodisias, the "second Aristotle," was nearly a contemporary of Sextus Empiricus. Following the tradition of Andronicos of Rhodes and the Aristotelians of the previous century, he applied himself to his studies of the Aristotelian text with philological scruple. Head of the Lyceum from 198 to 211, Alexander wrote a work on the soul and a great number of commentaries on Aristotle's logic, metaphysics, psychology, and anthropology. He attempted to liberate Aristotle's thought from the religious syncretism and Stoic tendencies that had been superimposed upon it, emphasizing logical nominalism and insisting on the autonomy of the world. Alexander explained the learning process by resorting to the concept of an active intellect that persisted at the basis of science, beyond the contributions of single individuals. Individual contributions, being closely tied to individual souls, would disappear along with the mortal souls.

Alexander's doctrine on the active intellect remained a point of reference for medieval and Renaissance Aristotelianism. The doxographical tradition was also very much alive during this century. The works of Sextus mentioned above were also doxographical in a certain sense as were several works by Aristotelian authors who referred back to Theophrastos in doxography. An example is the *Lives of Philosophers* (in ten books, the largest extant work from antiquity on the history of philosophy) by Diogenes Laërtios, who lived during the first half of the century.

From the Chaldean oracles to the Corpus hermeticum

Along with Aristotelian naturalism and Skeptic empiricism, the religious-mystic trend that had found its way into scientific tradition grew stronger around the beginning of the century. The so-called *Chaldean Oracles* had appeared at the end of the second century. Collected by Julian the Theurgist (also called the Chaldean), they make up a doctrinal body which is an intimate mixture of magic, mysticism, and Platonism. This century also witnessed the arrangement of the so-called *Corpus hermeticum*, a series of works that were almost surely written in Alexandria, Egypt, within those philosophical circles that were dominated by religious syncretism. The work contains seventeen *logoi*, in which the master (usually Hermes, often identified with the Egyptian god Thoth, and the god called *Trismeghistos*, or three times as large) reveals certain doctrines and spiritual directives to his disciples (usually Tat, Asclepios, and Amon). The contents are a mixture of Platonism, Aristotelianism and Stoicism, with some traces of Judaism.

Writings on magic, alchemy, and astrology were also circulated under the name of Hermes. In their original versions, these works probably came from a much earlier period. But it is difficult to make a clear chronological distinction, since the tendency to attribute theories and doctrines to ancient times and Oriental sages was not only a tradition in Greek culture, it also became a necessary feature in the realm of occult literature. When faith in the verifiable procedures that were typical of scientific research was lost, inspiration and revelation became the sources of knowledge and the monopoly of a few, who, detached from the world, were in a favorable position to receive the divine message. There was a great interest in Egypt, the land of the initiative cults; in Chaldea, the country of the astrologers; in Persia, the homeland of the Magi; and in India, the meeting place for the mystical and contemplative currents of the East.

Manichaeanism

Gnosticism had already directed much of its attention to addressing the problem of evil. The Persian Manes, who reportedly lived from about 215 to 274, explained the problem of evil by reproposing the Persian belief in two opposing principles, Ormuzd and Ahriman, good and evil, which were engaged in perpetual and indomitable struggle.

This struggle explained the stormy history of the world, even though the Manichaeans did not exclude a final triumph of the principle of good over evil, where evil would be totally annihilated.

Manichaeanism differed from Gnosticism on many points, especially when it opposed two powers on the same level, since the Gnostics believed that the demiurge, which was responsible for the creation of the world, was inferior to God. The differences were even more profound with respect to Christianity, which considered Manichaean dualism as not just a Gnostic-type heresy but a radical alternative. Harsh polemics arose between Manichaeans and Christians in those territories of the empire where Manichaeanism became popular.

Christian thought from Irenaeus to Tertullian

With the *Adversus haereses*, also known as the Ἔλεγχος χαὶ ἀνατροπὴ τηζ΄φεδονύμο΄γνώσεωζ (*The Unmasking and Confuting of False Gnosis*), by Irenaeus (a native of Asia Minor, thought he died at Lyon), the Christian polemic against Gnosticism surpassed the model set by the early apologetics with remarkable theological zeal. The work of Irenaeus was continued in Rome by Hippolytos (died c. 236) in his *Refutatio omnium haeresum* (or *Philosophoumena*). Hippolytos displayed a hostile attitude toward the doctrines of the Greek philosophers, which he placed together with mystical and astrological beliefs.

According to Hippolytos, the most recent developments in philosophy, science and even mysticism-permeated pseudo-science were just continuations of Gnosticism, from which the Christian doctrine should rigorously keep its distance. But the most radical refutation of classical tradition was expressed by Tertullian, who lived in Carthage from about 160 to 240. "What can Athens and Jerusalem, the Academy and the Church, the heretics and the Christians have in common? Our doctrine comes from the porch of Solomon, who taught us that God should be sought in simplicity of the heart. So much the worse for those who gave birth to Stoic, Platonic, or dialectical Christianity. We have no curiosity beyond Jesus Christ: we do not need to look beyond the Gospel." In his *Ad nationes* and his later *Apologeticum*, Tertullian, who was quite knowledgeable in both Greek language and culture and in all the apologetic literature of the previous century (written in Greek, even though there were works by authors from the western provinces of the empire), re-proposed that Christians return to their original condition and uphold Christian purity in the face of the degeneration of imperial society and its culture. His rigorism led him to embrace Montanus' theory that the second coming of Christ was imminent and one should prepare oneself through an ascetic refusal of the world, a separation of the body of the Church from social ties, and a refusal of any compromise with the worldly structure of the state. But in spite of enormous difficulties, the process that was bringing Christianity and classical culture closer together continued, and Montanism was soon condemned by the Church as a heresy.

However, Tertullian's ideas had considerable influence, especially in northern Africa. Cyprian of Carthage, for example, one of the first great organizers of ecclesiastical life and a prominent figure in the controversies of the first half of the century (he died a martyr in 258), was inspired by Tertullian.

The Christian school of Alexandria. Clement and Origen

The Christian school of Alexandria, which, under Panthenos, cultivated a wide range of scientific and religious interests during the second half of the second century, was of a very different nature. Clement, a teacher at the school between 190 and shortly after 200, unlike those who theorized a return to a primitive community of saints, held that sinners could best be converted not through a global condemnation of Platonism or Stoicism philosophy, but through realizing that Christianity's true enemy was Gnosticism. Clement dedicated almost all the books of his *Stromata* to confuting Gnosticism. Among Clement's works that have been lost were allegorical commentaries on the Bible and works of moral edification. Two extant works, the Προτρεπτίχός πρὸς Ἕλληναζ (*Exhortation to the Greeks*) and the Παιδαγωγός (*Tutor*), show his interest in morals and pedagogy.

The efforts that Clement made to reconcile the *humanitas* of late Platonic-Stoic pagan philosophy with Christianity is quite obvious in all these works. They were made in the belief that a confluence with the Christian doctrine would be better for the best traditions of classical thought than dissolving into Neopythagorean or Gnostic irrationalism.

Origen's position as the head of the Christian school of Alexandria from 203 to 232 was much more difficult, both on a personal and on a philosophical level. Origen was the author of allegorical commentaries on the Bible, apologetic works, such as Κατὰ Κέλσο' (*Against Celsus*), theological works, such as the four-volume Περὶ ἀπχδζ' (*On principles*), and writings of an ascetic nature, such as Ελζ μαρτ'ιο' προτπεπτιχάζ (*Exhortation to Martyrdom*). While Justin and Clement had demonstrated the compatibility of Christian and Platonic-Stoic morals, Origen attempted to incorporate Christianity into a speculative structure that had its justification in the more refined traditions of philosophical thought. Origen's bold theological work was often conducted with such broad-mindedness that some of his opinions were considered heretical because they approached Platonic and Stoic ideas too closely.

Neoplatonism. Ammonios Saccas, Plotinos, and Porphyry

In Alexandria, the great center of scientific life (Alexandrian mathematicians made some brilliant findings during this period, for example), but of metaphysical and religious speculation as well, a philosophical school with Platonic leanings was founded by Ammonios Saccas (who some say was also Origen's teacher) and flourished at the same time as Pantaenus' Christian school.

Plotinos, born about 204 at Licopolis, Egypt, was Ammonios' disciple for over a decade. He went to Rome in 244 and taught there until his death in 270. Among Plotinos' disciples was Porphyry, who organized his master's treatises into the Ἐννεάδεζ (*Enneads*), fifty four pieces grouped into six parts of nine treatises each. Although their form of Neoplatonism could be linked to previous Platonic thinkers (Plutarch and the middle Platonists), it also emphasized the metaphysical aspects of Plato's doctrine. For this reason, it was very selective in its choice of the traditions to which it referred. Its interest was focused on the *Republic*, the *Phaedo*, the *Phaedrus*, and the *Timaeus*, while the other Platonic dialogues were almost completely neglected. These texts furnished the philosophical "revelation" on which Neoplatonism was built—the "major mysteries"—while the

"minor mysteries," needed for a systematic treatment of the major mysteries, were drawn from Aristotle. On this foundation, Neoplatonism developed a complex hierarchical structure of the universe which was purely metaphysical in nature. The principle of the universe, according to Plotinos, was the One or Good (Plotinos uses the Platonic language of the *Republic* and the *Parmenides* indiscriminately). The One, the source of all being, allows being to flow from it because it superabounds. On the second level is the intellect (*Nous*) and the ideas contained in it, and on the third is the soul, which as the "soul of the world" orders the material world, imposing the shapes received from the sphere of ideas.

According to Plotinos and Porphyry, evil, like matter, was privation, not being. As a result, Neoplatonic doctrines had a substantial philosophical optimism which contrasted with those of Gnosticism, Manichaeanism, and in many ways, Christianity. Porphyry, for example, argues strongly against the followers of the new religion in his Κατὰ χριστιανῶν (*Against the Christians*), and proposes the tradition of the Delphic Apollo instead.

Neoplatonism's ideological ties with the society of its time did not end with the proposal of a philosophical religion that could validly stand up to the increasing penetration of Christianity. Neoplatonism, with its metaphysical hierarchy of the universe, offered a point of encounter between senatorial tradition and the imperial concept of the *res publica*. It is not an accident that many eminent senators were among Plotinus' disciples in Rome.

Neoplatonism was not a political philosophy but rather a great metaphysical system, much more refined and developed than many syncretistic theologies of the time. Porphyry's use of Aristotle, for example, in his Εἰσαγωγή (*Isagoge*), a commentary on the *Categories*, which, centuries later, gave rise to the controversy on universals, should also be seen in this light.

Late Christian apologetics. Arnobios and Lactantius

The third century was a time of great political and institutional upheaval, which was connected to the explosion of the great crisis in the Roman world, the premonitory signs of which had already been visible in the previous century. The empire, which had reached its maximum expansion under the Antonines, seemed unable adequately to resist the barbarians pressing at the borders with Germany, or Parthian and later Sassanian expansionism on the eastern borders.

The estate system, where the estates tended to become *villae*, that is, autonomous estates, was based on the progressive impoverishment of large sectors of the rural population, not only in Italy but in most of the provinces of the Empire. This situation fostered discontent and instability, and popular uprisings degenerated more and more often into episodes of banditry.

The function of the army proved to be increasingly more essential, both as a guarantor of the borders and as an arbiter in internal struggles. The so-called military anarchy (which began after the decline of the Severi in 235, and only ended definitely in 305 with Diocletian), strongly criticized by the conservative historians of the following century, had the characteristics of a class struggle and led to the transformation of the principality itself. The men who emerged in this situation, from Aurelian (died 270) to Diocletian, brought a new idea of central authority which took away many of the functions of the

Senate that were not simply representative, and aimed at more clear-cut forms of autocracy, following the example of the solar cults of Mithras. Diocletian's constitutional reform and the establishment of the tetrarchy toward the end of the century constituted both a restoration and an innovation. The last great anti-Christian persecution should also be looked at in this context. When Diocletian's experiment ended, the imperial authority followed the example of Gallienus, who had established a policy of tolerance in his time, and looked toward Christianity for the terms which would reconcile the old and the new, the unification of East and West, and the settling of the great ideological, political, and social conflicts. For this reason, an interest in Christian apologetics developed toward the end of the century. Of course there were those like the African Arnobios, a rhetorician at the time of Diocletian, who converted from an extremely fanatical paganism to an intransigent Christianity, which induced him to write "*luculentissimos libros adversus pristinam religionem*" (his seven-volume *Adversus nationes*) and to see the earthly world as simply an *ianua mortis*. But there were also those like Lactantius, who used elements of the tradition that attempted to reconcile Christianity and classical culture to lay the foundations for transforming Christianity into the official ideological base of the imperial state. Lactantius' own personal history is an example: a pagan, holder of an imperial chair in rhetoric, persecuted for a short time after his conversion, Lactantius became the tutor of Constantine's son as well as the official ideologist of the court.

MATHEMATICAL SCIENCES

Algebra in Greek mathematics

The last spark of ancient mathematics came not from geometry, but from algebra. Around 250, according to the most reliable critical reconstructions, the Alexandrian Diophantos wrote his *Arithmetica*, the only explicitly algebraic treatise that has survived from Greek mathematics. The work is a diversified collection of numerical problems, many of which can be translated into first and second degree equations (one can be expressed as a third degree equation). They are presented and solved, case by case, in specific terms, even though the implicit presence of some more general methods can often be seen. Diophantos' work is especially distinctive for the ingeniousness of the methods used to solve equations and systems of equations, and it shows a real virtuosity in computation. It also contains many arithmetical theorems that are not found in earlier authors although there is good reason to believe that it synthesizes previous knowledge.

It is quite difficult to conceive of Diophantos' work as a product of the development of Hellenic mathematics, since its abstract nature, its generality of approach, and its demonstrative rigor do not belong to that tradition. It can be much better understood in connection with the Babylonian-Alexandrian tradition already mentioned with reference to Heron. As a matter of fact, Diophantos' *Arithmetica* shares its non-systematic nature, its method of proceding by collections of problems, and its computational virtuosity, detached from any search for general principles, with Babylonian mathematics. The Greek influence can be seen in the fact that the problems, although specific, are never concrete, in the sense of finding solutions to effectively

practical problems. Another important characteristic of Diophantos' work is his use of an abbreviated, almost stenographic, form of writing, making him a forerunner of algebraic symbolism. With his work, we are moving from so-called rhetorical algebra, where the calculations are described in words, to so-called syncopated algebra, where a number of abbreviations are used. This type of algebra was later taken up and exploited by the Italian and French algebraists of the sixteenth century, who transformed it until by the seventeenth century it became authentically symbolic algebra.

Pappos, the author of *Μαθηματιχαὶ σναγωγαὶ* (*Mathematical collections*), lived in Alexandria at almost the same time as Diophantos. In his work, a number of fundamentals of Greek mathematics are reviewed, discussed, and reelaborated, with the addition of biographical, bibliographical, and historical data that provide invaluable information on ancient mathematics. Pappos also did original research on conics (introducing projective-type considerations), the theory of isoperimeters, and the theory of barycenters. He also invented the barycenter method for calculating areas and volumes of rotation, a method that was later unjustly attributed to Paul Guldin, who rediscovered it in the seventeenth century.

ASTRONOMY

The Astronomical doctrines of the church fathers

The diffusion of Christianity and the theoretical systematization of Christian doctrine by the church fathers, especially those in the East, brought up the problem of the structure of the world, with particular reference to the story of the creation in the first chapter of Genesis. But the new religion's opposition to classical culture, which was believed to have made too many compromises with pagan religion, left little room for considerations of a scientific nature, beyond allegorical interpretations of Biblical statements.

At the beginning of the century, Clement of Alexandria introduced the theory of a universe in the shape of a Biblical tabernacle with all its furnishings, a model that became very popular in patrology. In particular, the seven-stemmed candelabra represented the system of the seven planets with the sun in the middle.

Censorinus

Regarding pagan culture during the third century, we shall only mention the *De die natali*, an astrological compendium written by Censorinus around 238. It takes up certain Pythagorean themes and revives theories of celestial harmony that are based on the relationships among the distances between the planets. The work is a product of a Neo-Pythagorean cultural environment.

PHYSICAL SCIENCES

The progressive decline of scientific interests

Alexandria housed an extraordinary mixture of religions, philosophies, and sects during the third century, while the old scientific school was slowly fading. Christians, Jews, and pagans lived side by side, as Eastern cults and the cults of "Egyptian knowledge" grew. The impoverished schools of Plato and Aristotle still had a few, strictly conservative, followers. There were some Stoics and fewer Epicureans. An interest in science was very rare among the exponents of traditional philosophical thought. Neither the perspectives opened up by scientific research nor a desire to understand the physical world seemed to attract those who sought refuge from the crisis of the times in mysticism. A few principles of physical knowledge persisted in some of the philosophical theories of the period, so that we find speculations on acoustics and harmony by Neo-Pythagoreans and the Neo-Platonists.

APPLIED SCIENCES

The art of glass-making

The lack of sources from this period is an almost insuperable obstacle to research into the technology of the third century. The storm that was brewing and would later rage seems to have canceled all traces. However, it would be a serious mistake to pass hasty judgment and speak of a general decline in the arts.

The few sources available often report the contrary. Under the *Severi*, for example, the glass-making industry flourished, since glass was used not only to produce vases (both smooth ones and a variety of decorated types), but also for decorating houses (glass tables, etc.). First opaque, then transparent glass was widely used for windows in houses.

Around 220, the first mirrors also appeared.

Aurelian's wall

Except for written sources, there is not much tangible evidence of building from this period. The economic situation did not favor great civil projects, and Rome's urban expansion slowed down considerably. An eloquent indication of the uncertainty of the times is the huge wall, nineteen kilometers long, that Aurelian had built in 272 to protect the city. The wall, which took at least five years to build, made it possible to resist the attacks of Alaric, king of the Visigoths, a little longer, but did not keep him from conquering and sacking the city in 410.

300— 400

THOUGHT AND SCIENTIFIC LIFE

The alliance between Church and Empire. From the edict of tolerance to the Council of Nicaea

Diocletian's persecutions (304–305) were partly motivated by the desire to confiscate the property of the Christians, who controlled both chattel and land because of their rule that converts donate all their worldly goods to the community. They were also a final attempt to break the new religion by force. With the Edict of Milan in 313, Licinius and Constantine decreed that the confiscated goods be given back to the Christians and granted them the freedom to worship publicly. A few years later, Constantine adopted a policy that clearly favored the Christians. Bishops were placed in the courts, the anticelibacy laws were revoked, episcopal jurisdiction was recognized, and

the Church was allowed to possess property and to free slaves. The political counterpart of these concessions was a tight control over institutional structures and even doctrinal disputes. The "heavenly" emperor oversaw the ecclesiastical system, sanctioned juridical privileges for bishops and convoked local synods and councils.

In the long run, this attitude was aimed at reestablishing the empire's ideological and cultural unity, reorganizing consensus, and legitimizing authority on new grounds. The missionary work done among the Goths, who were defeated by Constantine in 322, had both religious and political implications (the conviction obtained through preaching could consolidate the success obtained with weapons). However, for the tactic to succeed, the Christian organization had to remain united. This explains why the emperor was so interested in the controversies about the Trinity, where theories, which admitted a subordination of the son to the father, following the teachings of Tertullian and Origen, were opposed to modalistic theories, according to which the "persons" of the trinity were nothing more than different forms of a single divine being. After a long debate, the subordinationist theories, defended by the deacon Arius, were defeated by those of Athanasios of Alexandria, who upheld the theory of the consubstantiality (*omousia*) of the Father and the Son, a theory that was half-way between subordinationism and modalism. The official condemnation of Arianism did not mark its end, however; it grew for several decades among the barbarian populations, as well as in the imperial court itself. Unending polemics were put forth by the orthodox; the twelve-volume *De Trinitate* by Hilary of Poitiers (died 366) was one of the most prominent of these.

Tendencies in Christian thought

The process of defining orthodoxy, the urgency of which was evident from the controversy surrounding the theses of Arius, took place at the highest levels. Involved in such definition was the problem of the differences and similarities between Christianity and Neoplatonism, on the one hand, and Christianity and Judaism, on the other. The school of Antioch, which tended toward a literal and historical interpretation of the Scriptures, opposed the already active theological-exegetic school of Alexandria; and an intellectual atmosphere also prevailed in Milan, the seat of the imperial court.

There were also rigorist tendencies, which looked upon classical tradition with suspicion. Jerome (c. 347–420), who, in 384; wrote the canonical translation of the *Bible* into Latin (the so-called Vulgate), reached the point of considering his love for the Latin classics such as Cicero and Virgil to be inspired by the devil, and suggested abandoning those authors because they were incompatible with Christianity. The figure of Jerome is also linked to those forms of asceticism and hermitage where flight from the cities and exemplary actions replaced active preaching in the context of an urban society.

The Christianity of Marius Victorinus (300–363), a teacher of grammar and rhetoric, was much different. He was converted to Christianity in 355 and helped bring Neoplatonic themes into Latin-Christian culture. Texts such as the partial Latin translation of the *Enneads* (perhaps done by Victorinus himself, but no longer extant) and, later, the commentary on Cicero's *Somnium Scipionis*, written by Macrobios (born c. 360) in a Neoplatonic key, began to circulate among the educated Christian followers of Victorinus in Milan. Thus Neoplatonism and Christianity met—and not only polemically—in the West as well as in the East, where the confrontation had begun to take place in Origen's time. Elements in the writings of Ambrose (340–397), bishop of Milan, especially in his *Sermones*, indicate that he read Plotinos and Porphyry, or was at least influenced by Neoplatonism. Neoplatonism also had a determinative influence on Augustine.

Developments in Neoplatonism

In any case, Neoplatonism remained a point of reference during this century for those who did not accept the new religion. It was at the core of an attempt made by the emperor Julian (361–363) to restore paganism by issuing an edict that prohibited all Christians from teaching the *artes*. The edict was retroactive, so that the Christians who occupied imperial chairs were forced to give them up. Julian's attempt proved ephemeral, however. A year after his death, the edict was revoked.

As for developments in Neoplatonism, a Syriac school, with Iamblichos (c. 250–325) as its main exponent, joined the Alexandrian center during this century. Iamblichos' doctrines were not always the same as Plotinos', however. The Syriac was much more open to theurgical practices and the revival of Pythagorean-Chaldean elements, as his *De vita pythagorica liber* and *De mysteriis liber* demonstrate. According to Iamblichos, those who philosophize speculatively do not reach the theurgical union with the gods that leads to the achievement of ineffable works.

Between the two great centers of Neoplatonic thought, the Syriac school was the first in which mathematics was used for purely metaphysical ends. Iamblichos wrote *In Nicomachi arithmeticam introductionis liber* and *Theologumena arithmeticae*. In these works he maintains that mathematics is the best place to start in order to reach the superior (magical-theurgical) knowledge that will truly unveil the most profound mysteries of reality. Mathematics should make it possible to understand the first basic aspects of the complex relationships that exist among the beings emanating from the One. About a century after great mathematicians such as Pappos and Diophantos had flourished, mathematical speculation was no longer involved in trying to discover new theorems nor in applying mathematics to limited areas of investigation, but instead was attempting to attain profound meanings that would fit into a metaphysical-religious view of the cosmos.

Attempts at interpreting the works of Aristotle in a Neoplatonic key, following the path already opened by Porphyry and taken by the Alexandrian Neoplatonics, encountered considerable resistence, especially from Themistios of Paphlagonia (c. 317–388), an expert in the works of Aristotle and Plato who lived at the imperial court of Constantinople. Although he was convinced of the profound unity, or rather the complementarity, of Aristotelianism and Platonism, Themistios rejected the speculative constructions of the Neoplatonics, whom he accused of not knowing how to distinguish between the search for God and the rigorous, i.e. scientific, study of human nature. It was in this perspective that Themistios wrote his commentaries on the paraphrases of Aristotle and Plato. We should also mention Clacidios (see ASTRONOMY 300–400), who influenced medieval tradition.

The Cappadocians

The period following Julian's failed attempt at a pagan restoration saw a revival of Christianization in the empire. On a doctrinal level, the Cappadocian fathers dominated in the East. Basil of Caesarea (330–379), Gregory of Nazianzus (329–389) and Gregory of Nyssa (c. 330–395) made some of the most conscious efforts to construct a Christian theology in their works. On many points, they refer back to Origen, although they come to quite different conclusions than Origen, who had developed a philosophical perspective that incorporated Christian doctrine. They sustained the self-evidence of the certitude of faith, the priority of God's word, and the limits of the intellect and human language (as Basil says, "the knowledge of the divine being consists only in the knowledge of its incomprehensibility"). However, they were remarkably open-minded toward pagan cultural tradition, Gregory of Nazianzus in a more rhetorical-literary way, Basil in a scientific way, and Gregory of Nyssa in a philosophical-metaphysical way. The combined efforts of the Cappadocian Fathers resulted in the preparation of the Council of Constantinople in 381, under the reign of Theodosius. The position taken on the nature of the Holy Spirit was analogous to that of the Council of Nicaea. A year before, with the edict of Thessalonica the emperor had imposed the Nicaean Creed. In 391, Theodosius issued a law that prohibited pagan sacrifices, already partially prohibited by another edict in 356. In Alexandria, which besides being an intellectual center also had a long tradition of Christian fanaticism, the patriarch Theophilos interpreted the edict as an invitation to raze the pagan temples. In the disorders that followed Theophilos' decrees, a considerable number of philosophers, scientists, and scholars left the city.

Christianity, sometimes imposing itself through persuasion, other times through tyranny, had become the official religion of the empire. The clergy, which displayed an increasing tendency to reproduce itself through professional formation in episcopal monasteries, began to structure itself hierarchically. An attempt was made to organize orthodox beliefs into a compact unit that would correspond to this institutional structure, as for example in the solutions to the controversies proposed in the councils. This was urgent inasmuch as the Church was being looked to for some kind of indication on how to mediate the expectations of the various classes that identified with Christianity. Doctrinal divergencies, exegetic conflicts, and theological disputes could produce centrifugal tendencies, provincial separatism, and rebellions of the marginal classes. This is exactly what happened in the African colonies that identified with the heresy of Donatus, and in the case of the Roman aristocrats who supported the heresy of Pelagius.

MATHEMATICAL SCIENCES

Serenos and Theon of Alexandria

No exceptional advances were being made in mathematics during this period. Serenos, who appears to have been of Egyptian origin, studied the plane sections of cylinders (recognizing them as elipses and establishing some of their properties) and demonstrated various propositions concerning the areas and volumes of various types of cones. Theon of Alexandria, a famous editor and commentator of Euclid and Ptolemy, set forth a system for extracting square roots that was quite similar to that which later became standard.

ASTRONOMY

Lactantius

Like mathematics, astronomy remained substantially extraneous to Christian thought during this century. The case of the polemicist Lactantius, who lived between the third and fourth centuries, is typical. In the name of a totally unrefined common sense, Lactantius rejected, among other things, the idea that the earth was a sphere and the concept of the existence and habitability of antipodes.

Chalcidius and Firmicus Maternus

Developments in pagan thought were more interesting. Toward the middle of the century, in the midst of an epoch of great Neoplatonic commentaries, Chalcidius translated a part of Plato's Timaeus into Latin and commented on it, producing a text that became one of the most important sources of knowledge on Platonic cosmology until the twelfth century. Chalcidius, who also used texts by Theon of Smyrna and Adrastos, upheld Heraclides on Ponticus' mixed system. He also alluded to the correspondence between humans and the universe, between microcosm and macrocosm, a concept which became very popular during the Middle Ages.

Along with Chalcidius' commentary, we should also mention Firmicus Maternus' great astrological treatise entitled Matheseos libri VIII, the largest astrological compendium of antiquity.

The Church and astronomy

Basil of Caesarea (330–379) reproposed the existence of "higher waters" in the tabernacle structure of the world, which separated and shielded the earth from the heat of the higher layers of the heavens.

Ambrose of Milan (c. 340–397) thought that it was in general not very profitable for a Christian to be concerned with topics like astronomy, which were extraneous to the body of knowledge needed for practicing the faith. Augustine (354–430) limited himself to rejecting those parts of scientific discourse, especially in cosmology, that were in open contrast with the Scriptures.

PHYSICAL SCIENCES

The indifference of the Church fathers toward physical science

Neoplatonism was destined to win out over Stoicism and Epicureanism and to swallow up even Aristotelian doctrine (as is evident in the commentaries on Aristotle), to the point that it became almost the only point of reference for those who did not accept the new Christianity. Various Neoplatonic thinkers took positions on specific questions in the philosophy of nature (especially polemicizing against orthodox Aristotelianism, Stoic determinism, and Epicurean materialism). Theories on light and vision, acoustics, harmony, etc., were all subjects on which the Neoplatonics could focus their attention. But they did not have the scientific spirit of the Alexandrian scientists or even of the Roman scholars of the previous centuries. The context remained one of a prevalently religious philosophy where matter was, above all, nonbeing. Not even the philosophy of the Church fathers, which, after many ups and downs, finally succeeded in prevailing over Neoplatonism during this century, contains anything more than an apologetic and manipulative use of a few bits of physical information. Christian writings of the time lacked any scientific information, and their disinterest in physical problems was evident. Their insistence on themes of faith, redemption, and salvation created a distance from physical considerations that was even greater than in the case of Neoplatonic mysticism.

LIFE SCIENCES

Christian medicine. St. Cosmas and St. Damian

Around the year 300, Cosmas and Damian, two Syrian physicians, were martyred and became the protectors of Christian physicians. Their medical activities and their martyrdom symbolized the beginning of a medicine that was based on caritas but soon became fixed into a religious dogmatism that was influenced by the idea that the salvation of the soul had priority over bodily health. Basil of Caesarea (330–379) founded and organized a large hospital at Caesarea sometime between 370 and 379, and a woman named Fabiola (fourth century) opened the first western hospital in Rome.

Veterinary medicine

Two works in Latin on veterinary medicine, the Mulomedicina Chironis and a treatise by Pelagoios, date back to the second half of this century. They were utilized and completed toward the end of the century by Vegetius Renatus (fourth–fifth century), a scholarly administrator who wrote a veterinary manual focused on the diseases of horses.

Byzantine medicine. Oribasios of Pergamos

The last great physician of antiquity, Oribasios of Pergamos, principal physician to Julian the Apostate, wrote works of notable importance. He lacked originality; and the classic period of Greek medicine and biology ended with his huge compilation, where scholarship replaced originality and direct contact with phenomena.

Nevertheless, this work of simple compilation was not yet subject to the deterioration in content, the doctrinal simplification, and the impoverishment of knowledge that characterized the entire Middle Ages.

APPLIED SCIENCES

The applications of the water mill

Classical technology left the Middle Ages an important heritage in the Vitruvian water mill. Roman engineers of the first century B.C. had transformed it into a true primitive engine. If the information that has come down to us is true, i.e. that Metrodoros introduced water mills into India during the Constantine period (between 312 and 317), the use of water mills finally became widespread throughout the territories of the empire and the East during the fourth century.

Thus water mills began to be more commonly used, but the real innovation lay not so much in the increase in number as in the variety of uses to which the water mills were put. This indicates that by the fourth century a clear idea of the water mill as a primary engine, and not only a mechanized grain mill, had finally developed. The variety of uses to which the mills were put is documented by Ausonios, who mentions water mills for cutting and levigating marble installed on a tributary of the Moselle in 379, and by the unknown author of the *De rebus bellicis*, written in 370, who describes a ship run by a paddle wheel, which is nothing more than an inverted water wheel. The many possible applications of water wheels were only partially exploited, however, in the general climate of profound decline that prevailed in the empire and affected all the arts.

The glass industry

The art of glass-making, which flourished during this period, was an exception to the general decline. The major centers of glass production were in Italy, in the north of France, and in the Netherlands. The Italian glass-making industry was characterized by the engraving on glass of illustrated scenes and by vases with reticulated designs. It flourished with the invention of free blowing, which probably took place during this period, following the other important innovation, blowing into molds, which was developed in the first century.

It is likely that continuous improvements that took place in glass-producing methods over the course of a few centuries gave the industry the particular boost that enabled it to prosper even under adverse conditions.

Handcraft production

The introduction of a loom for weaving cloth with patterns, which came to Europe from China, where it had been invented during the first century B.C., may indicate that the art of weaving still showed some signs of life. We should also remember that the first reliable reference to soap as a detergent goes back to 385; the physicist Priscianus speaks of it being used for washing hair. The soap-making industry must have prospered to a certain degree during this period, since soap was a common product in Europe by the ninth century.

Nordic shipbuilding

A profound renovation was taking place in Nordic ship-building techniques, which had not been influenced by ancient Mediterranean methods. A Viking craft found at Nydam, built during the fourth century, shows a certain similarity to Roman ships. There is not yet a true keel, but the shorter axis appears narrower and thicker. The craft is a typical mastless construction with overlapping sheathing and was moved by rowers.

400–500

THOUGHT AND SCIENTIFIC LIFE

The split between East and West. The crisis of the empire

When Theodosios died (395), the empire was divided between his sons Arcadius and Honorius, following a tradition that went back at least as far as Diocletian and did not intentionally mean to decree a break between East and West. But the events of the first half of the century made the separation fairly definitive, and confirmed an economic, linguistic, cultural, and in some ways even religious, division. Although both East and West were Christianized, there was a profound difference not only in their evaluations of specific questions but more generally in their very ways of living the religious experience.

We shall not go into the reasons for the relative stability of the Eastern empire as opposed to the rapid disintegration of the Western empire. From the earliest years of the century (Rome was sacked by Alaric's Goths in 410), the Western empire suffered the attacks of the barbarians, which eventually brought about the establishment of the so-called Roman-Barbarian kingdoms, first in the provinces (Africa and Spain) and then in Italy itself. Ravenna, the capital of the Western empire from 402 on, had become the capital of the Ostrogoth kingdom of Theodoric by the end of century.

As the structure of the State collapsed, the Church, especially in the figure of the bishop of Rome, seemed to be the only reliable institution that was able to mediate between the Latins and the barbarians, at least on a long-term basis. During this period, therefore, evangelization became the affirmation of Rome's prestige and its guarantee for survival. Episodes such as Patrick's evangelization of Ireland and the "conversion" to Christianity of many barbarian kings and all their subjects should be seen in this perspective.

The early Christian fathers

The most outstanding figure of Christian thought during this period was Augustine, who was born at Tagaste, Numidia, in 354 and died at Hippo in 430, while the walls of that African city were being besieged by the Vandals. We cannot in this limited space explore the complex ideas of the author of the *Confessiones* (400) and the *De civitate dei* (413–426) here. It will suffice to point out how the very steps in Augustine's difficult development signal themes that became central to his thought in later times. Augustine went from studying the classics (especially Cicero's *Hortensius*) to reflecting on the problem of evil; he went from an early adherence to Manichaenism, which was widespread in Africa, the Gauls, and even Italy at that time, to a probabilistic Skepticism of Ciceronian stamp; and he also went from an ever-broader use of Neoplatonic themes to mediating them with Christian thought. This path led Augustine first to become one of the most brilliant polemecists and Christian thinkers of all time and later to be a bishop who led his flock with "sullen energy," while his polemics against the Manichaeans were succeeded by those against the rigorist followers of Donatus, the increasingly more enflamed spokespeople for the need for social regeneration in many rural areas of Africa. Against the Donatists, Augustine called for order and unity, which the Church could not relinquish, especially at a time when the ecclesiastical hierarchy had progressively assumed civil power and was therefore responsible for administering it. This meant a strict obedience to the established authority. Egalitarian aspirations that upset the social order of the state must therefore be repressed, if necessary, according to Augustine thought, by force.

Such a step was not necessary, however, in dealing with another heresy which took shape around 410, the work of the Irish monk Pelagius and his disciples Celestine and Julian. Their ideas were especially popular among the intellectual Christians of the Roman aristocracy. In the face of this rationalistic deviation—which was as dangerous as Donatist rigorism to the image of the Church that Augustine defended, although it was not connected to the subversive revindications of the popular masses—the best weapon seemed to be a doctrinal confutation. Against the doctrine of free will and the rejection of Adam's sin as hereditary, which was especially convincing to those who, going back to Stoic themes, insisted on the *virtus* that human beings could use for their salvation, Augustine—who was afraid that such points of view, carried to the extreme, would cause a radical reduction in the value of redemption and the mediatory role of forgiveness, and consequently of the ecclesiastical hierarchy—developed his well-known theses of humanity as a *massa damnationis* and of predestination. His reflections in *De civitate Dei* were also tied to these themes. The work was written in the event of barbarian invasions, but its ample theorization on the subordination of the earthly city to the city of God made it the great model for the vindication of spiritual power over temporal power.

When Augustine died, the theological dispute over the doctrine of Pelagius, which had been condemned by the synod of Carthage in 418 and then at the Council of Ephesos in 431, abated. However, another doctrinal conflict soon arose, this time over the nature of Christ. The problem was to define the relationship between humanity and divinity in Christ intended as *logos*. The Council of Ephesos condemned the theory put forth by Nestorios, bishop of Constantinople, who saw two different natures, one human and one divine, in Christ. The opposing theory, upheld by Eutyches (c. 375–454), in which Christ had only one (divine) nature (the Monophysitic theory) was also condemned at the Council of Chalcedonia, and the theory according to which there were two natures in one person was recognized as orthodox.

These theological disputes widened the gap between East and West, and were often the cause of instability and disorder even within the Eastern empire. On a cultural level, especially in Greek-speaking culture, they were evidence of the continuous struggle that Christianity had to go through in order to explain its fundamentals in the philosophical terminology of Greek tradition.

The early Christian fathers and the exact sciences

Although the writings of the early Christian fathers were dominated by a liking for controversy, constant attention to theological detail, requests on the part of the ecclesiastical authority, and the need for a safe "orthodox" leader, there was complete silence on the question of the natural sciences. This did not stem from a mastery of the achievements of the

sciences, but rather from an indifference toward them. As Augustine said, "*in interiore homine habitat veritas.*" Although Augustine's appeal to interiority stimulated a wide range of ethical and religious doctrines focused on this particular aspect of his thought, it did not encourage a careful study of nature. The more general interest in problems regarding the divine and salvation, which became almost the sole concerns that Christians had, explains the scientific sterility of a movement that was extremely lively in other respects and certainly did not lack originality. Some fathers studied the philosophy of nature in order to combat atomism, which was considered blasphemous, others in order to comment on the Biblical account of the Creation. But natural processes as an object for autonomous study do not seem to have been of any interest. The complete devaluation of "secondary causes" with respect to God, the "primary cause," distracted attention from the study of natural processes. Mathematics was in a somewhat different situation; Augustine refers to it directly in his confutation of the Skepticism of the Academicians. But although on various individual questions Augustine showed remarkable intuition, these remained scientifically sterile during his time.

Neoplatonism. Mathematics and mysticism

The mathematician Theon of Alexandria was an outstanding figure in the Alexandrian school toward the end of the fourth century. His daughter Hypatia also dedicated herself to mathematics. However, our interest lies not in the particular results, but in the reinterpretations of classical works that took place in the light of the new philosophical interests. This explains the flourishing of commentaries on Euclid—an edition of the Elements was edited by Theon himself. Hypatia commented on the works of Diophantos and attempted to interpret the relationship between the One and the many in Plotinos' philosophy in arithmetical-geometrical terms.

A few years after the death of Hypatia, who was killed in 415 by a mob of Christian fanatics, several Neoplatonics organized a school in Athens in order to revive the tradition of the Academy. Proclos (410–485), who was its most illustrious scholarch, wrote commentaries on virtually all the Platonic dialogues. In his Στοιχείωσις νεολογιχή (*Elements of Theology*) and Εἰς τὴν Πλάτωνος νεολογίαν (*On Platonic Theology*), Proclos developed a broad inter-penetration of metaphysical and theological systems, emphasizing the theme, which had already been present in Iamblichos, the surpassing of philosophy by theurgy, the production of divinity. Thus there was a growing trend, which already existed in Eastern magical tradition, Gnosis, and the doctrines of Hermes Trismegistus, toward reevaluating the irrational. In Proclos, explicitly, the irrationality of the magical practices that were essential to theurgy was more important than metaphysical rationality.

Scientific rationality could still be a paradigm, however, for the rational organization of theology. Thus Proclos commented on Euclid's Elements, in the tradition of Theon and Hypatia and referred to their logical structure in the demonstrative organization of his theological treatise. Proclos' commentary is limited to the first book of the *Elements*, where the basic concepts of geometry are outlined, the part of the work that best lends itself to a philosophical reading. The preceding century's custom of writing commentaries on mathematical works continued after

Proclos as well. Marinos, Proclos' immediate successor at the school of Athens, for example, was the author of an introduction to Euclid's *Data*. In the next century, Eutocios of Ascalon wrote commentaries on the works of Archimedes and Apollonios. This usage indicates that the creative phase of classical thought had ended. The phenomenon was not only limited to mathematics, though it was accompanied by a gradual disinterest in demonstrative rigor.

The Dionysios who presented himself as a disciple of Paul, and whom Christian tradition has identified as one of the Areopagus converted in Athens by the apostle' preachings, flourished toward the end of this century. This identification gave the work of Dionysios the Pseudo-Areopagite the immense amount of authority that it enjoyed throughout the Middle Ages. In reality, Dionysios was a Neoplatonized Christian who took a number of themes from Plotinus, Iamblichos, and Proclos and gave them a fully Christian appearance. His themes of the celestial hierarchy, negative theology, ignorance as the highest form of knowledge, and evil as nonbeing became some of the favorites of Medieval Christian thought.

Commentators on Aristotle: Simplicios and Joannes Philoponos

Despite opposition, the tendency toward interpreting Aristotle's thought in a Neoplatonic key clearly prevailed, partly because it coincided more with the cultural needs of the epoch. The most important works in this current were the commentaries of Simplicios, the last representative of the Academy of Athens, where an immense amount of information on the history of philosophy and mathematics was preserved; the commentaries on the logical works written by Ammonios, who lived toward the end of the century at Alexandria; and the commentaries of Joannes Philoponos, a disciple of Ammonios and also a Neoplatonist. Philoponos took considerable interest in Aristotelian physics (it was in criticizing the doctrine according to which bodies supposedly fell at a speed proportional to their weight and the void did not exist that he outlined the theory of impetus, which was again taken up in the fourteenth and fifteenth centuries, opening the way to modern Galilean mechanics). The *Theologia*, a Neoplatonic work that has come down through the Arabs as a work of Aristotle, is also a product of this century.

The events in Philoponos' life are exemplary: his conversion to Christianity attests to the compatibility of Aristotle (re-read in a Neoplatonic key) and the new religion. Of course, a few renunciations were necessary in order to make the synthesis: polytheism and theurgy, as well as the Aristotelian theory of the eternity of the world, had to be given up.

Martianus Capella and the classification of the artes

Among the scholars of this century was the Latin writer Martianus Capella, who lived in Carthage at the end of the century and wrote an encyclopedic work entitled *De nuptiis Mercurii et Philologiae et de septem artibus liberalibus*. This arrangement of information was inspired by Cicero's concept of the *unito* between *eloquentia* and *sapientia*, symbolized by the marriage of Philology and Mercury. The liberal arts that Capella considers are grammar, dialectics, rhetoric, arithmetic, geometry, astronomy, and music. Little of what he says about these doctrines is original; what he omits with respect to Varro's famous classification is actually more interesting to our topic. There is no

place in "heaven," at Mercury and Philology's wedding, for architecture or medicine, which were excluded from the arts for being too closely tied to the human "hand," i.e. to manual work.

A clear and decisive devaluation of all forms of technology, which contaminated any discipline that needed to use them, had thus appeared by this time.

MATHEMATICAL SCIENCES

The epoch of the commentators

Mathematics seems to have played a leading role in Neoplatonic tradition. Proclos (410–485), a Neoplatonic philosopher and a director of the renowned Platonic Academy of Athens, gives an abundance of information as well as a very important historical outline of the evolution of Greek geometry in his commentary on Book I of Euclid's Elements. He was, however, more interested in the philosophical-speculative aspects of mathematics and did not make any real contributions to the discipline, except for an important discussion of Euclid's postulate of parallel lines. The discussion is accompanied by an attempt at a demonstration, based on another, more intuitive, postulate. In the prologue to his commentary, which is a valuable, if partial, source for the history of Greek mathematics, Proclos schematically described the development of Greek mathematics from Thales to Euclid. Plato is presented as the central figure of this history; Eudoxos was only "a disciple of Plato's friends"; Euclid was a convinced Platonic, and the real purpose of the *Elements* was the construction "of the so-called Platonic figures," i.e. the five regular solids.

Thus in mathematics as well as in philosophy, an epoch of commentaries on the classical texts had begun in the Greek-speaking world. Simplicios (the last head of the Platonic Academy) commented on Book I of Euclid's Elements; Asclepios of Tralles commented on Nicomachos' Arithmetic; Eutocios of Ascalon (born c. 480) commented on three Archimedean treatises and the first four books of Apollonios' Conics, which probably have only reached us in a Greek version, thanks to this commentary. The so-called Book XV of Euclid's *Elements* has been attributed to the Neoplatonic philosopher Damascios (c. 480–550), but is more probably the work of the mathematical school that formed around Isidoros of Miletus, the architect appointed by Justinian to reconstruct St. Sophia in 532, along with Anthemios of Tralles, the author of a treatise on burning mirrors that is interesting for its theory on conic sections.

Ammonios, son of Hermias, who lived in Alexandria between the end of the fifth and the beginning of the sixth centuries, also deserves mention. He divided mathematics into arithmetic, geometry, astronomy, and music, the famous quadrivium division that Martianus Capella (see THOUGHT AND SCIENTIFIC LIFE 400–500) and Severenus Bothius then introduced into Latin culture.

ASTRONOMY

Macrobius and the Egyptian system

Two important figures stand out in this century. Macrobius, who lived during the first half of the century, was a Neoplatonic scholar who introduced the exegesis of texts, common in the field of Christian literature, into profane culture as well. In particular, Macrobius mentions Heraclides of Ponticos' partially heliocentric system in his commentary on Cicero's *Somnium Scipionis*, attributing it to the Egyptians, as was always done when the stamp of a religious authority was needed. Since his commentary was widely circulated, this system was known as the Egyptian system for centuries.

Martianus Capella. Astronomy as a liberal art

The other important representative of this period is Martianus Capella. In the eighth book of his *De nuptiis Mercurii et Philologiae et de septem artibus liberalibus*, which was written around 470, Martianus describes the traditional astronomical doctrines of the Classical Age, and refers to Heraclides' theory on the systems of Mercury and Venus. He also places astronomy among the liberal arts, thus defining its role in the arrangement of knowledge in a way that was widely accepted and followed throughout the Middle Ages and beyond.

PHYSICAL SCIENCES

The preservation of classical science in the West

Even with Augustine of Hippo (354–430), the most important thinker of Latin patrology and one who was deeply influenced by Neoplatonism, Christian thought showed very little interest in physical questions (the attitude toward mathematics was different, especially toward music and acoustics, although these were usually thought of as arts rather than sciences). As in previous centuries, the role of scholars, compilers, and authors of encyclopedias and compendia was very important in preserving, though often in a schematic way, the great findings of earlier epochs.

... and in the East

Eastern scientific culture, which was economically, politically, and institutionally more stable, was able to maintain its ties with tradition more easily and systematically. The division of science into specialities, research done with a certain commitment and rigor, and scientific investigation all survived. But it was an age of examining traditional questions more thoroughly rather than one of original research. One of the most interesting innovations in dynamics, Joannes Philoponos' theory of impetus (see PHYSICAL SCIENCES 500–600), occurred in the environment of Aristotelian commentaries.

LIFE SCIENCES

This century saw no original contributions to the life sciences. Compilations were made in the Eastern empire and in northern Africa.

APPLIED SCIENCES

Roads and construction

The Barbarian invasions into the western part of the Roman Empire, this empire's precarious condition, and its fall in 476 caused a serious regression in technology. In agriculture, iron was used increasingly less often for making tools, until it almost disappeared. Although the road system lasted for quite a while, it was completely abandoned, and many parts of it became unusable. The high costs of repair and maintenance, which the empire had been able to afford, could no longer be sustained by the various communities.

A new type of road, more suitable for the new needs, appeared. A paving of pebbles or broken stone was laid on an easy-to-repair foundation of sand.

The paving in Roman cities deteriorated rapidly and was almost nonexistent in the newer cities. It reappeared in German cities in the twelfth century. Buildings constructed in England and France beginning in 400 were quite inferior to their Roman predecessors. Bricks were rarely used; in England, one of the few examples of brick construction is the arches of the basilica of Northamptonshire (670). Until the seventh century, windows were made with linen or perforated boards. Stone buildings prevailed everywhere.

The wool and glass industries

Although the wool industry was in a decline, it did not completely disappear. It was revived around 1000 in the Netherlands, mainly because of that area's geographical closeness to England, where the intensive selective breeding of sheep had continued in several ecclesiastical communities and better-quality wool was available.

The glass-making industry continued to prosper to a certain degree. Supplies of soda for glass production appear to have been regular in the Rhineland. In the ninth century, factories were moved to forest areas, signifying that potassium carbonate from plant ashes was being used. Glass then became less transparent. The Roman technique of blowing glass into molds disappeared, together with the arts of carving and painting on glass.

The spread of Roman artisanal technology

Quality wheel-turned earthenware continued to be produced in England and in several areas of Gaul, Italy, and the Rhineland, but vitrified glazes were not used in the West until the tenth century. Kilns remained fidentical to the Roman models. Bricklayers' tools stayed the same, but in carpentry, planes disappeared completely.

Roman techniques adopted in the Eastern empire

Roman artisanal traditions, interrupted in the West, were saved by the Eastern empire. The common brickmaking industry, the production of lead-base paint, the technique of blowing glass into molds, and various techniques of embellishment were all preserved in this way.

Preserving Roman traditions was very advantageous to the East, and several innovations were also added. A non-Vitruvian vertical mill that used a bucket wheel and was built around 470 has been discovered in Athens. It is the earliest known example of this uncommon type of mill, which required a direct, controlled water supply and was more efficient than the Vitruvian mill. The use of torches for street lighting goes back to 450; Antioch was supposedly the first city in the world to illuminate some of its streets.

Navigation in Byzantium and among the Nordic populations

The ancient marine tradition also disappeared with the barbarian invasions and was resumed by the Byzantines. In the North, large Irish skin-covered boats which could hold more than twenty people crossed the northern seas. The Irish were already good navigators by this time, and their expeditions probably reached Iceland, Greenland, and Newfoundland.

The preservation of classical science

The work of scholars who collected and preserved the findings of classical technology during this period has proved to be extremely valuable (for more information on this phenomenon, see the section on PHYSICAL SCIENCES).

The Feudal Middle Ages and Islam

(from the seventh to the twelfth centuries)

500–600

THOUGHT AND SCIENTIFIC LIFE

The Roman-barbarian kingdoms

The Roman-barbarian kingdoms were the political expression of a process of assimilation that took place between the traditional fabric of Roman society and the ethnic and political elements brought into it by the invaders. Although they came to the empire with extremely different traditions, the Germans set up quite similar state structures in Italy, Spain, and the various regions of Gaul. These structures united customs and laws characteristic of the various German groups with Roman institutions. Most of the barbarians spoke similar enough Germanic dialects, but they did not feel bound by any concept of national unity. As a matter of fact, even the populations that were the proudest of their origins had gradually absorbed the remnants of the nomadic and semi-nomadic cultures of northern Eurasia that they had encountered in the course of their migrations. The language of the vanquished culture, on the other hand, was imposed for political as well as religious reasons and constituted a unifying factor. During the early decades of the century, the rulers of the new states began to issue codes written in Latin, such as the Visigoths' *Breviarium Alarici*; the Ostrogoths' *Edictum Theodorici*; the *Lex romana burugundiorum*, which coordinated barbarian codes and Roman laws; and the *Lex burugundiorum* and *Lex salica*, which codified barbarian customs.

Although political power was concentrated in the hands of the German king and the Longobard soldier-farmers of the barbarian *exercitus*, the administrative framework of the individual regions remained predominantly Roman. The culture and the organizational structure of intellectual life were also Roman. As a matter of fact, in these states the Latins maintained a certain intellectual supremacy; and the way in which culture was elaborated remained the special patrimony of the Roman aristocracy, even though it had been forced to give up its traditional political supremacy and a large portion of its economic power. Thus intellectuals educated in the pure classical tradition preserved the juridical-philosophical-theological knowledge that had been produced from the meeting of Greek philosophical concepts, the concerns of the Church fathers, and the elaboration of Roman law. But this knowledge was not simply transmitted from its natural depositories to the new intellectual aristocracy that arose within the ecclesiastical institutions. Instead, a process of adaptation took place which soon involved intellectuals of barbarian origin, who were quickly assimilated by the organization of the Church. Within this context, culture, especially philosophical-scientific culture, was often the reflection of pedagogical and religious activities in a broad sense. For example, the most original work of Saint Boëthius, the *De consolatione philosophiae* (524), belongs to the Western tradition of introspective moral thinking that began with Augustine of Hippo's *Confessions*. On the other hand, medieval culture was much more strongly influenced by Botius' scientific-philosophical work than by his theological reflections, which directly followed Augustinian models. The author of two commentaries on Porphyry's *Isagoge*, as well as other treatises and commentaries on logic, Botius transmitted Greek philosophy and a part of Greek science to the Latin Middle Ages. The *logica vetus* that was based on the *Isagoge*, as well as on the *Categories* and the *De interpretatione* (the only extant parts of Botius' translation of the *Organon*) and on Bothius' treatises containing bits of Aristotelian logic filtered through Stoic logic and Neoplatonic commentaries, drew its choice of texts and problems largely from Botius.

Byzantium. The school of Athens. Persia

The religious antagonism and mutual diffidence that existed between the Romans (who were Catholics) and the Germans (who were often Aryans) undermined the delicate compromise that had been achieved in the Roman-barbarian kingdoms and kindled the ambition of Justinian, Emperor of Byzantium from 526 to 565, to reconquer the territories "that the Romans lost because of their indolence." In the designs of the Emperor, Byzantium represented the new Rome and would have the task of politically, religiously, and culturally reuniting East and West. The enormous efforts of the commission of jurists, presided over by Tribonian from 528 to 533, which culminated in the *Corpus iuris*, a collection of Roman and Byzantine laws and judiciary customs and commentaries on them, were part of this prospect.

Justinian's restoration of the authority of the Emperor as a Roman-Christian monarch involved a series of precise cultural choices. The Byzantine ideal was a deeply committed culture capable of theological subtleties. Scientific and philosophical works were also subjected to comment and exposition, in an imitation of the exegesis of sacred texts, following a method that was developed during the final decades of the previous century at the schools of Athens and Alexandria and that would later be typical of scholasticism. Like Bothius in the West, Eastern Greek thinkers such as Simplicios, Joannes Philoponos, Ammonios, and Damascios discussed and commented on the works of Plato, Aristotle, and Porphyry. They recorded valuable historical information about these thinkers, lingering on the scientific aspects of their work as well, while at the same time, in the Neoplatonic tradition, they maintained an interest in mathematics and commented also on the works of Euclid, Archimedes, and Apollonios. Perhaps the most original of the commentators was Joannes Philoponos, who harshly criticized some of the basic precepts of Aristotelian physics (see PHYSICAL SCIENCES 500–600).

One of the first victims of Justinian's cultural policy was the school of Athens. The last stronghold of pagan philosophy, it was closed in 529 by an imperial edict. A group of Greek philosophers, one of the foremost of whom was Simplicios, took shelter (529–533) at the court of King Chosroe of Persia, contributing to the diffusion of Greek culture in the East. The exiled Nestorians and Athenian emigrant philosophers were partly responsible for the flourishing of the Persian school of Jundīshāpūr, which had, under Chosroe's reign (531–579), the nature of an academy. The Nestorians contributed their knowledge of Greek works on medicine, while the Athenian philosophers brought the Neoplatonic theories that would influence Persian and Indian mysticism. Almost simultaneously, compilations and translations into Syriac of works by Plato, Aristotle, and Porphyry and the *corpus* of Pseudo-Dionysios were being done in Mesopotamia by Sergios of Resaina (died 536 in Constantinople). This laid the foundations for transmitting Greco-Roman philosophy and science to the Arab culture, a transfer which took place when the Moslems occupied the country.

Benedictine monasticism

One of the most important developments that followed the death of Theodoric in 526 was the gradual formation of monastic *curtes* on the Italian peninsula and elsewhere. Unlike the communities that had existed in great numbers throughout the Byzantine East for some time, the Western communities followed in the footsteps of the original nucleus created around 529 by Benedict of Nursia. They did not advocate isolation in places where religious souls could close themselves in meditation, but rather aimed at setting up centers to do active work for the collective good, to defend and aid those populations that had been terrorized by continuous wars and weakened by frequent famines.

Excessive rigorism was taken out of the Benedictine *regula* to make room for practical and intellectual work, which was alternated with devotions and works of charity—especially aid to the ill—in the daily lives of the monks. Monastic schools of medicine, (so important because this discipline had been excluded from the body of the seven liberal arts since the time of Martianus Capella) as well as hospitals developed out of this system.

Pedagogical responsibilities were transferred from the Roman free and individual educational system to the rule of the monasteries, which also had the task of forming novices. The strongest cultural accent was brought to monasticism by Cassiodorus, who founded the monastery of Vivarium at Scylacium, in Calabria, around 540. There, classical texts were collected, amended, and copied, forming an almost perfect library that was synthesized in encyclopedic compendia. The authors intended these compendia to be, above all, tools for a more thorough study of the Bible, and they belong to the body of manuals that were written for pedagogical purposes. Cassiodorus, in particular, in his *Institutiones divinarum et saecularium litterarum*, provided the Middle Ages with the division of the seven liberal arts into the *trivium* (grammar, rhetoric, and logic, or dialectics) and the *quadrivium* (arithmetic, music, geometry, and astronomy), drawing upon Vitruvius, Martianus Capella, and above all Bothius.

There was also an embryo of historical studies forming among the monks, as Cassiodoros' *Historia gothica* (553) shows.

Gregory the Great and the Church of Rome

While Italy was economically prostrated by the Greco-Gothic war (535–553), Justinian's political reunification of East and West proved to be short-lived. In 568, the Longobards invaded Italy; in 571 the Visigoths began their campaign to reconquer Andalusia. In Rome, the papacy assumed new powers and autonomy with respect to the political control of the emperor and the pretenses of eastern and African bishops. Gregory I, pope from 590 to 604, took the first steps toward political autonomy by offering his mediation to the Byzantines and the Longobards. He thus took on governmental responsibility in the areas not reached by imperial authority and organized the first of the great missions that left Rome for the purpose of converting the barbarian populations to Christianity and obedience to Rome. The mission to Britain (596) led by the monk Augustine—who was later named bishop of Canterbury—spread the elements of classical culture that had survived in the Latin-Christian world together with the Gospels among the still-pagan Anglo Saxons.

The universalistic stamp that Gregory placed on the political and cultural function of the church made the intellectual formation of the clergy of fundamental importance. Gregory discussed this problem various times in his pastoral writings and in epistles, in which he encouraged the study of the liberal arts, as long as their only aim was to clarify and explain the Scriptures. Although personally he was not at all ignorant of classical culture, Gregory was not one of its enthusiastic champions, as is obvious from his opinion that the ecclesiastics' practical use of Latin should not be inspired by the rules of the *artes* (i.e. grammar) which Donatus had set forth in the fourth century, but rather should follow the linguistic models contained in the Holy Scriptures.

MATHEMATICAL SCIENCES

The work of Boëthius

Although Severinus Boëthius (c. 480–524), one of the few learned figures of these years, did not make any original contributions to mathematics (the era of real research in the field had already ended some time before), he fully appreciated its value and worked to preserve the patrimony of the discipline. For this reason, he not only translated several works of Greek scientists into Latin—at least according to tradition—but he also wrote his own work, the *Institutiones arithmeticae*, a valuable contribution to the preservation and diffusion of mathematical knowledge at the beginning of the Middle Ages. It subdivides mathematics into four disciplines—arithmetic, music, geometry, and astronomy (the *quadrivium*). This division remained standard for many centuries and, along with the *trivium* (grammar, rhetoric, and dialectics), made up the typical curriculum for higher studies during the Middle Ages. Even Boëthius' Platonic-Pythagorean approach to arithmetic (a strong tendency to treat it as the "philosophy of numbers" rather than as an exact science or, even less, as a computational discipline) long influenced medieval mathematics. Other manuals on geometry and music which follow the *quadrivium* subdivision have come down to us under Boëthius' name, but they were almost certainly compiled at a later date.

The situation was not substantially better, however, in the eastern Mediterranean basin. After the Platonic Academy at Athens was closed by Justinian in 529, even the remarkable work of the commentators, which had made it famous, ceased. Byzantium became the center of Greek-speaking culture; mathematics continued to exist there in a substantially conservative form. The great texts of classical tradition were preserved and repeated, but no new contributions were made.

Practical arithmetic in the East and the West

The Akhmim Papyrus, which contains a manual on practical arithmetic, dates back to this period. It is especially interesting as a document, since it is the oldest extant treatise on applied arithmetic. As for its content, it shows no progress with respect to the knowledge of the classical period. It is of Byzantine origin and was written by a Christian. The diffusion of Buddhism from India into China revived the cultures of the countries converted to the new religion and also spread knowledge of Indian mathematics, where nine-numeral notation was already in use. Besides the notable development in practical arithmetic that took place in the Far East with the use of fractions, ancient local texts were recuperated and commented upon.

ASTRONOMY

Boëthius' compendium

Severinus Boëthius (c. 480–524) compiled a series of scientific and philosophical compendia written in Latin from classical Greek sources. One of these, which uses material drawn from Ptolemy's Almagest, is dedicated to astronomy.

Cosmas' synthesis

In a report on his travels, *Christian Topography*, written around 535, Cosmas Indicopleustes, a well-known traveler to the East and Africa, attempted to organize the astronomical theories of the church fathers by proposing a cosmological system in which the world had the form of a parallelepiped closed on top by a semi-cylinder. The earth was located at the base, and halfway up, the plane of the sky horizontally separated the earthly part from the heavens. There was also an enormous projection on the earth toward the north, behind which the heavenly bodies, moved by the angels, could hide during their movements, producing the succession of the days and nights. At the price of remarkable naïveté, Cosmas achieved one of the first syntheses between the Christian concept of the world and classical cosmology.

Joannes Philoponos' De opificio mundi

In the Byzantine world, where knowledge of the classical texts was preserved, but very little progress in astronomical studies was made, Joannes Philoponos (last decade of the fifth-second half of the sixth century), already well known as an energetic commentator on the works of Aristotle, expounded the classical astronomical theories of homocentric spheres and the sphericity of the earth in his *De opifico mundi*, in direct opposition to the positions sustained by the church fathers.

PHYSICAL SCIENCES

Byzantine science

Byzantine science produced figures of notable prestige, and the solidness of its research institutions (schools, libraries, etc.) contrasted with the chaotic situation in the West, where the preservation of classical knowledge was entrusted to people like Bothius and Cassiodorus, who were able to reconcile the old culture with the values that emerged within the Roman-barbarian kingdoms and where the monasteries became the only places where classical works were preserved and studied.

One of the foremost figures in Constantinople was Anthemios of Tralles (died c. 535), a mathematician, physicist, engineer, and architect. The extant fragments of his work display not only a remarkable knowledge of mathematics, but also a keen interest in questions of a practical nature. He appears to be the first scientist to have theorized the possibility of making burning mirrors (achieving what legend attributes to Archimedes). The hypothesis appears in his treatise on extraordinary machines. Anthemios rejected spherical concave mirrors and considered a suitable combination of plane mirrors adequate (in 1747, Buffon obtained the results that tradition attributes to Archimedes, by combining 168 plane mirrors). In keeping with Alexandrian tradition, Anthemios' optical and mechanical devices seem to be ruled on the one hand by practical objectives (construction, designs for war machines, etc.), and on the other by purely playful ones (amazing machines, automatons, etc.).

The theory of impetus

The figure of Joannes Philoponos (c. 490–570) stands out along with Anthemios. Philoponos, with his Neoplatonic leanings (he later converted to Christianity), profoundly revised Aristotelian doctrine. With respect to the doctrine of place, Philoponos belonged to a tradition of commentators that included Alexander of Aphrodisias. He took up the question again and maintained that three-dimensional space must be thought of as the natural place of all physical objects. Although this space could not be separated from the bodies that occupied it, it could be distinguished by thought, according to Philoponos. Space was immobile, both in its totality and in its parts, and thus satisfied Aristotle's requirements for place. There was also a strict analogy between local motion and qualitative change: when a body abandons a place, another comes to occupy that place; when a form perishes, another comes to take its place and mold the inert matter.

The ideas that scholastic thinkers later took up to formulate the theory of impetus were similar to Philoponos' theories. According to Philoponos, the medium was not at all the *motor conjunctus* of a missile, as Aristotle had believed. Instead, the mover was a certain immaterial force, a sort of incorporeal kinetic force (i.e. the impetus) that the thrower imparted to the missile at a particular moment during the act of throwing. Finally, Philoponos denied the assertion that a body fell increasingly rapidly in proportion to its weight. Contrary to Aristotle, Philoponos saw the resistance of the medium as simply a supplementary element, making the existence of motion in a void possible.

LIFE SCIENCES

The Persian medical school and the reawakening of Chinese culture

A flourishing Persian medical school, which brought together indigenous, Hindu and Syrian influences and had its roots in the classical medicine that the Nestorians of Edessa had introduced into the East, developed at Jundīshāpūr. After the Moslem conquest of the Sassanid kingdom in 651, it became one of the great sources of Arabic knowledge in medicine. In China, which was enjoying a cultural reawakening during this period, Tao Hung-ching (451–536), a Daoist philosopher, compiled a fundamental text of materia medica that was the basis of Chinese pharmacopea for centuries.

Aëtios of Amida

Latin and barbarian medical scholars began to compile manuals that had no original value, using the material that had survived from the Late Classical Age or had come indirectly from Byzantium. During Justinian's reign, medicine began to flourish in the regions of the Eastern empire and continued to be productive throughout the sixth century. This boom developed from the ancient tradition of classical medicine through Aëtios (c. 502–550), who wrote the τετράβιβλοι a vast compilation of ancient authors. But Atios was also an original observer and described the paralysis of the palatine velum. One of the great currents of Arabic eclectic thought, which became popular in the following centuries and passed on to Salernitan medical culture, was thus formed at Byzantium during the first half of the sixth century. The development of medicine was favored by the pres-

tige of the profession, which was often passed on from father to son, and by the simultaneous creation of hospitals.

Western monastic medicine

Western monastic medicine developed along with Byzantine medicine. Cassiodoros (c. 487–583) brought along his special interest in medicine when he took up life in the convent. He listed the rights and duties of medical practicers in an ordinance that he himself issued when he was the *magister officiorum* of the Gothic imperial court. Other edicts of Gothic-Longobard legislation also attest to the medical profession's interest in the social aspects of medicine and consequently in deontology. The monks of Scylacium read and copied classical works, while the Cassino Benedictines opened a hospice that became famous for its difficult and miraculous cures.

The creation of hospitals

In the West as well as in the East, the hospitals were seen as corollaries to the working of Christian charity. However, there was a brief revival of traditional Roman lay medicine. The reestablishment of a body of *comes archiatrorum* and the creation of lay hospitals at Benevento, Lyons (542), and Merida (580) were expressions of the latter phenomenon.

Alexander of Tralles: pathology and therapy

During this period, the scientific world was dominated by the figure of a great physician of the court of Byzantium, Alexander of Tralles (525–605), called the periodeute because he accompanied the emperor and Belisarius on numerous military expeditions. Alexander showed a keen sense of clinical experience, which even prevailed over his respect for the diagnostic outlines and curative maxims that were the legacy of Galenic tradition. Alexander was one of the last great figures to illuminate Alexandrian scientific culture before its fall under Persian (619) and Arab (642) occupation. The author of a treatise on internal pathology and a twelve-book work on therapy, Alexander's name is linked to various syndromes: Hepatic excess paludism, intestinal parasitism, and hemoptytic excess. He thought that the function of therapy should be to eliminate the pathogenic cause, and distrusted the treatment of symptoms, believing that it could be harmful even though it seemed beneficial. He had an objective spirit, and recommended ''acting as Aristotle says; that is, considering Plato, but also truth, your friend.''

APPLIED SCIENCES

Water mills

The diffusion of water mills caused the ancient mills run on animal power, to disappear, at least in urban centers. So widely used were water mills that during the siege of Rome in 537, the Goths cut off the aqueducts to prevent the mills from functioning. The cutting off of the aqueducts inspired Belisarius and his engineers to use the water of the Tiber, and floating mills were developed. Procopios tells us that ''...with the force of the water flow, all the wheels, one after another, turned independently, operating grindstones connected to them, and ground enough grain for the city.'' The use of floating paddle wheels became especially widespread in the Islamic world, where they powered grain mills, paper mills and other machines.

The heavy plow

During this period, the heavy plow, with its three basic elements—the coulter, the share, and the moldboard—appeared among the agricultural tools used by Slavic populations. A plow that may have been similar had already been mentioned by Pliny as early as the first century. This type of plow was developed because of the need for a tool that, unlike the simple plow, could also work the richer, but heavier, alluvial soil.

Architecture

The rare examples of famous architectural works built during the entire period preceding the second millennium all belong to Byzantine culture and were constructed during the sixth century. The basic principles of Roman architecture were followed: bricks were widely used, and large domes were built. However, Sassanid influences were also present. Uncentered vaults, which give buildings a light and harmonious aspect, are typical of Byzantine structures. The best example is the basilica of St. Sophia in Constantinople, built in 573.

The Byzantine Anthemios also deserves mention. Along with his vast theoretical knowledge in mathematics and physics (see the section on PHYSICAL SCIENCES for additional information on Anthemios), he was also very much interested in the practical applications of engineering and in architecture.

The technology of mosaics

The art of mosaics reached extremely advanced levels in Byzantium, even surpassing Roman mastery. A greater variety of materials and different sizes of tesserae were used, while metal oxides and salts—copper for blue and chromium for green—provided a wider selection of colors.

The textile industry

Under Justinian, silkworms were imported and bred extensively, perhaps in order to alleviate unemployment in the Byzantine textile industry. From that time on, silk-weaving flourished, and a state factory existed in Byzantium until the tenth century. Egyptian- and Sassanid-type looms were used. A wooden weaver's die has been discovered in a Byzantine necropolis of the sixth century, indicating that fabrics may have already been printed at that time.

Maritime transport

Ships already had triangular lateen sails during this period, although the first mention of them is only found in a Greek manuscript from the ninth century. The interpretation of a text by Procopios leads us to assume that they were used in Belisarius' expedition against the Vandals of Africa in 533.

600–700

THOUGHT AND SCIENTIFIC LIFE

The Visigoth kingdom. Isidore of Seville

Among the Roman-barbarian kingdoms, often troubled by grave political problems and profound crises, the Visigoth kingdom in Spain was rather exceptional. By 624, all of Andalusia had been reconquered from the Byzantines. At the same time that a satisfactory territorial unity was being reached, the results of the religious unification which resulted from King Recared's conversion from Aryanism to Catholicism in 591, the work of Leander, bishop of Seville, began to be felt.

During this period, Italy's cultural institutions declined, and Longobard's domination represented a continuity with the past. The defense of tradition was entrusted to the monastic schools and to the bureaucratic and juridical culture that persisted in the territories controlled by the Byzantines. Compared to the Longobard and Frankish kingdoms, the relative stability of the Visigoth kingdom, greatly influenced by the juridical and administrative traditions of the Roman system and dominated by the growing power of the ecclesiastical authority provided a notably restful cultural life. Some of the dominant themes of classical speculation and ancient science were preserved and readapted to the new needs.

The work of scholars, all of whom had an ecclesiastical education by now, was directed mainly towards utilizing the patrimony of classical thought and reducing it into easy-to-use encyclopedias or manuals that were suitable for educating the clergy, who now made up the educated class of Roman-barbarian society. The *Etymologiarum libri XX*, by Isidore (560–636), bishop of Seville, is typical. The work was written between 622 and 633, and was destined for exceptional success. One of the most widely read and diffused works of the Middle Ages, it is an example of how the intellectual patrimony of antiquity was reduced to a summary of ideas, equally useful for those who wished to study one of the classical disciplines and those who wished to dedicate themselves to ecclesiastical studies. The range of knowledge and reading had greatly diminished from Boëthius and Cassiodoros to Isidore of Seville. There was less knowledge of the classics, many references to science were full of gaps, and there was an acritical acceptance of sources. However, the quality of knowledge about medicine and natural science was good, presumably due to the flourishing Jewish schools in Spain, with their heritage of Platonic tradition.

Theology in the East

As in the second half of the sixth century, the exact sciences were cultivated very little both in the West and the Byzantine East during this century. Byzantine culture, too, produced compilations and commentaries, such as those inspired by the medical writings of Galen and Hippocrates, and the treatises on astronomy based on Greek sources, more than original works. The reasons were often theological; if the study of the heavenly bodies made it more possible to understand the mystery of the creation, then the study of human anatomy and physiology could be linked to the relationship between the body and the soul.

Theology was the main driving force of all cultural activity. In Byzantium, where there was ample access to a rich Christian tradition as well as the Neoplatonic speculative tradition, there emerged from the numerous theological debates trained thinkers, who often referred to the Church fathers and to Dionysios the Pseudo-Areopagite. Among the most representative of these was Maximos Confessor, who lived in Byzantium between 580 and 662. This was the period during which the Persians, traditional enemies of the empire in the East, were defeated (633), while in the West the conflict with the Pope was growing. Maximos, like other Eastern theologians, basically followed the philosophical pattern of Neoplatonism, which influenced not only Christian, but also Jewish and Moslem thought. Although Byzantine philosophy had very different features from Western philosophy, the great controversy concerning the mystery of the Trinity and the nature of Christ were certainly not overlooked. Maximos, for example, fought the Monothelitism of Sergios, patriarch of Constantinople, who denied Christs's human will and accepted only his divine will in order to sustain his impeccability.

Monasticism in the West

In places where the economy was prevalently agricultural and political systems were collapsing under the attacks of particularism, the spread of Benedictine institutions created centers of intensely productive life, where cultivation of monastic lands was alternated with study and mastering the rudiments of the liberal arts. As early as the first decades of the century, especially in the Gauls and in southern Germany, the network of abbeys, often linked by economic and administrative ties, had created a monastic culture that often made reference to classical tradition.

Monastic institutions which were independent of the Benedictine current were founded in Anglo-Saxon countries toward the beginning of the sixth century. They also participated in the reelaboration of the elements inherited from Greek and Roman culture. In almost exclusively pagan countries, Celtic monasticism arose; this current answered only to monastic rules, and not to the episcopal hierarchy of the church. Episcopal tradition was only revived in 596 under Gregory I, but after its initial success in the extreme southeastern part of England (the kingdom of Kent), the missionaries from Rome found themselves up against the tenacious resistance not only of the pagans, but also of the Celtic missionaries, whose rites had developed without any contact with the rest of the Catholic world and no longer conformed with Roman observances. The victory of Penda, pagan king of Mercia, who was allied with the Celtic-Christian king of Wales, over the Roman Catholic king of Northumbria, at first seemed to compromise all the work of Augustine of Canterbury and his successors. But Roman observance was clearly reaffirmed in 664 at the synod of Whitby. The way in which the Anglo-Saxon church was organized created a new model of obedience to Papal authority. While churches founded previously had obeyed the Pope, but had administered their own affairs within each Roman-barbarian kingdom, the new churches, like the Anglo-Saxon church from 597 on, the Longobards and Frisian churches in this century and the German church in the next, subjected themselves exclusively to a universal Roman Catholic Church discipline. Their example caused even the old churches of the West to establish closer bonds with Rome. Since political unity was lacking, the effects of this religious unity were felt in the organization of culture. Again the model was furnished by the Anglo-Saxon church. By 664, East Anglia already had its own Anglo-Saxon bishop. Ten years later, the Anglo-Saxon Deusdedit became bishop of Canterbury. He was succeeded by the Greek monk Theodore, an expert in Greek and Latin letters. Theodore and the African abbot Hadrian started a campaign of intellectual reform which gave quite a bit of space to the teaching of classical languages and the reading of the *auctores* in the framework of a religious education. The monks educated by Theodore and Hadrian diffused this type of teaching and practice throughout a network of monastical institutions that extended from Britain to Wales and from Ireland, which had already been converted by Patrick, to the first Scottish monasteries. As the Bible was read and explained, information was given on the Latin language, astronomy, and ecclesiastical history. The monk Adhelm, born around 640 at Malmesbury, Ireland and educated in the Canterbury school, is the most representative figure of this current. He often cites the classics in his works, which are of a strictly religious nature modeled after the writings of Augustine. These works came out of a teaching tradition that was quite similar to that found in the schools of the late Roman Empire, and was definitely successful, since by the end of the century the works of the Irish monks were being requested on the continent.

The Arabs and Islam

While Europe witnessed the consolidation of the Visigoth kingdom, the tribulations of the Frankish monarchy and the continuation of the complex power struggle in Italy among the Longobards, the Byzantines, and the Pope, an event was taking place that would prove to be decisive for the peoples of the Mediterranean basin, namely the birth of Islam and the Moslem expansion which followed the period in which Mohammed (died 632) preached. In 635, the Moslems conquered Syria, where the interpreters of Aristotle were flourishing (Severus Sēbōkht and George, Bishop of the Monophysite Arabs, translated several of Aristotle's works on logic into Syriac and commented on them). In 640 they conquered Egypt, and the African coast as far as Tripoli was conquered in 644. The rest of the Persian Empire was absorbed over the following years. In 670, the Arabs stormed all of northern Africa. Through these victories, The Koranic doctrine was diffused, and many of the traits of the vanquished cultures were assimilated. This was an extraordinary phenomenon, since the language that had to filter all of the new knowledge not only lacked any scientific and philosophical tradition, but was codified in a written form almost solely in the 633 edition of the Koran. The absolute ban against translating the sacred text of Islam, which was also the source of the juridical system, forced the subjected populations to learn Arabic. The language thus spread along the shores of the Mediterranean at the rate of Arab military conquests. Given the nature of the Koran, initial speculation about it combined religious, theological, and juridical considerations. Differences in interpretation gave rise to the first sects, a phenomenon that had reached ample proportions by the second half of the century (the struggle between the Shiites and the Sunnites is an example). Tradition has it that Hālid ibn Yazīd, an Umayyad prince who lived in Egypt in the second half of the century, was the first to encourage Greek sages to translate scientific works into Arabic. Toward the end of the century, the translating of scientific and philosophical works into Syriac was carried on in Mesopotamia by the disciples of Severus Sēbōkht, adding to the collection of works that had already been translated in the older Nestorian colonies.

MATHEMATICAL SCIENCES

Mathematics in China and India

The encyclopedic work of Isidore of Seville (c. 560–636) contains no indication that any progress was being made in the West to enlarge the extremely elementary mathematical knowledge of the times. In the East, on the other hand, the Chinese mathematician Wang Hsiao-t'ung, who lived under the first T'ang emperors, wrote a treatise of which only twenty problems are extant; however, some of them require the solving of cubic equations, which thus made their first appearance in Chinese mathematics.

In Byzantium, the mathematical skills needed for public administration in the areas of the public treasury and food supplies were common knowledge.

The most illustrious mathematician of this period was the Indian Brahmagupta (c. 598–663), who worked with first and second degree determinate and indeterminate equations, studied the so-called cyclical quadrilaterals, calculated the values of π and $\sqrt{10}$, and successfully solved problems in solid geometry and combinatorial analysis.

Hindu numerical notation and practical Chinese arithmetic

Around the middle of the century, the Nestorian Severus Sēbōkht made the monastery of upper Mesopotamia, of which he was bishop, into a center for translations and exchanges among Greek, Indian, and Arabic cultures. Through his Syriac adaptations, the Arabs were exposed to Greek astronomy. In one of Sēbōkht's writings, datable to 662, he mentions and appraises the Hindu nine-numeral notation, upon which our present system of numeration is based.

In China, Li Shun-feng, who lived under the T'ang dynasty in the second half of the seventh century, attacked problems in algebra and geometry, perhaps inspired by the increasing number of Hindu mathematical works that were being translated from Sanskrit. He also collected texts from the ancient Chinese arithmetical tradition and commented on them. Chinese mathematics, which was of a particularly practical nature, reached Japan through Korea in the wake of the missionary activity connected to the diffusion of Buddhism.

ASTRONOMY

Isidore of Seville. Astrology as a science

In his encyclopedia entitled *Etymologiarum libri XX*, Isidore of Seville (c. 560–636) not only discussed several questions of astronomy, he also came to terms, in a certain sense, with astrology. While Augustine had flatly condemned astrology (according to him, astrologers were nothing more than vulgar and blasphemous impostors), Isidore considered astrology a par-

tially legitimate science and made a distinction, between natural and superstitious astrology. Only the latter, with its pretext of reading prophesies in the sky, considering the twelve constellations of the zodiac capable of acting on the soul and the body, and predicting horoscopes and destinies of human beings from the course of the stars was to be condemned. Isidore accepted various astrological doctrines, such as the influence of the moon on the lives of plants and animals and the moods of humans, the close ties between astrology and medicine, and the influence of the constellation of Canis and comets. Other authors who wrote on natural science, compilers of encyclopedic collections, and scholars accepted analogous astrological doctrines. Similar concessions, on a larger scale, also appear in Isidore's treatise *De rerum natura*.

Nestorian culture

While Isidore's compendium of knowledge was gaining fame in Visigoth Spain, thanks in part to the prestige that its author enjoyed due to his ecclesiastical position, an interesting Syriac-speaking cultural environment was developing in Mesopotamia as a result of the expulsion of the Nestorian Christians from Byzantium. One of its most prominent figures was Severus Sēbōkht, who wrote the first treatise on the plane astrolabe around 650.

PHYSICAL SCIENCES

Science in the Roman-barbarian kingdoms

While in the previous century, Greek science had been passed down by the way of the "Ostrogoth renaissance" (Boëthius, Cassiodorus), in this century the Visigoth kingdom was the center of scientific culture. Of course, it was a small and impoverished culture that gave more attention to technical applications of scientific knowledge (construction, military machines, the art of navigation) than to its theoretical content. In physics, the encyclopedic work of Isidore of Seville (c. 560–636) deserves mention, although only occasional parts are important.

In Byzantine culture, investigations into the physical world were more systematic and concise, in a traditional sense. Here again, however, scholarly compilations, compendia, and encyclopedia expositions took the place of any original research.

LIFE SCIENCES

Isidore of Seville

The Etymologiae, compiled by Isidore of Seville (c. 560–636), capsulizes a good part of the knowledge of the ancient world in twenty books. The fourth book is dedicated to medicine, the eleventh to humans and monsters, and the twelfth to animals. The long quotations from ancient medical texts helped to connect monastic culture with Greek and Roman scientific tradition.

The compilation of the Hortuli

Monastic medicine began to acquire an interest in medicinal plants, and the Hortuli, a series of explanatory lists of the curative herbs, was compiled. In Byzantium and the area of the eastern empire medicine continued to flourish as it had in the previous century.

Paulos Aegineta, the last illustrious Byzantine physician

Paulos Aegineta (late seventh century), the author of a seven-volume *Epitome*, was the foremost physician of his time. The *Epitome*, an ample collection of the writings of ancient authors, corrected here and there on the basis of experience, was also translated into Arabic. The sixth book of the work is dedicated to surgery and shows remarkable progress in surgical techniques. Paulos Aegineta also worked in obstetrics, where his skill gained him the Arabic nickname al-Qawābiliyyi.

Public services

The imperial administration attended to the problems of welfare and public health through its great orphanage. A medical corps was formed in the army, and nurses had their own organization. The embryo of a hospital system existed, and women were admitted to the medical profession.

Chinese medical culture

In the Far East, Ch'ao Yuan-fang's treatise on the causes and symptoms of diseases, written in China in 610 at the behest of the imperial court, is the first indication of the original train of thought that was developing in medicine in that country. It later became part of a current that is still somewhat autonomous from the influence of European medicine. Medical knowledge spread from China to Japan thanks to the contacts the two countries had set up across the Korean peninsula, as well as to the work of the astronomer and physician Kwanroku, which was followed by students being sent to Chinese cultural centers in order to learn the medical profession.

In China, the emperor Kao Tsung, who took over the throne in 650, commissioned two successive updatings of Tao Hung-ching's medical texts, which brought the work to twenty books of text, twenty-five of illustrations, seven of explanations of the illustrations, and a one-volume index.

Sun Sumo, who died in 682 at a very advanced age, was considered the depository of medicine, and summarized all the medical knowledge of the times into thirty books.

St. Benedictus Crispus and the Commentarium medicinale

There was a complete lack of progress in the life sciences as such between 650 and 700. A treatise on the human constitution written by the Phrygian monk Meletios was dictated more by theological than by medical interests, and the *Commentarium medicinale* by Saint Benedictus Crispus, archbishop of Milan from 681 to around 730, is an elementary practical manual, in verse, on the use of medicinal plants for curing a series of illnesses.

Arabic expansion and the cultures of the East and the West

The Mohammedan conquests brought the Arabs in contact with both eastern and western culture. Almost everywhere, from Egypt to Persia, the practice of medicine was based on the *corpus* of doctrines derived from classical science, transmitted through various versions and adaptations. Medicine was the first discipline that the Arabs mastered, making use both of local schools, such as the Persian school at Jundīshāpūr, and of translations, especially those in Syriac, some of which, especially Galen's basic texts, had been made centuries earlier.

APPLIED SCIENCES

Arab town-planning

Mohammed began preaching around 610, and in 630, after much hardship, triumphed at Mecca. The Arabs poured out of their original territories. In 635 Damascus was Moslem; in 638 Basra and Kūfa were founded; Cairo was founded in 643; and Qairawān was founded in 670. These new cities often developed out of encampments set up to service a military and civil occupation. They generally followed the same layout as ancient cities, being split into two equal parts by a large street that was perpendicular to another. Even bathing establishments, quite common in Arab cities, were inspired by ancient models. For cities with no natural water supply, water was supplied by camel and donkey caravans.

Brick buildings up to seven stories high were erected in several important cities. Brick factories stretched outside the cities throughout the East, since brick-making was prohibited inside the city limits. The cities did not generally have roads for carts because in those areas in which the cities were located, vehicles could not compete with camels.

Arab manufacture

Several Arab cities became famous textile-producing centers. Fustāt, a suburb of Cairo, gave its name to fustian, and muslin was named after Mosul. But the most prestigious Islamic city was Damascus, which fell into the hands of the Moslems in 635. It became the center of Islamic military, economic, and cultural power under the Ommiad caliphate, before the advent of the Abassids, who moved the capital to Baghdad in 762. Damascus not only became an important textile-producing center (damask was named after it), but also acquired a solid reputation in the production of crucible-cast steel.

Following the ancient Indian method, round ingots, about thirteen centimeters in diameter and thirteen milimeters thick, weighing around one kilogram were produced, then hammered together in one or more directions to make blades that were then forged, cooled, and tempered. The blades of Damascus were in demand as far away as Europe, where they were copied and perfected, especially in their shapes.

In the East, steel weapons were shaped by crystallization, while in the West during this period steel was molded using a new technique developed by Burgundian blacksmiths. Strips of steel from the East were welded together and then cut, bent and forged until the desired shape was obtained. The Burgundian smiths' skill in decorating their weapons was so great that trade was established with the Arabs.

Windmills

The Arabs must be given credit for having introduced another basic engine, the windmill, into the Mediterranean. An Arab legend dates the construction of a mill that is turned by the wind to the caliphate of Omar I (ruled 634–644). The idea was supposedly suggested by a Persian slave and carried out under the orders of Omar I. In reality, a wind-turned vertical axis used for religious purposes already existed in central Asia. The windmill was common by the year 400, although its use was limited mainly to Persia and Afghanistan until the tenth century.

Military and naval technology

The Byzantines probably first used "Greek fire," a weapon that was widely used in combat in the second half of the century, during the attacks waged by the Moslems against Byzantium. The survival of the Byzantine Empire is even attributed to its use. According to tradition, Greek fire was invented by a Syriac, Callinicos, who supposedly settled in Byzantium around 670. It was made of crude petroleum from Asia Minor, which was transported to Byzantium in goatskins. It was also used in naval combat, where suitable flame-throwing syphons were installed on board ships in order to set fire to enemy ships.

Byzantine ships were no different from those used in ancient Greece. The most prevalent type was the dromond, with a single file of rowers protected by a roof, although dromonds with two files of rowers also appeared.

700–800

THOUGHT AND SCIENTIFIC LIFE

The effects of the Arab conquests

The second phase of Arab expansionism, which began under the Umayyad dynasty in the second half of the seventh century, culminated at the beginning of the eighth century when the Arabs penetrated as far as India and China and began pressing into Europe. Visigoth Spain was completely subjected in 714, and almost simultaneously, Byzantium was besieged (717–718). The enormous attraction of Islam from a religious point of view and the great number of conversions that followed the conquests, favored assimilation between the Arabs and the subjected populations. While the vanquished populations quickly acquired political equality with their conquerors, the Koran became the standard for juridical and religious life throughout a vast area that reached westward from the Arab peninsula as far as Spain, and eastward beyond the borders of Persia. As a consequence, theological and juridical debates became quite important in the Islamic world. During the first half of the century, the Muslim jurist Abū Hanīfa, (699–767) founded the first orthodox law school, where every subject was investigated from both juridical and religious points of view.

Throughout the eighth century, as well as later, Christianity and Islam existed in a permanent state of war. Arab attempts to conquer Europe were fiercely resisted, as exemplified in the Frankish victory at Poitiers in 732 and the resistance of the Isaurian emperors of Byzantium. This state of affairs made the relations between East and West much more difficult and increased Europe's tendency to turn progressively inward upon itself.

Byzantium under the Isaurian emperors. Iconoclasm

Once the Byzantine Empire had lost its territories in Syria, Mesopotamia, Palestine, Africa, to Islam and a large part of the Balkan peninsula had been taken over by Slavic and Mongolian populations, its tendency to easternize increased. As the imperial authority launched a vast program of political and military reorganization and radical reforms in the countryside, political-religious tensions increased. The Paulician movement grew out of an opposition to the wealth and lifestyle of the ecclesiastical hierarchies, especially in Anatolia. This religious movement called for a return to primitive Christianity and the abolition of the priesthood and also rejected the worship of sacred images. In 726, Emperor Leo III issued a decree forbidding the worship of images of saints, which he labeled as pagan, and ordering the images destroyed. Among its various consequences, the iconoclastic decree produced a revolt in Byzantine Italy and increased the autonomy of the Church of Rome. Iconoclasm also caused many theologians, in Byzantium and elsewhere, to take a stand. In any case, it marked a further split between Eastern and Western religious cultures.

The proselytism of the monks

Whereas the Arab conquests resulted in linguistic unification, in the West the Latin language became quite impoverished. At the beginning of the century, very few people knew how to read and write. Although those who controlled the chancelleries of the Roman-barbarian kingdoms came from the clergy, in many regions their cultural level was very poor. Monastic institutions remained an essential point of reference during this period. The cathedral schools of southern England flourished as a result of missionary activity in England and contacts with nearby Ireland. One of the most famous of these was the school of York, where Alcuin was educated. Venerable Bede (c. 672–735) led an active cultural life in England during this period. His education was based on the religious and profane texts brought from Italy and Gaul by Benedict Biscop, who founded the monastery of St. Peter in Wearmouth in 647. Bede's historical works, such as his encyclopedic treatise *De natura rerum*, are written in a remarkably pure Latin, and contain the great themes of Christian thought that Boniface soon carried to northern Europe. Bede's work is a product of the extensive proselytizing movement that raged throughout the world during this period, generating theological study, attempts and synthesis with or adaptation to local cultures, discussions, and debates (within the limits of the culture of the times, of course).

In 590, the Irish missionary Colomba founded the monastery of Luxeuil in the Gauls; monasticism quickly spread from there as far as Germany and Italy. In the eighth century, the Anglo-Saxon monk Wynfrith, who took the name Boniface when he became the first bishop of Mainz, led an evangelization of the Saxons. Wynfrith not only succeeded in his task, thanks to the protection offered him by the Frankish monarchy, and the environment of Roman episcopal leaders, but united Roman-Benedictine and Anglo-Saxon tradition as well. This established a link between the cultures of Irish and British monasteries. Finally, from 742 to 747, when the church of Neustria, whose clergy was showing signs of disciplinary laxity and an extreme lack of culture, was being reorganized, Boniface promoted a radical reform that not only subjected the Frankish episcopate to Roman authority, but also transformed the cultural environment that had developed in the monasteries of Britain founded during the course of the seventh century into episcopal and abbatial schools and institutions.

Theological speculation

Joannes Damascenus, who flourished in Palestine and died around the middle of the century, was one of the most illustrious figures in this field. He firmly opposed the iconoclasm promoted (as we have already seen) by the Isaurian emperors. In his [Greek here] (*Fountain of Knowledge*), Christian theology crowns an introduction to philosophy of Aristotelian inspiration and an analysis and condemnation of heresies. Significantly, he places Islam— on the rise at the time—among the latter. Philosophical argumentation was especially important now since Damascenus was no longer taking on fairly primitive forms of paganism, as the evangelizers of Germany had done, but rather a theological conception of the divine and the world that was acquiring a command of classical science and philosophy.

The rebirth of the sciences in Abbasid Baghdad

The two most significant events in the cultural life of the second half of the eight century were the promotion of a new culture on the part of the caliphs of Baghdad and the Carolingian renaissance. Although Arab civilization continued to exercise considerable influence from a religious point of view, in the areas of culture and politics it drew increasingly more frequently upon Byzantine and Persian traditions. A revolutionary movement that began in Persia overthrew the Ommiad dynasty, although the new rulers remained faithful to Mohammedan tradition. The caliphate fell into the hands of another Meccan family, the Abbasids, who were linked to the Persian revolutionary movement. While the political center was moved further east to Baghdad, in Mesopotamia, the western provinces of Africa and Spain began to acquire more autonomy under the rule of emirs who were formally emissaries of the caliph of Baghdad. New, lively cultural centers arose. Al Mansūr (ruled 754-775), the second Abbasid caliph (from 754) and the founder of Baghdad (762), sponsored the translation of a large number of philosophical and, especially, scientific texts from Syriac, Greek, and the languages of nearby countries such as Persia and India into Arabic. Countless linguistic as well as conceptual difficulties had to be overcome in translating these texts, since the Arab language was quite new to the in-depth treatment of scientific and philosophical topics and corresponding technical terminology had to be created for each discipline.

Hārūn al-Rashīd, the fifth Abbasid caliph (ruled 786-809), was one of the most prominent of al-Mansūr's successors. A patron of the arts and letters, he also favored a recovery of foreign, especially Greek, science. The effects of the Abbasid caliphs' cultural policies were soon felt in the individual sciences: astronomy, although often in the form of astrology, was actively cultivated, alchemy was revived, and the study of medicine flourished once more, thanks to numerous translations of medical works.

Charlemagne and the Palatine school

The Franks, mainly as a result of Charlemagne's victorious enterprises, came into contact with countries such as Italy,

Ireland, England, and Visigoth and Arab Spain, where native cultures had remained lively. From these lands came the first scholars who lived in the court; they were later joined by the Irish monk Alcuin. Charlemagne's political efforts to consolidate his conquests in the different countries of Europe developed into an attempt at unification that had legislative, juridical-administrative, and religious aspects, through the proclamation of various capitularies, the reorganization of the state bureaucratic structure, and the adoption of a single Latin-language liturgy and unification of the sacred texts into a single canonical version, respectively. Even before Charlemagne was crowned emperor, he promoted a vast cultural renovation in an attempt to fulfill the needs of the state chancellery and improve the administration of justice and the economy. This was done through the *Admonitio generalis* (789) the *Epistula de litteris colendis*, written to Baugulf, abbot of Fulda, between 794 and 797, and later (805) his instructions to the *missi dominici*. These actions promoted the institution and operation of schools in parish churches, cathedrals, and convents, with libraries and *scriptoria* for copying books (a small script hand, which was the basis of all of our present writing forms, was developed during this period). As Charlemagne wrote to the abbot of Fulda: "The priests are to set up schools of reading; psalms, notes, canto, computation, and grammar are to be taught in all the dioceses." Two didactic levels were provided for, one for elementary education and a higher one that had purely religious ends.

At the same time that theological and juridical debates in the East brought about the Qaraite split between the Jews and the Arabs, and another orthodox law school, the Mālikī school, was founded by Mālik ibn Anas (c. 715–795), Alcuin (735– 804), who had been educated at the school of York, was working (782–796) at the court of Charlemagne as an advisor and director of the cultural reform. He collaborated on the preparation of manuals for the reformed schools and the version of the sacred texts that would be used throughout the empire. The problem of reconciling pagan knowledge and Christian beliefs was solved by adopting the *trivium* and the *quadrivium*, considered as introductions to philosophy, at the top of which was knowledge of the scriptures. In accordance with Augustine and the Biblical precept of the *spoliare aegyptios*, the classical *auctores* were recognized as being unquestionably instrumental in the curriculum. Thus the Carolingian perspective was far different from the cautious attitude that Gregory I displayed at the end of the sixth century. This is attested to by Charlemagne's capitularies against medical superstition and in favor of teaching medicine, his introduction of a new system of weights and measures, and the promotion of studies to reform the calendar. The Palatine school, which Charlemagne founded in 782, was directed by Alcuin and soon became the meeeting place for the cultural elite that included Paulus Diaconus, the Latinist Peter of Pisa, the grammarian Paulinus of Aquileia, etc.

MATHEMATICAL SCIENCES

The art of computing

In his arithmetical works and chronicles, especially the *De temporum ratione,* Venerable Bede (c. 672–735) displayed a remarkable knowledge of mathematics, compared to the average level of the times. Bede also wrote the *Liber de loquela per gestum digitorum,* the only text that gives us information on the art of arithmetical computation carried out on the fingers. ''Finger reckoning'' was the only form of non-written computing that remained in use after the Roman abacus was invented.

In an attempt to find a solution to the calendar problem, which the Chinese worked on until the seventeenth century, the Indochinese astrologer Gotama Siddha translated a Hindu work on chronology, in which nine-numeral notation adapted from the Sanskrit was used, bringing the system to the attention of the scholars who lived at the T'ang court. The emperor Ming Huang assigned the study of the system to the astronomer and mathematician I-hsing (682–727), who unfortunately died before he was able to complete his task.

The Carolingian renaissance

Charlemagne's cultural revival, directed by Alcuin, also affected mathematics by introducing the *quadrivium* division. Alcuin (735–804) wrote didactic manuals, and mathematical knowledge proved to be essential to the administration of the empire.

In the meanwhile, Arab civilization's first findings in the field of mathematics began to appear. They presented several features that remained typical of Arab mathematics for centuries. One such feature was the association of astronomy with trigonometry, especially spherical trigonometry, usually found in a concrete form in the introductions to astronomical tables. As a matter of fact, the first two Arab treatises on trigonometry, one concerning spheres and the other on sines and chords, were written by the astronomer Ya 'qūb ibn Tāriq (flourished, Baghdad, second half of the eight century).

ASTRONOMY

Western astronomy. Venerable Bede

Venerable Bede (c. 672–735), in his *De natura rerum,* took up classical astronomical doctrines, integrating them with theories concerning meteorology and geography. His principal source of information for this work was Plinys' *Naturalis historia,* modified by numerous adaptations to the usual Christian orthodoxy. His *De temporum ratione,* dedicated to questions of chronology, is based on a remarkable amount of knowledge about the annual movement of the sun, the various periodical celestial phenomena, and the tides. Bede is an illustrious example of the type of culture that developed in Ireland and, at least to some extent, in England. Thanks to the mediation of Alcuin, this culture influenced the organization of Carolingian studies considerably.

Early developments in Arab astronomy

Around 773, in Baghdad, the capital of the Abbasid caliphate, the caliph al-Mansūr commissioned a transflation of the Hindu astronomical treatise *Siddhanta,* which dated back to the first half of the sixth century. The translation, executed by al-Fazari (flourished second half of the eight century), one of the first Arab astronomers, joined those works of Greek origin that were being translated into Arabic from their Syriac and Persian versions.

PHYSICAL SCIENCES

A century of transition

While natural science in the West during this period was wholly comprised in the work of Venerable Bede (c. 672–735), Alcuin (735–804) and Hrabanus Maurus (c. 776–856), the situation in the East was much more interesting. The situation in the Byzantine east and in the Moslem world contrasted sharply with the isolated West, which was troubled by severe economic and political crises that were only solved toward the end of the century with the Carolingian achievements. Already in the seventh century the Arabs, who lacked any consistent form of scientific culture of their own, had come into contact with an extremely rich cultural tradition, the survival of which had been guaranteed for the most part by the Nestorians. These scholars had translated not only the works of the great classical philosophers, such as Plato and Aristotle, but also those of the great scientists, such as Archimedes, Heron, and Claudius Ptolemy, from Greek into Syriac. Now the scholars were moving from Jundishāpūr, the Nestorian metropolis (see LIFE SCIENCES 500–600) to Damascus, the capital of the Umayyad caliphate. With the rise of the Abbasid dynasty (750), which transferred the capital to Baghdad, Hellenistic scientific culture was diffused throughout the Arab world, first in Syriac, then in Arabic, thanks mainly to the Nestorians.

Arabic science had not yet made any original findings. But by now Arab scholars were the guarantors of continuity in scientific tradition. Scientific and technical achievements arrived from China, India, and Persia, and the Arab world transmitted them to the West.

LIFE SCIENCES

The decline of western medicine

While the Arab world continued to expand its territories and enrich its culture, the spirit of research seemed to be asleep in the West. Only a small amount of the scholarly activity continued in the production of compilations based on secondhand sources. The *De natura rerum* by Venerable Bede (c. 672–735) belongs to this category. On the other hand, naturalistic works, such as bestiaries, herbaria, and lapidaries, which sometimes had medical import, began to be written. The bestiaries in particular were fablelike and didactic. The *Physiologus,* which may have been derived from a second century work and was translated into a number of languages, is a good example. It records about eighty animal species, including those mentioned in the Bible. Its descriptions of the animals are imprecise, and their habits are described anthropomorphically.

Arab culture

The most intense work in the field of medicine was being done by the Arabs. After they consolidated their conquests, especially in the East, they intensified their efforts to recuperate both Classical culture and the cultures of the occupied territories, especially medical-biological texts. Theophilos of Edessa (died 785), astrologer to al-Mahdi, the third Abbasid caliph, translated Galen into Syriac. Jibrā'īl ibn Bukhtishu of the court of al-Mansūr, began a medical tradition that was handed down from father to son. Abū Yahyā al-Batrīq, who also flourished at the court of al-Mansūr, translated Galen and Hippocrates, while his son, Abū Zakariyā Yahyā ibn al-Batrīq, together with a few other scholars, translated Latin as well as Greek texts. The latter is responsible for the Arab version of a pseudo-Aristotelian text, the *Sirr al-asrār (Secreta secretorum)*, which is full of physionomic observations and dietary rules. Other translators were commissioned by the Abbasid caliphs to render Pahlavi or Sanskrit medical works into Arabic. One of these scholars was 'Abdallāh ibn al-Muqaffa'.

Monastic medicine

In the West, the science of medicine remained alive only within the walls of the monasteries. It was reduced almost soley to practice, without any attemps at a thorough examination of the material that had been handed down by a fairly poor tradition. In the Far East, on the other hand, from Tibet to Japan, studies were flourishing and medical institution were being revived. The first Japanese hospital was founded in 758.

APPLIED SCIENCES

Agricultural techniques

In the West, especially in the Frankish area, signs of a reawakening of the applied sciences appeared after Charles Martel's victory over the Arabs at Poitiers in 732. The capitulary *De villis* is a valuable source of information on the agricultural tools of the period. The implements (coopers' axes, regular axes, gimlets, billhooks) all seem to be made for working wood.

It appears that the agricultural tools, especially those used for plowing, were usually made of wood. As a matter of fact, in documents from the Carolingian period, the blacksmith is placed in the same category as the goldsmith, described as a specialist in manufacturing unusual objects. The lack of sturdy tools obviously placed very strict limits on the capacities of individuals. A study of the names for the heavy, or moldboard, plow and the simple plow in German and Slavic dialects allows us to conclude that the moldboard plow was well enough known to have a special name in every language and that both tools existed simultaneously during the Carolingian age, although we do not know in what proportions. The Carolingian carruca was a plow with a wheeled forecarriage, an improvement that enabled the person operating the plow to control the depth of the furrow by pressing on the plow handles. The use of heavy plows brought about an evolution in the organization of peasant life.

The heavy plow was expensive to build and use, so that several peasant families had to join together in order to have one. This slowly brought about a new concept in dividing up the land, which was basically distributed proportionally to each person's contribution to pulling the plow. Another important innovation in agriculture was the adaption of the triennial crop rotations system. The first reliable indication that this system was in use dates back to 763; a manuscript written by St. Gall says: *"et in primum ver aratro iurnalem unum et in mense Junio brachare alterum et in autumno ipsum arare et seminare."*

Metallurgy techniques

While documents on agriculture tell us that iron was rarely used in farming, we learn from other sources that the art of metallurgy had a certain revival in the West, revealing itself in the techniques of casting bells. Since the fifth century, bells had been made with sheet iron; now for the first time cast bronze bells appeared.

Another sign of life in this area was the appearance of the first texts that refer, at least in part, to metallurgy. An example is the *Compositiones ad tingenda musiva* (Recipes for coloring mosaics), which contains descriptions of how to melt gold, silver, lead, copper, and tin ores and a discussion of metalworking. The word *brandisium*, which anticipated bronze for the alloy of copper, tin, and lead, also appears.

Another text is the *Mappae clavicula de efficiendo auro* (Little key to the formula for working gold), which contains other information as well (for example, how to make Greek fire).

Paper manufacturing

After an initial period of adjustment, which lasted until around 750 and coincided with the era of the largest conquests, Islamic civilization became the bearer of technical and scientific progress in the Mediterranean world.

The conquest of Samarkand in 753 taught the Arabs the Chinese techniques for making paper, which Chinese artisans had introduced there the year before. A paste made of fibers from rags, mullberry bark, grass, and hemp was drained in a piece of cloth. When most of the water was gone, a sort of fleece remained, which was dried to make a sheet that was later made smooth with alum or glue. The first paper factory was built in Baghdad in 793. Paper manufacture spread so rapidly through the Islamic world that papyrus and parchment soon disappeared.

Military techniques and the introduction of stirrups into the West

The technology of war was completely revolutionized by the introduction of stirrups, which became a part of the harness for horses. Stirrups originated in the Orient; the first mention of complete stirrups, indicating that they were already commonly in use, is in a Chinese text that dates back to 477. They may have reached Arabs through Persia toward the end of the seventh century, and they definitely reached western Europe by the first half of the eight century, though via an unknown route. It seems that Charles Martel, who lived during that period, had understood the inherent possibilities of using stirrups in combat. By giving the rider lateral support, stirrups transformed the horse and the rider into a compact fighting unit, capable of unprecedented force. Some attribute the Franks' victory over the Arabs at Poitier to Charles Martel's insight; he supposedly created a new kind of tactic by being the first to utilize the properties of stirrups.

The historical connotations of this innovation were extremely important: in western Europe, battles conducted on horseback took on increasing importance, and the need for riders brought about important changes in the social system.

800–900

THOUGHT AND SCIENTIFIC LIFE

The effects of the Carolingian cultural renaissance

After Charlemagne's death in 814, the effects of the cultural revival that he had promoted in the previous century were prolonged under the reigns of Louis the Pious (ruled 814–840) and later, Charles the Bald. The spirit of reformation survived in the great monasteries of St. Marin of Tours, Corbie, and Fulda, where Alcuin's disciples followed his teachings and applied his didactic methods. One of the most prominent of these was Hrabanus Maurus (c. 776–856), abbot of Fulda and later archbishop of Mainz, who earned the title of *primus Germaniae praeceptor* for his teaching activities. Hrabanus edited scholastic texts and encyclopedias, and dealt explicitly with education in his three-volume *De institutione clericorum*, where teaching was regulated by the system of the seven liberal arts. The historian and educator Einhard (c. 770–840), Charlemagne's biographer, and Lupus, abbot of Ferrières from 840 to 862, who promoted studies in literature, philosophy and science for the formation of the perfect Church scholar, also deserve mention.

Almost simultaneously, Agobard (769–840), archbishop of Lyon, was fighting prejudice and superstition. He particularly condemned ordeals and judgments of God, rejecting magical practices and giving naturalistic explanations for phenomena such as storms. The *Dicta*, by Candidus, Hrabanus Maurus's successor at Fulda (882), already present a dialectic demonstration of the existence of God, modeled after Cicero's *De natura deorum*.

These were the first signs of a revival in biblical exegisis and theological meditation, which opposed mysticism in the name of logic during the following centuries. At the same time, *dialectica* was also beginning to influence theological conflicts. These conflicts resulted from the various doubts that had arisen among ecclesiastical scholars. As a consequence, the imperial authority promoted a broad consultation of the most qualified Church scholars, as the opposing sides wrote pamphlets and letters on the controversial topics. The themes being debated were varied, ranging from the question of Christ's presence in the Eucharist, which Paschasius Radbertus, abbot of Corbie (c. 785–860) addressed in his treatise *De corpore et sanguine Domini*, to the relationship between body and soul, which involved Ratramnus of Corbie (died c. 868), among others. Ratramnus was also the author of a work on predestination, which kindled the controversy on that topic. Among those who took part were the monk Gotteschalk and Joannes Scotus Erigena, an Irish monk who moved to the court of Charles the Bald, where he wrote the often-condemned *De divina praedistinatione* against Gotteschalk.

The spread of the Corpus areopagiticum in the West. Joannes Scotus Erigena

Carolingian culture drew its theological doctrines from Latin patristic tradition (especially from Augustine of Hippo), although the Carolingian clergy were not totally ignorant of Greek patrology, which some British monks had read in the original, and of the Platonism in Boëthius' work. Irish monks had already

perfected their knowledge of Greek through direct contact with texts that were unknown to the continental schools of the same period. A large number of Irish teachers emigrated to the schools of the empire—to Rheims and Laon, for example—laying the foundations for a school of thought with a generally Platonic orientation. In 827, Michael II Balbos, the emperor of Byzantium, sent Louis the Pious a gift of a Greek code the contained the *Corpus* (four treatises and ten letters) ascribed to Dionysios the Areopagite, but in reality the work of a Christian writer who lived in Syria at the end of the fourth century. The Dionysian writings, solemnly kept in the Abbey of St. Denis in Paris from 827 on, were translated between 828 and 830 by the abbot Hilduin, who unconditionally accepted their attribution to the supposed disciple of Paul. Hilduin thus contributed to the legend that enchanced their diffusion throughout the Middle Ages, giving Christian *auctoritas* to an organic synthesis of Neoplatonic ideas. The work was soon associated with Macrobios' testimonies and Calcidius' Neoplatonic commentary on the *Timaeus*, as well. But interpreting the *Corpus* presented unsolvable problems for Western culture, which was still weak and fragmented and was venturing for the first time upon a Greek work that was a well-coordinated, but not very orthodox, attempt at interpreting Christianity in a Neoplatonic key. Hilduin's translation proved inadequate, and after several unsuccessful attempts, around 845 Charles the Bald entrusted the translation of the *Corpus* to Joannes Scotus Erigena. At the beginning of the second half of the century, Erigena added the Latin versions of Gregory of Nissa's *De hominis opificio* and Maximos Confessor's *Ambigua* to his version of the *Corpus*, creating a valuable source of reference for Platonically inspired theological speculation. The most important product of this current was the *De divisione naturae*, a dialogue written by Joannes Scotus himself. As in many of his works, he reiterated the function of reason; the conclusions reached by reason are the same as the truth of the faith. His constant appeals to the authority of Pseudo-Dionysios, Maximos, Gregory, Augustine, and many other Church fathers, however, did not suffice to prevent the condemnations that the ecclesiastical authorities pronounced and repeated in the face of the more daring parts of Erigena's thought. But even though contemporary theologians were suspicious of the novelty of meditations that reintroduced doctrines that had been forgotten or taken up in different ways in the West by Augustinian tradition, Scotus's works continued to be diffused until the twelfth century. They were a point of reference for philosophical speculations of Platonic origin, which also included a genuine scientific interest in nature.

Arab culture in the first half of the century

In Baghdad, al-Man'mūn, the caliph from 813 to 833, not only encouraged research into and translation of the classical texts, also being done by private citizens (for example, the three Banū Mūsā brothers), but also institutionalized it by founding the House of Knowledge, which included a library and, later (829), an observatory that competed in fame with the one built some time before by the Umayyads in Damascus. The beneficial effects of this cultural policy were soon felt in all the countries ruled by Islam. The scholars who made reference to the House of Knowledge, united by their common knowledge and use of Koranic language, came from every population and religious conviction. Although translations were still often done using and intermediate language, the texts were subjected to revision by experts, a caution that western Latin translators too often neglected to take. Ibn Batriq's translation of the *Timaeus*, done at the beginning of the century, figures prominently, since it introduced Platonic cosmology into the Arab world. Almost simultaneously, al-Nazzām (died 845) revived the idea of the potential creation of all humankind in Adam (a concept that the West owed to Augustine of Hippo). The *Theologia*, attributed to Aristotle but in reality written during the fourth century along with parts of Plotions' *Enneads*, and the *Liber de causis*, an exposition of Proclos' *Elements of Theology* arranged by theorems, were some of the more important works that reached Arab culture in translation from the Greek. These texts were all declaredly Neoplatonic and led the Arabs to a synthesis of Plato and Aristotle. But members of the Shiite movement often united the Neoplatonic concepts of divine emanation and illumination with ancient themes from the Persian cult of light and the religion of Mithras. This initiated a type of reflection that, when it did not lead to mysticism (as it did with Sūfism), developed a number of ties with science. This tradition flourished later in the West in the *metaphysica de lumine*, reappearing during the Renaissance and as late as Newton, with an increasing emphasis on scientific ideas.

Arab culture in the second half of the century. Al-Kindi's philosophical reflection

Baghdad became the most important cultural center of the East, both because of the original findings of the Moslem scholars and because of the activity in its centers of study. After the middle of the century, Thābit ibn Qurra (836–901) coordinated translations of works in mathematics and astronomy, and Hunayn ibn Ishāq (809–873) coordinated those in medicine and natural history. The latter, known as Joannitius to the Latins, researched and collected Greek codes and compared the Syriac and Arabic translations of texts as they were translated. Among his works is the *Cannon of Galen*, in which he lists 129 works by the Greek physician known to him, with a critical history of the Syriac and Arabic translations of each.

Al-Kindi, considered the first Arab philosopher in the proper sense of the term, also lived in Baghdad (where he died in 873). The specific intent of Arab philosophy, which was much freer than Western Christian thought in the areas of religion and theology, is already evident in Al-Kindi's work. He investigated nature and human beings, approaching them through the theory of knowledge, which he developed as a critical introduction to the various sciences, especially medicine, in the tradition of Galen. Al-Kindi's *Liber de intellectu*, which deals with the Aristotelian theory of intellect and was modelled after Alexander of Aphrodisia's interpretation, and his three-step logical method—division, resolution, and demonstration—were studied extensively in the West. Al-Kindi developed his method in a commentary on the theory of the *Posterior Analytics* that uses Euclidian geometry.

Byzantium from Leo of Thessalonica to Photios

In Byzantium, Leo the Iatrosophist, archbishop of Thessalonica, who lived under the reigns of Theophilos (ruled 829–842) and Michael III (ruled 842–867), began a job of reelaborating

past knowledge that displays very little originality but reawakened an interest in the classics and stimulated an impressive amount of copying and editing of ancient texts. In this way, the ancient works were transmitted to later generations and diffused beyond the borders of the empire by Byzantine scholars who were exiled because of the political-religious struggles that troubled the country.

The Byzantine world reached complete religious and cultural independence from the Western Church with the work of Photios (c. 820-891). Photios, like Leo in his time, gathered around him scholars who were dedicated to reading and discussing the classical texts. His most famous work, the *Myriobiblon*, contains reports on the books that his group of scholars read during their meetings. These ranged from theology to natural philosophy, law, and medicine. Photios' cultural interests were considered too profane by the more conservative factions of the religious cultural environment, and he was accused more than once of practicing magic and astrology. Nonetheless, Photios made quite an impact on culture, not the least in the revival of studies in logic, which he carried out on Aristotle's works from a perspective that conceded very little to Neoplatonic mysticism.

The cultural policy of the imperial authority gave a boost to the revival in philological, philosophical, and scientific studies. The regent Bardas reorganized higher learning in Byzantium in 863, placing Leo the Iatrosophist at the head of a university-type institution. Coinciding with this rebirth of Byzantine cultural institutions was a reform in Greek writing, which went from the large, all-capital uncial style to the small script that was already being used for chancellery notes and correspondence. This solved the problem of the high cost of books, which were by now exclusively written on parchment, since the Arab conquest of Egypt had made it almost impossible to get papyrus, the cheapest writing material in the ancient world. The new small writing took up less space and could be executed much faster than uncials.

The missionary work done by the two brothers Cyril (c. 827–869) and Methodios (c. 825–885) of Thessalonica among the Khazars of Crimea and the Slavs of Moravia and Pannonia also deserves mention. Cyril is credited with inventing the Glagolitic alphabet, from which the Cyrillic alphabet, still in use in Serbia, Bulgaria, and Russia, was later derived.

The end of the Carolingian age

In the Carolingian West as well, one of the most important events of the time was a reform in writing. The small Carolingian style, created in the *scriptoria* that Charlemagne had set up, slowly dethroned the various national forms of writing, a spontaneous result of the degeneration of non-epigraphic Romans forms. It outlived Charlemagne's ephemeral political system and later caught on as far away as Spain, England and Southern Italy. As for the degeneration of the Latin language outside the narrow circles of ecclesiastical scholars, it is important to note that as early as 813 the council of Tours recommended that the clergy translate their sermons into Romance or Germanic languages, so that the people could understand them. Twenty-nine years later (842), at the swearing-in ceremony that sanctified the alliance of the two cadet sons of Louis the Pious against the older brother of the Emperor Lothaire, Charles the Bald's soldiers gave their oath in the Romance language, while Louis the

German's gave theirs in German. The political division that followed favored this process even more during the second half of the century.

The work of Joannes Scotus Erigena ended in the efforts of the Carolingian renaissance. The politically divided Carolingian empire was already in crisis by 877 (the presumed date of the Irishman's death), while new Barbaric populations were pressing at the borders. The dissolution of linguistic and political ties met with the immobility of economic life (the difficult relations with Byzantium, for example) and the inborn weakness of the empire's institutions. The intellectual monopoly exercised by the ecclesiastical hierarchy affected cultural life, where the religious tones, already typical of the Carolingian reform, were accentuated. The *Exempla diversorum auctorum*, by Micon of St. Riquier, and the work of Adoard, a priest and librarian at an unknown French monastery, who used a number of Cicero's writings in a collection of examples from classical authors, probably both date back to the end of this century. But the work of grammarians dispersed in the various scholastic centers apart, theological and philosophical research did not die out. In England, Alfred the Great (849–899), King of Wessex from 871, gave an energetic boost to studies by having a great cultural patrimony translated into the Anglo-Saxon vernacular. Even more important was the work of the Benedictine school at Auxerre, which included Heiric (c. 841–876) and his pupil Remigius (c. 841–908), both of whom were influenced by Joannes Scotus' Erigena. Remigius, who had been a student of Hrabanus Maurus, taught in Paris during the last few years of his life, continuing the pedagogical and literary tradition of Alcuin. His teachings in Paris were supposedly the beginnings of the future Sorbonne University.

MATHEMATICAL SCIENCES

The beginnings of Arab mathematics

Having conquered territories with extremely varied cultures, the Arabs not only quickly assimilated the ideas of these cultures but also promoted their diffusion, by translating their basic works into Arabic. The Arabs brought the products of Greek mathematics and those of Hindu science together in the most prestigious capitals of their caliphates (Damascus, Baghdad, and Cordova), making an original synthesis of these two very different traditions and offering new contributions to the development of the various mathematical disciplines. Systematic translations were made of the scientific works of other languages, especially in Baghdad under the caliphs al-Mansūr (ruled 754–775), Hārūn al-Rashīd (ruled 786–809) and al-Ma'mūn (ruled 813–883). By around 773, for example, al-Fazārī had already translated the Hindu astronomical books of the *Siddhanta*, while in this first part of the ninth century, al-Hajjāj ibn Yūsuf (flourished c. 786–833) did the first Arabic translation of Euclid's *Elements*, which were then commented on by al-'Abbās. Not only the greats, such as Euclid, Archimedes, Appolonios, and Ptolemy, but even the minor authors were translated and commented upon by the Arabs, so that they came into contact with almost all the products of ancient science.

Muhammad ibn Mūsā al-Khwārizmī flourished at this time, during the twenty-year period of al-Man'mūn's caliphate (813–833). After an initial period of translating, al-Khwarizmī made a considerable number of original scientific contributions to Arab mathematics. His two major mathematical works are a treatise on arithmetic and one on algebra. They mark the explicit and conscious introduction of Hindu positional notation and the numerals used therein into the Mediterranean area, along with the use of several basic computational procedures and an interesting theory of equations. They also contain some purely Greek-style approaches to geometry, as well as discussions on the practical commercial and legal applications of algebraic methods. This interest in practical uses was an important new feature that the Arabs introduced into the way mathematics was thought about. It was favorably accepted by medieval Latin mathematicians and, as we shall see, even influenced the birth of modern mathematics. Al-Khwāruzmī also introduced the notion of the sine, taken from the Hindus and unknown to the Greeks, who dealt only with arcs and chords, into trigonometry. Our term algorism actually comes from al-Khwārizmī, which was Latinized into Alchorismus, while the word algebra comes from *Kitāb fi hisāb al-jabr wa l'muq bala,* the title of his treatise on the subject.

The philosopher, astronomer, and mathematician al-Kindī (c. 801–866) also flourished in Baghdad. He was the author of a four-volume work on Hindu numerals, and also wrote *De aspectibus*, a treatise on geometric and physiological optics that greatly influenced the development of those disciplines in the West during the Middle Ages. The *Liber karastonis*, on the conditions for the equilibrium of scales, and the treatise *De Geometria*, dedicated to various problems such as the measurement of spheres, the trisection of angles, and proportions, are attributed to the three Banū Mūsā brothers, who died during the caliphate of al Ma'mūn.

In Byzantium, meanwhile, Leo the Iatrosophist (see also THOUGHT AND SCIENTIFIC LIFE 800-900) also considered mathematics in the context of his encyclopedic interests. He started a movement of revival in this field as well, and we owe the survival of many of the scientific texts of the great classical authors, which have come down to us in manuscripts written during this period, to this movement.

Arabic translations and commentaries on the classics

The energetic work that the Arabs did in translating and commenting on Greek scientific texts, which added to the work of researchers and innovators in the fields of arithmetic, geometry and, above all, trigonometry, reached its apex in the second half of this century. Al-Māhānī, who flourished around 860, commented on Euclid and Archimedes, and attempted to solve the cubic equations—which Arab mathematicians called by his name—that would solve the Archimedian problem of dividing a sphere into two segments according to a given ratio. Halāk al-Himsī translated the first four books of Apollonios' *Conics,* while the Egyptian Ahmad ibn Yūsuf (flourished c. 912), known in the West by his Latinized name Hametus, wrote the *Liber de proportione et proportionalitate*, which made Menelaos' theorem known in the Latin world. Abū-l-'Abbās ibn Hātim al-Nairīzī (died 922), known in the West as Anaritius, wrote commentaries on the works of Euclid and Ptolemy.

Among al-Khwārizmī's pupils, one of the most prominent was Thābit ibn Qurra (836–901), who can be considered a typical Arab mathematician, in that he did both translations (he executed and directed translations of works by Euclid, Archimedes, Theodosios, Apollonios, and Ptolemy), and original research in arithmetic and, especially, geometry. In a treatise on the quadrature of the parabola and paraboloids, he actually made new findings in an area that Archimedes himself had already made famous.

Qusta ibn Luqa translated Diophantos, Aristarchos, and Heron's *Mechanics* and developed the use of tangents (*umbra versa*, used earlier by Anaritius and before that by al-'Abbas) and cotangents (*umbra recta*) in trigonometry.

The most ancient Arab documents using numerals written following the system of positional notation, which is our present system, date back to the years 874 through 888. Zero is indicated by a dot, the symbol that Arabic notation uses to this day.

ASTRONOMY

The Baghdad observatory

The Jewish astronomer Masha'allah (754–813), known for his observations and his skill as an astrologer, wrote a popular treatise on the plane astrolabe, an instrument that became widely diffused and was used in the West until the seventeenth century. In the East it is used even today to calculate ritual astronomical recurrences.

Mathematical and observational astronomy became popular once again in Baghdad, where, thanks to the patronage of the caliphs, the translators' school led by Hunain ibn Ishaq (809–873) translated Ptolemy's treatise on astronomy into Arabic, and the mathematician al-Khwarizmi (flourished early ninth century) used his own findings in trigonometry to compile astronomical tables that became very popular among the Arabs. The activities of the Baghdad and Tadmur observatories, where observational date was recorded for many years, were particularly directed toward the tasks of calculating the inclination of the ecliptic more accurately; measuring the degree of the meridian in order to calculate (following the method of Eratosthenes) the size of the earth; and preparing a huge map of the earth. al-Farghani (Alfraganus) (died after 861) compiled a manual on astronomy, which was translated into Latin in the twelfth century. The Latin version, entitled *Elementa astronomiae*, gave a considerable boost to the revival of studies in astronomy in the West and was only supplanted by the more modern works of Regiomontanus (see ASTRONOMY 1450–1500). Yahya ibn Abi Mansur directed the observatory of Baghdad and did a remarkable job of checking Ptolemy's data. The results of his investigation, along with others carried out under his direction, are gathered in the *Tabulae probatae*.

With the rise of new observatories, more instruments were built and treatises were written about them. 'Ali ibn 'Isa, an astronomer who was known as al -Asturlabi for the fame he acquired in building astrolabes, flourished in Baghdad around 830.

Western works

Perhaps the most interesting western *summa* from the years immediately following Charlemagne's reign (ruled 768–814) is the *De mundi coelestis terrestrisque constitutione liber*, erroneously attributed to Bede. The unknown author demonstrated the sphericity of the earth and discussed the order of the planets and the constitution of the sublunary sphere. In his *De mensura orbis Terrae*, finished in 825, the Irishman Dicuil used information gathered by travellers to explain the most elementary astronomical facts. Hrabanus Maurus (c. 776–856) attempted to reconcile classical astronomical theories with patristical ones in his encyclopedia *De universo*.

Arab astrology

Abu Ma'shar (Albumasar), who died a centenarian in 886, was long active at the 'Abbasid observatory. In the West, he is especially famous for his astrological works, the *Introductorium in astronomiam* and the *De magnis coniunctionibus*, which brought Arab judiciary astrology to the Christian world. These works enjoyed extreme prestige and were enormously popular up to the end of the sixteenth century. Abu Ma'shar contended that the world was created when the seven planets were in conjunction in the first point of Aires, and that the end of the world would come when the conjunction was repeated in the point of Pisces. The verification and prediction of this particular event were topics of an extremely lively discussion among astrologers throughout the Middle Ages.

The spherical astrolabe

The Persian al-Mahanni (died c. 880), beginning in 853, made a series of observations of solar and lunar eclipses and planetary conjuctions. These used in particular by the Egyptian astronomer Yuus to prepare the Hakemite Tables. Anaritius (died 922) worked on meteorology at the Baghdad observatory; he commented on Ptolemy and complied a treatise on the spherical astrolabe, in which he described the history, construction, use, and merits of that instrument compared with the other means of observation of the times.

Thabit ibn Qurra (836–901), very active at the House of Knowledge in Baghdad as a translator of works on mathematics and astronomy, also did important work in the field of astronomy. He made long and systematic observations of the sun, and used them to determine the tropical year. He took up the theory of the trepidation of the equinoxes put forth in the Hindu *Siddhanta*, and added a ninth sphere, the primum mobile, to the eight of the Ptolemaic system in order to account for this phenomenon. Both the theory of the nine spheres and the erroneous explanation of the precession of the equinoxes, which even Copernicus accepted, were diffused in the West through the translations of his works. Joannes Scotus Erigena (died c. 877), who flourished at the court of Charles the Bald, took up Heraclides of Pontos' theories in his *De divisione naturae*, in which he maintained that not only Mercury and Venus, but Mars and Jupiter as well, were heliocentric. This daring system was somewhat successful only until the diffusion of Arabic astronomy, which imposed the theories of Claudius Ptolemy.

PHYSICAL SCIENCES

Physics in eastern Islam

Al-Kindi (c. 801–866), who did research in meteorology, on the tides, and on the problem of specific weights, can be considered the first great Arab author in physics (see THOUGHT AND SCIENTIFIC LIFE 800–900). One of his most important works is a treatise on optics and the reflection of light.

Developments in technology and mechanics

Technological research, initially motivated by the need for a wide network of canals for water supplies, irrigation, and communication, flourished in Egypt and even in Mesopotamia at almost the same time that al-Kindī was conducting his studies in physics in Baghdad.

The developments in theoretical mechanics were quite interesting. Many books were written about lifting water and about water wheels, scales, and water clocks. The oldest work, which apparently appeared in 860, is the *Book of Artifices*, written by the brothers Muhammad, Ahmad, and Hasan, the sons of the great translator Mūsā ibn Shākir, who were themselves patrons of translations. The work describes over a hundred technical devices, about twenty of which have practical applicatins, for example containers for holding hot and cold water, permanent level wells, and water clocks. The Heronian and more generally Alexandrian tradition of technical devices intended to amaze and delight prevails, although the degree of theoretical knowledge is much higher here.

CHEMICAL SCIENCES

The mystical and magical nature of alchemy

For a long time, people practiced chemistry without even knowing it. Metals were extracted from ores, natural pigments and dyes were produced, ceramics and glass were made, and the natural desire to understand things soon drove people to inquire into the reasons behind this primitive technology, as well as into the physiological activities of plants and minerals.

Quite early on, however, due to the lack of investigative means and perhaps because of the Greek inclination toward metaphysics, this embryonic chemistry went through a phase of mysticism and magic, above all of Chaldean and Persian inspiration, which reached its apex in Alexandria and, later—although it aroused little interest—in Rome, during the first centuries of our era. This was the first phase of alchemy.

The practical aspects of alchemy

Only toward the end of the eighth and during the ninth centuries was alchemy effectively revived as an experimental science. It was still unaware of itself, however, until Roger Bacon (c. 1214–1294), and it continued to be influenced by magic and mysticism. It was during this period that the Moslems developed an interest in alchemy. It seems that Harrān, the Syriac city that saw a revival of its intellectual life during this early period of Islam, was the center of developments in alchemy. Harrān was already famous for the liberal nature of its culture; it was a meeting point and melting pot for Persian, Syriac, and Greek naturalistic currents, which created a favorable environment for the development of alchemy; it also favored this development in a practical sense. Harrān had a flourishing trade in precious metals, as well as in sulphur, borax and other substances commonly used by alchemists. In any case, whether the Arabs borrowed from Harrān or from the Greco-Egyptian alchemists of Alexandria, they soon had their own scholars. The first and greatest of these, Jābir ibn Haiyān, known in the West

as Geber, probably flourished during the last quarter of the eighth century. A *corpus* of theoretical and experimental works are attributed to him.

Geber's *corpus* contains about two thousand books, and was certainly not all written by him, or at least by him alone. Many fragments were probably added in later epochs. In any case, parts of the *corpus* contain descriptions of methods and equipment for numerous chemical and physical operations (especially for changing states, such as melting, distillation, sublimation, etc.). The properties of sal ammoniac, so surprising because of its volatility when heated, are described, and mercury is rationally defined as a metal.

Two texts, know as the *Compositiones ad tingenda musiva*, which some say dates back to an Alexandrian text of 650, and the *Mappae clavicula de efficiendo auro*, of which the first known manuscripts only date back to the year 1000, were written around the period between the eighth and ninth centuries. They are of a technological nature, and talk about colors, minerals, metals, and other topics. They do not contain any theoretical discussion, so are not alchemistic in nature, and have no relation to the Arab world. These two works are more thoroughly discussed in the section on APPLIED SCIENCES 700–800.

LIFE SCIENCES

Islam and the mediation between the West and the Indochinese East

Following the boost that the caliphs, and especially al-Ma'mūn, gave to the sciences, Islam's cultural presence emanated from Baghdad through its translations, its mediation between the Mediterranean and the Indochinese East, and the compilation of catalogues and encyclopedias that also dealt with biology and medicine. The *Paradise of Wisdom*, written in 850 by Alī Rabbān al-Tabarī, provided Islamic medicine with a summary of basic ideas which also contained a true philosophy of nature.

According to al-Tabarī, material changes resulted in changes in form, since matter was substance and its form an accident. The ideal state was equilibrium, and the practice of medicine should be based on standards of prudence and careful individual diagnosis. A casuistry that was new with respect to tradition arose. Until the time of Abu l-Qāsim al-Zahrāwī (c. 936–1013), surgery was cultivated very little, partly because of the strict Koranic ban on the dissection of cadavers. Ibn Massawaih (died 857) (Mesuë the Elder in Latin) returned to the old custom of dissecting monkeys in order to study anatomy. Ophthalmology also flourished, influenced by the numerous cases of trachoma and cataracts that appeared in Egypt and Syria, with Mesuë, who wrote a treatise entitled *Disorder of the Eye*.

In his *De medicinarum compositarum gradibus*, the philosopher al-Kindī (813–873) attempted to base the dosage of medicines on mathematical measurements. Works of a more naturalistic character also dealt with pharmacopea: the *Secretum secretorum* and the *De lapidibus*, both attributed to Aristotle, and 'Utarid's lapidary are examples. The Christian physician Sābūr ibn Sahl of Jundishāpūr compiled a twenty-two volume work on antidotes that dominated Islamic pharmacopea for almost four centuries.

An-Nazzām and the concept of evolution

An-Nazzam (c. 775–845), a Mu'tazilite philosopher, worked on the problems of natural science and outlined a concept of evolution that in part referred back to Augustine's theory of potential creation. All of the authors already mentioned also contributed to the diffusion of medical knowledge through their translations from the Greek and from the languages of nearby countries. The works of Hippocrates and Galen, as well as Aristotle's naturalistic treatises, were soon available in virtually complete Arabic versions.

In Byzantium, under Theophilos (ruled 829–842), Leo the Iatrosophist compiled a medical encyclopedia that contains a particularly well-developed section on surgery. In the West, Walafrid Strabo (c. 808–849) described a number of plants and their therapeutic properties in his *Hortulus*.

Abū Bakr al-Rāzī and Islamic medicine

The second half of the century was dominated by the work of the Iranian Abū Bakr al-Rāzī (c. 855–923), who later moved to Baghdad, where he became the director of the great hospital founded in 918 by Caliph al-Muqtadir. Al-Rāzī's work established the features of Islamic medicine: a tendency to synthesize in treatises and encyclopedias, structural ties with metaphysics and alchemy, a lack of direct anatomical knowledge, a wealth of clinical casuistry, and specific contributions, especially in the field of epidemiology. Besides his encyclopedic treatise *Kitab al-hāwī* (*Totum continens* in the Latin translation), the monumental body of writings by the master of Baghdad contains more than eighty other medical and naturalistic works and tens of other philosophical, metaphysical, logical, theological, and alchemical works. Altogether the *corpus* constitutes one of the major links between Medieval scientific knowledge and ancient science. Al-Rāzī's *Kitab al jadarī* (*De variolis et morbillis*) was an original contribution of the Middle Ages to medical knowledge, in a field that medicine again took up with the same enthusiasm at the end of the eighteenth century.

Treatises on medicine and zoology. Medicine in the convents

Hunain ibn Ishāq, the author of a treatise on ophthalmology and an introduction to Galen's *Ars parva*, the *Isagoga Joannitii ad Tegni Galeni*, also flourished in Baghdad at the same time as al-Rāzī. The latter work was put into a collection of medical works, called *Articella*, that was widely diffused in the West. The *Practica sive breviarium*, by the Christian physician Serapion the Elder of Damascus, was just as popular, especially the part on phlebotomies. The *Book of Animals*, by the philosopher al-Jāhiz, and the *Book of Plants*, by the Persian historian al-Dīnawarī (c. 815–895), are also of biological and naturalistic interest.

In Byzantium, Nicetas compiled a large collection of medical texts ranging from Hippocrates to Paulos Aegineta, as well as numerous panegyrics on the lives of the saints. Bertharius, the abbot of Montecassino from 857 to 884, composed two treatises, *De innumeris remediorum utilitatibus* and *De innumeris morbis*, that give us some insight into monastic medicine.

APPLIED SCIENCES

Agricultural techniques

Several agricultural techniques that would bring about some rather important changes began to be used in the West. The *Polyptychus Irminonis*, written at the beginning of the ninth century, tells of three plowing times: two before the winter planting and one before the spring planting. One third of the arable land was not cultivated, while one third was planted with wheat, and the other third with various legumes and vegetables. However, the rare documents from the period refer to large monastic or royal farms, which must have used regular triennial rotation, an idea that may have come from the Latin classics studied in the monasteries. Livestock was allowed to graze in the uncultivated spaces.

The large farms of the Carolingian provinces, which were at the vanguard in technology, had a remarkable range of equipment for grinding grain. The lords made their mills available to the peasants for a fee, and advantageous arrangement for the peasants since it freed them of a considerable amount of labor.

Descriptions of agricultural tools continue to be scarce. A passage from the polyptych of Sain-Maur-des-Fosses indicates that six oxen pulled a device for plowing. The number of animals suggests that the tool was powerful, but we know nothing about the harnessing system or whether a simple or a moldboard plow was used. The simple plow, which breaks up the soil without turning it over, requires a tedious operation of spading every four to six years.

The agricultural yields during the Carolingian age remained low, however, just double the amount of seed sown; in fact, the amounts available for comsumption were often considerably less than the amounts reserved for sowing.

Astronomical instruments

Around the year 800, Māssāllāh wrote a treatise on the astrolabe that was translated into Latin and became the most popular text on that instrument for many centuries. The astrolabe definitely dates back to Claudius Ptolemy, but the Arabs promoted it to such an extent that it became a universal instrument for measuring angles and calculating the positions of the stars. The Arab astrolabe consisted of a disc, graduated along the edge, and an alidade, or calibrated ruler with a sight. Harrān became renowned for designing and building different types of scientific, mathematical, and astronomical instruments (see ASTRONOMY 800–900).

The development of astronomical observations led to the creation of new precision instruments; the protable quadrant was known to the Moslems as early as 840.

The origins of chemical production

Al-Rāzī (c. 855–923), who flourished in the period between the ninth and tenth centuries, wrote a classification of the substances existing in nature. The work is especially remarkable because it is based on the properties of the substances and is totally free from the allegories and mysticism that permeate the works of Muhammad ibn Umayl, for example. Apart from the genius of Jābir and Rāzī, the reason Arab chemistry made such remarkable progress during this period was because of the importance of chemical products to the flourishing Arab industries, such as pottery, perfumes, and probably gasoline.

The Arabs became famous for the shiny colored enamels that they put on their earthenware. The enameled containers, many of which were fireproof, were particularly suitable for chemical experiments, while a growing number of chemicals also took part in the production process.

The perfume industry changed completely with respect to ancient times. The ancient method of making perfumes was to mix flowers and herbs with oils. The Arabs mixed flowers and herbs with water and distilled the mixture to get liquid perfumes. Cylindrical and conic ovens with rows of stills inside were invented to improve the distillation process.

Navigation and military techniques

The Arabs also brought technical innovations to shipbuilding. The historian al-Tabarī speaks of warships similar to galleys that were built by tying pieces of wood together, a method that the shipbuilders of Sīrāf were famous for. Coco trees were cut for boards and also provided the fibers that were spun and used to "sew" the boards together. In the field of mechanics, the *Book of Artifices* by the three sons of *Mūsā ibn Shākir (see PHYSICAL SCIENCES 800–900)* describes automatons similar to Heron's.

In 877, Pope John VIII gave the European navy a boost when he had dromonds built at Civitavecchia. Meanwhile, Nordic ships continued to evolve; a ship discovered at Goskad, in Norway, dates back to this century. It has a square sail, a single mainmast at the center of the ship, and rowing ports in its high sides. The ships later grew to have thirty to sixty rowers per side. The cock-boat, a large roundish Nordic mercantile ship with overlapping planking and a single square, sail may be descended from the Nordic warship.

The Byzantines made various innovations in military techniques. They invented a sort of grenade loaded with naphtha, saltpeter, and sulphur (a mixture very similar to gunpowder), and improved the Byzantine visual telegraph, a system of lanterns and fires used to transmit messages over long distances. The messages were beamed from hill to hill until they reached the beacon on the terrace of the great palace of Constantinople.

The art of navigation became highly developed in the countries of the Arab caliphates on the Mediterranean Sea, the Red Sea, and the Indian Ocean. The Europeans only began to be interested in this art somewhat later. The Portuguese prince Henry the Navigator (ASTRONOMY 1400–1450) contributed a great deal to Arab navigation on the open seas and to introducing these skills into Europe.

900–1000

THOUGHT AND SCIENTIFIC LIFE

The flourishing of Islamic civilization. Moslem science, the commentaries on Aristotle, and the work of al-Fārabī

Although on a political level an increased tendency toward fractionism brought about the creation of three large, inde-

pendent centers of Arab power (Asia, Egypt and the northern coasts of Africa, and Spain), the Arab world had a lifestyle and culture that were clearly superior to those of the Christian West, and even to the Byzantine Empire. Arabic was the language of all the subjected territories—which now extended from Gibralter to India, and from upper Egypt to Mongolia—and constituted a powerful vehicle of cultural unification and diffusion. Baghdad was still the liveliest center, but almost all the other cities that played important roles in the troubled political events of the times, even for short periods, competed with the capital of the 'Abbasid caliphate in promoting the progress of letters, science and the arts. Islamic knowledge was slowly developing its own traits. There was a typically Moslem science which, together with literature, included history, chronology, law, philology, and theology, united by their common foundations in the Islamic religion, as well as a lay science, which had its roots in the classical patrimony that had by then been completely mastered but was outside of Moslem tradition, both in content and in methods. The task of commenting on Aristotle, which had traditionally absorbed a large part of Arab speculation, was carried on in this century by al-Fārabī, Alfarabius to the Latins (c. 870–950). He accepted the position of the Pseudo-Aristotelian *Theologia* and in his *De concordantia Platonis et Aristotelis* attempted to reconcile the doctrines of Aristotle and Plato, and to harmonize classical philosophy with the religious ideas of Sūfism. He also developed a theory of motion that foreshadowed the famous theory of impetus; and the distinction between the essence and the existence of things, developed starting from Aristotle's observation that the idea of a thing does not necessarily imply its existence. This distinction was investigated in detail by thirteenth century philosophy and theology in the West, where it was known through the work of Avicenna (THOUGHT AND SCIENTIFIC LIFE 1000–1100).

A general interest in the question of establishing a harmonious relationship between faith and reason dominated both Moslem and Eastern Jewish thought. The Qaraite Jew Saiadia ben Joseph (882–942) attempted to reconcile rationality and faith, while his coreligionist al-Qirqisānī proposed applying scientific methods to philosophy, and in his commentary on the *Pentateuch*, even maintained that faith should be founded on reason. Simultaneously, the Moslem al-Ash'arī (c. 873–935) founded so-called Islamic scholasticism by attempting to rationalize faith.

Religious reflection and Islamic encyclopedism

During the second half of the century, the Buyid sultans, who in practice controlled the caliphate of Baghdad, became great patrons of the sciences, the letters, and the arts. 'Adud al-Dawlah (936–983) had a great hospital, built in Baghdad; which was finished in 979, and his son, Sharaf al-Dawla, had a private observatory constructed (observations were definitely made there around 988). The syncretism of Islamic authors, who by then had two centuries of monumental production behind them, created a need for encyclopedias that would synthesize and arrange knowledge, even in the light of specific research (especially in mathematics and astronomy). *The Keys of the Sciences*, by the Persian Muhammad ibn Ahmad al-Khwārizmī, a huge technical dictionary written in 976, and the *Kitāb al-fihrist, (The Catalogue)*, by Abū Ya'qūb al-Nadīm, finished around 990, are examples of works that satisfied this need. The latter is subdivided into "discourses," and is a collection of bibliographical and biographical information on the authors of hundreds and hundreds of works written in Arabic. These great encyclopedias required a classification of the disciplines that would make it possible to arrange their contents efficiently. The basic format was of classical, usually Aristotelian, origins. The division of the various disciplines into two large sections—Moslem science and other science (which included the sciences in a strict sense as well as philosophy, logic, music, and alchemy)—was a Moslem invention. The aims of the two sections were different: the former gave wisdom to the 'ulamā' who studied it, the latter to the al-faylasūf, i.e. the lay philosopher.

The *Epistles* of the Brethren of Purity, a politico-religious sect established at Basra in 983, are a case apart. The epistles (fourteen on mathematics and logic, seventeen on the natural sciences, ten on metaphysics, and the remainder on mysticism, astrology, and magic), reflect the eclectic inclination of the Brethren, who were convinced of the purifying function of science and were for this reason given the title of philosophers.

The cultural function of the Umayyads at Cordova

As early as 929, under the reign of 'Abd al-Rahmān III, the emirate of Cordova was on its way to great glory. When 'Abd al-Rahmān died in 961, he was succeeded by his son al-Hakam II, who ruled until 976 and continued his father's work of promoting culture. The contribution of Hasdai ibn Shaprut, the court physician, was of primary importance in this movement. Thanks to its university, Cordova could soon hold its own against Baghdad on a cultural level. The Caliph's library, which contained manuscripts from all the known world, brought together hundreds of thousands of works (the library's catalogue alone filled forty-four volumes).

Thanks to the patronage of Ibn Shaprut, a Jew, the city also became one of the major centers of Jewish culture. Jewish scholars began to be the mediators between the Moslem world and the Christian West. Gerbert of Aurillac, also known as Pope Sylvester II, most probably was educated in contact with the culture of Spain.

Limits of the Christian West. The Clunists

The conditions in the Christian West and the Byzantine East were much different from those in the Arab world. Both of the former were subjected to the pressures of new populations coming from the east or north (the incursions of the Normans, for example, who had already sacked Aix-la-Chapelle in 881, violating the tomb of Charlemagne, and in 911 installed themselves in the western regions of the Gauls, were particularly terrible). Although the new forces coming from these populations actually affected the traditional structures positively later on, for the time being Byzantium was far more able to resist and regenerate itself than were the remains of the Carolingian Empire. There the process of linguistic differentiation among the major European nationalities that had already led to the formation of the kingdoms of Italy, France, and Germany reached its completion, mostly in the first half of the century. On the one hand, Europe was organized under the political and social structures of feudalism. But behind this apparent uniformity were particularistic tendencies that were working to turn the most important feudal centers into just as many autonomous nucleii of economic, social, and political life. In this closed society, which was divided into hundreds of centers, the city was an island in a society with a rural structure, revolving around a feudal castle and the Carolingian *villae* system. But the feudal institutions that filled the void left by the collapse of the Western empire were the only bulwark against the invasions, and they gradually made it possible for a new type of productive community, able to resume permanent ties with the urban centers, to form. The pre-eminent authority of the bishops in the cities favored the perpetuation of an educational tradition that was almost always entrusted to the ecclesiastical schools. But there were also the so-called "royal" schools, those in Verona and Pavia, for example, where judges and notaries were educated. The culture transmitted in these schools was mainly ecclesiastical or juridical in nature. Generally a continuity in the teaching of the liberal arts and in the scholastic tradition that had its roots in the Palatine school, guaranteed the perpetuation of a predominantly literary tradition. There was not, however, a total lack of scientific and philosophical thought.

The most important cultural event of the first half of the century was the foundation of the Abbey of Cluny in 910, on the initiative of William the Pious, duke of Aquitaine. A reform of the monastic orders was launched from Cluny, thanks to the impassioned work of Odo (c. 879–942), the abbot from 927. The Benedictines in particular were summoned to the purity of the ancient order under the leadership of the abbot of Cluny himself. Odo also brought the monasteries of Monte Cassino and Subiaco under the reform rule. The monastic reform, which called the monks back to their religious and social duties, initiated a restoration of the Church, which in turn had repercussions on European cultural life, due to the role that the monks and their schools still played in it.

The resumption of studies in Aristotelian logic

In Byzantium, Constantine VII Porphyrogenitus (905–959), who was given the title of emperor in 912 but exercised effective power only from 945, commissioned the compilation of encyclopedic works, on which he himself collaborated. The works were to bring together the knowledge of the times on a wide variety of subjects—history, public administration, agronomy, veterinary medicine, medicine, and zoology. The ties that the imperial family of Byzantium set up with Emperor Otto II (972) and Prince Vladimir (988) of Russia through matrimony, were another important vehicle of civilization and culture.

The West had a period of relative stability under the empire of the Ottos. The work of Notger, who was educated at St. Gall and was then bishop of Liège from 972 to his death in 1008, shows signs of a reawakening in religious cultural activity. Notger was responsible for the intense cutural life that made Liège one of the most lively centers of the empire. In addition, during the second half of the century, monks like Ratherius of Verona and Odo of Cluny wrote works on dialectics and morals, although they were usually not very original. The writings of Abbo of Fleury, whose monastery was perhaps the most lively center of literary, philosophical and theological culture, gained a certain degree of popularity. There is reason to believe that the study of logic advanced at Fleury; as a matter of fact, the oldest manuscript of the *Prior and Posterior Analytics*, in the version attributed to Boëthius, comes from there, as does a treatise on categorical syllogisms that some attribute to Abbo. The Latin version of the *Corpus* of Aristotle's logical works, a more ample knowledge of which was transmitted to the Christian world through Arabic Spain, began to be formed during this period. The texts on logic, although few in number, were basically the only works available to theoretical research, and the schools naturally had to turn to them. There was an increasingly broader and more systematic use of dialectics and the practice of writing re-expositions and commentaries flourished.

Our picture of Western culture in the tenth century would not be complete without mentioning Gerbert of Aurillac, Pope Sylvester II from 999 to 1003. His cultural education was still relatively undefined, but already showed obvious signs of radical change. Gerbert was interested in philosophy, arithmetic, astronomy, and music. The various legends about his exceptional giftedness, his almost magical abilities, and his mysterious contacts with Arab scientists are not unfounded.

MATHEMATICAL SCIENCES

Greek methods in Arab mathematics

The work of the Egyptian Abū Kāmil 850–930 shows that the Arabs' familiarity with Hindu arithmetical methods did not keep them from carrying out mathematical research using more typically Greek methods.

Abū Kāmil dealt with various types of equations on different occasions and from many points of view. He examined the solution of quadratic equations, the multiplication and division of algebraic quantities, the addition and subtraction of radicals, and the algebraic treatment of pentagons and decagons. In these investigations, he quite often resorted to methods of algebraic geometry that had been used by the Greeks, set forth for example by Euclid. Abū Kamil's work was in turn commented on and developed by the mathematician and astronomer al-Imrānī (died 955), and the Latin translation of it profoundly influenced European mathematicians especially Leonardo Fibonacci.

In the area of translations, the Arab version of Pappos' commentary on Book X of Euclid's Elements deserves mention, since it is the only extant source of this particular work. The translation was done by Abū 'Uthmān, who lived in Baghdad under al-Muqtadir, the Abbasid caliph from 908 to 932.

The astronomers' contributions to mathematics

As was the case in Greece, Arab astronomers also contributed new findings to the development of mathematics. Besides commenting on Ptolemy's Almagest and the first book of Appolonios' Conics, Ibrāhīm ibn Sinān (908–946), the grandson of Thābit ibn Qurra, developed the most elegant method for the quadrature of the parabola prior to the invention of integral calculus. Another great astronomer was al-Battānī, known as Albategnius (c. 858–929) in the West after his famous work on astronomy was translated into Latin. Spherical trigonometry played an important part in his work, especially in his preparation of tables. He established several new theorems in that area, the most important of which was his theorem on the cosines of spherical triangles.

At about the same time the philosopher al-Fārābī wrote a treatise on music that was greatly superior to such works written in the West as De armonica institutione, by Regino (840–915),

abbot of Prüm, and the Dialogus de musica (926), by Odo (c. 879–942), abbot of Cluny.

In Byzantium, around 938, Hero the Younger wrote a treatise on land-surveying entitled Geodaisia; it was based on material from the classical period, especially Heron of Alexandria's Diopter.

Geometric curves and trigonometry

The translations of Apollonius' Conics and the commentaries on it, which, as we have seen, were executed during the course of the preceding one hundred years, reawakened an interest among Arab mathematicians in exploring that field more thoroughly.

Abū Ja'far al-Kāzin (died c. 965) solved the so-called al-Māhānī cubic equation by means of conics. Al-Kūhī, who lived in Baghdad under the first Buwayhids, approached problems connected with Archimedes' and Appolonios' works, which led him to solve and discuss the conditions of solvability of equations higher than the second degree. He also invented a compass for drawing conics. Al-Sijzī (c. 945–1020) studied the intersections of circles and conics, and geometrically solved the problem of trisecting an angle, using a procedure based on the intersection of a circle and an equilateral hyperbola.

The greatest mathematician and astronomer of the times, Abū al-Wafā' (940–998), also flourished in Baghdad. He wrote commentaries on Euclid, Diophantos, and Hipparchos and a treatise on geometrical constructions where, following a method used by Greek geometers, he attacked problems of the geometry of a fixed compass, the construction of regular polyhedrons and parabolas by points, and the solution of equations higher than the second degree. In spherical trigonometry, which was the natural complement of his work in astronomy, he introduced the secant and the cosecant, and studied the basic relations between the six trigonometric lines. He also proposed a new method for constructing tables and used it to obtain results with a very high degree of approximation.

The first reawakenings in the West

Simultaneously, the first signs of a renewed interest in the exact sciences began to appear in the Latin West, although this interest only reached very elementary levels and almost exclusively concerned practical applications. Hēriger. abbot of Lobbes in Hainaut from 990 to 1007, wrote about the abacus. Toward the end of the century, Hrosvitha, a Benedictine nun at the convent of Gandersheim in Brunswick, wrote religious plays in which references are made to scientific subjects and the first four perfect numbers are mentioned. Abbo of Fleury (c. 947–1004) wrote De numero et pondere super calculum Victorii, a commentary on the ancient computation tables of Victorius of Aquitaine (who lived in the first half of the fifth century). A treatise on multiplication and division by Joseph the Wise is mentioned twice in the letters of Gerbert of Aurillac (c. 940–1003), Pope from 999 under the name of Sylvester II. Gerbert, who was educated in contact with the Arab culture of Spain in the county of Barcelona, and taught at the school of Rheims, is the most prominent figure of the period in Western Christian culture. He wrote a treatise on the abacus, often dealt with scientific questions in his letters, and mentioned the ghubār numbers—

the Spanish form of the Arabic numerals, also mentioned in a code written in Logrono in 976. Since *ghubār* means dust, we can assume that the numerals were connected with the use of some kind of sand abacus; in addition, the abacus was reintroduced into the West during this period.

It is not an accident, as we have said, that Gerbert was in contact with Arab culture in Spain. In the following century as well, scholars such as Abraham ben Ezra (c. 1090–1164), Abraham bar Hiyya (1070–1136), and John of Seville (flourished mid-twelfth century), increasingly more skilful and rigorous compilers of mathematical subjects, were distinguished by the intellectual benefits they received from Arab-Spanish culture. They were the first to bring the contents of this culture to the attention of Europe, thus preparing the intellectual curiosity needed for the future success of the Arabic-to-Latin translations of classical works.

ASTRONOMY

The work of Al-Battānī

Al-Battānī (Albategnius) (c. 858–929) (see MATHEMATICS 900–1000), was one of the greatest Arab astronomers, observers, and calculators. With the data from his own observations, begun in 877 at Raqqa and continued at Baghdad, he revised astronomical data transmitted from antiquity as well as the information that had been gathered more recently by Islamic observers. He calculated new, more accurate values for the obliquity of the ecliptic, the annual precession, and the tropical year. He verified a yearly shift in the solar apogee following the order to the signs of the zodiac, and demonstrated the possibility of annular solar eclipses. He also extended the use of trigonometry in the field of astronomy. His results are collected in the treatise *De scientia stellarum*, which was widely diffused and respected in the West up to the end of the sixteenth century.

The Persian Albubather, a contemporary of al-Battānī, wrote the treatise *De nativitatibus*, a basic text of medieval astrology. It was joined by a flourishing of similar works, written by physicians and astrologers, while astronomical tables for calculating predictions for various localities were also compiled. Some of these were drawn up by the great observer Ibn Amājūr, who flourished from 885 to 933, and, for Yemen, by al-Hamdānī (died c. 946), who also wrote a historical and archeological work about his country that gives information on ancient Arabic science.

The Persian astronomer 'Abd al-Rahmān al-Sūfī (903–986), wrote *Book of the Fixed Stars*, in which he revived Ptolemy's catalogue of the fixed stars, revising all the data with extreme accuracy and prepared a map of the sky that was only surpassed by the celestial atlases of the eighteenth century.

The continued observations at Baghdad

In 988, Sultan Sharaf al-Dawla had his own observatory built at Baghdad. The instruments for it were made by the astronomer al-Sāghānī (died 990), while the scientific activity was directed by the mathematician al-Kūhī. The astronomer Abū-l-Wafā' (940–998), for a long time credited with the discovery of the third inequality of the motion of the moon, made observations there, wrote a manual modelled after Ptolemy's *Almagest*, and made considerable progress in trigonometry, which he used to compile his *Tables*. Other extremely accurate *Tables* were compiled in Baghdad by Ibn al-A'lam (died 985). While making observations at Ray in 994, the Transoxian astronomer al-Khujandī measured the obliquity of the ecliptic more accurately.

Al-Qabīsī (Alcabitius in Latin), who flourished around 970, wrote two astrological works, the *Liber introductorius ad magisterium iudiciorum astrarum* and the *Tractatus de coniunctionibus planetarum*.

Arab astronomers in Spain

Astronomy and astrology also began to be actively cultivated in Arab-occupied Spain, especially in Cordova under the caliphate of al-Hakam II (ruled 961–976). The Spanish Moslem Maslama ibn Ahmad Al-Moiriti (died 1007), who was born in Madrid and worked in Cordova, and the Arab-speaking Christian Rabi' ibn Zayd, bishop of Cordova and Elvira, were the first Europeans to deserve mention for their scientific contributions. The former published an improved version of al-Khwārizmī's *Tables*, a treatise on astrology and a commentary on Ptolemy's *Planisphere*. The latter wrote several astrological treatises in Arabic and a calendar dedicated to Caliph al-Hakam II. A work by an otherwise unknown author, Alchandrus (or some similar name), has come down to us through Arab sources. The work goes back to the end of the tenth century, and can be considered one of the oldest testimonies of Eastern influences on Latin astrological literature.

Gerbert of Aurillac's knowledge of astronomy was also derived from his contacts with Arab culture in Spain. He wrote a treatise on the astrolabe, built scientific instruments, and took an interest in astronomical questions.

PHYSICAL SCIENCES

Developments in optics

The study of physics increased in the Arab world during the tenth century. This trend continued in the following century, and had an inestimable influence on Western culture as well, through the complex mechanism of translating various scientific treatises from Arabic into Latin. There were both great technical developments and a flourishing reappearance of studies in optics. The research of Alhazen (Ibn al-Haytham, c. 965–1038), concluded the century. His *Optics* criticizes Euclid's and Ptolemy's theories on vision and develops a doctrine according to which the forms of objects perceived come to the eye and are elaborated by the lens (the so-called transparent body). Alhazen also investigated the propagation of light and colors, the problem of optical illusions, and the theory of reflection. He was undoubtably influenced by earlier traditions, but he found extremely original solutions to a number of questions. One example is the famous problem known as Alhazen's problem where, given a convex, spherical, conical, or cylindrical mirror, the point at which a ray coming from a given point would have to strike in order to reflect to another given point is calculated. The problem leads to a fourth-degree equation, which Alhazen solved using a hyperbola.

Alhazen considered light a kind of fire that reflected on the spherical edge of the atmosphere surrounding the earth. He furthered the study of refraction, and was also interested in phenomena such as rainbows and the halo around the moon. Several centuries later, the physical study of these phenomena became favorite topics for optical research in the West.

The position of physics in the classification of the sciences

One of the most original products of Arab thought between this century and the following one is a group of writings that are now called classifications of the sciences. The work of al-Bīrūnī (973–1048), a physician, astronomer, mathematician, physicist, geographer, and historian also belongs to this genre. Al-Bīrūnī's most important specific contribution to physics was the fairly accurate calculation of the specific weights of eighteen precious stones and metals, which he determined using methods indicated by Archimedes.

Toward the year 1000, when people in the West believed that the end of the world was approaching, the Arabs became the continuers of ancient traditions in physics. These ranged from static mechanics, dynamics, fluid statics, etc. to the theory of vision, geometrical optics, and acoustics, often involving a successful combination of mathematics and physics. Contributions of a more speculative nature, which were destined to influence the organization of physical knowledge considerably, also appeared. An example is the revival of the doctrine of the atomic structure of matter, which contrasted with the dominant Aristotelian concept accepted by Moslem theologians. Mesopotamia was the center of this rebirth of atomism. Esoteric sects, such as the Brethren of Purity, founded around 980, developed a sort of alternative culture to the orthodox physical image of the world. The Brethren of Purity compiled a huge encyclopedia, a part of which deals with natural phenomena, following Greek and Alexandrian tradition, especially concerning the formation of minerals, seismic movements, tides, and meteorological phenomena in relation to the spheres and the heavenly bodies. The Moslem orthodoxy opposed the ideas of the Brethren of Purity, and their works were often publicly burned. Nevertheless, their doctrine was quite widely diffused, and, by way of Spain, had a notable influence on science and philosophy in the Christian West.

CHEMICAL SCIENCES

The books of Abū Bakr al-Rāzī

Mysticism still completely pervaded the work of Muhammad ibn Umayl, but another work, best known in its later Latin translation as *Turba philosophorum*, appeared at the same time. It made an attempt to adapt Greek cosmological theories to Arabic science.

The books of Abū Bakr al-Rāzī (c. 855–923) probably belong to this period, although some contend that they were in part written by alchemists of a later era. This hypothesis is somewhat confirmed by the way in which chemical and scientific topics are generally treated, which seems far too advanced for al-Rāzī's time. Of his many books, the best-known is the *Kitāb Sirr al-asrār*, or *Secretum secretorum*, which reveals its 'secrets' by fur-

nishing a series of well-described practical procedures. For example, the process for causticizing sodium carbonate with lime is explained in great detail, and there is a description of how to use calcium polysulfate to destroy hairs. The book goes on to describe many of the caustic waters that constituted one of the basic discoveries of Arab alchemists, who, although they believed in the possibility of changing base metals into gold and silver, were still grounded in the results of experience.

Gunpowder

The discovery of gunpowder in China dates back to this period (around 919). It was used in the year 1000 as the chemical fire (huo yao) in explosive bombs. The Mongols probably brought chemical fire to the West, where it was described by Marcus Graecus (the first manuscripts date back to the twelfth century but most probably drew upon earlier texts). Berthold Schwarz, the monk of Freiberg, is only a mythical figure.

LIFE SCIENCES

Medical assistance

The organization of medical assistance in Baghdad was quite remarkable. Its supervision was entrusted to scholars such as Abū 'Uthmān, who directed the hospitals of Baghdad, Mecca, and Medina from 935, and Sinān Thābit (c. 880–943), grandson of Thābit ibn Qurra, who superintended the great hospital of Baghdad, which had been built in 918 at the behest of the caliph al-Muqtadir, and established an examination of certification to practice medicine for the physicians of Baghdad. The translators' school created by Joannitius translated Galen's *De anatomicis administrationibus*, preserving the work's integrity, since Books IX-XV have been lost in the Greek and are extant only in this Arabic version.

Islamic, Jewish, and Western medicine

The Moslems' tolerance toward other religious confessions and the exchanges that were established between the occupiers and the occupied populations, created new, tighter bonds between Islamic and Jewish medicine, and later, with Western medicine (Salerno and Montpellier). Isaac Israeli (died c. 955), the "prince of medicine," dominated in this field. He wrote classical works on fevers and uroscopy, as well as a *Guide of the Physicians*. The latter work displays another aspect of the medicine of the times—a strict deontology that was derived from Hippocratic-Galenic tradition and had humanitarian and even charitable ends. Constantine the African (1015–1087) transmitted this deontology to the West, along with the works of Moslem and Jewish physicians, beginning in Salerno (the first news of the activity of a medical center there dates back to this period).

In Byzantium, at the behest of Constantine VII (913–959), Theophanos Nonnos wrote an *Epitome de curatione morbum*, using Greek sources, and composed a prescription book of the most common drugs, the *Euporista pharmaca*.

In the West, in the wake of the cultural revival promoted by King Alfred the Great in England, the first medical works written in the vernacular appeared. The *Leech Book of Bald* and the *Lacnunga,* the two works in Anglo-Saxon, are of some interest to botany.

Al-Mas'ūdī's theory of evolution

Al-Mas'ūdī, a historian and geographer who died in Cairo around 957, composed an autobiographical work toward the end of his life, in which, among other things, he puts forth a theory that nature progressed from minerals to plants, from plants to animals, and from animals to human beings.

The development of pharmacology

Medical science during this period displayed a particular interest in pharmacology, and there was an increased number of works on materia medica, which often made various references to natural history. Between 968 and 977, Abū Mansūr al-Muwaffaq wrote the *Liber fundamentorum pharmacologiae,* which gives an elementary theory of pharmacology and gathers and classifies almost six hundred cures. In order to complete the work, the author made a number of journeys to Persia and India.

Al-Tamīmī, who lived in Jerusalem and Egypt and died in the last quarter of the century, wrote a guide on materia medica in which he also mentions experiences with drugs. Hasdai ibn Shaprut, a physician and influential person at the court of Cordova (see THOUGHT AND SCIENTIFIC LIFE 900–1000) and of al-Hakam II, with the help of the monk Nicola, called in from Byzantium for the purpose, translated the code of Dioscorides that Constantine VII had sent as a gift to the caliph. In Cordova, ibn Juljul wrote a commentary of Dioscorides, to which he added a supplement of prescriptions that the Greek physician had not known. Another text in materia medica was written by Donnolo, a Jewish physician who flourished in southern Italy during the second half of the century.

'Arīb ibn Sa'd, who flourished at Cordova at the time of Hasdai ibn Shaprut, wrote numerous treatises on gynecology and obstetrics, while the Persian physician 'Alī ibn al-'Abbas (end of the tenth century) composed the *Liber regius,* an encyclopedia that surpassed in part al-Rāzī's *Totum continens.*

Abū-l-Qāsim, the teacher of Arab surgeons

The physician Abū-l-Qāsim al-Zahrāwī (c. 936–1013), the greatest surgeon of Islam, flourished at al-Zahrā, near Cordova. He wrote a *Vademecum* (*al-Tasrīf*) which, when translated into Latin by Gerard of Cremona, gave him the fame that the Arabs, who were adverse to surgery for religious reasons, had difficulties according him. With his work, Islamic surgery became aware of its inferiority, which was due to a lack of sufficient knowledge of anatomy. The three books of the thirty-volume *Vademecum* dedicated to surgery have often been published, commented on, and illustrated separately from the rest of the work. They contain descriptions of the specific procedures for lithotomies, fistulae, hernias, trapanations, amputations, and cauterizations.

The hospital system. The military medical corps

Meanwhile, the medical profession continued to be organized in the Islamized countries, especially with respect to hospitals, which followed Byzantine models. New hospitals, divided into specialized sections, continued to be built in the large cities, usually near mosques. An orphanage and a library were often annexed.

A military medical corps was started, and in the following century attempts at providing medical assistance in the countryside with mobile health units were begun. Checks were made on professional competence, the basics of clinical teaching were established, and controls were placed on the selling of medicines.

The encyclopedias

During this period, a series of compilations, commissioned by the emperor, Constantine VII, were written in Byzantium. Constantine had gathered some of the best scholars of the time together in the city. Among the works they produced are encyclopedias on medicine, veterinary medicine, and zoology, which are of some interest to the life sciences.

APPLIED SCIENCES

Arab industry

This was the Golden Age of Islamic civilization. During this period, the arts that had developed during the previous centuries reached perfection, and the importance of the findings made as a result of the great Islamic technical-scientific efforts became obvious. In the field of agriculture, great projects were launched to restore the ancient dams; irrigation canals that had been abandoned for centuries were dredged; and paddle wheels mounted on pontoons were built on the great rivers to operate grain and paper mills.

The paper and bookbinding industries improved so much that as early as 900 in Baghdad, standard sizes of paper were available and special qualities were being produced for carrier pigeon service. Improvements were made in steel production in the centers of Damascus and Toledo and in the linen and wool industries in Egypt and Mesopotamia.

Ancient scientific and technical works were collected and translated in the Arab libraries. The library of Cordova held as many as six hundred thousand volumes in the tenth century.

European and Chinese agriculture

Compared to Arab growth, technical activity in the West appeared stagnant, even in the most advanced areas. However, an event of great importance for the future development of European agriculture deserves mention. A Frankish manuscript shows a wagon with shafts, pulled by horses wearing rigid collars. This illustration indicates that the complex evolution of draft harnesses had finally reached its conclusion.

Joseph Needham, world-famous authority on Chinese science, says that the collar was invented in China after the third century, and introduced into the West around 920. In reality, the horizontal instead of the vertical arrangement of breast-bands, connected in the middle by a belly-band, appeared in China during the period mentioned, but it is doubtful that the system gave rise to the modern harnessing system, which only developed when shafts were attached to the middle part of the breast-band and the breast-band effectively became a collar.

On the other hand, Needham's theory tends to be supported by the fact that shafts were commonly used in China by the beginning of our era, and were not at all diffused in the Roman world at the same time. Since both shafts and breast-bands existed in China, it is logical to believe that the evolution took place there before it occurred in the West. The use of this type of harness created a general increase in the traction power available for agriculture, as well as important changes in the system of transport. However, it was only adopted slowly, and did not immediately change the agricultural system in any significant way.

There is also evidence that the first reliable examples of horseshoes, found in the Yenisei region, also belong to this period.

1000–1100

THOUGHT AND SCIENTIFIC LIFE

Arab and Jewish culture. From Avicenna to Avicebron

At the beginning of the century, the fractionistic tendencies increased within each of the three states that had formed during the previous century. In Africa for example, which was ruled by the Fatmids, who founded Cairo as their new capital in 969, the Berber emirates soon stopped recognizing the authority of the caliph, while Maghreb was removed from Fatimid sovereignty by the Almoravids, who imposed their rule in the name of a more strict interpretation of the Moslem faith.

The foundation of a scientific academy, along with an astronomical observatory, in Cairo in 1005 gave a strong boost to the sciences. A great school of astronomy and mathematics was founded at the observatory (996–1020) at the behest of the Fatimid caliph al-Hākam. It lasted until the dynasty became extinct in 1171.

During the first years of the century, the poet Firdawsī (c. 935–1023), a native of Khurāsān, completed the Shāhnāma (Book of Kings), an epic poem about the Persian people. Besides the historical narration, it gives a broad view of the life of that people through the centuries. Almost simultaneously, the indefatigable al-Bīrūnī (973–1048) developed an interest in India, which created one of the most lively cultural exchanges between the Arab world and Indian civilization. A prominent figure in philosophy was Ibn Sīnā—Avicenna to the Latins (980–1037)— recognized by the Arabs as the third master of thought (the first two were Aristotle and al-Fārābī). His work united some of the essential elements of Aristotle's doctrine with themes from Neoplatonism and Moslem theological and mystical tradition, giving a considerable amount of attention to numerous topics of natural philosophy, such as movement, light, and heat. His Kitāb al-Shifā (Book of Healing), of which the sections on logic, physics, the doctrine of the soul, and metaphysics were especially well-known in the West, was the source of a theme that was debated for centuries, especially in Franciscan and sometimes even in Thomist circles. Important theological speculations were also made during this period, for example by Muhammad ibn al-Tayyib al-Bāquilānī of Basra (died 1013 in Baghdad), who attempted to introduce topics such as atoms and the void into Moslem scholasticism.

The Jew Salomon ibn Gabirol—Avicebron in Latin—who was born in Malaga in 1021 and died at Valencia around 1060, was a typical product of the multiracial environment of Spain. Different religious currents existed among the Jews, as well as among the Moslems: some adhered literally to the sacred text, while others formulated freer interpretations. Among Avicebron's works, the one that most influenced the scholastics was the Fons vitae, a treatise originally written in Arabic, but only extant in Ibn Falaquera's florilegium in Hebrew and the Latin translation by John Hispanus and Domingo Gundisalvo. The Fons vitae, which was first thought to be the work of a Christian or a Moslem thinker, had little success in Jewish circles, but gained considerable fame in the West because of the author's ability to reconcile Neoplatonic and Arabic themes with Jewish principles.

Culture and religion in Byzantium. Psellos and Cerularius

Byzantium, which had achieved economic prosperity, demographic growth, and cultural prestige under the Macedonian dynasty (starting with Basil I in 867), already showed the first signs of a political decline during the reign of Basil II (976–1025), although the period was full of great accomplishments, both military (the campaign against the Bulgars) and cultural (the increased prestige of Byzantium among the Russians). The harsh political struggles reemerged with the rise of the Comnenos dynasty in 1057. Philological and scientific tradition was, however, quite alive. Toward the middle of the century, a cultural revival in a Neoplatonic key occurred, centering around the figure of Michael Psellos (c. 1018–80). Psellos wrote historical works, treatises on medicine, and commentaries on Aristotle, and displayed an extremely vast knowledge of Hellenism, ranging from Greek orators and historians to mathematicians, such as Euclid and Diophantos and including the Neopythagorean and Neoplatonic philosophers. He also referred to the cultural traditions of Chaldea, Egypt, and Palestine, and even included the Hermetic works. This vast amount of research was presented with a rationalistic approach that was strongly hostile to Chaldeanism—that mixture of theurgy and occultism to which many Byzantine scholars still willingly made concessions. Psellos' critique did not even spare the mysticism of Michael Cerularius, the patriarch of Constantinople, under whose leadership the definitive schism between the Greek-Eastern Church (Orthodox) and the Latin Church (Catholic), then ruled by Pope Leo IX, took place in 1054. The reasons for the schism transcended the immediate cause (the question of legitimate authority over the dioceses of southern Italy), and the split soon became irreparable due to the ancient conflict of mentalities and differences in spiritual, cultural, and linguistic orientation between Rome and Byzantium. The Serbian, Bulgarian, and Russian churches, which had been founded under Byzantine influence, went with Cerularius.

In this atmosphere of anti-Roman sentiment and a rebirth of interest in religious debate, the conflict between the ecclesiastical authority and the rationalistic followers of Psellos, who were later harshly condemned in order to prevent an autonomous profane culture from developing, was virtually inevitable.

The West. The restoration of Europe

The weakening of the Arabs and the Byzantines was paralleled by a strengthening of Europe. The restoration of the monarchy, already begun with Otto I (ruled 936–973), considerably diminished the more degenerate aspects of feudal anarchy. The closed system of the *curtis* was opened up to more intense trade. Cities began to play a central role again, and strong ties were formed between the cities and the countryside. Technical innovations, especially in agriculture, favored considerable demographic growth. The fear of the year 1000 disappeared; institutions were reorganized, and cultural perspectives were widened.

While the fame of the Salerno medical school continued to grow, the jurists of Pavia elaborated two versions of the Longobard laws, the last barbarian codes, and transferred their attention to Roman law and its relation to canonical law, laying the foundations for the law school from which the university later arose.

At the end of the century (1088), Irnerio began teaching in Bologna, *alma mater studiorum*. The Normans extended their influence in Europe by occupying England (1066) and southern Italy, ending Byzantine domination (1071). Thus another possible contact between Rome and Constantinople was severed. The Latin West proceeded independently in its political, cultural, and economic revival, while Arab power began to vacillate. In 1055, the Seljuk Turks—nomads who had been hired as mercenaries by the caliphs of Baghdad in the ninth century and organized militarily by Seljuk around 1030—put an end to the Abbasid Empire, while the Pisans took the fabulous treasures snatched from the Moslems during the sacking of Palermo (1063) back to Pisa to build their cathedral. Venetian trade stretched even further eastward. In 1085, Alfonso IV of Aragon took Toledo from the Moslems. Pope Gregory VII (pope, 1073–85) claimed the absolute autonomy of the papacy in his *Dictatus papae*, initiating the struggle over investiture with Henry IV, and launched a vigorous project to reorganize the Church. Finally, in 1095, Pope Urban II proclaimed the first Crusade.

It was a fertile period for scholarly work. In 1056, Constantine the African (1015–87) of Carthage founded a center for Arabic-to-Latin translation at Monte Cassino. The physicians of Salerno became familiar with numerous medical works (see LIFE SCIENCES 1000–1100). A collaboration between scientific research and philology proved to be essential to the development of the discipline of medicine, as well as a model in a more general sense. At the same time, there was a considerable amount of renewed interest in the arts of the quadrivium, especially music, beginning with a codification written by the monk Guido of Arezzo (c. 995–1050).

Dialecticians and anti-dialecticians

While the major protagonist in cultural life was the Church, enlivened by the Cluniac movement, there were also lay persons who studied the liberal arts, in the prospect of taking public office or making a career of law. On the other hand, there was an increasingly keen interest in dialectics and rhetoric among the Church scholars, often at the expense of theology. Peter Damian (1007–72), an expert in dialectics, although he was busy governing the Church and reforming the monastic rule, complained that the problem of utilizing a hypothetical or categorical syllogism had become more important than defining and teaching the Truth. This was the reaction of a man who was theoretically close to the policies of Gregory VII and to

an intellectual movement that, following in the footsteps of Scotus Erigena, saw dialectics as the rational tool par excellence.

Anselm of Besate, called the Peripatetic because of his education in logic, wrote a *Rhetorimachia*, which already displays a sense of argumentative rigor, and Berengarius of Tours (c. 1000–1088) used dialectics to discuss specifically religious topics. Peter Damian was not alone in fighting these practices, since there were numerous supporters of the monastic reform who were uncompromisingly adverse to profane culture and complained that more weight was given to Boëthius than to the Holy authors, or even to pagan authors who were not, however, lacking in the spirit that teaches all truth. The work of Lanfranc of Pavia (c. 1005–89), abbot of Bec in Normandy, offered some reconciliation.

Anselm of Canterbury, Gaunilon, and Roscelin

In 1063, Anselm of Canterbury (c. 1033–1109) succeeded Lanfranc as the prior of the Abbey of Bec, and in 1078 became its abbot.

While from a purely political point of view, the struggle that he conducted from 1093 to 1106 against two kings of England concerning his investiture as archbishop of Canterbury was extremely important, his thirty-year period at Bec was the most important in a philosophical sense. These were the years in which he wrote the *Monologion* (c. 1070); the *Proslogion* (slightly later), which contains the famous ontological argument on the existence of God; and the *De veritate*. They were also the years of his polemics with the monk Gaunilon. Anselm's thought, which was permeated with Augustinian themes, was far removed from the extremism of the anti-dialecticians. In fact, the very structure of the *Monologion* and the *Proslogion* was an ulterior confirmation, in the eyes of Anselm's contemporaries, of the value of dialectics, since it was able to prove the existence of God without making recourse to the sacred texts. On the other hand, Anselm's interest in logic is attested to by the quarrel concerning universals, that is, notions that refer not to the single individual but a category of individuals.

The question of universals arose during the last years of the century, based on Porphyry's observations as transmitted by Boëthius. Countering the traditional position (Scotus Erigena, Anselm) that considered the universals reality and logic a *scientia de rebus,* Roscelin of Compiègne (c. 1050–1125) proposed a nominalistic solution, according to which logic was only a *scientia de vocibus*. His was a turn toward an attitude that looked at individual material objects as just that, instead of considering them only the shadows of eternal ideas as Augustine had.

The work of al Ghazzālī (died 1111) concluded the century. He reacted to the speculations of al-Fārābī and Avicenna and, in general, to the attempt at subordinating religious experience to philosophy, with his *Tahāfut al-falāsifa (The Internal Contradictions of Philosophy)* and *Ihyā 'ulūm al-dīn (Revivifying of the Sciences of the Faith)*. In the West, only the part of the *Contradictions* (called *Intentiones philosophorum*) where al-Ghazzālī expounds Avicenna's doctrine became known, so that al-Ghazzālī was thought to be a disciple of Avicenna. In reality, his position with respect to rational research had certain features in common with that of the anti-dialecticians in the West and with the ideas set forth by Michael Psellos' adversaries in Byzantium.

MATHEMATICAL SCIENCES

Arab mathematics

Optics had already, on occasion, proved to be a fairly fertile discipline for the development of mathematics in the ancient world, a characteristic it maintained in the Arab world as well. The famous *Optics*, by Ibn al-Haytham, or Alhazen (c. 965–1038), a work that would later instruct and inspire such scientists as Roger Bacon, Vitellione, and Kepler, contains a wealth of mathematical knowledge, also evident in other purely mathematical writings by the same author. Much of the knowledge is connected to the problems, the methods of treatment, and even the specific contents of Greek geometry (see PHYSICAL SCIENCES 900–1000).

A figure who is more famous for other aspects of his activity, but is also notably important for his studies in mathematics, is al-Bīrūnī (973–1048), the philosopher, historian, and traveller. He left a compilation on calculating the chords of a circle, which is also full of references to findings of Greek mathematics that have since been lost. While studying the construction of regular enneagons, he arrived at the solution of a third degree equation. In addition, he did important research in trigonometry, calculating the chords of large arcs, making a very close approximation of the value of π, and defining various properties of spherical triangles.

Ibn Sīnā, known as Avicenna (980–1037), also flourished in this period. In his native Bukhārā, Avicenna learned about Hindu numerals from the Arab-Egyptian Ishmaelite sect. He also wrote several fairly unoriginal works on geometrical and arithmetical topics.

Although Arab mathematics, taken as a whole, achieved a remarkable fusion of Greek and Hindu traditions, this synthesis was not always untroubled. It was during these years that al-Nasawī wrote the *Al-muqni' fi-l-hisāb al-Hindī (Satisfying on Hindu Calculation)*, which was theoretically opposed by the works of al-Karkhī (flourished c. 1000), a staunch supporter of the Greek methods, who even claimed ignorance of the Hindu methods, although his works include all of the Hindus' technical discoveries. Al-Karkhī made diverse discoveries in the theory of algebraic equations and also determined the sums of the first, second, and third powers of the first n natural numbers.

The continuity of Greek tradition

Arab mathematics continued to produce important works, such as the algebraic treatise by Abū-l-Fath al-Busti, a Persian poet and mathematician, who probably lived between 1036 and 1123, although some say he flourished at a much earlier date. The treatise subdivides the equations of the first four degrees into twenty-five classes, which are studied systematically. Fath also developed a system for calculating irrational radicals and wrote a new, annotated version of the work of Apollonius. This version, published in Florence in 1661, is particularly important to us since it is the only source for the contents of Books V–VII of the *Conics*, which along with Book VIII, have been lost in the original Greek.

It is interesting to note how the Byzantine Empire, although geographically close to the Arab world, was not influenced by it. The ancients (especially Euclid and Diophantos), continued to be studied and commented on without originality. Michael Psellos (c. 1018–1078) was a scholar of this sort.

ASTRONOMY

The caliphs' observatory in Cairo

Chronological studies were revived by the monk Bridferth, who wrote a *Compotus* in 1004.

At the behest of the Fatimid caliphs, a scientific academy with an astronomical observatory was established in Cairo, in 1005. It marked an energetic revival in astronomical observations and lasted until the end of their reign in 1171. The most important product of this revival was the *Hakemite Tables*, compiled by Ibn Yūnus (c. 950–1009) between 990 and 1007 and dedicated to the Caliph al-Hākim II, the protector of astronomers. They contain new data on observations of eclipses and conjunctions; more accurate values for the obliquity of the ecliptic, the longitude of the sun's apogee, the solar parallax, and the precession of the equinoxes; and information on the geodetic measurements carried out at the observatory of Baghdad.

Ibn al-Haytham (965–1038), known in Europe as Alhazen, also worked in Cairo. He became famous for his work in optics (see PHYSICAL SCIENCES 900–1000), but also did studies in astronomy and meteorology. He developed a theory on atmospheric refraction, and one on vision, discovered spherical aberration, studied various properties of light and colors, and invented the camera obscura. He was the first Eastern scholar to take an interest in the physical nature of the heavenly bodies and expressed interesting ideas on the nature of comets, the moon, and the planets, as well as the idea that the Milky Way was made of stars. Ibn al-Haytham's theories were widely known in Europe.

The astronomers of central Asia and Iran

The most important encyclopedic scholars working in this period in central Asia and Iran were al-Bīrūnī (973–1048), Ibn Sīnā (Avicenna, 980–1037) and Omar Khayyam (c. 1050–1123), famous for their works on astronomy, geography, mathematics, physics, medicine, and history.

In his *Chronology*, al-Bīrūnī gives a detailed description of the calendars of various peoples, while his *Cartography* and *Astrolabe* are dedicated to the problems involved in building astrolabes, celestial globes, and maps of the world. Between 1022 and 1024, al-Bīrūnī travelled to India; out of this experience, he wrote *India*, a monumental book on the history, the culture, and the science of that country. He also developed a series of trigonometric methods for accurately calculating distances on the earth, thus laying the foundations for modern geodesy (*Geodesy, Mas'ūdic Canon*). Al-Bīrūnī is best known for his astronomical encyclopedia, the *Mas'ūdic Canon,* dedicated to his protector, Sultan Mas'ūd. His measurements of the specific weights of precious stones laid the foundations for mineralogy as a science (*Mineralogy*), while his *Farmacognosia (Science of the preparation of Medicines)* perfected Avicenna's *Qānūn*.

Avicenna gained fame in Europe mainly as a physician and philosopher, although he also studied astronomy and mathematics. His commentaries on Ptolemy's *Almagest* and *Geography*, on Aristotle's *De Coelo* and *Physics*, and on Euclid, became very popular.

Omar Khayyam, who worked at the astronomical observatory of Isfahan, was a follower of Avicenna. He developed an original calendar system (more accurate than the Gregorian calendar). On the basis of his astronomical observations, he compiled the *Malikshāh Tables*, dedicated to his protector Malikshāh. He also did research in mathematics and other disciplines. Nonetheless, he is most famous in Europe and the United States as a poet. He is the unsurpassed author of quatrains with profound philosophical and lyrical meaning (the *rubā'īyāt*).

Al-Bīrūnī summarized the astronomical knowledge of the times in his *Mas'ūdic Canon*, and in the *Book of Knowledge*, where topics pertaining to astrology and mathematics are also discussed. He also wrote treatises especially dedicated to planispheres and astrolabes. Avicenna wrote a *Compendium* of the *Almagest* and a *Summa astronomica*, commented on Aristotle's *De coelo* and *Physics*, and made astronomical observations with the aid of an altazimuth-type instrument which he himself invented and built. The theologian al-Ghazālī (1058–1111) also composed an astronomical treatise on the nature of the stars.

The convents in the West

An interest in astronomy continued in the convents of the West. Contributions of some importance were made by Hermannus Contractus (1013–54), abbot of Reichenau, who wrote a treatise on the astrolabe; by Franco Liège (died 1083), superintendent (1047) of the cathedral school of St. Lambert, who wrote on computation and spheres, and may have invented the *torquetum*; by William, abbot of Hirsau from 1069 to 1091, who wrote a treatise on astronomy and built a famous model of the planetary system; and by Honorius Solitarius, a Benedictine monk, probably English, who flourished around 1080 and wrote a *De imagine mundi* using the works of Pliny, Isidore, and Bede, inspiring Walter of Metz to compose a famous poem in French with the same title, *Image du monde* (1246).

Arab-Spanish astronomers

In Spain, the Spanish Arab Abū-l-Qāsim al-Baghawi (979–1035), who flourished in Granada and is also known as a mathematician, was a famous compiler of astronomical tables, Ibn Said (1029–70) dedicated his attention to history, gathering considerable information concerning science, and to astronomy, becoming well-known as an observer. The results of his work, which was mainly done in Toledo, are used in the writings of al-Zarqālī (1029–89). Al-Zarqālī invented a new type of plane astrolabe (*saphaea Arzachelis*) and was both an observer and a calculator. He published the *Toledan Tables*, which were accompanied by a remarkable trigonometric introduction about the method used in the compilations, and were based on ancient material as well as the observations made in Toledo by a group of Arab and Jewish astronomers. The work, which was translated into Latin by Gerard of Cremona was extremely famous in the West and helped to spread the erroneous theory of the trepidation of the equinoxes. Al-Zarqālī was the first to demonstrate the motion of the solar apogee with reference to the stars.

The measurement of time and astronomical instruments in China. Su-sung's orrery

In ancient and medieval China, as Joseph Needham, Wang Ling and Derek J. de Solla Price (1960) have pointed out in *Heavenly Clockwork; the Great Astronomical Clocks of Medieval China*, Cambridge, Published in association with the Antiquarian Horological Society at the University Press, the Emperor's promulgation of the calendar was a right similar to that of coining money, and the acceptance of the calendar on the part of his subjects signalled their recognition of the imperial authority. These close ties between the calendar and the power of the State turned astronomical observation and the building of instruments for it into a highly protected monopoly of the imperial court during this period.

In 1090, the court astronomer Su-sung had a huge astronomical clock built at Kai feng, He nan province, the capital of the empire. It was the most complex orrery ever built prior to the one constructed by Giovanni Dondi in Padua (see APPLIED SCIENCES 1350–1400). The clock was closed in a tower and operated by a water wheel that rotated a celestial globe and an armillary sphere, which was located on a platform above the tower. At the same time, a series of figurines would appear outside to signal the hours and the quarter-hours to passers-by by means of sounds.

Because of the secrecy that cloaked the practice of astronomy, when the imperial experts in charge of the clock's operation were dispersed by the Ming invaders three centuries later in 1368, Su-sung's clock fell completely into disuse.

Chinese and Japanese records of the explosion of a supernova

Although no true tradition of astronomical studies of any importance existed in China after the fourth century B.C., one of the most interesting scientific and historical astronomical observations was made by Chinese and Japanese scholars toward the middle of this ninth century, during a period in which not much progress was made concerning knowledge about the heavens in the West either. It was the first recorded description of the explosion of a supernova, more than half a millennium before the novae of Tycho Brahe in 1572 and Johannes Kepler in 1604 (see ASTRONOMY 1550–1600 and 1600–1650).

Destined to remain an isolated event in Far Eastern culture, the reports tell of the sudden apparition of a star of extraordinary brightness in the constellation of Taurus, in the year 1054. Recent studies (1941) have made it possible to identify this event with the explosion of a supernova. The remains of its shell constitute the Crab Nebula (M1-NGC 1952), well-known to scholars as early as 1731, when the Englishman John Bevis first observed it. The residue of the supernova, the pulsar NP 0532, has also been found inside the Nebula (see GALACTIC ASTRONOMY, 1940–50).

PHYSICAL SCIENCES

Avicenna and the problem of motion

Studies in physics continued to flourish in the Arab world, not only in the field of optics (Alhazen, whom we have already mentioned), but also in dynamics.

The contributions of the physician-philosopher Ibn Sīnā (Avicenna, 980–1037) were particularly important in this field. He took up Joannes Philoponos' theory according to which a projectile continues to move after it has left the thrower's hand not because of the medium in which it is travelling (as Aristotle had maintained), but because of an unnatural force (or power) transmitted to it by the thrower. In a void, if that were possible, the aforementioned force would never be consumed, so that the motion would continue infinitely (see PHYSICAL SCIENCE 500–600).

The speed of bodies in a void—finite or infinite?

Toward the end of the century, the philosopher Ibn Bājja (Avempace) also defended Philoponos' theory, maintaining that the speed of projectiles, as well as that of freely falling bodies, would be proportional to the difference between the driving power and the resistance encountered. He also maintained that in a void as well (always if it were possible), the speed of a body would be finite, contrary to Aristotle's theory according to which that speed would be infinite.

An interest in physics began to arise in the West as well, closely related to the interest in logico-philosophical quarrels. As a discipline, its study was essentially based on Plato's *Timaeus*. But there were also those who opposed the new genre of studies. For example, the anti-dialectician Manegold of Lautenbach (c. 1030–1103) polemicized animatedly with the Platonic physicists, maintaining that Christians must give up their pretensions of understanding natural phenomena with the power of reason alone, since explanations of these phenomena should be based on the biblical story of Genesis.

CHEMICAL SCIENCES

The splendor of Arab alchemy

During the first part of this century, as often happened later, alchemy was dominated by one thinker—Ibn Sīnā (980–1037), known in the West as Avicenna, to whom works of a clearly later date have also been mistakenly attributed. Ibn Sīnā maintained that the mineral world could be divided into stones, fusible substances, sulphurous substances, and salts. He considered mercury a basic element of all fusible substances, including metals (which were transformed into mercury by fusion). He denied that alchemists could really effect transmutations, and admitted, only realistically, that alchemistic procedures could be used to tincture a metal, such as silver, so well that it might seem to be gold. The skepticism of a physician-philosopher like Ibn Sīnā, not expressed by other Arab alchemists, contributed a great deal to the evolution of alchemy in the Western world.

The *Rutbat al-hakīm* (*Sage's Step*), by Maslama al-Majriti, the first alchemistic book written in Spain, also belongs to this period. It records interesting and detailed chemical operations, including the formation of red mercury oxide by heating the metal (the same as Antoine Laurent Lavoisier's famous experiment), although it clearly states that the weight remained unchanged (see CHEMICAL SCIENCES 1770–1780).

Finally, al-Bīrūnī (973–1048) determined specific weights by using a pycnometer quite similar to the more refined types in use today.

Although Arab alchemy's period of greatest splendor ends at about this time, the Arab world can be credited not only with preserving the continuity of alchemy, but with restoring to it the valuable rule of experiment.

LIFE SCIENCES

Avicenna and the crystallization of ancient medicine

Islamic thought continued to flourish throughout the century, and during this period left its deepest imprint on the field of medicine and the natural sciences. Between the old and the new centuries, the great Avicenna (980–1037) "canonized" the features that Islamic medicine had assumed during the previous three hundred years. His synthesis of ancient medical knowledge was all-inclusive and systematic and was coordinated with a doctrine regarding symptoms that brought together the most recent experiences of Arab, Iranian, and Jewish clinicians.

Since medicine lacked any new, basic anatomical or physiological knowledge, it became dogmatized—the title of Avicenna's major work, *Qānūn*, even expresses this tendency—and justified itself by deferring to the authority of famous persons and illustrious works. Hippocratic and Galenic tradition tended to blend together, while theoretical medicine tightened its bonds with an Aristotelianism that was not always faithful to the original texts. While Abū-l-Qāsim's work on surgery even influenced Fabricius ab Aquapendente, Avicenna was read in Louvain until the seventeenth century.

Minor figures flourished along with the two great physicians: 'Arīb al-Qurtubī, an obstetrician and pediatrician, Ibn Wāfid and Ibn Jazla, pharmacologists, and 'Alī ibn Ridwān, a scholar of the endemic and epidemic diseases of Egypt.

Geometric optics and Ibn al-Haytham

An original physiological perspective was opened up with the optics of Ibn al-Haytham (Alhazen, c. 965–1038). The Cairo scientist adopted a theory of propagation by means of rectilinear rays to explain the phenomenon of vision as well as the reflection and refraction of light.

Ophthalmology began to develop with 'Ammār of Mosul, and 'Alī ibn 'Isā of Baghdad, the author of a treatise.

The Baghdad and Salerno medical schools

As the cultural and scientific supremacy of the Islamic world began to decline, new links were formed between Islam and the Latin West. The first Latin-Arabic glossary was compiled in Castile. The Baghdad medical school prospered with Ibn Jazla and Sa'īd ibn Hibat Allāh, and the Persian Zarrin Dast was a prominent figure in ophthalmology.

The vitality of the medical schools in Islamic Spain, especially Seville, can be seen in the example of Ibn Bajja (Avempace), as the premises were laid and the developments made for surpassing decadent Iranian and Mesopotamian medicine.

In the Latin West, Salerno appeared as a center for medical study and practice, evolving far beyond the primitive Benedictine hospice and the original lay school, supposedly started by a Greek, a Latin, a Jewish, and a Saracen teacher.

Gariopontus, a Longobard master who died around 1050, published an encyclopedic work called *Passionarius*, which broadened terminology by introducing a number of roots from the *sermo vulgaris.* Trotula, a woman physician who lived in Salerno between the later part of this century and the beginning of the following, published a work on pregnancy, giving birth, and care of the infant, which was widely read for many years. Alphanus, a Benedictine bishop of Salerno, published a treatise entitled *De quattuor humoribus corporis humani*, and carried out works of charity while the city was being besieged by Robert Giuscard. The *Civitas hippocratica* received a substantial contribution from the Islamic world with the arrival of Constantine the African, the *magister Orientis et Occidentis,* a Christian physician from Carthage who fled to Salerno because he was suspected of magic. Constantine brought Galen's *Ars parva* and Hippocrates' *Aphorisms* (ἀφορισμοί) to Salerno, laying the foundations for the rise of university medicine, which was sanctioned by several conciliar decrees in the following century.

APPLIED SCIENCES

Metallurgy

Beginning in the tenth century in the West, attempts to use coal in metallurgy were intensified, in order to make up for a lack of charcoal. The use of coal influenced the development of metallurgy. Although the documentation is not clear, it seems certain that coal was used at least in the initial steps of fusion.

Recent studies have shown that water power was used in metallurgical procedures beginning in the eleventh century. Documents tell of hydraulic hammers and bellows used in the western Alps and in Silesia. The increased draft and hammer power must have had a decisive influence on the modern evolution of furnaces, making it possible to obtain higher temperatures and work heavier blocks of iron.

This progress in metallurgical techniques was decisive for the European rural economy. Documents show that iron played a much more important part in the daily lives of the peasants at the end of the century than in Carolingian times.

Agriculture and the development of water power

During Carolingian times, plowing had been limited to only the easiest soil. In the eleventh century an important three-fold improvement in traction was made, which alone may have opened much more land to cultivation. It seems that better harnesses, such as the shoulder collar for horses, the frontal yoke for oxen, and shoes for both were developed for draft animals. However, horses did not yet completely replace oxen for plowing. Evidence exists that horses had been used for plowing for some time, although only along the northern coasts of Norway, while the first representations of horses plowing appear in a tapestry from the end of this century. Harnesses were improved as more leather was used when techniques such as oil tanning, or chamoising, and alum and vegetable tanning were introduced in leather-working.

Water power became widely diffused during this period, as documented in England in the *Domesday Book*, a property register drawn up in 1086 on the orders of William the Conqueror. The register records no fewer than 5,624 water mills in existence in three thousand British communities (an average of one mill for every fifty families) plus pestle mills for crushing ores and hammers operated by water wheels. Water wheels could be adapted by means of an intermediary mechanism—the cam-

shaft, which had already been known to Heron—that transformed the continuous circular motion of the wheel. In hammer mills, the cam acted on the handle of the hammer. The flywheel also appeared as a machine part between the eleventh and twelfth centuries, in a treatise by Theophilus the Monk.

With the variety of uses to which the mills were put, as well as their diffusion, medieval mechanization advanced considerably. The process continued in the following century, then stabilized until the fifteenth century.

The textile industry

The wool industry prospered, above all in the Netherlands, especially after the fulling machine was introduced. The fulling mill appeared in Normandy around 1086, causing the wool industry to move into the river valleys, where water power could be used. The result was to transform a predominantly town-based luxury industry into a rural industry, free from the laws that had greatly hindered its development.

Architecture

The building industry flourished once again throughout western Europe, influenced by Roman architecture. In France, the new Roman style could be seen in the Abbey of Bernay (1020) and the Abbey of Jumieges in Normandy (1040), which became models for European (especially English) architects.

In England, the abbatial church of Westminster and the chapel of St. John in the ''White Tower'' of London were built. They are typical examples of Roman architecture, with their barrel vaults, over thirty centimeters thick, the thrust of which is contained by massive walls.

Military technology

In the field of Western military technology, the diffusion of or reinvention of one-person crossbows, which brought about changes in warriors' armor, deserves mention. Although crossbows had been known to the Romans, they were not very widely used during the first part of the Middle Ages.

Sugar refining

The Arabs, with their knowledge of chemistry, developed a sugar-refining process that replaced the ancient method of pressing cane. At first, the Arabs mixed the juice with milk and allowed the mixture to decant, so that the impurities would settle along with the milk, but in the eleventh century they discovered that the juice could be extracted better by treating it with lime and ashes. After this liquid was filtered, sugar was obtained by evaporation and crystallization.

1100–1200

THOUGHT AND SCIENTIFIC LIFE

The twelfth century renaissance. Lay schools

The economic, political, and cultural revival of the Christian West which began in the second half of the eleventh century, reached its peak during the first decades of this century. Free cities flourished, especially in Flanders and northern Italy; and in England and France new states, such as the Norman kingdom of southern Italy on the Mediterranean basin, arose. The marine republics prospered. A common feature of these economic and social transformations was the increasingly important role of the cities, the trade and market centers, where merchants, artisans, and entrepreneurs emerged in contrast to the old feudal aristocracy. The organization of the free cities favored the establishment of a type of school that was not tied to ecclesiastical jurisdiction. The Salerno medical school was well established by the end of the eleventh century. It was now the turn of the lay law school in Bologna, which enjoyed universal prestige after Irnerio taught there, and played a leading role in the ideological polemics between the empire and the papacy. The study of law, traditionally included in the study of rhetoric, became autonomous; and the restoration of the Roman juridical tradition in the West not only proved to be in tune with the new political institutions, but deeply influenced the internal structure of the Church as well. After initial attempts by Burchard of Worms, Deusdedit, and Ivo of Chartres (died 1117) to compile canonical collections, the Church established its own autonomous law. The first code of ecclesiastical laws, the *Decretum* of Gratian was published in about 1139; and by 1150, the first commentaries had already appeared. The study of law, like that of logic, contributed to making language more precise in other branches of knowledge, as well.

The cathedral schools

The ecclesiastical institutions still played a dominant role in culture. Developments in the political struggle, especially the caesaropapism of the emperors and the theocratic theories of the papacy, required that the ecclesiastics acquire a more practical knowledge of the profane, which included logic and law. However, the development of city life had taken the traditional monopoly on culture away from the monastic schools to the advantage of the episcopal and capitulary schools which were usually located in the cities. The presence of scholars, who often exercised both curial and ecclesiastical functions in these scholastic communities, favored the formation of actual centers, with large libraries and adequate scholastic structures. It was in these cathedral schools that a rich philosophical and literary culture was developed, characterized by studies in law and a long tradition in the arts of sermonizing (grammar, dialectics, and rhetoric), as well as by subtle theoretical *reflection* combined with scientific research.

It was symptomatic that the most brilliant writers of this century were bishops (for example Hildebert of Lavardin, a scholar at Tours, and Gilbert de la Porrée, a teacher at Poitiers) or canonists (such as Hugh of Orleans), while others came directly from the French cathedral schools (for example Robert of Melun, William of Conches, and Bernard Silvestris). In England, the major cultural center, Centerbury, became the new episcopal center, where a group of jurists and canonists of Italian origin gave exceptional impetus to juridical studies. Later, Thomas Becket, bishop and chancellor of England, led the formation of a learned ecclesiastical elite, competent in both theological debate and the practice of the liberal arts.

Theology, logic, and philosophy. Paris and Chartres

The development of intellectual life varied, however, from place to place. While culture in the Italian cities was mainly juridical and medical, the cathedral schools of Chartres and Orleans were the centers of a literary and philosophical revival, Rheims and Laon maintained a solid scientific tradition, and the Parisian schools became increasingly noted for theological-philosophical speculation. A widespread familiarity with those texts that had permitted Platonic and Neoplatonic tradition to survive in the West and an increasingly broad and in-depth knowledge of Aristotle's *Organon* were unifying factors. Besides these purely philosophical documents, the scholastics of Paris and Chartres, along with those of Rheims, Laon, and Orleans, looked to the *auctores* of the great Latin literary tradition as models of *eloquentia* and the sources of scientific and moral, as well as rhetorical, *exempla*. The most important philosophical experiences of the first half of the century belong to this cultural context. For example, the teachings of Peter Abelard (1079–1142) in Paris, which were focused on commentaries on the classical texts of the *logica vetus* made some of the most original contributions to the question of universals, and the doctrine of the school of Chartres were centered around a revival of the cosmological themes of Platonic tradition. At Chartres, which had become famous as early as the end of the tenth century with the bishop Fulbert, and had contributed enormously to promoting the arts of the *quadrivium* and the *trivium*, the teachings (from 1114 to 1124) of Bernard of Chartres, a Breton and the chancellor of the cathedral, represented the richest source of literary culture in the Gauls during that time, according to John of Salisbury. At almost the same time, the school of St. Victor, against which Peter Abelard polemicized, was elaborating its doctrine.

Currents of religious spirituality. The Cistercians

The age of Abelard and the thinkers of Chartres was also an age of intense mystic fervor, not the least of which came from the various religious movements of popular origin that advocated the need to reform ecclesiastical life and often found themselves in open conflict with the hierarchy. Among these were the Cathars, the Poor of Lombardy, the Humiliati, and (later) the Waldenses. The need to reform the monastic movement was also felt among the ranks of the orthodoxy. One of the most famous attempts was made by the Cistercians, who came from the monastery of Cîteaux. Convinced that although dialectics and the profane sciences were useful in confuting heretical doctrines, they should always be subordinate to the faith, the Cistercians looked with suspicion upon the intellectual arrogance of the philosophers of Chartres, logicians such as Abelard, and, more in general, all those who cultivated mundane knowledge. This profound conflict reached its peak with the Council of Sens in 1141. As a result of the campaign waged by Bernard of Clairvaux (1091–1153), a great champion of Cistercian mysticism, several of Abelard's theological propositions were condemned, including various tracts of his *Scito te ipsum*. The doctrines of Gilbert de la Porrée (1070–1154), Bernard of Chartres' successor and later a teacher in Paris, fell to the same fate. Gilbert's renewed Platonism—much better articulated than that of his predecessor Bernard, and focused on a doctrine of the essences, the theological applications of which disturbed the ecclesiastical hierarchy—was the object of Bernard of Clairvaux's suspicions.

The Victorines reached a balance between mystical thought and profane culture. Hugh of St. Victor, who succeeded William of Champeaux in 1141, commented on Pseudo-Dionysios' texts and favored a revival of Augustinian mysticism, while profane culture is organically included in his *Eruditio didascalica*. Subsequently, Richard of St. Victor, prior and teacher of theology from 1162 until his death in 1173, followed Hugh's mystical-speculative path, although he did not reject an empirical and argumentative foundation. His writings, along with those of Godfrey of Saint-Victor (c. 1125–1194), had a notable influence on Franciscan mysticism. Around 1179, Walter of St. Victor, Richard's successor and disciple, condemned Peter Lombard and Peter of Poitiers, together with Abelard and Gilbert de la Porrée, in his *Contra quattuor labyrintos Franciae*.

Theological schools and the *Libri sententiarum*

Ecclesiastical censure and condemnation from the mystics did not stop the spread of Abelard's ideas—which thoroughly revised not only logic, but also theology and morals—or the development of the thought of Chartres. A brief commentary by Gilbert on Aristotle's *Categories*, the *Liber sex principiorum*, was also quite influential in the following century. In addition, Thierry of Chartres (c. 1100–1156), the younger brother of Bernard of Chartres, picked up a love for literary studies and an interest in the arts of the *quadrivium* from his brother's teachings. Theodoric's famous *Eptateuchon* is a valuable document on the cultural life of Chartres. William of Conches (c. 1090–1160) presented a broad synthesis of Chartrian themes in his treatise *Philosophia mundi* and in a commentary on the *Timaeus*.

The development of philosophical and theological reflection was part of a process of profound reorganization of the cultural institutions. Various centers of study grew, and the school of Bologna became a *Studium generale*, while the University of Paris was first organized around 1170, on the initiative of three schools, the cathedral school of Notre Dame, the school of St. Victor, and the abbatial school of Sainte Genevieve. The teachers and students gradually organized themselves into four faculties: the *inferior*, which contained the arts, following the formal arrangement of the *trivium* and the *quadrivium,* and the *superiores*, i.e. law, medicine, and, of course, theology. The first recognition from civil authority came from the King of France in 1180. At almost the same time, a university center was also arising at Oxford, especially after 1167 when English students were recalled to the homeland.

The new organization of these institutions required that a number of tools for study, adapted to the needs of scholastic life, be developed. On the other hand, the need for vast encyclopedic compilations that included all the elements indispensable to science, law, and theology was not new, as the success of Isidore of Seville's *Etymologiae* demonstrates. The influence of the heretical polemic, which forced the clergy to prepare themselves systematically in theology and learn the techniques of confutation, provided a powerful stimulus for the editing of the collections of *quaestiones* that had arisen from debates on biblical texts and interpretations of the Church fathers. The collections of juridical *quaestiones*, widely diffused in the law schools, furnished a model.

The first surveys of handy theological handbooks are found in the *Elucidarium*, attributed to Honorios of Autun, and the *Sententiae*, by Anselm of Laon. The school of Laon distinguished itself in the task of systematically organizing theological material and elaborating scriptural glossaries. While Robert of Melun (c. 1100–67), a follower of the Laon tradition, provided an actual manual with his *Summa sententiarum*, Hugh of Saint-Victor presented a more elaborate philosophical and mystical framework in his *Summa sententiarum* and *De sacramentis*. Various collections appeared in the second half of the century, including ones by Simon of Tournai, Peter Cantor, and Peter Comestor. All show the influence of Chartres and refer to Abelard's famous *Sic et non* as a paradigm. While lively debates arose around these collections, which often took up currents of thought that were not agreeable to the ecclesiastical hierarchy, a much more conciliatory line, respectful of the orthodoxy, was followed by Peter Lombard (c. 1095–1160) in his *Libri quattuor sententiarum*. This work was explicitly aimed at reconciling opposing interpretations of the Church fathers—which explains its success—and took a position that was very different from Porrée's or Abelard's. Peter Lombard's text was quickly popularized partly because of the work of his disciple, Peter of Poitiers (c. 1130–1205) whose *Sententiarum libri quinque* contributed significantly to the success of his teacher's ideas. The approval of the Fourth Lateran Council in 1215, after some debate, made Peter Lombard's work the compulsory text in faculties of theology in the West.

The translations from the Arabic

The propagation of Greco-Arabic science, mainly in the centers in Sicily and Spain, greatly widened the range of *auctores* available to Christian speculation. It was in southern Italy and Sicily that the logic of Adelard of Bath (c. 1090–1150), who mostly translated scientific works, came into contact with Arab culture. In Spain, the cathedral school of the city of Toledo became a true *collegium* of Arabic-to-Latin translators under the guidance of Raymond of Toledo, archbishop from 1126–1152. Some of the outstanding translators there were Domingo Gondisalvo, Aben Dahut (Avendeut to the Latins), and John Hispanus (whom some identify with Avendeut). In the space of thirty years, the school of Toledo translated Arabic scientific and philosophical writings into Latin, including al-Kindī's *De intellectu*, a good part of Avicenna's works, the philosophical works of al-Ghazzālī, and Avicebron's *Fons vitae*. In addition, the number of translations attributed to Gerard of Cremona bear witness to the presence of an efficient organization of translators who ventured not only upon Avicenna and al-Fārābī, but also upon Ptolemy's *Almagest* and several of Aristotle's works, such as the *Posterior Analytics*, along with Themistios' commentary on it, the *Physics*, and part of the *Meteorology*. The translators often made use of the combined efforts of an Arab and a Jew who translated into the Castilian vernacular, plus a scholar who translated into Latin. The oldest extant Latin-Arab glossary is in a Spanish manuscript of this century. In any case, the translators' task was arduous: not only did they have to acquire an in-depth knowledge of the Arabic language, but they also had to find their bearings regarding extremely complex questions and render a very wide range of technical terms into Latin. It is not surprising, therefore, that these translations were often literal and that sometimes, when the meaning of a word was not understood, it was simply transcribed in Arabic or Hebrew. In addition, some translations were also done directly from the Greek; an example is those of James of Venice, who in 1128 translated the *Analytics* and the parts of the *Organon* that were unknown to Boétian tradition (or the *logica vetus*). These became the principal source of the *logica nova*. Subsequently, Henricus Aristippus of Catania took a large number of Greek manuscripts from the library of the Byzantine monarch, Manuel I Comnenos, to Catania and, around 1156, translated the four books of Aristotle's *Meteorology*, as well as Plato's *Meno* and *Phaedo*, directly from Greek into Latin.

Arab, Jewish, and Christian philosophical culture in the second half of the century

The Moslem world, although torn by grave crises, was still full of philosophical and scientific thinkers of great intelligence. Ibn Rushd (1126–98), the Latin Averroés, lived in Cordova during the second half of the century. Although the climate was already intolerant, he followed his studies in law, medicine, astronomy, and mathematics with a broad philosophical synthesis that ranged from polemics against al-Ghazzālī (in his *Tahāfut al-tahāfut*, or *Destructio destructionis*), to commentaries on Aristotle. The latter greatly influenced the West in the following century, where Averroés became the commentator *par excellence*.

The Jew Moses Maimonides (1135–1204), a physician and Talmudist who in the last part of his life was the head of the local Jewish community in Cairo, also flourished in Cordova. In his work *Moreh nebukim* (or *Guide to the Perplexed*), he approaches Arabic science for religious ends, but without prejudice. Many Christian philosophers who were attempting to reconcile Aristotelian and Arab demonstrative rigor with Christian tradition regarded Maimonides' work with particular interest.

The West, in the meanwhile, was dominated by John of Salisbury (c. 1115–80), who was educated in France at Abelard's school and was connected to the cultural environment of Chartres, the episcopal seat he occupied during the last years of his life, after acting as secretary to the archbishop of Canterbury. His interests ranged from logic (*Metalogicon*) to politics (*Polycraticus*), and his work is rich in references to classical authors such as Cicero and Seneca. The strong emphasis on rationality that is found in the work of John of Salisbury can also be seen in the *Regulae de sacra theologia*, by Alain de Lille (c. 1117–1202), who taught in Paris at the end of the century and wrote poetry, libels against the heretics and a treatise on morals. Rational argumentation used against the heretics and a treatise on morals. Rational argumentation used against the heretics also plays a prominent role in the *De arte catholicae fidei*, a treatise written by Nicole of Amiens, a contemporary of Alain.

MATHEMATICAL SCIENCES

The work of the Arabic-to-Latin translators

The Arabs' contact with Latin culture, which basically took place in Spain and Sicily, made it inevitable for the ideas they had synthesized from Eastern cultures to be transmitted to the West, and even Greek science found its way into medieval Latin culture via this path. Just as the Arabs had established well-equipped centers for translating Greek and Hindu works into Arabic in the major cities of their caliphates, centers for translating Arabic works into Latin, be they Arabic versions of ancient classics or original works by Moslem scientists, began to be set up in the West. The first and most important of these was founded in Toledo by Raymond of Toledo, and others arose in Sicily and England during this period.

In these centers, illustrious translators like Gerard of Cremona (1114–87), Plato of Tivoli (flourished c. 1135), Adelard of Bath, and Robert Chester (who both lived around the middle of the century) put their names to the Arabic-to-Latin versions of Euclid, Ptolemy, and Theodosios, as well as to the most famous works of Moslem authors, during these years. The succession of translators went on for many years, reaching a sort of conclusion, almost a century later, with William of Moerbeke (c. 1215–86), the translator of Archimedes, among others.

Latin tradition and Arab influences

Despite the work of the translators, the recovery of ancient science in the West proceeded slowly. A *Practica geometriae Hugonis*, which probably belongs to this period, still contains vague and inaccurate information, and shows no traces of Greco-Arabic influence. It reflects only the pure and simple Latin tradition, which was soon to be suppressed by the return of the Euclidean speculative approach, made possible in turn by the translations.

Works that were in any way original were still Arabic; an example is the *Treatise on Perfect Compasses* by Muhammad ibn al-Husain, written between 1187 and 1195, which develops a theory of conics and describes an instrument for drawing all types of conics. Al-Husain also established a number of theorems on rectangular triangles and on congruous numbers, following research methods derived from Hindu mathematics.

ASTRONOMY

The Sanjaric Tables and the Malmudic Tables

In the Arab world, observational astronomy continued to develop with the production of increasingly accurate and extensive tables, profiting considerably from the progress made in spherical trigonometry and the perfection of instruments like the astrolabe and the *torquetum*. At Marv, al-Khāzinī published the *Sanjaric Tables*, which gave the positions of the stars for 1115–16. The *Malmudic Tables* were prepared by al-Asturlābī, a famous observer, theoretician, and maker of instruments, who died at Baghdad around 1139. The Spanish Jew Abrāhām bar Hiyya (Savasorda in Latin, 1070–1136), who lived in Barcelona and Provence, also compiled astronomical tables. He also wrote other astronomical works, including one on the calendar, in Hebrew, and collaborated extensively with Plato of Tivoli in the Arabic-to-Latin translation of numerous scientific texts. Abrāhām bar Hiyya is perhaps the oldest Hebrew-language scientific author, and one of the foremost intermediaries between Arabic and Latin cultures.

His example was followed by the philosopher Ibn Ezra, who was born in Toledo around 1090 and died in 1164 after travelling all over Europe. He wrote astrological treatises and works on the astrolabe and the calendar, eight of which were composed at Lucca in 1148. He translated al-Bīrūnī's commentary on al-Khwārizmī's *Tables* from Arabic into Hebrew.

Astronomy and astrology

That the close relationship between astronomy-astrology and medicine continued is clearly indicated by the works of two Spanish-Arab physicians: a treatise on the astrolabe, written by Abu al-Salt al-Doni, who died at Tunis around 1135, and and an exposition on astronomical knowledge useful to the practice of medicine, by al-Ainzarbi, who lived at Baghdad and Cairo and died around 1154.

Theoretical and observational astronomy

In the field of theoretical astronomy, the Persian al-Kharazi, who died at Marv around 1139, wrote a large treatise on *The Division of Spheres*. The work revives a doctrine of material spheres to explain the movements of the heavenly bodies, a theory that had already been mentioned by al-Khāzinī and Alhazen. Through the diffusion of Latin translations of Alhazen's works, especially the *De mundo et coelo*, this theory was accepted in the West.

The developments made in observational astronomy are an indication of the huge difficulties that the Ptolemaic system was having in explaining celestial motion as data were obtained with increasing accuracy. The task of revision was begun by Ibn Bājja (Avempace in Latin), who was born in Saragozza around 1105 and died at Fez around 1138, and carried on by the astronomer Jābir ibn Aflah (Geber, as he is known in the West) who lived in Seville and died around the middle of the twelfth century. Geber worked expressly on the theory of the planets; his work was continued by Ibn Tufail and Alpetragius.

The work of the translators

In the West, the Frenchman Raymond of Marseille made observations and compiled astronomical tables. His first table was on the moon, and the second adapted al-Zarqālī's *Toledan Tables* to the meridian of Marseille. As contacts between the Moslem, Jewish, and Christian worlds increased in frequency and depth, a period of extremely fervent activity in translating into Latin began. Through these translations, Western culture acquired a wealth of knowledge from Arab, and, to a lesser extent, European Jewish, civilization, while at the same time it recovered the classical Greek texts, once again by way of Arabic tradition. The historical conditions in Spain were such that the exchange was easiest there, and Toledo was the most active translating center. The translators often worked in pairs, using their skills in Arabic or Latin via a common intermediate language, which was often the local vernacular spoken by both.

In the same period in Italy, in the areas that had maintained contact with Byzantium—Venice, Ravenna, southern Italy and Sicily—some texts were translated directly from Greek into Latin.

The English philosopher and scientist Adelard of Bath (c. 1090–1150) translated al-Khwārizmī's *Tables*, as well as mathematical works, including the fifteen books of Euclid's *Elements*, from Arabic into Latin, introducing trigonometry into the West. Herman the Dalmatian, a student of Theodoric of Chartres,

Hugh of Santalla, and the Englishman Robert of Chester, also flourished around the middle of the twelfth century. They translated many Arabic works on mathematics, astronomy and astrology, and alchemy into Latin, often making original contributions to the various disciplines as well.

The dragoman John of Seville, a converted Jew, and Dominic Gondisolvi, archdeacon of Segovia, were the most prolific pair of Toledan translators. They flourished until after the middle of the century, translating a large number of texts in arithmetic, philosophy, medicine, and astronomy into Latin, including Alphraganus' *Compilatio*, al-Kindī's *De magnis coniunctionibus*, Albategnus' *Ceniloquium*, Alcabitius' *Libellus ysagogicus*, and many more, especially astrological texts, which were diffused along with medical texts in the West, as well.

Plato of Tivoli, who lived in Barcelona from 1134 to 1145, in collaboration with the mathematician Savasorda, translated mathematical and astronomical works from both Hebrew and Arabic into Latin. Among these was Claudius Ptolemy's *Quadripartitum*, translated in 1138, the first treatise by the great astronomer to be made available in the West.

At the court of Palermo, during the reigns of Roger II of Sicily (1095–1154) and William I (1154–66), Henricus Aristippus, archdeacon of Catania (died c. 1162) introduced copies of Greek philosophical and scientific works that he acquired in Byzantium into the West. Among these were Aristotle's *Meteorology*, which Aristippus himself translated into Latin, and Ptolemy's *Almagest*. The latter was translated around 1160, probably a collaboration between Aristippus and Eugene, "admiral" of Palermo under William I. The first Latin version of the basic text of Greek astronomy was not widely diffused in the West, and remained almost unknown.

Much more successful was the translation from the Arabic done by Gerard of Cremona in 1175 at Toledo. Gerard, who was born in Cremona around 1114 and died at Toledo in 1187, organized the largest translators' school of the Middle Ages in Toledo. Almost a hundred Latin versions of works originally written in Arabic, or translated into Arabic from the original Greek, on philosophy, logic, physics and mechanics, medicine and alchemy, and astronomy and astrology, came out of this school. Among the works on astronomy and astrology were those of al-Khwārizmī, Alfraganus, Thābit ibn Qurra, Geber the astronomer, Arzachel, Autolycos, Theodosios, Geminos, and Ptolemy.

The Almagest in the West

Just as Western culture was acquiring a knowledge of the *Almagest*, Arabic astronomers were developing a critique of the Ptolemaic doctrine of eccentrics and epicycles, and reviving the ancient theory of homocentric spheres that they found in Aristotle. Aristotle's thought was gaining in popularity, due to the translation into Arabic of his works and, especially, to the commentaries that were being written on them. A large part of the work of Arabic philosophers, Ibn Tufail and Ibn Rushd (Averroës in Latin), the commentator *par excellence*, for example, was dedicated to commenting on the Aristotelian texts.

The Spanish Arab Ibn Tufail, who was born early in the century and died at Marrakesh in about 1185, a philosopher and physician, commented on Aristotle's *Meteorology*. He taught Alpetragius, to whom he suggested a critique of the planetary theory in the *Almagest*, and was a friend of Averroes, who succeeded him in 1182 as physician to the sultan of Marrakesh. Ibn Rushd (1126–1198), a philosopher and a great physician, also wrote a treatise on the celestial spheres and a summary of the *Almagest*.

In Persia, Ibn Malka, philosopher, physician and astronomer, who flourished at Baghdad in the second half of the twelfth century, compiled new astronomical tables and dedicated a treatise to explaining why the stars are only visible in the night sky.

In Byzantium, John Camateros, who lived at the court of Emperor Manuel I (ruled 1143–80) and later became archbishop of Bulgaria, wrote two poems in Greek dedicated to astronomy and astrology.

In the West, Roger of Hereford (flourished c. 1180) wrote treatises on chronology and astrology and a *Theorica Planetarum*, and compiled astronomical tables for the meridian of Hereford.

Alpetragius

The Spanish Arab al-Bitrūji (Alpetragius in Latin) was born at Pedroche near Cordova and flourished at Seville around the end of the twelfth century. He wrote a *Theorica planetarum* in which he uses ideas suggested to him by his teacher, Ibn Tufail, in an attempt at explaining planetary motion that contradicts the Ptolemaic doctrine of eccentrics and epicycles and goes back to the Aristotelian theory of homocentric spheres. He imagined that each heavenly body was attached to a solid sphere, which received movement from the ninth, outermost sphere, the *primum mobile*. The system was characterized by the non-coincidence between the earth's axis and that of the various spheres, and the precession of the eighth sphere along with the rotation of its axis, which caused the ''trepidation'' of the equinoxes.

Maimonides

The Spanish Jew Moses Maimonides, who was born at Cordova in 1135 and died in Cairo in 1204, was well-known as a philosopher, physician, and commentator on the rabbinical texts. Besides explicitly condemning astrology, he also rejected the Ptolemaic theory inasmuch as it diverged from Aristotle's teachings. His student Joseph ben Judah ibn Sham'un (c. 1160–1226) shared his ideas.

PHYSICAL SCIENCES

Averroës: the limits of experience

With respect to the Arab world, we shall only mention that Ibn Rushd (Averroës, 1126–1198), the famous Aristotelian commentator, resolutely took a position against Avempace's explanation of motion which, we know, was inspired by Philoponos. It is interesting to note that the polemic was carried out in the name of a loyalty to concrete experience, which shows us that all bodies do, in fact, move within some medium. Therefore, if this medium should constitute an impediment to the continuation of movement, rather than its cause, as Aristotle maintained, we must conclude that all movement is non-natural. According to Averroës, to appeal to an immaterial force transmitted to moving bodies (especially projectiles) by their mover in order to explain their motion meant looking for the cause

of phenomena not in reality itself but in abstract entities. This explains Avempace's accusations of Platonism directed against him, and his need to revive Aristotle's primitive theory against Avempace's.

The school of Chartres

As for the Western world, first of all a rapidly growing interest in physics was developing among its most representative thinkers, although in a form that was still more philosophical than strictly scientific. The major physicist-philosophers of the epoch belonged to the school of Chartres, which was of clearly Platonic inspiration. Some of these were Theodoric of Chartres, William of Conches and Ivo of Chartres. Their Platonism, however, was essentially naturalistic, that is, open to an interest in the world of phenomena, which Plato's *Timaeus* had taught them to interpret rationally. This very openness induced these authors to attempt to integrate Plato's general concepts with others, advanced by Presocratic philosophers. For example, William of Conches integrated Plato with Democritos, arriving at a cosmological construction that was nearly materialistic.

However, the debate between Platonism and Aristotelianism that was so heated among Arab authors had yet to reach the school of Chartres. The fact is, in the twelfth century Aristotle was known in the West almost exclusively through his writings on logic, while his works on physics were much less well-known.

CHEMICAL SCIENCES

The first description of alcohol and cane sugar

The Arab world, which had made important contributions to the sciences and, in part, to alchemy during the few preceding centuries, was now seriously in crisis. Nevertheless, it transmitted a notable patrimony of information to the Western world, and along with it retransmitted the particular philosophy on which alchemy was based. Although this philosophy was directly derived from Greek Neoplatonism, for various reasons it took a sharp turn toward gullibility and charlatanism among Western alchemists and was less oriented toward mysticism.

The updated manuscripts of the *Compositiones ad tingenda* and the *Mappae clavicula*, which were partly from an earlier date and, in any case, not Arabic in origin belong to this century. The same is true for the *Liber ignium ad comburendos hostes*, by Marcus Graecus, which provides the first description of Greek fire and the use of saltpeter, later used in black powder. In *Mappae clavicula*, an anagram must be solved (it was solved by Marcellin Berthelot) in order to obtain the first description of a liquid, distilled from wine, that would catch fire. The liquid would later be called alcohol by Paracelsus in the sixteenth century, a word that is definitely derived from the Arabic, although its origins are not altogether clear. The *Liber ignium* also contains a description of how to prepare alcohol.

An exceptionally clear description of the extraction and purification of cane sugar was furnished by the Italian physician Matthaeus Platearius around 1150. Other explanations of how to make sugar are found in the *Mappae clavicula*. It is unusual that, although sugar had been used for a long time in the East, the Arab authors make no mention of it.

A remarkable text from this period is the *Diversarum artium schedula,* by the "humble servant of the servants of God," Theophilus the Monk, so-called to distinguish him from Theophilos the Greek alchemist. The work is in three volumes: the first is dedicated to the pigments used in painting, the second to glass-making and the third and largest to metal-working. Theophilus' descriptions are quite clear and instructive; his text on metallurgy is undoubtedly the best prior to the works of Vannoccio Biringuccio (c. 1480–1538) and Georgius Agricola, which appeared in the sixteenth century.

Latin translation of Arabic texts

The translations into Latin of Arabic texts, done in this century by Hugh of Santalla, Gerard of Cremona, and Robert of Chester (who is attributed, probably wrongly, with the translation of the alchemical works of Khalid ibn Yazid, one of the first Arab alchemists who lived during the second half of the seventh century) were very important, and bear witness to the transmission of Arab culture into the West. The *Liber sacerdotum,* which was compiled in Spain from Arabic texts essentially dedicated to producing metals similar to gold, also belongs to this period.

The *De anima in arte alchemiae,* which some (including Berthelot) believe is based on original texts by Ibn Sīnā, probably dates back to this century as well. The Latin text was definitely produced by a Spaniard, who often uses Spanish words instead of Latin ones in curious way (for example, *plata* for *argentum*). In any case, it is a fairly obscure text that contributed nothing of effective value to chemistry.

LIFE SCIENCES

Avenzoar and Averroës. Arab thought in medicine

Avenzoar (Abū Marwān 'Abd al-Malik ibn Abī l-Alā Zuhr, c. 1091–1162), who proclaimed the value of experience over tradition, flourished in Andalusia. He described new clinical syndromes and condemned the practice of surgery. His pupil, Averroës (Ibn Rushd, 1126–1198), returned to the compilatory-dogmatic position, and assembled, among other things, a large collection of Galen's works.

Hildegarde of Bingen's Physica

The school of Salerno produced a textbook, the *De aegritudine curatione,* made up of lessons and maxims of the various teachers, as well as the treatise *Regimen,* or *Flos sanitatis,* a manual containing common rules for hygiene and cures. An interest in naturalistic studies continued to be negligible. One of the few exceptions is the *Physica* of Hildegarde of Bingen, which contains lively descriptions of plants and animals. In 1140, Roger of Sicily granted the school of Salerno his protection and an official statute.

The influence of the Salerno school

Works and people moved out from Salerno into Europe. The *Liber Pantegni,* by Alī ibn Abbās al-Majūsī (died 994), and the *Isagogae* by Joannitius—one, a large treatise on anatomy, physiology, and pathology, and the other a basic outline of precedents—as well as the *Canon medicinae* and other analogous

treatises, reached Chartres, Montpellier, and Ireland. But even the last great authors of Salerno contributed to the growing influence of the *civitas hippocratica.* Foremost among them was Roger of Frugardo, a surgeon who initiated a new revival in surgery in the Latin West, partly because be was convinced of the importance of studying anatomy. The scholars of Salerno, although still loyal to Galenic tradition, very actively studied the anatomy of animals. Roger was one of the teachers who went from Salerno to the Languedoc province in the second half of the century, laying the foundations for the rise of the school of Montpellier in 1220. He was preceded by Rinaldo, Matthew Salomone "the honor and glory of Montpellier," and the oculist Graffeo, and accompanied by Arnald of Cremona, Bruno of Calabria, Louis of Reggio, Nicole of Florence, and Sylvester of Pistoia. The Greek- and Arabic-to-Latin translators were fairly active in the West, and medical and naturalistic works were acquired, including Avicenna's *Qānūn* and Alhazen's *Perspective.*

APPLIED SCIENCES

Agricultural techniques

Horses were almost always used in the place of oxen. The use of draft horses was profitable more for intensifying yield than for extending cultivation.

Using horses meant accelerating the working of the land considerably and plowing more frequently. Giving up using oxen also meant growing more oats, which seems to have brought about a more systematic use of triennial rotation. The farmers who decided to stop using oxen could improve the preparation of the soil—and consequently its fertility—considerably, reduce the fallow time, and raise their crop yields. Abandoning the yoke of oxen marked the advent of a more productive farming system. Toward the middle of the century, a four-wheeled *longa caretta* appeared, again pulled by horses.

Mechanical technology

Exchanges between the West and the Arab world increased. This phenomenon brought about a profitable exchange of techniques. The Arab method for manufacturing paper (beating the raw material—linen, cotton, straw, wood—and mixing it with water) was introduced into Europe around 1150, with an important innovation in the beating stage—the use of an elaborate mortar with a wooden pestle.

Around 1180, the windmill was introduced from the East. The first mills were built in Normandy and spread from there into England and the Netherlands. Unlike the Persian vertical-axis windmill, the European type had a horizontal axis.

Around 1125–33, tide mills appeared at the mouth of the Adoux. The industrial applications of this type of mill were scarce, however. In central Europe, the use of water power increased; for example, it was used in the silver and copper mines of the Harz mountains to operate winches, grindstones, bellows, and forges.

The Islamic world was a more fertile terrain for new inventions and was much more able than the West to recognize their value. This happened in the case of compasses and the distillation of alcohol. In 1190, news of floating magnets spread in the

West, and the compass may have been invented at Amalfi. But the Arabs were the ones to use it in navigating the Mediterranean Sea and the Indian Ocean, perhaps even as early as this century, but in any case long before Petrus Peregrinus described it as a rarity in the thirteenth century.

Chemistry

The same was true for the distillation of alcohol, which was probably discovered in southern Italy (Magister Salernitanus). The technique was immediately put to use by Arab chemists, especially for producing perfumes.

Nitric acid, which was distilled by Italian chemists as early as 1150 from a mixture of saltpeter and alum, was also utilized almost at once by the Arabs, who had considerably improved the arts of burnishing and assaying metals. The discoveries made in Italy, in any case, indicate that chemistry was being revived in that country, which had the best quality glass and enamels. Italy became an important center for the production of pottery, and thanks to improvements in enamels, chemists could work with materials that were resistant to strong acids. The discoveries in chemistry affected the arts, especially painting, which gained a number of new coloring materials, including cinnabar, introduced from the Arab world.

Building techniques

The gothic style made its appearance in the abbatial church of St. Dionysius, begun in 1140 near Paris. Gothic arches were used, even in the stone vault ceilings, in the cathedral of Canterbury, reconstructed in 1174. Until the end of the century, gothic arches were used beside windows and doors that had semicircular arches above them. For this reason this period is sometimes referred to as one of transition to the Gothic.

The works of Jordanus Nemorarius may belong to the end of this century; the most important is the *De ponderibus*, which contains several propositions in statics that inspired the new architecture, in particular the so-called concept of weight relative to the position of the body.

During the last years of the century, French architects studied a particular type of buttress called the flying buttress.

Some European innovations are known through Arab testimonies; one was the sternpost rudder, which appeared in the West during this period—around three centuries after its invention in China. The Arab traveller Ibn Jubair (1145–1217) wrote about it in 1184, while returning to Spain on a European ship that had two rudders that were "the legs on which it walked."

The Twilight of the Middle Ages

(the thirteenth to the fourteenth centuries)

1200–1250

THOUGHT AND SCIENTIFIC LIFE

Innocent III's cultural policy and the rebirth of Christian spirituality

The first half of the century opened with a theocratic experiment by Innocent III, the pope who claimed to be "placed by God above all peoples and kingdoms." But the supremacy acquired at the expense of the empire soon proved ephemeral. The crusade against the Albigensian heretics (1209–13) resulted in the annihilation of the feudality of southern France, to the complete advantage of the Capetian monarchy. Innocent's victory over Otto of Brunswick in the Battle of Bouvines (1214) marked the beginning of a historical era that saw the emergence of national monarchies and urban bourgeoisie. Even the Christian reconquest of Spain (Cordova was won back in 1236 and Seville in 1248, while the Arabs withdrew to the capital of the Moorish kingdom Garnata—now Granada) led to the rise of a new national state. The Fourth Crusade (1202–04), which took advantage of the weaknesses of the Byzantine Empire, proved more profitable to Venice than to the papacy. Nevertheless, Innocent's enterprise was immediately effective in a number of areas, especially cultural ones. Profane culture became subordinate to religious

culture, and the Church exploited the religious ferment of the times by reacting sternly to heretical tendencies. The establishment of an inquisition directed by pontifical legates who were immediately answerable to Rome is indicative. In France, mystical and eschatological themes appeared in the doctrine of Amalric of Bène, who drew upon Pseudo-Dionysios and Johannes Scotus Erigena. The University of Paris censured Amalric's theories in 1204, and, after his death in 1206, the Amalricians were subjected first to theoretical condemnation, then to physical repression. Analogous pantheistic themes ("God is everywhere"), developed by the theologian David of Dinant, were censured and finally condemned by the Fourth Lateran Council in 1215.

Franciscans and Dominicans

Other religious currents remained within the area of orthodoxy. The preachings of Francis of Assisi (c. 1182–1226) gave rise to the Franciscan order, which Innocent orally approved in 1210; written recognition came in 1223 from Pope Honorios III. By the time of its founder's death, the order was already disrupted by a conflict between the more conciliatory conventual friars and the strict spiritualists who were influenced by the prophecies of the Calabrian abbot Joachim of Fiore. While Francis' preachings were animated by the idea of a return to the poverty of primitive Christianity, the objective of Dominic

Guzman (c. 1170–1221) was to preach the Gospel among infidels and heretics. This gave rise to the Dominican order, which was recognized in 1215. A certain rivalry, even with regard to doctrines, soon arose between the two orders. The Franciscans, who were open to mystical experiences, were inspired by Augustine's Christian Platonism, while the Dominicans, heirs to the tradition of rigorous confutation of heretics, leaned toward Aristotelianism. Both Franciscans and Dominicans organized by provinces, each with its own *studium particolare*, or school for the monks. In the more important centers, *studia generalia* were established for preparing teachers for the provincial schools. These were also a part of the universities.

Translations from the Arabic and the Greek. Aristotelianism

While the Toledo center continued in its activity, the court of Frederick II in Palermo also became an extremely important translating center. As monarch, Frederick had created a centralized, absolutist state with a broad cultural policy aimed at integrating the most important ideas of Arab speculation into a strictly lay framework. The most famous translator at the court was Michael Scot (c. 1175–1235), who translated works by Aristotle, including *De coelo et mundo* and *De anima*, as well as Averroës' respective commentaries. Beginning in 1240, other works by Aristotle, including the *Nichomachean Ethics* and the *Rhetoric*, were translated in Toledo by Herman the German, who used al-Fārābī's glossaries, works by Avicenna, and Averroes' commentaries. As for translations from the Greek, new Aristotelian and Pseudoaristotelian works that completed the Greco-Latin versions of the *Metaphysics*, the *Nichomachean Ethics*, and other works that were already in circulation arrived from the East when Constantinople was conquered. The patrimony was further enlarged by the numerous Greek-to-Latin translations of works by Aristotle done by Robert Grosseteste (c. 1175–1253), bishop of Lincoln and a professor at Oxford, and his collaborators John of Basingstoke (d. 1252) and Nicholas the Greek (d. 1279). Grosseteste's own version of the *Nichomachean Ethics* was important, as were the translations of works by Ignatius of Antioch, Maximus Confessor, and Joannes Damascenus.

In Europe, Aristotle's name was often linked to Averroës', and the fact that the commentaries emphasized the gap between Aristotelian philosophy and religion created some confusion. Even though the much more conciliatory works of Moses Maimonides were also translated, the discomfort did not disappear, and as early as 1210 a provincial council banned all public and private teachings of Aristotelian physics and metaphysics in Paris. Cardinal Robert of Courçon renewed the ban, especially with regard to the University of Paris, in 1215. However, no objections were made to teaching Aristotelian logic, which had been considerably strengthened by the recovery of the entire *Organon*.

The universities. Paris

The importance of the universities continued to grow during this period, fostered by both the spirit—cultural and otherwise—of the new urban bourgeoisie, and the desire of monarchs like Frederick II and the kings of France and England to cultivate trained intellectual elites. New universities arose. The University of Cambridge joined the University of Oxford in 1209; the University of Padua was founded in 1222; and the University of Naples was created by Frederick II in 1224. The University of Paris, however, remained the most important cultural institution of the times, and all the teachings and debates that took place there had almost immediate repercussions throughout the Christian world. For this reason, the papacy made frequent attempts to control the orthodoxy of the school and to remove it from the influence of the French monarchy by conceding it special privileges. The Franciscans and Dominicans who presented themselves as an alternative to the secular teachers were also tools of this policy. The Franciscans established themselves in Paris in 1219, and in 1232 obtained a chair of theology. It was given to Alexander of Hales, who had already been a secular teacher in Paris in 1220. The *Summa theologica*, a work attributed to Alexander but probably not written by him alone, includes references to Arabic science and philosophy and shows the course that the Franciscans took when they first started teaching in Paris. Alexander was succeeded in 1238 by his Franciscan disciple John of Rochelle. Meanwhile, the Dominicans, who had settled in Paris in 1217, obtained their first chair at the university in 1229, followed by a second in 1231. There were secular teachers as well, such as the theologian William of Auvergne (1180–1249), who became bishop of Paris in 1228. He wrote such theological works as *De primo principio*, *De anima* and *De universo*, which reflect Arabic influences but also show a strong resistance to those particular developments that were in conflict with Christian doctrine. But the cautious openness of an Alexander of Hales or a William of Auvergne to Aristotle and Arabic culture got little response from the rigid intransigence of the pope. Gregory IX's letter to the theology teachers of Paris, dated July 7, 1228, is emblematic. It austerely recalled them to the tradition of the fathers, and censored them for having approached pagan culture too closely for the "spirit of blasphemous novelty." The University of Oxford fell under the control of the papacy. While the prevalent studies in Paris were dialectics and, in the midst of difficulties and suspicions, physics and metaphysics, attention in Oxford was mainly turned toward studying the *quadrivium* and, as a result, those scientific disciplines included a traditional Platonist-Augustinian perspective, often with reference to Pseudo-Dionysios as well. The scientific and philosophical work of Robert of Lincoln, mentioned earlier (a secular teacher who was the first Oxonian to teach in the local Franciscan school), figured prominently at Oxford. His doctrine of light had a considerable influence on the Franciscan teachers and, through them, on a great part of later English culture.

While Jewish, Arab, and Byzantine culture was no longer as lively as it had been in Avicenna or Averroës' time, the vitality of university studies in the Christian West was only one facet of a varied and articulated context that ranged from the great advancements made in the sciences—mathematics, and the revival of astronomical and astrological research (Grosseteste's interests are an example)—to the continued study of Roman law in Bologna and the further elaboration of canon law, solicited by the theocratic programs of the more intransigent popes, to the excitement of the first geographical explorations which produced the *Historia mongalorum* by Giovanni da Pian del Carpini (c. 1180–1252).

MATHEMATICAL SCIENCES

Trigonometry

Abū Ja'far, known as Nāsir al-Dīn at-Tūsī (1201–1274), a philosopher, astronomer, and mathematician, compiled astronomical tables of great repute and attempted to improve Euclid's theory of parallel lines. But his main contribution was a treatise on trigonometry, full of original ideas and important findings. It is to him that we owe the treatment of spherical trigonometry as a discipline independent of astronomy, as well as the systematic use of circular functions in the study of plane and spherical trigonometry.

The rebirth of European mathematics

By this time the work of translation and compilation that had been done in the preceding decades had laid the groundwork for producing original findings even in the West. The first results appeared in the *Liber abbaci* (1202) and the *Practica geometriae* (1220) by Leonardo Fibonacci, known as Leonardo of Pisa. The *Liber abbaci* marks the effective introduction of Arabic numerals—which the author had learned first-hand during his travels in the Mediterranean—in the West. It also contains a broad study of fractions and a rich series of applications of arithmetic to practical problems, especially those of a commercial nature, as well as a treatment of radicals and equations. (The first information on the Hindu numerical system dates back to an *Algoritmi de numero indorum*, translated from the Arabic around 825. Traces of Hindu numerals can also be found in the *Codex virgilianus* of 976. These only mentioned the numerals, however, without developing them.)

In the second work, Leonardo outlines rules for calculating areas and volumes, dividing figures, etc., and, indicative of the new mentality, even goes so far as to propose some improvements in the demonstrations made by the classical Greeks (Archimedes, for example). It is interesting to note how Fibonacci's arithmetical treatment, following an Arab-Hindu approach, and his geometrical treatment, after Greek models, in these two major works have a deep affinity with the spirit of Arab mathematics.

Jordanus Nemorarius (died 1237) also flourished in Europe during these years. He wrote on mechanics, geometry, and arithmetic, using alphabetical letters to indicate unknowns, foreshadowing symbolic algebra. He also developed some interesting considerations on the angle of contingence formed by a curve and a tangent of that curve.

A near contemporary was John of Holywood (c. 1200–1256), author of the *Tracatatus de sphaera mundi* (see ASTRONOMY 1200–1250) and of a short treatise, entitled *Algorithmus*, which, among other things, clearly states the rules for the principal operations of arithmetic.

ASTRONOMY

The introduction of Aristotelian cosmography and the Almagest into Europe

Latin translations of Aristotle's works, often inferior in that they were based on Arabic translations, had already begun to circulate in Europe in the second half of the twelfth century.

The phenomenon reached major proportions in the first half of the thirteenth century, partly due to the flourishing cultural and university centers in Paris, Oxford, Cambridge, Bologna, and Padua, along with Frederick II's ephemeral but fairly lively center in Naples, to mention only the most important. As for astronomy, the criticism brought against the *Almagest* in Spain by Ibn Tufail and his disciple, Alpetragius (flourished c. 1190), based on Aristotle's planetary system, became known outside Arabic culture through a translation done by Michael Scot (c. 1175–1235), a Scottish philosopher and astrologer who lived at the court of Frederick II from 1227 until his death (see THOUGHT AND SCIENTIFIC LIFE 1200–1250).

In these same years, the *Tractatus de sphaera mundi* by the Englishman John of Holywood, a short work that popularized the Ptolemaic system, became widely diffused. Born around 1200, John of Holywood studied at the University of Oxford and taught at the University of Paris from 1230. He died around 1256. The *Tractatus de sphaera mundi*, the first exposition of the *Almagest* written by a European, remained the basic treatise on elementary astronomy for almost four centuries, definitively replacing Pliny and Martianus Capella's works on the subject.

European scholars thus came into contact with both the *Almagest* and Aristotelian cosmography simultaneously and finally rose to the level of contemporary Arabic astronomy, which began to lose its position at the vanguard during these years. Compared to Alpetragius and Ibn Tufail's complicated constructions of homocentric spheres which, unlike epicyles and deferents constituted the actual material "packaging" of the planetary orbits, the simpler and more elegant Ptolemaic system resisted criticism and became the only acceptable model for Western astronomers for many centuries. It was the only system that faithfully represented planetary motion from a mathematical point of view, while Aristotelian cosmology, which had been harmonized with Christian theology, constituted the philosophical substratum.

The Church, meanwhile, took a defensive attitude toward the introduction of Aristotelian theories, partly because Pseudo-Neoplatonist and mystical theories were often smuggled in under their guise. In 1210, the provincial council of Paris prohibited the public or private reading of Aristotle's books on natural philosophy, as well as the commentaries on them. The same measures were taken at the University of Paris in 1215. We should not, however, underestimate the objective difficulties that the Church encountered in its attempt to reconcile some of the statements on astronomy found in the Holy Scriptures with the reawakening need for scientific verification. A new synthesis that could overcome the contradictions was needed; it was provided by Thomas Aquinas (c. 1225–74), who was in Cologne at the end of this first half of the thirteenth century in the retinue of Albert the Great (1193–1280). Albert himself, a man of remarkable culture and openness of spirit, was already working in that direction. Between 1245 and 1260 he commented on Aristotle's major works, attempting to integrate them with the Arabic commentaries written by Avicenna, al-Fārābī, and Maimonides.

Natural philosophy and astronomy. Robert Grosseteste

The Englishman Robert Grosseteste (c. 1175–1253), bishop of Lincoln and the first chancellor of the University of Oxford, a philosopher of nature best known for his optical theories, wrote works on astronomy which contain evidence of the first original European reawakening in this discipline. Unlike Aristotle, Grosseteste maintained in his *De generatione stellarum* that the stars were made of the four earthly elements. In his *De cometis et causis ipsarum*, he stated a curious theory, which is interesting in that it extended the laws of earthly phenomena to celestial phenomena. According to Grosseteste, comets are "sublimated fire" attracted by a force inherent in the stars, just as iron is attracted by a magnet and loses its earthly nature. *De sphaera* dates back to between 1215 and 1230, and *De motu supercoelestium* to after 1230. Both are bastardizations of Aristotle and Ptolemy. *De impressionibus aeris seu de prognosticatione*, which talks about tides and astrological influences, among other things, was probably written in 1249. Grosseteste's cosmological system, based on Ptolemy and Aristotle, is interesting because of the unifying force that it had on the theory of light. The universe is seen as a finite sphere within which light, the origin of corporeal matter itself, is reflected from the edges to the center, where the Earth is, and generates the nine celestial spheres and the four spheres of the prime natural elements along its way by a process of "rarefaction" and "condensation."

In a series of treatises—*Canon in kalendarium, Compotus, Compotus correctorius* and *Compotus minor*—Grosseteste pointed out several discrepancies between the calendar and observations of the full moons, the solstices, and the equinoxes. He proposed replacing Hipparchos and Ptolemy's measurements of the solar year, accepted by the Latins, with Albategnus's calculations. He also upheld the need to use accurate calculations to correct the errors in determining Easter. Grosseteste's project for reforming the calendar was based on three main points: making a more accurate calculation of the solar year, calculating the relationship between the solar year and the average lunar month, and using the findings of the first two points to more accurately determine Easter. Roger Bacon (c. 1214–94) carried out the project almost to the letter.

PHYSICAL SCIENCES

The problem of the cause of motion

The new century saw a surprising growth in the West of scientific research. In part it was still tied to general philosphical concepts, but in part it was now directed toward more specific topics. Western science continued to draw upon Arabic science, but would soon surpass it.

A typical example of the phenomenon was the transfer to the West of the Arab scholars' lively discussions on the role in motion of the medium (usually air) in which bodies move. Western "physicists," some of whom (Albert the Great, for example) resolutely took sides with Averroës's theory, and others, no less reputable (Thomas Aquinas, Roger Bacon, etc.), who enthusiastically defended Avempace's theory, resumed these discussions and elaborated them quite thoroughly. The investigation of Avempace and Philoponos' ideas that was begun during these years gave rise, in the fourteenth century, to the systematic elaboration of the so-called "theory of impetus," one of the major accomplishments of the Parisian school of physics.

Kinematics

Another important innovation introduced by Western science was the study of motion from a purely kinematic point of view, that is, without reference to its causes. The first treatise on the subject was the *Liber de motu*, written at an unknown date by the best-known mathematician of the first half of the century, Gerard of Brussels, who took up the heritage of Euclid's and Archimedes' kinematic ideas. It is significant that Gerard used the term *motus* in the same way that the term *velocitas* was later used. His ideas were revived in the fourteenth century, along with those of Avempace's Latin advocates.

Statics

In addition to dynamics and kinematics, statics also made considerable progress with the Latin scientists of this period. Again, this was made possible by the many suggestions that came from reading the Latin versions of Arabic and Greek works (usually retranslated from the Arabic).

The predominant figure in this field was the mathematician Jordanus Nemorarius, identified by some scholars with Jordanus of Saxony, general of the Dominican order from 1222 to 1237. He wrote several widely-read works that elucidated the important concept of *gravitas secundum situm* (i.e. the components of a force in a closed system), and made very intelligent use of the principle of virtual velocities. Some historians of statics, Pierre Duhem in particular, maintain that Nemorarius' writings had a considerable influence on Leonardo da Vinci.

CHEMICAL SCIENCES

The great encyclopedists

Even alchemy profited from the notable surge in European culture that took place in the course of the thirteenth century.

Contacts with the Arab world and the East in general intensified with the Crusades, and Marco Polo's journey definitively summed up and characterized this movement. Great encyclopedists, who gathered, elaborated, and compiled fairly complete data on the scientific knowledge of the times, appeared. These men were basically Aristotelians (even though Aristotle's natural philosophy had been banished) and usually expressed their ideas on the composition of metals in terms of the mercury-sulphur theory. On a theoretical level, these scholars still believed in the transmutation of metals, but they did not indulge in mystical speculations, foolish fancies, or allegories.

The first of these encyclopedists was Bartholomeus Anglicus, a disciple of Robert Grosseteste (c. 1175–1253). Around 1240 he wrote the *Liber de proprietatibus rerum*, a very successful book that continued to be used in the following centuries, since it tended to summarize all previous works.

Another Englishman, Richard Anglicus, who probably died around 1252, was the author of a *Correctorium alchymiae*. This work divided minerals into two classes: metals that originated from mercury (gold, silver, copper, iron, tin, and lead) on one hand, and those that originated from sulphur, viscous earth, and heat, along with the salts such as alums, vitriols, and orpiment, which had the same origins on the other.

LIFE SCIENCES

The Montpellier school

The last great masters of Salerno flourished during this period. Roland of Parma, author of a *Surgery* that was studied and commented on in other centers of medical studies, was one of the Salernitans who moved to Bologna and created a nucleus that in part converged in the university faculty. At the same time, others whom we have already mentioned followed the talented surgeon Roger of Frugardo to Montpellier.

The Languedoc city, which in 1204 passed from the sovereignty of the pontiff to the kings of Aragon and Majorca, saw the offical rebirth of the faculty of medicine with a papal bull, issued by Honofrius III, in 1220. Jewish physicians exiled from Spain flocked to Montpellier along with Spanish and English physicians. The environment was favored by an intelligent tolerance that allowed them to share their diverse experiences.

Frederick II and the growth of the medical sciences

Salerno, on the other hand, remained detached from the university that was founded in Naples in 1240 by Frederick II. But the same royal edict that created the university also entitled the Salerno school to issue licenses for the practice of medicine and authorized the dissection of cadavers. The Hohenstaufen ruler, who had a keen interest in naturalistic studies himself, displayed a spirit of observation unusual for the times in his *De arte venandi cum avibus*, a complete treatise on falconry.

Meanwhile, the life sciences found meaningful expression in the works of the Dominicans Albert of Bollstädt (Albertus Magnus, or Albert the Great, 1193–1280), Thomas of Cantimpré (c. 1204–1271) and Vincent of Beauvais (c. 1190–1264). Albert's teachings in Cologne and Paris kept alive questions on nature in general, and on phenomena and forms of life in particular. He dedicated his attention to zoology, botany, geography, astronomy, mineralogy, alchemy, and medicine, and wrote *De coelo et mundo* and *De natura*. Vincent of Beauvais' *Speculum naturae* is important for the variety of ancient and Arab sources it lists.

EARTH SCIENCES

The reawakening of naturalistic interests in Western culture

From the earliest times, human beings have been guided in many of their decisions by an empirical knowledge of geology, for instance in searching for the most suitable caves in which to live, or the best stones for making axes, lance tips, and arrowheads (flint, obsidian, jadeite, and pyrophyllite), and later for the colored clays with which to paint the splendid grotto pictures of the Upper Paleolithic. Mining activities and the great stone constructions also required skillful geognostic prospecting.

Indian and Middle-Eastern religious dealings touched upon questions of cosmology and terrestrial chronology as early as the second millennium B.C. In the Egypt of the Pharaohs, problems

of inundations, floods, the changing boundaries between the earth and the sea, and earthquakes were considered. The Assyrians remembered the flood as clearly as the Jews did.

In the areas of philosophical speculation and practical applications, and under the influence of geographical explorations, Greek and Roman civilization developed a knowledge of geology that was anything but negligible. On the basis of their observations of fossils, they hypothesized transgressions and regressions of the sea (Anaximander, Xenophon, Xanthos of Lydia, Herodotus, Xenocrates, Eudoxos, Ovid, and Tertullian); the earth was recognized as being spherical (Aristotle, Theophrastos, and Aristarchos of Samos), its size was estimated (Dicaearchos, Eratosthenes, Eudoxos of Cnidos, and Posidonios) and its climatic zones were outlined (Parmenides). Even the idea that the earth revolved around the sun found some supporters, such as Aristarchos of Samos, while Claudius Ptolemy codified the geocentric system. Earthquakes, volcanic eruptions, and tides were described and interpreted (Phytheas linked the tides with the position of the moon.) Among those who studied volcanic phenomena were two illustrious victims of scientific research, Empedocles of Agrigento, who died by falling into the crater of Etna in 433 B.C., and Pliny the Elder, who was suffocated by the vapors of the eruption of Vesuvius that destroyed Pompei in A.D. 79.

While Ptolemy's work in astronomy and geography left the legacy of the geocentric solar system, along with a vast amount of information, classical culture passed down another scientific error in the field of geology. Fossils, correctly interpreted as the remains of ancient organisms by many scholars, including Xanthos of Lydia, Herodotus, Eudoxos of Cnidos, and, later, Titus Livy, Gaius Svetonius Tranquillus, and Basil of Cesarea, were considered jokes of nature by Aristotle, Theophrastos, and Pomponius Mela.

Hellenistic scientific knowledge was transmitted to the West by way of the Arabs (Avicenna in particular), along with the false interpretations, which gave rise to theological, philosophical and scientific controversies until the birth of modern science. The decline of culture in the West also brought about a stagnation in the sciences. The last descendents of Roman scientific thought flourished in the seventh century in the Visigoth kingdom in Spain (Isidore of Seville) and among the Irish monks (Venerable Bede). (For information on the Greek and Roman authors mentioned here, see the relative sections on MATHEMATICAL SCIENCES, ASTRONOMY, and PHYSICAL SCIENCES, as well as THOUGHT AND SCIENTIFIC LIFE).

Arab expansion, the Islamic conquests in Europe, and the Christian reconquest reestablished the ties between ancient Greek thought and Western culture, which began its reawakening around the beginning of the twelfth century, when the first systematic treatments of both theoretical and practical scientific problems appeared.

Early interest in problems in the earth sciences

Beginning in the middle of the century, Western thinkers began to consider naturalistic questions and, in the field of the earth sciences, to study minerals and fossils. The writings of Alexander Neckam, who became a professor in Paris in 1213, mention many minerals and their properties.

Among the major works of the century were *De meteoris* and

De mineralibus et rebus metallicis (1260) by Albert the Great (1193–1280), who seemed to recognize the true nature of meteorites, and the writings of his disciple, Thomas Aquinas (1225–74). *De proprietatibus rerum*, by Bartholomeus Anglicus, was written in 1240, and shortly afterwards, Vincent of Beauvais (c. 1190–1264) wrote his *Speculum mundi*. Between 1220 and 1230, Arnold of Saxony wrote *De finibus rerum naturalium* and Thomas of Cantimpré wrote *De rerum natura*. Alfonso X of Castile (1221–84) had his *Lapidarius* translated into Spanish from the Arabic.

Journeys to the East

In 1245, Giovanni da Pian del Carpini (c. 1180–1252) travelled to the Mongol court at Karakorum. The events of his journey are described in his *Historia mongalorum*.

Oceanographic knowledge

Robert Grosseteste (c. 1175–1253), bishop of Lincoln, drawing in part upon ideas held by other thinkers of his time, hypothesized that the moon's influence on the tides was due to the light coming from the moon itself, which supposedly caused the water to swell and expand, mostly upward. Roger Bacon (c. 1214–94) also accepted this theory.

APPLIED SCIENCES

Agriculture

Cultivation began to be organized more rationally in the West. Spring wheat was introduced into the agricultural system and became widespread. Triennial rotation was instituted, so that the land lay fallow one year out of three. Quadrennial rotation was adopted during the following century.

Following the advancements made by cultivating barren land and forest areas came a phase of intensification, where attempts were made to reduce fallow land to a minimum. The first signs of quadrennial rotation appeared. Agricultural yields increased, reaching three to four times the amount of seed planted. This was still a modest result, although harvests in France were an exception.

Thirteenth-century English agriculturalists gave the following yields as normal: eight times for barley, seven for rye, six for legumes, five for wheat, and four for oats. In his treatise Walter of Henley wrote that land that did not yield more than three times the seed planted yielded nothing, an evaluation that attests to the progress made since the Carolingian period. It seems that methods of fertilization had not improved much, however. For example, a thirteenth-century rent contract drawn up near Paris required that the tenant fertilize the land "only once in nine years, during the fifth year." Surplus fertilizer was generally used on gardens, where the yields were more rewarding.

The progress made was due to increased diligence, improved tools for tilling, and the common use of horses for plowing. An agricultural literature that circulated beyond intellectual circles arose. Various manuals on 'housebondrie' (domestic economy), and the Fleta collection, another English work that describes a modern farm, deserve mention.

Metallurgy

Important advances in metallurgy were made during this period. The depth of furnaces for manufacturing iron was increased from three to four meters, and a special hole for removing the slag was built directly into the furnaces. Drafting was improved with the use of blowing machines operated by water power. This new drafting system, along with the use of more carefully selected carbon, made it possible to obtain higher temperatures and slag with less iron in it. A part of the iron contained enough carbon to fuse. Attempts to increase the carbon content and produce pig-iron began during this period, although they only became successful on a large scale in the fifteenth century.

Mechanical devices

A book of drawings made in 1235 by the French architect Villard de Honnecourt contains the first sketch of a hydraulically operated mechanical saw with spring-action return. However, the attribution of this invention to the architect is questionable, since Villard assembled his drawings after a long period of travelling all over the world.

The most important innovation of the period was the crankshaft, which made it possible to transform reciprocating motion into rotary motion. The earliest European references to the use of the pedal in weaving date back to the beginning of this century.

Technology and science

An original technical-scientific culture started to develop in the West. The first contributions were made by R. Grosseteste (c. 1175–1253), who developed the theory of the double refraction of light through a spherical lens, and was also interested in the practical use of lenses for enlarging small objects and bringing far-away ones closer.

The use of the flying buttress made it possible for Gothic architecture to develop fully in France. The cathedral of Amiens, with its forty-three-meter-high vault, stands as an example. Flying buttresses discharged the thrust of the central vault onto the ground, by way of the aisles. By placing buttresses in correspondence with each pillar of the central nave, the walls between the buttresses could be decorated with windows, since masonry was reduced to a minimum.

The most interesting example of Gothic architecture in England is the cathedral of Salisbury, built between 1237 and 1258. The stone covering was lightened by the use of ribs and a thin covering filling the spaces between them, while the downward thrust was discharged onto a series of columns and buttresses rather than onto the stone walls.

Fuels

During the thirteenth century, coal was picked up along the English coasts and carried to London on small barges. It burned so imperfectly that the smoke it released caused a number of people to die. The situation was so bad that Edward I (1239–1307) imposed the death penalty for anyone who dared to burn it. Coal only came back into use at the end of the fifteenth century, when refractory materials began to be used to build better fireplaces.

1250–1300

THOUGHT AND SCIENTIFIC LIFE

Religion and culture in the second half of the thirteenth century

In some ways, though, it was a glorious period: poetry, both scholarly and vernacular, was revived (this was the era of the "Dolce stil nuovo"), the Italian Gothic flourished, and the cathedrals of Orvieto, Strasbourg, Amiens, and Salisbury were constructed. An interest in making journeys and geographical expeditions developed. From 1261 to 1266, Brunetto Latini wrote his famous *Tresor*, a mandatory reference for the encyclopedic culture of the times, and in 1269 Peter Peregrine composed his first theoretical treatise on magnetism and the compass. Marco Polo undertook his great journeys into Asia between 1271 and 1292.

Aristotelianism and Averroism

The events that took place in the Islamic and Byzantine worlds directly influenced the diffusion of Greco-Arabic culture in the West. Many Arabic works were still being translated into Latin, but more important was the increase in the number of translations being done directly from the Greek texts. Bartholomew of Messina, who between 1258 and 1266 translated the *Magna moralia* and various Pseudo-Aristotelian texts, lived at the court of Manfred, King of Sicily. Elswhere, Aristotle's works on the natural sciences were being revised: Pope Gregory IX had named a commission especially for that purpose in 1231, and the Flemish-born Dominican William of Moerbeke (c. 1215–86) began translating from the Greek in 1260. Translations of the *Rhetoric*, the *Politics*, *De generatione animalium* and the *Meteorology* are attributed to him. While the translation of the *Politics* opened new perspectives to political theorizing in a period when states such as France and England were consolidating their institutions, the naturalistic works provided a model for scientists, researchers, and scholars. Besides the Aristotelian texts, William also translated works by Archimedes, Galen, Alexander of Aphrodisias, Themistos, Simplicios, and Proclos. With the addition of William's translations, almost all of Aristotle's writings were known, some directly, others through the Greek and Arabic commentaries. In 1252, the faculty of arts in Paris authorized the study of Aristotelian psychology, while three years later the study of all of Aristotle's works was permitted. As a result of the enthusiasm for Aristotle, considered the master *par excellence*, difficult texts such as the *Physics* and the *Metaphysics* were read, often through Averroës's commentaries.

Franciscans and Dominicans. Bonaventure, Albert the Great, and Thomas Aquinas

While the Dominicans began to look upon Aristotelian doctrines with more sympathy, the Franciscans remained more resolutely in the rut of Augustinian tradition. The Franciscan teachers in Paris (disliked as much by the secular clergy as by their Dominican rivals, as the polemics at the University of Paris around 1252 indicate) were at the forefront in the attacks against the teachers in the faculty of arts.

The Franciscan Bonaventure of Bagnoregio (c. 1217–74), who was a disciple of Alexander of Hales and taught theology in Paris from 1248 to 1255, was among those who avidly contested the "errors" of pagan philosophy and fit the Aristotelian themes that were most in line with Christian theology into traditional schemes. He became general of the Franciscans in 1256, and began teaching again in 1257 after the pope definitively sanctioned the right of regulars to occupy chairs in Paris.

But precisely while Bonaventure was writing his most famous works, including his celebrated *Itinerarium mentis in Deum* (1259), the Dominican teachers were elaborating a grandiose cultural program. While the secular professors, under direct pressure from the pope, were attempting, without success, to compile a purged edition of Aristotle's works, the Dominicans resolved to eliminate all traces of pantheism from the Aristotelian doctrine, and to replace the limited speculative groundwork provided by the Augustinian tradition with the much more articulated Aristotelian system as the new bastion of the Christian doctrine against pagans and heretics. Albert of Bollstädt (1193–1280) known as Albert the Great to his contemporaries, and Thomas Aquinas (1225–74) were prominent figures in this work of "Christianizing" Aristotle. Between 1245 and 1260, Albert wrote several original scientific works and an ample commentary on the Aristotelian *Corpus*, which contains traces of al-Fārābī, Avicenna, Moses Maimonides, and Averroës, as well as themes focusing on the Dionysian *Corpus* and the *Liber de causis* that were dear to the Platonizing speculation of early Parisian scholasticism. Thomas Aquinas, who was appointed to teach in Paris in 1259, later taught in Italy, again in Paris (1269–72), and finally at the University of Naples. He seems to have had less of an interest in scientific subjects than Albert, but dedicated himself to constructing a much more organic and rigorous theological and philosophical system. In collaboration with William of Moerbeke, Thomas commented on a large part of the Aristotelian *Corpus*, as well as on works by Proclos and the Dionysian *Corpus*. Although his commenrty *Super IV libros sententiarum* (1253) was very popular, he soon gained fame as a theologian with his *Summa contra gentiles* (1259–64). The work was written to demonstrate the truth of the Christian faith to the Moslems, who did not accept the authority of the revelation and thus required a persuasion that appealed to reason. This in turn created a need to define the boundaries of natural reason rigorously, a philosophical battle that Thomas conducted on several fronts. The years (1265–73) of the unfinished *Summa theologica* were the same years in which the debate was raging with the Averroists on the question of the unity of the intellect.

The faculty of arts. The Latin Averroists

The extended knowledge of Aristotle's work and its progressive absorption into Christian speculation appreciably influenced the organization of university studies. Teaching took place through lessons and debates. The lessons consisted in the reading (*lectio*) of a text, which in the faculty of arts was one of Aristotle's works, and in the faculty of theology was the *Bible* or Peter Lombard's *Libri sententiarum*. The teachers expressed their views in commentaries on the texts, while the debates consisted in the collective critical examination of various questions (*quaestio*) by a number of teachers who sometimes disagreed with one another. Within this institutional framework, the faculty of theology, which occupied the highest position in the ordering of studies, did not directly participate in the examination of questions in physics, metaphysics or morals, but dealt with more general topics, such as the relationship between reason and faith, philosophy and theology, and science and religion. The faculty of arts approached the study of philosophical and scientific disciplines autonomously and with specific aims.

As knowledge of Aristotle's works spread, a vast range of physical and metaphysical problems opened up, and teachers no longer limited their attention to dialectics alone. The Parisian *artistae* also claimed the right to practice philosophy freely. Philosophical commentaries on the works of Aristotle, Averroës, and other Arab authors had to be written keeping to the *littera* of the philosophers, while doctrines that seemed to be in contrast to the faith (such as the affirmation of the eternity of the world, the influence of the stars, the denial of providence, and the unity of the intellect) were presented in philosophy without any attempts at reconciling the philosophers' opinions with the truth of the dogma. The Latin theory of "dual truths" had its roots in this distinction of spheres. Positions where a preoccupation with strengthening philosophy prevailed at the expense of theology were not lacking. In fact, some *artistae* maintained that the only wise persons were the philosophers.

As early as 1267, Bonaventure attacked the Averroists who had infiltrated into the faculty of arts in Paris. The next year, Stephen Tempier, bishop of Paris, explicitly condemned fifteen theories, most of them inspired by Averroism. That same year, the Dominican Giles of Lessines wrote to Albert the Great, inviting him to confute fifteen theories taught "by the most eminent teachers of the Parisian schools." Albert, who had already written against Averroës' "theological errors," responded with his *De quindecim problematibus*. Almost simultaneously, Thomas Aquinas entered the scene with his pamphlet *De unitate intellectus contra averroistas*, which reflects his desire to make his position distinct from that of the Latin Averroists. On January 18, 1277, Peter of Spain (1210–77), Pope John XXI, issued an admonition to the Parisian teachers of the arts. On March 7 of the same year, Stephen Tempier issued a decree that prohibited and condemned 219 propositions, with an express interdiction against supporting them even as true doctrines from the point of view of reason but false from the point of view of faith. The theses condemned by the episcopal decree were of quite diverse origins: the most characteristic of Avicenna and Averroës' concepts, as well as several Thomist propositions, appeared together with Andrea of Cappellano's doctrines of courtly love. The exponents of Latin Averroism, such as Siger of Brabant (c. 1240–1284), Boëthius of Dacia, and Bernier of Nivelles were the hardest hit by the decree, although it even compromised Thomas's efforts.

Oxford: the disciples of Robert Grosseteste and Roger Bacon

The University of Oxford, a center of intense religious life dominated by the influence of the Franciscans, remained somewhat outside the disputes between the Augustinians and the Aristotelians, and became the center of a philosophical current that felt the effects of theological problems much less, and instead promoted the advancement of the arts of the *quadrivium*, especially Robert Grosseteste's *Metaphysica de lumine*. Grosse-

teste's teachings were directly carried on by his disciples Adam Marsh (died c. 1258) and Richard Rufus of Cornwall (died c. 1260). Thomas of York (died c. 1260), the author of a *Sapientiale* that contains almost all of Grosseteste's themes, was also an important exponent of this current, although the most prominent was Roger Bacon (c. 1214–94), one of Adam's disciples. Despite fierce opposition from his adversaries, Bacon, who was deeply convinced of the "operative" value of knowledge, outlined a broad program of philosophical, scientific and religious *renovatio* (especially in his three great works, the *Opus majus*, the *Opus minus*, and the *Opus tertium*). When his protector, Cardinal Guy Fucoldi (Pope Clement IV from 1265 to 1268) died, charges of heterodoxy were brought against Bacon. Tempier's condemnation of the doctrine of astrological determinism in 1277 furnished a pretext for condemning Bacon as well. Jailed in 1278, he was only released around 1292 and died shortly thereafter.

The Dominicans of Oxford also felt the effects of the condemnation of 1277. One of these was Robert Kilwardby (bishop of Lincoln when he died in 1279), who was also a commentator of Aristotle, although he was far from the declared Aristotelianism of Albert the Great and Thomas. In 1277, in the wake of the condemnation of Paris, Kilwardby in turn condemned the Averroist and Thomist texts. On October 29, 1284, the Franciscan John Peckham (died 1292), Kilwardby's successor in the bishopric, reiterated the condemnation of Aristotelian (and Thomist) ideas.

Thomists and anti-Thomists. Developments in political reflection

The end of the century saw an increase in the eschatological tendencies of Franciscan mysticism, which had already appeared with Bacon, and the flourishing of a Pseudo-Joachimite literature, which influenced the theologian Peter John Olivi, leader of the spiritual Franciscans, as well as the poet Jacopone da Todi and the mystic Ubertino of Casale. Both the Franciscan Matthew of Acquasparta (c. 1238–1302), a disciple of Bonaventure, and the secular teacher Henry of Ghent (c. 1217–1293), championed a return to Augustinianism. In France and Italy, numerous Dominican teachers, such as Bernard of Alvernia and John of Naples and with more originality, the secular teacher Godfrey of Fontaines and the Augustinian Giles of Rome (c. 1243–1316), who often disagreed with Henry of Ghent—defended Thomist theories, helping to make the doctrine of the *Doctor angelicus* the official doctrine of the order.

The polemic that involved Thomists and anti-Thomists also involved the political world. Thomas had put forth his mediative theses in his *De regimine principum*, completed after his death by his disciple Bartholomew (Tolomeo) of Lucca. A work with an analogous title was written in 1277–79 by Giles of Rome, who also wrote the famous *De ecclesiastica potestate* in defense of the supremacy of the papacy in 1302. But almost simultaneously, the French controversialist Pierre Dubois (c. 1250–1321) cleverly defended the sovereign Philip the Fair's policies, theorizing the right of the royal authority to intervene in matters of the Church. Along with Dubois, who was a disciple of Siger's lessons on Aristotle's *Politics*, were other French controversialists—the so-called royalists—who adapted the concept of the monarch as *legibus solutus*—which was the ideological foundation of Hohenstaufen policies for the jurists at the University of Bologna—to the needs of Philip's policies.

MATHEMATICAL SCIENCES

Campanus of Novara

The work of translating from the Arabic did not stop when the initial enthusiasm had worn off. During this period, Campanus of Novara (c. 1220–1296) translated the *Elements* into Latin, adding a very important commentary that made his the most popular version of Euclid's work until the Renaissance. Among other things, the commentary contained some considerations on the starred pentagon and the angle of contingency that, due to the widespread diffusion of Campanus's work, were much discussed in the following centuries.

The Arabic numerals and zero were also introduced into the Eastern Empire during this period, in the *Arithmetic according to the Hindus,* written by the Byzantine monk Maximos Planudes (c. 1260–1310).

ASTRONOMY

The Alfonsine Tables

Arabic astronomy on the Iberian peninsula was definitively abandoned and passed into the hands of the Christian king of Castile, Alfonso X, known as "el Sabio" (the Learned). Following the example of the caliphs, Alfonso established a center for astronomical studies in his court. His reign lasted for three decades, from 1252 to 1284. For some time, it had been obvious that new planetary tables were needed. The Ptolemaic system, on the other hand, with its complicated mechanisms, must have aroused some doubts in the king's mind. Legend has it that he said "If the Lord almighty had consulted me before beginning the creation, I would have recommended something simpler." In this spirit, the *Alfonsine Tables* were drawn up by Arab and Hebrew astronomers, including Ishak ben Said, known as Hasan, and Yehuda ben Mosheh Cohen, who worked for Alfonso in Toledo. The tables were finished in 1252, the year Alfonso ascended the throne of Castile. Although they were only published for the first time in 1483, a large number of handprinted copies, some of which are still extant, circulated much earlier. The *Alfonsine Tables* were considered the best available until the sixteenth century, supplanting the earlier *Toledan Tables.* Besides this initiative, Alfonso's circle of scholars also produced a number of treatises on astronomy. Twenty-six codes were later collected in the *Libros del saber de astronomia del rey Alfonso X de Castella,* published between 1863 and 1867. It is a rather mixed encyclopedia that added very little to the modest levels that astronomy had reached in the course of the thirteenth century.

The work describes the Ptolemaic system, with the variation of extending the theory on Mercury to the other planets. The theory locates the center of the equant half way between the earth and the center of the deferent, disregarding, however, the motion of the deferent along a small circular orbit. There is also an apparently anomalous drawing, which actually seems to anticipate Kepler's first law, of Mercury's orbit in the shape of a curve similar to an ellipse, with the sun apparently at the center. In reality, it is a sketch of the composition of the two circular motions described in the Ptolemaic theory on Mercury, the smaller being the motion of the center of the deferent and the larger the motion of the center of the epicycle along the deferent. The object at the center of the ellipse, which appears to be the sun, is the center of the first of the two motions and was added several centuries later.

Nasir al-Dīn at-Tūsī and the Observatory of Marāgha

Meanwhile, in Persia in 1258, the Mongol conqueror Hulagu Khan defeated the caliph of Baghdad. In 1259, following the advice of his vizier, Nasir al-Dīn at-Tūsī (1201–74), Hulagu Khan founded a great observatory at Marāgha, in northwest Persia, now Azerbaijan. It contained a 4.3 meter wall quadrant, armillary spheres, an azimuth circle, etc. The accuracy of the observations made there was only surpassed by the astronomers of the Samarkand observatory in the fifteenth century. A set of tables called *The Tables of Ilkhān,* or *Zij-Ilkhani,* which became well-known in Europe, were compiled on the basis of the *Marāsha* astronomers' observations.

One of al-Tūsī's colleagues at the Observatory of Marāgha, Najm al-Dīn 'Alī al-Katībī al-Qazwīnī (died 1277), wrote a treatise on philosophy in which he mentions "several philosophers"—he does not specify whether they were Greeks or Arabs—who believed in the daily motion of the earth. He considered the hypothesis indefensible, however, since all earthly motion was rectilinear and, consequently, the earth could not move in a circle. Another probable colleague of al-Tūsī's at Marāgha was Yūhannā Abū'l-Faraj (c. 1226–1286), known as Bar-Hebraeus (from Bar Ebraya—son of the Jew), who in 1279 wrote *The Ascension of the Spirit,* a course in astronomy written in Syrian that supports the theory of the material spheres. A similar system was upheld by Zakāriyā' ibn Muhammad ibn Mahmūd al-Kazwīnī (c. 1203–1283), author of a famous *Cosmography* (1275), and by Mahmūd ibn Muhammad ibn 'Umar al-Jagmīnī, author of a compendium on astronomy, about whom almost nothing certain is known, except that he flourished around 1300. In 1262, the Persian philosopher, physicist and physician Qutb al-Dīn al-Shīarzī (1236–1311) came to the Observatory of Marāgha and became al-Tūsī's best disciple. Among his works on astronomy are *Nihayat al-idrāk fī dirājat al-aflak (The Limit of Understanding of the Knowledge of the Heavens),* written in four books in 1281, *Ikhtiyārāt-i muzaffarī (Muazffari Heavens),* written in four books in 1281, *Ikhtiyārāt-i muzaffarī (Muzaffari Selections),* a logical summary of his previous work, written around 1304, and *al-Tuhfat al-shāhiyya fī'l-hay'a (The Royal Gift on Astronomy),* which mainly deals with astronomical mathematics and dates back to 1284. He also participated in collecting the observational material that was used to compile the *Ilkhane Tables.* Al-Shīrazī's corrections of the Ptolemaic system are technically quite interesting, and are based on the corrections suggested by his teacher, al-Tūsī.

Thomas Aquinas. The relationship between revelation and reason

During his stay in Italy, from about 1259 to 1268, Thomas Aquinas (1225–1274) studied the works of the Stagirite and commented on *De coelo.* He explicitly stated that although revelation was certainly one of the highest sources of knowledge, it was not opposed to human reason, which had to follow a different path to arrive at truth. In fact, revelation and reason came from the same source, although they were separate and distinct. His exposition of Aristotelian cosmography is accompanied by quotes that show Thomas's familiarity with the works of Simplicios, Plato, Ptolemy, and others.

During the same years, the mathematician and astronomer Campanus of Novara (c. 1220—1296) wrote a *Tractatus de sphaera* and a *Theoria planetarum,* a loose version of the *Almagest.*

Roger Bacon: science as experiment

In his *Opus majus,* Roger Bacon (c. 1214–94) follows Ptolemy and cites a considerable amount of data derived from Alfraganus's observations on the dimensions of the earth as compared to the fixed stars, the smallest of which was supposedly eighteen times larger than the earth. Contrary to scholastic tradition, he emphasized the serious scientific errors contained in the Holy Scriptures, such as in the episode of Joshua in the first book of Genesis, and in Jerome's statement according to which there should be twenty-two stars in Orion, nine of the third magnitude, nine of the fourth and the rest of the fifth magnitude, which is in disagreement with the seventh book of the *Almagest.* We should also mention that Bacon's information on the size of the ocean between Europe and Asia was indirectly quoted by Christopher Columbus in one of his letters from Hispaniola to the rulers of Spain in 1498.

PHYSICAL SCIENCES

Studies in optics

In the Latin world of the Middle Ages as in the Arab world before, the two branches of physics that aroused most interest were mechanics (with its three branches: dynamics, kinematics, and statics) and optics. The most important general contributions to optics were made by Robert Grosseteste (c. 1175–1253), bishop of Lincoln, who attempted to link all natural phenomena to a single principle, i.e. light, and John Peckham (c. 1220–92), archbishop of Canterbury, who indicated optics as the science *par excellence,* capable of revealing the highest mathematical and philosophical truths.

The major contributions of an empirical nature to this discipline were made by Witelo (born c. 1220), a scholar of Polish origin about whom very little is known, by Roger Bacon (c. 1214–94), an Englishman like Grosseteste and Peckham, who, among other things, conducted interesting research on lenses, with the same empirical-practical approach that he used in all the sciences, and by the German Theodoric of Freiberg, who recognized that the phenomenon of rainbows was due to the refraction of the sun's rays in drops of rain. Grosseteste's influence, as well as that of Ibn al-Haytham (see PHYSICAL SCIENCES 900–1000) can be seen in all three. Witelo's *De perspectiva* (one of the most widely-read books on optics of the entire Middle Ages) is particularly important, and was studied and discussed extensively during the Renaissance, up until Kepler's time (see PHYSICAL SCIENCES 1610–1620).

Magnetism

At this point, another group of physical phenomena that had aroused keen interest ever since antiquity—the phenomena caused by magnetic attraction—deserves brief mention. The famous *Epistola de magnete,* by Petrus Pereqrinus, or Peter Peregrine (Pierre de Maricourt), was written in 1269. The work marks an authentic turn in this area of research, both because its findings and the way in which the material is treated (using a method based essentially on observation and "a skillful use of the hands"). Peter Peregrine also anticipated a number of ideas that were later expounded by William Gilbert, at the end of the sixteenth century. The *Epistola de magnete* is one of the most important scientific treatises of the Middle Ages. At least thirty hand-written copies are known to exist, indicating that it must have had an enormous influence on successive developments in the theory of magnetism.

CHEMICAL SCIENCES

Vincent of Beauvais and Albert of Bollstädt

Another encyclopedist of this century was Vincent of Beauvais (c. 1190–1264), a Dominican and author of the *Speculum majus.* He is fairly important, though more for his accurate citation of a large number of authors than for his personal opinions. Particular attention is given to the alchemistic part of the *Speculum majus* in the *Speculum naturae,* written around 1250, which cites more than 300 authors, many of whom are not available to us elsewhere. Among the passages quoted are some from *De aluminibus et salibus,* a text that is attributed to al-Rāzī but was certainly written at a later date. The work does not display a particularly critical spirit: Vincent of Beauvais indiscriminately speaks of salts as products of the evaporation of water and of the burning of plants or bones, but in compensation, he seems fairly sure of the possibility of transmutating metals.

Less rich is the work of Albert of Bollstädt, better-known as Albert the Great (1193–1280), a philosopher and theologian (Thomas Aquinas' teacher) who worked in alchemy, although he had no practical experience. His five-volume work *De rebus metallicis et mineralibus* is especially interesting from an alchemistic point of view. Rather than citing other authors directly as Vincent of Beauvais did, Albert reelaborated the texts he used, contributing his own thoughts, if not his practical experience. Albert the Great also accepted the concept of transmutation, but was quite skeptical of those who claimed to have achieved it. A *Libellus de alchemia* is also attributed to him, although it mentions Arnald of Villanova (c. 1235–1313), who came after him, and Philip Ulstad, a sixteenth century alchemist. Thus the attribution is rather doubtful (unless later additions were made), even though the *Libellus* is included in his *Opera omnia* (Paris, 1890–99).

Alchemy as an experimental science. Roger Bacon

Even Roger Bacon (c. 1214–94), the theologian and philosopher, made some contributions to alchemy. Unlike Vincent of Beauvais and Albert the Great, he did not have absolute faith in the information passed down through the ancient texts (which was definitely what caused him all of his trouble outside natural philosophy), but he did have a strong enthusiasm for experimentation (as he states in the sixth part of the *Opus majus*), even though he did very little of it. This was the first time experimentation had been thought of as a universal method for research, an idea abstracted from its individual applications to a concrete problem. Bacon's definition of alchemy in his *Opus tertium* is characteristic of his ideas. Besides the *Opus minus*, which was definitely written by Bacon and contains passages on alchemy, a number of other texts on alchemy are attributed to him. The *Speculum alchemia*, the *Tractatus trium vertorium* and the *Breve breviarium* are definitely apocryphal (some doubts about the first aside), in that they are common alchemistic treatises and contain no trace of their presumed author's ideas.

Taddeo Alderotti (c. 1223–1303), a Florentine physician, should also be mentioned in connection with this period. He suggested cooling condensation coils with water in the alcohol distillation process.

LIFE SCIENCES

The growth of the faculties of medicine

Arnald of Villanova, a physician, astrologer and alchemist, flourished at Montpellier during this period. Open to the influences of the Salernitan tradition, Arnald was not opposed to polemicizing against the authority of the ancients in the name of his own experience. In this way, the clinical position won out over the dogmatic position at Montpellier, even though philosophical and theological ties continued to weigh negatively on scientific knowledge, distorting accurate scientific thought.

The Montpellier medical *universitas* continued to influence medicine for many centuries through several of Arnald's works—his *Breviarum practice* and *Regimen sanitatis* in particular—and those of other renowned teachers there—Giles of Corbeil's *De urinis* and Bernard of Gordon's *Lilium medicinae,* for example.

Elsewhere as well, studies in medicine and the medical profession tended to structure themselves around the *universitas magistrorum et discipulorum*, which, through official pontifical or royal patents, recognized and protected their integrity against the threats of charlatans and healers. Peter of Abano, Albert the Great, Giles of Corbeil, and J. Saint-Amand all went to Paris, where the faculty of theology had preceded the faculty of medicine, and taught there in the second half of the century.

Surgeons were excluded from the faculty of medicine, and came together in the college of St. Cosmas, founded around 1260 by Jean Pitard, while in the corporative "guilds" they were associated with the barbers. One of the most prominent surgeons was Lanfranc (died c. 1306), a refugee from Milan who went first to Lyon and then to Paris. His popular work, the *Chirurgia magna*, introduced the operating techniques of the Italian schools into France.

In Italy, the Universities of Bologna and Padua were opened up to the *universitas* of medicine. Taddeo Alderotti (c. 1223–1303), author of a manual on hygiene in the vernacular, inspired by Salernitan precedents, taught in Bologna, together with William of Brescia, the three Del Garbos (Bruno, Dino and Thomas), the two Varignanas (Bartholomew and William), Torrigiano di Torrigiani, and the renowned surgeons Hugh Borgognone and William of Saliceto. In Padua, in the tolerant climate of the Republic of Venice, a school that was open to the metaphysical and gnoseological problems of Aristotelianism began to flourish.

EARTH SCIENCES

Geological knowledge in the West and in the East

The most important figure in the field of geology was Ristoro of Arezzo, who in 1282 wrote his *On the Composition of the World.* He rejected the theory of spontaneous generation and supported the true nature of fossils, although he did not escape many of the beliefs of his times and appealed to the influence of the stars in a number of events.

Nasīr al-Dīn al-Tūsī (1210–74) studied the minerals of the Azerbaijan region. Toward the end of the century, Muhammad al-Qazwīnī wrote describing various aspects and wonders of nature.

Maritime explorations

In 1291, the Genoese navigators Ugolino and Guido Vivaldi set out with two ships in search of a sea route to India beyond the Strait of Gibraltar. Unfortunately, they never returned from their voyage.

The journey of Marco Polo

Toward the end of this century, the Venetian traveller Marco Polo (c. 1254–1324) dictated the tale of his journeys, the *Livre des merveilles du monde,* known in translation as *The Travels of Marco Polo,* which became very widely circulated. Marco Polo left Venice in 1271 and pushed forward across regions then unknown to Europeans (Pamir, Lop, Gobi) to the borders of China ("Cathay"). He stayed there for seventeen years (from 1275 to 1292), participating in numerous missions to Tibet, Cochin China, Yunnan, and Annam.

APPLIED SCIENCES

Industrial techniques

The climate in Europe seemed very open to Eastern inventions, evidence of its economical reawakening and its growing technical skills. All the machinery for the silk industry that was installed in Europe during this period—rotating reels, Chinese cage bobbins, pedal-operated horizontal looms with threading devices—was of Chinese origin. Winding rotators also appeared during this period. The only European invention may have been a type of silk-spinning machine that differed from the Chinese models. Other innovations that came from the East appreciably changed European technology. Ground and river transport was improved enormously by the diffusion of wheelbarrows and the use of sluice gates for river and canal regulation, both third century Chinese inventions. Devices activated by gunpowder—most likely some sort of rockets—were first reported in Europe in 1258, while Chinese information on gunpowder dates back to the tenth century.

New applications of water wheels and their repercussions on the agricultural economy

Water wheels for transforming rags into paper are first mentioned in a document from Xativa, near Valencia in Spain, dated 1238. In 1268, seven of these mills were in operation in Fabriano. In France the Moulin Richard-de Bas, the first of its kind, was built on the Dore river in Puy-de-Dome; the earliest news of it dates back to 1326. It has been preserved through the ages and still produces fine paper used by many contemporary artists.

Cloth fulling mills were widespread, especially in England, and increased the production of finished cloth and clothing considerably. The technological revolution brought about by the mechanization of fulling has been compared to the eighteenth century revolution in spinning and weaving (see APPLIED SCIENCES 1000–1100).

The tendency grew for the mills to be monopolized in a few centers of power, under lords and convents, preventing the peasants from using their traditional techniques and their own mills. In England, the story of the Abbey of Saint Albans in Hertfordshire is typical. A long dispute began in 1274, when the abbot Roger attempted to forbid the inhabitants of the village and the surrounding area to full their cloth by themselves (they pressed it with their feet), and forced them to bring the cloth and their clothing to the mill at the abbey. Many years later, in 1326, the people rose up against the monastery when the abbot demanded that all the peasants bring their grain to the abbey to be ground. Five years later, in retaliation, the abbot Richard Wallingford (c. 1292–1335), an illustrious scientist and inventor, had all the houses searched, confiscated all the grinding mills, and had the cloister of the convent paved with them, an act of humiliation for the townspeople. Fifty years later, in the peasant revolt of 1381, led by Wat Tyler, the people of Saint Albans destroyed the pavement of the cloister, an *ante litteram* example of a Luddite revolt against machines.

Optics

Roger Bacon (c. 1214–94), who studied the laws of refraction in order to design an instrument that could be used as a telescope, carried on Grosseteste's work. Bacon was the first after Claudius Ptolemy to attempt to draw a geographical map where the positions were determined by coordinates. Bacon's friend, Peter Peregrine (of Maricourt, in Picardy), described two different types of encased compasses, one floating, with a magnet, the other with a pivoted needle.

Spectacles were invented around 1286, before Bacon's death. Fra Gerolamo of Pisa testified to their exceptional importance when in 1306 he spoke of the invention as "one of the best and most useful arts in the world." The first lenses were convex and corrected presbyopia.

The first mechanical clocks

From a chronicle written in 1271, we learn that "clockmakers are attempting to construct a wheel that will make one complete revolution for each revolution of the equinoctial circle, but they are not succeeding in their task." In 1286, a mechanical clock built by a certain Bortholomew of the Cathedral of St. Paul in London was mentioned. The new clock of Canterbury, built in 1292, was of the same type, as was the first public clock, manufactured by Pierre Pipelart in 1300 in Paris.

The new mechanical clocks, described in detail by Richard of Wallingford (see ASTRONOMY 1300–1350), used a bar and rod movement connected to a mechanism operated by weights.

More than half a century passed, however, before this type of instrument, which made it possible to obtain a regular and accurate subdivision of the day into twenty-four hours, became more popular in the West than the traditional water clock, which showed hours of different lengths for the day and night, for the different periods of the year, and for varying latitudes (see APPLIED SCIENCES 1350–1400).

1300—1350

THOUGHT AND SCIENTIFIC LIFE

Changes in institutional structures

In 1300, Pope Boniface VIII celebrated the first Jubilee. His theocratic policy, theorized in the bull *Unam sanctam* of 1303, did not succeed in forcing Philip the Fair, king of France, into obedience. The conflict between Boniface and Philip was directly mirrored in the harsh polemics between the curialists and the royalists. The royalists defended the autonomy of state power against the curialists, and criticized the interference of the pontiff. The king of France chose his political and juridical counsellors, such as Philip of Nogaret, the creator and leader of the anti-papal policy, from among the royalists. Almost simultaneously, the ecclesiastical structures began to feel the need to reduce the pope's power in relation to the authority of the College of Bishops. In 1309, a few years after Boniface's death, the period of the "Avignon captivity" began. It lasted for almost seventy years and made the pope virtually a pawn of French politics.The other universalistic power of the Middle Ages, the Empire, was also in decline. After a final attempt at exerting real power within *christianitas*, made by Henry VII (ruled 1308–13), the imperial crown became increasingly more tightly bound to a limited area of influence and a particular dynastic structure. A precapitalistic structure, characterized by the flourishing of the great Italian, French, German, and Flemish trade centers, arose as the result of the sudden economical development that had occurred in the previous century. The rapid accumulation of mercantile capital caused the military and landed power of lay and ecclesiastical feudalities to be replaced by the new economical and political authority of the commercial and financial oligarchies. This happened not only in the Mediterranean regions, but along the Atlantic coasts as well, and even on the Baltic Sea (the German Hansa is an example). Mercantile capital soon became financial capital, as actual banking institutions, dislocated in the strategic points of commercial and productive activity, arose. The more advanced intellectuals were aware of these great political-social changes: their awareness came out of political discussions where the historical events themselves seemed to suggest topics for reflection and debate (the autonomy of civilian authority, its origins and meaning, the relations between political authority and the ecclesiastical hierarchy, the nature and function of state institutions, etc.). Doctrines extremely critical of the presuppositions and foundations of the twelfth century thought soon appeared in philosophical and scientific theories as well. In fact, the years in which Dante Alighieri wrote his *Convivio*, *De vulgari eloquentia* (1304–07) and *De monarchia* (1308–13) were the very same years in which John Duns Scotus developed his mature speculation and Ramon Lull composed his *Ars brevis quae est imago artis generalis* (1308).

Philosophy and theology from Duns Scotus to William of Ockham

The Scottish Franciscan teacher John Duns Scotus (c. 1266–1308) had a particularly strong influence on the Franciscan schools in England and France through both his teachings (at Oxford, in Paris, and, during the final years of his life, in Cologne) and his works, the most important of which was the *Opus oxoniense* (1300). Scotus' thought tended to considerably diminish Thomist optimism about the harmony between faith and reason, and to limit philosophy to those areas of knowledge that could be acquired through the strictly demonstrative procedures codified in Aristotelian logic. For this reason, it constituted a point of reference for those theologians and philosophers who still felt the effects of the condemnation of 1277 and the harsh polemics among Thomists, Averroists, and Augustinians in Paris. On the other hand, the complexity and ambiguity of many of the themes that the *doctor subtilis* dealt with allowed differing Scotist doctrines to develop. Thus Richard of Conington (died 1330) mixed Scotist and Augustinian themes in his *Quodlibeta*, Scotus' disciple John of Wottinghan drew on themes from Bonaventure, Robert Cowton insisted on a voluntaristic solution to the problem of divine prescience in his commentaries on the *Sententiae*, and Francis of Mayronnes, a disciple of Scotus from 1304 to 1307, took up the Scotist doctrine of essence and existence, emphasizing the realistic nature of ideas. Almost simultaneously, the English Franciscan William of Alnwick (1322) revised Scotist themes in an original way and with impeccable dialectical rigor in his *Quaestiones de esse intelligibili*.

Even in the Dominican schools, where Thomas's authority had become official, yet posed an obstacle to independent philosophical positions, there were still teachers who discussed the essential points of Thomism. James of Metz, who wrote a commentary on the *Sententiae* in 1295 and again in 1302, raised criticisms against Thomas that drew him a reprimand from Harvey Nedellec. Then, in 1314, Durandus of Saint Pourçain (c. 1275–1334), a representative of the Avignon Church, had to defend his right to the freedom of his own ideas before the hierarchy of the order, polemicizing first against Nedellec and then against John of Naples. The Franciscan Peter Aureoli, archbishop of Aix-en-Provence in 1321, took a similar position (1322). In his *Tractatus de principiis*, and in a commentary on the *Sententiae*, he criticized Bonaventurian as well as Thomist and Scotist doctrines. His resolute adherence to Aristotelian ideas led him to develop a clever doctrine that solved the problem of justifying the immortality of the soul, which, according to the rulings of the Council of Vienne (1311–12), had the

shape of the body. The Englishman Henry of Harclay (c. 1270–1317), a professor and later chancellor at Oxford, and even later Bishop of Lincoln, was also critical of the great masters of scholastic tradition. He was a strong adversary of the Dominicans as well as of the Scotists, and criticized the Scotist doctrine of *haecceitas*. A few years later, the same doctrine and its realistic roots were contested by the Franciscan William Farinier (1361). Polemical attitudes toward the great scholastic *auctoritates* can also be seen in the works written by the first Carmelite teachers in Paris, for example Gerard of Bologna (died 1317) and the Catalan Guy Terrena of Perpignan (died 1342) whose ontological and gnoseological doctrines were influenced by the ideas of the nominalist logicians. Any residues of realism were finally swept away by the critique of the doctrine of the reality of the *essentiae*, developed by William of Ockham (c. 1285–1349), a Franciscan teacher educated at Oxford from 1312 to 1318 and who taught there, commenting on the *Sententiae*, until around 1324. Taking up the logical problem of universals, Ockham outlined an organic concept of terms in his commentary on the *Sententiae*, as well as in his *Summa totius logicae*. His reduction of the problems regarding the reality of the *essentiae* to linguistic problems concerning the correct use of universal terms such as human being, animal, etc., marked both a revival of physical doctrines (in the *Summulae*) and a different way of stating the metaphysical problem of the relationship between God and the individual being.

The Ockhamist movement

In 1324, Pope John XXII (pontificate 1316–34) called Ockham to Avignon to answer for some of his theories that were suspected of heresy. In 1328, Ockham fled the papal court together with Michael of Cesena and other dissident Franciscan leaders who were in disagreement with the tyrannical power of the Avignon papacy. This made Ockham the theoretician *par excellence* of Franciscan mysticism and consequently the theoretician of the relation between civilian power and ecclesiastical authority in the anti-papal struggle led by the Emperor Louis the Bavarian.

Ockham's *Dialogus de imperatorum et pontificum potestate* and his libels and pamphlets against the pope were no less popular than his works on logic and physics. Around 1330, Ockhamism began to spread into the major European universities, from the English schools at Oxford and Cambridge to the faculties in Paris, even though, at least initially, it was often confused with nominalistic theories, such as those of Durandus or Aureoli. The reaction of the realists manifested itself in a rather violent way around 1335. Among the first adversaries, some of whom were Scotists and others supporters of traditional Augustinianism, was Walter Chatton, who confuted Aureoli's and Ockham's doctrines in his commentary on the *Sententiae*. Neither papal condemnation (1324–26) nor a ban (1340) in the statute of the Parisian faculty of arts against teaching Ockhamist philosophy stopped its diffusion. A new intervention on the part of Pope Clement VI in May of 1346 was of no use either.

In 1340, Adam Wodam, a philosophy teacher at Oxford, took up and emphasized the skeptical conclusions of Ockham's teachings. The idea of divine omnipotence, so dear to Ockham, was taken up by theologians such as Thomas Buckingham (flourished mid-fourteenth century), who reaffirmed the absolute contingency of the universe, which existed only by divine will. An analogous doctrine of contingency was developed by the Oxonian teacher Thomas Bradwardine (c. 1290–1349), who wrote a treatise entitled *De causa Dei contra Pelagium et de virtute causarum,* as well as numerous scientific works (on mathematics, mechanics, and astronomy in particular). Bradwardine, however, aimed for a theology based on a deductive method that was modelled after the mathematical method. The Franciscan Oxonian John of Rodington (c. 1290–1348), who united Ockhamistic themes with the revival of Augustinian tradition, polemicized with Bradwardine.

In Parisian circles, critiques of the doctrine of the *essentiae* were developed by the Augustinian Gregory of Rimini (died 1358), whose ideas were similar to Ockham's, even in his approach to physical problems such as time, space, and movement. In 1347, the Cistercian teacher John of Mirecourt, a lector of the *Sententiae* in Paris at the college of St. Bernard in 1345, was forced to repudiate forty theses contained in his lessons, and was condemned by the chancellor of the University of Paris, the Florentine Robert de' Bardi. That same year, the academic authorities also condemned Nicholas of Autrecourt, a teacher of the arts, who was forced to burn some of his writings before the teachers of the university. While Nicholas of Autrecourt represents the most radical product of Parisian Ockhamism, Jean Buridan of Béthune (1300–1358), a teacher of the arts and rector of the University of Paris in 1328 and 1340, took a substantially moderate position that drew upon Ockham's logical doctrines, but also upon Averroist tradition. Buridan did not aim for the total elimination of the doctrine of essences, and maintained, against Nicholas of Autrecourt, that one could start from the existence of one thing to demonstrate the existence of another.

Averroism

Buridan's interest in physics, and in the theory of impetus in particular, which he developed out of Philoponos' explanation, and the work of the Scotist Franciscans (Gerard of Odone and John Marbres revised the Aristotelian concept of space in an atomistic key; Nicholas Bonet formulated a theory of actual infinity and distinguished between mathematical time and physical time), demonstrate the lively interest in Aristotle's doctrines and the Greek and Arabic commentaries that was characteristic of the times. The Averroist tradition survived in the environment of Paris, where Siger's teachings were continued by John of Jandun (1280–1328), who collaborated with Marsilius of Padua (c. 1270–1342), on the famous *Defensor pacis,* in which the Aristotelian theory of the state is used to defend the rights of the lay state against the theocratic ideas of the papacy. Pope John XXII reacted strongly to this political Averroism. Marsilius and John were forced to flee Paris, and the papal legate Bertrand del Poggetto did not spare even Dante's poetry, which was publicly burned in 1329. The supporters of the pontifical monarchy, which included the Franciscan Alvaro Pelayo (who wrote *De planctu ecclesiae* around 1335), continued the intransigent polemics against Marsilius' doctrines. Simultaneously, attempts were being made, for example by the Irishmen Thomas of Wilton and Richard Fitzralph (c. 1300–60), bishop of Armagh, to include even Averroës in an orthodox-type culture. Fitzralph, however, disowned the ideas of his

youth when he became a dignitary of the Avignon court in 1335.

Padua was the center of a scientific and philosophical current that focused on a variety of Greek and Arabic texts. One of the prominent figures there was Peter of Albano, an advocate of Arabic and Jewish astrological and medical doctrines. Similar themes were being dealt with in Bologna at the same time. Between 1337 and 1347, Matthew of Gubbio developed an Averroist trend, while the Servite monk Urban of Bologna published a work about Averroës' commentaries on the *Physics*. There were also supporters of other currents, as well as polemics against Averroism.

Mysticism spread through Germany and Flanders after the Dominican Eckhart, who died in Cologne in 1328, revived the theme of the *unificatio* between humans and God, drawing upon Proclos, Pseudo-Dionysus, and Scotus Erigena. Speculative mysticism was revived by Johannes Tauler (c. 1300–61), Henry Suso (1300–65), and later by Jan van Ruysbroeck (1293–1381). At the same time, a classicistic trend, aimed at recovering the values of ancient Rome and often in conflict with the Avignon papacy, was becoming widespread in Italy, especially among the scholars. This tendency was very much present in Cola di Rienzo's election as the tribune of the revived *respublica romana* in 1347. It also influenced the works of Petrarch and Boccaccio.

During the early decades of the century, Richard Wallingford (c. 1292–1335) wrote (in Latin) about trigonometry and, together with Levi ben Gerson (1288–1344), an astronomer and mathematician who wrote several works on arithmetic, geometry, and trigonometry that display considerable originality and rigor, was the first European author to write on this subject. Wallington also ventured upon a strong criticism of Euclid's postulates.

The most original mathematician of this period was Nicholas Oresme (c. 1323–82). His *Tractatus de latitudinibus formarum* contains a foreshadowing of the Cartesian method of using coordinates to represent fluid forms, or functions, as they were later called. In his *Algorismus proportionum*, he proposed an original theory of proportions based on the use of fractional exponents, which was basically a theory of irrationals. Unfortunately, these works were not well enough known to influence the development of science as much as they might have.

ASTRONOMY

The culture of astronomy in the fourteenth century

Although scholars such as Roger Bacon, Robert Grosseteste, Albert the Great, and Thomas Aquinas approached the heritage of Ptolemaic and Aristotelian cosmology, more or less filtered through the Arabs, from an increasingly open and original perspective, even the best products of fourteenth century European astronomy (from which it is always difficult during these centuries to separate the component that should more properly be considered astrology) never reached those scientific levels that had characterized the previous century.

Jean Buridan and Nichole Oresme present two important exceptions (see ASTRONOMY 1350–1400).

Dante's cosmology

Dante Alighieri's *Divine Comedy* certainly provides a good example of the more advanced cosmological ideas that existed among scholars of this period. Written in the first twenty years of the fourteenth century, it drew upon many sources for its concept of the universe. Among these were Aristotle (through Saint Thomas's commentary on the *De coelo*), Brunetto Latini's *Li livres dou Tresor* (c. 1260–67), a scholarly thirteenth century encyclopedia of modest scientific importance, and especially the translation of al-Farghānī's *Elements of astronomy*, done by Gerard of Cremona in the twelfth century under the title *Liber de aggregatione stellarum*. Very little is said, and only fleetingly, about Ptolemy and his theory of epicycles (*Paridiso*, VIII, 2). There are ten celestial spheres, which, besides the eight for the planets and the fixed stars, include the primum mobile and the empyrean, where God resides. The nine inferior heavens, beginning with the primum mobile, are moved by angelical beings (Cherubs, Seraphim, etc.). In *Purgatorio*, XI 108, there is an allusion to the precession of the equinoxes, or rather the precession of the sphere of the fixed stars, the "circle in heaven that moves at slowest pace."

Dante's *Convivio*, presumably written between 1304 and 1307, in four books, contains almost the same notions on astronomy. The most obvious difference is in the explanation of sun spots, which according to the *Divine Comedy* are caused by the differing intensity of the moon's brightness, controlled by angelical beings; in the *Convivio,* they are supposedly caused by the rarity of certain areas on the moon that do not reflect sunlight well.

In the *Quaestio de aqua et terra*, a small work that probably dates back to 1320, Dante opposed the widespread belief that the sphere of the waters did not coincide with the sphere of immersed land, and consequently that their centers were different. More than a century and a half later, in 1498, while crossing in front of the current of the Orinoco River, which was blocking the way of his galleys, Columbus still believed that he had reached the highest point in the ocean, from which the water descended in order to reach the level of the terrestrial sphere.

Cecco of Ascoli: a martyr of science

The case of Francesco degli Stabili, known as Cecco d'Ascoli (1269–1327), is indicative of the extent to which astrological beliefs were spread along with scientific concepts in the fourteenth century. In his poem, *L'Acerba* (1324), Cecco spoke of the heavens, the beings that moved them, destiny, and the soul. He attacked the *Divine Comedy* and refuted the attacks it made against astrology. The work was judged heretical and its author was burned alive.

Jewish cosmology

Around 1340, in southern France, Levi ben Gerson (1288–1344), an astronomer of Jewish origin, wrote a *Sefer tekunah* (*Book of Astronomy*), in which he attempted to replace the Ptolemaic system with a system of forty-eight eccentric spheres, which was inspired by Aristotle. At the same time, he confuted Alpetragius' theory, in an attempt to rationalize Jewish cosmology (see ASTRONOMY 1100–1200).

The new physics and Aristotelian cosmology. Jean Buridan

Jean Buridan's philosophy of nature became popular in the north of France, at the University of Paris, in the years between 1330 and 1350. In his cosmology, Buridan did not hesitate to apply his doctrine of impetus to the theory of the celestial spheres, criticizing the doctrine of angelical beings that had come down from Aristotle. He advanced the hypothesis that at the moment of the Creation, God bestowed various impetus on the various celestial bodies, and that these moved indefinitely, without friction or resistance, not requiring the intervention of any intelligence.

His theory stirred up a lively controversy among philosophers and theologians that lasted for quite some time.

The work of Richard Wallingford

One of the greatest inventors of astronomical instruments in the West was Richard Wallingford (c. 1292–1335), abbot of St. Albans from 1326, the abbot who had the cloister of the abbey paved with grindstones confiscated from the peasants of the area (see APPLIED SCIENCES 1250–1300). Among his most original instruments and those which became most widely used were the *albion*, which he described in a treatise that Regiomontanus published in 1460, a sort of *equatorium* for calculating the positions of the planets, and the *rectangulus*, similar to the classical *torquetum*, consisting in four rulers that slid over one another.

Wallingford also had a grandiose astronomical clock with a mechanical movement which showed the motion of the sun, the moon, and the planets constructed in the Church of St. Albans.

PHYSICAL SCIENCES

The problem of method

The progress that took place in the natural sciences in the thirteenth century brought the problem of experimental method to the forefront, especially as far as physics was concerned. Roger Bacon had already been concerned with the problem, and although he attributed an essential function to empirical observation, he openly recognized that mathematics was no less important in the area of naturalistic research. Without mathematics, he wrote, "it is not possible to know any science." At the turn of the century, investigations into the subject were intensified and took on the aspect of extremely subtle discussions on the inductive method.

Scotus and Ockham: on the nature of physical laws

The positions taken in this debate by Johannes Duns Scotus (c. 1266–1308) and William of Ockham (c. 1285–1349), both Franciscans and Englishmen, and the heirs of the philosophical-scientific thought of Grosseteste, Peckham, and Bacon, deserve mention. Scotus maintained the need to make a clear distinction between causal laws and empirical generalizations. According to him, the former tell us why natural events occur, while the latter only tell us how they occur. Physical laws belong to the environment of metaphysics. When the particular sciences speak of laws arrived at inductively, in reality they only refer to generalizations made on observational data or to mathematical connections among those data, while the validity of the laws depends entirely on the principle of the uniformity of nature, which can only be guaranteed metaphysically. Ockham was even more radical. He stated that the only knowledge was knowledge gained from perceiving individual data and that the presumed causal relationship was only an association of these data, where at most one could speak of the immediate cause of a phenomenon, meaning the series of phenomena leading up to it. In other words, neither pure reason nor the inductive process allow us to go beyond the individual objects understood through perception. The value of the natural sciences lies in their ability to make us understand precise connections among phenomena, not to lead us to general causes that supposedly determine them.

In physics, Ockhamist empiricism contributed notably to shaking the faith that the general (philosophical) principles of the universe could be understood and favored instead the study of more limited areas of experience.

The two major centers of scientific research in the fourteenth century were the University of Paris and the University of Oxford, where teachers from the religious orders, especially Franciscans, taught together with lay teachers.

CHEMICAL SCIENCES

The Church's stand against the alchemists

The alchemists, who had not led an easy life for quite some time, found themselves in even greater difficulty after Pope John XXII, in his *Spondet partier* (1317), said: "The evil alchemists promise something they do not give.... They present a false metal for gold and silver, and trick you with a stream of words that mean nothing.... They coin false money.... We order that they all leave the country, the commissioners of gold and silver with them, and that they give all their real gold to the poor." In 1323, the general of the Dominicans threatened with excommunication any church members who practiced alchemy in any way and did not destroy their books within eight days. For his part, Dante put them all in the Inferno.

The work of Petrus Bonus

However, four books on alchemy, the *Summa perfectionis magisterii*, *De investigatione perfectionis*, *De inventione veritatis*, and the *Liber fornacum*, were written by a Spaniard right at the beginning of the century (c. 1310). We do not know whether it was for reasons of modesty or of cleverness that this person circulated the works under the name of Geber, but for lack of another name, the author is called the false Geber, or Pseudo-Geber. We know he was a Spaniard from Petrus Bonus of Ferrara's *Pretiosa margarita novella*, written in 1330, in which he is called Geber Hispanus. There are three principles of metals in Pseudo-Geber's books—sulphur, mercury, and arsenic—and transmutation is accepted along with the existence of the philosophers' stone and two elixirs (white and red, both basically made from mercury and sulphur). But the experimental part, which gives numerous directions for purifying salts, alkalis, and alums; for

obtaining solutions of metals; and even for extracting metals from ores, is the most important. Directions are even given for dry distilling vitriol (alum) to obtain sulphuric acid, and vitriol and saltpeter to obtain nitric acid (which it appears was produced in Italy as early as 1150). The addition of salt produced aqua regia, an acid that could dissolve even gold.

Bonus was more of a theoretician, and wrote an exhaustive description of the alchemistic theories of the times, which were still substantially based on the sulphur-mercury idea. Bonus's work can be dated with certainty, and for this reason serves as a valuable point of reference.

Alchemistic texts of doubtful authorship

Several works attributed to Arnald of Villanova (c. 1235–1313), who flourished as a physician during the second half of the thirteenth century, can be ascribed to this period. A great authority in his own field, Arnald appears as the author of texts such as the *Thesaurus thesaurorum*, the *Rosarius philosophorum*, and the *Novum lumen*, which are now considered apocryphal, or at least of fairly questionable attribution. If the works are his, then he was fairly active in the field of alchemy; however, if they are not, then no other evidence exists that Arnald of Villanova was ever an alchemist.

According to some, Arnald's real work in alchemy was destroyed by the Inquisition. In fact, the work attributed to him is banal and charlatanistic, unlike his usual work, and was probably written by another Arnold, who lived at Montpellier (where Arnald of Villanova taught from 1300) at the beginning of the fifteenth century.

Another name that appears, quite surprisingly, in the history of alchemy is that of Ramon Lull (c. 1235–1315), a Friar Minor who was born at Palma of Majorca and is well-known as a philosopher, theologian, and missionary who travelled extensively in order to convert the Moslems to Christianity. The texts attributed to him, published in the fifteenth century, were supposedly written between 1330 and 1333, definitely after Lull's death, and most probably were written even later. It is possible that Lull's original works, like those of Arnald of Villanova, were destroyed by the Inquisition. Among the works attributed to Lull are the *Testamentum duobus libris universam artem chymicam complectens* and the *Experimenta*. They are not very original, although they contain a number of good, detailed descriptions; the term *aqua fortis,* to indicate nitric acid, also appears for the first time.

LIFE SCIENCES

The improvement of dissecting techniques

There was new activity in the field of medicine. The dogmatism of Galenic tradition did not stop the first autopsies from being performed during this period, while the new scientific knowledge came into contact with Aristotelian metaphysics and gnoseology.

Mondino de' Liucci (c. 1270–1326) of Bologna published his *Anatomia* in 1316, providing the Italian and European universities with one of their basic texts for study. The work indicates that the author had participated in several dissections of cadavers during the preceding years. But the practice of dissection, which provided the foundations for studies in physiology as well as for the practice of surgery, remained within a traditional framework, and in more than one instance observations were improperly interpreted due to bookish, scholarly prejudices. In any case, dissecting techniques became better defined, and the dissection of cadavers was gradually favored by the fact that autopsies were required in some medico-legal situations. On May 27, 1308 (some sources say 1368), the Council of Venice authorized one anatomical dissection per year, although the dissector was not the *magister regens*, who watched the work of the autopsy reading a book from the rostrum, but a subordinate.

Among Mondino's many students in Bologna was a woman, Alessandra Galioni. Surgery continued to be excluded from the universities, despite illustrious figures such as Henri de Mondeville (1260–1320), Guy de Chauliac (c. 1300–1368), Lanfranc, and Bruno of Longoburgo, who in his *Chirugia magna* wrote: "*Ac operationes scarificationis et phlebotomiae noluerunt medici propter indecentiam exercere; sed illas bacteriorum in manibus reliquerunt.*"

A new epistemological concept

In Padua, the physician and philosopher Peter of Albano (1257–1315) opened a new epistemological perspective within the Aristotelian current by proposing a new, empirical demonstration of the cause through the effect, in place of the traditional demonstration of the effect through the cause.

The Black Plague

In 1347, a deadly epidemic of Black Plague entered Europe from the East via the communication centers of Genoa, Venice, and Marseilles. Rats and fleas spread the plague rapidly northward, from France to Germany, England, and Scandinavia. The more urbanized, more crowded areas provided especially fertile breeding places. It is estimated that more than one third of the European population was killed off by the plague between 1348 and 1353. The unstoppable spread of the epidemic was favored by the general conditions of poverty, filth, and promiscuity associated with the thriving Euro-Asiatic and Mediterranean trade and as Emmanuel Le Roy Ladurie has observed, by the extraordinarily cold and humid winter of 1347–48, which caused a large outbreak of broncho-pulmonary diseases, the best breeding ground for the most virulent strains of the plague.

EARTH SCIENCES

The origins of fossils

Giovanni Boccaccio attacked and correctly solved the problem of the origins of fossils in his *Filocolo* (c. 1340), where he deduced that the sea once extended over the earth.

Naturalistic notes and oceanography

Dante Alighieri's *Quaestio de terra et aqua* appeared in 1320, though he had already correctly explained the phenomenon of rain and the origins of meteoric waters in his *Divine Comedy*.

Giacomo Dondi (1298–1359) studied the salinity of the ocean (*Tractatus de causa salsedinis*) and the problem of tides, which he attributed to the combined effects of the sun, the moon, and the planets.

APPLIED SCIENCES

The crisis of technological progress in medieval civilization. Military science

During the golden age of the medieval technological revolution (eleventh–thirteenth centuries), a fortunate combination of the entrepreneur spirit, rationalistic and constructive attitudes, prolific trade and exchanges of ideas with the East, favorable climatic conditions (especially for agriculture), and relative international peace, had favored the development of an improved standard of living and a degree of technological creativity that remained unsurpassed for originality in the West until the coming of the modern age. This trend was inverted, however, beginning in the fourteenth century. Winters were harsh and rainy, endemic diseases spread, there was an almost uninterrupted state of war right in the heart of Europe (France and England were involved in the dismal Hundred Years War from about 1339 to 1453), the Church condemned rationalistic theories and mystic spiritualism spread (the staid rationalism of the reconciliation of faith and reason that had characterized Thomism was unable to affirm itself until the fifteenth century), and, finally, the epidemic of the black plague that quickly reduced the population of Europe by one third (see LIFE SCIENCES) raged around the middle of the century.

As Jean Gimpel has observed (*The Medieval Machine*, 1976), what occurred was a phenomenon that is typical of the decline of a civilization that has reached its peak of technological maturity; that is, interest was transferred from creating productive techniques to intensifying production that was mainly directed toward military ends (even in metallurgy, as we shall see in the following section). The treatises of the major engineers of this period reflect this trend. Guido of Vigevano made quite detailed drawings of revolutionary war wagons pulled by a system of windmill blades connected to the wheels by gears, while later engineers, up to Konrad Kyeser, Leonardo, and Francesco di Giorgio Martini, worked in the same direction, although they wre influenced by the new humanist interests as well (see APPLIED SCIENCES 1400–1450, 1450–1500).

The use of cannons

The developments made in metallurgy during the previous century were mainly applied to military science, which had been revolutionized by the introduction of gunpowder. According to Lynn White, Jr. (*Medieval Technology and Social Change*, 1962), gunpowder originated from regional discoveries, with occasional exchanges of technical improvements. The earliest reputable information on the use of cannons dates back to this period. Very small cannons, weighing no more than nine—eighteen kg., were used at Crecy in 1346. They were made by working metal, most probably forged iron, and rivets were used to fasten on the barrel.

Machines and instruments

Water power was used for drawing iron and operating silk machines, an innovation that came from China; the earliest information on the use of water power with silk machines in China dates back to 1310.

The idea became diffused almost immediately in the West. The inventor of the breaker, a device for beating flax, on the other hand, was Dutch.

The compass was improved during this period when the needle was fixed on a compass rose. In 1342, Levi ben Gerson (1288–1344) first described the Jacob's staff, an instrument commonly used by sailors during the Renaissance. In Italy, the single-masted, square-sailed Nordic cockboat with sternpost rudder was adopted as a mercantile ship, while galleys continued to carry lateen sails.

In 1344, Giacomo Dondi (1298–1359) built an astronomical clock with a mechanical movement in Padua. It was one of the first of its kind, along with the one in St. Albans, designed by Richard Wallingford (see ASTRONOMY 1300–1350).

1350–1400

THOUGHT AND SCIENTIFIC LIFE

Scientific developments during the late Scholastic period. Buridan's school and the calculatores of Merton College

The major exponent of Buridan's school was Nichole Oresme (bishop of Lisieux when he died in 1382), a teacher at the college of Navarra in 1356, and a teacher of theology in Paris in 1362. His reflections on the respective positions of the heavens and the earth in his *Livre du ciel et du monde d'Aristote* led Oresme to conclusions that were very similar to those later reached by Copernicus. No less important were his studies on the law of falling bodies, his research into pure and applied mathematics, and his development of the technique of *calculationes,* first introduced by Thomas Bradwardine (c. 1290–1349). Many of Oresme's works, even those on political and economic topics, were written in the vernacular, and he composed the first French paraphrases of Aristotle's *Politics* and *Ethics.* Albert of Saxony, also known as Albertus Parvus (c. 1316–90), a teacher at the faculty of arts from 1351 to 1362, also belonged to Buridan's school, where he conducted research in logic, mathematics, and physics in a period when attempts at reviving the Empire and the growing power of the German princes made Germany a country that looked to France and England as models for its culture. The University of Vienna was founded in 1365, the University of Erfurt in 1379, the University of Heidelberg in 1385, and the University of Cologne in 1388. Albert, nominated as the first rector of the new Viennese University, brought Buridan's influence into Germany with his polemics in defense of the new logic of the *nominales* or *terministae* Ockhamists against the *via antiqua* of the Thomists and Scotists, his unbiased commentaries on Aristotle's works (including those on ethics and economics), and above all his revival of the theory of impetus.

The Paris school also produced thinkers more closely tied to the conclusions of Ockhamist theology. Since many of these scholars ended up in the new German universities as a result of the difficulties arising from the Hundred Years War, the doctrines spread to Germany. One of the leading figures was Marsilius of Inghen (c. 1330–96), a student of Buridan, who was a professor in Paris from 1362 and rector in 1367 and 1371 and who moved to Heidelberg in 1386. Among other things, his *Quaestiones super IV II. sententiarum* reaffirms the possibility of metaphysically proving the existence of God, although he warns that this must not be considered a mathematical demonstration.

Bradwardine's influence, which can be seen in many of the thinkers just mentioned, was especially evident among those professors at Merton College in Oxford known as the *calculatores.* Unlike the exponents of Buridan's school, the Oxonian *calculatores* did not focus their attention so much on problems of dynamics as on questions of *intensio et remissio formarum,* that is, the quantitative increase or decrease of a given quality in relation to a fixed scale. Several works produced by the Oxonian circle—the *Liber calculationum* that circulated under the name of Richard Suiseth (whom some identify as the logician and physicist Richard of Swineshead) and the works of John of Dumbleton, who lived at Oxford from the middle of the century, for example—were soon in circulation on the continent as well. The refined logical and mathematical analyses of the Oxonian *calculatores* and the disciples of Oresme and Marsilius of Inghen became the targets of polemics from both Augustinians and cultivators of the *humanae litterae,* who were not sparing in their criticims of the naturalists' *sophismata* or of those whom they called *calculationes britannicae* (see PHYSICAL SCIENCES).

Popular reformist movements. Conciliatory doctrines

While the struggle begun by Michael of Cesena and the Franciscan movement ended in 1330 with the intervention of the pope, the end of the Avignon captivity (1378) did not mark the beginning of an epoch of peace for the Church. When only a few months after the Holy See returned to Rome it suppressed the reform movements protected by Louis the Bavarian, and with them the first schismatic tendencies, national forces and reformist currents banded together against the restored power of the pope. The great schism saw the revival of doctrinal opposition to papal authority and a rebirth of the doctrine of the supremacy of the council. While the Church, increasingly tied to worldly interests, lost prestige, and the empire often lacked effective authority over the great German princes ("a vain name without a subject," as Petrarch said), and while the kingdoms of France and England fought a bitter war (the events of which were closely linked to revolts of the feudal nobility and severe dynastic crises), popular malcontent expressed itself in a long series of peasant riots and insurrections (like the great French *Jacquerie* in 1358) and popular uprisings (such as the revolt of the Ciompi in Florence in 1378).

Toward the end of the century, the traditional need for spiritual renewal provided the religious groundwork for the revolutionary Lollard movement of the British lower classes. Its egalitarian theories came out of the preachings of John Wycliffe (c. 1330–84), an Oxonian theologian who drew on Bradwardine's theological determinism more than on his mathematical method and who championed a voluntarism that left very little room for mediation by the ecclesiastical hierarchy. Wycliffe's translation of the *Bible* into English (1380) together with Geoffrey Chaucer's *Canterbury Tales* (1387) remained a point of reference for written English. This along with his polemics against the pope, written between 1374 and 1384 made Wycliffe, who was excommunicated in 1377, the theoretician for a community of righteous people who on the one hand cultivated the direct reading of the sacred text and on the other aspired to genuine social reform. The idea of integrating the study of the sacred text and the liturgy into national tradition, together with a political and religious radicalism anti-papal polemics and egalitarianism, also animated the preachings of Jan Hus (c. 1372–1415) in Bohemia, where Prague was one of the mose lively centers of German culture. Much more moderate positions were taken up by the supporters of the Council, such as Peter of Ailly (1350–1420), chancellor of the University of Paris in 1389 and author of the *Imago mundi*, a sort of encyclopedic work that gives an outline of scientific knowledge at the end of the fourteenth century, and Jean Gerson (1363–1429), his disciple and successor as chancellor in 1395. The supporters of the Council looked with suspicion upon both the subtleties of the Parisian and Oxonian *magistri*, and the popular reform movements, but still felt a need to intervene in the crisis of the ecclesiastical institutions and culture. In 1391, Gerson requested the convocation of an ecumenical council.

Early Italian humanism

In Italy, where a precapitalistic-type society formed more quickly, a new type of intellectual figure appeared, much different from the traditional *clericus*, and in some ways the descendent of the lay jurist and the physician-philologist of the Salerno school. Like Petrarch and Boccaccio, these new intellectuals referred to the Greek and Latin classics not so much for theological-philosophical reflection or logical or physical research, but in order to regain the forgotten knowledge of ancient times. The contribution of the Byzantine scholars, who were leaving Constantinople under pressure from the Osmanli Turks, successors of the Seljuks in threatening the shaky Empire of the East, proved essential, as they came into Italy to teach Greek and spread the study of the classical authors. Manuel Chrysoloras (c. 1353–1415) began giving Greek language lessons in Florence in 1398. The following year, Coluccio Salutati (1331–1406) finished his *De nobilitate legum et medicinae*. Taking off from a statement made by a Florentine physician in a short treatise that medicine, and in general all knowledge of the physical world, was superior to knowledge about human reality, Salutati retorted that knowledge of the human world, i.e. rhetoric in the sense of the study of language and law, or the study of the civilized world, was preferable to knowledge about nature.

MATHEMATICAL SCIENCES

Fibonacci's followers in Italy

In Italy, the followers of Leonardo Fibonacci continued to develop computing (technics) for use in the practical sphere.

The best known representatives of this so-called abacus tradition were Paolo dell' Abbaco (1282–1374) and Prosdocimo de Beldamandi (1370–1428), who also wrote a treatise that contains the first sum of a geometrical progression. The professional activity of the "abacus shops" belonged to this tradition; the shops flourished above all in those cities where commercial activity was most intense and were practically institutes of consultation for the most varied economic activities. Some of the roots of the prodigious flourishing of algebra in Renaissance Italy lay in this very tradition of practical arithmetic, as we shall soon see.

ASTRONOMY

Texts on astronomy

Astronomy students everywhere were required to be familiar with the *Almagest*. At the University of Vienna, for example, where Albert of Saxony (died 1390)—who wrote *Super Aristotelis de coelo et de mundo* and was an adversary of Albert the Great and Thomas Aquinas—was rector from 1365, candidates for the degree of *Magister artium* prepared for their degrees using Campanus of Novara's *Theorica planetarum*, a work directly inspired by the *Almagest*.

Nichole Oresme: the debate on the infinity of worlds

Among those who worked in astronomy during this period, undoubtedly the most interesting was the Norman-born Frenchman Nichole Oresme (c. 1323–82), who taught at the College of Navarre in Paris from 1356. An avid opponent of astrology, translator and popularizer of the classics, and author of notable contributions to mathematics and economics, Oresme also wrote *Quaestiones de coelo*, *Quaestiones de sphaera*, and *Livre du ciel et du monde d'Aristote* between 1340 and 1377.

Although he almost always tends toward traditional cosmological solutions, Oresme's methods of attacking the corresponding problems show that he was an attentive and impartial critic of the cosmology of the times. Faced with his teacher, Jean Buridan's application of the theory of impetus to the motion of the celestial spheres, for instance, he recognized its "non-impossibility" and in fact spoke of the system of the heavens as a "mechanical clock" (a metaphor that remained famous), while at the same time he never abandoned the concept of a mover. In *Quaestiones de coelo*, he also sustained the substantial difference between the mechanics of celestial motion and that of bodies on the earth. As for the infinity of worlds, a much-discussed topic of the times, Oresme leaned toward the effective existence of one world only, but he also affirmed the possibility that God, in His omnipotence, may have created an infinite number. In discussions of the possibility that the earth itself moved, he had no doubts about the particular movement caused by the terrestrial sphere adjusting to changes in the earth's center of gravity, and he could not deny the possibility that the earth rotated around itself. Oresme also attacks astrology. In his *Proportiones proportionum*, on the basis of mathematical considerations he demonstrated the incommensurability of the relations among the motions of the celestial spheres. As a result, he denied that astrology could be a true science, since its previsions were based on calculations of incommensurable, or rather, irrational quantitative relations.

PHYSICAL SCIENCES

Oxford and Paris

The school of Oxford was headed by Thomas Bradwardine (c. 1290–1349), archbishop of Canterbury, who taught there for a number of years. The university also included a group of scholarly professors (Richard of Swineshead and several others), who taught at Merton College in Oxford toward the middle of the century and are generally known as "the *calculatores* of Merton College."

Ockham's influence on the Paris physicists as well as on Bradwardine and his *calculatores* was quite evident, though neither group can really be called Ockhamist, since neither accepted Ockham's philosophical positions in full.

Jean Buridan and the Paris school

In various works, including his famous *Quaestiones totius libri physicarum*, Jean Buridan took up Joannes Philoponos' theory, which had found various supporters among both Arab and Latin scholars in the eleventh, twelfth, and thirteenth centuries, and which even Ockham had shared in the fourteenth century. Buridan started from Philoponos' criticisms of two Aristotelian theories—first that once it had left the mover, the movement of a projectile depended on the medium (usually air) in which it was moving, and second that the acceleration of falling bodies was due to their attraction for their natural places—and maintained instead that all motion was due to impetus (Avicenna and Avempace's "moving force"), a force inside the moving body. In the case of projectiles, Buridan believed that the impetus was transmitted to them by the thrower at the beginning of the motion, and was slowly consumed by air resistance. In the case of falling bodies, natural gravity gave them a certain impetus at the beginning of the motion and gradually increased it as they continued to fall.

Buridan's contributions to the elaboration, or rather the re-elaboration, of the theory were that he arranged it systematically and formulated it with quantitative accuracy.

Geometric coordinates

Oresme and Albert of Saxony were also followers of the theory of impetus. But Oresme's greatest contributions are mainly to be found in his mathematical writings, where he introduced the idea of geometrical coordinates, using what would later be called the Cartesian method. His system allowed him to represent phenomena that changed in time graphically, opening the way for a study of kinematics that was, in a certain sense, modern. Albert further developed the theory of impetus, making remarkably interesting findings, especially in the area of falling bodies.

The British Sophists

Although the Oxford school had several traits in common with the school of Paris, it did not accept the theory of impetus. Oxford's major contribution to physics was that its scholars systematically applied mathematics to the descriptions of physical phenomena. For example, Bradwardine translated those laws of motion—both Aristotle's and Avempace's—that were usually accepted by the scientists of the times into algebraic equations

in order to demonstrate that none of them adhered strictly to the facts. The *calculatores* used methods very similar to Oresme's to describe the phenomena of motion, an approach that enabled them to make a clear distinction between dynamics, which considered the causes of motion, and kinematics, which studied its effects in time and space. They focused their investigations on kinematics, coming up with remarkable results, such as definitions of velocity (and, in a certain sense, of the concept of instantaneous velocity), uniform motion, acceleration, uniformly accelerated motion, etc. The definitions and theories they developed were later rediscovered by Galileo.

The subtlety of their investigations drew the admiration of many of their contemporaries, although it also provoked a certain amount of diffidence toward the Oxford school and the *calculatores* in particular. Petrarch even called them the "British Sophists."

CHEMICAL SCIENCES

The continuation of alchemy's dark age

The Church was not the only institution that looked with suspicion upon the activity of the alchemists. In 1380, Charles V of France, more realistic than the pope, banned alchemical research in his kingdom and prohibited even the mere possession of instruments for alchemical experiments. This interdiction, plus the intrusion of charlatans and speculators, caused alchemy to go through a dark period. Believers in alchemy became the brunt of satires (Petrarch, 1366) and were not thought well of (Chaucer, c. 1388). Not only was there no experimentation taking place, there was not even anyone—as at least there had been in the previous centuries—producing good texts.

However, around 1350, John of Rupescissa wrote *Liber lucis*, improved the production of alcohol as the quintessence of wine, extolling its therapeutic qualities (to the extent that some consider him the founder of medical chemistry), and maintained that a quintessence, i.e. a basic ingredient besides the traditional ones, could be extracted from anything. According to Marcellin Berthelot, Hortulain, who is attributed with the Latin translation of the *Tabula smargdina*, a mystical treatise on alchemy, also belongs to this same period (the historian Karl Christoph Schmieder places him about two centuries earlier).

LIFE SCIENCES

The work of hospitals

While scientific medicine contributed to the flourishing of universities as institutions, the organization of health services developed through the initiatives of both public authorities and the religious orders that worked in hospitals. But the hospitals had serious defects. They were so encumbered by superstitions—beliefs and practices that had no scientific basis—that they were more like depositories for the ill than places where they could be cured.

New forms of therapy

Medical practices were influenced by astrology and alchemy. Diagnoses were often schematical, based on dogmatic assumptions. Physicians had too much faith in pulses and uroscopy and disdained complete examinations. These attitudes arose from a science of pathology that was still founded on the theory of humors and temperaments. Therapy was based on substances extracted from plants—called "simples"—and was written up in herbaries or *hortuli*. But there were more complex medicines as well, such as theriac, a mixture of viper's meat and a number of other substances, which was widely used, although its origins are rather obscure.

The rulers of France and England were recognized as having the gift of the "healing touch." Those who suffered from certain diseases, especially adenitis and tubercular osteitis, went to be touched by the king in the hope of being healed.

Ophthalmology. The concept of contagion and epidemics

Among the specialized branches of medicine, ophthalmology flourished with the teachings of Barnaby of Reggio, who wrote *Libellus de conservanda sanitate oculorum*. The Arabic method of operating on cataracts was taught, and corrective lenses were first applied between the end of this century and the beginning of the next.

But perhaps the most important contribution to medical science during this period was the concept of contagion and epidemics, which the more attentive observers put forth in order to explain the calamity of the plague that was spreading throughout Europe (see LIFE SCIENCES 1300–1350). Curialti of Tussignano opposed the astrologers' belief that the conjunction of Saturn, Jupiter, and Mars on March 24, 1345 caused the plague, proposing instead direct and indirect causes (the contact of healthy persons with people who were sick and contact with infected clothing, respectively). In 1374, the Republic of Ragusa imposed the first quarantines on the crews of suspected ships.

EARTH SCIENCES

Oceanography

The *Catalan Atlas*, a very important work for navigating the more heavily travelled seas of the times, was published in 1375. It showed the distribution of the tides and listed the times of high tides in the English Channel and along the coasts of England.

APPLIED SCIENCES

Military technology

Cannons made by fusing together copper-base metals appeared at the beginning of this period, followed by large guns made of forged iron and the *ribandequin*, a sort of machine gun.

More traditional weapons were modified. Around 1370, the catapult was improved by the addition of a heavy steel arc that could only be bent with a winch-type device or the *cranequin*, a mechanism with a toothed wheel that had blocking teeth.

At the end of the century, a new construction technique was developed: cannons were made with iron rods placed parallel to the axis of the core and welded together over a form that was later removed.

The production of iron

Furnaces for producing iron were designed in such a way that they actually produced an alloy of iron, carbon, and other elements, i.e. basically pig-iron. Iron was in demand for making machine parts that had to be particularly sturdy. It was also used for constructing four-wheeled carriages, which, according to Lynn White, Jr., were quite common by this time.

Time in medieval society

Giovanni Dondi (1318–89), son of Giacomo Dondi (see APPLIED SCIENCES 1300–1350), constructed his "astrarium," an extremely complicated mechanical planetarium, between 1348 and 1364 in Padua. The device soon drew great admiration from all over Europe. Dondi left a very detailed description of the clock with fairly accurate drawings from which the Smithsonian Institute of Washington built a faithful reproduction in the twentieth century.

The central mechanism was operated by a weight and had seven quadrants attached to it that showed the movements of the five planets known at the time, the moon, and the primum mobile, along with another quadrant subdivided into twenty-four hours, two more for fixed and movable feasts, and one for nodes. However, Giovanni Dondi intentionally omitted any drawings of the supports of the "common" clock, the one that showed the hours, considering them too banal. While on the one hand, this indicates a certain widespread use of such mechanisms during that period, it also shows that common clocks were considered mere offsprings of the more complex astronomic clocks, to which the builders directed the greater part of their attentions.

As we have already seen in the section on APPLIED SCIENCES 1250–1300, it was during this period that mechanical clocks finally surpassed the water clocks of classical and Islamic tradition, which were based on the daily cycle of the sun. As far as we know, the first mechanical clock to strike twenty-four hours of equal length was the one built in the church of St. Gothard in Milan in 1335. This was followed by Giacomo Dondi's (1344), then by many others: Genoa (1353), Florence (1354), Bologna (1356), and Paris (1370).

As Lewis Mumford has observed, "...the regular striking of the bells brought a new regularity to the life of the workman and the merchant. The bells of the clock towers almost defined urban existence. Time-keeping passed into time-serving and time-accounting and time-rationing. As this took place Eternity ceased gradually to serve as the measure and focus of human actions.["][1] While the Roman Church accepted and adopted this implicit revolution, the Greek Orthodox Church, which was hostile to technological innovations and reluctant to make compromises, did not. Mechanical clocks were prohibited in Orthodox churches up until the twentieth century.

[1]Lewis Mumford, *Technics and Civilization,* New York, Harcourt Brace and Company, 1934, p. 14.

Humanism, the Renaissance, and the Reformation

(the fifteenth and sixteenth centuries)

1400–1450

THOUGHT AND SCIENTIFIC LIFE

Religious life and Late Scholasticism

The early years of the fifteenth century saw a revival of reform currents within the Church. The Western schism (1378) divided Christianity between those who supported the Roman pope and those who supported the Avignon pope. In 1409, the Ecumenical Council of Pisa attempted to end the conflict by deposing both popes and nominating a third, but the new pope was unable to impose his authority, so in the end there were three rival popes. The schism was only definitively resolved at the Council of Costanza (1414–18), where the currents favoring a reorganization of the Church on conciliar and federative bases came out victorious. A large number of theological and political articles foresaw the reform of ecclesiastical government through a council, but intentionally left no space for the type of radical criticism of papal power that Wycliffe had once made. Peter d'Ailly and Jean Gerson, the most prominent figures in ecclesiastical and university culture during this period, also emphasized their distance from Johann Hus's popular Bohemian movement. Hus was sentenced to death and burned at the stake in 1415. There were profound differences even on a theoretical level. Peter and Gerson supported Ockham's nominalism against Wycliffe's realism. According to Gerson, the realists were responsible for having intellectualized theology, depriving religious life of its more authentic meaning. At Costanza, the nominalists even maintained that the errors of the Hussites actually had their roots in realist rationalism. The latter did not hesitate to respond that Ockhamist voluntarism, with its clear separation of faith and reason, had contributed to radicalizing the polemics against the Church's worldly power.

The conflict between *reales* and *nominales* reached all the large university centers of Europe. Nominalism, which favored the autonomous development of the *artes*, but kept up rational research in the area of linguistic analysis and methodology without compromising on questions of a religious nature, was the prevailing tendency. Italy was not outside this great debate, which was at the same time theological, metaphysical, logical, and political, and which often influenced the individual sciences, sometimes providing bold, original solutions.

Between the fourteenth and fifteenth centuries, the teachings of Blasius of Parma and the physics of Paris and the tradition of the *calculatores* of Merton College were continued in Italy, despite Petrarch's condemnation of ''British sophism'' and the scholars' growing contempt of ''barbarian'' scholastic culture. Centers of philosophical-scientific study that in many ways remained detached from the typically Parisian debate between theology and science also existed, for example in Padua, where the faculty of medicine was dominated by an Aristotelian approach that was more inclined toward problems of medical biology and astrology than toward questions of mathematics and physics, and Bologna, where a school of medical anatomy flourished. The Bologna school considerably influenced the advances made in anatomy in the sixteenth century, especially after the first anatomical tables began to circulate around 1493.

Florence and the early humanists

At the beginning of the fifteenth century, the decline of the more typical structures of the Middle Ages reached its peak. We have already spoken of the grave crisis in the Church; the other universal institution, the empire, had also lost power and prestige as new states arose that were better able to stabilize their territorial unity, expand their financial and commercial enterprises and accelerate technological development. The creation of a central authority above all required the formation of a professional bureaucracy. As a result, studies became increasingly laicized. Protected by a monarch or a prince, they were no longer under any direct control from the ecclesiastical authority. This process took place both in countries such as France and England, where the monarch was able to create a solid centralized state, and in Italy, where the process of national unification was hindered by the establishment of a number of small states, ruled by lords or princes who were jealous of their autonomy. In a number of regions of Italy, commune experiments continued, and a class of wealthy merchants emerged, the nucleus of a new aristocracy that was much different from the warlike aristocracy of medieval tradition. This new aristocracy was interested not only in economic and political life, but in advancing the arts and letters, as well as scientific and technical research. The most important commune was Florence, which, as early as 1377, Coluccio Salutati called the ''illustrious city, the flower of Tuscany and the mirror of Italy, a rival of the very glorious Rome from which it is descended.'' The chancellors of the Florentine republic, which was theoretically modelled after Cola di Rienzo's Rome, or even after the republican Rome of the classical age, were, like Leonardo Bruni (c. 1369–1444), men of action, who were versed in Greek and Latin. They represented the highest authority in the city-state, and followed the theoretical teachings of Petrarch and Salutati.

Characteristics of early humanism

The Humanist-chancellors of Florence basically acted as mediators between the political institutions and the revival of classical studies. We should mention that the cultural current generally indicated as early humanism was not a unified school, insofar as the cultural attitudes of the scholars that were joined by a love for the classical world and a contempt for scholastic metaphysics were extremely varied.

In fact, echoes of Ockhamist voluntarism and Augustinian tradition were present at the beginning. Later, two themes emerged as central: the importance of the human being as an individual, and the resulting superiority of the will over the intellect. These themes developed in a more worldly framework than Augustinian tradition provided. It was a return to the classics through a recovery of Latin and Greek texts, with an added knowledge of the break that the Middle Ages had constituted. The uniqueness of the humanists' attitude toward classical culture lay not so much in ''an admiration or a love for antiquity, nor [in] a greater knowledge of antiquity, but consists rather in a well marked historical consciousness. The 'barbarians' were not barbarous because they had remained ignorant of the classics, but because they had failed to understand them as a historical phenomenon. The humanists, on the other hand, discovered the classics because they managed to detach themselves from them and comprehend their Latin without confusing it with their own Latin.'''

Thus commitment to civil life and careful philological research tended to merge. Beginning in 1417, Bruni himself translated Plato's dialogues and Aristotle's *Ethics* and *Politics* from the Greek. His model of the ideal republic was inspired by Aristotle's political theories. Manuel Chrysoloras taught Greek in Florence from 1398; Bruni, Pier Paolo Vergerio (1370–1444), and Guarino da Verona (1374–1460) all studied at his school. Vergerio and Verona both came from Venice, studied at Padua, and were later involved in the educational reform that viewed education as the formation of people who would be involved in the civil and political world, free from the worries of working at a job and intent upon attaining virtue and wisdom. Vittorino da Feltre (1378–1446) also played an important role in this pedagogical reform. Although he was not part of the Florentine group, da Feltre founded an experimental children's school in Mantua that fit the idea of a general education centered around rhetoric and philosophy and followed classical models.

Developments in humanism

Poggio Bracciolini (1380–1459) and Giannozzo Manetti (1396–1459) were prominent figures in the Florence circles. Bracciolini, who discovered a number of ancient works, including Lucretius' *De rerum natura*, criticized medieval asceticism, extolled human labor, and foresaw a unity of knowledge in function of civil life. Manetti developed the theme of human dignity in contrast to ascetic and pessimistic tendencies, identifying a theme that would later be taken up, with infinite variations, by a large part of humanist culture. We have already spoken of the pedagogical activities of Vergerio, Guarino, and Vittorino da Feltre; another important figure was Mapheus Vegius (1407–58), who dealt with pedagogical and moral themes in an attempt to reconcile classical culture with Christian tradition. The rapid spread of an interest in education attests to the extremely positive role that the cultivation of the classics played during this phase of humanism in picturing the cultured person as playing an active role in society.

Just as important was a new historical approach that can be seen, for example, in the works of Bruni and Bracciolini. An attempt was made to reassess the role of the supernatural in history in favor of an interpretation of events as seen through human emotions. This gradual secularization was taking place outside Florence as well, especially in the teachings and works of those scholars who moved about in the various cultural centers of Italy and contributed to spreading the new culture. Lorenzo Valla (1407–57) and Leone Battista Alberti (1404–72) are two typical examples. Valla taught from 1420 to 1433 at the University of Pavia, where a lively group of humanists held frequent critical and philosophical debates on Latin and Greek texts. In 1431 he composed *De vuluptate,* a detailed criticism of the first four books of Boëthius' *De consolatione philosophiae* (he critiqued the fifth one in his dialogue entitled *De libero arbitrio*). Disagreeing with both Aristotelianism and Stoicism, Valla maintained that the motive of all acts was pleasure (even religion tended toward a *divina voluptas*). Valla's hostility toward the medieval schools and their barbarian Latin is also apparent in the work. It was a hostility that was shared by many of the intellectuals of Pavia, including Maffeo Vegio, who was a friend of Valla's and appears as an interlocutor in the *De libero arbitrio.*

Almost simultaneously, in 1437–41, Alberti, born in Genoa but Florentine by adoption, composed his treatise entitled *Della famiglia*, in which he outlines the required education for the complete person, who would be integrated into public life and at the same time capable of *"bene et beate vivere."* Alberti's cultural and philosophical commitment is even more interesting, in that his intellectual activity ranged from architecture, painting, and sculpture to philosophy, pedagogy, mathematics, and engineering. On a philosophical level, Valla's teachings tended toward a revival of Epicurean themes, while Alberti's emphasized a resolute voluntarism, with touches of Stoic-type Christianity.

Controversies on astrology

Humanism brought about a revival not only of considerations on the human world, but of reflection on the natural world as well. Not only was there a new interest in the particular structures of the physical world and in technical devices for dominating them, but a new way of considering nature in its entirety was developed. Nature was no longer the scene of temptation and expiation, but rather an autonomous reality, often the residence or body of God. Along with this concept, magicians and scientists also developed an animistic perspective that saw the world as populated by genies and demons that humans hoped to learn about and dominate by means of occult practices.

These new pantheistic concepts of natural reality were united with traditional esoteric themes. Astrology is a typical example: in some cases it appeared as the embryo of a mathematical description of heavenly phenomena and in others as a tool for prophesying and intervening in the course of events. In any case, astrology was very popular among scholars, where the naturalistic and pantheistic currents already mentioned often merged with traditional Arabic ideas about the influence of the stars. Thus there were tendencies that ranged from Gerson's condemnation of astrological practices to the monk G. Ganivet's "astrologia physicata," in his *Amicus medicorum* (1431), which attempted to define the relations between celestial events and the care of the human body.

Developments in theological-political thought

The most important event in the life of the Church during this period was the Council of Basel, which began in 1431 and ended in 1449. The supporters of papal authority and those who supported the authority of the council confronted one another once again, and the problem of the relation between ecclesiastical and lay authority was again examined. *De concordantia catholica* by Nicholas of Cusa (1401–64), gave considerable support to the mediatory positions. Unlike the Council of Costanza, the Council of Basel reinforced the power of the pope (Nicholas himself later joined those who supported the pre-eminence of papal authority). The papacy also won a huge, though ephemeral, victory (it lasted twenty-three years), at the Council of Florence in 1439, with the signing of an edict that reunited the churches of the East and the West. It was an extremely critical period for Constantinople, which was under pressure from the Turks, and the reunification meant closer contacts between Western and Eastern scholars, marking an important turn toward a more direct knowledge of the great Greek works.

These papal victories did not prevent the debate on the legitimacy of the secular power of the popes from raging on. In 1440, in his *De falso credita et ementita Constantini donatione*, Valla corroborated the traditional perplexities concerning the so-called Donation of Constantine, using a compelling philological analysis. The importance of the work lay in the analysis itself, which constituted a model for all those who used philology as a weapon in theological and political debate. Two fundamental works, from a philosophical point of view, appeared that same year: *De docta ignorantia*, by Nicholas of Cusa (see ASTRONOMY 1400–1450), which openly attacked the "Aristotelian sect" and was full of speculative mathematical ideas that had a considerable influence on certain developments in that field; and the *Dialecticae disputationes*, by Valla, who, after a period in Pavia and a visit to the court of Alfonso of Aragon, had opened a school of rhetoric in Rome. Valla brought various criticisms against Aristotelian, Stoic, and Scholastic logic, proposing instead rhetoric, in which linguistic structures were studied in relation to persuasion. Valla definitely favored a science of language that would have a grasp on the civil and political world and would be able adequately to make use of Cicero and Quintillian's models. His position naturally had a considerable influence on developments in logic, and contributed to a devaluation of the formal-type studies of late Scholasticism. More generally, during the entire sixteenth century students of logic and rhetoric as well as politicians and theologians—both Catholic and Protestant—who were involved in religious debates also looked to Valla's model.

Byzantine scholars in Italy

Florence was chosen as the site for the Council of 1439, which was attended by the emperor of Constantinople (John VIII Paleologus) and the patriarch. Contacts between East and West progressively increased during the following years. Finally, when Constantinople fell to Mohammed II's Turks, all the major representatives of Eastern culture were forced to emigrate to the West, and Florence seemed to be one of the most suitable places to host the Byzantine scholars. It was natural for the city to become the most prestigious center in the rebirth of classical studies during this period. One of the most influential Eastern masters was George Gemistos Plethon (c. 1355–1452), who had attended the Council of 1439 and was an avid defender of ancient Greek theology against Latin Aristotelianism. Around 1440, he composed a work in Greek about the differences between Plato and Aristotle, which is a vindication of Plato against Aristotle. Plethon's Platonism did not coincide with the Athenian philosopher's thought, however, but rather with the Neoplatonism that was very much alive in the entire philosophical tradition of the Christian East. His interpretation of Platonism became typical of the Florentine environment and was a characteristic feature of most of Renaissance Platonism.

There were other Byzantine scholars, however, who defended Aristotle. There were, for example, George Scholarius (1405–72), Theodore of Gaza (1400–75), and George of Trebizond (1395–1486), who wrote *Comparationes philosophorum Aristotelis et Platonis*. The polemic created a conflict between Aristotelians and Platonists that dominated not only philosophy, but specialized scientific research as well. One of the first results of this polemic was a better knowledge of the two great Greek philosophers' works. A careful philological reconstruction was favored by theological interests and the taking of sides within cultural circles. The case of John Argyropoulos (1415–87), who was called to Florence in 1457 as a lector of Greek philosophy is typical; although he was a declared Platonist, he also taught Aristotelian thought. He compiled an edition of Plotinus's *Enneades* as well. Another figure deserving of mention is Cardinal Bessarion (1403–72), who, although he attempted to bring out the points that Plato and Aristotle had in common, insisted on the harmony between Platonic philosophy and the principles of true theology. He thus opened the way to a merging of Platonic (or Neoplatonic) tradition and the revived Christian culture. At almost the same time, Pomponius Laetus (c. 1425–97) revived an interest in Stoic and Epicurean doctrines at his Roman academy.

Nicholas of Cusa's thought developed during this same period. He added *De quarendo Deum* (1445), *De genesi* (1447), the *Idiota* (1450), *De visione Dei* (1453) and *De beryllo* (1458) to *De docta ignorantia* and his other great work of 1440, *De coniecturis*. The new works reiterated that we should not believe that we can succeed in fully understanding divine reality with our own intellect, and that we must proceed by simple conjectures instead. Although Nicholas's thought had different cultural roots from that of the Byzantine scholars, his speculative mysticism also contributed to a revival of the Platonistic view of the universe, in contrast to Aristotelian naturalism. Other works by Nicholas, such as his *De pace fidei* (1453), reflect the difficult times that Christianity was going through, especially with the fall of the millennial Byzantine empire and the confusion caused in the Mediterranean basin by Turkish conquest.

MATHEMATICAL SCIENCES

Mathematical culture and humanism

The fifteenth century, the typical century of humanism, marked the great reawakening of a literary civilization that found ideas for its profound resurgence in the rediscovery of classical values. Even the sciences (especially mathematics) profitted from the passion for rediscovering classical works that was characteristic of this century. Many ancient mathematical texts that had been lost during the Middle Ages were rediscovered, published, and translated, especially after the invention of printing. Even the resurrected Neoplatonism of the humanists contributed to arousing a keen intellectual interest in mathematics, especially in its theoretical and even its speculative aspects. An example in the work of Nicholas of Cusa (1401–64) is the importance given to reflections on mathematics and to philosophical considerations that made use of mathematical examples. In his most important mathematical work, *De quadratura circuli*, Nicholas considers the problem impossible from the strict point of view of equality.

Along with this speculative mathematics a much more elementary mathematics, concerned with concrete problems, survived as well. It was a direct descendent of the earlier abacus tradition, and was aimed at solving the arithmetical problems that arose in commerce, banking, and accounting. Several students of Fibonacci, Biagio Pelacani (died 1416) and Prodoximo of Beldomandi, flourished during these years.

It was also important that the best intellectuals recognized the educational value of teaching mathematics, along with the traditional value attributed to literary studies. Vittorino da Feltre (1373–1446), for example, gave mathematics a position of full standing in the curriculum of his educational institution for children.

Mathematics and the figurative arts

Even the extraordinary flourishing of painting and architecture had a direct influence on the development of mathematics: the desire to reproduce objects faithfully was such that artists were no longer satisfied with the resources offered by an intuitive perspective which had, for example, sufficed in Giotto's painting. So they attacked theoretical problems, and the result was a true science of perspective, founded on rigorous geometrical principles. The groundwork for this new science was laid by the *Trattato della pittura*, written by Leone Battista Alberti (1404–72) in 1436, while Alberti's *De re aedificatoria*, begun in 1445 and printed in 1485, introduced the use of multiple mathematical considerations to architecture. The first true treatise on perspective was *De prospectiva pingendi*, written by Piero della Francesca (c. 1420–1492) between 1470 and 1480. It remained unpublished (although it was quite well-known) until the beginning of the twentieth century. The *Libellus de quinque corporibus regularibus*, another important work on geometry written by della Francesca slightly later, had the same fate. Albrecht Dürer (1471–1528) continued this great tradition with his *Underweysung der Messung mit Zirckel und Richtscheyt in Linien Ebnen, und gantzen Corporen*, written in 1525.

Despite these stimuli, actual mathematical activity and research languished until the end of the century. Although the names of several talented scholars, such as George Peurbach and Regiomontanus (whom we shall return to later), can be mentioned, their interest lay more in other sciences, especially in astronomy, than in mathematics.

ASTRONOMY

The reflowering of astronomical studies in the fifteenth century

During this period, there arose a new and more productive interest in an experimental and theoretical knowledge of the heavens, especially in northern Europe. Professional philosophers and astronomers developed a series of theories and observations that provided the first impulse that was original and independent of Arab contribution which led to the great cosmological constructions of the following century. They also imposed a new observational rigor that would guarantee the scientific success of these constructs. Nicholas of Cusa, George Peurbach (1423–61), and Regiomontanus (1436–76) were the most illustrious representatives of this revival.

In Italy, even "judiciary astrology" produced an important figure in the astronomer and geographer Paolo dal Pozzo Toscanelli. In the Arab world, the last great cultivator of Eastern astronomical tradition was Ulugh Beg (1394–1449), a Mongolian Turk.

From Marāgha to Samarkand

In the East, between 1407 and 1416, the Persian mathematician al-Kāshī compiled tables and composed treatises in astronomy, improving the findings of al-Tūsī's school at Marāgha.

Around 1420, after travelling from city to city without finding a decent position, al-Kāshī reached Samarkand, in present-day Uzbekistan, where Ulugh Beg, the grandsom of Tamerlane, ruler of Transoxiana (western Turkestan) from 1407 and Emperor of Turkestan from 1447, welcomed him into the circle of scholars he was organizing in the city's *madrasa* (university). Samarkand was Ulugh Beg's capital and under him had become the East's major center for scientific and theological studies. Al-Kāshī's arrival and encounter with Ulugh Beg undoubtedly contributed to the great flourishing of studies at the Observatory of Samarkand.

Nautical astronomy

New observatories were founded in Europe and Asia, although they did not add anything substantially new in the way of construction or instrumentation. In 1418, Prince Henry the Navigator (1394–1460) of Portugal, having studied the navigational experiences of the Arabs, built an observatory at Sagres, near Cape St. Vincent, the southwesternmost point of continental Europe, for the purpose of compiling solar tables for use in navigation. The explorations that were being conducted along the African coasts made it necessary to improve existing nautical astronomy techniques and tables. This discipline became even more important in the second half of the fifteenth and during the sixteenth centuries, with the first voyages to the New World.

The Samarkand Observatory

Between 1420 and 1424, Ulugh Beg, whom we mentioned earlier, founded a large, well-equipped astronomical observatory at Samarkand. The Observatory, the remains of which were discovered in 1908, contained, among other instruments, a giant quadrant (a sextant, according to some) with a radius of 40.2 meters that must have made it possible to achieve an unprecedented accuracy. With the aid of this instrument, a catalogue of 1018 stars was compiled; it also contained astronomical tables called the *Zij Ulugh Beg*. The work of the Samarkand astronomers became famous throughout Europe and exerted a strong influence on developments in astronomy there.

While Islamic astronomical tradition experienced its last

period of glory at the court of Samarkand during these years, studies in astronomy and mathematics in Europe were carried on above all at the Universities of Padua and Bologna, and in Germany; simultaneously, the Paris school of the philosophy of nature was declining.

Around 1420, Dal Pozzo Toscanelli and Nicholas of Cusa met at the University of Padua, which remained one of the major European centers for the study of astronomy for many years. Toscanelli was a professional astronomer and astrologer who made several important celestial observations; Nicholas of Cusa was a philosopher and the author of a cosmological theory that was, in some respects, revolutionary.

Astrology and astronomy in Florence

Paolo Dal Pozzo Toscanelli (1397–1482) of Florence, a friend of Filippo Brunelleschi (1377–1446), to whom he gave lessons in mathematics and probably technical advice for building the dome of Santa Maria del Fiore (1420–34), was the major representative of "judiciary astrology" in Florence. As such, he was consulted by the governing council as to the appropriateness, on the basis of the astrological situation, of the most important decisions in affairs of state. In 1433 he observed a comet, also reported in Chinese annals and in Poland, and, using methods that anticipated those of Regiomontanus, illustrated its successive positions with respect to the fixed stars in a drawing that has since been shown to be extremely accurate.

Nicholas of Cusa's philosophical speculation and cosmology

Meanwhile, knowledge of the Greek language was growing, and Greek manuscripts were becoming more accessible. This fact proved to be quite beneficial to the sciences as well, especially astronomy, which could acquire a direct knowledge of Ptolemy's Sintaxis without the mediation and interpolation of Arabic translators and commentators.

Nicholas of Cusa (1401–64), a German prelate and philosopher born at Kues, Rhineland, cardinal in 1448, and bishop of Brixon in 1450 became interested in mathematics and astronomy during his stay in Padua, and in 1440 finished his De docta ignorantia, a work that presents, along with mystical concepts and metaphysical speculations, an extremely interesting theoretical cosmology, though generally speaking it has no experimental basis. The universe extends infinitely, "a sphere of which the center is everywhere, and the circumference nowhere"; consequently, the earth cannot be at its center. The earth is a "star that moves," despite the fact that we cannot perceive this motion, since there is no way to compare it with other bodies at rest. The sun is larger than the earth, which is in turn larger than the moon, "as we know by its shadow and by eclipses." There are inhabitants everywhere in the universe and, in fact, some of them are more perfect than humans.

It is not clear, however, whether the idea of the earth's motion meant for Nicholas that it revolved around the sun or around itself, or if it was a "motion" of the earth's surface toward the "center" of the world, as a result of gravity. The last hypothesis agrees more with the medieval notion of orogenetic phenomena and, in general, of variations in the earth's shape over time.

Apart from these theories which, in the light of more radical uses that Giordano Bruno later made of them, proved to be

forerunners of Copernicus's and Kepler's theories, Nicholas of Cusa's cosmology presents a number of obscure features and backward ideas, for instance the theory that the earth and the moon shed their own light, or that there were inhabitants on the sun who were spiritually superior to human beings. In reality, as Alexander Koyré has observed, we should not regard Nicholas of Cusa's concept of the infinity of the universe as having a precise astronomical meaning. In fact, he believed in the existence and motion of heavenly spheres, as well as the existence of a central region of the universe around which everything else moved. Nicholas also denied the explicative function of mathematics that is typical of modern science by the simple fact that his system was based on the "coincidence of opposites," which made circumferences equal to straight lines, small equal to large, and zero equal to infinity.

The decline of Arab astronomy

In 1449, Ulugh Beg was assassinated at Samarkand by a palace conspiracy led by his son. His circle continued its activity in the famous observatory and in other parts of Islam, through the work of al-Qūshji (died 1474), who moved to Istanbul, which became in 1453 the new capital of Ottoman Turkey. He can be considered the last in the line of Arab astronomers.

The work of George Peurbach

The Austrian George Peurbach (1423–1461) went to Italy around 1440, where he met the Ferrarese astronomer Giovanni Bianchini, author of a modified edition of the Alfonsine Tables. In his Theoricae novae planetarum, Peurbach essentially takes up the ideas of the Almagest, filtered through Sacrobosco's popularized version, though he also inherited the Arabs' concept of solid crystalline spheres, which replaced the Alexandrian astronomer's purely abstract geometrical construction. Among other things, Peurbach emphasized the fact that in Ptolemy, the movements of Venus and the three higher planets were in harmony with the movement of the sun, and also pointed out that Mercury followed an epicycle whose center described an oval orbit (already suggested in the thirteenth century codes of the Libros del saber). The Tabulae ecclipsium, another of Peurbach's works on the subject of astronomy, was published posthumously in 1514.

In 1456, various chronicles recorded the passage of an extremely large comet. This was Halley's comet, of which dal Pozzo Toscanelli, from the vantage point of his Florentine observatory, made a fairly accurate description, accompanied by drawings that indicated its successive positions.

PHYSICAL SCIENCES

Humanism and science

The contributions to the development of physics that can be traced back to the fifteenth century are much less important than those made by the great masters of the previous century. The most significant event for the culture of the new century was the remarkable increase in familiarity with Latin and Greek classical texts (the latter translated directly from the Greek rather than from earlier Arabic versions). As we know, the fifteenth century humanists were most interested in topics of a philosophical or literary nature, but their interests slowly expanded to include scientific works as well. The diffusion of these works contributed significantly to the general revival of mathematics, astronomy, physics, and biology that took place in the sixteenth century. The most obvious illustration of the meeting of *studia humanitatis* and scientific studies is found in the work of Giorgio Valla (c. 1430–1499), an expert in both. His two volumes entitled *De expetendis et fugiendis rebus opus* (published posthumously in 1501) deserve particular mention; they are an encyclopedic collection of excerpts from previously unknown writings by Hippocrates of Chios, Aristarchos of Samos, Heron, Archimedes, etc.

The spread of physical theories

Another fairly significant event that characterized the culture of this period (including science) was its gradual spread from the great centers of France and England, which had been the strongholds of medieval thought, outward to the rest of Europe, and Italy in particular. Physics spread in the same way, so that an increasing number of authors accepted the theory of impetus and the kinematics of the *calculatores*, a phenomenon that became part of the general reaction against Aristotelianism guided by a resurgent Platonism. We should note however, that although the theory of impetus evolved from a documented critique of Aristotle's explanation of motion, it still remained within the conceptual framework of Aristotelian physics, since it continued to be based on the belief that a force was necessary in order to keep a body in motion. It simply maintained that this force was inside the moving body, and not outside, as Aristotle had postulated.

Biagio Pelacani and the geometric method

In Italy, the most energetic defender of the ideas of the Paris and Oxford physicists was Biagio Pelacani of Parma (died 1416), who studied at Pavia, then taught in various universities on the peninsula (Pavia, Bologna, and Padua). In his numerous scientific writings, composed during the late years of the fourteenth and the early years of the fifteenth centuries, he clearly outlines the principles of Buridan's dynamics and the kinematics of the *calculatores* of Merton College. His use of the representative geometric method, also used by Oresme and the *calculatores*, also deserves mention.

Nicholas of Cusa: scales as instruments for research

Another harsh critic of Aristotle's explanations of motion and an avid defender of the theory of impetus was the German Nicholas of Cusa (1401–64), who lived for quite some time in Italy and is particularly well-known as a philosopher (of openly Platonic, or rather Neoplatonic leanings) and mathematician. Book IV of his *Idiota*, entitled *De staticis experimentis*, is dedicated to physics. The experiments he speaks about focus on the use of scales (with which he also wanted to demonstrate that air had weight), and are often vague, to the extent that some scholars believe he did not actually carry out some of them, but only planned them in his head. In any case, the work is valuable in that it shows that studies in physics were taking a new turn, one which would soon prove to be of major importance.

Leone Battista Alberti (1404–72), whose writings on architecture indicate a knowledge of mathematics and mechanics (see MATHEMATICAL SCIENCES 1400–1450), also displayed great interest in observing a variety of physical phenomena.

CHEMICAL SCIENCES

The birth of iatrochemistry

The infernal odor of sulphur that emanated from alchemic attempts at working magic led to further bans and confiscations. In 1404, Henry IV of England prohibited alchemic practices, and in 1418, the Great Council of Venice did the same. The danger of being accused of witchcraft, if not of heresy, and of ending up at the stake was still far more than theoretical.

Nevertheless, John of Fontana, a physician and military engineer, was able to shoot his diabolical multi-colored figures into space with rockets and black powder, frightening a large number of people, without being accused of witchcraft. Perhaps more caution was being taken in attributing supernatural powers to people.

Following suggestions made by Taddeo Alderotti and Joannes de Rupescissa, many physicians used not only *acqua ardens*, renamed *aqua vitae* for its curative powers, but other organic and inorganic products as well in the treatment of diseases. Their remedies—potent toxics such as compounds of antimony and mercury—were often worse than the diseases themselves, but iatrochemistry was incontestably born during this period, quite some time before Paracelsus.

Michele Savonarola (1384–1468), grandfather of Girolamo, used chemical remedies and carried out studies on the evaporation of mineral waters, which he describes in a book on mineral water cures. He may have written it *scripta manet* because the topic was safe—not many authors of works from this period are known; dates and names are often left vague and in doubt.

Notable progress was made in technology, especially in war science, where moral issues were not a problem. The quality of gunpowder was improved, progress was made in iron-working, and the fusion of pig-iron was finally accomplished.

LIFE SCIENCES

Early anatomical illustrations

Evidence of a new trend of thought geared toward concreteness, experimentation, and a direct relationship with classical antiquity can also be found in the life sciences. Anatomists were able to compare the Galenic text with observational data as prohibitions against dissecting cadavers were gradually lifted. Already in the previous century, Guido of Vigevano and Guy de Chauliac had combined Mondino's ideas with original evidence from autopsies. The practice of autopsies spread from Bologna, where the surgeon from Parma had studied, to Montpellier, Vienna, and Paris. A new style of anatomical drawing developed, which can be seen in a manuscript dated 1420 that contains the work of the English physician John Arderne.

Medical Aristotelianism

In Padua, medical Aristotelianism led to a considerable awareness of method and to speculative freedom.

Jacopo da Forlì (died 1413) and Ugo Benzi of Siena (c. 1360–1439), both professors of medicine in Bologna, investigated the difference between *resolutio* and *compositio dimostrativa* in their commentaries on Galen. They reduced the difference to the Aristotelian *qua* and *propter quid* demonstrations, contributing to detaching scientific reasoning from the pattern of the syllogism. In 1426, Guarino Veronese (1374–1460) rediscovered Celsus's works on medicine, which had been lost for more than five hundred years.

EARTH SCIENCES

Geographical discoveries

The era of great geographical explorations was inaugurated in 1416 when Henry of Portugal (1394–1460) founded a school of navigation.

Between 1427 and 1437, the Portuguese discovered the Azores and the Canary Islands, disputed by the Spanish, or more correctly, the Castilians. From this period on, new conquests followed rapidly: the Portuguese reached Cape Blanc in 1441 and the mouth of the Senegal River in 1445, and in 1446, Álvaro Fernandes landed on the shores of Sierra Leone after rounding Cape Verde.

APPLIED SCIENCES

Brunelleschi's dome

This was a period that decided the new direction of modern architecture. In 1415, Fillipo Brunelleschi (1377–1446) returned to Florence after a stay in Rome, where he had been inspired by the ancient monuments, and conceived his masterpiece, the dome of Santa Maria del Fiore. In 1417 he raised the tambour on four pillars to the height of the ''oculi,'' which let light inside.

His daring concept was harshly criticized by Lorenzo Ghiberti (1378–1455), the sculptor who made the famous doors of the baptistry. Ghiberti objected that without a keystone, the dome would not stay up, since the two meridian arch elements, separated along their axis by the vertical, had no point of support.

However, Brunelleschi saw the dome as a series of rings that fit inside one another. Each ring could be inserted into the conic-trunk bowl beneath it, which would be completely stable in that its elements were compressed by their own weight. The success of the project excited much amazement in Florence and in the world, since the dome was raised without scaffolding, buttresses, or arcs-boutants, in a void, as a series of crowns. The dome was finished in 1425, and later the lantern was added; according to Brunelleschi, its weight would only add to the stability of the vault.

Hydraulic engineering

The first Italian schools of engineering were also founded during this period, uniting architects, technicians, and scientists.

Of particular importance was the school established at the court of Sigismondo Malatesta in Rimini. Malatesta, a fancier of architecture, designed the fortifications of the Malatesta castle (begun in 1438), the construction of which he planned and personally directed.

In Italy, Giovanni Fontana wrote a treatise on aqueducts. In it, he proposed, among other things, a machine operated by hot air for raising water from wells. Fontana was one of the first supporters of the idea of replacing the gravity system of water supply with the pressure system.

In Holland, longer-lasting protection for dykes was developed in the form of barriers built with large piles which also acted as breakwaters. The dyke of Amsterdam was protected in this way. In 1435, the Dutch devised a ship with a large rake, called a rasper, for dredging their harbors.

The method of building canal locks was also improved by the use of devices that controlled the flow of water coming into and going out of the basin through central or lateral openings in the locks themselves. The same system, improved even more, was adopted in Milan in 1440, and is described by Leone Battista Alberti (1404–72) in his *De re aedificatoria*.

Iron-working processes

When water power began to be used to inject air into iron-ore reducing furnaces, it became profitable to build them, and they could be stoked with more fuel. The development process was complete when an iron that was richer in carbon was obtained. This development had its own specific notation: the furnace was no longer called a *Stuckofen*, but a *Flussofen*, i.e. the classical blast furnace, twelve to fifteen feet high with hydraulically-operated blowers. The change came about in the region of Liège and along the banks of the Rhine between the end of the fourteenth and the beginning of the fifteenth century, then spread into the Low Countries and the French provinces, and later into Germany and southern Europe.

Mechanical technology

The crank-and-connecting-rod system appeared at the same time as the blast furnace, probably as a result of the increasingly widespread use of water power. The system made it possible to convert continuous circular motion into reciprocating motion and *vice versa*, and was one of the most important mechanical acquisitions of the entire century. The first iconographical evidence of its use dates back to this period, and appears in the *Bellifortis*, a treatise on military engineering written by Conrad Kyeser (c. 1366–1405) in 1405. In Kyeser's work, the system appears on a hand mill that also has a flywheel for overcoming dead points.

The second important work of the German school following that of Kyeser also dates back to this period. More than a real treatise, it is a pad of notes and observations made by an unknown author and is called the *Anonymous Treatise on the Hussite Wars*. The manuscript is essential for an understanding of technology in the early decades of this century, especially in the field of mechanics. It contains the earliest drawings of a wood-boring machine and a machine for hollowing out tree trunks to be used as water pipes, as well as sketches for an early underwater outfit, with diving suit and breathing tube, preceding Leonardo's famous design.

If Kyeser's sketch of the crank-and-connecting-rod system still reflected some uncertainty, it is completely dispelled by the accuracy with which the system is illustrated here, as applied to a hand-and-foot mill. Another mechanical invention illustrated more tentatively in the Hussite manuscript is a machine for polishing precious stones, some elements of which (the pivot of the disc, wrapped with a cord moved by a crank) are recognizable in one drawing. In Italy, the first report on the crank-and-connecting-rod system was written between 1441 and 1458 by Mariano di Jacopo, known as Taccola.

A machine that connected the Archimedes' screw or cochlea to a windmill appeared for the first time in Holland in 1404. The machine was used in a drainage project, for pumping water out of polders where the canal sluice gates were not effective. (When the reclaimed land had been taken from the sea, it was in fact possible to drain off the water by taking advantage of outgoing tides.)

The first real spring-operated clocks began to appear around 1440–50.

Scientific works

One of the first scientific works on water distribution was the *Bellifortis*, mentioned earlier, although the techniques described mainly concerned military encampments. The entire third volume of the work is dedicated to hydraulics. The first two volumes contain descriptions of siege wagons, revolvers, and other military machines which were now much easier to produce, due to the more advanced systems for fusing iron. In 1409, Angelius prepared the first translation of Claudius Ptolemy's *Geography* into Latin. From this point on, the ancient circular world map began to disappear and be replaced by the new, spherical type of world map.

In 1414, an ancient manuscript containing the lost text of Vitruvius' *De re aedificatoria* (see PHYSICAL SCIENCES 100 B.C.–0) was discovered in a Swiss monastery.

The graphic arts

A printing house owned by Johann Gutenberg (1400–68) Johann Fust and Peter Schoffer was operating in Mainz in 1447. It is difficult to delimit the contribution that these printers made to the invention of the graphic arts, since similar experiments were being carried out simultaneously not only in various other parts of Europe, but in Korea, where a type foundry existed and a book printed with metal type was published in 1409.

The invention was in the air, favored by the general reawakening of inventive activity within the context of an expanding social structure. And it was urgent as well, since the skins of twenty-five goats were needed to produce a 200-page folio volume. The idea of type may have been imported from the East, or suggested by the experience of playing-card manufacturers, who existed in large numbers in the cities of southern Germany. They printed their cards by carving the figures in relief on wooden blocks.

The first printed works that can reliably be attributed to Gutenberg's printing house date back to 1448. The letters are composed and aligned so carefully that they must have been fused in a specially designed mold. Although the invention of type was essential, suitable inks and presses were just as indispensable.

Polydore Vergil, writing in 1499, attributes the invention of varnish-based ink to Gutenberg as well, but we should mention that the school of Flemish painters had already discovered that cooked linseed oil gave a good-quality, adhesive, fast-drying varnish. The invention of ink made it possible to use presses adjusted to operate faster.

Explosives

The manuscript known as *The Anonymous Treatise on the Hussite Wars* also contains a description of the method used to make gunpowder. In 1420, granular gunpowder appeared. This invention made it possible to achieve an even distribution of all the different ingredients throughout the powder. The evolution of gunpowder manufacturing techniques and the concomitant developments in metallurgy, explain the victories won by artillery and the improvements made in fortification systems.

Navigation

Progress was also made in navigation. In 1436, Andrea Bianco's *Atlas* carried *martelogio* tables, tables for calculating the distance sailed in the desired direction when the ship was tacking. The so-called *martelogio* method required the trigonometrical solution of triangles. It is believed that the method was already known in earlier times (Ramon Lull may have been alluding to it in his *Arbor scientiae*, therefore before 1295). However, along with nautical maps, it seems to have been the exclusive property of Italian navigators, so that the Italian sailor was the first technician to use applied geometry.

1450–1500

THOUGHT AND SCIENTIFIC LIFE

Traditional culture

The political and cultural developments mentioned earlier soon made the Italian cities mandatory reference points for scholars north of the Alps. But for this entire period and beyond, traditional European university culture was basically involved in the controversy between realists and nominalists and, more in general, in developing the ideas typical of late Scholasticism. In 1455, Gutenberg's *Bible*, the first printed book, came out. The advances made in the art of printing were initially put to the service of traditional theological and philosophical culture. The writings of Ockham and the Ockhamists had an especially wide circulation and were read in many scholarly circles. Only later, with the flourishing of the first great Italian printing centers, did the invention contribute to the development of humanist culture.

The Accademia Platonica of Florence

The times were extremely complex from a political point of view. The Western countries began to feel the effects of the fall of the Byzantine empire. The end of the Hundred Years War (1453) marked the beginning of a difficult period of adjustment for France and England, and the War of the Roses (1455–85) raged in France. A process of aggregation was taking place in Spain, while in Italy, with the Peace of Lodi in 1454, an equilibrium among the various states was reached that did not permit the type of centralization that had occurred in other European countries to take place. The interests of the papacy, which wanted to establish itself as both a strong state on the peninsula and the ideological center of Christianity, were antithetical to this process of non-centralization.

In many respects, however, Italy represented a reference point for the rest of Europe. Venice and Genoa still maintained a definite superiority in the area of trade, despite the increasingly insidious presence of the Turks in the Mediterranean. Lombardy and Tuscany continued to grow commercially, and Florence in particular became an extremely important banking center. The powerful Medici family, which had already asserted itself in the decade from 1430 to 1440, soon ruled all the institutions, though it had to fight harsh political battles to do so (the Pazzi Conspiracy of 1478 was typical). In this evolution, Florence was a sort of prototype of "many ancient communal cities [where] the center was gradually being shifted to the palaces of the great bankers.... In Florence, for instance, the center moved from the marketplace in front of the Palace of the *Signori* to the splendid palace of the Medici, the *Casa Medici*, a new architectural expression of a new function.

"In such a situation ... the rational task in hand ... intended to make use of classical theories as useful supports and as suggestions for a condition which had not yet found a sufficiently systematic treatment of its own.... Just as in physics Archimedes was considered more actual and more modern than Buridan, both Vitruvius and Plato appeared more relevant and modern than the theoreticians of the middle ages. It was held that to imitate ancient cities, in town planning as well as in political constitutions, was to obey both reason and nature."[1]

Already in the teachings of Chrysoloras, Plato's *Republic* had seemed to suggest the possibility of extending perfect geometrical order to human society. Interest in Plato as the master of rationality was later sharpened by constant contacts with Byzantine scholars. But the great development of Florentine Platonism came with the teachings of Marsilio Ficino (1433–99), at the academy he directed for twenty years, beginning in 1462 when Cosimo the Elder gave him a villa at Careggi in which to carry on his studies. Ficino, who knew Greek, began by translating the writings of Hermes Trismegistos, then from 1463 to 1468 translated Plato's dialogues. After that, he translated Plotinus, Iamblichos, Proclos, and the work of Pseudo-Dionysios. Thus Ficino approached the religious philosophy that was born among the Persians with Zoroaster and with Hermes among the Egyptians, nurtured itself with Orpheus among the Thracians, grew strong with Pythagoras among the Greeks and the Italians, and finally reached maturity in Athens with Plato. Ficino's Christian Platonism, which contains clearly humanist themes (the glorification of the beauty of the world, the call to the arts, the dignity and excellence of the human soul) and does not lack a keen interest in the study of natural phenomena, found its direct expression in his *Theologia platonica*, written in eighteen books between 1469 and 1474, and *De christiana religione*, written in 1474. Ficino's mystical tendencies are quite

evident and his animistic view of the universe is open to magical practices and astrology. Thus Ficino's perspective is very distant from that of early humanism, as it appeared in the products of Salutati's school or in the work of Alberti. But the Florentine Academy's speculative attitude was also different from that of Nicholas of Cusa especially in the philological scruple it displayed in approaching Plato's work directly in the original.

Developments in humanism outside Italy

As the *humanae litterai* flourished in Florence and in other Italian cities, humanist culture began to spread outside Italy as well, with lively polemics against the "subtleties" of the Scholastics. The most prominent representative of this current was Rodolphus Agricola (1443–85), who promoted the study of Greek and Latin literature in Germany. His *De inventione dialectica* had a strong effect on European culture at the beginning of the sixteenth century. In particular, Agricola reproached Scholastic logic for being inconclusive and supported the integration of rhetoric and dialectics.

The Accademia Platonica *in Florence; Pico*

Remarkable developments in art and literature, from Botticelli's painting to Poliziano's poetry, took place in Lorenzo the Magnificent's Florence. Studies in rhetoric and grammar were also being made, and careful philological research continued. A large part of the cultural activity, however, seemed to be an act of escape, avoiding those problems of civil life that had been the center of debate among early Humanists. The scholars, poets and artists that were emerging attempted, on the one hand, to keep alive the ideal of the universal human being as conceived by Alberti and, on the other, indulged in specialization, prefiguring a feature that would be typical of other phases of Western European intellectual life as well. Widespread, articulated philosophical reflection, with Ficino's Academy as its center, was also taking place. Giovanni Pico della Mirandola (1463–94), who was educated in Bologna but was also in contact with the speculative themes of Padua's Averroist tradition, was particularly connected to the Florentine school. He met the Florentine Platonists in 1484. Although he did not totally disown his previous Aristotelian and Averroist studies, after an exchange of letters with Ermolao Barbaro (1485), in which he discussed the value of grammar and philosophy, he revived the idea of a religious philosophy, in many ways similar to Ficino's. In 1486, Pico prepared nine hundred *conclusiones* that began with an allegorical explanation of the sacred text in an attempt to reconcile hermetism, cabala, Christian revelation, and philosophical investigation. Pico wanted to convoke a congress of scholars in Rome, but Pope Innocent V intervened to block the project. Like Ficino, Pico emphasized a unitary, animistic concept of the universe, even though by insisting on themes of a voluntaristic nature he was in open disagreement with astrology. In any case, the philosophies of Ficino, whose epistles were published in 1495, and Pico had a considerable influence on the development of humanism outside Italy. To get an idea of the close ties that existed between the entire European humanist movement and the Florentine Academy, we need only mention the figures of Johann Reuchlin (1455–1522), the great German humanist who was particularly influenced by Pico; John Colet (c. 1467–1519), the famous English scholar who had contacts with both Pico

---

[1] Eugenio Garin, tr. Peter Munz, *Science and Civil Life in the Italian Renaissance* (New York, 1969), 36.

and Ficino, from whom he drew his inspiration for a general reform of religion; and Jacques le Fevre d'Étaples (c. 1450–1537), who was raised on Florentine Platonism and initiated French humanism.

The traditional currents

We have already spoken of the conflict between the realists and the nominalists, and of the gradual dominance of the latter. The humanist thinkers were, of course, hostile to both currents and often proposed a Neoplatonist reform in contrast to scholastic tradition. Those Aristotelian thinkers who emphasized the autonomy of scientific research also became targets of the polemics. Ficino himself attacked them, observing that "the entire world is now invaded by Aristotelians, divided into two sects, the Alexandrists and the Averroists: the one maintains that our intellect is mortal, the other that it is unique in all humans; both equally destroy religion, particularly by denying any divine providence in the human world." Both currents were present in Padua and Bologna. In 1483, a Latin translation of Aristotle appeared in Venice, along with Averroës' commentary; both were done by Nicoletto Vernia (1420–99). The debates on the unity of the intellect alluded to in the excerpt from Ficino started as a result of the influence of Averroist doctrines. The bishop of Padua intervened against these debates in 1489, causing all the Aristotelians to become involved in the same philosophical battle in spite of the internal conflicts between the followers of Averroës and the followers of Alexander of Aphrodisias, as the refined Platonist-leaning humanists accused them all of barbarism. Even the term Averroist ended by losing its specific meaning and was increasingly used to indicate any heterodox Aristotelian.

The slow reawakening of interest in the sciences

We have already mentioned that, at least potentially, humanism contained lively stimuli both for a global reconsideration of nature and for specific research of a scientific and technical nature. However, in the heat of their raging polemics against the traditional currents (be they Parisian physicists, English *calculatores*, or Averroists) the humanists most often harshly criticized not only the way those thinkers did their scientific research, but even the investigation of the world of nature as such, reechoing Petrarch when he said, "I ask myself of what use is it to know the nature of wild beasts and birds and fish and snakes, yet be ignorant or not care about knowing the nature of human beings." This viewpoint has undoubtedly contributed to many historians' idea of humanism as a phenomenon that concerned the historical-literary sphere more than the actual scientific sphere. Such historians tend to see in the humanists' worship of antiquity something inimicable to the development of modern science. But the humanist idea of considering nature the direct expression of the divine being, which clearly came out in Neoplatonism, revealed an increasing openness to natural problems and consequently projected scientific research as an intellectual activity of unquestionable dignity. This is why it is impossible to reduce the complex currents of humanism to any single theme. It is true that if one compares the humanists' approach to problems in natural science with the approach of a Biagio Pelacani, one has the impression of a more limited view. The process that brought researchers working in the individual sciences, such as mathematics or medicine, closer to

the principal exponents of the new culture was anything but linear. However, by the end of the fifteenth century, scientists were ready to accept a humanist approach, since they realized that the research of Greco-Roman scientists was vastly superior to the work of the recent past. Translations of classical texts on scientific subjects, made directly from the originals and with a philological rigor equal to that of the best literary research or the exegesis of sacred texts, made a revival of scientific inquiry possible. An important figure in this area was Johann Müller, known as Regiomontanus (1436–1476), from the name of the city where he was born (Königsberg). He completed a rigorous Latin translation of the *Almagest*, begun by his teacher, George Peurbach, and an exposition of the Ptolemaic system entitled *Epitome in Ptolemaei Almagestum*, published posthumously in 1496 and greatly admired, both because of the astronomer's skill and because of his broad knowledge of the Greek texts. A prominent figure in the Italian circles was Giorgio Valla (c. 1430–1499), a professor at Pavia, Milan, Genoa, and Venice, who combined the *studia humanitatis* tradition with a considerable interest in science. A two-volume work, the *De expetendis et fugiendis rebus*, was published posthumously in 1501. It was an encyclopedic collection of ample extracts from scientific writings, unknown before, by Greek and Latin authors (Hippocrates of Chios, Aristarchos of Samos, Heron, Archimedes, Boëthius, etc.). Both Regiomontanus and Valla displayed an interest in mathematics, which would again be revived in the sixteenth century.

New cultural demands from the technical world

Another essential component of scientific life during this period was the range of diverse stimuli that came from advances in technology. Technicians, who acquired their skills from their work as artisans, seemed increasingly less able to attack and solve the problems posed by new developments in society. In many places (for example in certain European monarchies, in many courts of princes, and in the Florentine commune), the need arose for a new group of technicians able to deal with these problems in a rational, systematic way. However, the traditional approach of the universities, where theoretical ideas that were often quite bold were not given practical application, proved inadequate to the task of forming the new technicians.

In any case, the texts of the ancient scientists and engineers that the philologists were making available constituted a reference point. Although some authors were already known, a critical presentation of the works of Euclid or Archimedes was certainly a new occurrence not only for philology, but for applied science as well. Thus the figure of the Renaissance engineer slowly took shape, to emerge fully in all its potentiality in the cultural environments of the courts of the following century. One of the most culturally stimulating events of the fifteenth century was the blending of art, science, and technology that took place in painting and architecture. If, on the one hand, it was a component of the humanist movement, it also brought about remarkable developments in both technology and science (especially mathematics, with the first fundamentals of projective geometry). Without going into detail, we should at least mention the role played by the intuitions of Fillipo Brunelleschi (1377–1446) on the importance of perspective, and Alberti's interest in the same subject in his works on art and mathematics. *De prospectiva pingendi*, by Piero della Francesca (c. 1415–1492), an in-depth treatment of the new perspective (see MATHEMATICAL SCIENCES 1400–1450), also deserves special mention.

Piero also wrote *Libellus de quinque corporibus regolaribus*, on measuring regular polygons, spheres, polyhedrons inscribed in spheres, and solid figures in general. It is significant that Luca Pacioli (c. 1445–1517), the most important mathematician of this period, refers to this work in his *Divina proportione*, written in 1497. An increasingly more attentive, unbiased observation of natural phenomena also became a new component of scientific and technological research; this attitude can be seen toward the end of the century in Leonardo da Vinci's scientific writings. At least three currents converged in this tendency: that of the Parisian physicists (through the work of Pelacani), the humanist and Neoplatonic current that expressed itself in the relationship between human beings and the universe, and lastly a fruitful interrelation between the observation of natural phenomena and attention to technological developments.

Geographical discoveries, the growth of the Atlantic countries, and the crisis of Italian society

With respect to the influence of stimuli coming from technology, we should not forget that the last decade of the century was one of great geographical discoveries, of the search for a route to India, and the discovery of America. The impact of these discoveries on applied science was enormous and involved mathematics, geography, physics, astronomy, biology, and medicine. Besides, the conquest of the world beyond the Pillars of Hercules should be seen as a determining factor in a complex process of transformation that saw the importance of the Mediterranean countries decrease to the advantage of the countries on the Atlantic and the North Sea.

This process had repercussions in Italy and was a factor (though certainly not the only one) in the political and institutional crisis that exploded when Lorenzo the Magnificent died in 1494. This is not the place to examine the causes of that crisis in detail; however, it had very important effects on the entire development of culture in the Renaissance and on the spread of humanist culture throughout Europe. It was also a crisis of traditional values, for instance of Christian values at a time when the papacy was placing emphasis on worldly power, and of the new humanist values as well, which often proved to be extremely difficult to put into practice.

From this point of view, Girolamo Savonarola's preachings in Florence (1490–98) may be an extremely enlightening episode; and it is not an accident that during these years the artists and intellectuals involved in rediscovering classicism, Boticelli and Pico, for example, turned to political-religious experiences that focused on rigoristic, mystical and unworldly themes. Many humanists involved in the *rinovatio* of Christian culture in early sixteenth century Europe displayed the same tendencies, which were based on reformist stimuli, increasingly more intent on ending "the Babylonian captivity of the Church."

MATHEMATICAL SCIENCES

Mathematical works in print

The invention of printing, marked by the publication of the Bible in 1455, printed by Johann Gutenberg (1400–68), also gave a new thrust to mathematical research in the course of a few decades. It insured not only the diffusion of numerous ancient treatises, but also the more rapid circulation of new works.

Regiomontanus's treatise on trigonometry

One of the few authors of this period worth noting was Johann Müller, known as Regiomontanus (1436–1476). In 1464, he composed a famous work, the *De triangulis omnimodis*, although it was only published in 1533. It is the first real treatise on trigonometry written by a European, and it arose from needs in astronomical research (in which the author distinguished himself particularly). It reflects both classical Greek and Arabic influences, which the author was able to harmonize and blend together in an original way, making use of his vast cultural and historical knowledge. His originality as a mathematician can also be seen in his notes to Euclid's *Elements*, as well as in his copious scientific correspondence.

Practical arithmetic and its influences

It was certainly not an accident that the first mathematical work printed was a manual of practical arithmetic, the so-called *Treviso Arithemetic* of 1478, which preceded by four years the first printed edition of Euclid's *Elements*, published in Venice in 1482. The anonymous *Treviso Arithmetic* is a modest treatise, with exclusively practical aims, which teaches certain basic rules for carrying out operations, dealing with various commercial problems, and calculating the movable feasts of the religious calendar. However, in the course of a decade, this small treatise was followed by a number of similar ones, a sign that the need for this type of practical arithmetic was quite alive and widely felt. Nor was the phenomenon limited to Italy alone. In 1483, a similar *Arithmetic* was published at Bamberg, again addressed to solving various practical problems, but more extensive than the Treviso prototype in the amount of material concerning commercial affairs that it contained. In 1484 in France, Nicholas Chuquet (1445–1500) composed a treatise entitled *Triparty en la science des nombres*, the first mathematical work written in French. The influence of the Italian tradition is obvious, although the work is distinctly theoretical in nature. It contains a number of original points of view, and the first signs of a use of symbols that not too much later marked the transition to so-called syncopated algebra can be seen. It is significant that this important work—which, however, lacked any practical applications—remained unpublished until 1880.

In any case, the time was ripe for moving toward a theoretical understanding of problems arising from practical situations as well. The numerous algebraic treatises that began to circulate in Germany and that use an incipient symbolism (the so-called cossic characters, from the word *cosa* or *cossa* as the unknowns in equations were called, under the influence of the Italian nomenclature) are indicative of this phenomenon. A treatise by Johannes Widmann of Eger (c. 1460–1500), *Behende und hubsche Rechnung auff allen Kauffmanschafft*, written in 1489 and full of arithmetical and geometrical problems, is important in that + and − signs to indicate addition and subtraction appear for the first time.

The work of Luca Pacioli

But the most outstanding synthesis of mathematical studies of the entire century, and one that served as a starting point for future developments, was the *Summa de arithmetica, geometria, proportioni, et proportionalità* by Luca Pacioli (c. 1445–1517), published in Venice in 1494.

The work is actually a mathematical encyclopedia, in which the author compiled the arithmetical and geometrical knowledge of the period. It includes both the theoretical findings and a variety of practical applications of the arithmetic studied by Pacioli's contemporaries, mixed with copious philosophical-mystical digressions on the nature and properties of various numbers. Right along with these are explanations of methods for calculating interest and bookkeeping techniques, for instance various elaborations on the double-entry system (widely used at that time in the flourishing trade of the maritime republics of Genoa and Venice). A large section is dedicated to algebra, where symbols, some of them original, are used profusely. The geometrical part of the *Summa* is less significant, since it draws more directly upon preceding works.

ASTRONOMY

Regiomontanus and the popularization of the Ptolemaic system

In 1463, Johann Müller, known as Regiomontanus (1436–76), completed the *Epitome in Ptolemaei Almagestum*, a work begun by Georg Peurbach. Although it is clear that Regiomontanus wholly supported the Ptolemaic planetary theory, it was thought for a long time that he was the first to uphold the daily rotation of the earth, perhaps because of an incorrect interpretation of a work by Johannes Schoner (1477–1547), the *Opusculum geographicum* (1533), which has a chapter entitled *An Terra moveatur an quiescat, Johannis de Monte regio disputatio*. The error was indirectly reinforced by a letter written by the mathematician Georg Hartmann of Nuremberg (1489–1564) that cites an unpublished note by Regiomontanus, in which he supposedly stated that the motion of the stars varies *"propter motum Terrae."* This statement is not confirmed, however, by a comparison with other published writings of Regiomontanus. In reality, he confuted those who supported daily rotation before him, using the traditional arguments (birds in flight would remain behind, etc.), and explicitly maintained the theory of the earth's immobility at the center of the universe.

Also worth mentioning is the work that Regiomontanus did during this period to convince the king of Hungary, Matthaeus Corvinus, of the need for a reform aimed at using more accurate instruments in astronomy, such as corrected sundials and balance-wheel clocks.

Progress in observational astronomy

Following Nicholas of Cusa's death in 1464, his old colleague at the University of Bologna, Paolo dal Pozzo Toscanelli, contrived a simple astronomical device in the Church of Santa Maria del Fiore in Florence. A hole was bored in the dome at the base of the lantern. Light rays shone through this hole onto the floor of the cathedral, ninety meters below. Toscanelli was thus able very accurately to establish the day of the solstice, as well as local time and other important data for computing the ecclesiastical calendar. He calculated the exact date of the equinox in the same way and corrected the errors in the *Alfonsine Tables*, extending his corrections to the positions of the fixed stars and arriving at a more precise value of the precession of the equinoxes.

That same year, a comet appeared, and Regiomontanus made very accurate measurements of its course, calculating that its daily parallax did not exceed 6 degrees. He used Dal Pozzo Toscanelli's method, though its invention has until recently been wrongly attributed to him. The Florentine astronomer also wrote a description of the comet, which proved to be even more accurate.

At Cracow, in 1482, Albert of Brudzewo, or Brudzewski, wrote a *Commentariolum super theoricas novas*, which circulated among the students of the University of Cracow. It outlines Peurbach's popular work, emphasizing, among other things, how the center of the lunar epicycle must describe an oval orbit, similar to Mercury's, rather than a circle.

In 1483, the *Alfonsine Tables* were published, although a number of manuscript copies had been in circulation since the thirteenth century. The first drawing of the oval shape of Mercury's deferent, already mentioned in ASTRONOMY 1250–1300, appeared in the *Tables*.

PHYSICAL SCIENCES

Leonardo da Vinci, artist and scientist

Interest in nature reached a peak in the work of Leonardo da Vinci, which actually marked the transition to a truly new epoch. His art aside, Leonardo's exceptional genius manifested itself in engineering and anatomy in particular, although his research in the area of physics was also of undeniable importance. Indicative are his studies of motion, in which, starting from the theory of impetus, he arrived at several conclusions that seem to foreshadow the principle of inertia. He also studied inclined planes, the composition of forces, communicating tubes, and the camera obscura (pointing out its similarities to the eye) and discovered the analogy between light and waves. Nor were his reflections on method, in which he maintained that science must be founded on experimentation and mathematics, any less interesting.

Gregorius Reisch: the old science

In order to complete our sketch of fifteenth century physics, we must mention Gregorius Reisch, prior of a Carthusian monastery near Freiburg, who flourished in the second half of the century. He wrote a famous encyclopedia, the *Margarita philosophica*, fourteen editions of which were published between 1486 and 1583. It talks about all the sciences, from astronomy to biology, often in a chaotic fashion, mixing old hypotheses with new proposals for explaining various phenomena. The pages dedicated to optics, which explain the mechanism of vision and the nature of light, are particularly indicative. Although the author cites the names of Alhazen, Bacon, and Witelo, it seems that he basically agrees with the more traditional doctrines on the subject.

The enormous success of his work illustrates how the patrimony of old concepts continued to carry considerable weight in scientific culture. While a profound revolution of concepts and methods was already in sight, it cannot yet be said that science had emerged from its medieval phase.

CHEMICAL SCIENCES

Modest experimental contributions

Although this is not a period of many facts and names, there is the case of one important humanist, Nicholas of Cusa (1401–64), who suggested that scales be used systematically and planned an experiment—which he never carried out—for weighing soils, seeds, the plants obtained from them, and their relative ashes in order to find out how much of the soil transferred itself into the plant. This was the origin of Johannes Baptista van Helmont's experiment of 1648, which really involved the weighing of water (see CHEMICAL SCIENCES 1640–1650).

Aside from the relatively modest contributions of such scientists as George Ripley (c. 1415–90), and Philipp Ulstad (who taught medicine in Freiburg im Bresgau in 1500), it seems that preparations were tacitly being made during this half-century for the sudden reflowering of alchemy that took place in the Renaissance. The *Liber de arte distillandi de simplicibus*, by Hieronymous Brunschwig (c. 1450–1533), was written in 1550. It is a summary of the experiences of the preceding years and is an authoritative sixteenth century technical text.

LIFE SCIENCES

The publication of ancient works

A variety of events determined the history of the life sciences in the second half of the fifteenth century. New texts were published, there were critiques of several traditions, and exceptional people or those who were at least the precursors of future developments in scientific thought flourished. Theophrastos of Eresos' *Historia plantarum* was printed in Latin in 1483 and in Greek in 1497. In 1478, Dioscorides' *De materia medica* appeared for the first time in print, with numerous illustrations which, besides identifying the plants that the author wrote about, also helped satisfy the naturalistic interests of aristocrats and artists, as did the illustrations in other, similar works. The publication of Dioscorides gave rise—through the richly varied translations of the period—to a number of herbals in the vernacular. These works, all of a descriptive and utilitarian nature, differed in the plants they chose to include and the descriptions of their healing effects. Slowly, the search for more variety and originality developed into a true interest in science.

Pliny the Elder's *Naturalis historia* (1469), one of the first scientific works to profit from the new process of printing with type, gave rise to a lively debate. In 1492 (or, more probably, in 1491) Nicolò Leoniceno published a *De Plinii et plurium aliorum in medicina erroribus* in Ferrara, to which the humanist Pandolfo Collenuccio of Pesaro replied with a *Pliniana defensio*. Ermolao Barbaro joined the debate with his *Castigationes plinianores*, listing around five hundred errors made by the copyists and editors of Pliny's work and excusing the Latin naturalist (see LIFE SCIENCES 0–100).

The birth of pathological anatomy

Pathological anatomy was born with Antonio Benivieni (c. 1440–1502), whose *De abditis nonnullis ac mirandis morborum et sanationum causis* explained how the dissection of cadavers also served to determine the cause of death. For some other authors, an interest in medicine was linked to astrology and natural magic.

Nicholas of Cusa's work in medicine

Nicholas of Cusa (1401–64) demonstrated the medical importance of measuring the specific weight of urine and blood. He also made accurate observations of pulses and respiratory rhythms.

EARTH SCIENCES

Geographical discoveries

In 1488, Bartholomeu Dias rounded the Cape of Good Hope, in 1492 Christopher Columbus reached the island of Guanahani and discovered the New World, in 1497–99 Vasco da Gama, with the help of an Arab steersman, Ahmad ibn Mādjid, reached India by circumnavigating Africa, and between 1499 and 1501 Amerigo Vespucci conducted his explorations of America. The length of the equator was calculated with an error of only eighty kilometers with respect to its actual measurement.

In the course of 1492, before all these discoveries were made, Martin Behaim of Nuremberg constructed a globe that showed all the lands then known.

Geology, mines, and ores

There was a great deal of mining activity in Saxony. P. Schneevogel, a professor at Leipzig, wrote about mines, Grotterus (1473) told of rocks with marvelous powers, while the alchemists were all worked up about the philosophers' stone, which Thomas Norton wrote about in *The Ordinall of Alchimy* (1477).

The fall of Constantinople, the diffusion of the works of Plato and Aristotle in the West, and the invention of printing provoked a rediscovery and spreading of classical Greco-Roman thought. The world of naturalistic studies was reopened with humanism, while in the area of speculation, the Aristotelians prevailed and Avicenna's doctrines were defended. Nicholas of Cusa (1401–64), who was also interested in calculating the depth of the ocean, among other things, proposed a cosmology with an infinite universe. Leone Battista Alberti (1404–72), who worked on determining the depth of the ocean as well, also described the erosive action of the atmosphere, anticipating several ideas of modern hydrogeology. These two scholars opened an experimental period in the earth sciences, with its basis in accurate measurements rather than in speculation.

The *Philobiblon*, by Richard de Bury (1287–1345) was first published in 1473. It uses the word "geology" to describe a science, or law, of the earth, as opposed to "theology," a science of the divine.

With regard to cosmologies and cosmogonies, theoretical speculation ranged freely. For example, Gregorius Reisch's *Margarita philosophica* (Strasbourg, 1486) discusses the origin of metals, the causes of earthquakes, and the formation of springs, which supposedly came from underground vapors.

Oceanography and maritime explorations

In 1462, Roselli, a Catalan, published nautical maps of the North Sea. In the Mediterranean, *portolani* began to be used as the main source of the geographical and physical information needed for navigation. The first Italian *portolani* of the Mediterranean Sea were published in the course of the second half of the fifteenth century, probably in 1490.

In 1471, Portuguese navigators discovered the south equatorial current of the Atlantic, which flows east to west, in the Gulf of Guinea and provided the first descriptions of the phenomenon.

An interest in studying bodies of water in general, and oceans in particular, was revived and flourished in the Mediterranean area, especially in Italy. The time became ripe for early physical and physical-mathematical interpretations, rather than simple verifications of facts and philosophical-scientific speculation. Leonardo da Vinci (1452–1519) made some fundamental contributions in the area of theoretical (as well as experimental) hydrodynamics. He studied the motion of fluids from a physical point of view and attempted to explain ocean currents, demonstrating the absurdity of Aristotle's ancient ideas about sloping sea basins. He investigated the motion of waves, furnishing a striking example—the changes in the surface of a field of grain, due to the oscillation of the blades of grain moved by the wind. This same phenomenon probably suggested to Franz Joseph von Gerstner the first mathematical theory on ocean waves at the beginning of the nineteenth century. Leonardo also outlined the principle of the continuity of incompressible liquids and worked on hydraulic and marine technology, even designing underwater vehicles. In 1472, the Venetian Roberto Valturio also designed a submarine.

It is interesting to note that at this point, when science was about to take some decisive steps forward, the most strongly felt oceanographic problems of antiquity—the origin of salinity in the seas and the causes of the tides—both remained unsolved. The first could not be explained because of a lack of geological knowledge and, especially, because chemistry did not exist; the second remained unexplained because there was as yet no concept of gravity. While the positional and temporal relation between the tides and the stars had been known for some time, a wide variety of imaginative hypotheses were brought forth to explain the causes of the phenomenon.

APPLIED SCIENCES

Metallurgy and the mining industry

The demand for metals grew as cities developed and the circulation of money increased. Iron was needed for artisans' equipment and for plowshares, lead for roofs, and silver for coins. The demand for copper and iron increased even more with the advent of the cannon.

Advances in mining activities were apparent as early as 1450. In 1451, J. Funcken began to make use of the discovery that silver could be separated from copper ores with the aid of lead. According to John Ulric Nef, no other invention had such a stimulating effect on the development of the Central European mining and metallurgy industries. Mining techniques were improved, especially drainage methods, with the introduction of ingenious chains of buckets moved by water wheels or horses.

The great bankers of the period became increasingly more interested in mining activities. Large sums of money were invested in Bohemia, Tyrol, Saxony, Hungary, and Spain. In Bohemia, mining camps such as Joachimstal, Iglau, and Deutschbrod became actual cities, while other mining centers arose in Slovakia (where copper, silver, and lead were extracted), Styria, Carinthia, and Carniola (where calamine and silver were mined). In Germany, the mining and metallurgy industries grew so large that at the end of fifty years they employed almost a hundred thousand people. The activity reached its peak under Maximillian I and the Emperor Charles V, with the domination of the great financiers Fugger and Welser.

Even Italy participated in this growth. When rich deposits of alunite were discovered at Tolfa, the pope was convinced to exploit them in order to free Europe from its dependency on the Arabs. After two years, the Tolfa mines already employed seven thousand people.

Civil and military engineering

Leone Battista Alberti (1404–72), and, from 1446, Roberto Valturio (1405–75), worked at the court of Sigismondo Malatesta. Inspired and perhaps even guided by Sigismondo himself, Valturio composed a treatise on military science, *De re militari*, which he finished in 1455. The work was so extremely important that it was the first technical treatise to have the honor of being printed (1472).

In Milan, Francesco Sforza encouraged a similar type of work, though it was centered more around hydraulic projects than military works. The canal of Martesana, which connected Milan with Lake Como, was constructed between 1457 and 1460 under the direction of Bertola of Novate (c. 1410–75). A. Averlino, known as Filarete, and Aristotele Fioravanti of Bologna, who began the projects of the canal from Pavia to the Po, also worked in Milan. Fioravanti supposedly invented locks with movable doors, an invention attributed to Leonardo, while working on the canal project. In 1455, Fioravanti used a machine that he may have invented himself to move the 407-ton tower of the church of the Magione eighteen meters. The gears of the machine were made with worm screws.

Mariano di Jacopo, known as Taccola (c. 1381–1458), the Sienese Archimedes, flourished in Siena. He wrote a treatise on military machines, *De machinis libri decem*, and his contributions to this field were definitely remarkable. He was the first planner of fortifications and means for attacking them (mines and explosives) to have modern ideas.

The flourishing of the Italian schools of engineering reached its peak during this period. Under the rule of the Montefeltros, who were fond of surrounding themselves with military engineers, such as Luciano, and Fra Carnevale, Urbino became a major center along with Milan and Rimini. Franceso di Giorgio Martini (1439–1501) flourished in the Urbino environment. In 1469, after studying painting and sculpture under Angelino of Fiesole and architecture under Lorenzo di Pietro at Orvieto, Martini was appointed superintendent of the waters, aqueducts, and fountains of Siena. In 1477, he was called to Urbino by Frederick II of Montefeltro to build several fortresses; that same year he began his treatise on architecture, fortifications, mechanics and town planning, which he finished in 1480.

Martini's *Treatise on Architecture* owes much to Alberti and Filarete, especially with regard to town planning, where Martini adopts Alberti's idea of the rational city. Although the tenaille trace appears in his fortifications, they still do not make a clean break with medieval models. The most original part of Martini's work is the section on mechanics and hydraulics. It contains new applications of the crank-and-connecting-rod system, for instance to transmit the movement of a water wheel to a saw. The treatise also contains an application of sucking and pressing pumps. Martini shows a particular interest in worm-screw gears as well. His drawings are done with such clearness, accuracy, and good perspective that he seems to be the indisputable precursor of machine design.

Between 1478 and 1480, construction was begun on the Mon Viso tunnel, designed to connect the Dauphine with the Marquisate of Saluzzo. The tunnel, which is two meters high and two and one-half meters wide, was dug under the hills at an altitude of more than two thousand meters above sea level. The project was directed by Martino of Albano and Baldassare of Alpiano, and was greeted as one of the wonders of the times.

The use of mines to remove rock layers was recorded by a German Dominican of Ulm and refers to the construction of a road near Bolzano.

Guiliano di Francesco Giamberti, known as "the man from St. Gall" (c. 1445–1516), designed the citadel at Ostia, which was built in 1483. It marks an intermediate step between medieval and modern fortifications. Giamberti's machine designs show the strong influence of Martini, who was a friend of his and helped him defend Castellina in 1478 when it was being sieged by Giuliano della Rovere.

Between 1485 and 1508, Russian and Italian masters (Aristotele Fioravanti, Marco Ruffo, Pietro Antonio Solario, and others) worked on constructing the new walls of the Kremlin (2.3 kilometers total length and up to seventeen meters high) and the towers and fortifications inside, making the Kremlin one of the largest and mightiest fortresses in Europe at the beginning of the sixteenth century.

City plans and cartography

In 1452, Leone Battista Alberti finished his *De re aedificatoria*, which was printed in 1485. Its importance lies mainly in the new concept of urban esthetics it expresses. Alberti chose a radial-concentric plan for the city, with straight main roads and concentric circular streets.

Filarete reached almost identical conclusions with his sixteen-sided polygonal city. It had a large square in the center with sixteen streets going out from it, which passed through an equal number of squares.

In 1486, another treatise was printed, the Roman Vitruvius's *De re aedificatoria*, which had only recently been rediscovered. This work and Alberti's were very widely circulated and became essential parts of sixteenth century engineering and artistic culture.

The developments that took place in cartography also deserve mention. In 1459, Fra Mauro designed a splendid model of the Ptolemaic world map, while Nicholas of Cusa (1401–64) drew the first modern map of Germany.

Means of transport

Transport had been partially eliminated with the evolution of harnesses, though heavy four-wheeled vehicles continued to be difficult to handle. The first reputable drawing of a swivelling forecarriage appears in *Das mittelalterliche Hausbuch*.

With the first problem solved, there remained the problem of suspension. Leonardo da Vinci, before Girolamo Cardano (1501–76), described the suspension system that was given Cardano's name when it was perfected.

The work of Leonardo da Vinci

Leonardo worked in Milan from the spring of 1483 until 1499. There he met Francesco di Giorgio Martini, whose treatise he had read and annotated, and frequented Luca Pacioli (1445–1517).

During his stay in Milan, Leonardo worked on hydraulics, planning several canals which gave him a solid reputation as a hydraulic engineer. He also investigated new architectural problems when he worked on the project for the lantern of the Milan Cathedral. He found that he had to study arches made of stone ashlars, for which no adequate binding material existed. He attacked the problem by going to studies on the equilibrium of wedges.

Many of Leonardo's studies on machines have been traced back to this period. Among other things, the designs show a remarkable familiarity with basic mechanical devices such as gears, screws, pulleys, chains, cranks, and connecting rods. Leonardo also investigated the conditions of operation for these devices.

In Milan, Leonardo carried out military engineering projects. He also worked on problems of town planning, even though he was more drawn to the machines. His drawing of a pedal-operated lathe is quite remarkable, although the same can be said for all his drawings of machines. It is a clear representation, and is important not only because a pedal has been added, but also because it has a flywheel to send the pedal back up. The pedal lathe with inertia flywheel was very important to the future of later precision mechanics. It also allowed the cutting tool to be transferred from the hands of the operator to the lathe, fixed on a support bar, independent of the operator and capable of being directed and adjusted in advance.

Leonardo also thoroughly studied the problems of gears, including a system for designing and constructing them so as to decrease the attrition between them and consequently prevent them from wearing out too quickly.

Leonardo's manuscripts contain drawings and studies which anticipated machines and devices that were developed, in some cases, several centuries later.

Developments in ship-building. Voyages and cartography

Louis of Bourbon's seal of 1466 carries the first image of a three-masted carrack that can be dated with certainty. The carrack was the Mediterranean version of the round merchant sailing vessel of the North. The first step in the ship's evolution from the merchant ship was probably the addition of a second, smaller foremast. Then, in order to equilibrate the pressure that the sail on the bow exerted on the rudder, a third mast (the mizzen mast) was put on the quarter deck. If we examine this type of ship, we can see that we are at the end of one epoch and the beginning of another. Although very ancient, the ship is essentially modern. It could set sail in any wind, whereas Roman ships had to wait for favorable winds.

Fra Mauro's work in cartography was continued by Nicolaus Germanus, who drew a series of Ptolemaic maps in 1466. He made one improvement, however, by using a trapezoidal rather than a rectangular projection.

Spanish discoverers happily reached an island of the Bahamas, which they called San Salvador, on October 12, 1492. Christopher Columbus (1451–1506) had used his ingenuity and creativity to overcome the deficiencies of his instrumentation (see EARTH SCIENCES 1450–1500). Quadrants were of no use to him, since they could only be used to make observations on dry land; on the open sea the inevitable instability of the plumb line made them difficult to use.

In any case, more accurate measurements would have been purely a formality, since the latitude could not have been marked on a map. And the problem of establishing the longitude seemed completely unsolvable. So Columbus used lunar observations and conjunctions of the planets. Tables of solar declination were first printed in Venice in 1483, then in 1488, and again in 1492. Giovanni Bianchini's tables appeared in 1493.

1500–1550

THOUGHT AND SCIENTIFIC LIFE

The new historiography and the science of politics

During the early part of the sixteenth century, a series of uprisings shook Europe, affecting economic and political life, the organization of the religious community and the world of art, science, and technology. New problems arrived on the scene, giving rise to lively debates and new currents. The case of historiography is typical. Already in the fifteenth century the humanists had replaced the concept of the intervention of providence in the course of events with an analysis of the interests that motivated individual acts. The fact that the new historians were at the same time politicians (ambassadors, chancellors, leaders of cities, etc.) made them direct participants in the matters they were discussing. In many cases, this pragmatic historiography was unable to make the correct connection between persons and events, and at times deteriorated into a simple writing of memorials. But at other times, it took up the great themes of humanism (voluntarism, the immutability of nature, the contrast between virtue and success) in very original ways, and produced powerful analyses of historical-social reality. This was true of the work of Niccolò Machiavelli (1469–1527), who personally experienced the crisis in Florence following the rise and fall of Savonarola "the disarmed prophet." In addition to presenting a careful reconstruction of the most significant historical events of his time, he also developed a broad, in-depth view of history and politics.

The Prince was written in 1513, and the Discourses between 1513 and 1521; the Istorie Fiorentine were only published posthumously, in 1532. The works display, though in an unsystematic way, a rationalism that was not purely contemplative and that had the goal of working to establish our rule over known reality. Political realism, which brought the hostility of theologians and moralists upon Machiavelli, was a support for a revival of the humanist idea that nature, even human nature, could be coerced by human intelligence in order to direct the course of events toward goals that we ourselves could predetermine. This aspect made Machiavelli's work a model to accept or reject, in any case a fixed point of reference for historians, memorialists, and theoreticians and practitioners in the field of government, for the entire sixteenth century, as we can tell by the often-critical references that the other great Italian Renaissance historian Francesco Guicciardini (1483–1540) makes to the Prince and the Discourses.

European humanism. Sir Thomas More and Erasmus of Rotterdam

Despite the severe political crisis that afflicted Italy during the first half of the century, the Italian courts continued to be cultural reference points for all of Europe. While scientific research, technological innovation, and artistic experimentation all merged in the work of Leonardo da Vinci in the first half of the century, Renaissance art found a new center in Rome, where the popes undertook a vast project of restructuring the city. The great masterpieces of Michelangelo (the Sistine Chapel was finished in 1512) and Raffaello (the rooms of the Signatura were completed in 1518) were also done in papal Rome. But meanwhile, humanist art was spreading throughout Europe. One of the people responsible for this was Albrecht Dürer (1471–1528), the great German painter, who had ties with the Italian circles and wrote a treatise on geometry in order to familiarize painters with the new techniques of perspective.

With regard to intellectual activity, it should be noted that the study of the humanae litterae began later in other countries than it did in Italy, and soon merged with the Reformation. The early years of the sixteenth century constituted a brief period in which the new culture became widely diffused, without theological controversies coming directly into play, in France, Germany, and England. The experience of the Accademia Platonica of Florence provided a reference point, but the early sixteenth century European humanists kept their distance from the more distinctly heterodox and neo-pagan themes that were present in Italian culture.

The two major protagonists of European humanism during this period were Erasmus of Rotterdam (c. 1466–1536) and Sir Thomas More (1478–1535). Both belonged to the same current of humanist, anti-medieval, Platonizing Christianity that Colet had been a part of and that Ulrich von Hutten would follow, before his conversion to the Protestant reform. Erasmus, who moved from Louvain to England and from England to Italy, and later back to Holland, while the monarchs and princes of France, England and Germany contested him, soon became the most prestigious cultural figure in Europe. In 1504, he wrote the Enchiridium militis christiani, and in 1505 he published Lorenzo Valla's Adnotationes to the New Testament. In 1509, during his stay in England, he composed the Laus stultitiae, which was reprinted a number of times, and, on the occasion of the death of Julius II (1513), the prototype of the mundane pontiff for those who aspired to a renovatio of both culture and religious life Erasmus wrote his masterly satire Iulius exclusus e coelis. In all these works, Erasmus focused on the targets of his polemics: the excessive importance given to the rites of worship, fanaticism and superstition, the intellectualism of scholastic philosophy and theology, mystical escapism and authoritarian impositions. The just cause was that of the *bonae litterae*, which would enable people to rediscover the genuine spirit of primitive Christianity. In 1516, at Basel, Erasmus published editions of the writings of St. Jerome and of the New Testament, a revised and annotated Greek text, with a Latin translation that corrected the liberties taken in the Vulgate. That same year he wrote the *Institutio principis christiani*, composed for the future Charles V. The Englishman Sir Thomas More, a personal friend of Erasmus, also finished his *Utopia* in 1516. The Utopia is a strong criticism of English society at that time, with a proposition for an ideal state in which More mixes elements of medieval societal structures with communistic elements inspired by Plato's republic. Erasmus' and More's criticisms of the ecclesiastical hierarchy and scholastic culture contained a number of ideas that were taken up by the Reformation. In fact, at almost the same time, the Augustinian monk Martin Luther (1484–1546) proposed applying philological criticism to the sacred texts as the basis of a method for interpreting the word of Christ. From 1512 to 1517, Luther emphasized a voluntaristic concept of the nature of God, which came out of his education at Erfurt and was in agreement with the *via modernorum* of the late Ockhamists. In 1517, he published his famous ninety-five *Theses* against indulgences, and in just a few years (1517–19) the conflict with the ecclesiastical hierarchy and the authority of Rome sharpened. The financial bourgeoisie's growing hostility toward papal fiscalism, growing nationalistic feeling in the German-speaking countries, which expressed itself in a decided aversion for Romanism, and the social ferment that set peasant masses against landowners, urban proletariat against capitalists, and knights against the great feudal lords and the monarchy, were all connected to the Augustinian monl's rebellion. There was also the protest of lay intellectuals against the cultural monopoly of the clergy and the religious orders, which often looked toward the new reflections on the relationship between God, the world, and humanity that were typical of Renaissance thinkers for the theoretical basis. In 1519, Luther asked Erasmus to speak out in favor of the Reformation, but Erasmus refused.

Aristotelianism in Padua. Pomponazzi

By the end of the fifteenth century, Padua had become the most lively center of the revival in Aristotelian studies. It held this position throughout the sixteenth century, in contrast to Florence, the center of Platonism. The Averroist tradition in which Nicoletto Vernia (1420–99) had been a major figure in the fifteenth century was strong. Vernia's teachings were continued by his pupil Agostino Nifo (c. 1473–1538), who spread Averroism to the various Italian cities where he went to teach (Pisa, Bologna, Salerno, and Rome). In addition, the philological demand for a better knowledge of Aristotle's work, in contrast with medieval tradition, had a considerable influence in the Padua environment. Following the teachings of Ermolao Barbaro (1453–93), the Greek interpreters of Aristotle began to be considered superior to the medieval ones, leading to a closer examination of the great Greek commentaries. The revival of Alexander of Aphrodisias' work set off a polemic between Averroists and Alexandrists. Both currents had decidedly naturalistic leanings, and fought energetically against metaphysical-religious ideas, in the conviction that the object of philosophy was nature and the only way to investigate it was through reason. In this

battle, Averroism was still hindered by its concepts of an active reality (the intellect) separate from the individual and the resulting interpretation of matter as pure passivity. The Alexandrists, on the other hand, denied the possibility of a separation between form and matter and maintained that matter possessed an authentic activity and determination of its own.

The teachings of Pietro Pomponazzi (1462–1525), who taught first in Padua then in Bologna, were connected to the Alexandrist current. In his *De immortalitate animae*, written in 1516, Pompanazzi develops a detailed critique of Averroist, Platonist, and Thomist theories on the soul. The soul is *actus corporis physici organici* and consequently cannot live separate from the body. This and some of his other works (*De incantationibus* and *De fato*, written in 1520), display an attitude toward religion that was common even among the most avid Averroists, i.e. a complete separation between relevation and rational investigation, a separation that many Renaissance thinkers later resolved with the theory of dual reality.

The Reformation in Germany

With Luther, the debate on religious reform became a broad movement of European range that soon created institutional structures parallel to those of the organization that obeyed Rome. Reform, up to this point, idealistically pursued the spirit of protest against the increasingly worldly nature of the Church and the aspiration for a return to "primitive Christianity." These ideals had found numerous echoes in the cultural world, from William of Ockham and Marsilius of Padua to John Wycliffe and the debates at the Council of Costanza. This not the place to examine Luther's theological and political pamphlets written between 1520 and 1525; we shall only point out how his principle of "Christian liberty" denied the Church of Rome its teaching function, while his principle of "justification by faith alone" denied its function as the vehicle for forgiveness. Luther considered the Holy Scriptures as inspired, and rejected the tradition and interpretations of popes and councils and the commentaries of the doctors, leaving the interpretation of the Scriptures up to the reasoning of each individual believer, who was assumed to be aided by the Holy Ghost. It was this first affirmation of freedom of religion that led many German princes and their subjects to take a political position against the clergy and the Roman pope. This movement, which Luther basically greeted with pleasure, was joined by the revolt of the knights— the petty nobility that intended to advance the religious revolution with the use of arms. Ulrich von Hutten and Franz von Sickingen, in particular, exponents of the petty nobility who were linked to currents of the humanist movement, elaborated a plan for reforming German institutions that proposed to limit the power of the princes. This caused a split with Luther, whom they had joined at the time of his condemnation by the pope (the Bulls *Exsurge Domine 1520* and *Docet romanum pontificem 1521*), since the ex-Augustinian monk counted on the princes and their subjects to adhere spontaneously to the reformation. The revolt of the knights was put down in 1522. Von Sickingen fell in battle, and von Hutten was forced into exile. The peasant revolt that followed was even larger and more violent. The anti-feudal program of the revolt, which began in the Black Forest in 1524 and spread into the Rhineland, Franconia, and Thuringia, was written up in the *Twelve Articles of Memmingen*. A very lucid theoretician of the revolt was Thomas Münzer, whose position on the authority of the Holy Scriptures and whose program for political-social reform were both much more radical than Luther's. Fearing that the extremism of the peasant movement would alienate the sympathies of the ruling classes and interfere with the victory of the reform, Luther harshly condemned Münzer's doctrine, or "the spirit of Allstadt." In 1525, the peasant forces were demolished at Frankenhausen by the princes' army, exhorted by Luther. Münzer was tortured to death.

On a cultural level, the conflict between Luther and Erasmus of Rotterdam was even more significant. Erasmus, who initially did not let Luther pull him into the reform movement, finally came out in open disagreement with Luther's theories of devaluating works of charity and of justification by faith alone. He expressed his views in a work entitled *De libero arbitrio*, written in 1524. Luther replied with a harsh attack, *De servo arbitrio*, and Erasmus retorted with his *Hyperaspistes adversum servum arbitrium Lutheri* in 1525, defending the doctrine of free will and further criticizing Luther on the question of forgiveness. The polemics between the learned philologist and the combative German reformer revealed that the reform, which the humanists aspired to for moral reasons and conceived of as a slow and gradual development to be sought with the tools of reason and persuasion, had increasingly less to do with the German reformer's concrete political action, which united Germany's keenest aspirations of the moment with a mysticism and a rigorism that came from the medieval German mystical tradition.

The reformist movement in Germany after 1525 essentially developed along the lines laid out by Luther, with the support of princes and cities that adhered either out of conviction or because they were interested in getting their hands on the clergy's wealth. In 1525, the Grand Master of the Order of Teutonic Knights also converted, secularizing his wealth and forming the duchy of Prussia. The emperor Charles V soon took repressive action against him, but at the Diet of Spira, which the emperor convoked in 1529, the minority of Protestant princes united in resistance to imperial and Catholic pressures and drew up a *Protestatio* (from whence the name "Protestantism," used to indicate the entire reformist movement). The following year, the Emperor convoked a new diet at Augusta, inviting various religious groups to submit their professions of faith as the basis for discussion and an eventual reconciliation. The mediation failed, and the diet adjourned with an imperial document that was actually a declaration of war against the Protestants. The reformist document, the so-called *Confessio augustana*, prepared by Philipp Melanchthon, one of Luther's active collaborators, became the programmatic foundation of the new church. In 1534, Luther translated the *Bible* into German. This was a logical product of the theory of "Christian liberty," and an event of incalculable importance in German culture.

The Reformation outside Germany

While the Lutheran movement was suffering from bitter internal crises (Karlstadt's polemics, Münzer's radical movement, etc.), Martin Luther's ideas spread with remarkable speed in such Nordic countries as Denmark and Sweden, which from 1523 was independent of Denmark. An extremely original current within the Reformation formed in Zurich and later extended to Basel and Bern. It was headed by Huldreich Zwingli (1484–1531). Although it attempted to coordinate its action with that of the Lutherans, the Swiss movement took a considerably different stance from that of the Germans on doctrinal questions (for example, on the Eucharist) and in its attitude toward the Catholics, which was more radical among Zwingli's followers. Zwingli had a much more deeply rooted humanist education than Luther, as his *De vera eta falsa religione* (1525) and *De providentia Dei* both indicate. These works, unlike Luther's, still contain the humanist idea of a religious connection beyond the limits of Christianity as it existed historically, embracing even the scholars of the pagan world, who were also reached by the light of divine revelation, though to a lesser extent. On this basis, Zwingli emphasized the mystical theme of the subordination of human beings to God, and the need for divine intervention not only in salvation, but in understanding reality as well. Zwingli's defeat and death at the battle of Kappel (1531) did not mark the end of the Reformation in Switzerland; in fact, only a few years later the city of Geneva became the center of a lively reformist movement. Several Frenchmen, exiled for religious reasons, were prominent figures in this movement; two of them were Guillaume Farel (1489–1565) and John Calvin (1509–64). Calvin, who made his debut in 1532 with a commentary on Seneca's *De clementia*, gradually approached religious subjects (1534), and in 1536 the first edition of his *Institutio christianae religionis* was published in Latin. The work was later expanded and translated into French (the final edition was published in 1559). By 1540, Calvin's thought was fully developed. While on the one hand he was influenced by humanism and philosophy, on the other he espoused a radical religous mysticism, with predestination and forgiveness as the dominant themes, along with the idea that work was a human duty and material success a sign of divine predilection, almost a visible verification of predestination.

During the same period, the Reformation in England, came directly from the top, on the initiative of King Henry VIII. The crisis with Rome (which began in 1527) ended in 1534 with the Act of Supremacy, which established the king of England as the supreme head of the English Church. His first act was to suppress the convents and confiscate their wealth, which went to benefit the court and the royal administration as well as the petty nobility, merchants, and landowners. Thus the reformation became extremely popular among various ranks of society, all of whom shared a diffidence toward Rome and a traditional hostility toward the pretenses of the papacy. However, it created even further perplexities among those humanists who shared the need for a moral reform of the clergy, but who did not support an open break with Rome. Sir Thomas More's position was typical: he had favorably greeted Henry VIII's anti-Lutheran position, when the king wrote an attack on the "German Hercules," which earned him the appellative of "champion of Christ" from Rome. But More now criticized the religious policy of the king, who sentenced him to death (1535). This condemnation, like the conflict between Erasmus and Luther, marked one of the major breaks between humanism and the Reformation. It should, however, be added that unlike the reforms of Luther and Calvin, the English reform hesitated in defining its religious content. Henry VIII oscillated between preserving orthodoxy and making concessions to the evangelists, who were inspired by European reformist currents. This often led to the harsh repression of both Catholics (the papists) and Lutherans and Calvinists (the heretics).

In Italy, the Reformation involved neither a state apparatus nor great popular masses and was championed by those intellectuals who were most free in their theological and moral research. At the beginning, th predominant current followed the path laid out by Erasmus and turned to Charles V as the Christian prince *par excellence*. Its aim was to carry out a doctrinal and organizational reform of the Church in order to achieve a reconciliation with the Protestants and a return to Christian unity. The first center of Italian evangelism was Naples, where in 1535 the Spaniard Juan de Valdes founded a circle that was frequented by numerous scholars and persons of culture interested in the *renovatio*. The movement lasted until 1541. Indifferent to exterior forms of worship, Valdes cultivated a spiritualistic view of religion that allowed him to accept the principle of justification by faith, while at the same time professing homage to the Roman Church. Valdes' influence particularly affected a small circle of humanist nobles and aristocrats, which included Pietro Carnesecchi, Marc'Antonio Flaminio, Aonio Paleario, Gian Galeazzo Caracciolo, Vittorio Colonna and Bernardino Ochino. Some of them, starting from Valdes' premises, ended up breaking openly with the Church, sometimes adopting Calvinistic demands.

It should also be mentioned that a whole series of reformatory currents within Catholic orthodoxy were growing progressively stronger. These aimed at a reform of the Church that would reject the worldliness of the early Renaissance popes. The papacy's policy of power had already received a strong blow in 1527, with the sack of Rome (which also affected the developments of the Roman Renaissance). With the election of Paul III Farnese to the pontifical throne (1534–49), the Catholic reform currents rose from the base and reached the top. New forces entered the Catholic world, especially from Spain. In 1534, Ignatius Loyola founded the Jesuit order, which took up some of the themes of Catholic reform, merging them with a staunch struggle against heretics. This general renewal, however, had to deal with political pressures from the Catholic rulers, resistance from the more conservative clergy, protestant intransigence and the complex political structure of the church-state. For these reasons, it is not always easy to distinguish the features of the Catholic Reformation from those of the later Counter-Reformation.

Developments in humanism

We have already spoken of Erasmus and More and should also mention Juan Luis Vives (1492–1540), a Spanish humanist who was tied to both Erasmus and More by strong bonds of friendship as well as by common cultural ideas. His various writings include *De disciplinis* and *De anima et vita*, in which Vives carries on a fierce battle against Averroës, whom he considered a characteristic representative of medieval mentality. The culture of the scholastic tradition became the target of many other humanists, as well as many scientists, during this period. While Vives was attacking Averroism, equally harsh criticisms were being made against nominalism, Scotism, and Thomism, often focusing, as Vives did, on the need for a reform of dialectics and a radical critique of realistic metaphysics. In any case, the traditional currents were still very much alive and often tended to merge with the new tendencies. An illustrious Thomist, for example, was Thommaso de Vio Cajetan (1469–1547), from Gaeta, who wrote commentaries on Aristotle and Thomas. The major center of Scotist and Ockhamist tradition and the polemics between *reales* and *nominales* was the University of Paris. *Gargantua et Pantagruel* (1534) by François Rabelais (c. 1494–1553) contains echoes of the polemics between the humanists and the followers of traditional philosophies and expresses the need for a reform of religious life within the framework of an Erasmian-type tolerance. Its content is basically pedagogical, and it is full of philosophical ideas that are unsparing in their disagreement with the Parisian scholars and that project a reform of education even beyond humanist experiments.

The expansion of scientific research

The time of the great crisis of *christianitas* and the conflict between humanist culture and religious reform was also a time of great economic developments that stimulated a revival of artistic, philosophical, and scientific culture in general. It was a time in which capital investment found new outlets, the axis of world trade gradually shifted from the Mediterranean to the Atlantic, and the first conquests of the New World were made. This explains why new, urgent demands to make the means of production, communication, and even destruction more efficient continued to come to the technicians from all quarters. Without going into detail about the advances made in the applied sciences (see that section), it should be noted that the need, on a cultural level, for theories that would better meet the demands of technology (a need that had already been felt in the fifteenth century) was met to some extent in the work of the philologists, who made a wide range of ancient scientific texts available. Here we shall only mention the rediscovery of three great classical authors, a rediscovery that not only had an incalculable effect on the individual sciences (mathematics, mechanics, astronomy, physics, etc.), but was also a very important event in a more general sense, in that it led to philosophical theories and general discussions on the nature of science. The works brought to light were the *editio princeps* of the Greek text of Euclid's work (Basel, 1533), of which a Latin translation from the Arabic, done by Campanus of Novara, a thirteenth century scholar, had been published in 1482 in Venice, and a Latin translation by Bartolomeo Zamberti had been published in 1505; Memo's Latin translation (Venice, 1573) of the four books of Apollonius's *Conics* known at the time; and the first important Latin translation of several of Archimedes' works, done by Niccolò Tartaglia (Venice, 1543). Thus in about a decade (1533–43) most of the works of three great masters of Greek mathematics, long considered the model *par excellence* of scientific rigor, were made available. If we keep in mind that the specific needs of research were being successfully met by the work of the philologists in other scientific areas as well (for example, two complete editions of Galen's works in the original Greek came out, one in Venice in 1525 and one in Basel in 1538), we can see why scientific and technical research increased and specific areas of research multiplied during this period.

Magic and the philosophy of nature

Another characteristic feature of the scientific-technical culture of the epoch was a widespread interest in investigations into astrology, alchemy, and magic, a group of studies that during the sixteenth century were almost universally considered to be very closely connected to scientific research, while in the following centuries they were generally classified as non-scientific. These disciplines often presented aspects that were quite different from their medieval precursors. For example, ophthalmology was sometimes placed within a theological framework that attempted to reconcile it with Christian religion, and magic was presented as the apex of natural philosophy. These disciplines were also very different from one another. Alchemy contained a much stronger element of activism than astrology, while the scope of magic is very difficult to define, since a magician was generally speaking someone who could cause singular events to take place by acting more or less mysteriously on nature's occult forces (alchemistic procedures provided a valuable tool in this sense).

The thinkers who were reflecting on the sciences during the first half of the seventeenth century emphasized the difference between their own scientific approach and the pseudo-science of the magicians by defining the scientific method in various ways and by promoting various ideas on the verification of scientific theories. This was one of the features that distinguished the scientific revolution from Renaissance culture. Typical of the Renaissance were persons who were both magicians and physicians, such as Theophrastus Phillipus von Hohenheim (Paracelsus, 1493–1541), or magicians and mathematicians (and physicians), for example Girolamo Cardano (1501–76). An important exponent of magic and the philosophy of nature during this period was Heinrich Cornelius Agrippa of Nettlesheim (1486–1535), author of a *De occulta philosophia*, in which Christian revelation, cabala, and the Pseudo-Dionysian tradition are all reconciled in a typically Renaissance perspective of human domination over the world. Paracelsus wrote the famous *Paragranum* (first edition 1530), an *Opus paramirum* (1531), a large work on surgery (*Grosse Wundarznei*, 1537), an *Astronomia magna* (or *Philosophia sagax*, 1537), and many other pamphlets and treatises that made this physician, astrologer, and magician, admirer of Erasmus and enemy of the medieval authors—he publicly burned Avicenna's *Liber canonis* as a demonstrative act in Basel in 1527—the most prominent natural philosopher of the early sixteenth century. Girolamo Fracastoro (c. 1478–1553) was another important figure. He took up the Renaissance theory of the bonds of "sympathy and antipathy" between bodies but tended to interpret them in the light of the atomistic theory, revalued, often through readings of Lucretius, with new Epicurean tendencies. He also tended to criticize Pythagoreanism and astrology in his studies of both the human body (the microcosm) and the universe (the macrocosm).

The Reformation and intellectual activity

While Europe was torn by severe conflicts among states, and the first signs of religious strife appeared (the crisis in England when Henry VIII died in 1547, the religious war in Germany and the Franco-Imperial wars), the protestants strengthened their positions. In Germany, in particular, the Peace of Augustus (1555) sanctified the division of Germany into Catholic and Protestant principalities according to the principle of *cuius regio eius et religio*. The event was of inestimable importance in that it marked the end of the spiritual unity of Europe as *christianitas* (this is why nineteenth century thinkers like Hegel and Heine viewed this political recognition of Lutheranism as a key event

in the affirmation of the modern state). The most prominent intellectual figure in Germany was Philipp Melanchthon (1497–1560), Luther's main collaborator in organizing the doctrine. He wrote a Greek and Latin grammar, manuals for university and pre-university studies, and treatises on dialectics, rhetoric, physics, ethics, politics, and history, which became the foundation of German education. Luther's writings had contained original pedagogical ideas (including a discussion of the state's responsibilty in providing for the education of all its subjects). Melanchthon's reconciliation with the *bonae litterae* and scientific culture (within a basically Aristotelian framework) now laid the foundations for creating a class of functionaries and professors who would play a predominant role in German cultural life. However, many more radical Lutheran ideas were lost, and above all in the area of pedagogy.

Meanwhile, Calvin was making the city of Geneva the center of his reform (beginning in 1541), the year of the famous *Ordonnances ecclesiastiques*). Due to the zeal of the Calvinists, who continually aimed at making daily life, in all its manifestations, conform to their religious convictions, the "city of the elect" became a center for reformist currents in a number of countries. Especially in the Nordic countries on the Atlantic, which would later become the great rivals of the Catholic countries of the Iberian peninsula, the Protestant ethic became more and more closely tied to the economic activism of early capitalism. The climate in Geneva, which attracted the most brilliant exiles of the Catholic countries, where repression was growing increasingly harsh, was still quite far from an Erasmian spirit of tolerance. The episode of Michael Servetus (1511–53) is indicative. Servetus developed anti-trinitary theological ideas which led him, among other things, to confute the Galenic theory of the "three spirits" that resided in human beings. His anti-trinitarian positions attracted the hostility of the Catholic hierarchy, and he fled to Geneva to escape the Inquisition. But the doctrine expounded in his *De Trinitatis erroribus* also alienated the Geneva reformers, and in 1555, Calvin had Servetus sentenced to death for heresy and burned at the stake. Servetus's condemnation (even more than Luther's polemics against Erasmus or Thomas More's death sentence), dramatically emphasized the contrast between the reformers and the representatives of culture (Servetus was also a scientist) and caused many philosophers and scientists to lose interest in the themes of political and religious conflict. They claimed a sort of indifference and withdrew into extremely specialized research that was intentionally presented as not being threatening to religion or the wishes of the state authorities.

MATHEMATICAL SCIENCES

Paccioli's Divina proportione

Luca Paccioli (1445–1514) did not confine his activity to writing mathematical treatises, already quite remarkable in themselves, but also did a considerable amount of work in didactics, moving to various Italian cities as an enthusiastic teacher of the mathematical disciplines. During his stay in Milan, at the court of Ludovico the Moor, he became friends with Leonardo da Vinci (1452–1519), who in 1497 drew a series of splendid tables for his *Divina proportione*, published in 1509. In this work, Paccioli, as usual, alternated purely geometrical treatments, to which the treatise is dedicated, with a number of philosophical-speculative digressions that definitely reflect Platonist leanings.

However, mathematical interest during this period was still directed toward questions of arithmetic and practical geometry. Authors like Gregorius Reisch (died 1525), Adam Riese (1492–1559), and Peter Apian (1495–1552) in Germany, and Oranco Fine (1494–1555) in France, continued studying the arithmetical problems and geometrical constructions that had been developed at the end of the preceding century, making specific contributions of varying importance.

Practical arithmetic and theoretical mathematics

The work of the mathematicians just cited, which in a certain sense was minor, is definitely of interest because of the considerable number of solutions they came up with for various problems, which contributed to increasing the patrimony of mathematical knowledge available at the time. However, it is even more important for two other reasons. In the first place, these authors provided a link between practical arithmetic and theoretical mathematics that proved especially valuable in advancing the latter discipline. The practice of dealing with negative numbers, and soon, imaginary numbers, could be extended to academic mathematics because a certain change in mentality had come about through the work of the practical mathematicians, who made it common practice to operate on mathematical values even when that clarity of ideas about their nature that the rigors of classical mathematics seemed unable to give up was lacking. Only in this way, for example, could solutions for general cubic and biquadratic equations be developed. In the second place, algebraic symbolism, or the conventional signs introduced to indicate unknowns and various operations, began to appear more frequently in the works of these authors. This quickened the transition to syncopated and symbolic algebra, the two basic contributions of this century. Christoff Rudolff (c. 1500–1545) and Michael Stifel (c. 1487–1567) made contributions of this sort.

The solution of cubic and biquadratic expressions

The principal achievement of the sixteenth century was the discovery of a solution to general cubic and biquadratic equations. The Bolognese mathematician Scipione del Ferro (1465–1526) was the first to find the formula for solving cubic equations around 1515.

He kept is a secret, however, in order to use it to his advantage in the mathematical challenges that were so common at the time, and on which not only the scientists' prestige and fame, but also their possibilities of enjoying the protection of princes or republics and of holding university chairs rested. It was after he received a series of notices of mathematical challenges (containing problems that required solving a cubic equation) around 1530 that he turned his attention to finding a solution to the famous Niccolò Tartaglia (c. 1500–1557) equation, which he reached in 1535. Then, after much urging, he communicated it to Girolamo Cardano (1501–76), who published it in 1545 in his Ars magna, together with instructions on the method for solving biquadratic equations, discovered by his disciple Ludovico Ferrari (1522–65). Although Cardano gives credit to del Ferro, Tartaglia, and Ferrari in his work, he himself has generally been credited with the paternity of the discovery (even today the formula for solving cubic equations is called the Cardano formula). A violent polemic that lasted for several years broke out between Tartaglia and Cardano, who hid himself behind the interventions of Ferrari.

The value of this discovery of the solution to cubic and biquadratic equations went beyond its great mathematical importance, in that it was the first decisive advance over classical mathematics. Neither the Greeks nor the Arabs had been able to overcome the barrier posed by second-degree equations, and this demonstration that it was possible to surpass the exceptional achievements of the ancients was a powerful stimulus for all of Renaissance science.

Not only that, but from this time on geometry (which, according to tradition, was the paradigm of rigorous and perfect mathematical knowledge) ceased to be the leading discipline, even within academic-type mathematics, as algebra and the algebraic mentality began to predominate. With the passage of time, they would again merge with geometry, bringing new elements to it. Less than a century separates Cardano's Ars magna from Descartes' Geometrie, which introduced analytical geometry — the application of algebraic language to geometry (see MATHEMATICAL SCIENCES 1630–1640).

ASTRONOMY

Characteristics of astronomy in the sixteenth century

The new astronomy was born in the sixteenth century, although its gestation was long and difficult, and the work of one scholar alone was not enough. Nicholas Copernicus, Tycho Brahe, and the young Johannes Kepler took the first steps toward constructing a planetary theory and gave a new impulse to regular and constant observations of the heavens, while from a philosophical point of view, Giordano Bruno was the first to make a clear and decisive break with the centuries-old classical image of the closed geocentric system.

The science of the heavens and the concept of the world were not two separate issues in this first phase of modern astronomy. The lack of a solid physical base and of observational instruments more powerful than the human eye still left ample room for question and debate, which was often dictated by cultural and ideological positions. Paradoxically, some of the most lasting contributions to theoretical astronomy actually came from philosophical and mystical considerations, as in the case of Kepler. The Ptolemaic system was being investigated much more thoroughly, and progress was being made in mathematics. In addition, we must not forget the decisive importance to the birth of modern cosmology which the influence of Platonic and Pythagorean doctrines from Greece had on European scientists. There was also the widespread impression that "the world was small" following the geographical discoveries of the Portuguese and Spanish navigators. Astronomy, like the letters and the arts before it, was welcomed by the princes and sovereigns of Renaissance Europe much as it had been by the sultans and the caliphs of Islam in the past, and more recently at the court of Ulugh Beg. Astronomers were almost always practiced in astrology, which conferred a function to their role that went beyond mere scientific interest.

However, there were also princes, such as the Landgrave William IV of Hesse-Cassel and Fredrick II of Denmark, who seem to have been motivated by a totally unselfish worship of astronomy and who allocated huge sums of money to it. The island of Hevn, Brahe's "kingdom," is the most splendid and adventurous example of this phenomenon.

Alongside this flourishing of spontaneous initiatives of Renaissance patronage were the universities, for example those in Padua, Bologna, Prague, and Vienna, which also retained their role as leaders in the study of astronomy, although to a lesser degree than in the past. Their loss of influence was partially due to the diffusion of printing, which made texts more accessible and less expensive and therefore more readily available even outside academic libraries.

The first exposition of the heliocentric system

While the commentaries on Aristotle's De coelo and Sacrobosco's Sphaera along with the more recent commentaries on Ptolemy's Syntaxis by Peurbach and Regiomontanus were gaining fame in the universities, the planetary theory of Nicholas Copernicus (1473–1543) was forming in the cultural and scientific isolation of a diocese on the shores of the Baltic Sea.

If we are to believe the letter Copernicus wrote to Pope Paul III, preceding De revolutionibus, he began elaborating the new system in 1507, thirty-six years before the work was published in 1543. Around 1510–15, Copernicus gave several scholars a manuscript, entitled De hypothesibus motuum coelestium a se constitutis commentariolus, in which he presented the basic theses of the heliocentric theory, emphasizing its advantages over the Ptolemaic system.

In 1514, Copernicus was invited to the Lateran Ecumenical Council to work on a reform of the calendar, but he declined the invitation.

Meanwhile, Nicholas of Cusa's De docta ignorantia was published in Paris that same year, and in 1515, the first printed Latin edition of Ptolemy's Syntaxis was published in Venice.

The Church and the heliocentric theory

Meanwhile, the notoriety of the ideas presented in the Commentariolus began to produce some effects. In 1533, the German chancellor J. A. von Wildmanstadt, a friend and sincere admirer of Copernicus, gave a brief presentation of the Copernican heliocentric system to Pope Clement VII and his familiars in the Vatican gardens, sparking a great amount of interest.

The Roman Church's first reaction to the new theory was not hostile, but sharp attacks came from the Protestants. In a Tischrede, written in 1539, Martin Luther accused the "madman" who "wants to overturn the entire art of astronomy" by going against the Holy Scriptures (the usual example of Joshua), and later even Philipp Melanchthon turned against him.

Astronomy in Italy

Generally speaking, astronomy in Italy continued in the rut of Ptolemaic orthodoxy. Only a few, including C. Calcagnini, who presumably met Copernicus in Ferrara, or heard him speak during his visit to Cracow, diverged from the tradition and proposed several interesting variants that indicate a new interest in Preptolemaic Greek cosmology.

Perhaps before 1525, Calcagnini wrote a pamphlet entitled Quod coelum stet, Terra moveatur, vel de perenni motu Terrae (published posthumously in Basel in 1544), which talks about the daily rotation of the earth. Despite the ambitions of this small work, it shows a rather shallow knowledge of astronomy and, although it attributes other movements to the axis of the earth, its language is inexpert and often confused. However, there are those who see Calcagnini as a precursor of heliocentrism.

Several years later, around 1538, Girolamo Fracastoro (1478–1553) of Verona published his Homocentricorum sive de stellis, liber unus, in which, following the ideas of his recently deceased friend Girolamo della Torre, he proposed a system of solid homocentric spheres that drew upon Callippus and Eudoxos. According to this work, the axes of the spheres were perpendicular to one another, but other than that, the book contained very little else that was new. Thanks to an ingenious system of adjacent spheres placed between the spheres on which the celestial bodies moved, Fracastoro was able to explain various irregularities in motion (trepidation, etc.) and attribute the impulse given to the heavens to the primum mobile alone, in contrast to Eudoxus' system. He also made the interesting observation that the tails of comets (which it was then thought travelled below the sphere of the moon) always point away from the sun. Simultaneously, the same observation was made by the geographer Peter Bienewitz, better known as Peter Apian, in his Astronomicum caesareum, on a series of comets that appeared between 1531 (Halley's comet) and 1539. Fracastoro's work is also the first to mention the combined use of lenses, though not for astronomical uses. Apian compiled the first atlas of the constellations of the northern hemisphere (1536), which included the forty-eight described by Ptolemy plus the Coma Berenicis and the Canes venatici.

Copernicus's De revolutionibus orbium coelestium

Within a few months of a decisive encounter with Copernicus at Frauenburg in the summer of 1539, the mathematician and humanist Rheticus (1514–74) wrote a report on De revolutionibus that stirred up quite a commotion in the scientific world when it was published anonymously the next year in Danzig, under the title Narratio prima.

When it was re-published in 1541 in Basel, Rheticus' Narratio prima immediately became extremely popular, contributing to a better knowledge of the Copernican system in the scientific world. In his commentary on Peurbach's Theoricae novae planetarum, published in Nuremberg in 1542, Erasmus Reinhold (1511–53), a professor at Wittenberg, alludes to Copernicus as the "new Ptolemy."

The next year, Copernicus finally yielded to the urgings of Rheticus and his friend Tiedmann Giese, and gave permission for his opus magnum to be published. In the meantime, the trigonometric part, entitled De lateribus et angulis triangulorum, edited by Rheticus, was published (Wittenberg, 1542).

The anonymous introduction to the manuscript of De revolutionibus, now attributed with certainty to the Lutheran editor and theologian Andreas Osiander, states that the work's value is hypothetical and that it does not intend to give a real image of the world. Although De revolutionibus proved an editorial failure (not even all of the thousand copies printed were sold), and despite the excessive difficulty of the text (Galileo judged it almost unreadable), it signalled the birth of modern astronomy.

The struggle between geocentrism and heliocentrism

At this point, the hesitation with which, for another fifty years, scientists approached the heliocentric system—although appreciating its mathematical value and its practical utility—is understandable. While the objection that a solid physical basis for a kinematics of the heavens was lacking did not come up until Newton and Descartes, the problems arising from the resulting infinity of the universe and the dynamic effects of the earth's motion created considerable perplexity.

In 1543, the mathematician Francesco Maurolico (1494–1575) of Messina published a *Cosmographia* in Venice that alludes to the "human perversity" and the "absurdity" of maintaining that the earth turns on its axis, obviously in reference to *De revolutionibus orbium coelestium* (although some view it as simply an attack on Calcagnini's pamphlet, mentioned earlier).

PHYSICAL SCIENCE

Premonitions of the scientific revolution

Although the scientific revolution did not actually begin until the seventeenth century, we must acknowledge that the sixteenth century was a decisive phase of preparation for the later developments. It was a very complex phase, made possible by the merging of various cultural and social factors, including an increased interest in nature on the part of both Platonist and Aristotelian philosophers (those of both the Averroist and the Alexandrist currents), the rediscovery and vast diffusion of Greek scientific texts, published in good Latin versions and even in several modern-language translations, a new interest in practicality that emerged within society, or at least within its ruling classes, stimulating the inventive spirit of technicians in almost all fields (from production to mining, from transport to printing, etc.), and, as a result of these solicitations, precise problems that were posed to the scientists by civil and military engineers, who wanted to make effective improvements in their constructions.

It is a well-known fact that the science that underwent the most changes during this period was astronomy (the Copernican revolution, for example). But several advances of considerable importance were made in physics as well, both in studies of mechanical phenomena, where the mathematical side of physics where the observational side prevailed.

In outlining the profound transformations that took place in physics, we should begin with the progress made in the second of the areas just mentioned, since this will give us the opportunity to focus on one of the most significant innovations of the sixteenth century.

We have already mentioned that an increased interest in nature can be detected in all the major philosophical schools of the time. But now we must add that this new interest extended well beyond philosophical circles. There was a generalized curiosity about concrete phenomena as they appear to the common person, beyond the abstract formulae that medieval science had used to attempt to describe them. All fields of experience were investigated in order to uncover their innermost secrets and consequently to succeed in dominating the forces that lay behind natural processes. But attention was particularly turned toward exceptional phenomena, those that were out of the ordinary, miraculous in a certain sense, where natural forces seemed to manifest themselves more directly and explicitly.

Relationships between the occult and the experimental sciences

Only by keeping this particular climate in mind can we explain the extraordinary popularity of the occult sciences (astrology, alchemy, magic), which everyone, regardless of class, saw as providing the most useful tools for discovering the mysteries of reality. To the modern scholar, this may appear to be a totally illusory direction to take; on the contrary, however, it made some fairly remarkable contributions to the creation of a new scientific spirit. In fact, it was the particular curiosity provoked by the occult sciences that gave rise to a new type of observation, resulting in experimental science. Of course the process was very complex—the rising interest in science was often mixed with a strong dose of occultism in the intermediate phases—but its importance cannot be undervalued if we truly wish to understand the steps that led up to the scientific revolution.

This is especially true for experimental physics, which was a direct descendent of magic. This was the area of physics that dealt in the most mysterious phenomena, such as magnetic and certain optical phenomena; and nearly occult forces, such as the reciprocal sympathy or antipathy of the elements were constantly called up to explain them. But a trend toward experimentation could already be seen in the way these phenomena were observed. It appears in *De subtilitate* (1550), written by Girolamo Cardano (1501–1576) who, in his multiple activity as a physician, mathematician, astrologer, and magician, was one of the most typical examples of a scientist of this period.

The new attitude can also be seen in the *Magia naturalis* (first edition 1558, second edition, much enlarged, 1589) by Giambattista della Porta (1535–1615). Although it attempts to distance itself completely from "infamous magic" based on traffic with "unworldly spirits," it still contains many aspects of traditional magic, but in spite of all this, it contains several pages dedicated to actual scientific experiments.

Still another example is offered by *De magnete* (1600) by William Gilbert (1544–1603), usually considered a treatise on experimental physics. Under close examination, however, it appears to be, as Marie Boas says, the last important work on natural magic dealing with the study of the occult forces of nature.

The definitive leap from this type of experimentation for purposes of magic to scientific experimentation as such was only made at the beginning of the seventeenth century and above all consisted in substituting quantitative experiments (in some cases only mentally) for the essentially qualititive ones described by the authors of the above-mentioned texts.

It is a well-known fact that this new type of experimentation stimulated the building of suitable measuring instruments, which were invented and gradually improved as modern physics developed.

Physics and mathematics

Although mathematics was only applied to experimentation in the seventeenth century, transforming it from qualitative to quantitative, the determining influence of mathematics was felt as early as the sixteenth century in what, in modern terms, we might call theoretical physics, i.e. research in mechanics (dynamics and statics). We also know that the most famous fourteenth century physicists of Paris and Oxford had already made considerable use of mathematics in their studies of motion. Now however, an important step was taken along the path these physicists had opened when mathematics stopped being considered simply a useful tool for dealing with certain problems and became thought of as a necessary key for understanding the most profound realities of nature. This systematic recourse to mathematics characterized the work done in mechanics during this period, just as the connection with natural magic characterized studies in optics and magnetism.

However, for this understanding, the Renaissance physicists proclaimed their indebtedness to their readings of the classical texts, recently discovered and translated by the philologists, rather than to the heritage transmitted to them from medieval science. In fact, the appeal to mathematics as the principal vehicle for an understanding of nature constituted a break with medieval culture, where logic was seen as the basic vehicle. Thus there was an opposition between geometry and logic, between Euclid and Aristotle.

But how should this appeal to mathematics be interpreted? Some, influenced by Platonism and Pythagoreanism, justified it by maintaining that mathematics constituted the ultimate essence of reality. In their eyes, mathematics took on a mystical-magical significance and was a part, though on a more serious level, of the great current of occultism mentioned above. Others interpreted mathematics essentially as a tool, a rigorous language that was able to give clear coherence to human discourse and at the same time was superior to logic because it could describe not only the qualitative, but the quantitative aspect of reality.

Both tendencies can be found in many of the authors of the sixteenth century and the beginning of the seventeenth century. In Kepler, a mystical interpretation of mathematics prevails, while an instrumentalist interpretation, drawn from the works of Arichimedes, prevails in Galileo.

Niccolò Tartaglia's kinematics

Prominent in the field of studies in mechanics, which became prevalent in this climate, was Niccolò Tartaglia of Brescia (c. 1500–1557). In addition to being an engineer, Tartaglia is remembered as a mathematician for his contribution to the revolution of cubic algebraic equations and as a translator of Greek mathematical writings—in 1543 he published the first Italian translation of Euclid's *Elements* and the first Latin translation of several previously little-known works by Archimedes, for instance the famous treatise *On Floating Bodies* (see MATHEMATICAL SCIENCES 1500–1550).

His major writings on mechanics are *La nuova scientia* (1537) and *Quesiti et inventioni diverse* (1546). The first is particularly important; it is dedicated to ballistics, a science that Tartaglia is universally credited with having founded. In both works, Tartaglia still frequently used Aristotelian concepts of dynamics, but despite this, he began studying the kinematic aspects of

motion, rather than its presumed metaphysical nature, opening the way to Galileo's approach.

CHEMICAL SCIENCES

Leonardo da Vinci and chemistry

It was in the sixteenth century that a clearer tendency toward viewing experimentation as a tool for stimulating hypotheses and ideas rather than for confirming preconceived ideas emerged. The lesson of William of Ockham (c. 1285–1349) served to clarify the importance of experimental verification, perhaps with excessive severity concerning the possibility of arriving at principles through experience; and the physicists were the first to carry out this sixteenth-century scientific revolution that culminated with Galileo.

Leonardo da Vinci (1452–1519) figures among those who were interested in and wrote about chemistry and alchemy (*Codice atlantico*). Leonardo's interest in chemistry was largely tied to his interest in the materials he needed as an artist and military technician: his primers, plasters, and colors (which he produced himself and even attempted to extract from flowers, although with generally poor results), and the materials he used for building weapons. But he sometimes took a more direct interest in chemistry: besides making drawings for chemical apparatus (stills, kilns, mixing mills, refrigerators, etc.), he listed various recipes (for preparing nitric acid, or *aqua fortis*) and analytical tests (for finding copper in pyrite). He described and explained the dark cone inside a candle flame, and recognized that air played a part in feeding the flame, as well as in life.

Leonardo was hostile toward the alchemists. He compared the would-be inventors of perpetual motion to those who claimed to transform base metals into gold and made it clear that he did not share their belief that mercury was a component of all metals when he said: "The lying interpreters of nature say that quicksilver is a common ingredient in all metals and do not remember that nature varies her ingredients depending on the diversity of things that she wants to produce in the world."

The rise of technical chemistry

Works about technical chemistry characterized the progress in chemistry in the sixteenth century and can be considered the origins of industrial chemistry. They were largely published in the first half of the century. The *Bergwerkbuchlein*, dedicated above all to mineralogy and geology, and the *Probierbuchlein,* both written in the vernacular, appeared around 1510. The *Probierbuchlein* contained a large amount of information on chemical tests, introduced quantitative concepts and made abundant use of scales.

Two Italian works that came out in 1540 were widely circulated even outside Italy. The first was *De la pirotechnica libri X*, by Vannoccio Biringuccio (c. 1480–1539), which talks about the metals known at the beginning of the fourteenth century (gold, silver, copper, tin, lead, iron, and mercury), their metallurgy and alloys, as well as the semi-metals (arsenic and antimony), various ores (including some containing zinc, cobalt, and manganese), various chemical products and materials used for bombs, petards and the like. Biringuccio was among the first to note that the weight of iron increased (by eight to ten percent) during calcination in air, but he attributed this to an increase in density due to the elimination of "aqueous and aerial parts" from its pores. His major contribution was that he interpreted technical processes on the basis of experimental data rather than nebulous alchemistical abstractions. However, he was an empiricist and cannot be considered a pioneer of the experimental method. (More information on Vannoccio Biringuccio is given in the section on APPLIED SCIENCES 1500–1550).

The other technical work published in 1540 was the *Plichto dell'arte de' tentori*, by G. Ventura Rossetti, which gives detailed information on dyes for various textile fibers, mordants, etc.

A well-known work that was also translated into Italian was *De re metallica libri XII*, by Georg Bauer, known as Agricola (1494–1555), published posthumously in 1556 (see APPLIED SCIENCES 1500–1550). Agricola is also well-known for his *De natura fossilium* (1546), a very accurate work that has earned him the title of the father of mineralogy. It mentions petroleum and succinic acid (*sal succini*), a substance obtained from amber.

Iatrochemistry. Paracelsus and his followers

Another sort of technical chemistry was the chemistry that dealt in medicaments. In 1500, right at the beginning of the century, the *Liber de arte distillandi de simplicibus*, by Hieronymus Brunschwig (c. 1450–1512), appeared. It accurately describes the steps for distilling (really in steam currents) plant medicaments to obtain "waters." An enlarged edition appeared one year before Brunschwig's death.

The *Examen omnium simplicium medicamentorum, quorum in officinis usus est* (Rome, 1536), by A. Musa Brasavola, gives an idea of the status of pharmacy during this period. It describes numerous ancient medicaments extracted from plants, and a few more recent ones, including some derivatives of metals. But an essentially extractive pharmacy was now being challenged by preparatory pharmacy, championed by Philipp Theophrast von Hohenheim, or Paracelsus (c. 1493–1541), a physician, philosopher and chemist.

Disdainful of traditional medicine, which mainly used medicaments derived from plants, Paracelsus upheld the use of medicaments derived from minerals. He asserted that the true purpose of alchemy was not to obtain gold and silver, but to prepare medicines (*arcana*) from minerals, which would re-equilibrate what he guessed was an imbalance in the chemical activity of the diseased organism. His concept, which was based on intuition rather than on experimental data, was marred by certain assertions, for example that medicine rests on four columns—philosophy, astronomy, chemistry and virtue. However, by insisting on the importance of chemistry in medicine, Paracelsus was indisputably the founder of what he himself called "iatrochemistry," which would later be called chemotherapy. Paracelsus also coined a number of other words, such as "spagyric" as a synonym for chemistry, "alkahest" (a remedy for liver diseases), "archeus" (an immaterial principle, a sort of soul of the world), "zinc" (*Zincken*), etc.

Among his many works, most of them published posthumously, *De natura rerum* and the *Archidoxis*, written between 1525 and 1527, are of particular interest to chemistry. His views on the elements are unclear: originally, he admitted the four Aristotelian elements (earth, water, air, and fire), but he later substituted the sky for fire. He also maintained the exist-

ence of three "hypostatic principles"—salt (the principle of fixity), sulphur (the principle of combustibility) and mercury (the principle of fluidity). He did not, however, give them the status of actual elements, but of properties. In fact, he believed that there were as many kinds of sulphur, salt, and mercury as there were fruits. A certain kind of sulphur was to be found in gold, another in silver, a different one in lead, and a different one in iron, tin, and so forth."

Paracelsus is not to be particularly remembered as a chemical theoretician, nor as an exceptionally skilled experimenter, but as the person who reminded the physicians of the uselessness of blindly following tradition and of the need to use chemistry for preparing medicaments. His attempts at separating out the quintessence, that is, the active principle of medicines, are also important, in that they required processes of purification, to which he gave great importance. As a result, his followers tended not only to develop new medicines, but to refine their methods of preparing and purifying them. At the same time that this advanced the emergence of pharmaceutical chemistry, it also brought about rapid progress in experimental chemistry in general.

Some of Paracelsus' best-known followers in the course of the century were Oswald Croll (1580–1609), who isolated succinic acid from amber, Leonhard Thurneysser (1531–96), who went to Italy to produce gold from the base metals, J. Duchesne, or Quercetanus (c. 1544–1609) and Theodore Turquet de Mayerne (1573–1655), who prepared benzoic acid from benzoine and black mercuric sulphide.

LIFE SCIENCES

The new naturalism. Leonardo da Vinci and anatomical studies

Renaissance artists and scholars looked upon the human body and the morphological differences between animals and plants with new eyes. Italy was the cradle of this new naturalism, and its most important representative was Leonardo da Vinci (1452–1519), who was able to unite art with science. It seems that he dissected about thirty cadavers. His anatomical drawings are extremely beautiful and at the same time present a degree of accuracy never before achieved. Leonardo studied the flight of birds and the mechanics of the movement of the limbs of human beings and quadrupeds, was interested in fossils and guessed their true nature, and in his own way was searching for some systematic unity in the apparent diversity of living forms. However this immense task remained unfinished and, most importantly, influenced the scholars of the epoch very little. Leonardo da Vinci held a place in the history of the life sciences that was well outside the major currents of thought, and his contribution was only fully appreciated by later generations.

One only needs to glance at the anatomical tables in Magnus Hunt's *Anthropologium* (1501), still executed in the medieval fashion, very stylized and more symbolic than real, to understand how wide the gap that divided official anatomy from artistic naturalism still was. However, the philosophical and esthetic anthropocentrism of the Renaissance necessarily led to a better knowledge of the structure of the human body. As the anatomist Niccolo Massa (1485–1569) said, the human body was considered the supreme example of the perfection of nature. Galen's text, however, continued to be considered gospel. A process of intellectual liberation and revolt against this Galenic domination began to take place in the field of anatomy with Gabriele Zerbi and Alessandro Achillini (1463–1512). Although Zerbi taught Galenic anatomy, he insisted that those who wanted to understand the working of nature must not trust anatomical texts, but must rather observe nature as it presented itself to their eyes. Dissection however, remained a didactic procedure that served to illustrate texts, not to investigate inside the body.

The use of autopsies

In European cities, especially in Italy, autopsies were done for medical-legal reasons. The procedure opened the way for an anatomical concept of illness that was best expressed in the work of the physician and surgeon Antonio Benivieni (1443–1502) around 1507. Alessandro Benedetti, Volcher Coiter, and Jean Fernel (c. 1497–1558) later contributed to the development of pathological anatomy.

Traumatology, syphilis, and military fever

The first description of injuries caused by firearms is contained in Giovanni da Vigo's *Practica*, written in 1514. This new phase of traumatology and a raging epidemic of syphilis were the two most urgent problems in medicine for a large part of the sixteenth century.

Among the diseases considered new were scurvy, which occurred among the sailors on large sailing ships, and, above all, miliary fever (*sudor anglicus*), which appeared at the beginning of the century in England and invaded the European continent several times beginning in 1529.

Humanism and the revision of Arabic medical tradition

Another basic characteristic of medicine in the years between 1520 and 1540 was humanism, i.e. the revision of Arabic tradition and a return to the origins of Greek classical medicine. The first modern Latin translation of Hippocrates' works was published in Rome in 1525. In Venice and in other Italian cities, as well as in Basel, Paris, Lyon, Antwerp, and elsewhere, annotated translations of classical texts appeared, accompanied by tables, concordances, and marginal notes, and were shorn of the additions of Arab commentators.

Translating ancient works. Dictionaries of medicine and natural history

François Rabelais (c. 1494–1553) translated and commented on Hippocrates, as did Nicolò Leoniceno (1420–1524), and many others. Dictionaries of medicine and natural history were published (for example, the works of the publisher Estienne of Paris).

The medical-philological school of Ferrara worked on rediscovering the meanings of ancient scientific terms (Giovanni Manardo and others). Thomas Linacre (1460–1524), who translated Galen and founded the Royal College of Physicians in London, carried medical humanism from Italy to England.

Iatrochemistry and occultism

While physicians and naturalists were attempting to rediscover a rational, elegant, pure science in the distant past, there were also a number of original thinkers during this period who turned their attention to magic and the supernatural and who subverted the entire culture in an entirely original way. The most typical representative of this current of Renaissance thought was Paracelsus (1493–1541), the founder of iatrochemistry. A short time later, occultism found an exponent in Girolamo Cardano (1501–1576), who also gave a description of typhus (1536).

Descriptions of syphilis and its therapy

Epidemics of syphilis were wreaking havoc. The best clinical description of the disease was given by Girolamo Fracastoro (c. 1478–1553), who is best-known for his *Syphilis sive morbus gallicus* (1530), a poem on a subject that at first glance has very little that is poetic about it. Fracastoro not only gave the disease its name, but described it in an expert fashion. The severity of the syphilis pandemic led to a whole series of studies on the subject, done by Niccolo Massa (1532), Pietro Mattioli (1553), Ruy Díaz de Isla (1539), Gabrielle Falloppia (c. 1523–62), and others. Mercurial ointments and a decoction of guaiacum were introduced in therapy. Already the question had arisen of whether the disease was autochthonous to Europe or came from the New World.

The description of plants from the New World

Diseases were not the only thing that came from America. Naturalists were struck by the variety of plants and zoological specimens that arrived from the new continent. From 1525 to 1535 Gonzalo Hernandez de Oviedo y Valdes (1478–1557) published the first systematic description of the medicinal plants of Central America.

Botanical iconography

The decade 1530–40 saw the development of botanical iconography, especially with the *Herbarum vivae eicones*, by Otto Brunfels (c. 1489–1534), the first volume of which was published in 1530. Brunfels took his inspiration from nature, not from tradition, and his work marks the beginning of a new genre of herbals.

The reform of anatomy. The work of Vesalius

Dissections for didactic purposes and autopsies for judiciary reasons were practiced in many European cities, especially Bologna, Padua, and Venice. The reform of anatomy began with two works on the subject by Giacomo Berengario da Carpi (1460–1530), published in 1521 and 1522. The real break came with the great work of Andreas Vesalius (1514–1564), who taught at Padua, published the *Tabulae anatomicae* (1538) and prepared his innovative work. The advances made in anatomy and, above all, a certain state of mind that came from new attitudes toward the human body necessarily affected surgical practices. Mariano Santo da Barletta introduced median lithotomy, and recommended tying off arteries during surgery in order to combat hemorrhaging.

Many historians of science designate this period as the beginning of the history of modern medicine. In 1543, Vesalius, who at the time was a professor in Padua, published his major work, *De humani corporis fabrica*, in Basel. Vesalius' treatise, which came off the presses the same year that Copernicus' work was published, revolutionized the knowledge that people had about the structure of their own bodies. The work is important not so much for the number of discoveries, in the strict sense, that it contains, but for its demonstration of a new method.

The development of anatomical science

Although he was unquestionably the major anatomist of the sixteenth century, Vesalius was preceded by Berengario of Carpi, Charles Estienne (c. 1505–64) Giovanni Battista Canano (1515–79). Canano's *Muscolorum humani corporis picturata dissectio* (1541) was the first work to use copper engravings for anatomical illustrations. Canano and Estienne discovered the valves in veins. After Vesalius, discoveries in anatomy followed one another at an extraordinary pace. At Padua, Fallopius continued to do research that was inspired by Vesalius. The Padua school of anatomy reached its peak of glory with him. There were also distinguished anatomists in the other university cities, for instance Bartolomeo Eustachi (c. 1500–74) in Rome, Giulio Cesare Aranzo (c. 1529–89), and Conefanzo Varolio (1543–75) in Bologna, Giovanni Filippo Ingrassia (1510–80) in Naples, Giudo Guidi (1508–69) in Paris and Pisa, Jacques Dubois (1478–1555) and Estienne in Paris, Guillaume Rondelet and André du Laurens (1558–1609) in Montpellier, F. Platter in Basel, etc.

The progress made in understanding the anatomical structure of the ear is a good example of the advances made during this period. Achillini, Massa, and Ingrassia described the ossicles and how they work, Fallopius was familiar with the ear drum and the aqueduct. Eustachi and Coiter discovered other anatomical details that were fundamentally important to explaining how hearing worked. For his part, Francesco Maurolico (1494–1575) studied the optics of the crystalline lens and explained the origins of myopia and hypermetropy.

Progress in therapy; the founding of epidemiology

Therapeutics benefitted from the advances made in natural history, and materia medica was expanded. In 1546, Valerius Cordus (1515–44) published the first real pharmacopoeia. That same year, Fracastoro's major book, *De contagione et contagiosis morbis*, was published. It contained new ideas on the transmission of contagious diseases that can be considered the origins of a truly scientific study of epidemics.

EARTH SCIENCES

Geographic and oceanographic knowledge

The first half of the sixteenth century was characterized by a broadening of geographic knowledge, following the discovery of America, which was named in 1507 in honor of Amerigo Vespucci. In the thirty years from 1492 to 1522, the extension of the known world increased at a dizzying pace. The major

contributions were Christopher Columbus' first three voyages (1492, 1493, and 1498) to the New World, Vasco da Gama's explorations along the coasts of Africa and India in 1497–99, the explorations made by Vespucci and Vicente Yáñez Pinzón from 1499 to 1502 along the coasts of Brazil, Sebastian Cabot's voyages to Labrador and Newfoundland (1497–98), and the circumnavigation of the Strait of Magellan.

The first oceanographic observations on the open sea were made during these voyages. For example, on his third voyage Columbus noticed strong currents flowing westward off the northern coasts of South America. On the basis of this observation, he attempted to establish that the waters of the ocean flowed from east to west. Da Gama also recognized and studied the currents produced by the steady blowing of the wind as early as his first voyage (1492). Columbus supposedly recorded the existence of magnetic declination and recognized magnetic storms. Vasco Núñez de Balboa's discovery of the Pacific Ocean in 1513 and the sighting of several islands, including, it seems, the Philippines, and the circumnavigation of the globe (1519–22) by Ferdinand Magellan's expedition, definitively solved the problem of the shape of the earth (although it was still considered the center of the universe). In particular, the immense extension of the Pacific Ocean was verified, and the existence of an unknown *Terra australis* in the middle of it was hypothesized. Investigations of oceanic currents, the existence and importance of which the explorers had already understood, were intensified. In 1513, Juan Ponce de Leon furnished the first information on the currents of Florida.

In 1511, Pietro Martire d'Anghiera (1459–1526) published his *Oceani decas*, in which he describes various oceanic currents. The work includes the Gulf current, which he guessed was derived from the equatorial current deviated by the American continent.

Navigational systems and related information became increasingly refined. The first Dutch nautical maps marked with the North Sea routes and the routes along the northwestern coasts of Europe were probably published in 1518.

In 1524, Apian discussed the dimensions of the Earth in his *Cosmographicus liber*. In 1525, the physician Jean François Fernel (c. 1497–1558) established that one-fourth the length of the earth's meridian was equal to 10,011,000 meters. These studies foreshadowed an understanding of the exact shape of the earth, which was still thought to be perfectly spherical.

Geology

In the area of geological studies, Leonardus Camillus wrote about minerals and their origins in 1502 (*Speculum lapidum*) and in 1505 (*De mineralibus*, printed in Venice). He classified them according to their physical characteristics, but thought that they originated in the most improbable ways, in keeping with the ideas of the times. Ulrich Rülein von Kalbe wrote his text on ore deposits in 1505.

Geophysics

While Sebastien Münster of Ingelheim attempted to describe the entire surface of the earth in his universal cosmography (1544–50), in 1543 (the year of his death) Copernicus published his *De revolutionibus orbium coelestium*, in which he stated that the earth belonged to a heliocentric system.

In 1544, Georg Hartmann (1489–1564) discovered magnetic dip, an important part of the earth's magnetic field.

Leonardo da Vinci's observations and intuitions

Leonardo da Vinci (1452–1519) correctly answered the geological questions that were being debated at the time, maintaining, for instance, that fossils were derived from organic remains, springs were formed by meteoric waters, mountains were shaped by erosion, and the boundaries between the earth and the sea changed. Unfortunately, the manuscripts he wrote at the beginning of the sixteenth century, such as the Leicester manuscripts, remained unknown for many years.

Mining activity

Mining activity flourished in Bohemia, Germany, and Italy during this period. Various texts were written on mining laws, and geognostic observations, which two centuries later gave rise to the science of geology, were developed. But the genesis of minerals still remained a mystery, and the alchemists continued to argue about whether transmutation was possible or not. The first work by Georgius Agricola (1494–1555), which had a lasting influence on ideas about mining deposits, appeared in 1528. It outlined the idea that metalizing solvents took part in the genesis of metallic ores. In this same period, Paracelsus (c. 1493–1541) hypothesized that pumice was solidified sea foam.

It was still a time of the strangest and most varied conjectures, which were partly influenced by the ideas of the alchemists. Not only was the genesis of minerals the object of speculations that were often fantastic, but minerals were often attributed with magical, miraculous properties, such as the power of healing or, in cases of poisoning, of antidotal action through contact alone.

One of the most sought-after stones for these purposes was bezoar, which supposedly formed in the bodies of certain animals. Carbuncoli were said to have luminous powers, and to glow in the dark. Rösslin mentions them in his *Kreuterbuch* (1535). In 1546, Agricola published *De ortu et causis subterraneorum*, followed the next year by *De natura fossilium*, in which minerals are placed in the category of fossils, a common practice at the time.

The idea of a congealing liquid (*succus lapidescens*) that intervened in the formation of metals—a substance that today would be called a metalizing solution—appears in Agricola's works. He also gives a detailed account of geological knowledge beginning with the classical authors of antiquity and moving up to his time. For the origin of fossils, he devised the unfelicitous solution of a *materia pinguis* that supposedly formed them as it concretized. The study of mineral waters and hot springs attracted a considerable amount of interest during this period, as it already had in the past (we know that mineral waters were used for curative purposes from the earliest times). A book entitled *De balneis omnia quae extant apud graecos, latinos et arabos* was published anonymously in Venice in 1552; it probably contains the ideas of various authors, including those of Agricola himself and of previous scholars at the school of Padua who were experts on the Euganean hot springs (Michele Savonarola, Iacopo and Giovanni Dondi).

Many other authors of this epoch took an interest in minerals, and especially in gems. Rueus, in his *De gemmis aliquot* (1547)

and G. Cardano (1501–76) in his *De subtilitate* (1550) both talk about gems. Vanoccio Biringuccio (1480–1539) wrote about mines and extracting metals in his posthumously published work *De la pirotechnia* (1540).

The origin of fossils and their relation to the flood

In a famous letter written to the Veronese jurist Torello Saraina in 1517, Girolamo Fracastoro (c. 1478–1553), a professor of medicine, responded to questions he had been asked about fossils, stating that they were the remains of animals that lived in the past, and that they could not have come from Noah's flood, since in that case they would have been freshwater and not salt-water animals, and they would have been scattered on the surface of the earth rather than contained in rocks. For these reasons, Fracastoro concluded that the sea must once have covered what later emerged as land and mountains. The organisms lived in the depths of the sea, and the fossils that we observe are their petrified remains.

APPLIED SCIENCE

The art of mining and chemistry

Following the efforts toward universality of fifteenth century engineers and until new syntheses took place, the task of specialization began. Sixteenth century scholars and engineers interested in technological problems turned their attentions to specific techniques and attempted to popularize their procedures, stimulated and aided by the advent of printing.

For economical reasons, the arts of mining and metallurgy were among those activities that profited most from this effort of diffusion, especially in Germany. At the beginning of the sixteenth century, the first modest attempts appeared: two illustrated works, the *Bergwerkbuchlein*, a treatise on the extraction of ores, and the *Probierbuchlein*, a treatise on assaying. The *Bergwerkbuchlein* (the first edition of which probably dates back to 1510) is the oldest printed work dealing with the genesis of and the search for deposits of metals. The *Probierbuchlein* was probably first published in 1510, although some historians cite the Magdeburg edition of 1524 as the first. The texts are written in German and were not intended to be read by scholars, but rather by technicians and workers (see CHEMICAL SCIENCES 1500–1550).

Other techniques were not so widely discussed, except for the chemistry of distillation, described by the German Hieronymus Brunschwig (c. 1450–1512) in his *Liber de arte distillandi de simplicibus*, which appeared in Strasbourg in 1500. It is easy to see how research in mining and metallurgy stimulated these studies.

Chemistry and metallurgy

The interest in chemistry in Germany found an advocate in Paracelsus (1493–1541), a student at the mining school founded by the Fuggers. Paracelsus actually founded a new school of chemistry which, in its attempts to capture the "spirit" of liquids, stimulated other distillations. Andreas Libau, the author of *Alchemia* (1597), Johannes Baptista van Helmont and Johann Rudolph Glauber later came out of this school.

In Venice in 1540, Vanoccio Biringuccio (1480–1539) published his *De la pirotechnia libri X*, the oldest and most detailed manual of metallurgy. The first four chapters deal with the fusion of gold, silver, copper, lead, tin, and iron ores. They contain exhaustive descriptions of the amalgamation of silver, the flowing furnace, and the liquation process.

Biringuccio had travelled widely, especially in Germany, where he came into contact with the great German tradition of metallurgy. Before Pope Paul III engaged him as his founder and director of artillery, he worked for the duke of Parma, for the duke of Ferrara, and for Venice. His work, which appeared posthumously, was widely circulated. Biringuccio, along with Georgius Agricola (1494–1555) and Lazarus Ercker (died 1593), was a true authority on metallurgy. Esban Hesse first described the use of rolling mills in the iron-working process. Fortuno Imperato (c. 1550–1631) also took up Biringuccio's ideas in his *Historia naturale*, written toward the end of the century. Agricola's first great work appeared in 1546. It contained four essays, the third of which was *De natura fossilium*, the first systematic treatise on mineralogy.

Mechanical devices

German talent also manifested itself in other fields of research. Two important inventions by Renaissance artisans, the vice and the pocket watch, date back to this period. The vice appeared in Nuremberg around 1500, while the spring-operated watch with balance-wheel regulator was designed and built in Nuremberg around 1510.

A petition for a license to manufacture "real crystal mirrors," presented by the Dal Gallo brothers in 1507, attests to the progress made in the art of glass-making in Venice.

Textile machines

The imbalance between spinning and weaving within the textile industry was reduced by the invention of the pedal-operated spinner, one of the major innovations of the Renaissance. The flyer throstle, a widely-used invention of the fifteenth century, had the great inconvenience of being hand-operated, leaving the spinner only one free hand. The worker's seated position and continuous motion suggested the application of a pedal. The innovation is attributed to master Jurgen, a Brunswick bricklayer, in 1530. The use of the pedal made spinning much faster.

Engineering and hydraulics projects

The first Renaissance town planners flourished at the beginning of this period. Around 1500, Francesco di Giorgio Martini (1439–1502) published a series of drawings of ideal polygonal and fortified cities, with the streets laid out geometrically.

In 1546, Michelangelo built one of the major works of the Renaissance, the dome of St. Peter's in Rome. However, his structure presented no significant advance over Filippo Brunelleschi's dome on the Florence cathedral, built in 1420.

In hydraulic architecture, Benedetto da Missaglia's designs for the locks of the Paderno canal deserve mention. Leonardo had already built the first modern locks in Milan in 1497.

While canalization techniques were being perfected in Italy, techniques for draining *meer*, or internal lakes, were being developed in the Low Countries. The technique was new, totally different from embanking against the sea. Between 1540 and 1565 a total of 35,608 hectares of land were reclaimed in northern and southern Holland, Frisian, Gröningen, Sjaeland, and North Brabant, of which 1,349 hectares came from drained lakes. The new technique was used most of all in the seventeenth century, between 1615 and 1640, when out of 25,513 hectares reclaimed 19,060 came from draining internal lakes. The practice was important not only for agriculture, but for the boost it gave to technical developments in mills for drainage.

The greatest achievement of the hydraulic architects was the bridge of Notre Dame in Paris, where a caisson foundation was used.

Pistols for knights and fortified cities

The most important innovation of this period, however, came about in military technology. Around 1520, the wheel-lock pistol for knights was invented, probably in Italy. It was held in one hand only, although it was often quite large. All of the earliest portable firearms had been front-loading, and the powder charge would only burn if fire was applied directly, so that the knights who used them had to carry a slow-burning wick with them. It was exceedingly inconvenient for the knights to have both hands occupied, and the use of firearms was virtually impossible for them.

The invention of the wheel-lock pistol was the first step toward a completely mechanical firearm. The device had a spring under tension that was freed by a trigger, making a wheel with a toothed edge rotate against a flint. This produced sparks that lit the priming.

A second mechanical improvement in portable firearms came around 1525, when barrels began to be rifled. The combination of the rifled barrel and the wheel-lock firing device made it possible to construct accurate, manageable hunting weapons, although the cost limited their use to a few impassioned sharp-shooters.

The use of gunpowder and firearms in combat led to profound changes in the art of static defense. Vertical walls and towers of all thicknesses were vulnerable to projectiles and mines, and the larger and more complicated the defenses developed in the late Middle Ages, the more they limited the besieged in their offense, since they gave the attackers more protection from projectiles and sallies.

Fortified cities became the basis of the new defense system; they could both protect the civil population and hold thousands of soldiers and a large amount of artillery, materials, and equipment, besides water and food supplies. The new theory of fortifying cities and building citadels was illustrated by Albrecht Dürer in his *Etliche Underricht zu Befstigung der Stett, Schloss und Flecken*, published in Nuremberg in 1527.

That same year, M. Sanmicheli (1484–1559) used a new construction design on the Maddalena and Spain bastions in Verona. Uniform curtain-type walls were rejected and the city was surrounded by an intertwined system of regular and irregular polygonal defenses, where the angles formed by the outside walls were designed to give maximum crossfire concentration.

In 1529, Michelangelo followed the same basic rules for fortifying San Miniato.

Food and agriculture

During the fifteenth century, the only books on preparing food that were reproduced were the works of Marcus Terentius Varro, and Pietro de Crescenzi's treatise of 1471. The advent of printing, however, favored the spread of the art of more refined food preparation. One of the first kitchen manuals was the *Boke of Keruynge* (on the art of carving), published in England in 1508.

A few works on agronomy that showed some originality, even though their value was modest, began to appear. One of these, *The Boke of Husbandry* by the Englishman Fitzherbert, was a remarkably popular manual; eight editions of it were published between 1523 and 1550. In 1530, a work on botany, the *Herbarum vivae eicones* by Otto Brunfels (c. 1488–1534) of Strasbourg also appeared. The introduction of botanical gardens was of prime importance to agriculture. One of the most famous gardens was founded in Ferrara in 1528 by Alfonso I d'Este.

Meanwhile, a new product was added to the European diet: cocoa, in the form of bars and slabs, reached Spain in 1528 and slowly spread throughout Europe.

The nautical astrolabe

In 1533, Regnier Gemma Frisius (1508–55), a professor of mathematics at Louvain, explained the principle of triangulation, eliminating all measurements of distance except the base line. Gemma, aware of the errors the procedure was subject to when applied in the field, emphasized the importance of aligning the instrument accurately at both ends of the base line and of levelling it.

Another instrument for surveying that was important to navigation—the nautical astrolabe, a Hispano-Portuguese adaptation of the astronomer's astrolabe—was invented around 1535. It consisted of an alidade and a calibrated ring (the parts necessary for observation), and had no device for making calculations. It was used throughout the sixteenth century as an instrument for measuring the height of the sun and of the pole star.

1550–1600

THOUGHT AND SCIENTIFIC LIFE

The Counter-Reformation

The situation outlined in the last chapter occurred not only in those countries where Protestantism gained a hold, but in those dominated by the Catholic Counter-Reformation as well. In Italy, the Waldensians disbanded; Carnesecchi was sentenced to death in 1567; in 1542 Ochino, a preacher and general of the Capuchian order, began a long period of exile in the centers of the Reformation (which lasted until his death in Moravia in 1564); Caracciolo reached the Protestants in Geneva; and Seripando remained in the Catholic Church, where he became a cardinal and one of the most important exponents of the Counter-Reformation.

In 1542, the Inquisition was reorganized in Rome and a court

of the Holy Office was established. Its first targets were the theological, literary, and scientific works that were considered vehicles of heresy. The *Consilium de emendanda ecclesia*, a document of the commission that Paul III created in 1537 to reform ecclesiastical life, had already alluded to the need for instituting a censorship of books, and recommended banning the teaching of the *Colloquia familiaria*, one of Erasmus' most brilliant works. Erasmus' works began to be prohibited in 1551, when the struggle became more radical. In 1559, the first list of forbidden books appeared. Humanist culture began to feel the effects of the first Catholic restrictions, and a reconciliation with the various Protestant sects appeared increasingly unlikely. The problem of reorganizing the internal structure of the Church was still of prime importance, however. The basis of the Counter-Reformation movement was the Council, called by Paul III in November of 1544, which began in Trent on December 13. This is not the place to go into detail about the developments of the Counter-Reformation, just as we did not enter into the particulars of the Reformation. Nor can we follow the tortured progress of the Council. We shall only mention that those who championed a dialogue with the Protestants, such as Cardinals Reginald Pole, Jacopo Sadoleto, and Giovanni Morone, were defeated by the more intolerant and extremist current, which definitively gained the upper hand when Gian Pietro Carafa was elected pope (Paul IV, pope from 1555 to 1559) during a long recess of the Council (1552 to 1563). The last reforming tendencies within the Church were repressed, Cardinals Pole and Morone were tried, and the Protestant groups that remained on the peninsula were dispersed. The ecclesiastical body was pressed by the pope's insistent urgings into a position of total unreceptiveness to the demands that came from both the Protestant world and Renaissance culture itself. However, together with this repression, there was also a reorganization of both religious life and intellectual activity. The new religious orders that had come out of the Catholic Reformation during the first half of the century—the Theatines (1525), the Capuchians (1528), the Somaschi (1528), the Barnabites (1533), and the Jesuits (1534)—played a primary role in this reorganization. The most important of these orders was the Jesuits. Around 1550, after Ignatius Loyola had elaborated the *Formula instituti* (1539) and the first papal recognition was granted (1540), the order began to grow. By 1556, it numbered around a thousand. In keeping with its founder's military past, it called itself a company and became an actual war machine, directed first against the Moslems, and a little later against heretics. The structure of the company was rigidly hierarchical. The supreme leader was the Provost General, who was subject to the laws of the order and was responsible, by way of a special bond, only to the pope. This organization, which was new even to the Catholic world and was almost a church within the Church, was able, on a doctrinal level, to combine an optimism with respect to humanity with a pessimism with respect to the single individual. On the one hand, it accepted some of the more lively themes of humanist and Renaissance activism, which led to a lively conflict with the passive idea of salvation by pardon developed by the Franciscans in the late Middle Ages and taken up by the Reformists; on the other hand, it theorized the need for a more rigid authoritarianism, in complete agreement with the Counter-Reformist opposition to the Protestant interpretation of Christianity as spiritual freedom.

The intellectual debate: Aristotelianism and anti-Aristotelianism

In the midst of these theological-political controversies, with their often dramatic outcomes, the humanist movements were very much alive. The attacks against Aristotle and scholastic philosophy became even harsher during this period. In Italy, Marius Nizolio (1498–1576) taught the *humanae litterae* in Parma from 1549. His most important work, *De veris principiis et vera ratione philosophandi contra pseudophilosophus* (1553) is an analysis of the *Organon*, in contrast to which he proposes a philosophical knowledge based on a harmonic merging of logic and rhetoric. The work enjoyed remarkable success (Leibniz himself edited two editions, with explanatory notes and critiques). Nizolio's anti-Aristotelianism was echoed in France by that of Pierre de la Ramée (1515–72). Ramée's *Dialecticae partitiones ad Academiam parisiensem* was followed by a long series of manualistic and scholastic works, which led to a new current of studies that became popular not only in France, but in Germany, Holland, and England as well, as early in the first decades of the seventeenth century. The central theme of Ramism was the opposition of a "natural" logic to Aristotle's "artificial" logic.

Elements of anti-Aristotelianism also existed in Platonist and Neoplatonist thinking, which drew many of its ideas from Ficino and Pico. Here we shall only mention Francesco Patrizi (1529–97) who, like Plato, longed for an ideal republic in his *La città felice*, written in 1553. Patrizi's republic, however, had a number of features in common with Venice, which was fast becoming a center of free intellectual life and remarkable artistic flowering during this period. In his later works (*Discussiones peripateticae* and the treatise *Nova de universis philosophia* were written in the last decade of the century), Patrizi found the philosophical references that could best combat "Aristotelian blasphemy" in Platonism, emphasizing the theme of *prisca theologia* and a reconciliation of various doctrinal currents (Protestants included). But Patrizi's cultural policy remained a dead letter when the Thomists began to prevail within the Counter-Reformation.

Aristotelianism, despite the attacks against it, was still one of the foremost philosophies. In fact, the first four books of the *Physics*—of which three different Latin translations were published in 1554, 1558, and 1559—and Joannes Philoponos' commentaries on them, which reevoked a whole series of physical questions, including the theory of impetus, so dear to Late Scholasticism, were extremely popular. Padua was the most lively center of Aristotelianism. The debate on Aristotle's works involved a number of sciences (from logic to mathematics and mechanics, and from astronomy to biology and psychology). An interest in the concrete manifestations of nature, as opposed to Plato's ideas, dominated the continuing polemics between Averroists and Alexandrists. It also dominated the more moderate formulations of the doctrine of the intellect and the soul, which was modeled after Simplicio's commentaries, and the debates on the *Ethics,* the *Politics* and the *Poetics*, which undeniably influenced a great part of Renaissance essay-writing.

Anti-Aristotelian ideas were also present in the philosophy of nature. The *Physics* was subjected to various criticisms, and a concept of the universe was proposed that was more unitary in structure. The sublunary world and the world of the heavenly spheres were brought closer together, the features that humans had in common with other animals were examined, and the relationship between the body and the soul was studied. The universe was pervaded by animation and movement even where it appeared inert. Hidden causes and secret forces linked objects and phenomena, as Girolamo Cardano, the most famous magician, mathematician, and physician in all of Europe theorized in his *De rerum subtilitate* (1550) and *De rerum verietate* (1557), and Giambattista Della Porta (1535–1615) in the first edition (in four books, 1558) of his *Magia naturalis*.

Great developments in science

Aristotelianism, Platonism, and the philosophy of nature, although from different angles, all reawakened a growing interest in nature, no longer seen as an inferior reality, but as an object worthy of in-depth study. As a result, the profound connection between scientific investigation and philosophical speculation was emphasized in various areas. It manifested itself in both the mystical-religious notion that mathematics was inspired by "the divine Plato" or by Pythagoras, and in magic itself, as well as in the works of the Aristotelians most directly involved in scientific research.

We have already spoken of the rediscovery of the classics of science, promoted by humanists and philologists. Now, when new and more accurate editions of the original texts and Latin translations were available, the first versions in the vernacular began to appear. In 1543, Niccolo Tartaglia (c. 1500–57) published the first Italian translation of Euclid's *Elements* with an ample commentary. These translations were aimed at a larger public, beyond the traditional scholars, and later became a reference point for all research in the applied sciences that looked toward the classics for an adequate theoretical basis. Nor should we forget that exposure to these works, diffused via printing, which by now was clearly at the service of technical-scientific culture, brought up new problems, sometimes leading to huge cultural debates. It was not rare for one classical author to be contrasted with another. For example, Archimedes was often contrasted with Euclid, and Euclid, as an example of the richness of mathematical reason, with Aristotle, the creator of a sterile logic. The growing awareness of the multiplicity of models that antiquity proposed led the researchers to understand the importance of classical problems, although new paths were sought as well. An example of this was the flowering of research in arithmetic and algebra, with heuristic methods that often deviated from the Euclidean ideal. There was also the interest in the problem of areas and volumes, suggested by reading Archimedes, which led the mathematicians to develop the tool of calculus, apparently very distant from Greek geometrical rigor.

Two great scientific works that, in their respective fields, marked a revolt against the established authority, appeared in 1543: *De revolutionibus orbium coelestium*, by Copernicus (1473–1543) and *De humani corporis fabrica*, by Vesalius (1514–64). Their specific aspects aside (for these, see ASTRONOMY and LIFE SCIENCES 1550–1600), we should note that, although they referred to the teachings of the ancient masters (the influence of Pythagoreanism on Copernicus is especially evident), these scientists contested not only the authority of medieval tradition, but that of the celebrated scholars of antiquity—Aristotle and Ptolemy in astronomy, and Galen in medicine—as well. It became increasingly evident that science could not put any *a priori* limitations on its investigations, and that it had to be free to repudiate any theory, no matter how ancient and authoritative, whenever there was good reason to do so (more adherence to the facts, as the schools of anatomy in Padua and Bologna upheld; more "simplicity," as Copernicus maintained in the heliocentric theory he opposed to Ptolemy's geocentric theory).

The problem of freedom of scientific research with respect to the authority of the Scriptures also arose. For example, Copernicus' work implicitly conflicted with the teachings of the *Bible*. Although this fact escaped the exponents of official Catholic culture at first (in 1530, after reading a passage from Copernicus' work, Pope Clement VII urged the Polish astronomer to publish the work *in extenso*), it was grasped in Protestant circles. The Protestant theologian Andreas Osemann, also known as Osiander, wrote a preface to the first edition of *De revolutionibus* in which he emphasized that the new theory was only a mathematical hypothesis, thus introducing an "instrumentalist" interpretation of scientific theories that in some ways foreshadowed the interpretation later adopted in Catholic circles by Cardinal Bellarmino with respect to Galileo's position (see ASTRONOMY 1630–1640).

The Counter-Reformation and religious freedom

The Council of Trent finished its work on December 14, 1563. The task of reorganizing and moralizing the Church ended by defining the condition of residency for bishops, banning the accumulation of ecclesiastical benefices, reconfirming the condition of celibacy for priests, establishing control over preachers, and promoting ecclesiastical education through the founding of seminaries in every diocese. On a doctrinal level the Council rejected the theory of justification by faith alone, confirmed the number of sacraments according to tradition, reiterated the divine nature of the priesthood, and recognized the official interpretation of the Scriptures as the only valid version. In 1564, Pope Pius IV approved the conclusions of the Council and published them in the *Professio fidei tridentinae*, which was reduced to a more accessible, simpler form in the *Catechismo tridentino* (1566). In 1565, the pope proclaimed Thomas Aquinas a Doctor of the Church, and Thomism was on its way to becoming the official philosophy of the schools organized by the Catholic authority. A few years later, Pius V (pope, 1566–72) created a new tool of ideological pressure, the Congregation of the Index, to which he assigned the task of updating the list of prohibited books.

The sweeping restoration undertaken by the Church and its emphasis on authoritarianism (in particular, the authority of the pope was given priority over both doctrine and the government of religious life) coincided with the elimination of the last evangelistic groups in Italy. Thus those intellectuals who professed a spiritualism that did not fall within rigorous dogmatic formulations or Calvinist rigidity, especially after Servetus was burned at the stake, were dispersed. Perhaps the most prominent figure among the Italian exiles was Fausto Sozzini (1539–1604), who became the major organizer of the group of "Polish brothers" who spread rationalistic interpretations of religion and attitudes of tolerance in Poland (and elsewhere).

Developments in philosophical thought

The political events taking place during this period were particularly complex. While the Hapsburg monarchy acted as the right arm of the Counter-Reformation and simultaneously won a prestigious military victory against the Turks at Lepanto (Navpaktos, 1571), France was torn by religious wars from 1562 (the French Calvinists, the so-called Huguenots, were massacred on St. Bartholomew's Eve in 1572), the Low Countries were in revolt against Spanish domination from 1566, and England achieved relative stability under the reign of Elizabeth I. The study of political reality, which tended to structure itself scientifically, after the model of Aristotle's *Politics*, took place in this context. Undoubtedly the most important politician of the epoch was Jean Bodin (1530–96), who published his *Les six livres de la republique* in 1576 (it was republished in Latin ten years later). Its dominant theme was the idea of royal power as the only guarantee of order and peace, in contrast to the Counter-Reformist theories, upheld mainly by the Jesuits, that wanted the monarchy subject to religious authority (often to the point of advocating the assassination of politicians who deviated from behavior that was inspired by the dictates of the Catholic doctrine). As a result, Bodin and other cultivators of politics and law insisted on the need to tolerate the various confessions, in order to safeguard the bonds of political obedience.

The unbiased analysis of political reality, the revival of Machiavelli's ideas (Machiavellianism and anti-Machiavellianism turned up persistently throughout the seventeenth century and beyond), and the renewed interest in historiography often referred back to the Greek and Latin classics, especially the Aristotelian tradition. Aristotelianism was very much alive in studies of logic and the natural sciences as well. The *Quaestiones peripateticae*, by Andrea Cesalpino (1519–1603), a teacher of botany and medicine in Pisa and the papal physician in Rome, were written in 1571. Jacobus Zabarella (1533–89), a teacher in Padua from 1563 to 1589, wrote his work on logic in 1578. For both, Aristotle was the thinker who raised philosophy to its greatest heights. Cesalpino insisted that Aristotelianism had to be freed from the "commentaries of the barbarians," and leaned toward an unbiased openness to experimentation ("the true physician is one who unites experiment and reason"). Zabarella, for his part, developed an original doctrine of scientific medicine and, in the field of specific research, approached Alexandrist ideas, theorizing the union of body and soul and rejecting the theory of the unity of the intellect.

A few years earlier, the first two books of *De rerum natura juxta propria principia*, by Bernardino Telesio (1509–88), a broad attempt at revising and correcting the peripatetic philosophy of nature, had appeared (1565). They included extremely controversial ideas (Aristotle's doctrine, Telesio wrote, "was all founded on false principles"), especially when Telesio polemicized against the Aristotelian concepts of matter and form, replacing them with the notions of material mass and force. There are also references to authors before Aristotle, evident for example in the revival of the hot and cold theory, which goes back to Empedocles, or the idea of a material soul, which allowed Telesio to assert that all bodies are animated, thus reconnecting with Renaissance animism.

Reciprocal influences of science and technology. Cartography

While both the Aristotelian and the anti-Aristotelian philosophies of nature contributed to the development of specific scientific research (in mathematics, physics, medicine, etc.), the diffusion of the classics enriched the range of scientific problems being considered. The works of Euclid, Archimedes, Apollonios, Galen, Hippocrates, Pliny the Elder, etc. no longer circulated only in Latin and Greek, but in their relative Italian, French, English, and German translations. Another element of scientific progress was the enormous development of certain technologies, such as the technology of pumps, first used in mines and then in aqueducts; the techniques connected with the war sciences; and the first industrial technologies (for example, those in the area of textiles). The figure of the Renaissance technician moved away from that of the medieval master of the *artes mechanicae* and acquired new scientific and cultural dignity. While artists and scholars were astonished by the achievements of the new technology—from the aqueducts of London, Toledo, Bremen, and Gloucester to the terrifying wonders of the new guns—the humanist theme of human domination over nature was transformed from a literary theme into a progressive awareness that the exploitation of the forces of nature could be planned rationally.

Nor must we forget the profound interconnection among the needs of civil life, artistic research, problems in engineering, and scientific developments in various areas, from town planning to architecture to painting. Numerous ideas from Brunelleschi, Alberti, Piero della Francesca, and Leonardo da Vinci were rethought and taken up again throughout the sixteenth century. Studies in perspective led to further research in that field, which later evolved into projective geometry. Stereometry met the needs of engineers, architects, and town planners. For the first time, mechanical considerations were applied to problems in ballistics. Because of technological needs, fluid statics was one of the fields of most intense research.

Perhaps one of the most interesting fields of the entire sixteenth century was cartography. On the one hand, the new Renaissance town planning required the reproduction of city plans; on the other, geographical explorations—first the great trans-oceanic voyages, then the explorations of the internal regions of the New World, and finally the attempts at circumnavigating the globe—brought attention to the inadequacies of medieval cartography. The technique of drawing and printing more adequate maps was soon developed on a larger scale in centers such as Antwerp and Amsterdam. Cartography posed an important mathematical problem, that of reproducing a section of a spherical surface on a flat sheet. For this reason, the map made by Gerhard Kremer, known as Mercator (1512–94), in 1569 was quite important, since it introduced the famous projection that was later called the Mercator projection (see EARTH SCIENCES 1550–1600).

Geographical explorations, which brought about the discovery of new plant and animal species, contributed to a revival of zoology and botany. Typical was the curiosity displayed by the scholars regarding the "monsters" that the ships brought back to Europe. These explorations also posed new problems of a technological nature, concerning methods of navigation. And they added, at least indirectly, to new technical-scientific advances (in optics, the theory of magnetism, etc.). Nor should we forget that by placing Europeans in contact with other populations—although often only after brutal conquests and abuses, as in the Spanish conquests of Mexico and Peru, for example—they favored communication with cultures that were extremely different. Notions of tolerance and rationalism, the first signs of which can be seen in Rabelais's work, for example, benefitted considerably from these contacts.

Scientific culture was already extremely rich by around 1580. While astronomy was investigating the heavens, cartography made possible a better understanding of the earth, and medicine added to knowledge about the human body. The art of printing was a valuable ally for spreading information in all three fields (astronomical charts, geographical maps, and anatomical tables).

The Counter-Reformation and cultural policy. The role of the Jesuits

The energetic recovery of the ecclesiastical hierarchy led to a more diligent religious life and a more deeply felt mysticism among the humbler ranks of the clergy. In many intellectual circles, however, religious conformity was only a veneer, imposed by prestige and sometimes even by intimidation. But post-Tridentine Catholicism was unable completely to ignore the broad-minded development of Renaissance thought, the new ethical and political ferment and the great advances made in the sciences. The Catholic hierarchy's cultural policy was to attempt to conserve the basics of scholastic tradition and to integrate some of the more important themes of the new cultural currents into the old schemes. Within Scholasticism itself, preference was given to the *auctores* of early Scholasticism (Anselm, Thomas, Bonaventure), in a search for elements leading to an organic metaphysics that would have a connection with the new problems of the times. Nominalism (the preferred target of humanist polemics), with its overly subtle logical investigations, was rejected. St. Thomas Aquinas was the author who became the major reference point.

Another point that Counter-Reformist culture had in common with humanism was that the exponents of the Counter-Reformation also recognized the need to approach the sources of Christian doctrine directly. As a result, the Church promoted the study of the sacred texts, the works of the Church fathers, and the resolutions of the early councils, giving rise to "positive theology." As for the works of classical antiquity, their importance was not denied, but their value was seen as limited: they were proposed as models of literary style, but not of philosophical thought. Within these limits, they constituted—along with Thomist philosophy—one of the cornerstones of educational programs.

The ambitious Counter-Reformation program counted above all on the forces of the Jesuits, who were organized in the important teaching centers, such as the Spanish University at Coimbra and the Gregorian University, also a great center for scientific studies. In accordance with the general directives of the Counter-Reformation, the Jesuits accepted Thomism, generally speaking, although they did not hesitate to approve a few rectifications, as long as the dogma was left intact. The most famous teacher at Coimbra was Luis de Molina (1535–1600). Driven by a desire to find some kind of accord between free will and divine prescience in an anti-Protestant manner he arrived in his *Concordia liberi arbitrii cum gratiae donis* (1588), at conclusions that rival theologians considered close to the Pelagian themes that Augustine had fought against in his time.

The Jesuits displayed a remarkable openness toward scientific and moral problems. The company counted among its number many students of the social sciences and, at least at first, looked with benevolence upon the new developments in science. Their moral laxity was one way of attracting individuals that had strayed away back into the Church. An important figure in the realm of scientific problems was Father Clavius, who translated Euclid's *Elements* in 1574. In education, the Jesuits' strength lay in the large number of colleges they established in every Catholic country, in accordance with the norm codified in the *Ratio studiorum* of 1599. While popular and elementary education was entrusted to the orders, Jesuit colleges were founded explicitly for educating ranks of leaders. This explains how the company was able to exert an increasingly broad influence on political, social, and intellectual life.

Products and developments of Humanist culture

Apart from the concessions that the Counter-Reformation made to Humanist culture, the tradition of the *bonae litterae* remained alive in many centers. The three volumes of *Essais* (published in 1580 and 1588) by Michel de Montaigne (1533–92), who was educated in the Latin classics (such as Cicero, Terentius, Plautus, Virgil, Horace, and Ovid), came out of this tradition. Montaigne was considerably influenced by Sextos Empiricos' *Outlines of Pyrrhonism*, but while other authors referred back to ancient Skepticism in order to support their apologies for the revelation, which was contrasted to "weak" human science, Montaigne's goal was a direct knowledge of human beings and their world. This was in open disagreement with the great systems of both Aristotelian and Platonic inspiration of his time, and even with the Renaissance concept of humans as the microcosm and the center of the universe. Pierre Charon (1541–1603) was the foremost continuer of Montaigne's ideas.

The "libertine" movement also appealed to classical authors, especially the Epicureans and the Stoics; here the return to pagan authors had an openly anti-religious function. Although the main representative of the "libertines," Giulio Casare Vanini (1585–1619) published his major works in the second decade of the seventeenth century, the movement had its cultural roots in the final years of the sixteenth century. It actually arose as an open rebellion against the Counter-Reformation, inasmuch as it revived the theme of the "three imposters" (Moses, Christ, and Mohammed), which may have been of Islamic origins; borrowed the concept of religion as *instrumentum regni* from political realism; and recognized the so-called horoscope of religions, according to which the appearance of new religions depended on the position of the stars.

Philosophy of nature and Giordano Bruno

Several formulations that were organic and fairly rich in ideas came out of the philosophy of nature during this period: Telesio's entire treatise appeared in 1584, Giambattista Della Porta's *De humana physiognomonia* was published in 1586 (the second edition, in twenty books, of Della Porta's *Magia naturalis* came out in

1589), and some of the most important works of Giordano Bruno (1548–1600) date back to this period. A strong anti-Aristotelian attitude is evident in all these works. The reaction of the Peripatetics was not long in coming; in fact, in 1593 they succeeded in putting Telesio's work on the index of prohibited books.

Meanwhile, the natural philosophers became aware that their theories were not only quite often in contrast to the authority of the Stagirite, but that they rarely agreed with the growing dogmatic rigidity of the Counter-Reformation. However, they borrowed the doctrine of dual truth (sometimes learned through the thought of the Libertines) from Averroism, separating the religious level from the philosophical, as did Bruno, for example. The dominant themes were those of Neoplatonism and Magism, and often, especially in Bruno, a retrieval of authors like Epicurus, Democritos, and Pythagoras, which led to a revival of atomism and theories on the plurality of worlds.

Bruno is perhaps the most prominent figure. Without going into the specific doctrines of his Italian dialogues, which were written and published in England (*De la causa, principio et uno*, *De l'infinito universo et mondi*, and *Cena de le ceneri*, all written in 1584, mostly about metaphysical subjects; *Spaccio de la bestia trionfante* (1584) and *Degli eroici furori* (1585), mainly on ethics), or of the short Latin poems he published in Frankfurt (*Summa terminorum metaphisicorum*, *De mondae, numero et figura*, and *De immenso en innumerabilibus*, all written in 1591), we should note that Bruno, unlike Telesio, understood the revolutionary significance of Copernicus' doctrine. While he replied to Osiander that Copernicus "makes use not only of mathematics to guess, but of physics to demonstrate the motion of the Earth," he celebrated the discoveries (1582) of Tycho Brahe (1546–1601) on the nature of comets and the localization of their trajectories and, in Cusano's footsteps, theorized an infinite universe, maintaining that even Copernicus had to be corrected on this point. Bruno was aware of the conflict between the new cosmology and the Scriptures, but maintained that it was correct, from a philosophical point of view, to uphold the scientific doctrine, while he attributed to the Scriptures an essentially moral function.

In the *Cena de le ceneri* and other works, Bruno also attempted to defend the Copernican theory with mathematical arguments, but his mastery of geometry was inadequate. One of the features that distinguished Bruno's philosophical cosmology from Galileo's new astronomical concept can be seen here. It was not only for reasons of cultural policy, though these were paramount, did Galileo try to keep his defense of the Copernican system from being confused with Bruno's; Galileo's methodological approach to scientific work was very different from Bruno's speculation. The great debate on the new astronomy made it obvious that the protagonists of the so-called scientific revolution had very different attitudes from those of the Renaissance magicians and philosophers of nature. It was important that the new science tended more and more to present itself in opposition not only to the authorities (Aristotle, Galen, Ptolemy, etc.) and their followers, but to the magical-philosophical reflection that played such a large part in the culture of the fifteenth and sixteenth centuries.

In 1591, Bruno was reported to the Inquisition. The trial, after many alternating phases, ended with Bruno being burned at the stake in Campo de' Fiori on February 17, 1600. With this condemnation, the ecclesiastical authority, with Cardinal Roberto Bellarmino (1542–1621) one of its prominent figures, sanctified the opposition between the Counter-Reformist program and some of the most innovative ideas of Renaissance culture. In addition, Bellarmino not only censored thinkers whose ideas were still in many ways close to the intellectual themes of the Counter-Reformation (Bruno himself, and Campanella some years later), he also played a determining role in getting Copernicus's doctrine condemned and in the consequent warning that was issued to Galileo. The Counter-Reformation's attempts to channel new, culturally more important themes back into traditional schemes failed, however, not only with respect to Renaissance philosophical currents, but in the area of modern science as well.

MATHEMATICAL SCIENCES

Editions of and commentaries on the classics

It would be anti-historical to date the break with classical mathematics to this period. Although this break was undoubtedly prepared by the discoveries made by the Italian algebraists of this century, it only matured in the seventeenth century. In fact, those illustrious works of antiquity which had been rediscovered thanks to the enthusiasm of the humanists of the previous century, enjoyed their widest diffusion in printed versions during this very period. In 1533, a Greek edition of Euclid's *Elements* with Proclos' commentary, came out in Basel, while the first Italian translation of the work, done by Niccolo Tartaglia (c. 1500–57), was published in 1543. The first English translation appeared in 1570.

Important commentaries on Euclid's work can also be found in the Latin editions by Federigo Commandino (1509–75), written in 1572, and Christopher Clavius (1537–1612), written in 1574. Besides Euclid, the works of Archimedes, with editions by Tartaglia, Francesco Maurolico (1494–1575) and Commandino (including the famous edition of 1558), and of Appolonius, were also widely diffused. A Latin translation of the first four books of the *Conics* was published in 1537. In particular, studies of Archimedes' works on geometry and mechanics led to a series of studies on the calculation of areas, volumes, and barycenters, which in turn led to the first developments in infinitesimal calculus at the beginning of the seventeenth century. Luca Valerio (1552–1618) was among the most prominent scholars in this revival of stereometric studies.

But the most interesting ancient author in the eyes of the mathematicians of this period was Diophantos, the only classical author who dealt with those albegraic problems that deeply interested the mathematical world at this particular time. His work (the manuscript of which Regiomontanus had discovered in Venice a century earlier) was often studied and quoted. The original Italian translation was done by Rafaello Bombelli (born c. 1530), and the first Latin translation appeared in Basel in 1575. It is important to note that a knowledge of the classics of mathematics, and the commentaries and discussions on them, considerably influenced the cultural climate at some of the most famous philosophical and scientific schools of the times. For example, the Venetian mathematician and translator F. Barozzi greatly influenced the school of Padua.

Bombelli's work in algebra

Rafaello Bombelli published a famous *Algebra* in 1572, which became the basic text for study in that discipline up until Leibniz's time. In it, Bombelli very systematically continued the task of freeing mathematics from the domination of geometry and arithmeticizing it, after the example of Diophantos' work. Its many technical merits aside, Bombelli's work is important for two basic reasons. First, its use of abbreviations in mathematical language is so vast and systematic that it can unquestionably be considered the most perfect example of syncopated algebra, or algebra in which appropriate and well-codified abbreviations are used instead of ordinary mathematical language (so-called rhetorical algebra). Secondly, Bombelli (who introduced specific notations for imaginary units) succeeded in legitimizing the use of complex numbers in cubic and biquadratic equations, on the basis of the purely algorithmic justification that the sum of two conjugate complex numbers (as they are now called) always gives a real number. This was one of the first important innovative examples of the use of symbolism to overcome (at least practically) conceptual difficulties in mathematics.

Viète's algebra and geometry

The prolific research that took place in the field of algebra throughout this century (to the names already mentioned, we can add those of Jacques Peletier (1517–82), Robert Recorde (1510–58), Pedro Nuñes (1492–1578), Simon Stevin (1548–1620) and Adriaen van Roomen (1561–1615), all cultivators of this discipline) was in a sense crowned by the work of François Viète (1540–1603). His *Isagoge in artem analyticem*, published in 1591, contains notable theoretical contributions, such as its introduction of the law of homogeneity for comparing geometrical quantities. It also makes decisive advances in notational procedures. In particular, letters are used to indicate the coefficients of algebraic equations and, generally speaking, symbolic structure is used to its fullest extent. For these reasons, Viète's work marks the transition from syncopated algebra to real symbolic algebra and, thanks to this fact, is able to expound the general rules for treating and solving equations in an extremely systematic and rigorous way and in a form that in practice remained definitive.

Although algebra was the great conquest of sixteenth century mathematics, noteworthy developments also took place in other branches of mathematics. Although very little original research was done in pure geometry, some very interesting applications of perspective were developed. By this time, the science of perspective had passed from the hands of artists into those of professional mathematicians, and in 1558 Commandino published the first treatise on perspective ever printed. It was followed by the famous *Perspectivae libri sex* (1600) by Guidubaldo Dal Monte (1545–1607). Geometrical studies connected with perspective can also be found in several treatises on optics, for example the *Opticorum libri sex* (1613) François d'Aguilon (1546–1617), where stereographic projection was introduced and studied in detail. Trigonometry also advanced considerably following the development of research in astronomy. Tycho Brahe (1546–1601), like Copernicus before him, did important research in trigonometry, although scientists who were not involved in astronomy also did valuable work in the study of plane and spherical trigonometry. Besides making important theoretical contributions, discovering formulas and proving various theorems, these scholars distinguished themselves for the immense and patient work they did in computing tables that were valuable aids to astronomical calculations. François Viète and Rheticus (1514–76) are the most illustrious representatives in this field of research, which was first given the name trigonometry in *Trigonometria, sive de solutione triangulorum tractatus brevis et perspicuus*, a work by Bartholomaüs Pitiscus (1561–1613).

We should also mention that considerable research on the quadrature of the circle was done toward the end of the century, especially in France and Holland. Along with incorrect results and debates that were not always productive, these studies led to significant advances in calculating the value of π. In 1596, Ludolph van Ceulen (1540–1610) published a pamphlet which gave a development of π to thirty-five decimal places.

ASTRONOMY

The Tabulae prutenicae

We have already spoken of the hostile reactions of Martin Luther and Protestant theology to the new astronomy. In the *Initia doctrinae physicae*, published in Wittenberg in 1549 and 1550, Philip Melanchthon accused Copernicus of seeking something new at any cost out of vanity, and, at the same time, of copying Aristarchos of Samos. Erasmus Reinhold, although he continued to withhold his support of the system, published the *Tabulae prutenicae* (so called in honor of prince Albert of Prussia, for whom they were compiled), astronomical tables based on computations from *De revolutionibus*, in Nuremberg in 1551. The new tables soon became widely circulated and replaced all the previously-used versions. Similar to the tables in *De revolutionibus*, they were more accurate, giving approximate values to the second, while Copernicus' approximations were only to the minute. From 1556 on, the *Tabulae prutenicae* were used in England as well, where the science of the heavens began to be studied during these years, thanks above all to the work of John Dee (1527–1608), who made his house in Mortlake into a center for astronomical studies.

A certain John Field, also English, prepared an *Ephemeris* for 1557 entitled *Iuxta Copernici et Reinholdi canona*, in which, among other things, he supported the heliocentric theory. Dee, too, was convinced by the hypothesis, as he wrote in a preface to Field's work. However, the first published recognition of the Copernican system in England dates back to 1551, when the mathematician Robert Recorde (1510–58), in his *Pathwaie to knowledge*, alluded to the Copernican theory. He used very hesitant terms, but it seems he definitely believed in the physical truth of this theory, which he saw as a valid revival of astronomy in the footsteps of Aristarchos of Samos. Although Archimedes' *Arenarius*, which contains a description of Aristarchos' system, was only published after *De revolutionibus* (1544), it is possible that Copernicus studied a manuscript of it during his travels to Italy or Cracow.

Toward a verification of the heliocentric system

Copernican cosmology was not yet able to impose itself as a valid alternative to the Ptolemaic system. Both the inaccuracy of the data from the few observations of the heavens on which *De revolutionibus* was based and its highly geometrical and mathematical nature together with the difficulties mentioned earlier made it difficult for scholars to confirm it beyond its concrete applications in computing tables.

There were many who admired Copernicus' genius but did not share his ideas. Among these was Caspar Peucer, a professor at Wittenberg and Melanchthon's son-in-law, who published his *Hypotheses astronomicae* in 1571, "according to the observations of Copernicus and to the tables elaborated by him." However, he did not explain the heliocentric theory for fear of scandalizing beginners with its absurdity.

Meanwhile, opposition to Copernicanism had formed within the Catholic Church and was becoming increasingly more intransigent. While the Jesuit Christopher Clavius (1537–1612) limited himself to rejecting the theory in his *Tractatus de sphaera* (1570), a commentary on the homonymous work by Sacrobosco, the Italian Francesco Maurolico, who had already shown hostility toward it, attacked it violently in his *Compotus ecclesiasticus* (1575).

Among the more convinced adherents of the heliocentric theory during this period were Christopher Rothmann, the court astronomer for the Landgrave William IV of Hesse-Cassel and a friend of Tycho Brahe, with whom he carried on an important correspondence, and Rudolf Wursteisen, or Urstisius of Basel (1544–88), who, according to Galileo's *Discorso sopra i massimi sistemi*, held lessons on Copernicus in Italy. In 1573 in Basel, Wursteisen published his *Quaestiones novae in theoricas novas planetarum*, a commentary on the work of Georg von Peurbach, in which he mentions Copernicus twice ("a man of truly divine genius who has, in our century, attempted, and not without success, the restoration of astronomy"), without, however, mentioning his system.

In France, the mathematician Pierre de la Ramée, otherwise known as Petrus Ramus (1515–72), a professor of philosophy and rhetoric at the College Royal of Paris and an opponent of Aristotelianism, was the first to maintain the need for a radical reform of astronomy. In the second volume of his *Scholarum mathematicarum libri XXXI* (Basel, 1569), in which he states that astronomy must be founded on accurate observations and the use of logic and mathematics, abandoning all the theories of the ancients, he praises Copernicus's clever attempt to make astronomy depend on the motion of the earth, although he does not declare himself a Copernican. In a conversation with Tycho Brahe in Augsburg between 1569 and 1570, attested to by a letter Ramus wrote to Rothmann, he seems indirectly to have wished that the Greek principle of circular motion could be revised to give a mathematically more acceptable definition of the orbits based on the results of observations.

However, the first complete and original exposition (even graphically) of the Copernican system came from England. In 1576, at the end of his *Prognosticon everlasting*, Thomas Digges (c. 1543–95) son of the famous astrologer Leonard Digges, published a *Perfit description of the caelestiall orbes according to the most aunciene doctrine of the pythagoreans lately revived by Copernicus and by geometricall demonstrations approued*, where, as the title suggests, he shows himself to be an enthusiastic supporter of heliocentrism. Alexandre Koyré has recently demonstrated that, although it is not possible to see an anticipation of Bruno's theory of the infinity of worlds in this description of the Copernican system, as some would have it, the graphical representation of the sphere of the fixed stars, which is not limited by an external circle but extends "infinitely upward in a spherical shape" with stars filling up the corners of the page and where the "court of the heavenly angels" reigned, still remains a rather interesting example of how new, bolder concepts of the universe began to emerge in the wake of Copernicus's heliocentric theory.

Digges also wrote a very accurate description of the appearance of the nova of 1572, surpassed only by his contemporary Brahe's account. His contribution was published in the *Alae seu scalae mathematicae*, which came out in London the following year.

Tycho Brahe and observational astronomy

Meanwhile, the first observatory with a rotating dome was founded in 1561 at Cassel, by the Maecenas William IV, Landgrave of Hesse Cassel (1532–92). Among his collaborators were Rothmann, whom we have already mentioned, and the Swiss astronomer Jost Bürg, who installed a clock with a pendulum that served as a regulator in order to time his observations. Among the most important fruits of the Cassel observatory was a catalogue of the fixed stars, which was supposed to contain around a thousand stars but was never completed.

It was a period of European maritime expansion, with the related development of increasingly accurate compasses, clocks, and other observational and measuring instruments. The work of Tycho Brahe (1546–1601) came out of this climate of technical progress. More than a theoretician, Brahe can be considered the founder of modern observational astronomy.

Basing his studies on the results obtained by Eastern astronomers, mainly at Marāgha and Samarkand, with respect to both the methods of observation (the use of view-finders, the transversal method) and the preparation of instruments (accuracy increased with an increase in size), Brahe achieved the highest possible degree of accuracy in the field of naked eye astronomy. The largest instrument he used was a quadrant with a radius of 6.25 meters. Corrections for refraction and solar parallax enabled Brahe to surpass the degree of accuracy reached at Ulugh Beg's observatory at Samarkand for the first time in Europe.

Thanks to Brahe's new observations, the annual variation and inequality of the moon's motion was discovered, the exact value of the precession (50.3 seconds per year) was calculated for the first time, and a catalogue was prepared, which contained 788 stars and upon which J. Bayer later based his *Uranometria* (see ASTRONOMY 1600–1610). Starting with Brahe, accurate and constant observational data finally began to be available to astronomy, a *conditio sine qua non* for the later theoretical discoveries on the universe.

On the evening of November 11, 1572, Brahe was able to observe the most extraordinary heavenly phenomenon of those years, the sudden apparition of a new star near Cassiopeia, the same one that Thomas Digges described in England. Brahe published his first book, *De nova stella*, in 1573, the year after he had observed the star. Without taking a position on its nature, which he did not understand, he described the different phases of its light with marvelous accuracy and concluded that it did not have a parallax.

In 1575, after several years in various cities in Germany, Brahe went to Cassel, where the Landgrave William IV recognized his extraordinary talent. The following year, Brahe was offered the island of Hven, in Sund, near Copenhagen. He named the observatory that was built there Uraniborg. During these years, one of his achievements deserving of mention was his observation of the comet of 1557. This observation enabled him to establish that its orbit was at least six times farther away than that of the moon. This dealt the definitive *coup de grace* to the Aristotelian concept of the dichotomy of the changeable sublunary world *vs* the immutable celestial world, following the earlier conclusions drawn from the apparition of the nova of 1572.

The reform of the calendar

The old problem of adjusting the civil and religious (Julian) calendar to the actual length of the solar year, which had already been measured fairly accurately, was finally solved in Italy during this period. The plan for reforming the calendar was proposed by Luigl Lilio, or Aloysius Lilius, (1510–76), and the reform was launched in 1582 at the behest of Pope Gregory XIII Boncompagni (1502–85), after whom the Gregorian calendar was named. By that time, the deviation accumulated from the time of the Julian reform amounted to ten days, so setting the spring equinox at March 11, it was decided to "remove" the extra ten days from the year 1582. Thus Tuesday, October 4, 1582 became Friday, October 15.

In 1583 the Frenchman Joseph Justus Scaliger (1540–1609) published an important work on the calendar, *De emendatione temporum*, which contained a valuable comparative analysis of ancient chronologies in the light of the Copernican system. He also introduced the concept of the "Julian period" as a universal measure of chronology, equal to 7,980 years.

The Copernican system in Italy

While the philosopher Francesco Patrizi (1529–97) demonstrated his acceptance of the daily rotation of the earth, the Dominican Giordano Bruno (1548–1600), in a series of works on metaphysical and cosmological subjects written between 1584 and 1590, expounded a new view of the universe. Starting from Nicholas of Cusa's earlier intuitions, he took up the Copernican system and made it into a powerful guiding idea for future advances in astronomy. While he was certainly not the first person in Italy to declare himself openly as a follower of Copernicus, Bruno was undoubtedly the first to connect his system with a truly revolutionary interpretation of the universe. In his dialogues entitled *La cena delle ceneri* and *De l'infinito, universo et mondi*, published in London in 1584 and 1585, and in his Latin poem *De immenso et innumerabilibus* (1591), he expounded the theory of the infinity of worlds and the infinite extension of the universe. Our sun, located at the center of the planetary orbits, was nothing more than one of the many, infinite suns disseminated in the ocean of the heavenly ether. Condemned by the Inquisition for his heretical doctrines, Bruno died at the stake in a square in Rome in 1600.

The Tychonian system of the world

During the period he spent on the island of Hven, which lasted until 1597, Tycho Brahe continued to gather valuable data concerning his observations of the heavens. Simultaneously, he elaborated a planetary system with the earth at the center, surrounded by the orbits of the moon and the sun. The planets rotated around the sun, while the sphere of the fixed stars had the earth at its center. From the point of view of calculations, the so-called Tychonian system was exactly equivalent to the Copernican system. In his *De mundi aetherei recentioribus phaenomenis liber secundus*, printed in 1588 but published posthumously in 1603, Brahe presented an incomplete outline of his system of the world. It soon became known (a few copies of the book circulated among friends and correspondents before it was published) and was quite popular. Between 1589 and 1590, the Scotsman D. Liddell gave a course on it. The *Fundamentum astronomicum* by Nicholas Reimarus Ursus of Holstein, who had been to Uraniborg in 1584 and to Cassel in 1586, also appeared in 1588 in Strasbourg. This work ascribed the discovery of the Tychonian system to its own author, who claimed to have invented it and does not even mention Brahe's name. Ursus replied to Brahe's objections with his *De astronomicis hypothesibus* (Prague, 1597). A controversy arose that lasted until Ursus's death in about 1600, and was never resolved in favor of either, although the system is generally attributed (probably rightly so) to Brahe alone.

Astronomy and mysticism in Johannes Kepler

Meanwhile, Brahe's and Copernicus' successor, Johannes Kepler (1571–1630) began his work as an astronomer during this last part of the century. In 1595, in Tubingen, he published his *Prodromus dissertationum cosmographicarum continens mysterium cosmographicum*, known as *Mysterium cosmographicum*, written in collaboration with Maestlin. He immediately sent it to Brahe (who was then in Germany), and to Galileo (1564–1642), who answered, congratulating him and confessing that he was also a convinced Copernican.

In an attempt to find a law that would unite the motions of all the heavenly bodies harmoniously and describe the distribution of their orbits in space, Kepler proposed the famous geometrical construction of the five regular (Platonic) solids inscribed inside each other. Thus between the six planetary spheres that rotate around the sun, he inserted, beginning from the outside, a cube, a tetrahedron, a dodecahedron, an icosahedron, and an octahedron, respectively. Although the theory was obviously erroneous in its particulars and was marred by mystical assertions, his intuition of a law able to govern the distances of the planets from the sun was of paramount importance.

Astronomy in England

The new local tradition of astronomy continued in England. At the conclusion of a series of observational studies made between 1594 and 1597, Edward Wright (c. 1558–1615) published his *Certaine errors in navigation* (1599). William Gilbert (1544–1603) attributed the daily rotation of the earth to magnetic forces in his *De magnete*, although he did not recognize its motion of revolution. In *De mundo nostro sublunari philosophia nova* (published posthumously in 1651), he still stated that he was undecided between Brahe's and Copernicus's systems. Gilbert's magnetic theory later gave Kepler a valuable element for establishing the physical basis of his planetary model.

PHYSICAL SCIENCES

The work of Giambattista Benedetti

Two of the major Renaissance scholars of mechanical problems, besides Niccolo Tartaglia, whom we have already mentioned, were Giambattista Benedetti (1530–90) and Simon Stevin (1548–1620).

Benedetti's last work, the *Diversarum speculationum mathematicarum et physicarum liber* (1585), was his most important. In it, he continued Tartaglia's efforts to mathematize mechanics, concentrating especially on the mathematization of the theory of impetus. Without stopping to list all of Benedetti's many remarkable findings—for example those regarding falling bodies—we shall only mention the importance of his new general approach to research, already visible in his early works. In it, one can clearly discern the characteristic moment of transformation to a treatment which is distinguished by the constant comparison of ancient mathematicians—whose work had frequently just been rediscovered by human philologists and widely diffused by printing—with the traditional philosophers and by the total faith the author places in only those conclusions obtained *mathematica demonstratione et non aristotelica garrulatione, nor cum experientia ad sensum*.

Stevin: the rejection of perpetual motion

Stevin's most important work was entitled *De beghinselen der weeghconst* (*Elements of the Art of Weighing*, published in Flemish in 1586). Its most important contributions were to statics and hydrostatics and were motivated exclusively by Archimedian considerations, to which the author added a physical principle of considerable importance: the rejection of perpetual motion, which he considered physically impossible.

Without stopping to consider other contemporaty authors, such as Guidubaldo Dal Monte (1545–1607), who also studied statics within the Archimedian current, we can conclude that a link with the great Syracusan was a feature common to all the many Renaissance mechanicians.

While this link, on the one hand, confirmed modern science's strong dependency on Greek science, on the other, it brought to light the new active, operative, and technical nature of the theoretical research these mechanicians conducted. In a certain sense, their work shared this characteristic with the experimental research done by the first current of physicists mentioned earlier (the current that developed the themes of natural magic).

CHEMICAL SCIENCES

Science and magic

Giambattista Della Porta (1535–1615), who was especially active in the second half of the century and who touched on a variety of topics in science, curiosity and magic in his *Magiae naturalis sive de miraculis rerum naturalium libri IIII* (Naples, 1558), was one of the most interesting Italians of the epoch. The accusations of witchcraft brought against him proved salutary for a change. In the second edition, in twenty books, published in 1589, Della Porta omitted many of the magical themes and gave more space to a wide range of scientific topics. Among other things, the work contains descriptions of how to prepare sulphuric acid (*oleum sulphuris*) from sulphur, sodium carbonate (*sal sodae*), potassium carbonate (*sal tartari*), alum (*alumen*), and soaps (*smegmata sapones*). It describes various laboratory apparatus, such as retorts (*retortum vas*), refrigerating coils (*anguineus flexus*), various types of stills (even for fractional distillation), and the so-called Florentine flask. Della Porta's magic, however, no matter how it is looked at, even from the point of view of his accusers, was of quite a peculiar sort, no longer corresponding to Pliny's notion. As the title of his work clearly illustrates, it recognized a magic innate in natural occurrences.

Another scientist who dealt in magic was Girolamo Cardano (1501–1576), a physician, mathematician, and astrologer. His two works, *De subtilitate* (Nuremberg, 1550) and *De rerum varietate* (Basel, 1557), are both of interest to chemistry. Of the four Aristotelian elements, Cardano omitted fire, which he did not consider a substance. He considered three types of mixtures: *mistio* (combination), *drasis* (solution) and *coacervatio* (mechanical mixture). He denied the transmutation of metals and described various chemical apparatus and preparations. He explained the increase in weight of calcined lead (which he calculated at 1/30) as the exhalation of a *calor coelestis* and compared it to the supposed increase in a person's weight after death, caused by the loss of the soul. Julius Caesar Scaliger (1484–1558) found these explanations laughable and attributed the increase in weight to the assumption of air particles (1557).

The "secrets"

A. Piedmontese's *De secreti* (1555 or 1557), inspired the publication of other "secrets"—prescription-books of medicines, cosmetics, magical philtres, etc., especially in Italy. Among the Italian authors were Girolamo Ruscellai, I. Cortese, and Leonardo Fioravanti (c. 1530–88), a Bolognese physician and follower of Paracelsus, who also published *Li capricci medicinali* (Venice, 1568) and *Il tesoro della vita umana* (Venice, 1570), which describe many chemical and pharmaceutical preparations.

A true Italian secret, which Bernard Palissy (c. 1510–89) puzzled over at length, was how to produce the beautiful colored majolicas manufactured mainly in Faenza. Around 1560, he was finally able to begin manufacturing his *rustiques figulines*, which were soon widely admired and even requested by kings and queens, giving rise to the production of faiences in France as well. His writings do not reveal the secret, but only list the raw materials he used in his enamels, which he also contributed to formulating. He rose from being a simple glass-maker and potter to the highest levels not only of technology, but of science, and can be considered the most eminent French chemist of the sixteenth century. He opposed both alchemy and Paracelsus' iatrochemistry, and upheld the importance of both observation and experimentation. Like Vannoccio Biringuccio, however, he was basically an empiricist, and left very little room to theoretical ideas. In 1563, he published his *Recette véritable par laquelle tous les hommes de la France pourrent apprendre à multiplier et*

augmenter leurs trésors, in which he talks about agriculture, geology, mining, and forestry. In 1575, in Paris, he gave the first course in geology ever held in France. Finally, in 1580, he published his *Discours admirables sur la nature des eaux et fontaines* . . . . It included a *Traité de la marne*, in which Palissy maintained that mineral salts are necessary to plants and suggested that soluble salts and marle be added to manure (which the ancient Romans had already done).

The first treatise on chemistry

At the end of the century, the first systematic treatise on chemistry, the *Alchemia* (*Alchymia* in the second edition of 1606), by Andreas Libau (c. 1540–1616) was published. Through Jean Beguin (c. 1550–1620), this work inspired all seventeenth century French literature on chemistry. Although Libau was somewhat of a follower of Paracelsus, he criticized both Paracelsus' mysticism and his hermetism, as well as his complicated verbosity. Beguin methodically and clearly expounded the various parts of chemistry, from the apparatus to the manipulations, from preparations to descriptions of various substances (some of them new), and from some of the technological processes to simple analytical methods. The description of the *domus chymici*, accompanied by drawings, is particularly interesting. It includes a main laboratory with burners and stills, a preparation area, a room for water and sand baths, an area for crystallizations, a laboratory for assistants, a storage space for fuels, a museum, and a wine cellar, as well as covered walkways and gardens.

Along with the usual alchemistic preparations (transmutations of metals, "potable gold") are more properly chemical ones, some of them original, such as ammonium sulphate, lead nitrate (*calx plumbi dulcis*), lead acetate (*quinta Saturni essentia* or *saccharum plumbi*) and a distillate of it, acetone, as well as vinic ether (*oleum vitrioli dulce verum*), stannous sulphate (*vitriolum Jovis*) and stannous chloride (*spiritus sublimati*), which was later called *spiritus fumans Libavii*, although Arnald of Villanova had already described its preparation three centuries earlier (see LIFE SCIENCES 1250–1300).

Libau made no distinction between organic and inorganic substances. He believed that both were combinations of salt, sulphur, and mercury. He began to differentiate among gases, recognizing two besides air: one inflammable, which existed in mines, and the other capable of putting out fire. Libau did special research on mineral waters, and is considered one of the pioneers of qualitative analysis because of the analytical assays, both wet and dry, that he did on them.

Publications on technical chemistry also flourished in Italy during these years. The best known of these are *De metallis* (1596) by Andrea Cesalpino (1519–1603), which contains, among other things, observations on the crystallization of various substances, a description of alum production at Tolfa and an attempt at explaining the increase in the weight of lead during calcination, and *Dell'historia naturale libri XXVIII* (1599) by Ferrante Imperato (c. 1550–1631), which discusses topics in mineralogy, metallurgy, etc., some of which Biringuccio and Georg Agricola had already considered. Particularly interesting are the chapters on naphtha, bitumen, graphite, and precious stones.

LIFE SCIENCES

Problems in physiology

In a theological work written in 1553, Michael Servetus (1511–53) described the pulmonary circulation of the blood, which Matteo Realdo Colombo also studied. This clever anticipation remained unnoticed, however.

Physiology was still dominated by Galenic thought. Even though Jean Fernel (1497–1558) had just coined the term, the discipline did not yet take shape as an independent branch within the biological sciences. Fernel made the last great effort to construct a physiological system of classical inspiration. Andreas Vesalius (1514–64) contradicted Galen when it was not a question of completely subverting the physiological doctrine that had been transmitted through the ages, but he did not yet dare to challenge the actual foundations of Galenic physiology.

Clinical medicine

Vesalius was the first to diagnose aneurysm of the aorta in a living person. Clinical medicine advanced thanks to the new bedside method of teaching. Giambattista de Monte's teachings in Padua are an important example. The new tendency expressed itself in an interesting series of fairly faithful descriptions of clinical cases. Amato Lusitano should also be mentioned as a typical representative of this particular genre of medical literature, which involved detailed observations of the symptoms and the course of diseases.

Herbals and bestiaries

The history of the natural sciences during this period was marked by the publication of various herbals and bestiaries, which contained accurate morphological descriptions and, above all, beautiful iconography. An attempt at devising a rational botanical nomenclature can already be seen in the work of Leonhard Fuchs (1501–1566), who described various plants from America. Hieronymus Bock (1498–1554) took an interest in the flora of his country and provided descriptions that were based more on his personal experience than on tradition. Pierre Belon (1517–64) wrote works on the nature of fish (1551) and birds (1555). By emphasizing the homology that exists between the human skeleton and that of birds, Belon took the first step toward the modern notion of comparative anatomy. Guillaume Rondelet (1507–66), a naturalist and physician, studied marine animals; he also had the first anatomy amphitheater built in Rome (1566). Botanical research in England began with William Turner. *De differentiis animalium* (1552), by Edward Wotton (1492–1555) represents a turning point in zoology, in that the author attempted systematically to define the similarities and differences in the external morphologies of various living creatures. The naturalist Conrad Gesner (1516–65) wrote an encyclopedic work that summarized all the ancient and new knowledge in the fields of botany, zoology, and mineralogy. His five volume *Historiae animalium* (1551–87), in which animals are grouped following Aristotle's classification, is an inexhaustible gold mine of information.

Discoveries in anatomy

Many researchers moved into the field opened up by Vesalius's work. Thanks to both their personal talents and the use of productive methods, they made spectacular progress in understanding human anatomy. Gabriel Falloppius (1523–62) described the female genitalia and the placenta (1561). Bartolemeo Eustachio (1524–74) described the adrenal glands, the heart valve, and the tube that bears his name, studied the structure of the teeth, and accurately observed the first and second dentitions (1564). He also observed the *ductus thoracicus* (thoracic duct) in the horse, but this observation was forgotten until Gasparo Aselli (1581–1626) did work on it.

Julius Caesar Aranzi studied the anatomy of the embryo, taught that the circulation of blood in the fetus was separate from that of the mother, and recognized the importance of pelvic deformities in obstetrics (*De humano foetu,* 1564). Constanzo Varolio (1543–75) published original research on the anatomy of the optical nerve (1573), while Volcher Coiter advanced comparative osteology (1575). Andrea Cesalpino (1519–1603) took the first step toward discovering the circulation of the blood.

Exotic plants: tobacco, coffee, and tea

Botany during this period was characterized by the importation of exotic plants into Europe. Garcia da Orta simultaneously compiled the first good compendium on the materia medica of the American Indians and the first European manual on tropical medicine (1563). Nicolas Monardes published an important treatise on the medicinal herbs of the New World, which contained a description of tobacco (1565). That same year, Jean Nicot introduced tobacco into France, from whence it soon became known in many other European countries. In 1580, Prospero Alpino brought coffee into Europe, and Maffei imported tea. The botanical works of Cristobal Acosta, Mathias de L'Obel, and Pietro Mattioli, which mostly concerned medicinal herbs, also deserve special mention.

The development of pathological anatomy

The physician Guillaume de Baillou (1538–1616) defended the old Hippocratic doctrine against Arabic tradition, and also gave excellent descriptions of "pestilential fevers." It is to him that we owe the classic descriptions of croup (1575); whooping cough, or pertussis (1578); measles, or morbilli; and scarlet fever, or scarlatina. He also coined the term rheumatism. A pioneer in pathological anatomy, Baillou was the first to observe adhesive pericarditis complicated by dropsy. The physician Girolamo Mercuriale (1530–1606) published several treatises on skin diseases and gymnastics. M. Donati gave the first description of a gastric ulcer in 1586.

The battle against superstition: psychology and psychiatry

This period is particularly interesting from the point of view of the history of psychology and psychiatry. In the name of science, the German physician Johann Weyer (1555–88) launched a campaign against superstition and the persecution of witches. In his *De praestigiis daemonum* (1563), he gives a rational psychopathological explanation of several phenomena generally attributed to magic. Felix Platter (1536–1614) dedicated his attention to the psychological treatment of the mentally ill (1570). In 1575, the Spanish physician Juan Huarte de San Juan published his *Exémen de ingenios para las ciencias*, the first modern attempt at systematically studying the functions of the brain and establishing a link between psychology and physiology.

In his *De humana physiognomonia* (1586), Giambattista Della Porta (1535–1615) attempted to establish the relations between the psychological characteristics and the external morphology of an individual.

The natural sciences and agriculture

Andrea Cesalpino defined the basic criteria of botanical taxonomy (1583). Prospero Alpino (1553–1617) described the plants of Egypt and studied the fecundation process in plants (1592). He also made important findings in the field of nosology and described epidemic diseases that were particularly serious in northern Africa.

The advances made during this century in the field of the natural sciences found their technical applications in agriculture (Olivier de Serres, 1539–1619). Andrea Bacci, a pioneer in balneology (*De thermis*, 1571), also published a natural history of wines (1596).

Animal anatomy

In zoology, Thomas Moufet wrote an elegant work on insects, and Carlo Ruini (c. 1530–98) composed a treatise on the anatomy of the horse, equine diseases, and their cures. The illustrations in Ruini's book, which was published in 1598, are as good as those in Vesalius's work on human anatomy. It is also worth noting that this was the first book entirely dedicated to the anatomical structure of non-human living creatures. The great French artist and geologist Bernard Palissy (c. 1510–1589) recognized the affinity between fossils and living forms.

Plastic surgery and gynecology

The period seems to have been dominated by considerable advances made in surgery. Georg Bartisch (1535–1606) founded operatory ophthalmology, and Gasparo Tagliacozzi (1545–99) introduced rhinoplasty, which until then had been confined to popular medicine, into official medicine. In 1581, François Rousset published the first treatise on the Caesarean section. A treatise by William Clowes (1540–1604), a very important military surgeon, illustrated the situation of medicine in England at the time.

EARTH SCIENCES

Cosmography and cartography

A geographical description of the earth was given by Francesco Maurolico (1494–1575) in his *Cosmographia* (1558) and by Giovanni Paolo Gallucci in his *Theatrum mundi et temporis*, written in a cosmological-apologetic vein. The first geographical map by Gerhard Kremer, known as Mercator, (1512–1594) appeared in 1553.

In 1569, Mercator published a huge eighteen-sheet world map, the first example of a Mercator projection map. The *Theatrum orbis terrarum*, by Abraham Ortelius, the first general atlas, appeared in 1570. It was accompanied by a commentary by the author entitled *Thesaurus geographicus*. In 1578, Mercator revised the maps and text of Ptolemy's *Geographia*, restoring them to their original Alexandrian form.

Mercator's atlas

Another work by Mercator, the first great atlas of the world, entitled *Atlas*, appeared in 1585. A few years later (1590), the Scotsman Edward Wright improved Mercator's cartographic method. In 1594, an enlarged edition of the *Atlas* appeared, published, in part posthumously, by Mercator's son Rumold. Ocean depths were first indicated in cartographic representations in maps of Holland beginning in 1585.

Geographical explorations

In an attempt to complete the exploration of the Americas, Sir Martin Frobisher set out in 1576 in search of a northwest passage from the Atlantic Ocean to the Pacific, north of the Canadian coast, but was blocked by ice. Willem Barents attempted a passage from the Atlantic to the Pacific along the northern coasts of Asia in 1594, but was unsuccessful. He did, however, discover the Spitsbergen islands.

Oceanography

In Norway, Olaus Magnus (1490–1557) attempted to make various measurements of the depth of the ocean off the Norwegian coast. Some of his measurements were successful, while others were not, which reintroduced the whole problem of measuring ocean depths and what instruments to use.

In any case, in his *Historia de gentibus septentrionalibus* (1555), Olaus Magnus advanced the idea that the bottom of the ocean followed a course that was linked to that of the land above sea level and that the depth of the sea was proportional to the height of the adjacent land above sea level.

William Cunningham's work on tides, *The cosmographical glasse*, appeared during this period (1559). Although the author still supported the old idea that the light of the moon made the water expand, causing the phenomenon of the tides, the book contained detailed maps, including the times of the high tides for every day of the month, calculated for a number of English ports as well as various ports along the European coasts.

It became increasingly important to have adequate instruments available for measuring ocean depths (a topic that Nicholas of Cusa had already brought up). Jacques Besson in 1567 proposed devices that made it possible to measure the time it took for a weight to reach the bottom. It had already become obvious that enormous errors could result from measuring the length of a cable that was dropped, and new methods were being sought. Nicholas of Cusa's suggestions and Puehler and Besson's attempts were only the first rudiments for solving this problem, which was actually quite simple.

Between 1574 and 1578, William Bourne (flourished 1565–83) published his *Treasure for traveilers* and *A regiment for the sea*, two treatises on a wide range of topics. He attacked problems of navigation and also gave interesting interpretations of the effects of the sea on coasts and on the transportation of sediment by currents and tides. He imagined that the waters of the ocean moved under the influence of a tide force, just as water in a tank moves and oscillates when it is shaken. As we can see, these ideas were quite modern and anticipatory. On the basis of data gathered by navigators, Bourne also gave valuable descriptions of ocean currents, recognizing that the wind also helped cause them. Bourne believed that salinity basically came from minerals dissolving, favored by the jolts provoked by earthquakes.

Between 1575 and 1587, the French cosmographer André Thevet (1502–90) made a number of cognitive and critical contributions to oceanic cartography and oceanography, with his *Cosmographie universelle* and his *Grand insulaire*. These treatises also contained discussions on marine flora and fauna. Some of Thevet's experience was direct, since he made numerous exploratory voyages, even crossing the Atlantic. In 1588 he also wrote about the *terra australis*.

Minerals and fossils

Systematic treatises on various kinds of minerals and stones were accompanied by collections that brought together rocks, minerals, and fossils, plus a variety of strange objects, some forged by swindlers who tried to pass them off as marvels. The second edition of Rosslin's work (1551) lists eighty-five metals, stones, and gems. A lapidary by Christopher Encelius (1517–86), a student of Luther and a friend of Agricola, appeared that same year under the title *De re metallica*, with a preface by Melanchthon. A work by Georgius Agricola (1494–1555) with the same title was published in 1556.

Works on stones and minerals were written by Johann Mathesius (*Sarepta oder Bergpostill*, 1564) and Olaus Magnus, who mentioned metallizing emanations (1567), while J. Kentmann spoke of stones that were generated by thunder in his *Nomenclaturae rerum fossilium* (1565). During this period, the idea that metals grew and vegetated was supported by Pietro Andrea Mattioli (1579), and opposed by Aubertus (1575) and Quercetanus (1575). At the request of Pope Sixtus V, Michel Mercati illustrated the fossil collection of the Vatican.

The origin of stones was still the topic of a number of treatises, such as Cesalpino's *De metallis* (1596), while more generalized works by Cardin (*De rerum varietate*, 1581) and Giordano Bruno (*De l'infinito universo et mondi*, 1584) gave a pantheistic cosmological interpretation of the phenomenon (see THOUGHT AND SCIENTIFIC LIFE).

In 1584, the Kamennyj Prizak, an organization for identifying and exploiting natural resources, was founded in Moscow. Ferrante Imperato (c. 1550–1631), in his *Dell'historia naturale libri XXVIII* (1599) and G. Meyer (1594) both supported the idea that minerals vegetated.

Bernard Palissy's scientific work

The *Discours admirables de la nature des eaux et fontaines*, by Bernard Palissy, one of the precursors of chemistry, was written in 1580. He recognized the real nature of fossils and correctly solved the problem of the origin of springs. Palissy believed that the salinity of the ocean was of mineral origin and came from the waters of rivers washing over land that was above sea-level. Palissy's scientific ideas spread quickly among a vast public, partly because the author was so widely discussed and well-known and because he gave private lessons in geology in Paris in the circles where the themes of the Reformation were being debated (see CHEMICAL SCIENCES).

Geophysics

In 1583, Galileo (1564–1642) discovered the isochronism of the pendulum, a phenomenon that presumed the existence of gravity, as did his experiments on falling bodies (c. 1590).

In 1600, William Gilbert (1544–1603) described the earth as a huge magnet. With a small magnet appropriately mounted on a box called a *terrella*, he simulated and studied the properties of the magnetic earth (which differed from the *terrella* only in size).

Gilbert was undoubtedly the founder of the study of magnetism as a modern science (the phenomenon of magnetic attraction had been known since the most remote times, and compasses had already been used in navigation for quite some time). It was therefore natural and obvious for Gilbert to guess that magnetism (or something similar to it) caused the tides. This idea was even accepted by a number of scholars in later periods. But the first rudiments of the concept of gravitational attraction, the major initial contributions to which were made by Stevin and Johannes Kepler (1571–1630), began to be developed when this analogy with magnetic attraction was made.

APPLIED SCIENCES

Food-producing techniques

New methods were developed in agriculture, and the products for consumption were selectively grown. In England, irrigated fields and the cultivation of vegetables for the market were innovations of this period. Tasser's *Hundreth good points of husbandrie*, a work on gardening, grafting, the cultivation of hops and the arts of caring for plants and raising chickens and cattle, appeared in 1557. One of the most important agronomical treatises in France, *L'agricolture et maison rustique*, was published; the first edition, in Latin, came out in Paris in 1554 under the title *Praedium rusticum*. In Spain, tobacco began to be grown for non-therapeutic purposes in 1558. Some practical progress was also made in methods of grinding grain. Girolamo Cardano (1501–76), in his *De subtilitate*, described a sort of hand sieve for sifting flour.

The cultivation and consumption of new vegetables, typical of the Americas and the Far East, was introduced into Europe. Spain was the first to experiment with growing potatoes (no later than 1570) and corn, which the Spaniards called *panizo de las Indias* because of its similarity to millet (*panizo*, in Spanish). Corn spread to France, to Italy, and later to the Balkan countries.

Iron-working and metallurgy

The production of iron continued to increase thanks to the use of blast furnaces and to further improvement in drafting techniques—the wooden box bellows, which replaced the old leather bellows in larger furnaces—devised by Hans Lobsinger in Nuremberg in 1550.

The fusion of iron was mainly used for making iron and steel cannon barrels. The old method of assembling pieces of forged iron was abandoned, but no complete descriptions exist of the methods used to manufacture molds for iron cannons or how they were cast, so that in practice the method must not have been much different from that used in casting bronze. In 1586, Andrej Čohov made the largest cannon of the times, which weighed 40 tons. *De re metallica*, the most important work of Giorgius Agricola (1494–1555), appeared in 1556. Because of its richness of detail and the clarity of its many illustrations, *De re metallica* has been called one of the great monuments of technology.

Machines for drainage projects and for operating blast furnace bellows required a constant supply of water power, but the energy needed was not always available at the site. One solution to the problem was the use of valve rods, which could transmit the energy of a water wheel for a distance of up to several kilometers. This system was in use in Erzgebirge around 1555.

Benvenuto Cellini (1500–1571) mentioned the use of a press for coining money, although it was probably not a fly press, which we find for the first time in 1548 among the machinery acquired by the King of France for a new coining procedure. These machines, which were installed in the Louvre palace in 1552, included rolling mills for reducing the plates, drawing benches for regulating their thickness, circular punches for making discs and fly presses for coining. This use of rolling mills in mints, following Esban Hesse's description of their application in iron-working, is particularly interesting, and even suggests that techniques of working and mechanically shaping metals developed along lines that were very similar to those used in coining metals.

The *Beschreibung allerfürnemisten mineralischen Ertzt- und Berckwercksarten*, a treatise on the most important types of ores and metal compounds, appeared in 1574. It was written by Lazarus Ercker (c. 1530–94) who, like Biringuccio and Agricola, was an authoritative sixteenth century popularizer. It had very little to do with mining activities or fusing procedures, but gave more accurate instructions for assaying and as such supplemented Agricola's work.

The growing experimentation with methods of assaying and evaluating ores and metals was a contributing factor in the development of quantitative chemistry.

Hydraulic projects

Giacomo da Vignola (1507–73) applied the theoretical ideas of Leonardo da Vinci and Benedetto of Missaglia to three modern locks near Bologna.

In 1550, construction began on the canal of Brussels, the first of any importance constructed outside of Italy in the sixteenth and seventeenth centuries. It was built under the direction of Jean de Locquenghien (1518–70). The canal connected Brussels with Willebroek, on the Rupel. It was thirty kilometers long and had four locks to compensate for the ten-meter difference in level between Brussels and the Rupel. The canal was completed in 1561.

Topography and cartography

In cartography and surveying, the work of Regnier Gemma Frisius (1508–55) was finished by A. Foullon (1513–63), an attaché to the house of the King of France and a scholar of mathematics. Foullon was the first to describe a plane table, which he called a *holometre*, in 1551. The basic characteristic of surveying with a plane table was that positional lines were plotted directly on a piece of paper fastened to the table, just as they were sighted. It was very easy to draw a plan or a map using the plane table, since no mathematical knowledge was necessary. As a result, surveyors, particularly land-surveyors, soon became very numerous.

The most important invention for the advancement of car-

tography, and for its effects on mapping navigational routes, was the cylindrical isogon projection with increasing latitudes, devised by Gerhardus Mercator (1512–94) in 1569. The nautical maps used up until then, the so-called flat maps, had been indicated by mathematicians as the leading cause of disasters at sea. Among other inconveniences, the flat map progressively distorted east-west distances as one moved further away from the equator. As early as 1537, the Portuguese mathematician Pedro Nuñes (1492–1578) had analyzed the causes of the error. However, there was nothing that could replace the flat map prior to the invention of maps based on the projection devised by Mercator, who in his youth had been a student and assistant in mathematics of Gemma Frisius, as well as a skillful builder of instruments and globes. The foremost feature of the Mercator projection was that it gave correct directions between any two points. Ships' routes could be found by simply placing a ruler on the map. However, the Mercator principle was not adopted immediately, and only came to the attention of cartographers and sailors in a practical form toward the end of the sixteenth century.

Topographical surveying acquired its most important instrument, the theodolite, in 1571. It was invented by the Englishman Leonard Digges, who gave the name *theodolitus* to the directional quadrant, or circular disc, with alidade, and suggested that it be combined with another topographical instrument, the sighting rule that rotated on a vertical scale and could be used to measure heights. Digges died in 1558, leaving his invention in the form of notes, which were published under the title *Pantometria* by his son, the mathematician Thomas Digges, in 1571.

Progress in the area of machinery and mechanical technology

Francesco di Giorgio Martini's research into the more efficient utilization of water power was continued in 1579 by Jacques Besson, who wrote *Théâtre des instrumens mathématiques et méchaniques*. Besson described a horizontal water wheel without gears, which exploited the reaction effect of water on curved blades.

In 1561, Ekerhardt Baldewin, clockmaker to William IV of Assia, built a planetary clock that was a worthy successor to Dondi's. His innovations were a transmission with worm screws, cylindrical bearings, and universal joints.

The printing press was further transformed in 1550 when Danner of Nuremberg replaced the wooden screw with a copper one. This innovation made it possible to obtain clearer printed images.

In 1561, H. Spaichel of Nuremberg invented a new type of lathe with a tool rest. The story of this machine is exemplary: in order to keep it from becoming known beyond the walls of Nuremberg, and because of the opposition of other master metal turners, the Municipal Council of Nuremberg decided in 1569 to destroy the model. On the whole, the corporate environment was hostile to technical progress and, as we shall see, Spaichel's case was not an isolated event. In London, in 1589, William Lee (ca. 1562–ca. 1610) introduced a knitting machine that he had invented for manufacturing stockings.

The most important treatise on mechanical engineering of the century, *Le diverse et artificiose machine*, by Agostino Ramelli (1531–after 1591) was published in Paris in 1588.

In the area of hydraulic machinery, however, the prize goes to the Dutch engineers. In 1584, Simon Stevin (1548–1620) obtained a patent for his invention of a type of piston pump for draining polders and harbors. Four years later, he obtained another patent for a "high capacity drainage pump" that could lift four times as much water. Besides these practical achievements, Stevin also did original theoretical studies on statics and hydrostatics, developed in his important treatise on mills, entitled *Van de molens*. Stevin's work was continued by Cornelis Corneliszoon, who in 1597 obtained a patent for a type of drainage mill that had a crankshaft operated by a windmill, which alternately powered two pumps.

Mechanization took place in other areas as well. An example of primitive mechanical sifting was illustrated by F. Venanzio (1551–1617) in his *Machinae novae*, published in Venice in 1596. A lantern pinion struck a lever that in turn shook an inclined triangle covered with different types of more or less tightly-woven cloth. At Danzig, in 1586, Anton Möller invented a loom that anyone could operate, since the person only needed to push a bar or a lever in order to make it run. We should, however, mention that resistance was still strong to attempts at replacing skilled labor with machines. The use of Möller's loom led to a number of riots in Leyden, and the Dutch government was forced to limit its use. As we have already mentioned, the resistance came above all from corporate circles.

An innovation that later revolutionized transportation passed almost unobserved in England, where wooden tracks were laid to carry coal from the mines of Wollaton to the River Trent.

Civil constructions

Two bridges deserve mention as major achievements of fluvial architecture: the Santa Trinità bridge in Florence, built in 1567, and the New Bridge in Paris, begun in 1578, the foundations of which were dug with box caissons and laid more than three meters under the rock bed.

In 1578, Andries Vierlingh (1507–79) of Brabant wrote his *Tractaet van dickagie*, a work on the construction of dikes that was never published. It is important in that for the first time it explains the methods used during the sixteenth century for building not only dikes, but flood-gates as well.

Philobert de l'Orme's work on architecture, which describes "new inventions for building well at a low cost" (1561) deserves mention. In it, the author illustrates various methods for building curved roofs with short pieces of lightweight wood, eliminating the need for chain beams.

City planning and the protection of embankments

Italian architects toward the end of the century were involved in town planning projects. Domenico Fontana (1543–1607) was working in Rome under Pope Sixtus V; Giulio Savorgnan designed a plan for the city of Palma Nova that was then carried out (1593); Buonaiuto Lorini published his designs of fortified polygonal cities in 1592 in Venice. The problem of flooding was particularly urgent. Antonio Lupicini (c. 1530–98) wrote a *Discorso sopra i ripari del Po e d'altri fiumi che hanno gli argini di terra posticcia* (1587), in which he examined the causes of erosion of the banks in detail and described his remedy: a circular structure made of poles placed one or two meters apart, connected crosswise to form mats. One end would be attached to the shore and then the mat would be unrolled and laid over the point of erosion.

The step foundation of the Rialto Bridge in Venice, designed by Antonio da Ponte, also deserves mention.

Scientific instruments

As the seventeenth century approached, the most interesting developments took place in the area of scientific measuring instruments, especially with Galileo (1564–1642). Devices for making calculations were of the utmost importance in a period when mathematical symbolism and techniques of arithmetic were primitive and much more difficult to acquire than skills in taking measurements. Various types of slide rules were devised for calculating the capacity of barrels and the range of guns.

The most important mathematical instrument was undoubtedly the sector, a hinged, graduated ruler that used the theory of similar triangles to make a wide variety of computations possible. Galileo and his assistant, Marcantonio Mazzoleni, worked on perfecting the sector during this decade. The odometer was also redesigned during this period, as a magnetic compass and recording devices were added.

Between 1593 and 1597, Galileo invented the thermometer, "a glass instrument with air and water." He was probably inspired by reading Heron's *Pneumatica*, which contained a description of the principle on which the air thermometer is based (see PHYSICAL SCIENCES 100 B.C.–0). Heron's work had been translated into Latin in 1575 by the mathematician Federico Commandino.

The first idea for a telescope also dates back to this period. In his *Magiae naturalis libri XX*, published in Naples in 1589, Giambattista Della Porta (1535–1615) talks about a combination of concave and convex lenses to improve long-distance vision.

The Seventeenth Century

1600–1610

THOUGHT AND SCIENTIFIC LIFE

The rise of the bourgeoisie and the development of scientific institutions outside the university

The structure of European feudal society changed over the course of the seventeenth century as a result of the rise of the bourgeoisie and the consolidation of absolute monarchies. In France, after the Treaty of Vervins and the Edict of Nantes (1598) had put an end to the religious wars, the restoration of the monarchy offered political and religious guarantees that created a climate favorable to the development of industry and trade, and consequently of scientific studies. Commercial development, protected by the first traces of mercantilism, and the advances made in the arts and crafts, favored the bourgeoisie. Most of the scientists of this period came from the bourgeoisie, or else from the petty nobility. Another prime factor in this process was colonial expansion, which brought solid economic prosperity to England and Holland. They enlarged their fleets and founded the powerful companies of the East Indies, which organized expeditions involving large numbers of technicians and physicians. Simultaneously, the accumulation of capital stimulated industry (the wool industry in England and the silk industry in France). The Gobelins factory was founded in 1602.

The establishment of new universities continued to be a trend. Paul V founded the Sardinian University of Cagliari in 1603; the University of Giessen was inaugurated in 1607, a rival to Marburg; Gröningen acquired a school of higher learning in 1614 and Strasbourg, in 1621. In the course of the century, however, scientific activity was no longer all absorbed by the universities and the professors there. The major scientists of the epoch did not teach from the heights of university chairs; the new science sought its own place at the outskirts of official science, creating its own original means of expression. Scientific societies were created, a particular form of private correspondence was developed, and scientific journalism was founded. At the beginning of the century, in the New Atlantis, Francis Bacon foresaw the creation of a center for scientific research and an organization for the collective study of natural phenomena. The first Accademia dei Lincei was created in 1609 in Rome, on the initiative of Prince Federigo Cesi. Among its members were Galileo and Della Porta.

New libraries were founded. The Bodleian Library, which opened at Oxford in 1602, and the Biblioteca Ambrosiana, opened in Milan in 1609, were especially important. Parallel to the taste underlying Baroque art was a predilection for private natural history collections, which brought together every imaginable "curiosity" from the animal, vegetable, and mineral worlds, especially "extravagances of nature," such as fossils and monsters.

Philosophy and utopia: Bacon and Campanella

European philosophy was dominated by the Italian Tommaso Campanella and the Englishman Francis Bacon. While Bacon can be considered one of the most radical innovators of philosophical-scientific thought of the time, Campanella (1568–1639) essentially remained tied to the old culture (i.e. to both Telesio's philosophical naturalism and some of the themes of the Counter-Reformation), although he displayed a lively interest in concrete problems of nature and tenaciously opposed Aristotelianism.

A comparison between Campanella's view of a utopian society as outlined in his Città del Sole, written in 1602, and the one sketched by Bacon in the New Atlantis fragment mentioned above, is fairly indicative in this respect. Campanella's view is inspired by magical-astrological ideas and oscillates between a defense of individual freedom and the hope of a renewed theocracy. Bacon's vision describes the perfect society as entirely founded on science and technology and working to advance them for the good of humanity.

One of Bacon's most interesting works, Temporis partus masculus, dates back to 1608–09. In it, the attack against Aristotle takes on a new tone, becoming a defense of modern things against all past doctrines. Bacon maintains that truth is the daughter of time and that science, no longer directed toward constructing a general view of the universe, but rather toward describing and studying individual phenomena, must be sought only in the light of nature, and not in the "shadows of antiquity."

MATHEMATICAL SCIENCE

Algebra and applied geometry

Typical sixteenth century problems still dominated the first years of this century. In 1608, for example, Peter Rothe reiterated the principle that the number of roots of an algebraic equation could not exceed the degree of the equation itself, and in 1600 Guidubaldo Dal Monte (1545–1607) published his treatise on perspective, mentioned earlier, in which, among other things, he demonstrated that the central projection of a system of parallel straight lines was a bundle of concurrent straight lines.

ASTRONOMY

Theory and observation in the first half of the seventeenth century

After Tycho Brahe's death in October 1601 in Prague, his assistant Johannes Kepler (1571–1630), named imperial mathematicus in November of that year, succeeded in procuring the valuable collection of observational data that the Danish astronomer had compiled. In Kepler's hands, Brahe's data proved to be the key to discovering the basic laws of the new astronomy.

In the course of a few decades, heliocentrism grew from a genial, avant-garde hypothesis into a complete, coherent theory that was able, both mathematically and physically, to explain all the observations made up until then. It also explained those observations that resulted from the recent invention of the telescope, especially Galileo's, that were constantly being proposed to the scholars.

Bayer's Uranometria

In Augsburg, in September of 1603, Johann Bayer (1572–1625) of Rain, Germany, published his *Uranometria*, a folio volume that contained descriptions of the constellations and introduced the modern use of Greek letters to indicate the stars. On the basis of data gathered by the Dutch navigator P. Dirckszoon Keyzer (Petrus Theodori), Bayer added twelve constellations of the southern hemisphere to the 48 constellations already described by Claudius Ptolemy, the positions of which were given in Tycho Brahe's catalogue. This was the last important contribution to observational astronomy prior to the invention of the telescope.

Practical astronomy and the telescope

A telescope with two lenses—one concave and one convex—that gave an image magnified seven times, was presented at the Frankfurt exposition in 1608. The instrument aroused the curiosity of the visitors, and a few days later (on October 2), Hans Lippershey (1587–1619) asked the states-general of the Low Countries for a thirty-year license to manufacture simple and binocular models. The Dutch government gave two of Lippershey's instruments to the king of France, and in April 1609 similar ones could be bought in Paris from the better lens manufacturers. In the summer of that same year, in England, Thomas Harriot (1560–1621) directed a telescope toward the moon and began to draw lunar maps, even before Galileo. A number of Dutch telescopes were also sold in Italy in 1609. On August 21 of that year, Galileo invited the Senate of Venice to examine a telescope on the bell tower of St. Mark's that he had built, following the Dutch models. The *perspicillum*, as he called it, was a huge success, and three days later Galileo gave the Senate one that magnified nine times. Several months later, thanks to the Dutch invention, Galileo made the first revolutionary discoveries of modern observational astronomy, which he presented in his *Sidereus nuncius*.

Although Galileo confessed, in a letter to Kepler dated August 4, 1597, written in reply to receiving a copy of the *Mysterium cosmographicum,* that he had adopted Copernicus's doctrine some years earlier, he in fact continued to teach the Ptolemaic system, at least until 1610; and in a treatise, extant in a manuscript dated 1606 (*Treatise on the Sphere, or Cosmography*), he publicly came out against the motion of the earth, justifying his position with the traditional arguments: rotation would cause the earth to disintegrate, clouds would remain behind with respect to the earth, etc.

The heliocentric theory found its most courageous and committed supporter during these years in Kepler. Between 1600 and 1606, on the basis of data gathered by Brahe, Kepler completed the study of the orbit of Mars, which had been a serious obstacle for every astronomer before him. However, economic difficulties and conflicts with Brahe's heirs, who claimed rights to Kepler's findings, deferred their publication. Meanwhile, in 1604, Kepler published his *Astronomiae pars optica*, which contains a theory of vision, reflection, and refraction that has remained a classic. It also outlines the specific theoretical principles of Galileo's telescope.

In 1604, a nova appeared in the constellation of Ophiuchus, in coincidence with a great conjunction of Jupiter, Saturn, and Mars, which only happens every 800 years. The star, known as "Kepler's nova," caused a huge uproar in the scientific world, and Galileo dedicated three courses of lessons (now almost all lost) to studying it. Two years later, Kepler published his *De stella nova in pede Serpentarii*, in which he demonstrated that the nova was located in the "immutable" region of the fixed stars.

Kepler's first two laws

The work that contained Kepler's studies on the orbit of Mars, *Astronomia nova, seu physica coelestis tradita commentariis de motibus stellae Martis ex observationibus G.V. Tychonis Brahe,* came out in 1609. It became the cornerstone of modern astronomy, not only because it provided the basis for a complete heliocentric theory of the heavenly bodies, but also because for the first time a "physical" explanation was given, even though it did not match the rigor and precision of Newton's system.

From the point of view of the history of science, part of the interest of the *Astronomia nova* lies in its detailed (and reliable) description of the long and controversial heuristic process that led to the final writing of the work. The difficulties in correctly interpreting the orbit of Mars, which required abandoning the dogma that the motion of heavenly bodies was circular, a notion rooted in cosmology from its origins, explains why Kepler actually discovered the so-called second law, or the law of areas (the line joining the sun and the planet, the radius vector, covers equal areas over equal periods of time) first (in 1604). Only later did he discover the first law, which states that the planetary orbits are ellipses with the sun in one of the foci. He arrived at this conclusion after first reintroducing the *punctum aequans,* which Copernicus had rejected in the name of the uniformity of circular motion, and then guessing that the orbit of Mars was oval, retaining that to be a more probable figure since, like the circle, it had only one focus. The theory of elliptical motion was, for the time being, extended only to Mars, but Kepler suggested that it might be generalizable for the other planets, including the earth.

The other extremely interesting idea put forward in the *Astronomia nova* was a hypothesis on the existence of a magnetic-type force, emanating from the sun, that governed the motion of the planets. With respect to Copernicus, who attributed planetary motion to the center of the earth's orbit (which did not coincide with the sun), Kepler took an important step forward. The sun, which was at one of the centers of the elliptical orbits of the planets, was not only the geometrical focus of those orbits (even in Ptolemy the motion of the planets had depended geometrically on the sun), more importantly, it was their physical cause.

The *Astronomia nova, seu physica coelestis tradita ...,* the findings of which it seems even Galileo never took seriously into consideration, encountered only moderate success in Germany. In England, on the other hand, it became immediately well-known and appreciated, especially by Harriot.

PHYSICAL SCIENCES

Scientific optics

Mechanics and optics, the two main branches of physics at the time, took decisive steps toward reaching truly scientific levels during the first decade of the seventeenth century. There was one big difference between them, however. The research in mechanics done by Galileo (1564–1642), which followed the definitive break between medieval and modern mechanics, did not lead to the publication of any books (this took place only a few decades later), but simply gave rise to the exchange of a number of important scientific letters (which at least make it possible to date the research). Research in optics, on the other hand, led Johannes Kepler (1571–1630) to publish his *Ad Vitellionem paralipomena* (1604)—recalling Witelo, the thirteenth century scholar (see PHYSICAL SCIENCES 1250–1300)—one of the most important works in the reestablishment of this discipline.

The telescope

Meanwhile, notable progress was made in technical optics. The construction of the first telescope, which was able to make far-away objects visible, took place in Holland in 1608. The telescope was primarily used for military purposes, to discern the enemy's movements before they were visible to the naked eye. The curious device was not constructed according to the laws of geometrical optics (those known at the time), but was simply a product of the ingenuity of skillful artisans. For this reason, it aroused almost no interest in scientific circles.

In 1609, Galileo made several substantial improvements and elevated it for the first time to the rank of a true scientific instrument, using it for his historical observations of the moon, the planets and the sun. The methodological discussions that soon arose out of these observations are especially interesting from the point of view of physics. Was it or was it not right to trust those things that one saw with the aid of the telescope but that even the sharpest vision could not verify? In other words, should it be maintained that the telescope effectively potentiated the organ of sight, or should the findings that it led to be rejected as pure optical illusions? The suspicion that the results obtained using devices of this type were illusory lasted for a long time, though not so much with respect to astronomical observations as to observations made with microscopes.

The interest in magnetism

Although mechanics and optics were the two basic branches of physics during this period, other types of phenomena began to be studied in a more systematic way as well. This was especially true of the phenomena connected with magnetism, which became a very popular subject following the publication in 1600 of *De magnete*, by William Gilbert (1544–1603) (see EARTH SCIENCES 1550–1600). Galileo, who held the English scientist in the greatest esteem, also studied these phenomena on a number of occasions. Kings and princes also attempted to enrich their collections with examples of increasingly larger and stronger magnets.

CHEMICAL SCIENCES

Observational wealth and theoretical poverty

The state of chemical knowledge at the beginning of the seventeenth century is well-represented by the works of the Alsatian Benedictine monk Basilius Valentinus, which were published in the first decade of the century: *De microcosmo* (1602), *Tractatus de rebus naturalibus et supernatura libris metallorum et mineralium* (1603) and *Currus triumphalis antimonii* (1604). On the one hand, these works show the richness of observational data that was available to the scholars of this discipline, and, on the other, they illustrate how unscientific the levels reached in the conceptual elaboration of these data still were. Valentinus was a devotee of mystical alchemy as well and wrote *Les douze clefs de philosophie*, published in Latin, German, and French, on that subject. Although he recognized the four Aristotelian elements (earth, water, air, and fire), he insisted more on the three alchemistic principles (salt, sulphur, and mercury). He displayed a sound knowledge of even uncommon metals, such as bismuth and cobalt, and was an expert on mines, recognizing that there was a gas in them that was different from air. He knew of various processes for purifying gold, including amalgamation, and was aware that *aqua regia* (*Scheidewasser*) could be used to separate gold from silver. He described the preparation of fulminating gold, silver chloride, and various compounds of copper, mercury, arsenic, and especially antimony. He was also very familiar with the preparation and properties of the three most important mineral acids—hydrochloric acid, sulphuric acid, and nitric acid—as well as acetic acid. Valentinus had also learned techniques for making glass, including adding manganese dioxide to make it clearer. And he knew that the distillation of wine and beer produced alcohol (*aqua vitae*), to which he added hydrochloric acid to get ethyl chloride, and sulphuric acid to produce "a marvelous medicine" (possibly ether).

The alchemistic tradition

Of course Valentinus's work contained a lot of information and considerations that were of a more strictly alchemistic nature. Thus the precipitation of copper from a solution of one of its salts with the addition of iron was a transmutation; a philosophers' stone could transform 30,000 portions of silver into gold; saltpeter contained two elements, fire and air, which prevailed over the other two, water and earth, as well as a "subtle spirit"; in the combustion of alcohol, sulphur and mercury separated and while the first burned, the other volatilized and turned back into a gas.

The terms alchemy and chemistry were still almost interchangeable. An *Alchymia*, written by Nicolas Guibert (c. 1547–1620), which denied the possibility of transforming metals, was published in 1603; the *Basilica chymica*, by Oswald Croll (c. 1580–1609), a major vehicle for the diffusion of Paracelsus' teachings, came out in 1609. Johann Hartmann (1568–1631) was nominated professor of *chymiatria* at Marburg in 1609 and can be considered the first professor of chemistry.

LIFE SCIENCES

New descriptions of plants and animals

The study of animals and plants was still in its descriptive phase. Collectors and encyclopedists expanded the knowledge of the living world and, above all, systematically reorganized ancient knowledge. The physician and naturalist Ulisse Aldrovandi (1522–1605) left a Museum of Natural History and an enormous collection of literary and iconographic documents on animals, plants, and minerals to the University of Bologna. Only three of his volumes on ornithology (1599–1603) and a volume on insects (1602) came out during his lifetime, but publication of his extensive works did not stop with the author's death. Nine more volumes were published between 1606 and 1648 by his widow and a number of scientists. Although Aldrovandi's encyclopedia is open to criticism both for its editorial heaviness and for a certain lack of originality and critical spirit, still it is a remarkable collection of ideas and was used as a reference for a long time.

The evolution of anatomical investigations

The University of Padua was the world center for anatomical research. During this period, the chair of anatomy was occupied by Hieronymus Fabricius ab Aquapendente (c. 1533–1619), who did an important study on venous valves (1603) and wrote a treatise on embryology (1604).

His magnificent atlas of comparative anatomy was only a valuable manuscript until his description of the embryological development of chicks was published posthumously in 1621.

His students came from everywhere. One of them was William Harvey (1578–1657), the English scientist who recognized that Fabricius da Aquapendente's research, especially his description of venous valves, was the true starting point for understanding the circulation of the blood.

In the Italian, Swiss, German, French, and Spanish universities, anatomical research continued following Andreas Vesalius' example and methods.

Anatomical and sensory structures

Gaspard Bauhin (1560–1624) described a number of anatomical structures that were previously unknown, especially the ileocecal valve (1588). The physiology of sight and hearing began to be understood better when the anatomical structures of the eye and the ear were interpreted on the basis of recent physical findings. Johannes Kepler (1571–1630) applied the principles of the new dioptrics to the eye (1604), in particular describing the refraction of light rays by the cornea and the crystalline lens, and the inversion of the image formed on the retina, which he considered the true photo-receptive organ.

Guilio Casseri (1561–1616) of Piacenza published several findings on the organs of speech and the hearing apparatus in humans and domestic animals (1600–01). The work is considered remarkable because of the method of comparative physiology developed in it.

Clinical medicine, surgery, and gynecology

Organizational works were developed in the field of clinical medicine. In this current, Felix Platter (1536–1614) published a classification of diseases based on their clinical symptoms in 1602, while the surgeon Wilhelm Fabry (1560–1634) compiled a monograph on the classification and therapy of burns.

Surgeons were becoming increasingly bold. In 1602, Florian Mathis performed the first operation to remove a foreign body from a stomach.

In the course of 1609, Louise Bourgeois (1563–1636), a midwife in Paris, published a manual on gynecology and obstetrics which described the face-forward presentation of the fetus and how the umbilical cord comes out.

EARTH SCIENCES

As far as we know, the word "geology" was first used in the modern sense in 1603. Ulisse Aldrovandi (1522–1605) used it in his will to refer to the collections and works that he had amassed in this field of natural science.

Cartography and geological observations

The century of great geographical discoveries brought about an enormous increase in knowledge about the earth. Cartography and geographical descriptions slowly approached scientific accuracy. Geological observations, on the other hand, remained sporadic and were mainly focused on immediate practical problems, such as the exploitation of ore deposits. Various volcanic eruptions and the effects of earthquakes were described, and fossil remains were reported—for example, the bones that D. de Avalo y Figueroa found near Tarija in Bolivia (1602). However, when it came to theorizing, fantasy and reckless speculation (often the products of ancient and medieval beliefs) reigned.

The influence of the sixteenth century naturalists, such as Girolamo Fracastoro and Bernard Palissy, who, for example, had correctly guessed the origin of fossils, was quite limited. Treatises on metals and their origins seem to have had better luck, but numerous absurdities persisted in this area, as well; even Agricola was not exempt from them. Experimental science was on the verge of replacing the speculative method, but geology only benefitted from the new scientific tendency somewhat later.

The formation of precious stones

It was commonly believed at the time that the stars influenced the formation of precious stones—or that they were of vegetable or animal origin. Magical or at least supernatural powers were also attributed to gems, and even those who did not think that the powers came directly from the stones, believed, like Anselmus De Boodt (c. 1550–1632), that they were exerted by God or other spirits through the stones (*Gemmarum et lapidum historia*, 1609). The main problem was whether or not precious stones and metals, especially the noble metals such as gold and silver, grew with time from base materials and reached maturity, which was the prevailing notion among the alchemists of the time.

The inside of the earth

The belief in a central fire in the earth, which came from observing volcanic eruptions, was almost universally accepted. It held a privileged position in the cosmogonies of the period.

The phenomenon of tides

The works of Simon Stevin (1548–1620) on the equilibrium of fluids, a prelude to the later important work of Bernoulli, Euler, etc., the founders of hydrodynamics, on which theoretical oceanography is based, dates back to the beginning of the seventeenth century. His *Hypomnemata id est de cosmographia de praxi geometrica, de statica, etc.*, which among other things, contains the rudiments of the famous equation of the equilibrium of a heavy liquid, was published in 1605.

Johannes Kepler (1571–1630), in his *Astonomia nova* (1609), was the first to explain the phenomenon of the tides as an effect of the gravitational pull of the moon on the water of the oceans.

APPLIED SCIENCES

The fusion of iron

Iron was becoming one of the most widely used metals. Once it could be cast, technologies related to its use also began to grow. New methods of producing steel were experimented, and the number of patents increased throughout the seventeenth century. In 1608, Anton Zeller of Aschhausen suggested manufacturing steel by carburization, a process that involved heating iron bars in a ''box'' with beechwood charcoal. In another process, iron bars were immersed in a bath of fused pig iron, raising its carbon content and turning it into steel. This method, like the others that followed, was on the whole quite ineffcient. A plentiful substitute for wood charcoal had to be discovered before efficiency could be achieved.

The search for a substitute for wood charcoal

The problem of replacing wood charcoal with mineral coal as a fuel was complicated by the fact that both mineral coal and the iron had to go through various stages of transformation before the metal was ready in the form of bars, and in each of these stages the mineral coal could damage the iron. The use of mineral coal began indirectly—in the beer-making industry—possibly as a result of efforts made to replace wood charcoal. When mineral coal was used to dry malt, necessary for certain fermentations, it gave the beer a bad taste; and since wood was charred to obtain charcoal, Hugh Platt (1552–1608) had the ingenious idea of refining coal to eliminate the scoriae.

Platt also invented a way of producing blocks of coal dust in the shape of bricks, in order to purify hearth fires in homes. This opened the way to the manufacture of coke a few decades later.

Techniques of land reclamation

In Holland, techniques for draining land continued to be improved. In 1602 the states-general granted Cornelis Corneliszoon (1562–1639) a patent for a pump that appears to have been a primitive sort of centrifugal pump.

In 1608, work was begun to drain the Beemster, the largest internal lake in the northern part of the Low Countries. Jan Adriaanszoon Leeghwater (1575–1650) was placed in charge of manufacturing and constructing the water mills for the project. He developed a system for raising water in several stages. Although this technique had already been known in the previous century, Leeghwater was the first to utilize it to its fullest extent.

A new loom

In weaving, probably during the first decade of this century, Claude Dangon invented a loom that made it possible to increase the number of leashes from 800 to 2400.

Scientific instruments

The use of scientific instruments as indispensable aids in research began to spread during the early years of the seventeenth century. This marked the beginning of a long and profitable union between the worlds of practice and theory, a union that became characteristic of the modern scientific-technical era. The most important instrument of this decade was the telescope built by Galileo (1564–1642) in 1609, an improvement on the already existing Dutch invention (see ASTRONOMY).

The proportional compass

In 1607, Galileo published *Le operazioni del compasso geometrico e militare*, which described the construction and use of the proportional compasses (Galileo called it the geometrical compass) invented by Guidubaldo Dal Monte, and studied by Galileo from 1597. The instrument, which was also used to determine the caliber of firearms, became very widely known. When Baltesar Capra plagiarized it in 1607 in his *Usus et fabrica circini cuiusdam proportionis*, Galileo brought charges against him. The ensuing trial resulted in Capra's condemnation and an order that all copies of his pamphlet be destroyed.

Treatise writing

In the field of mechanical engineering, the *Novo teatro di machine et edificii*, by Vittorino Zonca (1568–1602), a work that contains information important to the history of technology, was published in 1607. Zonca was the first author, for example, to mention anti-friction metals, suprisingly ignored by all the other scientists who wrote about metallurgy and mechanics. He was the first to note that all metals except bronze wore down when they rubbed against steel. Zonca's *Novo Teatro* also contains the first illustration of the cloth-napping machines of the time, which had a roller covered with cards that rotated while the cloth passed underneath in the opposite direction.

Meanwhile, Olivier de Serres (1539–1619) published his second treatise of importance on agriculture, entitled *Théâtre d'agriculture et des champs*, in Paris in 1600. In 1609, the Dutch West Indies Company introduced tea into Europe.

A large work on metallic antimony, the *Currus triumphalis antimonii*, traditionally attributed to Basilius Valentinus, appeared in 1604.

1610–1620

THOUGHT AND SCIENTIFIC LIFE

The Counter-Reformation and religious intolerance

The Reformation and the Counter-Reformation had an enormous influence upon education, literary and artistic activity, the intellectual climate and lifestyle, and the situation of science in all the countries of Europe during this period. Baroque art spread a particular worship of the sumptuous, the broken line, research into perspective, and emphasis.

Putting the new architectural style into practice presented various technical problems, and baroque art left its characteristic mark on publications in physics, medicine, and the natural sciences as well. Those who were curious about nature were able to extend the range of their research thanks to the invention of new scientific instruments (lenses, telescopes, etc.). It was these instruments that made the transition from qualitative to quantitive observation (and sometimes even experimentation) possible.

The Reformation was undeniably successful in appealing to reason rather than to the established authority and in opening new roads for pedagogy. Protestantism introduced the democratic spirit into education and reformed teaching methods. This process had already begun as early as the sixteenth century, but only matured with Jan Amos Komensky, or Comenius (1592–1670).

The Catholic Church was forced to take regulatory, repressive measures. The Jesuit order organized and watched over scholastic teaching in most European countries. In France, the abbot Pierre de Berulle, later a cardinal, founded the Congregation de l'Oratoire, which was dedicated to teaching. Thanks to its flexible, tolerant policy, the order had an important influence on the development of science. Although the Counter-Reformation stimulated the development of mathematical research and the improvement of conditions for the ill in hospitals (the apostolate of St. Vincent de Paul, the writing of St. François de Sales), it often denied freedom of scientific thought and maintained outdated Aristotelian positions.

The controversy over the Copernican system

The foundations of one of the most important controversies of the modern age, the relationship between the discoveries of science and the dogmas of faith, were laid during this period. The conflict started from the cosmological problem of whether or not the Copernican system was compatible with the teachings of the Bible. The most convinced and authoritative supporter of compatibility, the position avidly upheld by all "progressive" Catholics (Prince Cesi for example), was Galileo. He wrote his *Lettere copernicane* (1613–15), which were not published but circulated among a large group of friends, on this subject. The Roman Church came out against the Copernican doctorine in 1616.

The theory, if considered a physical truth and not purely a mathematical hypothesis, was declared incompatible with the Catholic faith, and the books that maintained the contrary were condemned. Since the *Lettere copernicane* were private, they did not specifically fall under the condemnation. Galileo was warned not to pursue the question any further.

In 1616, Campanella hurredly composed an *Apologia pro Galileo*, in which he attempted to demonstrate the innocuousness of the heliocentric theory. It is interesting to note that Bacon, on the contrary, did not grasp the importance of the new astronomy at all, and consequently did not feel it necessary to intervene in favor of the Copernican system.

In 1615, the "libertine" thinker Giulio Vesare Vanini (1584–1619) published a volume of dialogues on natural philosophy entitled *De admirandis naturae reginae deaeque mortalium arcanis*. It created a real scandal, partly because the text was very different from what the faculty of theology in Paris had authorized the author to write. Vanini was forced to flee the city and was arrested in 1618. In 1619 he was sentenced to have his tongue cut out and be burned at the stake. However, "libertine" thought did not die with him; in fact it spread throughout France, becoming a target of attack for many philosophers of the times.

MATHEMATICAL SCIENCES

Continued fractions

The problem of dealing with infinity (in the dual sense of analyzing the infinitely small and the infinitely large, which culminated in the creation of infinitesimal calculus) was destined to become a central question during this century, and in a certain sense characterized it. The problem was already foreseen in the *Trattato del modo brevissimo di trovare la radice quadra delli numeri*, published by Pietro Antonio Cataldi (1552–1626) in 1613. In the mathematical tradition, of the University of Bologna Cataldi introduced and developed the infinite algorithm of continued fractions, starting from the rules approximated by ancient logistics.

The invention of logarithms

A new instrument of calculation, the logarithm, with its developments and applications, was also created during this decade. In 1614, John Napier (1550–1617) published his *Mirifici logarithmorum canonis descriptio*, in which he introduced logarithms as an expedient for making numerical calculations easier. Calculations using the new units, which Jost Bürgi (1552–1632) may have anticipated independently, became very widespread, especially after the valuable contributions that Henry Briggs (1561–1630) made with his *Arithmetica logarithmica* (1624). Meanwhile, the first logarithmic tables were compiled. The most important ones were Napier's (1614), for sines and tangents of angles of the first quadrant, and Briggs' tables of decimal logarithms, calculated in 1617 and 1624. Bürgi's tables of antilogarithms (calculated, it seems, between 1603 and 1611), appeared in Prague in 1620, while Edmund Gunter (1581–1626) constructed the first rule for making logarithmic calculations in 1623.

The study and improvement of this new branch of mathematics continued throughout the following decades. Francesco Bonaventura Cavalieri (1598–1647) illustrated the numerous applications of logarithms in plane and spherical trigonometry, astronomy, and practical geometry in his *Directorium generale uranimetricum* (1632) and his *Centuria di varii problemi* (1639). The first convergent series of logarithms was given in *Geometria speciosa* (1659) by Pietro Mengoli (1625–86), while Nicolaus Mercator (c. 1620–87) proposed a new method for calculating logarithms, based on the procedure of convergent series, in his *Logarithmotechnia* (1668).

ASTRONOMY

The controversy over the new discoveries made by Galileo and Kepler with the telescope

In March, 1610, Galileo's twenty-four-page octave pamphlet *Sidereus nuncius* came out in Venice. In it, in simple persuasive, elegant Latin, he announced a series of surprising discoveries made with a telescope. The pamphlet immediately drew heated attacks and enthusiastic consent from scientific circles. The moon's aspect was similar to that of the earth, furrowed by valleys and full of very high mountains (a phenomenon that had already been observed by Thomas Harriot); the Milky Way was made up of an enormous number of stars; these stars were infinitely more numerous than those visible to the naked eye (as could be seen, for example, in the consellations of Orion, Pleiades, Hyades, and Praesepe); and Jupiter was circled by four moons that rotated at different speeds (which Galileo named the *Medicea Sider*, after his protectors). It was this last, unexpected phenomenon that disoriented the scholars, who in some cases refused to accept the existence of the moons even when faced with factual evidence. A certain Martinus Horky published a *Brevissima peregrinatio contra Nuncium sidereum* in Modena in 1610, maintaining among other things, that Galileo mistook simple spots on the lens for the so-called "planets" of Jupiter.

Kepler, who had already received a copy of the *Sidereus nuncius* from Galileo, answered with an open letter to the author, published a short time later in Prague (1610). Although he pointed out that Giambattista Della Porta (1535–1615), in his *Magia naturalis*, had already described the mechanism of the telescope with concave and convex lenses, and that he himself had already provided an exhaustive geometrical explanation of simple *perspicilla* in his *Astronomiae pars optica*, he nevertheless, expressed total enthusiasm and support for the recent discoveries. After the imperial *mathematicus* backed Galileo, many of those who had opposed him changed their position. The first of these was Clavius, who was soon followed by a number of other Jesuit scholars. In Frankfurt, the following year, after he too had been able to use a telescope, Kepler published the findings of his own observations, in his *Narratio de observatis a se quatuor Iovis satellitibus erronibus*. His findings fully confirmed those of Galileo. Three years later, in 1614, the German astronomer Simon Mayr, or Simon Marius (1570–1624), also known for being the first to mention the Great Andromeda Nebula, published his *Mundus Iovialis* in Nuremberg. In it, he calculated tables for the Medicean satellites and gave them the names: Io, Europa, Ganymede, and Callisto, names they bear to this day.

Meanwhile, Kepler's *Dioptrice* came out in 1611. Among other things, it contained an exposition of the theoretical principles of the astronomical telescope (also called the Keplerian telescope) and a description of it. Unlike Galileo's model, Kepler's gave an upside-down image, but in return it achieved a greater degree of magnification.

Sunspots

The phenomenon of sunspots had been observed by astronomers before, but a satisfactory explanation of them had not yet been provided. In 1611 and 1612, the phenomenon was simultaneously studied with the use of the telescope by Christoph Scheiner (1575–1650), Thomas Harriot, Johannes Fabricius (1587–1615) (the son of David Fabricius, who in 1596 discovered the variable luminosity of Mira Ceti), and Galileo, who claimed priority for the observations in several letters, published in 1613. The letters also contain Galileo's first explicit statement in favor of the Copernican system.

Galileo and the Copernican system

A short time later, the Church began its attack against the Copernican system. The first blow came on May 26, 1616, when the Holy Office issued an injunction against Galileo to not "defend or support" the Copernican doctrine, in that it was "completely contrary to the Holy Scriptures," and placed *De revolutionibus orbium coelestium* by Nicolaus Copernicus on the Index until it could be corrected and expunged.

De revolutionibus orbium coelestium remained on the Index until 1822, but as early as 1620 corrections were made in the references to the Holy Scriptures in the preface by Pope Paul III, and in the points where the earth was called a planet. But the revisions had little effect. No editor would republish the expurgated version of Copernicus's text. It was too difficult and was drastically outdated following the findings of Tycho Brahe and Kepler. After the second Dutch version came out in 1617, it was only reprinted twice: in 1854 in Warsaw and in 1873 in Torùn.

The harmony of the universe

Meanwhile, pursuing his plan to unify the science of the heavens under one great law (which he had begun in his *Mysterium cosmographicum*), Johannes Kepler finished his *Harmonice mundi* in 1618 and published it the following year. It contains the formulation of his famous third law: the square of the time it takes the planets to describe their orbits is proportional to the cube of their mean distance from the sun. It was on the basis of this law that Isaac Newton, almost half a century later, discovered the elements for stating the law of universal gravitation. In the *Harmonice mundi* Kepler also described a study on the musical properties of the planets. Higher notes were assigned to the planets closer to the sun, which moved at a higher speed. Between 1618 and 1621, Johannes Kepler also published the *Epitome astronomiae Copernicanae*, in three parts, which extended the first two laws of the *Astronomia nova* to the other planets, the moon, and the satellites of Jupiter.

PHYSICAL SCIENCES

Kepler's organization of optics

Galileo and Kepler continued to be the two major protagonists in research in physics. In 1611, Kepler published his *Dioptrice*, in which he applied geometric optics (already expounded in 1604 in the *Paralipomena*) to the study of lenses and combinations of lenses, making a true theory of the telescope possible.

Although Galileo continued his astronomical observations (using telescopes built with increasing accuracy), he also extended his investigations to a progressively wider range of phenomena.

Galileo's defense of the Archimedean principle

We have already mentioned Galileo's keen interest in magnets; he displayed the same degree of interest in the phenomenon of bodies floating in water. It was a well-known fact that the accepted explanation did not hold for those cases in which a body such as a thin shaving of gold or a small bit of ebony remained on the surface even though its specific weight was greater than that of water. Galileo's pamphlet on the subject, *Discorso intorno alle cose che stanno in su l'acqua o che in quella si muovono* (1612), was not able to give a scientifically valid explanation of the unusual fact, which was caused by capillary action, still an unknown phenomenon to Galileo. Nevertheless, the work is of major scientific interest for the rigor of its arguments supporting Archimedes' principle.

The Aristotelian Ludovico delle Colombe wrote a scathing attack against it, which stands as evidence of the desperateness with which traditional science tried to uphold its theories. *Risposta alle opposizioni del sig. Ludovico delle Colombe*, written by the monk Benedetto Castelli (1577–1644), one of Galileo's disciples and collaborators (although it was corrected, and mostly written by Galileo himself), came out in Florence in 1615.

CHEMICAL SCIENCES

Developments in iatrochemistry and the new atomism

The influence of Paracelsus could still be felt in Jean Beguin's *Tyrocinium chymicum* (1610), which describes many pharmaceutical preparations, some of them new—for example, ammonium polysulphide (*oleum sulphuris*), later called *spiritus sulphuris Beguinii*—for the use of students. A later edition described a double decomposition reaction (with antimony chloride, mercury, and antimony sulphide) in terms of "sympathy." In the field of pharmaceutics, the publication of the *Pharmacopeia londinesis* (1618), the first national pharmacopea, deserves mention.

Changes took place in iatrochemistry. In his *Epitome naturalis scientae* (1618) and *De chymicorum cum Aristotelicis et Galenicis consensu ac dissensu liber I* (1619), Daniel Sennert (1572–1637) alludes to the atomic theory, which he later developed more fully. He speaks of atoms (*atomi*, *atoma corpuscula*, *minima naturae*, *corpora indivisibilia*) of the four elements, and of second-level atoms, i.e. molecules (*prima mixta*), which constitute compounds or are formed by the association of the atoms themselves. This general revival of atomism, both in philosophy and in the sciences, deserves particular mention.

At first, Sennert believed in transmutation, but he later acknowledged the erroneousness of it, at least in some cases: boiling water becomes aqueous vapor not air. However, he maintained that the transmutation of metals had been proved. In reality, a few years earlier (1609–13) A. Sala (1576–1637) had stated that copper was present in the solutions of the sulphate from which it precipitated with the addition of iron and that it was attracted by the iron, liberating sulphuric acid.

Finally, we should mention Hamerus Poppius's observation (in his *Basilica antimonii*, 1618) that antimony increased in weight during calcination, even though fumes were released. This fact was universally accepted only after Antoine Laurent Lavoisier's antiphlogistic reformation.

LIFE SCIENCES

The use of the microscope

Some extremely important discoveries in the life sciences were made during this decade, although their consequences were not felt until sometime later.

In 1610, Galileo (1564–1642) used his spectacle as a microscope and observed the minute structure of the eye of an insect. Although N.C. de Periese (1580–1637) had already examined cheese mites under a microscope in 1612, the method was not really appreciated by naturalists. The first biological illustration executed on the basis of observations under a microscope did not appear until 1625, in the *Melissographia* of the Accademia dei Lincei of Rome. It was an embellished image made by Francesco Stelluti (1557–1651).

The *Apiarium*, by Francesco Cesi (1585–1630), published that same year, contained a description of the minute structures of bees. Other images obtained with the first microscopes were later published in Stelluti's *Persio tradotto* (Rome, 1630).

Harvey's discovery

The other great discovery of this period was that of blood circulation. It came about gradually, and so is impossible to date accurately. William Harvey began his public lessons on anatomy in London in 1616. It was probably during these lessons that he first expounded his ideas on the function of the heart and the movement of blood (around 1619).

The principles of the new mechanics as stated by Galileo and the spread of the experimental method also had repercussions in the field of medicine and in the other life sciences.

The development of medical instruments

In 1612, Santorio Santorio (1561–1636) published the first description of the *pulsilogium*, an instrument for measuring the frequency of pulse beats and their variations. That same year, he conducted several experiments with an open-air thermometer that he himself had built. He also made an attempt to measure the body temperature of people who were healthy and people who were ill. The first edition of Santorio's major work, *De medicina statica*, appeared in 1614. It marked a new era in the history of research in nutrition and represented the birth of systematic quantitative experimentation in the biological sciences. Through the regular use of scales, Santorio discovered and studied imperceptible transpiration. In his *Commentaria in primam fen primi libri Canonis Avicennae* (1625), Santorio described his hygrometer, a number of open air thermometers and several new surgical instruments. He is credited with inventing the trocar, and recommended its use in thoracic and abdominal paracentesis. Santorio's work anticipated the medical doctrine that was later called iatrophysics. Although Santorio established the methodological foundations of this new science, he always remained faithful to the precepts of Galen.

Iatrochemistry

Although a true science of iatrophysics only developed later, iatrochemistry was already achieving full development. The physicians of the northern countries were particulary inspired by the ideas of Paracelsus.

In *Basilica chymica*, which appeared in the course of 1608, Oswald Croll (1580–1609) proposed medical applications for a series of chemical substances.

The most typical practitioner of iatrochemistry, Johannes Baptista van Helmont (1579–1644), compared respiration with combustion around 1610. Several years later, van Helmont drew attention to the chemical activity of gastric juice and bile in the digestive process.

Therapy

Practical achievements were also realized in medical treatment. A publication by Cesare Magati (1579–1647) about a conservative and particularly effective method for treating firearm wounds (1616), the successful removal of a gallstone by W. Fabry in 1618, and the publication of a treatise on naval medicine by John Woodall (1556–1643), who recommended lime juice against scurvy (1617), all deserve mention.

EARTH SCIENCES

Fossil findings

In 1613, J. Tissot announced that he had found the remains of the giant Teutobochus, the legendary king of the Cimbrians, in the Dauphine, near the Castel of Langon. In reality, they were the remains of the Deinotherium, a proboscidean that had become extinct during the Pleistocene.

Three years later, Fabio Colonna (1567–1650) stated that the so-called glossopetrae were not snakes' tongues but sharks' teeth. In his *Osservazioni sugli animali acquatici e terrestri* (1616), he also distinguished between marine and terrestrial fossil shells.

The growth of metals

Georg von Löhneyss (c. 1552–1625), among others, reiterated the idea that metals grew inside the veins of rocks. He stated his views in an essay (*Bericht vom Bergkwerck*, 1617) that also talked about mines.

In 1613, Galileo (1564–1642) conducted an experiment that had been in the making for some time, and that Nicholas of Cusa had already attempted. He succeeded in calculating the weight of air, which he found to be 1/460 that of water. Galileo also invented a thermoscope, or air thermometer, which Brancani described in his *Sphaera mundi* (1620).

Measuring the earth's meridian

Willebrord Snell (1591–1626) of Leyden performed the first triangulations for the purpose of geodetic surveying. He calculated one-fourth of the earth's meridian as equal to 9,660,000 meters. He communicated his findings in the *Eratosthenes batavus; de terrae ambitus vera quantitate a Willebrordo Sneillio suscitatus*, published in 1617.

APPLIED SCIENCES

Hydraulic architecture and town planning

Additional progress was made in Holland in hydraulic architecture and town planning. The huge, ambitious project for concentrically extending Amsterdam was begun in 1610. The new bastions, which had just been finished in 1593, were demolished and replaced by a canal. Two other canals were then dug parallel to it. A fourth canal, ten kilometers long, was built around the other three in 1658, with a new line of bastions. Small radial canals, which crossed the main concentric ones, formed a spiderweb configuration.

The automatic loom

The introduction in London of an automatic loom, known as the bar loom or Dutch machine, deserves mention. It was a ribbon loom that could weave four to six ribbons simultaneously.

Scientific instruments

In his *Dioptrice* (1611), Johannes Kepler (1571–1630) expounded on the theory of astronomical telescopes, which had numerous advantages over Galileo's. The instrument was only built later, however, in 1630, by Christoph Scheiner (1575–1650) (see ASTRONOMY 1610–1620).

Another scientific instrument that was especially important for medicine was the *pulsilogium*, invented by Santorio Santorio (1561–1636). It was a direct application of Galileo's laws on pendulums. Santorio described it in his *Commentaria in artem medicinalem Galeni*, written in 1612 (see LIFE SCIENCES 1610–1620).

A triangulation instrument that made it possible to read unknown distances and angles without making any calculations was also invented in 1617.

Treatise-writing

In 1612, Antonio Neri (c. 1576–1614) published his *L'arte vetraria*, a classic on the technological aspects of glass production in the sixteenth century. The situation of glass-makers in Great Britain was changed radically in 1615 by an edict prohibiting the use of wood to fuel kilns. A modern-type covered crucible was quite probably already in use during this period. In kilns for making "green" glass, for example, the sulphurous compounds produced by the combustion coal would have reacted with some of the lead particles of the vitrifying mass, forming lead sulphide, which is black. In this case, open crucibles were probably used.

In 1612 and 1613, Simon Sturtevant, who may have been Dutch, and John Rovenzon published several treatises on metallurgy in which they proposed the use of blast furnaces fueled with mineral coal; however, their methods seem not to have been effective. William Ellyott and Mathias Mersey were more fortunate. In 1614, probably inspired by the results achieved in glass-making, they invented a system for using coal as the fuel in preparing steel from fused iron. Meanwhile, the *Bericht vom Bergkwerck* by Georg von Löhneyss (c. 1552–1625), another treatise on metallurgy that contains important observations on mines and on the organization of the mining industry, was printed (1617).

1620–1630

THOUGHT AND SCIENTIFIC LIFE

Institutions at the time of the Scientific Revolution

The introduction of the Baconian inductive method, the Galileian experimental method and later, Cartesian rationalism among the first advances made in researching laws (instead of the essence of things) characterized the birth of the Scientific Revolution.

However, these trends did not immediately lead to changes on an institutional level. These occurred only gradually, becoming evident through both an increase in the absolute and relative number of scientists and a progressive abandonment of Latin in favor of national languages. Debates on scientific problems no longer took place only in universities or closed circles, but entered the literary salons as well with the result that they often became mundane events.

The Jardin des Plantes was inaugurated in Paris in 1626. It later became the Museum d'Histoire Naturelle, one of the most important scientific centers of the world. The botanical garden at Jena was founded in 1629.

The great scientific-philosophical battles

Francis Bacon's masterpiece, the *Instauratio magna*, was published in 1620. It was sent to press in an incomplete form and contained only a general plan for the *Instauratio* and summaries (presented as aphorisms) of the second part (*Novum organum*). In 1623, Bacon published his *De dignitate et augmentis scientiarum*, the reelaboration of a shorter work of the same title which he had written in 1605. In the *Novum organum*, the disagreement with Aristotle is essentially expressed as a radical critique of Aristotelian logic which Bacon deemed affected and sterile, proposing instead an inductive logic that would allow new truths to be discovered.

Galileo's *Saggiatore*, which clearly explains the famous distinction between primary and secondary qualities—later one of the pillars of the so-called mechanistic concept—was published in 1623.

The first great work by Pierre Gassendi (1592–1655), *Exercitationes paradoxicae adversus aristotelicos*, written in an empirically skeptical tone, appeared in 1624. The author accepted the Skeptics' criticisms of metaphysics, but made an attempt not to let them encroach upon all areas of human activity. Several years later, Gassendi made a thorough study of Epicureanism and became one of the most authoritative supporters of the atomistic theory.

In 1628–29, René Descartes (1596–1650) wrote his first important work, *Regulae ad directionem ingenii*. It was, however, never finished and was only published posthumously in 1701. The first version of the *Meditationes de prima philosophia* also dates back to 1629–30. Descartes continued to work on it for several more years and published it in 1641.

MATHEMATICAL SCIENCES

The attitude toward ancient science

Francis Bacon (1561–1626) published his *Novum organum* in 1620. In it he opposes the deductivistic approach of traditional science and, after a vigorous defense of the experimental method, proposes rules for what could already be called inductive logic. The work is particularly interesting because it shows that new paths were being sought in the field of logic, especially for dealing with the problems of scientific knowledge. Mathematics, which in the following decades was found to be an ever more effective instrument for scientific discourse, able to play a privileged role, benefitted greatly from this critique of Aristotelian formal logic.

This break with tradition, however, was not always controversial, as it was with Bacon's attitude toward Aristotle. In particular, it was joined to a certain veneration for the classics of scientific thought if not of philosophy. For example, a new edition of Diophantos' *Arithmetic*, edited by Claude-Gaspar Bachet de Méziriac (1581–1638), appeared in 1621, evidence that, even in this century, there was a continued interest in the classics, which was characterized by translations, commentaries, and "divinations," or attempts to reconstruct the contents of lost parts of ancient works by conjecture. Divination reached such proportions that it could almost be considered a trend. One of the most talented scholars in the field, and one who took advantage of the occasion to make new discoveries, was Vincenzo Viviani (1622–1703), to whom we owe, among other things, a reconstruction of book V of Apollonius' *Conics* (1659) and an edition of Euclid's *Elements* (1690).

ASTRONOMY

The Astronomia danica

While the Copernican system was fast becoming a symbol of the claims of the new science against academic and Aristotelian obscurantism, the system developed by Tycho Brahe turned out to be the one most able to reconcile the evidence of the facts with the needs expressed by tradition and the Church for seventeenth century scientists.

In 1622, the Danish astronomer Christian Severin, otherwise known as Longomontanus, after Longberg, his place of birth, Tycho Brahe's old assistant at Uraniborg and in Prague, published the *Astronomia danica*, a compendium of his teacher's planetary theory. The work contains, among other things, the results of studies Longomontanus made on the motion of the moon. Despite its many defects and the competition of Johannes Kepler's *Tablae Rudolphinae*, the *Astronomia danica* enjoyed considerable success and was reprinted in 1640 and again in 1663.

The Tabulae Rudolphinae

Johannes Kepler's last great contribution to astronomy was the *Tabulae Rudolphinae*, which he finished in 1623. Hot off the press, they were presented by the author himself at the Frankfurt book fair in September 1627. They were the result of Brahe's observations at Uraniborg and Kepler's meticulous mathematical elaboration of the data. For more than a century, they remained an indispensable tool for studying the planets and the fixed stars. Besides the tables of the planets, they contain Brahe's catalogue of 777 fixed stars, which Kepler brought to 1005, refraction and logarithm tables, used here for the first time in astronomy, and a list of the cities of the world with longitudes relative to the meridian of Hven.

The first Keplerian telescope

The era of the telescope, which was based mainly on the theoretical foundations laid by Kepler, began in the field of practical astronomy. The *Rosa ursina*, written by the Jesuit Christoph Scheiner, which contains the first description of a telescope using the Keplerian optical system (with two convex lenses, unlike Galileo's, which had a concave eyepiece), came out in 1630. As early as the middle of the century, the Keplerian telescope surpassed Galileo's model, mainly because it widened the field of vision. During this period, Scheiner also developed a helioscope for observing the sun. Research with this instrument led to the discovery that sunspots appear to rotate at different speeds, depending on their latitudes.

PHYSICAL SCIENCES

The Church and the new science

Despite the condemnation of the Copernican theory, pronounced by the Holy Office in 1616, Galileo continued to believe that the powerful Catholic Church could be convinced to take a more open position toward the new science. He felt encouraged by the election in 1623 of Cardinal Maffeo Barberini to the papacy. Barberini (Urban VIII) was a bright, talented man who was undeniably sensitive to the needs of the culture. Galileo dedicated his *Saggiatore*, which was published in 1623, to the new pope. The work is specifically concerned with the study of the nature of comets, but in reality it also discusses a great number of questions concerning philosophy and physics having only indirectly to do with astronomical problems. The author's vigorous argument against the pseudo-scientific methods of the Aristotelians is very effective. He upholds the rights of experimentation and mathematics with extraordinary energy and admirable precision. The work is a masterpiece, even though Galileo's explanation of the nature of comets is incorrect.

The microscope

Meanwhile, Galileo's scientific interests continued to multiply, constantly embracing new areas of experience. While he tenaciously continued his investigations of the heavenly world, he put an equal amount of energy into figuring out why the telescope could be used in a so-called inverse way (as he had already discovered) "in order to see minute things close up," i.e. as a microscope. In 1624, he sent an example of his microscope to Prince Cesi, accompanying it with a letter in which he explained how the device worked and its merits. He also worked in applied mathematics, in magnetism (trying to build increasingly stronger magnets), and in hydraulic engineering (the Galilean Benedetto Castelli later dedicated special attention to this discipline).

The interests of the Galileans

Galileo was able to communicate his enthusiasm to many of his disciples, as is evidenced by the many letters that he exchanged with Italian and foreign scholars. A general reawakening of the spirit of observation soon pervaded all of Europe, favored by various complex factors of a cultural and social nature. In the course of several decades, the effects of this revolution

of knowledge were felt in all areas of civil life.

Among the many scientific publications of the decade, special mention should be made of the *Philosophia magnetica* (1629) by Niccolo Cabeo (1585–1650). Cabeo continued Gilbert's work, repeating and probing further into his experimental investigations, although he rejected the theory that the earth was a huge magnet.

CHEMICAL SCIENCES

The experimental method and the crisis of alchemistic principles

In his *Philosophiae naturalis adversus Aristotelem libri XII* (1621), Sebastian Basso put forth ideas developed independently, though often similar to those of Daniel Sennert. He revived the theory of the Pre-Aristotelian atomists and denied the transmutation of the elements. He proposed five elements in contrast to Aristotle's four: spirit (or mercury), oil (or sulphur), salt, earth (*faeces* or *caput mortum*), and phlegm. These joined in second order groups, third order groups, etc. There was no void, instead there was the ether.

Another iatrochemist with original ideas was Angelus Sala (1576–1637), a student of pharmaceutical chemistry and an avid opponent of the universal remedy (*res absurdissima*), potable gold, and the like. He wrote a number of publications between 1608 and 1637, full of interesting observations, such as the fact that silver chloride turns black in the sun. He demonstrated that the oil of vitriol obtained from iron or copper sulphates was the same as that obtained from sulphur. He also described the preparation of ammonium chloride and phosphoric acid and was the first to mention producing alcohol from grains.

Another champion of the experimental method was Francis Bacon (1561–1626). Chemistry found its place in the multitude of his activities, especially in its connection with physics. From his numerous calculations of specific weights, Bacon concluded that the theory of the four elements was false, since the most dense of them, earth, was less dense than many metals. He was fairly skeptical about the atomic theory. Although he distrusted the alchemists, he believed that the transmutation of metals was possible. He believed that the transformation of lead into silver was more probable than its transformation into gold, since gold was denser. He supported the idea of the conservation of mass and observed that some reactions were exothermic.

Tommaso Campanella (1568–1639) also believed that the transmutation of metals was possible (1623), but rejected both Aristotle's four elements and the three principles of the alchemists (1638).

LIFE SCIENCES

The publication of Harvey's work

William Harvey's *Exercitatio anatomica de motu cordis et sanguinis in animal'ibus* appeared in 1628 in Frankfurt. It was a work that would revolutionize the biological sciences and medicine, both because of the theoretical and practical significance of the discovery of blood circulation and because of the experimental method that was behind the discovery. All this represented the triumph of the quantitative method and of the mechanistic idea in physiology. In a book written that same year Giovanni Colle (1558–1631) already predicted blood transfusion.

The fundamentals of ophthalmology

Kepler's ideas on the physiology of sight inspired Christoph Scheiner (1575–1650), who gave an experimental demonstration of the role of the retina, correctly explained the function of the crystalline lens, and described adjustment reflexes in his *Oculus* (1619). Thus scientific ophthalmology was born. In 1624, Fabricus Hildanus, (Wilhelm Fabry) used a special magnetic instrument to extract a fragment of iron from the eye of a patient.

EARTH SCIENCES

The fossilization process

Francesco Colonna, again writing about fossils, distinguished various ways in which they were preserved (*Historia lapidum figuratorum*), revising the conclusions he had already reached concerning petrified shells, which he divided into terrestrial and marine types.

Andrea Chiocci, in his *Museum Calceolarii*, described a large collection of fossils that belonged to the Museo Calceolari in Verona.

Variations in the earth's appearance

In his two books (published in 1625), Nathanael Carpenter (1589–1635) described the changes that take place in the shape of the earth's surface as continuous and gradual. *Sylva sylvarum*, a posthumous work (1627) by Francis Bacon (1561–1626), dealt with various oceanographic problems, such as tides, waves, currents, ocean depth, salinity, and the color of the sea. While studying refraction in 1627, René Descartes (1596–1650) formulated one of the first physical explanations of rainbows.

APPLIED SCIENCES

Research in hydraulics

One of the most urgent problems that the society of the times demanded scientists and technicians to solve was that of harnessing water, both to protect them from frequent flooding and to improve irrigation. The most valid contributions to this problem in Italy again came from Galileo and his school. A disciple of Galileo, Benedetto Castelli (1577–1644), was the most prominent Italian scholar in hydraulics in the first half of the century. His research on the subject marked the beginning of a truly new phase in the art of land reclamation. In 1623, Castelli was appointed by the Grand Duke of Tuscany to study solutions to the constant flooding of the land between the Serchio and the Morto rivers. In 1625, Urban VIII named him assistant to the superintendent of general drainage in Tuscany, in the zone between the Po and the Reno. In this way, he was able to make theoretical and practical observations, which he applied to solving many concrete problems, giving rise to a rational hydraulic technology. The results of his studies are summarized in his *Della misura dell'acque correnti* (1628), a masterpiece of richness of thought in this art.

Galileo also did work in hydraulics. In January, 1631, he gave the Grand Duke his opinion on a plan that several engineers had presented for checking the frequent inundations from the river Bisenzio. Although the letter that contains his evaluation is only a few pages long, it is a jewel of scientific accuracy and technical skill.

The art of printing

The first important innovation in the type of printing press that had been in use for over a century was made during this period. Willem Janszoon Blaeu (1571–1638), the Dutch cartographer, once a student of Tycho Brahe, redesigned the hose frame of the galley press, which was then named the Dutch press.

Scientific instruments

During this decade, the use of the microscope (to which Galileo dedicated several very accurate studies in 1623–24) became widespread, producing surprisingly interesting results in the field of biology.

Another instrument that found an increasingly wide variety of uses (especially in medicine) was the thermometer. According to Vincenzo Viviani (1622–1703), Galileo had already during the final years of the sixteenth and the beginning of the seventeenth centuries designed instruments "of glass with water and air to distinguish, through changes in heat and cold, variations in temperament in certain locations." In 1612, Santorio Santorio (1561–1636) had built a graduated clinical thermometer to measure his patients' temperatures (see LIFE SCIENCES 1610–1620).

In 1620, Francis Bacon (1561–1626) described a thermometer similar to Galileo's in his *Novum organum*. Another was described in 1624 by Jean Leurechon (1591–1670), who was the first to refer to it by the name "thermometer." Although these instruments were still very imperfect, they proved quite useful and became increasingly popular.

Treatise writing

A treatise containing a new system for using steam, *Le macchine diverse del Signor Giovanni Branca, cittadino romano, ingegniero architetto della S. ta Casa di Loretto*, by Giovanni Branca (1571–1645) was published in 1629. Giambattista Della Porta (1536–1605) and Salomon de Caus (1576–1626) had already advanced ideas on the subject, but Branca's system was original. According to his design, a jet of steam emitted by a heated bronze container would strike the blades of a wheel with enough force to make them turn, a motion that could have a number of applications. This was the first example of a steam turbine engine, an idea that had to wait 150 years before it found a practical application.

1630–1640

THOUGHT AND SCIENTIFIC LIFE

The spread of scientific ideas

In 1630, the physician Theophraste Rénaudot opened the Bureau d'Adresses in Paris, where he examined patients for free. It was an open challenge to the faculty of medicine, both theoretically and practically. The following year, *La Gazette* was founded. Thus the year 1631 saw the birth of journalism; besides being a vehicle for correspondence, journals became one of the most important channels for the spread of scientific information.

The contributions of academies and universities were also important to the spread of knowledge, despite the fact that very few authors of this period ever worked professionally as university teachers, and the university environment often showed a strong resistance to new ideas. While the old universities were enlarged, many more were founded during this decade, including the Athénée of Amsterdam in 1631, and one in Utrecht in 1636. In 1632, King Gustav Adolf created the University of Dorpat in Estonia, and a *studium generale* was founded in Trnava, Slovakia, by Cardinal Péter Pázmány. In France, the Académie française was created in 1635 by Cardinal Richelieu, who five years later established the Imprimerie nationale.

Scientific institutions were founded in the New World as well. In 1636, fifteen years after the Pilgrim Fathers arrived in Massachusetts, John Harvard helped found the college that later bore his name. It was proclaimed a university in 1650. The first printing house in North America was set up in 1639 in Cambridge, Massachusetts. In February 1632, Galileo published his *Dialogo sopra i due massimi sistemi* (*Dialogue concerning the two chief world systems*). He was immediately brought to trial in the court of the Holy Office in Rome. The opponents of the Copernican system and of Galileo charged that the *dialoghi* were "detestable and more dangerous to the Church than the writings of Luther or Calvin," and Galileo was found guilty (see ASTRONOMY 1630–1640). After Galileo's abjuration (1633), Descartes, who had just finished writing his *Traité du monde ou de la lumière*, which leaned toward the Copernican view, decided as a precaution not to send it to press. It was published posthumously in 1664. In 1632–33, Descartes had also composed *De l'homme*, an organic work directly connected to the *Traité du monde*. It too was only published several years after the author's death (in Latin in 1662, and in French in 1664).

The work of Mersenne

However, ecclesiastical censure and the precautions of scientists did not keep the new ideas from being discussed, at least in scholarly circles. The work of Father Marin Mersenne (1588–1648), a philosopher and physicist who created a private scientific academy in the Minorite Convent in Paris, is emblematic. Mersenne's role was anything but negligible, thanks especially to the regular correspondence he kept up with scientists and scholars in a great many countries, from Descartes, who wrote to him from Holland, to Pierre de Fermat (1601–1665) and Christiaan Huygens (1629–95).

As early as 1623, Mersenne supported the idea that the earth moved. He published his *Méchaniques de Galilée* in 1634, and five years later published *Nouvelles pensées de Galilée*, paraphrasings of the Pisan scientist's works. His *Harmonie universelle*, written in 1636–38, contained many Galilean ideas, but Mersenne, like G.P. de Roberval (1602–75), the leader of the Mathématiciens de Paris, did not directly defend the Copernican concept, since he could not find satisfactory physical proof for it in Galileo's work. There were instances of misunderstandings with Galileo himself (three of Mersenne's letters went unanswered) and with

Evangelista Torricelli (1608–47) and Francesco Bonaventura Cavalieri (1598–1647). But on the whole Mersenne—who was personally concerned with problems in physics (acoustics, hydraulics, falling bodies, etc.), although he mainly stimulated debates and discussions—carried out the important task of spreading Galileo's ideas widely, making it possible to overcome the negative effects of the condemnation of 1633.

The genesis of Hobbes's program and developments in Cartesian and Galilean thought

While the ideas of Galileo and Descartes were spreading throughout Europe, (Galileo's *Discorsi e dimonstrazioni matematiche intorno a due nuove scienze* was published in Holland in 1638), though with considerable difficulties, the thought of Thomas Hobbes (1588–1679) was maturing in England.

In 1636–37, Hobbes developed a plan for a grandiose systematical work entitled *Elementa philosophiae*, which would deal with matter, human beings, and—a new aspect in comparison with Descartes—the state. The work was thus to include a physics, an anthropology, and a politics, all interconnected by a rigorously unitary deductive network. This order was not followed, however, when the work was published.

Almost simultaneously (1637), Descartes' *Discours del la méthode* was published as the introduction to three scientific works, the *Dioptrique*, the *Météores* and the *Géométrie*. Although the contents of the three works are quite interesting, the methodological approach described in the preface was much more important. Descartes outlined in detail four canons (the rules of evidence, analysis, synthesis, and enumeration) on which the search for truth must be based. But in the last instance, these rules all led to a single rule: it was not right to accept any idea as truth that was not perfectly evident.

MATHEMATICAL SCIENCES

The exceptional event of this decade was the great encounter between algebra and geometry brought about by the creation of analytical geometry. The event was not, however, unexpected, and the great flourishing of systematical treatises on both geometry and algebra during the early years of this period actually appears to have been an explicit preparation for this new discipline.

The organization of algebra

Albert Girard (1595–1632) had already published his *Invention nouvelle en l'algebre* (1629), a work that carried on the tradition of Viète and introduced the use of parentheses, along with various technical findings such as a description of the method for breaking down a polynomial into factors. This current of refining formal and symbolic tools also included English authors, such as Thomas Harriot (1560–1621) and William Oughtred (1575–1660), who besides developing the theory of equations and relative techniques, introduced new symbols, such as $>$ and $<$ for inequalities, and the symbol $\times$ for multiplications. Almost as though to crown this current, the Frenchman Pierre Hérigone (flourished 1634) in his *Cursus mathematicus*, demonstrated a full understanding of the importance of the purely formal, abstract and completely explicit treatment of questions that symbolism made possible (1634).

The birth of indivisible geometry

Even more significant were the results obtained in geometry. The *Geometria indivisibilibus continuorum nova quadam ratione promota,* by Francesco Bonaventura Cavalieri (1598–1647), came out in Bologna in 1635, after more than a decade of incubation. It conceived of geometrical figures as made up of the sum of their indivisibles (lines in the case of plane figures and planes in the case of solids). This allowed the introduction of a general principle for squaring, which states, in substance, that the areas (or volumes) of two geometrical figures are to each other as their indivisibles are to one another, when these are obtained by properly cutting the figures along certain straight lines or planes. Cavalieri's principle already contained the basic idea of integral calculus, even though it was limited to cases where squaring was done through comparisons with known areas and volumes.

Another precursor of the method was Johannes Kepler (1571–1630), who had already solved the problem of calculating the areas and volumes of various figures in his *Nova stereometria doliorum vinariorum* (1615) by reducing them to infinitesimal elements, although he did not arrive at a general method. Among the opponents of the method (which in a certain sense led to the use of actual infinites and infinitesimals) was Paul Guldin (1577–1643), who attacked it in his *Centrobaryca* (1635–41). Evangelista Torricelli (1608–47), on the other hand, not only applied it, obtaining brilliant results, but also formulated a remarkable generalization of it. John Wallis (1616–1703) should also be counted as a continuer and, in fact, the last great cultivator of the method of indivisibles.

Galileo

The first edition of *Discorsi e dimostrazioni matematiche intorno a due nuove scienze,* by Galileo (1564–1642), appeared in Leyden in 1638. Besides laying the foundations of dynamics, this treatise contains a number of concepts that were important to the birth of infinitesimal calculus. For instance, it contains the idea of instantaneous velocities and the "compound" approach to the relations between the continuum and the indivisibles that make it up. These were described as its non-quantum parts that were, however, different from zero.

Analytic geometry

The *Dioptrique,* the *Météores,* and the *Géométrie,* prefaced by the *Discours de la méthode,* by René Descartes (1596–1650), were published in Leyden in 1637. The *Géométrie* laid the foundations of analytic geometry and made the first important contribution to it. Using the method of coordinates, the study of geometric quantities could be translated into algebraic problems, and information on various types of curves could be gathered by examining their equations. The needs of the new science also induced Descartes to do research on the theory of equations.

Its great technical value aside, the invention of analytic geometry was also quite important conceptually. In a certain way, it marked the end of the primacy of geometry which, at least in academic mathematics, had seemed sacred since the time of the Greeks. Algebraic methods were finally endowed with a scientific dignity equal to that of the purest geometric methods, and in addition they proved to be even more flexible and general. Only this union of algebra and geometry could make the geo-metric representation of a huge class of functions possible, a necessary step in the genesis of infinitesimal calculus. The study of the problems of tangents and areas that Kepler, Cavalieri, Torricelli, and Gilles Personne de Roberval (1602–75) conducted using purely geometric methods, were only partial forerunners of the new method. In particular, the reciprocal nature of derivations and integrations—without which integral calculus would have died out after the first squarings, which could be done with direct methods, were computed—could only become apparent on an analytical level.

The basic ideas of analytic geometry, in particular the method of coordinates, were also discovered by Pierre de Fermat (1601–65) independently of Descartes. In his lifetime, Fermat published just one brief dissertation. Otherwise he only communicated his findings in a series of letters to scholars of the time. This fact makes the dating of his discoveries uncertain; however, from his correspondence with Roberval, it is possible to conclude that he had devised analytic geometry, and in particular, written his unpublished memoir *Ad locos planos et solidos isagoge,* which contains the most complete exposition of the method, by 1637.

Fermat is also remembered as the founder of the theory of numbers, since he began the systematic study of the properties of natural numbers, and as one of the precursors of infinitesimal calculus, because of his treatment of the problems of determining the maxima and minima of curves and of squaring higher-order parabolas and hyperbolas.

Projective geometry

Along with the birth of analytic geometry, the high point of geometry was marked by the beginnings of projectivity. In 1639, Gérard Desargues (1593–1662), who in 1636 had already published a study on perspective in which he presented the method of axonometric projection, published a brief treatise on conic sections, which established principles of projective geometry and demonstrated theorems on quadrangles inscribed in conic sections and homological triangles. However, the obscurity of his language and an excess of new terminology prevented the work from exerting a strong influence during its time, despite the fact that it was written with practical intentions.

Among the few who were able to understand and appreciate it were Descartes and Blaise Pascal (1623–1662), who that very same year wrote his first essay containing the theorem of a hexagon inscribed in a conic section (the *esagrammum mysticum*).

ASTRONOMY

The second phase of Galileo's trial

On November 15, 1630, Johannes Kepler died at Regensburg (Bavaria). A short time later (1631), his old teacher Michael Mastlin (1550–1631) also died. Although he had continued to publish his popular *Epitome astronomieae,* which supported the Ptolemaic system, for didactical purposes, Mastlin had initiated his disciple into the Copernican doctrine, and remained his sincere admirer and friend until his death.

Galileo's conflict with the Inquisition and the Aristotelians also continued during these years. In February 1632, in Florence, Galileo published his *Dialogo sopra i due massimi sistemi.* He had finished it three years before and was only able to obtain an imprimatur for it on the condition that the book undergo revision by Father Nicolò Riccardi, a friend of Galileo's. In reality, Galileo openly sided with Copernicus, reproposing a theory of the tides which later proved to be false (see EARTH SCIENCES 1630–1640), as proof of the Copernican theory. This time (October 1632–April 1633), the Inquisition came down with a heavier hand than in 1616. Under the threat of torture (territio verbalis), a provision in the sentence of his trial, Galileo was forced, on June 22, 1633, to abjure the Copernican doctrines (officially proclaimed heretical), and his book was placed on the Index. However, a prohibited copy of the Dialogo sopra i due massimi sistemi del mondo reached Strasbourg, where the work was translated into Latin and published in 1635. From there it became widely circulated throughout Europe. When Marin Mersenne (1588–1648) informed the philosopher René Descartes (1596–1650) of Galileo's condemnation in 1633, Descartes, who had finished his great cosmological treatise Traité du monde, in which he upheld the Copernican theory, decided not to have the treatise published, so that he would not incur the sanctions of the Inquisition as well. But neither Galileo's condemnation nor the deferment of the publication of Descartes' Traité was able to stop the spread of Copernicus' ideas, which gradually gained precedence over the Aristotelian-Ptolemaic and hybrid systems.

Observational astronomy in France and Denmark

In Paris, on November 7, 1631, the philosopher Pierre Gassendi (1592–1655) observed a transit of Mercury across the sun through a telescope. The following year, he published his Mercurius in sole visus, in which he explained the phenomenon acknowledging Kepler's planetary theories. This was the first effective observation of the transit of an inferior planet in front of the sun; Kepler's Mercurius in Sole, written in 1609, actually talked about sunspots. In the controversy around Galileo, Gassendi took Galileo's side (on July 12, 1625, he had already written a letter to Galileo declaring his Copernican convictions). Gassendi was joined during this period by Ismaël Boulliau (1605–94), a Calvinist converted to Catholicism, who became one of Galileo's most avid supporters and one of the few astronomers of the time to accept Kepler's theories. His Philolaus, which shows him to be a follower of Copernicus, though not yet of Kepler, was published in Amsterdam in 1639.

Meanwhile in 1632, Longomontanus inaugurated the construction of a fixed astronomical tower in Copenhagen, which King Cristian IV was having built. However Longomontanus died in 1647 before the observatory was finished.

Developments in research in England: Horrocks, Crabtree, and Gascoigne

During these years of transition from the generation of Kepler and Galileo to that of Christiaan Huygens and Giovanni Domenico Cassini, the greatest contributions to astronomy came from England, where the use of the telescope was becoming increasingly more widespread. The most important discoveries were made by Jeremiah Horrocks, who died very young, at the age of twenty-two or twenty-three. In 1635, thanks to his friend William Crabtree (1610–44), Horrocks discovered Kepler's

Tabulae Rudolphinae, and from that moment on became a convinced disciple of Kepler. Noticing a gap in the tables, he predicted a transit of Venus in front of the sun for December 4, 1639. He and Crabtree were the only people in Europe to observe and report it. A communication that Horrocks wrote about the experience (Venus in Sole visa) is contained in Johannes Hevelius's Mercurius in Sole visus (1662).

Horrocks also made fairly accurate measurements of Venus's disc and the radius of the earth's orbit, and above all provided a valid theory of the elevation of the moon, marking an important step forward with respect to Kepler. He also accepted the physical basis of Kepler's cosmology and emphasized the attraction that the planets exercised on one another. Possibly influenced by Galileo's Dialogo sopra i due massimi sistemi del mondo, Horrocks linked celestial dynamics to the principles of falling bodies, maintaining that the planets had a tendency to fall toward the sun or oscillate freely about it, as the pendulum bob does about its mean position. The result of the combination of this motion with the force of the sun created their elliptical motion. Many of Horrocks' writings were destroyed during the civil war (1642–46); a large part of the remaining ones were published by John Wallis in 1672–73 under the title Opera posthuma.

Between 1638 and 1640, another English astronomer and a friend of Horrocks and Crabtree, William Gascoigne (1612–44) invented a micrometer with cross wires, an extremely useful device for making observations. It fell into oblivion when Gascoigne died and was reinvented in France by Adrien Auzout (1622–91) and Jean Picard (1620–82). Utilizing the property of the Keplerian telescope (with a convex eyepiece), that made it possible to superimpose cross-wires on the focal point of the lens, the micrometer consisted of a reticle of cross hairs (initially very simplified) that allowed the telescope to be aimed more accurately. Gascoigne was also the first to apply collimators to telescopes, but these too disappeared when he died, to be reinvented later (around 1666) by Robert Hooke and Christopher Wren.

PHYSICAL SCIENCES

The ecclesiastical condemnation and Galileo's great works

The decade opened with the complete defeat of Galileo's program. The condemnation passed against him in 1633 demonstrated unequivocally that the Catholic Church did not intend to open up to the demands of the new science, preferring to stay faithful to traditional culture. However, this did not prevent the Mechanics, a work that Galileo wrote at the turn of the century, from being published in Paris in 1634, in a French translation by Marin Mersenne. Although it is very brief, this treatise is of considerable importance, both because of its clarity and elegance and because it presents the first correct statement of the principle of virtual work.

Galileo's great work, the Discorsi e dimostrazioni matematiche intorno a due nuove scienze, came out in Leyden in 1638. Picking up and completing research begun in his youth, Galileo outlined, with exemplary demonstrative rigor, the first basic steps of a truly scientific treatment of dynamics.

Cartesian mechanism

During the final years of the previous decade, René Descartes (1595–1650) had begun to take an interest in the great problems of physics. However, news of Galileo's condemnation had kept him from publishing his lengthy treatise on the subject. Instead, in 1637, he published his *Dioptrique* and which, along with the *Géométrie*, were intended as concrete examples of the possibilities opened up by the new way of conducting cognitive investigation laid out in his *Discours de la méthode*. The most interesting scientific innovation presented in these short treatises was the law of refraction, contained in the *Dioptrique*. The same law had already been discovered some years earlier by the Dutchman Willebrordus Snellius (1591–1626).

The originality of Descartes' treatment lies in the fact that it was fundamentally aimed at basing science on clear and evident principles, from which the particular laws that ruled various phenomena could then be deduced with mathematical rigor. Experimentation was given a much less important role than Galileo had attributed to it; it was no longer a real "proof" but a simple "confirmation" of results already obtained, or at least obtainable, by purely rational means. The superiority, if one can call it that, of Cartesian physics with respect to Galilean physics lay in a rigorous philosophical concept (mechanism). In other words, the Frenchman, unlike the Italian, was no longer satisfied with rejecting Aristotelianism to make modern dynamics possible, but proposed replacing it with a new philosophy that included a mechanical explanation of all natural phenomena.

Specific research

Meanwhile, research into particular areas of physics was conducted in the Galilean spirit in Italy. The best example is *De motu gravium solidorum* (1638) by Giovanni Battista Baliani (1582–1666), a treatise dedicated to studying the phenomenon of impact (the focus of a number of scientific debates during those years).

There were also a number of scholars in France who conducted their scientific investigations with more of an interest in particular phenomena than in a general concept of the world. One of these was Pierre de Fermat (1601–65), who deduced the laws of light reflection and refraction from the so-called principle of the shortest path (the path that light rays travel in the shortest amount of time in two or more media with different refraction indices, and that does not correspond to the shortest geometrical distance), an idea that was highly criticized.

CHEMICAL SCIENCES

Hypotheses and experiments on the composition of air

Joachim Junge (1587–1657), the German Bacon, had a lot in common with his Londonese contemporary, although he had a sounder knowledge of chemistry and more modern ideas. He supported the atomic theory, denied transmutation, and correctly interpreted the reaction between copper sulphate and iron, demonstrating that the iron dissolved while the copper precipitated. He had a good understanding of zinc, and knew that zinc and copper produced bronze. He also recognized that air

played a certain role in combustion, although Jean Rey, who published his *Essays sur la recherche de la cause pour laquelle l'estain et le plomb augmentent de poids quand on les calcine* in 1630, seems to have had clearer ideas on the subject. Rey came close to the truth when he deduced that air contained heavier and lighter parts and that during calcination a partial compensation took place between the loss of weight from the exhalation of fumes and a weight gain from air, which, thickened by the heat, adhered to the "lime." He also mentioned a possible experiment for weighing air, which Galileo actually performed, determining its specific weight (in *Discorsi e dimostrazioni matematiche intorno a due nuove scienze*).

An iatrochemist who was harshly critical of Paracelsus was Fabrizio Bartoletti (1588–1630), famous for having isolated lactose, the sugar in milk, in its crystalline form (a discovery that was only published in 1633).

LIFE SCIENCES

The organization of zoology. The hypothesis of spontaneous generation

John Jonston (1603–75) summarized the zoological knowledge of his time in his *Taumatographia naturalis* (1632), and later in his *Historia naturalis* (1650–53). Although it is not terribly original, Jonston's encyclopedia of natural history still has the advantage of being more organized and easier to read than the works of Conrad Gesner and Aldrovandi. This explains the extreme popularity that the work enjoyed up until the time of Linneaus.

The *Insectorum sive minimorum animalium theatrum*, by Thomas Moffett (1553–1604), contained new zoological observations. Even though the work was begun in the sixteenth century, it was only published much later, in 1634. From many points of view, it stands as evidence of the difficulties that the naturalists of the era found themselves faced with when dealing with a world as varied as that of insects. Among Moffett's many errors, his unfailing belief in spontaneous generation deserves particular mention. For example, he believed that bees were born from the decaying cadavers of bullocks. This is an opinion that must always be kept in mind when weighing the importance of the attacks launched against ideas of this type on the nature of generation in the second half of the seventeenth century.

EARTH SCIENCES

Metals and metallography

The *Mineralogia*, written by the Modenese Jesuit Bernard Cesi or Caesius, (c. 1581–1630) and printed in Lyons in 1636, contains a systematic exposition of contemporary knowledge about minerals. In 1632, the physician Edward Jorden (1569–1632) supported the theory of the fermentation of minerals in *A discourse of natural bathes and mineral waters*. Etienne de Clave's *Paradoxes,* on "petrifying seeds," was published in Paris in 1635. A treatise on the art of metals, *El arte de los metales*, by Alvaro Barba, director of mines in Potosi, South America, came out in Madrid in 1640. The Prizak, an organization for exploiting the metals and natural resources of Siberia, was founded in 1637.

Oceanography and explorations

Between 1631 and 1632, captains Luke Fox (1586–1635) and Thomas James made further attempts at finding a northwest passage that would make it possible to communicate between the Atlantic and the Pacific avoiding Cape Horn and the Strait of Magellan. Bathymetric measurements were also taken on these expeditions, and the first ideas on the existence of an underwater shelf that was less deep than the actual ocean floor were formed. While measuring the depth of the ocean floor was difficult and not important to navigation, bathymetric calculations on the shelf could prove very useful.

Galileo and the tides

The *Dialogo sui massimi sistemi del mondo* appeared in 1632. In it, Galileo criticized Kepler's gravitational ideas about tides, hypothesizing that, on the contrary, the tides were simply caused by the effects of the earth's revolution (today we know that this factor does modify the distribution of the tides, although it does not cause them). Although these ideas were only made public in 1632, they may date back to as early as 1595. In any case, Galileo not only formulated the rotational hypothesis, but demolished all the old ideas that heat, either from the sun or the moon, was the cause of the tides. He also introduced other concepts that were later expanded and developed and which remain valid to this day. For example, he believed that the differences in the times of the high tides from place to place and the differences in range were due to the different geometry of the various basins. He knew that the frequency of oscillation depended on depth. From there at least a rudimentary concept of resonance was immediately arrived at. As for the variations in the range of the tides in the course of a month, Galileo attributed them to the action of the moon, which supposedly interfered with the earth's rotation.

Descartes and the tides

Several years after Galileo's *Dialogo* came out, Descartes proposed a totally different idea on the cause of the tides in his *Principia philosophiae*. Descartes's concept was completely revolutionary and found numerous supporters even many years later. He introduced the notion of an ether, a sort of invisible material that filled all of space. As the moon rotated around the earth, it compressed the ether surrounding it, generating a pressure that caused the tides when it was transferred to the surface of the ocean. When in 1658 Robert Boyle (1627–91) observed that when the moon transited over the meridian, variations in atmospheric pressure could be noted with a barometer (in fact, Boyle was the unknowing discoverer of atmospheric tides), Descartes's hypothesis seemed definitively confirmed.

APPLIED SCIENCES

Land reclamation projects

Work began on draining the Great Level between the Nene River and the Norfolk plateau in England in 1630. The marsh area covered about 124,000 hectares, seven-tenths of the total area of land reclaimed in the Low Countries between 1540 and 1690. Projects for reclaiming the Great Level had been advanced at the end of the sixteenth century, and Humphrey Bradley had submitted a treatise on draining the Fens to Lord Burghley in December of 1589. Given the expanse of land involved, this can be considered the largest reclamation project of all times. Cornelis Vermuyden (c. 1590–1677), a Dutchman whom Charles I had already contracted in 1621 to drain approximately 28,000 hectares of the Hatfield reserve in Yorkshire, was named director of the project. Vermuyden outlined the principles on which he based his work in *A discourse touching the Drayning the Great Fennes*, written in 1638 and published in 1642. The plan was based on a purely gravitational system, since, after a series of surveys, Vermuyden reached the conclusion that the entire surface was above the maximum sea level, a fact that Bradley had already noticed. Vermuyden also had the advantage of being familiar with Dutch technology, which was practically unknown in England.

Town planning

In France, Jacques Lemercier (1585–1654) planned the model city of Richelieu, Indre-et-Loire, which is almost the only example of Renaissance town planning.

Scientific instruments

Further advances were made in the field of scientific instruments during this decade. In 1632, Jean Rey invented a liquid thermometer that was based on the expansion of water, rather than air, an innovation with respect to the thermometers built by Galileo and Santorio Santorio. René Descartes (1596–1650) demonstrated that in order to make light reflect or refract at a certain point, lenses or mirrors with parabolic or hyperbolic curvatures had to be used instead of spherical lenses. The first quantitative measurements of the physical and mechanical properties of metals were also made during this period, beginning with density. Marin Mersenne (1588–1648) made quantitative measurements of the toughness of metals. His *Harmonie universelle*, published in Paris in 1636, contains descriptions of his experiments on their resistance, and also gives the densities and natural frequencies of semi-spherical bells made of various types of metals and alloys.

Naval engineering

The construction of galleys reached its peak during this period. War galleys generally had three decks and reached a lenth of sixty to seventy meters. *The Soveraigne of the Seas* (English), built in 1637, and the *Couronne* (French), built in 1638, are typical examples. The first had seventy-two cannons, and the second one hundred. They were no bigger than the ships built three or four centuries earlier, while mercantile ships were much larger.

1640–1650

THOUGHT AND SCIENTIFIC LIFE

Jansenism

The *Augustinus*, by the Catholic bishop Cornelius Jansen (1510–76), was published posthumously in 1640. Directly inspired by the Augustinian theory of grace, the work was written with the intention of introducing a concept into Catholicism that was in some ways similar to Calvinism and was in complete antithesis to the Pelagian trend manifested among the Jesuits, especially in the work of their philosopher Luis de Molina (1535–1600). The *Augustinus* gave rise to Jansenism.

The debate on Cartesian thought

René Descartes sent his *Meditationes de prima philosophia,* which he had begun writing in 1620–30, to press in 1641. Mersenne circulated it among various philosophers and theologians and then brought together the objections that were raised against it. A prominent critic was Pierre Gassendi (1592–1655), who attempted not so much to demonstrate the falsity of Descartes's metaphysical theses as to confute the proofs he used. In his critique of Descartes's proofs of the existence of God, Gassendi echoed the controversial themes that were characteristic of the "libertine" culture of the century, such as an emphasis on the antimonies of all theological thought on the problem of evil and wrong.

Hobbes also raised his objections in 1641, along with Gassendi. A refugee for the entire decade because of his pro-monarchy, anti-parliamentary positions, Hobbes was living in Paris, in close contact with Mersenne and Gassendi, with whom he shared several uncertainties about Cartesian metaphysics. From 1640 to 1651, Hobbes dedicated his attention to writing his *De corpore,* and in 1642 he published *De cive,* in which the third part of his grandiose project is developed. The same material was reelaborated and expounded again in 1651 in the *Leviathan.* In 1637, taking up ideas from a work of 1640 that was never sent to press, but which had circulated widely in manuscript form, he published the *Elements of law, natural and politic.* In three treatises on optics, two in Latin and one in English, written between 1642 and 1646, Hobbes again took a decidedly anti-Cartesian stand.

Descartes's most systematic work, the *Principia philosophiae naturais,* in four books (the first of which is on philosophical subjects, the others on physics) came out in 1644. It clearly outlined the features of Cartesian mechanism: the entire extended world, that is, all the natural world, including the organic world, must be explained on the basis of laws that consider only the "real" properties of beings, i.e. their primary qualities. The French translation of the *Meditationes,* edited by the Duke of Luynes, came out three years later. In it, Descartes presented the various objections, divided into seven different parts, with his respective answers beside them. Descartes's last great work, the *Traité des passions de l'âme,* in which he attempted to present the phenomena that were linked to the relations between body and soul in a systematic fashion, was published in 1649. Almost simultaneously, Gassendi finished his "restoration of Epicureanism" (*epicurea anastasis*), which he expounded in 1647 in his *Commentarius de vita, moribus et placitis Epicuri,* and in 1649 in *Animadversiones in decimum librum Diogenis Laertii.*

Gassendi explicitly tried to provide the theoretical foundations of the new science, which is why his teaching was so popular. He believed that Catholic orthodoxy and Epicurean doctrine were reconcilable. He also worked with a number of ancient and modern problems in physics, such as the correct formulation of the principle of inertia, falling bodies, and questions concerning

astronomical observations. For Gassendi, the ideas of Galileo, of whom he declared himself a fervent follower and admirer, were the crucial test for the scientific-philosophical controversies of the time. But the characteristic features of the method of Galileo and the Galileans escaped Gassendi, if for no other reason than because of his declared hostility toward infinitesimal methods in mathematics. This hostility led him systematically to charge with "vanity" any presumed demonstration that was founded on the calculation of indivisibles.

MATHEMATICAL SCIENCES

The apogee of the method of indivisibles

While the first steps were being taken in analytical geometry, the Cavalierian method of indivisibles reached its peak and progressively declined. In 1643, Francesco Bonaventura Cavalieri reformulated the principle of indivisibles in a more synthetic and rigorous manner in his *Exercitationes geometricae sex,* and outlined a number of applications. A year later Evangelista Torricelli published his *Opera geometrica* (1644), which explained a procedure for constructing the tangents to plane curves based on the composition of two movements, and presented important quadratures (of cycloids and cochleas). His premature death prevented Torricelli from perfecting and publishing a number of discoveries that he had made, and they remained in oblivion for centuries. Besides his many findings obtained by applying the method of curved indivisibles (a generalization of Cavalieri's method), he arrived at the notion of definite integrals, and in certain special cases, used the principle of inversion that connected the problem of tangents to the problem of the areas closed by a curve, i.e. the link between derivation and integration.

After Torricelli, the Galilean school declined from a mathematical point of view. Although they enjoyed their due fame at the time, scholars like Giovanni Alfonso Borelli (1608–79), Michelangelo Ricci (1619–82), and Vincenzo Viviani (1622–1703) represented the end of a great tradition, and did not belong to the current of the new mathematics.

The method of indivisibles was cultivated by prominent figures, such as P. de Fermat, Blaise Pascal and Gilles Personne de Roberval, in France as well. Roberval also had animated conflicts with Torricelli and even tried to claim authorship of the method of indivisibles (which he had applied in ingenious quadratures) and contended with Torricelli for priority in the use of the kinematic method for determining the tangents of curves. Despite the fact that Roberval's pretensions were unfounded, it is important to note that the French school, by conceiving of indivisibles as quantities with the same number of dimensions as the continuities they formed, took a further step toward the formulation that would later be typical of integral calculus, where elements of area or volume, rather than lines or surfaces, were considered as the infinitesimal elements that made up plane figures and solids.

ASTRONOMY

Descartes's theory of vortices

René Descartes's *Principia philosophia* came out in Amsterdam in 1644. It contained, among other things, a cosmological theory that remained very popular in France throughout the seventeenth century, and which was finally opposed by Isaac Newton's gravitational theory. The work essentially expounds the ideas of Descartes's unpublished treatise, *De mundo*, although this time an ingenious, though rather specious, solution protected the author from any possible clash with the theologians: the earth, in fact, did not move freely in space, but was carried around the sun in a vortex of matter, with respect to which it could be said to be at rest. On the basis of the rationalistic principle of the impossibility of a vacuum, Descartes conceived of the universe as filled everywhere with matter. God had put this matter in motion at the beginning, and it had given rise to an immense number of vortices. The solar system was one of these vortices, and the same structure was duplicated everywhere there was matter. A system of spherical elements tended, on the one hand, to move away from a common center in its motion around it, while a very fine dust moved toward the center of the vortex and lodged there, forming a star.

Although Decartes's theory is interesting from a philosophical point of view, it did not give a reason for the elliptical shape of the orbits, which Johannes Kepler's magnetic theory was much better able to explain.

The theoretical work of Ismaël Boulliau

In the following year, 1645, Ismaël Boulliau (1605–94) published his *Astronomia philolaica*, one of the most important treatises written during the period between Kepler and Newton, and one which had important repercussions, especially in England. Boulliau replaced Kepler's second law with a theory that attempted to reconcile the ellipticity of the orbits with the uniformity of circular motion. The ellipse was seen as a section of an oblique cone, with the blind focus, i.e. the one not occupied by the sun, located on its axis, while the angular velocity measured parallel to the base of the cone was constant with respect to its axis. The theory, which seems artificial from a geometrical and physical point of view, still offered several advantages in computing planetary tables and gave values that were quite close to the true ones.

Hevelius's observatory

During this period, Johannes Hevelius set up an observatory at Danzig, which for a short time, before the observatories of Paris and Greenwich were founded, was the most important in the world. Located on the roof of Hevelius's house, it was enlarged beginning in 1644 with the addition of a platform and two turrets, one of which revolved. Hevelius built his own telescopes, which he wrote about in 1679 in his *Machina coelistis*. Convinced by Christiaan Huygens's discoveries about the usefulness of long-focus lenses, he set up instruments of gigantic proportions (up to twenty meters long), the use of which required the aid of several assistants and winches. In 1647, Hevelius published his *Selenographia*, a magnificent folio with maps of the moon, which he himself engraved. Many of the names he gave to the formations of the lunar surface (mountains, craters, etc.) are in use even today. His study of the phenomenon of libration is particularly interesting and detailed.

In 1645, Pierre Gassendi entered the College Royal in Paris as a professor of mathematics (or rather, astronomy). He expounded the Copernican system in his *Leçon inaugurale* and in his *Cours*, published in 1647, although he wisely adhered to Tycho Brahe's model.

PHYSICAL SCIENCES

The publication of Descartes' principles

Sixteen forty-four was a particularly important year in the history of physics. René Descartes' *Principia philosophiae* was published in four books; beginning with the first book, which mainly dealt with philosophical topics, it expounded the general principles of Cartesian physics (the nonexistence of the vacuum, the constancy of the quantity of motion, the infinite speed of light) and outlined Descartes' theory of vortices.

That same year, in his *Opera geometrica*, Evangelista Torricelli (1608–47) published a short treatise entitled *De motu gravium naturaliter descendentium*, which contained some of the basic developments of Galilean dynamics.

Torricelli's experiments

Again in 1644, in a famous letter to Michelangelo Ricci, Torricelli described his classic experiment with a barometer, conducted the previous year. The invention of the barometer, as Vincenzo Antinori writes, changed the course of physics just as the telescope affected astronomy and Volta's battery affected electrology. It goes without saying that Torricelli's vacuum was interpreted as an experimental datum that conflicted with the first principle of Cartesian physics mentioned above.

Despite this and other contrasting evidence that it encountered in the area of experimentation, Cartesian physics began to draw the attention of scientists, mainly because it was able to provide a new global view of the universe. The most serious obstacle that it encountered was not the experimental data that did not easily fit into the system, but the blind mistrust it engendered in the official culture, especially in the ecclesiastical hierarchy. This is why it did not penetrate into Catholic France, whereas it quickly obtained widespread acceptance in Holland.

In fact, the first work written explicitly to spread and defend the new physics was composed by a Dutchman, Henri Duroy, or Regius (1598–1679). His *Fundamenta Physices*, which was published in 1647, expounded Descartes's ideas not only faithfully, but with such a dogmatic certainty that it provoked the disapproval of the French philosopher himself.

The problem of the vacuum

Discussions of the vacuum multiplied. Blaise Pascal (1623–62) attacked the problem in a purely experimental way, in the spirit of Galilean rather than Cartesian physics. In 1647, he published an important work on the subject entitled *Experiences nouvelles touchant le vide*. Not only did the author succeed in confirming Torricelli's experiment, he also made a new and very important discovery that is still known today as Pascal's principle.

Pascal dedicated two other interesting treatises to the study of fluids. *Traité de la grande expérience de l'equilibre des liqueurs et de la pesanteur de la masse de l'air* was written between 1651 and

1654, but only published in 1663, one year after the author's death. Considering the basic contributions that Torricelli and Pascal made to liquid mechanics, we can actually say that this discipline was born and established as a science during these very years.

As for advances in optics, we shall only mention that Jan Marcus Marci von Kronland discovered the diffraction of light in 1648. This important phenomenon did not, however, come to the attention of physicists until it was rediscovered by Isaac Newton, who conducted experiments on it.

CHEMICAL SCIENCES

The French atomists

Two currents—one philosophical-speculative and the other experimental—can be clearly distinguished in this period. The first was animated by René Descartes (1596–1650), who maintained the existence of elementary, though not necessarily indivisible, particles, and the nonexistence of the vacuum in his *Principia philosophiae* (1644). The essential feature of matter was not its weight, but its volume, so that fire was also matter. Although chemistry does not owe much to Descartes, the fact that he emphasized the importance of volume, at the same time that Evangelista Torricelli (1608–47) was measuring atmospheric pressure with his barometer (1643), seems to have paved the way for Robert Boyle's research (see CHEMICAL SCIENCES 1660–1670).

Illustrious French atomists who also belonged to the first of the two currents just mentioned were Johann Chrysostom Magnenus (c. 1590–1679) and Pierre Gassendi (1592–1655). According to Magnenus, who was a follower of Democritos and the author of a *Democritus riviviscens sive de atomis* (1646), there were three elements (water, earth, and fire), made up of atoms that were imperceptible to our senses, even though they did have dimensions. The atoms of all elements were equal to each other and joined in various ways to form the *mixta corpuscula*. The vacuum did not exist. Gassendi essentially revived the ideas of Epicurus (341–270 B.C.). He expounded his theories in his *De vita et moribus Epicuri* (1647). Only atoms and the void existed; atoms could have a number of forms and were in constant motion. They joined together to form first *corpuscula*, then *concretunculae* and *moleculae*, which then united, generating the various bodies. There were also cooling atoms that impregnated bodies to cool them, and heating atoms, which were the same as lighting atoms. The transmutation of metals through a rearrangement of their atoms was possible, just as an elixir for long life was possible.

Experimentation and quantitative techniques

The experimental current included Johannes Baptista van Helmont (1579–1644), a more rigorous follower of Paracelsus. His major work *Ortus medicinae, id est initia physicae inaudita* (1648) touches on various topics in medicine, chemistry, physics, philosophy, etc. Although he was not exempt from some of the misconceptions of the times (the philosophers' stone, the universal remedy, the *alkahest*, which he interpreted as a universal solvent), van Helmont has two important points in his favor: he often resorted to quantitative techniques (using scales, ther-

mometers, and gravity bottles or pycnometers), which led him to formulate a clear statement of the law of the conservation of matter; and he conducted accurate studies on gasses (in fact, he even coined the word gas, from *chaos*, distinguishing several of them, even though not all of them corresponded to individual chemical entities distinct from one another. He identified *gas ventosum* (air), *gas carbonum* (from the combustion of coal), *gas sylvestre* (in certain mineral waters and from the reaction of acids with limestone), *gas musti* (from alcoholic fermentation), *gas sulphureum, gas pingue* (intestinal gas), and others. Besides various chemical preparations, he described ingenious experiments, although he sometimes interpreted them incorrectly. Thus a tree which increased considerably in weight, growing in watered sand, the weight of which decreased very little seemed to demonstrate to him that water was transformed into wood. In the same way, the alkaline fusion of sand, which forms fused glass that appears to be water, from which he reobtained silica by neutralizing the alkali, induced him to maintain that water could become earth and vice versa.

LIFE SCIENCES

The beginnings of histology

In 1645, Marco Aurelio Severino (1580–1656), a surgeon in Naples who was very courageous in performing operations, published *De recondita abscessuum natura libri octo*, the first modern manual of surgical biology. Two years later, he printed an important work entitled *Zootomia democritea, idest anatome generalis totius animantium opificii*, where the term *anatome* was intended to mean *resolutio in indivisibilia* (ἀνὰ ἄτογιε) rather than *dissectio* (ἀνὰ τογή). The work put a whole new perspective on anatomical investigations, and was a real manifesto of histology in which efforts were made to penetrate deeply into the minute structures of organs. Severino proposed to establish a certain unity of conformation among animals, beyond their specific differences.

In the field of comparative morphology, the study conducted in 1640 by Nicholas Tulp (1593–1674) on the orangutan, in which he emphasizes the profound similarities between this monkey and human beings, deserves mention.

Macroscopic anatomical discoveries

Anatomical research as inspired by Vesalius also made considerable progress during this period. Johann Georg Wirsung (1600–43), for example, described the path of the pancreatic duct in humans (1642), while Jean Pecquet (1622–74), a student, discovered the thoracic duct, the canal leading to the superior *vena cava* (*canalis Pecqueti*), in a dog, and studied its relation to the lacteal vessels (1647).

Clinical descriptions of several infectious and deficiency diseases

Urbanization and the beginning of the industrial revolution worsened the health situation of city dwellers, especially children. For the first time, physicians were able to recognize the clinical features of some of the endemic diseases that raged at the time. Following the classic descriptions of diphtheria and whooping cough provided by French physicians, an English district doctor, Francis Glisson (1597–1677), described rickets (1650). The diseases and remedies of the natives of tropical countries began to be known in Europe to some extent. J. Bontius (1598–1631), a Dutch physician who lived in Indonesia, gave an excellent report on the medical history of that part of the world in his *De medicina indorum* (1642). The book contains the first mention in the West of beri-beri, as well as a description of cholera.

The introduction of quinine

The quinine tree, the bark of which was used by the American Indians as a febrifuge, was introduced into Spain at the beginning of this decade. The first work on the use of quinine powder for treating malaria, written by Pedro Barba, appeared in 1642. Willem Piso (1611–78) pointed out the effectiveness of ipecacuanha against dysentery (1648).

Surgical practices

Two remarkable events in the practice of surgery deserve mention. Jean Riolan (1577–1657) described trapanation of the sternum for pericardial effusion, while Marc Aurel Severino used frozen mixtures (snow with salt) as a surgical anesthesia.

Iatrochemistry and iatrophysics

Iatrochemistry reached its apogee toward the end of the decade, with the publication of Johannes Baptista van Helmont's *Ortus medicinae* (1648). This treatise, important also from a purely chemical point of view, indicated the author's keen interest in biology, in that chemical explanations of numerous physiological phenomena are given. It also introduced gravimetry in urine analysis. Van Helmont did the first quantitative experiment in plant physiology, demonstrating that the growth of plants is directly proportional to the expenditure of substances contained in the soil and the air.

In 1648, René Descartes revised his notes on his treatise *De homine*, where he presented his mechanistic theory of life. He accepted William Harvey's theory on blood circulation, but gave his own explanation of how the heart worked. On his part, Harvey explained his theory in more detail and answered the objections that Jean Riolan the younger had advanced (1649). The Neapolitan Giovanni Alfonso Borelli (1608–79) published his *Delle cagioni delle febbri maligne della Sicilia negli anni 1647 e 1648*, which contained the first description of an iatromechanical system. Borelli applied the Galilean method to the fields of medicine and physiology. His major work (*De motu animalium*) did not appear until over a year after his death, but Borelli's ideas influenced medical thought from the middle of the seventeenth century on.

EARTH SCIENCES

Meteorological instruments

The enthusiasm felt in Ferdinand II's Florence for the new experimental science led in the course of a few years to the creation of the basic instruments of meteorology. Benedetto Castelli invented the rain gauge in 1639. The perfection of the alcohol thermometer with a closed stem and bulb in 1641 on the initiative of the Grand Duke himself crowned several decades of experimentation with thermoscopes, beginning with Galileo's and Santorio Santorio's models built in 1579 and 1611 respectively. In 1646, the instrument evolved to its final form, giving rise to a much more well defined thermometer. For at least a century, numerous copies almost all with a fifty-degree scale, were built, and spread throughout Europe, long retaining the name of Florentine thermometers. Their excellent quality is attested to by tests that have been run on the many examples that still exist.

In 1644, five years after Gaspare Berti conducted the first experiment with water, the mercury barometer had also evolved. Torricelli clearly pointed out its principal utilization in measuring atmospheric pressure. The Frenchman Marin Mersenne (1588–1648) went to Italy that year to learn how to use the new instrument. Between 1646 and 1647, all the major French scientists became interested in the barometer. Among them were Descartes and Pascal, whose brother-in-law Floriu Périer—possibly on a suggestion made by Mersenne—carried out his famous measurement of the barometric level on the Puy-de-Dôme. Beginning in 1649, barometric readings were taken daily for one to two years at Clermont, and in Stockholm where Descartes was.

British naval power and oceanography

Meanwhile, a truly modern oceanographic consciousness was being formed through an initial collaboration between Gresham College and the navy and navigation offices of Greenwich. The original impetus for this collaboration probably came from essentially practical considerations, since all sorts of environmental information was needed for navigation at a moment when England was becoming the greatest naval power in the world. This collaboration, which was later extended to other institutions and scholars and finally coordinated by the Royal Society of London, brought extremely important geographical and physical contributions to the study of the oceans. One of the problems that most interested the British navy was the hydrological situation of Gibraltar, the port of entry for English penetration into the Mediterranean, and the first studies of modern English oceanography were focused there. In 1646, John Greaves (1602–52) noted the continuous flow of Mediterranean waters through the Strait of Gibraltar. Since he could not see any outflow toward the Atlantic, he guessed that the water balance must be maintained by water flowing out via underground paths, through porous bottom rock.

APPLIED SCIENCES

Atmospheric pressure

It was in this decade that the experiment was conceived and conducted that led to the demonstration that air is a substance, though a rarefied one, and that its evacuation creates a vacuum. This in turn led to the first important application of the principle in the invention of the barometer. By 1641, fountain builders in Florence had noticed that a suction pump could not raise water higher than thirty-two feet (10.33 m), and that above this height the tube remained empty. Ancient wisdom had coined the motto that "nature has a horror of the vacuum," but when Galileo was consulted concerning the phenomenon observed by the fountain builders, he only replied that the horror of the vacuum ceased above thirty-two feet, an answer that obviously could not have satisfied him, but that his subsequent death left without a sequel. Evangelista Torricelli (1608–47) and Vincenzo Viviani (1622–1703) collaborated to discover the cause of the phenomenon, which was atmospheric pressure, counterbalanced by the height of the water as observed. In 1643, using a tube that was closed at one end and filled with mercury, Torricelli verified that the height of the column of mercury (0.76 m) corresponded to the ratio between the densities of water and mercury. The discovery of the existence of atmospheric pressure aroused considerable general curiosity, as well as the interest of scientists. In 1647, Floriu Périer (1605–1702), Blaise Pascal's brother-in-law, performed his barometric experiment on the Puy-de-Dôme, which Pascal reported in his *Récit de la grande expérience de l'equilibre des liqueurs.*

The way was open for using atmospheric pressure in machines. Its first application was in improving suction and press pumps for water supplies.

The Briare canal

In France, the so-called Briare canal joining the Loire with the Seine, was finally finished in 1642. It had been planned in 1604 by Huguss Cosnier. Altogether it had required building dozens of locks, due to the considerable difference in level. The work, which was well on its way in 1611, was suspended that year because of economic and political difficulties and was only resumed in 1638, nine years after Cosnier's death. The enterprise immediately proved very profitable, to the extent that in the second half of the century it yielded an annual profit of twelve percent on the capital invested.

Scientific instruments

The progress made during the preceding century in the field of scientific instruments continued at a rapid pace. In 1640, William Gascoigne (1612–44) built a micrometer that could be adapted to the eye-piece of a telescope. In 1645, A.M. Schyrlaeus de Rheita (1597–1660) invented the movable eye-piece for terrestrial telescopes. Around 1650, systems of multiple lenses began to be used in microscopes.

An important advance in the development of optical instruments was made in 1647, when Johannes Hevelius (1611–87) described a way of simultaneously grinding a number of small lenses. He used a sort of vertical lathe with a disc of wood, on which the small pieces of unworked glass were mounted, attached to the chuck.

The American chemical industry

The American chemical industry can be said to have started in 1635, when John Winthrop, of Boston, began to produce saltpeter for manufacturing gunpowder.

Treatise-writing

In 1641, Jan Adriaanszoon Leeghwater (1575–1650), who had directed the project of draining the Beemster, published his *Haarlemmermeerboek*, in which he described a plan for draining the Haarlem *meer*.

In Paris, in 1640, Alvaro Barba (c. 1569–1640) published *El arte de los metales*, a work that essentially talked about how fusion was practiced in the gold and silver mines of the New World. In 1641, Johann Schroeder gave the first accurate description of how to prepare semi-metallic arsenic, in his *Pharmacopoea medicochymica*. In 1644, in Paris, Marin Mersenne (1588–1648) published his *Cogitata physico-mathematica*, in which he listed the densities of various metallic materials.

In Amsterdam, in 1646–50, Johann Rudolph Glauber (c. 1604–70) published his *Furni novi philosophici*, one of the most important collections of chemical and technological information published in the seventeenth century. The work contained the discovery that lead was resistant to sulphuric acid, a finding that was used in the manufacture of sulphuric acid in 1746.

1650–1660

THOUGHT AND SCIENTIFIC LIFE

The gradual laicization of scientific research

As the personal lives of many scholars illustrate, scientific research became increasingly more laicized in nature, despite the fact that scholastic education remained under the rigid control of the ecclesiastical authority. Hobbes, for example, became aware that scientific research was, for the most part, independent of religious authority, and during this decade elaborated his *De motu, loco et tempore*, a large manuscript (that remained unpublished) in which he defended Galileo's *Dialogo*. In 1651, he wrote his *Leviathan*, in which he expounded his materialistic doctrine on nature and human society.

The institutional situation of research itself was also changing. The idea that the state could profit from scientific discoveries spread, and the possibility of politically and economically exploiting scientific research in all areas—from colonial expansion to the growth of the various professions—began to be taken into consideration. Problems of social organization and public administration also became objects of scientific analysis, increasing the interest of those responsible for government, production, and commerce in the exact and natural sciences, with a view toward achieving an objective understanding of the world. The scientists mainly depended on public financing for a living. The natural environment for innovative, anti-conformist thinkers had become the court rather than the university, as the traditional scholastic institutions closed themselves in a sterile Aristotelianism. Even in those faculties where theology and philosophy did not play a decisive role, such as medicine, the universities of France, Austria, Spain, and the Nordic countries were too often tied to a formalism that had no possibilities for future development; and more attention was given to safeguarding ancient privileges than to promoting true scientific progress. The conservative academic front was not homogeneous, however; and in several Italian and English universities, such as

Padua and Bologna, and Oxford and Cambridge, the experimental method rapidly took hold.

Ties among the various scientific communities of Christendom were strengthened by journeys and the exchange of letters among the *savants* of Western Europe, despite the economic, political, and religious crisis of which the Thirty Years War was the most obvious sign. Hopes declined, however, for a possible reciprocal enrichment through contacts be*tween the West and the Far East, especially after the Jesuits' attempt to insert themselves actively into the social and cultural fabric of the Orient had failed.

The needs of rationalism and mysticism

The conviction that nature was intelligible and the confidence in the power of reason was fully shared by supporters of both Empiricism and Cartesian rationalism. There was also a productive interaction taking place between science and philosophy. Once the way, i.e. the method, was discovered, be it experimentation or critical reasoning, one needed only to follow it scrupulously, without losing oneself in purely speculative reflection. This rationalistic faith often led to a deism that existed outside the constituted religions, or to an atheism that some "libertine" scientists were barely able to conceal.

However, this victory of rationalism, far from being definitive, provoked reactions of skepticism and mysticism. Some scientists distanced themselves from the rational method because it did not seem to possess the magical key for deciphering the secrets of human destiny. The conversion of Blaise Pascal (1623–62) to Jansenism and his retreat to Port Royal (1654) was an indicative episode. In contrast to the Jesuits, who wanted to impose a formalistic, rational, and efficient education within the context of a Catholic society, and to the Protestant pedagogues and free thinkers who entrusted themselves unconditionally to reason, the Jansenists set out in search of a mystical inner truth. On the other hand, at the same time that the ideas of Galileo, Descartes, and Hobbes were spreading, many scientists, especially in the Nordic countries, were attracted by the teachings of alchemists and mystics, followers of Jakob Böhme, or members of the Rosicrucian sect.

Russia remained relatively untouched by these Western European trends. However, a reform of the Church did take place (with the publication of the *Nomocanon*, 1650–53). The Patriarch Nikon led a campaign against superstition and favored more liberal policies. In 1654, the first Russian school of medicine was founded; it was also the first scientific-technical school ever established in that country.

The consolidation of the mechanistic concept

De corpore and *De homine*, the two parts that were still missing in the system outlined by Hobbes, came out in 1655 and 1658. The first section of *De corpore* also contained a long exposition of Hobbes's ideas on logic. The general theme of Hobbes's philosophy was that the surrounding world could be explained as an exclusively material system: everything that exists is corporeal, and as a result God must also be thought of as corporeal. With respect to the problem of knowledge in particular, he rigorously supported the conventionality of both logic and scientific law (as well as ethico-political law). Gassendi's *Syntagma philosophicum*, which was intended as a systematic exposition of the Epicurean concept of the universe, came out posthumously

in 1659. Both Hobbes and Gassendi, although in ways that differed from Descartes, defended the mechanistic idea, according to which the true reality of the natural world was of a geometrical-mechanical nature. As a result, any real explanation of phenomena could only be based on the primary qualities of objects. Mechanism seemed to become an inevitable presupposition for all serious scientific research.

The controversy between Jansenists and Jesuits

Meanwhile, great events were taking place in the area of Catholic thought. In 1653, Pope Innocent X, incited by the Jesuits, condemned five propositions of Cornelius Jansen's *Augustinus*. The followers of the late bishop replied that those propositions were not included in the incriminated text. The controversy between Jansenists and Jesuits became increasingly heated, and in 1657 Blaise Pascal intervened with his famous *Provinciales*, which contained scathing attacks against the Jesuits. The Jesuits in turn had Pascal condemned and unleashed the repression of the French government against the author and his supporters.

Despite his involvement in this struggle, Pascal continued to pursue his philosophical-scientific interests. Two of his most interesting works on methodology, *De l'esprit géométrique* and *De l'art de persuader,* along with a number of ingenious studies on the infinitesimal properties of the cycloid (the so-called *roulette*), were written in 1658.

MATHEMATICAL SCIENCES

The properties of the cycloid

The last great work that systematically made use of the method of indivisibles was John Wallis's *Arithmetica infinitorum* (1655), which contains a remarkable calculation of the value of π as the product of an infinite number of factors. Wallis also applied his method to the study of curves such as the cycloid, the cissoid, and the conchoid. Blaise Pascal made detailed studies of the cycloid; after having discovered a number of its properties, he made it the object of a mathematical challenge in his *Lettres de A. Dettonville contenant quelques-unes de ses inventions de géométrie trie* (1659), in which he expounded his findings and illustrated the new way in which he obtained them using the method of indivisibles.

The reflowering of formal logic

The revival of studies in formal logic during this century was marked by the publication of the *Logica hamburgensis* (1638) by Joachim Junge (1587–1657), a very important work that even Leibniz admired. The German Cartesian Johann Clauberg (1622–55) sent his *Logica vetus et nova* to print in 1656. In it, for the first time, logical themes are embellished and weighted down with a series of psychological and gnoseological considerations.

John Wallis also wrote a valuable *Institutio logicae*, which appeared the same year (1658) that Pascal most probably composed his important *De l'esprit géométrique* and *De l'art de persuader,* In a certain sense, Wallis' work can be considered a prelude to *La logique ou l'art de penser*, which became famous under the name *Logique de Port-Royal*, published in 1667 by Antoine Arnauld (1612–94) and Pierre Nicole (1625–95). Although considerable space was given to formal questions, the work also contained approaches inspired by the Cartesian theory of ideas, along with a number of methodological considerations. The *Logica* (1662) by Arnold Geulincx (1624–69) was also written in the Cartesian spirit. The interest in formal logic continued even beyond these years of particular enthusiasm.

Thomas Hobbes' logic

In his *Leviathan* (1651), Thomas Hobbes expounded his nominalist and conventionalist gnoseology: language consists of names of things and names of names, joined by predicates. Although the names of names, i.e. the universals, exist only in the mind, they play a central role in the formation of judgments. For Hobbes, truth was analytical and the product of correct reasoning about the names. In his *De corpore* (1655), he also very rigorously outlined his logic, which was based on the resolutive-constituent method. The method consisted in resolving a given set of ideas into its constituent parts, or prime principles, and then recomposing it analytically. Hobbes applied the method above all in his theory of the social body.

The calculation of probability

The *Tractatus de ratiociniis in ludo aleae*, by Christiaan Huygens (1629–95) appeared in 1657. While Galileo, Pascal, and Fermat, working with gambling games, had attempted to outline a theory on the applications of combinatorial analysis, Huygens based his on the principle of the relationship between favorable and possible cases—the basis of the so-called classical theory of probability—and deduced several theorems from it.

Convergent series

Pietro Mengoli was the first to obtain convergent series of logarithms in 1659. The theory of convergent series later became a feature of the English school, with the work of James Gregory (1638–75) and Nicolaus Mercator, while Newton's research on infinitesimals was directly connected to it. Convergent series made it possible for mathematicians to broaden the concept of functions and begin to analyze new functions that could not be directly calculated with algebraic operations but could be represented and calculated using series of powers.

Giovanni Domenico Cassini in Bologna

In 1650, the head chair of astronomy at the University of Bologna, which had been left vacant in 1647 at the death of Francesco Bonaventura Cavalieri, was assigned to Giovanni Domenico Cassini (1625–1712). Cassini was chosen over Giovanni Alfonso Borelli (1608–79), who was suspected of excessive sympathy for the moderns, i.e. Galileo and Copernicus.

Variants in Kepler's second law

A period of intense critical activity in theoretical research took place between 1650 and 1670, while the physical groundwork of Newtonian cosmology was being developed. The work of Ismaël Boulliau, who in the 1640s had already proposed a revision of Keplerian dynamics, led to a flourishing of studies in that area, especially in England. In London in 1651, Vincent Wing published his *Harmonicon coeleste*, which basically followed the French astronomer's modifications. Several years later, in his *In Ismaelis Bullialdi astronomiae philolaicae fundamenta inquisitio brevis* (Oxford, 1653) and *Astronomia geometrica*, published in London in 1656, Seth Ward, a professor of astronomy at Oxford, proposed a theory that was intended to oppose Boulliau's but retained its refutation of the nonuniformity of motion of revolution. No longer resorting to oblique cones, the motion of a planet was uniform with respect to the blind focus of the ellipse (which acted as the equant), and as a result the line that connected this focus to the planet had a constant angular velocity. Boulliau replied with his *Astronomiae philolaicae fundamenta clarius explicata* (Paris, 1657), in which he proposed an alternative theory, according to which the equant was not fixed, but oscillated about a focus.

G.B. Riccioli's Almagestum novum

One of the foremost Italian exponents of the opposition to the Copernican system was Giovanni Battista Riccioli (1598–1671), who published his major work, the *Almagestum novum astronomiam veterem novamque complectens* in Bologna in 1651. The work contained a new planetary theory, similar to Tycho Brahe's but with several variants and a description of the lunar surface. Riccioli change some of Hevelius's nomenclature, and his version became definitive. Craters, mountains, and valleys were named after astronomers and philosophers, while the dark regions, called seas, were given imaginary names, such as *Mare imbrium*, *Oceanus procellarum*, etc.

Pierre Gassendi: the lives of the great astronomers

In 1653, Giovanni Domenico Cassini supervised a reconstruction of the sundial, originally built by Egnatio Danti in 1575, in the Church of San Petronio in Bologna. Like Riccioli, Cassini leaned toward a planetary system for the most part similar to Tycho Brahe's. He proposed his own model in 1659.

Meanwhile, in France, Pierre Gassendi published a work on the lives of modern astronomers, *Tychonis Brahei.... N. Copernici, G. Peurbachii et J. Regiomontani.... vitae* (Paris, 1654). He mentions Galileo's condemnation in the chapter on the life of Nicolaus Copernicus but does not go into any detail about it. Gassendi also accepted Galileo's erroneous theory that the tides were proof that the earth moved.

Christiaan Huygens and observational astronomy

The Dutchman Christiaan Huygens, together with his brother Constantijn, manufactured lenses that were more powerful than any ever before made, as well as telescopes, which he used to make several important observations. In March 1655, he discovered the largest of Saturn's satellites, Titan (the sixth out from the planet), and established its period of rotation as sixteen days. The next year, he published his findings in the pamphlet *De Saturni Luna observatio nova*, to which he added an anagram, according to a custom of the period (which Galileo also observed), that concealed the announcement of a second discovery, a correct description of the nature of Saturn's so-called arms. These were nothing more than a thin, flat ring that did not touch the planet and that was inclined on the ecliptic. The complete theory of Saturn, together with a calculation of the period of Mars (of which the author also drew maps) and the discovery of the nebula of Orion, was outlined in Huygens's *Systema Saturnium* (1659). He arrived at the theory of the ring on the basis of René Descartes's theory of vortices (see ASTRONOMY 1640–1650).

On June 16, 1657, Huygens presented his first pendulum clock, which he described the next year in his *Horologium* (not to be confused with his *Horologium oscillatorium*, written in 1673), to the states-general. Besides its fundamental importance in rigorously determing time in astronomical observations, the pendulum clock also constituted a radical advance from a theoretical point of view, in that it was now possible to establish a valid universal scale of astronomical time based exclusively on mechanics.

Huygens must also be credited with another discovery of considerable importance to the development of observational astronomy. He found a focal plane in the Keplerian telescope where it was possible to superimpose an object on the image without producing distortions. Even though William Gascoigne in England had already exploited this characteristic to apply a cross-wire micrometer to the telescope, his discovery remained unknown until 1666.

PHYSICAL SCIENCES

Optics: laws and technical limits

In the field of optics, the discussion between Pierre de Fermat and the Cartesians on the law of refraction (which the latter denied could be derived from the so-called Fermat principle) continued with the same intensity as before. Research in technical optics, aimed at improving the techniques of cutting and polishing glass for lenses, also continued. However, the difficulties encountered here were so great that the idea of turning to another type of telescope, using a mirror (i.e. the reflecting telescope) began to be considered. Niccolò Zucchi (1586–1670) described this type of telescope quite accurately in his *Optica philosophica* (1652–56).

The vacuum and pneumatic machines

The most important progress made during this period in the area of building scientific instruments was connected with studies on producing a vacuum. In 1652–55, Otto von Guericke (1602–86) invented his famous pneumatic machine, and began to conduct the first experiments with it, soon creating an enormous uproar. He did not provide a detailed description of the experiments until almost fifteen years later; however, they were reported as early as 1657 by Gaspar Schott (1608–66) in his *Mechanica hydraulico-pneumatica*. Robert Boyle (1627–91) took an interest in the subject, and was able to improve von Guericke's pneumatic machine and conduct new experiments with it. He wrote up his findings in 1660, in a work entitled *New experiments physico-mechanicall touching the spring of the air and its effects*.

Huygens's research in mechanics

Christiaan Huygens (1629–95) began to do research in physics during this decade and soon became one of the foremost figures in the field.

He began by considering problems in mechanics. In 1656, he wrote an interesting work on impact between elastic bodies (*De motu corporum ex percussione*) and in 1659 a work on centrifugal force (*De vi centrifuga*). Both were only published posthumously, but Huygens further developed the ideas put forward in them in several important scientific memoirs that were printed after 1660. In 1657, Huygens built the first pendulum clock in order to solve the problem of determining longitude on the open sea. He described it in a short treatise (*Horologium*), published in 1658. That Huygens's clocks did not solve the problem of longitude was certainly no fault of his.

CHEMICAL SCIENCES

The work of Johann Rudolph Glauber

Experimental chemistry definitely broke away from medicine with the work of Johann Rudolph Glauber (c. 1604–70). Glauber was not, in fact, a physician, but mainly practiced chemistry, and not only pharmaceutical chemistry. His work contains traces of the metallurgy tradition and, in particular, of Georgius Agricola, Vannocchio Biringuccio, and Lazarus Ercker. Most of his writings (from 1646 on) are collected in his *Opera chymica, Bücher und Schrifften* (1648–59), the most important book of which is *Furni novi philosophici oder Beschreibüng einer neuerfundenen Destillirkunst* (1646–50). The work also includes his *Pharmacopoea spagyrica* (1654). As we can see, Glauber wrote his texts in both Latin and German, a sign of his desire to be both practical and a scientist at the same time.

Among the compounds he described is *sal mirabile* (sodium sulphate, decahydrate, or Glauber salt), which he discovered in a mineral water in Austria, but also prepared from sea salt and sulphuric acid. Glauber determined the loss of weight when the salt was dried out and suggested using anhydrous salt as a dehydrating agent. Glauber extolled *sal mirabile* for its extraordinary therapeutic powers, and sold it in many of his pharmaceutical preparations. By melting it with coal, he obtained a compound (the sulphide) that freed sulphur when it was mixed with acids.

Glauber also verified that salts were formed by an acid and a base, and prepared a number of them. He made acute observations on the shapes of crystals and on various chemical reactions (the preparation of manganates and permanganates; the formation of ammonia compounds of silver and copper salts). He described various mining and metallurgy procedures (see APPLIED SCIENCES 1650–1660), and worked on preparing colored glasses, some of which he made for analytical purposes. He suggested various analytical assays, both dry and wet. He worked on distilling wood and other organic materials, and obtained three liquid fractions from the distillation of coal. He also obtained crude extracts of the alkaloids of opium and *nux vomica* and crystallized grape sugar and honey. He synthesized, in their crude states, acetone and acrolein. He noticed that when cotton was boiled with alkalis it became shiny (mercerized cotton), a process that was perfected by John Mercer (1791–1866) in 1844.

LIFE SCIENCES

The use of the microscope and epigenetic embryology

The use of the microscope in the field of the life sciences had its first triumph during this period. In an embryological monograph written in 1651, William Harvey outlined the results of his research on the development of eggs, partly done with the aid of a microscope. Harvey preferred Aristotle's epigenesis to the preformation theory and postulated that all living creatures came from eggs. He thus proposed to search for the egg in mammals. Pierre Borel (1620–89) observed the scabies mite, as well as the irritating mechanism in nettles, under the microscope. He also provided important information on the first steps of microscopy, and recommended the use of the microscope in human anatomy. In 1656, Giovanni Alfonso Borelli (1608–79) observed several corpuscles in blood and chyle, later named white corpuscles or leucocytes.

That same year, Borelli and Marcello Malpighi (1628–94) both became professors at Pisa, Borelli of mathematics and Malpighi of theoretical medicine. Their encounter was particularly important, in that Malpighi accepted the iatromechanical ideas of the mathematician, physiologist, astronomer, and physicist Borelli, one of the most representative figures of post-Galilean science, and as a result directed his research in microscopic anatomy along particularly productive paths. Borelli transformed his house in Pisa into a scientific laboratory, where Marcello Malpighi and Aubry of Lorraine, among others, worked under Borelli's direction. In 1657, Malpighi, who was definitely one of the most ingenious innovators in the field of medicine, discovered the spiral fibers of the heart.

The discovery of red corpuscles

During these same years, Jan Swammerdam (1637–80) discovered the red corpuscles of the blood, although he did not make his observations public. Use of the microscope also extended into the domain of pathology. Athanasius Kircher (1602–80) wrote a number of extensive volumes on the various branches of the natural sciences. These were always a sort of *summa* of the topic being considered, in which Kircher attempted to bring together all the knowledge of his time. He attributed the plague to the presence of 'little animals' that lived in the blood and discovered some of the bacteria in milk (1658).

The development of normal and pathological macroscopic anatomy

At the same time this research in microscopy was being done, considerable progress was taking place in normal and pathological macroscopic anatomy as well. François de la Boë (1614–72) described the tubercule as the typical anatomopathological formation of consumption, while Nathaniel Highmore (1613–85) described the maxillary sinuses that today bear his name, though in reality they were discovered by Giulio Casseri. In 1659, the Swede Olaf Rudbeck (1630–1702), a student at Uppsala, proved the existence of lymphatic ducts, that is, a vascular system of a type that was previously unknown.

In 1660, Thomas Bartholin (1616–80), a professor of anatomy at Copenhagen, confirmed Pecquet's discovery and found the thoracic duct in humans (see LIFE SCIENCES 1640–1650).

In 1653, Bartholin and Rudbeck published their theories on the lymphatic ducts. Although Bartholin's report came out in May, slightly anticipating Rudbeck's, the latter was given credit for the discovery.

Anatomical research on glands received a decisive boost following the publication of Francis Glisson's monograph on the liver (1654), and the *Adenographia* (1656), written by his colleague Thomas Wharton (1614–73). Glisson recognized the glandular nature of the liver and stated that bile was formed in this organ and not in the gall bladder. Wharton discovered the secretion duct of the submaxillary gland.

In 1658, Johann Jacob Wepfer (1620–95) described the arterial hexagon at the base of the brain, and demonstrated that apoplexy was often caused by cerebral hemorrhage.

EARTH SCIENCES

The origins of metalliferous deposits and the genesis of springs

The origins of ore deposits and the genesis of springs were the two major questions that engaged the attention of scholars in the earth sciences during this period. Various vitalistic theories continued to circulate on the production of precious stones and metals. The chemist Johann Rudolph Glauber (1604–70) clearly tended toward this interpretation (in 1652), and Jean-Baptiste Duhamel (1623–1706) also attributed the formation of fossils to a *succus lapidescens* (*De meteoris et fossilibus*, 1660). Johan Joachim Becher (1635–82) believed that metals were deposited by vapors formed by sea water penetrating underground and coming into contact with the central fire (1653). B. Mazotta thought that some stones were formed by thunder (1653).

Becher, among others, also believed that springs were the result of water that penetrated into the depths of the earth, turned into steam and then condensed. Others, such as Athanasius Kircher (1602–80), in his *Iter extaticum celeste* (1660), thought that the waters of the ocean rose directly from the depths of the earth, due to a variety of causes.

Tides, currents, and ocean depths

The *Geographia generalis*, by Bernhardus Varenius (c. 1622–1650), published in 1650, was an important work that reported a variety of concepts and interesting data on oceanography. Varenius attributed a primary role in the creation of ocean currents to the wind (although he also attributed the origin of tides to the wind). He recognized the Indian Ocean currents, which were created by monsoons as well as the currents that flowed along the coasts of Chile and Peru as among those caused by the wind.

Although the problem of measuring ocean depths had still not been satisfactorily solved, Varenius stated that the depths were variable and might even be greater than one German mile. Meanwhile, the English school developed a new interest in the study of the tides. Contributions were made by various scholars, including Jeremiah Horrocks (c. 1619–41), Robert Moray (1608–73), and Christopher Wren (1632–1723).

The first organic effort to understand the atmosphere

In 1655, the Grand Duke Ferdinand II, struck by the phenomenon of condensation on the outside of a cup filled with a cold drink, immediately had several examples of a rudimental hygrometer built in order to measure humidity in the air by the amount of water that condensed in a certain amount of time. On September 1, when, for the first time, no condensation was produced the Grand Duke, reacting as a true synoptic meteorolgist, noted that a rather strong wind was blowing from the Southwest. As a matter of fact, during those same years, he was already setting up what would today be called a synoptic network, that is an organization for observing the weather conditions with identical instruments in more than one place at the same time. He had thermometers (1653) and later (1657) barometers, hygrometers, and anemoscopes distributed to around a dozen stations, including Paris, Osnabruck, Innsbruck, and Warsaw. His correspondents were charged with recording the observations on special forms and sending them to Florence. Thus the famous Tuscan or Florentine for Medicean network was set up, one of the most significant events in the history of meteorology. It operated regularly until the events of 1667, which overwhelmed even the Accademia del Cimento. Readings continued to be taken only at Vallombrosa and Paris. A small part of the data collected was published, but most of it—it is unclear even how much—is conserved in libraries and archives, and as of 1977 still awaited thorough study.

APPLIED SCIENCES

The pneumatic machine

The experiments conducted by Evangelista Torricelli Vincenzo Viviani, and Blaise Pascal were utilized by Otto von Guericke (1602–86), burgomaster of of Magdeburg and a physicist, who invented the first pneumatic pump shortly after 1650. This "machine for making vacuums" was very important to physics, although it was no less important in the history of technology. In 1654, von Guericke, called to Regensburg for business concerning his office, performed an experiment that remained famous before the Prince of Auerberg. A piston placed inside a vertical cylinder that was closed at the lower end was held in the upper part of the cylinder by a cord. The cord was passed through a pulley and held, via connecting rods, by twenty people. A container, which had previously been emptied using a pneumatic pump, was connected to the bottom of the cylinder. When a tap connecting the container with the cylinder was opened, part of the air from the cylinder went into the container, and the pressure of the air outside the piston became greater than the pressure inside the cylinder, pushing the piston down. Twenty people were not able to hold it. This was the first example of a cylinder and piston motor system that operated under the effects of atmospheric pressure. That same year, Otto von Guericke conducted the famous hemisphere experiment in Magdeburg (two pairs of horses pulling in opposite directions were not able to pull apart two tightly-fitting hemispheres inside of which a vacuum had been created). Gaspar Schott (1608–66) described these experiments in 1657.

After von Guericke, Robert Boyle (1627–91) and Christiaan Huygens (1629–95) also worked on the vacuum pump. In 1658, Robert Hooke (1635–1702) built the first machine following Boyle's description in the *New experiments physico-mechanicall touching the spring of the air and its effects.*

Meanwhile, the variations in atmospheric pressure continued to be observed. In 1660, von Guericke studied fluctuations in atmospheric pressure using his great water barometer, and a sudden fall in pressure enabled him to forecast a storm. In Italy, Giovanni Alfonso Borelli (1608–79) began taking systematic barometric readings in 1657–58.

Naval engineering

The frigate, a ship that was lighter than the galley and was armed with no more than forty cannons, appeared during this era, although it did not yet have the design of those built a century later.

The pendulum clock

Christiaan Huygens, who was working on vacuum pumps during this period, was also at the center of another great invention, the pendulum clock (*horologium*) of 1658. Galileo had suggested to the states-general that a pendulum, with a recording mechanism for counting the oscillatons, be adopted as an instrument for measuring time. Huygens must have known about this idea, although he did not know what kind of escapement Galileo had in mind. As a result, the escapement of his pendulum clock still utilized the bar mechanism, which was considerably inferior. However, the fact that Huygens substituted the driving weight with a spring and balance wheel in his clock was decisive. Galileo's pendulum only had an amplitude of five to six degrees, which notably eliminated the error caused by imperfect isochronism, although he was not aware of this. When Huygens published his *Horologium*, he definitely knew that the amplitude affected isochronism, but he only discovered that a true isochronic pendulum had to oscillate on a cycloidal arc in 1659, and his findings were only published several years later.

The clavichord. The first coffee shop

The clavichord (1659), the progenitor of the harpsichord, which was in turn the precursor of the piano, was invented in this decade.

Coffee, which had been known since the ninth century in the Islamic world, slowly became popular in Europe during the first half of the seventeenth century. The first coffee shop was opened in London in 1652. Chocolate, obtained by mixing cocoa with sugar according to a recipe that was jealously guarded by the Spanish, also became well-known during this period.

Scientific-technical research

Johann Rudolph Glauber, who had already published his *Furni novi philosophici* in Amsterdam between 1646 and 1650, furnished the best description of the fusion process and of the composition of the metal used in mirrors, an alloy of copper, tin, and arsenic, in his *Opera mineralis*, another treatise written in 1651. In 1656, he published his *Prosperiatatis Germaniae pars prima*, which contains the results (also quite interesting for their practical uses) of his research on alloys. He was ahead of his time in proposing both surrogates (wine and vinegar obtained from various fruits), and war gasses (mineral acids), which he suggested spraying or shooting against the Turks.

In the field of optical research Niccolò Zucchi (1586–1670), in his *Optica philosophica* (1652–56), was the first to advance the idea of a reflecting telescope (see PHYSICAL SCIENCES 1650–1660).

1660–1670

THOUGHT AND SCIENTIFIC LIFE

The founding of the major scientific academies

The experimental method, based on the notion of the repeatable experiment collectively controlled by a group of scholars, asserted itself with the foundation of the Accademia del Cimento. Giovanni Alfonso Borelli, Vincenzo Viviani, and Francesco Redi, heirs of the Galilean approach who shared the same methodological principles, joined together in Florence around Prince Leopold, where they built and used new apparatus (thermometers, hygrometers, barometers, etc.). The results of their experiments were published in 1667 in the *Saggi* of the Academia. Although the Academy had a very short life—from 1657 to 1667—it helped keep alive in Counter-Reformation Italy an interest in scientific research solidly based on expeimentation and was a center for scholars all over Europe.

A short time later, a group of German naturalist physicians, who had been meeting in Schweinfurt since 1652 under the presidency of Johann Lorenz Bausch, founded the Academia Naturae Curiosorum, officially recognized by Leopold I and renamed Academia Leopoldina when it moved to Halle in 1677. As early as 1657, Henri Louis Harbert de Montmor organized a true scientific academy at his home in Paris, although it never received official recognition.

In 1660, a group of English scientists meeting at Gresham College promoted the creation of a society for studying nature according to Baconian precepts. The president of the association, Robert Moray, obtained the protection of King Charles II, and with a decree dated July 15, 1662, the Royal Society of London was officially formed. Thanks to the importance of some of its members, among whom were Robert Boyle and Robert Hooke, the institute soon gained worldwide fame, even though its economic and institutional situation was initially quite weak. Both Locke and Newton belonged to it, and under Newton's presidency it became one of the most important academies of the era. The first secretary was H. Oldenburg, a cultural leader and prominent scientific researcher, who corresponded widely with the most outstanding European scientists and expanded the exchange of scientific opinions. By 1665, the society had its own journal, *Philosophical Transactions*, essentially the first journal of a strictly scientific nature. Actually, the *Journal des Sçavans*, which appeared in Paris in January, 1665, preceded it by two months, but its founder, Denys de Sallo, published such severe critiques that publication was soon suspended. It began to be published again in 1666, under the direction of the abbot Jean Gallois.

In France, at the height of the *grand siècle*, i.e. under the absolutism of Louis XIV, scientific and cultural life was totally controlled by the minister Jean Baptiste Colbert. Colbert transformed the free society of Parisian *savants*, which had formed around Mersenne during his time, into an official institution, the Académie Royale des Sciences. Founded on December 22, 1666, established in the king's library, and with Jean-Baptiste du Hamel as secretary, the academy still had neither a statute nor a precise program. Its first tasks attest to a preoccupation above all with collective activity and responsibility (the findings of research

were made public by the perpetual secretary, while the author remained anonymous). A short time later, several scientific associations were created in the French provinces (for example, Pierre-Daniel Huet's Académie de Physique at Caen).

In this way, the absolutist state exhorted the application of new techniques, offered more or less substantial financial help to scientific societies, and demanded that it be informed about the inclination of the research, especially when it could be dangerous, either theoretically (undermining theological dogma, for example) or practically (for example, experiments on performing blood transfusions from dogs to people).

The situation of scientific institutions in Austria and Spain, where the spirit of the Counter-Reformation reigned, was rather unusual. The Jesuits completely controlled higher education and scientific activity. New universities were founded at L'vov (1661) and Innsbruck (1669). The Croatian University later developed out of an academy founded at Zagreb in 1669. Other new universities were inaugurated at Kiel (1665) and Lund (1666).

The triumph of Cartesianism

On a philosophical level, Descartes' thought met with increasing success in all the most important centers of European culture. The publication, shortly after 1660, of various writings that the French philosopher had left unpublished, stands as proof of this trend, as does the fact that the first work of Baruch Benedictus de Spinoza (1632–77) was dedicated to an exposition of Cartesian philosophy (*Principia philosophiae cartesianae*, 1663). The publication of *La logique ou l'art de penser*, also known as *La logique de Port-Royal*, which came out anonymously in 1667, written by the Jansenite Pierre Nicole (1625–95) and Antoine Arnauld (1612–94), is another confirmation of the popularity of Cartesian thought.

However, Cartesianism still left various problems open, such as the metaphysical problems connected with the *res cogitans* and *res extensa* dualism, the gnoseological problems concerning the criteria of rational evidence, and the specific problems around mechanism (the continuity or discontinuity of matter, the admission or denial of atoms and the vacuum, the theory of impact or reciprocal action, etc.). Many of these problems were attacked by thinkers who started from Cartesian concepts but developed increasingly autonomous and original solutions. Spinoza is a typical example. In 1665, he finished writing his *Ethica ordine geometrico demonstrata*, begun (possibly) in 1661, which contains interesting critical ideas on Descartes' dualistic doctrine. Like other authors, Spinoza also participated in the exchange of correspondence that had become common among both English and continental philosophers. The problems that Cartesianism had left open provided the topics of much of this correspondence.

MATHEMATICAL SCIENCES

The basis of infinitesimal calculus provided by Newton

It was during this period that Isaac Newton (1642–1727) began to mature his theories on infinitesimal calculus, although he did not publish them. The *Tractatus de quadratura curvarum*, in which he first developed the method of fluxions, was probably written in 1665, but it was only published in 1704 as an appendix to the *Opticks*. Besides considering a function variable with respect to a certain quantity (usually time), called the fluent, he considered the speed of its variation with respect to variations in that quantity. That speed of variation was called the fluxion of that fluent, and corresponded to what was later called the first derivative of that function with respect to the independent variable (time for example). Fluxions of fluxions, or successive derivatives of the same function, could obviously be given, indefinitely. Newton's notation, when the fluent was indicated as x, was x for the first fluxion, ẍ for the fluxion of the fluxion, etc. Newton studied how fluxions could be determined in various cases, as well as some of their basic properties.

In 1669, Isaac Barrow (1630–77), in admiration of his disciple's genius and above all of the new mathematical ideas he was developing in the area of infinitesimals (it was just then that Newton presented him with his *Analysis per aequationes numero terminorum infinitas*, still unpublished but written about three years earlier, which contained remarkable findings on convergent series and on the sums of infinite series, and a number of theorems on the calculus of fluxions), surrendered his chair at the University of Cambridge to Newton. The results of the young scientist's investigations after his appointment to the chair at Cambridge are included in the short treatise *Methodus fluxionum et serierum infinitarum* (probably written in 1671), in which he attacked the problems of finding fluxions and fluents, bringing together and arranging the studies he conducted for years on derivation and integration, and applying them to numerous problems. His investigations on the interpolation of a function also appeared for the first time in this treatise. This work, like the previous ones, remained unpublished until the controversy with Leibniz broke out. It was published posthumously in 1736.

The beginnings of Leibniz's work

Gottfried Leibniz (1646–1716) also rose to prominence during this period, though not strictly in the field of mathematics. In 1666, he graduated from the University of Altdorf with a *Dissertatio de arte combinatoria*, which already contained the germs of some of his principal ideas. In particular, it included the concept that all knowledge could be reduced to a composition of elementary ideas, which could be mastered by algorithmic means when the appropriate system of symbols governed by explicit laws could be devised. On the one hand, this idea led Leibniz to lay the foundations of mathematical logic; on the other, the great valorization of symbolism implicit in it also became the strongest point of his actual achievements in the field of mathematics.

ASTRONOMY

Newton's mechanics

The controversy over the work of Ismaël Boulliau continued during this decade as well, especially in England and Italy. In London in 1661, Thomas Streete published his *Astronomia carolina*, which followed a recent hypothesis put forth by Boulliau on the equant. The *Hypothesis astronomica nova* (London, 1664) by Nicolaus Kaufmann (Mercator), and the *Astronomia britannica* (London 1669) by Vincent Wing, were along the same lines.

Simultaneously, the theoretical premises of Newtonian celestial mechanics were being developed by Robert Hooke (1635–1703), the architect Christopher Wren (1632–1723) and Isaac Newton (1642–1727) himself.

The reflecting telescope

Wrongly believing that it was impossible to eliminate chromatic aberration in lenses, in 1668 Newton invented a new type of fairly powerful, small telescope based on the principle of reflection. Light rays from an object at an infinite distance were reflected by a parabolic mirror, creating a real image on the focal plane. The instrument proved to be one of the most valuable devices for observational astronomy prior to modern times.

Giovanni Alfonso Borelli's theoretical astronomy

Ismaël Boulliau's most important critic in Italy was the Jesuit Giovanni Alfonso Borelli, the author of *De motu animalium* and member of the Accademia del Cimento. He outlined his theory of planetary motion—which was critical of the positions of the French astronomer and returned to Kepler—in two volumes dedicated to specific topics: *Del movimento della cometa apparsa il mese di dicembre 1664*, published in Pisa in 1665 under the pseudonym Pier Maria Mutoli, and the *Theoricae mediceorum planetarum ex causis physicis deductae*, which came out in Florence the following year. Borelli's work is of great historical importance, both because it is the first and most famous evidence of Kepler's influence in Italy, and because, while it emphasized the Galilean principle of the uniformity of nature and the extension of earthly laws to celestial mechanics, it went beyond Galileo, in that it attributed centrifugal force to circular motion. The author came to write the *Theoricae* following a series of observations he made between 1665 and 1666 in the castle of San Miniato near Florence with a telescope built by the Roman optician and astronomer Giuseppe Campani.

Observational astronomy in Europe

In Danzig, in 1668, Johannes Hevelius published his *Cometographia*, in which he maintained that the orbits of comets were parabolic. This theory had already been proposed by Borelli in a letter to the Grand Duke Leopold of Tuscany in 1666. On the nature of comets, Hevelius agreed with Galileo's theory that they were condensations of planetary exhalations.

In France, Adrien Auzout (1622–91) and Jean Picard (1620–82) began to carry out observations in 1666, using telescopes with micrometers with cross-wires for the first time on the continent.

In 1667, Ismaël Boulliau established the periodicity of the variable star Mira Ceti in his *Ad astronomos monita due*, and explained the phenomenon as the rotation of a semi-luminous body. The period he calculated, 333 days, is only two days longer than the average period verified today.

Between 1667 and 1670, the architect Claude Perrault built an observatory in Paris that soon became the most important in Europe, along with the one at Greenwich (1675). On April 4, 1669, Giovanni Domenico Cassini (1625–1712) went to Paris, invited by minster Colbert to collaborate on the construction of the observatory, and two years later became its active director.

From 1664 to 1668, Cassini had made a series of observations with Campani's long-focus telescope, and in 1668 he published the *Ephemerides bononienses mediceorum siderum*, which enabled him to complete his rigorous determination of longitudes. On the basis of these, in 1675 Olaus Römer (1644–1710) was able to demonstrate that light had a finite velocity. During his Italian period, Cassini also calculated the period of rotation of Jupiter as nine hours fifty-six minutes (1664) and that of Mars as twenty-four hours and forty minutes (1666).

Meanwhile, his collaborator Geminiano Montanari (1633–87) was working in Bologna. In 1662, Montanari had finished a detailed lunar map, thirty-eight centimeters in diameter, for Cornelio Malvasia's ephimerides of 1661–66. He is particularly famous for his construction of large lenses, and his name is linked to observations of the variable star Algol (β Perseus), which he reported in his "*Sopra la sparizione de alcune stelle et altre novità celesti*," published in 1671 in the *Prose de' signori accademici Gelati*.

PHYSICAL SCIENCES

The Saggi di naturali esperienze

The Accademia del Cimento, founded in Florence in 1657, was one of the most effective driving forces of Italian scientific life in this epoch; but its brief life ended in 1667. That same year, the famous *Saggi di naturali esperienze*, by Lorenzo Magalotti (1637–1712) came out. The work was later translated into English and Latin, and had an enormous influence on all of European culture.

Impact between bodies

Giovanni Alfonso Borelli (1608–79), one of the most active members of the Academy, published a work, entitled *De vi percussionis*, in 1667. It dealt not only with the phenomenon of impact (between in-elastic bodies) but with many other problems of mechanics as well, achieving remarkable results on all the topics considered.

The problem of impact between bodies continued to arouse keen enthusiasm among physicists. The theoretical aspects were studied by Christiaan Huygens (who had already considered them in the preceding decade), and by the English mathematicians John Wallis (1616–1703) and Christopher Wren (1632–1723). The experimental aspects were studied by Edme Mariotte (1620–84), who wrote a *Traité de la percussion ou choc des corps* on the subject. It was published posthumously in 1717.

Fluid mechanics

Robert Boyle did the most important research in fluid mechanics during this period. His studies led him to determine experimentally the famous law of inverse proportionality between the volume and the pressure of a gaseous mass at a constant temperature. His *Paradossi idrostatici*, published in 1666 with the prime intention of confuting the arguments that the traditionalists used to attempt to oppose modern science's explanations of the phenomena caused by atmospheric pressure, deserves special mention.

Other areas of experimentation that attracted the attention of physicists were the phenomena of capillarity, thermic and electrical phenomena, etc. But the two basic fields on which research was focused continued to be mechanics (both solid and fluid) and optics. The most remarkable advances of this decade were made in the second of these two disciplines.

Diffraction and double refraction

A great treatise on optics, the *Physico-mathesis de lumine, coloribus et iride*, written by the Italian Francesco Maria Grimaldi (1618–63), a scholar and extremely industrious researcher, came out posthumously in Bologna in 1665. His most original contribution was the discovery of the phenomenon of diffraction. He also discussed in detail the general problem of the nature of light, but he did not take a univocal or well-defined stance on it.

The *Micrographia*, by Robert Hooke (1635–1703), which was full of experimental findings and perceptive theoretical considerations, also came out that same year. The author essentially focused his interest on microscopic observations (he described some of the properties of thin laminae).

In 1669, Erasmus Bartholin (1625–98) discovered the phenomenon of double refraction, which he reported in his *Experimenta crystalli islandici disdiaclastici, quibus mira et insolita refractio detegitur*. His explanation was naïve and unacceptable, but the discovery proved to be very important and was the source of numerous debates.

Newtonian optics

Meanwhile, optical research found a new follower in England, in Isaac Newton (1643–1727), who soon proved to be exceptionally talented in the field. Newton was introduced to this field of investigation by his teacher, Isaac Barrow (1630–77). From 1665 to 1667, when he was not yet twenty-five years old, Newton withdrew to a small estate at Woolsthorpe to escape from a terrible plague that was raging throughout England. There he was able to concentrate wholly on his studies, and he succeeded in elaborating the nucleus of all his most important physical discoveries (besides his mathematical ones), from the unequal refrangibility of light rays to universal gravitation. When he returned to Cambridge toward the end of 1667, Barrow gave his treatise, *Lectiones opticae et geometricae*, which he had composed during this period, to Newton to revise. Barrow's work came out in 1669–70, but meanwhile Newton had already presented a memoir to the Royal Society, considerably surpassing his teacher's ideas.

CHEMICAL SCIENCES

Robert Boyle and the birth of atomistic chemistry

Progress was made during this decade by Robert Boyle (1627–91), who was the first to have a modern view of chemistry, considering it an autonomous science, independent of both alchemy and medicine. Although he maintained some alchemistical ideas and did not overlook alchemy's therapeutic applications, Boyle studied chemistry as such, using rigorous experimental methods. He is considered the founder of pneumatics because of his famous research on gasses, which among other things led him to state the law on the inverse proportionality between pressure and volume.

In *The sceptical chymist* (1661), and in other minor works, Boyle supported an atomic-corpuscular theory which admitted primary (atoms) and secondary corpuscles. He clearly defined the modern concept of elements, and maintained that there must be more, and different, elements than Aristotle's. He reiterated the concept of affinity, vaguely outlined by Albert the Great and Johann Rudolph Glauber, which was later taken up by Isaac Newton (1643–1727) and, especially, by eighteenth century chemists. The theory—which obviously lacked the concept of atomic weight—was not based on those quantitative ideas that were advanced in the nineteenth century. Boyle also did studies on combustion, calcination, and respiration, phenomena in which he recognized that a component of the air participated (although he was unable to isolate it). He discovered methyl alcohol in the aqueous distillate of wood. He also made notable contributions to chemical analysis, which he even named. Among his inventions were test papers, including litmus papers. He wrote his *Experiments and considerations touching colours* in 1664 (see APPLIED SCIENCES 1660–1670).

Acids and bases: qualitative notions

The influence of Paracelsus was still very much felt in Germany. François Dubois, or de le Boë, Sylvius in Latin, (1614–72) thought that illnesses were caused by an imbalance between acid (pancreatic) and basic (bile) juices, and treated them on this basis. He also recognized the analogy between respiration and combustion. He had the first university chemistry laboratory built at Leyden, where he taught from 1658 to 1672. Laboratories at Oxford, Altdorf, and Stockholm followed.

Sylvius's theories were introduced into Italy by Otto Tachen, or Tachenius (c. 1622–after 1699). In his *Hippocrates chimicus* (Venice, 1666), Tachenius stated very clear ideas on the composition of salts: "*omnia salsa in duas dividuntur partes, in alcali nimirum et acidum.*" He classified soap as a salt, litharge as an alkali, and silica as an acid, and recognized that there were stronger and weaker acids and bases. His descriptions of several Venetian industrial activities, such as the manufacture of soap and its sublimate, are very interesting.

An attempt at theorization: Johann Joachim Becher

The first attempt at formulating a theory of chemistry was made by Johann Joachim Becher (1635–82). His most famous work, *Physica subterranea* (1663) was reprinted in 1703 and became known to Georg Ernst Stahl (1660–1734), who drew from it the inspiration for his phlogiston theory. Becher believed that there were three types of "terrae": mercurial, vitreous, and combustible (*terra pinguis*). The more *terra pinguis* a body contained, the more easily it would burn. In 1665, Niclaus Steensen, or Steno (1638–86), briefly stated the law of the constancy of crystallic angles.

LIFE SCIENCES

A new idea on the mucous membranes

A number of anatomical and physiological discoveries were made during this decade. The two sciences were still connected, and physiology was seen as "animated anatomy." The old humoral pathology was destroyed by the anatomical discoveries of Konrad Victor Schneider (1614–80), a professor at Wittenberg, who provided a new concept of the mucous membranes (1660). He demonstrated that there was no direct connection between the nasal cavities and the brain, and that rhinitis could not be described as a cerebral cold. The mucous flow did not come from the brain matter, but from the mucous membrane in the nose.

Microscopic anatomy. The cell

Niclaus Steno (1638–86) described the parotid duct in 1661. Anatomy profited above all from two research methods: post mortem vascular injections of various substances and the fine separation of tissues and their examination under the microscope. Using the former method, Frederik Ruysch (1638–1731) made several good anatomical preparations (now preserved in Leningrad) and discovered the valves of the lymphatic ducts (1665).

Malpighi's work revolutionized the knowledge of human, animal, and plant microscopic anatomy. In 1660, while he was a professor in Bologna, he discovered the alveolar structure of the lungs and the capillary vessels in frogs. Thanks to these two findings, the complete mechanism of blood circulation, already postulated by William Harvey, was understood. Malpighi's discoveries, contained in a letter written and addressed to Borelli, were published in 1661. His research on the structure of the tongue and on the organ of feeling in the skin (1665), his description of the red corpuscles (1666), his splendid studies on the silkworm (1669), and his early discoveries in plant anatomy also deserve mention.

The physicist and naturalist Robert Hooke (1635–1703) improved the microscope and observed the cell walls in a fragment of cork. He used the term cell for the first time in his *Micrographia* (1665). His work also contains the results of microscopic examinations of numerous invertebrates (descriptions of the eyes of a dragonfly, the sting of a bee, the radula of a snail, etc.), as well as observations made on several plant and animal fossils. In Palermo, in 1664, G. Odierna (1597–1660) published a monograph on the microscopic structure of the eye of a fly. These were not just observations made with the aid of a crude optical instrument; procedures were used that made the actual microscopic dissection of small organs possible. By dissecting a silkworm under the microscope, Malpighi discovered the tracheae with their stigmata and their air sacs. When he found these organs in other insects as well, he guessed that they functioned as lungs.

The same year in which Malpighi's work on the silkworm appeared (1669), Jan Swammerdam's *Historia insectorum generalis* was also published. In addition to a classification of insects, the work contains some basic information on their development.

Muscular physiology

In 1663, Swammerdam showed his friends an experiment on the contraction of a muscle immersed in water. The experiment proved that the muscle worked without changing its volume. William Croone (*De ratione motus musculorum*, 1664), Niclaus Steno (in 1664 and 1667), and Thomas Willis (in 1760) proposed several new physicochemical ideas on muscular activity. Motor fibers were postulated as the primordial elements of the animal organism. In antithesis to Descartes's ideas that the motor power of muscles did not come from an intrinsic force but from the flame of the heart, Willis (1621–75) believed that the power originated from a chemical process in the muscle tissue. Meanwhile, the muscular nature of the heart was recognized by Steno and Richard Lower (1631–91), among others.

Critiques of Cartesian anatomy and physiology

Descartes's *De homine* was published for the first time in 1662, in a Latin translation, followed by the publication of the original French text two years later. Niclaus Steno, in a famous speech on the anatomy of the brain made in Paris in 1665, criticized Cartesian anatomy and physiology and made important clarifications on the structure of the central nervous system. Willis assigned precise functions to the various regions of the brain, and proposed a new classification of the cranial nerves.

Progress from the experimental method

Lorenzo Bellini (1643–1704), a representative of iatromechanics, rediscovered the renal canaliculi that Bartolomeo Eustachio had already observed but had not understood (1662). Physiology also progressed thanks to the experimental method. Regnier de Graaf (1641–73) discovered the pancreatic ducts and obtained pure pancreatic juice (1664). Richard Lower and Robert Hooke showed that blood became red as it passed through the lungs (1667). In a letter to Steno, Francesco Redi (1626–67) outlined the results of experiments done by injecting air intravenously into dogs, foxes, and other animals. When he regularly observed fatal caisson disease (bends), Redi concluded (in contrast to the opinions of the ancients) that the vessels of warm blooded animals could not normally contain air. In 1669, Robert Boyle (1627–91) used a vacuum pump on blood, demonstrating that gasses were present, but only in solution.

The best examples of the experimental method are furnished by Redi's work. After numerous experiments on the poison of vipers, this Florentine physician demonstrated that the substance did not take effect if it was administered orally (1664). In addition, he showed that the worms in decomposing meat came from flies' eggs. He used the control group method, and confuted the theory of spontaneous generation. In his *Esperienze intorno alla generazione degli insetti* (1668), he made morphological observations on lice and other insects.

Intravenous injections and blood transfusions

Although he did not know about Christopher Wren's experiments mentioned in relation to the previous decade, as early as 1664 the German physician Johann Daniel Major (1634–93) anticipated intravenous injections as a technique in human therapy. He did experiments on dogs, published a mongraph on the subject (1667), and finally on March 21, 1668, performed an injection on a patient. In almost the same period, another German physician, Johann Sigmund Elsholtz (1623–88), did the same type of experiment. In February, 1665, Richard Lower performed a blood transfusion from one dog to another. In 1667, Jean Baptiste Denis (1643–1704), a Parisian physician, made a transfusion of animal blood into a human. He believed that it would cure madness, but his patient died.

Gynecology and obstetrics. Medical statistics

In the field of gynecology and obstetrics, François Mauriceau (1637–1709) arranged all the knowledge known at the time and published it in the form of a manual, entitled *Traité des maladies des femmes grosses et de celles qui sont nouvellement accouchés*, which for a long time remained the standard text on the art of obstetrics. In 1670, Hugh Chamberlen demonstrated the forceps, an instrument that had been invented by his ancestor, Pierre Chamberlen (1572–1626), but the shape of which had been kept a secret. The work of John Graunt (1620–74) marked the beginning of medical statistics.

EARTH SCIENCES

The nature of fossils

A fundamental work, although it did not become immediately known, was published in 1669. Niclaus Steno (1638–86) wrote *De solido intra solidum naturaliter contento dissertationis prodromus*, in which he sketched a geological picture of Tuscany, investigated the causes of orogenesis, and affirmed the real nature of fossils, distinguishing between marine and continental fossils and drawing paleographic conclusions on the ancient boundaries between the earth and the sea. Besides recognizing that some fossils belonged to extinct forms, Robert Hooke (1635–1703) suggested that they might be used to stratigraphic ends. At the end of the decade, Agostino Scilla (1629–1700) recognized the true nature of fossils and compared them with analogous modern organisms in his *La vana speculazione disingannata del senso* (1670).

Steno wrote other works in the field of paleontology that deserve mention. In 1667, he described the anatomy of the head of a shark, and explained the real nature of the so-called glossopetrae, which turned out to be the teeth of selachii. In 1666, Morin signalled the remains of fossil tortoises near Paris. John Narborough described the abundance of fossil oysters in Patagonia in 1670. Johann Daniel Major (1664) fought against the notions, then popular, that fossils were of an inorganic nature.

The great eruption of Etna

The eruption of Etna in 1669 was a geological event of immense proportions and tremendous consequences. Lava reached Catania and flowed out into the sea, joining the island of Castello Ursino to the mainland, while solid matter, sand, and ashes formed a huge double cone, called Monti Rossi. The eruption was described in chronicles and reports, and knowledge about it was extremely significant to the later development of ideas in the field of geology.

The nature and origin of ocean currents

Kircher's work of 1665, *Mundus subterraneus*, which we have already mentioned, also contained a map of various ocean currents, although their circulation was described in a fantastic manner. Reminiscences of past concepts, even Aristotelian ones, run throughout the work. The water of the oceans supposedly flowed in continuation through the center of the earth, entering at the North Pole and probably coming out at the South Pole. Hypothetical communications between other seas were explained by secondary subterranean circulations, for instance between the Baltic and the Norwegian sea, the current that supposedly caused the phenomenon of the maelstrom.

Other hypotheses on the origin of ocean currents were advanced by Georges Fournier (1667), who believed that the large amount of evaporation that took place in the equatorial zones compared to the high levels of rainfall in the colder zones had a great effect on currents. Since the level of the sea must remain constant, movements of water were generated across oceans to balance it. In 1663, Isaac Vossius (1618–89) suggested, in his *De motu marium et ventorum*, that the North Atlantic current moved clockwise.

Oceanographic problems were studied intensely from a purely physical and experimental point of view by the London school, especially by Robert Moray, Robert Boyle, and Robert Hooke. During the same period, Blaise Pascal (1623–62) stated that pressure was equally distributed in all directions.

Tides and the measurement of salinity in the ocean

Robert Moray (1608–73) directed most of his attention to studying tides. In 1661, he reported exceptionally high tides on the coasts of the western islands of Scotland. It is interesting to note that until Moray's time, only the times of high and low tides and the range of the rise were observed. The curve of the level was not followed, since the phenomenon was not recorded, but only observed. It was Moray who noticed its sinusoidal form when instruments (tide gauges) were developed that could record the evolution of the level continuously for months or years. In this sense, Moray was a great innovator, as can also be seen from the series of oceanographic initiatives promoted after 1660 by the Royal Society, thanks to the work of Moray and Henry Oldenburg (c. 1618–77). In this climate, an oceanographic campaign was promoted in the Mediterranean (1661–62), under the leadership of the Count of Sandwich. The Strait of Gilbraltar was, as usual, a major area of investigation. What is most surprising is the extraordinary modernity of the theme of the research: the measurement of salinity, pressure, currents, and tides. It was probably the first time that salinity was measured systematically, although the chemistry needed to discover its exact nature had not yet been developed. The measurements were made following the ideas of Robert Boyle, whose contribution to scientific progress during this period was enormous. Salinity was determined by comparing the weight of a volume of salt water to the weight of an equal volume of fresh water. The system of evaporating the water and weighing the residue had proved to be too inaccurate, both because of the hygroscopicity of the salts and because of the losses that took place during evaporation. But it seems that Boyle also experimented with a system of precipitation, treating sea water with a solution of silver in aqua regia (the now well-known Mohr-Knudsen reaction based on the precipitation of the chlorides with $AgNO_3$). The saltiness of the water was then measured not only on surface samples but on samples taken from deep water as well, using devices that Robert Hooke and Boyle himself worked to construct. These instruments were also designed to measure the depth of the ocean with the best possible accuracy, solving an age-old problem.

Measurements of salinity in the depths of the ocean definitively established that even deep water was salty; in fact, it was often saltier than surface water, demolishing once and for all the old Aristotelian idea that salt water was formed by evaporation caused by the heat of the sun. And these results were no longer inventions of the mind or philosophical deductions, but data gathered using the experimental method.

Boyle also recognized the analogy between sea salt and the salt that was found in terrestrial deposits, as well as that found inside organisms.

Boyle made other studies of water temperature (the first verifications that deep water was generally colder than surface water), currents, and the effect of the wind on waves and on the mixing of the waters.

John Wallis (1616–1703) and Childrey made contributions to the study of tides in 1666 and 1670, respectively. Although Wallis's work was still based on Galileo's rotational theory on the genesis of tides, it contained a large amount of observational data. Childrey pointed out new relationships, such as the different ranges of the tides in syzygy and in quadrature, as well as the perturbing effect of the wind, which causes higher levels on the coasts toward which it blows and lower levels on the other side.

Hooke, Huygens, and the earth sciences

Robert Hooke was also one of the pre-eminent scientists of the epoch in various areas of the earth sciences. His contributions to meteorology were the construction of barometers (even water barometers) and the study of atmospheric pressure (1663). Designing instruments for measuring ocean depths led him to study the speed at which heavy bodies fell in the ocean and consequently to consider problems of resistance in the medium. In his *Discours of earthquakes*, written in 1668, Hooke also considered the influence of earthquakes in the formation of underwater upcasting and buckling, of volcanoes in the formation of islands, and of earthquakes in the formation of waves at sea (tidal waves).

Problems of navigation also interested physicists and astronomers. The need for systems that would make it possible to determine a ship's position on the open sea was especially strongly felt. Christiaan Huygens (1629–95) proposed measuring longitude on the open sea with highly accurate pendulum clocks (1661).

APPLIED SCIENCES

Military and civil architecture

Military and civil architecture underwent a great boom, especially in France. The year 1667 consecrated the fame of Sébastien le Prestre de Vauban (1633–1707), who directed the siege of Lille and later the fortification of the city. Considered one of the greatest military engineers of the time, Vauban's skill lay above all in applying the principles formulated in the sixteenth century and adopting logical, definitive forms. From the defenses of Florence and Verona, which he had visited, he drew useful deductions for the most effective way of defending squares by lowering bastions, decreasing dead angles, developing low casemates etc. The novelty in his concept of defense was that it envisioned a unit that would form a continuous boundary, an idea that led him to build and rebuild a remarkable number of citadels and fortifications of cities.

In 1666, work began on the Languedoc canal to connect the Mediterranean with the Garonne, and consequently with the Atlantic Ocean. The idea was not new, but it was Pierre-Paul Riquet de Bonrepos (1604–80) who finally drew up the plans. First, a difference in level of 196 meters above the Mediterranean had to be overcome in order to reach the Garonne at a point located, at its average water-level, 60.4 meters above Naurouze. One hundred locks had to be built over a total distance of 241 meters in order to compensate for the difference in level. The project, finished in 1681, was rightly considered the greatest feat of European civil engineering prior to the nineteenth century (see APPLIED SCIENCES 1680–1690).

In England, architects and town planners were busy working on the plans to rebuild London after the famous fire that destroyed it. In 1666, various architects, including Christopher Wren (1632–1723), Robert Hooke, Peter Mills, and John Evelyn (1620–1706) submitted plans for the project.

In 1661, Samuel Morland (1625–95) obtained a royal concession for the exclusive use of his invention for lifting water to a certain height by mechanical means.

Admiral William Penn (1621–70) developed regular codes for communicating between ships. At the end of the eighteenth century, these codes were transformed into a flag system very similar to that still in use today.

Glass working

In 1662, Christopher Merrett (1614–95) published the English version of Antonio Neri's book *L'arte vetraria* (1612), and for the first time described the ''bench'' with the long flat arms where the glass-blower rotated the blowing tube in order to make a round container. This advancement with respect to a small wooden board tied to the glass-blower's thigh took place between 1575 and 1662. Merrett also described a coal kiln for glass-making, after an edict of 1615 had prohibited the use of wood.

Lens grinding

In 1665, Christiaan Huygens proposed a machine for grinding lenses that would mechanically rotate the implement or glass disc that was being worked.

Paper manufacture

The Hollander paper making machine (breaker) which mangled rags in an oval container that contained a cylinder with blades on it, was invented in 1670. The breaker produced more pulp in one day than eight stampers could in a week.

The hardening of steel

In 1664, Robert Boyle (1627–91), whom we mentioned in CHEMICAL SCIENCES as one of the founders of modern chemistry, published his *Experiments and considerations touching colours* in which, among other things, he discussed the important phenomenon of the sequence of colors in hardening, noting that the succession of colors indicated a change in the structure of the steel. The following year, Robert Hooke developed a theory on the hardening of steel on the basis of colors in his *Micrographia*: hardening and tempering depended on different amounts of vitrified substances dispensed in the steel.

Treatise-writing

In 1662, in Cologne, G. A. Böckler published his *Theatrum machinarum novum*, which described, among other things, a mill

that ground grain with a system very similar to that of modern mills.

The *Saggi di naturali esperienze* came out in 1667. In it, Lorenzo Magalotti (1637–1712) described many experiments conducted by the Accademia del Cimento, of which he was the secretary, during the ten years of its life (1657–67). Particularly important is the description of the so-called Florentine thermometer, a major step forward in the development of thermometry. The characteristic feature of the new thermometer was that it was closed, so that it was not affected by variations in atmospheric pressure. It was this feature that was responsible for its popularity in Europe.

Carriages

More carriages began to be built as roads were improved. A millennium of stagnation, which only ended around 1400, had passed between the fall of the Roman Empire and the fifteenth century. Carriages from this epoch were already evolved: a model from 1669 has glass doors and a metal spring suspension.

1670–1680

THOUGHT AND SCIENTIFIC LIFE

Philosophical debate: Spinoza and Malebranche

Blaise Pascal's *Pensées* were published posthumously in 1670. The profoundly religious climate of the Jansenists can still be felt in this work although Pascal's intent by then went far beyond the anti-Jesuit theme of his *Provinciales*. His goal was the much more demanding one of defending Christianity in the face of the needs of the ''rationalist'' century. But the very premise of his defense was an acknowledgment that faith could never be guaranteed by reason, because faith contained an irremovable factor of risk.

Scholars of the rationalist century were particularly well-trained in critiquing the Scriptures, in natural law, and in ethical-political thought. In 1670, in the Hague, Spinoza anonymously published his *Tractatus theologico-politicus*, which gave rise to a myriad of letters and polemical libels. A large part of the work was dedicated to showing how the Bible should be interpreted and explained as a historical human document, elaborated by authors who often were not the persons to whom the various books are attributed. All things considered, the Scriptures seemed to be a tool for inducing obedience to moral laws. At the same time that he took up Hobbesian political theory and criticized it, Spinoza, who was also sensitive to preserving the natural rights of individuals, elaborated a rationalist theory of the individual and the state that made a distinct separation between reason and religion. A short time later (1672), natural law was amply theorized by the German Samuel Pufendorf, in his *De iure naturali et gentium*, which also contained a defense of religious freedom and tolerance. Finally, with the *Histoire critique du Vieux Testament* (1678), the Frenchman Richard Simon launched a genre of Biblical criticism that rigorously applied the canons of philology and history to readings of the Bible.

Meanwhile, philosophical reflection on the themes of Cartesianism culminated in 1674 with the first three books of the *Recherche de la vérité* (the complete edition, in six books, came out in 1675) by Nicholas de Malebranche (1638–1715). The work proclaimed the need to resort to God in order to explain the relations between the two substances, thought and extension. The publication of Spinoza's *Ethica* upon the author's death (1677), on the other hand, introduced the theory—already anticipated in some of Spinoza's earlier works—that thought and extension were not two "substances," but two "attributes" of the only "substance": *Deus sive natura*. While Malebranche's philosophy was intended to be, and was, profoundly Christian, proposing to demonstrate that the worst difficulties of modern thought could be solved by appealing to a divine being that transcended the world, Spinoza's philosophy belonged completely outside the traditions of Christianity. This is why the theologians (Jewish, Catholic, and Protestant) were so hostile toward the doctrine of the "atheist Spinoza," and why the more moderate rationalist thinkers were hesitant with respect to it (the attitude of Gottfried Wilhelm Leibniz (1646–1716), who denied any connection between his thought and that of the *Ethica*'s author, was typical.) These very features, together with others, explain Spinoza's later influence on the Enlightenment, especially in Germany, while he even left his later mark on Hegelian philosophy.

Classicism and the spread of scientific institutions

The return to a classical taste in literature and art, in contrast to the excesses of the Baroque, also played a special role in the diversified philosophical-cultural debates of this decade. The classical models provided paradigms of order and harmony that contributed to the affirmation of a rationalistic ideal. But this reference to the classics took on a political and cultural significance in those areas, such as ethics and law, where the idea of classicicity was in antithesis to the obscurity of the ethics of positive religions and to the desegregation of the state brought about by the religious wars.

A lively picture of the cultural and scientific life of the epoch is given in the letters of Madame de Sévigné and in Molière's comedies. The image of the social conditions of physicians, scientists of the court, and intellectuals is particularly realistic. In this respect, the severe criticism in Molière's *Malade imaginaire* is an extremely vivid document of the intolerance that was felt toward the medical profession, which by then was crystallized into a rigid formalism.

Many cultural initiatives of the previous years also reached maturity in this decade. The "Cartesian" physicist Christiaan Huygens (1629–95), protected by Colbert, dominated the Académie des Sciences of Paris. The orator Jean Baptiste du Hamel was still the powerful secretary of the Academy, which was now working on systematic anatomical descriptions of animals (Claude Perrault, Jean Pecquet (1622–74), etc.), astronomical observations (Adrien Auzout, Jean Picard (1620–82)), studies in levelling and land surveying, solicited by those working on the castle of Versailles, the new sumptuous royal palace of the Roi Soleil (Jean Picard, Phillipe de La Hire), and hydrodynamics (for installing majestic fountains in the gardens of the royal villas). In 1675, Colbert asked the members of the Academy to prepare a *summa* of the technical knowledge of the times.

The Royal Society of London was very active as well. At Oldenburg's death (1677), Robert Hooke (1635–1703) became the new head, and physical and chemical research in particular were increased. In 1672, Isaac Newton (1642–1727) was elected to membership, and a few months later submitted his famous memoir entitled *A new theory about light and colours*. In 1686, Newton presented the results of his mechanical-astronomical research, originally entitled *De motu corporum*, and later *Philosophiae naturalis principia mathematica*, to the Royal Society. Newton was nominated president of the society in 1703. The famous astronomical observatory at Greenwich was built in 1675, and three years later another went into operation in Paris.

MATHEMATICAL SCIENCES

The work of Huygens. Leibniz's first work on infinitesimals

In 1673, Christiaan Huygens published his *Horologium oscillatorium*, in which, in addition to his basic considerations on mechanics, he illustrated the properties of evolutes and involutes, thus offering the first treatise ever published on this important topic. Huygens combined his interest in elementary and analytical geometry (in which other contemporaries, such as Frans van Schooten, Jr., Jan Hudde, Phillipe de La Hire, and John Wallis also distinguished themselves) with a number of studies in algebra and infinitesimal analysis as well. We should remember that it was Huygens who directed Leibniz—who in the meantime had gone to London, where he consolidated his friendships with various English scientists—toward the problems that led him to discover infinitesimal calculus. Huygens was also among the first to appreciate the value of this invention, accepting it and contributing his authority to its diffusion.

Upon his return from London, Leibniz passed a productive three year period (1673–76) studying in Paris, during which time he made a thorough study of the works of Descartes and Pascal. He also began to achieve the first notable results in differential and integral calculus, and in particular established his important notational system in this field. In fact, although the infinitesimal calculus that he developed was the same as Newton's calculus of fluxions and fluents in content, it was Leibniz's notation that gave his approach supremacy in the development of the new calculus, in that it provided a flexible and powerful symbolic tool. Leibniz used the differential coefficient dx/dy to indicate the derivative of x with respect to y, and the notation $\int(x)dx$ to indicate the integral, notations that are still used today because of their great formal efficiency.

During this same period, taking up an idea proposed by Pascal, Leibniz designed a machine (effectively built later) that would carry out the four operations and extract square roots. As a matter of fact, Pascal had built the first machine for making arithmetical calculations in 1642–44.

ASTRONOMY

The parallaxes of Mars and the sun

In the course of 1671, Saturn's ring disappeared, as Christiaan Huygens had forecast in 1659 in his *Systema Saturnium*.

In the range of the astronomical and geodetic activities promoted by the Académie des Sciences, one of which was the completion of the Observatory of Paris in 1672, particular mention should be made of the expedition (1671–72) to Denmark, directed by Jean Picard, for the purpose of more accurately determining the position of Tycho Brahe's Observatory at Uraniborg, by then in ruins, in order to better utilize the large amount of data left by the Danish astronomer. Another important expedition (1671–72), led by Jean Richer (1630–96), went to Cayenne to observe an opposition of Mars at the same time as Giovanni Domenico Cassini (1625–1712) observed the same phenomenon in Paris. The purpose of the project was to determine the parallax of Mars and indirectly the parallax of the sun, since the distance between the two points of observation was known, in order to calculate the distance between the sun and the earth, a fundamental astronomical datum. The sun's parallax was calculated at 9″.5, which corresponded to a distance of 21,600 radii of the earth. This was a better approximation not only with respect to Aristarchos's ancient figure, and to Johannes Kepler's conjecture, but even to the calculation set at 10″ by Edmund Halley (1656–1742) in 1672, using the conjuctions of Venus. While Picard returned to Denmark in 1672 together with Olaus Römer (1644–1710), a member of the Académie whose scientific research soon produced conspicuous results, Richer, in the course of the work at Cayenne, discovered that a pendulum clock adjusted to strike the seconds in Paris ran about two and a half minutes slow per day in the new location of observation, which was 5° latitude north of the equator, so that the stem had to be shortened about three millimeters. Richer explained the phenomenon correctly, supposing that the earth did not have the shape of a sphere, but of an ellipsoid of gyration. His explanation was confirmed by Huygens, Halley, and Newton, although Cassini rejected it. Richer's discovery marked the beginning of modern geodesia and, after further tests, was a basic element in favor of Newton's theory.

The *Horologium oscillatorium*

In the area of improving observational instruments, Olaus Römer, at the beginning of his stay in Paris and independently of Adrien Auzout and Jean Picard, improved the telescope micrometer, while around 1672, the Frenchman Giovanni D. Cassegrain (1625–1712) invented a type of reflecting telescope that was later named after him. On instruments in general, Huygens published his *Horologium oscillatorium* in Paris in 1673. Of particular importance to the history of astronomy, it contained the formula for the simple pendulum that linked the period of oscillation to the lenght of the pendulum and, above all, to the acceleration of gravity, as well as a determination of the centripetal force necessary to maintain a body in uniform circular motion.

Universal gravitation

Meanwhile, Johannes Hevelius published the first part of a great work on astronomy, *Machina coelestis*, in 1673 in Danzig. The second part came out in 1679, and the third, posthumously, in 1690, with the title *Prodromus astronomiae*. The work, in which the astronomical instruments used are also amply illustrated, was the first in the history of astronomy to use coordinates with respect to the equator instead of the ecliptic, as

had been done up until then, in giving the right ascension and the declination of the stars. It was also the last great product of naked-eye observational astronomy, in which Hevelius was so skilled that he obtained results comparable to those achieved with the aid of telescopes.

Cassini's division

In 1675, Giovanni Domenico Cassini discovered the line that separates the two rings of Saturn, later named Cassini's division. In the course of his observations in 1671 and 1672, Cassini had already discovered two of the planets's satellites; he discovered the other two in 1684.

The Greenwich observatory

Jonas Moore called upon John Flamsteed (1646–1719), who had been in contact with Isaac Barrow (1630–77) and Newton, to London to build a great observatory. The site chosen was the hill of Greenwich. The institution of the observatory was decreed on June 22, 1675 by Charles II "for the progress of the art of navigation," and it was founded on the following August 10.

Römer's determination of the velocity of light

On November 22, 1675, Olaus Römer presented a memoir entitled *Démonstration touchante le mouvement de la lumière* to the Académie des Sciences. In it, he recapitulated his extensive observations of Jupiter's satellites; and, on the basis of the retardations or advances observed in the eclipses of the satellites with respect to the calculated values, depending on whether the earth was moving toward or away from Jupiter, he reached the conclusion that the velocity of light was not infinite, but had a greater value than what had already been calculated. Thus a question that had long been debated in physics ended, and a new element assumed a fundamental role in astronomical knowledge.

Flamsteed's catalogue

In Nuremberg, in 1676, G. Vitali (1624–98) of Capua published a huge catalogue of the stars that Brahe and others after him had observed (*Absolutissimae primi mobilis tabulae*), giving both the equatorial and the ecliptical coordinates of each of them and prefacing the work with an extremely extensive table of transformations of the coordinates. That same year, John Flamsteed settled in Greenwich, where he continued his observations. When the Royal Society, under the presidency of Newton, published the results of these observations against Flamsteed's will, a heated debate arose between Flamsteed, Newton, and even Halley. In an attempt to avoid clashes with the great astronomers of his time, Halley decided to turn his attention to studying the less well-known austral sky. For this prupose, he spent eighteen months in 1677–78 on St. Helen's, the southernmost English colony. The result of his observations are collected in the *Catalogus stellarum australium*, published in London in 1679, which lists 341 stars. His observations of the transit of Mars across the sun in 1677 made Halley realize the opportuneness of studying the transits of Venus as well, while his preparatory work on the *Catalogus* brought the changes in brightness of the heavenly bodies to his attention.

Jean Picard began publishing the *Connaissance des temps*, an ephemeris that is still being published today, in Paris in 1678. In 1679, Cassini brought together the results of his own observations of the moon and those of his predecessors in a tiny *Carta della Luna*, twenty inches in diameter. That same year, Richer, using pendulum measurements, was the first to discover that gravity decreased with latitude.

PHYSICAL SCIENCES

The affirmation of Cartesian mechanism

We have already mentioned that Cartesian physics began to spread in the Low Countries even before the middle of the century, while it met with a number of difficulties in penetrating into French scientific circles. Things began to change several years later, and the situation virtually reversed itself in the decade between 1670 and 1680. The determining factor in this reversal was the publication in 1671 of the *Traité de physique* by Jacques Rohault (1620–75), an extensive, serious and extremely clear manual which, although it adopted a rationalistic approach to Cartesian physics, showed a considerable openness to experimentation, correcting to a certain degree the dogmatism of previous rationalists.

Rohault's work was enormously successful and had an immediate and profound impact on French culture. By this time everyone had recognized the substantial validity of the concepts it defended, and in the final years of the seventeenth century and the beginning of the next, Cartesianism ended up becoming the ''official physics'' of France, blocking all efforts to make Newton's ideas welcome there.

The nature of light: the corpuscular theory

As we have already mentioned, Newton presented his important memoir on optics, ''A New Theory about Light and Colours,'' to the Royal Society in 1672. Because of the rigor with which the results of many new experiments are presented and discussed in this work, it is exemplary from a methodological point of view. The author still seems to tend toward a wave concept of light, although his later research convinced him more and more that it must be of a corpuscular nature. This latter theory was sharply contested by Robert Hooke and Christiaan Huygens.

The finite velocity of light

In 1675, starting from several accurate observations of Jupiter's satellites, the Danish astronomer Olaus Römer (1644–1710) calculated the velocity of light (see ASTRONOMY 1670–1680). His findings were very important in that they showed that light had a finite velocity, challenging one of the general principles of Cartesian physics.

Experiments on the vacuum

The same year that Newton's memoir came out, Otto van Guericke published a detailed report, entitled *Experimenta nova magdeburgica de vacuo spatio*, on the experiments he had conducted, beginning in 1654, with the famous pneumatic machine. His work is further evidence of how fast the turn toward rigorous experimentation was spreading among the physicists of the era.

Scientific instruments took on primary importance in this phase of research.

Huygensian dynamics

One of Christiaan Huygens's major works, the *Horologium oscillatorium*, was published in 1673. It was a magnificent synthesis of mathematical treatment and empirical observation. Its guiding spirit was obviously Cartesian, although it was enriched by an interest in the concrete that demonstrates the author's close ties to the Galilean school (he even emphasized how useful his invention could be for measuring longitudes). The work is divided into five parts, two of which (the second and the fourth) are particularly interesting for the study of mechanics, in that they study the descent of weights and their movement along a cycloid and the determination of the center of oscillation for any compound pendulum.

Hooke's law

Meanwhile, the problem of the motion of the earth and the planets, and the causes that determined it, reemerged in all its importance. Newton was studying it and had already conceived the theory of gravitation, although he had not yet published anything systematic on it. In 1674, Robert Hooke published a remarkable essay on the topic, later included in his *Lectiones Cutlerianae* of 1679. This work contained the famous law on elasticity *ut tensio sic vis*, that is, the proportionality between force and deformation, known even today as Hooke's law.

In 1679, Hooke wrote to Newton asking his opinion on the planetary theory. The correspondence that followed stimulated Newton to state his theory more precisely (see ASTRONOMY 1670–1680).

CHEMICAL SCIENCES

The analogy between chemical and biological oxidation

Combustion and respiration, now recognized as analogous processes, were assiduously studied by two English physicians, Thomas Willis (1621–75) and John Mayow (1640–79). They acknowledged that air contained a substance that fed both processes via a heat-producing reaction. Willis maintained (1670) that ''sulphureous particles'' in the fuel united with ''nitrous particles'' in the air to produce flames (he spoke of ''nitrous particles'' because saltpeter notoriously activated fire). The heat of blood was of the same nature. Mayow was more scientific; after dealing with respiration in his *Tractatus duo* (1668), he returned to the subject, and others, in his *Tractatus quinque medico-physici* (1674). He also drew a parallel between saltpeter and a ''nitro-aerial spirit,'' and explained the increase in weight of calcined antimony as the absorption of particles of this spirit. The heat was the result of a violent conflict between the ''nitro-aerial particles'' and the ''sulphureous particles.'' Mayow described experiments with candles and mice under a bell-jar over water, and concluded that in both cases the air was deprived of its nitro-aerial and elastic particles, so that it became unsuitable for supporting combustion or respiration, while its ''elasticity'' (pressure) decreased. Although these experiments went back as far as Philon of Byzantium and Heron of Alexandria, and were repeated by Johannes Baptista van Hemont and others, Mayow studied them quantitatively, and can thus be considered a precursor of Lavoisier and Lomonosov.

New compounds and elements

Several interesting substances were obtained during this period. In 1672, Pierre Seignette (1660–1719) prepared potassium sodium tartrate (Rochelle salt). Around 1675 (1669 according to some), Hennig Brand (flourished c. 1670) obtained phosphorus by evaporating urine, calcinating the residue and distilling it; in 1676, Johann Kunckel (1630–1703) obtained it in a fairly pure form ("Kunkel phosphorous").

The famous *Cours de chymie*, by Nicolas Lemery (1645–1715), which came out in various editions and translations (the Italian one appeared in 1700), was published in 1675. It opposed alchemy and upheld the atomic theory. It considered the use of scales important, and clearly and accurately described various preparations, with the addition of personal observations. In Italy, C. Lancilotti published his *Guida alla chimica* in Modena in 1672.

LIFE SCIENCES

Investigations with the microscope

Antoni van Leeuwenhoek (1632–1723) of Holland contacted the Royal Society of London and in 1673 began publishing, in the form of letters, his important discoveries with the microscope. He described red corpuscles, Infusoria (1673), and the *animalcula seminis* (spermatozoa), observed by his disciple Johan Ham (1650–1723). With the aid of a simple, but relatively powerful, microscope, he also succeeded in observing bacteria (1683).

Embryology

In 1673, Marcello Malpighi (1628–94) published his *De formatione pulli in ovo*, a work that systematically followed and studied the embryonic development of chicks using methods of microscopy.

Botany and comparative anatomy

This was a particularly important decade for the history of botany. In London, Nehemiah Grew (1641–1712) published a book entitled *Anatomy of vegetables begun*, in which he uses the term and concept "comparative anatomy." Grew's major work on the structure of plants came out in 1682. Between 1675 and 1679, Malpighi also made several important studies of the microscopic structure of plants.

On the basis of these studies, flowering plants were divided into two large groups, monocotyledons and dicotyledons. The Académie des Sciences of Paris launched a program in botanical research planned by Denis Dodart (1634–1707), who provided external descriptions of plants using phytochemical analysis.

In 1667, Robert Hooke described the fungous disease of plants.

Animal anatomy and physiology

Animal anatomy and physiology continued on the wave of the previous century. Claude Perrault (1613–88) and Joseph

Guichard Duverney (1684–1730) performed dissections of a large number of different animals and compared their respective structures. They saw many tiny machines in the fine structures of these different organisms. This iatromechanical approach was especially popular in Italy with Lorenzo Bellini (1643–1704), Giorgio Baglivi (1668–1707) and above all, Giovanni Alfonso Borelli, whose major work, *De motu animalium*, appeared posthumously, in 1680–81.

The ovarian follicles

In 1672, Regnier de Graaf (1641–73) observed that the ovarian foillicles of mammals were normal organs linked to reproduction (Graffian follicles). He believed that these follicles were the eggs of mammals, an error that was later corrected by Karl Ernst von Baer (see LIFE SCIENCES 1820–1830).

The physiology of respiration

In 1674, John Mayow (1640–79), a physician at Oxford, taught that air contained a particular substance, *nitrum aeris*, necessary to respiration and combustion. The addition of this substance supposedly transformed venous blood into arterial blood.

The concept of irritability

Physiologists began to discuss a concept that was new in many ways: the concept of irritability, which Francis Glisson defined as a basic property of living fibers.

Clinical medicine

Altogether, the nature of medicine became more scientific as more and more physicians understood that the art of healing had to be based on new ideas. Some physician-anatomists, such as Claude Perrault, refused to practice a completely conjectural art. In England, Thomas Sydenham (1624–1689) promoted a return to Hippocratic observational medicine. He created the modern clinical concept of illness and gave exemplary descriptions of dysentery, scarlet fever, which he distinguished from measles, and many other acute diseases. He also provided a good description of gout, a disease from which he himself had suffered. In 1672, Thomas Willis described creeping paralysis and *dementia praecox*. John Mayow published the first observations on mitral stenosis in 1674, while Lazare Rivière (1589–1655) gave a description of aortic stenosis.

EARTH SCIENCES

Fossils and the Flood

Martin Lister (1638–1712), an expert on living mollusks, considered fossils *lapides sui generis*, generated as such, and, although he recognized their analogies with living forms, pointed out that, case for case, their composition was the same as that of the rock in which they were found (*Historiae animalium Angliae tres tractatus*, 1678, and previous works). Robert Plot (1640–96) in 1674 also supported this opinion, rather than the idea that fossils were the remains of organisms, as some believed. Paolo Boccone (1633–1704), on the other hand, was of the latter opinion regarding glossopetrae. Among those who maintained that there was no direct connection between Noah's flood and fossils was Georg Wolfgang Wedel (1645–1721) in his *De conchis saxatilibus* (1672).

The alembic theory on springs and the water balance

Herbinius, among others, supported the alembic theory in his *Dissertationes de admirandis mundi cataractis* (Amsterdam, 1678).

The first serious attempt to determine the amount of rainwater that fell and calculate whether it was enough to feed springs was made by Pierre Perrault (1608–80) in 1674 (*De l'origine des fontaines*, published anonymously). He was able to demonstrate that the amount of water that the Seine carried in a year was barely equal to one-sixth of the rainfall in its hydrographic basin over the same period of time.

The foundations of physical geography

Bernhardus Varenius's *Geographia generalis* (1672 edition revised by Isaac Newton) laid the first foundations of physical geography and established a division of geographical phenomena that was the basis for distinguishing between the branches of geography up until modern times.

The dawning of geophysics

The astronomer Jean Richer (1630–96) discovered the difference in range of the oscillation of a pendulum between Paris and Cayenne (see ASTRONOMY 1670–1680), anticipating an understanding of the variability of g (the acceleration of gravity), on which modern gravimetric measurements are based (1672). Slightly earlier (1666), Isaac Newton had begun to lay the groundwork for his theory of gravitation, which was also very important to geophysics, in that gravimetry was based on it.

Between Gibraltar and Greenland

English explorations of the Mediterranean and Gibraltar continued, especially with the work of Richard Bolland (who invented depth gauges and current meters, among other things), and Henry Sheeres, whose research was later reported in his *Discourse of the Mediterranean Sea and the Straight of Gibraltar* (1703).

In 1671, Friedrich Martens, after a voyage to the Spitsbergen islands, described the hydrology of the oceans around Greenland and Iceland, and made the first observations on the formation of marine ice.

APPLIED SCIENCES

The art of spinning and weaving

In both France and England, inventors were attempting to mechanize the operations of spinning and weaving. In 1678, Richard Dereham and Richard Haines patented a hand-operated machine that turned six to one hundred spindles. That same year, M. de Gennes, an officer in the French navy, described a loom for producing linen cloth without the help of an artisan.

Glass working

The *Ars vitraria experimentalis* by Johann Kunckel (1630–1703) appeared in 1679, and that same year the author was invited to Berlin by the Elector of Brandenburg to become the director of glass factories. Kunckel's work was the first to describe the manufacture of glass made with potash, which was first obtained from clay.

English architecture: Wren's contribution

In 1673, Christopher Wren began constructing the dome of St. Paul's in London, the first great architectonic project of the English Renaissance. Although Wren was very famous as an astronomer and a scholar of geometry, it appears that he did not apply his knowledge of statics to his work as an architect. At Oxford, Wren had met the mathematician John Wallis, who had studied the theory of loaded beams and collaborated with Robert Hooke, famous for the law on the elastic behavior of materials that bears his name, among other things. One of Wren's sketches for the dome of St. Paul's, however, suggests that he had in mind Galileo's theory of catenary curves, according to which the ideal shape for a uniformly thick arch, bearing only its own weight, was a catenary, i.e. the curve described by a cable or a chain freely hanging from both ends. The dome of St. Paul's is a milestone in the history of English architecture in that it was the first dome ever built in that country. Wren's design for the dome may have been structurally derived from the domes of St. Peter's and the Cathedral of Florence, which he studied and which were strong enough to hold the heavy stone lantern that crowned his work.

Wren also designed the astronomical observatory at Greenwich, construction of which was begun in 1675 (see ASTRONOMY 1670–1680).

In the area of hydraulic machines, Samuel Morland (1625–95), an English mathematician and inventor, patented a piston pump in 1674. The following year, he succeeded in lifting water from the Thames to a height of more than eighteen meters.

The enlargement of the castle of Versailles

In France, the work begun during the previous decade to enlarge the castle of Versailles continued, under the direction of J. Hardoin-Mansart (1646–1708), who was named the king's architect in 1675. During the final years of the century, Mansart also built the dome of the chapel of Les Invalides, the most majestic dome in Paris. The system of pumps that regulated the playing of the fountains in the gardens of Versailles is a masterpiece of French hydraulic engineering in the second half of the seventeenth century.

Improvements on the pendulum

Decisive progress was made in the theory and improvement of the pendulum during this decade. In 1673, in Paris, Christiaan Huygens published his most important treatise on instruments for measuring time, *Horologium oscillatorium*, in which he described the theory that a truly isochronic pendulum would have to oscillate along a cycloidal arc. Three years earlier, William Clement had invented the anchor escapement, which made it possible to shorten the arc of the circle described by a pendulum, so that it approached a cycloidal arc. In 1673, Joseph Knibb, a clock-maker in London, built the first clock that showed fractions of a second, for James Gregory (1638–75), an astronomer at the University of St. Andrews. Christiaan Huygens described his balance-spring in 1675, but Robert Hooke claimed priority for its invention. It was Hooke's balance-spring with an oscillation of around 120 degrees that became commonly used.

Edward Barlow invented the rack-and-snail striking mechanism in 1676.

Microscopes

Robert Hooke had already illustrated a compound, extendible microscope in his *Micrographia*, and another was built by Eustachio Divini (1620–95) in 1668. In 1676, John Christopher Sturm (1635–1703) designed an even more complex model, equipped with interchangeable lenses so that the proper enlargement could be obtained each time. The common goal of all of these microscopes was to reduce the aberrations created by increased enlargement. The results achieved were not very satisfying, however; in fact, the most important microscopic discoveries of the period were made by Antony van Leeuwenhoek (1632–1723), who made his observations using a simple microscope. He described his findings in reports sent to the Royal Society beginning in 1673.

1680–1690

THOUGHT AND SCIENTIFIC LIFE

Scientific life in Germany and Russia

The people of the German-speaking countries lifted themselves back up after the disastrous wars. Scientific life was also reorganized during this (sometimes slow) rebirth. The first scientific circles were formed as well. The *Acta Eruditorum* began to be published in Leipzig in 1682. Although the editors of this journal (Otto Mencken and Wilhelm Ernst Tentzel) were not famous, they soon counted figures of great prestige all over Europe among their collaborators (for instance, Leibniz). Significantly, these German intellectuals, Leibniz first among them, had an extremely cosmopolitan outlook, which was often translated into hopes of a lasting peace among the populations of "Christianity" after the devastations of the religious wars.

In Russia, Czar Peter the Great, in power from 1682, conducted a Westernization campaign in his country, often forcing the stages. A member of the Parisian Académie des Sciences and an admirer of the scientific and technical achievements of the English and the Dutch, the Czar imposed Western rationalism, science, and technology upon the Boyards and the Russian clergy.

Developments in natural philosophy: Newton

One of the principles of Descartes's mechanism was that all natural phenomena, including those of the organic world, could be explained with the laws of mechanics. However, the French philosopher's program encountered a number of difficulties on this point, with respect to the organic world. With this situation in mind, it is easy to understand why the work of Giovanni Alfonso Borelli (1608–79), *De motu animalium*, published posthumously in 1680–81, excited so much interest among scholars in both biology and philosophy. The author's ingeniousness in providing quite clear mechanical explanations not only of muscle contractions but of the performance of the major organic functions as well (such as the circulation of blood and digestion), stood as a confirmation of the unity of scientific knowledge and the primary position that physics held in it. The work thus made a notable contribution to the success of that science.

But the major contribution to natural philosophy came some years later with the publication of Isaac Newton's *Philosophiae naturalis principia mathematica* (1687). It succeeded in presenting all of mechanics in a systematic form almost as perfect as Euclid's geometry, and at the same time it was able to unify celestial phenomena with terrestrial phenomena through the law of gravitation. It is true that the work aroused considerable controversy, especially among the Cartesians, who accused Newton of returning to prescientific notions because of the obscure and mysterious nature of the idea of gravitation. But these criticisms were not enough to place the great system in crisis nor to keep Newtonianism from being viewed throughout the following centuries as the most perfect form of mechanism.

Around the same time that Newton was working on his *Principia*, Leibniz elaborated a new spiritualistic philosophy. It proposed to resolve the difficulties of Descartes's metaphysics by denying the substantial nature of extension and recognizing instead that all reality is made up of single, unextended substances (monads), each of which is able to evolve from one state to another, but not to go outside itself. On the one hand, this philosophical innovation enabled him to salvage and develop the Cartesian concept of innate characteristics; on the other, it led him to reshape virtually all the sciences (from physics to biology to psychology) in a highly original way.

Rationalism and deism: Bayle and Locke

The increased successes of scientific and philosophical research, meanwhile, suggested a new theme for reflection: the role of science in society in general. A very important work in this regard, *Pensées diverses écrites à un docteur de Sorbonne à l'occasion de la comète qui parut au mois de décembre 1680*, was published by Pierre Bayle (1647–1706) in 1682. While the common people and many theologians had interpreted the phenomenon as a sign of God's anger, Bayle gave a purely scientific explanation of it and seized the occasion to criticize the belief that God would resort to such trivial means to announce His punishments. For Bayle this was a superstitious belief that denied true religion.

Reflections on the enlightening role of science again brought up the problem of relations between modern thought and religious beliefs. How was it possible that scientific truth was able to obtain the almost unanimous consent of scholars, while religions continued to pass off as absolute truth a variety of different dogmas and to quarrel among themselves with unmitigated anger? These were the questions that stimulated the new culture's interest in religious problems. Two substantially complementary currents of thought thus arose: one based on a profound tolerance in contrast to the sectarian struggles among various religious confessions, and the other on a search for a religion without dogmas, to be the rational nucleus of Christianity. The first of the two currents found a supporter and a theoretician of the highest order in John Locke (1632–1704), who wrote his famous *Epistola de tolerantia* (1689) on the subject. The work is generally considered a true masterpiece in this genre of philosophical-religious literature.

MATHEMATICAL SCIENCES

Leibniz's first essays on infinitesimal calculus

The first works on infinitesimal calculus were published during this period. Gottfried Wilhelm Leibniz promoted the foundation of *Acta Eruditorum* (1682), in which both he and his friends and pupils published a number of important contributions. One of the contributors was Ehrenfried Walther Tschirhaus (1651–1708), who, after having learned the principles of differential calculus from the master and believing that he had derived his own formulation of it, published an article on determining the tangents of curves in *Acta* in 1683. This induced Leibniz to compose a memoir defending his right of paternity to the method. Thus one of Leibniz's most important memoirs, *Nova methodus pro maximis et minimis itemque tagentibus*, appeared in *Acta* in 1684. In it, the author systematically presented his ideas, his notations, and his findings on infinitesimal calculus. This was the first publication that dealt with the new algorithm; Leibniz wrote others during the years following.

While Leibniz continued confidently and consciously to expound his calculus and defend it, Isaac Newton did not do the same. He published his famous *Philosophiae naturalis principia mathematica* in 1687, but the work spoke only of the "method of the first and last reasons," without explicitly dealing with the method of fluxions and fluents, or with the methods of analytical geometry. The author was probably afraid of arousing suspicion in many of his readers if he openly resorted to methods that were too new. As a result, although he had definitely used the new algorithm to arrive at the results he was presenting, he reformulated and demonstrated them using the methods of traditional geometry and algebra.

ASTRONOMY

Halley's comet

The apparition of the comet of 1680, which was observed in Paris by Giovanni Domenico Cassini and Edmund Halley, followed by those of 1682 and 1684, revived an interest in studying these heavenly bodies, especially the shape of their orbits. Among those who worked on the problem were Jakob Bernoulli (1654–1705), who published a memoir in Basel in 1681 and one in Amsterdam in 1682, Isaac Newton, and later Halley, who applied Newtonian methods to determine periodicity (in a work published in 1705). Astrology was also strengthened by the apparition of the comet, and old discussions on the validity of astrological theories were resumed. The astronomer, physicist and philosopher Geminiano Montanari (1633–87), a convinced supporter of Galileo's doctrines, contributed to these discussions with his *Astrologia convinta di falso* (Venice, 1685).

The plurality of worlds

Bernard le Bovier de Fontenelle (1657–1757), of Rouen, for many years the perpetual secretary of the Académie des Sciences in Paris, published his *Entretiens sur la pluralité des mondes* in 1686. This work, which was immensely popular, theorized that life was not a prerogative of our planet, but could be present elsewhere.

Hevelius's catalogue

Johannes Hevelius died in Danzig on January 28, 1687. His *Catalogus fixarum stellarum*, which contained data on 1,500 stars, was published that same year. Hevelius made naked-eye observations, although his data competed in accuracy with the information his contemporaries obtained with the aid of telescopes.

Newton's Principia

In 1687, Newton published the first edition of his *Philosophiae naturalis principia mathematica*, which presented a definitive arrangement of classical mechanics. The one solution which had been found to fit both the problem of the motion of earthly bodies and of heavenly bodies, which he applied to the more general case of the law of universal gravitation, became of fundamental importance to astronomy, since Kepler's laws could also be applied to it. Thus Kepler's laws were not only confirmed by observation, but supported by the fact that Newton's differential equations of motion could be deduced from them. In fact, Kepler's empirical laws, considered in the light of Newton's generalization, provided a proof of the law of universal gravitation *more geometrico*. A new world opened up before the eyes of mechanicians, astronomers, and mathematicians. Phenomena such as the precession of the equinoxes, the inequalities of motion of the planets and satellites, the shape of the earth (and of Jupiter), the trajectories of all the heavenly bodies, and the tides, which for so many centuries had challenged human intelligence, now found their law, which descended mathematically from a single principle. Only light appeared to rebel against being confined to formulas.

In 1688, Father Marco Vincenzo Coronelli (1650–1718), of Ravenna, reduced the stupendous terrestrial and celestial globes that he had constructed in Paris in 1683 for King Louis XIV, from a diameter of fifteen feet to three and a half feet, and published them.

The ether and the planets

In a memoir in the *Acta Eruditorum* of 1689, Gottfried Wilhelm Leibniz (1646–1716) attempted to explain planetary motion by resorting to the idea of a fluid around the sun, which in a sense transmitted the sun's force of attraction to the nearby bodies that were immersed in it. Christiaan Huygens made recourse to a moving ether to explain the force of gravity, with explicit reference to Descartes's vortices, in his *Discours de la cause de la pesanteur*, published as an appendix to the *Traité de la lumière* in Leyden in 1690, after the author had left Paris (1681).

Olaus Römer left Paris, in 1687, and set up an observatory in his own home, where he constructed the first transit instrument, using a telescope mounted on the meridian plane and equipped with illuminated cross-hairs and a clock that struck the seconds. His equipment was completely destroyed in the Copenhagen fire of 1728.

PHYSICAL SCIENCES

Universal gravitation

The most important turn of events in physics, not only in this decade but in the entire second half of the century, came as a result of Isaac Newton's theory of gravitation.

In 1684, the great English scientist provided the first summary exposition of his theory in a memoir entitled *De motu corporum*. A short time later, he gave a more ample and systematic explanation in his *Philosophiae naturalis principia mathematica* (1687). The very title of the work shows that the author was thinking of René Descartes's *Principia;* however, Newton's approach was distinctly different from that of the French philosopher in that he made much broader use of mathematics and introduced a remote force (gravitational attraction) into science, which the Cartesians considered magical, not mechanical.

The *Principia* is subdivided into three books: the first, purely mechanical in its subject matter, is dedicated to the study of motion determined by central forces; the second deals with fluid mechanics; and the third is an application of the general findings arrived at in the first book to the study of celestial motion. The three books are preceded by a very important exposition of the basic concepts and principles of dynamics. In the second edition, published in 1713, Newton added his famous *Scholium generale*, a profound philosophical statement, to the end of the work. The third book opens with the no less famous *Regule philosophandi*, which outlines Newton's idea of the inductive method.

The diffusion of Newtonian science

The work was a great success. Besides the second edition, just mentioned, a third, which was virtually identical to the second, came out in 1726, within the author's lifetime. In the eighteenth century, Newtonian physics was considered the most valid patrimony of all natural science. Only France contested this acknowledgment and contrasted Cartesian physics to the Newtonian system.

The controversy on living forces

In addition to the controversy between Cartesians and Newtonians, a new controversy arose during this decade. In a certain sense it was even more important, and it caused quite a stir in physics for several decades. Gottfried Wilhelm Leibniz (1646–1716) started it, when he published a memoir entitled "Demonstratio errori memorabilis Cartesii" in the *Acta Eruditorum* of 1686. It aimed at confuting the Cartesian principle of the conservation of motion (mv), and replacing it with a new principle, the conservation of living forces (which Leibniz identified as the product of m x v2). The interesting fact is that the Newtonians took sides with the Cartesians in this quarrel, so that it was transformed into a controversy between Newtonianism and Leibnizianism.

As to more specific research in physics, we shall only mention that in 1686 Edmund Halley (1656–1742) succeeded in developing a fairly reliable formula linking the altitudes of various localities with the atmospheric pressure measured in them. The altimetric formula was one of the first interesting practical applications of the abundant research of seventeenth century physicists that began with Torricelli's discovery of the barometer.

CHEMICAL SCIENCES

The popularization of chemistry

C. Lancilotti's *Guida alla chimica* reawakened an interest in chemistry in Italy, possibly because it was written in the vernacular. Various editions were published, and it was followed by the *Vaglio chimico* (Venice, 1682), the *Farmaceutica antimoniale* (Modena, 1683), the *Farmaceutica mercuriale* (Modena, 1683), and other works, all by the same author.

The study of chemistry was made more accessible in Germany as well. The *Chymia philosophica perfecte delineata*, a collection of the lessons that J. Barnel (1641–86) gave in Padua, was published posthumously in 1689, in modern German. Georg Ernst Stahl (1660–1734) read it in manuscript form when he was fifteen, and learned it almost all by memory.

On the composition of air, water, and crystals

In his *Magisterium naturae et artis* (Brescia, 1684), Francesco de Lana (1631–87), a chemist, physicist, and inventor, described, among other things, various curious experiments, such as the transformation of (humid) air into water. L.M. Barbieri, drawing most of his ideas from John Mayow (1640–79), underlined the importance of air—precisely its "nitro-aerial spirit," the "elastic part"—in combustion and life, in his *Spiritus nitro-aerei operationes in microcosmo* (Bologna, 1680).

In contrast, Giovanni Alfonso Borelli (1608–79), the founder of iatromechanics, maintained in his *De motu animalium* that air was necessary to life, but not because of its nitrous nature. The relationship between saltpeter and the oxygen in air was still unclear; Olaus Borrichius (1626–60), who in 1680 described saltpeter as a salt "*particulis flabelliferis praegnantem,*" guessed that oxygen was present in potassium nitrate. Borelli, criticizing Johannes Baptista van Helmont's tree experiment, explicitly says *aquae moleculae non sunt corpora prima simplicia, sed sunt congeries, compositae ex atomis primis elementaribus* (see CHEMICAL SCIENCES 1640–50).

Another iatromechanician was Domenice Guglielmini (1655–1710), who anticipated René Just Hauy's theory by a century in his *Reflessioni filosofiche dedotte dalle figure de' sali* (Bologna, 1688), when he formulated the hypothesis that artificially prepared mineral and salt crystal resulted from the juxtaposition of elementary cells that had the same polyhedral shape as the solid of cleavage of the crystals themselves.

The distillation of coal

Johann Joachim Becher (1635–82) obtained a tar from the distillation of coal (already described by Johann Rudolph Glauber) which he described accurately (1681), while John Clayton (1657–1725) obtained a gas, methane, and recognized its flammable nature (1688). These were the origins of two discoveries that aroused extreme practical interest in the nineteenth century.

LIFE SCIENCES

Systematics: the great classifications

A basic text of biological mechanism, *De motu animalium*, by Giovanni Alfonso Borelli (1608–79), came out posthumously in 1680–81.

This decade marked the rise of systematics and comparative anatomy. In his *Methodus plantarum nova* (1682), John Ray (1627–1705) described thousands of plant species, classifying them on the basis of a new taxonomic system.

Pierre Magnol (1638–1715) had the idea of compiling a natural classification of plants based on one organ, the calix. His *Prodromus historiae generalis plantarum in quo familiae plantarum per tabulas disponuntur* (1689), in which he considered all the characteristics of each plant, was the first attempt at a simple and rational classification into families. A systematic description of species of parasitic animals was undertaken by Francesco Redi (1684).

Comparative anatomy

Nehemiah Grew studied the comparative anatomy of the alimentary canal in various mammals, birds, and fish. His essay is original in that it makes a detailed examination of just one organic system using the comparative method (1681). Raymond Vieussens (1641–1715) published a *Neurographia universalis* (1684), which contained the first description of the *centrum ovale*, and represented a remarkable step forward in understanding the structures of the brain, the spinal cord, and nerves.

In 1686, John Ray published Francis Willoughby's *De Historia piscium*. Willoughby's and Ray's zoological system was still inspired buy Aristotle, although their use of several characteristics of internal anatomy marked an important innovation. Ray made a distinction between fish that breathe with fins and marine vertebrates that breathe with lungs.

EARTH SCIENCES

The beginnings of crystallography

The phenomenon of double refraction in calcite (Iceland spar), discovered in 1669 by Erasmus Bartholin, was now studied from a physical point of view by Christiaan Huygens, who established the law of the propagation of light in that mineral (1690), and through it attempted to discover the intimate structure of crystals. In 1684, Francesco de Lana (1631–87) anticipated Linnaeus's theory on the genesis of crystals as saline bodies. In 1688, Domenico Guglielmini (1655–1710) anticipated René Just Haüy's work of 1784 on the structure of crystals.

The question of the genesis of metals continued to be debated throughout this period, and various authors spoke of exhalations, which could also develop outside veins of metal in the form of poisonous vapors. One of these was Johan Joachim Becher (1635–82) in his *Physica subterranea* (1681).

The water balance

In 1686, the famous physicist Edme Mariotte (1620–84) made an improved study of the water balance of the Seine and its basin above Paris to check whether it was possible that springs were fed by rainwater. His conclusions were strongly affirmative.

T. Burnet's "sacred theory of the earth"

The Anglican theologian Thomas Burnet published a text, in Latin in 1681 and in English in 1684 (*Telluris theoria sacra, The Sacred Theory of the earth*), that became widely popular. It was written to support the Scriptures, and advanced a fantastic story about the earth. A large part of the work is dedicated to facts that are presumably connected with Noah's flood, the most catastrophic event that ever struck our planet from the time of its creation.

Newton and the tides

In his *Philosophiae naturalis principia mathematica* of 1687, Isaac Newton (1642–1727) produced unequivocal, clearly documented proof that the tides are produced by the gravitational attraction of the sun and the moon, following the original attempts made by Simon Stevin and Johannes Kepler. Since this force of attaction (Newton's law) is directly proportional to the mass and inversely proportional to the square of the distance (the lunar-solar tide force, on the other hand, is inversely proportional to the cube of the distances), although the mass of the moon is much smaller than that of the sun, it produces tides that have almost double the range of solar tides because it is much closer.

Edmund Halley (1656–1742), who collaborated intimately with Newton on this work for a time, also studied the tides, as well as the relations between the wind and currents (see PHYSICAL SCIENCES 1680–1690).

In navigation, attempts were made to apply the new concepts of physical mechanics to regulating ships' movements, continuing (and partly rediscovering) Leonardo da Vinci's work. The initiatives taken by the Republic of Venice contributed considerably to the study of these problems.

APPLIED SCIENCES

Papin's steam machine

When vacuums began to be used to draw energy from atmospheric pressure, the idea of using the explosive force of gunpowder in these kinds of engines arose. In 1680, Christiaan Huygens designed a gunpowder machine in which the explosion was produced in a cylinder underneath a piston that was raised by a counterweight. The gasses from the explosion went out through safety valves, and air pressure pushed the piston down. However, some gas remained in the cylinder after the explosion, and a satisfactory vacuum could not be produced. Denis Papin (1647–1712), who assisted Huygens in these experiments, was intrigued by the problem. To keep boilers from eventually exploding, he invented the safety valve, and in 1681 became famous for his invention of the digester, or steam pressure cooker. However, most of his attention went to the problem of finding a substance that did not leave a residue. He found the solution in 1690, by building "machines in which water, with the aid of moderate heat, and at a low cost, would produce that complete vacuum that could not in any way be obtained using gunpowder." Papin's first steam machine was built in 1690 and consisted in a vertical tube six centimeters in diameter, with a rod and piston. The tube served the triple function of boiler, motor cylinder, and condenser. Papin's machine had no practical use, but he was the first to use steam to move a piston in a cylinder, and to indicate the work cycle that was later incorporated into Newcomen's machine.

Bridges and canals in France

The Languedoc canal, joining the Atlantic and the Mediterranean, was inaugurated in 1681 (see APPLIED SCIENCES 1660–1670). Promoted by the minister Colbert, it had been started in 1666. The project was directed by Pierre-Paul Riquet de Bonrepos (1604–80), who died seven months before the work was inaugurated. His son, Jean-Mathias Riquet, directed the final work of completion, which lasted until 1692. The canal was considered the prototype of all the great navigable canals that were constructed after it in other European countries.

J. Hardoin-Mansart (1646–1708) designed the Pont Royale in Paris in 1685, in collaboration with Jacques Gabriel IV (1630–86). In contrast to the original plan, the bridge was built with caisson foundations filled with pozzolan and lime, a combination that formed a natural cement in water and set quickly, even underwater. The Pont Royale was the first example of the use of caissons in France.

Assaying ores

Some progress was made in assaying as well. In practice, the most valuable assay for iron ores was the forge test. Lazarus Ercker (c. 1530–94) suggested a magnetic test for iron ores (see APPLIED SCIENCES 1550–1600).

C. Schindler was the first assayer to provide a method of assaying iron ores that was based on their fusion with a reducing material, producing a regulus of cast iron. He described the method in his *Der geheimbde Müntz-Guardein und Berg-Probierer*, published at Frankfurt am Main in 1687.

Treatise-writing

Cornelis Janszoon Meijer, who had presented a project for draining the Pontine marshes to Innocent XI in 1676, published the results of his studies in *Del modo di secare le Paludi Pontine*, in 1683.

In the second edition of the *Ars vitraria experimentalis*, dated 1689 (the first edition was published in 1679), Johann Kunckel (1630–1703) recorded another technological advancement made by German glass-makers: opalescent glass was manufactured by adding calcined bone or horn.

In 1683, J. Moxon (1627–1700) published his *Meckanick Exercieses*, in which he described the technique of fusing letters of type.

In 1686, Robert Plot (1640–96) published his *Natural history of Staffordshire*, which contains the first detailed description of the procedure of the complete case hardening of iron bars soldered together in order to produce steel for further working. The furnace was similar to a baker's oven, with a hearth about sixty centimeters wide and heating boxes with a capacity of more than five hundred kilograms of iron in bars up to one hundred and fifty centimeters long. The hardening process was done with wood charcoal alone, and lasted from three to seven days.

1690–1700

THOUGHT AND SCIENTIFIC LIFE

Deism: Locke and Toland

In 1690, 1692, and 1704, John Locke wrote a second, a third, and a fourth letter on tolerance (this time in English), in order to further develop the topic he had introduced in his *Epistola* of 1689, and to answer the many objections that traditional theologians had raised against it.

The first attempts at constructing a rational religion that would be the true nucleus of all religions dated back several decades (in particular to the so-called Platonic school of Cambridge, which depended on the university and flourished toward the middle of the century). The years between the seventeenth and eighteenth centuries saw a revival of these attempts, especially in England, and the group of doctrines that went down in history under the name "deism," as opposed to the "theism" of positive religions, was systematically developed. The most important work was *Christianity not mysterious* (1696) by John Toland (1670–1722). It was obviously inspired by Locke's writings, but definitely surpassed them, in that it proclaimed the need to follow rational truth alone, even in religious matters. Even from the Bible, Toland emphasized, we can only accept what is most credible to human reason.

The first edition of Pierre Bayle's *Dictionnaire historique et critique* came out almost simultaneously (1695). In it, historical criticism seems to have become increasingly corrosive in its attitude toward the old "noble" traditions.

Bayle's work and that of the English deists notably influenced the development of thought in the Enlightenment.

Empiricism and rationalism: Locke and Leibniz

Locke's position marked a considerable change with respect to the main current followed by all the earlier rationalists. Although he had been deeply influenced by Descartes, Locke maintained that the philosopher's essential duty should not be the search for an absolute criterion of truth, but rather the investigation of the many processes of our minds. Nor did the author simply stop with the processes that gave rise to scientific knowledge (privileging it a priori with respect to every other form of knowledge), but investigated the formation of all ideas, scientific and otherwise. He thus arrived at the famous distinction between simple and complex ideas and a critique of the idea of substance. His masterpiece on this subject, *An essay concerning human understanding* (1690), is full of extremely piercing philosophical analyses and is an important landmark in the history of modern empiricism.

On the rationalist front, Gottfried Wilhelm Leibniz very diligently continued his research aimed, as we have already seen, at building a new pluralistic metaphysics that would be able to overcome the major difficulties of the Cartesian system. In 1695, he published a work on the subject entitled *Système nouveau de la nature et de la communication des substances*. Despite the importance of this work, it should be noted that neither this, nor any other of Leibniz's systematic works, contains his entire metaphysics. This, however, can be constructed in its entirety, from his correspondence and various unpublished fragments.

The spread of scientific-technical knowledge. The development of institutions

While critical thought was turned toward great metaphysical speculations (the prototype of which had become the Cartesian system), an interest in scientific and technical knowledge, especially the traditional mechanical arts, was growing. Bayle declared that the wealth of humanity could be greatly increased by the natural philosophers' interest in the trades. Locke recognized that arts that were very useful in life derived from mechanics, practiced by men without formal educations and he reprimanded the Cartesians for talking a lot about machines but never producing much that was useful. Even Leibniz compared "the vague and sad discourses" of the scholars to the "unwritten knowledge" of those who conducted technical activities. A rapprochement between scientific-philosophical theorization and technical-scientific practice seems often to have been difficult. On the other hand, however, while the scholars turned to technical studies, the technicians also appealed to the knowledge of the scholars. This phenomenon is abundantly evidenced by the increase in publications on the pure and applied sciences written in modern languages rather than in Latin. It was a trend that developed in the course of the eighteenth century, especially where the work of Enlightenment thinkers was most intensive.

The development of institutions during this decade also deserves mention. The Regia Accademia dei fisiocratici (Royal Academy of Physiocrats) was founded in Siena in 1691, and in 1693 the University of Halle was inaugurated, and the Biblioteca Casanetense was created in Rome. On January 22, 1699, Louis XIV established the Académie des Sciences in the Louvre, imposing strict rules and reserving the right to elect its members. The relations were coordinated by Bernard Le Bovier de Fontenelle (1657–1757), who succeeded du Hamel in 1697. Fontenelle was by then famous for his *Entretiens sur la pluralité des mondes* (1686) and his *Histoire des oracles* (1687), among other things. On the advice of Leibniz, Frederick I created the Société des Sciences in Berlin in 1700, an institution that was destined to have a great future.

MATHEMATICAL SCIENCES

The earliest developments of the new calculus

Although it had just been created, intense research on the new calculus was soon being done. Leibniz himself furnished important findings in differential calculus in his memoirs and letters, while others were made by his first followers, among whom were Jakob (1654–1705) and Johann (1667–1748) Bernoulli. Not only did they correspond with Leibniz; they understood his calculus and quickly developed a series of remarkable results. They were the first in the long line of great analysts who characterized the following century in particular, and who all belonged to the Leibnizian current. Jakob Bernoulli also introduced the term integral calculus (1690), while Leibniz first used the term "function" in mathematics in an article written in 1692. In 1690, Michel Rolle (1652–1719) published a *Traité d'algèbre*, which was full of new perspectives on solving equations and contained the important differential theorem that still bears his name. Almost as though to crown this process, the *Analyse des infiniment petits pour l'intelligence des lignes courbes,*by Guillaume François de L'Hôpital (1661–1704) appeared in 1696. It basically contained the concepts that de L'Hôpital had learned from Johann Bernoulli about the new methods of calculus. It was the first true manual of infinitesimal calculus and, especially, of differential calculus, and came out only fifteen years after Leibniz's first publication.

ASTRONOMY

Halley's research

In 1691, again advancing ideas that he had already developed in 1677–78, Edmund Halley predicted that Venus would transit in front of the sun in May, 1761. In 1716, he returned to these same thoughts and even devised a plan for observing the phenomenon.

Halley's calculation of the amount of heat radiated by the sun at different latitudes of the earth's surface in each season considerably advanced climatology, with respect to both its object and its methods.

Coronelli's globes

In his *Epitome cosmografica*, published in Cologne in 1693, Marco Vincenzo Coronelli (1650–1718) described his procedure for manufacturing globes of the earth and of the heavens, which he had built for Louis XIV.

The problem of telescopes

A great obstacle to the spread and improvement of observational instruments was the problem of chromatic aberration. Attempts were made to obviate it by building both reflecting telescopes and telescopes with extremely long-focus lenses. David Gregory (1661–1708) vaguely presented the idea of achromatism in lenses in his *Catoptricae et dioptricae elementa* (Oxford, 1695).

Scientific expeditions

In 1695, Giovanni Domenico Cassini returned to Bologna from Paris in order to determine the exact time of the equinox using the giant gnomon in the church of San Petronio, for the purpose of checking the secular adjustment of the Gregorian calendar that was to take place in 1700. In 1698, Edmund Halley was appointed director of an expedition which was organized by the British Admiralty into the Atlantic Ocean, in order to make geographical, exploratory, geodetic, and astronomical observations. The ship *Paramour* was placed under his command, even though he was a civilian.

PHYSICAL SCIENCES

The nature of light: the wave theory

One of the most important works in the history of optics, Christiaan Huygens' *Traité de la lumière* appeared in 1690, right between the last two decades of the seventeenth century. Its main theme was the nature of light. The author first attempted to confute the corpuscular theory (by then openly supported by Newton in various special memoirs), and then to demonstrate— using substantiated physical and mathematical considerations—how all phenomena known at the time could be explained by interpreting light as a wave motion of the ether. Although Huygens's theory contained many points that were anything but obvious (for example, his concept of the ether is quite unclear), his treatment was definitely exemplary, considering the epoch.

The corpuscular-wave controversy

Beginning in 1690, the year in which the *Traité de la lumière* was published, physics was forced to acknowledge that it was faced with two conflicting theories on light, both of which possessed full scientific dignity: the corpuscular theory, linked above all to the name of Newton, and the wave theory, linked mainly to the name of Huygens. In the centuries that followed, both underwent considerable reelaboration (the wave theory, in particular, was profoundly modernized in the nineteenth century, especially with the work of Augustin Jean Fresnel), but the mark that Newton and Huygens had placed upon them was indelible. Each time that one of the two has seemed "definitively" defeated, some previously unknown experimental phenomenon has been discovered, giving it new credibility, until today it is believed that in reality light possesses both corpuscular and wave characteristics.

The affirmation of experimental science

Despite all the controversies we have mentioned, the interest in physics continued to grow among increasingly wide ranks of the population, as the remarkable success of the *Entretiens sur la pluralité des mondes*, a work of scientific popularization published in 1686 by Bernard Le Bovier de Fontenelle (1657–1757), shows. By this time, everyone very clearly understood the radical difference between the sterile theological controversies of the past and the extremely productive ones of modern science. This understanding prepared the way for the advancement and victory of the spirit of the Enlightenment.

However, we should not let excessive optimism make us forget that, in reality, some of the controversies of the epoch hid antagonisms between populations and conflicts of interest that were anything but purely scientific. This was undoubtedly true of the controversy between the Newtonians and the Leibnizians regarding infinitesimal calculus which exploded at the beginning of the eighteenth century. And in this period the same was true of the conflict between the Cartesians and the Newtonians, a controversy that involved all aspects of physics.

We have already mentioned that the decade 1670–80 saw the defeat of the reactionary forces in France who were opposed to Cartesianism. The Cartesian current became increasingly more victorious during the last decade of the century. The main protagonists of this victory were Fontenelle, who was elected the secretary of the Académie des Sciences in 1699, and made it into the major bulwark of Cartesian science, and Pierre-Sylvain Regis (1632–1707), who in 1690 published a three-volume work, *Systèmes de philosophie, comprenant la logique, la metaphysique et la morale*, in which Descartes's physics was given authority by being set in a sumptuous logical and metaphysical framework.

Even English scientific circles felt the need to update themselves on Cartesian physics and its most recent developments, as evidenced by the publication, in 1697, of the English translation of Rohault's famous treatise of 1671. However, the work was edited by the Newtonian Samuel Clarke, who, even though he praised the text that had been translated, added various notes and clarifications expressly formulated to induce the reader to recognize the superiority of Newton's physics over that of Descartes.

CHEMICAL SCIENCES

The decline of alchemy

German scientific texts progressively abandoned the use of Latin. All the works of Johann Kunckel (1630–1703), the most important of which is the *Collegium physico-chymicum experimentale, oder laboratorium chymicum*, which appeared posthumously (1716), were written in the vernacular. Alchemy still played an important role. For example, air and water were considered elements, and the work says that metals are generated in the bowels of the earth from water being heated with the "universal seed." In certain respects, however, Kunckel distanced himself from alchemy, for example when he denied that sulphur was an ingredient in metals and when he asserted that the *Alkahest* was nonsense ("*alles Lugen ist*"). He was even modern when he worked with quantitative experiments, such as determining the concentration of nitric acid by saturating it with silver, evaporating it, and weighing the residue of silver nitrate. He determined the amount of silver in solution by using common salt to precipitate it as a chloride.

Quantitative experiments were described by others as well. S. Boulduc attempted to determine the quantity of acid in distilled vinegar by neutralizing it with potassium carbonate, evaporating it, and weighing the residue of potassium acetate (1698). Wilhelm Homberg (1652–1715) applied the method to various acids, wrongly concluding that all acids differed from one another only in their water content (1699). After making measurements of the density of air, together with Edme Mariotte (1620–84), Homberg deduced that it depended on temperature and humidity as well as on pressure (1698).

LIFE SCIENCES

The reproductive mechanism and plant sexuality

Through his studies on plant lice, A. van Leeuwenhoek found that they were viviparous (producing living young, not eggs) and discovered the phenomenon of parthenogenesis (development of the egg without fertilization). Almost simultaneously, Rudolf Jakob Camerarius (1665–1721) furnished several experimental proofs of plant sexuality, and Joseph Pitton de Tournefort (1656–1708) published the *Elements of Botany*, which contained a clear and simple method for classifying plants.

The return to Hippocratism

De praxi medica by Giorgio Baglivi (1697) presented a typical example of the new trend in medical science. While Baglivi defended the ideas of the school of iatromechanics, he proclaimed the need for a return, as far as clinical medicine was concerned, to Hippocratism, i.e. to an observation of illnesses that was not marred by doctrinal prejudices. The works of Thomas Sydenham became models for the new way of describing diseases and the unbiased study of environmental pathogenetic factors.

Malpighi's *Opera posthuma*, which included his famous *Risposta* to the objections that Giovanni Girolamo Sbaraglia has raised against rational medicine, appeared in 1697. Some modern scholars consider the *Risposta* Malpighi's true scientific testament.

EARTH SCIENCES

Fossils, the flood, and the history of the earth

The problem of the link between fossils and Noah's flood had already caught the attention of scholars during the preceding decades. Now, however, it took on particular importance in the wake of numerous discoveries of fossil bones. Among these were Thomas Molyneux, for what was later called the Irish megaceros deer (1697), and Ernst Tentzel for the fossil elephant from near Gotha, which he correctly identified. There were general works that spoke of fossils as well, mostly connecting them with Noah's flood. William Whiston (1667–1752) supported that idea in his *New Theory of the Earth* (1699), a work full of fantastic and sometimes extravagant ideas on the history of the earth. John Woodward (1665–1728) also took the diluvialist position in his *Essay toward a natural history of the Earth* (1695), although he recognized the real nature of fossils. Edward Lhuyd (1660–1709) on the other hand, an antiquarian and collector, was of the opposite opinion, supporting the idea that fossils originated from seeds scattered in the rocks. His *Lithophilacii britannici iconographia* (1699), which presented this view, was an extremely influential work. Lhuyd also rejected the common ideas on the Flood, as did John Ray (1627–1705), one of his correspondents, who was uncertain whether fossils were of organic origin or not (see his *Miscellaneous discourses concerning the dissolution and changes of the world*, London, 1692, 1693, and 1713, and other writings).

De generatione conchitarum, a paleontological work by Johann Jakob Scheuchzer (1672–1733), in which fossils are considered *lusus naturae,* appeared in 1697.

Earthquakes and volcanoes

The general works mentioned in the previous section also discussed the causes of earthquakes and eruptions, often advancing fantastic ideas. R. Hooke linked earthquakes and volcanic eruptions but did not reach a definite conclusion as to their causes, although he seems to connect them to the work of underground sulphurous vapors. Martin Lister (1638–1712) had attributed eruptions to the burning of underground sulphur, and Nicolas Lemery (1645–1715) attempted to test this theory experimentally in 1700.

Gottfried Wilhelm Leibniz (1646–1716) indicated the central fire as the prime cause of many events in the first draft (1693, *Acta Eruditorum*) of his *Protogaea,* published posthumously in 1749, which presented an idea of the development of the earth's history that agreed with the Biblical texts.

In 1693, R. Buckley discovered the Giant's causeway in Antrim county, Ireland, a characteristic basalt formation. S. Foley and T. Molyneux described it in detail in 1694 and 1698.

Artesian waters

In 1691, Bernardino Ramazzini (1633–1714) advanced an interpretation of the phenomenon of artesian waters in the area of Modena (fed by the Apennines), although he was unsure whether the water originally came from rain and snow or from the sea (*De fontium mutinensium admiranda scaturigine tractatus physio-hydrostaticus*).

Minerals

In the area of prospecting, for minerals, Martin Lister's suggestion to the Royal Society of London that a mineralogical map be drawn up deserves mention. It was actually done much later.

The privateer Dampier and oceanography

Although William Dampier (1697) was a privateer and sometimes pirate, he published a detailed and succinct report on the winds, storms, currents, and tides in the equatorial zones, performing a remarkable service to navigation and geographical knowledge.

Urban Hjarne (1641–1724) linked the variations in the shorelines of Scandinavia to the motion of the ocean's waters.

APPLIED SCIENCE

Sources of energy

Water and wind continued to be the major sources of energy throughout the seventeenth century, as they had been in previous centuries. To serve the growing needs of workshops that continued to enlarge (some of them already employed several hundred workers), gears were invented to exploit this energy. But the possibility of utilizing a new kind of energy—thermic energy—arose.

Denis Papin's idea found its first practical application when Thomas Savery (c. 1650–1713), a military engineer, invented the steam pump during the last decade of this century. Steam was condensed to create a vacuum, which sucked water up as high as atmospheric pressure would allow. The water was then collected and lifted even higher by applying maximum steam pressure. Savery obtained his most important patent in 1698 for a "new invention for lifting water and producing motion in any type of factory, by means of the force exerted by fire." In 1699, he demonstrated his pump to the members of the Royal Society.

Innovations in glass-working

Almost all the industries were able to increase their production, although this was due more to better internal organization and an increased number of employees than to significant technical innovations. One innovation of the period was Lucas de Nehou's method for molding glass, which was introduced in the Saint-Gobain workshops in France during this decade. The method made it possible to produce larger sheets of glass than ever before.

Construction

Remarkable construction projects were carried out during this period, especially in France and England. Some of the most beautiful squares of London, such as Grosvenor Square (1695), Berkeley Square (1698), Red Lion Square (1698), etc., were developed during the last few years of the century.

The first serious attempt to state a theory on arches for bridge building dates back to 1695: Philippe de la Hire (1640–1718) published his *Traité de mécanique,* which contained the "theory of the smooth wedge."

Scientific instruments

Special attempts were made to improve the accuracy of scientific instruments. Clocks in particular were perfected when precious stones began to be used as almost frictionless, non-wearing pivots. The idea was introduced by Nicolas Faccio (1664–1753), a Swiss who settled in London in 1687.

In the field of practical scientific instruments, Robert Boyle, in his *Medicina hydrostatica* (1690), suggested using hydrostatic scales to make densimetric checks on rainwater, wine, *aqua vitae*, cider, beer, etc.

The English fleet

At the end of the seventeenth century, the English mercantile fleet, already the largest in the world, counted almost 3,400 ships and a capacity of 300,000 tons. This was doubled during the next fifty years, and reached a capacity of 1.5 million tons by the end of the eighteenth century. Among the technical improvements were steering wheels, which were introduced on English ships in 1705.

Treatise-writing

Some of the most important works on hydraulic architecture came out in Italy during this decade. In 1696, Cornelis Janszoon Meijer published his most important book, *L'arte di rendere i fiumi navigabili in varij modi con altre nuove invenzioni, e varij altri segreti*, a work that offers a picture of the status of several branches of technology, especially hydraulic technology, in Italy at the time. In 1697, Domenico Guglielmini (1655–1710) published his *Natura dei fiumi*. In 1699, Giovanni Battista Barattieri published his *Architettura d'acque,* which brought together essays that had been published separately in 1656 and 1663.

The gap between scientific and technological competence at the end of the seventeenth century

From the point of view of Western humanity's progressive domination over nature and the world, the end of the seventeenth century respresents an important milestone, but at the same time, a halting point, as the physicist and historian of science Kurt Mendelssohn has so accurately observed in his *Science and Western Domination*: "The 'century of genius', as it has often been called, had set the stage, but the action of the drama to be performed on it was as yet unknown.' "[1] In fact, the laws of nature had begun to unveil their secrets more clearly, and the quantitative method, refined with the techniques of infinitesimal calculus, had already provided the mathematical view of reality that would remain substantially unchanged during the following centuries. However, the gap between theoretical knowledge and practical application was still quite large. At this point, a decisive step needed to be taken: machine power had to be exploited for the mass production of work in the service of humankind.

The century that was about to begin opened with an emphasis on research aimed at the optimum exploitation of power (the steam engine, for example). Only with this premise could intellectual domination over nature gradually extend to the concrete technological and industrial achievements that would lead to the effective domination of human beings over nature and the world.

---

[1] Mendelssohn, Kurt, F.R.S., *Science . . .* , London, Thames and Hudson, 1976, p. 104.

The Eighteenth Century

1700–1710

THOUGHT AND SCIENTIFIC LIFE

Philosophical debates and developments in scientific culture. Leibniz

During this decade, the sciences began to be taught outside the traditional centers of Western Europe: In America, Yale College was founded in New Haven, Connecticut in 1701; and in Russia, as a part of his vast westernization plan, Peter the Great created the first great Russian hospital, and in 1707 had a medical school opened. At the same time, the cultural scene in the German-speaking countries was particularly active. Leopold I founded the University of Breslau in 1702, and almost simultaneously the nucleus of the future Academy of Science was being formed in Berlin.

The most prominent figure in Berlin was Gottfried Wilhelm Leibniz (1646–1716), who was involved in organizing scientific projects and in philosophical debates. Feeling a need to respond to the empirical arguments in Locke's *Essay*, which had become extremely popular, Leibniz prepared his *Nouveaux essais sur l'entendement humain*, in which he defended innatism. However, when he received news of Locke's death in 1704, he decided not to publish the work. He continued working in various areas of science; for example, in 1705 he wrote a treatise on biology entitled *Considérations sur les principes de vie et sur les natures plastiques.*

Deism and English culture

During these same years, Locke's ideas were the center of a huge debate in England, where reactions to the publication of Toland's work (1696) were also very strong. Locke denied even indirect paternity for Toland's ideas, while Samuel Clarke

(1675–1729) attempted to give a rational proof of the truth of Christianity in his Boyle lectures of 1704–05. This effort to respond to Toland's theories on a rational basis is evidence that rationalistic needs were being felt even in those circles where tradition was defended. The first great investigations of classical economy and the early theories of moral sentiment arose at almost the same time as Clarke's Christian deism: *The Fable of the Bees, or Private Vices, Public Benefits*, by Bernard Mandeville (c. 1670–1733), came out in 1705, and "A Letter Concerning Enthusiasm," by Lord Ashley, Earl of Shaftesbury (1671–1713), was published anonymously in 1708.

MATHEMATICAL SCIENCES

The controversy over infinitesimal calculus

The conflict between Gottfried Wilhelm Leibniz (1646–1716) and Isaac Newton (1642–1727) over priority in the invention of infinitesimal calculus began in 1699. Up to this time, there had been no reason for conflict or rivalry over the discovery of the new calculus, but in 1699 a certain Fatio de Dullier published a pamphlet in London that created a controversy between the two mathematicians. Besides the two principal contenders, a number of other major and minor figures intervened in the debate over a period of more than two decades. The Royal Society, of which both Newton and Leibniz were members, became very much involved in the debate. A commission was appointed to resolve the conflict by examining a series of documents. The commission, probably maneuvered by Newton, who was president of the Royal Society from 1703, decided in favor of the English author.

This delayed controversy, provoked by others and soon fed by nationalistic political conflicts, does not appear to have been historically justified. In one sense, infinitesimal calculus cannot even be called a true invention, given the many long and detailed developments that led up to it. If, however, we consider only the strictly algorithmic aspect, we cannot deny that Newton and Leibniz actually devised the new calculus. But it is clear that the two great figures reached their conclusions via different paths, independently of one another. As to strict chronological priority, Newton was the first to formulate the rules of the calculus, but he did not publish anything, so the first publications came from Leibniz. Much more importantly, Leibniz was fully aware of the potential and the full scientific legitimacy of the new formal tool.

The development of infinitesimal analysis

Despite the bitterness of the conflict over infinitesimal calculus, which was carried on by Newton after Leibniz's death in 1716 and continued by the followers of both even after Newton's death in 1727, the new calculus immediately flourished all over Europe. From a mathematical point of view, it can be said that the eighteenth century was the century of infinitesimal calculus. Even though truly new ideas appeared quite rarely (with the important exception of the concept of partial derivatives and the consequent flourishing of the theory of differential equations), the application of the new methods led to an enormous mass of both theoretical and applicative findings. This was above all the result of the productive relationship that soon developed between analysis and rational mechanics. Besides furthering technical developments in both areas, this relationship appeared to the scholars of the times as a brilliant confirmation of the program of rationalization that they were promoting in all areas of learning.

The Swiss German school immediately distinguished itself in promoting the progress of analysis, as the mathematical dynasty of the Bernoulli family began. Johann (Jean) Bernoulli (1667–1748), who left a huge patrimony of findings concerning various questions of infinitesimal calculus in trigonometrical research, in the area of integrations, ordinary differential equations, and series, became a prominent figure during these years. The great Leonhard Euler came out of the same school a short time later.

The new calculus also reached Italy: Luigi Guido Grandi (1671–1742) was the first to apply the Leibnizian algorithm to a squaring problem (1703), and in 1707 Gabriel Manfredi (1681–1761) published his *De constructione aequationum differentialium primi gradus*, the first Italian work dedicated to pure integral calculus. Manfredi's work contained a valuable arrangement of the theory of differential equations. Jacopo Riccati (1676–1754) also made a number of contributions to the same theory, while Giulio Carlo Fagnano (1682–1766) formulated several important theorems on integral calculus.

It is also important to remember that Newton himself published some remarkable mathematical works during the first years of the century. His *Enumeratio linearum tertii ordinis* (which appeared as an appendix to the *Opticks* in 1704) can be considered the beginning of a general theory of curves, while his *Arithmetica universalis* (1707) was a collection of lessons given by Newton, particularly on the theory of algebraic equations.

ASTRONOMY

The calendar and ephemerides

In 1700, construction was begun on the Berlin Observatory, where astronomical observations began to be made six years later. The first director of the observatory was Gottfried Kirch, who died in 1710, one year before construction was completed. That same year, the Protestant states of Germany adopted the Gregorian reform (except for the Gregorian calculation of Easter, which was only accepted in 1775) by eliminating the last eleven days of February. Denmark and the Low Countries had also conformed in 1700. Governments were becoming increasingly aware of the practical importance of problems connected with determining time and the motion of the stars, and of circulating the relative findings. In 1702, the French government became the first to begin publishing annual ephemerides. The "Connaissance des temps" continues to be published even today; until 1759 it contained only the ephemerides, but in 1760 the new editor, Joseph Jêrome de Lalande (1732–1807) initiated the tradition of including astronomical information as well (see ASTRONOMY 1750–1760).

New instruments

It was in this decade that important new observational instruments were invented or improved upon. One of the most significant of these was the meridian circle, designed and built by Olaus Römer (1644–1710) when he returned to Denmark. Until the beginning of the eighteenth century, the best instrument for measuring the positions of the planets and determining their distances as they passed over the meridian was the mural quadrant, invented and used by Tycho Brahe. With it, the meridian arc was described with a maximum approximation of about 5 seconds, while the error in determining the time of passage of a planet due to inaccuracies in aiming the instrument sometimes amounted to seven to eight seconds. Römer's transit instrument, or meridian telescope, like the modern versions, was a telescope mounted orthogonally on an axis aligned in an east-west direction, which could rotate around the axis while remaining on the meridian plane at all times. This instrument was more accurate than the previous one, since it was built around an axis worked on a lathe; it was further improved when Römer applied a micrometer with illuminated cross-wires to it. This was one of Römer's last efforts before his death in 1710. Eighteen years later, a fire destroyed his observatory and, along with it, examples of this and other instruments that he had built.

The orbits of the comets

Meanwhile, one of the greatest achievements ever was about to be made in theoretical astronomy. In his *Astronomia cometicae synopsis*, published in 1705, Edmund Halley (1656–1742) stated that, using Newton's theory of universal gravitation, he had calculated the orbits of several comets and discovered that the comets that had appeared in 1531, 1607, and 1682 were in reality the same comet, the orbit of which had a period of between seventy-five and seventy-six years. He predicted that the comet would reappear in 1758, which it did. This greatly helped to demystify those heavenly bodies, whose apparently sudden and unpredictable apparition had been the source of a superstitious terror ever since antiquity. Halley's discovery proved these fears to be totally unfounded. Halley also made an important contribution to practical astronomy during this decade by proposing that the motion of the moon be determined by the way in which it moves with respect to known stars. He explained this method in the second edition of Street's *Astronomia carolina*, which he edited in 1710.

PHYSICAL SCIENCES

Newton's Opticks

The first great work in physics of the century, *Opticks, or a tractatus of the reflections, refractions, inflections and colours of light*, by Isaac Newton (1642–1727), appeared in 1704. It was in three books: the first two, similar in approach to Newton's *Principia*, were aimed at demonstrating that optics should essentially be treated mathematically, although the treatment should be based on numerous accurate experiments. The third book, on the other hand, took a completely different approach: first of all, because it consisted of a certain number of queries, quite independent of one another (later added to and partly modified in the various editions that came out during the author's lifetime); secondly, because Newton dealt with purely physical subjects as a physicist and not as a mathematician; and finally because in discussing these topics, he used methodological arguments that were very interesting from a philosophical point of view. The most characteristic aspect of Newton's treatment is that

he expressed deep perplexities about many of the questions he approached, especially the problem of the composition of matter. He displayed an openness to various solutions, and although he tended to favor some of them over others, he was careful not to present any of them as definitive. In fact, he carefully and objectively outlined all the arguments for and against each of them.

Research in acoustics and electricity

Experimental research on sound phenomena grew during this decade. In 1705, the English theologian and physicist William Derham (1657–1735) was able to prove that wind affected the velocity of the transmission of sound. Accurate investigations carried out by various scholars, in particular the English physicist Francis Hauksbee (c. 1666–1713), led to a final consolidation of Otto von Guericke's discovery that sound was not transmitted in a vacuum. To test this finding and perform other interesting experiments, Hauksbee made significant improvements on Guericke's vacuum machine, building the first two-cylinder pneumatic pump in 1709.

This great English experimenter also did a number of experiments on electrical phenomena, especially on discharges in gasses (1706). Nor should we forget that it was Hauksbee who introduced the use of glass rods, electrified by being rubbed with a piece of wool, into electrical research.

CHEMICAL SCIENCES

Iatrochemistry, the last frontier of alchemy

Alchemy essentially declined between the end of the seventeenth and the beginning of the eighteenth centuries, while iatrochemistry was still in its prime. In their attempts to isolate the active ingredients in medicinal herbs and to prepare synthetic remedies from minerals that they hoped would reestablish the altered chemical balance in diseased organisms, the iatrochemists laid the foundations not only of pharmaceutical chemistry and biochemistry, but of the chemical sciences in general. A typical figure of this period was Wilhelm Homberg (1652–1715), who was basically an iatrochemist, although he was not immune to the fascination of alchemy. In fact, in 1709 he claimed to have changed silver into gold. But he was also interested in chemistry as an experimental science. The best-known result of his experiments was the discovery of boric acid (*sel narcotique du vitriol*), which he obtained in the form of a sublimate in 1702 by heating borax with ferrous sulphate.

The phlogiston theory

In 1703, Georg Ernst Stahl (1660–1734), another iatrochemist, took the first step toward formulating the famous phlogiston theory. He had his teacher Johann Joachim Becher's *Physica subterranea* republished, prefacing it with a long introduction entitled "Specimen beccherianum," in which he outlined the basics of his theory, already alluded to several years earlier and developed over the course of the next twenty-five years. The term *phlogiston*, which Becher and others had already used, and which Aristotle had used to mean flammable, was here used to mean combustible (*brennlich*). According to the theory, combustible substances and metals that became "lime" when they became red-hot in the presence of air lost a combustible principle during combustion or calcination. This principle was the phlogiston. The theory either ignored or devised awkward explanations for the fact that the weight of metals increased, rather than decreased, during calcination. Although the theory was wrong, it stimulated research on oxidation phenomena and the quantitative aspect of chemical reactions. It was the first attempt to summarize chemical knowledge from a unitary theoretical point of view.

LIFE SCIENCES

The development of organisms and spontaneous generation

The debate over the generation of living creatures and how to interpret the development of organisms was one of the central themes of scientific interest during this century. In his *Dialoghi sopra la curiosa origine di molti insetti*, published in Venice in 1700, Antonio Vallisnieri (1661–1730), a student of Marcello Malpighi and a continuer of Francesco Redi's work, denied that the insects that caused plant galls were produced by spontaneous generation, as Redi had believed, admitting this one exception to the general rule that no living species could originate from another living species. In a later work entitled *Considerazioni ed esperienze intorno alla generazione spontanea dei vermi ordinari del corpo umano* (1710), Vallisnieri demonstrated that the law of generation from eggs also held true for intestinal parasites. He later placed these works into a preformational framework in his *Istoria della generazione*. The debate over generation extended to pathology as well: the French naturalist Nicholas Andry (1658–1742) attributed diseases to eggs in the atmosphere that penetrated into the body. For this reason he was nicknamed *homo vermiculosus*.

In 1706, the Bolognese explorer and naturalist Luigi Ferdinando Marsigli (1658–1730) sent to the Académie des Sciences in Paris a memoir describing flowers that protruded from branches of coral. Marsigli's observations were repeated by Jean André Peyssonel (1694–1759), who concluded that the projections were not flowers, but tiny animals.

The influence of mechanism and vitalism

The conflict between mechanism and vitalism proved to be of fundamental importance not only in biological research, but in practical medicine as well.

On the one hand, the life sciences were increasingly influenced by the methods and basic concepts of the new science of motion and adopted its tendency toward a mechanistic cosmology. In his *Fundamenta medicinae* (1703), Friedrich Hoffmann (1660–1742), a professor at Halle and court physician in Berlin, stated the principles of a humoral system, in which organic functions depended on the circulation of the humors (blood, lymph, and nerve fluid). The flow of the humors was in turn caused by the ingestion and assimilation of food. Living matter possessed an intrinsic motility; diseases were caused by external agents, which passed through the digestive tract and damaged the spinal cord, creating specific disturbances; the disturbances basically consisted in changes in the "tone" of the fibers of the organism under the influence of abnormal nervous stimuli. Hoff-

mann's notion of tone opened the way to neural pathology, which in the last quarter of the century also attributed the various forms of irritability observable in living organisms to the nervous system.

The premises and method of mechanistically inspired medicine were further defined, with remarkable and unusual balance, in the *Institutiones medicae*, by Hermann Boerhaave (1668–1738), the greatest exponent of iatrophysics. The work was published in Leyden in 1707 as a textbook for students of physiology, since nothing of the kind existed. Boerhaave distanced himself from the iatrochemical views of his teacher, Franciscus Sylvius de le Böe, and followed the physical and mechanistic ideas of Giovanni Alfonso Borelli, Lorenzo Bellini (1643–1704), and Giorgio Baglivi (1668–1707), since he was aware of the conceptual inadequacies of eighteenth-century chemistry, which was still entirely based on a presumed "affinity between bodies." Boerhaave's *Institutiones* is divided into five sections: *oeconomia animalis*, *de morbis*, *de signis*, *de sanitate tuenda* and *de morborum curatione*. According to the author, the body is made up of solids immersed in humors (solidism). Since the soul is immaterial, it does not influence the body. The balance of the solids, or state of health, is determined by the external action of the air and the internal action of the humors. Air and humors can cause diseases if they change. The Dutch physiopathologist's fame spread throughout Europe: among his pupils were Albrecht von Haller, Julien Offroy de la Mettrie, and Carolus Linnaeus.

On the other hand, the growing influence of the physical disciplines (mechanics in particular) on the life sciences was contrasted by a vitalistic and animistic current that not only claimed that the vital processes were qualitative in nature, but even connected them with immaterial and, in some ways, conscious factors. In 1706, the German physician and chemist Georg Ernst Stahl (1660–1734) published a memoir entitled *Disquisitio de mechanismi et organi diversitate*. He followed it with a *De vera diversitate corporis mixti et vivi* and the large work *Theoria medica vera* (1707). Together, these three works present the theses of biological animism. According to the author, life depends on an incorporeal entity, the soul, which acts on the body through circulation, secretion, and excretion. Disease is the result of a malfunction in the soul. The academic discussions between Hoffmann the mechanist and Stahl the animist soon became open rivalry.

Specialization in medicine

While the two currents just mentioned competed in the field of general medical issues, a phenomenon that would soon have very significant consequences was taking place on another front: the tendency was growing for medicine to be subdivided into specialized branches, which often corresponded to modern divisions. The fields of occupational pathology and hygiene were created when a work by Bernardino Ramazzini (1633–1714), *De morbis artificum* was published in 1700. In this work, the author described the major illnesses caused by almost fifty different job activities, including saturnism, the syndromes caused by handling antimony and tin, and mercurialism. He also proposed preventive measures. The importance of Ramazzini's *De morbis* would be understood in the following century, in the light of industrial labor pathology, although silicosis, another widespread occupational disease, was already being studied by eighteenth-century English doctors. Another special discipline, otology, which combined aspects of anatomy, physiology, and medicine, was created with *De aure humana* (1704), a work by Antonio Maria Valsalva (1666–1723).

EARTH SCIENCES

Geological knowledge at the threshold of the eighteenth century

At the beginning of the Age of Enlightenment, geological knowledge, certainly not alone among the sciences, was characterized by an unusual interweaving of accurate observations with explanations that can definitely be called fantastic. The scientific heritage of the Middle Ages, and of classical antiquity as transmitted through the Middle Ages, was rapidly disintegrating, although it still exerted quite a bit of influence, especially on broader, more general problems, such as those of cosmology, which were directly connected to geology. One of the issues often debated was the coherence of scientific ideas with the Bible. Although the principle of a non-literal interpretation of the Bible had already been introduced with the affirmation of the Copernican system, most scholars still compared each new scientific explanation with the literal version of the sacred texts.

One of the most widely debated problems was that of the origin of fossils. Although Leonardo da Vinci, (whose works were still unknown), Girolamo Fracastoro (1517), Bernard Palissy (1580), and others had already clearly supported the theory that fossils were the remains of organisms that had once lived in the sea and had later become petrified, most scholars still believed that they were concretions formed directly within the rock as a result of the most varied and completely improbable causes. Aristotle's error continued to influence scientific thought.

Edward Lhuyd listed over a thousand types of fossils and attributed their existence to an *aura seminalis* in his Lithophylacii *britannici ichlographia* (1699), while Karl Niklaus Lang (1670–1741), in his *Historia lapidum figuratorum Helvetiae* (1708), provided excellent descriptions of the fossils and minerals that he had gathered together in a museum of natural history in Lucerne. The idea of the inorganic origin of fossils finally ended with a tragicomic episode in 1726. However, good descriptions and treatments of fossils were being made even at this time, for example David Spleiss's work on the mammal remains found around Stuttgart, and the various publications of Johann Jakob Scheuchzer (1672–1733), a physician in Zurich. In 1708, in his *Piscium querelae et vindiciae*, Scheuchzer declared that he was abandoning the idea of the concretional origin of fossils in favor of the Flood theory. In doing so, he proclaimed himself a follower of John Woodward (1665–1728), with whom he carried on an active scientific correspondence. Scheuchzer produced a large number of paleontological works; during this particular period he wrote a description of fossil plants (1709).

Cartography

In 1700, Guillaume Delisle (1675–1726), considered the first modern cartographer, began publishing geographical maps that quite accurately showed the coasts of the continents.

The introduction of the seismoscope

In 1703, Jean de Hautefeuille (1647–1724) built a rudimentary seismoscope that used drops of mercury. Though very primitive, it was probably the first instrument ever built in the West for studying earthquakes. It gave the direction and the intensity of shocks.

Marsigli's oceanography

Fundamental to the scientific research of this period was the work of Count Luigi Fernando Marsigli (1658–1730). At the beginning of the eighteenth century, after an adventurous life, Marsigli resumed the oceanographic investigations that he had begun some years earlier. His work was particularly intensive between 1706 and 1707. He measured depth, currents, temperature, and salinity, mostly in the waters of the Mediterranean south of France. He also made some original contributions in the area of instrumentation.

Marsigli made a number of synthetic deductions in interpreting his findings. One of his most important observations, which Marsigli was the first clearly to point out after intensive explorations in the Gulf of Lions, was the existence of the continental shelf, evidently an underwater continuation of the continental areas above sea level. The shelf was clearly distinct from the ocean floor and was separated from it by a steep slope, where depth increased rapidly, although it did not reach unmeasurable levels as the ancients had believed. On the basis of his observations, Marsigli deduced that the southern European mountain chains and the Mediterranean had the same origins and that the rock formations of the ocean floor were just like those found in nearby parts of the continent, except for the deposits of sediment.

As for meaurements of temperature and salinity, Marsigli used instruments based on modern concepts. He also used quite accurate determinations of specific weight to measure salinity.

Although Marsigli's research in marine thermometry was still done using instruments that were not really adequate, he found that deep-water temperatures in the Mediterranean were constant, as were surface temperatures. In addition, he found that the temperature generally fell as the water got deeper. These findings were extended a priori to the oceans, and later observations confirmed their validity as generalizations.

APPLIED SCIENCES

Progress in metal working. The cupola furnace

Innovations in the field of metallurgy, which came, in part, from the need to provide military weapons, assumed a leading role at the beginning of the new century. The experts concentrated above all on the problem of utilizing coke in the fusion process. A new type of furnace, called the cupola furnace and used for refusing pig-iron, was adopted in 1702 and partially solved the difficulties. The cupola furnace was an improved kind of reverberatory furnace in which the iron did not come into direct contact with the coal. Coke could conveniently be used to produce pig-iron, while the cupola furnace made it possible to produce a much better quality iron. Thanks to the use of the cupola furnace, instances of accidental explosions of cannon on English ships were drastically reduced.

In 1709, it seemed that Abraham Darby (1677–1717) had made a major advance when he succeeded in melting iron ore with coke, but it was later discovered that his success was due to the nature of the peat found near Coalbrookdale, which contained an unusually low percentage of sulphur. This explains why, during the early period of fusion with coke, the technique was limited to the region of Coalbrookdale and a small number of nearby furnaces.

In 1704, Christopher Polhem (1661–1751) built a factory for working iron and other metals at Stjärnsund, Sweden. The plant was designed to make use of water power in all the possible steps of production, limiting manual labor to the finishing work. Polhem's factory was designed with a logical, codified division of all the productive processes in mind and stands as evidence of the progress that had been made in methods of organizing work. Polhem's system of labor achieved the highest level of perfection ever reached in an ironworking industry that was based solely on the use of wood charcoal, water power, and human and animal power.

The mining industry

The demands of the ironworking industry, especially after the introduction of coal into the ironworking process, stimulated the mining industry enormously. But its expansion seemed to be insurmountably limited by the type and quantity of power available. Mining and industrial experts were perfectly aware of the need to find new sources of power, since the costs of water and animal power were becoming prohibitive. Anyone who could produce a pratical steam-powered engine for removing water from mines would be assured of great success. Inventors immediately began to make the first feeble attempts. In 1702, Thomas Savery (c. 1650–1713) published *The Miner's Friend*. In 1705, Denis Papin (1647–1712) attempted to improve upon Savery's engine, abandoning his own earlier plan of condensing steam under a piston in order to obtain driving power from the effects of atmospheric pressure.

The porcelain manufactories of Meissen

Developments in other technological sectors were less important. Johann Friedrich Böttger (1682–1719), an alchemist at the court of the Great Elector of Saxony, in collaboration with the mathematician Ehrenfried Walter von Tschirnhaus (1651–1708), worked on improving the techniques for manufacturing porcelain. In 1707, he obtained an extremely hard red stoneware by adding a flux (alabaster or marble) to the clay of Saxony, the melting point of which was too high. Using a suitable white clay (kaolin), Böttger finally obtained a mixture that could be used to make white porcelain. Thus a technique that had escaped European potters for generations, as they futilely attempted to discover the secret of manufacturing Chinese porcelain (imported to Europe as early as the sixteenth century), was finally developed.

The first porcelain factory was opened in Meissen, near Dresden, in 1710, more than fifty years before those at Berlin and Sevres.

Various technologies. Marturin Jousse's L'art de charpenterie

Other improvements in various branches of technology took place during the early years of the eighteenth century.

Some progress was made in agricultural tools when Jethro Tull (1674–1741) invented a new type of sower in 1701. In the area of scientific instruments, the first two-cylinder pneumatic pump was built by the English physicist Francis Hauksbee (c. 1666–1713) in 1709. Meanwhile, Daniel Gabriel Fahrenheit (1686–1736) devised his thermometer scale, which was only described in 1724 in *Philosophical Transactions*.

A treatise on carpentry, *L'art de charpenterie*, by Maturin Jousse, was published in Paris in 1702. It contained the first technical description of a mill, complete and detailed enough to serve as a guide for building windmills, which were still a valuable source of energy. Jousse's work is also famous because Diderot drew most of the information for the entry on mills in his *Encyclopédie* from it.

1710–1720

THOUGHT AND SCIENTIFIC LIFE

Also important was the development of the Paris and London academies of science, which continued their activity of centralizing a large part of scientific research. The Paris Academy investigated the novelty and usefulness of inventions, and descriptions of several hundred machines and inventions approved by the Academy were published. In 1716, the Academy adopted a new set of rules, increased the number of its members, and generally reorganized. There were other active institutions as well: the Royal Society of Sciences was founded in Uppsala in 1710; the University of Bologna was restructured in 1711, incorporating the Academy of Sciences that had been founded in 1690; and the Spanish Royal Academy was founded in 1713 in Madrid, under the auspices of Philip V, who officially approved its statute in 1714. Specialized institutions were enlarged, and new one were created. The teaching of surgical medicine was improved, and new anatomy amphitheaters were built. As to technical schools, the Bridge and Road Corps, which brought together engineers at the service of the King of France, was created in 1716; the technical schools of Prague were opened in 1717. In Paris, the Jardin des Plantes was reorganized: its direction was no longer entrusted to physicians alone, and it was transformed into a center for research in botany, zoology, anatomy, and mineralogy that served as a model for many institutions throughout the world.

A further sign of progress was the establishment of huge libraries. Two of the most important were the Biblioteca Lancisiana, created in Rome in 1711, and the Biblioteca Nazionale, established in Turin in 1719.

MATHEMATICAL SCIENCES

The first treatises on the calculus of probabilities

There was a noteworthy revival in studies on probability during this period. *Essay d'analyse sur les jeux de hazard*, by Pierre Raymond de Montmort (1678–1719), appeared as early as 1703. *Ars conjectandi* by Jacques Bernoulli (1654–1705) was published posthumously in 1713 by his nephew Nikolaus. It was a wide-ranging work which contained, among other things, Bernoulli's theorem and a statement of the so-called law of large numbers. Abraham de Moivre (1667–1754), who in 1711 had established the theorem of compound probabilities, published *The Doctrine of Chances* in 1718. In it, he developed a theory of random error that contained the idea of probable error and the normal curve of probability. De Moivre presented further findings on the calculus of probabilities in 1730, in his *Miscellanea analytica*.

This development of studies in probability took place as the theory of probabilities surpassed the limited range of problems that Pascal, Fermat, and Huygens had considered in the early research of the previous century. In their time the area of inquiry barely went beyond gambling games. The application of the calculus of probabilities to physics (it became one of the great mathematical tools of that science in the second half of the nineteenth century) could not have been foreseen this early, but the theory of probabilities was already beginning to be applied to new problems, such as actuarial and insurance calculations (establishing mortality tables, etc.) and juridical questions (the reliability of evidence, judicial errors committed by courts, etc). In general, this branch of mathematics was not yet approached with sufficient rigor and accuracy, so that its use was confined to those human sciences which, due to their very vagueness, seemed better to lend themselves to a treatment that was not yet very precise. This lack of precision was caused by the fact that the basic concepts underlying the calculus of probabilities had not yet been clarified. It was thought of as belonging to the realm of *evidentia mathematica* or to that of *evidentia moralis* by different authors at different times. This situation of uncertainty persisted even much later as can be seen in the demonstration of the theorem of inverse probability published in the *Essay Towards Solving a Problem in the Doctrine of Chances* (1763, posthumous) by Thomas Bayes (1702–61) and in Pierre Simon de Laplace's essay on the probability of causes. It was only overcome at the beginning of the nineteenth century by Laplace.

ASTRONOMY

The problem of longitude

The problem of determining longitude, for which a satisfactory solution had never been found, preoccupied navigators so much that in 1714 the British Admiralty was induced to offer a prize of 20,000 pounds sterling to anyone who could find a method for determining longitude at sea within half a degree, which is equal to thirty nautical miles. In 1716, the Duke of Orleans, the regent of France, also promoted a competition.

The shape of the earth

While the problem of the position of a traveller moving along the surface of the earth was still far from being solved, a controversy broke out over the very shape of the earth. The flattening of the earth, first suspected by Jean Picard, was determined only after a long period of hesitation. For the moment, since the only measurements made of meridian arcs were too close to one another, it was not possible to determine whether the meridian degrees were longer in the south or in the north of France, that is whether the earth was flattened at the equator

or at the poles. In 1718, in an attempt to solve the dilemma of whether the earth was a lemon or an orange, as someone said, Jacques Cassini (1677–1756), along with Giacomo Filippo Maraldi (1665–1729) and Phillipe de la Hire (1640–1718), completed the calculation of the meridian begun by Picard and continued by Cassini's father. The Paris meridian arc was measured for eight degrees, five minutes between Dunkerque and Collioure. On the basis of this measurement, Cassini wrote two articles (1713 and 1718), which were published in the book *De la grandeur et de la figure de la Terre* (1720, reprinted in 1723). In them, Cassini expounded in detail his opinion that the earth was elongated at the poles (a spheroid). His theory was harshly criticized by the Newtonians.

The proper motion of the stars

That same year, Edmund Halley made a very important discovery. While comparing the positions of the stars given in the old catalogues with those listed in the newer ones, he found that some of them, such as Aldebaran, Sirius, and Arthur, had definitely moved in the heavenly vault. This discovery destroyed one of the oldest myths, i.e. that the stars were immobile (the ancients had even believed that they were attached to a single sphere). It also opened the way to later research in stellar dynamics.

PHYSICAL SCIENCES

The spread of Newtonianism

This and the following decades saw the spread of Newtonianism throughout Europe as well as the revelation of its dual nature: what can be called a physical-mathematical component, on the one hand, and an essentially experimental component on the other. The first was an expression of the need to organize natural phenomena within mathematically elaborated structures and to consider the rigorously deductive nature of these elaborations as the sole guarantee of truly scientific study. Its goal was basically to reduce all of "true" physics to rational mechanics. The second, on the other hand, was an expression of the need to link physics to observational data, gradually increasing the amount of data available and attempting to clarify inductively rules of reciprocal relationships. The goal here was to establish, as the only authentic guarantee of truly scientific research, that no empirically unverifiable element be introduced in describing the data.

Newtonianism and Cartesianism

A look at the history of physics shows that the two currents of Newtonianism and Cartesianism persisted throughout the century, each of them giving rise to interesting research and important findings. Here it is sufficient to observe that these developments brought to light the basic conflict between the two schools. While the Cartesians attempted to construct physics around a general (philosophical) concept of matter and its properties, the Newtonians wanted to found the science of nature solely on observation and calculation. Newton's famous motto *hypotheses non fingo* became the slogan of the Newtonians in this long-lasting polemic, not, it should be understood, in the sense that they wanted to refuse as a matter of principle all hypotheses on the nature of phenomena, but in the sense that they were only willing to accept an explanation on two conditions: (1) that it be formulated in terms that were mathematical or capable of being restated mathematically, so that it could be the basis of precise calculations, and (2) that it refer to a limited range of phenomena, so that it could be subjected to experimental verification and the effective development of the phenomena could be observed as accurately as possible.

Among those who contributed the most to diffusing Newtonianism in Europe was William Jacob van's Gravesande (1688–1742), who wrote *Physices elementa mathematica experimentis confirmata sive introductio ad philosophiam newtonianam* (1719–20), a treatise that became one of the most popular works on physics of the epoch.

France put up the strongest resistance to the penetration of Newtonian science. This situation changed only after 1730, partly due to the work of Voltaire and Enlightenment philosophers in general and partly because of the support that Newtonian science received from the evidence of the facts.

Fahrenheit's thermometer

One of the most valuable additions to experimental physics during this decade was made in the field of thermometry. Around 1714, the German-born Dutch scientist Gabriel Daniel Fahrenheit (1686–1736) built an alcohol thermometer (which he himself later converted to mercury) that was exactly graduated according to the scale that even today bears his name. In 1730–31, the Frenchman René Antoine de Réaumur (1683–1757) and the Swede Anders Celsius (1701–44) did further research on building thermometers and proposed two more basic thermometer scales.

CHEMICAL SCIENCES

The concept of affinity and the use of symbology

The recognition of chemistry as a science in and of itself had already led to the accumulation of new experimental data; these were now being organized in an attempt to interpret them better.

In 1718, Étienne-François Geoffroy (1672–1731), known as L'Aîne, presented his *Table des Différens rapports observés en chymie entre différentes substances*, a table of the affinities between various acids and alkalis or metals. Although it contained many errors and gaps, the table gave a general picture of chemical reactions. It was especially remarkable in that it concretized and delimited the concept of affinity and used symbols for the various substances (still basically derived from alchemy), a practice that Geoffroy considered very advantageous. Both the concept of affinity and the use of symbology were adopted by chemists, many of whom published increasingly accurate and wider-ranging tables of the same type in the decades that followed.

LIFE SCIENCES

Physiological experimentation: digestion

The methodology of the life sciences tended to grow out of a synthesis of rigorous observation and physiological experimentation. Examples of this trend can be found in the work of the naturalist René Antoine de Réaumur (1683–1757), as well as in that of Lazzaro Spallanzani in later decades. In 1712, Réaumur described the phases in the process of regeneration of the legs of crayfish after partial or total amputation. The following year, the author began an experimental study of the digestive processes, which had been the object of theoretical controversy during the preceding century. He fed several pigeons meat encased in perforated metal tubes and found that the meat was broken down even when it was not ground. This showed that the stomach performed a chemical as well as a mechanical function. That the chemical function prevailed over the mechanical was confirmed by the physician Jean Astruc (1684–1766) in his *Mémoire sur la cause de la digestion des aliments*. The decisive experiments were conducted in 1780 by Spallanzani, who succeeded in extracting gastric juices from the stomachs of geese and turkeys using small sponges, connected to a string, which he made the animals swallow. The gastric juices he obtained dissolved various types of food *in vitro*.

The debate on the generation of living organisms continued. In 1714, a *Dissertatio de generatione fungorum*, by Luigi Ferdinando Marsigli, listed probative arguments against the theory that fungi formed spontaneously from decomposition. This helped convince those mid-century scientists who upheld the theory of spontaneous generation to begin using the microscope.

The spread of epidemic diseases

The English physician Griffith Hughes (1707–79) brought news of a serious disease, "black vomit" (yellow fever), to a Europe already plagued by the spread of exotic diseases that had arrived from Africa and the Americas through trade with the colonies. Several cases of this disease had already been observed almost a century earlier on the island of Guadaloupe by the French missionary Jean Baptiste du Tertre (1610–87). The new disease hit Lisbon (1723), Cadiz (1730), and Philadelphia (1791) in epidemic form. An American physician, Josiah Clark Nott (1804–73), hypothesized that mosquitoes were involved in spreading the disease: the insect that carried it, a mosquito of the genus *Aedes*, was discovered in 1881 by Carlos Juan Finlay y de Barres (1833–1915).

Other syndromes and diseases belonging to this new and exotic pathology that menaced Europe in the eighteenth century were *enfermedad del garotillo* (diphtheria), localized in Spain and Italy in the seventeenth century; whooping cough, which in two successive epidemics (1749–64) killed forty thousand children in Sweden; and scarlet fever and influenza, which spread in pandemic form in 1767. However, the most widespread and serious diseases were still smallpox and malaria. Vaccination was the most important advance of the century in the area of hygiene.

The reform of medical studies

The tendency for all the natural sciences to turn toward the experimental method, the new ties between the life sciences and the physical sciences, and the birth of numerous specialized disciplines in medicine all made it necessary to update the teaching of these sciences both inside and outside the universities. In his *De recta medicorum studiorum ratione instituenda* (1715), Giovanni Maria Lancisi (1654–1720), a physiologist, surgeon, botanist, scholar, and founder of the medical library later dedicated to him, confronted the problem of reforming medical studies, hoping to connect them to chemistry and the use of new techniques (microscopes and thermometers). His teaching in Bologna and Rome was inspired by the same ideal. The reform of medical studies continued to be of major importance until the French Revolution, when mandatory clinical practice was instituted. It was connected both to a tendency for hospitals to pass out of the hands of religious orders and into the control of the state and to the beginnings of scientific research based on clinical findings. Thus the great currents along which medical-scientific thought progressed in the following years were formed.

Transformism and the theory of the fixity of species

Transformism and the theory of the fixity of species, two opposing views on the relations between living forms, were in the air. Besides the conviction, of Platonic origin, that there was a gradual transition from one organic form to another, morphological observations of plant species in the great botanical gardens also contributed to establishing the transformist hypothesis. In 1716, the botanist Jean Marchant (1650–1738) discovered two new species of Euphorbiaceae, also known as mercuries, that differed from the typical species in the disposition and dentation of their leaves. The new species were able to transmit their characteristics to their offspring, inducing the author to guess that all plant species could produce varieties, some of which would perpetuate themselves and become new species. These were the first timid signs of a concept of limited transformism that was historical in nature and not purely idealistic like the concept of the ''chain of being.''

EARTH SCIENCES

The return to the problem of the origin of springs

One of the problems that had long been a source of lively scientific debate, the origin of springs, was solved definitively during this period. In a lecture held at the University of Padua in 1715, and published under the title *Lezione accademica intorno all'origine delle fontane*, Antonio Vallisnieri (1661–1730) stated that the springs of water that gushed out of mountainsides came from rainwater and melted snow that had penetrated into the ground and was resurfacing. He maintained that the amount of water that the soil absorbed from rain and snow was enough to account for the water produced by springs. He thus confuted the ancient idea that springs were a product of the condensation of underground steam, formed by the evaporation of ocean water that had penetrated underground, or into underground cavities, and that came into contact with the central fire or with heat inside the earth. This was also called the alembic theory, because the mechanism was the same as that used in artificial distillation, which produced fresh water from salt water.

Descriptions and illustrations of fossils

Various monographs from this period contain valuable illustrations of fossil remains. In 1717, Giovanni Maria Lancisi (1654–1720), physician to Clement XI, published the *Metallotheca vaticana*, in which he included drawings that had been made by Michel Mercati (1541–93) and stated that fossils were formed by the influence of the stars. Friedrich Mylius illustrated *Memorabilia Saxoniae subterranee* in 1704–18; Johann Jacob Baier (1677–1735), in his *Oryctographia norica*, drew belemnites and Ammon's horns (ammonites), which he compared to present-day nautiluses and distinguished from false fossils or *ludi naturae*. In 1714, Cotton Mather (1663–1728) announced the discovery, made in 1705, of gigantic bones (mastodons) found near Albany, New York. In 1712, Johann Jakob Scheuchzer (1672–1733), a convinced diluvialist, presented his *Herbarium diluvianum*, illustrations of fossil plants from various localities, to the Académie des Sciences; and in 1716 he published an organized catalogue of paleontological collections. In 1718, William Stukeley (1687–1765) signaled gigantic fossil bones discovered in Newark, England. These were later found to have belonged to the reptile Teleosaurus.

Edmund Halley

Edmund Halley (1656–1742), the astronomer, dealt with oceanographic problems in four works that appeared between 1687 and 1715, in which, for example, he considered the water balance of the Mediterranean. In 1715, he proposed to determine the age of the oceans on the basis of how fast the salinity level changed. He developed clear ideas on the problem of water exchange, and can be considered a precursor of the concept of permanence of a substance in the ocean, a basic problem in modern oceanography in the period just after 1950. In addition, Halley outlined a theory on the general circulation of the atmosphere.

The longitude-measuring competition

At the beginning of the eighteenth century, as English colonial power grew, there was an increasing demand for research to find the most economical and safest routes for navigation. In 1714, Queen Anne proclaimed a public competition for finding a method of measuring longitude on the open sea, an extremely important factor in at least partially determining a ship's position (see ASTRONOMY 1710–1720).

APPLIED SCIENCES

The steam engine

It is ironic that in 1705 Denis Papin decided to work on improving Savery's engine and abandoned his promising project for condensing steam under the piston of a cylinder in order to draw power from atmospheric pressure. Papin's idea was independently pursued by Thomas Newcomen (1663–1729). Historians generally agree that although Newcomen made a deal with Savery, this does not mean that he got his idea from him. Newcomen's Swedish collaborator, Marten Triewald, has explained that Newcomen had to make a contract with Savery when he built his first steam engine in England in 1712 because of patent of Savery's was still in effect. The 1712 engine had a boiler that produced steam which acted between atmospheric pressure and the vacuum that could be obtained using a direct contact cold water condenser. It was the first steam engine widely used for removing water from mines.

Mechanical and chemical industries

Toward the end of the decade, a new process for producing sulphuric acid was introduced in England. It consisted in burning sulphur and saltpeter together under bell jars.

In 1715, George Graham (1673–1751) invented a modified anchor escapement for clocks, called the "Graham frictional rest escapement."

Johann Friedrich Böttger directed the porcelain factory founded at Meissen in 1710 until his death in 1719. He was succeeded by J.G. Herold, who made considerable improvements in Böttger's technique bu substituting the calcareous flux that Böttger had used with a feldspathic mineral.

Most probably it was in this decade that John Lombe (1693–1722) and Thomas Lombe (1685–1739) introduced the silk-twisting machine into England and improved upon it. Silk extracted from reels was passed onto bobbins below, guided by wire vanes. When the bobbins and vanes turned, the thread was twisted as it was wound onto the bobbins.

In 1713, the Swiss inventor Mariz introduced a method for making cannon by turning both the inside and the outside from a single solid piece. In this way, a precise, controlled thickness could be obtained.

Public works

The problem of repairing roads became increasingly important as the demands of trade grew, and the need to reorganize road networks according to an overall plan was felt in the more advanced countries. What happened in France was typical: in 1715, all the roads and bridges were placed under the control of a single Directeur-général des ponts et chaussées, and the following year a corps of engineers was placed at his disposal. The École Polytechnique, which incorporated the École des Travaux Publiques, was expressly created in 1767 to train engineers for this corps.

In Russia, during the early years of the eighteenth century, work was begun on a series of canals and water works between the Volga and the Don and in the water system of Vyšnevoloc, between the Baltic Sea and the Volga.

Scientific apparatus and treatise writing

In the area of scientific instruments, the only significant development was the alcohol thermometer built by Daniel Gabriel Fahrenheit. It was accurately graduated according to the scale that Fahrenheit had devised during the previous decade.

Among treatises on engineering, two works by Henri Gautier (1660–1737), the *Traité des chemins* (1715) and the *Traité des ponts* (1716), deserve mention. They describe the proportions usually followed for support pillars, thickness, and span in masonry bridges.

1720–1730

THOUGHT AND SCIENTIFIC LIFE

Scientific academies in Europe and America

The Academy of Sciences in St. Petersburg, planned by Peter the Great in 1721, was established immediately after his death, in December, 1725. It soon counted some of the major Western scientists, such as Leonhard Euler (1707–83), Daniel Bernoulli (1700–82), and the astronomer Joseph Nicolas Delisle (1688–1768), among its ranks. Together with the already established Academies of London, Paris, and Berlin, it became the center of assiduous and continuous activity. Particularly important were the competitions promoted by these academies, which helped clarify basic problems on both theoretical and practical levels.

In addition, numerous astronomical observatories operated within the environment of the academies and universities (especially important were the English observatory at Greenwich and the Paris Observatory). The publication of the first modern star catalogues stands as evidence of a notable collective effort.

A new institution also arose in the new world. On the initiative of Benjamin Franklin (1706–90), a group of American scientists met in Philadelphia and formed the Leathernapron Club (1727), more familiarly called The Junto, which can be considered the original nucleus of the future American Philosophical Society. It was a true academy of science, even though it did not call itself one.

The spread of Leibniz's thought and the work of Wolff

While the controversy over deism, which also involved Newtonian scientific and methodological thought, continued in England (George Berkeley's *De motu*, which harshly criticized many of the premises of Newton's system, was written in 1721), the most significant cultural event in Germany was the publication (1720–23) of four works on metaphysics, ethics, politics, and physics by Christian von Wolff (1679–1754). These were written in German but were followed, beginning in 1729, by further systematical treatises on the same subjects written in Latin. Wolff, whose thought explicitly went back to Leibniz and who expounded the various topics in manual form, with a definite didactical intent, was a mandatory point of reference for the diffusion and discussion of Leibniz's thought.

Like English deism before it, Wolff's doctrine constituted a step forward in the development of the themes of the Enlightenment. Typical Wolffian themes were the negation of the possibility of miracles and the demonstrability of the existence of God as the "architect of the universe," starting from the rational evidence of the order of nature. These themes reappeared extremely frequently, often as the target of polemics, in the German Enlightenment (and not only in Germany: viz. Voltaire's *Dictionnaire philosophique*).

French and Italian culture

In France, although no systematic philosophical works were written, there was a flourishing of investigative activity. The *Lettres persanes*, by Charles de Secondat, Baron de Montesquieu (1689–1755), was written in 1721. In 1726, François Marie Arouet (1694–1778), who later changed his name to Voltaire, went to England, where he remained for several years, studying the works of Locke, Newton, and the deists. After 1730, he contributed a great deal to introducing English thought into French culture through a series of works of great literary value.

A certain cultural revival also took place in Italy. The University of Turin was reopened on a new site, the Papal University of Camerino was founded in 1727, etc. But altogether Italy remained outside the early Enlightenment movement, and more connected to Humanist-Renaissance tradition.

The first edition of the *Scienza nuova*, entitled *Scienza nuova prima*, by Giovanni Battista Vico (1668–1744), came out in 1723 (a second and third edition, considerably enlarged, came out in 1730 and 1744). That same year, publication was begun on the monumental work *Rerum italicarum scriptores*, in twenty-four volumes (the last volume came out in 1738), written by Ludovico Antonio Muratori (1672–1750) in collaboration with various scholars of the epoch. The *Istoria civile del Regno di Napoli*, by P. Giannone (1676–1748), also appeared in 1723. In it, a historical approach to the study of human institutions was seen as a weapon of political struggle for freeing the state from ecclesiastical power.

MATHEMATICAL SCIENCES

The peculiarities of British mathematics

British mathematics remained almost completely outside the main current of European mathematical research, which was, as we have seen, essentially based on infinitesimal analysis. In fact, as a result of the nationalistic battle over the discovery of infinitesimal calculus, British mathematicians refused to use the Leibnizian algorithm. This proved to be more than slightly detrimental to their own research in analysis, which was conducted, at least initially, using only traditional methods, even when topics that mathematicians on the continent were developing with the new procedure were being considered. This tendency can be seen among the major authors of the period, such as Brook Taylor (1685–1731), a Newtonian; his *Methodus incrementorum* (1715) contains the famous series that bears his name and describes the method of integration by parts. In 1720, Colin Maclaurin (1698–1746) published his *Geometria organica*, in which he applied Cartesian methods to the study of the generation and properties of many types of algebraic curves of various orders. In 1730, James Stirling (1692–1770) published his *Methodus differentialis*, which contained remarkable findings on the theory of series and the method of finite differences. In any case, adhesion to Newton's fluxion method was not a totally static situation; in fact, the sometimes subtle and pertinent criticisms brought against it, such as those in the *Analyst*, published in 1734 by the philosopher George Berkeley (1685–1753), were exactly what forced mathematicians to refine the fluxion method. The results of this process can be seen, for example, in the *New Treatise on Fluxions* (1737), by Thomas Simpson (1710–61), and even more in Maclaurin's *Treatise on Fluxions* (1742), which was characteristic in that it expounded the fluxion method using only procedures of classical geometry. Maclaurin's treatment incontestably affirmed, against all the diffidence of the critics, the rights of the new calculus. The treatise also contained a number of original discoveries, including the famous "Maclaurin series."

ASTRONOMY

Halley's observations

When John Flamsteed died in 1717, Edmund Halley was nominated Astronomer Royal and Director of the Observatory of Greenwich. Unfortunately, he found the observatory empty: Flamsteed's widow had taken away all the instruments that had

belonged to her husband and would not accept Halley's offer to buy them. Only in 1722 was Halley able to acquire a six-foot meridian telescope, built by Robert Hooke about twenty years earlier. With it, he began to make daily observations of the moon as it passed over the meridian, in order to calculate the true right ascension and compare it with the figures in the tables. For some time, Halley had believed that in order to compile fairly accurate lunar tables that would be useful in calculating longitudes on the open sea, one would only need to observe the position of the moon every day for eighteen and a half years. After that, when the line of nodes of a complete revolution about the ecliptic had been computed, the cycle of variations would be repeated with the same values and in the same order. Despite his advanced age, Halley was able to complete the entire cycle two years before his death.

The main reason for the bitter enmity that had existed between Halley and Flamsteed lay in the events surrounding the publication of Flamsteed's *Historia coelestis britannica*, which was finally published posthumously, in 1725, at the expense of his estate. The three-volume work, which contained, among other things, the stellar, lunar, and planetary observations that Flamsteed had made between 1676 and 1705, along with his very celebrated catalogue of 1,884 stars, had first been published, against his will, in 1712 by the Royal Society, under the direction of Isaac Newton (1642–1727), with a preface by Halley. Flamsteed had bought up 300 of the 400 copies printed and burned parts II–VI (387 pages), so that only the hundred that had already been distributed were saved (one exemplary is now in the Bologna Observatory).

Stellar aberration

In 1726, George Graham (1675–1751), an English mechanician who two years earlier had become a member of the Royal Society, invented the compensation pendulum. A builder of astronomical instruments, Graham had also constructed a huge mural quadrant, which James Bradley (1693–1762) began to use in December of 1725 to observe the star Draconis, suspected of having quite a large parallax. Comparing his observations of December 3 and 17, Bradley noticed that the star had moved, but in the opposite direction from what he had expected and with quite a large yearly amplitude. He then noted that all the stars described an annual ellipse, where the minor semiaxis varied according to the latitude of the star and the major semiaxis was 20″.25 (the presently accepted value is 20″.49). In 1728 Bradley himself furnished an explanation for this apparent motion, which had also been independently observed by others, including Gabriel Manfredi, who in 1729 was the first to use the term aberration. The movement of the stars across the celestial vault during the course of the year was due to a combination of the velocity of the light coming from the stars and the velocity of the earth in its orbit. Thus the first direct proof that the earth moved around the sun finally came in 1728, and the idea was no longer just a reasonable hypothesis.

PHYSICAL SCIENCES

The nature of force

The early years of this decade saw a rekindling of the quarrel that had arisen some time before concerning the nature of force.

While Isaac Newton, in accordance with René Descartes's ideas, had affirmed that force should be measured by the quantity of motion, Gottfried Wilhelm Leibniz (1646–1716) had believed that it should be measured using so-called living force (or kinetic energy). The event that reawakened the controversy was the conversion of van's Gravesande from the Newtonian-Cartesian side to the Leibnizian side. Many physicists of the epoch participated in the debate, taking sides in favor of one position or the other.

Physics and mathematics

A very subtle advance, important not only for its scientific interest, but for its philosophical aspects as well, was also made thanks to van's Gravesande's work. In 1724, he published a short treatise, *De evidentia*, in which he intended to define the difference between mathematics and physics. Modern scholars have pointed out that van's Gravesande defended a position taken up several years later by David Hume (1711–76). Inspired by the ideas put forth in van's Gravesande's work, another Dutchman, Petrus van Musschenbroek (1692–1761), upheld the full autonomy of physics with respect to mathematics (1730–40). He even maintained that physics could do without mathematics, since its laws could accurately and without exception represent all properties revealed by the senses.

Experimental research

Several important steps were made by experimental physics during the last years of this decade. In 1726–28, the astronomer James Bradley (1693–1762) discovered the aberration of light coming from the fixed stars and was able to provide a satisfactory explanation for it. In 1729, Pierre Bougeur (1698–1758) began conducting research in photometry and came up with a number of remarkable findings. That same year, Stephen Gray (c. 1696–1736) discovered the difference between conductors and insulators while studying electrical conduction.

CHEMICAL SCIENCES

Discoveries on the nature of living organisms

Chemistry became increasingly more distinct from iatrochemistry. Georg Ernst Stahl's major work, *Fundamenta chymiae dogmaticae et experimentalis* (1723), written in a mixture of Latin and German, was dedicated to chemistry. The work is interesting in that, besides providing an accurate description of the chemical procedures and apparatus of the period and of the major chemical substances known at the time, it contains a well-developed presentation of the phlogiston theory.

Pharmaceutical chemistry was the subject of a number of specialized publications, such as B.G. Capello's *Lessico farmaceutico* (1728). Increasingly accurate studies of the components of living organisms led Giacomo Bartolomeo Beccari (1682–1766) to discover gluten in wheat flour in 1728. This was the first proteic substance of plant origins to be found. Although some attribute this discovery to Francesco Maria Grimaldi (who supposedly made it in 1665), Beccari was undoubtedly the first to find the richest source of gluten and guess its importance in nourishment, especially after he discovered casein, the analogous substance in milk, in 1767.

In the area of metallurgical chemistry, Johann Friedrich Henckel (1678–1744) succeeded in isolating zinc—which Georgius Agricola had considered an "imitation of tin and silver"—in 1721. However, he did not announce his discovery until 1743.

LIFE SCIENCES

Animalculism and ovism

Two tendencies existed within the preformationist current: one that saw the miniature organism as preexisting within the male seed (animalculism), and the other that saw it as contained in the female seed (ovism). The seventh (and last) volume of the *Arcana naturae ope exactissimorum microscopiorum detecta*, by Antony van Leeuwenhoeck (1632–1723), was a valuable contribution to the animalculist preformation theory.

Surgery during the eighteenth century

Surgery as a profession—excluded from the universities in the twelfth century and relegated to the level of artisanal guilds—finally achieved academic dignity. The Académie Royale de Chirurgie was founded in 1731 on the initiative of Georges Mareschal (1658–1738) and François de la Peyronie (1678–1747), and soon gained respect in scientific circles, thanks to the advanced level of its debates. Surgeons were finally allowed to wear the official toga. In 1741, a college of surgery was opened in Montpellier, and other cities followed suit over the next few years. In 1742, the College in Paris reacquired its ancient status. In the course of the century, the range of surgical interventions was widened, and the first specializations were created. Skin sores, abscesses, phlegmons, firearms wounds, tumors, hernias, anal fistulae, and hydroceles were all treated.

Vertebral surgery and orthopedics also arose with the work of Nicholas Andry (1658–1742), and Jean Andre Venel (1740–91). John Hunter (1728–93) performed a ligature of the femoral artery, while vescical calculus began to be treated by professional surgeons. Besides the surgeons already mentioned, Pierre Joseph Desault (1744–95), Antoine Louis (1723–92), and Jean Louis Petit (1674–1750) became prominent figures in France. In England, besides Hunter, there were Alexander Monro (Secundus) (1733–1817), James Douglas (1675–1742), Percival Pott (1714–88), who described the form of vertebral tuberculosis that was later called Pott's disease, and William Cheslden (1688–1752). An important Italian surgeon was Antonio Scarpa (1747–1832), and famous Germans were Lorenz Heister (1683–1758) and August Gottlob Richter (1742–1812).

Otolaryngological surgery and surgical odontology were born and developed during this period, and surgeons gradually replaced midwives.

Historiography in the life sciences during the eighteenth century

Bacon's idea of a history of knowledge that would include the history of scientific thought was also taken up in the life sciences. The example set by Daniel Le Clerc (1652–1728) of Geneva, the 'father of the history of medicine,' whose work on the history of medicine was reprinted a number of times between 1696 and 1729 was followed by numerous physicians. Above all there was John Freind (1675–1728), whose large historiographic work came out in 1725. Later, in the course of the century, came the Belgian Nicolas Eloy (1714–88), the Frenchman Antoine Portal (1742–1832), the German Kurt Sprengel (1766–1833), and the Italian Rosario Scuderi (1767–1806). The major scientists also did work in historiography: Hermann Boerhaave (1668–1738) edited ancient and Renaissance texts, Giovanni Morgagni (1682–1771) wrote commentaries on Celsus and a biography of his teacher Antonio Maria Valsalva, and Albrecht von Haller (1708–77) published his monumental *Bibliotheca botanica, anatomica, chirurgica, medicinae practicae* in 1771–79.

Experiments in animal, plant, and human physiology

Other investigations—mostly physiological—based on experimentation and the quantification of results, were made in addition to the prevalently morphological investigations, based on rigorous observation, already being conducted. This resulted in strengthening of the methodological foundations of the life sciences. In 1727, the physiologist Stephen Hales (1677–1761) published *Vegetable Statiks*, the first volume of his *Statical Essays*. The second volume, entitled *Haemostatiks*, came out in 1733. In volume one, the author approached various problems of animal and plant physiology with physical methods and measuring procedures. For example, he calculated the ratio between the water absorbed by the roots of a plant and that given off by the leaves, thus determining the amount of plant transpiration as well as the speed of rising sap and its pressure. He also guessed that plants absorbed material substances from the air.

In the second volume of the *Statical Essays*, Hales attacked several problems regarding circulation using the same methods, he had used in his earlier studies. He demonstrated that blood pressure was different in arteries than in veins, that it fluctuated in the systole and the diastole, and that its average value was characteristic for any given animal species. Hales's contribution to promoting observation and quantification in plant and animal physiology was very great. Human physiology was also quantified, following the example set by William Harvey. Daniel Bernoulli (1700–1782), a professor of anatomy in Basel from 1733 to 1751, defined the work done by the heart as the product of the weight of the blood expelled from the ventricle times its systolic displacement.

EARTH SCIENCES

Geography and geology

Observations on the shape and formation of mountains were systematically arranged in an important publication by Antonio Vallisnieri, *De' corpi marini che sui monti si trovano* (1721, second edition 1728). In it, the author maintained that fossils did not come from Noah's flood, but belonged to organisms that had once lived in the same places where they were presently fossilized, when those places had been covered by the sea. These theories had, at the time, a profoundly innovative influence, in that they reinforced those currents of thought that tended to minimize old myths in the name of reason. Vallisnieri also discussed the disposition of fossil layers in the regions of Friuli,

Vicenza, Verona, Emilia, Romagna, Tuscany, and other areas. He emphasized the length of time it must have taken for these deposits to form, compared to the brief episode of Noah's flood. John Woodward, however, continued to maintain that there was a link between fossils and the biblical flood.

The last of the inorganic interpretations of fossils

The last supporters of the inorganic origin of fossils date back to this period. Among them was Johann Bartholomeus Beringer (c. 1667–1738), who remained famous for the cruel practical joke played on him by his students. A professor at Würzburg, he discovered some artificial fossils that his students had made and planted without his knowledge. They reconfirmed his belief that fossils were a joke of nature, and in 1726 he published his *Lithographia Wirceburgensis*—today considered a bibliographical rarity—in which he expounded his ideas, accompanying them with numerous illustrations. But a short time later, he found a "fossil" with *Vivat Beringerius* written on it. This forced him to reconsider his conclusions and attempt to find and destroy all the copies of his work.

Another gross paleontological error dates back to the same period. In 1725 and 1726, Johann Jakob Scheuchzer described the discovery of *Homo diluvii testis*, in reality an ancient giant salamander (later named *Andrias scheuchzeri* in his honor) which he mistook for the remains of a child who died in the Mosaic flood. Here again, enthusiasm for a particular theory caused the scholar to be deceived, even though he was a physician and certainly had an adequate knowledge of anatomy.

Fossils were observed by Luigi Ferdinardo Marsigli, who drew a map of the deposits of fossil fish on Monte Bolca, near Verona (1724); by Giuseppe Monti, who stated that the fossils found in the hills around Bologna corresponded to species that populated the Indian Ocean rather than to those that lived in the seas of Italy (1729); and by Hutchinson, who illustrated John Woodward's collections in his *Principia mosaica* (1724), although he rejected Woodward's diluvian ideas. Giacomo Bartolomeo Beccari (1682–1766) found a microscopic chambered fossil shell in the sands of the beaches of Romagna that Linnaeus named *Nautilus beccarii* in his honor. It was really a foraminifera, although for quite some time paleontologists interpreted it as a microscopic cephalopod.

Observations on rocks and minerals

In 1728, John Woodward wrote a systematic treatise on minerals (*Fossils of All Kinds Digested into a Method Suitable to Their Relation and Affinity*). Giacinto Gimma (1668–1735), in his *Della storia naturale delle gemme, delle pietre e di tutti i minerali, ovvero della fisica sotterranea* (1730), also offered an extensive bibliographical review of the subject. B. Holloway made a practical application of lithological studies in 1723 when he searched for a new tanning clay by following the continuum of layers of *craie* (chalk) between Oxford and Cambridge, already observed by Woodward.

The physics of the oceans

The first recorded observations made in marine optics date back to 1729, and were carried out by Pierre Bouguer (1698–1758), the same scholar who conducted interesting studies on the shape of the earth during the Peru expedition of 1736. Marsigli's *Histoire physique de la mer* was published in 1725, following a previous monograph written in 1711. The later work was the decisive compendium of the founder of modern oceanography; besides containing a number of remarkable thoughts on marine biology, currents, density, color, and temperature, it summarized some of Marsigli's most important contributions to the progress of oceanography. He was the first to take correct readings of deep-water temperatures and to note the differences between the ambient of surface waters and that of deep waters. Although he used the same reagents that Boyle had already used to analyze water (spirit, sal ammoniac, and oil of tartar), he definitely knew about the precipitation that occurred when silver was dissolved in aqua fortis (AgCl, known as cerargyrite). However, for determining salinity, he originally preferred to evaporate the water; later, fearing losses and alterations, he began measuring specific weight using suitable hydrometers.

Measuring ocean depths

The problem of making a true, accurate measurement of depth was not yet resolved, and many scholars puzzled over it, following old ideas that went back as far as Nicholas of Cusa and Leone Battista Alberti, or else devising new systems. One of the new instruments was invented by Jean Théophile Desagulier (1683–1744) in 1727, probably after an idea proposed by Stephen Hales (1677–1761). In operations on the open sea, the only universally used system was a weight (usually a sixteen-kilogram cannon ball) attached to a hemp rope. The system gave good results when the depths being measured were not very great, but for abyssal depths, the results were still quite unreliable.

APPLIED SCIENCES

Metallurgy

No significant progress was made in iron metallurgy, although the first real study on ferrous metals and the relative processes for working them, *L'art de convertir le fer forgé en acier et l'art d'adoucir le fer fondu*, by René Antoine de Réaumur (1683–1757), was published in 1722.

Mechanical industry

Meanwhile, a number of new inventions appeared in the textile industry, marking the beginning of the technological revolution that changed the face of the industry and started the industrial revolution. During this decade, mechanization took place mainly in the primary and preparatory phases of the textile industry. In 1721, Henry Browne invented a machine for stamping hemp. A device with cylinders driven by a water wheel for scutching and beating flax, invented by David Donald in 1727, met with immediate success. The following year, James Spalding invented a machine for cleaning flax. Research was also being done to build a mechanism that would weave automatically and, above all, simplify those models that required about two weeks of specialized labor every time it was necessary to set up the drawboys and cords on the loom. The first important step was made in 1725 by Basile Bouchon, who designed a mechanism

that would automatically choose the cords to be drawn. Falcon improved upon the mechanism in 1728 by increasing the number of needles.

In the area of agricultural machinery, the sifter, a Chinese invention, was introduced into Europe in 1727.

Important developments were also made in precision mechanics. George Graham invented the cylinder escapement in 1721, and in 1726 James Harrison invented the bimetallic pendulum, which compensated for the effects of changes in temperature on the length of the pendulum.

Reflecting telescopes began to be built on a large scale at the end of the decade.

Instruments and treatise writing

The different types of scientific instruments that were being invented and the demand for increased accuracy led to the creation of specially equipped workshops. More and more of them were established with time, and they became the best schools for engineers and specialized laborers, whose work proved indispensable to the dawning industrial revolution.

Antonio Stradivari, who since 1670 had been producing the most famous, most perfect violins ever made, died in 1725. His construction techniques, continued until the middle of the eighteenth century by his two sons, Francesco and Omobono, remained a family secret and have never been discovered.

In the area of treatise writing, *La science des ingénieurs* (1729), by Bernard Forest de Belidor (1693–1761), deserves mention. It contained, among other things, the elementary rules of statics for checking the stability of walls, calculating the resistance of beams, and other problems.

1730–1740

THOUGHT AND SCIENTIFIC LIFE

The Enlightenment in England

The Enlightenment movement began in both England and on the continent at this time. Although the movement displayed different characteristics in each country, there were unifying features that transcended the various traditions and cultures, giving a homogeneous aspect to the battle of the *philosophes* to laicize philosophical, political, and scientific culture. Among these features were a faith in reason, which was seen as a tool for clarifying all human problems; a firm conviction that knowledge should be diffused among the largest number of people possible; an optimistic concept of human destiny; and a rational belief in progress. In England, the debate on natural religion provided a strong stimulus; the most important works of the decade were *Christianity As Old As the Creation* (1730), by Matthew Tindal (c. 1656–1733) and *A Treatise of Human Nature* (1739–40) by David Hume (1711–76). Hume's work was in three volumes: the first took up Clarke's arguments and concluded the research of the deists, maintaining a rationalistic concept of God and morality, while the second led to a new phase in empiricism, with the express intent of introducing the experimental method of reason into the treatment of moral subjects.

The Enlightenment in France and Germany

While Wolff's philosophy, which bore the stamp of Enlightenment thought but was essentially moderate in tone, dominated in Germany, in France, from the very beginning, the Enlightenment assumed a polemical zeal unknown in other countries. Montesquieu's *Considérations sur les causes de la grandeur des romains et de leur décadence,* a work on a historical subject, but with strong political implications as well, came out in 1734. It met with immediate and widespread success. That same year, Voltaire published his *Lettres philosophiques*—exalting English philosophy, science, and political institutions—in French (it had appeared in English the year before). Immediately condemned as liable to provoke a licentiousness most dangerous to religion and the order of civil society, it was enormously successful. Five editions were published in 1734, and five more between 1734 and 1739. In 1738, Voltaire published *Les éléments de la philosophie de Newton,* which contrasted Newtonianism with Cartesianism, while the Marquise of Châtelet began translating Newton's *Principia,* with the help of the mathematician Alexis Claude Clairaut (1713–65). The translation came out in 1759. Francesco Algarotti's *Newtonianismo per le dame* stands as further evidence of the interest in Newton's philosophy that existed on the European continent.

The academies and the Enlightenment

The universities and academies played a contradictory role in the bitter polemics that accompanied the spread of Enlightenment thought. On the one hand, they were naturally inclined to respect the traditions that supported the socio-juridical structure, while on the other, they were becoming increasingly more open to innovative and revolutionary tendencies, especially in those countries, such as France, where the rationalistic approach of the Enlightenment merged with the demands of a bourgeoisie whose opposition to the old social institutions was growing.

Following the restructuring of the University of Copenhagen in 1732, the University of Göttingen, the famous Georgia Augusta University, was opened in 1737 with a solemn ceremony. In 1735 and 1736, the Académie des Sciences in Paris organized two great scientific expeditions to measure the meridian arc: one at the equator (south of Quito, in Peru), and the other in Lapland. Together, they confirmed Newton's theory that the earth's spheroid was flattened at the poles, a finding that gave his theory a victory over Descartes's. *Savants* from every discipline participated in these expeditions: mathematicians, geographers, and naturalists. In fact, the two expeditions made observations in natural history as well as geodetic measurements; it was on this occasion that Charles Marie de La Condamine, a veteran of Peru, introduced cautchouc and curare into Europe. Entire fields of science and technology profited enormously from expeditions of this type. A new way of organizing, coordinating, and directing explorations to distant places was also developed.

One great enterprise of this type was the expedition organized between 1733 and 1743 in Russia, known as the Second Expedition of Kamchatka, or the Great Northern Expedition. The numerous participants explored vast areas of Siberia, the Far East, and Kamchatka, as well as the coast of the Arctic Ocean and the eastern coast of Russia. An enormous and varied amount of historical, geographical, ethnographical, botanical, zoological, and astronomical data was collected. A number of discoveries were made, the geographical coordinates of many locations were calculated using astronomical methods, and accurate geographical maps were drawn of zones that had been unknown up to then.

The many advances made in the various scientific disciplines (especially in botany, which had been revolutionized by the Swedish scientist Linnaeus) also affected scientific academic institutions: a number of new botanical gardens, professional scientific medical societies, university museums of natural history, and so forth, were established during this period.

MATHEMATICAL SCIENCES

The affirmation of Newtonianism on the continent

An obvious symptom of the difference in approach between British mathematics and mathematics on the continent was that the British often drew upon the classical procedures of geometry to solve infinitesimal-type problems and justify the infinitesimal methods themselves, while the opposite tendency prevailed on the continent, where differential and integral calculus continued to be widely applied in the study of geometry.

A prime example of this continental tendency can be found in the *Recherches sur les courbes à double courbure* (1731), by Alexis Claude Clairaut (1713–65), the first study on the infinitesimal geometry of curves not in the same plane. An illustrious analyst, still remembered by the well-known type of differential equation that bears his name, Clairaut also distinguished himself for his courageous and incisive defense of Newtonianism in a France that was still very much tied to Cartesian physics. His *Théorie de la figure de la Terre, tirée des principes de l'hydrostatique* (1743), marked a decisive step forward in continental science's acceptance of Newtonian perspectives in physics.

Clairaut's decision to take a position was part of a process of accepting Newtonian mechanics to which Pierre Louis de Maupertuis (1698–1759) also made significant contributions. In a famous essay written in 1732, de Maupertuis discussed the differences between Descartes's and Newton's cosmological systems, and in 1736 he directed a scientific expedition to Lapland to measure a degree of the meridian arc. The findings of the expedition, which were published in his *Sur la figure de la Terre* (1739), supported Newton's theory that the earth was flattened at the poles, a valuable contribution to the affirmation of Newtonianism in France.

Similar in many ways to de Maupertuis was Georges Louis Buffon (1707–88), who is famous as a naturalist but was actually an encyclopedic genius. He started out with research on the calculus of probabilities (1733), was accepted into the Académie des Sciences of Paris in the mechanics section in 1734, and in 1740 translated Newton's *Methodus fluxionum;* he later contributed to the birth of geology as well as to the development of the biological sciences. In his *Essai d'arithmétique morale,* written in 1777, he appeared as one of the most refined supporters of the theory of probability as the basis for "moral certainties," and indicated the theory as the typical tool of intelligibility in all the not-absolutely-exact sciences.

Preludes to non-Euclidean geometry

With respect to geometry, we must not forget that the most famous work on Euclid's postulate of parallels—*Euclides ab omni naevo vindicatus* (1733), by Girolamo Saccheri (1667–1733)—

was published during this decade. From antiquity, through the Middle Ages, and even into the Renaissance, attempts had been made to prove this postulate. While, on the one hand, Saccheri's work belonged to this tradition, on the other, it distinguished itself from all the rest in that it was the first step in the process that led to the formulation of non-Euclidean geometry. Instead of trying to prove the postulate directly, Saccheri attempted a proof via *reductio ad absurdum*, that is, by attempting to reach a contradiction starting from the two possible negations of the postulate. By using this method, he established, in practice, the first *corpus* of non-Euclidean theorems. And, although he wrongly believed that he had reached the *reductio ad absurdum* proof he was looking for, he opened the way to further research by other mathematicians, which had as its primary result the gradual understanding that this type of proof by *reductio ad absurdum* was impossible. From this point on, the recognition of geometries other than Euclidean as legitimate gradually matured (around 1830).

ASTRONOMY

The first sextant

Technical progress continued to be made in nautical astronomy. In 1731, John Hadley (1682–1744) invented the reflecting octant, the first sextant-type instrument, to measure the position of the heavenly bodies from ships. Robert Hooke had already conceived the idea for this instrument nearly sixty years earlier, and Newton himself had explained the principle to Edmund Halley; but Hadley was the first actually to build the instrument (even though some historians give priority to the optician Thomas Godfrey [1704–49] of Philadelphia). A controversy over precedence, in which Halley came to the support of his friend Hadley, arose almost immediately. John Harrison, a very skilled clockmaker who had begun working several years earlier on building a precision chronometer for calculating longitudes that would work perfectly on ships, perfected his invention in 1736. After a successful trial on a voyage to Lisbon and back, Harrison was awarded the prize of 500 pounds sterling and urged to improve his invention. Although he continued his work, later chronometers were not tested until 1761.

The earth is flattened

The event that most characterized this decade was the finding of a solution to the age-old question of the shape of the earth. After Jacques Cassini's measurements of 1718, discussion and debate had grown continuously, reaching its peak in 1733 when Pierre Louis Moreau de Maupertuis (1698–1759) intervened with his *Discours sur la figure des astres*. Works by Louis Godin (1704–60), Charles Marie de La Condamine (1701–74), Alexis Claude Clairaut (1713–65), and Pierre Bouguer (1698–1758) appeared almost simultaneously. The Académie des Sciences decided to resolve the question by sending an expedition to Peru to measure a three-degree arc of the meridian near the equator. Just after this expedition, which included Godin, Bouguer, and La Condamine, had left (May, 1735), Maupertuis demonstrated that a comparison of those findings with measurements of another meridian arc near the North Pole would be decisive, and offered to carry out the task himself. Thus a second expedition left for Lapland on April 20, 1736. Besides Maupertuis, it included Clairaut, Charles Camus (1699–1768),

Pierre Charles Le monnier (1715–99), and the abbot Renauld Outhier (1694–1774). Anders Celsius (1701–44), professor of astronomy at Uppsala, joined the expedition on behalf of the Swedish government. On August 20 of the following year the expedition returned to Paris, after having measured more than one degree of the meridian at plus sixty degrees latitude. The work of the Peruvian expedition, however, which was charged with measuring an arc three times as long, together with other tasks, lasted for about ten years. When it was finished, a comparison of the length of meridian arcs at minus two degrees, plus forty-seven degrees, and plus sixty degrees latitude proved that the earth was flattened at the poles.

The St. Petersburg Observatory

In 1727, an observatory was set up at the Academy of Sciences in St. Petersburg. Joseph Nicolas Delisle (1688–1768), invited from Paris by Czar Peter I, was appointed director. Systematic observations of astronomical and meteorological phenomena, laboratory experiments on light diffraction and refraction and photometry, and research on the physics of the atmosphere of the planets, as well as research in geodesy, geography, cartography, and the history of science were all carried out there. Daniel Bernoulli, Leonhard Euler, and (from 1741) Mikhail Vasilievich Lomonosov participated in the work of the observatory. The findings were published in the editions of the Academy and in Delisle's *Mémoire pour servir à l'histoire et au progrès de l'astronomie, de la géographie et de la physique* (1738). Delisle's work also contained a description of his mercury thermometer, divided according to a 150-degree scale, on which Anders Celsius based his thermometer of 1742 (see EARTH SCIENCES 1740–1750). Delisle also described his method for studying the atmospheres of heavenly bodies, which was based on the diffraction or refraction that could be observed in them.

Equatorial mountings

It was in this epoch that the first equatorial mountings, which would later become indispensable to modern astronomy (especially after the introduction of photography) appeared. They were descendents of astronomical rings, of Tycho Brahe's equatorial armilla, and of the parallactic machine (or parallactic telescope) described and used by Christoph Scheiner and Giovanni Domenico Cassini. According to James Short (1710–68), the first equatorial mounting was built in England, where a description of it was also published in 1749.

In reality, Vayringe, an artisan from Longuyon who lived at Luneville (where he later became a professor of physics at the Academy of the Duke of Lorraine) built several of them around 1735, as Joseph de Lalande (1732–1807) later testified. Lalande owned one of the original models, which, although it was undated, he estimated as having been built before 1737.

PHYSICAL SCIENCES

The two types of electricity

Research into the nature of electricity continued with good results during these years. Gray discovered the important phenomenon of electrostatic influence, or induction. Meanwhile, in 1733, the Frenchman Charles François Du Fay (1698–1739)

advanced the theory that there were two distinct types of electricity, which he called "vitreous" and "resinous." This dualistic theory was taken up by the abbot Jean Antoine Nollet (1700–70), in contrast to the unitary theory proposed by Benjamin Franklin (1706–90).

Euler's Mechanica

Two very important works in the field of mathematical physics deserve mention: Leonhard Euler's treatise, *Mechanica, sive motus scientia analytice exposita*, which was also fairly significant from a methodological point of view in that it maintained that the principles of mechanics were necessary truths; and *Hydrodynamica, sive de viribus et motibus fluidorum comentarii*, a treatise by Daniel Bernoulli (1700–82) that marked the beginning of theoretical hydrodynamics. Bernoulli advanced several basic hypotheses that were later used by nineteenth century physicists developing the so-called kinetic theory of gasses. Two other important memoirs by Euler also date back to 1738–39: one on the nature of heat, which maintained that heat "consisted in a certain movement of the small particles in bodies," and one on vibrating strings, which proposed to formulate "a new theory of music" (the velocity of sound was also experimentally determined during these years).

CHEMICAL SCIENCES

The first chairs of chemistry

In the schools, chemistry became an autonomous discipline during this period. Although its teaching was generally assigned to physicians who had a particularly sound knowledge of the natural sciences, it became progressively distinct from the teaching of medicine. In fact, one of the first great didactic works in chemistry, the *Elementa chemiae* (1732), was written by the most famous physician of the epoch, Hermann Boerhaave (1668–1738). The work presents a concise outline of chemical knowledge at the beginning of the century. Boerhaave criticized both alchemy, with its futile attempts at finding the philosophers' stone, and iatrochemistry, which by definition was limited to substances connected to medicine. He did not completely accept the phlogiston theory and can thus be considered a mediator between Georg Ernst Stahl and Antoine Laurent Lavoisier.

Chairs of chemistry were also established in Italy. The first one, at the faculty of medicine in Bologna, was assigned to Giacomo Bartolomeo Beccari (1682–1766) in 1737. In the course of the century, other chairs were established in Padua (1749), Pisa (1757), Parma (1769), and Pavia (1770).

Logical research

Logical experimentation gradually replaced more or less random experimentation in chemistry. Starting from research in plant physiology, Henri Louis Duhamel du Monceau (1700–82) studied the nature of salts. He was the first to use the term "base" to indicate those substances that combined with acids to form salts and to make a clear distinction between soda and potassium.

Experiments with minerals led Georg Brandt (1694–1768), a professor at Uppsala, to discover cobalt in 1735. He succeeded

in isolating it in 1743. These were the first instances of the important chemical and mineralogical research done by the glorious Swedish school.

LIFE SCIENCES

Linnaeus and the Systema naturae

Carolus Linnaeus (1707–78) began his academic activity in Uppsala in 1730, when he replaced Olof Rudbeck as public lecturer and became the curator of the city's botanical garden. In the face of multiplying plant species and genera, he became more and more convinced that the basic problem of taxonomy for living species was still unresolved. The *Historia plantarum*, written in 1686 by John Ray (1627–1705) had introduced the idea of species—groups of individual plants with similar seeds—into botany; and the concept of genus—groups of all related species was introduced in the *Eléments de botanique ou méthode pour connaître les plantes* (1694) by Joseph Pitton de Tournefort (1656–1708). Tournefort had also recognized the importance of the disposition of petals and flowers in classifying plants. In 1735, Linnaeus published his *Systema naturae, sive regna tria naturae systematice proposita per classes, ordines, genera, et species* in Leyden, where he had gone to meet Hermann Boerhaave and attend his courses. It was the first of thirteen editions of the work that were published during the author's lifetime. For the second of the three kingdoms of nature, the plant kingdom, Linnaeus proposed a new method of classification, the sexual system, based on the number of stamens (monandria, diandria, etc.), their arrangement (monadelphia, diadelphia, etc.) and the sexuality of the flowers (hermaphroditic, monoecious, dioecious, poligamous). While still a student, Linnaeus had written a poem on the marriage of plants, in which he maintained—in the vein of such prestigious botanists as Nehemiah Grew (1641–1712) and Rudolf Camerarius (1665–1721), but in contrast to most other authors—that plants reproduced just like animals, via female and male organs. His scientific trips to northern Sweden and his experience at the botanical garden of Uppsala convinced him that his youthful intuition was correct. In Leyden, in 1735, he met the Dutch naturalist Johann Friedrich Gronovius (1690–1760) and showed him a copy of his manuscript, twelve folio pages in all. Gronovius and the wealthy Scottish physician Isaac Lawson offered to finance its publication. The year after the *Systema naturae* came the *Bibliotheca botanica* and the *Fundamenta botanica*, an outline of the rules to be used in studying plants. Linnaeus returned to his homeland in 1738, after having established contacts with the scientific circles of London and Paris.

The establishment of entomology as an autonomous science

Entomology became an autonomous field of study in the life sciences. René Antoine de Réaumur, already known for his experiments on the regeneration of limbs in shrimp and on gastric juices, published the first volume of his *Mémoires pour servir à l'histoire des insectes* in Paris in 1734. The work was completed in 1742; it reported the author's observations of the lives of caterpillars and butterflies, entomophagous insects, and cicadas and bees. There was a growing interest in morphology: studies were made not only of the structures of humans and mammals, but of small animals as well. Studies in comparative morphology were linked with ethological and ecological observations.

The physiology and pathology of nerves and muscles

Physiologists took a growing interest in the functional properties of nerves and muscles, since these were an important common point of comparison for iatromechanicians, animists, and vitalists. Giovanni Alfonso Borelli's *De motu animalium* was republished in Naples in 1734, with two appendices by Jean Bernoulli (1667–1748): *De effervescentia et fermantatione* and *De motu musculorum*. In the second of these, Bernoulli applied differential analysis to the study of muscle contraction. Another important work on the subject, *The English Malady* (1735) by George Cheyne (1671–1743), came out during the fourth decade of the century. In speaking about muscular contraction, the author rejected any speculation about essences and postulated that the fibers were elastic due to the attraction between their various parts.

Venereal disease

The French syphilologist Jean Astruc, of the Montpellier school, wrote an important work on venereology, *De morbis venereis libri sex*, in 1736. This field of study had gained importance after Girolamo Fracastoro's preliminary work, as the diseases became more widespread and severe. Astruc summarized all the existing knowledge on lues, including information on visceral and congenital forms. He believed that the agent of the disease could "remain hidden in the blood for a number of years." As to the nature of the agent, the prevailing opinion was that it was a microscopic worm.

EARTH SCIENCES

The formation of rocks and mountains

Attempts to give a systematic explanation of the formation of mountains and the rocks that formed them, two problems that were then considered indistinguishable from one another, basically focused around the question of whether the main causal agent was fire or water. Naturalistic theories were still always compared with the *Bible*.

Antonio Lazzaro Moro (1687–1764), of Venice, wrote his *De' crostacei e degli altri marini corpi che si truovano su' monti* in 1740. In it, he disagreed with Thomas Burnet's and John Woodward's diluvian hypotheses and attributed the formation of mountains and plains to frequent underwater eruptions. It seems the idea was suggested to him by the eruption of 1707, which created one of the Santorin islands in the Aegean Sea, and by the eruptions of Monte Nuovo and Vesuvius that Gaspare Paragallo wrote about in 1737. Moro was the first to introduce the distinction between primary and secondary mountains: the first being formed by non-stratified rocks, and the second consisting in layers of various natures, one on top of the other; in some places secondary layers even covered primary mountains. The time scale for the formation of the world was proportional to the six days of creation in the Bible. On the other hand, in 1734, M. Colonna emphasized the importance of water in his *Histoire naturelle de l'univers*. Considering water, air, earth, and fire the four basic elements, he believed that the sand in the ocean had crystallized from water, partly because it looked like granules of glass under the microscope. He also believed that mountains grew in time, like plants, but more slowly, no more than a hundred paces in a thousand years, since the phenomenon could not be seen, even when watched very carefully.

Paleontological systematics

The first edition of the *Systema naturae*, by Carolus Linnaeus (1707–78), came out in 1735. It introduced binary Latin nomenclature as the rule for taxonomy and made a distinction among the mineral, vegetable, and animal kingdoms. Binary Latin nomenclature had already been used by other authors; one of these, Jacob Theodor Klein, discussed living and fossil echinids, grouped under a common systematic arrangement, in his *Naturalis dispositio echinodermatum* (1734), which also contained thirty-six tables. In 1739, Giovanni Bianchi (1693–1775), more commonly known by his Latin pseudonym, Plancus, observed foraminifera from the Adriatic coast under the microscope. Like his predecessors, he believed them to be microscopic ammonites that corresponded zoologically to the well-known fossils. Ammonites and other fossils had been described in detail by Jacob Breyn (Breynius) in 1732. In 1737, Spada, the parish priest of Grezzana, near Verona, asserted that the tertiary mollusks of that area had nothing to do with the biblical flood. Johann Georg Gmelin (1709–55), who had observed a number of mammoth remains on the banks of the Lena River in Siberia, described them in 1733. He maintained that they were elephants and ox-like animals which for some reason had fled northward, where they died of cold and hunger or had been drowned in floods that carried their bones to the places where they were found.

Geodetic measurements

The two expeditions conducted to measure an arc of the meridian (see ASTRONOMY 1730–1740) confirmed that the earth was flattened at the poles, as Newton had anticipated.

Nicolas Louis de Lacaille (1713–62) and César François Cassini de Thury (1714–84) made another measurement of a meridian arc in France in 1740. Their findings led to the conclusion that the error that had caused the controversy between the Cartesians and the Newtonians was due to the inaccuracy of previous measurements.

Hales and the depth gauge

In 1733, Stephen Hales devised a depth gauge using the principle of the compression of a given volume of air by a liquid submerged in the sea and therefore subject to increased pressure as the depth increased. A colored liquid left a mark corresponding to the depth reached, which could be determined when the device was pulled up. The same principle was later used by Lord Kelvin. Measurements carried out using this system (reported by Hales in 1754) compared well to those obtained using a ballasted line; however, errors could be introduced because of the time it took to pull the device up. Hales also attempted to describe the course of the ocean's depth, on the basis of the available data, though some of his descriptions were very hesitant (see EARTH SCIENCES 1720–1730).

Developments in hydrodynamics

This was a decisive decade in studies in hydrodynamics, especially with the work of Daniel Bernoulli (1700–82) and Leonhard Euler (1707–83). In 1733, Bernoulli established some of the basic principles of fluid dynamics in his *Hydrodinamica, sive de viribus et motibus fluidorum commentarii*. In 1740, the French Academy promoted a decisive study on the tides, in which Colin Maclaurin (1698–1746), Bernoulli, and Euler all participated. Newton's old theory was refined and completed by introducing the variable factor of the positions of the heavenly bodies that caused the tides. The spheroids that represented sea level perturbed by the tide force were determined on the basis of the positions of the moon and the sun. In particular, Bernoulli formulated equilibrium theory of the tides (*Traite sur le flux et reflux de la mer*, 1740), and Euler (*Inquisitio physica in causam fluxus ac refluxus maris*, 1740), discussed the formulation of the tide forces and for the first time emphasized the importance of their horizontal component in generating the tides. In 1755, Euler also provided the very famous hydrodynamic equation on which much of the theoretical development of modern oceanography is based.

APPLIED SCIENCES

Mechanical industry

During this decade, mechanization reached the textile industry. The major achievements came from three great pioneers: Lewis Paul, John Wyatt (1700–66), and John Kay (1704–80). Kay took one of the most important steps toward devising an automatic loom when he invented the flying-shuttle, which he patented in 1733. He was already known for having patented a machine for twisting wool yarn in 1730, but it was his shuttle mechanism that made him famous. With the shuttle, a single person could weave fabric of any width much faster than before.

The invention of the flying-shuttle considerably increased the disproportion between the operations of spinning and weaving. Before, from three to five spinners were needed to provide yarn for a single weaver, and the imbalance became even greater now. Paul and Wyatt provided a solution in 1738 when they built the first truly mechanical spinning machine. It had a roller traction system with a series of paired cylinders, where each pair turned faster than the previous one.

Other innovations in the textile industry during this decade were a mill for beating flax, which appeared in Holland in 1735, and a machine for beating wool using small wooden boards moved by cams mounted on a wheel, invented by Kay in 1733.

In 1734, Humphrey Perrott of Bristol patented "A Furnace [which] is contrived in a new Manner with Artificial Draughts to it, whereby to Force the Heat or Fire the sooner to perform its office." This was probably a cone-shaped glass-making furnace, which concentrated all the air currents into a single ascending movement.

Scientific instruments

Important progress was also made in precision mechanics. At the beginning of the decade, George Graham designed the first precision dilatometer, which was considerably improved upon during the following decades. This instrument made it possible to determine experimentally the now well-known coefficients of dilatation for solids.

Various types of octants with sights were also invented between 1730 and 1732. In 1739, John Harrison built his first marine chronometer (government authorities urged him to continue his research and improve the instrument).

Meanwhile, starting in 1737, the first telescopes built by Georg Friedrich Brander (1713–83) began to be used in Germany.

Treatise-writing

The treatises that appeared during this decade were not up to the level of the achievements made. In 1731, Jethro Tull published his *Horsehoing Husbandry*, a treatise on using horses for hoeing in agriculture. In 1738, Schlüter published a manual on metallurgy. An impressive enterprise of this period was the four-volume *Architecture hydraulique*, written by Bernard Forest de Belidor. Publication was initiated in 1737 and completed sixteen years later, in 1753.

1740–1750

THOUGHT AND SCIENTIFIC LIFE

The Enlightenment and enlightened despotism

Although many of the works of the *savants* of the Age of Enlightenment lack originality, there was definitely an interest in stepping up research and popularizing scientific knowledge. Enlightenment, or *Aufklarung*, meant above all, as Kant said in 1784, the attempt of human beings to overcome the inferiority to which sin had relegated them, that is, their incapacity to use reason in a way that was free from tutelage. Science was increasingly viewed as a means of reaching that freedom. However, almost all scientific research was carried on under the auspices of monarchs who considered themselves "enlightened." The support that these sovereigns, such as Frederick II of Prussia, Catherine II of Russia, etc, gave to science was a part of their general programs for reform, which were aimed at modernizing their state structures, overthrowing old institutions (especially ecclesiastical ones), and favoring trade in order to revive the economy. As soon as Frederick II took the throne in 1740, he began enlarging the Academy of Berlin (which in 1743 changed its name to Königliche Preussische Akademie der Wissenschaften) and generally promoting its activity. He brought scholars such as Maupertuis (1698–1759), Euler, and Lagrange (1736–1813) to the Academy, while the other European monarchs competed to create scientific academies and societies. Thus in 1741, the Collegium Curiosorum, a private society that had gathered in Stockholm around Linnaeus (1707–78) in 1739, became the Royal Swedish Academy, and the Royal Danish Academy of Sciences was established in 1742 under the protection of King Christian VI. In Belgium, that same year, the old Académie des Curieux de la Nature was renamed the Leopoldine Academy in honor of its protector.

The Antica Accademia dei Lincei was reestablished in 1745. A climate of optimism reigned in scientific circles. Research was honored and sometimes rewarded. While in the past most scientists, except for physicians and engineers, had been amateurs, at least as far as income was concerned, scientific work was now becoming more of a professional job. Science was better able to meet the demands of technology and the economic interests of society. For example, the invention of the Leyden bottle and the improvements of several electrical machines made it possible to conduct spectacular experiments in literary salons and in the laboratories of the naturalists. Jean Antoine Nollet electrified an entire company of guards before the king of France. The demonstration that electrostatic sparks were the same as lightning, carried out by Benjamin Franklin (1706–90) in the American colonies, had remarkable repercussions.

The effects of enlightened despotism and the progressive laicization of higher learning could not be ignored. State control gradually replaced ecclesiastical control.

Prominent among the new universities were the College of Philadelphia, founded in 1740, the University of Erlangen, the University of Santiago de Chile (1743), and the College of New Jersey (1746), which later became Princeton University. An École des Ponts et Chaussées was established in Paris in 1747, under the direction of Jean Rodolphe Perronet (1708–94).

What later became known as ''the reign of Buffon'' began in the field of the natural sciences. In 1739, Georges Leclerc, Comte de Buffon (1707–88) was appointed director of the Jardin du Roi, which became a model for scientific institutions throughout Europe.

The great philosophical-cultural battles

Even though many exponents of the new culture, including Voltaire, were working on popularizing Newton's ideas, imported English empirical philosophy was not the only current on the continent. In France, for example, Cartesian tradition was still alive and was even followed by the Jesuits. Although in the seventeenth century the Jesuits had been opposed to Descartes's ideas in the name of Aristotelian philosophy and science, Cartesian physics now seemed to be the only weapon against Newton's philosophy of nature. The fate of the Jesuit Ruggiero Giuseppe Boscovich (1711–87), who was forced to leave his teaching post at the College of Rome when he became one of the major promoters of Newton's ideas in Italy, was typical. The traditional Cartesian mechanistic-materialistic current reappeared in the radical themes of the materialists, who extrapolated the more daring concept of the human-machine from the Cartesian idea of the animal-machine. *L'homme machine*, by Julien Offray de La Mettrie (1709–51) was published in 1745. The ties between the Enlightenment and empiricism grew progressively closer. In Germany, there was a rebellion against the predominance of Wolff, brought about by a desire to emphasize experimentalism rather than abstract reasoning. Outstanding figures in this rebellion were Alexander Gottlieb Baumgarten (1714–62) and Christian August Crusius (1715–75). In 1742, Baumgarten taught courses in esthetics at the University of Frankfurt on the Oder. He considered esthetics a philosophical science of sensitivity, which, as such, was aimed at understanding the individual as opposed to logic, which was aimed at understanding the universal. He collected the contents of these lessons into a famous volume, entitled *Aestetica*, which was published in 1750. For his part, Crusius believed in the extralogical nature of existence; it was on this idea that he based his confutation

of the Wolffian theme that the world was sustained by a necessary order or a preestablished harmony (as Leibniz had surmised). His main work on the subject was entitled *Entwurf der nothwendigen Vernunftwahrheiten* (1745). In France, Etienne Bonnot de Condillac (1715–80) adopted a decisively anti-metaphysical (and therefore anti-Cartesian and anti-Leibnizian) position, which led him to advance not only empirical, but openly sensist theories. His first important work on the subject was the *Essai sur l'origine des connaissances humaines*, written in 1746. In England, Hume once again expounded his position on the question of understanding, this time in a simpler form than in his *Treatise* of 1739–40. His *Philosophical Essays Concerning Human Understanding* (1748), unlike his previous work, met with widespread and total success. That same year, in Geneva, Montesquieu published his masterpiece, *L'esprit des lois*, which reunited all the studies that he had conducted since 1730 on the constitutions of Europe. The work was very widely circulated.

MATHEMATICAL SCIENCES

The work of Leonhard Euler

Two figures stand out distinctly over all the others in eighteenth century mathematics: Leonhard Euler and Joseph Louis Lagrange (1736–1813). Euler carried out his most important work in the quarter-century that began in 1741. In 1727, he had left his native Basel to succeed Daniel Bernoulli (1700–82) as a member of the Academy of Sciences in St. Petersburg, and the extent of his mathematical genius had immediately become obvious in the capital city of imperial Russia. There he did decisive research in integral calculus (introducing Euler's first and second integrals, as they were later called, in works of 1731), the theory of series, the theory of partial differential equations (distinguishing them for the first time from ordinary derivatives and demonstrating the theorem of the inversion of derivatives for them in 1734), and the theory of isoperimeters. In 1741, he was invited by Frederick II to move to Berlin, where he became director of the science class at the Academy of Prussia (from 1746 to 1753, which partially coincided with the period in which Maupertuis was president of the Academy). Although he did not teach from a university chair, his influence was felt in almost all of European mathematics, and even opened completely new perspectives. For example, in his *Methodus inveniendi lineas curvas* (1744), he approached classical problems of isoperimetric and geodetic curves with methods that foreshadowed the calculus of variations.

An indication of Euler's stimulating influence on mathematics during his time can be found in his relationship to the work of Jean le Rond d'Alembert (1717–83). Besides his great accomplishments in the field of mechanics, d'Alembert made some important contributions to integral calculus and the solving of differential equations, which were published in the memoirs of the Academy of Berlin between 1746 and 1748. In 1747, in studies on the problem of vibrating strings, he established and solved the famous partial differential equation that bears his name, and which he discussed with Euler and Nicholas Bernoulli (1687–1759). In addition, in 1746 d'Alembert provided the first correct formulation of the basic theorem of algebra, outlining an imperfect demonstration which Euler improved upon in 1751.

ASTRONOMY

Studies of comets

On the basis of observations made of the comets of 1742 and 1744 at the St. Petersburg Observatory, Joseph Nicolas Delisle and Georg Wolfgang Krafft proposed graphic methods for determining their orbits, while in Berlin Leonhard Euler (1707–83) developed several analytical methods which Johann Heinrich Lambert later perfected.

In order to explain the shape of the most important comets, Euler introduced the idea of the 'repulsive force of the sun' in 1746 and outlined a mechanical theory on the shape of comets. Thus celestial mechanics was born and began its rapid development. After Euler finished his work, Mikhail Vasilievich Lomonosov put forth his theory on the appearance of comets (1753).

Celestial mechanics

After Leonhard Euler had completed his analytical treatment of the problem of two bodies and established the ten integrals of the problem of n bodies, he became the first to make a general detailed analysis of the problem of planetary perturbation, in his Theoria motuum planetarum et cometarum, published in 1744. Isaac Newton, in his Principia, had already defined the question perfectly, emphasizing that Jupiter's effect on Saturn during the conjunction of the two planets was 1/211 that of the sun on Saturn, and was therefore not at all negligible. This initiated a series of industrious studies into the "three-body problem," which concerned not only the sun and one planet, but a third body for the non-negligible influence that it exerted due to either its mass or its proximity. In particular, the sun-earth-moon complex was considered an especially important system because of the practical problems (such as that of calculating longitude) connected with the motion of the moon. After having founded the method of the variation of constants, published in an initial work of 1746, Euler developed his research in other works published in 1749, 1750, and 1752. Meanwhile, in 1747, Alexis Claude Clairaut and Jean le Rond d'Alembert (1717–83) presented the results of their research on the three-body problem, in particular on the motion of the moon around the earth, perturbed by the sun, to the Académie des Sciences in Paris. Unaware of Euler's first treatise, which came out almost simultaneously, they followed another path, which was still, however, based on calculating variations in the coordinates. Clairaut's method can be found in his Théorie de la lune, which the Academy of St. Petersburg awarded a prize in 1750 and published two years later. D'Alembert described his method in the three-volume Recherches sur différents points importants du systeme du monde, published in 1754–56. Among other things, this work contained the first treatment of the problem of determining the motion of a planet perturbed by its satellites. All these early works dealt only with circular or almost-circular orbits, i.e. those of the planets and their satellites, and not those of comets. Clairaut was the first to tackle this latter problem, around a decade later, from 1759 to 1762.

Nutation

In 1748, James Bradley reported that he had discovered nutation, a phenomenon that caused a periodic alteration in the apparent position of the stars. Newton had predicted it as a product of the theory of attraction, and various astronomers, including John Flamsteed and Olaus Rmer, had looked for it. Bradley had already started making observations in 1727 in an attempt to find it and finally discovered it after about nine years, or half a period. News of the discovery was already widespread when it was published, since Bradley had communicated it privately to French astronomers.

The first accurate map of the moon

In 1748, Johann Tobias Mayer (1723–62) attempted to draw a map of the moon, based on a number of accurately measured specific points, but his plan for a map in 25 sections was cut short by his untimely death. He only completed a 19.5 centimeters in diameter map, which was published posthumously in 1775. Even though this was only a small part of the plan Mayer had outlined, it was the first accurate lunar map. It was used to calculate longitudes at sea.

PHYSICAL SCIENCES

Fluid mechanics

The findings and methods outlined by Daniel Bernoulli in his Hydrodynamica (1738) led to further investigations during this decade. In 1742, Colin Maclaurin published a treatise that critically analyzed the possibility of basing a study of the motion of fluids on the principle of the conservation of living force. Together with Maclaurin's work, which was entitled Treatise on Fluxions, Johann Bernoulli's Nuova idraulica, dated 1743, also deserves mention. It contained general equations for solving problems in hydraulics. In 1743, Alexis Claude Clairaut (1713–65) presented his Théorie de la figure de la Terre tirée des principes de l'hydrostatique, a work on hydrodynamics. In it, he attempted to formulate the laws of fluid mechanics in a manner that would be compatible with the various hypotheses on gravitation, within the context of the debates that were going on between Neocartesians and Newtonians.

The research of Jean le Rond d'Alembert (1717–83) was focused on a critical analysis of Daniel Bernoulli's theories. In 1744, d'Alembert published a treatise on the equilibrium and motion of fluids (Traité de l'équilibre et du mouvement des fluides).

Rational mechanics and d'Alembert's Traité de dynamique

In 1743, d'Alembert published his most famous work, Traité de dynamique. Mechanics, according to d'Alembert, was a rational science par excellence, due to the fact that its principles were necessary truths. The laws of motion should not be based on forces inherent in bodies, because these forces were "obscure, metaphysical entities, which could only throw shadows on a science that in itself was clear." Instead, one had to start from the "principle of equilibrium, along with that of the force of inertia and compound motion." These principles were "essentially different from one another," and the general laws of motion for all types of systems could be deduced from them, so that "all the problems of mechanics" could be solved. In this area, d'Alembert criticized the foundations of Eulerian mechanics.

Those who maintained that "the accelerating or retarding force is proportional to the (differential) element of velocity" were basing their ideas on "that vague and obscure axiom according to which the effect is proportional to the cause." It was not a question of proving that this principle was a necessary truth; in reality, the available proofs were not at all convincing. Nor was it possible to trust it as a contingent truth, according to d'Alembert, this would cause "the certainty of mechanics to fall into ruin and reduce it to nothing more than an experimental science." Therefore according to d'Alembert, the expression of the basic relationship between the accelerating force ø, the time t, and the velocity u as

$$ø = du,$$

was not a necessary or contingent truth, but a definition of ø.

The principle of least action

Pierre Louis de Maupertuis (1698–1759) continued his work in mechanics and in 1744 stated the famous principle of least action in a memoir entitled *Accord de différentes lois de la nature qui avaient jusqu'ici paru incompatibles*. He took up the principle again and developed it in a later memoir, written in 1747. Today, the scientific value of this principle is indisputable, though Maupertuis presented it within a philosophical-theological framework that aroused many doubts among his contemporaries. He maintained that all laws of movement were founded on it, so that one could even say that God himself imposed it at the time of the creation. This idea started a quarrel that involved almost all the major scientific figures of the time. Even Voltaire entered the controversy with a treatise that was strongly satirical and disrespectful of Maupertuis.

The Leyden bottle

The construction of the so-called Leyden bottle, built in 1745 by both E. Georg von Kleist (died 1748) and Petrus van Musschenbroek (1692–1761), working independently and without any knowledge of one another, occupies a special place in the history of electricity. Perfected considerably over the following years, this device gave research in electricity a huge boost, since large discharges could be obtained with it.

The electrical fluid according to Franklin

In 1747, the American Benjamin Franklin elaborated his famous theory of electricity, according to which electricity was a single fluid that existed in the proper amount in all bodies, in their natural state. Bodies became electrified when there was an excess or a lack of this fluid in them. He put forth this theory and various experimental findings (including the "power of points") in a series of letters on electricity (1747–54). The subjects of Franklin's investigations and his early findings were mainly concerned with the problem of the analogy between light and electrical phenomena, and with various questions about the Leyden bottle.

Developments in optics

The research done in optics during this decade, especially on the nature of light, was fairly important. Here physicists were divided into two camps: those who supported the cor-puscular theory of light and those who supported the wave theory. Leonhard Euler, who published a number of memoirs on refraction and achromatism in 1746–47, belonged to the second group (see PHYSICAL SCIENCES, 1770–1780).

CHEMICAL SCIENCES

The discovery of platinum

The Frenchman Guillaume-François Rouelle (1703–70), a professor at the Jardin du Roi, succeeded in creating an active school of chemistry in which ample space was given to both theoretical considerations and demonstrative experiments. Antoine Laurent Lavoisier and Joseph Louis Proust were both students of his. In 1745, Rouelle published several important memoirs on the nature of salts.

Around this time, platinum was discovered in a number of gold and silver mines in South America; it soon proved to be very valuable, even for making some of the apparatus used in chemical and physical experiments. Antonio de Ulloa (1716–95) introduced platinum into Europe in 1736 and wrote an accurate description of it in 1748. The first experiments done on it were reported in 1750 by William Brownrigg (1711–1800) and William Watson (1715–87), and Henric Theophil Scheffer (1710–59) made a more detailed study of its nature in 1752.

The first large chemical industries

The most important event in the history of chemistry during this decade was the rise of the first large chemical industry, for the production of sulphuric acid. Toward the end of the decade, other industries—for extracting sugar from beets for example—were developed. From this point on, chemistry proved to be the science most directly tied to the development of industry. This was obviously a result of its indisputable ability to devise new techniques that could immediately be used to increase production.

These industries are further discussed in the section APPLIED SCIENCES, 1740–1750.

LIFE SCIENCES

Parthenogenesis, the theory of ovist preformation and regeneration

The naturalist Charles Bonnet (1720–93) of Geneva demonstrated parthenogenesis in plant lice (1740) by keeping a specimen in isolation from the time of its birth. The discovery, announced to the Académie des Sciences in Paris by René Antoine de Réaumur (1683–1757) and immediately echoed by Albrecht von Haller in Berne, stimulated interest in the debate on generation and development. It also seemed strongly to support the ovist side in the theory of preexistence of germs. The ovists believed that the adult organism existed in miniature inside the egg, and attributed an indirect role to the sperm in the process of fecundation. According to Bonnet, the sperm initiated cardiac motion within the fetus inside the egg and contributed a certain number of nutritional molecules. In his *Considérations sur les corps organisés* (1762), he added the idea of *imboîtement* to the ovist preformation theory (i.e. the adult organism was already formed inside the egg, along with its ovary, inside of which all the individuals that would later be born from it also already existed).

The polemic on generation was joined by the debate on regeneration. This in turn was linked in various ways to the "monster controversy"—the debate over what caused teratomas—that arose at the Académie des Sciences during those same years. Some scholars considered these formations to be congenital while others believed them to be accidental.

Abraham Trembley (1710–84), an amateur French naturalist, in his *Mémoire pour servir à l'histoire d'un genre de polypes d'eau douce*, reported the curious regenerative properties of Hydra tissues. This little polyp could be divided into segments, and each segment would regenerate a complete Hydra. Two polyps could be grafted together; and a Hydra could be turned inside out just like the finger of a glove. Trembley's experiments were confirmed by Bonnet, Réaumur, and Lazzaro Spallanzani (1729–99), while Voltaire doubted that the Hydra was an animal. Their explanations divided the preformationists and the epigenetic minority even further. Bonnet advanced the hypothesis that "germs" were contained not only in the ovaries, but in other parts of the body as well, and that the regenerative processes originated from germs with varying potentialities.

The experiments conducted by the naturalist John Tuberville Needham (1713–81) also had widespread repercussions. In *An Account of Some New Microscopical Discoveries Founded on an Examination of the Calamary and Its Wonderful Milt-Vessels* (1745), Needham maintained that Infusoria reproduced *via* spontaneous generation. Formative particles were diffused everywhere, and supposedly united under the effects of a vegetative force to form miniscule living beings, similar to those observed by Leeuwenhoeck. His theory was based on findings that he claimed to have obtained from experiments on plant and meat Infusoria which had been exposed to high temperatures: after a certain period of time, microscopic living beings appeared. Otto Friedrich Müller (1730–84), the major micrographer of the times, and Buffon, about whom we shall speak in a moment, sided with the spontaneists; the opponents included Bonnet, Réaumur, and Spallanzani, who strongly confuted Needham's affirmations.

Another contribution to the debate on these issues came from observations of hereditary phenomena. In his *Vénus physique* (1745), the philosopher, mathematician, and naturalist Pierre Louis de Maupertuis (1698–1759) contested Bonnet's ovist ideas by citing the case of a hereditary anomaly (polydactylism) that was passed down by females and males alike. Maupertuis also formulated the idea of accidental variations in living species. He attributed the creation of new species to these variations.

Extremist mechanism and transformism in living organisms

At the beginning of the century Georg Ernst Stahl defended an extreme concept of vitalism, which was animistic in nature. This was contrasted by an equally extremist mechanism in *L'homme machine*, written in 1748 by the physician Julien Offray de la Mettrie (1709–51), who three years earlier had published an *Histoire naturelle de l'âme*. According to the author, all the faculties and properties of human beings—including the rational soul, which René Descartes substantially distinguished from corporeal extensions—could be explained as mechanical processes.

Human beings were machines; the human body was made up of a huge number of organic molecules joined together; the soul was the activity of the mind. The life of the parts did not depend on the life of the organism; in fact, there were parts of the body that survived and moved after death.

Analogously, a transformist concept of living nature appeared in contrast to Linnaeus's theory of the fixity of species, and new views on the age and history of the earth challenged biblical geology, which was presupposed by fatalism. The first three volumes of the *Histoire naturelle, générale et particulière avec la description du cabinet du roi*, by George Louis Buffon (1707–88), appeared in 1749. Projected as a description of the three kingdoms of nature, the work came out of a collaboration between Buffon and other naturalists. It was written from a philosophical, as well as an empirical, point of view. Some of the assertions—for example those concerning the age of the earth, which was estimated at 17,000 years, and others that compared the similarities between humans and monkeys—led to conflicts between the author and the faculty of theology at the Sorbonne. In the fourth volume (1753), Buffon retracted his statements, but he later reiterated his geological ideas in his *Époques de la nature* (1773). He also stated his belief in the transformation of animal and plant "families." Publication of the *Histoire* continued until 1804; volumes XXXVI–XLIV appeared posthumously.

With his *Traité de la structure du coeur, de son action et de ses maladies* (1749), Jean Baptiste Senág (1693–1770) inaugurated cardiology as an autonomous discipline within medical pathology.

EARTH SCIENCES

The role of fire and water in the origin of mountains

The problem of whether fire or water was the cause of mountain formations continued to be debated. Major contributions were made by Benoit de Maillet (1656–1738) in France. In Italy, Giuseppe Cirillo Generalli defended the ideas advanced by Antonio Lazzaro Moro during the previous decade and upheld the importance of the action of streams and rivers in wearing away mountains. Generelli was also aware of how comparisons with biblical texts were often abused to support theories that otherwise could not be satisfactorily explained by the observed facts. A Dominican monk, he openly opposed the attempts of those scholars who looked to the Bible for support for theories that would not stand alone.

Fossils were thought to be the remains of species that still existed, and research on the analogies led Donati to probe the bottom of the Adriatic in order to examine the similarities between its form and that of the sub-Apennines. He found that the mud and sand being deposited there was the same as that which formed the reliefs at the edges of the Apennines. This supported the theory advanced by Moro and his predecessors that the sea had once extended over land that was presently above sea level. In Chile during the same period, Antonio de Ulloa found marine shells in rocks as far as five leagues from the coast and one hundred meters above sea level; he considered them proof of the biblical flood. In 1742, Jose Turrubia attributed the fossils of Spain to Noah's flood.

De Maillet's work *Telliamed*, the title of which was the author's name spelled backwards, was published posthumously in 1748. Its transformist ideas, which went against what seemed to be the common evidence of the facts, aroused strong reactions. The work, which also contained cosmological and evolutionary theories, advanced the idea of a cyclical transformation of planets into burning stars and cool, habitable lands.

From observations of a cliff that was slowly emerging from the sea, and on the basis of his grandfather's recollections, de Maillet saw a confirmation of the old idea that the sea had at one time covered land that was now above sea level. He believed that all rocks came from the sea and attributed the relief of the land to the effects of the water, giving even more importance to the role of the sea than the later Neptunists did.

From the gradual lowering of the water level in wells near the sea (the rate was three inches per century, as his grandfather had also observed) de Maillet calculated the length of the earth's history, which on the basis of various considerations he established at two thousand million years. He also calculated that human history was very old and had begun many thousands of years before. The distinction that de Maillet made between primary and secondary mountains was very interesting as well; he believed that the former were created by the sea and the latter by erosion. De Maillet developed his ideas between 1715 and 1716, but they only became universally known when his work was published.

The first volume of the *Histoire naturelle*, by Buffon (1707–88), entitled *Théorie de la Terre*, was published in 1749. It aroused great interest because of its concept of a central fire and its negation of the biblical flood, which brought many attacks against it (see LIFE SCIENCES 1740–1750). The work was the first to outline the various topics that constitute the science of geology, which the author elaborated in later works. The idea of a central fire in the earth, with a molten core enclosed by a solid crust, can also be found in Leibniz's *Protogaea* of 1749.

Maps

Several geological maps of this period deserve mention. Giovanni Arduino (1713–95) drew a map of northern Italy, which was published in 1744; Christopher Packe (1686–1749) made one of England in 1743; and Jean-Étienne Guettard (1715–86) made one of France in 1746. France was the first country in the world to have a cartographic representation of its national territories that indicated the lithological makeup and mining resources.

Geodesy

In 1743, Alexis Claude Clairaut (1713–65) published his *Théorie de la figure de la Terre tirée des principes de l'hydrostatique*, a work on the shape of the earth, which he believed was determined by a sort of water balance of masses that were subject to the centrifugal force of rotation and the effects of gravity. Clairaut later made other major contributions to geodesy; earlier he had participated in the French expedition to Lapland to measure the meridian arc.

The measurement of deep-water temperature

During one of the numerous exploratory voyages made during this period, Henry Ellis (1721–1806), at the suggestion of Stephen Hales, the great promoter of oceanographic studies, collected a large body of data on surface and deep-water temperatures (1749). Since the thermometers that Luigi Ferdinando Marsigli (1658–1730) had used had by then been lost, measurements of deep-water temperature were made on large samples of water collected from the desired depth in insulated (or almost insulated) containers that were pulled up very quickly so that no heat exchange could take place, or with thermometers imbedded in spheres of insulating material (resin) that were left at the desired depth for a number of hours so that the temperature equilibrated, then quickly pulled up and read. With this system, the young Horace Benedict de Saussuve (1740–1799) took readings in Swiss lakes and found that the bottom temperature stabilized at around four degrees centigrade. Given that this was the temperature at the point of maximum density of fresh water, various hypotheses were formulated regarding temperature of ocean depths. This brought up a problem that remained almost completely unsolved until the voyage of the *Challenger* in 1872. However, Ellis noted that the temperature decreased with depth and extended Marsigli's findings in the Mediterranean to the oceans. William Wales (c. 1734–98), William Bayly (1737–1810), and Constantine John Phipps made similar observations in the years that followed.

The Celsius scale and Franklin's lightning rod

In 1742, on the basis of Josep Nicolas Delisle's 150-degree scale, Anders Celsius (1701–1744) proposed the centigrade thermometer scale, which became extremely important in numerically defining the thermic conditions of the atmosphere and the sea (the Fahrenheit scale dated back to 1714, and the Réaumur scale to 1731). In 1749, Benjamin Franklin (1706–90) demonstrated the physical nature of lightning and pointed out that it was an electrical phenomenon. Following these studies, Franklin invented the lightning rod (which it seems had already been used for some time in other countries, based on empirical principles). In France, just a few years later (1752), Louis-Guillaume le Monnier (1717–99) carried out independent studies on atmospheric electricity and in particular on the electrical nature of lightning. The Englishman John Freke and the German Johann Heinrich Winkler had also studied the electrical nature of lightning as early as 1746–48.

APPLIED SCIENCES

Iron working

Important discoveries and innovations were made in the field of metallurgy during this period. Mining work benefitted from an invention by Carlisle Spedding. With his device, a handle was turned to rotate a toothed flint and steel wheel, producing a constant spray of sparks. Besides illuminating the work, it indicated the presence of "flammable air" when the sparks became brighter. The production of coke was improved when closed brick ovens replaced the old open piles. In 1742, Thomas Bolsover found that a billet made by melting or welding together a thick sheet of copper with a thinner sheet of silver could be rolled and flattened without the metals separating. However, the most important discovery was made by Benjamin Huntsman (1704–76), an instrument-builder, who in 1742 finally found a method for producing molten steel. By adding a suitable flux, he was able to melt carburized steel rods in closed clay crucibles. Coke was used to provide the necessary heat for the operation. Crucible steel made it possible to increase production and, consequently, decrease costs. A description of the ironworking methods used at the time is given in the *Political testament*, written by Christopher Polhem (1661–1751) in 1746.

In 1745, the first (bituminous) coal mine was opened in Richmond, Virginia. The deposit had been known to exist as early as 1673. Other deposits were discovered in Ohio in 1755. Anthracite was not found in the United States until 1790.

The textile industry

Mechanization of the other phases of the textile industry proceded during this decade. In 1748, Lewis Paul invented two wool carding machines: the first had a fixed card with parallel strips of a special carding fabric, and was known as the rectangular machine; the second had a hand-operated cylinder with the carding fabric attached to it, and was known as the crank machine. That same year, Daniel Bourn, of Leominster, patented a rotary carding machine, in which four cylindrical cards, turned by hand or by a water wheel, worked against one another.

The chemical industry

The first large chemical industry was founded in 1740, when Joshua Ward (1685–1761) began the industrial production of sulphuric acid at Richmond, near London. The acid, a very important chemical substance, was obtained by heating sulphur with saltpeter in iron capsules and condensing the vapor produced by the reaction in large glass balls. In 1746, John Roebuck (1718–94) and Samuel Garbett (1717–1805) replaced the glass balls with lead chambers, and large-scale production was begun in Birmingham and at Prestonpans, Scotland (production in Italy began in Milan in 1799).

Meanwhile, in 1747, Andreas Sigismund Marggraf (1709–82) succeeded in extracting sugar from beets, obtaining an important natural organic substance that was the same as cane sugar. The process soon gave rise to new industries. Marggraf's disciple, Franz Achard (1753–1821), developed the discovery and opened the first European sugar refinery in Silesia in 1801.

The optimum composition of gunpowder was again revised, and in 1742, after much consideration, the saltpeter-sulphur-carbon ratio was set at 6:1:1.

Roads, canals, and scientific instruments

Great roads and networks of navigable canals were extended. Jean Rodolphe Perronet (1708–1794), director of the École des Ponts et Chaussées, introduced important innovations in bridge building techniques (bridges could be built with much thinner pillars and arches than before). In England, new navigable canals were dug and port services were greatly improved.

In the area of scientific instruments, the Leyden bottle, which was independently built by Ewald Georg von Kleist and Pieter van Musschenbroek in 1745 deserves particular mention (see PHYSICAL SCIENCES 1740–1750).

Treatise writing

In 1742, Mikhail Vasilievich Lomonosov finished writing his most important work on the basics of metallurgy, or the mining industry, *Pervye osnovanija metallurgii, ili rudnyh del'*, an encyclopedic treatise that was first published in 1763. One of the most important works on shipbuilding, the *Traité du navire, de sa construction et de ses mouvements*, by Pierre Bouguer (1698–1758), was written in 1746. It was one of the most widely read, studied, copied, and translated theoretical and practical texts on shipbuilding in the eighteenth century. In 1744, John Theophilus Desaguliers (1683–1744) published his *Experimental philosophy*, which, among other things, described Newcomen's method for calculating machine power. In 1749, Jean-Baptiste Bourguignon d'Anville published a map of Africa, which became the starting point for African explorations.

1750–1760

THOUGHT AND SCIENTIFIC LIFE

The Encyclopédie *and the context of French culture*

In the second half of the century, publication began on the *Encyclopédie, ou dictionnaire raisonné des sciences, des arts et des métiers par une société de gens de lettres*. For several years, almost all the major authoritative thinkers of eighteenth-century France collaborated in the writing of the *Encyclopédie*. The promoters of the initiative were Denis Diderot (1713–84) and Jean le Rond d'Alembert (1717–83). In 1750, Diderot published 8,000 copies of a *Prospectus* that outlined the editorial features of the work and presented the general plan. The first volume, which contained d'Alembert's famous *Discours préliminaire*, came out in 1751. The second and third volumes came out in 1752 and 1753. Publication of the work immediately led to harsh conflict with the clergy and ruling authorities. A number of attempts were made to interrupt it, and the encyclopedists formed a sort of *parti des philosophes* that battled openly against obscurantist and repressive tendencies.

At this same time, other extremely important works were also published in France, making that country the center for European culture. (Meanwhile French gradually replaced Latin as the language of scientists and scholars.) In 1750, Jean Jacques Rousseau (1712–78) made his debut with *Si le rétablissement des sciences et des arts a contribué à épurer les moeurs*. Voltaire wrote his *Précis du siècle de Louis XIV* in 1753, and his *Essai sur les moeurs et l'esprit des nations* in 1756; these were later recognized as the first works of modern historiography. In 1754, Condillac published his *Traité des sensations*, which although it gave rise to various polemics—for example from Buffon, who claimed that Condillac had plagiarized much of the work from him—gained wide acceptance to the extent that it was considered the basic text of sensism. In 1758, Claude Adrien Helvetius (1715–71) published his *De l'esprit*, which reiterated the gnoseology of sensism, but also presented a new view of the moral world. The work caused such a scandal that it was immediately condemned by the archbishop of Paris, the pope, and the French Parliament. The *philosophes* feared that the indignation against Helvetius would also involve the *Encyclopédie*, and that same year d'Alembert broke with Diderot. However, strong critiques of traditional culture were still made, some against religion and others that focused on the traditional metaphysics of Descartes and Leibniz. An example is Voltaire's *Candide* of 1759.

The Enlightenment in England, Germany, and Italy

As far as England is concerned, we shall only mention that during this decade even Hume focused his interest on ethical, religious, and historical problems. He wrote numerous essays defending natural religion and several basic works on the history of England.

In German culture, a transition was made from the first phase of the Enlightenment, which was dominated by Wolffian rationalism, to the second, where interest was more centered around human, historical, religious, ethical, and political problems. Both the spread of so-called popular philosophy (a current of thought essentially aimed at bringing Germany up to date on all the great questions of European thought and converting people to religious tolerance) and the publication of several serious works that defended natural religion, were indications of this trend. The works included the *Abhandlungen von den vornehmsten Wahrheiten der natürlichen Religion* (1754), Hermann Samuel Reimarus (1694–1768), and two early essays by Gotthold Ephraim Lessing (1729–81) entitled *Das Christentum der Vernunft* (1753) and *Über die Entstehung der geoffenbarten Religion* (1753–55). The first philosophical-scientific works of Immanuel Kant (1724–1804) also appeared during this decade. The most important one, the *Allgemeine Naturgeschichte und Theorie des Himmels* (1755), was based on Newton's cosmology; however, it introduced a great innovation in that it maintained that the order of the heavens was not imposed by an architect of the universe, but could be fully explained by simply referring to the general laws of nature.

Meanwhile, the Enlightenment spirit began spreading through Italy as well. As in France, it essentially focused on economic-juridical political, and religious problems. The two major centers of the new culture were Naples and Milan. The first works that appear to have been obviously inspired by Enlightenment thought were *Della moneta* (1751), Ferrando Galiani (1728–87) and *Meditazioni filosofiche sulla religione e sulla morale* (1758), by Antonio Genovesi (1712–69). Genovesi was appointed the head chair of economy at the University of Naples in 1754.

The Enlightenment and institutions

By this time the Enlightenment was the voice of a revolt not only against the old metaphysics, but against the old political and social structures as well. Even the conservative forces, which more than once attempted to form a coalition against its advance, were well aware of this tendency, as were the more enlightened monarchs of Europe. Even though these rulers were often influenced by some of the more ardent demands advanced by the *philosophes* and attempted to make them their own (though at the same time moderating them), they nevertheless feared a certain radicalism in them. Despite the difficulties that arose within the camp of reason, however, the new culture continued to spread unremittingly. Its advance was favored by the cosmopolitan attitude of the scientists, who easily moved from one country to another and, fought over by the reformist monarchs, became the directors of one institute or another. As to the new institutions, the British Museum in London, established in 1753 around Hans Sloane's library and collection, became an important center. Meanwhile, a Society of the Sciences was created in Göttingen, thanks above all to the work of the scientist Albrecht von Haller (1708–77). An Académie des Sciences Utilitaires was founded in Erfurt in 1754, and the Académie des

Sciences de Bavière opened in Munich in 1759. In 1757, the Academia Reale delle Scienze was established in Turin; it began as a private association of scientists, including Lagrange. In the history of higher education, the creation of the University of Moscow in 1755, on the initiative of Mikhail Vasilevich Lomonosov (1711–65), and the restructuring of the universities in Austria, both deserve mention. In America, King's College, which later became Columbia University, was founded in 1754.

MATHEMATICAL SCIENCES

The calculus of variations

While Euler dominated the European mathematical scene from Berlin, Joseph Louis Lagrange (1736–1813), the scholar who would share with him the title of most prominent mathematician of the century, came onto the scene in Italy. Cofounder of the future Academia delle Scienze di Torino in 1757, Lagrange began publishing important works on the calculus of variations and the integration of differential equations (two areas in which he did particularly important work in the future as well) in the memoirs of the academy (*Miscellanea taurinensia*).

Lagrange got his inspiration for laying the foundations of the calculus of variations, one of his major claims to fame, from Euler's *Methodus inveniendi lineas curvas*. Even the name "calculus of variations" was coined by Euler when, inspired by the appearance of Lagrange's works, he returned to the subject in three memoirs between 1764 and 1767.

Topology and the theory of numbers

Another branch of mathematics that began with the works of Leonhard Euler was topology. In fact, the considerations that he used in demonstrating his famous theorem on the relationship between the number of faces, vertices, and edges of an ordinary polyhedron, were basically of a topological nature. Numerous memoirs on the theory of numbers and important publications on eliptical integrals were other conspicuous contributions made by the great mathematician during his extremely productive Berlin period (1756–1766).

Euler's treatises on infinitesimal calculus

Our brief outline would be seriously lacking if we failed to mention the great treatises that Euler wrote, beginning with his *Institutions calculi differentialis* (1755) and ending with his three-volume *Institutiones calculi integralis*, which came out between 1768 and 1770, after he had returned to St. Petersburg as director of the Academy of Sciences toward the end of his life. The treatises are important not only because they were extremely popular and exerted a great influence all over Europe, but also because they stand as evidence of the effort at systematic arrangement and rational, unitary coordination of knowledge that was an essential part of the Enlightenment mentality and its ideal of rationality. Other sciences as well (especially mechanics) reflected this typical need of the epoch.

Other developments: algebra and logic

In any case, infinitesimal calculus was not the only area of mathematical research during this period. In 1750, Gabriel Cramer (1704–52) published his *Introduction à l'analyse des lignes courbes algébriques*. In it he applied algebra to the study of plane curves and to a particularly rigorous and exhaustive classification of them. There were also interesting developments in logic, the formal components of which were considerably attenuated. In 1747, for instance, Christian August Crusius (1712–75) published a work on logic that presented it not as a formal discipline but as a ''theory of reason.'' This same tendency was reflected in the very title of the *Vernunftslehre...* (1756) by Hermann Samuel Reimarus (1694–1768), possibly the most popular work on logic of the era. In the area of research somehow related to mathematics, but not directly mathematical in nature, the publication in 1758 of the *Histoire des mathématiques*, by Jean Etienne Montucla (1725–99), deserves mention. This work can be considered the first true history of mathematics; it covered a wide range of topics from an authentic historical perspective. It came out of the tendency to produce vast historiographical syntheses that was a particular characteristic of eighteenth century culture.

ASTRONOMY

The austral skies

Nicolas Louis de Lacaille (1713–62) went to the Cape of Good Hope to observe the stars of the austral hemisphere and in 1752 published the first fairly extensive catalogue (1,935 stars). He also composed new constellations, which appeared in his *Coelum australe stelliferum*, published posthumously by Maraldi in 1763. From the Cape, he coordinated his observations with others made by Joseph Jérôme de Lalande in Berlin in order to calculate the moon's distance from the earth. Kolb and Wagner had used the same method in 1727, although their results were very poor. They obtained a parallax of 67 minutes 33 seconds, while almost a decade earlier Edmund Halley, with another method, had calculated it at 57 minutes 18 seconds. Lacaille and Lalande found the parallax to be 57 minutes 14.8 seconds (present value: 57 minutes 02.45 seconds).

The reform of the calendar

In 1752, England and Ireland updated their calendar by adopting the Gregorian reform, suppressing September 3–13, and beginning the year according to the modern method (January 1 instead of March 25, the Incarnation). These reforms (especially the second, which eliminated January, February, and twenty-four days of March from 1751) created a considerable amount of discontent and made the promoters, such as Lord Chesterfield and the astronomer James Bradley, extremely unpopular.

Mayer's table and the problem of distances

1752 was an important year for solving the problem of longitudes. In Göttingen, Johann Tobias Mayer published his *Novae tabulae motuum solis et lunae*, which were better than any previous tables because of the accuracy of the observations on which they were based. Enthusiastic friends and colleagues convinced Mayer to let them contact the Royal Society of London, and in January 1755, Mayer delivered the tables and the manuscript that described the method of lunar distance for determining longitude to the first Lord of the Admiralty. However, the discussion on awarding the prize got bogged down, and a decision was not made until 1765, after Mayer's death. The prize was awarded to Mayer, Leonhard Euler (on whose theories Mayer's tables were based), and John Harrison (who had built a chronometer). Thus the important problem of determining longitudes at sea was finally solved. The use of chronometers came later with respect to Mayer's method, which was first published in 1767 in the Nautical Almanac.

Mayer also made an important contribution to the development of stellar astronomy. By generalizing Halley's discovery and comparing his own measurements of star positions against those taken by Römer more than half a century earlier, he discovered the proper motion of eighty stars, which he collected in the first catalogue of its kind (1760).

The first cosmological theories

The first modern cosmological theory also appeared during this progress-filled decade. At the time, it was based merely on speculation and only began to be proved by the observations made by Frederich Wilhelm Herschel at the end of the century. In 1750, Thomas Wright (1711–86) published his first work on the structure of the universe. In it he stated that the stars were not distributed uniformly in all directions, i.e. in spherical symmetry with respect to earth, but in a very thick but finite layer. Immanuel Kant (1724–1804) took up this idea and supported it in his *Allegemeine Naturgeschichte und Theorie des Himmels*, published anonymously in 1755 and initially unrecognized. In this work, Kant also advanced his famous nebular hypothesis on the origin of the solar system, later taken up and developed by Laplace, and for the first time affirmed that nebulae were other stellar systems, unresolvable because of the inadequacy of the instruments used to observe them. As far as star clusters and galaxies were concerned, this affirmation proved to be prophetic.

Achromatic objectives

The technical development of telescopes, which astronomers were impatiently awaiting, took a decisive turn in 1756 when John Dollond (1706–61) began building the first achromatic objective, using a technique based on Euler's theory of achromatism. Dollond patented the objective two years later but ran into problems concerning the priority of his discovery. Without a doubt, others had worked on the problem simultaneously and even before, but excellent, compact achromatic telescopes without irridescent images only became available throughout the world beginning in the following decade, thanks to the work of Dollond and his sons.

The return of Halley's comet

We have already mentioned (ASTROMONY 1700–1710) Halley's prediction that the famous comet today known as Halley's comet would return in 1758. Later Alexis Claude Clairaut corrected Halley's calculations and specified that it should appear during the first months of 1759. The actual appearance of the comet in the skies of Paris in March of that year was welcomed with much enthusiasm, and the event was considered a great victory of reason over superstition.

Photometry

The research done in photometry during this period, described in two classical works published in 1760—*Traité d'optique*, by Pierre Bouguer (1698–1758) and *Photometria sive de mensura et gradibus umbrae*, by Johann Heinrich Lambert (1728–77)—was very important to the further development of astronomy and physics. Lambert also introduced the word "photometry" to indicate the new discipline, which found numerous applications in astronomy and physics.

The history of astronomy

Publication of the *Histoire des mathématiques*, by Jean Étienne Montucla (1725–1799), a history of all the physical-mathematical sciences, including astronomy, was begun in 1758. The second edition of the work, edited by Joseph Jérôme de Lalande, came out in 1799–1802.

PHYSICAL SCIENCES

The generalization of mechanics according to Euler

Several extremely important developments took place in mechanics during the early years of this decade, thanks to Leonhard Euler's research on the motion of all types of mechanical systems, both discrete and continuous. For the first time the equations that are often called Newton's equations were written out in their most general form. Thus the more than half a century of theoretical research that began with the statement of Newton's second law was formally synthesized for the first time, thanks to Euler's mathematical work. The synthesis, according to Euler, contained all the principles that could lead to an understanding of the motion of all bodies, no matter what their natures.

In 1755, Euler published the results of his studies on the equilibrium of fluids (*Principes généraux de l'état d'équilibre des fluides*). He analyzed, again in a very general form, the behavior of a fluid under the effects of all types of forces. That same year, Euler formulated the basic equations of hydrodynamics, which he put forth in a memoir on the general principles of the motion of fluids (*Principes généraux du movement des fluides*). He pointed out the difficulties involved in reaching a complete understanding of the motion of fluids. These difficulties were rooted not in some inadequacy in the principles of mechanics, but rather arose from unresolved problems in the area of mathematics.

Developments in d'Alembert's investigations

In 1752, taking up several theories that had been advanced in the previous decade, Jean le Rond d'Alembert published an essay on fluid mechanics (*Essai d'une nouvelle théorie de la résistance des fluides*). In it he attempted to reduce all problems concerning the resistance of fluids to general laws on equilibrium. The second edition of d'Alembert's *Traité* appeared in 1758, considerably enlarged with respect to the 1744 edition and with the addition of interesting methodological observations.

Boscovich's Theoria

An important work by Ruggiero Giuseppe Boscovich (1771–87), entitled *Philosophiae naturalis theoria redacta ad unicam legem virium in natural existentium*, came out in 1758. A third edition of the *Theoria* appeared in Venice in 1763, edited by Boscovich himself. The work's importance lies above all in the fact that the author did not accept the corpuscular theory of matter, defended (though with reservations) by Newton and adopted by all the major scientists of the epoch (Euler, for example). Instead, Boscovich proposed a "dynamistic" theory, according to which matter was made up of unextended physical points that exerted a force of attraction or repulsion on each other, depending on their reciprocal distances. Matter, Boscovich thought, was made up of subtile, unextended, and perfectly indivisible points. Many points existed in a very small space, though the number was finite, and one could say that the universe did not consist of a void disseminated between matter, but of matter disseminated in the void and floating in it.

Atomism and dynamism

One of the most basic controversies of the century, one that involved not only physics, but philosophy as well, arose over the issue of atomism versus dynamism. In his *First Metaphysical Principles of the Science of Nature* (1786), Immanuel Kant (1724–1804) supported the dynamistic concept, accusing those who upheld the corpuscular theory of introducing absolutely unverifiable entities (atoms) into physics. The dynamistic theory was later adopted (at the end of the eighteenth century and early in the nineteenth century) by the followers of *Naturphilosophie*.

Franklin's theory of electrical atmospheres

In 1751, Benjamin Franklin presented a detailed outline of his theories on electrical phenomena in a work entitled *Opinions and Conjectures Concerning the Properties and Effects of the Electrical Matter*. In this work, Franklin described the electrical fluid as a form of matter not normally perceived by the senses. It differed from common matter, he said, in that the parts of the latter mutually attract, while those of the former mutually repel each other. But, though the particles of electrical matter do repel each other, they are strongly attracted by all other matter. This implied an attempt to reduce macroscopic interactions to microscopic attractions and repulsions. According to Franklin, the mediation between macroscopic and microscopic occurs because of electrical atmospheres dislocated on the surface of bodies; the presence of these atmospheres electrifies the bodies. The atmosphere theory presented a number of difficulties, which Franklin attempted to resolve in 1755, though he never really succeeded. Especially difficult were the problems involved in explaining why the atmosphere did not disperse, as well as those encountered when Franklin's theory was applied to the repulsion observed between bodies with negative charges. The static interactions invoked by Franklin in his theory of fluids contrasted with the explanation provided by Jean Antoine Nollet (1700–70) in his *Lettres sur l'électricité,* published in 1753.

Despite the various difficulties it presented, Franklin's theory opened the way to the theories advanced by Aepinus (1724–1802), Henry Cavendish (1731–1810), and Charles Augustin Coulomb (1736–1806). Another important step in this direction came with a treatise written by Giovanni Battista Beccaria (1716–1781) in 1753, *Dell'elettricismo artificiale e naturale*. Beccaria attempted to present an organic outline of all the experiments on electricity known at the time, plus several fairly interesting new ones, in the light of Franklin's unitary theory.

That same year, Jean-Baptiste Le Roy (1720–1800) published his *Mémoire sur l'électricité*, in which he disagreed with Nollet and supported Franklin's ideas. The static interactions of Franklin's theory were interpreted from a dynamic point of view by various scholars, including John Canton (1718–72), who, in 1753, published a memoir in which he maintained that bodies emitted and absorbed electrical fluid. This idea led him to abandon the notion of electrical atmospheres. Beccaria's theories began to be circulated with the publication of his *Lettre sur l'électricité, par le R.P.J.B. Beccaria* in Paris in 1754.

Aepinus's theory

Franz Ulrich Theodosius Aepinus's major work, *Tentamen theoriae electricitatis et magnetisimi* was published in St. Petersburg in 1759. It elaborated a theory of electrical fluid that differed from Franklin's on a number of points. Not only did Aepinus eliminate the concept of electrical atmosphere and introduce a new repulsive form between particles of ordinary matter, but above all, he mathematically connected his thoughts on magnetic phenomena with those on electrical phenomena. As he himself stated, his intention was to consolidate the theory of magnetism rather than that of electricity. Within this context, the *Tentamen* emphasized the need to give an internal cohesiveness to all knowledge on these phenomena by using mathematical deductions. At the same time, Aepinus took an interest in problems in experimental physics, directing his attention to pyroelectricity and the Leyden bottle. He published his findings in 1758 in his *Mémoire concernant quelques nouvelles expériences électriques remarquables*, a work that gained him much authority and prestige. The diffusion of the ideas advanced in the *Tantamen* encountered some difficulties during the first few years after its publication. When the English scholar Robert Symmer (c. 1707–63) again attempted to take up and modernize Charles François Du Fay's old two-fluid theory in 1759, a lively controversy arose in which numerous scholars intervened. Meanwhile, the first electrostatic machines were developed. The one built and distributed by the English mechanician Jesse Ramsden (1735–1800) deserves particular mention.

CHEMICAL SCIENCES

Studies on gasses and the theory of caloric

New metals continued to be discovered. Following the discoveries of cobalt and platinum, nickel was discovered in 1751 by the Swede Axel Fredrik Cronstedt (1722–65).

The study of gasses or "airs," i.e. pneumatology, which had counted Johannes Baptiste van Helmont, John Mayow (1640–79), and especially Robert Boyle among its devoted followers in the seventeenth century, was neglected during the first half of the eighteenth century, except for the research done by Stephen Hales. Joseph Black (1728–99), primarily known for his caloric theory, brought pneumatology back into vogue with a series of experiments on alkaline and alkaline-earth carbonates

done between 1752 and 1756. He pointed out that these substances, when treated with acids and some when exposed to heat, emitted a gas that was different from atmospheric air, i.e. carbon dioxide ("fixed air"). This gas was also produced in the combustion of coal, in respiration, and in fermentation, and was therefore present in small quantities in the atmosphere. It was unable to support combustion and respiration, and transformed caustic alkalis into their relative carbonates and then into their bicarbonates. Black also considered the quantitative aspect of his experiments, which led him to question the phlogiston theory.

In 1756, the Russian Mikhail Vasilievich Lomonosov (1711–65) presented a series of arguments against the phlogiston theory, but for a long time his experiments and writings remained little-known in the West.

LIFE SCIENCES

Hygiene and the "rules of life"

The opening of the first public baths in Liverpool around the middle of the century indicates that an understanding of the problem of public health had become widespread. In England, limits were placed on the consumption of alcoholic beverages, and books on the "rules of life" became fashionable. Investigations were made into the state of hospitals, and demands were made to improve prisons. John Howard (c. 1726–90) published a report on the state of prisons in 1777 and a history of the major lazarettos of Europe in 1789.

The causes of muscular contraction

An essay on the vital and other involuntary motions of animals (1751), by the Scottish physician Robert Whytt, reopened the debate on the causes of muscular contraction. In his work, Whytt proposed a theory that lay between Hermann Boerhaave's mechanical explanation (according to which muscular fibers contracted because of pressure exerted on them by the nervous fluid) and Stahl's animistic concept (according to which all local reactions depended on a single vital principle. Motion depended on the soul, but on its sensitive, rather than its rational faculty. Whytt's ideas were criticized by Johann August Unzer (1727–99), who based his denial that the brain had anything to do with muscular motility on the movements of a decerebrated frog. In his *De functionibus systematis nervosi commentatio* (1784), Unzer formulated another new explanatory hypothesis: neuromuscular automatism, supported by the law of the conservation of living force and parallel to the faculty of voluntary response to peripheral stimuli. In any case, the distinction between irritability and contactility, made by Albrecht von Haller, one of the major physiologists of the century, was the real prelude to the concept of autonomism. In a memoir entitled *De partibus corporis humani sentientibus et irritabilibus* (1753), Haller distinguished irritability from contractility— the specific irritability of the muscles—and from sensitivity, denying that irritability depended on sensitivity, as Stahl had stated. Irritability existed in all parts of the organism. His formidable scholarship and ability to synthesize led von Haller to begin publishing his *Elementa physiologiae corporis humani*, a treatise in nine volumes that replaced Boerhaave's *Institutiones*, in 1759. The most important chapters were those on the mechanics of respiration, osteogenesis, and circulatory dynamics. Haller's *Elementa* dominated European medicine until Johannes Peter Müller's *Handbuch der Physiologie des Menschen* appeared in 1833 (see LIFE SCIENCES 1830–1840).

Binary nomenclature in animal and plant classification

The first volume of the tenth edition of Carolus Linnaeus's *Systema naturae* came out in Stockholm in 1758 (the first edition dated back to 1735). It was followed by the second volume in 1759. "Binary nomenclature," proposed by Linneaus himself in his *Philosophia botanica* (1751), simplified and rationalized animal and plant classifications, which before that were based on long and necessarily inaccurate descriptions. Now each species was designated by two names: the first, the genus, was common to all similar species, and the second belonged to that particular species. The "diagnosis," which listed the characteristics of the species, was different from the binary designation. There were many species, as many as the infinite being had created, and there were sudden variations, although these did not affect the specific structure. Plant classification was based on the disposition of the reproductive organs (the sexual system). Species were divided into genera, families, and orders; the subspecific categories were of little importance. Animals were divided into six classes: mammals, birds, amphibians, fish, reptiles, and worms. There were 1,767 plant genera in the last edition of the *Genera plantarum* (1789–91), while the first edition (1737) contained 995. Linnaeus considered his classifications "natural"; in reality, however, they were still based on an arbitrary choice of reference point.

Epigenesis v. preformation

Epigenesis took precedence over preformation with the *Theoria generationis* (1759), written by the embryologist Caspar Friedrich Wolff (1734–94). According to Wolff, the structure of the adult organism was formed through successive transformations in the embryonic structure. In chick embryos, which Wolff used for his studies, for example, the blood vessels developed from primitive vacuolar formations. In a later work, entitled *De formatione intestinorum* (1768), Wolff showed that the intestine was formed from a sheath of tissue, and from this concluded that embryogenesis was supported by a *vis essentialis*. In the nineteenth century, while reading Wolff's *Theoria*, Karl Ernst Baer grasped the central idea of his comparative embryology, the development of types (see LIFE SCIENCES 1820–1830).

In 1750, Rudolf August Vogel (1724–74) began publishing the *Medizinische Bibliotec* in Erfurt. The *Journal de Medecine* in Paris (1754), the *Giornale della medicina* in Venice (1763), and the *Raccolta di osservazioni* in Pavia (1788) followed.

EARTH SCIENCES

The earth's crust

Some basic geological ideas were developed during this decade, and in the following decades, debates on them gave rise to modern geology. Studies were made to discover some sort of order in the composition of the earth's crust, generally considered total chaos and often thought to have remained unchanged since the time of creation. In general, the idea prevailed that fossil layers came after crystalline formations because they were found above them. Discussions about fossils focused on their relationship to Noah's Flood and other catastrophic events, since by this time their real nature as the remains of organisms had been accepted.

The nature of basalt rock

In 1752, Jean-Étienne Guettard presented a memoir entitled *Sur quelques montagnes de la France qui ont été volcans* (published in 1756) to the Académie Royal des Sciences in Paris. In it, he maintained that the outcropping basalt rocks in the Auvergne were of volcanic origin. The problem of the origin of basalt later became the basis of a controversy between the Neptunists and the Plutonists, beginning in the first decade of the nineteenth century. Guettard's discoveries did not become very well known during his lifetime; he even had to claim credit for them twenty-eight years later. In 1752, he also compared Canada and Switzerland on the basis of geological information sent to him by Canadian experts.

Stratigraphic systems

In a series of works written between 1751 and 1754, Giovanni Tarqioni Tozzetti (1712–83) opposed Buffon's ideas on the underwater origins of valleys, and attributed their formation in the Apennines to fluvial and torrential erosion. He also stated that the remains of organisms (nummulites, thought to be corals; elephants) came from species that had once lived in the same regions. He then proposed a stratigraphic system for the region of Tuscany, subdividing it into three orders. His major work, in six volumes, *Viaggi fatti in diverse parti della Toscana per osservare le produzioni naturali e gli antichi monumenti di essa*, was published between 1751 and 1754.

Georg Wolfgang Knorr's paleontological tables, *Lapides diluvii universalis testes*, with an explanation by Joahnn Ernest Walch (1725–78), were published between 1755 and 1778. A work by Johann Gottlob Lehmann (1719–67), *Versuch einer Geschichte von Flötzgebirgen*, which can be considered the first to have actually laid the foundations of a stratigraphic classification (in part because of the notoriety it achieved and the influence it had on later scholars), appeared in 1756. Lehmann divided rocks into primary, secondary, and tertiary types. The first, crystalline rocks without fossils, were formed when the world was created and before the creation of animals and were the original elements of the earth's crust. Next came secondary rocks, which were deposited in the water together with debris from earlier rocks following universal catastrophies such as Noah's Flood; these contained both plant and animal fossils. Tertiary rocks, formed following localized catastrophes, including volcanic eruptions and floods, and partly by Noah's Flood, also contained fossil remains. In his *Oratio de generatione metallorum a terrae motu* (1757), which talked about metals and earthquakes, Mikhail Vasilievich Lomonosov (1711–65) took into account gradual and imperceptible effects on the earth's crust as well as catastrophic events and deduced the geological nature of past phenomena by comparing them with present ones. He was basically applying the concept of actualism, later developed by Charles Lyell (see EARTH SCIENCES 1830–40). Lomonosov also distinguished between external and internal processes with respect to the earth's crust; both of these were responsible for its continual transformation.

Also in 1757, Emmanuel Mendes Da Costa made a systematic study of various rocks and, in particular, described granite, the eruptive nature of which was unsuspected at the time.

The most important stratigraphic work of this period, which among other things proposed the names that are used even today to indicate the major subdivisions of the earth's history, was published in 1759. In his *Osservazioni sulla fisica costituzione delle Alpi Venete*, Giovanni Arduino distinguished between primary, secondary, tertiary, and quaternary orders of rocks. His ideas came from the experience he acquired as mines expert for the Republic of Venice and from his stays in the Tyrol and Tuscany. Two categories of rocks belonged to the primary order: chalky-micaceous fossil stones, which together with granite were the earliest rocks with no marine animal fossils, and calcareous and areno-micaceous rocks, which very rarely contained relics of marine animals and were associated with vitrescent (eruptive) rocks. The secondary order consisted of layers of marble and rocks of a calcareous nature, with abundant remains of marine animals, along with vitrifiable rocks that were different from those of primary mountains. Calcareous and arenaceous stones and petrified or loose gravel formed from fragments of marine animal shells and bits of rock from the disintegration of primary and secondary mountains belonged to the tertiary order. Vitrescible stones were black and either very hard or spongy (compact and pimply basalts, respectively). The quaternary order was made up of gravel, sand, mud, and earth, which flooded the plains and was carried by rivers and streams. The accuracy and validity of Arduino's subdivisions were a product of both his remarkable abilities of observation and synthesis and the fortunate circumstance that he did his research in the Veneto region, where a fairly complete stratigraphic range exists within a relatively small area. The local stratigraphic subdivisions of Saxony, where Lehmann did his studies, were not as suitable for generalizing.

The physical characteristics of sea water

Important studies on the physical characteristics of sea water were begun during this period. In 1753, William Watson studied processes for desalting sea water. In 1757, Henry Cavendish (1731–1810) invented thermometers for use in the ocean. John Canton (1718–72) experimentally determined the compressibility of water, which until then had been considered uncompressible (his results were published in 1762).

John Michell (1724–93) made a study of waves and tides in 1760, following the Lisbon earthquake of 1755. His study of the effects of the earthquake led him to deduce theories recording the mechanisms and causes of this phenomenon, and by analogy, the propagation of tidal waves. Immanuel Kant, who also did work in physical geography, hypothesized in his *Allgemeine Naturgeschichte und Theorie des Himmels* (1755) that the tides caused a very slight slowing up in the velocity of rotation of the earth. A description of the morphology of the oceans, written by Phillipe Buache (1700–73), appeared in 1752. Buache also made correlations between the morphology of the ocean bottom and corresponding land above sea level.

APPLIED SCIENCES

The mechanical industry and precision instruments

Precision mechanics underwent its most intensive phase of development during this decade. This formed the basis for further improvements in measuring and surveying instruments, with extremely important practical consequences for exploration and navigation. In 1755, John Dollond (1706–61) invented the achromatic objective by grinding lenses with different indices of refraction and placing them one over the other in such a way as to obtain an objective with no chromatic aberration. Dollond's son, Peter (1730–1820), built a telescope using the objective.

John Harrison finished his fourth chronometer, which was successful in navigational tests, in 1759. A copy made for the Board of Longitude performed equally well. Toward the end of the decade the octant evolved when the length of the radius was decreased and the length of the arc increased. The sixty-degree sector was subdivided into 120 parts, and the new instrument was called a sextant.

Construction

In the field of construction theory, Leonhard Euler (1707–83) elaborated a mathematical expression for the critical load under which a long, thin column became unstable. Although this was an important step in the study of the critical behavior of materials, it was not immediately applied. Meanwhile, building concepts became increasingly bolder. In 1757, the engineer Hans Ulrich Grubenmann (1709–83) built a double-arched beam bridge over the Rhine at Schaffhausen. Grubenmann would have liked to build the bridge with a single 110-meter span. He later successfully did this, building a bridge with a 120-meter span, rising to a height of thirteen meters, over the Limmat at Wettingen.

That same year, the architect Jacques-Gabriel Soufflot (1713–80) began building the church of Sainte-Geneviève in Paris, using non-traditional methods that drew the hostility of conformist builders. He had iron bars sunk into grooves in the mortar at joints and in the course of the masonry, wherever there was danger of breakage. His technique was a precursor of reinforced concrete, even though the iron was still not completely immune to corrosion. The first attempts to produce cement go back to 1756: John Smeaton (1724–92) obtained one type by cooking lime and clay together. Smeaton also constructed the third Eddystone lighthouse in 1759.

The textile industry

Two patents in the area of spinning deserve mention: in 1754, James Taylor patented ''an engine to be worked either by men, horses, wind, or water for spinning cotton wool into yarn''; and Lewis Paul patented his spinning machine in 1758.

Naval engineering

By 1750, the English mercantile fleet had already reached a capacity of 610,000 tons, and the Nelsonian frigate (twenty-eight to fifty-six cannons) was devised for the navy. The *Victory*, Nelson's flagship at Trafalgar, was built in 1765 and armed with 100 cannons.

The spread of technological culture

The results of John Smeaton's research on the output of water wheels and windmills came out in 1759 in the Royal Society's *Philosophical Transactions*. The academies, which had been one of the first products of the scientific movement, began to have a direct influence on technology, as they organized the systematic collection and publication of data that illustrated the status of the various sectors of technology. The program of the Royal Society also included a historical reconstruction of "all sorts of curious and useful techniques from every country." In carrying out this program, however, the Royal Society was far surpassed by the Académie des Sciences, and even more so by the work of Denis Diderot (1713–84). The Royal Society's *Descriptions of the Arts and Crafts* was overshadowed by Diderot's *Encyclopédie*, the first volume of which was published in 1751.

In this way, industrial processes were made known to the educated classes and became a component of the culture of the epoch. However, it was not just a question of a social awareness of factors that had an ever-increasing influence on daily life. Scientific awareness of these processes was an essential element for their progress. In effect, just as this awareness was reaching the educated classes, the industrial processes were becoming so complex that the uneducated were increasingly unable to maintain the overall view of their jobs that is an essential factor for progress.

In this context, the *Eléments de l'architecture navale ou traité pratique de la construction des vaisseaux*, Henri-Louis Duhamel du Monceau (1700–82), published in 1752, deserves mention.

1760–1770

THOUGHT AND SCIENTIFIC LIFE

New aspects in education

A significant step was taken in the process of laicizing education when the Jesuits were expelled from a number of European countries (Portugal in 1759; France in 1762; Spain, the Kingdom of Naples, and the Ducate of Parma and Piacenza in 1767–68). The order was finally suppressed by Clement XIV in 1773. The defeat of the Jesuits meant more freedom in teaching Newton's philosophy of nature, which was the scientific model for new theorizations, even in economics, politics, and morals. While new universities were founded and famous centers were restructured, new types of schools arose that attempted to make a more direct link between theorization and the concrete observation of phenomena. Veterinary schools (the first prototype of which was the École de Lione, founded in 1762, and later the École de Alfort, founded in 1765), mining academies (at Selmecbànya in Hungary (1763), Freiberg in Saxony (1765), etc.), and new botanical gardens (Vienna, Cambridge, Saint-Vincent, etc.) were created. This tendency was very much in harmony with the practical spirit of certain entries in the *Encyclopédie*, which became the basic text for educating new generations of Europeans.

Scientists were often bold explorers, too. A number of scientific expeditions, especially out of France and England, were organized to observe the passages of Venus across the sun in 1761 and 1769. Scientific exploration became popular in other

countries, as well. Peter Simon Pallas (1741–1811) travelled to Siberia and the Caucasians (1768–74); Carsten Niebhur was cartographer and sole survivor of a Danish expedition into Arabia, Persia, and Mesopotamia (1763–67); James Bruce led an expedition to Ethiopia (1768–73); Louis Antoine de Bougainville journeyed from Tahiti to Java (1766–69); and Captain James Cook made his famous voyages (beginning in 1762). These expeditions not only made important contributions to various sectors of scientific research and stimulated scientists toward a more unbiased observation of nature, they also aroused a lively interest in populations and countries with extremely different religions, cultures, and customs. One typical product, on the level of intellectual theorization, was the moral relativism that constituted one of the *philosophes'* strongest weapons in their critique of traditional European institutions.

The French Enlightenment: Rousseau and d'Holbach

Diderot tenaciously proceeded in his preparation of the remaining ten volumes of the *Encyclopédie* (publication was suspended for political reasons in 1759, and d'Alembert had withdrawn his collaboration in 1757). Finally, in 1766, after the Seven Years War was over, the French government tacitly consented to its distribution. The ambitious work was completed with eleven volumes of *plates*, which appeared between 1762 and 1772. Jean Jacques Rousseau, of Geneva, had also collaborated on the *Encyclopédie* until 1757, when an open conflict with d'Alembert had broken out and he had cut off all ties. In 1761, Rousseau published his *Nouvelle Héloïse*, and in 1762 his *Émile* and *Le contrat social*, which outlined his points of dissent with the *philosophes*. In opposition to the encyclopedists, Rousseau proposed a notion of reason which he considered more profound, as well as a more direct contact with nature. Manifest in his writing was the influence of literature about the "noble savage," who was not a slave to civilization and its abstractions and could therefore understand authentic natural truths. A short time later, from 1766 to 1770, Rousseau wrote his famous *Confessions*, which had a considerable influence upon the Romantics. In 1764, Voltaire finished his *Dictionnaire philosophique*, while Paul Henri d'Holbach (1723–89) published his *Système de la nature*, which heightened the attack against religion and replaced Voltaire's deism with a clearly materialistic atheism.

Developments in philosophy and politics

Of particular importance in European culture was the rediscovery of Spinoza, who for around a century had been condemned as blasphemous. Lessing, in particular, studied him and acknowledged his extreme importance in a pamphlet entitled *Über die Wircklichkeit der Dinge ausser Gott* (1762). Kant, meanwhile, continued his research, dealing in problems that were increasingly more philosophical in nature. In 1766, he published his *Träume eines Geistersehers*, in which he argued not only with the Swede Swedenborg, but with the metaphysics of Wolff and Crusious as well. His dissertation *De mundi sensibilis atque intelligibilis forma et principiis* came out in 1770; it clearly outlined the basic ideas of the new theory of knowledge which he later developed in 1781.

Meanwhile, groups of Italian Enlightenment thinkers, stimulated by French and English culture, intensified their work. The Società dei Pugni was founded in 1761–62 in Milan; its very name intended to express the full freedom of discussion that its members were able to enjoy. It was promoted by Paolo Verri (1728–97), assisted by his brother Alexander (1741–1816), Cesare Beccaria (1738–94) and others. All of them were intensely interested in problems of civil life, scholars of economic questions, and writers of ethical and political essays. Their journal, Il Caffè, was modelled after the English Spectator, and came out every ten days from June 1764 to May 1766. In 1764, Beccaria published his small volume Dei delitti e delle pene; the work was soon extremely well-known throughout Europe, despite condemnation by the Church, which placed it on the index of forbidden books.

MATHEMATICAL SCIENCES

Mathematics and the encyclopedism of the Enlightenment

This attention to a wider range of philosophical, historical, logico-critical, cultural, and even popularizing (in the higher sense of the word) points of view, in addition to technical developments and strictly mathematical conceptual perspectives, was not by any means extrinsic; on the contrary, it intimately responded to the spirit of scientific (and mathematical) research of the epoch. We have already mentioned the prolific interweaving of the philosophical disputes and technical research that surrounded analysis in the British Isles, and the interference of ethical and juridical (besides socio-economical) problems in the development of the calculus of probabilities. The philosophical and theological controversies that arose around the principle of least action are discussed elsewhere.

On this note, it would be unnatural to isolate the purely mathematical contributions of a figure like d'Alembert without considering only how closely they were related to problems in mechanics and the broader cultural context in which his scientific research took place. For example, various articles that he published in the Encyclopédie, ou dictionnaire raisonné des sciences, des artes et des métiers were of considerable mathematical importance. Likewise only in this perspective can the importance of the Lettres à une Princesse d'Allemagne, which Leonhard Euler began publishing in 1768, be fully appreciated. This work presented a sort of culturally unitary synthesis of all the scientific knowledge of the times. We have already mentioned the many faceted work of scientists as Maupertius and the naturalist Buffon.

Another author who displayed this multiplicity of interests common to Enlightenment scientists was Johann Heinrich Lambert (1728–77), who in 1759 published Die freye Perspective, a work intended for painters and artists as well as for mathematicians. In 1764, he published a Neues Organon, in which logical themes are accompanied by various logical and methodological considerations, using a Leibnizian-type approach. In 1765 in his De universalioris calculi idea disquisitio, Lambert very decisively took the same approach, which Gottfried Ploucquet (1716–90) had taken the previous year in his Methodus calculandi in logicis. Lambert's work came out the same year in which he presented his famous, purely mathematical demonstration of the irrationality of π. This proof is a worthy companion to other findings of a strictly mathematical nature that Lambert made,

such as those on continuous fractions. In 1766, he composed his Theorie der Parallel-Linien (published posthumously in 1782), in which he very ably pursued the work in non-Euclidean geometry begun by Saccheri's Logica demonstrativa (see MATHEMATICAL SCIENCES 1730–1740).

ASTRONOMY

The cosmological problem

Johann Heinrich Lambert (1728–77) published his Cosmologische Briefe during this period, in which he not only guessed the existence of the galaxy, but also maintained that it was part of an order of four systems: the planetary system, the star system (cluster), all of the clusters (galaxy) and all of the galaxies. Lambert's hypothesis has been effectively proved to be true in our century.

The passages of Venus

The rare phenomenon of the passage of Venus across the sun, which happens twice in eight years every 121.5 or 105.5 years, took place in 1761 and 1769. Expeditions were organized for the occasion in order to obtain the best possible observations of the phenomenon. The findings could then be used to calculate the sun's parallax (thus giving the distance of the earth from the sun), on the basis of Halley's method as perfected by Joseph Nicolas Delisle. More than 1,200 people from various parts of the world took part in these expeditions. In 1761, 120 stations were set up, and in 1769 there were 150. France, England, and Russia sent the largest number of observers. Delisle, who had returned to France from St. Petersburg in 1747, was the general director of the project. The elaboration of the data gathered during this part of the eighteenth century was completed by Simon Newcomb in 1890, and gave the value of the parallax as 8.79 s (present value, 8.794 s). During the observations of the passage of Venus in 1761, Mikhail Vasilievich Lomonosov (1711–65), using Delisle's method, noticed effects of atmospheric refraction on Venus, thus discovering that it had an atmosphere. This was the first astrophysical discovery of the eighteenth century. In the years that followed, Lomonosov took a considerable interest in astrophysical problems, and in 1762 he made a detailed outline of the goals and duties of this science, which he named ''astrophysics.''

The Titius–Bode law

In 1766, Johann Daniel Titius (1729–96) published his law on the distances of the planets from the sun, which he formulated as $0.4 + 0.3 \times 2'' = D$, where substituting $-\infty, 0, 1, 2, 3 \ldots$ for n gives the distance D from the sun, expressed as a function of the distance of the earth. The law did not become well-known until it was diffused by Johann Elert Bode in 1772.

Lalande's Astronomie

Joseph Jérôme Lalande gathered all the discoveries and methods of astronomy from antiquity to the present, and published them in 1764 in his Astronomie, a classical treatise first in two, then in three volumes. Three editions of the work came out up to 1792, and it was reprinted in 1966.

PHYSICAL SCIENCES

Euler: solid mechanics and popularization

The beginning of the decade was marked by the publication of Leonhard Euler's *Theoria motus corporum solidorum seu rigidorum* in 1760, a work on solid mechanics. It described the results of Euler's research on rigid bodies, mostly done between 1758 and 1760. In his research, Euler made a critical analysis of the knowledge gathered by various scholars up to then, and his findings contributed to a profound conceptual revision in eighteenth-century mechanics. His mathematization of the problems of the mechanics of rigid bodies led him to determine the equations for the general motion of bodies with forces acting upon them.

One of the most significant works of the scientific Enlightenment, Euler's *Lettres à une princesse d'Allemagne* (in three volumes, 1768–72), began to come out at the end of the decade. A high-class popularization, it clearly and precisely attacked all the major problems of physics at the time and advanced and defended the positions that the author took on them in his more scientific works (defense of the wave theory of light, the corpuscular theory of matter, etc.). It was inspired by a firm conviction that science should not be the monopoly of specialists, but rather the patrimony of all educated persons and a basic factor in the progress of civilization.

Problems in fluid mechanics

In several works, Jean Charles de Borda (1733–99) denied the possibility of solving all the problems of fluid mechanics using only the theories known at the time. He strongly emphasized the inadequacies in elaboration that a number of studies on the living force in fluids, done with a Newtonian approach, presented. Borda's *Mémoire sur l'écoulement des fluides par les orifices*, published in 1766, spoke particularly to these inadequacies. It focused on those problems of hydrodynamics that concerned "losses of living force."

Pyroelectricity

Investigations into the properties of tormaline, analyzed in detail by Franz Ulrich Theodosius Aepinus from 1756, and by John Canton (1718–72), gave a considerable boost to research into electrical phenomena. A work by Aepinus on the subject, entitled *Recueil de différents mémoires sur la tourmaline*, appeared in 1762, proclaiming pyroelectricity as a basic field of study. Research on pyroelectricity was given a prominent place in the *History and Present State of Electricity, with Original Experiments*, by Joseph Priestley (1733–1804), the first edition of which was printed in London in 1767. The work was above all a comparative analysis of the various theories on electrical phenomena, with special reference to the ideas of Benjamin Franklin, Jean Nollet, and Aepinus. Priestley's experiments led him to discover the conductivity of coal and metallic salts, and to guess that electrical force depended on the inverse of the square of the distance.

The nature of electricity

Meanwhile, debates on the nature of electricity continued. In 1766, the anatomist Giovanni Francesco Cigna (1734–90), of Turin, carried out several interesting experiments on the charges of conductors. In 1767, Giovanni Battista Beccaria (1716–81) attempted to interpret Cigna's experiments theoretically, presenting his ideas in a printed letter addressed to Benjamin Franklin. Allessandro Volta (1745–1827) criticized these ideas in his first work, *De vi attractiva ignis electrici ac phaenomenis indepentibus* (1769), written in the form of a letter addressed to Beccaria. It introduced the important concept that is today called potential.

CHEMICAL SCIENCES

The preparation of hydrogen

Research continued to be done on gasses. Henry Cavendish (1731–1810) prepared hydrogen from the reaction of diluted acids on certain metals and, in 1766, defined it as a substance in its own right, different from other gasses. He collected it with apparatus that he himself invented, and studied its properties, often subjecting it to quantitative experiments. He maintained that it was liberated by metals and thus identified it as the phlogiston.

The concept of "equivalent"

Quantitative methods began to be diffused. In his analysis of the water of Rathbone Place, one of the first quantitative analyses ever, Henry Cavendish used the term "equivalent." In the area of quantitative analytical techniques Antoine Baumé (1728–1804) introduced the areometer, an instrument for determining the amount of alcohol in a hydroalcoholic mixture on the basis of density, in 1768.

The Dictionnaire de chymie

The first chemical dictionary, in the modern sense of the term, came out of the land of the encyclopedists: the *Dictionnaire de chymie* was published by Pierre Joseph Macquer (1718–84) in 1766 and translated into several languages. The Italian translation was done by Giovanni Antonio Scopoli (1723–88), who included notes and several writings by Allessandro Volta (1745–1827). It should be mentioned that Macquer was an avid defender of the phlogiston theory, an indication of how deeply that doctrine was rooted in the scientific ideas of the epoch. In fact, Macquer considered the increase in weight observed during the calcination of metals as sure proof of the phlogiston theory. During the process, the metal supposedly lost the phlogiston, which he believed had negative weight. However, the French chemist Tillet had already expressed some doubts about this in a memoir published in 1762, in which he observed that the loss of phlogiston could not sufficiently explain the particular increase in weight that occurred in the calcination of lead.

LIFE SCIENCES

Medical semiotics and pathological anatomy

Medical semiotics arose as a branch of medical knowledge distinct from traditional nosography. The new discipline, which was closely tied to clinical medicine, proposed to distinguish the physical symptoms of disease syndromes. It was founded by the Austrian clinician Leopold Auenbrugger (1722–1809), whose *Inventum novum ex percussione thoracis humani* (1761) minutely described the percussion technique and how it could be used in diagnosing various diseases. The work was translated into French and Italian, but was greeted with diffidence; the percussion technique was only popularized in 1808 by Jean Nicolas Corvisart des Marets (1755–1821).

Decisive developments also took place in pathological anatomy. Like semiotics, it presupposed an assimilation of the most general features of the mechanical method in the life sciences, where the spatial, temporal, and quantitative aspects of observed phenomena, as well as their physical concomitances, were no longer being ignored. Pathological anatomy proposed to correlate systematically the anatomical lesions and functional disorders connected with various disease syndromes. Studied from a new localistic point of view as well, syndromes ceased to be a casual set of symptoms and became definite entities on the structural and functional levels. This new trend was reflected in *De sedibus et causis morborum per anatomen indagatis*, by Giambattista Morgagni (1682–1771), a professor of theoretical medicine. The work came out the same year as Auenbrugger's *Inventum*. The most important sections were those concerning cardiovascular infections, paralytical syndromes caused by circulatory disorders, and tubercular lesions. Between its first edition and 1825, the work was republished seven times and translated into every major European language.

Controversies over botanical taxonomy

Although botanical taxonomy was simplified and rationalized by Linnaeus' sexual system and binary nomenclature, it was still the object of controversies and polemics, as it was accused of artificiality. A system that mirrored all the natural affinities and divergencies among the large groupings was needed. The *Familles des plantes*, by the botanist Michel Adanson (1727–1806), was the first work that attempted to base distinctions among plant groups on a natural criterion.

The attempt was resumed in 1789 by Antoine-Laurent de Jussieu (1748–1863). Another scholar of plant systematics and a skillful geographer of plants of this period was Peter Forskal (1736–63). He participated in the scientific expedition organized by the king of Denmark, which left Copenhagen in March in 1761. The scientists travelled to Constantinople, Malta, Smyrna, Egypt, and the Red Sea, where Forskal successfully conducted his research on botany. Other studies in botany, especially on the sexual propagation of plants were made by Joseph Gottlieb Kölreuter (1733–1806). In 1760, he conducted crossbreeding experiments similar to those done by Linnaeus and laid the foundations for understanding hybridation and its importance in the modes of plant sexuality. His findings are described in his *Vorläufige Nachricht von einigen das Geschlecht der Pflanzen betreffenden Versuchen und Beobachtungen* (1761), to which he made three successive additions (1763, 1764, and 1766). However, his research on the fecundation processes was limited by the scant technical possibilities of microscopic research at the time.

Lazzaro Spallanzani and spontaneous generation

Lazzaro Spallanzani made a decisive contribution to the controversy over the generation of living organisms, with respect to the views on spontaneous generation of this period, with his *Saggio di osservazioni microscopiche concernenti il sistema della generazione de' signori Needham e Buffon* (1765). Based on direct observations through the microscope and refined experimental techniques, the Italian naturalist was able to demonstrate that, in samples of meat identical to Needham's, the protozoa were born from tiny eggs, i.e. from cysts, and that the germ cells entered the meat from the air. Spallanzani's *Saggio* was published in French in 1769, and was edited by Needham himself, who accompanied it with notes in defense of his own theory. Spallanzani replied with the first volume of his *Opuscoli di fiscia animale e vegetale* (1776). In it, he above all denied that the prolonged effects of heat acted on the so-called vegetative force rather than on the protozoa cysts contained in the specimens. Voltaire wrote to Spallanzani: "Your book will live on because it is founded on experimentation and reason." In the *Opuscoli*, Spallanzani also distinguished between Infusoria and molds, and described the shape and movements of the former, which he recognized as animals. He pronounced himself in favor of the ovist theory in the form proposed by Bonnet, and confuted the *aura seminalis* theory. His research on the digestive effects of gastric juices confirmed the Italian author's experimentalism as one of the most authoritative expressions of biology of this century.

In his essay *De la nature* (1776), Jean Baptiste Robinet (1735–1820) once again proposed the Platonic idea of the great chain of being: science needed only a single prototype to explain the variety of organisms that were found in nature. Robinet's transformism, however, was speculative rather than historical. It was a guide to scientists' reason and fantasy, not an overview of the history of living things.

EARTH SCIENCES

The search for the causes of changes in the earth's appearance

A number of lists and illustrations of fossils belonging to the different orders recognized in stratigraphy were published during this decade. Stratigraphic subdivisions multiplied as observations were extended to different countries, while the first systematic treatises in mineralogy based on the characteristics of the components of known rocks were written. However, the causes of the events that formed the various layers of the earth's crust were still uncertain and obscure, and, except in rare instances, planations were vague or totally fantastic. John Michell's *Conjectures Concerning the Causes and Observations upon the Phenomena of Earthquakes* (1760), mentioned earlier, attributed the causes of the great earthquake that devastated Lisbon in November, 1755, to the effects of underground vapors that formed in caves and in cracks in the earth's crust when sea water or rivers came into contact with the central fire. He also believed that the irregularities and folds in rock layers were the result of the same forces, which were often accompanied by earthquakes. In 1761, Antonio De Ulloa (1716–95), who had found the univalve fossil Lamellibrancia at the Guanca Velica mine in Peru at an altitude of 4,330 meters made accurate deductions concerning the climate in ancient times, which must have been milder, and about the rock, which was originally ocean mud. For the rocks to have been lifted to such great heights, the upheavals had to have been much different from those occurring nowadays.

That same year, Georg Christian Füchsel (1722–73) published his *Historia terrae et maris ex historia Thuringiae per montium descriptionem erecta*, which unfortunately remained unknown for sixty years. In it, he increased Lehmann's stratigraphic distinctions to five for the Harz mountain area and Thuringia and maintained that the earth had always been subject to phenomena similar to those of today. He also explained that layers could be distinguished by the organic fragments they contained, as well as by their mineralogical composition. Series of rock layers (formations) provided a representation of the epochs in the earth's history. The rocks were originally deposited horizontally, and later became inclined as a result of earthquakes and oscillations of the ground. Unfortunately, Füchsel's unusually modern ideas did not have any influence on his contemporaries. In 1763, Mikhail Vasilievich Lomonosov (1711–65) published an important work in which he developed his ideas on the processes that worked on the earth's crust, already outlined in his work of 1757. His studies were quite influential in Russia and in other European countries, where numerous translations of his works were published. John Tuberville Needham (1713–81) wrote that mountains were formed by the action of underground fire and volcanoes.

New ideas on the origin of basalt

In 1765, Nicolas Desmarest (1725–1815) presented a paper to the Académie Royale des Sciences (published in 1771) in which he associated the basalts of the Auvergne with present-day lava. His observations of veins of basalt in porphyry (trachyte) and veins of porphyry in granite, led him to deduce the eruptive nature of both. This interpretation was not taken into consideration by adherents of the Neptunist school in subsequent years. In 1769 Giovanni Arduino explicitly recognized the existence of ancient volcanoes at Chiampo near Vicenza, and compared their eruptive activity with that of Vesuvius and Etna.

Arduino's assertions and those of some of his contemporaries (for example, Alberto Fortis [1741–1803], who described the Montegalda hills, also in the area of Vicenza, in 1765) seemed incredible to the geologists of the time. Jean-Étienne Guettard modified his opinion somewhat and defined the basalt he studied in southern France as a vitrescent rock that was formed by crystallization in a water-like fluid rather than in a molten fluid. This explanation anticipated some of the ideas of the Neptunists.

Paleontological studies

In North America, George Croghan discovered the remains of mastodons in Kentucky in 1765. Buffon studied them and concluded that animals larger than any of the largest known elephants had once existed and had inhabited North America and Europe, where similar remains had already been found.

Daniel Charles Solander (1763–82) and Gustavus Brander (1720–87) rejected John Woodward's idea that shells in the tertiary clays of Barton, in Hampshire, had been deposited by the Biblical flood. They illustrated the fauna of these clays in detail (1766), and claimed that most of the species were different from any that were presently living.

The two-volume *Dictionnaire universal des fossiles propres et des fossiles accidentels* by Élie Bertrand (1712–55) was published in 1763. It was the basic treatise for geology students during the eighteenth century.

Scientific expeditions

On August 25, 1768, the *Endeavor*, commanded by Captain James Cook (1728–79), set sail from Plymouth to observe the passages of Venus across the sun from the South Seas. This and later voyages undertaken up to 1779 completed the geographical exploration of the earth, except for the extreme polar regions. English power in the Pacific was also established through Cook's discoveries. In 1769, under the auspices of Catherine II, a Russian expedition left for Siberia, also to observe the transit of Venus. One of the participants in the Russian endeavor was the geologist Peter Simon Pallas (1741–1811), sent from Berlin in 1767.

Cook's expedition made important findings in oceanography as well. During the first explorations of the Pacific islands, especially at Tahiti, Cook and the expeditions's astronomer Charles Green, noted and described tides with a particularly small range (and considerable daily variations). The systematic observation of the phenomenon (small tide ranges had already been noted on islands in the central and southern Atlantic, and Newton had studied the problem) definitively confirmed that ocean tides were relatively small and that they were more or less exalted on the continental coasts due to the effects of coastal geometry (shallow waters). On the occasion of the second voyage (1772–75), William Wales (c. 1734–98) and William Bayly (1737–1810), who had gone along to do research, invented a tide gauge and repeated Cook's and Green's earlier observations, confirming them entirely.

APPLIED SCIENCES

Watt's engine

A new type of blowing machine appeared in 1762. Some time later, John Smeaton introduced blowing cylinders. However, the driving force was still supplied by water power. Coke only won out against wood charcoal when the cylinders began to be powered by steam engines.

In 1763, James Watt (1736–1819), who had been appointed Mathematical Instrument Maker to the University of Glasgow, was able to repair a model of Newcomen's engine, succeeding where a London instrument builder had failed. From the experiments he carried out, Watt deduced that the cooling effect of the water that was let in on every active stroke was the cause of so much steam being consumed. These reflections led Watt to his first important conclusion, that there was latent heat in the condensation of steam. Watt's other great invention dates back to the beginning of 1765, when he came up with the idea of a separate condensor that would fulfill his needs. The principle of maintaining the cylinder at the highest temperature possible led to the idea of using steam instead of atmospheric pressure to push the piston down.

In 1769, Watt obtained a patent for "A new Method of Lessening the Consumption of Steam and Fuel in Fire Engines."

In 1769, the Frenchman Nicolas Joseph Cugnot (1725–1804) applied the steam engine to a three-wheeled vehicle that could move at a speed of four kilometers per hour.

Textile machines

Spinning was revolutionized by James Hargreaves (1720–78) and Richard Arkwright (1732–92). Hargreaves invented the spinning jenny, a hand-operated spinning machine, in 1764 (patented in 1769). In 1769, Arkwright patented his water-powered spinning machine, a wooden frame with four bobbins on top onto which the yarn passed.

Considerable progress was also made in the production of porcelain; in 1768 the English chemist William Cookworthy (1705–80) obtained the first patent for manufacturing hard china.

Mechanics, construction, and instruments

Simultaneous to the development of the first engines and progress in the spinning industry came developments in precision mechanics and machine tools. In 1765, the Duc de Chaulnes (1714–1769), an academician in France, built two dividing machines, one for circles and the other for straight segments. From 1760 to 1770, Jesse Ramsden (1735–1800) worked on the first dividing machine that could be used on an industrial scale. Then in 1770, Ramsden invented two screw-cutting lathes. In 1768, for the first time, Jacques de Vaucanson (1709–82) adapted Thiout's principle to machines for large-scale production, designing lathes and drills with screw-operated carriages.

In 1764, P.M. Trésaguet (1716–94) invented a new method for building roads by first laying large stones that would separate the surface from the road-bed. His system allowed the road surface to drain properly without the excessive convexity that made the older roads dangerous.

James Small (flourished 1763–84) built the oscillating Scottish plow in 1763, and in 1767 introduced into Scotland the Rotherdham plow, an oscillating plow with a curved moldboard that turned the sod very well. The major innovations in the plow were basically a new arrangement of the parts that made up the main frame and the use of a triangular plow instead of a quadrilateral one, the most important change in plow design in the eighteenth century.

As to progress in the field of scientific instruments, Jesse Ramsden built his famous electrostatic machine in 1770. It was the device most widely used for producing large quantities of high tension electricity for almost a century.

From 1764 to 1781, in Nuremberg and Augusta, Georg Friederich Brander (1713–83) published a series of little books that gave detailed descriptions of the instruments he had invented. These included a compound microscope and a hydrostatic weighing unit.

Treatise writing

The *Nautical Almanac* containing astronomical tables calculated for Greenwich, was published in 1767. In 1769, James Cook (1728–79) determined the longitude of the northern point of Tahiti, arriving at an estimation that was within a minute of the present value.

In the area of treatise writing, we shall only mention the *Photometria* (1760), by Johann Heinrich Lambert (1728–77), which long remained the basic text for this new discipline. The final volumes of Diderot's *Encyclopédie* appeared in 1766; the eleven volumes of illustrated plates completing the work also came out between 1762 and 1772.

The plates gave considerable attention to designs for machines and to works of engineering in general, an interest which contributed a great deal to stimulating that type of activity.

1770–1780

THOUGHT AND SCIENTIFIC LIFE

Progressivism and power politics

Two political events, in a certain sense antithetical to one another, had immediate and profound repercussions on the cultural scene: the first division of Poland (1772) and the American Declaration of Independence. The agreement between Russia, Prussia, and Austria to appropriate some of the most important provinces of the Polish state showed all impartial observers (including the admirers of Frederick II, Catherine II, and Joseph II) that the three enlightened monarchs, despite their sympathies for the new ideas of the philosophers and scientists, had no difficulty at all in pursuing pure power politics. This clearly proved the impossibility of establishing an authentic, lasting alliance between the progressive movement and monarchic absolutism. On the other hand, the Declaration of Independence and the victorious war of the North American colonial rebels against the English troops aroused sincere enthusiasm among European progressives: the Declaration of Independence inspired the French revolutionaries of 1789.

The decline of the English and French Enlightenment

Voltaire wrote his *Questions sur l'Encyclopédie* in 1770–72, and his *Histoire de l'établissement du christianisme* in 1777, while the polemics on religion continued, and the *philosophe* front split into deists and atheist-materialists. But by this time the prolific output of the French Enlightenment was beginning to decline. After Hume's death in 1776, in reaction to his skepticism, a current of thought that is generally known as the philosophy of common sense arose in England with Thomas Reid (1710–96) as its central figure. Following a solidly critical approach, Adam Smith (1723–90), from his early studies in moral philosophy (which dated back to 1759), began developing an economic analysis that reached its peak in *An Inquiry into the Nature and Causes of the Wealth of Nations* (1776). The work was an important milestone in the research on economic policy that became very important in Scotland during this period. This type of research was pursued in France as well and continued to be the basis for research in England throughout the following century.

The Italian Enlightenment continued along the path opened in the preceding years. It was characterized by a prevalence of historico-economic and ethico-political interests. In 1773, Paolo Verri published his *Meditazioni sull'economia politica* and *Idee sull'indole del piacere e del dolore*.

Lessing and Kant; Sturm und Drang

In 1777, Lessing published an outline of his most important philosophical work, *Die Erziehung des Menschengeschlechtes* (the complete edition came out in 1780). In it, he maintained that the search for truth was more important than processing it, and outlined a new concept of history, conceived of as ideal order.

Between 1774 and 1778, he edited the publication of *Fragmente eines Anonymus* (a work by Reimarus), which can be considered the most openly anti-Christian work of the entire German Enlightenment. Although he started from a concept based on Wolffian principles, as Lessing explained in his introduction, Reimarus manifestly carried it to completely heterodox conclusions, undeniably demonstrating the intolerance of the new German thought for Wolff's moderation.

Meanwhile, a new movement—*Sturm und Drang*—arose, soon attracting young Germans who were interested in literature and philosophy. It exalted themes of a clearly irrationalist stamp but at the same time maintained a number of ties with the Enlightenment. Johann Wolfgang Von Goethe belonged to it in his youth, and its influence can be seen in both his play *Götz von Berlichingen* (1773) and his epistolary novel *Die Leiden des jungen Werthers* (1774). The *Sturmer* and Enlightenment thinkers shared a vehement condemnation of the defects of the old society, an interest in nature, and substantially laical tendencies. They differed in that the *Sturmer* introduced the category of genius in place of enlightened rationality. This is why Lessing enthusiastically condemned their ideas on esthetics, an important area for him as well as for the supporters of the new current.

The spread of the new culture

Meanwhile, the cultural revival was spreading to an increasingly large number of countries. This was clearly evidenced by the creation of new institutions, including the *Académie royale de Belgique* in Brussels (1772), which had begun as a private literary society, and the University of Trnava, in Slovakia (1770), which later became the University of Budapest. Others were the University of Münster in Westfalia (1773), the Society of the Sciences in Prague (1774), and finally the University of Palermo (1779) and the Academy of Sciences in Lisbon (1779).

In 1772, J.C. Senckenbert left a substantial fortune to a foundation for the advancement of the sciences in Frankfurt. The Royal Society celebrated its hundredth anniversary in 1776, and for the occasion created the Copley Medal, an award for the authors of the best scientific research.

MATHEMATICAL SCIENCES

Lagrange's work and the flourishing of studies in algebra

Joseph Louis Lagrange moved to Berlin in 1769 as director of the science classes at the Academy there. His important works on the theory of numbers began to appear in the memoirs of the Academy between 1768 and 1771. A decade of prolific studies and findings, especially in analysis, followed: series development, partial differential equations, the theory of singular integrals, and indeterminate analysis, in addition to developments in the calculus of variations, which he had basically intuited in earlier work done while he was still living in Turin. His prodigious work and the many contacts he established with the major scholars of the time (not only mathematicians), make him one of the major scientific figures of European culture.

While Euler's style of research was marked by a great boldness and richness of mathematical imagination, which often lacked the rigorous critical control of a strictly logical approach, Lagrange (although he did not yet feel the explicit need for rigor that characterized the following century), was much more attentive to the demands of deductive rigor. He was also the most typical representative of the analytical mentality, in the sense that he applied methods of infinitesimal analysis in areas where it had previously been considered extraneous; for example, he made wide use of the infinite algorithm in continued fractions in the fields of arithmetic and algebra. In algebra, he worked on the problem of approximating roots in a numerical equation using continued fractions. He also attacked the general question of solving literal equations, expressing his conviction that it was impossible to solve general equations beyond the quartic, and that the problem of resolvability depended on a study of the coefficients of the equations.

Lagrange's profound interest in algebra (the famous series that bears his name appeared in 1770 in a work written on the theory of equations) corresponded to the emphasis that was being put on research in this discipline, which actually flourished in a number of places during this very period. The *Meditationes algebricae*, by Edward Waring (1734–98), a work in which the concept of transformation stood out amid a wealth of themes, came out in 1770.

In 1779, Etienne Bezout (1730–83), who in 1764 had proposed the method for eliminating the unknowns from equations that is named after him, published his *Théorie générale des équations algébriques*, in a certain sense the first extended manual on the subject, which was full of original findings.

ASTRONOMY

The School of Celestial Mechanics in St. Petersburg

Of particular importance was the work done by the School of Celestial Mechanics in St. Petersburg, founded by Leonhard Euler in the Academy of Sciences of that city. The *Theoria motuum Lunae*, compiled by Euler and his students Johann Albrecht Euler (1734–1800), Wolfgang Ludwig Krafft (1743–1814), and Anders Johan Lexell (1740–1784), was published there in 1772. This so-called "second theory of the moon" proposed a new method for calculating all the irregularities of the moon's motion, which was seen as a rotating system of coordinates. These ideas were later developed in the work of George William Hill and Ernest William Brown in the nineteenth and twentieth centuries (see ASTRONOMY: THE SUN AND THE SOLAR SYSTEM, 1920–1930).

Frederich Wilhelm Herschel

Astronomical production and discoveries were in the midst of a period of stagnation when Frederich Wilhelm Herschel (1738–1822) appeared on the scene. Herschel dominated the following decades and founded modern astronomy. It was in this decade, after having read a popularizing book by Ferguson, that he felt a desire to observe the skies himself. He began building increasingly larger and better telescopes, with which he completed his first celestial review in 1775 and began his second in 1779.

Messier's catalogue

The celestial bodies outside the solar system began to be identified more accurately. Charles-Joseph Messier (1730–1817) distinguished between clusters and nebulae without stars and compiled the first catalogue of both types of formations. He discovered most of them himself while exploring the skies for comets, of which he was a passionate observer. In 1771, he published a description of forty-five celestial objects.

The Berliner Jahrbuch

Publication of the Berliner Jahrbuch, which contained accurate annual ephemerides and astronomical news, was begun in 1776. Besides improved ephemerides, methods of calculation were perfected, and new geometrical data on the solar system were obtained.

Improvements were made in guessing and calculating the orbits of comets with the use of a new method elaborated by Heinrich Wilhelm Olbers (1758–1840) in 1779. Meanwhile, Joseph Louis Lagrange (1736–1813) determined the mass of Venus using a combination of two methods, the precession of the equinoxes and the inequalities in the sun's motion. This was the first time that the mass of a planet without satellites was calculated. In fact, the classical method for determining the mass of a planet consisted in applying Kepler's and Newton's laws to the motion of the satellite or satellites.

PHYSICAL SCIENCES

New developments in fluid mechanics

Following the important theoretical studies of the preceding decades, and the requests that Turgot presented to the Académie des Sciences in 1775, a monograph entitled *Nouvelles expériences sur la résistance des fluides* came out in Paris. Published in 1777, it was a product of the work of researchers such as Jean le Rond d'Alembert (1717–83), Marie-Jean Caritat (Marquis de Condorcet) (1743–94) and Charles Bossut (1730–1841), in collaboration with mathematicians such as Adrien Marie Legendre (1752–1833) and Gaspard Monge (1746–1818).

The prevalent experimental nature of the research did not, however, overshadow the role played by theory in directing it; in fact, the monograph affirms that "the art of interrogating nature by means of experiment is very delicate," and that "no science exists without reasoning, or rather without theory." In the discussion of the results, substantial reference is made to progress in Newtonian mechanics, with particular reference to knowledge acquired on the resistance of fluids. The first edition of the *Principes d'hydraulique*, by Pierre du Buat (1734–1809), which pointed out that the problems connected with the resistance of fluids to the passage of solid bodies implied a reelaboration of the available theories, was published in 1779.

The questions that emerged in these studies led to further investigations in the area of resistance and friction. In 1781, the Académie des Sciences promoted a competition on the problem of friction in the construction of machines. It was on this occasion that Charles Augustin de Coulomb (1736–1806) presented his theory of simple machines (*Théorie des machines simples en ayant égard au frottement de leurs parties et à la raideur des cordages*).

Euler's Dioptrics

The three volumes of Euler's *Dioptrics*, a wide-ranging work that systematically expounded the ideas that the author had elaborated over a number of years of research on this important topic, were published between 1769 and 1771. The work contains, among other things, a very important finding made almost twenty years earlier, i.e. the demonstration that refraction and dispersion in lenses were not proportional to each other. This implied the possibility, which Newton had denied, of building achromatic objectives. (As we have already said, John Dollond had succeeded in producing them in 1758, after Euler had supported the possibility of eliminating chromatic aberration in 1747.) An achromatic objective was built with two lenses, one concave and the other convex made of two kinds of glass with different indices of refraction, for example, crown glass and flint glass.

Cavendish's theory of electrical phenomena

Henry Cavendish's "An Attempt to Explain Some of the Principal Phenomena of Electricity, by Means of an Elastic Fluid," an important work in the physical sciences, appeared in 1771 in *Philosophical Transactions*, the journal of the Royal Society. Cavendish's theory was similar to Franz Ulrich Theodosius Aepinus's, although it differed in extent and accuracy. Cavendish himself pointed out the differences and emphasized the importance of mathematical reasoning in reaching a more thorough understanding of electrical phenomena. According to the author, the electrical fluid was "matter," even though it was of a type different from ordinary matter. The particles that made up the fluid repelled each other and attracted particles of ordinary matter, while the latter also repelled each other. Cavendish's analysis focused on discovering the types of forces acting among the particles, and was developed via successive generalizations of the problem according to which interactions were of the form

$$\frac{1,}{r^n}$$

where $1 < n < 3$.

According to Cavendish, a body is in the natural state when it is saturated with electrical fluid, with saturation being a condition of equilibrium between attractive and repulsive forces. With respect to the natural state, a body can be overcharged or undercharged with fluid, in which case attractions and repulsions among different bodies will occur on a macroscopic level. Cavendish's theory was general in that it attempted to establish all possible distributions of the electrical fluid inside bodies and on their surfaces, and analyzed situations in which different bodies were connected by "canals" along which the fluid could flow. Important in Cavendish's thought are several considerations in which the notion of potential is expressed more clearly than in the letters of Allessandro Volta, to Giovanni Battista Beccaria, which were published in 1769. (see PHYSICAL SCIENCES 1760–1770).

Volta's electrophorus

In 1775, Alessandro Volta's research on electrified bodies led him to build a machine called the "perpetual electrophorus" and to conduct a series of investigations on the capacity of conductors. Volta's letter of 1778 to de Saussure was entitled

"Osservazioni sulla capacità dei conduttori elettrici" and laid the foundations for modern electrometry. His investigations developed through a progressive clarification of the relations that existed among the notions of charge, capacity, and tension and led Volta to build the so-called condensor electroscope in 1782.

The spread of studies on electrical and magnetic phenomena

Publications of studies on electrical phenomena and magnetic phenomena multiplied during this decade, following trains of thought that contributed to a critical revision of concepts advanced in their times by scholars such as Benjamin Franklin and Nollet. Beccaria's treatise entitled *Elettricismo artificiale*, a work influenced by the theories of Aepinus, was published in 1772. A translation of the work into English, entitled *A Treatise upon Artificial Electricity*, appeared in 1776. The third edition of Joseph Priestley's *History and Present State of Electricity, with Original Experiments* was printed in 1775 (see PHYSICAL SCIENCES 1760–1770). The spread of studies on electricity was marked by a progressive abandoning of the theories of effluvia and a growing articulation of theories regarding the remote effects exerted by forces of an electrical nature. The works of Aepinus and Cavendish contributed to the mathematization of electrical theories.

Coulomb's investigations into magnetism

Going back to Petrus van Musschenbroek's research (*Dissertatio de magnaete*) and to Aepinus's great work, Charles Augustin de Coulomb published his findings on the best way to produce magnetic needles (*Recherches sur la meilleure manière de fabriquer les aiguilles aimantées*). According to Coulomb, all the hypotheses on magnetic vortices had to be rejected because they were contrary to the principles of mechanics. An explanation of magnetic phenomena could only be found by going back to the "attractive and repulsive forces of the kind treated by gravitational and celestial physics." This memoir of Coulomb's, together with his research on elasticity (*Recherches théoriques et expérimentales sur la force de torsion et sur l'élasticité des fils de métal*, 1784), laid the premises for establishing the basic laws that even today are called Coulomb's laws (see PHYSICAL SCIENCES 1780–1790).

Thermology and investigations into gasses

The most important work of this decade on problems of thermology was a posthumous treatise by Johann Heinrich Lambert entitled *Pyrometrie*. It was dedicated to the study of the expansion of air and to questions connected with heat radiation, and was a point of reference for various types of thermological research until the beginning of the nineteenth century. The research done by Joseph Priestley on the properties of "different types of air" and the *Recherches sur les modifications de l'atmosphère* (1772) by Jean André Deluc (1727–1817) also deserve mention.

Frisi's Cosmographia

An important treatise in two volumes by Paolo Frisi (1728–84), entitled *Cosmographia physica et mathematica*, was published in 1775–76. The work brought immense fame to its author, who was made a member of the most famous European academies (the Royal Society of London, the Académie des Sciences, etc.).

CHEMICAL SCIENCES

The discovery of nitrogen and studies of its chemistry

Important research on gasses was done in the 1770s. New gasses were discovered, while others were studied more thoroughly. But it is above all important to remember that it was in the climate of this research that the ideas of Antoine Laurent Lavoisier (1743–94) matured.

Nitrogen was isolated almost simultaneously (1772) by Daniel Rutherford (1749–1819) and Carl Wilhelm Scheele (1742–86). Joseph Priestley (1733–1804), who was a philosopher as well as a chemist, discovered nitrogen protoxide (1772–73) and characterized nitrogen oxide. Noting that the contraction in volume was proportional to the breathability of the air (i.e. to its oxygen content), Priestley developed a method for measuring the "freshness" of air; the method became more accurate when Felice Fontana invented the eudiometer in 1774. He also obtained but did not isolate carbonic oxide, which Joseph Marie Lassone (1717–88) had already prepared in 1776 by heating zinc oxide with wood charcoal. The question of the chemical identity of carbonic oxide, which was confused with hydrogen, was finally resolved in 1801 by William Cumberland Cruikshank (1745–1800), who named the compound gasseous carbonic oxide; the name was later changed to carbonic oxide.

Chlorophyll in photosynthesis

While studying the properties of carbon dioxide, Priestley found that plants regenerated air that had become stale due to combustion or respiration. He thus laid the groundwork for the discovery of chlorophyll photosynthesis, which Malpighi had predicted in 1671 and which was finally announced by Jan Ingenhousz (1730–99) in 1779.

The discovery of oxygen

During the same period (1772–74) Antoine Laurent Lavoisier gathered proof that in the combustion of carbon, phosphorous, and sulphur, as well as in the calcination of metals, the increase in weight was caused by the absorption of air (or of a gas contained in it) and was proportional to the volume of air absorbed.

However, oxygen was not discovered by the French chemist alone; the discovery was published independently by both Carl Wilhelm Scheele and Joseph Priestley in 1775. Scheele obtained oxygen (1770–73) by heating several metallic oxides and treating pyrolusite with sulphuric acid. Priestley obtained it (1774) by heating red mercury oxide in the absence of air, using the sun's rays, concentrated by a large lens. Both noted that the gas supported combustion and respiration better than air. Lavoisier repeated Priestley's experiment a few months later and concluded that heating a metallic oxide produced the metal and an air that was particularly pure.

The composition of air

While pursuing these studies and checking, completing, and coordinating similar experiments done by various scholars including the Italians Giovanni Battista Beccaria (1716–81), Giovanni Francesco Cigna (1734–90) and Fontana, Lavoisier conducted his classic experiments on combustion, respiration, and the composition of acids. He observed that in the combustion of phosphorous or a candle, as well as in the oxidation of metals and the respiration of animals, only the purest part (from one-quarter to one-sixth) of the air was consumed. The part that remained was unable to support combustion, oxidation, or respiration. He concluded that atmospheric air was a mixture of about one part of *air vital* and four parts of *air méphitique*, or *mofette atmosphérique*. The first gas was present in all acids, and could therefore be called the *principe acidifiant* or *oxigine* (later modified into *oxygène*). The second, which was incompatible with life, was called *gas azotique*, and later *azote*. Gasseous oxygen contained the oxygen base combined with the matter of fire, light, or heat (*calorique*); the fuel combined with the oxygen base, producing an acid, while the caloric was liberated. This occurred in combustion, oxidation (slow combustion), and respiration. This, briefly, was the hypothesis that Lavoisier opposed to the phlogiston theory.

Other gasses discovered during this period were chlorine and sulphydric acid (Scheele, 1773 and 1777), and methade, which Allessandro Volta (1745–1827) discovered in 1776 at Angera, on Lago Maggiore.

New elements and inorganic compounds

In addition to the enormous advances made in the chemistry of gasses, some progress was also made in mineral and organic chemistry. In 1770–71, Johan Gottlieb Gahn (1745–1818), a collaborator of Scheele's, discovered the phosphoric acid in bones. In 1774, he isolated manganese from pyrolusite (after Scheele, that same year, had shown that pyrolusite contained a metal other than iron) and, together with Scheele, discovered barite. Scheele prepared phosphoric acid (1774), arsenic acid (1775) and molybdic acid (1778) and in 1778 discovered molybdenum, which Peter Jacob Hjelm (1746–1813) isolated in 1782.

The discovery of urea and picric acid

Carl Wilhelm Scheele was also interested in organic chemistry. He isolated various organic acids by precipitating them as calcium salts and displacing them with sulphuric acid. He obtained tartaric acid (1770) and uric acid (1776) using this method. In 1771, Peter Woulfe (c. 1727–1803) obtained picric acid by treating indigo with nitric acid. The substance dyed wool and silk a lovely color of yellow and can be considered the first artificial organic dye. Urea was discovered in urine by Hilaire-Marin Rouelle (1718–79) in 1773, though it was only isolated in 1799 by Antoine François de Fourcroy (1755–1809) and Louis Nicolas Vauquelin (1763–1829).

LIFE SCIENCES

The smallpox pandemic and the prevention of epidemics

A pandemic of smallpox descended on all of Europe in 1770, heightening the spread of the disease, which had already been dramatic in the preceding years. By this time, four out of five people had contracted the disease and had either died or carried permanent marks of it on their bodies. An awareness of the

need for a public commitment to fighting epidemics was reawakened in European countries. In France, the study of the problems connected with the prevention and hygiene of epidemic diseases was entrusted to the Royal Society of Medicine, founded in 1776 by Joseph Marie François de Lassone (1717–88) and Félix Vicq d'Ayr (1748–94). In 1776, L.S. Le Brun published a treatise on epidemic diseases. The time was ripe for the discovery of vaccination.

"Neural pathology"

The activity of nerves and muscles continued to be a focus for both physiologists and pathologists. By the middle of the century, two currents had emerged. One, of Stahlian inspiration, attributed muscle contractility to a vital principle located in the nerve centers (and consequently to the nerve centers themselves); the other, after Haller, believed that contractility was a characteristic of the muscle tissue. A current that, without falling back on the ideas of Stahlian animism, insisted on the importance of the nerve centers in the genesis of diseases, developed in eighteenth-century pathology. In his *First lines of the Practice of Physic* (1776), William Cullen (1710–90), a pathologist from Edinburgh, attributed to the nervous system in general the same function that Friedrich Hoffman had attributed to the spinal cord in pathology. The pathogen acted directly on the nerves, causing a reaction: this was the origin of contractions and spasms and of relaxation and atony.

John Brown (1735–88), a student of Cullen's who later became an adversary, organized and corrected Cullen's views in his *Elementa medicinae* (1779). According to Brown, irritability was an expression of life itself, but stimuli were needed to sustain it. Diseases were caused by an increase or decrease in stimulation. Giovanni Rasori (1766–1837), an Italian chemist, was converted to Brown's views and translated the *Elementa* into Italian (1792). Rasori maintained that medical knowledge had to be reconstructed in the light of the new principles of neural pathology. During the last decade of the century, research on electricity furnished neuromuscular physiology with new hypotheses and tools; but traditional views were still being echoed in 1796 by Samuel Thomas von Sömmerring (1755–1830). In his *Organ der Seele*, von Sömmerring affirmed that the *sensorioum commune*—the centers that received sensory stimuli and transformed them into movement impulses—were located in the cerebral vescicles, and that the humors contained in the ventricular cavities were identified with animal spirits.

The naturalistic vitalism of the Montpellier school

Stahl's nonmaterialist vitalism—which had taken on the characteristics of a true animism—was joined by a naturalistic, and in some authors, declaredly materialistic, vitalism that came out of the Montpellier school. This tendency was given an organic formulation in the *Nouveaux éléments de la science de l'homme* (1778) by Paul-Joseph Barthez (1734–1806), after earlier statements by Théophile de Bordeu (1772–76). The soul ruled spiritual actions, while a "vital principle" inherent in the single parts of the organism was the source of life. The vitalism of the *montpelliériens* also contrasted with biological mechanism, represented by Hermann Boerhaave and Friedrich Hoffmann, and above all by Julien Offray de La Mettrie, in that it recognized a qualitative principle of life. The psychiatrist Phillipe Pinel

(1745–1826) and the anatomist Marie-François Bichat (1771–1802) gravitated around the Montpellier school along with Barthez. Barthez insisted that he could only see "one of the follies of the century" in the success achieved by Bichat's work.

The classification of plants and plant physiology

In a work on the study of the family Ranunculaceae, the French botanist Antoine-Laurent de Jussieu applied the natural principle of plant classification anticipated in the work of Michel Adanson. The author demonstrated the existence of common traits in all Ranunculacaea flowers, despite differences in shape, structure, and symmetry. In his *Genera plantarum* (1789), de Jussieu attempted a natural classification of all plants, dividing them into acotyledons, monocotyledons, and dicotyledons, and subordinately, into hypogynous, perigynous, and epigynous (and dycotyledons into apetalous, monopetalous, and dialypetalous).

Once the quantitative method of the new mechanics had been assimilated through the work of Stephen Hales, the work of Jan Ingenhousz (1730–99), brought plant physiology to the same heights of methodic experimentation reached in animal physiology in the works of Lazzaro Spallanzani. In his *Experiments upon Vegetables* (1779), Ingenhousz showed that green plants produced oxygen during the day and carbon dioxide at night and that this phenomenon was caused by the effects of light during the day.

Basing his ideas on the discoveries of Antoine Laurent Lavoisier (1743–94), Ingenhousz later demonstrated (1798) that plants got their carbon from the carbon dioxide in the atmosphere. In a work entitled *Mémoires sur l'influence de la lumière solaire* (1782), Jean Senebier (1742–1809) also showed that some plant carbon was taken from carbonic acid dissolved in water. Other important findings in plant physiology, concerning the fecundation mechanism in plants, were made by Christian Konrad Sprengel (1750–1816). In his *Das entdeckte Geheimnis der Natur im Bau und in der Befruchtung der Blumen* (1793), Sprengel first affirmed the biological interdependence of flowers and insects in fecundation and the special importance of the color and smell of the flowers, both new concepts for the natural sciences at the time. The value of Sprengel's work was later recognized when Charles Darwin studied the conditions of impollination in orchids.

Andrei Timofeevich Bolotov (1738–1833), a Russian botanist and agronomist, conducted studies on plant fecundation and reproduction. Reechoing the research of Carolus Linnaeus, Rudolf Camerarius, Joseph Kölreüter, and Sprengel, Bolotov supported, developed, and spread the theory of sexual differentiation in plants, which at the time was opposed or ignored by most botanists. He considered the union of the male and female sexual elements an indispensable premise for the fecundation and development of plants that grew from seed. He also emphasized the importance of the quantity of pollen for the purpose of fecundation, the widespread diffusion of cross-pollination in nature, and the function performed, from this point of view, by the wind and insects. Bolotov is considered the founder of scientific pomology.

Animal transformism

Mainly basing his work on Buffon's writings, the Russian zoologist Nathanael Gottfried Leske published a pamphlet in Leningrad, entitled *Von Abartung der Thiere* (1775), in which he developed transformist ideas using arguments that were more feasible than Buffon's. He also maintained that the environment had an effect on organisms.

The psychic structure: hypothesis and therapy

The last decades of the century saw new diagnostic and therapeutic ideas come into psychiatry. These were suggested by new currents such as phrenology, the theory of animal magnetism, and a view of the causes, structure, and classification of mental illness that was much different than in the past.

EARTH SCIENCES

The birth of the Neptunist school and the progress of historical geology

While there were continuing observations regarding the existence of extinct volcanoes in different parts of France and Italy, the major center of geological research shifted from Paris to Freiberg, in Saxony, where the School of Mines (Bergakademie) was founded in 1765. While teaching at this school, Abraham Gottlob Werner (1750–1817) formulated the Neptunist doctrine, according to which almost all rocks had originated underneath the waters of the ocean. Thanks to the author's fame, the doctrine spread quickly throughout the scientific world of the period, supplanting even the correct deductions of various geologists on the eruptive nature of basalts. Werner also proposed a stratigraphic system that was widely acclaimed and used in Europe and America, even though it was less accurate than Arduino's.

Among those who affirmed the existence of extinct volcanoes were Ignaz von Born (1742–91) for Bohemia, John Strange, the British ambassador to Venice, for the Euganean Hills near Padua (1775) and other parts of the Venetian State (1778), Barthélemy Faujas de Saint-Fond (1741–1819), the first professor of geology in France, for the Auvergne (1778), Jean Louis Giraud-Soulavie (1752–1813) for the same region, and Alberto Fortis (1741–1803) for the area of Ronca, near Verona (1778). William Hamilton (1730–1803) observed live volcanoes, at Etna, Vesuvius, and later (1776–79) the Phlegraean Fields. Dieudonné Gratet de Dolomieu (1750–1801) visited Ponza in 1778.

In 1774, Werner published his *Von den äusserlichen Kennzeichen der Fossilien*, a work in which he proposed a mineralogical system that was very valuable because it gave practical indications for identifying minerals. This was the first of Werner's few but important works. In 1778, he wrote the first systematic treatise on geological formations, the *Kurze Klassifikation und Beschreibung der verschiedenen Gebirgsarten*, which was published ten years later. It divided formations into primitive rocks (*Urgebirge*)—granite, gneiss, mica schist, porphyry, basalt, and primitive limestone without fossils; sedimentary or stratified rocks (*Flözgebirge*)—limestone, sandstone, graywacke, coal and clay, often fossiliferous; volcanic rocks—lava, ash and pumice, which were further divided into real and pseudo-volcanic rocks (the latter deriving from the combustion of coal beds, sometimes fossiliferous); and finally alluvial or transported rocks (*aufgeschwemmte Gebirge*)—gravel; sand and clay, which sometimes contain fossils. The rock categories were not clearly separate, and gradual transitions often occurred from one to the other.

Werner's ideas spread quickly thanks to the many students he had at the Bergakademie. As mentioned earlier, Werner believed that rocks originated under water. He felt that his theory was supported by further observations in the field, to the point that he definitively concluded that even basalt came from the water. He believed that volcanoes were a very limited phenomena that had only existed in the more recent periods of the earth's history.

During this same decade, Peter Simon Pallas presented the first findings from his Siberian explorations in St. Petersburg (1777). He distinguished Primitive, Secondary, and Tertiary mountains. He also wrote detailed reports on a number of mammoth remains that had been found in that immense territory, even though earlier amazing discoveries had already been made, such as the complete rhinoceros, with its skin and hair intact, found in 1771 in the glaciers not far from Jakutsk. Pallas reproached his predecessors and contemporaries for often having extended conclusions that were based on the study of very restricted areas to the entire planet. In 1780, Giraud-Soulavie noted the differences in fossil content in the various geological formations into which he divided the stratigraphic succession of southern France.

Modern crystallography and mineralogy

In 1773, the Swede Torbern Olof Bergman (1735–84), famous for his contributions to oceanographic chemistry, printed a work on the crystalline forms of calcite and other minerals, anticipating René Just Haüy's hypothesis on the elementary structure of crystalline forms (Jean-Baptiste Romé de L'Isle [1736–90] did analagous research in 1772). In 1777, Bergman published quantitative analyses of gems, applying the principle of chemical composition as the basis for mineralogical distinctions. Axel Fredrik Cronstedt (1722–65) used this principle for classifying minerals in his *Versuch einer neuen Mineralogie* (1775). But the most important work in this field was done by Haüy, who applied calculus to the study of crystals and devised the theory of regular decrement and primitive, or elementary, forms, thus relating external morphology to the intimate structure of crystals (*Extrait d'un mémoire sur la structure des cristeaux de Grenat*, written in 1780 but published in 1782).

Geological cartography

A great mineralogical atlas of France, by Jean-Étienne Guettard, was published between 1778 and 1780. Along with Buffon, Guettard was the major scientific authority in the field of geology in Paris. Collaborating with him were Antoine Laurent Lavoisier (1743–94) and Antoine-Grimoald Monnet (1734–1817), who contributed the final thirty-one folios to the sixty-folio work. The atlas, accompanied by a number of geological sections, was very important and made France the leader among European countries in thoroughness of study and regional geological illustrations. A geological map in color, possibly the first ever published, appeared in 1775. It showed a part of Saxony and was drawn by Gottlieb Gläser.

Studies of the alpine region

The highest and most complex mountain range in Europe, the Alps, became the object of systematic explorations when Horace Bénédict de Saussure studied it from 1779 to 1796.

Buffon's history of the earth

George Louis de Buffon's voluminous work, *Époques de la nature* (1778), had a considerable influence on contemporary thought. In it, the author affirmed that the earth originated as a molten body which, just like the other planets, had been broken off of the sun by the impact of a comet, and had slowly cooled to the present temperature over a period of at least 74,800 years. Buffon illustrated the succession of living beings that he believed had populated the earth during different epochs and outlined a prototype of the earth's geochronology, based, however, on fantastic considerations. All the geological knowledge available to science at the time appeared in Buffon's work.

Oceanographic chemistry

An important advance in the ocean sciences during this decade was the determination of the chemical makeup of salinity, and consequently the birth of oceanographic chemistry. The pioneers in this field were Antoine Laurent Lavoisier, the famous French chemist, and Torbern Olof Bergman. Lavoisier described the precipitation of cerargyrite when sea water was treated with silver dissolved in aqua fortis, but, while only qualitative analyses had been made in the past, the French chemist attempted to discover the ingredients of salinity and determine their amounts by weight. Lavoisier made his first analysis of sea water in 1772; in water collected at Dieppe, Lavoisier found 1.7 percent salt (evidently the water was diluted). He evaporated the water and separated the salt with alcohol (taking advantage of the low solubility of NaCl in alcohol), and found that 1.2 percent was "*sel marin à base d'alkali fixe de la soude*" (NaCl), while the rest was a mixture of "*sel marin avec Epsom et Glauber sel*" (Mg and Na sulphates), "*terre calcaire*" ($CaCO_3$) "*et sélénite*" ($CaSO_4$). Thus, for the first time, salinity was found to be made up of various ingredients.

The Swedish chemist Bergman began to study sea water around 1779, after a period in which he analyzed minerals, in particular the composition of mineral waters. During this epoch, when Paracelsus's iatrochemistry was evolving into pharmacology, mineral waters were of great interest to medicine and, consequently, to the newly emerging science of chemistry. Bergman found that sea water contained brine (NaCl), salty magnesia, and gypsum. He also noted the alkaline reaction of sea water.

The thermometric determination of currents

The decade was extremely productive in other branches of oceanography as well. In particular, Benjamin Franklin's research, first (in 1733) on the effects of oily surfaces in subduing wave motion, and especially (1775–85) on determining currents, was conducted during this period. In particular, Franklin studied the Gulf Stream current beginning as early as 1769–70. During later crossings of the ocean, by taking systematic temperature readings, he was able unequivocally to determine that the temperature of the Gulf Stream was higher than that of the rest of the ocean mass. In fact, the spatial distribution of the current could be followed by measuring the temperature of the water. When extended to other cases, this phenomenon proved to be quite valuable to navigation, since the zone being crossed could be determined on the basis of water temperature. This led to theories of thermometric navigation, which were based on the fact that temperatures varied where there were currents, as well as in other cases, for example when there were variations in depth, in distance from the coast, etc. These theories were advanced by John Williams (1750–1815) and Alexander von Humboldt (1769–1859), among others. The true structure of the ocean's waters, which were composed of different bodies that mixed and interacted, began to be perceived (see EARTH SCIENCE 1810–1820).

Laplace's theory of the tides

The work of Pierre Simon de Laplace (1749–1827) began in 1775, and continued until he died in 1827, at the age of seventy. His work provided essential and exceptional contributions to hydrodynamics in general, and in particular to the study of wave motion and the theory of the tides (Laplace's dynamic theory of the tides, which supposed that the ocean covered the entire globe).

Explorations in the polar seas

The expedition into the north polar seas made by Constantine John Phipps in 1773, with the ships Racehorse and Caracass, deserves mention in the area of explorations. Captain Phipps, in search of routes in the Arctic Ocean, came within 1,000 kilometers of the North Pole. During the expedition, which was also an experiment in navigation through ice, temperature readings were taken using Cavendish's new special recording thermometers. It was discovered that in the deeper waters of the polar region the temperature rose slightly instead of falling. Analagous observations were made in the austral waters by William Wales and William Bayly during James Cook's expedition of 1772–75. During Phipps's expedition, depth measurements were also made, and some samples of the ocean bottom were taken. The interest in studying glacial oceans led to numerous experiments on the phenomena of freezing and melting and to research on the salinity of ocean ice, a subject on which opinions were still confused and often differed. In this respect, the research done by Edward Nairne (1726–1806) in 1776 and Charles Blagden (1748–1820) in 1778 deserves special mention.

The composition of air

Definitive progress was made during this decade in understanding the composition of air. The basic components were recognized and separated: nitrogen in 1772 by Daniel Rutherford and Henry Cavendish; oxygen in 1772 by Carl Wihelm Scheele and, independently, in 1774, by Joseph Priestley; and carbon dioxide in 1775 by Joseph Black (see CHEMICAL SCIENCES 1770–1780 for a more thorough discussion of these events).

APPLIED SCIENCES

Mechanical industry

In 1774, using a new method for boring the cores of cannon, John Wilkinson (1728–1808) also succeeded in producing the first internally-turned cast-iron cylinders completely suitable for use in steam engines. The following year, John Smeaton was able to double the yield of Newcomen's engine, giving it its maximum level of perfection. In 1776, a steam engine was used in Wilkinson's blower for drafting furnaces. This was the first time (Cugnot's vehicle aside) that such an engine had been used for purposes other than pumping water.

The first ''velocipede'' was built in Paris in 1779.

In 1779, Samuel Crompton (1753–1827) invented the self-acting mule by combining the beams of the water-powered loom with the moving carriage of the spinning jenny. The machine made it possible to produce any kind of yarn.

The chemical industry

In 1773, von Weigel introduced the principle of continuous countercurrent operation in the processes for distilling alcohol and spirits. The following year, Jacques François Demachy (1728–1803) published the first systematic account on the use of the hydrometer to test the strength of alcohol. The hydrometer was invented by Robert Clarke, and was commonly used in England from 1746.

Construction

In 1773, Charles-Augustin de Coulomb (1736–1806) formulated a theory of the strength of materials. A few years earlier, in 1770, Emiland Marie Gauthey (1732–1806) had designed and built a machine for testing materials that provided accurate data for designing and constructing buildings. The use of this machine had made it possible to resolve the controversy concerning the reconstruction plan for the Church of Sainte-Geneviève, in which the architect Jacques-Gabriel Soufflot (1713–80) had used iron rods sunk into the mortar.

In 1779, the first iron bridge in history was completed, spanning the Severn near Coalbrookdale, at a point that was later named Ironbridge. The access structures, which were subject to compression, were made of cast iron. The bridge was designed by Abraham Darby III (1750–91). The Brombery Canal, which connected the Oder and the Vistula, was completed between 1772 and 1775.

Scientific instruments

In 1770–74, John Harrison built one of the first precision scales. Built for Henry Cavendish (1731–1810), the scales used the triangular fulcrum for the first time. These were the scales that the great chemist used for his important research. At the same time, a new precision instrument for navigation, Borda's circle, named after its inventor, Jean-Charles de Borda (1733–99), a navigator and astronomer, was built by the French engineer Étienne Lenoir in 1772. In 1778, William Green (1725–1811) published his Description and Use of the Improved Reflecting and Refracting Telescopes and Scales for Surveying, in which he described the ''stadimeter.'' Publication of The Atlantic Neptune (a collection of surveys made on 1,600 kilometers of the Atlantic coast of North America) by James Cook (1728–79), J.F.W. Des Barres (1722–1824), and Samuel Holland (1728–1801) was begun in 1777. Cartographic techniques that represented geographically the structural and spatial relationships of data elaborated in other sciences were developed. The relationship between longitude and magnetic phenomena was illustrated in maps that showed magnetic declination. In 1770, Johann Heinrich Lambert (1728–77) traced isogonic lines—i.e. lines connecting places that have the same magnetic declination—on the surface of the globe. Lambert's work was a direct continuation of the studies made by Edmund Halley on the Atlantic and other oceans around 1700 (see ASTRONOMY 1690–1700).

As to actual scientific apparatus, the perpetual electrophorus, the prototype for the electrostatic induction generator, described by Allessando Volta (1745–1825) in 1775, deserves special mention. It was an extraordinary success, and in a very short time numerous models in varying sizes were being manufactured and distributed in every civilized country in Europe. Meanwhile, Jean André Deluc (1727–1817) established the rule for measuring altitude using a barometer.

Treatise writing

In the area of engineering, the first English text on bridge building, *The Principles of Bridges*, by Charles Hutton (1737–1823), came out in 1772.

Musical instruments

The spinnet (or harpsichord), which had prevailed since 1660, was replaced by the piano.

Armaments

In 1779, Henry Shrapnel (1761–1842) began experimenting with hollow spherical projectiles filled with musket balls and explosive charges. The shell later became oblong.

1780–1790

THOUGHT AND SCIENTIFIC LIFE

The fall of the ancien régime

The decade opened with the American colonies gaining their independence (1781, military victories; 1783, Peace of Versailles) and the birth of the United States of America, welcomed by many Continental (and even English) intellectuals as a model for the new rational state.

New intellectual societies were founded: the American Academy of Arts and Sciences in Boston (1780), the Società italiana della scienza in Rome (1782), the Royal Irish Academy in Dublin (1782), the Royal Society of Edinburgh and the Hunterian Museum in Glasgow (1783), the Asiatic Society of Bengal in Calcutta (1784), the Svenska Akademie in Stockholm (1786), the Sociétié philomatique in Paris (1788), and the Linnean Society in London (1789). The University of Bonn was inaugurated in 1784, but Göttingen University, where literary and scientific activity was very lively, remained the major German university. Although English science and technology counted such figures as Black, Cavendish, Priestly, and Watt among its ranks, and Italy had Galvani, Volta, and Spallanzani, Paris was the center of European scientific life, partly because science in the French capital was connected with solid institutions, such as the Académie des Sciences, the Jardin des Plantes, and the Observatoire Astronomique.

But French society was going through a serious economic, social, and political crisis during this period, which led to the convocation of the Estates-General (nobility, clergy, and the third estate), called for by an assembly of notables in 1788. With the consent of the monarch, Louis XVI, the Estates-General met at Versailles in May of 1789, and, on the initiative of the Third Estate, became the Constituent Assembly a short time later. Passing over problems of public finance, the assembly aimed at formulating a new constitution. The Bastille was taken in July 14. On August 5, 1789, feudal privileges were abolished, and between the twentieth and twenty-sixth of the same month, the Declaration of the Rights of Man and of the Citizen was passed. In October, the king and the Assembly were forced to move from Versailles to the capital.

The last of the Enlightenment thinkers

The Enlightenment had opened up new areas of investigation, uncovered unsuspected connections between the various aspects of philosophical and scientific research, and conducted important cultural and civil battles in a number of countries. The fruits were now being gathered in various areas, especially in reflections on politics, morals, and history. Here we shall only mention, as an example, the enormous influence that the monumental *History of the Decline and Fall of the Roman Empire* (1781), by Edward Gibbon (1737–94), had on historians, writers, poets, and scholars of religion and law.

In France as well, the great flourishing of Enlightenment writers had ended. Their worthiest heir was Marie Jean Caritat, Marquis de Condorcet (1743–94), who published and commented on the complete works of Voltaire. Attached to d'Alembert, Voltaire, and especially Turgot, Caritat had already become well-known at a fairly young age for his various writings on mathematics and philosophy, which had earned him the position of perpetual secretary of the Académie des Sciences. In 1782 he entered the Academy of France. When the revolution broke out, he became an active participant, taking on high positions of responsibility.

In Italy, the writers associated with the group known as "Il Caffè" continued their activity. Pietro Verri published some of his most interesting philosophical-cultural writings around 1780: *Elogio del cavalier Isacco Newton* (1778), *Elogio di Bonaventura Cavalieri* (1779), and *Opuscoli filosofici* (1781). His famous *Eligio del signor d'Alembert* came out posthumously in 1786. In 1787, Verri wrote his *Memorie appartenenti alla vita e agli studi del signor Don Paolo Frisi*. Prominent among the Enlightenment thinkers of the Neapolitan group was Gaetano Filangieri (1752–88), whose masterpiece in eight volumes, *La scienza della legislazione*, began to be published during this period (the first seven volumes came out between 1880 and 1885, the last posthumously in 1891).

German culture and Kantian critique

Meanwhile the center of philosophical thought in Europe shifted to Germany, where a very active artistic and literary culture was also flourishing. In 1780 Johann Wolfgang Von Goethe read his *Urfaust* to Duke Carl August of Weimar and Prince August of Gotha, and in 1781 Schiller finished his *Die Räuber*. The first edition of Kant's *Kritik der reinen Vernunft* (1781) marked the completion of the turn in his thought that had already been outlined in his dissertation of 1770. In 1787, a second edition of the work came out, in which there were a number of changes with respect to the original. The following years were extremely productive for the great philosopher. In 1783 he published his *Prolegomenza zu einer jeden künftigen Metaphysik, die als Wissenschaft wird auftreten können*, which together with the previous work provided one of the major expositions of Kant's theory of knowledge. In 1784, he wrote a brief essay entitled *Beantwortung der Frage: was ist Aufklärung?* His *Grundlegung zur Metaphysik der Sitten* came out in 1785, which together with the *Kritik der praktischen Vernunft* of 1788 outlined the basics of Kantian ethics. In 1786, Kant wrote a brief work on the fundamental metaphysical principles of the science of nature that was very important for the philosophy of science. His so-called third critique, the *Kritik der Urteilskraft* came out in 1790. In 1786, a work by Karl Leonhard Reinhold (1758–1823) entitled "Briefe über Kantische Philosophie" had appeared in the *Deutscher Merkur*, contributing immensely to the circulation of Kant's critique. In 1789, Reinhold published his major work, *Versuch einer neuen Theorie des menschlichen Vorstellungsvermögens*, a thorough study of Kantian gnoseology.

Finally, several interesting works by Gottlob Ernst Schulze (1761–1833), Salomon Maimon (1754–1800), Johann Georg Hamann (1730–88), Johann Gottfried von Herder (1744–1803) and Friedrich Heinrich Jacobi (1743–1819) appeared during the last few years of this decade and the early years of the next; although they raised various objections to Kant's doctrine, they indirectly increased its notoriety.

MATHEMATICAL SCIENCES

Analysis, "the mother of all sciences"

The spread of the mentality and methods of analysis into all mathematical discplines, a tendency that could already be seen in the work of Joseph Louis Lagrange, had by this time become a general phenomenon. In geometry, for instance, Gaspard Monge (1746–1818)—who was already known for the introduction and study of orthogonal projection and who later gained his greatest fame as the founder of descriptive geometry—did important work in infinitesimal geometry concerning the theory of curves not in the same plane and the infinitesimal properties of areas during this decade. In fact, the use of analytical and infinitesimal techniques in geometry was so important in Monge's work that he was able to coordinate findings he had put forth in various memoirs and bring them together in one large work, to the fourth edition of which he gave the expressive title *Analyse appliquée à la géométrie* (1809). It is also significant that infinitesimal methods were used not only in dealing with new questions, which could really not be solved without them, but were also applied in reproposing classical problems. In 1782, for example Simon-Antoine-Jean Lhullier (1750–1840) published the first part of a remarkable theory on isoperimeters, in which the use of differential methods made it possible to rediscover classical results and extend them considerably. Lhullier also received the prize, offered in 1784 by the Berlin Academy (at the suggestion of Lagrange, who was also a judge), for providing a rigorous definition of mathematical infinity, to help eliminate the conceptual vagueness that surrounded the use of this notion in analysis. This was a particularly important symptom of the new critical needs that were emerging and that would be confronted in the following century.

The analytical reelaboration of mechanics

On the other hand, analysis itself benefitted from this union with other disciplines. New problems and new openings for research were proposed. It is fairly obvious that analysis would be applied to mechanics (we shall return to this in a moment), but it is significant that it was also used in slightly more collateral disciplines, such as astronomy. For example, Adrien Marie Legendre (1752–1833) introduced the polynomials that bear his name in a memoir of 1782 on the shape of the planets, and Pierre Simon de Laplace (1749–1827) presented his differential equation in a memoir on the theory of Saturn's ring (1787–89).

It is no surprise that the major fruits of the collaboration with analysis were gathered in mechanics, not only because analysis itself was closely linked to mechanics from its beginning (e.g. Isaac Newton), but also because the construction of rational mechanics as a rigorously mathematical discipline had been the ambition of all Enlightenment science and a projection of its ideal of "rationalizing" nature. Among the major victories of analysis that came from research done in mechanics during this period was Laplace's introduction of potential functions, together with the formulation of the well-known operation and the completion of the techniques of spherical analysis.

The crowning achievement of this great effort, which came about thanks to the contributions of such scientists as Leonhard Euler and Jean le Rond d'Alembert, came again from Lagrange. After moving to Paris in 1787 upon the death of his famous protector, Frederick II, Lagrange as an honorary pensioner of the Academy there, published his *Méchanique analytique* (1788), a great methodical, and unifying synthesis that brought all mechanics down to the principle of virtual work and d'Alembert's principle. But perhaps even more significant was the fact that it explicitly put forth and carries out the intention of resorting to "neither constructions nor geometrical and mechanical arguments but only algebraic operations subject to a regular and uniform development. . . . Those who love analysis will," says the author, "with pleasure, see mechanics become a new branch of it and will be grateful to me for thus extending its field."

ASTRONOMY

The systematic exploration of the skies and the discovery of Uranus

From the contributions of Galileo and Newton up until 1781, astronomy developed through the great efforts of mechanicians and analysts such as Leonhard Euler, Alexis Claude Clairaut and Jean le Rond d'Alembert, who attempted to restructure a model of the universe that was no longer simply geometrical, but was dominated by inertia and gravitation, and the elements of which were thought to be fully known by this time. As for observational astronomy, its goal was mainly to check out more accurately increasingly large amounts of data which were to be subjected to analysis and inserted into the Newtonian system.

However, an actual revolution began to take place in 1781 with Frederich Wilhelm Herschel (1738–1822), a German-born scientist who lived almost all his life in England. His main intention was to make a systematic exploration of the heavens beyond the areas already known and to classify the stars according to their magnitudes, a particularly pressing problem at the time. He built his own instruments, gradually increasing their power. From 1774 on he built new telescopes; the last one, which was put into operation in 1789, had a focal distance of around twelve meters and a parabolic mirror that was 1.47 meters in diameter.

On the night of March 13, 1781, while exploring the constellation of Gemini in the vicinity of ⁿGeminorum, he discovered a body that was definitely larger than the fixed stars around it. Herschel thought that it was a comet, but calculations made by Anders Jean Lexell of the Academy of Sciences in St. Petersburg showed that its orbit was almost circular. It was, therefore, a new planet in the solar system—Uranus, as Johann Elert Bode named it—the sixth to add to the five already known since remote times. The enormous interest that the new discovery aroused and the work of preparing tables on Uranus's motion, which brought to light several anomalies, led, via purely theoretical considerations, to the prediction of the existence of Neptune as well as its place in the celestial sphere (1846).

If the discovery of Uranus was due to chance and good luck, Herschel studied other large problems concerning the nature of heavenly bodies and the physical dimensions of the universe scientifically and carried out his observations following a plan that he developed over a period of forty years. He was particularly interested in optical doubles, the determination of stellar parallax in order to measure the distance of stars from the earth, the proper motion of stars, the nature of nebulae and the dimensions of the galaxy. In 1783, for example, he began making observations to measure stellar parallax, i.e. the angle with the star at its vertex and its sides passing through the extremities of the earth's diameter, so that he could calculate the star's distance from the earth trigonometrically. However, instrumental difficulties invalidated Herschel's results.

The birth of galactic astronomy

In 1783, Herschel began making systematic explorations of the heavens to determine quantitatively the distribution of the density of stars throughout the galaxy in order to establish the shape of the universe, starting from hypotheses previously formulated by Thomas Wright, Immanuel Kant, and Johann Heinrich Lambert. This was Herschel's most ambitious project in his long career as an observer; he wanted to confirm by observing the heavens in all directions the hypotheses that his predecessors had reached simply through conjecture. Assuming that the stars were uniformly distributed in space and that the instrument he was using—a telescope with a focal distance of approximately six meters—would reach the outer limits of our galaxy, he proposed to count the number of stars up to the eighth magnitude that passed across the field of the telescope in a given amount of time. Since the stars were uniformly distributed throughout the cone that was proportional in height to the cube, the number of stars counted in any given direction should give the depth of the galaxy in that direction.

The nebula catalogues

The enormous number of observations that were required to carry out the entire plan forced Herschel to reduce its scope and limit his investigations to specially selected directions. Nevertheless, he was still able to establish that the galaxy had

the shape of a very flat disc and that the distribution of the stars was much denser along the plane of the galactic equator and sparser in the direction of the poles. However, in 1818, he was forced to acknowledge that his instruments were not adequate to enable him to calculate the ratio between the diameter of the equatorial plane and the polar axis of the galaxy. While conducting his systematic explorations of the heavens, Herschel also encountered a large number of nebulae (by the end of his work, he had catalogued around 2,300), which he continued to observe with passion. During that same period, in 1784, the French astronomer Charles-Joseph Messier studied and catalogued 103 nebulae, although he remained in doubt as to whether they were star clusters or clouds of gasses. Herschel, who had been able to resolve a number of them into star clusters thanks to his telescopes, maintained that nebulae were, in fact, all of the same nature, and that the remainder of them could not be resolved because the instruments used to observe them were too weak with respect to their distance. In 1785, he explicitly acknowledged the plurality of galaxies, which had been postulated by Wright, and maintained that they too were islands of universes similar to the Milky Way. Only later in his life did Herschel recognize the existence of unresolvable gasseous nebulae.

The history of astronomy

Publication of the *Histoire de l'astronomie* in five volumes, by the French astronomer Jean Sylvain Bailly (1736–1793), which gave an abundant and detailed account of the history of astronomy up to 1781, was begun in that year. The work was complemented by the *Bibliographie astronomique avec l'histoire de l'astronomie depuis 1781 jusqu'à 1802*, by Joseph Jérôme de Lalande, published in 1803.

PHYSICAL SCIENCES

The rationalization of mechanics

The *Essai sur les machines en général*, by Lazare Nicolas Carnot (1753–1823), a systematic exposition of mechanics that looked at the concept of force in a critical light, was published in 1783. Carnot's arguments, influenced in some ways by English empirical thought, emphasized the notion that force works in imperceptible degrees and is not apt to produce sudden variations. Carnot's reflections on force as an obscure, metaphysical concept that could only be interpreted as the cause of motion on a didactic level were expressed most explicitly in a work of 1803 on the basic principles of mechanics (see PHYSICAL SCIENCES: MECHANICS 1800–1810).

In 1784, the Memoirs of the Royal Berlin Academy of Sciences and Literature printed a famous late eighteenth-century text on mechanics entitled *Lucrèce newtonien*, by George-Louis Lesage. In it, Lesage again proposed ideas on the mechanist theory of gravitation that he had already expounded a number of times before. Lesage's theory hypothesized an infinite number of gravitational particles—or ultramundane particles—that could travel through space at extremely high speeds. A body that was very far away from the remaining matter would be acted upon by Lesage's particles equally from all directions, and would consequently remain in a state of rest. If another body was nearby, however, then a sort of gravitational shadow would fall upon the first body, provoked by the shield that the second body formed. Consequently, the first body would be pushed toward the second by the gravitational particles that hit its unshielded part. The variation in the resulting force, according to Lesage, was inversely proportional to the square of the distance. However, gravitational fluid did not explain the dependence of force on mass, a problem that plagued every attempt to elaborate an atomistic and mechanistic explanation of gravitational interaction.

The masterpiece of the process of rationalizing mechanics, *Méchanique analitique*, by Joseph Louis Lagrange (1736–1813), appeared in 1788. In it, Lagrange used a rigorous mathematical treatment to unify mechanics, according to a plan that he himself summarized as follows: "To reduce the theory of mechanics and the art of solving the associated problems to general formulae, whose simple development provides all the equations necessary for the solution of each problem."

Given the close ties between Lagrange's rational mechanics and the development of analysis in mathematics (see MATHEMATICAL SCIENCES 1780–1790), the elaboration of the concept of potential, already introduced by Lagrange and developed by Pierre Simon de Laplace (1749–1827) with his second partial differential equation (1782), should also be mentioned in connection with this plan.

Inquiries into the nature of heat

Important advances were made in the field of calorimetry, thanks above all to the research of Laplace and Antoine Laurent Lavoisier (1743–94). The two authors wrote the classical monograph "Mémoire sur la chaleur" in 1783. Along with a large amount of experimental data on specific heats, the work maintained that there was a certain equivalence between two different theories on heat.

According to the first theory, which was similar to Lavoisier's theories, thermic phenomena were caused by the effects of a caloric fluid, while the second theory claimed that they were caused by the motion of the molecules of matter. On the basis of the second theory, Laplace and Lavoisier wrote that heat was the living force resulting from the nonobservable motions of the molecules of a body. If two bodies with different temperatures were placed in contact with each other, the two scientists stated, the increase in the living force of the cooler body would be equal to the decrease in that of the warmer body, and this increase would continue until the quantities of motion exchanged were equal, at which time the temperature of the bodies would have reached a uniform state.

According to Laplace and Lavoisier, the two models would be reconciled, in that the explanations that could be drawn from one could be translated into explanations that could be drawn from the other. It is interesting that in his search for a more rational reduction of thermology to mechanics, Laplace adopted the fluid theory in the following years, and decisively rejected the models based on molecular motion as the cause of heat. In any case, the memoir of 1783 was definitely a turning point in calorimetric investigations and remained a basic text until the beginning of the nineteenth century. In 1786, *Philosophical Transactions* hosted a memoir entitled "New Experiments upon Heat," written by Benjamin Thompson (Count Rumford) (1753–1814), which expounded the results of a number of observations on heat conductivity in a vacuum, in water, and in mercury. Although it was inspired by economic and military interests—heat conductivity measurements were made to determine the fabric best suited for the uniforms of the Bavarian army—and Rumford still made the analogy between heat and the electrical fluid, his memoir initiated a wave of experiments that were increasingly more frequently aimed at contradicting caloric interpretations.

The physical properties of gasses

In 1787, the French scientist Jacques Alexandre Charles (1746–1823) discovered that oxygen, nitrogen, hydrogen, carbon dioxide, and air all expanded equally if heated from 0 degrees to 100 degrees Celsius (at a constant pressure). His findings, which remained unpublished, were taken up and developed fifteen years later by Joseph Louis Gay-Lussac, who explicitly mentions his indebtedness to his predecessor.

In 1789, the Italian Allessandro Volta, who had already written about his research in electrology, began a long series of studies on vapor pressures. These led him to important findings on the partial pressures of the components of gaseous mixtures. Volta's findings were included in a more general context in the so-called Dalton Law, established by John Dalton in 1801.

Electricity and magnetism

Research on electricity and magnetism during this decade was dominated by Charles Augustin de Coulomb's theoretical and experimental studies, the results of which were published in seven long memoirs between 1785 and 1789.

The first three works are dated 1785, and are entitled, respectively, *Construction et usage d'une balance électrique fondée sur la propriété qu'ont les fils de metal d'avoir une force de torsion proportionnelle à l'angle de torsion*; *Mémoire où l'on détermine suivant quelles lois le fluide magnétique ainsi que le fluide électrique agissent soit par répulsion, soit par attraction*; and *De la quantité d'électricité qu'un corps isolé perd dans un temps donné. . . .*

The first memoir contains a description of the famous device that Coulomb built to measure what he called the basic law according to which electrified bodies repel each other. According to this law as Coulomb stated it, the repulsive force between two electrified spheres with the same type of electricity is inversely proportional to the square of the distance between the two spheres.

In his second memoir, Coulomb extended his observations to the interactions between spheres with positive and negative charges, as well as to magnetized bodies. By making an analogy between the behavior of electrical fluids and the behavior of magnetic fluids, which he confirmed by experiment, he was able to state that magnetic fluid acts by attraction or repulsion, in a way that is directly proportional to the density of the fluid and indirectly proportional to the square of the distance between its molecules.

In the third memoir, Coulomb examined the phenomenon by which electrified bodies gradually lose their charge as a result of the physicochemical conditions of the environment.

In 1786, Coulomb presented a fourth memoir describing the properties of the electrical fluid, which stated that "in conducting bodies, the fluid, having reached a state of equilibrium, distributes itself over the surface of the bodies without penetrating into the interior.'' This fundamental property for studies in electrostatics was again taken up in Coulomb's fifth memoir, entitled *Sur la maniére dont le fluide électrique se partage entre deux corps conducteurs mis en contact, et la distribution de ce fluid sur les differentes parties de la surface de ce corps* (1787). At this point, Coulomb's investigations unequivocably confirmed the generalization of the inverse-square law of variation which became the basis of mechanistic systems.

The formal analogy between Coulomb's laws and Newton's law on gravitational attraction constituted, at the time, a reasonable motive for attempting to reduce all observable interactions to explanations based on mechanics. We might also mention that this remained a reasonable motive throughout the nineteenth century, to the extent that it represented, though often through rather complex theoretical mediations, a point of reference for notions of electromagnetism acting at a distance.

Coulomb's sixth memoir, which continued with the theoretical considerations put forth in the fifth, appeared in 1788. Here Coulomb discussed various theories on the electrical fluid, with special attention to Aepinus's theory, which maintained that there was only one fluid, and to the theory ''already proposed by many physicists'' according to which there were two fluids. Coulomb supported the two-fluid theory, although he emphasized that this was not because he wished to show the true causes of electricity but because he wanted to present the results of calculations and experiments in the most concise way possible.

In 1789, Coulomb presented his seventh memoir, entitled *Du magnetisme*. According to Coulomb, Aepinus's hypothesis on the magnetic fluid had to be modified to take into consideration new experimental data. Coulomb's revision of Aepinus's theory saw the magnetic fluid as tightly bound to each molecule of the magnetized substance, although it was relatively free to move from one end of the molecule itself to the other. Thus each molecule had magnetic polarity and behaved just like a tiny magnetized needle.

Siméon Denis Poisson completely reelaborated Coulomb's ideas in a mathematical form beginning in 1881 (see PHYSICAL SCIENCES: ELECTRICITY 1810-20).

We should also mention the new science that was called experimental acoustics, initiated during this decade by Ernst Chladni (1756-1827). In 1787, Chladni published a famous memoir on the vibrations of plates, in which he explained the acoustical figures that even today are called Chladni figures. The figures were obtained by covering a metal plate that was fixed at the center and free on the edges with fine sand or lycopodium powder and making it vibrate by drawing a violin bow across the edge. He damped the vibrations by placing a finger on a point that then became a node. In this way, Chladni proved that there were fifty-two figures for a rectangular plate, forty-three for a circular one and almost one hundred for other shapes.

CHEMICAL SCIENCES

The synthesis and electrolysis of water

The time was ripe for proving that water was not an element, but a compound of hydrogen and oxygen. In experiments begun in 1781 and made public in 1784, Henry Cavendish synthesized water from the two gasses. He demonstrated that they combined in a ration of two-to-one by volume and that the reaction did not cause any change in weight. Antoine Laurent Lavoisier repeated the experiment with Jean-Baptiste Meusnier de la Place (1754–93) and proved that decomposing water with a hot iron produced hydrogen. A few years later, just after Volta had invented his battery (1800), Anthony Carlisle (1768–1840) and William Nicholson (1753–1815), and then Volta, William Cumberland Crikshank (1745–1800), and others, succeeded in decomposing water by electrolysis.

In 1738, Cavendish demonstrated that striking a spark in a mixture of ''dephlogisticized air'' (air without oxygen, i.e. impure nitrogen plus noble gasses) in the presence of water produced nitric acid. Martin Van Marum (1750–1837) obtained ozone in 1785 by striking an electric spark in oxygen. In 1787 Jacques Alexandre Charles (1746–1823) studied the reactions of gasses at a constant pressure to variations in temperature and established the law that is known as the Charles-Gay-Lussac law. In 1788, Charles Blagden (1748–1820) wrote on super-cooling water and the effects of solutes on the freezing points of solutions. He established Blagden's law, which stated that the drop in freezing points was proportional to the concentration of the solute.

New uncommon elements

By this time, the concept of element had been established with sufficient clarity, and analytical methods had been improved enough, so that new, fairly uncommon elements began to be discovered. Tungsten, or wolfram, was discovered by Carl Wilhelm Scheele in 1781 and isolated, under his guidance, by the Spanish chemist Faust d'Elhyuar y de Suvisa (1755–1833) in 1783. Tellurium was discovered by Franz Müller, Baron von Reichenstein (1740–1825) in 1782 and isolated in 1798 by Martin Heinrich Klaproth (1743–1817). Uranium and zirconia were discovered by Klaproth in 1789 and isolated in the nineteenth century. Titanium was discovered by William Gregor (1761–1817) in 1789 and isolated after more than a century. Strontium was discovered by Adair Crawford (1748–95) and Cruikshank in 1790.

Organic and inorganic chemistry

Progress was also made in organic chemistry. Scheele discovered lactic acid and mucic acid in 1780, hydrocyanic acid in 1782, citric acid in 1784, malic acid in 1785, and gallic acid and pyrogallic acid in 1786. He also discovered various esters in 1781, acetaldehyde in 1782, and glycerine in 1783–84. Only in 1784, with the work of Torbern Bergman (1735–84), was a clear distinction made between organic and inorganic substances.

Lavoisier's new chemistry required more modern terminology and accurate popularization. The *Méthode de nomenclature chimique, proposée par MM. de Morveau, Lavoisier, Berthollet & de Fourcroy,* was published in 1787. In 1789, Lavoisier published his *Traité élémentaire de chimie,* which can rightfully be considered the first modern text in chemistry. In the second edition of his *Disquisito de attraction electivis* (1783), Bergman published a much more thorough table of affinities than any of his predecessors, listing affinity values at room temperature and high temperatures as well.

LIFE SCIENCES

Phrenology

While Johann Kasper Lavater (1741–1801) was transforming physiognomy into a logical study of the features of the human face, the physician Franz Joseph Gall (1758–1828) founded a new discipline, phrenology, the study of cranial protuberances, believed to be the skeletal characteristics corresponding to parallel developments in the cerebral "organs" where specific functions and attitudes were located. According to Gall, there were thirty-one specific functions, localized in the same number of zones in the brain. The high scholastic performance of students with bulging eyeballs, for instance, led him to surmise that the cerebral memory organ was located behind the eyes. Forced to leave Vienna in 1802 because of the supposed materialistic nature of his doctrine, Gall travelled through Europe for five years spreading the ideas of phrenology and accumulating a good number of disciples. Although phrenology fell into disrepute because of its unilateral affirmations, it left a lasting and valid legacy in the notion of cerebral localization.

Animal magnetism

Another new current was the theory of animal magnetism, devised by Franz Mesmer (1734–1815). According to Mesmer, there was a reciprocal influence between the heavenly bodies, the earth, and animated bodies. This influence supposedly came from a continuous, universal fluid that could receive, propagate, and communicate all motion and that had properties that were similar to magnets. Forced to leave Vienna by the physicians of the city, who were hostile to his hypnotic practices, Mesmer moved to Paris, where he became extremely popular in worldly circles. However, in 1784 a royal commission of study, which included Benjamin Franklin, Jean Sylvain Bailly (1736–93), Antoine-Laurent de Jussieu, and Lavoisier, judged that there was no reliable proof of the existence of animal magnetism.

Biochemistry and biophysics

Lavoisier's studies on animal heat, begun with the help of Pierre Simon de Laplace (1749–1827), demonstrated the chemical nature of the phenomenon, thus revealing the genesis of the oxidation process. Biology tended to establish increasingly close ties with all the inorganic disciplines.

EARTH SCIENCES

The affirmation of Neptunist theories and the rise of Plutonist ideas

The idea that most known rocks originated in the depths of the ocean, developed by Abraham Gottlob Werner, became accepted by the great majority of scholars, who attempted to apply Werner's stratigraphic classification, based on the division into crystalline, sedimentary, and alluvial rocks to the various regions where they were doing research. During this same period, the first doubts about and then opposition to Neptunist theories were advanced, mainly by James Hutton (1726–97) of Edinburgh.

Werner, a professor at the Bergakademie of Freiberg in Saxony, completed his theory on the origin of rocks from the crystallization of material within the oceanic waters, which he had first presented in 1777. In 1787, Werner visited Scheibenberg Hill in Erzgebirge, where he observed a layer of basalt with a layer of sandstone underneath and a layer of altered basalt in between; this led him to conclude that both rocks belonged to the same system and, consequently, that basalt also originated from the sea.

Werner's stratigraphic system, which was detailed and complete for the zone that he studied, was considered valid for all areas and was accepted as the basis for a universal system. Difficulties arose when zones with a very different stratigraphic history were explored, especially those where the relations between eruptive and sedimentary rocks were more evident.

In 1785, Hutton read his *Theory of the Earth* (published in 1788), an event that can rightly be considered as marking the birth of Plutonist thought. (Meanwhile, that same year, Werner completed his *Kurze Klassifikation der verschiedenen Gebirgsarten*.) The Scottish geologist strongly emphasized the stratigraphic irregularities that so often appeared in the geological succession and maintained that forces from inside the earth must have repeatedly lifted and folded the layers, which had originally been laid down horizontally. The frequency with which eruptive rocks passed through sedimentary rocks in many parts of Scotland led him to believe that these internal forced acted through the pressure exerted by great masses of molten rock, which intruded into the rock above. Each irregularity, therefore, marked a revolution in the earth's history, which could not be thought of as continuous and uniform within geological time. The retreat of the waters of the ocean was caused by the Earth rising, not by a decrease in the volume of water, which Werner believed evaporated into extraterrestrial space. Volcanic eruptions were ventings of the underground fire, and kept earthquakes from happening and the Earth from rising locally.

In 1790, in his *Observations on Granite*, Hutton stated that granite, earlier believed to be the original, primitive part of the earth, was more recent even than some parts of the Alpine schist, which was stratified and therefore itself not original. The internal forces acted constantly, and neither signs of a beginning nor the prospect of an end to the earth's history could be seen. Hutton even explained that the lithification of sediments was due to the effects of internal heat. Attacks from the Irish Wernerian Richard Kirwan pushed Hutton to gather further proof for his ideas by visiting the Alps and extensively travelling in Scotland and England. His conclusions were therefore based on studies of a wider range of regions than the Neptunist leader ever made.

Even prior to Hutton's opposition, a student of Werner's, Johann Carl Voigt, interpreted the basalt of Scheibenberg as an ancient lava flow. In the debate that followed, Werner reiterated his position on the aqueous origin of basalt, explaining that volcanic eruptions, which he maintained were very limited, came from the combustion of huge underground coal deposits. Meanwhile, William Hamilton (1730–1803) affirmed the volcanic origin of the Giant's Causeway, the columnar basalt formation on the island of Staffa in Antrim county (1790).

John Michell (1724–93) presented a stratigraphic report on England, with observations that were more complete than those contained in Werner's system.

Alpine ascents

Ascents continued to be made to the major peaks of the Alps. Mont Blanc was first climbed by Jacques Balmat and Michel-Gabriel Paccard in 1786. Horace Bénédict de Saussure dedicated his attention to research on the glaciers, making surveys on Mont Blanc (he climbed to the peak with Balmat in 1787), Monte Rosa (1789) and the Little Matterhorn. Hostility toward the more inaccessible regions, the 'accursed mountains,' common among his contemporaries, began to soften. The first volume of de Saussure's *Voyages dans les Alpes* came out in 1786. The fourth and final volume was published in 1796.

Progress in mineralogy

The second edition, in four volumes, of the *Essai de crisallographie*, written by Jean-Baptiste Romé de l'Isle, came out during 1783 with the title *Crisallographie, ou description des formes propres à tous les corps du régne minéral*. The essay contains a statement of the law of constancy of the dihedral angle between the faces of crystals, which Romé de l'Isle discovered using a goniometer built by his student Carangeot.

In his *Essai d'une théorie sur la structure des cristeaux appliquée à plusiers genres de substances cristallisées* of 1784, René Just Haüy perfected the theory of the elementary parts that make up crystals and emphasized the fundamental importance of crystallography in mineralogical research and study.

The fifth and last volume of George Louis Leclerc, Compte de Buffon's last work, *Histoire naturelle des minéraux*, was completed in 1788, just before the author's death. An essentially descriptive and speculative epoch, but one in which the natural sciences had always had direct philosophical implications, ended with Buffon's death. At the same time, Werner became a leader for those scholars who emphasized direct observation of the facts and systematic study of the phenomena observed.

Paleontological studies on vertebrates

Alberto Fortis described elephant bones found at Romagnano, near Verona, during this period, and S. Volta wrote his *Ittiologia veronese* (1788), a treatise on the famous fossil deposit of Mount Bolca. In 1789, the viceroy of Buenos Aires sent to Madrid a complete megatherium skeleton, found in the alluvial pampas near Lujan. A short time later, more of these huge skeletons were discovered in other parts of South America.

The mechanics of wave motion

Theoretical work continued. In 1781, Joseph Louis Lagrange (1736–1813) introduced the concept of velocity potential, which was important to certain mathematical developments concerning wave motion. In 1786, Lagrange made a thorough study of the problem of wave propagation and established the famous formula $c = \sqrt{gh}$, which defined the velocity c of a long wave simply as a function of the depth h of the sea (g was known to stand for the acceleration of gravity).

In 1782, Richard Watson (1737–1816) gave a unitary interpretation to thermometric observations of the sea, based above all on new findings made by Constantine John Phipps and William Wales. He attributed thermic stratification to the interaction between superficial heating and cooling over the course of the seasons in temperate zones. Temperature measurements were commonly made at sea at the time, both for navigational purposes, following Benjamin Franklin's suggestions (e.g. Charles Blagden's observations along the American coasts in 1781), and to study the temperature gradient and solve the problem of abyssal temperatures (Horace Bénédict de Saussure's studies in the Mediterranean in 1780).

In 1785, an expedition sponsored by the French government and led by Jean François de La Pérouse (c. 1741–88) set out with two ships to explore the northern Pacific. After three years the explorers disappeared, possibly in a shipwreck.

APPLIED SCIENCES

Metallurgy: the adoption of coke and puddling

In the decade between 1780 and 1790, the metallurgy industry was definitively freed from the use of wood charcoal, which had forced the foundries to move to mountain areas where water power and wood were abundant. The price of wood charcoal, which had risen eighty percent in two centuries because of extensive deforestation, was also about to cause the industry to go into crisis.

The adoption of pit-coal coke, which Abraham Darby II (1711–63) obtained by dry distillation around the middle of the century, had partly solved the problem. However, the cast iron produced with coke was too fragile to use in making homogenous iron because of impurities it contained.

The solution came in 1784 with the puddling process, invented and patented by Henry Cort (1740–1800). This process made it possible to convert raw pig iron into forgable iron in a reverbatory furnace, still using coke as the fuel. The pig iron was first fused in a pig bed, then stirred with an iron rod until the carbon and the other impurities were eliminated through combustion. Peter Onions had made the same discovery a year earlier. Cort ensured his fame by combining puddling with the use of grooved rollers (patented in 1783) similar to those that had long been used in rolling and cutting iron, but shaped in such a way that the bars produced had a circular cross section. Later, this new type of rolling mill made it possible to produce bars of any shape section, a feature that opened the way to producing rails directly in the rolling process.

Forging works moved to the great coal fields, as the foundries had already done. Thus a gradual concentration of the various branches of the metallurgy industry occurred; they merged into a single block in the following century.

There was an enormous increase in the production of iron, which in Great Britain, rose from 20,000 tons in 1720 to 70,000 tons in 1788. Connected to this boom was a proportional increase in the production of coal, which went from 4,500,000 tons in 1750, to 6,000,000 tons in 1770 and 10,000,000 tons in 1780 (again in Great Britain).

In the case of France, the figures speak of only 700,000 tons in 1789, which rose to 5,000,000 tons in 1798 with the annexation of Belgium.

The search for more power from the steam engine

The success of James Watt's steam engine was immediate, since it was considerably more powerful than Newcomen's; however, its use was limited to pumping water in mines, salt works, distilleries, iron works, and all other cases where the alternating motion of a balance could be used. The need to adapt the steam engine to a wider range of uses had already led James Pickard, an Englishman, to patent a device in 1779 that transformed the reciprocal motion of Newcomen's engine into rotary motion. In 1781, Watt introduced a similar innovation into his steam engine. The quest for more power also induced Jonathan Carter Hornblower (1753–1815) to build a compound expansion engine, where the steam used in one cylinder then passed into another, in 1781.

But the best results in this area were again achieved by Watt, who in 1782 was able to double the power of his engine by making the steam work alternately on both faces of the piston (the double-acting engine).

In order to get the most power from the steam introduced into the cylinder, Watt devised two regulators: the first (1784) automatically stopped the steam coming in when the piston reached a certain point in the stroke, while the second (1788) automatically regulated the amount of steam that entered the cylinder according to variations in stress on the piston. The principle of retroaction, on which the second valve was based, found one of its first and most stimulating applications in this instance.

The use of rotary motion

The steam engine was first used to operate machines in the metallurgy and textile industries (the forging hammer in the Wilkinson factory in 1783 and spinning machines in 1785). In order to centralize power sources, the various branches of the textile industry, which corresponded to the various phases of production (carding, spinning, weaving), began to merge.

From 1781 on, carding and spinning were done in the same workshop (the machinery was patented by Richard Arkwright but the patent was later revoked under pressure from the spinners of Manchester).

Most of the technical innovations of the previous decades had been made in spinning. Figures from 1788 show that no fewer than 20,000 of James Hargreaves's spinning jennies were in operation at the time. Progress was also made in weaving, thanks to the invention, by Edmund Cartwright (1743–1823), of the automatically operated loom, the first in a series of looms that, over the next twenty years, led to a tenfold increase in the amount of fabric produced. The era of the great textile factories, which in Manchester alone increased from two in 1782 to fifty-two in 1802, had begun.

The relationship between chemistry and the textile industry

Bleaching and dyeing procedures, which were extremely important to the manufacturing industry, underwent radical changes during these years, as they began to benefit from discoveries made in the chemical sciences. As a substitute for the traditional methods of bleaching (exposure to sunlight in special bleaching fields and the use of diluted sulphuric acid), Claude Louis Berthollet (1748-1822) introduced a mixture of chlorine and alkali that was less corrosive and bleached better (1789). Only in 1799 was Charles Tennant (1768–1838) able to produce a low-cost whitening agent in powder form that was less irritating.

New means of locomotion

The physical properties of gasses and the possibilities offered by the steam engine stimulated the invention and perfection of new means of locomotion. In 1783, Michel Joseph (1740–1810) and Étienne Jacques de Montgolfier sent up the first hot air balloon, designed by Jacques-Alexandre-César Charles (1746–1823). That same year Pilâtre de Rozier (1756–85) became the first person to ride in a balloon: his "montgolfier" rose to a height of twenty-four meters. Two years later, together with the marquis of Arlands, he flew across Paris, while John Jeffries (1744–1819) and Jean-Pierre Blanchard (1750–1809) flew across the English Channel.

Attempts, still on the experimental level, to use steam as the driving force for a land vehicle multiplied during this decade. In 1784, with the help of William Murdock (1754–1839), Watt obtained a patent for building a steam vehicle. He was followed in 1786 by William Symington (1763–1831) who built a three-cylinder vehicle. In the United States, John Fitch (1743–98) built the first steamboat in 1787 by connecting a Watt engine to a series of oars. (It operated on the Philadelphia-Burlington route in 1790.) In 1789, Oliver Evans (1755–1819), who worked on high-pressure steam engines, obtained a patent for a steam-propelled vehicle.

Iron: the new building material

A new and extremely important advance in the construction field was the incorporation of forged iron into masonry as a reinforcement. In 1780, the architect Jacques-Gabriel Soufflot (1713–80) adopted the technique to reinforce the vault of the dome of the Church of Sainte-Geneviève (construction of which was begun in 1757), later the Pantheon, in Paris. Cut stone was still the primary material being used in bridge construction, however. In 1791, Jean Rodolphe Perronet completed the Pont de la Concorde, begun four years earlier using stones taken from the ruins of the Bastille (see APPLIED SCIENCES 1740–1750).

However, in Coalbrookdale, where one of the major extracting and metallurgy complexes in England was being built, John Wilkinson and Abraham Darby III built a cast iron bridge as early as 1779. The bridge is still in use as a footbridge.

1790–1800

THOUGHT AND SCIENTIFIC LIFE

The organization of the sciences in revolutionary France

From 1789 to 1793, the revolutionaries' policy toward science was characterized by criticism of the official institutions and suspicion of the men who represented them. Criticisms were above all directed at the Observatoire Astronomique and the Académie des Sciences, which, on August 8, 1793 was decreed closed by the National Convention. That same month, the Paris Observatory was renamed the Observatoire de la République and received new statutes calling for parity among all the astronomers working there. The director, Jacques Dominique Cassini (1748–1845), resigned. The Jardin des Plantes had already, in June, been transformed into the Muséum d'Histoire Naturelle, with parity among the researchers, on the basis of a plan which they themselves had elaborated and presented to the legislature three years earlier.

On September 20, 1792, after having voted in the new constitution, the Constituent Assembly was replaced by the Legislative Assembly, which was in turn succeeded by the Convention. The sentencing to death of Louis XVI and Marie Antoinette (January and October, 1793), preceded by the proclamation of the republic, deeply affected French political life. At the same time, the country was being threatened by the forces of the legitimist coalition. Following the defeat of the Girondin bourgeoisie, the Jacobin democrats, led by Robespierre, confronted the war-like situation and launched reforms that benefitted the artisan and working classes. Within this context, a mobilization of scientists was also called, in 1793, on the initiative of the Committee for Public Health: the leaders were Claude Louis Berthollet (1748–1822), Jean Antoine Chaptal (1756–1832), and Gaspard Monge (1746–1818).

The École Polytechnique was born out of these experiences in October, 1794. It offered a three-year course of serious scientific preparation for youth who wanted to pursue military or engineering careers. The vital force at the new school was Monge, who imposed high standards on the teaching of science there. The major representatives of French science during the following decades came out of this school. The École Normale Supérieure for training teachers was also founded in October, 1794. In October, 1795, the Institut National des Sciences et des Arts was formed, on the initiative of the Directory. The first class corresponded to the old Académie des Sciences (which took back its old name in 1816 when the Institut was completely reorganized by a royal ordinance of March 21). The universities established in other countries for the purpose of training future army technicians in the sciences (the Josephinum of Vienna, the Frederich Wilhelm Institut of Berlin, the Imperial Medico-Military Academy of St. Petersburg, etc.) were often modelled after the rival French Institute, at least to the extent allowed by ideological and political differences.

On the one hand, the events of the Revolution tended to relegate theoretical problems to a position of secondary importance, focusing instead in the larger problems of the country's political life. Also, revolutionary positions did not always correspond on both the political and the philosophical and scientific levels, so that the political theories advanced by the Jacobins, for example, often proved to be quite moderate philosophically. On the other hand, many scientists placed their knowledge at the service of the revolutionary government, for example Lazare Carnot (1753–1823), who organized the revolutionary army, and Monge, who organized the powder factories and cannon-foundries to defend revolutionary France.

Some of the same scientists—Lavoisier (1743–94), Laplace (1749–1827), Condorset, and Monge—worked on an extremely important project: the establishment of the universal system of measures on which the decimal metric system was based.

The crisis of the Enlightenment; philosophy in Germany

We have already spoken of the organization of scientific life and university education. Condorcet was also active in this process; he presented a project concerning education to the Legislative Assembly in 1792. Although it was never approved, his plan was the basis for all succeeding reforms in the French school system. In 1793, with the Jacobins in power, Condorcet abandoned public life and lived as a recluse. In eight months, he elaborated his famous Esquisse d'un tableau des progrès de l'esprit humain, which was only published after his death in 1794 (he was arrested and died in prison, a possible suicide). Antoine Laurent Lavoisier was also sentenced to death in 1794. In 1793, when all the fermiers généraux were decreed under arrest, Lavoisier turned himself in, perhaps trusting that his fame as a scientist would save him, or counting on the support of other scientists. His hopes were in vain, however, and he was guillotined.

Anti-revolutionary literature, modelled after the Reflexions on the Revolution in France, by the Irish-born English conservative Edmund Burke (1729–87) appeared all over Europe during these years. Some scholars who had previously followed Enlightenment ideas also took up anti-revolutionary positions: of the two Verri brothers, for instance, Pietro, who had enthusiastically supported the Revolution, continued to defend it (though with considerable reservations) even when Jacobin ideas triumphed, while Alessandro rapidly underwent a complete political and cultural reversal.

After the fall of the Jacobins (the Thermidor coup d'etat of July, 1794) and the rise of the Directory (October, 1795), the old themes of sensism were once again taken up, although the ways in which they were approached and the solutions found to basic problems were much different. This was the beginning of the idéologues movement, one of whose major representatives was Antoine Louís Destutt de Tracy (1754–1836); his most important works came out at the beginning of the nineteenth century.

In England, the first edition of An Essay on the Principle of Population as It Affects the Future Improvement of Society, by the preacher Thomas Robert Malthus (1766–1834), was published anonymously in 1798 (a second, much larger edition came out in 1803).

Meanwhile, in Germany, Kant pursued his activities, although less intensively than in the preceding decade. In 1794, he published Die Religion innerhalb der Grenzen der blossen Vernunft, which provoked such a denunciation by the ruling authorities that King Frederick Wilhelm II ordered Kant to stop writing about the relationship between reason and religion. In 1795, Kant published Zum ewigen Frieden, and in 1797 his last systematic work, the Metaphysik der Sitten.

Meanwhile, a new philosopher, Johann Gottlieb Fichte (1762–1814) emerged. In 1792, Fichte anonymously published his Versuch einer Kritik aller Offenbarung; it was so obviously influenced by Kant that some believed he had actually written it. In 1794, Fichte again defended the old philosopher, but by that time he was rapidly moving away from his critical position. That same year, Fichte published his Grundlage der gestamten Wissenschaftslehre, which was of a clearly idealistic nature. This was the first step toward a new philosophical perspective that soon became established throughout Germany.

Shortly thereafter a student of Fichte, Friedrich Wilhelm Joseph von Schelling (1775–1854), took another step in the same direction. In his Ideen zu einer Philosophie der Natur (1797) and even more in his three works Von der Weltseele (1798), Erster

Entwurf eines Systems der Naturphilosophie (1799), and *System des transzendentalen Idealismus* (1800), he outlined a speculative physics that was antithetical to Enlightenment science and, in general, supported a concept of the world that was clearly inspired by romanticism.

MATHEMATICAL SCIENCES

New institutions of higher learning in France

The foundation of new institutions of higher learning in Paris during the Revolution, i.e., the École Normale Supérieure and the École Polytechnique, provided a totally new model for the development of scientific research. The push toward research shifted decisively from the academies to these new schools, where research was associated with highly specialized and rigidly selective teaching. With regard to mathematics, the École Polytechnique was extremely influential; both schools favored the intensification of contacts among scientists and the use of their teaching activity for research (Joseph Louis Lagrange, Pierre Simon de Laplace, and Adrien Marie Legendre, for example, taught simultaneously in these schools).

Preludes to new research

This fact, furthermore, led to a flourishing of important treatises. While on the one hand these followed the already-mentioned tendency towards writing treatises, they also quite often fulfilled very precise needs connected with teaching. This was the case, for example, with Legendre's *Éléments de géométrie*, which came out in 1794 and was later reprinted and reedited a number of times. The same was true for Lagrange's *Leçons élémentaires sur les mathématiques données à l'École Normale en 1765*, *Théorie des fonctions analytiques* and *Leçons sur le calcul des fonctions*, which first appeared in the records of the École Normale sessions and were later published separately. In other cases, however, the treatises were works of synthesis, which attempted to collect scattered results in fields that were, by then, rich in content. This occurred, for example, in the case of Legendre's *Essaie sur la théorie des nombres* (1798), which, in a certain sense, completed and organized the research of an entire epoch of studies in the field. This work also presented original findings, such as the "law of reciprocity" for quadratic remainders. Lagrange's *De la resolution des équations numériques de tous les dégrés* (1798) was the same sort of attempt: despite its title, it is more a collection of separate findings than an actual treatise.

Particular mention should be made of the *Réflexions sur la métaphysique du calcul infinitésimal* (1797), an essay by Lazare Carnot (1753–1823). The need to clarify the general principles of this type of calculus, a concern of the following century, is already apparent in this work.

These works, as we have already mentioned, in some ways constituted a *summa* of eighteenth-century mathematical knowledge, and provided almost a natural ending to an epoch of great enthusiasm for research. It is also historically very interesting that in the same year the two treatises just mentioned appeared in France, other works that opened brand new perspectives in the same areas appeared in other parts of Europe. In 1798, Karl Friedrich Gauss (1777–1855) began publishing his famous *Dis-quisitiones arithmeticae* (finished in 1801), which was the basis for almost all nineteenth-century research on the theory of numbers. In 1798, in his *Teoria generale delle equazioni*, Paolo Ruffini (1765–1822) demonstrated the theorem that general equations higher than the quartic could not be solved algebraically. His results were rediscovered and utilized several years later, giving rise to new branches of mathematical research.

Gaspard Monge's *Géométrie descriptive* was connected to his teachings at the École Normale. The new discipline originated from the study and solution of problems in military engineering (for this reason it was kept secret until 1794), and flourished above all in the course of the following century, after Monge's work was finally published and became known throughout Europe.

The relationship between Monge's scientific research and military applications was characteristic of a trend that saw a significant part of mathematical research being done in military schools beginning in this decade (especially in France). Several famous mathematicians were also more or less involved in military careers and, in general, took on political and administrative responsibilities. During this particular period, for example, Monge and Jean Baptiste Fourier (1768–1830) both participated in the Napoleonic expedition to Egypt, and Napoleon later appointed them to administrative positions. Even more significant was the work of Laplace, who held a number of offices both during the Revolution and as an adviser to Napoleon. When Napoleon fell, Laplace pursued his political activity at the service of the Bourbons.

In concluding our survey of some of the most important mathematical treatises that appeared toward the end of the century, we cannot neglect to mention the vast *Traité du calcul différentiel et du calcul intégral*, by Sylvestre-François Lacroix (1765–1843); in a certain sense, this work can be considered a typical example of eighteenth-century treatise writing in the field of analysis, with all of its positive aspects as well as its defects (the first edition came out in 1797–1800).

ASTRONOMY

Units of measure

Following a recommendation from the Académie des Sciences and the approval of the French National Assembly in 1791, it was decided to adopt the meter, equal to one ten-millionth of a meridional quadrant of the earth, and its decimal multiples and sub-multiples, as the unit of measure for length. Units of measure for area, volume, and weight would be derived from these. It was further decided to verify the measurement of the meridian arc on the basis of the findings made in Peru by Pierre Bouguer and Charles Marie de la Condamine in 1735–44. The task of checking the length of the meridian arc between Dunkirque and Barcelona, already partially determined in previous campaigns, was entrusted to the astronomers Pierre-François-André Mechain and Jean Baptiste Delambre (1749–1822). The work was begun in 1792 and was carried out in the midst of innumerable difficulties, most of them due to the political situation in France and Spain; it was only completed after several years. The two astronomers were aided by

Jean Charles de Borda, who provided two valuable measuring instruments for the campaign, the repeating transit and the bi-metallic rule for directly measuring the base line, and by the mathematician Adrien Marie Legendre (1752–1833), who gave the formulas for calculating triangles drawn on the ground. A quarter of the meridian was found to be equal to 5,130,740 toise, so that a meter was equal to 0.513,074 toise. The measure was adopted by the Assembly on April 7, 1795.

Geodetic measurements, around which a large portion of astronomical and geographical activity was centered during the eighteenth century, also brought attention to the importance of gravimetric determination.

Cosmology founded on mechanics

The *Exposition du systéme du monde*, by Pierre Simon de Laplace (1749–1827), a classical popularizing work on astronomy, which also contained an interesting summary of the historical development of this science and was written without resorting to mathematics, came out in Paris in 1796. The work also expounds (in note number seven) the cosmological theory usually referred to as the Laplace-Kant hypothesis. In his *Allgemeine Naturgeschichte und Theorie des Himmels* of 1755, Kant had proposed the idea that the solar system emerged from an original chaos, in which nuclei of matter were formed due to forces of attraction. These nuclei were later animated by circular motions. Basing his idea upon three well-known characteristics of the solar system that, up until then, held true without exception (all motion of both revolution and rotation was direct, i.e. from west to east; the eccentricity of the orbits was very small; and the inclination of the orbits with respect to the ecliptic was minimal), Laplace maintained that a single cause had produced the three phenomena characteristic of all the planets by acting upon a single, original nebular material that extended throughout all space that is currently occupied by the solar system, i.e. beyond the orbit of Uranus, the furthest of all the planets known at the time. The nebula was animated by a motion of rotation, and within it the planets were formed by matter cooling and contracting with a consequent increase in angular velocity that enabled first one ring, then others (as can be seen in the system of Saturn), to pull away from the outside edge of the nebula when centrifugal force became greater than the attractive force of the central mass. Satellites later formed in the same way from the rotating matter of the rings. These were also animated by direct motion and lay close to the plane of the ecliptic. The most serious difficulties in Laplace's theory arose when the furthest satellites of Jupiter and Saturn, which had retrograde motion, and the satellites of Uranus, which orbit on a plane almost perpendicular to the ecliptic, were discovered. In more recent times, considerations of a mechanical nature have shown that, in reality, the distances between planets are too great and the sun's rotation is too slow for Laplace's theory to be true.

Around the end of the century (1799), Laplace published the first and second volumes of his *Traité de mécanique céleste*, which was only completed in 1825. In this great theoretical work, he arranged astronomy analytically in the same way that mechanics had already been arranged, above all in the work of Joseph Louis Lagrange, by collecting and thoroughly reelaborating, in a unitary manner, the many contributions already published on the mechanics of celestial bodies.

PHYSICAL SCIENCES

Rational and celestial mechanics

In 1796, in his *Exposition du systéme du monde*, Pierre Simon de Laplace outlined the notion of reducing all experimental data collected by scientists to explanations founded on mechanics. Permeated by the Englightenment spirit, this work sanctioned the need to generalize Newtonian mechanics in order to arrive at the "analytical relationship between particular facts and a general fact" that, according to Laplace, constituted a theory. In this sense, the "simplicity of nature" did not, in Laplace's opinion, derive from the fact that natural laws were easily understandable, but from the fact that they were in perfect agreement with all phenomena. Laplace also put forward the theory of the original nebula, or the Laplace-Kant hypothesis in this same work (see ASTRONOMY 1790–1800).

An important step in the direction of study indicated by Laplace's program was made by Jean Baptiste Fourier (1768–1830) in 1797. He generalized a basic theory of the so-called momentum of forces. The momentum of forces was the product, with its sign changed, of the force times the "virtual velocity" of the point on which the force was applied. Fourier expounded this theorem, along with its applications to very general systems and its correlations with theorems on levers, in his *Mémoire sur la statique, contenant la démonstration du principe des vitesses virtuelles et la théorie des moments*.

The first two volumes of Laplace's *Traité de mécanique céleste* came out in 1799. The laws of mechanics, Laplace wrote, were first of all laws of an experimental nature. Force could only be measured in relation to the effects it produced, and it was wrong to try to understand the nature of forces, which "is and will always be unknown." It was therefore necessary to elaborate a mathematical theory of the laws obeyed by the effects of forces. Referring to the eithteenth century theories that were based on the great law of nature proposed by Isaac Newton, Laplace summarized his program of research. He proposed to present from a single point of view those theories, scattered throughout a large number of works, which together, embracing all the effects of universal gravitation on the equilibrium and motion of the solid and fluid bodies that make up the solar system and analogous systems scattered throughout the immensity of the sky, form celestial mechanics. Astronomy thus became a general problem of mechanics, which could be and had to be solved by making accurate observations, perfecting analysis, and rejecting all forms of empiricism. It was, according to Laplace, a question of formulating a "theory of the system of the world" that was based on the laws of equilibrium of a point "stressed by any number of forces acting in any direction whatsoever," on the "laws of motion of a system of bodies, in all the possible mathematical relations between force and velocity." All of this implied a deeper knowledge of differential equations and the means of solving them, and a rigorous application of mathematical methods to the general problem of the interactions between spheroids and points inside them, on their surfaces and at any place in space outside them.

The problem of the caloric

The debate on the nature of heat, which had been opened very specifically by Laplace's and Antoine Laurent Lavoisier's memoir of 1780, was furthered during this decade by Count Rumford. In 1792, Rumford published the sequel to the research he had begun in 1786 and, in a communication entitled "Experiments upon Heat," dealt with questions of the heat conductivity of textile fibers in relation to the presence of "particles of air" between them. Here again Rumford used, in a qualitative way, language that was based on quantities of heat. However, in his famous later work, the "Inquiry Concerning the Source of the Heat which is Excited by Friction," published in 1798, Rumford made a global criticism of this type of language and of the physical meanings underlying it, posing the following questions: "What is Heat? Is there such a thing as an igneous fluid? Is there anything that can with propriety be called caloric?" In stating these questions, Rumford made reference to his observations of increases in temperature due to friction in metals. An explanation of these phenomena was possible within a fluidist context only if the hypothesis according to which the heat capacity of metals depended heavily on the size of their fragments was true. However, through a series of measurements of specific heat, Rumford demonstrated that this dependency was not observable: therefore the large quantities of heat produced by friction did not originate from the production of metallic fragments, and another explanation had to be found. Rumford suggested the following theory: "It is hardly necessary to add, that anything which any *insulated* body, or system of bodies, can continue to furnish *without limitation*, cannot possibly be *a material substance*; and it appears to me to be extremely difficult, if not quite impossible, to form any distinct idea of anything capable of being excited and communicated in the manner the Heat was excited and communicated in these experiments, except it be MOTION." Rumford finally took the side of the mechanical-wave theory on heat phenomena during the early years of the nineteenth century (see PHYSICAL SCIENCES: THERMOLOGY, 1800–1810).

The controversy over animal electricity

There was, during this decade, a profound innovation in the area of electrical phenomena. In 1789, while investigating the effects of electricity on the nerves of animals, the abbot Luigi Galvani (1737–98) had discovered the famous muscle contractions of frogs. After repeating his observations a number of times, he communicated them in 1791 to the Academy of Sciences in Bologna, in a memoir entitled *De viribus electricitatis in motu muscolari animalium*, in which he suggested the possiblity of interpreting the phenomena as a manifestation of a new type of electricity, animal electricity. The memoir was republished in 1792 with notes and a long introduction by Giovanni Aldini.

As soon as he heard about Galvani's discoveries, Allesandro Volta set about carefully examining them. His studies led him to confirm the existence of the phenomena, although he rejected Galvani's interpretations of it. He sought various proofs to show that the frogs' contractions were caused not by a presumed animal electrical fluid, but by the metal arc with which the contractions were detected. A long polemic ensued, stimulating both authors continually to seek new proofs to support their respective theories. In 1797, Galvani published what was actually a monograph in the form of letters to Lazzaro Spallanzani (1729–99), his five *Memorie sull'elettricità animale*. The research that Volta conducted in order to refute Galvani's interpretation led

him to discover the famous laws, still named after him, on the differences in potential produced in contacts between heterogeneous metals. In 1798, the year of Galvani's death, the Académie des Sciences of Paris appointed a commission to repeat Galvani's experiments and report their results. The commission included Charles Augustin de Coulomb and Alexander von Humboldt (1769–1859), who defended Galvani's theory in a treatise published in 1797, supporting it with numerous experiments. In 1799, on the basis of the same laws, Volta finally succeeded in building his famous battery, opening the way to electrodynamics, a new and extremely important branch of electronics. He announced his achievement in a letter dated March 20, 1800 to Joseph Banks, president of the Royal Society (see PHYSICAL SCIENCES: ELECTRICITY, 1800–1810). Despite the extraordinary success of Volta's invention, Galvani's name was not forgotten: in 1802, a Galvanic Society was formed in Paris for the explicit purpose of studying animal electricity. Ironically, the electricity generated by the battery was actually called galvanic electricity, to distinguish it from the electricity generated by the electrostatic machines that were built in the eighteenth century. Galvanism particularly attracted the attention of the physicists and philosophers of the romantic *Naturphilosophie* current, who were interested in any new phenomenon that came out of the conceptual framework of Enlightenment science.

CHEMICAL SCIENCES

Lavoisier's law

Almost as though to crown the important discoveries made during this century—discoveries that first of all definitively liberated chemistry from the fetters of alchemy and then, with the application of quantitative methods, succeeded in defeating the contrived phlogiston theory, making chemistry an exact science--Immanuel Kant (1724–1804) had written in 1786 that chemistry could never be more than a systematic art or an experimental doctrine. It could never become a science in the proper sense of the term, because its principles were merely empirical and consequently did little to explain the fundamentals of chemical phenomena since they were not suitable for the application of mathematics. Despite the eminently empirical nature of the science, chemists were able to formulate several general laws, foremost among them being the law of the conservation of matter. Although it was known as Lavoisier's law, in reality it had been more or less recognized by others (including Mikhail Vasilievich Lomonosov as early as 1756), and was reiterated, more than stated, by the French chemist.

Stoechiometry

Jeremias Benjamin Richter (1762–1807) is quite rightfully considered the founder of stoechiometry, with his *Anfangsgtünde der Stöchymetrie oder Messkunst chymischer Elemente* (1792–94). As early as 1792 he stated, in a nebulous form, the laws of equivalents. In 1797, Joseph Louis Proust (1754–1826) stated the law of definite proportions; it was only confirmed, however, in 1808, after a long period of intensive work. The law contrasted with the opinion of Claude Louis Berthollet (1748–1822), who believed that a continous variation in the relations between two elements was possible. Today we know that this is possible in some cases, and we divide compounds into daltonian (or daltonides) and non-daltonian (or berthollides), terms later introduced by Nikolai Semyonovich Kurnakov.

Antoine Laurent Lavoisier died under the guillotine in 1794. His ideas were spread in Italy by Vincenzo Dandolo (1758–1819) and Luigi Vincenzo Brugnatelli (1761–1818) in Pavia, Giovanni Antonio Giobert (1761–1834) and Constanzo Bonvicino (1739–1812) in Turin, and Giovanni Valentino Fabbroni (1752–1822) in Pisa.

The discovery of more new elements

New elements continued to be discovered, bringing the number of those known to thirty-four, though not all of them had yet been isolated by the end of the century (no more than fifteen had been known previously). In 1794 the Finn Johan Gadolin (1760–1852) discovered yttrium, which Carl Gustaf Mosander isolated in 1843; this marked the beginning of one of the most unusual chapters in inorganic chemistry, the discovery and isolation of the rare earths. Louis Nicolas Vauquelin discovered chrome in 1798 and isolated it in 1799: in 1798 he discovered beryllium, which was isolated independently by Antoine Alexandre Bussy (1794–1882) and Friedrich Wöhler (1800–82) in 1828. Alessandro Volta arranged the known metallic elements into the electromotive force series in 1795.

In 1796, Wilhelm August Lampadius (1772–1842) obtained a substance that he maintained was a compound of sulphur and hydrogen; but in 1802, Charles Bernard Desormes (1777–1862) and Nicolas Clement Desormes (1779–1841) identified it as CS_2 (carbon disulphide), prepared by passing a current of sulphur over red-hot carbon.

LIFE SCIENCES

Aid to the mentally ill

The clinician Phillipe Pinel (1745–1826), director of the Salpêtrière asylum and author of the famous *Nosographie philosophique* (1798), also published a treatise entitled *Traité medico-philosophique de l'aliénation mental.* Pinel's work was the source of a new explanatory and curative attitude toward mental illness. Each symptom was interpreted as the effect of an organic or functional alteration; in this were included social and educational causes. The treatment of the patient in the psychiatric hospital became, in consequence, a "moral treatment." In diagnosis, the doctor was expected to know the "history of human intelligence." Pinel divided mental illnesses into the categories of mania, melancholy, insanity, and idiocy. The Italian physician Vincenzo Chiarugi (1759–1820) stated analogous concepts in his *Della pazzia in genere e in ispecie* (1793); simultaneously, a movement to reform institutions for psychiatric care was spreading to Switzerland, Germany, Austria, and England. Everywhere the mentally ill were being freed from their chains.

Electrical phenomena and experimental physiology

The study of several aspects of the activity of living organisms—especially the functioning of nerves and muscles—was at this time being linked to research in electrical phenomena, which had become increasingly important from the 1740s

on. In the life sciences, the connection with research on electricity served to bring certain problems, which animistic vitalism had given preconceived and unverifiable solutions, onto the experimental terrain. We have already mentioned the discoveries of Luigi Galvani (see PHYSICAL SCIENCES, 1790–1800) and the different interpretation that Allessandro Volta gave to them, as well as the lively controversy that ensued between the two scientists. Here it is sufficient to add that the experiments which Volta used to demonstrate that the contractions of a frog's muscles could originate outside the tissues of the animal did not succeed in convincing Galvani. The debate spread; the controversy regarding animal electricity began. It ended in 1844 with the demonstration given by the physiologist Carlô Matteucci of the existence of electrical phenomena that were concomitant with the vital processes and indentical to those occuring in inorganic nature (see LIFE SCIENCES, 1840–50).

Epidemiology and prophyaxis

The hygienist Johann Peter Frank (1745–1821), a professor at Pavia and director of the Allegemeines Krankenhaus of Vienna from 1795, published his six-volume *System einer vollständigen medizinischen Polizei* between 1779 and 1819. In it, he discussed problems of hygiene and public health and outlined the responsibilities of public officials. The European states—France and England above all, but Venice and the Kingdom of Naples as well—financed epidemiological research and decreed prophylactic measures, often opposed by the population, who saw them as a threat to retail trade and inveterate customs. A pioneer in antitubercular prophylaxis was the Neapolitan chemist and anatomist Domenico Cotugno (1736–1822).

The classification of human beings. Craniometry

The first works in physical anthropology came out during this period. A work by the anatomist Pieter Camper (1722–89), entitled *Sur les différences que présente le visage dans les races humaines* and published posthumously in 1791, established the bases for craniometric measurements, improving upon conventions and techniques already proposed by Louis Jean Daubenton (1716–1800). In 1790, Johann Friedrich Blumenbach (1752–1840), *magister Germaniae*, published a large work on comparative craniometry, the *Collectionis suae cranoriorum diversarum gentium illustratae decades*, in which he compared the shape and size of the outside of the skull in individuals of various races. Blumenbach also formulated a classification of the races in a memoir entitled *De generis humani varietate nativa liber* (1798). Transformist hypotheses were again affirmed in the *Zoonomia*, by the physician Erasmus Darwin (1731–1802): more complex organisms had developed out of a primal organic filament in response to the ever-growing needs imposed on life by the external environment. Darwin's transformism was historical and not purely speculative in nature; it was closer to Maupertius's theories than to those of Robinet.

Vaccination

In 1796, Edward Jenner (1749–1823) innoculated a human being with cowpox lymph matter, in order to provoke a slight infection that would protect the individual from contracting smallpox. The experiment came from Jenner' observation that people who caught cowpox never got smallpox. The vaccination became known throughout the civilized world, in the wake of Jenner's report on the results of his experiments, *An Inquiry into the Causes and Effects of the Variolae Vaccine.* In Italy, Luigi Sacco (1769–1836) credited Jenner's procedure in his *Osservazioni pratiche sull'uso del vaiuolo vaccino come preservazione del vaiuolo umano* (1800).

EARTH SCIENCES

The stratigraphic systems of Abraham Gottlob Werner and Peter Simon Pallas

In the heated polemic between the Neptunists and the Plutonists, the latter theory prevailed, although the stratigraphic system proposed by Abraham Gottlob Werner, leader of the Neptunists, continued to dominate. The Saxon scholar completed his system in 1796 by adding the category of transition rocks between the categories of primitive and sedimentary rocks. This new layer contained the first fossil remains. Werner's five basic divisions were now primitive rocks (crystalline and without fossils), transition rocks, stratified rocks, alluvial rocks, and volcanic rocks. Among Werner's many students at Freiberg were Christian Leopold von Buch (1774–1852), Alexander von Humboldt (1769–1859), and Jean François D'Aubuisson de Voisins. Around the end of the century, von Buch visited the Alban Hills, together with Scipio Breislak (1748–1826), and went to Vesuvius, which Breislak had described. He was able to observe the eruptive nature of basalt lava and other rocks, including "feldspathic porphyry," which, as a Neptunist, he believed had crystallized out of the waters of the ocean. He was impressed by the immensity of the volcanic forces and began to harbor serious doubts about the validity of the Neptunist interpretations. Von Humboldt began his travels in South America during the same period (1799).

Scholars studying the volcanic regions of Italy adhered to the Plutonist theory. Lazzaro Spallanzani (1729–99), who visited the Kingdom of the Two Sicilies (1792) to observe active and extinct volcanoes there, deserves particular mention. He also maintained that the Euganean Hills (near Padua) were of volcanic origin. In France, many scholars recognized the volcanic origin of the basalts of the Auvergne (Faujas de Saint-Fond and Dieudonné Gratet Dolomieu (1750–1801) confirmed their interpretation during this period). In general, however, the stratigraphic relations among volcanic rocks were overlooked at the time.

In Paris, the Conseil des Mines was founded in 1793 in order to promote a survey of mining resources in France. This institution became a center for initiatives in geological research, as did similar state mining and geological services in other countries. Among the envoys of the Conseil des Mines was Dolomieu, who in 1791 collected in the Southern Alps samples of the calcium-magnesium rock that was later named dolomite in his honor.

One of the great scientific authorities of this period was Peter Simon Pallas (1741–1811), whose stratigraphic system, perfected in 1794, was not very different from Werner's, although it did provide better explanations for the position of crystalline rocks in mountain chains and their relationship to stratified rocks. According to Pallas, crystalline rocks gradually became less inclined the further they were from the nucleus of the chain, because they were deposited while the sea was gradually withdrawing and the mountains were rising as a result of volcanic forces. Pallas believed that these forces came from the combustion of great masses of pyrite that formed inside the sediments as a result of the degradation of primary mountains.

Horace Bénédict de Saussure's *Voyages dans les Alpes*, completed in 1796, was the first large geological work written about that mountain chain. In it, the author maintained that shell limestone must have been formed by mechanical deposits and not by crystallization from the waters of the primitive ocean, and that the folds in it were not from the same period as its formation, but were made later, either by thrusts from below or by folds that occured in the layers, which were originally horizontal and in a plastic state. De Saussure hesitated to attribute the main cause of the thrusts to a central fire in the earth, since he felt that such a theory did not have sufficient basis in observation.

The birth of stratigraphical paleontology

William Smith (1769–1839), inspector of coal mines in England, carried out observations in the field aimed at designing canals for transporting the fuel. In 1796, after recognizing the regular succession of layers in the regions where he travelled and noting their varying fossil content, Smith proposed using fossils to identify the stratigraphic position of fossiliferous strata in the rock series (guide fossils). Although the importance of this discovery was not immediately recognized, Smith's system was soon universally applied, as it is even today. Smith also demonstrated the importance of the circulation of underground water through porous rocks.

The beginnings of comparative paleontology

The numerous discoveries of fossil vertebrates led to increasingly precise studies. In 1796, Georges Cuvier (1769–1832) presented a memoir comparing fossil elephants to living species, showing that the two were different. During the previous year, Cuvier had recognized that the South American megathere belonged to the Edentata.

In 1799, Johann Friedrich Blumenbach (1752–1840) named the gigantic mammoth *Elephas primigenius.*

Oceanography: hypotheses and applied research

Between 1790 and 1810, Richard Kirwan (1733–1812) performed several analyses on samples of sea water, contributing to a refinement of the methods of analysis and improving the definitions of the ingredients of salinity. In 1795, Jens Rathke began a scientific study of the problems of fishing in the Norwegian Sea. Jean-Claude de Lamétherie (1743–1817) published *La théorie de la Terre* in 1797, in which he maintained that circulation in the oceans was caused by colder, fresher surface water flowing from the poles toward the equatorial zones, while saltier, heavier water flowed in the opposite direction deeper down. A year later, Benjamin Thompson (Count Rumford) (1753–1814) criticized the basis of this hypothesis.

An expedition led by Alexander von Humboldt went to the austral oceans from 1799 to 1804, for the purpose of visiting the Spanish colonies and South America.

APPLIED SCIENCES

Steel and the new alloys

The immediate diffusion of Henry Cort's process (in 1790 four-fifths of English blast furnaces operated on coke) led to an increase of 300 percent in the production of pig iron in about fifteen years (it rose from 50,000 to 150,000 tons between 1785 and 1800). Steel and copper also became common materials in the manufacture of tools and the creation of new alloys. Steel, obtained by melting in crucibles in a reverberatory furnace, a method invented and perfected by Benjamin Huntsman (1704–76), found its "capital" in Sheffield.

Copper was mainly used for making alloys (the so-called "Sheffield" plate, made of silver and copper, and brass, of which Birmingham industries produced 1,000 tons in 1795) and, from 1797 on, as a metal for coining (the first copper penny made its appearance).

Iron and cast iron began to be used, though still in small amounts, in the area of civil engineering. Although traditional materials continued to be used, French (in the footsteps of Jacques-Gabriel Soufflot) and English architects (John Soane (1753–1837) in the Bank of England building) adopted wrought iron and cast iron for support columns, reinforcements, framework, and putrelles. In 1791, the Committee of the Association of English Architects recommended the use of iron, especially in public buildings.

In 1793, an attempt to build a metal bridge over the Wear at Sunderland failed. Built with sections of cast-iron joined to wrought-iron arches, the bridge fell as soon as the framework was removed. In 1795, Thomas Telford (1757–1834) built the second iron bridge over the Severn, at Buildwas. The first suspension bridge, held up by iron chains, was built in 1800 by James Finley (c. 1762–1828) in Pennsylvania.

Devices for the steam engine

The steam engine, construction of which was dominated by the Boulton & Watt Company until 1800, continued to evolve through a series of improvements to increase its yield. The major innovations were the steam pressure indicator, which James Watt added to his own engine in 1794, and a sliding valve that regulated the steam entering the cylinder, invented by William Murdock, a collaborator of Watt's (also 1794). Murdock also studied lubricating blends for reducing the friction between the piston and the cylinder. By 1800, when Watt's patent ran out, the Boulton & Watt Company had built 496 engines: 164 for operating pumps, 25 for blast furnaces and 308 as prime motors for machines.

The automatic spinning machine

The mechanization of the textile industry proceeded without any particular innovations, with the exceptions of the first carding machine, built by Edmund Cartwright in 1792, which gave very disappointing results when put into operation, and the automatic spinning machine, invented by the American Samuel Slater (1768–1835) in 1790 but still unknown in Europe. The first incidents of rebellion against machines by workers who were afraid of losing their jobs took place during this period: in Manchester, in 1791, weavers destroyed a factory with 400 looms.

Costs and markets in the textile industry

Cotton began to surpass linen and wool in production and consumption. In 1785 the English imported 58,000 quintals of raw cotton, while in 1800 the amount rose to 182,000 quintals. However, the percentage production of cotton, wool, and linen was still seventeen percent, forty-two percent, and forty-one percent respectively in 1800; by around 1830 it had become forty-one percent, twenty-six percent, and thirty-three percent.

The main reason for the success of cotton was the cotton gin, invented in 1793 by Eli Whitney (1765–1825) and perfected in 1796 by Hogden Holmes. This machine very easily removed the seeds from the cotton ball, making it convenient to cultivate even short-stapled cotton, the only kind that would grow in the inland regions of America.

Whitney also used the American system (in reality introduced by the Frenchman Le Blanc in 1785) of producing interchangeable machine parts; thanks to this method, he was able to negotiate a contract with the American federal government for mass-producing muskets.

Chemical production

The application of Antoine Laurent Lavoisier's research in France to industry created a chemical industry that was the first instance of technology born out of theoretical science. The French chemical industry was developed for military purposes and to provide for scarcities in important raw materials. The most urgent problem, the synthetic production of soda (originally imported raw from Egypt or extracted from the ashes of marine algae), was solved by Nicolas Leblanc (1742–1806), who in 1791 discovered the chemical procedure for preparing it from common salt. Leblanc's process made the price of soda in France drop from ninety to ten francs per quintal, inducing the Convention to confiscate the patent, which was only used in 1823 by James Muspratt (1793–1886) in Liverpool and in 1825 by Charles Tennant in Glasgow.

The first aerial reconnaissance

Military needs in France stimulated attempts at long-distance communication and aerial observation. In 1793, Claude Chappe (1763–1805) invented the optical telegraph for transmitting the orders of the revolutionary government. The French army was outfitted with a fleet of balloons, the first of which was used for reconnaissance in the battle of Fleurus in 1794.

Developments in road networks

The roads of France were admired throughout Europe, both for their paving and for their extension (40,000 kilometers of roads already built or under construction), but they fulfilled military needs more than the needs of the national economy. The opposite was true in Great Britain, where local administrations began constructing a huge network of roads for commercial traffic and the transport of industrial products.

The availability of food and farm machinery

Although James Cooke had built a mechanical seeding machine in 1782, followed in 1786 by Andrew Meikle's threshing machine, many of the steps in agriculture only began to be mechanized toward the end of the decade. The stimulus came from a decrease in labor power, which was shifting toward the nascent industries, and from food needs, which were being felt in both continental Europe and in England. The need to regulate agricultural production induced Albrecht Thaer (1752–1828) to found an agricultural institute in Hanover in 1802, for the specific purpose of making the first attempts to establish quantitatively the food needs of livestock. He was followed in 1803 by von Voght, who set up a similar school in Holstein. In the course of the year 1800, Joseph Boyce mechanized reaping by attaching a series of scythes to a circular disc, while Robert Salmon and Smyth improved Cooke's seeding machine.

The Nineteenth Century

1800–1810

THOUGHT AND SCIENTIFIC LIFE

Research during the Napoleonic decade

The Napoleonic decade (which culminated in the creation of the empire in 1804) affected scientific life not only in France, but in all of Europe. When Bonaparte became a member of the "first class" of the Institut de France, which corresponded to the former Académie des Sciences, in December 1797, he used it as a tool for his policy on science, which was based on national prestige and a bold open-mindedness toward all new disciplines that were likely to have technological applications. In 1802, the requirement that each class present an annual report, with the suggestion that they consider the possible practical applications of theoretical work, was reestablished. In 1803, a "second class" was created in French literature, corresponding to the disbanded (1793) Académie Française; its basic task was seeing to the ideological formation of the new ruling class. The former "third class" of moral and political science was split into two classes, one for languages, antiquity, history, and politics in relation to history, and one for the fine arts. The editing of historical reports on the progress of science from 1789 to 1808 was promoted; these reports were to constitute a documentation of the history of civilized society and scientific thought. In November, 1801, Allessandro Volta (1745–1827) spoke at the Institut in the presence of Bonaparte, who then offered a prize of 60,000 francs for important research in the field of electricity. In 1807, the emperor promoted the construction, at the École Polytechnique, of a larger Voltaic battery than any ever before built. Prizes were also offered to encourage technological innovation: two tenth-anniversary prizes were offered in industry and agriculture to celebrate the coup d'état of 1799.

A faculty of the sciences was established at the imperial university in 1808, although scientific teaching still had a minor place in Napoleonic secondary schools. Meanwhile, an international scientific circle, which included Joseph Louis Gay-Lussac (1778–1850), Arago (1786–1853), and Alexander von Humboldt (whom Bonaparte suspected of being a Prussian spy) formed around Berthollet and Laplace in the Parisian suburb of Arcueil. The Arcueil Society promoted avant-garde research, while the Institut de France operated within the realm of official science.

Napoleonic restructuring (especially the Kingdom of Italy of 1805) exerted a positive influence on academic life in Italy. The Accademi degli Inquieti in Bologna became the Istituto Nazionale Italiano della Repubblica Cisalpina (1802–10). The Istituto Nazionale Italiano in Milan was established in 1797; the Istituto Nazionale in Naples was created in 1799, although it was destined to have a short life; the Ateneo Veneto was established in Venice in 1810, etc. Many of these institutions continued to function under new names during the Restoration, and were centers for the intellectual elite during the nineteenth-century Risorgimento.

The ideologists and Napoleonic politics

The differences in the development of French and German thought, which had already become evident by the end of the previous century, grew more pronounced during the early years of the nineteenth century.

The *Rapports du physique et du moral de l'homme*, Pierre Jean Georges Cabanis's most famous work on the relationship between body and spirit, came out in 1802. It firmly rejected the old separation between the physical and the spiritual world, maintaining that without a precise understanding of the organs of the body, it was impossible to understand how thought, will, and passion develop. Meanwhile, Destutt de Tracy's *Éléments d'idéologie* in four parts: metaphysics (1801), grammar (1803), logic (1805), and will (1817), began to be published. Although the work was inspired by Condillac's sensism, it approached the research of ideas with a methodicalness and an adherence to concrete observational data unknown to eighteenth century philosophers. But the movement of the ideologists, which had enjoyed considerable authority in the Institut de France during the years of the Directory, fell into disgrace under Napoleon, who became increasingly interested in cancelling all ties with the *philosophes* of the Enlightenment and in reaching an accord with the Catholics.

Romanticism

Cultural life in Germany was particularly intense, often in dramatic contrast to the country's political, economic, and social backwardness, as compared to England and France. This was the time of Goethe's *Faust* (the first part was completed in 1808); it was also the time of the romantics, and consequently of a vindication of sentiment over intellect, tradition (even religious tradition) over the "light" of reason, and intuition over mathematical-mechanical knowledge. The romantic current had in some ways been prepared by the *Sturm und Drang* movement, which flourished between 1770 and 1780, although romanticism had a much wider range and was much more radically anti-Enlightenment. It had its beginnings in the work of the two Schlegel brothers, Friedrich (1772–1829) and August Wilhelm (1767–1845), directors of the review Athenaeum, published in Berlin from 1798–1800, and in the philosopher-poet Friedrich von Hardenberg, alias Novalis (1772–1801), an active collaborator on the review. The philosopher-theologian Friedrich Ernst Daniel Schleiermacher (1768–1834) also made a significant contribution to the rise of romanticism; he championed the complete autonomy of religion—based on sentiment—with respect to scientific knowledge.

Novalis's concept of the world, which he called magical idealism, was strongly influenced by Fichte's idealism, although it did not identify itself as such. Fichte also undoubtedly contributed to the development of the romantic mentality, especially with his later thought, particularly his famous speeches to the German nation (*Reden an die deutsche Nation*, 1807–08), which made a major contribution to the recovery of a Germany invaded by Napoleon's armies. Schelling's philosophy also belonged to the general current of romanticism, from which it above all drew a new attitude toward nature, especially developed in the speculative physics mentioned earlier.

Hegelian idealism

Georg Wilhelm Hegel (1770–1831) also took up some romantic themes, although his philosophy—basically rationalistic, but with a rationalism unlike that of the eighteenth century—was completely outside the patterns of pure romanticism.

The distance between Hegel and Fichte or Schelling is evident in the *Phänomenologie des Geistes* (1807), which as Hegel himself stated in the introduction of his *Wissenschaft der Logik* (see THOUGHT AND SCIENTIFIC LIFE 1810–1820) expounds "the movement of consciousness, from the first crude opposition between itself and the Object, up to absolute knowledge. This process goes through all the forms of the *relation of thought to its object*, and reaches the *Concept of Science* as its result." Here Hegel was attempting to grasp the concentration of phenomena, or temporary manifestations of the spirit, with respect to the result, which also included all movement that led up to it. Just as the final product of the process included all the preceding instants, according to Hegel, the results were constantly present in the various moments of the process.

MATHEMATICAL SCIENCES

The work of Gauss

Some of the previously mentioned events continued to affect the field of mathematics during this period. This was true, for instance, of the editing of Joseph Louis, Comte de Lagrange's treatise, two of which were published during this decade. As we have already mentioned Karl Friedrich Gauss's *Disquisitiones arithmeticae* was published in 1801, bringing immediate fame to its author.

The profound connection between mathematics and the physical sciences, which had already characterized the end of the eighteenth century, became even more evident. The names of such famous mathematicians as Pierre Simon de Laplace, Siméon Denis Poisson (1781–1840), Jean Baptiste Fourier, and George Green (1793–1841) ended up achieving greater prominence in the physical sciences (especially mechanics) than in the field of pure mathematics. Furthermore, it is well known that Gauss, the *princeps mathematicorum* himself, dedicated more of his attention to the physical sciences than to pure mathematics. From 1801 to 1817, he did most of his work in astronomy (he was given a chair in astronomy at Göttingen in 1807); from 1821 to 1825 he did geodetical research; and he studied magnetism from 1831 to 1841 in collaboration with Weber.

The beginnings of mathematical physics

However, research in physics spurred the better mathematical minds on to findings in pure mathematics as well. For example, Adrien Marie Legendre presented the method of least squares (which Gauss had already discovered, but did not publish until 1809) in 1805, in the appendix of a memoir on comets. Gauss himself published a number of important mathematical findings in his *Theoria motus corporum coelestium* of 1809. This is not surprising; it was during these years that the volumes of Laplace's *Traité de mécanique céleste* (1799–1825) began to come out, and at this particular moment, it was astronomy that gave the most impetus to mathematical research, leading to results that were, at times, important enough to be taken up and coordinated in separate publications. This was the case, for example, for Laplace's *Mémoire sur différents points d'analyse*, published in 1809. In it, the author illustrated the major findings in pure mathematics made in the course of his research in celestial mechanics and in the calculus of probabilities. These mathematical discoveries and applications of purely mathematical techniques on the fringes of research connected to physics increased even further, as we shall see, in the course of the following decades.

On the other hand, however (and this was true for at least the first half of the nineteenth century), the scientists who made the greatest contributions in this respect retained the title of pure mathematicians, even when they worked on questions of mechanics or physics. For example in 1809 the foundation of the faculties of sciences in Paris asked Poisson to take over the chair of pure mathematics there.

Mathematical physics gradually began to take form as an autonomous discipline in the second half of the century and was in a sense, the result of a development that actually began during these years. After the application of analytical methods to mechanics, acoustics, optics, the theory of elasticity, the heat theory, electricity, magnetism, and electromagnetism had established a *corpus* of problems, techniques, and areas of investigation, where the ties between certain mathematical tools and certain physical questions were particularly direct and effective, a discipline arose as a specialized branch of mathematics, to deal with these problems from a different point of view from that of the analyst or the pure geometrician.

ASTRONOMY

The Lilienthal meeting and the discovery of Ceres

Friedrich Wilhelm Herschel's discovery of Uranus in 1781 was seen at the time by many astronomers as a confirmation of the empirical law formulated in a Pythagorean and Keplerian spirit in 1776 by Johann Daniel Titius (1729–96) and later taken up by Johann Elert Bode. The law linked the relations between the distances of the planets from the sun, taking the distance between the earth and the sun as the unit of measure for a geometrical progression times two, where the first term was three, four was added to each product, and the final result was divided by ten.

The average distance of Uranus was actually 19.2, which was fairly close to the value of 19.6 obtained with the Titius-Bode law. However, there was a fifth place in the series, between Mars and Jupiter, where no planet was known to exist. In 1800, Franz Xaver von Zach (1754–1832), director of the journal *Monatliche Korrespondenz*, gathered a group of astronomers at Lilienthal to organize a systematic search for the missing planet. On January 1, 1801, while making observations for a huge star catalogue (published in 1814), Giuseppe Piazzi (1746–1826) discovered an eighth magnitude "star" that by the following evening had shifted in position with respect to the fixed stars. Although Piazzi had not attended the Lilienthal meeting, he sent von Zach his data on the position of the object. Unfortunately, the news was late in reaching von Zach, and the body had moved too close to the sun to be visible. Piazzi was forced to suspend his observations due to illness, and the heavenly body, the asteroid Ceres, seemed destined to be lost again in the skies right after it had been discovered. However, despite the small amount of information available from the short arc of the orbit observed by Piazzi, Karl Friedrich Gauss (1777–1855) was able to establish its probable position in the constellation Virgo, where it was effectively discovered on December 7, 1801. Thus the century opened with a new discovery made, thanks to an observer's cleverness, and an impressive confirmation of the possibilities of calculus. Gauss generalized his method in a memoir of 1809, the *Theoria motus corporum coelestium in sectionibus conicis solem ambientium*, in which he also put forth the method of least squares, later systematically used to check errors, not only in astronomy, but in all other fields where observation was used. Gauss drew much of the material for his *Theoria* from the discovery of other planetoids made between 1802 and 1807, mainly by the astronomer Heinrich Wilhelm Olbers (1758–1840), already known for having proposed an easier method for calculating the orbits of comets in 1797.

Double stars

While the *Historie céleste française*, a catalogue of more than fifty thousand stars compiled by numerous collaborators under the direction of Joseph Jérôme de Lalande (1732–1807), was published in France in 1801, in England, Frederich Wilhelm Herschel continued his systematic exploration of the skies. He focused on double stars in particular, hoping to be able to measure stellar parallax more easily by comparing the data on both stars, basing his calculations on the assumption that the apparent pairing of two heavenly bodies in the sky was only the effect of perspective, caused by the very small angular distance between the two stars. However, his observations convinced him that the pairing was real in many cases, and that both stars orbited around a common center of gravity. This proved the validity of Newton's law, even for distant points in

space. From changes observed in the proper motion of several bright stars, including Procion and Sirius, Friedrich Wilhelm Bessel (1784–1846) assumed basing his conclusion on purely theoretical calculations that they too must be double stars.

A study of the proper motion of fourteen stars, later extended to thirty-six, and the supposition that the sun, as a star, also possessed proper motion, led Herschel to consider the problem of determining the apex of solar motion, i.e. the direction in which the sun, together with its system, was travelling. He solved it in 1805, indicating the apex in the direction of ρ Herculis. The fourth volume of Pierre Simon de Laplace's *Traité de mécanique céleste*, which can be considered conclusive as an organic treatment of the topic, also came out in Paris in 1805. The treatise summarizes all the activity of eighteenth-century astronomers and mathematicians that was directed toward resolving the great problems of the mechanics of the solar system in such a way as to make the theory coincide so closely with the observations that empirical equations would no longer be necessary. The work seems actually to fulfill this desire for absolute rationalization, to the extent that even the anomalies of Jupiter and Saturn, which seemed to compromise the theory of gravitation, end up becoming the most convincing proofs of it.

Johann Franz Encke (1791–1865) turned his attention to observing comets, attempting to define their periodicity, in particular by following the motion of a comet with a very short period of 3.3 years (Encke's comet). He attempted to give an explanation of the delays in the comet's appearance by attributing them to phenomena of friction with the cosmic ether, which he believed braked the motion of the comet as it passed near the sun.

PHYSICAL SCIENCES: MECHANICS

Programs for extending mechanics

The process of further rationalizing mechanics and extending its laws to all natural phenomena assumed a prominent role between the end of the eighteenth and the beginning of the nineteenth centuries. This process, which was carried out via an increased application to mechanics of the deductive methods elaborated by mathematics, in turn produced a wealth of problems for mathematical investigation and particularly benefitted from the cultural impulses and transformations that emerged as a result of the French Revolution.

In 1801, Charles Augustin de Coulomb applied the Newtonian concept of mechanical phenomena to a study of the theoretical and experimental correlations between the resistance of fluids to bodies in motion and the forces of the velocities of the bodies. The results of the study were reported in a memoir entitled *Expériences destinées a déterminer la cohérence des fluides et les lois de leur résistance dans les mouvements très lents*.

In 1803, Lazare Nicolas Carnot published his *Principes généraux de l'équilibre et du mouvement*, a reflection of the fundamentals of mechanics. Going back to several considerations already made in 1783 in his *Essai sur les machines en général*, Carnot maintained that the notion of force was "metaphysical and obscure." In his opinion, such notions had to be redefined, in order to connect them not so much to the hypothetical prime causes of motion and mechanical phenomena as to the quantities of motion impressed upon physical systems. Carnot again set himself the goal of establishing the proper relationships between mechanics and experimentally observable quantities. The *Principes* contains an exposition of Carnot's theory on kinetic energy in collisions and the notion of geometric motions, a particular form of the notion of virtual displacement.

Louis Poinsot's *Eléments de statique* and the third volume of the *Traité de mécanique céleste* were also published in 1803. In his third volume, Pierre Simon de Laplace elaborated the "particular theories of celestial motions" with an accurate mathematical analysis of the perturbations provoked by gravitational interaction. He proposed further improvements in the theory of perturbations in 1803. The fourth volume of the *Mécanique céleste* appeared in 1805: it dealt with different aspects of the theories on the satellites of Jupiter, Saturn, and Uranus, and of the theory on the motion of comets. Laplace made another very important contribution to the development of mechanics in 1806, when he published his *Théorie de l'action capillaire*, an appendix to the fourth volume of the *Mécanique*. In it, he considered several problems linking the laws of mechanics to studies on the structure of matter and on the interactions between molecules and particles within the caloric. Within Laplace's program, it became increasingly more obvious that connections were being made to reduce thermology to a branch of mechanics, through the progressive mathematization of a theory on the interactions between the caloric and matter. Thomas Young (1773–1829) also studied fluids and discussed several of their aspects in *An Essay on the Cohesion of Fluids* (1805).

Poisson's brackets

It was Siméon Denis Poisson (1781–1840) who succeeded in carrying a basic aspect of dynamics to a high level of abstraction, introducing the algorithm known even today as Poisson's brackets. Based on several considerations made by Joseph Louis Lagrange between 1808 and 1809 (integrated into the 1811 edition of his famous treatise entitled *Mécanique analytique*), Poisson analyzed the properties of the prime integral with the arbitrary constant a, which is a function of the coordinates u_j, q_j, e and t, and given another prime integral with a similar constant b, Poisson showed that:

$$\frac{\partial b}{\partial u_j}\frac{\partial a}{\partial q_j} - \frac{\partial b}{\partial q_j}\frac{\partial a}{\partial u_j} = \text{a constant,}$$

or:

$$(b, a) = \text{a constant,}$$

where

$$(b, a) = -(a,b), \qquad (a, a) = 0.$$

These studies of Poisson's influenced the research of William Rowan Hamilton (see PHYSICAL SCIENCES: MECHANICS, 1830–40) and played a primary role in Paul Adrien Maurice Dirac's formulation of quantum mechanics.

The attraction of ellipsoids

James Ivory's *On the Attraction of Homogeneous Ellipsoids*, also published in 1809, deserves particular mention. In it, Ivory attempted to reexamine earlier studies made by Colin Maclaurin, Adrien Marie Legendre (1752–1833), and Pierre Simon de Laplace on the attraction between spheroids of revolution, and credited Laplace with having carried this type of physico-mathematical research to levels of extreme generalization. However, Ivory believed that an even higher degree of generalization was necessary and outlined a program of research that he carried out over the course of the following years.

PHYSICAL SCIENCES: THERMOLOGY

The wave-corpuscle dilemma

During the early years of the century, physical studies on heat phenomena were marked by a conflict that went back to the eighteenth century. This conflict had to do with the problem of choosing between a wave and a corpuscular theory of heat and light, and it appeared above all in the complicated relations that developed between thermology and optics and continued to grow until it was overcome, around the second half of the nineteenth century, by a radical transformation of several fundamental concepts, which later led to the emergence of the new thermodynamics.

In 1801, Thomas Young developed a description of heat phenomena within a wave context in his Bakerian lecture "On the Theory of Light and Colours," in which he proposed a common line of study for both optical and heat phenomena. In his opinion, heat was caused by vibrations in the ether, provoked by molecular oscillatons. According to Young, this concept of heat was confirmed by experimental results, especially considering the observations of Fredrich Wilhelm Herschel (1738–1822), Humphrey Davy (1778–1829), and Count Rumford. In 1800, Herschel had observed unusual heat effects that could be attributed to invisible radiation. He thus faced the problem of verifying by experiment whether this form of heat radiation was the same as light radiation or not. His communications of 1800 did not solve the problem, although they did lead to a wave of innovative studies on infrared radiation. Herschel's memoir on refraction, *Experiments on the Refrangibility of the Invisible Rays of the Sun*, deserves special mention. The debate that arose from it spread rapidly to include such scholars as John Leslie, Johann Wilhelm Ritter, and William Hyde Wollaston (1766–1828) and quickly surpassed the understanding of radiation that had previously been reached through the work of Carl Wilhelm Scheele and Marc-Auguste Pictet. Young made reference to theories expounded by Davy in his *Essay on Heat, Light, and Combinations of Light* (1799). According to Davy, there was no caloric, and heat effects were due to a particular type of motion, which he called "repulsive motion." The different forms of motion (gravitational, mechanical, and repulsive), according to Davy, were linked through laws of mutual conversion that ruled all natural transformations. It was within this context that Davy described his famous experiment on melting ice with friction. Davy and other scientists viewed this experiment as convincing proof against the materiality of the caloric and in favor of wave theory. Rumford's work to disprove the corpuscular theory was dictated by scientific, technological, and commerical motivations and went back to 1786. Rumford's main theory concerned the non-materiality of the caloric, which he elaborated in a particularly explicit form in 1804 in *An Enquiry Concerning the Nature of Heat, and the Mode of Its Communications.* According to Rumford, instead of studying the caloric, it was necessary to analyze the invisible mechanical operations that take place when bodies are heated or cooled. Heat, in his opinion, was a product of the rapid oscillations produced in the aetherial fluid, vibrations that were isochronous with respect to the material particles that made up bodies.

Models of the caloric

Programs of research related to the theory of the materiality of the caloric were particularly favored in the physical sciences during this period. In a memoir of 1802 (*Quelques remarques sur la chaleur, et sur l'action des corps qui l'interceptent*), Pierre Prevost proposed a particle theory of the caloric that contained qualitative considerations of a probabilistic nature.

According to Prevost's model, the caloric was an imponderable substance made up of a great number of elementary parts moving at a very high speed in no particular direction within physical systems that were in thermic equilibrium. If a physical system were perturbed in such a way that a gradient of heat was created inside it, then a flow of igneous particles tending to eliminate the gradient and return the system to its equilibrium would be generated.

Besides development of models such as this, intense experimental activity was also conducted, especially by French researchers. In 1802, Joseph Louis Gay-Lussac observed heat dilation in gasses with remarkable precision, publishing his results in a work entitled *Recherches sur la dilatation des gaz et des vapeurs*. The *Lectures on the Elements of Chemistry*, by Joseph Black, which urged further investigation of specific heats and latent heat, came out posthumously in 1803. The early works of John Dalton (1766–1844), one of the founders of the atomist theories, also favored the corpuscular concepts. Between 1799 and 1802, Dalton carried out studies on the heat capacity of gasses, which he published in 1802. In the following years, Dalton's primary interest lay in chemical research, though he did not neglect the need to develop theories on heat. He returned to them in the pages of *A New System of Chemical Philosophy*, written in 1808.

Differential equations and models

Another line of research emerged around 1804 with the studies that Jean Baptiste Biot (1774–1862) conducted on differential equations describing the heat conductivity of solids. Jean Baptiste Fourier also did work on the same topic and in 1807 presented a memoir entitled "Recherches sur la théorie de la chaleur" to the Académie des Sciences. Fourier's memoir laid the groundwork for an alternative to Laplace's program of reducing thermology to a chapter of mechanics. In 1806, in his analysis of capillarity (see PHYSICAL SCIENCES: MECHANICS, 1800–1810), Laplace introduced the interaction between the caloric and matter, and initiated the process of mathematizing the corpuscular model of the caloric.

Laplace again intervened between 1809 and 1810 in the question raised by Biot regarding differential equations concerning heat conductivity. He attempted to apply mathematical formalism to increasingly explicit and detailed hypotheses on the structure of the caloric and the mechanism of heat radiation.

PHYSICAL SCIENCES: ELECTRICITY

Coulomb's interactions

A basis for research on electrical and magnetic phenomena was provided at the beginning of the century by Charles Augustin de Coulomb's laws on the interaction between electrical charges and between magnetic poles. These laws established that the interactions were inversely proportional to the squares of their distances and therefore provided a reasonable basis of reducing electrical and magnetic phemomena to mechanics. In his memoirs, published between 1785 and 1789, Coulomb had applied these laws to models of electrical and magnetic fluids, attributing the inverse square interaction directly to the particles of these fluids (see PHYSICAL SCIENCES, 1780–90). However, Coulomb had always maintained that he did not intend to provide an explanation of "the true causes of electricity" with these models, but "to present in the most concise way the results of calculations and experiments." Coulomb perfected his theories in 1801, in a treatise entitled *Détermination théorique et expérimentale des forces qui ramènent différentes aiguilles, aimantées à saturation, à leur méridien magnétique.*

Contact electricity

Another essential point of reference for the development of knowledge in this area was furnished by the results of Allessandro Volta's research on contact electricity. This research led to the construction of piles and batteries, as Volta himself pointed out in his famous memoir of 1800, published in Philosophical Transactions under the title "On the Electricity Excited by the Mere Contact of Conducting Substances of Different Kinds." That same year, William Nicholson used Volta's discoveries to study electrolytic decomposition. He presented his early observations on the subject in *Account of the New Electrical and Galvanic Apparatus of Sig. Alex. Volta, and Experiments Performed with the Same.*

In 1803, Vasily Vladimirovich Petrov completed his *Izvestie o gal'vanivol'tovskih opytah*, in which he reported his experiments on galvanic and voltaic phenomena. He also used the battery to study the electrolysis of various substances.

Electrochemical investigation progressed very quickly, as attested to by Humphrey Davy's Bakerian lectures of 1806 and 1809. The first of these conferences, published in 1807 under the title *On Some Chemical Agents of Electricity*, had repercussions throughout Europe. Besides presenting a critical summary of the experimental data gathered by various researchers and by the author himself, this Bakerian lecture was a serious attempt to correlate information on electrochemistry within a unitary framework and was accepted as such by physical scientists in France and England.

Oersted's concepts

During this same period, other impulses of a theoretical and experimental nature came into the debate on the nature of electrical and magnetic fluids, linking these fluids and their properties to studies in optics, thermology, and chemistry. In the case of scientists such as Hans Christian Oersted (1777–1851), this intertwining of problems was considered via the mediation of a philosophical concept of nature that was influenced by German thought and saw the ultimate origin of various phenomena in a conflict of interchangeable forces. In a book written in 1808 on the science of the general laws of nature Oersted emphasized this concept (see PHYSICAL SCIENCES: THERMOLOGY, 1830–40).

Wave and corpuscular optics

The science of optics was visibly affected by the contradictory situation that developed out of the debates on wave and corpuscular theories (see PHYSICAL SCIENCES: THERMOLOGY, 1800–1810). In 1802, Thomas Young (1773–1829) formulated a general principle, based on the wave nature of light, to explain phenomena of interference. In his memoir *An Account of Some Cases of the Production of Colours Not Hitherto Ascribed*, Young stated the principles of interference, using the example of two systems of waves moving on the surface of a lake and forced to flow together, one on top of the other, in a narrow canal. Also in 1802, Young carried out his classical experiment on interference, where sunlight was made to pass through two pinholes in an opaque screen. A wave interpretation of the results of his observations with this device enabled Young to provide the first quantitative values of the length of light waves, and to extend the wave explanation to optical phenomena in thin plates. Young's reflections, which he often expressed unclearly, drew rather strong criticisms. Even his *Course of Natural Philosophy and the Mechanical Arts* (1807) did not succeed in arousing serious scientific interest in the wave theory of light.

The major impulses for the development of optics came, in fact, from research into corpuscular theories, which went back to the hypothesis of emission contained in eighteenth-century Newtonian physics. It should be mentioned, in this respect, that, using corpuscular theories, scientists were effectively able to formulate the problem of polarization, even though they could not provide an adequate explanation for it. An explanation was finally given during the period 1810–20 by physicists using the wave theory, although this theory had to undergo profound modifications in order to provide this explanation.

In 1808, Étienne Malus (1775–1812) attacked the problem of double refraction, which the Académie had proposed as the theme of a competition. Malus's memoir, entitled *Théorie de la double réfraction de la lumière dans les substances cristallisées*, was published in 1810 and was a serious attempt to find a corpuscular explanation for polarization phenomena. According to Malus, particles of light had a sort of polarity. While the poles were mixed quite homogeneously in natural light, in light subjected to reflection and double refraction phenomena the poles lined up along privileged directions. During this same period, Pierre Simon de Laplace proposed a laborious and complex theory on double refraction that was also based on a corpuscular model of light.

Toward 1810, Jean Baptiste Biot (1774–1862) and David Brewster conducted studies on the relations between polarization and refraction, maintaining an attachment to Newtonian concepts, which were still widely preferred over those proposed by Young. Despite all the obstacles standing in the way of a satisfactory corpuscular explanation of optical phenomena, there was a general hope that these obstacles could be overcome through a mathematization that would not place the basic hypotheses in question.

CHEMICAL SCIENCES

The laws of gasses and Dalton's atomism

Some important laws on gasses and chemical combinations were stated between 1800 and 1810. In 1801, John Dalton (1766–1844) formulated the law of partial pressures for the components of a gaseous mixture. In 1803, William Henry (1775–1836) determined the effects of pressure on the solubility of gasses in liquids. In 1808, Joseph Louis Gay-Lussac (1778–1850) gave the law of combining gasses in terms of rational and simple relations of volume. This last law was particularly important, and inspired Amedeo Avogadro (1776–1856) to discover the principle which bears his name. Meanwhile, in 1803, Dalton stated the law of multiple proportions, which applied to two elements that could combine in more than one way, as well as his atomic theory, which overcame the resistance of those who feared that the concept of atoms was too metaphysical in nature. Unlike the old atomism, Dalton's theory maintained that all atoms of a given element were identical. In 1805, Dalton published the first table of atomic weights; between 1803 and 1810, he proposed a system of chemical symbology.

The discovery of new elements and the electrochemical preparation of elements

By this time, almost all the common elements had been discovered, and the less common ones began to be discovered. In 1801, Charles Hatchett (1765–1847) discovered columbium, which Heinrich Rose (1795–1864) later renamed niobium; Anders Gustaf Ekeberg (1767–1813) discovered tantalum in 1802; in 1803, cerium was discovered by Jöns Jacob Berzelius (1779–1848) and Wilhelm Hisinger (1766–1852), and independently by Martin Heinrich Klaproth; William Hyde Wollaston (1766–1828) discovered palladium in 1803 and rhodium in 1804; and Smithson Tennant (1761–1815) discovered iridium and osmium in 1804.

William Hyde Wollaston's observation (1802) of black lines in the solar spectrum opened the way to the discovery of new elements in outer space, but it was sixty years before chemists were able to use spectroscopy profitably. In 1808, Gay-Lussac and Louis-Jacques Thénard (1777–1857) obtained sodium and potassium by chemical means and prepared potassium amide for the first time.

The Voltaic battery, on the other hand, was utilized much sooner. Besides providing a stimulus to research to find out how it worked (chemical theory, as opposed to the contact theory), it gave rise to electrochemistry, a new branch of chemistry. One of the applications of electrochemistry was electrolysis, which enabled Humphrey Davy (1778–1829) to isolate potassium and sodium in 1807, and calcium, strontium, and barium in 1808. He also isolated magnesium (which Joseph Black had already recognized as an element in 1755) from magnesium oxide (*magnesia usta*). Antoine Alexander Bussy (1794–1882) is credited with having obtained it in the pure state in 1830. Gay-Lussac and Thénard also isolated boron by electrolysis during this period.

The great development of inorganic chemistry

Despite the uncertainty in which it still moved (for example, the formula for water had not yet been positively determined, and it was only between 1809 and 1810 that Gay-Lussac, Thénard, and Davy put forward the hypothesis that chlorine was an element and therefore did not contain oxygen), inorganic chemistry was more advanced than organic chemistry. Researchers in inorganic chemistry not only analyzed the compounds present in nature—calcium fluoride was discovered in tooth enamel by Domenico Lino Morichini (1773–1836) in 1802 and by Gay-Lussac in 1805, and calcium and magnesium phosphate were identified in bones by Antoine François de Fourcroy (1775–1809) and by Vauquelin in 1805—but also prepared new ones: Sodiumbicarbonate was discovered by Sigismund Friedrich Hermbstaedt (1760–1833) and independently by Valentine Rose (1762–1807) in 1801: sulphur chloride was discovered by Thomas Thomson (1775–1852) in 1803; and Sulfocyanic acid by Robert Porrett (1783–1868) in 1802.In 1806, Charles Bernard Desormes and Nicolas Clément-Desormes (1779–1841) discovered the catalytic role of nitrogen oxide in the reaction that produced sulphuric acid in lead chambers. It was partially on the basis of these results that Berzelius later formulated his theory of catalysis.

In 1809, Amadeo Avogadro introduced the concept of relativity for acids, i.e. a substance that is acid with respect to another may be alkaline with respect to a third and propounded a theory regarding the chemical affinity between any two substances based on their acidity.

Organic chemistry. Courses in chemistry

Progress in organic chemistry, on the ther hand, was almost exclusively limited to isolating new natural compounds: allantoin was isolated by Francesco Buniva (1761–1834) and Louis Nicolas Vauquelin in 1800; morphine was discovered by Armand Séguin (1767–1835) and Pierre Joseph Pelletier (1788–1842) in 1804 (published in 1814) and by Friedrich Sertürner (1783–1841) in 1805 (published in 1806); asparagine was discovered by Pierre-Jean Robiquet (1780–1840) and Vauquelin in 1806; mannite was isolated by Joseph Louis Proust in 1806; hematotoxylin was discovered by Miche Eugène Chevreul (1786–1889) in 1808; nicotine was isolated by Vauquelin in 1809; and glycyrrhizin was isolated by Robiquet in 1809.

In 1805, John Bostock (1773–1846) differentiated between the albumins in gelatine and in salivary mucus. In 1807, Thénard distinguished between ethers and compound ethers (esters). Pierre François Boullay (1777–1869) did research on the same subject between 1801 and 1810, preparing various alcohol esters with phosphoric, acetic, and arsenic acids.

An important observation for the future development of biological chemistry was made by Berzelius (1807) when he discovered that meat juices contained lactic acid.

Four important texts were published during this period: *A System of Chemistry* (1802), by Thomas Thomson; *A New System of Chemical Philosophy* (1808–27), by Dalton; *Lehbuch der Chemie* (1808–18), by Berzelius; *Système des connaissances chimiques et de leurs applications aux phénomènes de la nature et de l'art*, by Antoine François de Fourcroy.

The invention of the oxy-hydrogen blow-pipe by Robert Hare (1781–1858) in 1801 also deserves mention.

LIFE SCIENCES

Histology

Marie-François-Xavier Bichat (1771–1802), the dominating figure at the beginning of the century, died at the age of 31, after having published his *Traité des membranes*, *Recherches physiologiques sur la vie et la mort*, and *Anatomie générale*, works that contributed some very stimulating ideas to morphology and the theory of life. A vitalist who differed with Stahl, a morphologist who was distrustful of the microscope, a supporter of "function" who was disinterested in the chemical bases of life, Bichat applied the tradition of critical sensism of Condillac (1715–90) and Cabanis (1757–1808), going beyond sterile antitheses and promoting a successful system of instruction that lay between the structuralist and physiological currents. His general anatomy, founded on the identification of twenty-one "tissue systems," provided a solid basis for the anatomy of organs and systems, which was compromised by humoralism and arbitrary descriptions of presumed elementary microscopic entities. The histology and cytology of the following decades would have been impossible without these premises. Bichat and Karl Friedrich Burdach (1776–1847) were the first to use the term biology to designate the science of the general properties of organisms. Its use was sanctioned in 1802 by Gottfried Reinhold Treviranus (1776–1837), with publication of his *Biologie oder die Philosophie der lebenden Natur für Naturforscher und Aerzte*.

Evolution in relation to the environment

In the natural sciences, the century opened with a revival of George Louis de Buffon's and Erasmus Darwin's evolutionary hypothesis by Jean Baptiste de Lamarck (1744–1829). In his *Discours d'ouverture du cours de zoologie de l'an VIII* (1800), Lamarck claimed that spontaneous generation was actually possible and affirmed that living individuals, not species, existed, in that Carolus Linnaeus's presumed fixed species were subject to variations as the environment changed. His *Philosophie zoologique* of 1809 specified that the evolution of organisms was caused by an "inner feeling," or a sort of perception of the problems that the habitat posed to the living being, and consequently to the organism, which pushed the organic fluids toward the parts of the body where they were needed.

Experimentation

The French school of physiology preceded the German school, partly because the fertile ideas proposed by Antoine Laurent Lavoisier with his work on the chemical origins of animal heat (for example, in his *Mémoire sur la combustion en général* of 1777 and his *Expériences sur la respiration des animaux* of 1780) were developed in France. The foremost figure was François Magendie (1783–1855), whose *Précis élémentaire de physiologie* came out in 1816 and was published a number of times with important additions. Magendie's authoritativeness was confirmed by his admission to the Académie des Sciences in 1821 and to the College of France in 1831. The asystematic nature of his research, which was later corrected by Claude Bernard (1813–78), manifested itself in his toxicological work on the poison extracted from nux vomica (isolated by Pierre Joseph Pelletier (1788–1842) in 1818 and given the name strychnine.

In Germany, the *Recherches sur le système nerveux in général et sur celui du cervau en particulier* (1808), by Franz Joseph Gall, reiterated the phrenologists' that the study of cerebral anatomy and physiology had to come before any statements of a psychological nature. Among those who escaped "nature philosophy" was the naturalist and physiologist Karl Asmund Rudolphi, appointed to the chair of physiology in Berlin in 1810. Among his students was Johannes Peter Müller (1801–58), the leader of German biology during this century.

In plant physiology, the most important work was the *Recherches chimiques sur la végétation* (1804), by Nicholas Théodore de Saussure (1767–1845). The controversy of the previous century between the botanical vitalism of Johann Gottschalk Wallerius (1709–85) and the physicistic approach of authors such as Jean Sénebier and Jan Ingenhousz, was not yet resolved: de Saussure referred to the scholars of the second group and demonstrated that the nutritive process in plants took place through the assimilation and successive transformation of inorganic substances contained in humus, as well as through the decomposition of carbon gas and water.

Semiotics

René Théophile Laënnec (1781–1826), who worked for a short time at Nantes, then in the great Parisian hospitals, and was a teacher at the College of France, formulated and applied the principle of objective, systematic, and logical observation in clinical practice. Semiotics took form as the intermediary between clinical medicine and pathological anatomy; it was founded on the conviction that alterations in organs were the most frequent and observable indicators of localized diseases. Disease syndromes should not be defined by "symptoms," which were always necessarily somewhat subjective, but by certain and, as far as possible, physical signs. In his *Traité de l'auscultation médiate et les maladies des poumons et du coeur* (1819), Laënnec explained the application of his diagnostic invention, i.e. auscultation using a stethoscope, in his research on thoracic diseases.

The comparative method

The five volumes of the *Leçons d'anatomie comparée* by Georges Cuvier (1769–1832), which extended the comparison of analogies and structural homologies to the entire world of zoology and constituted the first complete work in the field, came out in 1800–05. Cuvier started from organic, but not vitalistic, premises, and proposed two basic criteria for anatomy: the correlation of the parts of an organism to each other—in both a functional and a formal sense—and the subordination of characteristics. Above and beyond these two criteria stood the principle of the conditions of existence: every organism is connected to the environment in which it lives. Thus Cuvier's zoological classification, unlike Antoine-Laurent de Jussieu's botanical system, had a predominantly physiological bent. In 1807, Cuvier placed the "essence of the animal" in the nervous system. A debate arose between Cuvier and Lamarck, who were colleagues at the Institut de France. Their differences were fueled by both personal conflicts and the contrast between Cuvier's structuralist position and Lamarck's transformist ideas.

In 1806, André Marie Duméril (1774–1860) published his *Zoologie analytique ou méthode naturelle de classification des animaux.* Duméril, named professor of comparative anatomy at the

Museum of Natural History in Paris in 1802, had collaborated in the writing of the first two volumes of Cuvier's *Leçons d'anatomie comparée*, and later made notable contributions to enlarging the collections belonging to the museum's chair of reptile and fish zoology, especially the herpetology collection, which became one of the richest in the world and gave rise to the creation of the first herpatarium in Europe. From 1834 to 1854, Duméril published his *Erpétologie générale*, which described 1,311 species of reptiles; in 1856 he published his *Ichthyologie analytique ou essai d'une classification naturelle des poissons à l'aide de tableaux synoptiques*, and four years later his *Entomologie analytique.*

Two scientific currents developed in Germany. At first tending to oppose each other, they later became partially integrated. The first referred back to "nature philosophy," and the second was created by Alexander von Humboldt (1769–1859). "Nature philosophy" inspired a prolific debate on the relationship between embryology and comparative anatomy: i.e. whether embryology could be considered "a transitory comparative anatomy"—an expression coined by the French embryologist Antoine Serres (1786–1868)—or whether the morphological characteristics of the adult being were already visible in the early phases of embryonic development and presided over its further development, as Karl Ernst von Baer (1792–1876) maintained. Von Humboldt became well-known at the beginning of the century with the first publication of his scientific findings from his long journeys (1799–1804) in South and Central America. His *Essai sur la géographie des plantes accompagné d'un tableau physique des régions équinoxiales*, dedicated to Johann Wolfgang von Goethe, came out in 1805. His *Ansichten der Natur, mit wissenschaftlichen Erläuterungen* followed in 1808. A unitary but diversified concept of the physical universe, where organic and inorganic nature corresponded to each other and together determined the personality of various natural environments, emerged in von Humboldt's work. He thus provided a very significant middle point, both historically and conceptually, between Goethe's ideas and late eighteenth century French science (especially Antoine Laurent Lavoisier's chemistry), posing an obstacle to the "nature philosophers'" often-negative influence on the new German science. Louis Tredern's dissertation on the development of the beak, the bile duct, the intestine, and the limbs in the chick embryo, with numerous illustrations, was published in 1808.

Speculative biology provided another germ for morphological reflection during this period with the vertebral theory of the skull. Proposed by Goethe (1749–1832) and Lorenz Oken (1779–1851), it was developed by Richard Owen in his *Archetype and Homologies of the Vertebrate Skeleton* (1848) and later confuted by Thomas Henry Huxley in a work on the theory of the skull in vertebrates (1858).

Homeopathy

The first edition of the *Organon der rationellen Heilkunst*, by the German physician Samuel Hahnemann (1755–1843), the first complete exposition of the homeopathic method, was published in 1810. The work was a collection of rules for understanding diseases, patients, how remedies worked, and the choices and criteria for administering them. The title of the volume was changed for the second edition; it was no longer a text on rational medicine, but rather a work on the art of healing. In the *Organon*, Hahnemann, who had studied regular medicine, had many years of therapeutic practice, and had consistently applied himself to the study of chemistry, took up, developed, and organized the law of similitude, already discovered by Hippocrates, rediscovered in the Middle Ages, and taken up by Paracelsus and his followers. Hahnemann established a close link between the disorders that a patient displayed (disease syndrome) and disorders that were experimentally produced by administering a given animal, vegetable or mineral substance (experimental syndrome). *Similia similibus curantur* was the slogan of homeopathy, which defined itself as a therapy that consisted in giving the patient small doses of a substance that, administered in large doses to a healthy organism, produced symptoms analagous to those observed. Another characteristic of the method was specified in the definition: the medication was to be administered in very small (infinitesimal) doses, obtained through successive dilutions. The third major characteristic of homeopathy was the inseparable association between the disease and the patient: the diseased state observed in an individual appeared in manifestations that depended on the particular reactions of that particular individual, so that the homeopathic remedy also had to be individual.

EARTH SCIENCES

The decline of Neptunism

The most important change in geological theories during this period was the conversion of Abraham Gottlob Werner's main and most direct disciples to Plutonism. In 1802, while visiting the volcanoes of the Auvergne, Christian Leopold von Buch reached the conclusion that the rocks of those extinct volcanoes, both basalt and trachyte lavas (domite), were of eruptive origin.

Jean François d'Aubisson de Voisins adhered to Plutonist explanations, as did another friend of Werner's, Christian Samuel Weiss, who informed his teacher of his observations in southern France. Werner, however, was not willing to change his theories without first personally checking out the facts that seemed to contradict them.

Alexander von Humboldt returned from his voyages to South America, and studied volcanoes and earthquakes in particular, using Plutonist interpretations to explain them. Meanwhile, in Edinburgh, the home of James Hutton, founder of the Plutonist school, Robert Jameson (1744–1854), another of Werner's pupils, established the Wernerian Natural History Society in 1808. However, he soon converted to Plutonism, following his discussions with John Playfair (1748–1819), a professor of natural philosophy in the same city and author of an *Illustrations of the Huttonian Theory of the Earth* (1802). Jameson is remembered for not having exaggerated the importance of volcanic phenomena and for having recognized that glacial activity was responsible for having carried large erratic masses far from their rocks of origin. (Von Buch, for example, maintained that the blocks of Alpine rock found on top of the Jura mountains had arrived there as a result of tremendous explosions.) Jameson's ideas were also supported by the experiments of James Hall (1761–1832), who obtained crystallized masses by melting and cooling various volcanic rocks.

The Wernerian system was still used by some scholars, however: William Maclure's geological map and illustrative memoir of the United States were products of the application of Neptunist ideas to North America. Even in studies of volcanic regions, there were some convinced Neptunists, such as Ermenegildo Pini, who also studied the Phlegraean region near Naples and who entered into a controversy with Scipione Breislak. To his credit, Pini was the first to explain that the holes made by pholas in the columns of the temple of Serapide at Pozzuoli must have been the result of variations in the ground level, which caused the ruins of the temple to become partially submerged and then to reemerge from the water.

The Geological Society of London was founded as an independent institution in 1807. Its goal was to bring together English geologists and promote scientific debates, at times via publications.

Systematic studies of stratigraphy

By this time, remarkably detailed and accurate scientific contributions were being developed for various regions. Particularly important were the *Corsa pel bacino del Rodano* (1806), by Giuseppe Marzari-Pencati (1779–1836), another Neptunist who converted to Plutonism, the stratigraphic system of the Paris basin proposed by Cuvier and Alexandre Brongniart (1770–1847) in 1808, which used fossils for stratigraphic correlations, completed in 1810 with a geological map and profiles, and the studies made in Belgium (1808) by Jean-Baptist de Omalius D'Halloy. Various scholars, including Brongniart from different parts of France (1810), particularly emphasized the discovery of deposits that had formed in fresh water.

Invertebrate paleontology and paleobotany

The possibility of using fossils for stratigraphic correlations, which was generally recognized by this time, gave rise to a lively flourishing of paleontological research. The works of Jean-Baptiste de Lamarck (1744–1829) on the tertiary shells of the Seine basin (1802–09) deserve particular mention. Elsewhere we have noted the evolutionary theories of this scholar, which contrasted with those of Cuvier, who was a convinced supporter of the fixity of species. Lamarck found confirmation for his evolutionary ideas in his paleontological discoveries. We should also remember that a number of scholars studied plant fossils: for example, in 1804 Ernst Friedrich von Schlotheim (1764–1832) llustrated imprints of plants found in Germany.

On the origin of mountain chains

His observations of the tremendous effects of volcanic forces and of the widespread existence of volcanoes induced von Buch to formulate the hypothesis that the origins of mountains could be attributed to the activity of heat inside the earth. Heated vapors melted deep masses of rock and caused them to intrude into the overlying rock, lifting it up into a dome. If the molten mass reached the surface and solidified, maintaining its original form, a particular type of hill without a crater was formed; an example was the Puy-de-Dôme in the Auvergne. If, on the other hand, there was an explosion, the top part collapsed and a volcano with a crater was formed. Von Buch expressed his ideas on the subject in his *Geognostische Beobachtungen auf Reisen durch Deutschland und Italien*, which appeared in 1809, and developed them into his great "crater upheaval" theory that was so important in later geological thought. Several authors once again proposed the theory in a more modern form between 1950 and 1970.

Practical applications of geology

Geological studies began to be utilized for practical ends in areas other than mining research, with which they had always been associated and from which they had actually originated.

Works directed by William Smith in England, such as the draining of the Prisley Bog in 1801 and the reactivation of the spring at Bath, which had gone dry because of a crack, which Smith blocked, deserve particular mention. This was the beginning of hydrogeology. (It should be noted that Lamarck's work of 1802 entitled *Hydrogéologie* was really a treatise on dynamic geomorphology.)

In 1807, Hans Conrad Escher of Zurich engineered a correction in the course of the Linth in Glarona canton; when the work was a success, the Diet honored him with the title "von der Linth" (of the Linth).

Geological theories

In 1802, Playfair (1748–1819) published his *Illustration of the Huttonian Theory of the Earth*, spreading the theories of Hutton and even converting Jameson (1774–1803), a follower of Neptunism as well as the founder of the Wernerian Natural History Society in Edinburgh and author of a work entitled *Mineralogy of the Scottish Isles* (1800), to them. In 1802, G.B. Alberto Fortis (1741–1803) published his *Mémoires pour servir à l'histoire naturelle et principalement à l'oryctographie de l'Italie*, which brought together the observations he had made mainly in Dalmatia, around Verona, Trieste, and Naples, and on Etna and Vesuvius.

Oceanographic voyages

Several important oceanographic voyages took place during this period: William Scoresby (1789–1857) led a polar expedition to explore zones of the Arctic Ocean, making magnetic, meteorological, and oceanographic measurements (investigating in particular the make-up of ice), as well as biological observations of both plankton and whales.

François Peron (1775–1810) commanded a French fleet that explored the austral seas and Oceania between 1800 and 1804 and gathered a vast number of zoological specimens. Peron included illustrations of the collections in his report on the voyage, published upon his return. Between 1803 and 1806, Adam Johann von Krusenstern (1770–1846) guided a Russian expedition to the Arctic, giving special attention to the seas of Siberia. Among other things, measurements of deep-water temperatures were made (using six thermometers).

Gerstner's wave theory

In the area of theoretical and experimental research, the wave theory (1802) of Franz Joseph von Gerstner (1756–1832) deserves particular mention. Von Gerstner became the first to give a mathematical explanation of wave profile with his very original model, which started from orbital motion (completely different from wave motion). The motion of the ocean's surface was clearly distinguished from the motion of the particles of water, in agreement with the ideas of scientists studying hydro-

dynamics at that time, especially Pierre Simon de Laplace. Modern views have considerably modified Gerstner's ideas; however, he deserves recognition as the first to have outlined the problem and consequently to have provided a stimulus for further research.

The density of water

In 1805, Thomas Charles Hope (1766–1844) conducted experiments on the expansion of water and on phenomena of congealment (it was still believed that the maximum density of sea water occurred at around four degrees centigrade, the same as for fresh water). Hope was a professor at the University of Edinburgh, where he continued the geology lessons, initiated by his predecessor Joseph Black (1728–99), within the course in chemistry.

Beaufort's wind scale

In 1806, Francis Beaufort (1774–1857), who had participated in the naval campaigns in Spain and who later carried out very delicate assignments in America and in the Aegean Sea (such as surveillance against pirates), proposed his empirical scale for evaluating wind speed. The scale, divided into twelve units, proved very useful. With some modifications, it is still in use today. Beaufort's scale rated wind speed on the basis of its effects, originally on the sails of ships and later on more common objects that were also affected by the wind.

APPLIED SCIENCES

The effects of the continental blockade

The Napoleonic wars contributed decisively to increasing the demand for raw pig iron and iron. The production of first-casting pig iron, which was also used for manufacturing guns, rose from 150,000 to 350,000 tons in Great Britain, while iron, widely used for producing agricultural building and military implements, reached 250,000 tons in 1806. The Auserlitz bridge was built of cast iron in 1804, and in 1806, the architect Beranger covered the hall of the Trade Exchange in Paris with an iron and cast iron dome, forty meters in diameter.

Continental Europe also developed a steel industry in 1802, thanks to the Swiss industrialist Johann Conrad Fischer, who developed a method similar to Benjamin Huntsman's and made it known in Austria and in France. The English had kept Huntsman's process secret. The continental blockade, mutually applied by France and England in 1806, stimulated the creation of national industries on the continent, aimed mainly at filling the gaps created by the abrupt halt of importations (in France, which remained without a supply of sugar cane, a procedure for extracting sugar from beets was revived and perfected). The fall of Napoleon forced the governments of continental Europe to adopt rigid protectionist policies in order to reduce the effects of English competition. The result was a decrease in the application of new modernization techniques.

Gas: a more economical source of illumination

Experiments conducted to obtain good coke showed that carbon produced a flammable gas during distillation. As early as 1797, Philippe Lebon, a teacher of mechanical engineering at the École des Ponts et Chaussées, had begun studying the gas produced by heating wood; in 1790, he had already patented his "thermolamp," a retort connected with tubes to fan-shaped burners. Using Lebon's findings and the new circular burner for liquids invented by Pierre Ami Argand (1750–1803). William Murdoch succeeded in installing a lighting system in his own house. He had already set up a gas lighting system at the Boulton & Watt foundry in 1798.

An interest in gas lighting began to develop among industrialists during this decade, since it could be used to extend work time into the evening and night hours. In 1807, a large factory in Manchester adopted the new system of lighting, and it was shown that the use of 900 gas burners resulted in a savings of 2,350 pounds sterling over the expense of lighting by candles.

Ground and ocean traffic and steam

James Watt's steam engine was brought to a high level of structural perfection combined with good yield at the beginning of the new century. It is estimated that a maximum of around fifty horsepower was produced. The drive to increase the engine's power led experimenters to overcome Watt's distrust of high-pressure steam. Independently of one another, Oliver Evans (1755–1819) and Richard Trevithick (1771–1833) built two different types of high-pressure engines, for extracting ore, the former in Philadelphia in 1801 (patented in 1804), and the latter in Cornwall in 1800. The use of high-pressure steam resulted in more horsepower per unit of weight. The new type of engine did not replace Watt's, but its smaller size made it attractive for use in means of transport. In 1801 and 1803, Trevithick experimented with steam carriages on roads, introducing the smokestack to increase draft. In 1804 he designed a locomotive that would pull a load of ten tons over a railway approximately fifteen kilometers long. The locomotive completed the run at a speed of eight kilometers per hour, demonstrating that there was enough friction between the iron wheels and the rails to make steam traction possible.

Steamboats were soon used as towboats for transporting passengers and goods. In 1801, commissioned by the war ministry, William Symington (1763–1831) built a motor especially designed to turn the paddle wheels of a boat. It was installed on the towboat *Charlotte Dundas*. Robert Fulton (1765–1815), who had watched the *Charlotte Dundas* being tested, used a Boulton & Watt engine on the steamship Clermont, which in 1807 travelled 240 kilometers up the Hudson River, on the New York-Albany route, in thirty-two hours.

Interchangeable parts and mass-produced machines

The rapid development of the steam engine and the construction of machines were made possible by the existence of instruments and procedures for manufacturing accurately finished machine parts. The technicians of the Boulton & Watt factories and a number of engineers of the school of Joseph Bramah (1748–1814) and Henry Maudslay (1771–1831) made decisive contributions in this area. In 1797, Maudslay invented a screw-tapping lathe, which was built three years later. Maudslay also devised strict standards for parts such as screws, bolts, and nuts, which made it possible to mass-produce them and

rendered them interchangeable. A number of machine tools, made completely out of iron, were designed during this decade, beginning with the drilling and broaching machines of Marc Isambard Brunel (1769–1849), Eli Whitney's milling machine of 1802, and Bramah's mechanical planing machine, also invented in 1802.

The most important innovation in machines for the textile industry was the silk loom, designed by Joseph Marie Jacquard (1752–1834) in 1801, which solved the problem of selecting the warp (up to then done by an apprentice). By 1820, 11,000 of these looms were operating in France.

In 1801, Robert Hare (1781–1858) introduced the use of the oxyacetylene torch.

1810–1820

THOUGHT AND SCIENTIFIC LIFE

The foundation of the University of Berlin: a new ideal of scientific knowledge

The fall of Napoleon (Leipzig, October, 1813; Waterloo, June, 1815), the *Congress of Vienna*, the Holy Alliance (September, 1815), and the policies of the Restoration had serious repercussions on scientific and cultural life. In every country, either for reasons of internal politics or because of direct or indirect pressure from its allies (the cases for France and Italy, respectively), controls over culture, science, and the arts were very strict, even though an anachronistic return to the culture of the *ancien régime*, longed for by many, was not possible.

The situation of the Kingdom of Prussia was particularly interesting. Having emerged from the long struggle against Napoleon in military glory, but with considerable damage as well, it was still founded on the personal union of two unconnected territories (Eastern Prussia with its center at Königsberg and Western Prussia with its center in Berlin) and was torn by numerous political and social conflicts. But Prussia was slowly becoming industrialized. While the bolder economists were drawing up the first plans for joint customs agreements with the other Germanic states, Prussia became the potential rallying point for the German petty states and a natural opponent of the Hapsburg Empire. Definitely an essential component of this complicated evolution was the educational policy promoted by Frederick William III. In 1810, the University of Berlin was born out of a project formulated by Wilhelm von Humboldt, the director of the cultural and educational section of the Ministry of the Interior. Philosophers such as Schleiermacher and Fichte, who was also the first rector, were called to teach there. Foreseen by Kant and theorized by some of the major exponents of romantic idealism, the Berlin university was founded on a new ideal of scientific knowledge, characterized by a recognition of the supreme principles of all disciplines, the methodological consciousness of the scholar, and teaching, in the Fichtian manner, viewed as the mission of the teacher. It was symptomatic that physics and biology were closely modelled after the humanist disciplines, which, at least initially, were considered more prestigious and culturally dignified in Berlin. Two decades passed before German "nature philosophy" and the agnosticism of researchers such as Johannes Peter Müller and Heinrich Gustav

Magnus opened the way to the generation of Helmholtz, Du Bois-Reymond, and Virchow. The authoritative mediation between science and humanistic culture provided by Alexander von Humboldt (1769–1859) (Wilhelm's brother) and his intervention from his influential position as royal advisor in favor of the young scientists, proved essential during these early years of the university's existence.

Humboldt's project embraced other institutions of higher culture as well, especially the Academy, which was seen as the seat of pure research and "a community with the task of submitting the work of each individual to the judgment of all." The Academy was renovated in 1812 and rebaptized the Preussiche Akademie der Wissenschaften.

The development of Hegelian philosophy with regard to scientific thought

Hegel occupied an important position in the newly reformed German academic world. Although he had been a convinced admirer of Napoleon ("spirit on horseback," and "the Sunday of life" at the time of the *Phänomenologie*), Hegel quickly became aware of the new situation created in Germany by the fall of the Napoleonic Empire and the ephemeral state structures connected with it. From 1812 to 1816, while staying in Nuremburg (he left Jena at the time of Napoleon's victory), he composed his *Wissenschaft der Logik*, by then far removed from enthusiasm for the French Revolution and at the same time opposed to involvement with the romantic movement. The *Encyklopädie der philosophischen Wissenschaften im Grundrisse* of 1817 was a product of Hegel's later stay in Heidelberg, where he was a university professor. In it, he gave considerable space to the philosophy of nature, for which many critics harshly reproached him, and which scientists such as Hermann von Helmholtz (1821–94) considered the fruit of a pretension to construct nature a priori, building it dialectically from reason. This was the source of the accusations made against Hegel of having replaced the image of the world given by the natural sciences with a purely arbitrary construction and, even more seriously, of having, at least indirectly, hindered the progress of science itself. While one should take into account the fact that Hegel often criticized some of the institutional aspects of science as it was usually taught in Germany (as well as contemporary avant-garde science), it is necessary to emphasize Hegel's purely rationalistic aspiration to a unity of knowledge, which had little to do with the contempt with which certain of his followers regarded scientific thought. In 1818 Hegel was called to the University of Berlin, where, under the protection of the major authorities of the Prussian state, he soon became one of the most authoritative scholars of the epoch.

The rebirth of the academies. French thought during the Restoration

Changes in the academic world did not occur only in Prussia. For example, the universities of Gand and Liege in Belgium were inaugurated in 1815, and the University of St. Petersburg in Russia was established in 1819. One peculiarity of French cultural development, in this area as well, ways the compromise between ultra-legitimist tendencies and the innovative currents that were the heritage of the French Revolution and Napoleonic policy. The radical cancellation of the revolutionary and Napo-

leonic past created profound imbalances everywhere, from life in the countryside to the urban situation. This explains why, in cultural as well as in political and administrative life, the cadres that had been formed under Napoleon or even had a Jacobin past often remained in their positions. The roles played by Pierre Simon de Laplace (made a peer by Louis XVIII in 1816) and Jean Baptiste Fourier are good examples. But while efforts were being made to effect policies that were as open and tolerant as possible toward the individual, policies concerning science and the institutions had partly to give in to legitimist demands and partly to favor, wherever possible, a tacit compromise in order to reduce ideological conflict as much as possible. Thus a royal decision of March, 1816, reinstituted the Académie Française and the Académie des Sciences, restoring to the former its privileges and priority of rank. A number of research centers began to complain of a lack of funds. The École Polytechnique, closed in April of 1816 following student demonstrations, was reopened a few months later (September, 1816), though restructured with a new set of rules that eliminated any residue of the Enlightenment spirit. It now provided a prototype for specialization in scientific research, completely detached from philosophical problems. Tendencies toward specialization had already surfaced during the preceding century; in fact, in more than one case powerful trends toward the autonomous development of given scientific disciplines had appeared. But for the thinkers of the Age of Enlightenment and for those of the Napoleonic era who still aspired to the Enlightenment ideal of an overall rational concept of nature and humanity and perceived the need for a political and cultural battle in order to gain it, specialization had not led to rigid separations between the various disciplines, nor to the "theoretical neutrality" of science that researchers (not only in France) often invoked from this time on.

The movement of the ideologists had lost all its weight in French cultural life: Destutt de Tracy was forced to publish his *Commentaire sur l'"Esprit des lois" de Montesquieu* in an English translation in the United States. After the fall of Napoleon, the movement futilely awaited a return to liberalism in France, and, faced with the triumph of reactionary forces, could do nothing more than once again shut itself into a position of sterile opposition. The major philosopher of the ideologist circle, Marie François Maine de Biran (1766–1824), left the current in order to reestablish Catholic spiritualism in France. His work took on the aspect of a refined investigation of the human soul.

Claude Henri de Saint-Simon (1760–1825) emerged in a completely different environment. He supported a utopian renewal of society, no longer based on political revolution, as the Enlightenment thinkers had envisioned it, but on a radical reorganization of production and the distribution of wealth. Saint-Simon's movement later changed and attempted to generate a new Christianity (as a matter of fact, the title of his last work was *Nouveau christianisme*, published in 1825).

The spread of romanticism in Europe

The influence of spiritualist currents on militant scientists was no more than a sign of the spread of romanticism (often critical, if not of science itself, then of any form of rationalistic philosophical reflection that made concrete reference to the sciences) from Germany into the rest of Europe. The extraordinary success of the short work *De l'Allemagne* (1810), by Anne-Louise-Germaine Necker de Stal-Holstein, known as Madame de Staël, was certainly indicative. Romanticism drew upon both the disillusionment of many revolutionaries and the sense of political frustration that spread along with the Restoration, especially in intellectual circles. In reevaluating the past, the romantic movement was open to a return to tradition: for example, historians began to judge the Middle Ages and consequently the historical role of religion, positively. Often the heir of certain demands that had already existed during the Age of Enlightenment (for example in Rousseau), romanticism adopted a critical (with respect to the institutions) and fundamentally democratic tradition that led many intellectuals to oppose the policies of the Holy Alliance, which they considered carried out "over the heads of the people" and, more generally, to contest the official culture of the Restoration.

English thought: Bentham, Ricardo

In England, a leader in simplifying its productive institutions, the scene was extremely lively on a number of levels, from socio-economic ideas to philosophical, scientific, and political thought. The English system had not experienced the direct impact of Napoleon. England's domination of the seas had enabled its economy to survive the interruption of trade relations with the continent; the effects of technological innovation had transformed the means of production; and the class struggle was much more advanced than on the continent. England was, therefore, on a cultural level, the depository of the great tradition of classical economy. Of course there were a number of cultural experiences of a romantic stamp there, too, but the so-called utilitarian movement escaped the wave of romanticism, and instead remained closer to eighteenth-century thought. The major representative of utilitarianism was Jeremy Bentham (1748–1832), who published *A Table of the Springs of Action* in 1815. Despite the harsh attacks launched against it by the romantics, the utilitarian current succeeded in exerting a strong influence over ethical and political thought in England and became one of the pillars of English liberalism. In the area of political economy, David Ricardo's *The Principles of Political Economy and Taxation* came out in 1817.

MATHEMATICAL SCIENCES

The classical theory of probability

Pierre Simon de Laplace's treatise *Théorie analytique des probabilités* provided a formulation of the theory of probabilities that was in many ways definitive. For the first time, the calculus of probabilities was given the satisfactory analytical treatment that it had long been lacking, even to the point of providing the opportunity for the development of new concepts in analysis (such as the important theory of generating functions, contained in this treatise). Two years later, Laplace gave a complete exposition of the conceptual principles of the so-called classical or a priori doctrine of probability in his *Essai philosophique sur les probabilités*. With these premises, the theory of probability was able to make enormous contributions to physics (in the second half of the century), even though for sometime precedence was still given to its applications in the human sciences.

As late as 1837, in his *Recherches sur la probabilité des jugements*, Siméon Denis Poisson investigated the possibilities of applying the calculus of probabilities to juridical and political questions.

Meanwhile, the first signs of new currents in mathematical research began to appear. Some of them were still extrinsic and of a practical nature, such as the Analytical Society, founded in Cambridge in 1812 by George Peacock (1791–1858), Charles Babbage (1792–1871), and John Frederick Herschel (1792–1871) for the purpose of introducing Leibniz's notation and techniques of infinitesimal calculus into Great Britain. The work of this society was successful, partly because of the important contribution made to the cause by illustrious and influential scientists such as William Whewell (1794–1866) and George Biddel Airy (1801–92). Other signs were of a conceptual nature. In 1811, for example, Jean Baptiste Fourier (1768–1830) provided the first rigorous definition of the concept of convergence of a series, a prelude to the work of conceptual refinement and increased logical rigor that began a short time later with Cauchy and continued for the entire century.

Cauchy and the need for rigor in analysis

Augustin Louis Cauchy (1789–1857) began publishing his first significant contributions during these years; after an important memoir on algebra (1812), in which he laid the foundations for the theory of substitutions and determinants, he read his fundamental "Mémoire sur les intégrales définies" at the Institut de France (1814, published in 1825). This memoir marked the real beginning of his research in the field of integration and, above all, laid the basis for the theory of functions of a complex variable and their integrals to which he made essential contributions during later years as well. His work as an analyst, however, was important not only because of the numerous technical findings he made, but also because of his program of maintaining the greatest possible conceptual and deductive rigor in elaborating theories and demonstrations to which he scrupulously adhered. In this sense, he can be considered the initiator of the concept of the need for rigor in analysis, which led to there being practically no nineteenth-century mathematicians who did not feel committed to furthering conceptual analysis and increased logical rigor while pursuing more properly technical research in the discipline. This resulted, as we shall see, in a great number of concepts gradually being defined in detail. The impulse culminated at the end of the century in the great problem of research into the fundamentals of mathematics.

Returning specifically to infinitesimal analysis, we should note that (as its name states) it had long been focused around the poorly defined concept of the infinitesimal, approximately conceived of as a sort of infinitely small quantity, different from zero but in many cases practically, and even theoretically, negligible. For several decades, analysis had been able to develop even in the absence of an adequate definition of this concept, but between the end of the eighteenth and the beginning of the nineteenth centuries, a number of technical difficulties arose, especially within the theory of series, that pointed up the impossibility of proceeding without more accurate definitions of the concepts. These were supplied, almost all at once, by Cauchy, when he advanced and carried out his program to reduce all the fundamental concepts of analysis to a single cardinal one, i.e. the concept of limits. It is true that other mathematicians, such as Jean le Rond d'Alembert, Simon-Antoine-Jean Lhuillier, and Sylvestre-François Lacroix, had already used the notion fairly extensively in their work; however, they had not established its central, "foundational" role in analysis. Even more importantly, they had not established the necessary and sufficient conditions for the existence of the limit itself (which Cauchy did). Furthermore, the idea of limits was always linked more or less directly to representations or intuitions of a geometrical or mechanical nature in the works of the earlier authors, while Cauchy gave it a totally abstract and formal definition (the same that later remained definitive in manuals on analysis). On the basis of Cauchy's definition, not only could one do without references to geometry or physics, but other concepts such as those of derivative, differential, integral, sum of a series, continuity of a function, and infinitesimal, could be defined as well.

Cauchy recorded this enormous task of conceptual clarification, prepared through reflections developed during these years, in three famous treatises published in the following decade: the *Cours d'analyse de l'École Royale Polytechnique* (1821), the *Résumé des leçons données à l'École Polytechnique sur le calcul infinitésimal-Analyse algébrique* (1823), and the *Leçons sur le calcul différentiel et le calcul intégral* (1829).

We must, however, remember that this same need for rigor was brilliantly resolved ahead of its time, in this very decade, by Bernhard Bolzano (1781–1848). In 1817, Bolzano rigorously formulated the notions of limit and continuity of a function, and demonstrated what is today called the Bolzano-Weierstrass theorem, although, at the time, no one was aware of his achievements. Bolzano's research remained unknown until more than half a century later, when it was rediscovered by Hermann Hankel (1839–73); but by then his findings (like the theorem just mentioned) had already been made independently by other mathematicians.

ASTRONOMY

Bessel's Fundamenta astronomiae

After he had finished editing James Bradley's observations, publication of which was completed in 1805, Friedrich Wilhelm Bessel sent to press his *Fundamenta astronomiae* (1818), an outline of new and higher standards for *astronomical observations* and instrument building. He introduced a calculation for correcting errors in instruments and observations, applying the *theory of errors* to the second case.

In 1830, Bessel began publishing annual tables (the *Königsberg Tables*) that made it easier to correct observational errors. These tables became a consistent part of all astronomical annuals. Bessel's works stimulated research in the theory of errors, and also found widespread applications in geodetic and gravimetric measurements (see PHYSICAL SCIENCES: MECHANICS, 1820–1830).

Fraunhofer's lines

In 1814–15, Joseph von Fraunhofer (1787–1826), an instrument builder and a student of Pierre Louis Guinand, examined the spectrum produced by sunlight passing through a prism with the telescope of a theodolite and detected the presence of the characteristic dark lines that now bear his name. In reality, William Hyde Wollaston (1766–1828) had already observed

these lines and described them in the Philosophical Transactions of the Royal Society in 1802; however, Fraunhofer succeeded in examining them more closely and determined the positions of approximately seven hundred of them (he designated the major ones, beginning from the red end, with the letters of the alphabet from A to H).

He also demonstrated that they always maintained the same arrangement with respect to each other and concluded that the sunlight must be absorbed by some unknown substance where the dark lines appeared. His hypothesis was later proved to be correct, when it was found that a gasseous substance placed in front of a continuous spectrum of light will absorb light at exactly the same frequences it emits as spectral lines when it is reexcited. Von Fraunhofer also spectroscopically analyzed the light coming from several heavenly bodies, pointing out similarities and differences in a comparison of Venus, the moon, and the stars with the sun.

Astronomical spectroscopy

From the very beginning, the spectroscope was used as an instrument for astronomical investigation: the improvements that Gustave-Robert Kirchoff (1824–1887) made in it around 1860 rendered it indispensable to astronomy, as observational techniques were radically transformed through its use. This marked an enormously important turn of events in the development of astronomy: the birth of astrophysics.

Astronomical tables and lunar motion

The impression made by Pierre Simon de Laplace's treatise and the conviction that it was possible to arrive at values in astronomical tables without resorting to observation, induced the Académie des Sciences to promote a competition for "the formation, by means of theory alone, of lunar tables." The prize was awarded in 1820 to a memoir submitted by two Italian astronomers, Francesco Carlini and Giovanni Antonio Plana (1781–1864), who published their *Théorie du mouvement de la Lune* in 1832. Alexis Bouvard pointed out the difficulties involved in establishing an orbit for Uranus that corresponded with observational data. This was the first indication of perturbations in the motion of Uranus.

The history of astronomy

The French edition of Ptolemy's *Almagest*, prepared by Nicolas Halma with comments by Jean-Baptiste Delambre, was published in 1813. It marked the beginning of the critical publication of classical works in the history of this science. Delambre published his major work, *Histoire de l'astronomie*, between 1817 and 1827.

PHYSICAL SCIENCES: MECHANICS

Developments in mathematical physics

Three major areas of investigation were opened at the beginning of the second decade of the nineteenth century: research concerning the theory of the oscillation of pendulums, a critical analysis of Pierre Simon de Laplace's program, and research on the theory of non-elastic impact. The *Lehrbuch der Astronomie*, by Johann von Bohnenberger (1765–1831), which reexamined and applied some of Huygens's theorems on the oscillation of pendulums, was published in 1811 at Tübingen. Bohnenberger's treatise proposed ideas that were examined in greater detail by numerous mathematical physicists, in particular Frederich Wilhelm Bessel (see PHYSICAL SCIENCES: MECHANICS, 1820–1830).

In 1812, returning to theories he himself had already put forth in 1809, and to some of Joseph Louis Lagrange's observations on the attraction of spheroids, James Ivory investigated Laplace's findings and developed his own ideas in a work on mathematical physics entitled *On the Grounds of the Method Which Laplace Has Given in the Second Chapter at the Third Book of His "Mécanique céleste" for Computing the Attractions of Spheroids of Every Description*. In Ivory's opinion, the Laplacian method gave theorems that were limited to a particular class of spheroids. It was therefore necessary to analyze mathematically the entire set of problems on which Laplace's method was based in order to broaden knowledge of the basic laws of mechanics. With a second memoir of 1812 (*On the Attractions of an Extensive Class of Spheroids*), Ivory repeated that the mathematical difficulties encountered in attempting to solve these problems came from the need to clarify the real foundations of Laplace's method and of the equivalent method, which he himself outlined. In the fifth volume of the *Mécanique céleste*, Laplace explicitly acknowledged Ivory's contribution.

Considerable progress was made in solving problems concerning phenomena of non-elastic impact when mathematical methods, such as those developed by Lagrange in his *Theory of Functions,* were applied to them.

As the process of rationalizing mechanics continued, interests of an applicative nature also developed. Indicative of this trend was the publication in 1810 of three volumes edited by Jean Baptiste Delambre (1749–1822) entitled *Base du système métrique décimal*. The work included the classical *Expériences pour connaître la longeur du pendule qui bat les seconds à Paris* (1792) by Jean Charles de Borda (1733–99) and Jacques Dominiques Cassini (1748–1845).

Siméon Denis Poisson presented a mathematical discussion of perturbations in relation to problems of elasticity in his *Mémoire sur les surfaces élastiques* of 1812. A lengthy debate on the subject developed between Poisson and Jean Baptiste Fourier. Works on the applications of the laws of mechanics to the oscillations of pendulums that deserve mention include a treatise by Jean Baptiste Delambre (1749–1822) entitled *Astronomie théorique et pratique* (1814), *Sur la longeur du pendule à secondes* by Laplace (1816), and Poisson's *Sur les oscillations du pendule composé* (1816).

Gaspard de Prony's *Leçons de mécanique analitique données à l'École Polytechnique*, a panorama of physico-mathematical developments in the laws of motion, appeared in 1815.

Laplacian determinism

Regarding Laplace's basic theory on the role of the laws of mechanics in the learning process, the famous passage from his *Essai philosophique sur les probabilités* (1814) deserves to be quoted in full: "We ought then to regard the present state of the universe as the effect of its anterior state and as the cause of the one which is to follow. Given for one instant an intelligence which could comprehend all the forces by which nature is animated and the respective situation of the beings who compose it—an intelligence sufficiently vast to submit these data to analysis—it would embrace in the same formula the movements of the greatest bodies of the universe and those of the lightest atom; for it, nothing would be uncertain and the future, as the past, would be present to its eyes. The human mind offers, in the perfection which it has been able to give to astronomy, a feeble idea of this intelligence. Its discoveries in mechanics and geometry, added to that of universal gravity, have enabled it to comprehend in the same analytical expressions the past and future states of the system of the world.''

In 1816, Laplace formulated a new theory on the velocity of the propagation of sound waves in gasseous media (see PHYSICAL SCIENCES: THERMOLOGY, 1810–1820). Correcting Isaac Newton's treatment of the problem, he demonstrated the following proposition: "The velocity of sound is equal to the product of the velocity given by the Newtonian formula times the square root of the ratio between the specific heat of the air at constant pressure and the specific heat of the air at constant volume." This was one of the most brilliant successes of the systematic application of mathematical methods, as developed in the *Mécanique céleste*, to classes of phenomena studied with the use of heat theories.

The credibility of Laplace's plan to reduce thermology to mechanics was further reinforced when his formula was verified experimentally; it was no accident that Laplace gave ample space to this new theoretical elaboration in the fifth volume of the *Mécanique céleste* (see PHYSICAL SCIENCES: MECHANICS, 1820–1830).

In 1818, following this line of research, Siméon Denis Poisson presented a general theory on the velocity of the propagation of sound, while Laplace examined its applicability in the case of sounds propagated in fluid media.

The theory of oscillations

Progress was made in the theory of oscillations and the applications of this theory to problems in geophysics. In 1818, Laplace published his *Mémoire sur la figure de la Terre*, which discussed questions regarding certain parts of the general theory that was expounded in the first volumes of the *Mécanique céleste*.

A memoir written in 1819 by Jean Baptiste Biot (1774–1862), *Sur la longeur du pendule à secondes observée a Unst, la plus boréale des îles Shetland*, deserves mention.

A classical memoir by Thomas Young, entitled *Remarks on the Probability of Error in Physical Observations . . .* , appeared toward the end of the decade. It attested to an increasing interest in the need for a rigorous evaluation of errors in measurement. It marked the beginning of the theory of errors, which became increasingly important in physics and in all the experimental sciences.

PHYSICAL SCIENCES: THERMOLOGY

The problem of specific heats

On the basis of the increasingly urgent need for accurate data on values for specific heats—data that was most important from a technological as well as a scientific point of view—and correct theories on heat phenomena in general, the Académie promoted two special competitions in experimental and mathematical physics in 1811. The first was won by a monograph entitled *Sur la détermination de la chaleur spécifique de differens gaz* (1813), written by François Delaroche and Jacques Etienne Bérard. The experimental data that the two researchers gathered using the equipment in the laboratories of Claude Louis Berthollet (1748–1822) were great advances. The best European theoreticians used the new values measured by Delaroche and Berard for many years, until the findings of Henri Victor Regnault (1810–78) were published (see PHYSICAL SCIENCES: THERMODYNAMICS, 1840–1850). However, theoretical interpretation was rendered particularly difficult by the fact that the values measured by Delaroche and Berard suggested a dependency between specific heat and pressure.

Fourier's program

The second competition of the Académie was won by an important memoir by Jean Baptiste Fourier, *Théorie du mouvement de la chaleur dans les corps solides*, presented in September, 1811. In this memoir, Fourier generalized the theories that he had already advanced in 1807, theories that were later brought together in his great work on thermology, which was published in 1822 (see PHYSICAL SCIENCES: THERMOLOGY, 1820–30). The core of Fourier mathematical reflections lay in the decisive stand he took against the Laplacian pretense of reducing the theory of heat to a branch of mechanics. Since heat phenomena developed independently of the initial conditions of a physical system and moved in a particular direction, tending toward an equilibrium in temperature, they could not, according to Fourier, be explained by the principles of motion. Instead, they required a new theory, independent of mechanics and essentially founded on the internal coherence of the deductive system. This theory was valid not because it applied to physical interpretations made using models of the nature of heat, but because it led to "equations of the phenomenon" that were numerically calculable and could be confirmed by experiment. Fourier explicitly stated in 1822 that this depended, in the final analysis, on the fact that all nature was connected mathematically and the rules of mathematical analysis perfectly reflected this connection. Fourier's considerations on the essentialness of investigations into the nature of heat and on the resulting need to elaborate equations that would describe heat phenomena are cited by Auguste Comte (1798–1857) in the first lesson of his *Cours de philosphie positive* (1830). According to Comte, Fourier's physics stood as proof of the theories that science should not presume to discover the intimate causes of phenomena, but should limit itself to formulating "the most important and most accurate laws of the phenomena" themselves. In this, wrote Comte, lies the "eminently positive philosophical nature" of Fourier's scientific work.

Interactions between heat and matter

However, another line of research different from Fourier's, was developed thanks to the work of François Delaroche, Pierre Prévost (1751–1839). Pierre Simon de Laplace, and Siméon Denis Poisson. Poisson in particular debated the problem of whether the modes of interaction between heat and matter were valid, an aspect that Fourier's research tended to neglect.

Following Delaroche and Bérard's observations, new research that contributed to progress in understanding the parameters of gasses was done. In 1815, Pierre Louis Dulong (1785–1838) and Alexis Thérèse Petit (1791–1820) began a series of exper-

iments to perfect the techniques of measuring the dilatation of thermometric liquids. This research became the basis for Dulong's and Petit's great work on the specific heats of solids, described below.

The mathematical rationalization of thermology in the context of Laplace's program was advanced in 1816 when Laplace himself elaborated an explanation of the velocity of the propagation of sound perturbations in matter in the gasseous state. Laplace's considerations enabled definitions of some of the macroscopic parameters of gasses to be formulated in terms of the interaction between heat and matter, and there seemed to be enough of rational connection between theory and experiment to confirm the existence of "latent heat." Laplace's theory was based on a mechanical representation of material molecules, which were seen as centers of emission and absorption of particles of the caloric.

The development of caloric theories profoundly influenced chemistry. This was particularly evident in the memoirs published by Amedeo Avogadro (1776–1856) in 1816 (*Memoria sul calorico specifico de' gaz composti*) and in 1817 (*Continuazione e fine della memoria sul calorico specifico de' gaz composti*). In them, Avogadro returned to themes he had expounded in a paper of 1811 (see CHEMICAL SCIENCES, 1810–1820), which contained his proposition on the number of molecules per unit volume.

John Herapath began his studies during this same period. He suggested analyzing gasses as statistical molecular units and viewing heat as a result of disordered movements of the material particles. Temperature, from this point of view, could be defined in terms of molecular mechanical quantities. The Royal Society did not take Herapath's work seriously into consideration, and it was published in the Annals of Philosophy without arousing the interest that it undoubtedly merited.

In 1817, Fourier confronted the problem of heat radiation (which was becoming increasingly important in Laplace's publications) in two memoirs entitled *Note sur la chaleur rayonnante* and *Questions sur la théorie physique de la chaleur rayonnante*. According to Fourier, what was needed was a mathematical theory of heat radiation that depended on only a few general empirical principles and was completely autonomous with respect to the physical interpretations suggested by supporters of the caloric theory, such as Prevost.

The Dulong–Petit law

The final years of the decade were characterized by two experimental research projects on specific heats, one conducted by Pierre Louis Dulong and Alexis Thérèse Petit, and the others by Nicolas Clément-Desormes (1779–1841) and Charles Bernard Desormes (1777–1862).

In 1819, Dulong and Petit published the data they had collected on the specific heats of solids and sent to press their classical work *Sur quelques points importants de la théorie de la chaleur*. This work contained a statement of the law according to which "the atoms of all simple bodies have exactly the same capacity for heat." A work by Clément-Desormes and Desormes on the expansion of gasses, containing a precise evaluation of the relationship between specific heats at constant pressure and at constant volume, was published that same year. This relationship confirmed the validity of the theoretical approach suggested by Laplace in 1816 regarding the speed of sound in gasses.

PHYSICAL SCIENCES: ELECTRICITY

Electrostatic potential

During the second decade of the century, the divergencies between Laplacian programs and those influenced by other currents of thought became greater. In the area of investigations into electricity and magnetism, the studies of Poisson and Hans Christian Oersted were characterized by completely different approaches, while in the field of optics the battle between wave and corpuscular theories became radicalized.

The theoretical results developed by Poisson and presented in a two-part memoir—the first part dated May and August 1812 and the second, September 1813—were exceptionally important.

Introducing a concept of potential deduced from the analogous notion provided by Laplace's equation of second partial derivatives in mechanics, Poisson analyzed the distribution of electricity on conducting spheres (see PHYSICAL SCIENCES, 1760–1770 and 1770–1780 for the concept of potential as developed by Allessandro Volta and Henry Cavendish). Referring to Laplace's analysis of attraction between spheroids, Poisson maintained that the distribution of electricity on the surface of an ellipsoid had to be such that the sum of the partial derivatives of a certain function that varied as the inverse of the distance would be zero inside the ellipsoid. This was the same as affirming the constancy of potential within the conductor. Poisson's analysis led to a general conclusion that could be extended to conductors of any shape. However, Poisson was aware of the difficulties involved in making this extension using the formulas on the attraction of spheroids. Availing himself of a demonstration originated by Laplace, Poisson elaborated a discussion that partially repeated results already obtained by Charles Augustin de Coulomb. In fact, he demonstrated that the total electrical force exerted by a conductor with surface charge density δ on a point outside the conductor but taken to be infinitely close to it was equal to $4\pi\delta$.

Oersted's conflict

Oersted expounded completely different ideas in 1813 in his *Recherches sur l'identité des forces chimiques et électriques*, in which he maintained that light and heat were the results of an electrical "conflict." Although this approach was qualitative, it enabled Oersted in 1820 to shake the faith that the physical sciences had once again established thanks to Poisson's work in electrostatics —in the possibility of reducing even electrical and magnetic phenomena to manifestations of motion.

The ether and Fresnel's coefficient

Thanks to the flexibility of Augustin Jean Fresnel's research programs, explanations were provided in wave terms for several experiments conducted by Arago on the aberration of light in optically dense media. In 1818, Fresnel's interpretation was published in the *Annales de Chimie et de Physique* in the form of a letter to Arago, and the subject of the ether, with all its problems, emerged. As Fresnel wrote in this memorable letter, if one accepts that the earth transfers its motion to the ether surrounding it, and allows the propagation of light, then several optical phenomena can be explained, though others, such as the aberration of stars, become impossible to explain. In order to resolve this contradiction, according to Fresnel, the following condition must be imposed upon the ether: "Suppose that the ether passes freely through our globe and that the velocity communicated to this subtle fluid is but a small part of that of the earth." This condition led to Fresnel's definition of the coefficient of partial drag, which even today bears his name.

CHEMICAL SCIENCES

Avogadro's hypothesis

In 1811, Amedeo Avogadro specified the difference between atoms ("elementary molecules") and molecules ("integrating molecules"), and stated the hypothesis that equal volumes of gas at equal temperatures and pressures contain the same number of molecules. Avogadro's hypothesis, today recognized as a principle, was opposed by the major scientists of the epoch, including John Dalton and Jöns Jacob Berzelius, and was practically ignored until 1858, when Stanislas Cannizzaro demonstrated its enormous value. Ampère also realized the need to distinguish between atoms and molecules.

The laws of constancy of the product of specific heat times atomic weight (1919), proposed by Pierre Louis Dulong (1785–1838) and Alexis Thérèse Petit (1791–1820), and the law of isomorphism (1919–20), stated by Eilhard Mitscherlich (1794–1863), were helpful in establishing the atomic weights of elements. In 1815–20, Joseph Louis Gay-Lussac drew the first solubility curves for several salts in water and gave the composition of five nitrogen oxides.

Iodine and other elements

In the field of inorganic chemistry, new elements were discovered: iodine was discovered in 1811 by Bernard Courtois (1777–1838). Its properties were studied by Gay-Lussac and, independently but simultaneously, by Humphrey Davy, who claimed the right of priority and distinguished between iodine and hiodic acid, stating that iodine had properties similar to those of chlorine and that iodine was to hiodic acid as chlorine was to (hydrochloric) acid. Other discoveries were selenium (Berzelius, 1817); cadmium (Friedrich Stromeyer [1776–1835], in 1817); and lithium, discovered by Johan August Arfvedson (1792–1841) in 1817 and isolated by Davy in 1818. New compounds were also prepared: barium peroxide (Gay-Lussac and Louis-Jacques Thenard in 1811); nitrogen trichloride (Pierre Louis Dulong, 1811); chloric acid (Gay-Lussac, 1814); nitrogen iodide (Gay-Lussac, 1814), the production and explosivity of which had already been observed by Charles Bernard Desormes; hypophosphoric acid (Pierre Louis Dulong, 1816); and hydrogen peroxide (Thenard, 1818).

New natural and artificial compounds

In organic chemistry new natural and artificial compounds were discovered. Among the first were cystine, isolated by William Hyde Wollaston, in 1810, xanthine, isolated by Alexander Marcet (1770–1822) in 1817, and maleic acid, isolated by Jean-Louis Lassaigne (1800–59) in 1819. Of particular interest was the discovery of naphthaline in coal tar by William Thomas

Brande (1788–1866) in 1819, and independently by Alexander Garden and numerous alkaloids, most of which were isolated by Pierre Joseph Pelletier and Joseph Bienamié Caventou (1795–1877) between 1818 and 1820 (strychnine, brucine, quinine, colchicine, etc.). Among the compounds obtained artificially, but not analyzed, were phosgene (Davy, 1811) and alloxan (Luigi Vincenzo Brugnatelli, 1818). Michel Eugène Chevreul's work on fats and soaps, done in 1810 and published between 1813 and 1823, deserves particular mention.

Biochemistry

An interesting biochemical observation was made by William Prout (1785–1850) when he discovered that the excrement of birds and reptiles contained high levels of uric acid.

In 1818, Nicholas Théodore de Saussure (1767–1845) advanced the important theory that the transformation of starch into sugar in the presence of malt was accelerated by the gluten in the malt. It was also pointed out during this period that certain metals performed the same function in certain reactions between gasses (Thenard, 1813; Davy, 1817; Döbereiner for platinum, 1820; and Angelo Bellani [1776–1852] in 1824).

Analyzing organic substances

In 1811, Gay-Lussac and Thenard proposed the first efficient general method for the analysis of organic matter and analyzed a number of organic substances. They discovered the relationship between carbon, oxygen, and hydrogen in sugars that gave rise to the new name carbohydrates. They also demonstrated that hydrocyanic acid did not contain oxygen; and when Davy showed that hydrochloric acid also did not contain oxygen, Antoine Laurent Lavoisier's theory that that element was characteristic of acids was disproved.

Gay-Lussac also greatly simplified analysis (and at the same time increased accuracy) by introducing the volumetric (titrimetric) method in 1817. With this technique, weighing (which was always difficult and often inaccurate) was not necessary; instead comparisons were made between solutions with known concentrations, i.e. containing quantities of substances that had been weighed once and for all.

Optical activity

Jean Baptiste Biot (1774–1862) recognized the optical activity of certain crystals, such as quartz (1812), certain liquids, such as essences and sugar solutions (1815–18), and some gasses.

Treatise writing

Two important treatises appeared during this period: the *Traité de chimie élémentaire*, by Thenard (first edition 1813–16; sixth edition 1834–36) and the *Handbuch der theoretischen Chemie*, by Leopold Gmelin (1788–1853), an attempt at unifying chemistry that came out in 1817. In the fifth edition, Karl Kraut eliminated the part on organic chemistry from Gmelin's text, creating the *Handbuch der anorganischen Chemie*.

LIFE SCIENCES

Fixity of species and evolutionism

In 1817, Georges Cuvier published *Le règne animal*, which proposed a new zoological classification divided into four groups—vertebrates, molluscs, articulates, and zoophyta—based on the four structural forms of the organisms. Living and fossil species were brought together for the first time. Cuvier's morphology emphasized the physiological aspects, the nervous system became the central element on which the system was built. In his *Histoire des poissons* (1828–33), the group became a "type," a concept that was later applied even to the lowest taxonomical units (genera, species). After the publication of the *Histoire naturelle des animaux sans vertèbres* (1815–1822), by Jean Baptiste de Lamarck, the polemic between Lamarck's evolutionism and Cuvier's fixity of species became more heated. In his *Discours sur les révolutions du globe* (1825), Cuvier formulated the hypothesis of successive catastrophes to explain the extinction of certain species and the origins of others.

Phylogeny and ontogeny

An important systematic contribution to plant morphology was the *Théorie élémentaire de la botanique* (1813) by Augustin Pyrame de Candolle (1778–1841), who conceived of species in the Linnaean sense, distinguished cellular plants from vascular plants, and viewed morphology as the study of structural symmetries separate from considerations on the functional properties of the organs. Cuvier's physiological morphology and de Candolle's symmetrical-structural morphology established themselves as partially antithetical concepts; Karl Ernst von Baer leaned toward structural morphology. In 1811, under the influence of German "nature philosophy," Johann Friedrich Meckel (1781–1833) formulated the principle that ontogeny followed the stages of phylogeny. This was later expressed as a law by Ernst Heinrich Haeckel (see LIFE SCIENCES, 1860–1870).

In his *Dissertatio inauguralis sistens historiam metamorphoseos, quam ovum incubatum prioribus quinque diebus subit* (1817), Christian Pander (1794–1865) presented his discovery of two membranes ("serous" and "mucous") in the chick embryo. In 1827, von Baer established the existence of a third ("vascular") membrane (see LIFE SCIENCES, 1820–1830).

Other authors limited themselves to extending anatomical and systematic knowledge. Karl Asmund Rudolphi described a number of new species of parasites in his *Entozoorum synopsis* (1819), and continued to believe that they were created by spontaneous generation. Progress in descriptive morphology was favored by voyages and geographical explorations: Robert Brown (1773–1858) followed Matthew Flinders's expedition to Australia and Tasmania from 1801 to 1805; François Peron (1775–1810) and Charles-Alexandre Leseur were sent by the Institut de France to visit the austral territories (1800); Joseph Dalton Hooker (1817–1911) travelled to the Antarctic in 1839–43, gathering valuable samples of marine flora and fauna; and Jean-René Quoy explored the austral seas and territories (1826–29).

EARTH SCIENCES

The predominance of von Buch's tectonic theories: the upheaval of volcanoes

Christian Leopold von Buch's observations of the effects of volcanic forces and his discovery that many volcanic hills were aligned, together with his knowledge of volcanic islands, such as the Canaries (which he believed could not have been formed by normal eruptions beginning at the bottom of the ocean, since the lava would have solidified immediately when it came into contact with the water) led him to formulate his theory that vertical forces gave rise to mountain chains. This theory is generally called the upheaval theory of volcanoes (*Erhebungs Cratere*).

Von Buch applied this explanation to the origin of all mountain chains, and in particular to the Alps, which he believed were formed by forces exerted by masses of volcanic rock, working in much the same way as those that created volcanic islands, but which never reached the surface because of the strong resistance of the thick overlying rock. Thus the principle of vertical or radial thrust in the formation of mountains, which later became generally accepted, was stated during this period.

In the same context, Scipione Breislak, in his *Introduzione alla geologia* (1811), gave considerable emphasis to the internal thrusts provoked by contractions in the earth as it cooled.

Cuvier's "revolutions"

By this time, the dispute between the Neptunists and the Plutonists had abated, and in 1819, Christian Keferstein wrote a summary of its development.

Interest was mainly focused on the earth's geological history, documented by the simultaneous development of stratigraphic and paleontological research. Georges Cuvier's *Recherches sur les ossements fossiles des quadrupèdes* was published in 1812. In the "Discours préliminaire" to the first volume, the founder of comparative anatomy maintained that the succession of fauna and flora that populated the earth in the past and that suddenly disappeared from one stratigraphic layer to the next, was evidence of a series of "revolutions," or cataclysmic events, that every so often wiped out life over certain large areas. Since the fixity of animal and plant species was indisputable, the only explanation for the remains of organisms in layers following each of these revolutions, one of which may have been Noah's Flood, was the creation of new populations.

Vertebrate paleontology

Cuvier had a very thorough knowledge of paleontology, especially vertebrate paleontology, of which he can be considered the founder. His famous "law of organic correlation" enabled him to make sensational discoveries based on findings of just a few fossil fragments, such as the fossil marsupial found in the gypsum of Montmartre, which he recognized by its jaw bones. Cuvier isolated the skeleton from the gangue in the presence of his colleagues, who were highly skeptical until the marsupial's pelvis bones were uncovered, providing overwhelming proof that Cuvier's interpretation was correct. He worked mainly with the fossils of mammals and reptiles, which were sent to him from all over Europe thanks to the reputation of his studies and the important offices he held under both Napoleon and Louis XVIII. One of the remains that Cuvier analyzed, discovering its true nature, was the giant salamander *Andrias scheuchzeri* (1814), which Johann Jakob Scheuchzer had discovered in 1725 and identified as *Homo diluvii testis*, and which Pieter Camper, in 1787, had thought was a giant lizard. One of Cuvier's particular paleontological merits was that he definitively demonstrated that the remains he studied belonged to species that had become extinct at various moments of the earth's history and were no longer found among present-day fauna.

Cartography and regional geological illustrations

Stratigraphic studies were accompanied by geological maps and illustrative memoirs. The Huttonian principles of violent upheaval prevailed in England as the explanation for tectonic events. Thomas Webster supported this theory for the Isle of Wight (1816), William Daniel Conybeare (1787–1857) and William Buckland (1784–1856) for Ireland (1813), and Conybeare and William Phillips for England and Wales (1822). John McCulloch, on the other hand, was more cautious with respect to Scotland (1811–21).

In France, catastrophic theories were especially influenced by the work of Cuvier and Alexandre Brongniart. The stratigraphic illustrations of Karl Georg von Raumer for the area of Paris (1815); those of Marcel de Serres for the lacustrine limestone of France (1818) those of Jean-Baptiste de Omalius D'Halloy for the Paris basin (1813–16); and for Belgium and nearby areas of Germany, those of Ami Boué (1794–1881), which attempted to parallel the formations of Scotland with those of the Paris basin (1820), all deserve particular mention in this context. A good correlation between the Paris basin and southern England was proposed by Thomas Webster in 1814.

William Smith began his stratigraphic map of England in 1812 and completed it three years later, while George Bellas Greenough's geological map of the same region, considered a true prototype in the field, appeared in 1819.

Another significant event of 1819 was the appointment of William Buckland, already a professor of mineralogy, to a teaching position in geology at Oxford. He thus became the first teacher of this discipline in the famous English university. His position on the relationship between geology and the Bible is very well known: he affirmed that geological discoveries provided proof of the truth of the sacred texts, which he interpreted literally.

Paleontological monographs on invertebrates and plants

Jean-Baptiste de Lamarck's *Histoire naturelle des animaux sans vertèbres* appeared between 1815 and 1822. In this work, Lamarck described and reiterated his evolutionary ideas and his belief in the effective importance of the modifications that animals went through in response to the enironment, which he saw as the basis of transformations in fauna through the ages of the earth's history.

An extremely valuable malacological monograph was the *Conchiologia fossile subappennina* (1814) by Giovanni Battista Brocchi (1772–1826), which illustrated the fauna of the Emilian, Piedmontese, and Tuscan Tertiary. Other important monographs were, for example, those of James Sowerby (1812), completed by his son James de Carle Sowerby, and *Die Petrafaktenkunde* (1820, with appendices of 1822 and 1823) by Ernst Friedrich von Schlotheim (1764–1832). James Parkinson (1755–1824) completed his *Organic Remains of a Former World*, an illustration of fossil plants, in 1811.

Crystallographic optics and systematic mineralogy

The relationships between interference colors in thin sections of minerals and birefringence were recognized in 1812 by David Brewster. In 1820, John Frederick Herschel (1792–1871) described the dispersion of the optical axes in biaxial crystals.

Finally, in 1818, Brewster attempted to merge his optics with crystallographic optics, although he was unable to make them correspond completely due to imperfections in Haüy's crystallographic system, which he was using as a basis. Beginning in 1815, Weiss studied the systematics of crystalline forms and recognized the vectorial nature of their properties. Johann Nepomuk von Fuchs (1815) and François-Sulpice Beudant (1817) conducted the first studies on isomorphism in minerals, further developed by Eilhard Mitscherlich (1794–1863) between 1818 and 1820. In 1817, the chemist André Laugier recognized the similarity between aerolites and the native iron that Pierre Simon Pallas had brought back from Siberia. Carl Friedrich Christian Mohs devised his famous empirical scale of the hardness of minerals, still in use today, in 1820.

The ocean floor

The first deep-sea research and investigations into the nature of the ocean floor were begun during this period. The first direct study of the ocean bottom was conducted in 1818: John Ross succeeded in bringing up muddy sediments from a depth of 1,800 meters using a "clamp" that he himself had designed. Among other things, the study revealed the presence of life on the ocean bottom.

The navigator-scientists

Explorative expeditions into the oceans were gradually extended. Alexander von Humboldt, already known for his explorations (in 1799 he had set out on a five-year voyage around the world), described the presence of fairly low abyssal temperatures even in the tropics, an effect of water coming from the poles (1814) (see LIFE SCIENCES, 1800–1810). In 1818, in the course of a great expedition to the polar region, John Ross (1777–1856) succeeded in measuring temperature to a depth of 2,000 meters. He found that the temperature rose slightly with depth, a trait that was later recognized as characteristic of the polar seas.

Ross's and Buchan's polar voyages included programs for measuring temperatures (using Six thermometers), specific weight, currents, tides and depth. From a navigational point of view, their goal was to discover a northwest passage between the Northern Atlantic and the Pacific, and gather general information on conditions in the Arctic. Meanwhile, the Russians were investigating a northeast passage: between 1815 and 1818, Otto von Kotzebue navigated the Arctic and reached the Pacific through the Bering Strait. The Russian expeditions proceeded with the great polar campaign of 1823–26, in which various oceanographic measurements were again made, enabling Emil Lenz to draw some important conclusions. In any case, the two great polar expeditions to the northwest and the northeast led to an understanding of the thermic regimen in the deep waters of the polar seas. Ross and Kotzebue's early assumptions that the waters of the ocean depths were isothermic (even earlier, Peron had believed that they were frozen), were later modified, although they remained valid within a certain approximation for the polar zones.

In 1813, Thomas Young made important contributions to the tide theory.

The chemists of the sea

But possibly the most significant and definitive contributions to the progress of oceanographic knowledge during this decade came from the chemists of the sea. Heinrich Link (1767–1851), Christian Heinrich Pfaff, and von Lichtenberg made normal systematic measurements of salinity, especially in the Baltic Sea. John Murray (1818) and Alexander Marcet (1819) also did some very important research. Murray conducted the most complete analysis of the epoch, establishing a number of analytical methods as well. In Murray's studies, the composition of salinity was determined using quantitative data on sodium and magnesium muriate (chloride), sodium, magnesium and potassium sulphate, and potassium and magnesium carbonate. Marcet, on the other hand, determined salinity using data on the amounts of substances that precipitated from sea water, concentrated by evaporation, in the form of silver muriate (chloride), barium sulphate, calcium oxalate, and magnesium phosphate. Marcet established a relationship between the sum of these precipitates and the weight of the residue after evaporation, and between the latter and density (measured with extreme accuracy, to the fourth or the fifth decimal). Marcet postulated that the ratios among the ingredients of salinity remained constant, anticipating one of the most important concepts of modern oceanographic chemistry. He also definitively pointed out the dependency between density and temperature; as a result, the fact that density depended on both salinity and temperature became universally accepted. Marcet examined water from all over the world: from the Atlantic, the Arctic, and the Mediterranean. He also worked on methodology and instrumentation, designing bottles for gathering samples, for example. He began his work on the chemistry of the sea as early as 1807, when he analyzed water samples from the Dead Sea.

APPLIED SCIENCES

Industry for domestic needs

In England, the experimental phase of gas production that distinguished the preceding decade blossomed into a phase of its industrial production. After various attempts were blocked by continental governments and the English Parliament, the German Frederic Albert Winsor (1763–1830) finally succeeded in founding the Chartered Gas-Light and Coke Company in London in 1812. He was aided by the decisive contribution of Samuel Clegg, the elder, (1781–1861), a student of James Watt and William Murdoch, who solved the problems connected with the distribution and transport of the new raw material. Clegg invented the watertight gas main (1811), the closed water gasometer, the valves, and the meter that is still used today. In 1813, the Westminster Bridge was illuminated with gas lights, and in 1816, London had about forty kilometers of lighted streets.

The gas industrialists made up one of the sectors of business that was most aware of technological developments. There was considerable interest in the new method of illumination on the continent as well. Between 1811 and 1816, Wilhelm August Lampadius (1772–1842) conducted extremely successful experiments in Freiberg. The Royal Museum of Philadelphia was illuminated by gas in 1816. The New York Gas-Light Company was founded in 1823. The advent of gas for lighting and heating reinforced the tendency to centralize the distribution of raw materials for domestic use. A centralized water supply had already been perfected by James Peacock (1738–1814) as early as 1791.

Safety in the mines

Most of the progress in mining techniques took place in the area of illumination. The frequency of explosions caused by pit-gas (grisou) induced John Wilkinson and John Hodgson (1779–1845) to found the Sunderland Society for the Prevention of Accidents in Coal Mines in 1813. In 1815, Humphrey Davy (1778–1829), invited to collaborate in the work of the society, announced that he had discovered "absolute safety" for miners: a lamp made with a cylinder of metal mesh around a naked flame. At the same time, George Stephenson (1781–1848), who in 1812 had modified the mine pump so that it could hoist coal up from a slope, devised a series of "safety lamps" based on similar principles. Davy's lamp was so successful that it was made compulsory in France in 1823. The problem of ventilating the pits, which was more serious in Belgium than in England, since the Belgian mines were deeper, was partially resolved by using mechanical ventilators (air pumps); the first one was used by John Buddle in 1807 in the Hobburn mine.

New machines and machine tools

The use of machines with rotary and reciprocal motion led the gradual replacement of wood with iron in their construction, beginning with the parts that were most subject to friction, and consequently to wear (connecting rods, balance-beams and pins). The resulting demand for metalworking machines led Richard Roberts to build an automatic back-geared lathe and, in 1817, a metal-planing machine. The following year, Eli Whitney introduced a small milling machine.

The range of machines in the textile industry increased. In 1813, William Horrocks invented a new mechanical loom, the basis for a standard loom perfected by Roberts ten years later. The tendency was no longer to limit mechanization to a single operation, but rather to extend it to the entire productive cycle. In 1814, Crigton succeeded in directly connecting machines that performed consecutive operations, joining a beamer and a scutcher.

The revolt against the machines

The gradual replacement of a number of manual operations with mechanical ones, especially the introduction of a new machine for cutting cloth, increased the workers' distrust of machines. Spontaneous local rebellions evolved into the planned destruction of machines (the Luddite riots, 1811–12).

The high-pressure engine devised by Richard Trevithick (1771–1833) began to replace Watt's after its yield was improved during this decade. In 1812, Trevithick built the Cornish boiler, which made a more efficient use of fuel. In this way, the steam engine reached an output of 100 horsepower. The high-pressure engine changed the way in which the steam engine was used. The generator could be made smaller and no longer required fixed installation. In addition, the piston and the boiler were parallel, allowing a horizontal arrangement that

made for better use of rotary motion. The vertical engine began to disappear in 1825, when Taylor and Martineau introduced the use of the horizontal cylinder in London. The steam generator thus proved to be the most flexible and economical. In 1820, there were 60 engines in Birmingham; by 1829, the number had risen to 169.

Railways: the debate over traction

The increase in the price of feed for horses, a result of the Napoleonic wars, led to experimentation with new locomotives. Even though the first models, invented by Blekinsop (1811) and Hedley (1813), were better than Trevithick's as far as consumption and yield were concerned, they were still uneconomical. In 1814, George Stephenson began testing his steam locomotive. The central topic of discussion at the time, however, was what sort of road to use with steam-traction vehicles. Although Trevithick's trials with locomotives on railways had been successful and had demonstrated that the combination was convenient, there was still a certain lack of trust in the possibility of making wheels stay on rails. In 1812, Blekinsop used toothed wheels that meshed with a rack fixed to the rails, while in 1815, Hedley attempted to improve adherence by increasing the number of wheels to eight.

Steam propulsion

The success of Robert Fulton's enterprise was an incentive for building new steamboats. The first commercially successful European steamer was the *Comet*, which began operating on the Clyde River in 1812. The *Ferdinand I*, built for the King of Naples (the first steamship on the Mediterranean), and the *Savannah*, the first sailing ship to cross the Atlantic using steam as an auxiliary power source, were launched in 1818 and 1819, respectively. The first Atlantic crossing using steam alone was made by the steamship *Curaçao* in 1827. The paddle-wheel steamboat *Aaron Marby* made the voyage from England to Paris in 1822.

Road conditions and artificial canals

The growth of industrial production brought to light the problem of transport. The English road network, already built with trade needs in mind, had to bear heavy commercial traffic. The pace of road-building was not adequate to the needs of trade and industry during this phase of rapid development: a load of wood normally took a year to reach the coast from inland and the average speed of postal coaches did not exceed eight kilometers per hour. The tendency among English engineers was to substitute the French-style heavy stone-block road foundation with a surface of gravel, clay, and stone chips over a foundation that allowed adequate drainage.

In 1819, John L. McAdam (1756-1836) published *A Practical Essay on the Scientific Repair and Preservation of Roads,* in which he proposed a system for building a new type of road surface. The "waterproof macadam" road prevailed until the appearance of automobiles.

The reconstruction of the Cenes Road (the Sempion Road had been opened in 1805), projected by Napoleon, who had planned to have fourteen imperial arteries radiating out from Paris, was completed in 1810.

These improvements in the road network, however, did not diminish the tendency to utilize existing waterways and construct artificial ones to improve transport. Despite the large investments required for these fluvial and maritime projects, the Saint Quentin canal, connecting the North Sea with the English Channel, was opened in 1810, and in 1817, De Witt Clinton began digging the Erie Canal, which would connect New York with the Great Lakes, covering a distance of 590 kilometers.

Stress on materials

Systematic studies began to be made of the mechanical behavior of elastic solids under given types of stress. An early treatise, *Mémoire sur la flexion des verges élastiques courbes*, by Claude-Louis-Marie Navier (1785–1836), appeared in 1819. Navier futher developed the theory of elasticity together with Augustin-Louis Cauchy and Siméon Denis Poisson, and published his research in a *Mémoire sur les lois de l'équilibre et du mouvement des corps solides élastiques*.

1820–1830

THOUGHT AND SCIENTIFIC LIFE

Constitutional movements and nationalistic ferment

Demands for political freedom from liberals, claims from nationalities that had not been respected by the Congress of Vienna, and resistance from the members of secret sects (Filippo Buonarotti's organization, the Carbonari, etc.) all created opposition to the legitimist program of the Restoration. The disorders that had originated in Vienna exploded in the Spanish riots of 1820 and had widespread repercussions throughout Europe (revolts in the Kingdom of the Two Sicilies, agitation in the Lombardy-Veneto region, the Greek insurrection against the Turks, etc.). Meanwhile the Spanish colonial empire in the Americas collapsed. The armed intervention of England and France against Turkey in support of the Greek rebels brought about a definitive crisis in the system of the Holy Alliance (1829–1830).

The first congress of the Gesellschaft Deutscher Naturforscher und Aerzte, established for the purpose of fostering the "knowledge of its members" met in 1822; it was organized by Lorenz Oken (1779–1851), who the year before had participated in a similar congress of Swiss naturalists in Berne. Other congresses were held in Halle (1823), Würzburg (1824), Frankfurt (1825), Dresden (1826), and Munich (1827). The congress in Berlin in 1828 was presided over by Alexander von Humboldt and Martin Heinrich Lichtenstein. It was through annual congresses such as these in Germany and in Italy, that scientists more or less directly became a part of the historical national revival movement.

Scientific journals

New scientific periodicals were created in order to increase and speed up communication among researchers. Two of these were the *Rendiconti sui progressi della fisica e della chimica*, founded by Jöns Jacob Berzelius in 1821, and the *Journal für die reine und angewandte Mathematic*, founded by August Leopold Crelle (*Journal Crelle*) in 1826. The rise of these publications confirmed the growing need for an increasingly widespread circulation of scientific ideas, a need that could no longer be fully satisfied by the academies created in the seventeenth and eighteenth centuries.

At the University of Berlin, von Humboldt took advantage of his prerogative as a member of the Académie des Sciences to hold a course in physical geography. Held in 1827, the course described the observations von Humboldt had made during a voyage to South America in 1799–1804. Physics and biology continued to advance in the cultural environment of the Prussian capital; Rudolf Virchow wrote in 1893 that the university was slowly passing "from a metaphysical to a naturalistic phase." Humboldt's course later became the nucleus of the *Kosmos*, and ample encyclopedia of naturalistic information (1843–58).

The Royal Asiatic Society of Great Britain and Ireland was founded in 1823 to study the nature and civilization of the various regions of Asia. The Hungarian National Academy was inaugurated in Budapest in 1825.

Alternatives to the hegemony of Hegel

During this entire decade, German philosophy was dominated by Hegel (who died in 1831). The one work that the great philosopher published during this period was dedicated to an outline of the philosophy of law (*Philosophie des Rechts*, 1821), in which Hegel returned to and expanded the idea of the "objective spirit," already developed in the *Encyklopädie* Hegel's attack in his new preface, against Fries, a disciple of Kant and a colleague at the University of Berlin, who in 1817 had participated in the student uprising at the university, was symptomatic. Much of Hegel's distrust of the populism Fries supported came from his hostility toward the romantic movements, which contraposed vehement individual or collective feelings ("the pap of the heart") to the light of reason.

The lessons on the history of philosophy and on the philosophy of art, religion, and history that Hegel later taught at the university (they were collected and published by his students), were extremely well-received by the young people, giving rise to a school that, despite internal divisions, exerted a profound influence on European thought for most of the century.

Arthur Schopenhauer (1788–1861) attempted in vain to oppose Hegel's dictatorship in Berliner culture. He had published a work entitled *Die Welt als Wille und Vorstellung* in 1819, and in 1820 he qualified for a teaching post at the University of Berlin. But the students stopped coming to his lessons, and he was forced to suspend his course. Even editors lost interest in him and refused to publish his writings.

However, the one German university that remained almost completely outside Hegel's influence was Königsberg University (in western Prussia), where the chair that had been occupied by Kant was taken over in 1809 by Johann Friedrich Herbart (1776–1841), who supported a pluralistic metaphysical realism. Herbart had already published several remarkable works in previous years, but his major works came out during this decade. These included the *Psychologie als Wissenschaft neu gegründet auf Erfahrung, Metaphysik und Mathematik* (1824–25), and the *Allgemeine Metaphysik nebst den Anfängen der philophischen Naturlehre* (1828–29). In 1833, after trying in vain to be called by the University of Berlin, Herbart transferred to the University of

Göttingen. In 1835, he published his famous *Umriss pädagogischer. Vorlesungen*, which marked a decisive step in the development of pedagogy in the nineteenth century.

"Positive philosophy"

Hegelian philosophical thought began to spread to France around 1820, thanks above all to the work of Victor Cousin (1792–1867), who had met Hegel in Germany in 1818. The German philosopher went to Paris in 1826 to return the visit. Although not a very original thinker, Cousin was one of the most popular professors among the Parisian students and the cultured circles of the city. During the reign of Louis Philippe, he became one of the most authoritative figures of the regime (he was appointed rector of the university as well as Minister of Education). Auguste Comte (1798–1857), on the other hand, a close contemporary of Cousin's, remained completely outside official philosophical culture. He held only temporary teaching positions at the École Polytechnique where he had studied, never achieving permanent status. However, while Cousin exerted his influence exclusively in academic circles and for a limited time only, Comte's influence was felt in much wider circles and was incomparably longer-lasting. From 1817 to 1823 he worked with Saint-Simon, sharing many of the same ideas. However, in 1824 a major break occurred between the two, basically as a result of Comte's refusal to follow his teacher in his dreams for a religious reformation of humanity.

Comte developed the first elaboration of his philosophical system in 1823–24. On April 2, 1826, in his own apartments, he inaugurated a course in positive philosophy before a chosen audience of illustrious scientists. Although he was unable to continue beyond the first few lessons due to a serious breakdown, the course constituted the original line of his positive philosophy, published several years later in six volumes. The problems that Comte considered were both cultural and social, in that, in his opinion, a reorganization of knowledge (in an articulated form but not dispersed into specialized areas of research, independent of one another) was the basic condition for a profound moral reorganization of post-revolutionary society.

MATHEMATICAL SCIENCES

Pure mathematics and physical research

The accent that was placed on critical problems and on the need for rigor had by this time become a prelude to a rapid revival in studies in pure mathematics, once again a product of research in physics, as had already been foreseen. Jean Baptiste Fourier's *Théorie analytique de la chaleur* was published in 1822; it carried investigations that the author had begun as early as the beginning of the century to their fullest development and marked the stable, systematic introduction of trigonometric series into analysis. This was important not only from a technical point of view (the versatility of Fourier's series development of a function was well known), but conceptually as well. In fact, the possibility of expressing functions that were not strictly analytical in an analytical form, which Fourier's series allowed, contributed a great deal to clarifying the concept of real variable, and led to a definition of the concept that did not depend on the existence of analytical expressions that could represent it.

The other great example of pure mathematical research connected to problems in physics was the *Essay on the Application of Mathematical Analysis to the Theory of Electricity and Magnetism*, published by George Green (1793-1841) in 1828. It introduced the term "potential" to designate the function Laplace had considered in relation to the theory of gravitation.

The work of Abel and Jacobi

The revival of pure mathematics, however, was represented above all by Niels Henrik Abel (1802–1829), who in the course of a few years produced works of exceptional value. In 1824, independently of Paolo Ruffini (1765–1822), he composed a memoir in which he began treating the problem of the algebraic insolubility of general equation higher than the quartic. Two years later, he obtained extremely important results in his efforts to discover which algebraic equations could be solved by radicals and, finally, in a masterly memoir of 1829, completely and definitively resolved this difficult question, which had resisted the efforts of the greatest mathematicians for centuries. From 1826 to 1829, at the same time he was working on the problem of the resolvability of algebraic equations, Abel was also studying the no less vast and complicated problem of expressing a given integral with known functions. In his *Untersuchung über eine allgemeine Eigenschaft einer sehr verbreiteten Klasse transzendenter Funktionen* of 1826, he made considerable progress in this direction, by formulating a basic generalization of the theorem of the addition of elliptic functions. In the wake of this achievement, he went on, in works of 1827, 1828, and 1829, to create what can effectively be considered a new theory of elliptic functions, establishing their dual periodicity, providing formulas for addition and muliplication, resolving the consequent problem of division, and so forth. Abel's work was characterized by great logical rigor and an effort towards maximum generality, making it one of the best examples of the critical revision of analysis mentioned earlier. It is also significant to remember that Abel may have been the first to grasp Augustin Louis Cauchy's suggestion of developing rigor in analysis and that, in particular, he was the most adamant in expressing dissatisfaction with the lack of rigor in the theory of divergent series, making some of the first efforts to correct it.

Meanwhile, Karl Gustav Jacobi (1804–51) had independently developed the foundations of the theory of elliptic functions. In 1829, he published his *Fundamenta nova theoriae functionum ellipticarum*, a collection of his own contributions to this branch of analysis. The work also managed to touch upon almost all branches of mathematics.

The foundations of projective geometry

While these brilliant developments were taking place in analysis, geometry was also being restructured on new foundations. It was during this period that a distinction was being made between the projective and the metric properties of figures, creating the true basis for projective geometry. In 1822, Jean Victor Poncelet (1788–1867) published his *Traité des propriétés projectives des figures*, in which he very ably developed the method of transformations as a tool for geometrical research and recognized the particular nature of the projective properties of figures that rendered them different from the metric properties—i.e. that they remained unaltered by projection and section. He also introduced the concepts of infinite lines and planes, cyclical points and the theory of polarity. Between 1824 and 1827, Joseph Diaz Gergonne (1771–1859) arrived at a general formulation of the principle of duality in projective geometry, which had already been partially outlined by Poncelet.

Toward an infinitesimal geometry of surfaces

Other important seeds for later developments in geometric research were also planted during these years: in 1822, Karl Friedrich Gauss introduced the theory of conformal transformations and, for the first time, used curvilinear coordinates on a surface. In 1827, he published an important memoir, *Disquisitiones generales circa superficies curvas*, in which he presented several of his most important contributions to geometry (including the concept of total curvature of a surface), opening the way to the infinitesimal geometry of surfaces and laying the foundations of modern differential geometry.

Developments in analytical geometry

That same year August Ferdinand Möbius (1790-1868) published *Der baryzentrische Calcül*, in which he introduced barycentric coordinates and studied various significant geometrical transformations, such as collineation, laying down the lines for research that he developed significantly over the following decades. The first volume of the *Analytisch-geometrische Entwicklungen*, by Julius Plücker (1801–68), came out in 1828; in it, Plücker made systematic use of homogeneous coordinates and of the line and plane coordinates that he had already introduced two years earlier. He developed highly innovative points of view in analytical geometry, especially in the theory of *algebraic curves*.

Non-Euclidean geometry

In geometry, as in analysis, this flourishing of technical research and new trends, was accompanied by investigations into the foundations of geometry; in this case, the investigations took the form of research on the postulate of parallel lines, and led to the creation of non-Euclidean geometry. Gauss had already arrived at a basic formulation of this type of geometry in 1817, but he never published anything on it. In 1825, Franz Adolf Taurinus (1794–1874) published two volumes containing important steps toward the explication of non-Euclidean geometries, but they passed unnoticed. Finally, in 1829–30, Nikolai Ivanovich Lobachevski (1792–1856) published the ideas of non-Euclidean geometry, which he had already verbally communicated to his colleagues three years earlier. These were the first works ever printed on non-Euclidean geometry. Lobachevski expounded his theories in even more detail and developed them further in later writings that appeared in 1835 and 1836–38. János Bolyai (1802–60) did research at the same time as Lobachevski, but independently; in 1832, he published an essay on non-Euclidean geometry as an appendix to a didactic work written by his father. Bolyai viewed non-Euclidean geometry more in the sense of an absolute geometry, in that the accent was placed on building a geometry independent of the postulate of parallel lines, rather than starting from a negation of it. Adrien Marie Legendre (1752–1833) published an important memoir on the theory of parallel lines in 1833, presenting his latest findings on the topic.

The historical importance of non-Euclidean geometries is well enough known that there is no need to go into it here. We shall only mention that by providing effective examples of mathematical theories in which there were no contradictions, even though they contained a series of theorems that contrasted with ordinary geometrical intuition, the developers of these geometries launched a process of looking at mathematics in an abstract and formal way that grew increasingly stronger in the second half of the century. They also challenged the way in which demonstration in general, and mathematical demonstration in particular, had been viewed for thousands of years. Demonstration had always been conceived of as a reduction to the evidence, or rather as a logical procedure by which the truth of a certain proposition could be traced back to the original truth of certain other propositions, which were accepted as being evident, or true in themselves (the axioms and postulates of various mathematical theories). Following the crisis of evidence brought about by the development of non-Euclidean geometries, demonstration became a simple formal procedure of reduction to postulates, profoundly changing the way in which even the nature and function of the axiomatic method was perceived. However, these final results were achieved only after non-Euclidean geometries were shown to be non-contradictory through the discovery of their Euclidean models, which occurred only after 1868.

In 1826 in Berlin, August Leopold Crelle (1780–1855) founded the famous *Journal für die reine und angewandte Mathematik*, known as the *Journal Crelle*, under direct encouragement from Abel, who published many of his most important findings in it. Originally, it was a German mathematical journal, though it soon rose to international fame. In any case, Joseph Diaz Gergonne had founded the *Annales de mathematique* in France as early as 1810.

ASTRONOMY

Catalogues of double stars

The German Friedrich Georg von Struve (1793–1864), a professor of astronomy and director of the Observatory of Dorpat (now Tartu, USSR), who beginning in 1813 continued Frederick Wilhelm Herschel's research, published the first catalogue of 795 double stars in 1822. This was followed by two more catalogues, which listed the positions of the components of more than three thousand double stars (measured with a micrometer) and some of their characteristics. In the 1822 volume, Struve also published his first estimations of stellar parallax, including that of α Aquilae, which he evaluated at 0.181″ (present value: 0.198″). This was the first calculation of the distance of a star from the earth. Struve's work on double stars and triangulation in the Baltic (1816–1831) demonstrated that cartography and navigation could benefit from the use of astronomical catalogues, tables, etc., only if these were based on new and more accurate observations. This induced Czar Nicholas I to have a new astronomical observatory built at Pulkov, near St. Petersburg. Construction was completed in 1839, and Struve was the first director.

On August 25, 1822, Frederick Wilhelm Herschel died at Slough, where he had set up his own observatory. The work in astronomy that he had conducted for almost half a century in collaboration with his sister, Caroline Lucretia, was continued by his son John Frederick Herschel (1792-1871). With Herschel, astronomy had undergone enormously important qualitative and quantitative transformations. Besides the improvements he made in instrumentation, which enabled him to achieve such an abundance of remarkable results, he also extended the boundaries of astronomical knowledge from the restricted environment of the solar system to the furthest reaches of space, given an absolutely new and physically concrete vision and dimension to the planets, double stars, nebulae, the galaxy, the motion of the fixed stars and the cosmos.

On September 11, 1822 with a decree confirmed by Pius VII on the twenty-fifth of the same month, the Congregation of the Holy Office abrogated the condemnation of Copernican theories that had been issued on March 5, 1616. The repeal was prompted by a request from the Roman clergyman and astronomer G. Settele, who in 1820 had presented the second volume of his *Elementi di ottica e di astronomia* to the same Congregation for the *imprimatur* (first volume, Rome, 1818; second volume, in which the heliocentric theory was expounded, January 1821. Following the deliberation of the Holy Office, Copernican works were taken off the *index librorum prohibitorum* in the 1835 edition.

Pierre Simon de Laplace died in 1827. Publication of his *Traité de mécanique céleste*, begun in 1799, had been completed in 1825. The last part of the work was a collection of the developments that his theories and discoveries had undergone during that period, including his research onthe figure of a rotating fluid in equilibrium, on which he had worked all his life and which was connected to his formulation of a cosmological theory of the original nebula. In his work on the problem of attractions, Laplace introduced two analytical tools for studying physical problems: Laplace coefficients and potential functions.

In 1829, at the Observatory of Königsberg, using the most perfect astronomical instruments of the epoch, in particular the extremely accurate precision heliometer built by Joseph von Fraunhofer, Friedrich Wilhelm Bessel began measuring the parallax of 61 Cygni, which led him to determine its distance from the earth.

The measurement of degrees

The nineteenth century was characterized by a notable flourishing of measurements of degrees and studies in gravimetry in a number of countries throughout the world. In Germany, Karl Friedrich Gauss and Friedrich Wilhelm Bessel did work in this area. Most meridian arcs were measured in India (1800–73, George Everest: 23 degrees 50 minutes), in the USA (during the thirties, from Texas to Minnesota: 23 degrees, data elaborated by John Fillmore Hayford in 1906–9) and above all in Russia (1816–55, Karl Ivanovich Tenner, Fredrich Georg Struve: 25 degrees 20 minutes). Struve's and Tenner's measurements, which were made on an enormous arc that went from the Danube to the Barents Sea, were more or less accurate. Research was also done in celestial mechanics and gravimetry (for more details see PHYSICAL SCIENCES: MECHANICS, 1829–30).

PHYSICAL SCIENCES: MECHANICS

Elasticity and hydrodynamics

A new stimulus to the progressive rationalization of mechanics was provided by studies of the properties of elastic media and work on problems in hydrodynamics. This stimulus arose not only from questions that remained unanswered in the gradual process of mathematizing the laws of motion, but also from new interests that emerged as scientists studied the role of the ether as the seemingly ideal elastic medium for the propagation of optical phenomena.

The *Mémoire sur lois de l'équilibre et du mouvement des corps solides élastiques*, by Claude-Louis-Marie Navier (1785–1836), appeared in 1821. In this memoir Navier laid the foundations for a modern theory of elasticity. According to Navier, the bases of such a theory were closely tied to hypotheses on the interaction among molecules.

Molecular action

The interaction among molecules, in Navier's opinion, was caused by two types of forces—one attractive and the other repulsive—produced by heat. The result, as Navier wrote, was that between any one molecule M and any molecule M′ among those near it there exists an action P that is the difference between these two forces. The mathematical form of this resultant action was not describable, so that the equations of elasticity had to be studied with the presupposition that molecular actions decrease rapidly with distance according to an unknown law. In 1822, Navier adopted an analogous though more detailed model in a discussion outlining the theory of hydrodynamics. In this research, Navier assumed as a principle that, thanks to the motion of a fluid, the repulsive actions of the molecules are increased or decreased by an amount proportional to the velocity with which the molecules themselves move closer together or further apart from each other.

Siméon Denis Poisson also took an interest in the theory of elasticity, and attempted to define the form of the function that would describe molecular interaction, attributing an exponential behavior to it. A rather bitter polemic arose in the pages of the ''Annales de Chimie et de Physique'' on the subject.

Beginning in 1822, Augustin Louis Cauchy (1789–1857) carried out a program of mathematical research on the problems of the finite deformations of continuous media. These analyses also played a fundamental role in the development of fluid mechanics as it related to theories on the ether.

Volume five of the Mécanique céleste

From 1823 to 1825, the science of motion in general was dominated by the publication of the fifth and last volume of the *Mécanique céleste*. It was published in three parts constituting a single text. The first part, dated 1823, contained the results of Laplace's findings on the shape of the earth and its motion of rotation, as well as the statement of an essential part of his theory in a chapter entitled *L'attraction et la répulsion des sphères et lois de l'équilibre et du mouvement des fluids élastiques*. Here Laplace used mathematical analysis in an attempt to reduce the laws of gasses and heat to purely mathematical explanations. In particular, he defined temperature as a function of the density of free heat in casual motion in the free space between material molecules and formalized the discussion of latent heat, linking it to the speed of sound in gasses, which he had already studied in 1816. The second part of the volume appeared in 1824; it examined the oscillations of liquids covering planets and the motions of planets and comets. In the last part, dated 1825,

Laplace returned to the motions of satellites. One of the characteristics of the fifth volume is the attention that Laplace gave to the history of the progress of mechanics, which he saw as the history of the development of knowledge about the universe.

The research of Bessel, Gauss and Abel

The first works of Friedrich Wilhelm Bessel, Karl Friedrich Gauss (1777–1855), and Niels Henrik Abel (1802–29) on oscillations and the theory of pendulums appeared between 1823 and 1824. Evidence of the amount of attention given to these problems in connection with questions of geophysics can be found in Jean Baptiste Biot's report, *Troisième voyage entrepris in Italie et en Espagne pour la détermination de la figure de la Terre, dans les années 1824 et 1825*, and in Laplace's communication entitled *Sur la réduction de la longueur du pendule au niveau de la mer* (1825). Ivory also analyzed the problem of the shape of the earth in the same context in a series of memoirs published in the *Philosophical Magazine*. The most important work in this line of studies was a monograph presented by Bessel in 1828 entitled *Untersuchungen über die Länge des einfachen Secundenpendels*. Bessel regarded the experiments he described and the calculations made in connection with them within the context of an experimental verification of the principle according to which the acceleration of bodies was always the same. As Bessel wrote in 1832 (see PHYSICAL SCIENCES: MECHANICS, 1830–1840), the basic problem was that Newton's system was not mathematically necessary. Its real existence as a system of nature could not be absolutely demonstrated. It could be established with a certain degree of probability, which depended upon the accuracy of experiments.

The end of the decade was marked by the statement of the principle of least constraint according to the form elaborated by Gauss. After analyzing d'Alembert's principle of least constraint and Maupertuis's considerations on the principle of least action, Gauss, in a paper entitled "Über ein neues Grundgesetz der Mechanik" (1829), stated the proposition that the sum of the products of the squares of the deviations of each material point with respect to its free motion times the mass of each point is minimal in real motion.

PHYSICAL SCIENCES: THERMOLOGY

Critical temperature and caloric models

The debate on the nature of heat remained unresolved. It was no accident that Biot, in the second and third editions of his *Précis élémentaire de physique expérimentale* (1821–24), did not take a definite position on the subject, tending instead to evaluate the various caloric models as more or less equivalent.

In 1822 Fourier's major work on the theory of heat, the *Théorie analytique de la chaleur*, was published in Paris. In the *Théorie analytique*, Fourier reproposed, in a general form, his ideas on the impossibility of reducing theories of heat phenomena to the laws of motion. His divergences from the hypotheses on which Laplacian-type programs were based became increasingly more evident. In fact, between 1823 and 1824, Fourier emphasized that thermology was a new branch of mathematical physics, and that its purpose was to calculate the effects caused by heat and not to individualize the nature of the forces upon which those effects depended.

Nevertheless, models played a dominant role that could not easily be eliminated in favor of an increased mathematization of theoretical devices. The help that they provided in the application of physics to chemistry was evidence of their value as interpretive tools, even though, for example, in his memoir *Nuove considerazioni sulle affinità de' corpi del calorico calcolato per mezzo de' loro calori specifici, e de' loro poteri refringenti allo stato gazoso*, after having made use of both the notion of caloric and the wave theory, Amedeo Avogadro advised setting aside these theories, which could only continue to be shrouded in many obscurities and uncertainties, and returning to the formula and to its applications.

The driving power of heat

The investigations of Nicolas Leonard Sadi Carnot (1796–1832) on heat engines made their appearance during this period on a scene dominated by the relationships among thermology, technology, and the basic laws of motion. Carnot's famous *Réflexions sur la puissance motrice du feu, et sur les machines propres à developper cette puissance* was published in 1824. Although he harbored serious doubts on the validity of the principles of the caloric theory—doubts which he developed in notes published posthumously (1878)—Carnot made use of some of the official points of the theory, particularly the ideas on the indestructability of the caloric. On this basis, resorting as little as possible to mathematical discussions and using qualitative language, Carnot stated a general theory of heat engines and cyclical transformations. In this context, Carnot analyzed several propositions regarding the fact that "wherever a difference of temperature exists, driving power can be produced." The new thermodynamics of the second half of the century was based on these analyses, after William Thomson Kelvin (1824–1907) and Rudolf Julius Clausius (1822–88) "rediscovered" the *Réflexions*.

Heat engines and heat

The study of heat machines was given a boost in England by a memoir written by Davies Gilbert on the yield of steam-operated devices, entitled *On the expediency of Assigning Specific Names to All Such Functions of Simple Elements... And Also Some Observations on the Steam Engine...* (1827). Gilbert reproposed several theoretical evaluations already published in his entry on "Steam" in the third edition (1797) of the *Encyclopedia Britannica*, and, for information on yield, suggested referring to engineers' tables, which numerically correlated "the pounds lifted to the height of one foot" with the consumption of "one bushel of coal."

Given the spread of these interests, it is not suprising that, based on the hypothesis that heat is motion, Carnot wrote in one of his notes that according to some ideas on the theory of heat that he had formed, the production of one unit of driving power required the destruction of 2.70 units of heat.

These apparently solely technological considerations did not have any echoes in the academic physical sciences. An 1827 edition of Joseph-Louis Gay-Lussac's text on physics, *Leçons de physique á la faculté des sciences de Paris*, stated the impossibility of finding reasonable motives for choosing among the various hypotheses on the nature of heat. In this situation, opinions on the non-essentiality of caloric models and the need to develop applications of more refined mathematical methods to thermology began to emerge in various scientific circles. Jean Baptiste Fourier stressed this notion in 1829, in a memoir entitled *Remarques générales sur l'application des principes de l'analyse algébrique aux équations transcendentes*. According to Fourier, the study of the theory of equations would clarify physical problems.

PHYSICAL SCIENCES: ELECTRICITY

Oersted's ideas

In 1820, investigations on electrical and magnetic phenomena underwent a radical transformation, thanks to the observation of strange new electromagnetic properties and the consequent need to reelaborate previous theoretical systems. This transformation completely revolutionized nineteenth century physics, both because a unifying theory for electrical, magnetic, and optical phenomena eventually emerged from it, and because this new theory led to a series of developments in the idea of field, which found their earliest synthesis in relativistic theories. In this sense, 1820 marked the beginning of a wave of critical analyses of mechanics and its foundations.

The memoir that opened the debate on the nature of electromagnetic phenomena was Hans Christian Oersted's *Experimenta circa effectum conflictus electrici in acum magneticum* (1820). Oersted experimentally demonstrated that an electrical current flowing along a conductor was able to deflect a magnetic needle located near the conductor. He maintained that this deflection was caused by "effects" of a conflictual nature that took place in the conductor and in the space surrounding it. These effects, in Oersted's opinion, generated spiral motion that acted on the magnetic particles of the matter and had a very wide range of activity in space.

The few pages in which Oersted supported his ideas aroused an enormous amount of interest, since the effects that he demonstrated could not immediately be explained by Newtonian-type interactions. Therefore, not only did the theories that Poisson had developed for electrostatics seem to be called into question; even the general foundations of mathematical physics, bound to attractive and repulsive forces that travelled in straight lines, were challenged, since they were unable to provide an explanation for the spiral motion of the magnetic needles in Oersted's experiments.

Ampère's electrodynamics

In this situation, the premises of which could already be seen in the different approaches taken by physical research in the preceding decades, but which only became explicitly important at this time because of the strange interactions observed in the area of electromagnetism, the organization of programs of study became even more complex. English scientists assumed a critical position in the face of research based on Laplacian-type physical theories. The early reflections of Michael Faraday (1791–1867) offer a particularly good example of this tendency. Debates based on the divergences between Poisson's works and the works of André Marie Ampère arose among French scientists. From these debates, in which attempts were also made to find a possible common basis in Newtonianism, sprung electrodynamics, which emphasized the reducibility of magnetic phenomena to electrical phenomena. This was an enormously important

step, the workability of which was founded on a hypothesis regarding the electrical nature of magnetism, elaborated by Ampère. Poisson, on the other hand, in his famous communication *Sur le magnetisme en mouvement* (1826), attempted to extend the mathematization that he had developed in the preceding decade for electrostatics to the new experimental data. In order to do this, he analyzed a model in which magnetic phenomena were linked to the behavior of the elementary parts of two fluids, the austral and the boreal, in terms of Newtonian interactions. Meanwhile, Ampère directed his investigations against both this and Oersted's approaches.

Ampère also developed his research along Newtonian lines, but, being influenced by Fourier's ideas, he criticized the prejudices that led scientists to accept hypotheses on magnetic fluids, and linked electrodynamics to both propositions whose empirical contents could be rigorously controlled and to a refined mathematization. However, his numerous writings on electrodynamics, published between 1822 and 1827, all rested on the above-mentioned hypothesis, according to which, as Ampère himself wrote in 1822, magnetic phenomena "fell under the law of attraction of two electric currents so long as it was assumed that, on the surface and in the interior of a magnet, there were as many electric currents in places perpendicular to the magnet as there could be lines forming closed curves without crossing one another."

The most complete exposition of Ampère's electrodynamics is contained in his great *Mémoire sur la theorie mathématique de phénomènes électrodynamiques uniquement déduite de l'éxperience*, dated 1823 and published in 1827. In this work, Ampère declared that, guided by the principles of Newtonian philosophy, he had reduced the phenomenon observed by Oersted to forces that always act along the straight line connecting the two particles between which they are acting.

After having thus negated Oersted's spiral theory, and citing Fourier, Ampère maintained that the equations of electrodynamics were "the expression of the facts." Consequently, although one had to recognize the equivalence between his hypothesis and Poisson's fluidistic theory, one could not deny, wrote Ampère, that the electrical nature of magnetism was coherent with his own mathematical formulation and was "founded on the body of facts." In addition to Ampère's writings, a memoir by Faraday on electromagnetic motion and the theory of magnetism, published in 1822, Biot and Savart's memoir on the magnetization of metals caused by "moving electricity," published in 1823, and Félix Savary's *Mémoire sur l'application du calcul aux phénomènes electrodynamics* (1823), all deserve particular mention.

The concept of potential and Ohm's law

The investigations of George Green (1793-1841), who in 1828 published an *Essay on the Application of Mathematical Analysis to the Theories of Electricity and Magnetism*, proceedeed along a line of research that was based on *Poisson's* works in mathematical physics. By means of a general theorem on the integrals of surfaces and volumes and a correct formalization of the concept of potential, Green developed a very modern theory, which was taken up many years later when Kelvin had it published in Crelle's mathematical journal (*Journal de Crelle*).

Contributions to the development of experimental research

were made by Thomas Johann Seebeck (1770–1831) and Georg Simon Ohm (1789–1854). Seebeck's communication on his observations of thermoelectric phenomena (*Magnetische Polarization der Metalle und Erze durch Temperatur-Differenz*) appeared in 1822, and in 1826 Ohn publicized experimental proofs of the law on electrical currents that bears his name even today. He also correlated the intensity of electrical currents in metallic conductors to their resistances and their differences in electrical potential.

In 1827, Leopoldo Nobili (1784–1835) resumed Galvani's studies on animal electricity, experimentally demonstrating the existence of a "frog current," which he wrongly attributed to thermoelectric effects. In order to carry out these experiments, he built the famous astatic galvanometer that still bears his name; he first mentioned it in an article in the *Annales de Chimie et de Physique* in 1828.

The elastic ether

Developments in optics were no less important than those related to electrodynamics. An exhaustive outline of these developments was given in an important memoir by Augustin Jean Fresnel, published by the Académie in 1826. The memoir provided the basis for a series of researches on the optical and electromagnetic ether. It also provided a strong stimulus for studies on the mechanics of elastic solids, which were begun by Navier, Cauchy, and Poisson (see PHYSICAL SCIENCES: MECHANICS, 1820–1830).

Fresnel's intent was to reduce optics to mechanics through the formulation of a wave-type dynamic theory of light. His theory viewed the ether as an elastic medium, assuming that it was made up of interacting particles. The problem, therefore, was to provide a model of the elastic ether, introducing special hypotheses, in addition to the general laws of motion, in order to describe the type of forces that operated among the ethereal particles. As to the discrete ether, Fresnel moved away from a Huygens-like theory of a continuous medium and raised serious questions within the theory of elasticity, in that he was forced to adopt ad hoc hypotheses to exclude the possibility that the discrete medium could transport longitudinal waves.

Cauchy confronted this situation beginning in 1830, still within the context of mathematizing an ether model based on the assumption of central, Newtonian-type forces. The serious problems that arose began to be solved in the following decade, thanks to George Green's critiques of the excessively restrictive nature of Fresnel's and Cauchy's mechanical models of the elastic ether. According to Green, the ether was rigid, not elastic, eliminating the possibility of longitudinal waves (see PHYSICAL SCIENCES: ELECTROMAGNETISM AND OPTICS, 1830–1840).

CHEMICAL SCIENCES

The first organic synthesis: urea

In 1827, on the basis of his long experience in the laboratory, Jöns Jacob Berzelius asserted that a special vital force intervened in the formation of organic compounds (i.e. compounds that came from living organisms), and that consequently they could never be obtained in the laboratory. However, just a year later, almost as if to refute this assertion, Friedrich Wöhler (1800–

82) succeeded in synthesizing urea by heating ammonium cyanate. Berzelius, however, did not yield, maintaining that urea was a borderline case between inorganic and organic substances. But the great barrier was down, and organic substances began to be produced increasingly frequently.

The growth of the list of elements

Inorganic chemistry continued to hold its predominant role, and important contributions were made by Friedrich Wöhler, who obtained aluminum (1827) and beryllium (1828) by electrolysis, and Berzelius, who isolated silicon (1823) and thorium (1828). Nils Gabriel Sefström (1787–1845) discovered vanadium in his laboratory (1829–30), and in France Antoine Jérôme Balard (1802–76) discovered bromine (1825–26). Among the inorganic compounds discovered during this decade were potassium ferrocyanide, discovered by Leopold Gmelin in 1822, sodium pyrophosphate, discovered by Thomas Clark (1801–67) in 1827 and metaphosphoric acid, discovered by Berzelius and Johann Friedrich Engelhart (1797–1837) in 1827.

In 1827, Heinrich Gustav Magnus (1802–70) discovered the first compound $[Pt–NH_{34}][PtCl_4]$, Magnus's green salt, and also worked on the additive compounds of $PtCl_2$ with ethylene, later described by William Christopher Zeise. Magnus was also the author of a rule of electrochemistry, according to which the deposition of a metal by a solution of one of its salts takes place at a given potential.

From iatrochemistry to pharmaceutical chemistry

Among the new organic compounds isolated from plants were coumarin, discovered by Heinrich August von Vogel (1778–1867) in 1820, solanine (R. Desfosses, 1821), caffeine, discovered by Friedlieb Ferdinand Runge (1795–1867) in 1821, codeine (Robiquet, 1822), alizarin, isolated by Robiquet and Jean Jacques Colin (1784–1865) in 1827, aspartic acid (August Arthur Plisson, 1827), and salicin (Leroux, 1829). The discoveries of salicin, the first known glucoside, and the first alcaloids, which took place during this and the preceding decades, marked the triumph of chemistry in pharmacology. The advances made in this field were partly due to the accurate analytical work of Jean Baptiste Dumas (1800–84) and Pierre Joseph Pelletier, who carried out a number of elementary closed-tube analyses of alkaloids (1823). Amadeo Avogadro, on the other hand, gave the elementary formulas for ethyl alcohol, ether and urea in 1821.

The concept of isomerism

The discovery that silver cyanate—Wöhler—and silver fulminate—Justus von Liebig (1803–73)—had the same elementary composition led Joseph Louis Gay-Lussac to formulate the idea (1824) that atoms could arrange themselves in more than one way in a molecule to form different compounds. Berzelius initially rejected Gay-Lussac's hypothesis, although he later accepted it and named it isomerism.

Benzene and aniline

Gay-Lussac's name is also linked to the centesimal alcoholometer (1824), which gave the concentration of alcohol in a water–alcohol mixture based on its density.

The discoveries of benzene by Michael Faraday (1791–1867) in 1825 and aniline by Otto Unverdorben (1806–73) in 1826 were of enormous interest, both theoretically and practically. In 1826, Unverdorben also discovered guaiacol, prepared by the pyrolysis of guaiacum wood.

Progress in biological chemistry

Important work on biliary acids was done by Gmelin (1826) and by William Prout, who found hydrochloric acid in gastric juices (1823) and divided the substances that make up foods into three classes: carbohydrates, fats, and proteins. An important new figure on the biochemistry scene was Liebig, who in 1829 discovered ippuric acid in horses' urine.

The emergence of physical chemistry. Treatise writing

Physical chemistry became increasingly well-defined as an independent branch of study. In 1822–23, Charles Cagniard de la Tour (1777–1859) indicated the existence of a critical temperature for gasses, above which they could not be liquified in any way, even when highly compressed.

In 1824, Germaine Henri Hess (1802–50) stated the law of constant heat summation for heats of reaction. Henri Dutrochet (1776–1847) conducted the first quantitative experiments on osmosis (1826–28); he was also the first to observe (1826) the motion of particles suspended in a liquid, described more in detail by the botanist Robert Brown (1773–1858) in 1828 and now commonly called Brownian motion.

The following year, Thomas Graham (1805–69) stated the law of the rate of diffusion of gasses. In 1829, Augustin Pierre Dubrunfaut (1797–1881) introduced the osmotic process for eliminating the salts from molasses.

The most famous treatises that appeared during this period were the *Traité de chimie appliquée aux arts,* by Jean Baptiste André Dumas (the first volume was published in 1828), and the *Handbuch der analytischen Chemie* (Heinrich Rose, 1829).

LIFE SCIENCES

The articulation of semiotics

René Théophile Laënnec's lesson became popular in France and elsewhere: syndromes became diseases through the definition of their symptoms and their connections with specific structural lesions. Descriptions such as "irritation" and "inflammation" gave way to a varied and articulated pathogenesis and a rigorous semiology. The supremacy of French medicine was partly due to the organization of its hospitals and to clinical practice, which was made compulsory for all medical students by the university reform passed by the Convention. Pierre Fidè Bretonneau (1778–1862) shared Laënnec's criteria when he asserted that diagnosis and therapy should be founded on a triple goal: etiological, anatomapathological, and clinical specificity. Bretonneau applied this methodology to his descriptions of diptheria and typhoid fever.

The objections raised by Jean le Rond d'Alembert at the end of the eighteenth century against those who professed to apply the beneficial prognosis of preventive variolation as verified on large numbers of people to singles cases were again brought up

to oppose the use of the statistical method in medicine. It was Lambert Adolphe Jacques Quételet (1796–1874) who overcame these difficulties, giving rise to true biomedical statistics. The progress of organic chemistry, especially the synthesis of urea, influenced the development of the semiological approach in pathology.

During the third decade of the century, noteworthy progress was made in blood transfusion in both laboratory animals and in humans, especially in cases of hemorrhaging after childbirth. There was considerable progress as well in pharmacology, with the discovery of salicyline, an effective remedy for acute rheumatism of the joints. The substance was first extracted from willow bark by Felice Fontana (1730–1805), and later synthesyzed by Adolphe Wilhelm Hermann Kolbe (1818–84) as salicyclic acid. The gamut of chemical medicines was enlarged with the addition of cloral, several bromides, antipyrin, and phenacetin.

Botany and paleontology

Alexandre Brongniart (1770–1847) made some important contributions to botany and paleobotany during this period. His essay *Reserches sur la génération et le développement de l'embryon des phanérogames* (1827) gave the first description of the structure and development of pollen. His *Prodrome*, published in the *Grand dictionnaire d'histoire naturelle* (1828), created the basis for his later system of classification; that same year he completed his *Histoire des végétaux fossils*. It has been said that these two works were to paleobotany what Linnaeus's *species plantarum* was to the classification and nomenclature of present-day plants.

Embryological classes

In 1827, the Estonian embryologist Karl Ernst von Baer communicated to the Academy of St. Petersburg that he had succeeded in observing the ovicell in the ovary of a mammal (a dog). The following year, von Baer published the first volume of his *Entrickelungsgeschichte*, which provided the premises for comparative embryology and instituted a productive correlation between comparative anatomy and later organic science. Cuvier's "type" became the embryological class, which von Baer identified in the particular relationship of the parts to the whole. The second volume of the work came out unfinished in 1837, when the author had returned to the philosophy of nature. He later entered into a passionate polemic against Charles Darwin, whose views on nature he attacked from a theological point of view.

François Magendie formulated the law of the function of medullary nerve roots in 1822. Although he may not have been aware of analogous researches by Charles Bell (1774–1842), he actually clarified their meaning.

Structural unity

Jean Baptiste de Lamarck's activity declined after he finished his *Histoire naturelle des animaux sans vertèbres* in 1822; Étienne Geoffroy Saint-Hilaire (1805–61), an anatomist and comparatist who advocated the "unity of plan" by which all living things were made, took over as Cuvier's antagonist. Expounded for the first time in his *Philosophie anatomique* (1818), this concept presided over Geoffroy Saint-Hilaire's successive works, in which he attempted to demonstrate how invertebrates and vertebrates shared a like structure, with the exception that invertebrates lacked a circulatory system. Cuvier replied that the various groups had only their "animality" in common. Geoffroy Saint-Hilaire also seemed to place more of an emphasis on morphology than on phylogeny. In 1825, Cuvier accused Geoffroy Saint-Hilaire of seeing "the embryo of vertebrate animals in the worm" and of "making the classes originate out of one another." The controversy culminated between February and April of 1830 at the Académie des Sciences, and its echoes even reached as far as Goethe. This conflict of the comparatists prepared the ground for the birth of Charles Darwin's theory of evolution.

Meanwhile, completely opposite notions invested embryology, where Johann Friedrich Meckel, Antoin Serres and Martin Heinrich Rathke maintained that "a universally valid idea—plan, scheme, type—exists for the different classes and governs the early formation of the various parts" (Rathke). Von Baer's concept of development was different. In his *Entwickelungsgeschichte*, he maintained that the general characters of a type appear early in the embryo and direct the successive phases of evolution. Von Baer intended type in a morphological sense as the "positional relationship among the parts of an organism." There were four types, which he defined as periperical or radial, metameric or longitudinal, massive and vertebrate. Comparative embryology originated with von Baer, and together with comparative anatomy defined the essential differences between the zoological categories. Von Baer's research on the eggs of mammals was also important (Baer vescicles). He was a convinced animalculist.

EARTH SCIENCES

Opposition to catastrophism

Jean Léopold (Georges) Cuvier's interpretation of the earth's history as a series of catastrophic upheavals predominated throughout the scientific world as a valid explanation for the succession of different fauna and flora in the superimposed geological formations of a particular region. The implications of attributing the organisms that followed each "revolution" to new populations led to philosophical and theological controversies. For those who supported the fixity of species, these new species could only have come from two sources: migrations from regions that had not been devasted (Cuvier supported this hypothesis) or from a new creation of plants and animals. Cuvier gave a complete exposition of the catastrophic system in 1825 in his *Discours sur les révolutions du globe*, an elaboration of his *Discours préliminaire* of 1812.

Some, such as Louis Constant Prevost (1787–1856), opposed the catastrophic theory from the very beginning. The most lively polemic on the subject took place between Cuvier and Étienne Geoffroy Saint-Hilaire (1805–61) at the Académie des Sciences, beginning in 1820. When Geoffroy Saint-Hilaire, who had had the occasion to observe crocodiles in the Nile during the Napoleonic campaign in Egypt, studied the "crocodile" (Teleosaurus) of Caen, already examined by Cuvier, he asserted that it could be a link between reptiles and mammals. The greatest opposition to this theory came from Charles Lyell (1797–1875), who had met Cuvier in Paris in 1829. In 1830, Lyell began publishing his *Principles of Geology*, in which he discussed the causes that acted to modify the earth's crust and maintained that an analogy existed with present activity, in contrast to the succession of sporadic cataclysms proposed by Cuvier.

The application of catastrophism and attempts to enumerate the upheavals suggested a comparison between the successions of rocks and their fossils in regions distant from each other. In 1823, Alexandre Brongniart wrote a memoir comparing the rocks of the area about Vicenza to those of the Paris basin; his *Tableau des terrains qui composent l'écorce du globe*, which divided sedimentary layers into forty-five groups, appeared in 1829.

Alexandre's son, Adolphe Brongniart (1801–76), studied fossil plants. Anselme-Gáetan Desmarest (1784–1838) outlined the paleontological history of crustaceans (1822), and William Buckland examined and illustrated the fossils of the previous epoch in his *Reliquiae diluvianae*, which appeared in 1823.

Variations that had occurred in the landscape during our epoch were illustrated in 1825 by Count Fossombroni, prime minister to the Grand Duke of Tuscany. He compared the course of the Arno River as described by Strabo and as shown on ancient maps with the present situation.

Criteria for dating tectonic events

The lessons held by Alexander von Humboldt in Paris in 1827–28 gave the broadest contemporary presentation of the geology of the entire earth, and at the same time went beyond the more rigid Plutonist positions.

The greatest innovation of the period came from Jean Baptiste Élie de Beaumont (1798–1874), who believed that movements in the earth's crust were caused by the contraction of the globe and explained the presence of crystalline rocks in the nuclei of mountain chains as the effect of huge longitudinal faults. On the side of the fault that moved downward, sedimentary rocks surfaced, and on the side that rose, crystalline rocks, which were older and located further below the surface, appeared. The age of the dislocations, which were caused by catastrophic events, could be determined on the basis of the newer layers involved in the movement, as well as the oldest layers formed after the event, and therefore untouched by it.

This was the basic criterion used by geologists from this time forward to date tectonic events. Among the observations that led to a reexamination of the structure of mountain chains, those of André Brochant De Villiers (1772–1840) in 1816 on the non-central position of granite in the Alps, and analogous observations made by Louis François Ramond De Carbonnières (1775–1827) in the Pyrenees, deserve particular mention.

In 1821, Christian Leopold von Buch refused to acknowledge that the "granite" (monzonite) of Predazzo in the Trentino region, discovered by Giuseppe Marzari-Pencati, who observed its metamorphic effects on the limestone of the Trias, was recent. Von Buch defended his own ideas against the evidence of the facts.

Symmetry in crystalline forms

Studies in physical crystallography were given a significant boost by the work of two scholars: Carl Friedrich Naumann (1797–1873), who in his *Grundriss der Krystallographie* (1825) and other later writings used inclined axes of symmetry to study trimetric crystals, and Johann Friedrich Hessel, who, in the entry "Krystall" in Gehler's dictionary of physics, considered the microscopic symmetry of all crystalline forms, demonstrating that they could all be categorized into thirty-two classes of symmetry.

The mathematical and physical study of crystals was well-established by this time.

The internal heat of the earth

Regular and continuous measurements of the internal temperature of the soil were begun in 1828 in the mine shafts of Prussia and Saxony. Similar observations were made in provinces and regions of England, as well as in other countries. A thermometer located twenty-eight meters under the ground in Paris has given a reading of 11.8 degrees Celsius since 1780.

Geological and paleontological periodicals

Like many geologists of his time, Heinrich Georg Bronn (1800–62) journeyed to northern and central Italy, Switzerland, and southern France to study the Alps and the Auvergne (1824–28). In 1830, in order to fulfill the growing need to speed up scientific communication and to bring together writings on geology and paleontology that were scattered throughout the publications of the various academies, Bronn, together with von Leolinart, founded the *Jarbuch für Mineralogie, Geognosie, und Petrefaktenkunde* in Heidelberg. This initiative, which also had the purpose of promoting scientific debate, was soon joined by the *Lethea Geognostica*, which had the same goal of uniting and spreading knowledge in the field of geology, but was oriented more towards problems regarding the history of the organic world.

The need for communication had become so strongly felt because of the increased number of scholars who worked in this discipline. Periodicals tended to make the results of studies in various countries quickly available beyond the boundaries of academic meetings.

Treatise in geology

In 1822, William Daniel Conybeare (1787–1857) published his *Outlines of the Geology of England and Wales*, which contained illustrations of the Primary, Secondary, and Tertiary geological formations of England and Wales, with the typical fossil associations and their state of the conservation in the various sediments.

Paleontological discoveries

In 1822, Gideon Algernon Mantell (1790–1852), together with Charles Lyell, visited the stone caves of Tilgate Forest, where interesting paleontological discoveries of typically tropical fauna and flora had been made, revealing the deltaic nature of the deposit.

Fox's observations

In 1830 (perhaps even a few years earlier), Robert Were Fox (1789–1877) observed the phenomenon of spontaneous electrical polarization in the lead deposits of Cornwall. Almost exactly a century later, this observation gave rise to a geoelectrical method of prospecting for ores.

Studies on Vesuvius

In 1825, in collaboration with Niccolò Covelli (1790–1829), Teodoro Monticelli (1759–1845) published a *Prodromo della mineralogia vesuviana,* an unfinished treatise on Vesuvius and its eruptions.

New oceanographic voyages

A voyage into the seas of the Antarctic, led by Thaddeus Bellinghausen, ended in 1821 (it had begun in 1819). Other voyages of circumnavigation were undertaken by the *Beagle* (known for Darwin's voyage) and the *Adventure* between 1826 and 1836. Between 1826 and 1829, Jules Dumont d'Urville made oceanographic voyages in which he gathered a large amount of data, especially on temperature. However, because of errors made in using the apparatus on board and numerous instrument failures, d'Urville was led to believe that deep-water ocean temperatures never went below about four degrees Celsius, giving rise to controversies that were only definitively resolved after the expedition of the *Challenger,* even though contemporary opinions contrary to d'Urville's were advanced. In fact, Emil Lenz correctly interpreted a number of deep-water temperature measurements between 1823 and 1826. Recognizing that there was no physical law that would prevent abyssal temperatures from dropping below four degrees Celsius, Lenz made accurate studies on the physical conditions of the oceans, also studying their circulation.

Hydrological experimentation

Several important experimental findings were made during this decade. In 1827, Jean Daniel Colladon (1802–93) and (Jacques) Charles-François Sturm (1803–55) determined the velocity of the propagation of sound in water. In experiments conducted in the lake of Geneva, a value of 1,435 meters per second was measured. Heinrich Wilhelm Brandes (1777–1834) encouraged the use of tephigrams and meteorological bulletins as auxiliaries to navigation as early as 1820. In 1828, John Augustus Lloyd determined that there was a difference in water level between the Atlantic and the Pacific oceans at the isthmus of Panama.

The contributions of Fourier, Navier, and Merian

Valuable theoretical results were presented in the *Théorie analytique de la chaleur,* a basic treatise on mathematical physics by Jean Baptiste Fourier (1768–1830) that was extremely important for the future development of marine thermics and problems of microstructure, diffusion, etc. In 1823, Claude-Louis Navier (1785–1836) introduced the famous hydrodynamic equations (later named after him) that perfected those of Euler. In 1828, Johann Rudolf Merian put forth his theory on seiches (valid even today), which analytically established the eigen periods of oscillation for water in a basin.

APPLIED SCIENCES

English supremacy

The political stability that followed the Restoration favored the rapid development of productive forces, especially in industries that were not dominated by military needs. England had the advantage in this situation. Since its industrial structure was already much stronger than that of other European countries, it could now make the most of the opening of extra-European markets.

The production of iron, which had fallen off after 1815, began to rise again until 1835 (the production of pig iron in England reached a record one million tons); however, it was conditioned by the transition from coal to coke and by the fact that as blast furnaces grew larger, neither steam nor water compressors could produce enough air to reach the high temperatures required. In 1829, James Beaumont Neilson (1792–1865) of Glasgow devised a new method, which consisted of introducing preheated air into the furnace. The inflammable gasses of the blast furnace were used to preheat the air needed. In this way, there was less dispersion of heat and a more economical use of fuel. In addition, a direct connection was made between the mining industry and the metallurgy industry, since coal could now be used directly in the furnaces. Furnaces also became larger, and the shape went from quadrangular to cylindrical.

The preparation of new alloys of steel was tied to scientific experimentation. In 1822, Michael Faraday (1791–1867) obtained chrome and nickel alloys of steel in the laboratory; in 1823, Hans Christian Oersted (1777–1851) chemically produced a small quantity of aluminum. However, this discovery was not immediately exploited industrially.

Large factories

The mechanization of the textile industry basically depended on the spread of the mechanical loom, the primary machine of the industry. In England alone (excluding Scotland), the number of looms rose from 2,500 in 1813, to 12,150 in 1820, and to 45,000 in 1829, reaching 85,000 in 1833. The dimensions of the phenomenon transformed it from a simple quantitative increase into a qualitative increase of extraordinary importance; this was the birth of great modern industry. The mechanization of the entire productive cycle was finally reached during this period. In 1821, Liéven Bauwens and Didelot patented a flat silk comber; in 1824, Charles Danforth of Massachusetts built a cotton spinning machine; and in 1825, Philippe de Girard (1775–1845) developed a mechanical wool comber, preceding Platt and Collier by two years. The need to achieve mechanical continuity between the various operations without human intervention led to the development of machines that performed more than one operation. The tendency to connect the various operations and the machines that performed them gave manufacturing the structure and the aspect that characterized it for the entire nineteenth century: a single building held the various machines, which were arranged in series around a single central motor. The first great factories were the cotton mills of Manchester. In fact, cotton was the industry that experienced the greatest development during these years. It took advantage of the low purchase price of the raw material (447,000 quintals were imported in 1821) and the large market for its product, which was essentially more accessible than the French product.

Urban concentration

The English work force, historically privileged with respect to European workers by the lack of ties constricting its mobility, poured into the new sectors of employment. In 1833, the cotton industry alone absorbed 237,000 workers in England.

Their total independence, by this time, from fixed energy sources (wood and waterways) and the need to reduce the incidence of transport costs favored the concentration of the various industries in commercially privileged centers. The convergent process of urbanization of large masses of peasants (farm hands and share-croppers) gave English industrial centers the structures upon which the large modern cities later modelled themselves.

In any case, the rapid influx of inhabitants into the rather small urban centers led to modifications in the traditional structure of houses, as buildings began to develop vertically. Techniques were adapted to the new needs, and by this time the method of metallic reinforcements was almost always used. Toward the end of the decade, the first angle irons for construction work were laminated in England.

In 1824, Joseph Aspdin (1779–1855) devised a new type of cement by calcinating mixtures of limestone and clay at temperatures high enough to synthesize them. The new product was so similar in aspect and consistency to Portland stone that it was named after it. Concrete made from Portland cement was used by Marc Isambard Brunel (1769–1849) to fill the riverbed over the tunnel under the Thames, construction of which was begun in 1826.

The steam locomotive

By this time, the advantages of rail transport had been recognized in England. Immediately after the Stockton–Darlington line was built (1825), work was begun on the Liverpool–Manchester railway. The great brick-and-stone Sankey viaduct, which had nine arches, approximately twenty-two meters high and with spans of fifteen meters, was built on this line.

However, the use of locomotives did not progress at the same rate as the growth of railways. Even after the successful trial runs of George Stephenson's first engines on the Stockton–Darlington line, the directors of the Liverpool–Manchester branch were leaning towards a railway that used inclined planes and cable tractions on uphill stretches. Only the high cost of fixed steam installations convinced them to experiment with locomotives. Meanwhile, after the success of his *Locomotion No. 1* (1825), Stephenson incorporated two important innovations. He installed a tube boiler, designed by Marc Sequin (1786–1875) in 1828, and connected the piston of his cylinder directly to the wheels of the locomotive. *The Rocket* won a race at Rainhill in 1829; the following year, in collaboration with his competitor *Rennie* (1791–1866), Stephenson founded the first locomotive factory. In 1830, research into stabilizine rails led to the adoption of wooden ties set into a roadbed of crushed stone. By the end of the century, there were twenty-six locomotives operating in England, while two had been exported to France and four to America.

The daguerreotype and Braille

Joseph Niepce (1765–1833), who had exposed photographic plates in 1827, founded the Société Photographique in 1830, together with Louis Jacques Daguerre (1787–1851). In 1830, Louis Braille (1809–52) devised a form of writing for the blind.

Intensive agriculture and diet

Agricultural production also had to adjust to the new needs of the industrial society. The work of the large schools that had been opened at the end of the preceding century—the Hohenheim Institute, founded by von Schwerz in 1818, and the Agricultural School of Roville, created by Matieu Dombasle in 1819—finally became tangible. Farmers no longer used markets only to rid themselves of surplus products, but began to produce for the market. In England, where more efficient methods of production accompanied intensive cultivation, the diet of large sectors of the population changed: the consumption of potatoes became common, and meat was no longer considered a luxury. Vacuum canning of products became widespread and ice replaced salt for preserving fish.

Thanks to the machines introduced into agriculture, especially harvesting machines—the best one was built by Patrick Bell (1799–1869) in 1826—English farmers were able to achieve a fifty-percent higher per-acre production than French farmers.

1830–1840

THOUGHT AND SCIENTIFIC LIFE

The new situation in Europe. Scientific associations and institutions

The forces of the Restoration suffered a serious defeat in France with the events of July, 1830 and successive Orleanist monarchy. Essentially the product of a new compromise and destined in the long run to disappoint the expectations of political forces that were initially neutral or favorable, the new regime was the result of the victory of moderates who were not aware of escalating social problems. These were the typical bearers of a culture that in some ways went back to the now classical themes of liberalism, but in other ways was open to innovative currents, as long as they were compatible with the specific needs of the constitutional monarchy. On the level of cultural organization, figures who were both popular and authoritative, such as Victor Cousin, were able to lend a certain prestige to the scientific and didactic policies of a monarch such as Louis Philippe, who was on the throne not by divine right, but by the force of bayonets.

On an international level, the "July monarchy" was not the only sign of the decline of the Holy Alliance. Liberal reforms were being made in a number of countries such as Switzerland and England. England and France established new foreign policies, and it became an international principle to reject the so-called right of intervention. Even in the area of revolutionary thought, following the failure of the uprisings of 1831, secret sects like the Carbonari were replaced by nationalist associations modelled after the Giovine Italia, founded by Mazzini in 1831.

The renovation of scientific institutions in England, still in the vanguard from an economic point of view and extremely advanced technologically but in some ways increasingly isolated culturally from the continent, was particularly significant. In 1830, in his *Reflections on the Decline of Science in England*, the mathematician Charles Babbage (1791–1871), reflecting on the causes of the scientific decline in England, pointed to the inactivity of the Royal Society, whose golden age seemed long past, the traditionalism of university life, and the purely artisanal origins of technology, which made the importance of scientific progress a less tangible factor in the organization of economical and social life than in other places.

However, Babbage's critique was also political. It included stabs at the state for both its absenteeism with respect to reforms and its absolute lack of any program for financing research projects, which were left to the ingenuity of artisans or, if the projects were more complicated and costly, to the private initiative of factory-owners, mineowners, etc.

The long war with France and the successive isolationist policy had also contributed to severing England's contacts with French scientific culture and even with German culture. The first signs of recovery came mainly from the Scottish universities, which were open to the influence of continental, and especially German, science. Premises established during the preceding decades, such as the creation of efficient new institutions (the Royal Institution in 1801, the Linnean Society in 1805, the Geological Society in 1807, the Astronomical Society in 1820), favored the formation of a reformist scientific opinion that was antagonistic to the conservatism of Oxford and Cambridge. In particular, demands were made for a revival of the Royal Society, for abolishing the rights of the Anglican Church over the universities, for more emphasis on scientific subjects in the schools and, finally, for contributions from the treasury for research. In 1831, these efforts at reform came together in the establishment of the British Association for the Advancement of Science. The work of the society was not in vain. A reform of university teaching was approved, new centers of research such as the Royal College of Chemistry (1845) and the Government School of Mines and Science (1851), were created, and research programs were launched.

While countries with long scientific-cultural traditions such as England were profoundly renovating their academic structures, other countries that had entered into more direct relations with the rest of Europe established new universities. For example, the University of Athens was founded in 1832.

Comte's Cours de philosophie positive

Important happenings in philosophical thought during this decade were the development of Auguste Comte's positive philosophy in France and the crisis of Hegelian philosophy in Germany.

The first volume of Comte's *Cours de philosophie positive* was published in 1830, followed by the second volume in 1835, the third in 1838, the fourth in 1839, the fifth in 1841, and the sixth and last in 1842. Comte's famous classification of the sciences, his law of the three stages, and his description of sociology as the supreme science, able to provide the "positive" basis for a rational reorganization of society, were all expounded in detail in these six volumes. Comtian positivism was in some ways a direct heir of Enlightenment thought (faith in science, a completely lay concept of humankind and the world, etc.), though in other ways it was radically innovative (its profound interest in the historical development of humanity, even in its primitive phases, as well as in the social problems arising out of the industrial revolution).

The conviction that the only valid knowledge in modern times was scientific knowledge; the notion of science as a public concern; the relative autonomy of mature theoretical research; the applicability of a criterion for discriminating between theories worthy of further investigation and programs to be abandoned; and finally the conviction that a necessary condition for the progress of humanity in the broadest sense was the progress of knowledge, meaning scientific knowledge: these were the essential features of Comte's thought. They made him the first positivist thinker and the source of the school of thought that became one of the dominant currents in European philosophy in the second half of the century, even though this school's ideas often assumed different characteristics than those of its founder.

The crisis of Hegelianism

Hegel had staunchly opposed various cultural and religious currents, such as Pietism, romantic liberalism, and the historical school of law represented by Friedrich Karl von Savigny (1779–1861), also a professor at the University of Berlin. It was in the polemic against Savigny that Eduard Gans (1797–1839), one of Hegel's most brilliant disciples, especially distinguished himself. Gans was actually hired as a professor at the faculty of law in Berlin to oppose the historical school's influence on the students. Another Hegelian, the theologian Philipp Marheineke (1790–1846), a professor in the faculty of theology, polemicized instead against those who, in the wake of Schleiermacher, upheld the predominance of feeling over reason in religion. Thus while a fairly large current, later called the Hegelian right, tended to interpret the master's thought in a clearly conservative way, indirectly supporting Fries and the other theoreticians of romantic liberalism, still what neither the government nor the liberals saw, at least one man saw as early as 1833 (as Friedrich Engels wrote in 1888). This was the poet Heinrich Heine (1797–1856), for whom Hegel was "the Philip of Orleans of philosophy," founding, or better arranging a new eclectic regime, in which everything and everyone was given a precise place. But for Heine the path of history had not come to an end, and "naturally" had to pass from philosophy to politics.

Heine's work, *Zur Geschichte der Religion und Philosophie in Deutschland,* which violently criticized Christianity, appeared in 1834. Almost simultaneously, a "leftist" tendency in the interpretation of Hegel sprung up. Two practical and political problems most impassioned the youth of the period—constitutional reform and the religious question—but because censure was still extremely severe in Germany, in contrast to France and England, the second topic was the most bitterly debated, often in connection with the numerous religious conflicts of a multiconfessional country, while state control prevented any debate on the first. The initial factor that led to the split in the Hegelian school came from theology: the first volume of *Das Leben Jesu,* by David Friedrich Strauss (1808–74), was published in 1835. Strauss's use of the Hegelian method, which he closely combined with Enlightenment tradition and with the empirical tools of historiography (philology, critiques of sources, etc.), led the Hegelian right, as the interpreters of the equivalence between rationality and reality as a legitimization of the existing *status quo,* to assign to the theologian Bruno Bauer (1809–82), a student

of Marheineke's, the task of blasting Strauss in the *Jahrbücher fur die Wissenschaftliche Kritik,* the official journal of Hegelianism. In 1836, Strauss published the second volume of his review, and by 1837 the break was definitive. The *Hallische Jahrbücher,* which became the organ of the left, began coming out in January, 1838. In 1839, Bauer himself clamorously passed to the left, while the first polemical works of Ludwig Feuerbach (1804–72), another collaborator of the Halle *Annals,* appeared during these same years (1839–39).

Philosophy in Great Britain and Italy

The philosophical scene in Great Britain and Italy was much less polemical. A predominant figure in Great Britain was William Hamilton (1788–1856), a professor at the University of Edinburgh from 1836, who among other things provided his students with the first information on recent German philosophy. (Personally, Hamilton considered himself a supporter of Kantian-type ideas.) A project for reforming the Baconian inductive method, proposed by William Whewell (1794–1866), was also particularly significant. In his *History of the Inductive Sciences* (1837) and *The Philosophy of the Inductive Sciences* (1840), Whewell applied a typically British train of thought to an extremely articulated philosophical framework, which also contained Platonic and Kantian elements. (It is interesting to note the recent interest in this author in twentieth-century philosophy, especially in the context of inductive logic and the logic of scientific discovery, in the modern sense.) Inspired by Whewell, John Stuart Mill (1860–73) undertook a systematic study of logic, which culminated in 1843 in a work entitled *A System of Logic, Ratiocinative and Inductive.*

The figure of Antonio Rosmini-Serboti (1797–1855), the champion of a declaredly spiritualist current that nevertheless reflected a certain Kantian influence, emerged among Italian philosophers. His *Nuovo saggio sull'origine delle idee* was published in 1830, and his *Antropologia* in 1838. But, in opposition to spiritualist metaphysics, a movement inspired by eighteenth century Enlightenment thought arose. Its program was to assign philosophy the primary task of investigating the concrete—economical, political, scientific-technical, etc.—phenomena that historically gave rise to civilization. The driving force behind these studies was the journal *Il Politecnico,* founded in 1839 by Carlo Cattaneo (1801–69).

The congresses of Italian scientists

Italian scientists also attempted to find their place in European scientific life by holding scientific congresses. The promoter of the first meeting, held at Pisa in 1839, was Prince Carlo Luciano Bonaparte, nephew of Napoleon and a naturalist and scholar, who had come into contact with the ferments and tendencies of foreign scientific life during his many study trips to England, Germany, and France. Later congresses were held in various cities of Italy, from Turin (1840) to Venice (1847). Pope Gregory XVI, however, refused the invitation extended to him and prohibited his subjects from participating in the meetings.

Nevertheless, the congresses took place fairly frequently, and flourished among the scientists and scholars of the various states of the peninsula.

MATHEMATICAL SCIENCES

The emergence of a new concept of algebra

One of the most salient events around 1830 was the gradual transformation of the very notion of algebra. Traditionally, it had been considered the branch of mathematics that dealt with the study of literal equations, but by now this task seemed fairly limited by the results of impossibility that had been reached. The demonstration of the algebraic insolubility of general equations higher than the quartic provided by Ruffini in 1798 had gone practically unobserved at the time; but by now this concept was completely accepted, having been reconsidered in Niels Henrik Abel's work of 1824, and by August Leopold Crelle (1780–1855) in 1826, and then finally established in a perfectly rigorous manner by Abel's theorem of 1828. Of course the study of this type of equation did not end here. In fact, the modern theory of algebraic equations, or the study of the conditions of resolvability of particular classes of equations, actually arose at this time. Although as early as 1829, Jacques Charles Sturm (1803–55) had presented a memoir to the Paris Academy in which he put forth his famous theorem on the resolution of numerical equations, the novelty was that considerations of an abstract and structural nature now emerged in treatments of these problems. This could already be seen in the writings left by Evarist Galois (1811–32), who (developing ideas that Ruffini and Lagrange had already glimpsed and Abel had outlined) associated with every algebraic equation a certain number of permutations of its coefficients, and brought the question of its resolvability by radicals down to the study of certain structural conditions that the relative group of permutations had to possess. Thus the new theory of groups, the first branch of modern abstract algebra, was born out of the old theory of equations. This occurred much later, however. Galois's intuitions, which were only barely outlined, remained unpublished until 1846, and even then were seldom understood and appreciated. The heritage of the young French mathematician could only be considered completely assimilated toward the end of the 1860s. This acceptance took place mainly as the result of the efforts made by Camille Jordan (1838–1921) to clarify and develop Galois's ideas.

However, the abstract tendency in algebra also developed independently of the theory of equations. The *Treatise on Algebra*, a work by George Peacock (1791–1858), came out in 1830. The first volume dealt with common numerical algebra, called arithmetic algebra, while the second proceeded to build a so-called symbolic algebra. This was arrived at by applying a fairly nebulous principle of permanency of equivalent forms, through which, in essence, Peacock postulated the permanency of the validity of the formal laws of operations as defined for the common classes of numbers, without the restrictions placed upon them in those applications. In this way, the fairly precise concept of symbolic algebra as a theory of operations defined by formal properties that were independent of considerations on the range of objects to which they were applied and, in particular, were free of any reference to numerical interpretations began to emerge in the discipline. This was clearly a prelude to the abstract algebraic mentality that characterized English mathematics in the following decade.

The birth of the analytical theory of numbers

Analysis and geometry mainly underwent developments of a technical nature during this period. Deserving of mention among these, with regard to analysis, were the creation of the theories of elliptical (1837) and curvilinear coordinates (1838) by Gabriel Lamé (1795–1870), as well as the systematic application of methods of infinitesimal analysis to arithmetical questions. This was the birth of the analytical theory of numbers, the foundations of which were laid out in several works published in 1839 and 1840 by Peter Gustav Lejeune Dirichlet (1805–59). Dirichlet thus took up the great arithmetical tradition of Karl Friedrich Gauss, whose works he knew very well. A large body of new knowledge came out of Dirichlet's application of these methods. Dirichlet was one of the most distinctive representatives of the already cited quest for rigor in analysis. Having complained of a number of imperfections in the theory of trigonometric series, he offered ways of correcting them. He later extended his investigations to series of spherical functions (1837), and in that endeavor was inspired to introduce the concept of absolutely convergent series. In 1829, he had already provided the definition of the notion of function in the form that then remained definitive, while in the area of the theory of potential, he later stated and applied the principle that is named after him and that gave rise to numerous studies in mathematical physics.

The development of projective and analytical geometry

Geometry was mainly characterized by the development of projective research. The work of Michel Chasles (1793–1880), who in 1829 presented his *Aperçu historique* to the Academy of Brussels, stands out in this field. In addition to a profound historical reconstruction of the principles of the science of geometry, the *Aperçu historique* presented a number of new results in the theory of quadrics and developments in projective geometry obtained through a systematic application of the principles of duality and homography. The memoir also contained the complete theory of involution. Chasles's work during the following decades also provided fundamental contributions to projective geometry (for example, he created the theory of the characteristics of conic systems).

Among the principles that the geometers of the period utilized most frequently was the principle of continuity, with which Karl Georg von Staudt (1798–1867) obtained new metric and projective properties of second order curves in 1831. In 1832, Jakob Steiner (1796–1863) published his *Systematische Entwicklung der Abhängigkeit geometrischer Gestalten von einander*, in which he reconstructed projective geometry by successive generations, beginning with extremely simple figures and presenting new theories which, in the following decades, led him to very productive results. With the work of these authors, the task of determining the peculiarities of projective properties with respect to metric properties, initiated by Poncelet, can be considered finished. By this time, projective geometry possessed an abundance of procedures and principles of its own that clearly distinguished it from metric geometry.

One factor that should not be overlooked is the appearance during this period of a need to axiomatize branches of mathematics other than geometry, first expressed in *The Mechanical Euclid* by William Whewell (1794–1866), published in 1837.

This tendency was in some ways a part of the general aspiration to increased logical rigor, already mentioned several times before, and it began laying the cultural foundations for the general reevaluation and restructuring of the axiomatic method that characterized the final decades of the century.

Of particular importance to German scientific life during this century were the seminars held at the major universities, where the teachings of the greatest scientists focused on contemporary questions in research. The first of these seminars was founded in 1834 in Königsberg by Karl Gustav Jacobi.

Meanwhile, in France, Joseph Liouville (1809–82) founded the *Journal de mathematiques pures et appliquees* (1836), which soon became quite authoritative. Liouville directed the journal for forty years.

ASTRONOMY

The measurement of stellar distances

At the beginning of the century, Joseph von Fraunhofer had built an instrument for accurately measuring small angles. It consisted of a telescope with the lens cut in half along the optical axis, so that two images of the same object could be obtained by changing the position of the two halves with respect to each other. The instrument was called a heliometer, since it could be used to measure the apparent diameter of a celestial body such as the sun. The heliometer, no longer in use today, had already been designed, using two whole lenses, by Pierre Bouguer in 1748, and built by John Dollond, who cut the negative lens in two. Karl Friedrich Gauss had used Fraunhofer's first model in 1814. The second model had helped Friedrich Wilhelm Bessel reach one of the most coveted goals of antiquity: that of measuring the distance of a star. The method for determining this distance (measuring the change in position of a star observed successively from the two extremities of the earth's orbit) is so simple that no particular author is known for it. However, the displacement that had to be measured was so small, even for the closest stars, that it had been impossible ever to obtain an appreciable value. The measurement required a high-precision instrument and an able observer, the combination was finally reached with Fraunhofer's heliometer used by Bessel. His and other accurate measurements, made by several astronomers between 1822 and 1838, provided the first exact evaluations of stellar parallax. In 1837, Friedrich Georg Struve obtained the value of 0.125 seconds, and in 1838 the value of 0.26 seconds for α Lyrae (present value: 0.121 seconds). In 1838, Bessel obtained the value of 0.314 seconds for 61 Cygni (present value 0.299 seconds), while Thomas Henderson (1798–1844) obtained the value of 0.91 seconds for α Centauri (present value 0.76 seconds). These discoveries began to provide a scale for the stellar world, which had been given a structure by the speculations of Wright, Kant, and Lambert, and by Frederick Wilhelm Herschel's observations.

Halley's comet

Halley's comet reappeared in August of 1835, and was simultaneously observed by numerous astronomers. After calculating the centrifugal velocity of the masses away from the nucleus of the comet, and on the basis of methods elaborated by Leonhard

Euler in 1746, Bessel was able to determine the trajectory of the particles of the comet and the shape of its tail, thus laying the foundations of the mechanical theory on the shape of comets.

Maps of the moon and Mars

Meanwhile, progress continued to be made in the physical study of the solar system. Two enthusiasts, Wilhelm Beer, a banker and brother of the musician Giacomo Meyerbeer, and Johann Heinrich von Mädler, an elementary school teacher, founded a small observatory in Berlin, in which they compiled one of the best maps of the moon. It was surpassed only by the work of Johann Friedrich Schmidt (1825–84) in 1878. In 1837 they also drew the first complete map of the planet Mars. It was based on their own observations, which they began in 1830, and which they published in 1839. It established the point of departure for longitudes on Mars (Sinus Sabaeus) that is still used today. In 1849, while studying the changes in the planet's white polar ice-caps, they found that both reached their maximum point of constriction just after the summer solstice, and from this concluded that they must be made of gradually melting snow.

Shooting stars

Important research began to be conducted on shooting stars, about which opinions were still very confused. At the beginning of the century, Alexander von Humboldt had observed that the meteors of a given shower all seemed to emanate from a single point (radian). Simultaneous observations had also made it possible to determine distances and velocities. In 1836, Lambert Adolphe Quételet (1796–1874) published the results of his research comparing the number of shooting stars observed on different nights of the year. He pointed out the presence of several maximum points that corresponded to swarms. Then, in 1837 and 1839, he compiled the first catalogues of the most extraordinary apparitions. In 1835, Dominique-François-Jean Arago (1786–1853) recognized the periodicity of the November swarm of shooting stars (Leonids), already suspected by Heinrich Wilhelm Olbers (1758–1840) who, immediately after the shower of 1833, connected it with the shower of 1799.

The periodicity of shooting stars of August (Perseids) was recognized two years later by Quételet. All these discoveries supported the evidence that shooting stars were caused by small to very small bodies moving around the sun in orbits that the earth passed through every year; these bodies generally moved in swarms, with maximum concentration in a particular zone that the earth encountered only every certain number of years. These results led to Giovanni Virginio Schiaparelli's great discovery of 1866 of the cometic origin of falling stars (see ASTRONOMY 1860–1870).

New observatories

An astronomical observatory was built at Pulkovo, in Russia, in 1839. Thanks to the extremely modern and original configuration of its instruments, and to the organization of observational methods and their elaboration, projected by Friedrich Georg Struve, director of the observatory, it soon became the acknowledged world capital of astronomy. Astronomers from all over the world, including some Italians, such as G. V. Schiaparelli, trained there.

At about the same time in the United States, construction was begun on the Harvard Observatory in Cambridge, Massachusetts, largely financed by a private donation. While the Pulkovo Observatory was specifically set up to direct work in geodetics and nautical astronomy, and only later began to deal with astrophysical research, the Harvard Observatory focused primarily on astrophysics from the very beginning.

PHYSICAL SCIENCES: MECHANICS

The Hamiltonian function

During the fourth decade of the century, mechanics became progressively better articulated, according to forms dominated by new problems that opened up in various sectors of physics. The research programs responsible for this were mainly characterized by a growing use of increasingly more powerful mathematical formalisms. The major works of William Rowan Hamilton (1805–1865), for example, came out of an analysis of the relations between dynamics and optics that was simultaneously aimed at elaborating unitary mathematical methods and centered around problems of an interpretive nature. The essay in which Hamilton examined these relations in order to build a "deductive mathematical optics" that would be as internally coherent and as fomally correct as Lagrangian mechanics seemed to him to be was published in 1833. In this essay, significantly entitled *On a General Method of Expressing the Paths of Light, and of Planets, by the Coefficients of a Characteristic Function*, Hamilton basically focused his attention to two problems. First of all, in his opinion, it was essential rigorously to put in mathematical form all knowledge on optical phenomena in order to eliminate the "imperfections" that still existed in optical theories. Then, affirmed Hamilton, regardless of whether the Newtonian or the Huygenian theory on the nature of light is adopted, it remains a fact that the laws of optics can be regarded as objects of a separate study, independent of wave or corpuscular interpretations. The basic problem, according to Hamilton, came from the need to analyze the principle of least action. For this purpose, Hamilton introduced a characteristic function in his essay of 1833.

Coriolis's theorems

During these same years, other scientists obtained significant results following lines of research that were already under way. In 1835, Gaspard Gustave de Coriolis (1792–1843) developed the idea of the composition of accelerations in a memoir on the equations of relative motion of systems of bodies (*Sur les équations du mouvement relatif des systèmes de corps*). Working in the area of dynamics, and especially pressed by the need to study the technical problems of driving mechanisms using rational mechanics, Coriolis presented his second theorem in 1835, after having demonstrated an earlier theorem on the dynamics of relative motion in 1831. Louis Poinsot conducted a similar program, and in 1834 published an important memoir on the motion of a solid around a fixed point. Coriolis's and Poinsot's works were basic references for the planning and building of Foucault's gyroscope (see PHYSICAL SCIENCES: MECHANICS, 1850–1860). Friedrich Wilhelm Bessel's famous report on experiments concerning the force with which the earth attracts various types of bodies, presented in 1832 and aimed at confirming experimentally "the law according to which the masses of earthly bodies are proportional to the attractions to which they are subject," also deserves mention in this context.

While a second edition of Siméon Denis Poisson's *Traité de mécanique* came out in 1833 (the first edition was published in 1811), another problem arose out of the difficulties caused by the fact that the perturbations in the oscillatory motion of bodies moving in gasseous media could not be deduced from the principles of hydrostatics. These difficulties were implied by the discussion that Poisson attempted to develop in 1821 in a memoir on the motion of a pendulum in a resistant medium (*Mémoire sur le mouvement d'un pendule dans un milieu résistant*), where a molecular model was used to describe the generation of vibrations in the gasseous medium in which an oscillating body was immersed. These subtle problems, which only found solutions with the work of George Gabriel Stokes (see PHYSICAL SCIENCES: MECHANICS, 1850–1860), were also considered by George Green (1793–1841), in his *Researches on the Vibrations of Pendulums in Fluid Media* (1833).

The equations of Hamilton and Jacobi

The most interesting aspect of the process of mathematizing mechanics, however, was the development of Hamilton's program, already mentioned, which led to the canonical equations of motion and to what are still known today as the Hamiltonian functions. These results appeared in two important essays published by Hamilton in 1834 and 1835 and entitled *First Essay on a General Method in Dynamics* and *Second Essay on a General Method in Dynamics*.

Using methods already outlined in his treatise on mathematical optics, Hamilton attempted to "reduce the complete problem of the dynamics of conservative systems to the study of a single function and the development of a single equation." According to Hamilton, the kinetic energy T of a dynamic system could be expressed as:

$$T = U + H,$$

where H was the so-called Hamiltonian function.

A "characteristic function" V was then introduced, which depended on the initial and final coordinates of the system, as well as on H, and which satisfied the differential equation:

$$\frac{\delta V}{\delta H} = t.$$

Given V, a "principle function" S could be defined as:

$$V = tH + S,$$

where:

$$S = \int_0^t (T + U)\, dt.$$

Then, as Hamilton reiterated in 1834 in a memoir entitled *On the Application to Dynamics of a General Mathematical Method*

Previously Applied to Optics, all mathematical dynamics could be reduced to the determination and study of this function. In effect, the variation of S led to both the Lagrangean equations and the canonical equations of motion. The main problem, therefore, was to elaborate methods for calculating S. A good part of Hamilton's activity from 1836 on was concentrated on developing these methods. A short time later, Karl Gustav Jacobi (1801–51) developed a generalization of Hamilton's theory (see PHYSICAL SCIENCES: MECHANICS, 1840–1850).

An extremely important part of mechanics was concerned with theories on the optical ether; for a discussion of this issue, see PHYSICAL SCIENCES: ELECTROMAGNETISM AND OPTICS, 1830–1840.

The decade ended with a significant work by Poisson on the resistance of media to motion, entitled *Recherches sur le mouvement des projectiles dans l'air.*

PHYSICAL SCIENCES: THERMOLOGY

Wave theories of heat

Early in the 1830s, notions that heat and light could be explained by studying oscillatory-type motions captured the interest of many scientists who, although they were no longer satisfied by caloric theories, still did not want to abandon explanations based on mechanical models. This was, for the most part, the attitude expressed in a work by Thomas Thomson (1773–1852) entitled *An Outline of the Sciences of Heat and Electricity* (1830), in two communications by André Marie Ampère (1775–1836) on heat and light (*Idées de M. Ampère sur la chaleur et sur la lumière,* 1832, and *Note sur la chaleur el sur la lumière considérées comme resultant de mouvements vibratoires,* 1835), and in a memoir by Macedonio Melloni (1798–1854), *Observations et expériences relatives à la théorie de l'identité des agents qui produisent la lumière et la chaleur rayonnante* (1835).

The ideas advanced by Melloni, who later became a harsh critic of caloric models, were at first rejected by French scientists; however, after they had obtained recognition and prizes in England, they exerted a considerable influence on science throughout Europe.

Poisson's theory

The Laplacian program had not, however, lost its force, despite the criticisms brought against it from more than one quarter. In fact, Siméon Denis Poisson reelaborated and generalized it in 1836 in his monumental *Théorie mathématique de la chaleur.* According to Poisson, Fourier's mathematical elaborations constituted a particular case of the Laplacian equations on the interaction between the caloric and matter, rather than an alternative to the modellistic approach. After Laplace's initial generalization of Biot's differential equations, the modellistic approach had been further perfected from a mathematical point of view. Poisson insisted on this point, maintaining that the development of the mathematization of the laws on heat was in no way independent of a physical interpretation of the hypothetical mechanisms of emission and absorption of particles of the caloric by molecules of matter. In fact, in Poisson's opinion, mathematics was nothing more than a correct and appropriate language for translating already-acquired empirical knowledge into formulas. The mathematical theory of heat thus had the

goal of deducing, by means of a rigorous calculation, all the consequences of a general hypothesis on the communication of heat founded on experience and analogy. These consequences were therefore merely a transformation of the hypothesis itself, to which the calculation took away nothing and added nothing. It was from this point of view that Poisson considered the application of the calculus of probabilities to thermology. In Poisson's theory of 1836, this application played an even more distinct role than in Laplace's models. The systems of material molecules and particles of caloric that, according to Poisson, constituted gasses, were analyzed in terms of the average values of microscopic parameters. Poisson justified this probabilistic approach by the irregularity of molecular motion and the large numbers of microscopic physical events that took place in each infinitesimal fraction of volume and time.

Faced with the growing opposition of programs of study that were so different from one another, scientists tended to consider the various models as equivalent. An example of this attitude can be found in the *Cours de physique de l'École Polytechnique* (1836), a treatise on physics by Gabriel Lamé (1795–1870).

Mossotti's molecular mechanics

One of the major monographs in mathematical physics of the first half of the century, a memoir by Ottaviano Fabrizio Mossotti (1791–1863) entitled *Surs les forces qui régissent la constitution intérieure des corps, aperçu pour servir à la détermination de la cause et des lois de l'action moléculaire,* was published in Turin in 1836. Mossotti was fully convinced that all of physical science had to be reduced to the laws of motion. To this effect, he proposed to investigate further the study of molecular mechanics, in order to make it a source of knowledge equal to celestial mechanics. With molecular mechanics, according to Mossotti, it would be possible to arrive at a general theory of physical phenomena that would unify the interactions analyzed by optics, thermology, and electricity into a single law of nature. This law would also contain a term representing gravitational interaction. However, any mathematical expression of this law implied the knowledge of several parameters concerning the structure of the ether. Mossotti wrote that further studies along these lines were being made in physics. He analyzed the conditions that existed in a dielectric subject to electrical action, and discovered what was later called the Causius–Mossotti equation. Mossotti's ideas, which were accompanied by a brilliant mathematical system, exerted a strong influence on European science.

The Clapeyron equation of gasses

While the physical sciences were accumulating the results of such varied lines of research, toward the middle of the century, the reflections inspired by the growing interest in application and technology led to a revival of Nicolas Leonard Sadi Carnot's ideas. It was in this perspective that Benoit-Paul-Émile Clapeyron (1799–1864) published his important *Mémoire sur la puissance motrice de la chaleur* in 1834. Clapeyron, who took a critical position with respect to Laplace's and Poisson's hypotheses, maintained that in order to rationalize the mass of information available on heat and gas phenomena, it was necessary to refer to Carnot's *Réflexions.* In his opinion, Carnot's theories were correct; they just had to be formulated in proper mathematical language. Clapeyron introduced the classical equation of gasses, which connected the temperature t, the volume V and the pressure p in the equation:

$$pV = R\,(273 + t).$$

In the same context, it was again Clapeyron who utilized graphs of pressure and volume to represent the transformations of a gas. The same memoir of 1834 contains a statement of the equivalency between heat and work. Despite the authoritativeness of the journal in which Clapeyron's work was published, it aroused no real interest until it was published in a German journal in 1843. It should, however, be noted that the process that gave rise to modern thermodynamics in the following decade was already quite well advanced by this time. Certain considerations of a philosophical nature concerning the mutual transformability of natural agents, as well as improvements in the methods for evaluating experimentally the yield of steam engines contributed to this growth. Regarding the convertibility of natural agents, we should mention the analyses conducted by Hans Christian Oersted (1777–1851), who as early as 1817 had considered the observable relations between the compression of fluids and the development of heat. It was in this same perspective that, in the winter of 1839, under the direction of Oersted, Ludwig Colding (1815–88) began the experiments that would lead him to formulate the principle of equivalency in relatively correct terms. As to the diffusion of statements on the relation between technical determinations of the yields of steam engines and their scientific implications, a treatise by Marc Seguin (1786–1875) entitled *De l'influence des chemins de fer et de l'art de les construire* (1839), deserves particular mention.

PHYSICAL SCIENCES: ELECTROMAGNETISM AND OPTICS

Faraday's theory

Ampère's new electrodynamic theory raised two orders of problems. On the one hand, it manifested itself as a mathematical device that was able to describe correctly many phenomena caused by the interactions between electrical currents and magnets and, as such, was accepted by a growing number of scholars. On the other hand, however, it was founded on a very particular hypothesis that tended to reduce magnetism to a series of effects caused by hypothetical electrical currents circulating at a molecular level. Although this hypothesis was surprisingly modern, it encountered the open skepticism of the majority of scientists. For example, Ampère's model was accepted by researchers such as Lord Kelvin only in the second half of the century. Even Michael Faraday, at the beginning of the 1820s, was critical of this model and of all physical theories that presumed, on the basis of very little experimental knowledge, to explain the more profound nature of magnetism. However, in the winter of 1831, Faraday accepted Ampère's notion, generalized it, and from this perspective elaborated the first two parts of his famous *Experimental Researches in Electricity*. The reasons that Faraday made this choice—and set out in the direction of a revolutionary transformation of the physical sciences in the area of electrical, magnetic, and optical phenomena—lay in the possibility of interpreting induction phenomena (which Faraday had systematically observed) on the basis of a sort of principle of symmetry. According to Faraday, who was strongly influenced by the hypothesis that all powers operating in nature were mutually transformable, electrical and magnetic phenomena found their "true identity" in induction. Electromagnetic action, in his opinion, was of a single nature, which could manifest itself on an empirical level sometimes with electrical effects, and other times with magnetic effects. In addition, this manifestation took place in the space outside the electrified and magnetized bodies in a way that could be represented by lines of force, the geometrical configuration of which depended on the distribution of the bodies and was not reducible to Newtonian interactions. There was no other known power according to Faraday, that manifested itself in the same direction that characterized the mutual action between an electric current and a magnetic pole: this action was tangential, while all other forces acting at a distance operated along straight lines.

The complexity of the problems that arose from this alternative to models based on interactions at a distance was without precedent in the history of the analysis of electrical and magnetic phenomena. In fact, it is not surprising that Faraday's ideas were often considered obscure by his contemporaries. Faraday's physics focused on space as the place where electromagnetic events occurred, thus approaching the theories of Oersted and putting the role played by the physical medium that allowed the realization and propagation of these events in a new light. It was not possible, from Faraday's point of view, to accept those explanations that referred to fluids interacting mechanically in a physically inert space, just as attempts to formulate mathematical models of the ether as an elastic medium able to provide for the propagation of perturbations with the speed of light could no longer be valid. In fact, the medium to which Faraday attributed a fundamental role was a medium made of matter itself within a space crossed by lines of force.

Faraday was also inspired to elaborate this innovative concept of the physical world by his basic researches in electrochemistry. In this area of study, he experimentally established the empirical laws that even today bear his name and, taking advantage of suggestions made by William Whewell (1794–1866), coined the terminology based on ions, catons, anions, etc. His studies comparing electromagnetism and electrochemistry led to experimental investigations on dielectric constants and phenomena of polarization, and reinforced the credibility of the theory of action by contact. All these problems were first solved in the great memoir that constitutes the eleventh series of his *Experimental Researches*, published in 1838. This work in turn opened up new questions that led Faraday to examine critically the concept of material particles. According to him, an understanding of the generality of interpretations of electromagnetism implied rejecting the explanations previously elaborated in the physics of Ulrich Theodor Aepinus, Henry Cavendish, and Siméon Denis Poisson. In his opinion, these authors' physics necessarily dealt with the category of action at a distance, and this category had become a real obstacle to the further development of knowledge on the subject.

The new investigations showed that this obstacle had to be removed, and that new theories could arise only if they were based on the idea of action by contact between adjoining particles. Faraday believed that, in all cases, ordinary induction was an action of adjoining particles consisting in a sort of polarity, rather than an action that took place at an appreciable distance between particle or masses; and, given this, then the clarification and the foundation of such a truth were of maximum importance for further progress in the study of the nature of electrical forces.

It was in this context of a radical revision of the physical sciences that Faraday introduced lines of electric force, together with lines of magnetic force, for the explicit purpose of "throwing light on the operations of nature." In the process of doing so, Faraday involved not only ether and fluid mechanics, through his critique of action at a distance, but the very supremacy of mechanics, to the extent that he began to speak of the need to reduce all phenomena to effects of the power of electricity. Faraday's research dominated the decade, even though other scholars such as Karl Friedrich Gauss (1777–1855), William Snow Harris (1791–1867), Claude Pouillet (1790–1868), W. E. Weber (1804–1891), Emil Lenz (1804–65) and Leopoldo Nobili, conducted extremely important studies and investigations during those same years. Toward the end of the decade, Faraday became influenced by the reflections of Ottaviano Fabrizio Mossotti (1791–1863) on the opportuneness of developing a molecular physics that would theoretically unify electrical and gravitational phenomena (see PHYSICAL SCIENCES: THERMOLOGY, 1830–1840), and made his critical revision of the concept of particle mentioned earlier.

In fact, while lines of force and the symmetry between electricity and magnetism made it possible to draw an early electromagnetic picture of field, the problems of gravitational interaction and of the "essence of the phenomenon of transmission" of physical events, which Faraday attempted to solve through a reformulation of the concept of matter, still remained open.

Ether and light

While Faraday made the greatest contribution to the development of electromagnetism, George Green and his criticisms against the approach of Augustin Louis Cauchy (1789–1857) laid the foundations for a turn of events in the progress of optical theory, the consequences of which played a leading role in nineteenth-century physics of the ether. Thanks to the researches of Augustin Jean Fresnel, Thomas Young, Claude-Louis Navier, Poisson, and Cauchy, an understanding of wave optics led to the development of fluid mechanics and the mechanics of elastic solids. Cauchy himself, in fact, in a series of steps in 1830, 1836, and 1839, formulated a molecular theory of the ether that united the new knowledge of mechanics with the explicative needs that arose from the wave approach to optical phenomena. However, in a treatment of this sort, the problem inevitably arose that an elastic medium is always able to transmit two spherical waves, one transversal and the other longitudinal. Green suggested eliminating the problem by adopting an ether that was so rigid that the velocity of propagation of longitudinal waves was rendered practically infinite. But this implied that different lines than those used with Cauchy's theory be followed in putting the model into mathematical form, as Green explicitly maintained in his memoir of 1838, entitled *On the Laws of the Reflexion and Refraction of Light at the Common Surface of Two Non-Cristallized Media*. Using Lagrangean mechanics, wrote Green, the restrictive assumptions made by Cauchy could be overcome.

As a matter of fact, although it referred to a molecular model of the optical ether; Green's discussion was general enough to be used with a continuous ether.

CHEMICAL SCIENCES

The distillation of tar

In-depth studies of tar, which led to the discovery of numerous organic compounds and, consequently, to rapid advances in organic chemistry and important practical achievements, were made by various scientists during this period. Karl von Reichenbach (1788–1869) obtained paraffin and oil of vaseline from petroleum tar (1830–31), and creosote from wood tar (1832). In the course of 1834, Friedlieb Ferdinand Runge isolated from coal tar a substance he called Kyanol, which basically remained the industrial source of aromatic compounds until the end of World War I (1918). It was later identified as aniline, which Otto Unverdorben (1806–73) had obtained in 1826 from indigo (anil), observing its great ability to produce crystalline salts with acids. By oxidizing this substance, Runge obtained a dye, aniline black, but the discovery was not developed.

Runge also obtained quinoline (Leukol), phenol (*Karbolsäure*), and pyrrole from tar. Jean Baptiste Dumas discovered anthracene, a polycyclic hydro-carbon (1832), from which his pupil Auguste Laurent (1807–53) obtained anthraquinone by oxidation (1835). Laurent later (1837) found other polycyclic hydrocarbons in tar (pyrene, chrysene), from which he prepared various nitro-derivatives.

The development of organic synthesis

French chemists obtained important nitrate compounds: H. Graconnot (1780–1855) extracted xylodine from starch (1833), and Théophile Jules Pélouze (1807–67) obtained nitrocellulose from paper and cotton (1838). These were the first hesitant steps in the field of actual organic synthesis. Numerous other organic compounds were discovered: some were found in natural products or obtained by transforming them, others were prepared in the laboratory. Among these were amygdalin (Pierre-Jean Robiquet and Antoine François Boutron-Charland, 1830), styrene (J. F. Bonastre, 1830), carotene (Heinrich Wackenroder, 1832), narceine (Pierre Joseph Pelletier, 1832), codeine (Robiquet, 1832), atropine (Mein, 1831), furfural (Johann Wolfgang Döbereiner, 1831), creatine (Michel Eugène Chevreul, in 1832), methyl alcohol (Dumas and Eugène Melchior Peligot, 1834), thebaine (Thibauméry, 1835), xanthophyll (Jöns Jacob Berzelius, 1837), cholic acid (Eugène Demarcay, 1838), toluene (Pelletier and Philippe Walter, 1838) and salicyclic acid (J. S. F. Pagnestecher, 1838), which Rafaelle Piria (1814–65) also obtained during that same year from its relative aldehyde. Among the products obtained in the laboratory were chloral and chloroform (Justus von Liebig, 1831), urethan (Dumas, 1833), ethyl mercaptan (William Christopher Zeise, 1834), pyruvic acid (Berzelius, 1835), acetylene (Edmund Davy, 1836), phthalic acid (Laurent, 1836), quinone (Alexander Woskresensky, 1838), and trichloroacetacid (Dumas, 1838). Empirical formulas were found for various alkaloids (Dumas and Liebig, 1831), acetone (Dumas, 1831), a number of terpenes, including camphor and menthol (Dumas, 1832), chloroform (Dumas, 1834), and acetaldehyde

(Leibig, 1835). Particularly extensive were Eilhard Mitscherlich's studies on benzene (1833–35), Ludwig Clamor Marquart's studies on the dyes in flowers (1835), and Liebig and Friedrich Wöhler's studies on uric acid (1837–38). In 1834, Carl Julius Fritzsche (1808–71) found that granules of starch produced a blue color when combined with iodine.

The concept of radicals

A memoir by Wöhler and Liebig on the benzoyl radical appeared in 1832, opening a path in what Wöhler called "the obscure forest of organic chemistry." The following year, Robert John Kane (1809–1890) pointed out that alcohol ether and some esters contained the same radical, which was first called ethereal, then ethyl. In 1836–37, Kane synthesized mesitylene from acetone and sulfuric acid, a reaction that later proved to be important in understanding the structure of benzene.

The appearance of modern chemical terminology

By this time, it was clear that elementary, or brute, formulas had little meaning in organic chemistry; Berzelius introduced the expressions isomer, polymer, empirical formula, and rational formula (1830–33).

Laurent advanced the theory of nucleus (1834–35), which viewed hydrocarbons as the basis of the various organic compounds. Justus von Liebig proposed tracing all organic bases to ammonia (1837), according to a theory later developed (1848) by August Wilhelm von Hofmann (1818–92).

The critique of the dualistic theory and the laws of electrochemistry

Several advances in general chemistry should also be mentioned. In 1838, Liebig stated that acids were compounds in which one or more atoms of hydrogen could be replaced by metals. In 1839, Liebig and Dumas openly attacked Berzelius's dualistic, or electrochemical, theory, at least as far as organic chemistry was concerned, since for example, an atom of chlorine (negative) could replace an atom of hydrogen (positive) without changing the nature of the compound. In 1839, while studying elimination reactions, Charles Frédéric Gerhardt (1816–1856) formulated the theory of residues, according to which the parts eliminated from the reactant molecules united to form a simple inorganic molecule, such as water, while the residual parts combined with each other according to a process called copulation. Unlike the radicals of the dualistic theory, the residues did not have a real existence, but were simply conventional ways of illustrating a reaction. Liebig's and Dumas's critiques did not, however, bring about the decline of electrochemistry; in fact, it was during these very years that Michael Faraday stated the laws of electrolysis (1834), while in 1836 John Frederic Daniell (1790–1845) discovered that the voltage drop in electrolytic cells was due to the deposition of copper on the zinc plate and designed the battery that still bears his name.

In 1835–40, Faustino Jovita Malaguti (1802–1878) studied mucic acid, demonstrating its dibasic nature, and prepared some of its derivatives; one of these was ethylic ester, the first crystallized ester on record. Together with Alexandre Eduard Baudrimont (1806–80), Malaguti demonstrated that cystine contained sulfur (1837), and with his observation that not all colorless liquids are equally transparent to light rays, anticipated the law of photochemistry, stated by John William Draper (1811–82) in 1841.

Enzymes: vitalistic theory and chemical theory

Theories began to be formulated regarding what Berzelius called catalytic force (1835) and which, with remarkable intuition, he attributed not only to chemical, but to biological catalysts, which Willy Küne later called enzymes (1878). Liebig attributed this force to a transfer of molecular motion that accelerated the reaction, more than to simple contact action, which was Berzelius's hypothesis (1836–37).

As several enzymes were discovered—e.g. pepsin, by Theodor Schwann (1810–1882) in 1836 and emulsin, by Friedrich Wöhler and Liebig in 1836–37—discussions about the fermentation processes and, in particular, alcoholic fermentation arose on all sides, while two theories on their nature, the vitalistic theory and the chemical theory, began to be formed.

In 1836, Charles Cagniard de la Tour (1777–1859) discovered that yeast was made of microorganisms. This observation, already made by Antony van Leeuwenhoek as early as 1681, seemed indirectly by Schwann's finding that heating to high temperatures blocked fermenting activity (1837). Despite other authoritative confirmations, Berzelius denied that yeast was made of living cells. Liebig, who was more cautious, maintained that, even if yeast were really made up of microorganisms, fermentation still had a chemico–mechanical basis, as Antoine Laurent Lavoisier had already hypothesized. Everyone seemed to be skirting the truth, which was only discovered in 1897 by Eduard Buchner (see CHEMICAL SCIENCES, 1890–1900), although quite a while earlier Giovanni Valentino Fabbroni had probably come the closest of all. In 1787, he had considered yeast a vegetal–animal ferment of principle similar to gluten, i.e., a protein. In this same context, it should be mentioned that the word protein was coined in 1838 by Gerardus Johannas Mulders (1802–80).

Analytical chemistry: titration

Little progress was made in inorganic chemistry: Thomas Graham's research on phosphates (1833–34) and the discovery of lanthanum (1839) by Carl Gustav Mosander (1797–1858) deserve mention. In 1839, Marc Antoine Gaudin (1804–80) made the first synthesis of rubin by fusing alumina with chromium oxide. In 1836, Christian Friedrich Schönbein (1799–1868) attributed the phenomenon of passivation of metals to the formation of a surface film of oxide.

In the field of analysis, James Marsh (1794–1846) discovered the extremely sensitive assay for detecting arsenic that still bears his name (1836). By introducing the volumetric analysis of silver in a nitric solution with sodium chloride (1832), Joseph Louis Gay–Lussac introduced the concept and practice of titration into chemical analysis, although the French term titre (titer) had been known since the beginning of the century, following the pioneering work of François-Antoine-Henri Descroizilles (1751–1825). In 1833, Michael Faraday produced magnesium from melted magnesium chloride by electrolysis. In 1834, Antoine Jérôme Balard (1802–76) prepared hypochlorous acid.

In 1832, Justus von Liebig founded the *Annalen der Pharmacie*, which symptomatically changed its name to *Annalen der Chemie und Pharmacie* in 1840, and in 1873 was renamed *Annalen der Chemie.*

LIFE SCIENCES

Physiology and microstructure

This was the decade of Johannes Peter Müller, (1801–58), Theodor Schwann (1810–82), and Agostino Bassi (1773–1856), and the period in which the use of the microscope in biological research became established.

In 1833, Müller was called to Berlin to the chair of his teacher Karl Asmund Rudolphi. The first volume of his *Handbuch der Physiologie des Menschen,* which soon replaced Albrecht von Haller's *Elementa phisiologiae,* came out that same year. Müller was a vitalist, open to the influence of "nature philosophy" and the naturalistic ideas of Johann Wolfgang van Goethe; however, in addition and above all, he was one of the most rigorous and open-minded experimenters that the history of biology had ever known. Physiology, Haller's *animata anatome,* became with Müller the science of the properties and phenomena of organic bodies, of plants and of animals, as well as of the laws according to which their effects manifested themselves. A correction of animistic vitalism through a need for logic and definitions, later accentuated by his students, was evident in Müller's program.

A physico-chemical anti-vitalism was also active in France; its major document was the *Recherches anatomiques et physiologiques sur la structure intime des animaux et végétaux* (1824), by René-Joachim-Henri Dutrochet (1776–1847). The publication of Müller's *Handbuch* took seven years. The work was divided into eight sections: organic fluids; the motion of fluids and the vascular system; the transformations of fluids and organized parts; the physics of nerves; movement, voice, and language; the senses; psychic life; reproduction; and development. A group of very young researchers gathered around Müller, and by the 1830s this circle had already begun to produce the major figures of German biology of the century. Meanwhile their teacher left yet another lasting mark with his histological research on tumors.

Progress in microscopy. Cell theory

Decisive progress was made in microscopy. The naturalist Giovanni Battista Amici (1786–1863) built objectives that were free from chromatic aberration and introduced the technique of immersion. In this way, Amici was able to observe how pollen grains formed the pollen tube that penetrated into the style to the ovule in *Portulaca.* Amici sent Mattius Jakob Schleiden (1804–81), who doubted his results, one of his microscopes, so that he could make "more accurate observations." Amici's results were also confirmed by Hugo von Mohl (1805–72), W.F. Hofmeister (1824–77) and Robert Brown. Brown discovered that vegetable cells had a nucleus. Schleiden, in turn, demonstrated that the nucleus was involved in cell reproduction (1838). At this point, a conceptual change of profound significance to the natural sciences and biology took place: cell theory was born, and with it, living things were reduced to a basic common structure, the cell, center of a well-defined property, growth. Cell theory in this sense was formulated in the *Mikroskopische Untersuchungen über die Uebereinstimmung in der Struktur*

und dem Wachsthum der Thiere und Pflanzen (1839), by Theodor Schwann, a disciple of Müller who worked in Berlin and then in Louvain. Schwann still believed in the *generatio aequivoca* of cells from the amorphous material contained in the intercellular spaces of tissues (cytoblastema), and regarded the cell as the formation from which all other elementary anatomical structures were derived. Schwann's cytogenetic law was modified during the 1850s by Robert Remak (1815–65) and Rudolf Virchow (1821–1902) (see LIFE SCIENCES, 1850–60).

Félix Dujardin (1801–60) doubled the power of the microscope by adding a system of prisms and lenses. His microscope became the prototype for all modern light condensers, and Dujardin became a pioneer in microphotography (1852). He also gave the name "sarcode" (1835) to "living jelly," which he defined as insoluble in water and the physical foundation of life, preceding organization. In addition, he was the first biologist to dare to affirm that sarcode also had to exist in the more highly differentiated tissues of the human body. In 1846 von Mohl rebaptized Dujardin's great intuition "protoplasm," a term inspired by Purkyně (1839). Another contribution of Dujardin's was the discovery of Trematoda larvae and Cestoda larvae, and of their dependency on an intermediate host (1837–38). He attempted to confirm his finding by making various animals swallow cysticerci.

Morphofunctional and pathological principles

While conducting experiments in London on decapitated newts, the Scottish physician Marshall Hall discovered the existence of reflex movement, which did not depend on the integrity of the connections between the encephalon and the medulla (1832–33).

Müller reported analogous experiments in the second section of his *Handbuch* (1834), but maintained that some reflexes required the intervention of the brain centers in order to take place. Physiology and pathology proceeded together to build the new morphofunctional principles of biology. Pathology made its contribution by both observing diseased tissues and researching disease agents. William Beaumont (1775–1853) conducted research on digestion and, in collaboration with chemists, established the definite presence of free hydrochloric acid in the stomach. In 1834, he published his *Experiments and Observations on the Gastric Juice and the Physiology of Digestion.*

Microorganisms and "fermentations"

During this decade Agostine Bassi, a physician from Lodi, Italy, did important work beginning the systematic quest for the invisible causes of diseases, and blazed a new path in the science of minute—microscopic—life. A two-volume work by Bassi, entitled *Mal del segno calcinaccio o muscardino,* demonstrated that the cause of the silkworm disease known as *mal del segno,* or muscardine, was a parasite that could be seen under the microscope. He formulated the hypothesis that most contagious diseases were caused by microorganisms. This hypothesis was again taken up by Friedrich Gustav Henle (1809–85) in his *Pathologische Untersuchungen,* and later by Louis Pasteur and Robert Koch, a disciple of Henle's. Investigations into minute life proceeded in other directions as well: J. Payet found the agent of trichinosis (1833), while Johann Lucas Schönlein (1793–1864) discovered the ringworm fungus. But microbiology, as it could actually be called by this time, also established unforeseen connections with the chemistry of fermentations. In 1837, via indirect paths, Schwann, Charles Cagniard de la Tour (1777–1859), and Friedrich Kützing (1807–93) reached the conclusion that yeast, the agent of alcoholic fermentation, was a living, cellular structure (Schwann). Schwann also affirmed that fermentation was linked to the metabolic power of the cell, and that it took place inside the cell. Justus von Liebig violently attacked these ideas in the *Annals of Pharmacy* (1839). Fermentation, putrefaction, and slow combustion, according to Liebig, were caused by organic substances breaking away from the control of the living organism's vital force. Meanwhile, following in the footsteps of Beaumont, Schwann found a substance, "present in small quantities" in gastric juice, which he likened to a fermentation and called "pepsin." This was the agent of proteic digestion, active in acid environments, which was later isolated by Ernst Wilhelm von Brucke and crystallized (1930) by John Howard Northrop.

Renal physiology and pathology

Richard Bright (1789-1858), a clinician at Guy's Hospital in London, conducted his studies and observation of nephropathies during this decade. Bright's activity ranged over a somewhat longer period of time, from 1827 to 1843, and succeeded in uniting edema, the presence of albumin in the urine, circulatory troubles and a rise in the haematic count in the urine all to a single factor, i.e. that chronic nephritics suffer a noticeable enlargement of the left ventricle. Clinical data stimulated physiological research: in 1842 William Bowman (1816–92)—a histophysiologist who later became an ophthalmologist—succeeded in demonstrating that urine was the super-filtration of blood plasma by the membrane of the Malpighian bodies of the renal cortex. In 1844, Karl Friedrich Ludwig (1816–95) hypothesized that even the tubules were involved in the formation of urine. Meanwhile, research into pigments and the bile salts of urine progressed with the work of Leopold Gmelin (1788–1853) (1830), and Max Joseph von Pettenkofer (1828–1901).

Insect systematics

In 1830, Pierre André Latreille (1762–1833) took over Lamarck's position at the Museum of Natural History in Paris, where the first chair of entomology was created. Latreille dedicated most of his attention to the taxonomy of insects. He attempted to define the higher categories and bring some kind of natural order to the somewhat chaotic state in which entomology found itself at the beginning of the nineteenth century. He created a whole intermediate hierarchy, and for the accuracy of his work, his contemporaries nicknamed him the "prince of entomology." He had already published a *Précis des caractères génériques des insectes disposés dans un ordre naturel* in 1796, and in 1831 he published a *Cours d'entomologie*.

EARTH SCIENCES

Actualism, or the principle of uniform causes in geology

Charles Lyell's *Principles of Geology*, which stated the law of

uniformitarianism, or actualism, was published in 1830–33. According to this theory, the factors that had caused the various changes in the earth's crust over the course of its history were still in operation at the present time, obeying the same physical and chemical principles. The accumulation of their effects over the enormous, inestimable duration of geological time sufficed to explain the great modifications that had taken place in the earth's aspect, even though observing the final effects, our reason might lead us to believe these changes had been sudden and violent, almost as though they had occurred over short periods of time, such as the span of human existence. This anti-catastrophic theory was presented to scholars in a complete documented form that supported a creed that was in contrast to the diluvian and cataclysmic theories, which at the time seemed more in accordance with the Bible, since one of the floods could have been the great flood. The opposition of English Protestant circles repeated what James Hutton's *Theory of the Earth* had stated fifty years earlier, when it proclaimed that the earth's history appeared to be without beginning and without end. The documentation provided by Lyell in his *Principles*, however, was so convincing that the law of uniform causes, or actualism, became the major animating principle of the geological sciences and the basic belief upon which geologists based their research.

The study of stratigraphy benefitted most of all, and in 1835 Lyell himself proposed the divisions of the Eocene, Miocene, and Pliocene for the Tertiary (to which Heinrich Ernest Beyrich (1815–96) added the Oligocene in 1853).

In 1836, Adam Sedgwick (1785–1873) proposed the Cambrian period for the Paleozoic Era, while Roderick Impey Murchison (1792–1871) proposed the successive Silurian. In 1839, Murchison and Sedgwick distinguished the Devonian, and in 1841 Murchison separated the Permian from the Carboniferous, which William Daniel Conybear (1787–1857) had already distinguished as a period in 1822. In 1834, the German Friedrich August von Alberti (1795–1878) introduced the Triassic in his stratigraphic system, thus completing the periods of the Mesozoic. The other periods of this era were the Jurassic, established by Alexandre Brongniart and Alexander von Humboldt (1829) and the Cretaceous, distinguished by Jean-Baptist Omalius D'Halloy (1783–1875) in 1822.

Lyell also recognized that the differences between metamorphic and sedimentary rocks were due to later transformations (as Hutton had already guessed) and coined the term metamorphism. Lyell found a confirmation of his theories in the journeys he made to the continent, especially to France, where he visited the volcanoes of the Auvergne, studied by George Poulett-Scrope (1797–1876) from a vulcanological point of view; the Vincenza region of Italy (where richly fossiliferous tertiary strata alternate with volcanic spills); and to Vesuvius and Etna. He did not hesitate to call attention to the stratigraphic use of fossils as evidence of the immense duration of geological time.

The equilibrium of rocky masses, later known as isostasy, was mentioned as one of the possible causes of variations in the earth's crust in a letter that John Frederick Herschel sent to Lyell in 1837.

In 1833, Gideon Algernon Mantell (1790–1852) finished a work on the geological conditions of Sussex and southeast England that he had begun in 1822.

Paleontology: the controversy over evolution

Ideas about the evolution of organisms, which considerably influenced interpretations of the succession of flora and fauna in the stratigraphic series of the soil, were still contradictory during this period. Many illustrious geologists supported the fixity of species; besides the work of Jean Léopold Cuvier, who died in 1832, William Buckland's *Geology and Mineralogy Considered with Reference to Natural Theology* (1836) also deserves mention. In 1837 Edouard Lartet (1801–71) discovered the jawbone of a fossil primate and founded human paleontology. In 1835, Félix Dujardin (1801–60) recognized the unicellular nature of Foraminifera. Upon his return from the voyage of the *Beagle*, Charles Darwin (1809–82) expounded his theory that atolls originated from the sea level rising and slowly submerging volcanic islands, on which coral colonies formed. This constituted one of the proofs against von Buch's upheaval theory.

Research on ancient glaciation

An explanation of the origins of the huge erratic blocks found scattered throughout the Alpine regions, far from the nearest outcroppings of similar rocks, began to be formulated: in 1834, Jean de Charpentier (1786–1855) suggested that they had been carried to their present location by ancient glaciers. His research was followed by that of Louis Agassiz (1807–73), the famous paleoichthyologist, and Karl Schimper, and reference began to be made to a glacial era in a recent period of the earth's history.

Further progress in crystallography and mineralogy

In 1832, Carl Friedrich Naumann (1797–1873) recognized the coincidence between optical and crystallographic axes; he perfected his studies in 1835, resolving the problem for monoclinic crystals as well. In 1837, James Dwight Dana (1813–95) wrote the first edition of his *System of mineralogy*; in 1838, in a *Treatise on Crystallography*, William Hallowes Miller (1801–80) proposed the crystallographic symbols that are still in use today.

The method of studying thin cross-sections under the microscope, invented by William Nicol, was also applied to rocks and fossils (e.g. Witham's observations of fossil wood, 1831).

Instruments for oceanic measurements

A new interest arose in bathymetric measurements and the instruments needed to make them, probably as a result of experiments conducted by Jean Daniel Colladon and (Jacques) Charles François Sturm. Partly because of the diversity of results that came out of various expeditions, it became clear that the measurements made with mechanical leads were subject to serious error. Some measurements falsely gave fantastic depth readings, up to 15,000 meters and more, leading to totally erroneous interpretations of the morphology of the ocean floor. As a result, more reliable methods were sought. In 1833, William Henry Talbot (1800–77) suggested setting off charges of explosives on the ocean bottom so that the depth could be measured from the time it took for the sound wave to reach the surface. In 1836 Urbain Dortet de Tessan proposed using an ecographic system for detecting the bottom. Unfortunately, the technology of the epoch was not yet advanced enough to build the device he proposed. In 1832, going back to very old principles, Edward

Massey built an instrument that was lowered to the bottom, then freed; buoyancy tended to make this instrument rise to the surface, and a calibrated propeller attached to a set of gears made it possible to measure the length of its path as it came up. However, this instrument also proved inadequate, since it rarely reascended vertically. Meanwhile Matthew Fontaine Maury (1806–73) continued to use a lead with a hemp line in his explorations. To avoid distortions, other than loss of time when it was hoisted, the line was cut as soon as the bottom was detected, and the depth was calculated by measuring the remainder. Using his own measurements together with data collected by others, Maury compiled the first bathymetric map of the Atlantic Ocean. Although it was obviously inaccurate, it gave an early overall view of depth distribution in that ocean. Since by this time it was also a well-known fact that deep water thermometric readings were distorted by pressure (as d'Urville had observed during the voyage of the *Astrolabe,* in 1850 François Walfredin (1795–1880) proposed protecting thermometers from pressure in order to determine the correct temperature; and using the difference in readings between protected and unprotected thermometers to determine the depth. This system was finally realized using Negretti and Zambra's reversing thermometers.

Other achievements in the area of instrumentation and experimentation also deserve mention. The first recording tide gauge, in the modern sense of the word, was built for the port of London by Henry Robinson Palmer in 1831. Also in 1831, Michael Faraday (1791–1867) proposed measuring the speed of ocean currents by measuring electric potential. This idea came directly out of Faraday's discoveries in electricity (the water of a current was a conductor that moved within the earth's magnetic field, so that the difference in potential could be measured perpendicular to the direction of movement). Faraday's attempts were unsuccessful due to the inadequacy of his instruments; however, Faraday's principle has been widely used for measuring currents in the second half of the twentieth century.

New exploratory voyages

In 1832–36, the ship *Beagle,* commanded by Robert Fitzroy (1805–65), made its voyage carrying Charles Darwin around the world. Between 1839 and 1842, James Clark Ross (1800–62) made his famous exploration of the Antarctic waters with the *Erebus* and the *Terror.* Famous French expeditions were those of the *Venus* (1836–39) and the corvettes *Astrolabe* and *Zelé* (1837–40).

Rennell and ocean currents

A work by James Rennell (1742–1830) on ocean currents appeared posthumously in 1832; among other things, it contained a remarkably detailed map of the currents of the Atlantic Ocean, although there were some imperfections on the eastern side, where some boundaries of the circulation were marked incorrectly. In any case, Rennell was the first to make a clear distinction between wind currents (drift currents) and other currents (stream currents), today called gradient currents, which he correctly attributed to the different grades produced in the free surface of the water when it encounters obstacles in its path of motion. This happens, for example, when a current originally produced by the wind flows toward a coast and accumulates there. It then flows away from the coast in a direction that is no longer related to wind direction. It was later discovered that qualities intrinsic to the water—variations in density—can also cause some differences in level, and consequently currents, without the preliminary action of the wind.

The hydrodynamic theories of Coriolis, Green, and Russell

In the field of hydrodynamic theories, Gaspard Gustave de Coriolis (1792–1843) discovered the quantitative relation that took geostrophic action, which induced deviations in the motion of all types of currents, into account (1835). In 1839, George Green published an important work on waves (*Note on the Motion of Waves in Canals*), in which, among other things, he demonstrated that the velocity of short waves was proportional to their length. A short time before (1834), while studying the propagation of waves in canals, John Scott Russell (1808–82) had provided the equation for solitary waves. Even today these waves are used as models for studying wave propagation; they exist in reality, and are particularly visible (and important to navigation) in canals.

Studies on tides

An extraordinary development of studies on tides took place during this decade, especially among English scientists. As early as 1830, John William Lubbock (1803–65) had returned to Laplace's theory and elaborated it with tide tables. He also recommended setting up a large network of tide gauges in order to determine the characteristics of the tides in different places. Then, in 1833, Lubbock and William Whewell (1794–1866) made further studies on tides, and finally, between 1833 and 1839, Whewell came out with numerous works that gave the characteristics of the tides in a large number of ports, together with the distribution of range and phase. A synthesis of this information enabled Whewell to show the cotidal configuration of the ocean (later developed by others), after he had introduced the concept of cotidal lines (lines connecting points that have simultaneous high tides), and to develop a theory on ocean tides.

Gathering meteorological data

Geographical expeditions, especially polar expeditions, required adequate meteorological knowledge, while at the same time they made it possible to gather huge amounts of data (Fitzroy's treatise on nautical meteorology, mentioned earlier, is a good example). Meanwhile, navigation tended to develop through regular intercontinental voyages along fixed routes. These required ready information on the meteorological situation, obviously possible only after the introduction of the radio in navigation, though scientists were already at work on the problem. Edward Becher, William Marsden (also known for subdividing the oceans into zones still widely used to represent various hydrometeorological situations), and William Reid were the major promoters of the need to collect synoptic meteorological data on an international level and to exchange information gathered by ships as well. Reid even made an agreement with Kruesenstern to collaborate with Russia on the exchange of data. Reid also studied hurricanes and researched quantitative, though empirical, relations on the phenomenon. William C.

Redfield (1789–1857) also studied the effects of hurricanes, as well as the currents and tides of the North Atlantic. The study of optical phenomena in the atmosphere received an important contribution with the rainbow theory, described by George Biddell Airy (1801–92) using physico-mathematical concepts.

APPLIED SCIENCES

Transport and railways

In 1830, one-sixth of English foreign trade took place with the United States, and more than three-quarters of the cotton worked in the United Kingdom came from America. Liverpool, the center of this trade, had pressing transport problems caused by the insufficiency of canals and the slowness of boats. The success of the Liverpool-Manchester railway line, officially opened on September 15, 1830, touched off a railroad fever not only in England, but all over the world. That same year, the first trunk lines were opened in France and the United States. The first railway in Germany was opened to traffic in 1835, and the first one in Italy in 1839 (Japan had to wait until 1872).

The first steam locomotives in Russia were built by M.E. and E.A. Čerepanov, with the collaboration of F.I. Švecov, in the Nižne-Tagil' factory in 1833–1836. The first Russian passenger railroad was inaugurated in 1837; it connected St. Petersburg with Carskoe Selo. The following year it was extended to Pavlovsk and covered a distance of twenty-five kilometers.

The world railway network expanded by leaps and bounds until the end of the nineteenth century, by which time all the important areas of the world with possible geographical links were connected by railroad. Different types of steam engines were developed and perfected along with the development of railroads.

The Rocket, built by George Stephenson, had reached a speed of forty-four kilometers per hour, the *North Star,* built in 1837, exceeded ninety-six kilometers per hour.

It was during this period that Robert L. Stevens (1787–1856) introduced the T-section rail into railroad construction. It was soon adopted everywhere.

The first English railway boom ended in 1837. In two years—from 1835 to 1837—eighty-eight railway companies were founded, with a capital of around seventy million pounds. During the same period, seventy-one mining companies, twenty banks, and eleven insurance companies were also founded. Only the railway companies survived the depression that followed the boom.

The development of railways helped industrial development reach maturity in England, while in other countries it provided the thrust for take-off. Belgium, which had coal and iron ores and was on one of the major routes of continental transit, got its industrial start in railroad construction. The Brussels-Malines line, inaugurated in 1835, carried more passengers than all the lines in England.

The entrance of Louis Phillipe's France onto the industrial scene was slower than that of Belgium, but it also took place as the result of railways: the first line—eighteen kilometers at St. Étienne—was planned by a French engineer, Marc Seguin, and built with ample help from English technicians and laborers.

One of the most important lines, the Paris-Rouen run, was promoted by an English diplomat. The engineers, locomotives and capital were also English.

Railway lines in which the traction was provided by pneumatic systems placed between the rails were also studied and built, especially when the route included relatively steep slopes. Eugène Flachat published a study of locomotives that could go up these slopes, and this led to the abandonment of pneumatic systems.

The iron-working industry and manufacture

The growing demand for iron and coal was partially satisfied by several innovations in iron-working, as well as by David Mushet's discovery of the coal and iron basin of Scotland, the first in a series of discoveries of deposits of easily-treated ferrous ores. Technical innovations included the rapid diffusion, in England and abroad, of the method devised by James Beaumont Neilson (1792–1865) (in Baden, inflammable gasses were already being used to heat air by 1832); the success of the circular furnace, known as the Scotch hearth, especially after improvements made by T. Oakes in 1832; and the modification of the puddling furnace by Joseph Hall's patent for calcinating slag (1839). Calcined slag was used to line the sole of the furnace; it combined with the burden to form carbonic oxide, which was released from underneath the surface of the metal, producing an apparent effervescence. The new method made it possible to accelerate the process of fusing iron, so that it was effectuated in a single step, while in "dry" puddling the metal had to be agitated slowly to expose it to action of the atmosphere.

In America, the iron industry evolved especially rapidly after 1838, when David Thomas built the first anthracite furnace at Catasanqua, Pennsylvania. Thomas applied a patent belonging to W. Geissenheimer, a Lutheran pastor who in 1833 had discovered that burning anthracite in the presence of hot air would fuse iron.

Important innovations were made in the field of non-ferrous metals as well. In 1832, G.F. Muntz patented a type of bronze that was three parts copper and two parts zinc. In 1833, Hugh Pattinson (1796–1858) patented a new process for separating silver from lead, based on the discovery that when fused argentiferous lead was cooled until it crystallized, crystals of pure lead would form first. The process could be repeated to remove four-fifths of the original lead.

Five years after the opening of the Liverpool-Manchester line, James Nasmyth (1808–90) moved to Manchester. Railroads and the textile industry had created a huge need for machine tools (lathes, drilling machines, reamers, planing machines, and slotters). The push toward mechanical engineering was considerable, and the standardized machine tools that Nasmyth was able to build (and which he listed in a catalogue) did not require particularly qualified workers, but, as he emphasized, only "well-chosen workers."

In 1836, Nasmyth invented the shaper, for producing flat surfaces, and in 1839 he designed and built the first of his famous steam hammers. The hammer made it possible to forge very heavy pieces more quickly and economically, enabling Nasmyth's factory to begin producing railroad locomotives. Joseph Whitworth (1803–87), a pupil of Henry Maudslay and a contemporary of Nasmyth's, also moved to Manchester in 1833. By 1834, Whitworth had already proved his great talent by building a gauge that could calibrate yard-long specimens to within one-millionth of an inch. The adoption of machine tools in factories answered the well-defined needs of the entrepreneurs. The capitalist entrepreneuring bourgeoisie was putting all of its energy into lowering the cost of production and placing increasingly large quantities of low-priced products on the market, in order to pay for the cost of the machines and increase profits. This caused the bourgeoisie to exploit the work of the laborers to the utmost. By this time, the labor supply was plentiful, since many farmers were abandoning the countryside, a situation that was created after farms were developed on a capitalistic scale. The great mass of workers exposed to the conditions of factory work, gradually evolved into a worker proletariat, with an emerging consciousness of its own. This consciousness gradually developed into a class consciousness opposed to the bourgeoisie. The establishment of solidarity among the workers was first translated into spontaneous, unorganized uprisings; however, these fueled an organized movement that finally revealed itself in class struggle.

The development of industry through the construction of machines connected with the building of railways was not a phenomenon peculiar to England alone. Belgium not only built the locomotives it needed, but, between 1830 and 1840, exported machines of all types to various countries of the continent, including Holland, Germany, and Russia. Meanwhile, the steam engine continued to evolve. In France, in 1836, Joseph Farcot patented a control system for steam pressure that used a governor.

England's major export commodity, however, was still cotton, and the English textile industry proved itself capable of stimulating and assimilating the inventive process. For example, the comber invented by Phillipe de Girard (1775–1845) passed immediately into England, where it was sold as the Wordworth comber thanks to improvements made upon it by a certain Wordworth. Among the most important patents were those for two automatic spinning mules, presented in 1830 and 1834 by Richard Roberts and James Smith, respectively, and the screw stay for pressing linen, which was later used with silk as well (Lawson and Westly, 1833). In 1839, Kenworthy, together with Hornby, patented the web sizing machine, and, together with Buloogh, patented an automatic ironing device, or temple, that could hold cloth taut across its width.

Agricultural technology

An important factor in English industrial supremacy was the efficiency of its agriculture. Even though there were laws prohibiting the importation of wheat and production of it increased, in the course of thirty years (from 1811 to 1841) the number of farm families involved in cultivating it dropped from one-third to one-fourth of the population. This meant that, with respect to other countries, England had many more laborers available for industry, and especially for its mills and mines.

The array of machines available to the English farmer increased during this period with the introduction of new types of hoes that had several rows of blades and could hoe more than one furrow of seeded land at a time; rotary machines for cutting roots; and Armstrong's zig-zag harrow (1839). The most important invention of the period, however, came from the

United States: Cyrus McCormick (1809–84) devised a reaper that was pulled instead of being pushed, and employed a knife and cutter bar. He patented this machine in 1831. McCormick's reaper became popular in England during the years following the Great Exhibition of London of 1851.

Food conservation was also improved with the invention of a type of refrigeration based on the expansion of volatile liquids, patented by Jacob Perkins (1766–1849) in 1834, and L.W. Wright's patent for producing ice by compressing air, cooling it to room temperature and successively expanding it by cooling.

Road construction

Besides a rapid and efficient exchange of raw materials, large-scale production required the speedy distribution of its finished products. Along with the lively intensification of city traffic, a second phase began in the development of road construction, especially in France. In 1820, De Sassenay had demonstrated that a good pavement could be obtained by spreading and levigating natural asphalt mixed with tar. The method was used in 1835 to pave the Place de la Concorde in Paris and was later employed in other French and English cities. In 1837, De Coulaine discovered that when natural asphalt was heated it became a powder, and that the powder could be spread and compressed with a steam roller to make a hard, impermeable pavement. In 1838, Polenceau invented a heavy compressor for that purpose. This type of road, particularly suitable for wagons and horses, was used until the end of the century, when it was replaced by the tarred macadam road. The first tarred macadam road was built in Nottinghamshire in 1832, and macadam was used on the streets of Nottingham in 1845.

Electrical technologies

In the area of technologies that were more closely connected to advances in scientific research, several discoveries and achievements that proved, with the passage of time, to be of exceptional practical importance deserve special mention: in 1832, Joseph Henry (1797–1878) discovered the phenomenon of self-induction; he perfected electromagnets, by building electrically isolated primary wire coils, and invented the relay, which was the basis of the later development of the telegraph. The electromagnetic electric motor, built by Moritz Hermann von Jacobi (1801–1874) in 1838, also appeared during this period.

A new type of electric motor that had a rotary shaft and a collector with friction contacts was built in 1834 by Jacobi. In 1838, a boat with one of these motors installed in it navigated the Neva River at a speed of 2.8 kilometers per hour. In 1838, after he had become a Russian citizen, Jacobi put the finishing touches on the technical procedure for galvanoplasty, which became widely used in the manufacture of coins and typographies. In 1840, he published his *Gal'vanoplastika*.

It was during these years that the telegraph, the first non-visual system for the almost-instantaneous transmission of messages, was introduced. Many attempts had been made prior to 1830, but in 1832, an American artist, Samuel F.B. Morse (1791–1872), invented the telegraph and a telegraphic alphabet. That same year, in Russia, Pavel Silling presented a non-printing, electromagnetic telegraph with two wires and a needle, the different positions of which corresponded to a signal code.

Morse's first telegraphic installation began operating in 1857. The attenuation of the signal, which made it inaudible after thirty to forty kilometers, was regulated by the use of a relay, which provided energy for another circuit of the same length, and so on. Between 1840 and 1850, long telegraph lines were set up, while the laying of the first transatlantic cable was completed in 1865.

1840–1850

THOUGHT AND SCIENTIFIC LIFE

The Hegelian left. Feuerbach

The years between 1840 and 1850 were historically significant, not only politically, but culturally as well, with German philosophy playing a leading role. At the beginning of the decade, this was primarily due to the young Hegelians of the left. The *Hallesche Jahrbücher für Deutsche Wissenschaft und Kunst* had become a focal point, mainly as a result of the enthusiastic work of Arnold Rüge (1802–80), a Hegelian educated at Jena. The original theme of the *Jahrbücher,* as Rüge explicitly theorized, was a defense of Protestant liberalism. The theoreticians of the Catholic restoration were its immediate targets; and the journal, which in 1838 attacked Heine for having over-exalted the French Revolution against reformism, aspired to be an autonomous third force, independent of both the reaction and the revolution. But the brutality of the reaction against the journal was destined to spoil this utopian idea of mediation. By 1839 the *Jahrbücher* had become the object of criticisms from Savigny's followers, from the pietists of Hegstenberg, and from the ultramontane Catholics; and the young Hegelians were severely hindered in their university careers. Several events complicated the situation still further: The elderly Schelling, whom Ludwig Feuerbach had attacked harshly, was called to fill the chair that had been Hegel's. Bauer was converted. And Strauss and Rüge further defended their positions. In 1840, the romantic Frederick William IV took the throne in Prussia and appointed a fanatical anti-Hegelian, Johann Albrecht Eichorn (1779–1856) Minister of Education. In 1841, in a courageous article in the *Jahrbücher,* Bauer openly accused the government of having taken a position against Hegel's teachings in philosophy. In 1842, he was expelled from the university, and his right to teach was revoked. In January, 1843, after three years of furious polemics, the *Jahrbücher* was forced to cease publication. Meanwhile, however, primitive Protestant liberalism had engaged in a fundamental struggle with Christianity, the Hegelian system and method was being criticized from a number of points of view, and, with Feuerbach's work, the left had found a philosophical base that was no longer only young-Hegelian, but in part even anti-Hegelian and antispeculative. Feuerbach's *Das Wesen des Christentums* had come out in 1841, and his *Grundsätze der Philosophie der Zukunft* appeared in 1843. By this time, through a critical study of the phenomenon of religiosity, Feuerbach had succeeded in reversing Hegelianism: the infinite no longer realized itself in the Hegelian absolute spirit, but rather in human beings and only in human beings.

Scientific socialism

Karl Marx (1818–83) and Friedrich Engels (1820–95) began their philosophical work as supporters of the Hegelian left. At first, Marx viewed the Hegelian left as a party with a concept that could bring about real advances in philosophy's struggle against the world. From May, 1842, to March, 1843, he collaborated on the *Reinische Zeitung*, initially a liberal-leaning journal that soon moved to a more radical position (thanks to the commitment of Moses Hess and of Marx himself). A more thorough investigation of the relationship between method and system in Hegel, a gradual understanding of the limits of the Hegelian left, and contact with Feuerbach's thought led Marx to develop an extremely original body of thought, which he outlined in such famous works as the *Kritik des hegelschen Staatsrechts* (1843), the so-called *Oekonomisch-philosophische Manuscripte aus dem Jahre 1844* (unpublished) and "Zur Kritik der hegelschen Rechtsphilosophie, Einleitung" which appeared in the *Deutsch-Französischer Jahrbücher*, published by Marx and Rüge, and outdid even Feuerbach's reflections in abstractness and basic speculativeness. His contact with Parisian Communist circles (1844) and his collaboration with Engels led Marx to criticize his friends in Berlin. Initiating a collaboration that lasted until Marx's death, Marx and Engels published *Die heilige Familie,* subtitled *Kritik der Kritischen Kritik* (A Critique of Critical Critique), in open disagreement with the Hegelian left, in 1845; and in 1845–46 they composed *Die deutsche Ideologie,* which was particularly directed against Strauss, Bauer, and Feuerbach. Their Paris contacts and the constitution of the League of Communists inspired Marx and Engels to write the *Manifest der Kommunistischen Partei,* published in February, 1848, on the threshold of the fall of the July monarchy in France and the revolutionary storm that swept Europe.

A fairly lively current of political philosophers had also arisen in France, the theoretical and practical work of which had in a certain sense been prepared by the diffusion of the works of Saint-Simon. Among the new philosophers of this current were François Marie Fourier and Pierre Joseph Proudhon. Their ideas had considerably influenced more advanced thinkers all over Europe, including Marx, who had emigrated to Paris for several years. But the socialism elaborated and propagandized by this current mainly appealed to the need for universal harmony and justice and was, as a result, of an irremediably utopian character. This was the radical difference between it and Marx's and Engels's scientific socialism, and it was the basis of Marx's intense polemic against Proudhon, author of the *Philosophie de la misère,* which Marx criticized in his *Misère de la philosophie* of 1847.

Positivism

As we have seen, one of the basic themes of Auguste Comte's philosophy was social problems, which he believed could be solved by the creation of a new positive science, sociology. Despite this programmatically scientific approach, the founder of French positivism abandoned himself to utopistic concepts after 1844 and fantasized founding a religion of humanity (1849) that would become a trustworthy new guide for resolving the moral and political crisis that plagued the modern world. But the majority of scholars who had admiringly received the ideas developed in the *Cours de philosophie positive* refused to follow the philosopher in this new initiative, and the church he founded survived only as a curious little group of fanatics lacking any link whatsoever with historical reality.

Comtian philosophy had been especially favorably received in England, where positivism had been viewed as the direct continuation of eighteenth-century empiricism. This very interpretation, however, explains the complete refusal with which even Comte's most sincere English friends greeted the final (religious) phase of his thought.

Among these, John Stuart Mill, a philosopher and economist and author of a monograph on the French thinker entitled *Auguste Comte and Positivism* (1865), deserves particular mention. During this decade, Mill published various works, including *A System of Logic, Ratiocinative and Inductive* (two volumes, 1843), already cited earlier, and *Principles of Political Economy* (two volumes, 1848).

In Italy, true positivism only became widespread at a later date. The closest thinker to this current was Carlo Cattaneo, mentioned earlier, who for the time being, however, dedicated most of his attention to studies in economy and political action. In 1844, he published an essay entitled *Notizie naturali e civili della Lombardia.* Philosophical circles continued to be dominated by Catholic thinkers, prominent among whom were Rosmini Vincenzo Gioberti (1801–52), and Raffaello Lambruschini (1788–1873). While Lambruschini worked primarily on problems in pedagogy—he founded and directed the periodical *Guida dell'educatore* (1836–46)—Gioberti became directly involved in the great metaphysical problems and in political action. Among the most important writings he published during this decade were *Degli errori filosofici di A. Rosmini* (1841), in which he accused Rosmini of subjectivism, *Del primato morale e civile degli italiani* (1842–43), and *Il gesuita moderno* (1846–47), which contained a harsh criticism of the powerful order of the Jesuits, who had progressively become the backbone of Catholic conservatism.

Scientific institutions, academies, and the foundation of new universities

The Physikalische Gesellschaft, which united a group of young scholars, physicists, and biologists from the school Heinrich Gustave Magnus and Johannes Peter Müller (1801–1858) under the banner of experimentalism, was established in Berlin in 1845. The teachers' agnostic, positive approach to the basic problems of scientific research was gradually replaced by a tendency toward speculation, which grew out of the experimentism itself and followed the example of Alexander von Humboldt.

On July 23, 1847, Hermann Ludwig von Helmholtz (1821–94) read a memoir on the conservation of force before the Society of Physics.

Among the democrats who participated in the Berlin riots of 1848 was Rudolf Virchow (1821–1902), who edited a report on the environmental causes of a typhus epidemic in Upper Silesia, denouncing the responsibility of the government.

Feliciano Scarpellini, who in 1795 had founded a physico-mathematical academy in Rome that pronounced itself—in the face of considerable discussion and protest—the direct continuer of the *Academia Linceorum,* died in 1840. Gregory XVI decreed the suppression of the Academy, but in 1847, Pius IX definitively reinstated the *Academia Pontifica dei Nuovi Lincei,* promulgating a new statute for it.

In 1841, the prototype of the American state university was created in Ann Arbor, Michigan, while more large universities established by private patronage continued to spring up: Cornell University, was founded by Ezra Cornell and Andrew White in Ithaca in 1865; Johns Hopkins University was set up by Johns Hopkins in Baltimore in 1876; and the University of Chicago was created in 1890 with the support of John Davison Rockefeller. The Saxon Society of the Sciences was inaugurated in 1846 in Leipzig, and the Irish University was opened in 1849.

MATHEMATICAL SCIENCES

The spread of the abstract mentality and the creation of quaternions

Abstraction, which originally affected algebra in particular, began to extend into other branches of mathematics as well. In 1844, Hermann Günther Grassmann (1809–77) published the first edition of his theory of extension, the *Ausdehnungslehre*, which presented a geometrical calculus that did not use coordinates and was based on operations defined in a purely formal, symbolic, and extremely general manner. This work proved to have points of view and procedures in common with the theory of quaternions, which William Rowan Hamilton (1805–65) was beginning to elaborate during this same period, and it foreshadowed the salient characteristics of vector analysis. However, even after the new improved and enlarged edition was published in 1862 and Grassmann had written various articles about it, the work did not gain the recognition it deserved.

The first edition of the *Geometrie der Lage*, by Karl Georg von Staudt (1798–1867), came out in 1847. It marked a new phase in the history of this discipline. From this point on, the acknowledged difference between projective geometry and metric geometry was replaced by the complete liberation of projective geometry from any reference whatsoever to metric geometry. As a matter of fact, this work developed geometry using purely projective methods, without the aid of figures and adhering to a very rigorous formal approach.

The following year, in his university lessons, W.R. Hamilton began systematically to develop the theory of quaternions, which he had already glimpsed in 1843 while attempting to represent points in space algebraically. (In 1837, he had proposed an arithmetic theory of complex numbers as pairs of real numbers, a theory that in many respects remained definitive.) The importance of the theory of quaternions, besides its intrinsic value as a powerful technical device (its versatility in physical applications was fully understood by James Clerkwell, for example), also lay in its purely algebraic and abstract approach, exemplified, for instance, in the particular introduction of a non-commutative operation of multiplication.

This led to a surpassing of the very prospects of Peacock's symbolic algebra, which only contemplated the extension of the same laws that held true in numerical algebra, and to an acceptance of operations that obeyed different laws, even though they continued to apply to quantities. Vector and matrix algebra were further concretizations of this same type of theory. Hamilton's *Lectures on quaternions* came out in 1853, and his larger *Elements of quaternions* was published posthumously in 1866. Other British mathematicians, such as Peter Guthrie Tait (1831–

1901), also wrote treatises on the subject of quaternions that later became classics.

Early treatises in mathematical logic

Possibly the most important, and in some ways unexpected, result of this new algebraic mentality was the birth of mathematical logic, in many respects a new discipline that had undergone virtually no significant developments since Leibniz had made his predictions regarding it. *Formal Logic*, by Augustus De Morgan (1806–71) and *The Mathematical Analysis of Logic*, by George Boole (1815–64), the two works that initiated modern mathematical logic, came out in 1847. In them, logic was viewed as a special algebraic calculus, which could be interpreted as a description of basic inferential processes, but was established and operated on the basis of pure formal rules governing the use of symbols. A more complete and detailed development of this program was provided by Boole in *An Investigation of the Laws of Thought* (1854). Thus one might say that mathematical logic arose as a special branch of algebra; as a matter of fact, not only was this type of study called the algebra of logic for several decades, but, in effect, the treatments developed by the scholars of this discipline can in many respects be considered authentic algebraic theories. The name Boolean algebra, for example, still exists in current mathematical terminology to designate certain types of reticular structures that Boole first presented in his studies on logic. The algebra of logic was developed over the following decades, especially with the work of William Stanley Jevons (1835–82), Charles Sanders Peirce (1839–1914) and Friedrich Wilhelm Karl Ernst Schröder (1841–1902).

It is interesting to observe that De Morgan's and Boole's studies intentionally made reference to problems of probabilities. Interest in the theory of probabilities (at this point directed more toward its applications than its mathematical development), was also confirmed by the publication of the *Lettres sur la théorie des probabilités, appliquée aux sciences morales et politiques,* Lambert Adolphe Quételet (1796–1874), which predicted the applicability of the theory of probabilities to social phenomena. In addition, the first edition of the famous and highly successful *A System of Logic, Ratiocinative and Inductive*, by John Stuart Mill, appeared in 1843. It gave considerable space to so-called inductive logic, and, in that context, considered problems of probability.

Important results were also registered in other branches of mathematics during this period, and the foundations were laid for theories that were destined to bring about important advances. In 1844, Joseph Liouville (1809–82) demonstrated the existence of infinite transcendental numbers, and around 1846 Peter Gustave Dirichlet (1805–59) conducted a series of important investigations on the theory of potential, which, as we have already mentioned, led him to discover the principle that bears his name.

The birth of hyperspace geometry

Among the new theories that appeared, special mention should be made of a memoir by Arthur Cayley (1821–95), *Chapter in the Analytic Geometry of* n *Dimensions* (1843), which led to research into the theory of hyperspaces. Decisive contributions to these studies were made by Georg Friedrich Bernhard Riemann, with his *Über die Hypothesen, welche der Geometrie zu*

Grunde Liegen of 1854, and later by Eugenio Beltrami, with his *Teoria fondamentale degli spazi a curvatura costante* of 1868 (see MATHEMATICAL SCIENCES, 1850–1860).

The circumstances in which the term topology first appeared in 1848 also deserve mention. It was introduced in the title of a work by Johann B. Listing (1808–82) to designate the particular qualitative properties of the geometrical figures studied in what was called *analysis situs*.

In 1850, Barnaba Tortolini (1808–74) founded the *Annali di scienze mathematiche e fisiche,* which in 1857 became the *Annali di matematica pura e applicata,* successively directed by Enrico Betti (1823–92), Francesco Brioschi (1824–97), and Angelo Genocchi (1817–89).

ASTRONOMY

Geodetic measurements

Measurement of the earth became increasingly more accurate. In 1841, by critically reexamining the more important geodetic measurements available, Friedrich Wilhelm Bessel was able to calculate the metric elements of the earth's ellipsoid, which were internationally used until they were replaced in 1909, by John Fillmore Hayford's calculations, which are still in use today. In 1843, Otto Wilhelm von Struve began an accurate determination of the difference in longitude between Pulkovo and Altona. In 1844, the first attempt to determine longitude via the instantaneous transmission of the time by telegraph was made between Washington and Baltimore. The first effective determination (between Washington and Philadelphia) was made two years later, followed by about a hundred others of the same sort. The extreme value of this method, however, was only revealed much later, when it became possible to apply it to ships, using the wireless telegraph.

The Doppler effect

A physical effect that proved to be of enormous importance to astronomy was recognized and defined during this period. In 1842, Christian Doppler (1803–53) demonstrated that the frequency of sound waves varies as the source moves closer or further away with respect to the observer. He also predicted that light waves would behave in the same manner. In 1848, Armand Hippolyte Fizeau (1819–96) experimentally proved Doppler's conclusion for light, thus opening the field of research on stars and their radial motion based on the positions of their spectral lines, which were discovered by Joseph von Fraunhofer.

Solar prominences and the solar cycle

Meanwhile, two issues called attention back to the sun. During the eclipse of July 8, 1842, astronomers were surprised to observe (though not for the first time) pink protuberances appearing in several points around the disc obscured by the moon. Discussions and arguments regarding their reality and nature arose among those who considered them optical illusions, clouds of solar gasses, or even mountains on the moon. These debates made the next eclipse, in 1851, an eagerly awaited event.

In addition, in 1844, Heinrich Schwabe (1789–1875), a pharmacist and amateur astronomer who had observed the sun continuously from 1826 to 1842, discovered that solar activity was cyclical, repeating itself approximately every ten years. A

more accurate period, 11.1 years. was obtained eight years later by Rudolf Wolf (1816–93), then director of the Observatory of Berne, through a critical examination of all the observations of sunspots made from Galileo's time onward. His statistics expressed solar activity in empirically derived numbers, called Wolf numbers. Johann von Lamont (1805–79) discovered the relationship between this cycle and the new cycle of variations in the earth's magnetism.

The discovery of Neptune

A great discovery was made in this decade, crowning the enormous structure of celestial mechanics that had been founded by Newton and perfected by Pierre Simon de Laplace using the contributions of the greatest mathematicians of two centuries. Since 1831, it had been noticed that the motion of Uranus indicated the existence of a perturbational planet beyond it. Research on the features of this perturbational planet had been independently undertaken by two mathematicians, Urbain Jean Le Verrier (1811–77) in France and John Couch Adams (1819–92) in England. In September, 1846, after reporting the conclusions of his calculation to the Academy, Le Verrier wrote to the Observatory of Berlin indicating the point in the sky where the new planet should be found. The astronomer J. Gottfried Galle (1812–1910) began looking for it, using the Berlin Academy's new, as yet undistributed, star chart, and on the night of 24 September he discovered the planet. At the suggestion of Dominique François Jean Arago, it was named Neptune. On 10 October William Lassell (1799–1880) discovered Triton, the first of the two satellites of Neptune known today. Adams had also finished his work at the same time, and in the summer of that year, on the basis of the first hand-written results, James Challis (1803–82) had begun the search with the Cambridge telescope. He even saw the perturbational planet on the nights of 4 and 12 August; however, since he did not immediately analyze his observations, the planet went unnoticed.

The beginnings of astrophysics

While classical astronomy seemed to have outdone itself with the discovery of Neptune, interest in the physical study of the heavenly bodies continued to grow. But the astronomers of the old school looked down upon these new interests, and many of the shared Le Verrier's feelings and ideas, when he complained that time and intelligence were being wasted on diversions that he considered useless. Thus it is common to find that many discoveries went unnoticed and that new techniques were looked upon as little more than amateurish games.

In the first category was the discovery of the spiral structure of several galaxies that still appeared as simple nebulae. This was made by William Parsons, Earl of Rosse (1800–67) in 1845 with his huge 1.83 meter-diameter telescope. In the second category were the first attempts to photograph the heavenly bodies.

Astronomical photography

Photography may be said to have originated among astronomers. After having made several attempts with Laplace and Étienne Louis Malus (1775–1812), Arago followed the experiments of Joseph Nicéphore Niepce (1765–1833) and Louis Jacques Daguerre (1787–1851) and, on 7 January 1839, finally announced the big news to the Academy of Sciences (see APPLIED SCIENCES, 1840–1850).

The original discovery of the solvent action of hyposulphites on silver halides (i.e. the well-known process of photographic fixing) had been made by John Frederick Herschel, the son of William and an astronomer in his own right, who in 1839 was also the first to coin the term photography. After Daguerre's first failure in photographing the moon, various astronomers made attempts with daguerrotypes. In 1840, John William Draper (1811–82), a physicist from New York and the father of Henry Draper, later a famous astronomer, became the first to photograph the moon fairly successfully. Ten years later William Cranch Bond (1789–1859) and George Mathews Whipple obtained excellent results at Harvard, using the thirty-eight-centimeter refractor there. On 2 April 1845, after an attempt made three years earlier by the optician Lerebours, Armand Hippolyte Fizeau and Jean Bernard Léon Foucault (1819–68) obtained good images of the sun, which from then on was photographed with increasing success, both because of its luminosity and because it was one of the objects most suited to the slow speed of photographic material. In 1851, in Scandinavia, August Ludwig Busch (1804–55) obtained the first photographs of the protuberances and of the internal corona.

PHYSICAL SCIENCES: MECHANICS

The Hamilton-Jacobi formalism

Mechanics developed along the following lines during the period between 1840 and 1850; a generalization of the method already suggested by William Rowan Hamilton; a reformulation of fluid dynamics; a return of mechanics to the pinciple of the conservation of energy; and a series of physico-mathematical researches concerning the structure of ether.

The first line took up Hamilton's ideas of the preceding decade, and led to two distinct programs of study. In 1842-43, Karl Gustav Jacobi taught a course at the University of Königsberg that drew its inspiration from Hamilton's concepts and had the goal of both extending and formally simplifying them. The result Jacobi achieved was an extremely rigorous formal scheme that is still known today as the Hamilton-Jacobi system for classical mechanics. In this system, it is possible to write the canonical equations of motion in a very generalized form and to proceed to a geometrization of the principle of least action, which Jacobi himself considered capable of ridding the principle of all its "metaphysical causes." Jacobi's lessons were not published until 1866. They appeared in a volume entitled *Vorlesungen über Dynamik,* which was based on a transcription edited by Rudolf Friedrich Clebsch.

The algebra of quaternions

Hamilton also attempted to link his early writings on dynamics to his new formalism through the study of quaternions. In 1844, at the suggestion from John Frederick Herschel (1792–1871), he applied the new mathematical method to the "problem of any number of bodies that attract each other according to Newton's law" and communicated his results in a memoir of 1845 entitled *On the Application of the Calculus of Quaternions to Some Dynamical Questions.* Further developments in Hamilton's researches—which especially attracted the attention of scholars because of the non-commutative nature of the operation of multiplication in the calculus of vectors or of quaternions—came in 1847, with a memoir entitled *On the Application of the Calculus of Quaternions to the Theory of the Moon.*

Hydrodynamics

The general problems that arose within research on the mechanical structure of the ether provided new incentives for the further development of fluid dynamics (see PHYSICAL SCIENCES: ELECTROMAGNETICISM AND OPTICS, 1840-1850). The "Notes on Hydrodynamics," published by George Gabriel Stokes and Lord Kelvin (1824–1907) in the Cambridge and Dublin Mathematical Journal between 1847 and 1849, played an important role in this sense. The subject also involved the creation of the formal tools that enabled Kelvin to attempt a mathematization of a part of Michael Faraday's ideas in 1847 (see PHYSICAL SCIENCES: ELECTROMAGNETISM AND OPTICS, 1840-1850). A subtle polemic on hydrodynamics and the mechanics of elastic solids was also being waged. It contrasted the program proposed by Stokes with that of James Challis (1803–1882), a champion of the necessity for a fluid concept of ether. The polemic began with a memoir written by Challis entitled *On the Analytical condition of the Rectilinear Motion of Fluids* (1842), and continued animatedly until 1845, with a communication by Stokes entitled *On the Theories of the Internal Friction of Fluids in Motion, and of the Equilibrium and Motion of Elastic Solids.*

The conservation of force

In 1847, Hermann Ludwig von Helmholtz (1821–94) published a memoir on the conservation of force (*Über die Erhaltung der Kraft*). According to Helmholtz, the problem of the natural physical sciences consisted in reducing natural phenomena to unchangeable attractive and repulsive forces, the intensity of which depended solely on distance. The resolution of this problem, he felt, was the condition for a complete understanding of nature. On the basis of this notion of scientific knowledge, Helmholtz stated a generalization of the Leibnizian principle of the conservation of living force, linking it to the assumption that it was not possible, through any combination of natural bodies, contunually to produce force starting from nothing, and suggesting that it was applicable to all branches of the physical sciences. As a matter of fact, in his memoir of 1847 he did not limit his reflections to dynamics, but extended them in order to exemplify their applicability to thermoelectric and electromagnetic phenomena.

In December, 1850, Stokes presented a long monograph on the internal friction of fluids in relation to pendular motion to the Philosophical Society. This work solved many of the problems that had plagued mechanics, and especially hydrodynamics, for decades and provided a formal clarification of the equations describing the dynamics of fluids.

PHYSICAL SCIENCES: THERMODYNAMICS

The equivalence of heat and work

A profound revision in the concept of thermology took place in the period between 1840 and 1850. It led to the formulation, toward the end of the century, of the two principles that constituted the foundation of the new dynamic theory of heat, or thermodynamics. This revision was the result of a merging of various research programs, very different from one another but sharing the theme of the reciprocal conversions of thermic and mechanical energy. For this reason, it had many aspects in common with the critical processes launched by Helmholtz in mechanics and by Faraday in electromagnetism.

In 1841, Julius Robert von Mayer (1814–78) sent an early work on the reciprocal transformations of forces to *Annalen der Physik,* but it was refused because it contained a number of inaccuracies. A second version was printed in 1842 under the title "Bemerkungen über die Kräfte der unbelebten Natur." Using a language that was related to the philosophical terminology of nature philosophy and basing his ideas on Pierre Louis Dulong's experimental data on specific heats, Mayer stated that heating a given weight of water from zero degrees to one degree corresponds to the fall of an equal weight from a height of approximately 365 meters. Comparing this result with the performances to the best steam engines, he pointed out, showed how small a part of the heat applied to the boiler is actually transformed into motion or the lifting of weights. Colding reached similar conclusions during this same period. However, he followed the suggestion of Hans Christian Oersted (1777–1851), who asked that he not make his considerations on the equivalence of heat and work public, since they were not yet confirmed by a sufficient amount of empirical data.

In 1843, Colding presented a memoir entitled *Nogle Saetninger om Kraefterne,* where, independently of Mayer, he developed the consequences of the experimental relationship observable between the mechanical work needed to overcome the friction between solids and the resulting heat.

The Joule effect

The work of James Prescott Joule (1818–89) in England proceeded on a much broader and more reliable experimental base. In 1841, Joule presented a paper entitled *On the Heat Evolved by Metallic Conductors of Electricity and in the Cells of a Battery during Electrolysis.* The essay affirmed the existence of a proportionality between the heat manifested in conductors and the square of the strength of the electrical current flowing through them. This was the so-called Joule effect. A further step in the direction of the principle of equivalence was taken by Joule following other classical experiments, described in 1843 in his *On the Calorific Effects of Magneto-Electricity and the Mechanical Value of Heat.* Here Joule maintained that the heat needed to raise the temperature of a pound of water one degree Fahrenheit was equivalent to the work needed to lift a weight of 838 pounds to a height of one foot. Karl H.A. Holtzmann also obtained results somewhat similar to Joule's in a work of 1845 on heat and the elasticity of gasses and vapors (*Über die Wärme und Elasticität der Gase und Dämpfe*).

Another advance toward the new theory of heat phenomenon was made with the famous experiments that Joule outlined in his *On the Changes of Temperature Produced by the Rarefaction and Condensation of Air* (1845). As a result of these experiments, the author was able to state that no variation in temperature occurs when air is allowed to expand in such a way that mechanical power is not produced. That same year, Joule formulated the principle of equivalence in a general form, even though he resorted to a contrived model of molecular structure (*On the Existence of an Equivalent Relation Between Heat and the Ordinary Forms of Mechanical Power*).

The problem of models that would connect heat phenomena to the principles of mechanics was raised in a brilliant monograph by John James Waterson (1811–83) entitled *On the Physics of Media that are Composed of Free and Perfectly Elastic Molecules in a State of Motion.* Waterson's work contained a correct treatment of the statistical mechanics of gasses but was rejected by the Royal Society. Brief excerpts appeared in 1846 and 1852, but the entire essay was only published in 1892 by Lord Rayleigh (1842–1919).

Regarding the mathematization of the theories proposed by the French school of physics, interesting developments emerged from Lord Kelvin's study on the formal analogies between equipotential and isothermic surfaces. The results of the study were published in 1842 in a memoir entitled *On the Uniform Motion of Heat in Homogenous Solid Bodies, and its Connexion with the Mathematical Theory of Electricity.* Kelvin's research, which was greatly influenced by his studies of Jean Baptiste Fourier's works, led to the rediscovery of several theories that had already been demonstrated by Michel Chasles, Karl Friedrich Gauss, and George Green. It was Kelvin who reprinted Green's monograph on potential (see PHYSICAL SCIENCES: ELECTRICITY, 1820–1830).

Absolute temperature

The possibility of applying the ideas that Nicolas Léonard Sadi Carnot had developed in 1824 was put forth by Lord Kelvin in 1848 in an important memoir, *On an Absolute Thermometric Scale Founded on Carnot's Theory of the Motive Power of Heat, and Calculated from Regnault's Observations.* This early definition of the notion of absolute temperature was of fundamental importance to a theoretical reorganization of the theories on heat phenomena, even though Kelvin expounded it within the context of caloric theory. The definition also had the merit of reintroducing Carnot's ideas into the official physical sciences. Kelvin further pointed up the value of these ideas in 1849 when he linked them to Joule's principle of equivalence. However, in Kelvin's opinion, the consequences of Carnot's axiom and Joule's principle constituted the source of irresolvable contradictions unless the theory of heat phenomena underwent profound revision. In fact, according to Kelvin, if one accepted Carnot's axiom, then Joule's principle was contradicted by all cases where heat flowed through a conductor that had a different temperature at each of its ends. When thermic action was consumed in the conduction of heat through a solid, he wrote, something must become of the mechanical effect that it should produce. A perfect theory of heat would necessarily require an explanation of this phenomenon yet no such explanation could be given at that time.

For this purpose, he elaborated a model regulated by mechanical interactions, following a plan of research that he later abandoned in favor of energetic programs.

The mechanical equivalent of heat

In addition to the contradictions denounced by Kelvin, all the experimental data on the properties of gasses obtained using the most accurate laboratory apparatus available at the time entered into the formation of thermodynamics. The laborious experimental research conducted by Henri Victor Regnault (1810–78) and Heinrich Gustav Magnus (1802–70) deserves particular mention. The decade ended with two basic memoirs for the new science of heat and gasses: *On the Mechanical Equivalent of Heat* (1850), by James Prescott Joule and *Über die bewegende Kraft der Wärme,* by Rudolf Julius Clausius. While Joule's memoir marked the end of the open skepticism with which his previous communications had been met by a large part of the scientific world, Clausius's work laid the foundations for resolving the contradiction indicated by Kelvin a few months earlier. According to Clausius, Kelvin's fundamental objection could be eliminated provided that one reject, once and for all, the idea of the so-called total heat contained in bodies and accept instead, without reserve, the general hypothesis on the mechanical nature of heat. Heat was not a substance, but consisted in a movement of the basic parts of material bodies. From this point of view, Carnot's axiom, as stated by Clausius, was completely valid upon the condition that all reference to the constancy of the quantity of heat be eliminated from it.

PHYSICAL SCIENCES: ELECTROMAGNETISM AND OPTICS

The debate over continuity versus discontinuity

The contrast between Michael Faraday's program and the series of programs in continental mathematical physics that continued to be based on action at a distance became particularly intense during this decade. However, it did not bring about a crisis in the comprehensive development of experimental and theoretical knowledge about electromagnetism; on the contrary, it acted as an incentive to further investigations. Meanwhile, studies in optical phenomena were dominated on the one hand by Faraday's attempts to reconduct light vibrations to oscillations of lines of force and on the other by Stoke's mathematization of the luminiferous ether.

In 1844, in a letter to *Philosophical Magazine,* Faraday examined the problem of the notion of material particle. As we mentioned earlier, this problem arose within Faraday's physics to the extent that it geometrized the space in which electromagnetic events took place. Therefore, it had to answer the question of whether or not material particles could exist apart from a continuous space. Faraday's letter of 1844 said no, maintaining that the common notion of material particles was an abstract concept that, within the context of action by contact, had to be replaced by a continuous material, which Faraday envisioned as follows: "Hence matter will be continuous throughout and in considering a mass of it we have not to suppose a distinction between its atoms and any intervening space. The powers around the centres give these centres the properties of atoms of matter; and these powers again, when many centres by their conjoint forces are grouped in to a mass, give to every part of that mass the properties of the matter." This definition of matter as continuous and space as the place where electromagnetic phenomena occur naturally led Faraday to confront openly the question of the ether. In a short communication entitled *Thoughts on Ray-Vibrations,* published in 1846, Faraday admitted the possibility of doing away with the ether. In the face of a theory of action by contact within a

material continuum, the ether was a sort of useless duplicate of matter. What now assumed a central position in Faraday's eyes was the velocity of propagation in the continuum of the vibrations that constituted not only electromagnetic phenomena, but all radiant action as well as gravitational interactions. Thus, the role of the finite velocity of propagation of physical events came into play in the context of a general quantitative theory of field. In Faraday's reflections was a deep sense of the need to interpret the new phenomena that he himself had brought to light with his patient work of experimental research. In 1845, he was able to observe in his laboratory the rotation of the plane of polarization of light in the presence of a magnetic field. That same year, he also confronted the problems of diamagnetism. What we wish to emphasize here is that, to the extent that it attempted to explain new phenomena, Faraday's program was forced to develop in directions that increasingly diverged from the traditional explicative scheme, which had triumphed with Ampère's electrodynamics and seemed to represent a victory for the concept of action at a distance.

Weber's electrodynamics

We have already mentioned the productiveness that resulted from the contrast between this program and the "continental" one. The latter, in fact, led, during this period, to extremely valuable works, such as the memoir published by William Weber (1804–91) in 1846, the first in a series of communications known under the general title of *Elektrodynamische Massbestimmungen.* In this memoir, Weber examined Ampère's electrodynamics, observing that it was incomplete and was still based on somewhat unreliable data. In Weber's opinion, there had been a real neglect of electrodynamics, despite the advances made by Faraday and Gauss's work on magnetism (1840), the studies of Joseph Henry (1797–1878) on induced currents (1842), and the work of Franz Ernst Neumann (1798–1895) on the quantitative formulation of Lenz's law (1846).

In order to overcome this situation, Weber proposed an accurate experimental reexamination and reformulation of the laws of electrodynamics, aimed at providing a systematic completion of the theory outlined by Ampère. Basic to Weber's effort was a fundamental distinction between electrodynamic and electromagnetic action. According to Weber, the former included two types of forces acting between the molecules of ponderable bodies. The motion of electrical fluids—and here Weber adopted the suggestion of Gustav Theodor Fechner (1801–87) on current as the result of two streams of electrical charges flowing in opposite directions—provoked interactions between the molecules, and these interactions were "the cause of all Galvanic and electrodynamic phenomena." Thus there were forces, still completely unknown, with a narrow range of action, and electrodynamic forces at a distance that acted among the molecules of different bodies according to the inverse of the square of the distance, producing the phenomena upon which Oersted and Ampère's physics were built. In addition to electrodynamic action, there was electromagnetic action: according to Weber, there were again interactions between molecules of two different bodies; in this case, however, one body had electrical fluid flowing through it, while the other was a container of magnetic fluids. Weber wrote that this distinction between the different actions had to be maintained until Ampère's general hypothesis was able completely to replace the oldest hypothesis on magnetic fluids.

The first part of Weber's lengthy memoir contained a description of the experiment with which he intended to demonstrate the correctness of the electrodynamic laws, while the second part was dedicated to a theoretical interpretation of those laws. Here Weber expounded his famous formula that included both Coulomb's and Ampère's laws: the force acting between two charges, e and e', with variations in their relative distance r is equal to:

$$\frac{ee'}{r^2} \left[1 - \frac{1}{c^2} \left(\frac{dr}{dt} \right)^2 + \frac{2r}{c^2} \frac{d^2r}{dt^2} \right] ,$$

where the constant c is proportional to the conversion factor between the units of charge. (In fact, Weber took both electrostatic and electrodynamic forces into consideration.) This constant played a critical role in nineteenth-century theories. The constant is a velocity, equal to the speed of light multiplied by $\sqrt{2}$. It was originally interpreted physically as the relative velocity between two charges that did not exert any influence on one another, and no correlation was found between it and the velocity of a perturbation crossing the elastic ether. Only during the following decade did the question begin to attract the attention of theoretical physicists and mathematicians, James Clerk Maxwell and Georg Friedrich Bernhard Riemann in particular.

The profound differences between Faraday's and Weber's approaches were characteristic of the contradictory development of electromagnetism in the second half of the century. Already a line of research aimed at the development of a mechanistic reduction of optical and electromagnetic phenomena, beginning with Ampère and perfected by Weber, could be seen contrasted by another line that began with Hans Christian Oersted and, via Faraday's generalization, matured in Maxwell's treatment of the electromagnetic field and Hendrik Antoon Lorentz's studies.

The theories of Kelvin and Stokes

An essential contribution to the establishment of a relationship between Faraday's and Maxwell's physics came with a brief but important essay, *On a Magnetic Representation of Electric, Magnetic, and Galvanic Forces*, published by Lord Kelvin in 1847. Making reference to Faraday's researches, Kelvin defended the idea that for every problem connected with the distribution of electricity in conductors and to the forces of attraction and repulsion exerted by electrified bodies, there corresponds a problem of the theory of elastic solids. The key for correctly interpreting this idea, according to Kelvin, was Faraday's discovery of the interaction between magnetic fields and polarized light, while the mathematical tool that made it possible to elaborate its correct expression was the set of general equations for the equilibrium of elastic solids formulated by George Gabriel Stokes (1819–1903). Using these equations, Kelvin was effectively able to discuss, in terms of an uncompressible elastic solid, the forces acting between a magnet and an "ideal unit" of magnetism, as well as those acting between an element of current and a unit of magnetism. What characterized this analysis (which, however, Kelvin did not elaborate) was the introduction of the infinitesimal rotation of an elastic solid.

Stokes' influence on Kelvin was considerable to the extent that it involved the relationships established between optics and electromagnetism in the field of studies on the luminiferous ether (see PHYSICAL SCIENCES: MECHANICS, 1840–1850). In 1845, following his disagreement with Challis, Stokes criticized Augustin Jean Fresnel's hypotheses in a memoir entitled *On the Aberration of Light*. It was on that occasion that Stokes stated his famous hypothesis that the earth and the planets carry with them a part of the ether, so that the ether near their surfaces is at rest with respect to them, while its velocity changes as it moves away from these surfaces until, at not very great distances, it is at rest with respect to space. Stokes' theory was perfected in two successive communications of 1846 and 1848, both entitled *On the Constitution of the Luminiferous Ether*, in which the medium that transported the light waves was interpreted as a fluid with respect to the motion of the planets and the heavenly bodies in general and as a particular elastic solid with respect to the "small oscillations that constitute light."

Fizeau's experiment

Optics at the end of the decade was characterized by the experiment conducted by Armand Hippolyte Fizeau (1819–96) on the speed of light, and by the publication of a work by Macedonia Melloni (1798–1854) on heat radiation.

Using a method that was not based on astronomical constants, Fizeau employed two telescopes, placed 8,633 meters apart, to observe the passage of a ray of light through the openings in a toothed wheel moving with a known angular velocity. Fizeau's results were made public in a communication of 1849 entitled *Sur une expérience relative à la vitesse de propagation de la lumière*. Melloni, whose experiments had by 1842 already explained several aspects of ultraviolet radiation, proposed the validity of a wave concept in that area as well. The work in which the Italian scientist summarized his ideas and laboratory observations, *La thermochrôse, ou la coloration calorifique*, was published in Naples in 1850.

With respect to the properties of the spectrum, the contribution made by Christian Doppler (1803–53), who in 1842 stated the principle that bears his name and explained the influence on the spectrum of the relative motion between the light source and the observer, deserves mention (see ASTRONOMY, 1840–1850).

CHEMICAL SCIENCES

The emergence of biochemistry as an independent science

Biochemistry emerged as a discipline in itself, thanks above all to the schools of Jean Baptiste Dumas and Justus von Liebig. In 1841, Dumas and Jean-Baptiste Boussingault (1802–87) published *Essai de statique chimique des êtres organisés*, in which they drew a parallel between animal and vegetable metabolism. In 1840, Liebig published his *Die organische Chemie in ihrer Anwendung auf Agrikultur und Physiologie*, and in 1842 his *Die Thierchemie oder die organische Chemie in ihrer Anwendung auf Physiologie und Pathologie*. In the first work, developing concepts that had already been expressed by Richard Kirwan (1733–1812) in 1794 Nicholas Théodore de Saussure in 1804, Liebig asserted that plant nutrients consisted of inorganic substances (water, ammonia, mineral salts). In contrast to the humus theory, he stated that manure did not act directly through its organic components, but through the inorganic products of its decomposition. The experiments on chemical fertilizers that Liebig conducted between 1845 and 1849 established the premises for the fertilizer industry. In the second work mentioned above, Liebig maintained that animal energy came from the reaction between the components of foods and oxygen in the atmosphere. The iron in red corpuscles played an important part in this process, since it was alternately oxidized and reduced. Beside the "respiratory" foods, there were the "plastics," which produced blood and the various organs. Even though Liebig believed in a vital force, he maintained that alkaloids, proteins, etc. would one day be synthesized. Although they were not lacking in errors, Liebig's theories represented an important contribution to the progress of animal and plant biochemistry. Liebig made important experimental contributions to biological organic chemistry as well. Worthy of mention are his researches on creatine and creatinine (discovered in urine by Max Joseph von Pettenkofer [1818–1901] in 1844), and the discovery of tyrosine, sarcosine, and inosynic acid (1846–47). In 1846, his student Julius Bodo Unger discovered guanine in guano. Two well-known biochemical reagents were introduced in 1849: the reagent discovered by Hermann von Fehling (1812–85) for reducing sugars and the reagent of Eugene Auguste Millon (1812–67) for proteins.

Also in the area of organic chemistry, Théophile Jules Pélouze (1807–67) partially synthesized a glyceride, tributirrina (1843), and glycerophosphoric acid (1845). Adolph Wilhelm Hermann Kolbe (1818–84), who achieved the total synthesis of trichloroacetic acid, from which Henri Melsens (1814–86) had already obtained acetic acid, fully realized that acetic acid, which up to that time was known only as a product of the oxidation of organic material, could also be prepared by synthesis from the elements.

Other natural substances

New natural products were isolated: Nicolas Théodore Gobley discovered lechthin (1846–47), Gustave Bourchardt isolated fructose (1847), and G. Merck (1825–73) discovered papaverin; Augustin Homolle obtained amorphous digitalin (1845), a sufficiently pure mixture of the active components of digitalis, from which C. Adolphe Nativelle later prepared crystallized digitalin (1869). Auguste Laurent, who had isolated isatin in 1841, noted that natural bases, unlike artificial ones, were optically active (1844).

The beginnings of the dye-making industry

The varied researches initiated in 1843 by August Wilhelm von Hofmann and his pupil, Charles Blachford Mansfield (1819–55), on the products of the distillation of coal tar led to the preparation of new organic compounds, such as anilides and sulphanilic acid (Charles Frédéric Gerhardt, 1845–46), and opened the way to the great organic chemical industry. In 1845 Rudolph Christian von Böttiger (1806–81) obtained styphonic acids, which possessed explosive properties, from the oxidation of wood and resin with nitric acid.

Thomas Anderson (1819–74) studied the components of bone oil, an industrial byproduct of the dry distillation of bones in the production of animal carbon, and succeeded in isolating and distinguishing several fairly important azotate bases: lutidine, collidine, picoline (1849) and, later, pyridine. He maintained that these substances constituted a homologous series, as derivatives of ammonium. He correctly assigned pyrrole, first obtained by Friedlieb Ferdinand Runge from coal tar in 1834, the formula $C_4H_{15}N$. He also obtained petitine, which Gerhardt showed to be n-butylamine, from bone oil.

Classical organic reactions

Various classical methods of organic chemistry were discovered: the preparation of aromatic polynitroderivatives with fuming nitric acid by Henri Étienne Sainte-Claire Deville (1818–81) in 1841, the reduction of aromatic nitroderivatives to ammines (the reduction of nitrobenzene to aniline) with ammonium sulphide by Nikolai Zinin (1812–80) in 1842 or with zinc and hydrochloric acid (Hofmann, 1845); the benzadinic transposition of hydrazobenzene to benzadine (Zinin, 1845), the first diamine in the diphenol series; the preparation of halogen derivatives of acids by Auguste Cahours (1813–91) in 1846 and of aromatic compounds (Hofmann, 1848); the preparation of carboxylic acids from nitriles by Jean Baptiste Dumas, Faustino Jovita Malaguti (1802–78), and Félix Le Blanc (1813–86); Edward Frankland (1825–99) and Kolbe in 1847; the transformation of primary ammines into alcohols with nitrous acid by Rafaelle Piria in 1848; and the preparation of isocyanic ester by Charles Adolph Wurtz (1817–84), which led to Wurtz's discovery, in 1849, of aliphatic ammines.

Compounds that would later be of great practical use were also prepared: nitrocotton (Christian Friedrich Schönbein, 1846), nitroglycerin (Ascanio Sobrero 1846), and propylene (J.W. Reynolds, 1849). In 1841, Antoine Jérôme Balard (1802–76) prepared oxamic acid.

Theoretical advances: the formulation of the theory of types

Theoretical organic chemistry took its first steps forward. After Dumas had linked the melting point of fatty acids to the number of carbon atoms in the molecule (1842), his assistant Gerhardt generalized the observations of them (1845) and recalled attention to homologous series in 1842 (the concept of homology went back to J. Schiel), classifying organic substances into families, each of which contained the same number of carbon atoms. In 1841, after he had isolated cacodyle oxide from Cadet's arsenical liquid and prepared a number of derivatives of it, Robert Wilhelm Bunsen (1811–99) demonstrated the existence of organic radicals (although he called them "molecules"), which, in his opinion, proved the extension of Jöns Jacob Berzelius's dualistic theory to organic chemistry.

In an attempt to look at organic and inorganic chemistry from a unitary point of view, Gerhardt and Laurent reconducted a number of organic and inorganic compounds to a water "type" (1843–46); in 1848, Gerhardt announced his "antidualistic" or "unitary" theory. The close ties between organic and inorganic chemistry were confirmed by the discovery of the first organometallic compounds, the zinc alkyls (Frankland, 1848–49).

Allotropic forms of known elements

Among the achievement in inorganic chemistry during this period were the discovery of two new elements in the rare earths group, erbium and terbium (Carl Gustav Mosander, 1843), and two allotropic forms of already-known elements, ozone (Schönbein, 1840) and red phosphorous (Anton Schrötter, 1847). In 1842, Eugene Melchier Peligot (1811–90) isolated metallic uranium, which Martin Heinrich Klaproth had already discovered as early as 1789 in pitchblende. Two years later, Carl Claus (1796–1864) discovered ruthenium and obtained it in the pure state, then pointed out the affinities between the elements of the Ru-Rh-Pd and Os-Ir-Pt groups, later used by Dimitri Ivanovich Mendeleev in compiling the periodic table. In 1846, Sainte-Claire Deville prepared nitrogen peroxide.

In 1847, Lambert Heinrich Babo (1818–99) stated the law that bears his name, according to which the fall in vapor pressure of a solvent upon the addition of a non-volatile solute is proportional to the molar concentration.

Photochemistry and thermochemistry. Treatises

Between 1840 and 1841, John William Draper (1811–82) and Germain Henri Hess (1802–50) stated the two well-known laws of photochemistry and thermochemistry. From 1845 to 1853, Pierre Antoine Favre (1813–80) and Jean Thiebaut Silbermann (1806–65) carried out calorimetric tests on the heat of combustion of carbon, sulphur, and phosphorous, demonstrating that it was different for the different allotropic forms of the elements.

In 1857, Sainte-Claire Deville introduced the term "thermal dissociation" and studied the equilibrium between the substance that dissociated and the products of the dissociation.

Karl Remigius Fresenius (1818–97) published manuals on qualitative analysis—with a subdivision of metallic cations into analytical groups (1841)—and on quantitative analysis (1846), both of which were printed in numerous editions and translations. Hermann Franz Kopp (1817–92) published his lengthy *Geschichte der Chemie* in 1843. In 1847, Liebig and Kopp began publication of the *Jahresbericht über die Fortschritte der Chemie*, a continuation of Berzelius's work. The *Cours élémentaire de chimie à l'usage des facultés des établissements d'enseignement secondaire, des écoles normales et des écoles industrielles*, by Henri Victor Regnault (1810–78), a work that became quite popular and was translated into several languages, was published between 1843 and 1849. And finally, the first edition of Liebig's famous *Chemische Briefe*, a work that became the source of an extremely animated debate a few years later (see THOUGHT AND SCIENTIFIC LIFE, 1850–1860), came out in 1844.

LIFE SCIENCES

Experimental physiology

This was a productive decade for physiology, which acquired an entire generation of young researchers coming out of Johannes Peter Müller's Berlin school and François Magendie's Paris school. These new scientists were outstanding for their experimentalism, their biophysical and biochemical approach, and their interest in methodology. Emil du Bois-Reymond (1821–94) joined with researchers from other schools and disciplines to form the Physical Society in Berlin. On 23 July, 1948, Helmholtz read to the Society the memoir in which he discussed the conservation of force and upheld the conservation of energy in all natural processes. The second concept had already been stated in 1842 by the physician Julius Robert von Mayer (1814–78), who reiterated it in 1845 in a pamphlet that addressed the problem of organic motion in relation to metabolism. But the accomplishments of Meyer, who lacked a solid experimental and mathematical background, were only recognized after James Joule and Helmholtz had completed their work (see PHYSICAL SCIENCES, 1840–1850). In 1850, on the threshold of the following decade, Helmholtz announced the invention of a device for observing the fundus of the eye in a living subject, the ophthalmoscope. In Königsberg, Brücke began his studies on the physiology of the sensory organs, which he continued in Vienna, where he was one of the founders of the Second Viennese Medical School together with Carl von Rokitansky (1804–78), Ferdinand von Hebra (1816–80), and Jošef Hyrtl (1810–94). In 1848, du Bois-Reymond published the first volume of his *Untersuchungen über tierische Elektrizität*, which openly disagreed with Carlo Matteucci (1811–68), author of the *Saggio sui fenomeni elettrici negli animali* (1840) (see LIFE SCIENCES, 1790–1800). But the two authors came to the same conclusions, though du Bois-Reymond's method was more refined, in their studies of the laws of the "frog current" (as Leopoldo Nobili had called it; see PHYSICAL SCIENCES, 1820–1830), obtained by connecting a point on a transversal section with a point on a longitudinal section of a muscle. In 1840, du Bois-Reymond was named professor of physics at the University of Pisa by the Grand-Duke of Tuscany. He later demonstrated that in the course of stimulation, an electrical perturbation, which showed up as a negative oscillation on a galvanometer, was produced along the nerve. Educated in the prestigious school of Müller—"a man who bore the stamp of the extraordinary," as he described his teacher in a commemorative speech—du Bois-Reymond in turn cultivated a number of disciples: Albert von Bezold, Ludimar Hermann, Hermann Munk (1839–1912), and Edward F. Pflüger (1829–1910), who successfully continued his teacher's electrophysiological research. The development of Karl Ludwig (1816–95), on the other hand, took place outside the Berlin school; in 1842, he published his *De viribus physicis secretionem urinae adiuvantibus*, in which, among other things, he sustained the hypothesis that the renal tubules participated in the formation of urine, as mentioned earlier. In 1849, Ludwig left Marburg for Zurich, after he had begun important studies on the mechanics of circulation and built the kymograph (1846), the first of a series of devices for measuring various physiological processes. In 1865, called to succeed Ernst Heinrich Weber (1795–1878) in the chair of physiology in Leipzig, Ludwig returned to Germany, where he set up a school that was attended by numerous European and American scholars.

The examination of perception. The concept of reflex. Animal glycogeny

Meanwhile, Weber's group represented the most important circle of physiological studies, besides Müller's Institute, in the German universities during this period. From his *Wellenlehre* (1825) to the *Adnotationes anatomicae et physiologicae* (1851), Ernst Heinrich Weber founded significant parts of human physiology on physico-mathematical bases and obtained numerous experimental results that indicated that the increase in intensity of a stimulus that was able to provoke a perceptible change in sensation always maintained the same ratio with respect to its initial intensity. In 1854 Weber demonstrated the inhibitory effect of the vagus nerve on cardiac contraction. His brother, Eduard Friedrich Weber, in an article of 1846 in the physiological dictionary of Rudolf Wagner (1805–64), provided a new description of muscular physiology and defined the elasticity of muscle tissues. Wilhelm Eduard Weber (1804–91), the well-known physicist, also worked on applying physical principles to physiological processes. Another important current of German physiology was led by Jan Evangelista Purkyně (1787–1869), a professor at Breslau from 1823 and director of one of the two existing physiological laboratories in Germany during the third decade of the century (the other, in Fribourg, was directed by M.J. Schultze). From 1820 to 1840, Purkyně developed a method for microscopic investigation that shifted the focus in the examination of several vital phenomena from the level of the organism to the tissues. Together with his pupil, Gabriel Valentin (1810–83), he discovered ciliary motion. J.N. Czermak (1828–73), inventor of the laryngoscope (1856), also came out of Purkyně's school. Other scholars who left lasting traces of their work were L.W. Bischoff (1807–82) of Giessen; R. Wagner of Göttingen, who was involved in a bitter disagreement with his pupil, G. Meissner (1820–1901), over priority for the discovery of the tactile corpuscles. But Müller's and Ludwig's school's brought positive ideas that already existed elsewhere to a higher level of consistency, from the use of the microscope to the quantitative method and the connections between physiology, physics, and chemistry. In Great Britain, the vitalistic tradition was still strong among several authors, the representatives of "natural theology," which had its origins in John Hunter's (1728–93) and Charles Bell's studies on animal mechanics (1828). Bell was the discoverer of the motor function of the anterior spinal roots. However, a distinctly experimental tendency emerged in neurophysiological studies in the English school with Benjamin Collins Brodie (1783–1862), and Marshall Hall. Hall was the author of approximately 150 memoirs, those dedicated to reflex action being particularly inportant.

The future of experimental physiology, however, lay in the Berlin and Paris schools. In the 1840's, two pupils of Magendie, Moritz Schiff (1823–96) and Claude Bernard (1813–78), began to show their talent. Schiff, who moved to Paris from Göttingen, from whence he later occupied the chair of physiology at Geneva, worked on cardiac innervation, experimental thyroidectomy and the functioning of the digestive tract. But the advance of physiological research in France was above all linked to the development of Georges Cuvier's comparative morphology, i.e. the dependence of function on form. In 1844, Bernard split with Magendie, whom he succeeded in 1848 at the College de France. By this time, the destinies of clinical physiology, subordinated to medical needs, and experimental physiology, which was actually the true "scientific medicine," or "experimental medicine," had become separate, and Bernard was the leader of the new current. In the fifth decade of the century, he began a broad and varied series of researches, which ranged from the digestive functions of gastric juice and the pancreas to hepatic glycogeny, from nervous vasoconstriction and vasodilation to the paralysis induced by curare. The "diabetic puncture" of the fourth ventricle topped off this first phase of his work with a sensational experiment.

Pierre Flourens (1794–1867) entered the Academy of France in 1840 after having succeeded Cuvier in the office of perpetual secretary of the Academy of Sciences. In a work entitled *Examen de la phrénologie* (1842), he reiterated the anatomical and physiological method that had led him to identify with the functions of the various parts of the brain.

Physiological chemistry

Many important works that marked the passage from organic to physiological chemistry came out during this decade. The *Essai de statique chimique des êtres organisés*, by Jean Baptiste Dumas (1800–84) and Jean Baptiste Boussingault (1802–87), was published in 1841. Justus von Liebig reiterated Antoine Laurent Lavoisier's theory (see CHEMICAL SCIENCES, 1840–1850), recognizing the existence of a metabolism of tissues, with the production of nitrogenous wastes as well as carbon dioxide and water. Liebig expressed the need for quantitative analysis through complex reconstructions of the matter-energy balance: he created a "desk-top biochemistry" that corrected the original errors by actually amending the addenda of organic economy. His work on soil fertilization, in which he demonstrated the untenability of the humus theory—according to which plants developed by absorbing substances from decayed plants and animals—was also important.

New nosographical definitions. Surgical anesthesia

New nosographical definitions were developed in pathology and clinical medicine: Acute anterior poliomyelitis, or Heine-Medin disease, was described by Jacob von Heine (1800–79), rheumatic heart disease by Jean Baptiste Bouillaud (1796–1881), acute nephritis from mercury poisoning by William Charles Wells (1757–1817), sciatica by François Valleix (1807–55), whooping cough by Frédéric Rilliet (1814–61) and Antoine Barthez (1811–91), sclerotic pericarditis by N. Chevers, exothalmic goiter by Robert James Graves (1796–1853), and epithelioma by Adolf Hannover (1814–49). At Guy's Hospital in London (500 beds, founded in the eighteenth century), Thomas Addison (1795–1860) followed the nephrological research of Richard Bright (1789–1858) and furnished the description of a syndrome that was characterized by a bronzing pigmentation of the skin and muscular asthenia and that had an unpromising outcome (1849–55). Autopsies of eleven cases enabled Addison to connect the disease with a bilateral lesion of the adrenal gland. The Swiss physician Jean Louis Prevost (1790–1850) identified and described iodine deficiency goiter (1849). In surgical practice, the administration of ether (1846) and chloroform (1847) for the purpose of general anesthesia was introduced: the American surgeon Henry Jacob Bigelow (1818–90) spoke of it as a "capital event in the annals of surgery." The Russian surgeon Vassili A. Basov (1812–1879) reported the suture of a gastric ulcer in experimental studies on digestion (1842).

Publication of the *Archiv für patoligische Anatome und Physiologie und für klinische Medicin*, directed by Rudolf Virchow, began in 1847. The platform of the new periodical was contained in the statement: "Practical medicine as applied theory, theoretical medicine as pathological physiology." Pathological physiology was, in turn, identified with biology, by which was meant science of life in general and of human beings in particular.

The Humboldtian concept of science based on the interaction of natural forces

The first volume of *Kosmos: Entwurf einer physische Weltbeschreibung*, by Alexander von Humboldt (1769–1859), came out in 1845. It was a synthesis of the scientific knowledge of the times, from biology to geology, from anthropology to astronomy—a historico-philosophical integration of the "progressive development of the idea of the universe." The work was born out of the author's journeys to South America, central Asia, and the major European countries, but the nucleus consisted in the text of sixty lessons on topics in physical geography held by Humboldt in Berlin between 1827 and 1828. Humboldt's concept of science, based on the idea of the "concatenation of things" and the "interaction of natural forces," was backed up by the author's exceptional experience and came to represent an effective counter to Hegel's and Schilling's philosophy of nature in Germany. Rudolf Virchow saw in Humboldt the "second founder of the University of Berlin," a person who was able to give science standing within the culture and as culture. The successive volumes of the *Kosmos* came out in 1847, 1850, and 1858 and contributed to moving beyond mechanism and agnosticism in the scientific field, while reinforcing the defenses against the invasion of philosophically derived rationalism. Humboldt's work helped strengthen the ties between the life sciences and the inorganic disciplines. By living in both Paris and Berlin, Humboldt was able to become the interpreter, in German scientific circles, of the currents that Antoine Laurent Lavoisier, heir to the generation of Lamarck, Cuvier, and Comte, had developed in French science at the end of the eighteenth century.

Animal systematics

Félix Dujardin, in his systematic studies (1841), eliminated otifera, algae, small worms, flagellata, and tentacolates from the old infusoria group (already recognized by Leeuwenhoek), endowing the group with its present connotation. In 1845, Dujardin published his *Histoire naturelle des helminthes ou vers intestinaux*. He was also a precursor of the protoplasmatic notion of cells.

Plant systematics

The *Grundzüge der wissenschaftlichen Botanik*, by Mattias Jacob Schleiden, which came out in 1842–43, brought the science of plant life up to a level of rigorous, inductive methodology. Schleiden saw embryology, viewed in the light of Kaspar Friedrich Wolff's theory of metamorphosis (see LIFE SCIENCES, 1750–1760), as the criterion for morphological description, freeing morphology from the chains of dogmatism. Cryptogamae were raised to the level of phanerogamae in systematics, even though knowledge on their development was still incomplete. Systematics acquired a number of important contributions during this period. In his *Genera plantarum secundum ordines naturales disposita* (1836–40), Stephen Endlicher (1804–49) divided the plant kingdom into Thallophytes and Cormophytes, and systematically identified the conifers. Alexandre Brongniart (1801–76) insisted on the distinction between phanerogamae and cryptogamae and reordered the botanical garden of the Museum of Natural History in Paris on the basis of that distinction. John Lindley (1799–1865), in *The Vege-*

table Kingdom, proposed an alternative to Linnaeus's botanical system. Joseph Dalton Hooker (1817–1911), one of the major systematicians of all times, began publication of his *Flora antarctica* in 1844. Between 1844 and 1860, three thousand new species were classified; Hooker listed 16,000 more new species in his *Flora of British India* (1872–97).

In the area of zoology, the work of Carl Theodor von Siebold (1804–85) and Friedrich Stannius on arthopodae and of Michael Sars (1805–69) on echinodermatae deserve mention. Tangible progress was made in entomology, which became more articulated. In particular, von Siebold became the first to decribe protozoa as organisms in themselves, in his *Lehrbuch der vergleichenden Anatomie der wirbellosen Tiere* (1848). This supported the theory proposed by Schleiden and Theodor Schwann in 1839 that all organisms were made up of cells.

EARTH SCIENCES

The glacial era

By this time, the idea of an ancient glaciation was generally accepted, although authors differed as to its extent. Jean de Charpentier theorized that it was a larger-scale version of the phenomena that could still be observed in the highest parts of the Alps and other mountain chains (1841). Louis Agassiz, on the other hand, maintained that an ancient glacial cap had extended over Northern Europe and came to similar conclusions for North America (1847). Charles McLaren (1842) saw glaciation as the possible cause of a lowering of the sea level as large masses of water remained on the continents in the form of ice, with a return to the original level as the ice melted. James David Forbes (1809–68) discovered ancient moraines in Great Britain (1845), which contributed to the spread of the glacier theory. Arnold Henri Guyot (1807–84) established a large collection of erratic masses from North America. Nicolas Léonard Sadi Carnot and John Thomson made obervations on the movements of present-day glaciers (1849), while J. Adhemar (1842) gave cosmic explanations for glacial phenomena in a theory that was later elaborated by James Croll.

Coral reefs

The question of how coral atolls and reefs were formed, a problem that received considerable attention during this period, was connected to the problem of variations in sea level, possibly in concurrence with glacial phenomena. In 1842, Charles Darwin published his famous work on *The Structure and Distribution of Coral Reefs,* in which he reiterated the link between the submersion of volcanic islands and the formation of atolls. James Dwight Dana, who had visited the coral islands of the South Seas during Charles Wilkes's expedition (1839–41), supported the same idea.

Orogenetic theories

The American brothers William Barton Rogers and Henry Darwin Rogers noticed the wave-like folded tectonic structure of the Appalachians (1842) and attributed it to the infusion of molten material deep underground. In 1847, Charles Babbage (1792–1871) observed that sediment was particularly thick in correspondence to mountain chains. Jean Baptiste Élie de Beaumont and Alexander von Humboldt studied the relationships between magmatic injections, earthquakes, and orthographic projections, though they were not able to arrive at a satisfactory, acceptable theory. In 1845, de Beaumont published his *Leçons de géologie pratique*, and the following year Humboldt completed his *Voyage aux régions équinoxiales*. Dana also connected *magnetism* and orogenesis (1847–49).

Milne and Forbes on earthquakes

Observations on the effects of *earthquakes* and the position of *epicenters* and suggestions for instrumental measurements were published by David Milne Home and James David Forbes (1809–68) between 1841 and 1844.

Stratigraphic problems and regional geology

A number of studies of regional geology, with particular reference to stratigraphic problems appeared during this period. Particularly important among these were the studies of Switzerland published from 1846 to 1848 by Hans Conrad Escher von der Linth (1807–72); on the formations of North America published in 1843 by William Edmond Logan (1798–1875), which made him a pioneer in the geology of the Pre-Cambrian; the research conducted in Russia by Roderick Impey Murchison and his separation of the Permian from the Carboniferous (1841); the paleozoological studies of the Russian Karl Frantsovich Rouillier; the completion of a geological map of France, with explicative notes (1841), by Élie de Beaumont on the Venetian Alps (1850), which included paleontological studies on the fossil fish of Monte Bolca and geological descriptions of the formations between the Carnia and the Adige. Ebenezer A. Emmons (1799–1863) recognized the age of the fauna of the Taconic system (1844), but his discovery aroused virtually no interest. Alcide Dessalnes d'Orbigny (1802–57) proposed the basic paleontological subdivisions for the Cretaceous of France (1842–43), and defined "plane" in geology (1850).

In 1840, Leopoldo Pilla (1805–48) wrote a treatise on geology entitled *Studi di geologia ovvero conoscenze elementari della scienza della Terra*.

Primitive industry and vertebrate paleontology

In a work entitled *L'industrie primitive*, Jacques Boucher de Crèvecoeur de Perthes (1788–1868) maintained that the human race was very old, contemporary with the mammoth remains discovered at Abbeville. Agassiz's monumental *Recherches sur les poissons fossiles* was published between 1833 and 1834; it included considerations on the ichthyofauna of Monte Bolca, near Verona. Agassiz continued the work of Georges Cuvier, who had skimmed over the study of fossil fish like his teacher Agassiz showed himself to be a catastrophist and creationist. The *Zoologie et paléontologie générales* by Paul Gervais (1816–79), appeared in 1848–50. Georg August Goldfuss (1782–1848) completed his *Petrefacta Germaniae*, written between 1826 and 1844; it included fossils of all kinds and even illustrated new forms. In 1844, Gideon Algernon Mantell (1790–1852) published *The Medals of Creation*, a richly illustrated paleontological work.

Debates on wave motion

Numerous important studies on waves, from both theoretical and an experimental points of view, were made during this decade. A number of disputes and controversies also arose, creating a geat deal of confusion in identifying the parameters of waves, which were specified during the next few decades. Several fundamental contributions deserve mention. George Biddell Airy (1801–92) in 1842 and George Gabriel Stokes (1819–1903) in 1845 and 1847 studied the mathematical expression of wave profile and of the associated particle orbits, making substantial modifications, though mainly of a formal nature, in Gerstner's earlier theory. Starting from the hydrodynamic equations, Airy provided the first theory for small-range sinusoidal waves in the open sea. Stokes perfected and extended Airy's ideas, studying waves with finite ranges and arriving at a trochoidal profile, but by way of Gerstner's parametric equations, but by superposing sinusoids of different lengths and ranges. Stokes did not impose the form of orbital motion *a priori* as Gerstner did, but deduced it from the wave profile by solving the hydrodynamic equations (the Euler-Navier-Stokes equations) and Laplace's equation. Stokes's concepts could also be applied to oceans with limited depths, as well as to internal waves. Increasingly accurate successive approximations could always be made: these were problems that were, to some extent, resolved a century later, with the use of electronic calculators.

In 1844, John Scott Russell studied the formation of waves and defined "wave germs" as the rippling capillary waves that form as soon as the wind blows with any intensity. His work provided the premises for studies of turbulence and the energy exchanges, on a micro- and mesoscale, between the atmosphere and the ocean.

New oceanographic instruments

New oceanographic instruments were designed and built. In 1842, G. Aimé invented a new type of marine thermometer, based on the new concept of "inversion." This was the instrument that finally solved the problems involved in measuring temperature even at great depths. It was modified and perfected by Negretti and Zambra, who by 1857 were already preparing protected thermometers for Robert Fitzroy. In 1845, Aimé invented the first propeller current meter.

In 1844, with a driving bell that he himself invented, Henri Milne Edwards (1800–85) made the first modern (i.e. instrumental) underwater explorations, with immersions off the coast of Palermo.

Biological and chemical research in the Mediterranean

In 1842, Edward Forbes (1815–54) made an expedition to the eastern Mediterranean and the Aegean, mainly for the purpose of biological research. Large amounts of data were gathered, but Forbes arrived at the erroneous conclusion that the ocean depths were azoic below 300 fathoms (600 meters). This assertion, which reflected extremely ancient beliefs, was not rejected due to Forbes's great authority, and it generated numerous perplexities until the voyage of *Challenger*, despite the fact that almost simultaneously Aimé found organisms at a depth of about

1,800 meters off the coast of Algeria. During this period, Aimé carried out a long series of investigations for the French government in the waters of the Mediterranean, including measurements of current, tides, temperature, etc. He correctly hypothesized that cold waters formed during the winter and sunk to the bottom due to their density.

Advances were also made in the chemistry of the ocean waters by J. Usiglio, who worked on methods of analysis (1849) and carried out a number of analyses of the waters of the Mediterranean. He was especially interested in bromine, and defined salinity, quantitatively identifying the following components (by this time correctly defined according to the symbology of modern chemistry): $NaCl$, $NaBr$, KCl, $MgCl_2$, $MgSO_4$, and $CaSO_4$.

Other instruments for meteorology and geophysics

The aneroid barometer, which presented definite practical advantages over the mercury type, was invented by Vidi (1804–66) in 1848.

In his *On the Variation of Gravity on the Surface of the Earth* (1849), Stokes provided the first representation, with a relative theoretical discussion, of the distribution of the acceleration of gravity over the earth's surface. In 1842, Humphrey Lloyd (1800–81) built a sensitive magnetometer for determining the earth's magnetic field, and organized a series of observatories for obtaining simultaneous measurements.

APPLIED SCIENCES

Mechanical technology

The Great Exhibition of London in 1851 provided an overview of the technical activity that had taken place in the decade from 1840 and 1850. England proved to be the most important industrial country and won prizes in almost all the categories. Further progress had been made in the area of machine tools, thanks mainly to the work of Joseph Whitworth, who attempted to standardize machine production, providing increased precision. Whitworth achieved these results by using an extremely accurate metrology and machines such as the caliber, which enabled him to reach a degree of precision impossible before then. As early as 1841 he proposed the adoption of a constant angle between the two planes of screw threads: Whitworth threading was universally adopted in mechanical technology until 1948. In 1842, he built an automatic, steam-operated planing machine, on a patent of 1835, and in 1843 he constructed a lathe incorporating improvements that he had patented in 1839—these included a transversal feed mechanism with guide screws. Thanks to Whitworth's machine tools, it was finally possible to produce machines mechanically, with an accuracy of construction that was unthinkable in manual execution.

However, the delay with which Whitworth's machines entered into the English market is significant. This was probably related to the absence of mass production and the abundance of labor. In general, English technology showed itself particularly attentive to absorbing techniques that made savings in capital, rather than savings in labor, possible, since England did not suffer from a scarcity of labor-power. This explains the strong impression that the so-called American work system—the method of interchangeable parts that Samuel Colt (1814–62)

applied in the production of revolvers—made on the English public at the Exhibition. Two other machines also attracted considerable attention: the sewing machine, invented by Elias Howe (1819–67), a Massachusetts mechanic (patented in 1846), which eliminated the last obstacle for the textile industry—manual needlework—and the reaper of Cyrus McCormick (1809–84), an improved model of threshing machine that the inventor had already patented in 1834. In 1842, Captain (later General) de Beaulieu theoretically expounded the advantages of rifling the insides of guns. The first rifled cannon, built by Giovanni Cavalli (1808–79) in Sweden, was adopted by the Piedmontese army in 1846.

In the area of steam engine construction, important contributions were made by Ernst Alban (1791–1856) with his *Die Hochdruckdampfmaschine* of 1843, in which he expounded his principles on high-pressure steam machines, and by John Penn (1805–78), who installed an oscillating steam motor where the connecting rod was directly attached to the shaft of the paddle-wheel in the boat *Black Eagle*.

Metallurgy and industrial chemistry

In 1847, the Gold Rush swept the United States after rich gold deposits were discovered in California. Innovations in mining set the pace, however. In 1844, John Buddle introduced a system of slow and continuous hauling with a cable, while at Hadwell, in Durham, the "main and tail" of pulling full wagons and returning empty ones was used. Working conditions did not noticeably improve, even though Mauseler perfected the first safety lamps for mines and ventilators began to be widely used.

Chemical research led to inventions, discoveries, and patents in all fields, from metallurgy to the refrigerating industry. In nonferrous metallurgy, the most important patents pertained to methods for separating copper from silver using sulphuric acid (1840) and for purifying tin ore (cassiterite) with sulphuric and hydrochloric acid (1840), a process for extracting copper that involved the use of sodium sulphate in the melting furnace and then crushing the mixture in water (James Napier, 1846), and finally, a method for producing mercury that was similar to the process used for producing gas for lighting (1847), invented by Andrew Ure (1778–1856).

Important advances were also made in the field of explosives, when Christian Friedrich Schönbein (1799–1868) discovered nitrocotton, an explosive substance formed by the action of nitric acid on cotton, and Ascanio Sobrero (1812–88) discovered nitroglycerin, both in 1846.

Other achievements worthy of mention were: the vulcanization of rubber, obtained by Charles Goodyear (1800–60) and Thomas Hancock (1786–1865) with molten sulphur (in 1840 and 1843–44, respectively), by Alexander Parkes (1813–90) with sulphur chloride (1846), and by Stephen Moulton with lead hyposulphite (1847) (the following year, Moulton founded one of the first large rubber factories in England); the patent by Thomas Clark (1801–67) of a process for softening hard water with milk of lime, and his soap method for determining the hardness of water (1841); and the process for mercerizing cotton (1844), devised by John Mercer (1791–1866). Justus von Liebig (1803–63) conducted the first experiments with non-natural fertilizers (see CHEMICAL SCIENCES 1840–1850).

Food technology

Advances were made in the food industry when Stephen Goldner patented a method for cooking foods in a bath of calcium chloride, which made it possible to cook the foods at temperatures higher than the boiling point of water (1841). A patent was also obtained by John Gorrie (1803–55) for a cold-air refrigerating machine for commercial use (1849).

In 1840, Ferdinand Schichau (1814–96) invented a steam machine for distilleries, sugar mills, and toasting chicory, which was used as a coffee substitute.

Urban services

Providing services essential to urban life, such as the distribution of water and gas, was made possible during this period (James Nasmyth also contributed in this area when in 1840 he devised a double-spouted, wedge-shaped cock). The production of tubing advanced when, in 1846, D.Y. Stewart patented his system for producing iron tubes that were fused vertically instead of horizontally in 1846. Methods of extraction were improved as the hydraulic plunger, invented by Joseph Michael Montgolfier in 1797, entered into use in 1840. Reservoirs were built everywhere, the most daring of which involved the construction of the thirty-five meter high Zola canal dam in France in 1843. Thomas Clark worked on purification and patented his system for softening water, mentioned earlier, in 1841.

The use of gas became convenient thanks to the introduction in 1840 of a new type of burner (the so-called atmospheric burner), in which air was introduced into the net of gas just below the combustion area. The first crude incandescent lamps were built and patented in 1841, and the arc lamp was patented in 1845.

Photography

One technological discovery of this decade that was destined for great future development was photography. The first experiments, conducted by Joseph Nicéphore Niepce and Louis Jacques Daguerre, dated back to 1830, but it was only during this particular decade that truly remarkable results were finally obtained. It should be mentioned that daguerrotypes were first used mainly in astronomy (see ASTRONOMY, 1840–1850).

Manuals

The growth of industry obviously increased the demand for manuals dedicated to each individual technology. Examples of these were the *Handbuch der Eisenhüttenkunde,* by Karl Johann Karsten (1841), the *Practical Treatise on the Manufacture and Distribution of Coal-Gas,* by Samuel Clegg (1841), and two English encyclopedias on construction: *An Encyclopedia of Architecture,* by Gwilt, published in 1842, and *An Encyclopedia of Civil Engineering,* by Edward Cresy, published in 1847.

1850–1860

THOUGHT AND SCIENTIFIC LIFE

The reorganization of Europe

The first years of the second half of the century were char-acterized by great constructive endeavors: the progressive creation of national unity in Germany, the economic and colonial establishment of the second empire in France, the liberal and imperialistic development of England, the internal transformations in czarist Russia, and the unification of Italy. There was undeniably a profound gap betwen the revolutionary prospects of 1848 and the political developments which followed. The ruling classes gradually adapted themselves to the new times, and national movements relied more on the initiative of chancellors than on the push for democracy. Industry continued to develop at an ever-increasing pace. Germany and Great Britain were in the lead during this decade, though they were soon joined by the United States.

European economic production rose constantly throughout this decade and the years following it. In the 1860s, the construction of railroads, already fairly widespread during the fifties, gradually connected manufacturers with their markets, reinforcing the structures of large industry. In correspondence to this capitalistic development was the movement toward proletarian organization. In the course of just a few years this movement began to surpass Proudhon's prospects of a simple reform of economic structures in the direction of cooperativism. The work of Marx and Engels was a driving force in this process (the First Workers' International was founded in London in 1864; it was dissolved in 1876 because of conflicts between Marxists and the anarchist followers of Bakunin).

The culture of this decade directly reflected the historical changes that followed 1848; while the Romantic spirit declined, science, which during the previous decades had slowly prepared for its rebirth, now gained a wide cultural dominance and was restored to its supremacy, influencing the principal manifestations of life and knowledge. It became increasingly clear that positivism was the most adequate cultural movement for this scientific revival.

The controversy over materialism in Germany

Although it never reached high philosophical levels, the so-called controversy over materialism (*Streit des Materialismus*) in Germany succeeded in exerting considerable influence on German culture, and provoked a new interest in the general concepts of science (especially biology) among a vast public. The first edition of the *Chemische Briefe,* by Justus von Liebig (1803–73), in which the author stated his conviction that a knowledge of nature would lead to a knowledge of God, had come out in 1844. In 1852, Jacob Moleschott, a Dutch professor of physiology at the University of Heidelberg, published a volume in response to Liebig's letters on chemistry entitled *Der Kreislauf des Leben. Physiologische Antworten auf Liebigs chemische Briefe.* Influenced by Feuerbach, Moleschott maintained that nature did not depend on any divine intervention. That same year, Karl Vogt (1817–95), a professor of zoology at Geneva, where he had been forced to emigrate after the failure of the 1848 revolution in Germany, published another lively volume, *Bilder aus Tierleben,* in support of materialism. At the congress of German physicians and naturalists, held in Göttingen in 1854, the supporters of a materialistic view openly aligned themselves against the supporters of a spiritualistic-religious view, who counted Rudolf Wagner (1805–64) among their leaders. The clash was violent and forced the protagonists to radicalize their positions. Wagner maintained that faith was wholly independent of science, and even went so far as to declare that, in questions of religion, one should emulate the humble, sincere faith of the coal vendor. Vogt, who as a political emigrant did not intervene in the congress, immediately published a satirical response on the "faith of the coal-vendor" and the natural sciences, entitled *Köhlerglaube und Naturwissenschaft*. Two other works in defense of materialism came out in 1855: one, the *Kraft und Stoff*, by the physician Ludwig Büchner (1824–99), vivaciously polemical, and the other, *Neue Darstellund des Sensualismus*, by Heinrich Czolbe (1819–73), somewhat calmer but just as firm.

The publication of *Kosmos, Entwurf einer physischen Weltbeschreibung* (1845–58), a four-volume work that claimed to be an outline of a physical description of the world and is considered by some as the first great book of German positivism, also contributed to the transformation of the German cultural environment. Marx and Engels, and even Joseph Dietzgen, polemicized at length against "vulgar materialism," but the reaction against the materialistic, positivistic trend, especially in literary circles, did not usually lead toward a renewed form of materialism, but rather toward a revival of spiritualistic themes. On the one hand, Schopenhauer's philosophy was rehabilitated; on the other, slightly later, a lively interest arose in the writings of the Dane Sören Kierkegaard (1813–55), which were directly aimed against Hegel's metaphysical rationalism but in reality could be used against scientific rationalism as well.

Philosophical and scientific thought in England, France, and Italy

In France and England, too, the need to introduce a new "positive" methodology into the treatment of philosophical problems grew increasingly stronger, inspired by the same method that was producing such successful results in scientific research. In order to achieve this end, French philosophy had only to reconnect itself with the first great works of Comte, omitting those of his last phase. In England, the movement of philosophy toward science evolved along two lines, one of which, initiated by Mill, was clearly empirical in nature, while the other viewed philosophical thought as a mere generalization of scientific thought.

The most illustrious representative of this second type of positivism was Herbert Spencer (1820–1903), who wrote *The Development Hypothesis* (1852) and *Principles of Psychology* (1855) during this period. Spencer supported a generic evolutionism, which led Charles Darwin (1809–82) to list him among the precursors of his theory. Darwin himself must be credited with the revival, on a strictly scientific level, of the theory of evolution, already hypothesized by Lamarck around fifty years earlier. In fact, Darwin published his *On the Origin of Species*, which contains a fully developed exposition of his evolutionary concepts, in 1859. This work, which was destined to revolutionize biology, was greeted by many thinkers of the epoch as the forerunner of a grandiose scientific and cultural renewal. Marx was an open admirer of Darwin; and for Engels, Darwin's theory constituted one of the most significant theoretical developments of the century, of incalculable importance to the entire organization of knowledge. The debate on the theory of evolution soon involved the most visible exponents of the culture, both radicals and traditionalists. Famous encounters immediately took place in England, such as the 1860 clash between Bishop Samuel Wilberforce and the combative Thomas Henry Huxley (1825–1895). But even the Spencerian concept profited from the rapidly growing popularity of Darwin's theory: Spencer took a moderate position on such key questions as the origin of human beings and the collocation of the human species among the other animal species, while Darwin, for his part, did not hide his affinity for a skeptical and rationalistic attitude in many instances, although proclaiming himself an agnostic as far as religion was concerned.

In Italy, on the other hand, the philosophical scene continued to be dominated by more-or-less orthodox spiritualistic currents, inspired by the thought of Rosmini and Gioberti. Among the noteworthy adherents of Italian spiritualism was Terenzio Mamiani (1799–1885), known above all as a patriot and statesman.

The founding of new universities

The positivistic approach naturally tended to favor the founding and strengthening of *scientific institutions*, often emphasizing the duty of state structures to intervene in the rational planning of this sector. In 1854, Switzerland created a new center for research and scientific teaching in the form of the Federal Poytechnical Institute of Zurich, an institution that was open to theoretical research and avant-garde technological applications. Particularly important were the researches in mathematics, physics, and chemistry that were carried out there. The Dutch National Academy was established in 1854, and the Norwegian National Academy in 1857. The first universities in India were opened at Calcutta, Bombay, and Madras in 1857, during a period in which the British administration was attempting to increase its prestige in reaction to the diffidence of the masses and the open hostility of the exponents of indigenous cultural tradition. In 1859, Spain reorganized its entire system of higher education, placing the universities under the control of a ministry and dividing the kingdom into ten districts, on the basis of the ten existing universities. That same year, the University of Manila was inaugurated in the Phillipines, which at the time were under Spanish domination.

Scientific life in the United States

The renovation of universities and research institutes had by this time become a worldwide phenomenon, and the United States was a leading participant in this process. The National Institution for the Promotion of Science, was created in 1840 in Washington with a structure that was a cross between an academy and an association and included a wide representation of state and federal officials. It had a short life and little effectiveness, but two other institutions, also created during the first half of the century, proved to be more vital. These were the Smithsonian Institution (1846) and the American Association for the Promotion of Science (1848). The first originated from a bequest, by the English scientist James Smithson to the American government, for the creation of an institute that would further "the increase and diffusion of knowledge among men." After a long debate on the appropriateness of accepting the bequest of a foreign citizen—a reflection on the growing political and economic isolationism of the United States—the government in Washington founded the institute and entrusted its direction to the physicist Joseph Henry of Princeton University. Henry proved himself an energetic organizer and promoted a vast program of research and publication that drew considerable praise both in America and in Europe. Despite all this, however, the limits of the institute were implicit in its private origins.

The American Association for the Advancement of Science, modelled after the analogous English society, was more incisive in public life. It grew out of the American Association of Geologists, which shortly after its founding in 1840 had admitted into its ranks naturalists who were not geologists with the intent of directing the efforts of all those working in the field of the physical and natural sciences toward common ends. Meetings were promoted in various cities throughout the United States, and the association's membership doubled over the course of a decade. In particular, the association fought for the creation of a national scientific academy, the establishment of adequate means of spreading information, and public subsidies for research.

MATHEMATICAL SCIENCES

The problem of infinity

The second half of the century opened with the appearance of several particularly significant works. The *Paradoxien des Unendlichen*, by Bernhard Bolzano, came out posthumously in 1852. In it Bolzano foreshadowed the ideas on infinity that became characteristic of the Cantorian theory of sets and, in particular, emphasized that infinite sets could be placed in biunivocal correspondence with parts of themselves. This work did not have adequate circulation.

The work of Bernhard Riemann

In 1851, Georg Friedrich Bernhard Riemann obtained his university degree with a dissertation on the principles of a general theory of functions of a complex variable. In this work he conducted the studies on complex variables that constitute one of his major contributions to science, together with his closely related researches on surfaces of least area (a particularly significant step in these researches was his 1857 memoir on the theory of Abelian functions). Three years later, he presented a memoir (published posthumously) on the possibility of representing functions of a real variable by trigonometric series. In it, he made use of a particularly rigorous concept of integral that is known today as Riemann integral. That same year, he became a university teacher with a dissertation entitled *Über die Hypothesen, welche der Geometrie zu Grunde liegen* in which he expounded the necessary and sufficient conditions for defining a space with any number of dimensions.

In addition to the considerable technical value of this memoir for the geometry of hyperspaces and differential geometry, its importance with regard to the problems of the fundamentals of geometry should be emphasized. In fact, it demonstrated that Lobachevsky's non-Euclidean geometry held for bi- and tridimensional varieties with negative constant curvature, while a new non-Euclidean geometry held for those with positive constant curvature, where there were no parallel straight lines and space was finite, though unlimited. This memoir came out posthumously in 1867, and inspired Hermann Ludwig von Helmholtz

(1821–94) to compose his essay *Über die Tatsachen, die, der Geometrie zu Grune liegen*, which corroborated Riemann's ideas with considerations of a physical nature.

New points of view in geometry

Remaining for a moment longer in the field of geometry, we should mention that in 1851 Joseph Alfred Serret (1819–85), published his famous differential geometry formulas. But the most important event, besides the presentation of Riemann's memoir, was perhaps the publication of Arthur Cayley's *Fifth Memoir on Quantics* (1859), which demonstrated that any metric property of a plane (or solid) figure could be viewed as a particular case of its relation to a fixed conic (or quadric), called the absolute. A return was made to the case of traditional geometry by assuming, respectively, that a conic breaks in two points and a quadric reduces to a conic. In this way, projective geometry, after having distinguished itself from metric geometry and becoming established as a totally autonomous discipline, seemed actually capable of swallowing up metric geometry. The importance of this possibility in the investigation on the fundamentals of geometry later appeared in two essays by Felix Klein (1849–1925) entitled *Über die sogenannte nicht-euklidische Geometrie* (1871–73), which showed how the appropriate choice of an absolute could give rise to non-Euclidean metrics (see MATHEMATICAL SCIENCES, 1870–80).

Other disciplines also received very significant contributions: in 1851, Ludwig Otto Hesse (1811–74) published a memoir that introduced the concept of the determinant of second derivatives of a homogeneous function. The importance of this concept was recognized by assigning it the well-known name "Hessian determinant." That same year, Joseph Liouville (1809–83) defined the notion of geodetic curvature.

In 1856, Ernst Eduard Kummer (1810–93), in the course of his various attempts to demonstrate Fermat's great theorem, which he began in 1837, created the theory of ideals. Constantly inspired by Gauss, Kummer, together with Dirichlet, can be considered the most significant exponent of arithmetical research during this period, thanks to the depth and richness of the results he obtained in that field, although the work he did in other sectors of mathematics, for example in geometry, was quite important as well.

In 1854, Giusto Bellavitis (1803–80) published his *Teoria delle equi pollenze*, in which he anticipated vectorial calculus and the theory of transformations for reciprocal radius vectors, already outlined in 1847 in a memoir by William Thomson Kelvin on the distribution of electricity (see PHYSICAL SCIENCES: ELECTROMAGNETISM AND OPTICS, 1840–1850).

ASTRONOMY

The popularization of astronomy

Up until the first half of the nineteenth century, the study of astronomy had been limited to specialists and restricted circle of scholars. Few people of average or lower education had any knowledge of it, and when they did, they often ended up going so deeply into the subject that they quickly joined the ranks of the specialists, even when they did not become professionals. Around 1850, an increased popular interest began to be shown in astronomy; it became more and more widespread and is still growing even today. This interest was part of a larger movement of the general population in the direction of an expanded awareness in which an understanding of the position of human beings in the universe played a fundamental role. In the 1850s the general populace was only just emerging from the astronomical Middle Ages. This movement had been favored by several occurrances that recalled the general attention to the great discoveries made by the scientists of the past two centuries. Most important of all was Urbain Jean Le Verier's discovery of Neptune, which to the public, still ignorant of the great advances in celestial mechanics, seemed almost superhuman. In 1851, another event of not only scientific, but popularizing value took place in Paris. Jean Bernard Léon Foucault (1819–68) became the first to demonstrate the rotation of the earth with his famous pendulum experiment (see PHYSICAL SCIENCES, 1850–1860).

Kosmos, in which Baron Alexander von Humboldt (1769–1859) provided a sometimes wordy but always accessible and intelligible view of the world, came out between 1845 and 1862. It was followed, between 1855 and 1859, by Astronomie populaire by Dominique-François-Jean Arago (1786–1853). These two works, in seven and four volumes respectively, were translated into all the major languages and became rapidly diffused, even, and above all, among non-specialists. A short time later, the works of Nicolas Camille Flammarion achieved an even wider popularity. In 1861, at the age of only nineteen, Flammarion published La pluralité des mondes habités, the first of around fifty popularizing works, the major part of which concerned astronomy. These included the famous Astronomie populaire, which appeared in 1880.

Galactic astronomy

The application of statistical methods to the elaboration of star catalogues made it possible to determine the direction of motion of the solar system in the galaxy (Friedrich Wilhelm Argelander, 1837; Friedrich Georg Struve, 1842); to study the structure of the Milky Way (Johann Heinrich Mädler, Mârian Albertovich Kovalsky Struve, 1847–1860); to discover the laws of distribution of stars in space (John Frederick Herschel, Struve, 1847); and to elaborate the theory of the rotation of the galaxy and a method for determining the major directions of stellar motion.

Photographs of the sun

With the introduction of wet collodion photographic plates in 1851, astromonical photography came out of its experimental phase. The first photographs of the stars were obtained in 1857, along with excellent plates of the moon, with which Warren de la Rue (1815–89) was even able to obtain a stereoscopic effect in 1859 by making use of the phenomenon of libration. In 1858 de la Rue also began to make a regular and systematic daily record of the sun's image, taking over three thousand photographs at the Observatory of Kew up until 1872, when a new, very lengthy series was begun at the Observatory of Greenwich. As early as 1865, his material enabled him to conduct important statistical studies of variations in solar activity and the distribution of sunspots. In 1855, Christian Henry Peters (1813–90) discovered that the motion of sunspots was irregular when he found discrepancies in the periods of the sun's rotation as deduced from various spots. In 1859, Carrington pointed out

that the sun does not turn as a rigid body, but with an angular velocity that decreases gradually from its equator to its poles.

The chemical composition of the sun

While celestial mechanics continued to become more consolidated, mainly through a series of works published by Le Verrier between 1855 and 1876, Gustave-Robert Kirchhoff (1824–87) and Robert Wilhelm von Bunsen (1811–99) made a discovery that led to the birth of astrophysics, a discipline to which the new astronomical researches of the previous decade, had already opened the way. In 1859, while comparing the spectrum of the sun with a sodium spectrum obtained in the laboratory, Kirchhoff found that the two bright lines in the spectrum of the element coincided with the black lines in the solar spectrum. He thus discovered that the lines of a given element can be seen in emission or in absorption, depending on the temperature of the light source emitting the continuous spectrum that is intercepted by the element. Thus through a comparison of the lines in the spectrum of the sun or of other heavenly bodies with those obtained in the laboratory for known elements it was possible to discover what elements were present in the heavenly bodies. Together with Bunsen, Kirchhoff later inverted and identified the potassium, strontium, calcium, and barium lines in the solar spectrum, in addition to the sodium lines.

PHYSICAL SCIENCES: MECHANICS

The conservation of energy

For the physical sciences in general and for mechanics in particular, the second half of the century opened under the sign of energy. Although the term energy had already been suggested during the early years of the nineteenth century by Thomas Young, who proposed using it to indicate the product of the mass of a body times the square of its velocity, the central role that the mechanical energy of systems now came to assume was based on changes that had taken place in mechanics itself, thanks to the principles of conservation formulated by various scientists between 1840 and 1850. In fact, the principle of conservation of energy found applications in the analysis of both heat phenomena (see PHYSICAL SCIENCES: THERMODYNAMICS AND STATISTICAL MECHANICS, 1850–1860) and electromagnetic phenomena (see PHYSICAL SCIENCES: ELECTROMAGNETISM AND OPTICS, 1850–1860), and led to various reflections on the relations between this science and others.

This ferment brought about innovations in scientific thought from the very beginning of the decade. In his Principes de mécanique fondées sur la cinématique, published in 1851, Adhémar de Saint-Venant (1797–1886), referring back to the French school of Lazare Carnot, maintained that the notion of force was a derived rather than an original notion of mechanics. In his opinion, it was essential to give priority to a rigorous definition of the concept of mass, a definition that he stated in terms of the ratio between two numbers, in order to separate the notion of mass from what was generically thought of as "quantity of matter." In this sense, Saint-Venant anticipated Ernst Mach (see PHYSICAL SCIENCES: MECHANICS, 1860–1870).

An alternative to Saint-Venant's program was elaborated by Frederick Reech, founder of the so-called *école du fil*. In 1852 this program, which was explicitly founded on a notion of force in the sense of pressure and traction, found its manifesto in Reech's course on mechanics (*Cours de mécanique d'après la nature généralement flexible et élastique des corps*). Here, the fundamental dynamic event was seen as the lengthening of a cord, and the term force was no longer used to designate as the cause of motion, but rather the effect of the causes that were generally called pressure and traction.

Leçons sur la théorie mathématique de l'elasticité des corps solides, by Gabriel Lamé (1795–1870) was also published in 1852 in Paris. In it, Lamé developed a program that corresponded to Fourier's mathematical physics and was very critical of Laplace, Poisson, and Cauchy.

Foucault's experiment

In applied mechanics, Jean Bernard Léon Foucault (1819–68) became the author of a famous experiment for physically demonstrating the earth's motion of rotation. In 1851, Foucault published a memoir describing the experiment, which was based on the rotation of the plane of oscillation of a huge pendulum hung in the Pantheon in Paris. The experiment lasted for months. The cord was sixty-seven meters long, and the sphere of the pendulum weighed twenty-eight kilograms. An experiment of this type had already been anticipated by Poisson, who had thought it impossible to carry out.

At the same time, Foucault successfully applied the studies of Coriolis and Poinsot to the construction of the first gyroscopes.

Kelvin's program

In England, the interest in mechanics found in William Thomson (Lord Kelvin) a scientist who was able to elaborate a general program aimed at satisfying the need for a unitary explanation of all natural phenomena.

Referring back to the ideas of Joule and Carnot on the one hand, and on the other basing his reflections on Helmholtz's considerations regarding the principle of conservation of forces, Kelvin explained the problem of energy sources and their utilization. As he observed in his short but fundamental memoir of 1852, *On a Universal Tendency in Nature to the Dissipation of Mechanical Energy*, "an absolute waste of utilizable mechanical energy" took place in every phenomenon where heat was allowed to pass from one body to another. This waste, however, could not be thought of as a destruction of energy, acording to Kelvin, but as a transformation of the energy itself. Nevertheless, the dissipation of utilizable energy would eventually lead to physical conditions prohibitive to organic life on our planet. The influence on Kelvin of the new thermodynamics, which he helped develop, is evident (see PHYSICAL SCIENCES: THERMODYNAMICS AND STATISTICAL MECHANICS, 1850–1860). The principle of dissipation of mechanical energy, the importance of which was judged fundamental for the sciences of motion in Helmholtz's famous conference of 1854 on the interaction of natural forces, brought up the problem of the origin of universal energy transformation. In Kelvin's opinion, this problem could be solved by referring to the potential energy of gravitation, which in the pages of a communication of 1856 (*On the Origin and Trans-*

formations of Motive Power) he defined as "the basic antecedent" of "all motion, heat, and light presently existing in the universe."

Rankine's energetics

The English school represented by William John Rankine (1820–72) examined the new ideas on energy transformation within a structure that, although it underwent modifications, influenced various currents of European scientific thought. In 1855, Rankine published his *Outlines of the Science of Energetics*, in which he strongly supported the trend of establishing mechanics as a rigorously deductive science, based on conjectures on the "state of things" as opposed to the idea of elaborating physical theories based on models, which Rankine judged subjective, giving the hypotheses on electrical fluids as examples. Models, according to Rankine, should be used with caution so that scientists would not attribute to them "the authority that belongs only to facts."

As to hydrodynamics, Hermann Ludwig von Helmholtz's studies of 1858 on the integrals of hydrodynamic equations that express whirling motion (*Über Integrale der hydrodynamischen Gleichungen, welche der Wirbelbewegungen entsprechen*) deserve mention. Despite the complexity of the mathematics used by Helmholtz, during the next decade these studies became the point of departure for new ideas on the structure of matter that eventually replaced the old atomistic concepts.

Also deserving of mention are the physico-mathematical researches on gravitation conducted during this period. Referring back to the analyses on fluids elaborated between 1840 and 1850, James Challis (1803–82) attempted a formal discussion of gravitational interaction in two important communications entitled *A Mathematical Theory of Attractive Forces* (1859) and *A Theory of Molecular Forces* (1860).

PHYSICAL SCIENCES: THERMODYNAMICS AND STATISTICAL MECHANICS

Kelvin's dynamic theory of heat

The first three parts of William Thomson Kelvin's *On the Dynamical Theory of Heat* were published in March, 1851. In this monograph, which can be considered the first nineteenth-century treatise on thermodynamics, Kelvin stated the two famous fundamental principles, or propositions, of thermodynamics, which he attributed respectively to James Prescott Joule and to Nicolas Léonard Sadi Carnot and Rudolf Julius Clausius. In December of that same year, Kelvin presented both the fourth part, which concerned several experimental and theoretical consequences of Joule's observations on gasses, and the fifth part, *On the Quantities of Mechanical Energy Contained in a Fluid in Different States, as to Temperature and Density*. In this essay, he stated that "... the aggregate value of the mechanical effects produced, [by a body that undergoes a heat transformation] must depend solely on the initial and final states of the body, and is therefore the same, whatever be the intermediate states through which the body passes, provided the initial and final states be the same." Although he judged the contributions of other scholars positively, Kelvin made critical allusions to the hypotheses expounded by William John Rankine (1820–72) and Clausius, and gave a negative evaluation of the method followed by Julius Robert von Mayer to reach the principle of equivalence. The sixth part of the monograph, dedicated to thermoelectric currents, was ready in 1854. It contained a recapitulation of the *Fundamental Principles of General Thermodynamics*, which is extremely interesting in that Kelvin formally developed the principles following a line of research that led to profound divergencies with Clausius. If a physical system is subject to a complete cycle of perfectly reversible operations, and if $H_t(i)$ stands for the quantity of heat absorbed or emitted by the system during the operation at temperature $t^{(i)}$, W is the total work and J is Joule's equivalent, then, according to Kelvin, the two principles of thermodynamics can be expressed by the two relations:

$$W + J \left(H_t + H_{t'} + \ldots + H_{t^{(n-1)}} + H_{t^{(n)}} \right) = 0$$

$$\frac{H_t}{t} + \frac{H_{t'}}{t'} + \ldots + \frac{H_t^{(n-1)}}{t^{(n-1)}} + \frac{H_t^{(n)}}{t(n)} = 0$$

It was the formulation of the second principle of thermodynamics that set off the above-mentioned debate between Kelvin and Clausius, which in many respects characterized the development of the notion of irreversibility and the possibility of providing a physical interpretation based on the calculus of probabilities.

Clausius's Q/T function

In 1855, Clausius intervened in the problem posed by the mathematization of the second principle, publishing a memoir on a modified form of the second fundamental "theorem" of the mechanical theory of heat in Poggendorff's *Annalen*. In it, Clausius maintained that although the method he had first used in his 1850' memoir had sufficed to demonstrate the compatibility of the first and second principles, it could not clarify "the real nature" of the second principle. In Clausius's opinion, a mathematical function had to be found that would quantitatively express the differences among the various physical transformations. He sought to find the law that would express the transformations as mathematical quantities, so that the equivalence of any two transformations would be obvious from the equality of their values. The mathematical function that Clausius proposed for this purpose was similar to Kelvin's, in that it was given by the ratio between the quantity of heat Q and the temperature T. According to Clausius, if we accept that the "value of equivalence" of a transformation connected to the passage of a quantity Q from a temperature T_1 to a temperature T_2 is represented by:

$$Q \left(\frac{1}{T_2} - \frac{1}{T_1} \right),$$

then it is possible to demonstrate that in the case of perfectly reversible processes the relation

$$\int \frac{dQ}{T} = 0$$

is always valid. Taking into account the fact that in the case of irreversibility "the algebraic sum of all the transformations that

occur" can only be positive, the above relation constitutes, as Clausius wrote, "the analytical equation of the second fundamental theorem."

Rankine's function

During these same months, Rankine attacked the problem of irreversibility with an approach that was linked to the ideas of energetics (see PHYSICAL SCIENCES: MECHANICS 1850–1860), introducing the so-called metabatic function, with which he attempted to reduce the considerations regarding the privileged tendency of heat phenomena to the principles of energetic mechanics. In the context of this approach, Rankine also reexamined several of his previous reflections on molecular vortices, which contributed to the affirmation, especially in English science, of research programs aimed at elaborating alternatives to the typically atomistic models that lay at the base of Clausius's formulations.

Krönig's model

The debate on the second principle and models was fueled by the publication in 1856 of a brief memoir on the elements of a theory of gasses, the *Grundzüge einer Theorie der Gase*, by the physicist August Krönig (1822–79). The model of gasses proposed by Krönig was based on an assumption that the author himself summarized as follows: "The path of every atom of gas must be irregular, so that it escapes calculus. However, according to the calculus of probabilities, a complete regularity can be considered instead of this irregularity." It therefore became possible to elaborate an initial microscopic specification—i.e. in terms of parameters that referred to the molecules—of the macroscopic parameters used to define the state of a gasseous system. A few months later, Clausius generalized Krönig's simple probabilistic model in a communication entitled *Über die Art der Bewegung, welche wir Wärme nennen*. In the study of gasses, Clausius wrote, a distinction had to be made between conclusions that could be deduced from certain general principles and those that presupposed "a particular type of molecular motion." In this case, he explicitly chose the second route, since the calculus of probabilities made it possible to introduce numerous simplifications and to "disregard all the irregularities that derive from an imperfect gasseous state." Thus out of the mathematical generalization of Krönig's model arose the model for the perfect or ideal gas, in which the molecules have equal velocities in relation to a certain average value, possess complete equiprobability with respect to the possible directions of motion and are characterized by constant numbers of collisions with the walls of the container holding them. Both pressure and temperature could then be correctly expressed as a function of the number, the mass and the square of the velocity of the molecules, and an answer could be given to a question of the following sort: "How far, on the average, can a molecule travel before its center of gravity enters into the sphere of action of another molecule?" Making drastic assumptions of a probabilistic nature, Clausius answered this question by introducing the new concept of mean free molecular path. It became easy to calculate that a molecule would rarely complete a path that was significantly longer than the mean free path. For example, out of N molecular paths, only $0.000,045\ N$ of them are as long or longer than ten times the mean free path. This new notion enabled Clausius to completely invalidate the objections raised by several scientists that, given the high molecular speeds of gasses, all phenomena of interdiffusion among them, contrary to all empirical evidence, should produce explosions.

The reaction against these probabilistic models was very strong. Critics maintained that the properties of matter in the gaseous state could not be explained by attributing them to hypothetical atoms, but rather had to be deduced from the general principles of mechanics.

Maxwell's dynamic theory of heat

A further development of the molecular and probabilistic model came at the end of the decade from James Clerk Maxwell (1831–79), who in September, 1859, presented a memorable communication entitled *Illustrations of the Dynamical Theory of Gasses*. In this communication, published in 1860, Maxwell stated that he wanted to base his researches "on strictly mechanical principles." In his opinion, the problem was to derive mathematically all the possible consequences of the atomistic hypothesis on the structure of matter, in order to check their validity. This validity, observed Maxwell, did not depend so much on a consistency among all the deductible consequences as on their experimental confirmation. The decisively innovative step taken by Maxwell was his introduction of a distribution of molecular velocities around a mean value, which made it possible to calculate "the average number of particles whose velocity lies between certain limits," after a great number of collisions between a great number of "equal spherical particles in a perfectly elastic vessel." This surpassed Clausius's assumption that the model for gasses could be analyzed in terms of molecular velocities equal to a mean velocity, and at the same time raised several fundamental problems for kinetic theory with regard to the physical significance of the second principle of thermodynamics. The function of distribution, which Maxwell found by drawing mathematical analogies with the calculus of distribution of errors, possessed a special property, according to him: "In the cases we are considering," he wrote, "the collisions are so frequent that the law of distribution of the molecular velocities, if disturbed in any way, will be reestablished in an inappreciably short time...." Maxwell's function, in short, represented a privileged condition of matter in the gaseous state, toward which physical systems necessarily tended.

Various possibilities of interpretation emerged and were analyzed during the following decade by Clausius, Ludwig Boltzmann, and Maxwell himself. Through this analysis, the analytical demonstration of the second principle, already attempted by Clausius, was generalized, and one of the basic notions of modern physics, the concept of entropy, was introduced.

PHYSICAL SCIENCES: ELECTROMAGNETISM AND OPTICS

Theories on field and action at a distance

The second half of the century opened with Faraday's statement of a program for a general theory of field that would include gravitational interaction as well, and Maxwell's first attempt to give a mathematical formulation to the physical concepts of action by contact. At the same time, the analysis based on action at a distance was being further refined in physics, with particular attention to the properties of local equations. Both points of view took into consideration the role of the speed of light. Great progress was made in the technological applications of the laws of electromagnetism, thanks to the transatlantic cable projects in which Kelvin also participated. In the field of optics, advances were made in researches on flourescence, spectroscopy, and radiation in general.

Relations between electricity and gravity

Michael Faraday's report *On the Possible Relation of Gravity to Electricity,* which expressed his hope of "establishing, by experiment, a connection between gravity and electricity," was published in 1851. He had to report that the results of his experiments were negative, in that they did not reveal in gravitational interaction anything that corresponded to the "dual or antithetical nature of the forms of force in electricity and magnetism." But, wrote Faraday, this lack of empirical proof did not suffice to negate his belief in the connection between these forces, which was based on "the long and constant persuasion that all the forces of nature are mutually dependent." Faraday reproposed the same ideas in relation to physical space in series XXV of his *Experimental Researches*, published that same year with the title *On the Magnetic and Diamagnetic Condition of Bodies.* The mutual transformations of natural forces, according to Faraday, implied that space in itself possessed a magnetic relation that would someday reveal itself to be of the utmost importance. The answers that Faraday was forced to give at this point, in response to the hostility with which his ideas were received in many circles, were precise and polemical. In a work of 1852, *On the Physical Character of the Lines of Magnetic Forces,* he explicitly accused the so-called Newtonian programs of presenting an obstacle to the progress of knowledge. The physics of action at a distance, according to Faraday, was incapable of going beyond "the pure and simple facts." Faraday repeated this same criticism in a memoir of 1857.

Weber's constant

Wilhelm Eduard Weber's systematic program of measuring the fundamental electromagnetic quantities, led to the determination of the numerical constant c, which appeared in the basic formula of Weber's monograph of 1846. Together with Rudolf-Hermann Kohlrausch (1809–58), Weber published the results of this measurement in 1855. As we have already mentioned, Weber did not interpret this value in the same way that Maxwell later did. Meanwhile, a growing number of observations were indicating that a perfect conductor should produce periodical oscillations of electrical current with a velocity of propagation equal to the speed of light, i.e. Weber's constant divided by $\sqrt{2}$. Gustave-Robert Kirchhoff (1824–87) reached a similar conclusion, observing that this velocity was numerically equal to the speed of light. However, his communication of 1857 on the movement of electricity in wires (*Über die Bewegung der Electricität in Drähten*), in which he expounded these considerations, did not contain any reflections on the importance of such a coincidence. Kirchhoff's greatest interest, with respect to the analyses he conducted during this period, published in a second memoir of 1857 on the movement of electricity in conductors (*Über die Bewegung der Electricität in Leitern*), lay in the possibility of writing the so-called local equations for a continuous conducting medium. By means of an equation of continuity, he connected the density of current with the density of free charge and also arrived at the wave equation that described the propagation of the above-mentioned oscillations.

Once again, the problem of the speed of light did not emerge in all its importance, partly because Kirchhoff's analysis dealt mainly with large linear circuits, the mathematical treatment of which was of great practical value during this period, especially in light of its applicability in the engineering of undersea cables.

In 1858, Georg Friedrich Bernhard Riemann presented a memoir on electrodynamics (*Ein Beitrag zur Elektrodynamik*) in which a central role was played by the assumption that the velocity of propagation of electrodynamical action was finite, and that velocity and the velocity implicit in Weber's works were equal. However, Riemann's memoir contained several mathematical imperfections, and as a result was retracted.

Faraday's lines of force

In England, meanwhile, interest in the concepts advanced by Faraday found valid and clarifying support in Maxwell's mathematics. Maxwell's first memoir on electromagnetism was published in 1857 under the title *On Faraday's Lines of Force*; it contained a discussion that was substantially based on a hydrodynamic model of action by contact. The influence of Kelvin's and Stokes's mathematical physics on Maxwell was evident. But Maxwell's investigation went beyond the limits of these influences, mathematizing the consequences of an analogy based on the flow of an incompressible fluid in a resistant medium. According to Maxwell, only an incomplete geometrical model could be built using Faraday's lines of force. It needed to be completed by achieving generality and accuracy, but avoiding the dangers that arise from a premature theory that intends to explain the causes of electromagnetic events. In Maxwell's analogy of 1857, the lines of force became a complex set of tubes with an imaginary fluid flowing through them. However, according to Maxwell, this analogy did not represent the reality being studied, but rather served as an object for mathematically analyzing the consequences of the "purely geometrical idea of the movement" of this fluid. The most important results obtained by Maxwell with this hydrodynamic treatment concerned the mathematization of one of the most obscure concepts of Faraday's physics, that of electrotonic state. According to Faraday, the electrotonic state was a particular condition of bodies, the only observable effects of which were the variations connected with the formation of induced currents within closed conductors in the presence of variable magnetic fields. Maxwell discussed the electrotonic state by referring back to Helmholtz's conjectures on the possibility of explaining electromagnetic induction by means of considerations on implied energy transformations and the principle of conservation. In this way, he arrived at a result that, in modern notation, indicated the possibility of dealing with the electrotonic state in terms of vector potential. In Maxwell's opinion, deducing the consequences of the hydrodynamic analogy had not so much the goal of providing a physical theory of electromagnetic phenomena, as of elaborating a formal concept that was based on action by contact and demonstrating its equivalence to the theories of action at a distance. The task

of examining the usefulness of such an approach was up to the physicists, Maxwell commented.

Meanwhile, Faraday continued his battle in favor of action by contact. In 1857, he published a work entitled *On the Conservation of Force*, soliciting an explanation for the more profound nature of gravitational phenomena, which he believed could not remain isolated from the general context of electromagnetic phenomena. And this sort of investigation implied the determination of the velocities of propagation of all events.

Communication via cable

In relation to the development of the theories of electromagnetism and concepts of field, it is impossible to ignore their vast application to technical problems. In 1853–56, William Thomson Kelvin (1824–1907) solved the mathematical difficulties connected with problems of oscillating currents and laid the groundwork for a mathematical theory on the transmission of electrical signals via cable. In these studies he collaborated with George Gabriel Stokes, and toward the end of the decade the first electrical signals were transmitted via transatlantic cable. Kelvin directly contributed to planning and setting up the cable project (see APPLIED SCIENCES, 1850–1860).

Researches in experimental optics

This decade opened with several important experimental researches in the field of optics. A confirmation of the wave concepts and of the validity of explanations of the type derived from Fresnel's drag coefficient (for more information, see PHYSICAL SCIENCES: ELECTRICITY, 1810–1820) was provided in 1851 by Armand Hippolyte Fizeau's measurements of the velocity of light through a mass of water that was mobile with respect to the earth's surface. The results obtained by Foucault, who determined the velocity of light in optically different media, such as water and air, also supported the wave concepts. Foucault found that the speed of light in water was about three-quarters its speed in air.

In a primarily experimental memoir published in 1852 concerning a possible reinterpretation of several phenomena already observed by Brewster and John Frederick Herschel, Stokes laid the foundations for modern investigations on fluorescence.

Kirchhoff's spectroscopy

In 1854, Gustave-Robert Kirchhoff moved to Heidelberg, where he began working with Robert Wilhelm Bunsen (1811–99) and carried out a program of experimental research that culminated in 1859 with the construction of an instrument that proved to be of fundamental importance for the physical sciences—the spectroscope. The optician Karl August von Steinheil contributed decisively to its construction; its light source was the flame of a Bunsen burner.

Also in 1859, Kirchhoff published two memoirs in which he provided an initial interpretation of Fraunhofer's spectral lines and attempted to formulate an explanation of the relations betwen the emission and the absorption of light and heat. These investigations made it possible to begin a systematic study of the optics of the sun's atmosphere (see ASTRONOMY, 1850–1860).

In the second memoir, Kirchhoff stated a law, valid for all physical systems, according to which "for waves of the same wavelength at the same temperature, the ratio between emissive power and absorptive power is the same for all bodies.''

The interpretation of this law later became the focus of all blackbody radiation physics.

Vacuum tubes

Another extremely important field of optical investigation opened up in the second half of this decade. Thanks to the possibility of obtaining high vacuums by using Geissler's new pumps, many scientists were able to study in detail the strange phenomena observed during electrical conduction in matter in the gaseous state under conditions of extreme rarefaction. Between 1858 and 1859, using Geissler's technique and several high vacuum devices used by Anders Jonas Ångström (1814–72) to observe gas spectrums, Julius Plücker (1801–68) began studying cathode rays. These studies were very important in that they brought to light problems regarding the structure of matter and led to developments that made it possible, at the end of the nineteenth century, to determine the ratio between electron charge and mass.

CHEMICAL SCIENCES

The continued development of organic synthesis

This period marked the beginning of the golden age of organic synthesis. In 1850, in experiments on α-nitronaphthaline Rafaelle Piria discovered the reaction of simultaneous reduction and sulphonation of aromatic nitroderivatives with sulphites, and obtained naphthionic acid, which became an important intermediate for azoic dyestuffs. Also in 1850, August Wilhelm von Hofmann found the general method for synthesizing aliphatic ammines based on the ammonolysis of halogenated compounds. Charles Frédéric Gerhardt obtained various anhydrites of organic acids and acetylsalicitic acid (1852–53), which was placed on the market in 1899 as an effective pharmaceutical under the name of aspirin. In 1852, Alexander William Williamson (1824–1904) devised a method for preparing aldehydes and ketones based on the pyrolysis of the calcium salts of carboxylic acids, but it was only put to practical use in 1856 by Piria and Heinrich Limpricht (1827–1909). Cesare Bertagnini (1827–57), a pupil of Piria's, discovered the disulphitic compounds of the aldehydes, which could be separated out in the pure state (1853). After synthexizing cyanamide with Stanislaus Cloëz in 1851, Stanislao Cannizzaro (1826–1910) discovered the reaction that was named after him (1855), i.e. the dismutation of aromatic aldehydes into alcohols and acids in the presence of strong alkalis. Gerhardt and Luigi Chiozza (1828–89) prepared various amides in 1853–56; Pierre Jacques Béchamp (1816–1908) discovered the method for reducing nitrobenzene with iron and acids, a new industrial process for preparing aniline, in 1854; and in 1855 Charles Adolphe Wurtz (1817–84) devised a synthesis, named after him, by which an aliphatic chain could be continued indefinitely by causing alkyl iodides to react with sodium. Marcellin Berthelot (1827–1907) was also quite active in the area of organic synthesis during this period, synthesizing ethyl and isopropyl alcohols (1853–55), formic acid (1855), methane (1856), methyl chloride and methyl alcohol (1857), and various aromatic hydrocarbons (some of which he discovered). He also made important con-

tributions to the chemistry of fats, sugars, and terpenes. In 1856, Bertagnini synthesized cinnamic acid, and Heinrich Debus (1834–1916) prepared glyoxal and glyoxalic acid and discovered a new heterocyclic nucleus, that of the glyoxalines or imidazoles. In 1858, Victor Dessaignes (1800–85) synthesized malonic acid.

The first synthetic dyestuff

The year 1856 marked the creation of the first true synthetic dyestuffs, the aniline dyes. In the course of purely empirical attempts to synthesize the quinine alkaloid from allyltoluidine in addition to the already well-known aniline black, William Henry Perkin (1838–1907), Hofmann's young assistant in London, obtained mauvine, a substance that dyed silk a magnificent bright red-violet—from the chromic ozidation of aniline. He immediately patented it and placed it on the market. He was soon followed by Jacob Natanson, who that same year prepared fuchsine, which Hofmann and Emanuel Verguin studied and prepared, using the most advanced methods of the times, from 1856 to 1859 in France (Magenta red). The structure of these dyestuffs was only definitively understood in 1876 by Otto Philipp Fischer (1852–1932) and Emil Fisher (see CHEMICAL SCIENCES, 1880–1890).

Aniline dyestuffs took a new direction and were more widely developed after the discovery of the diazocompounds by Johann Peter Griess (1829–88), 1858–60. Between 1859 and 1860, Adolphe Kolbe synthesized salicylic acid from sodium phenate and carbon dioxide, introducing a method that later found diverse applications. Rudolph Fittig (1835–1910) obtained pinacol by reducing acetone with sodium in 1859, and in 1860 discovered pinacolinic transposition.

The theory of types

Notable contributions were also made to theoretical organic chemistry and, simultaneously, to general and physical chemistry. A first attempt at a unitary systematization of organic chemistry was made by Gerhardt in 1853 with his theory of ''types,'' i.e. typical substances from which all others could be derived. To the water type (proposed by Williamson) and the ammonia type (proposed by Justus von Liebig and later by August Wilhelm von Hofmann), he added the hydrochloric acid type and the hydrogen type and attempted to formulate a theory that did not conflict with the theory of radicals.

The tetravalency of the carbon atom

After the concept of valency had been clarified, primarily through the work of Edward Frankland and Friedrich August Kekulé (1829–96), Kekulé and Archibald Scott Couper (1831–92) postulated the tetravalency of carbon and the ability of carbon atoms to link with each other in chains and even complex cyclical systems, making it necessary to resort to structural formulas, rather than rational ones based on radicals (1857–58).

The principle of greatest work. The kinetic theory of gasses

In 1854, Hans Peter Thomsen (1826–1909), and slightly later Berthelot, stated the principle of greatest work, according to which reactions spontaneously occur in the sense in which heat is produced.

In 1857, Rudolf Julius Clausius (1822–88) stated the kinetic theory of gasses. In 1858, after reiterating, on the basis of Amedeo Avogadro's writings, the difference betwen the molecular and atomic weights of elements, Cannizzaro reproposed the method of determining molecular weight from vapor density and suggested a method for finding atomic weights. In order to explain the anomalies found at various temperatures, Cannizzaro, and later Kopp and Kekulé, suggested thermic dissociation, already anticipated in 1838–39 by Armand Bineau D'Aligny and Dmitri Ivanovich Mendeleev.

Colloids, spectrophotometry, and electrolysis. Ozone

Between 1850 and 1852, Francesco Selmi (1817–81) conducted pioneering research on colloids. In 1852, August Beer (1825–63) stated the basic law of spectrophotometry, which regarded the proportionality between the concentration of a colored solution and light absorption. Between 1853 and 1858, Johann Wilhelm Hittorf (1824–1914) conducted studies on the velocity of ions in electrolysis. In 1856, Pierre Antoine Favre began the studies that led him to a verification of the validity of conservation of energy—stated in the first principle of thermodynamics by Hermann Ludwig Helmholtz (1821–94) in 1847—and recalculated the mechanical equivalent of the calorie, which he found to be 4.13 joules, not much different from the present value.

In 1856, Thomas Andrews (1847–1907) demonstrated that ozone from whatever source derived was not a compound but oxygen in an altered or allotropic form. In 1860, together with Peter Guthrie Tait (1831–1901), he discussed the volumetric relations between oxygen and ozone, preparing the way for the researches that led to the establishment of O_3 as the formula for ozone.

New techniques in the laboratory and in industry

Several chemists began to take an interest in war chemistry. In 1854, Lyon Playfair (1818–98) proposed using incendiary gasses (phosphorous in carbon disulphide) and poisonous gasses (cacodyl cyanide) in the Crimean War, but his suggestions were not followed. Robert Wilhelm Bunsen and Leon Schischkoff analyzed the chemical reactions involved in the explosion of gunpowder (1857).

Several advances in laboratory instrumentation also deserve mention, namely the burette with pincers, invented by Karl Friedrich Mohr (1806–79), and the Bunsen burner (1857). New processes and improvements in the field of industrial chemistry and metallurgy are discussed in the section on APPLIED SCIENCES.

The first chemical-pharmaceutical firms sprang up during this decade: the Carlo Erba company was established in Milan in 1853 and the E.R. Squibb company in Brooklyn in 1858. In 1851, Gustave Chancel (1822–90), inventor of the sulphorimeter, a device for determining the purity of sulphur, published a *Cours élémentaire d'analyse chimique* and in 1855, with Gerhardt, a *Précis d'analyse chimique qualitative*, printed in a number of editions.

LIFE SCIENCES

Agassiz and creationist morphology

This was the decade of *On the Origin of Species* (1859), when problems of morphology became fundamental to the life sciences, and in fact to culture as a whole. Charles Darwin (1809–1882) published *On the Origin of Species* in 1859. The 1,250 copies of the first edition were sold in a single day. Darwin had intensified his work in 1858 when the naturalist Alfred Russel Wallace (1823–1913) sent him an essay entitled "On the Tendency of Varieties to Depart Indefinitely from the Original Type," which mirrored many of the views Darwin had already developed. Darwin's evolutionary theory was opposed by Georges Cuvier's structural morphology and by the creationistic morphology of Louis Agassiz (1807–73), a Swiss-born scientist who lived in the United States. A paleontologist and author of a five-volume *Recherches sur les poissons fossiles* (1833–43), Agassiz still attributed much more importance to ontogeny and embryology than to phylogeny and the study of the evolution of forms. Change, according to Agassiz, was a "growth cycle." Another anti-evolutionary paleontologist was Richard Owen (1804–92), who applied the notions of "archetype" and "homology" to the study of fossil reptiles, distinguishing two orders—*ichthyopterygia* and *Sauropterygia*—in the Enaliosauria of William D. Conybeare (1787–1857). Thomas Henry Huxley (1825–95) presented Owen, a supporter of the vertebrate theory of the skull, with a precise empirical confutations in his *On the Theory of the Vertebrate Skull* (1858). All this controversy was a sign of the maturation of the new areas of enquiry of which Darwin became the interpreter.

Charles Darwin, the origin of species and natural selection

Naturalistic metaphysics, invoked by Agassiz and others to justify morphological fixity, found itself in difficulty with Darwin. During his five-year voyage in South America and the Pacific islands (1831–36), Darwin had found a succession of similar species living at various latitudes and had noted similarities between fossil species and living species, as well as between continental and insular species. Darwin coordinated these facts, which contradicted the fixity of Linnaean tradition, with the two postulates of the struggle for existence and natural selection, to which he added the inheritance of acquired features. Living beings varied because of changes in environmental conditions; the modifications were transmitted to their descendents and, if useful, aided the individuals in their struggle for existence, favoring the victory of the "fittest." This was the elementary mechanism by which living species could originate from one another. In *The Descent of Man* (1871), Darwin added sexual selection, which he had already mentioned in *On the Origin of Species*, to natural selection, which was described in the fourth chapter of *On the Origin of Species*. A return to the "very complex" laws of the variability of organisms is explicit in Darwin, while in numerous passages he mentions the principle of the correlated variation of organs, already formulated by Karl Ernst von Baer. Even in the case of domesticated animals, Darwin said in *Variations of Animals and Plants Under Domestication*, "It is an error to speak of man 'tampering with nature'... If organic beings had not possessed an inherent tendency to vary, man could have done nothing. He unintentionally exposes his animals and plants to various conditions of life, and variability supervenes...." Thus the distinction between species and variety ceased to have meaning, or was at least diminished, while the polemic against the creationism of authors such as Louis Agassiz and Robert Chambers (1802–71) led Darwin to accentuate the casualistic, anti-theological aspect of his theory of the evolution of living species. The influence of Darwinism spread quickly and extensively: in England, Thomas Henry Huxley and the geologist Charles Lyell (1797–1875), whose *Principles of Geology* had provided necessary support to the constitution of Darwin's theory, were converted to Darwin's ideas. In the United States, Darwinism achieved success with the botanist Asa Gray (1810–88). In Germany, Fritz Müller (1821–97) and Ernst Haeckel (1834–1919) enthusiastically echoed the themes of *On the Origin of Species.* In France, which was divided for and against Darwin, Paul Broca (1824–1880) and Albert Gaudry (1827–1908) supported the new ideas.

Charles Frantsovich Roullier (1814–1858), a professor at the University of Moscow who since 1841 had been working on a historical study of the organic world, maintained that the phylogenetic development of living beings and plants depended on variations in the surrounding environment. His ideas were expounded in detail in his *Žizn' životnyh po otnošeniju k vnešnim uslovijam* (*The Life of Living Beings in Relation to Environmental Conditions*, 1852).

Cell pathology

Virchow, Bernard, and Helmholtz also made major contributions during this decade. Cell pathology and evolutionary theory provided the biological disciplines with primitive concepts, directions of investigation, and systematic criteria. Rudolf Virchow's lessons, which were preceded by years of experimental and theoretical work, came out in 1858 under the title *Die Cellularpatologie in ihrer Begründung auf physiologische und pathologische Gewebelehre.* It is the method of natural science, he wrote in an article published in the *Archiv* in 1849, that makes it possible for us to formulate scientific questions. Virchow's method focused on giving precedence to experience over hypothesis; in his work on pulmonary infarction (1846), in which he demonstrated the priority of emboly over phlebothrombosis, Virchow reproached the anatomophysiologists of the schools of Paris and Vienna, such as Jean Cruveilhier (1791–1874) and Karl von Rokitansky, for jumping from hypothesis to hypothesis. Betwen 1852 and 1853, an explanation of the phlogistic process in physiopathological terms and at cell level gradually took shape: phlogosis became a particular metabolic manifestation of the cell. New methodological trends arose against the "ontologism" of clinicians and semeioticians—i.e. their tendency to establish disease syndromes without investigating their origins—and the anatomism of Rokitansky's Viennese school, which was opposed by the functional approach. In an article of 1855 in *Archiv,* entitled "Die Cellularpathologie," Virchow stated the new cytogenetic law: *omnis cellula e cellula,* as opposed to the Schwannian theory of *generatio aequivoca,* according to which cells were formed by an amorphous blastema present in the tissues. The ideas of Robert Remak (1815–65), a student of Müller and from 1859 professor of anatomy in Berlin, developed along similar lines. After his first memoir of 1852, in which he affirmed that cells were generated by preexisting cells by means of division, Remak published his *Untersuchungen über Entwicklung der Wirbeltiere* in 1855, and his *Über die embryologische Grundlage der Zellenlehre* in 1862. In these works, cytogenesis by division was supported by numerous examples taken from observations of embryonal tissues, but in the last work mentioned, the endog-

enous formation of new cells was reintroduced for certain pathological cases. Virchow, however, remained faithful to the ideas expressed in his article of 1855, and in his 1858 work applied the cellular principle to a wide range of problems: from the classification of tissues to the physiopathology of blood and of the nervous system, from dyscrasia to degenerative processes, and from phlogosis to tumors. Simultaneously, a new vitalism, founded on mechanics and the causal determinism of life phenomena, was also taking shape.

Quantitative physiology and psychophysics

The first part of Hermann von Helmholtz's *Handbuch der physiologischen Optik,* publication of which continued until 1867, came out in 1856. In it, Helmholtz, who had been called to Bonn from Königsberg in 1855, united morphology, physics, and physiology into a single conceptual synthesis, the best example of which can be found in the chapter on the accommodation mechanism.

The "psychophysics" of Gustav Theodor Fechner (1801–87), though it was rooted in a complex spiritualistic ideology, was also characterized by an effort to create a quantitative physiology starting from esthesiology. After having outlined a program for future studies in his *Zend-Avesta* (1851), Fechner began experiments on thought, sight, and distance sensations, which led him to formulate the law of the logarithmic ratio between the increase in intensity of a stimulus and the increase in intensity of the sensation produced by it. In his *Elemente der Psychophysik* (1860), Fechner recognized Ernst Heinrich Weber's priority (1846) for these results (the Weber-Fechner law).

The inner environment

During these same years, at the College de France and the Sorbonne, where he was called to the chair of physiology in 1854, Claude Bernard expanded his researches even further, attempting even more subtle experiments, and abandoned Magendie's fragmentariness for an initial attempt at synthesis, which he committed to copy-books full of notes and reflections that translated into generalized concepts such as the inner environment of the organism. His method, like that of his teacher, focused on vivisection and demonstration. In 1851, Bernard demonstrated the existence of the vasomotor nerves. In 1857, he isolated glycogen and, with irrefutable arguments demonstrated that oxidation and combustion could not take place in the lungs. In 1856, his long series of experiments on the paralysis caused by curare led to the hypothesis that the toxin acted on the ends of the motor fibers. In 1858-9, he elucidated the effects of nerves on the secretory organs. Bernard's "experimental physiology" was based on the unity of the organism, while Virchow's "pathological physiology" was oriented toward the elementary unity of the cell and a concept of the organism as an aggregate. Schwann, Bernard, and Virchow attempted to define the characteristic properties of life: metabolism and growth for Schwann; metabolism, reproduction, and excitability for Virchow; and creation, or the predominance of function over structure, for Bernard. The problem considered by Darwin—the stability or instability of the species—was different; but an evolutionary interpretation of life was reflected in above-mentioned scientific circles as well.

Clinical pathology

Scientists developed new disease descriptions and diagnostic methods for use in semeiotics, pathology, and clinical medicine. Descriptions were furnished for alcoholic polyneuritis (Magnus Huss, 1807–90), tabetic locomotor ataxia (Guillaume Duchenne de Boulogne, 1806–75), and chronic forms of rheumatism (Jean Charcot, 1825–93).

The nosological and semeiological study of thoracic diseases progressed, and the first counts of corpuscles in the blood, the first determinations of the acetone level in urine, and the first larygoscopic examinations were made. The works of Karl Wunderlich (1815–77) made it possible to conduct a systematic study of the relations between body temperature and disease states (1856). The Scottish physician Alexander Wood (1817–84) developed the subcutaneous method for giving therapeutic injections (1853).

Armand Trousseau (1801–67), who in 1852 took over the chair of clinical medicine at the Hôtel-Dieu in Paris, initiated clinical teachings that combined traditional nosography and diagnostics with numerous elements drawn from physiopathology, physiological chemistry, and microbiology.

EARTH SCIENCES

Cartography

In 1857, Jacques Babinet introduced homalographic projection, his name for the direct pseudocylindrical projection devised by Karl Brandan Mollweide (1774–1825). It was used to represent planispheres and hemispheres.

The catastrophistic theory and evolution

The theory that huge cataclysms had repeatedly destroyed the earth's flora and fauna, which had been recreated by successive repopulations, a theory that already counted numerous opponents, was gradually disproved as studies progressed. In 1858, Heinrich Georg Bronn (1800–62) pointed out that numerous fossils occurred in rock formations that were both earlier and later with respect to a presumed catastrophe. Charles Darwin provided an explanation for the succession of various organisms throughout geological history in his famous On the Origin of Species (1859), in which he formulated the theory of evolution, based partly on paleontological observations. Darwin especially emphasized the discontinuity of the paleontological information contained in fossil strata, which was caused not by catastrophes but by the common processes of erosion, sedimentation, and diagenesis, in accordance with Charles Lyell's actualistic principles.

The entire interpretation of paleontological succession thus acquired new foundations, and stratigraphic studies were soon being based on the new views. Often contested for non-scientific reasons, the theory of evolution soon proved quite useful in paleontology and was favorably greeted by geologists. The search for the "missing links" in the phylogenetic series stimulated scientific expeditions and excavation projects for the purpose of recovering vertebrate remains, campaigns of the type that are still under way today. The remains of Neanderthal man, described by Johann Carl Fuhlrott in 1858, were discovered in 1856, and in France, Edouard Lartet placed the miocenic Dryopithecus in the direct lineof human ascendency (1858).

Research in stratigraphic and regional geology

Stratigraphic studies continued to develop rapidly. The significant contributions made by every civilized country to historical geology are too numerous to list, so we shall mention only a few as examples:

Roderick Impey Murchison, who in 1855 succeeded Henry Thomas de La Beche as general director of the Geological Survey in Great Britain, studied the Silurian period, which he considered the oldest fossiliferous period; however, in 1852 Adam Sedgwick demonstrated that the Cambrian period was even older. Joachim Barrande (1799–1883) and Jules Marcou (1824–98) researched the American "Taconic System" (1860), and Barrande examined the Silurian in Bohemia, about which he began publishing a vast descriptive monograph in 1852. Karl Frantsovich Rouillier (1814–58) examined the conditions under which the deposits of the Carboniferous period were formed in Russia (1857). Albert Oppel and Friedrich August Quenstedt (1809–89) devised biozonations based on ammonites and conducted stratigraphic researches on the Jura region, mainly in Germany, while Alcide d'Orbigny dedicated his attention to the period Cretaceous in France and its fauna. In 1853, August Heinrich Ernst Beyrich (1815–96) distinguished the Oligocene period as a period in itself. Important regional treatments were given in Ebenezer Emmons's American Geology (1855) and the Geologie der Schweiz (1851–53), by Bernhard Studer (1794–1887). Studer pointed out the absence of a single axial crystalline zone in the Alps, demonstrating instead the existence of various "central" masses made up of crystalline rocks separated by sedimentary soil.

The work of Antonio Stoppani (1824–91) on Alpine geology (from 1856) and that of Hermann Wilhelm Abich (1806–86) on the Caucasus (1856–59) also deserves mention. During that same period, Karl Eugen von Mercklin wrote his Palaeodendrologikon Rossicum (1855), a work dedicated to fossil wood. In 1859, Abramo Massalongo became the first to use photography to illustrate fossil fish and fossil plants (of the Verona area). In 1855, Andrew Ramsay illustrated the glaciation of the Permian. Important studies on the glaciers of the Quaternary were made by Robert Chambers (1851) and John Tyndall (1857). Quenstedt wrote his Geologische Ausflüge in Schwaben and his Der Jura in 1857.

In 1856, Gérard Paul Deshayes (1796–1875), a zoologist and paleontologist, became president of the Geological Society of France, which he had helped found in 1830.

Seismology

Robert Mallet (1810–81) carried out important researches on the earthquake that took place in Naples in 1857. In 1859, he proposed a seismic scale (perfected in 1862).

Petrographic microscopy: vulcanology

In 1851, Henry Clifton Sorby (1826–1908) became the first to study thin cross-sections of rocks under the microscope. In 1858, he published a work entitled On the Microscopical Structure of Crystals.

General problems in petrography were investigated by Luigi Palmieri (1807–96) with respect to Vesuvius (1865), and by Wolfgang Sartorius von Waltershausen, who studied Etna and discussed the origins of magma (1853). In 1854, Charles Lyell visited the Canary Islands and the Madeira Islands.

The exploration of the ocean floor

Matthew Fontaine Maury (1806–73), one of the pioneers of modern oceanography, published his famous treatise *The Physical Geography of the Sea and Meteorology* during this period. This work can be considered the first real treatise on oceanography, since it presents and discusses, often in abundant detail, the fundamental topics of the discipline. Currents, temperature, tides, morphology, and meteorological events are described on the basis of Maury's direct experience (from his numerous explorations at sea) and observations made by other earlier and contemporary authors. There is a special section describing the characteristics of the gulfstream current. Maury also made a general bathymetric map of the Atlantic Ocean, possibly using all the data then available; although it contains a number of numerical errors (a product of inaccurate surveying methods), it actually gives a fairly correct representation. On other topics as well, such as the effects of the wind on currents, Maury's text reflects the inadequacies of the times, and the influence of controversies that were still raging on various topics can often be seen.

By this time, practical research at sea had assumed "industrial" aspects as far as bathymetric measurements were concerned: this was the epoch during which the first undersea telegraphic cables were laid, and knowledge of the ocean depths and of the nature of the ocean floor was no longer the speculation of scientists but important data for engineers. The first device for examining samples of the ocean bottom was created by John Mercer Brooke (1826–1906) around 1855. It was nothing more than the canon ball used as ballast in cable soundings with holes drilled in it. When it was dropped to the bottom, the holes filled with silt (or remained empty if the ball landed on rocks), which could be brought to the surface and examined. Occasional samples of the ocean floor had already been taken in the past, for example by John Ross in the polar campaign of 1818, and even by Marsili (and probably, at lesser depths, by others in the past as well). But by this time, systematic determinations were necessary. Even the direct exploration of the depths became a coveted goal.

The Origin of Species *and oceanography*

Charles Darwin's fundamental work *On the Origin of Species*, which appeared in 1859, stimulated new oceanographic studies, especially in the field of biology. Forbes's assertions that the ocean depths were azoic had still not been disproved. Brooke had shown that there were numerous shells of foraminiferae and skeletons of other organisms in the mud of the ocean floor; however, Jacob Whitman Bailey correctly interpreted them as the remains of forms that lived on the surface of the ocean and whose skeletons sank to the bottom when they died.

The birth of dynamic meteorology

William Ferrel (1817–91) studied the effect of the earth's rotation on the motion of fluids (air and ocean currents). Later studies, made in 1858 and 1864, enabled him to arrive at the famous quantitative relation between the barometric gradient and wind speed. These were the premises of dynamic meteorology, which had its major development in the work of Vilhelm Bjerknes (1862–1951).

One of Ferrel's major works, *Essay on the Winds and Currents of the Ocean*, published in 1856, contains a complete, rigorous treatment of the forces acting on a particle of air, as well as a representation of the general circulation of the air, deduced mathematically from well-defined hypotheses. The work also makes use of Coriolis's force to explain the curvature of the paths followed by air around centers of high and low pressure. In 1852, Christopher Buys–Ballot (1817–90) drew synoptic maps that used horizontal and vertical cross-hatching to indicate areas where the temperature was below or above average and arrows and arcs to represent the wind at a particular point and its eventual variation over the course of an entire day.

Synoptic networks

The astronomer Urbain Le Verrier (1811–77), who was deeply interested in meteorology and designed projects for reorganizing meteorological activity in France, took a decisive step toward the realization of a synoptic program for practical ends. Le Verrier was charged with the task of determining whether the storm that seriously damaged the Franco-English fleet in the Black Sea on 14 November 1854 was the same one that had hit a good part of the Mediterranean during the preceding few days. In just a few weeks Le Verrier succeeded in obtaining information on the weather conditions in more than 250 localities for the period 13–16 November 1854. He was convinced that the events he was examining were caused by a phenomenon with specific features that had moved from one place to the other. On 16 February 1855, he proposed gathering meteorological observations via telegraph so that storm warnings could be issued. The Emperor gave his consent, and by 19 February Le Verrier was able to present a 10:00 AM synoptic map for the day. This was the direct progenitor of the ones universally in use today. As a matter of fact, the present-day synoptic network developed straight out of Le Verrier's project, without any breaks in continuity.

The extraordinary success of this initiative changed the course of meteorology. It was acknowledged that the weather could be forecast with a fairly high degree of accuracy by extrapolating the movement of various meteorological parameters, especially isobaric configurations. At this point, practical meteorology took precedence over theoretical meteorology.

APPLIED SCIENCES

The steel industry

The technological developments of this period were greatly influenced by the Crimean War. When, on 26 March 1854, England and France went to war against Russia, the memory of the destruction of the Turkish fleet during the previous November was fresh in the minds of the ship builders. The idea of armoring ships came from the French, who used three armored vessels in the attack on Kinburn (not far from Odessa). Although the ships were hit some seventy times, they were not broached. In 1855 the English government, relying on its heavy industry, also turned to building armored vessels. Sir John Brown (1816–96) founded the Atlas Steel and Spring Works at Sheffield to produce boilers, armor, etc. In 1860, after a visit to a French naval base, Brown proposed rolling the armor instead of forging it. By the time he died, Brown had armored three quarters of the English fleet.

Just as war was being declared, Sir Henry Bessemer (1813–98), experimenting with a cast-iron cannon of his own construction, perfected a patent for imparting rotary motion to an oblong projectile. News of Bessemer's experiments reached Napoleon III, who offered to finance them. Bessemer built a small model of his cannon out of steel and sent it to Napoleon. His idea was to replace cast iron with steel; however, the cost of steel was still prohibitive at sixty pounds per 1,000 kilograms. Returning to England, Bessemer continued to run experiments in his small foundry, until he noticed that a jet of air had decarbonized two bars of pig iron lying at the edge of the furnace. This clue led him to the discovery (patented 17 October 1855) that blowing air into a closed cupola would remove the impurities in molten iron. In December Bessemer obtained a patent for what was basically the Bessemer converter.

The first Bessemer steel, however, was very brittle due to its phosphorous content. The process was only successful when valuable Swedish ore from Dannemora, which contained almost no phosphorous, was used. On August 11, 1856 Bessemer presented a paper on manufacturing malleable iron and steel at the annual congress of the British Association for the Advancement of Science.

The process invented by Bessemer had already been anticipated by William Kelly (1811–88), an American who had begun working on it in 1847. Kelly had noticed that when a current of air was passed through crude pig iron in the refining furnace, the iron became much hotter if it was not covered with wood charcoal than if it was covered. The carbon in the pig iron was acting as fuel and could therefore be eliminated with a jet of air. The novelty in Kelly's "boiling" process was that the temperature was raised—without the addition of fuel—by the rapid combustion of the carbon contained in the pig iron. However, when Kelly announced his method at the beginning of the 1850s, he was taken for little less than a madman. He was therefore extremely surprised in 1856 to learn that an Englishman had obtained a patent for a process very similar to his own.

Bessemer, in fact, also encountered obstacles. The first reaction of the Sheffield industrialists was negative, and the foundry that Bessemer, in partnership with others, set up there was not at all welcomed. The price of his product was little more than the seven pounds per 1,000 kilograms, which was the cost of 1,000 kilograms of good Swedish iron.

The repercussions on industry, especially ship building, were enormous. The shareholders of the Great Eastern, the first large iron ship, begun in 1854 by Isambard Kingdom Brunel (1806–59), were traumatized, as was John Ramsbottom, superintendent of locomotives of the London and North Western Railway, when Bessemer suggested he use steel rails. However, the first steel rails in the world were laid in the station at Crewe just a few years later, on 9–10 November 1861.

The iron industrialists had hardly recovered from the shock when in 1856 Frederick Siemens (1826–1904), a German-born English citizen, obtained a patent for the regeneration of waste heat. The regenerative principle was first applied in the manufacture of pig iron, in a stove devised by Siemen's associate E.A. Cowper. The Cowper stove was built with towers for preheating the blast. The regenerative principle became a standard feature of open-hearth steel production. While the heat for reaching operating temperature in Bessemer's converter was generated in the process, in Siemen's open-hearth furnace it was supplied from outside.

Arms and explosives

After the outbreak of the Crimean War, Sir William Armstrong (1810–1900), a hydraulic engineer, was given the task of designing underwater mines for use against the Russian ships anchored in the port of Sebastopolous. His mines were successful, and he was commissioned to build cannon. Armstrong built his first forged-iron breech-loading gun, which soon replaced bronze and cast-iron front-loading cannon.

During this same period, technicians in Russia were also producing for the war. K.I. Konstantinov worked on perfecting rockets; called to direct the St. Petersburg rocket factory in 1850, he devised a number of measures for improving the quality of rockets and ensuring safety in production. Range and accuracy were increased, rockets were made to last longer, and premature explosions were almost totally eliminated.

A Swedish engineer living in St. Petersburg, aware of the explosive properties of nitroglycerine, attempted to employ it in these wartime operations, but failed. His son, Alfred Bernhard Nobel (1833–96), who returned to Sweden in 1859, discovered that nitroglycerine could be made to explode with a detonator. However, the first idea for using nitroglycerine for wartime purposes went back to Nikolai Nikolaevich Zinin (1812–80). In the United States, Benjamin Hotchkiss (1826–85) invented the pressure fuse.

Chemistry and the mining industry

An accidental discovery made by William Henry Perkin (1838–1907), then an eighteen-year-old student at the Royal College of Chemistry, gave rise to the industry of synthetic organic products mentioned in the section on CHEMICAL SCIENCES. These products found immediate use in the textile industry, which, during this period, had automatized the combing process. In 1850 Greenwood and J. Warburton invented the comb presser for silk; in 1856, de Jongh, an Alsatian, invented an automatic comber, which was perfected by S.C. Lister (1815–1906) in 1877. In France, Quinson patented a circular comber, and in 1859, Lister and Warburton invented a silk comber.

The mining industry profited considerably from research in progress. A series of patents concerning nonferrous metals were issued during this decade. In 1850, Alexander Parkes proposed a method for extracting silver, based on the observation that when lead and zinc were melted together and then cooled, any silver present became concentrated in the zinc. The following year, Karl Friedrich Platner (1800–58) proposed extracting gold from arsenopyrites. In 1854, Henri Etienne Deville Sainte-Claire (1818–81) elaborated a method for producing aluminum chloride and sodium, the reaction of which produced aluminum. And in 1850, Henry Hussey Vivian (1821–94) modified copper mining by bringing calcination directly into the fourth phase of the process. In 1859, John Glover (1817–1902) designed the tower that is named after him, which considerably improved the lead-chamber method of producing sulphuric acid by recovering the nitrous gasses.

Fuels

In the area of fuels, the most sensational discovery was made in the United States, where the competition began to favor petroleum, despite the efforts of James Young (1811–83), who in 1850 obtained a patent for producing kerosene by distilling coal, and in 1856–57 manufactured lubricants from bituminous substances. In 1855, Benjamin Silliman (1779–1864), a professor of chemistry at Yale, described the general properties of crude petroleum. Influenced by Silliman's report, George H. Bissel (1821–84), founder of the Pennsylvania Rock Oil Company, commissioned Edwin Drake (1819–89) to conduct a search for petroleum. On 28 August 1859 (a fundamental date in the history of energy production), Drake put the first oil well into operation; it had been drilled using iron drilling bits with steel flangings. Despite improvements made by P.P. Fauvelle, materials that were able to break through rock formations only became available after 1860.

Transport and communications

In November 1851, Moscow and St. Petersburg were connected by a two-track railway built under the direction of P.P. Mel'nikov and N.O. Kraft, it was the longest (650 kilometers) and most important railroad line in Russia.

The Crimean War had cut off Great Britain's supply of Russian hemp, so that production of jute in India was stepped up. Tea and coffee also came from India, and the demand for them grew as the British standard of living rose. Despite the negative effects of Perkin's discovery, which dealt a severe blow to the Indian cultivation of indigofors, the stream of trade increased. The only thing that was needed was a trade route that was shorter than sailing around the coast of Africa. Ferdinand de Lesseps (1805–1894) promoted an initiative to solve this problem. Despite the fierce opposition of Great Britain, his project for cutting the isthmus of Suez was accepted in 1856 with the consent of Said, viceroy of Egypt, and the subscription of funds from France. The cutting of the isthmus was begun on April 25 of that same year with the founding of Port Said. The canal was opened on 17 November 1869.

The development of telegraphic communications was also extremely important. In 1842, B.S. Jacobi had directed the laying of underground cables for the St. Petersburg-Carskoe Selo telegraph line, and the so-called printing telegraph, with a device for automatically transcribing messages, was invented in 1850. The first transatlantic cable was laid in 1858, under the direction of William Thomson Kelvin. The enterprise received worldwide praise, and there was great disappointment when, just a few weeks after the first signals had begun to be transmitted, the cable suddenly broke down. Kelvin re-examined the problem, and in 1868 directed the laying of another, improved cable, which was not damaged by the sea. Alphonse Beau de Rochas

also investigated the topic, and in 1859 published a *Théorie mécanique des telegraphes sous-marins*.

Technical literature

Technological publications increased with the expansion of industry and the rise of specialized schools for training technicians and engineers.

Two tendencies in this area are particularly worth mentioning: first, the most interesting inventions were usually patented immediately in order to make the most profit on them (the first patent for an airplane, in the modern sense, was issued on 2 May 1857 to Félix du Temple de la Croi; however, nothing was ever built under it); and, second, the most recent scientific discoveries were more and more frequently used in the construction of new devices, so that technical literature naturally ended up merging with more properly scientific literature (and the specialized treatise writing of each science and discipline automatically became divided into two distinct branches, the pure science and the applied science).

1860–1870

THOUGHT AND SCIENTIFIC LIFE

Developments in American scientific life

During the first half of this decade, the United States was faced with a grave crisis, the secession of the southern states. The Civil War did not, however, stop the country's scientific and technological progress. In fact, this war, which in certain respects—such as the extreme importance assumed by military technology—foreshadowed the great wars of the twentieth century, and actually favored the organization of scientific research. The efforts of the American Association for the Promotion of Sciences to create a national scientific academy were crowned with success. ''Science without organization lacks power'' became a popular slogan in a society shaken by the events of the war. In was in this climate that in 1863 the Senate voted to establish the National Academy of Sciences, on the basis of a project presented by Charles Henry Davis, Joseph Henry, and Alexander Dallas Bache, with the support of Louis Agassiz (1807–73).

With the end of the war and the new problems presented by the expansion westward (railway networks, problems regarding the rational exploitation of the frontier), the American scientific world tended above all to develop applied science. The Academy itself had, among other things, the task of responding to any requests from government bodies in the way of studies, investigations, and experimentation.

Science in Russia

The organization of scientific activity, once centered exclusively around the Academy of St. Petersburg, was accelerated in Russia as well, especially after the abolition of serfdom in 1861. The secondary education system had been broadened to meet industrial and military needs by facilitating admission to the secondary schools from the grammar schools. New universities were founded (including the Institute of Railway Engineering in St. Petersburg, begun as early as 1810, and a school of higher technical studies in Moscow, started in 1832); and the Academy stopped serving a scholastic function, remaining the most important center of scientific activity in the country and one of the major centers of the world. As in Western European countries, scientific life gradually became more animated with the establishment of associations. The Society of Naturalists had already been formed in 1805, and the Society of Geographers had been created in 1845. These were now joined by societies of mathematics (1864) and chemistry (1868). In 1868, Russian naturalists and physicians met for the first time in a national congress in St. Petersburg.

The number of university students also increased appreciably after the abolition of serfdom (1861) and the beginning of a phase of more rapid industrial development. By the second half of the century, Russian science had become an authoritative presence in the areas of mathematics, chemistry, and embryology. In contrast to what was happening in the United States at the same time, however, a lack of governmental institutions kept a rich technology from developing out of this situation. The geographic explorations promoted by the Academy of Sciences in St. Petersburg contributed greatly to the progress of scientific knowledge.

In Poland, the University of Cracow, located in the part of the country that was under Austrian rule, became the center of Polish science in 1869 (the city was also the headquarters of an academy of sciences).

As far as other countries were concerned, the universities of Iasi and Bucharest were opened in Rumania (in 1860 and 1869, respectively), the University of Belgrade was founded in Yugoslavia (1863) and the University of Tokyo was established (1867) in Japan, quickly filling the scientific-technological gap between itself and the West.

The International Red Cross was founded in 1864 on the initiative of Jean Henri Dunant of Geneva (although the first overtures were made by Ferdinando Palasciano, a physician in the Bourbon army, in 1861).

The triumph of positivism in France

While the United states and Russia remained on the outskirts of scientific-philosophical reflection (and Europe often had a rather vague, inaccurate idea of the huge advances that were being made in these countries on economic and technological levels), this particular decade was characterized by the diffusion of positivism in Europe.

Three scholars—Emile Littré, Joseph Ernest Renan (1823–92), and Hippolyte Taine (1828–93)—contributed significantly to this diffusion. Littré is mainly remembered for a subtle essay on the founder of positivism entitled *Auguste Comte et la philosophie positive*, written in 1863, and for having founded the *Révue de philosophie positive* in 1867. Renan wrote one of the most popular books of the epoch, which also created the greatest scandal among religious believers, *Vie de Jésus* (1863). Taine wrote *Les philosophes français du XIX siècle* (1857), *Philosophie de l'art* (1865), and *Sur l'intelligence* (1870). The immediate success of all these works indicated that the authors knew quite well how to interpret the cultural needs of the French bourgeoisie, or at least its lay faction. Not without

reason, they also received considerable recognition from the French public after the fall of Napoleon III.

Of the French scientists who were most influenced by positive philosophy we shall only mention the chemist Marcelin Berthelot (1827–1907), a friend of Renan's. The cultural atmosphere of positivism is also reflected, in a certain sense, in the work of the physiologist Claude Bernard (1813–78), author of the *Introduction à l'étude de la médecine expérimentale* (1865), one of the most important texts of nineteenth-century philosophy of science.

English philosophy. Spencer's evolutionism

Herbert Spencer published several works in England and soon achieved great fame. His *First Principles* was published in serial form between 1860 and 1862 and as a book in 1867; his *Principles of Psychology* (1870–72) was a reelaboration of an 1855 work of the same title. Publication of two other very important works, *Principles of Sociology* (1877–96) and *Principles of Ethics* (1879–93), was begun during the next decade. The Spencerian system was remarkably successful, both because it was presented as "the philosophy of evolutionism" and because it seemed able to reconcile the needs of science with the needs of faith.

However, a need for other philosophical concepts was also beginning to be felt in university circles. This gave rise to a large movement to assimilate the great German philosophies of the first half of the century. In 1865, James Hutchison Stirling (1820–1909) published a work entitled *The Secret of Hegel* (1865).

Germany. The "return to Kant"

The situation in Germany was more complex. On the one hand, the positivistic atmosphere was spreading rapidly among scientists, while on the other, a lively rebirth of metaphysics was taking place among philosophers. The positivism of the scientists reflected their need no longer to allow themselves to be sucked into the bitter disputes between materialists and spiritualists as they had been in the previous decade. They saw their escape in an agnosticism, which exalted the value of scientific knowledge but at the same time assigned insuperable boundaries to it. The position of the physiologist Emil DuBois-Reymond (1818–96), synthesized in his *Über die Grenzen der Naturerkenntnis* (1872), a famous discourse on the limits of understanding nature, was typical from this point of view. The "return to Kant" (*zurük zu Kant*), championed by prestigious scientists and philosophers, such as the physicist-physiologist Hermann von Helmholtz (1821–94) and the philosopher Friedrich Albert Lange (1828–75), was also essentially an attempt to satisfy the same need. The title of Lange's major work, *Geschichte des Materialismus und Kritik seiner Bedeutung in der Gegenwart* (1866), is indicative of this trend.

The philosophers did not feel that they were in a position to confute the numerous objections brought against "nature philosophy" and Hegelianism by the scientists, for example by von Helmholtz. They preferred instead to resort to new forms of spiritualism, based on the affirmed existence of a non-mechanical order in the universe, separate from the mechanical order studied by science. The major representative of this neo-spiritualism was Rudolf Hermann Lotze (1817–81), whose most important work, *Mikrokosmos*, came out in three volumes between 1856 and 1864. Lotze published two other significant works during the next decade: *Drei Büche. der Logik* (1874) and *Drei Bücher der Metaphysik* (1879). Among Lotze's students at the University of Göttingen was the greatest German logician of the end of the century, Friedrich Ludwig Frege (1848–1925).

Italian positivism. Ardigò

Spiritualism declined in Italy, supplanted by neo-Hegelianism and, several years later, by positivism. The major representative of neo-Hegelianism was Bertrando Spaventa (1817–83), who was named a professor at the University of Naples in 1861. The most illustrious supporter of positivism was Roberto Ardigò (1828–1920), who became a professor at the University of Padua in 1881. Ardigò's first important work was a speech on Pomponazzi, which he read in 1869 while still a priest: it marked his estrangement from Catholicism and his approach to positivism.

MATHEMATICAL SCIENCES

The birth of algebraic geometry

This decade marked the beginning of a particularly productive period in geometry. In 1860, Rudolf Friedrich Clebsch (1833–72) began an important series of memoirs in which he very effectively applied algebra to a variety of questions in geometry, such as third- and fourth-order curves, cubic surfaces, and the stationary planes of intersection of two algebraic surfaces. A short time later, in memoirs of 1863 and 1864, he inaugurated the as yet practically unexplored field of the geometrical applications of transcendental functions, developing a general theory on the application of elliptical functions to the study of rational and elliptical curves. Then, between 1865 and 1868, he successfully dealt with various problems connected with the representation of surfaces on a plane.

In 1862, Luigi Cremona (1830–1903) published his *Introduzione a una teoria geometrica delle curve piane*, in which he reconstructed the general theory of algebraic plane curves in a geometrical form. The following year, he published the first rigorous and exhausitive memoir on the birational geometric transformations of plane figures (still called Cremona transformations), the theory of which constituted the transition from projective to algebraic geometry. In fact, Cremona's work is usually designated as the beginning of the so-called classical period of algebraic geometry in the sense of the study of the properties of geometric bodies by means of algebra, a study that can even be thought of as having begun with the creation of analytical geometry. More specifically, algebraic geometry was linked to the development of investigations into the functions of complex variables, so its true precursors can be found in the works of that sort written by Abel and Riemann, followed by the work of the English school (Cayley) and the German school (Clebsch). A turning point was Cremona's discovery of birational transformations (between planes and between linear spaces), after which algebraic geometry tended to characterize itself as the study of the invariant properties of birational transformations. In this new capacity, it was above all developed by the Italian school, which besides Cremona, counted among its ranks Corrado Segre (1863–1924), Eugenio Bertini (1846–

1933), Guido Castelnuovo (1865–1952), and Federigo Enriques (1871–1946).

Developments in topological research

In 1863, August Möbius (1790–1868) published his first important works on topology in which he studied the elementary transformations that could be used to make the infinitesimal elements of two figures mutually correspond. In the years that followed, he produced other topological results in the theory of polyhedrons. In this way, topology, or in a certain sense the qualitative study of figures, began to take shape as an autonomous discipline within geometry. Topology had already been anticipated by Gottfried Wilhelm Leibniz and Leonhard Euler and had been studied seriously in works by Karl Friedrich Gauss (the first person to express, by means of an integral, the whole number that represented the reciprocal concatenation of two closed curves and remained unchanged when they were subjected to continuous deformations without breaks or crossings) and Georg Friedrich Bernhard Riemann (he studied surfaces that illustrate the behavior of functions of a complex variable—now called Riemann surfaces). We have already mentioned the work of Listing, while Klein in Germany and Enrico Betti (1823–92) in Italy were the main contributors to this discipline, after Möbius.

The studies of Dedekind and Georg Cantor on the continuum were also of topological interest (see MATHEMATICAL SCIENCES, 1870–1880).

New directions in geometry

Geometricians who had already been active in the past also contributed to opening new paths. In 1864, Michel Chasles expounded the theory of the characteristics of conic systems and of the appropriate numbers for defining them, thus inaugurating numerative geometry. Following an interval of around twenty years during which he turned his attention to physics, Julius Plücker returned to geometry with his *Neue Geometrie des Raumes* (1868), in which he reformulated solid geometry using the straight line as the generative element of spatial figures.

In 1868, in his *Saggio di interpretazione della geometria non euclidea*, Eugenio Beltrami (1835–99) proposed the first Euclidean model of non-Euclidean (hyperbolic) geometry, thus opening the way to the admission of the relative non-contradictoriness of non-Euclidean geometry. Up to this point, non-Euclidean geometry had only proved non-contradictory in practice, but there were no formal guarantees that it shared the characteristic non-contradictoriness of traditional geometry. Other models were later proposed by Klein, Poincaré, Hilbert, and others.

Studies in the geometry of hyperspaces were advanced with Beltrami's *Teoria fondamentale degli spazi a curvatura constante* (1868) and A Cayley's *Memoir on Abstract Geometry*.

With regard to other branches of mathematics, the remarkable studies on the theory of surfaces of least area conducted in 1869 by Hermann Schwarz (1843–1921) deserve particular mention. That same year, Elwin Bruno Christoffel (1829–1900) introduced the symbols (named after him) for the study of transformations between differential forms.

One initiative that definitely should not go unmentioned is the publication of the *Bollettino di bibliografia e storia delle scienze matematiche e fisiche*, founded in 1868 by Balthasar

Boncompagni (1821–94). This bulletin was an extremely important tool for the history of science and continued to be published until 1887. Also in 1868, Clebsch and Karl Gottfried Neumann founded the *Mathematische Annalen,* which became one of the most authoritative journals in the world.

ASTRONOMY

The birth of astrophysics

After having developed in embryonic form through the discovery of Fraunhofer's lines, the first photographs and the first researches on the nature of the sun, astrophysics was finally born. The determining discovery was made by Gustave-Robert Kirchhoff and Robert Wilhelm von Bunsen, after which the new science was immediately applied to the study of the sun and the stars. The study of the sun developed rapidly. During the eclipse of July 18, 1860, two expeditions, one led by Angelo Secchi (1818–78) to the Desierto de las Palmas on the Spanish Mediterranean coast and the other led by Warren de la Rue (1815–89) to Rivarbellosa on the Atlantic coast of Spain, were organized to photograph protuberances. Upon comparison of the photographs, it was noted that the protuberances were the same and were equally distributed with respect to the dark limb of the moon, and it was concluded that they definitely belonged to the sun, thus resolving the question affirmatively. The study of proturberances developed considerably. On August 18, 1868, after having observed them through the spectroscope during a total eclipse, Pierre Jules Janssen (1824–1907) decided to look for them without an eclipse, by blocking out most of the diffused sunlight with the spectroscope. In this way, he could observe the objects in one of the lines in which they emitted the most light, for example the red hydrogen line. His attempts were successful, especially when he oriented the slit of the spectroscope tangentially to the disc of the sun. Meanwhile, Joseph Norman Lockyer (1836–1920) achieved the same results in Europe. Thus the protuberances could be observed at the limb of the sun at any time.

The discovery of helium

The possibility of conducting chemical analyses of the sun, and the discovery that protruberances belonged to the sun, along with repeated observations of them, aroused the first real interest in the study of the physical constitution of the sun and its varied activity. This was quite different from the attitude of the preceding epoch, when studies had stopped as soon as the few directly observed phenomena had been examined. The enthusiasm reached its peak in 1868, when Lockyer discovered an unknown element in the sun. Only in 1895 was it found on the earth and given the name helium.

Stellar spectra

No less important were the discoveries relating to the stars that were made possible by use of the spectroscope. As soon as he received news of Kirchhoff's discovery, William Huggins (1824–1910) built a spectroscope for observing the stars and discovered in them many of the elements already found in the sun. In 1864, he also discovered that several nebulae (many of which appeared in John Frederick Herschel's last catalogue, published that same year, which contained 6,245 non-stellar objects) gave a spectrum with luminous lines only. It was noted almost immediately that not all stellar spectra were alike. A Secchi, who directly examined around four thousand of them between 1862 and 1868, divided them into four major groups, laying the foundations for modern spectral classification.

Theoretical astrophysics was also born during this period. In 1869, Jonathan Homer Lane (1819–80) developed the first star model, stating that the sun and the stars were gasseous spheres in a condition of hydrodynamic equilibrium.

Dark companions

In 1862, while trying out a new objective, the famous telescope-builder Alvan Clark (1804–87) and his son discovered the companion of Sirius. Bessel had already predicted its existence as early as 1844, on the basis of variations in the proper motion of Sirius, and Christian August Peters (1806–80) had calculated its apparent orbit in 1851. That same year, Arthur von Auwers (1838–1915) predicted the orbit of the companion of another star, Procion, which was only effectively observed in 1896 with the great refractor at the Lick Observatory. A new area of research—the search for invisible stars—was thus born.

The Astronomische Gesellschaft

At the end of August, 1863, twenty-six astronomers meeting at Heidelberg founded the Astronomische Gesellschaft, an international astronomical society that six years later gave rise to the first great precision star catalogue, which contained around 120,000 stars and was compiled in sixteen observatories. The catalogue, identified with the initials of the society (AGK), was followed by the AGK_2 and AGK_3 catalogues, compiled in order to deduce proper motions from a comparison of the position of stars in various epochs. Other valuable initiatives of the Astronomische Gesellschaft were the editing of an annual astronomical bibliography (*Astronomischer Jahresbericht*), begun in 1899, and an individual history of variable stars (*Geschichte und Literatur des Lichtwechsels der veränderlichen Sterne*).

The origin of falling stars

In 1866–67, Giovanni Virginio Schiaparelli (1835–1910) made the important discovery that falling stars were of cometary origin, when he found that the Perseids (of August) came from the comet 1862 II and the Leonids (of November) came from 1866 I. This discovery was important not only because it solved a complex problem, but because it contributed to defining the new tasks of the most ancient of the sciences. For the first time, in fact, an essentially physical result (the nature of comets) was reached starting from purely mechanical observations and data, and the existence of a solid link between classical astronomy and the nascent astrophysics—which from this time forward constituted an essential component of the new astronomy—was demonstrated.

PHYSICAL SCIENCES: MECHANICS

Critiques of the Laplacian programs

The necessity of founding scientific knowledge on mechanics, and consequently the general validity of Laplacian type programs, was looked upon from a particularly critical perspective during this decade. This perspective had its roots in the debates on the concept of force that had already been under way for decades and, in turn, provided a premise for the great analyses of mechanics that emerged during the last three decades of the nineteenth century. Nevertheless, it must be remembered that the critique of mechanics as a privileged cognitive form was not a process that took place solely within purely mechanical studies. It had various points of reference in the development of all the sciences of nature and in philosophical reflections on this development.

In 1861, Antoine Cournot (1801–77), a French mathematician and philosopher, raised certain questions regarding the presumed scientific purity of mechanics. In his treatise on the concatenation of the fundamental ideas of science and history (*Traité de l'enchaînement des idées fondamentales dans les sciences et dans l'histoire*), he attempted to point out that several of the fundamental notions of the discipline were not rigorously scientific but were instead clearly anthropomorphic in nature. This was especially true, according to Cournot, of the notion of force and was exemplified by its inseparable link with well-known muscular functions (the lifting of weights, pressure on a surface, etc.). Cournot's critique, which was partially based on investigations that he himself conducted on the calculus of probabilities, ultimately arrived at anti-deterministic positions and merged with a philosophical concept of a spiritual nature, in contrast to the positivist thought that was popular among many of the scientists of the time. It was a sign of the revival of spiritualistic themes in French scientific culture (see THOUGHT AND SCIENTIFIC LIFE, final decades of the nineteenth century).

Ernst Mach (1838–1916) started out from very different positions. In his compendium of physics for physicians (*Compendium der Physik für Mediziner*) of 1863, he still adhered to a rigorous approach aimed at reducing all scientific explanations to the principles of mechanics. However, the researches he conducted between 1862 and 1872 on the physics and physiology of optical phenomena and acoustic phenomena convinced him of the sterility of that approach, which he first criticized in a memoir of 1868 on the definition of mass (*Über die Definition der Masse*). In this memoir, Mach attempted a global reconstruction of mechanics by means of three experimental propositions and two definitions. The first definition stated the concept of mass as follows: the inverse of the product, changed in sign, between the mutually induced acceleration of every pair of bodies is called their ratio of mass.

It followed, according to Mach, that the term "force" was reduced to the pure and simple product of mass, defined as above, times acceleration. This brief memoir of 1868 already laid the groundwork for the program that Mach developed over the following decades.

Analyses of the concept of force

The debate on the fundamental notions of mechanics expanded toward the end of the decade. In 1869, in *Nature* magazine, Peter Guthrie Tait (1831–1901) published an article entitled "On Force" that fueled the disputes already under way. According to Tait, it was necessary to make a distinction between the subjective nature of every sensation of force or tension and the real properties of nature. On the basis of this distinction, it

became clear, in his opinion, that the notion of force was purely an intellectual construction and not at all an objective entity. With these affirmations, Tait did not intend to deny the validity of explanations regarding the laws of motion—which he in fact studied in collaboration with William Thomson Kelvin—but rather to criticize their materialistic context, within the realm of a vast cultural battle that led him to coauthor *The Unseen Universe*, with Balfour Stewart in 1875 (see PHYSICAL SCIENCES: MECHANICS, 1870–1880). Closely related to all this was Tait's polemic against Clausius's and Boltzmann's lines of physico-mathematical research (see PHYSICAL SCIENCES: THERMODYNAMICS AND STATISTICAL MECHANICS, 1860–1870 and following decades).

The influence of developments in physiology on analyses of the concept of force is attested to by the importance assumed by the reflections of Emil Du Bois-Reymond (1818–96). In the introduction to his researches on animal electricity (*Untersuchungen über thierische Elektrizität*, 1848–60), he defined force as a rhetorical instrument of our brain and an abstruse product of that irresistible tendency toward personification that is impressed in us.

Mechanics, the nucleus of physics

The spread of these critical attitudes that attempted to eliminate the anthropomorphic aspects of the basic notions of mechanics was, however, paralleled by an increase in investigations on the possibility of arriving at a more fundamental rationalization of the laws of motion, with particular reference to the laws governing gravitational interaction. This intricate set of problems was the focus of several important communications, including James Challis's *On the Principles of Theoretical Physics* (1861), *On Newton's Foundations of All Philosophy* (1863), and *On Fundamental Ideas of Matter and Force in Theoretical Physics* (1866), Émile von Keller's work on the cause of weight and the effects attributed to universal attraction (*Mémoire sur la cause de la pesanteur et des attribués à l'attraction universelle*, 1863), and the contribution of Paul Émile Lecoq de Boisbaudran (1838–1912) to the theory of gravity (*Notes sur la théorie de la pesanteur*, 1869).

Given the very nature of the problem of the role of mechanics in the physical sciences, it should be mentioned that its role was also discussed in relation to physics, namely with regard to the development of electromagnetic theories, statistical mechanics, and reflections on the structure of matter. See PHYSICAL SCIENCES; THERMODYNAMICS AND STATISTICAL MECHANICS; ELECTROMAGNETIC THEORIES; and RADIATION AND MATTER for the period from 1860 to 1900 for a discussion of these connections.

PHYSICAL SCIENCES: THERMODYNAMICS AND STATISTICAL MECHANICS

The "heat death" of the universe

From 1860 on, the application of statistical methods to questions that emerged from the need to find an exhaustive interpretation of the two principles of thermodynamics and their relation to the laws of motion created fundamental differentiations among the programs that had sprung up during the preceding decade. William Thomson Kelvin's *Physical Considerations Regarding the Possible Age of the Sun's Heat*, which returned to the theme of the "heat death" of the universe, in light of the principle of dissipation of energy, stated in 1852, appeared in 1861. Given the natural tendency of physical phenomena expressed by the second principle, Kelvin affirmed that, if the universe were "a single finite mechanism," then this mechanism would stop when "the consumption of the potential energy within the material universe" was reached. A memoir written by Clausius in 1862 on the concept of inner work dealt with the same theme, but from a very different point of view with respect to the kinetic-molecular interpretation. According to Clausius, it was essential to investigate the second principle of thermodynamics further by means of specific hypotheses on the distribution of molecules in bodies. If it were true that by this time a rigorous mathematical demonstration of the form assumed by the second principle had been reached, it was also just as true, Clausius wrote, that the principle still had "an abstract form that is difficult to grasp with the mind." Basing his treatment, directed at "researching the precise physical cause" on which the second principle depended, on a molecular model, Clausius proposed mathematically representing the action of heat on molecules by means of a new value, which he called the disgregation of the body and by means of which he could define the effect of heat simply as a tendency toward an increase in disgregation. According to Clausius, this new function Z could be defined as:

$$Z = Z_0 + A \int \frac{dL}{T} ,$$

where A is the heat equivalent of a unit of mechanical work L, T is the temperature, and Z_0 is an initial, indeterminate value of the disgregation function. The problem of linking the two expressions $\int dZ = 0$ and $\int dQ/T = 0$ through circular processes immediately arose. In a second memoir written that same year, Clausius examined the irregularity of molecular motions that these mathematizations presupposed, disagreeing with several scholars, including James Clerk Maxwell. According to Clausius, mathematical physics should not be afraid of taking into account irregular motions. But this did not mean limiting studies only to the formal problems that arose: Clausius praised the formal elegance of Maxwell's communications of 1859 (see PHYSICAL SCIENCES: THERMODYNAMICS AND STATISTICAL MECHANICS, 1850–1860) but criticized its incompleteness and lack of internal consistency, asserting that Maxwell's discussion was based on implicit hypotheses that were not generalizable in an acritical way.

Laws and models

This progressive mathematization of kinetic models began to be contrasted by a series of very precise programs toward the middle of the decade. One of these was the program that William Rankine continued to develop after his conversion to energetics. In two communications, published in 1864 and 1865 and entitled respectively *On the Hypothesis of Molecular Vortices* and *On the Second Law of Thermodynamics* Rankine reiterated that models were not an indispensable part of thermodynamics. The two

laws of this science, according to Rankine, were nothing more than "abbreviated expressions of the experimental facts," and their link with mechanics should be sought by analyzing energy conversions. The answer was not to exclude all ideas of a modelistic nature, but to evaluate the various hypotheses with extreme caution and, in any case, to recognize the superiority of the molecular vortices model over all other hypothetical forms in relation to energetics. Even more hostile to modellistic approaches was Gustav Adolphe Hirn (1815–1890) in his *Théorie mécanique de la chaleur* (1865). According to Hirn, a mechanical theory of heat phenomena had to be completely free of any contrived hypotheses on the nature of heat and instead had to link the phenomena themselves "to the most elementary principles of mechanics and translate them into mathematical formulae."

Entropy

Despite the rules obtained with the use of kinetic models, a great part of the scientific world opposed them because of the various programs for reducing the physical sciences to mechanics and the different notions of the relations between physical theories and the objective world. In his memoir of 1862, Clausius wrote that he was aware that his ideas concerning the heat contained in bodies were different from those presently accepted. He also noted that in the course of the years, he had become increasingly convinced of the fact that scientists must not give too much weight to ideas that were, in part, founded more on habit than on scientific grounds. In his communication of 1865 (*Über verschiedene für die Anwendung bequemen Formen der Hauptgleichungen der mechanischen Wärmetheorie*), he explicitly introduced the innovative concept of entropy. Commenting on the equation:

$$\int \frac{dQ}{T} = S - S_0 ,$$

Clausius wrote: "If we wish to indicate S with an appropriate name, we can say that it is the 'content of transformation' of the body, in the same way we say that the quantity U is the 'heat and work content' of the body itself. Nevertheless, since I believe that it is better to draw names for quantities like these that are so important to science from the ancient languages (. . .), I propose indicating the quantity S with the name 'entropy' of the body, from the Greek term ἡ τροπή, transformation. I have intentionally formed the word entropy in such a way as to render it similar to the word 'energy,' since both these quantities are so closely related in their physical meaning that a certain similarity between their names seems useful." According to Clausius, the concepts of energy and entropy made it possible to state two principles that contained "the fundamental laws of the universe: 1) the energy of the universe is constant, and 2) the entropy of the universe tends toward a maximum.'

The quantity of entropy, according to Clausius, was given by the sum of the quantities previously elaborated starting from the hypothesis that heat was a form of molecular motion: the "transformation value" of the heat content of a system and the "disaggregation" of the system. The treatment of entropy now had to be extended to include all the effects of heat radiation.

Dynamics and probability

Two important contributions to the development of the kinetic model were published by James Clerk Maxwell and Ludwig Boltzmann in 1866. In his *Dynamical Theory of Gases*, Maxwell attempted to respond to the criticisms that Clausius had brought against him by approaching the mathematical treatment of the distribution function from a different point of view. In order to accomplish this, he based his arguments on both the results of experiments described in his memoir *On the Viscosity or Internal Friction of Air and Other Gases* (1866), and an emphasis on applying the calculus of probabilities to the molecular model. The "exceptional position" of the second principle was the central theme of Boltzmann's work on the mechanical significance of the two principles of the heat theory (*Über die mechanische Bedeutung des zweiten Hauptsatzes der Wärmetheorie*). By mathematizing the concept of temperature and making various probabilistic assumptions on the irregularity of molecular motion, Boltzmann attempted to demonstrate that the second principle could be deduced from the principle of least action, just as the first was a consequence of the "principle of living forces." The reductionistic approach that characterized Boltzmann's early works is evident. The same tendency can still be seen in the communication of 1868 (*Studien über das Gleichgewicht der lebendigen Kräfte zwischen bewegten materiellen Punkten*) in which Boltzmann presented an initial formulation of the equipartition of energy, though it gradually disappeared over the following two decades, replaced by a probabilistic approach.

The vortex-atom

The debate on the physical meaning of the second principle and the role of kinetic models became even more heated toward the end of the decade. Kelvin intervened resolutely in 1867 with his *On Vortex-atoms*, explicitly attacking the hypotheses on which the work of Clausius and the other "atomists" was founded. The real atoms, according to Kelvin, had to be analogous to the vortex structures in a perfect fluid, which von Helmholtz had studied from a hydrodynamic point of view (see PHYSICAL SCIENCES: MECHANICS, 1850–1860). In Kelvin's opinion, a hydrodynamic model had to replace "the monstrous hypothesis" that the "rash" analyses of some chemists had deduced from Lucretius's thought. From this point of view, wrote Kelvin, it was necessary to reconsider all the studies of D. Bernoulli, Herapath, Joule, Krönig, Clausius, and Maxwell. Similar theories appeared in William Rankine's memoirs of this same period: *On the Thermal Energy of Molecular Vortices* (1869 and 1870), and *On the Thermodynamic Theory of Waves of Finite Longitudinal Disturbance* (1870). The hydrodynamic model, according to Rankine, was the only one that permitted a study of the relation that necessarily exists between the laws of elasticity and heat of all substances (gasseous, liquid, or solid) and those of the wave propagation of a finite longitudinal perturbation. In particular, wrote Rankine, it was necessary to analyze the propagation of adiabatic waves and the conversion of the energy of these waves into energy of molecular agitation.

Numerous questions arose at the end of the decade around

the second principle of thermodynamics and the possibility of attributing a precise physical meaning to the concept of entropy through the use of models. The complexity of the situation derived from the fact that these problems did not fall within the realm of one particular branch of the physical sciences. In fact, they involved all the relations between the calculus of probabilities, the principles of mechanics, the laws of thermodynamics, and the use of hydrodynamic models in interpreting electromagnetic phenomena (see PHYSICAL SCIENCES: ELECTROMAGNETIC THEORIES, 1860–1870), as well as the very concept of matter in philosophy and science.

PHYSICAL SCIENCES: ELECTROMAGNETIC THEORIES

Retarded potential

This decade was dominated by two great memoirs written by James Clerk Maxwell on the electromagnetic field and the meaning of the speed of light. With these memoirs, Maxwell provided the first mathematical representation of Faraday's notions on electrical, magnetic, and optical phenomena and emphasized the difficulties involved in further extending this scheme to include gravitational events. During these same years, the programs connected with the studies of Ampère, Weber, and Kirchhoff were further developed in several works by Ludwig Valentin Lorenz (1829–91). Hendrik Antoon Lorentz later referred back to these works, especially with regard to the problem of "retarded potentials" (see PHYSICAL SCIENCES: ELECTROMAGNETIC THEORIES, 1890–1900).

Light as an electromagnetic phenomenon

In 1861, in a memoir on the theory of elasticity of homogenous bodies with constant elasticity (*Mémoire sur la théorie de l'élasticité des corps homogène à l'élasticité constante*), Ludwig Lorenz, inspired by ideas that went back to Oersted's philosophy, elaborated a notion of retarded potential that he applied in the context of Kirchhoff's theory in the years that followed. In 1866, Lorenz's research, which was based on the conviction that analogies existed between light and electricity, led to the recognition of the finite propagation of electrodynamic action. In fact, in a work published in 1867. (*Om identiteten af Lyssvingninger og elektriske Stromme*), Lorenz recognized that local equations were the basic expressions of electromagnetic interaction and linked the latter to the vibrations of light, thus surpassing some of the very premises he started from and arriving at the concept of action by contact.

The electromagnetic theory of light

In 1861, the same year that Ludwig Lorenz introduced retarded potentials, Maxwell published the first two parts of his communication *On Physical Lines of Force*, the final sections of which came out in 1862. Just as Lorenz developed Oersted's concepts through a mathematical revision of Weber's and Kirchoff's theories, so Maxwell developed Faraday's ideas by surpassing Kelvin's analyses and his own geometrical conclusions in 1857. According to Maxwell, the so-called Newtonian system was unsatisfactory, while the geometrical model of 1857 should not be taken as a representation of the reality of the field but merely as a tool to aid the imagination. His idea ran as follows: Electromagnetic phenomena are due to the existence of matter in certain conditions of motion or pressure in each part of the magnetic field and not to a direct action at a distance. The substance that produces these effects could be a certain part of the ordinary matter, or else an ether associated with the matter. The mathematical development of this hypothesis implied a new model, which Maxwell drew from the developments that had taken place in the physical sciences in England during these years and which involved the introduction of molecular vortices. However, as he himself pointed out, the vortex model did not reflect and mode of connection existing in nature and served only insomuch as it was conceiveable in mechanical terms and easily investigated. The hypothetical vortex structure that supported the field was thus a provisional structure, even though it was advisable to seek experimental confirmation of it, for example by attempting to render observable the physical dimensions of the vortices. However, Maxwell did not enter into these considerations on vortices in detail, and his model was exposed to the criticisms brought against it by scholars such as Challis, who in 1861 published an article entitled *On Theories of Magnetism and Other Forces, in Reply to Remarks by Professor Maxwell*. In any case, these criticisms did not minimize the fact that, although he provisorily resorted to hypotheses on molecular vortices, Maxwell succeeded in establishing the electromagnetic nature of optical phenomena and identifying the optical ether with the electromagnetic ether. The numerical value of the speed of light played a privileged role in Maxwell's considerations and was the principal result of his memoir of 1861–62. "I have deduced from this result," Maxwell observed *a propos*, "the relation between the statical and dynamical measures of electricity, and have shown, by a comparison the electro-magnetic experiments of M.M. Kohlrausch and Weber with the velocity of light as found by M. Fizeau, that the elasticity of the magnetic medium in air is the same as that of the luminiferous medium, if these two coexistent, coextensive, and equally elastic media are not rather one medium. . .The velocity of transverse undulations in our hypothetical medium agrees so exactly with the velocity of light that we can scarcely avoid the inference that *light consists in the transverse undulations of the same medium which is the cause of electric and magnetic phenomena*." In the same memoir, on the basis of a modelistic analysis of interactions between "cells" of the medium and layers of "particles" constituting electricity, Maxwell stated the possibility of "displacement currents," linking them to electrical displacement. This displacement, wrote Maxwell, did not amount to a current, since when it reached a certain value it remained constant, but it was the beginning of a current, and its variations constitute currents.

Maxwell's equations of electromagnetic field

The next step in James Clerk Maxwell's program involved definitively moving beyond the connection between electromagnetic theory and the theory of elasticity that had been established with the model of molecular vortices. This step, which was later amply discussed in the pages of a treatise written in 1873, was taken in two classical memoirs of 1865 and 1868, entitled respectively *A Dynamical Theory of the Electromagnetic Field* and *On a Method of Making a Direct Comparison of Electrostatic with Electromagnetic Force, with a Note on the Electromagnetic Theory of Light*. In the 1865 memoir, Maxwell established that the equations of field were valid, independent of any mechanical model of the ether. In his opinion, they constituted a consistent representation of experimental data. Secondly, Maxwell stated the impossibility of establishing a theory of field that would explain gravitational, as well as electromagnetic, phenomena. Just when he had succeeded in providing a rigorous formal structure for electromagnetic events, he was forced to reduce the range of Faraday's program so as not to fall into paradoxical consequences regarding the energy of the gravitational field.

Maxwell affirmed the separation between equations of electromagnetic field—not as yet expressed in vectoral form—and mechanical terms. While the phenomenon of the induction of currents was deduced from their mechanical effects thanks to von Helmholtz and Thomson (Lord Kelvin). Maxwell followed the inverse order, and deduced mechanical action from the laws of induction. Naturally, in expressing the theory and its fundamental equations, Maxwell used a language related to mechanics. Nevertheless, he specified that any statement that used mechanical terms had to be taken as an illustration and not as an explanation of the real connections of the field. Only in one point was field theory inseparably linked to mechanics; namely where energy was concerned. Writing about the energy of the field, Maxwell said that all energy was mechanical energy, whether it was in the form of movement, or of elasticity, or in any other form whatsoever. The energy of electromagnetic phenomena was mechanical energy, the only problem being ascertaining where it was found. Maxwell's answer to the last question was explicit. While the old theories viewed energy as localized in bodies, it was now seen as existing in the field, and therefore in all space and in all matter. It was also possible to demonstrate— except in the case of gravitational interactions—that the dependency of electrical and magnetic interactions on the inverse of the square of the distances was not the starting point for the electromagnetic theory, but its point of arrival.

All of these radical innovations were lucidly and consistently expounded in the communication of 1868. But Maxwell's theory still contained an unresolved problem, the premises of which already existed in the earlier theories of the optical ether, i.e. the difficulties caused by the longitudinal waves that the universal medium should be able to support on the basis of the common laws of elasticity. And, as a matter of fact, Maxwell was unable to deduce the laws of refraction and reflection, which had deep roots in the very idea of the ether and implied going beyond these theoretical difficulties.

So, despite the force of Maxwell's new theory, it contained reasonable grounds for the skepticism that it provoked in numerous scientific circles, both in those associated with the class of action at a distance, and those that developed programs founded on Faraday's ideas. This skepticism also found ample justification in the fact that Maxwell's so-called "dynamic theory" was not generally deduced from a precise mathemization of the mechanical structure of the ether and therefore could not be interpreted on the basis of the main propositions of the theory of motion. Maxwell attempted to respond to these flaws in his *Treatise on Electricity and Magnetism* of 1872 (see PHYSICAL SCIENCES: ELECTROMAGNETIC THEORIES, 1870–1880), but they definitely continued to pose an obstacle to the acceptance of his ideas for years to come.

Space and geometry

To the extent that the electromagnetic theory extended theoretical analysis to space, seen as the basic physical location where the events studied by the theory took place, it is important to remember that the problem of geometry in relation to space began to pose itself with growing insistence during these years. In 1868, von Helmholtz's communication on the basic hypotheses of geometry (*Über die Tatsachen, die der Geometrie zu Grunde liegen*) sanctioned the formation of an opposition to the idea that the validity of Euclidean geometry was based on *a priori* truths. The geometrical nature of extension, which was common to all real bodies, was an experimental fact, according to Von Helmholtz; and we must remember that the considerations for Von Helmholtz, emerged from a context that was closely related to investigations into non-Euclidian geometries. On the other hand, Riemann's treatise on the basic hypotheses of geometry (*Über die Hypothesen, welche der Geometrie zu Grunde liegen*), in which the importance of non-Euclidean geometry is derived from the fact that Euclieean space and Lobachevsky's and Bolyai's space can be shown to be particular cases of generalized space, also appeared (posthumously) in 1868.

By 1870, William Kingdon Clifford, the English-language translator of Riemann's works, took into account the connections between geometry and physics. These connections later became the object of violent disagreements.

PHYSICAL SCIENCES: RADIATION AND MATTER

The black body

The problems related to the structure of matter entered directly into the theoretical considerations of physicists during this decade, after having long primarily studied in the area of the chemical sciences. Various factors caused this particular turn of events which was full of important consequences for physics between the end of the nineteenth and the beginning of the twentieth centuries. Among these factors were new findings regarding the spectra of various radiations, the discovery of new elements through the refined spectroscopic techniques developed toward the end of the previous decade and the manifestaton of unanticipated properties of matter in states of extreme rarefaction (for example in vacuum tubes).

In a conference held in Heidelberg in 1861, Gustave Robert Kirchoff proposed a problem that would engage the efforts of some of the major theoretical and experimental physicists of the time for years: the problem of the "black body." A black body, according to Kirchoff, was a physical system that would absorb "all the rays" that hit it and could be characterized by a coefficient of absorption equal to one. The determination of the properties of these bodies implied the existence of a relation between absorption and emission that did not depend on the nature of the physical systems being considered, but was rather a function of their temperature and the wave-length of the radiation. Thus the fundamental problem was to find the form of this function.

Spectroscopy

The discovery of new elements had been made possible by Kirchhoff's and Bunsen's spectroscopic technique. Of these, the identification of cesium (1860), rubidium (Bunsen, 1861), and thallium, (William Crookes, 1861) and the observation of the spectrum of indium (Ferdinand Reich and Hieronymous Richter, 1863) deserve mention. The complexity of the spectra of these elements raised perplexities regarding the elementariness usually attributed to them, leading to the first conjectures that the heavier elements might be aggregations of lighter ones, while the problems connected with finding the causes of the various spectral components became increasingly urgent. Thus, in 1863 for example, the German physicist and chemist Hermann Kopp suggested the existence of extremely simple forms of matter with very low atomic heats (*Über specifische Wärme starrer Körper, und Folgerungen bezüglich der Zusammengesetzheit s.g. chimischer Elemente*). This overlapping of chemistry and physics was the source of questions that called for a deeper investigation of theoretical knowledge on the structure and origin of spectra, which involved making radical revisions in the very foundations of the physical sciences.

Cathode radiation

In 1869, Johann Wilhelm Hittorf (1824–1914), a teacher of physics and chemistry, published a memoir on the conduction of electricity in gasses (*Über die Elektricitätsleitung der Gase*) affirming on solid experimental grounds the rectilinear propagation of rays emitted by a cathode in vacuum tubes. With his researches, Hittorf, a student of Julius Plücker (see PHYSICAL SCIENCES: ELECTROMAGNETISM AND OPTICS, 1850–1860), carried cathode radiation physics to a crucial point: from this point on, the investigations of Eugen Goldstein, Varley, and Crookes studied the nature of cathode rays, and a new contradiction arose regarding the validity of the corpuscular or wave interpretations of the phenomena observable in connection with electrical discharge in very rarefied gasses.

Meanwhile, a new series of problems began to emerge in relation to the possibilty that cathode radiation might be the carrier of electrical charges and to the observability of interactions between this new type of radiation and the electromagnetic field, as well as to the observability of effects caused by a particle structure of the rays themselves.

CHEMICAL SCIENCES

From the Karlsruhe Congress to the periodic table of the elements

The first International Congress of Chemistry was held in Karlsruhe at the beginning of September, 1860. Stanislao Cannizzaro took advantage of the occasion to publish the ideas of Amedeo Avogadro together with his own method for determining atomic weights. From 1860 to 1865, Jean Servais Stas (1813–91) carried out accurate determinations of atomic weights in order to test Prout's principle, according to which they were all multiples of the atomic weight of hydrogen. He found that they were not exact multiples but for the most part were close. This strange but irrefutable result led him to state *"il faut croire qu'il y a quelque chose là-dessous."*

John Alexander Newlands (1837–98) placed the various elements in order of their increasing atomic weights and proposed the theory of "octaves" (1864). The previous year, Alexander Émile de Chancourtois (1820–1886) had presented his "tellurian table" of the elements, but it had not met with much success. In 1869, after the publication of similar tables by Alexander William Williamson in 1864 and William Odling (1829–1921) in 1865, Dimitri Ivanovich Mendeleev (1834–1907) and, independently, Julius Lothar Meyer (1830–95), published the famous periodic system. Although it still contained uncertainties and gaps, these were eliminated in the course of the following decades. Even the elements recently discovered through the use of spectroscopic analysis (introduced by Robert Wilhelm Bunsen and Gustave Robert Kirchhoff [1824–87] in 1860) found their rightful places in the periodic table. These newly discovered elements included cesium and rubidium (Bunsen and Kirchhoff in 1861), thallium (William Crookes in 1861), and indium (Ferdinand Reich and Hieronymous Theodor Richter in 1863).

In two publications, one of 1864 (in Norwegian) and one of 1867 (in French), Cato Guldberg (1836–1902) and Peter Waage (1833–1900) stated the well-known law of mass action. The law had been glimpsed by various other authors (Carl Friedrich Wenzel, 1740–93; Claude Louis Berthollet; Alexander William Williamson; Faustino Jovita Malaguti; John Hall Gladstone, 1827–1902; Ludwig Ferdinand Wilhelmy, 1812–64; Marcellin Bertholet; Leon Péan de Saint-Gilles; Augustus George Harcourt, 1834–1919; and William Esson), but Guldberg and Waage were the first to state it clearly, following an accurate investigation of both its experimental and its mathematical aspects. In 1869, Bertholet and Emile-Clement Jungfleisch (1839–1916) stated the law of distribution of a solute between two immiscible solvents. Thomas Graham (1805–69), in particular, studied colloidal solutions (1861–65); Graham is also responsible for having named the colloids.

The law of mass action

From 1869 to 1876, Thomas Andrews (1847–1907) studied the liquefication isotherms of carbon dioxide and other permanent gasses, discovering their discontinuity below a certain critical temperature. Above this critical temperature, the gas will not liquefy, no matter how much it is compressed.

Raoul Pierre Pictet (1846–1929) and Louis Paul Cailletet (1832–1913) applied Andrews' studies to the liquefication of oxygen and nitrogen (1877).

Theories on structure; bonds; benzene

Enormous progress was made in organic chemistry during this period, on both a theoretical and a practical level. By now the tetravalence of carbon had been ascertained and the carbon atom had been compared to a tetradedron—Alexander Butlerov (1828–86), Friedrich August Kekulé, and Emanuele Paterno (1847–1935). The custom of writing formulas according to the theory of radicals or types was being gradually abandoned, and modern structural formulas were beginning to be developed by Archibald Scott Couper, Butlerov, Lothar Meyer, Joseph

Loschmidt (1821–1895), and above all Alexander Crum Brown (1838–1922).

It was Butlerov who recognized the importance of determining the arrangement of compounds. He introduced the name "structure" (1861) to indicate the mutual bonds between atoms and stated that it was the structure, together with the composition, that determined the physical and chemical properties of compounds. He found a wide range of data for developing his structuralist theory in Gerhardt's theoretical work and was satisfied with representing arrangement by means of simple formulas, without entering into the problem of determining the position of the atoms in space, still a rather difficult endeavor at that time.

The existence of unsaturated compounds (for which Hofmann proposed the ending -ene) was partially explained by Meyer and Kekulé by means of "holes" or "free affinity units," until Loschmidt in 1861, and even more clearly Emil Erlenmeyer (1825–1909) in 1862 postulated the existence of double and triple bonds. The term valency was only introduced in 1868 by Carl Hermann Wichelhaus (1842–1927)).

Two hexagonal formulas were now proposed for benzene, one with three alternating double links (Kekulé, 1865) and the other quinoid (James Dewar, 1867). However, neither explained the existence, recognized during this same period, of only three isomers—ortho, para, and meta—of the di-substitution derivatives. While the diagonal formula proposed by Albert Karl Claus (1838–1900) in 1867 and the prismatic formula advanced by Albert Ladenburg (1842–1911) in 1867, both seemed to account for these isomers, neither of these structures contained double bonds, and thus did not escape criticism. In the wake of Kekulé's formulas for benzene, Carl Graebe (1841–1927) proposed formulas for naphthaline and anthracene (1868) and Wilhelm Körner (1839–1925) advanced one for pyridine (1869). Vladimir Vasilevich Markovnikov (1837–1904) stated the rule that bears his name on the direction of addition of water and halogen-hydric acids on double ethylene bonds.

New organic preparations

Noteworthy among new organic reactions were the hydrogenation of primary amine nitriles (O. Mendius, 1862); the preparation of homologs of benzene from halogen derivatives and sodium (Rudolf Fittig, 1864) and phenols by the alkaline fusion of sulphonic acids (Lucien Dussart, 1864); and the first organic derivatives of tetravalent sulphur, the sulphoxides, prepared by Aleksandr Mikhailovich Zaitsev (1841–1910) in 1866.

Natural compounds: cocaine, glumatic acid, and the nucleic acids

Isolation of natural compounds and syntheses were extremely numerous during this period. Among the first were cocaine (Albert Niemann, 1860), choline (Adolph Strecker (1822–71), 1862), serine (E. Cramer, 1865), glutamic acid (isolated by K.H. Ritthausen (1826–1912) in 1866) and the nucleic acids (Johann Friedrich Miescher (1844–95), 1869). Nicorinic acid (C Huber, 1867) and barbituric acid (Adolf von Baeyer, 1864) were obtained from natural compounds. In 1864, Cannizzaro began the researches on the structure of santonin that he and his collaborators pursued for many years, right up to the beginning of the twentieth century. The definitive structure of santonin was only established in 1929 by Robert Downs Haworth's English school. From 1860 to 1871, Jean-Baptiste Boussingault studied nitrogen fixation in plants and realized that the amount of fixed nitrogen increased even in the absence of atmospheric nitrogen. Jean-Jacques Shloesing (1824–1919) attributed the cause of this phenomenon to a microbic action. Boussingault had already demonstrated that the ammonia present in rainfall was not enough to satisfy the fixed nitrogen requirements of plants.

Acetylene from carbide; uric acid and indole

Particularly important syntheses were those of acetylene from calcium carbide (Friedrich Wöhler, in 1862) and formaldehyde from methyl alcohol (Hofmann, 1868). New synthetic compounds were diphenyl (Fittig, 1862), hexamethylentetramine (Butlerov, 1860), acetacetic ester (Anton Geuther, 1863), diphenylamine (Hofmann, 1864), Schiff's bases (U. Schiff, in 1864) and α-naphtol (Johann Peter Griess, in 1867). From 1860 to 1864, Baeyer worked on uric acid and its derivatives, making important contributions to information about these compounds. In 1865, he began his researches on indigo, from which he obtained indole, among other things. For the time being (1869), his attempts to synthesize indigo were only partially successful.

The chemistry of dyestuffs: the azoics, induline, and alizarin

The preparation of new series of organic dyestuffs underwent important developments during this period. In 1861 François-Zacharie Roussin (1817–94) discovered alizarin black. In 1863, Carl Alexander Martius (1830–1920) obtained the first azoic dyestuff of practical interest, Bismarck brown, from the action of nitrous acid on m-phenylendiamine. Also in 1863, Heinrich Caro (1834–1910) and John Dale discovered the indulines, an important class of azine dyestuffs. In 1866, Johann Peter Griess carried out the diazotization of aniline, preparing various salts of diazobenzene and describing copulation with another molecule of aniline. In 1867, Charles Lauth (1836–1913) prepared Paris violet. Carl Graebe and Karl Liebermann (1842–1914) synthesized alizarin from 1.2-dibromoanthraquinone (1868). In 1868, these same two scientists in Germany and William Henry Perkin in England prepared it with the more economical process of fusion and alkaline oxidation of sodium β-sulphoanthraquinone.

Oximes and carbylamines. The determination of molecular weights

In 1865, Wilhelm Lossen (1838–1906) discovered hydroxylamine, an important reactant of the carbonyl compounds, which was only isolated in the pure state in 1894 by Cornelius Adrian Lobry de Bruyen (1857–1904).

In 1867, Hofman discovered the carbylamines, or isonitriles, and the identification reaction for the primary amines based on their formation. That same year, he isolated formaldehyde for the first time from the oxidation of methyl alcohol. Hofmann also described (1868) the method for measuring vapor densities and consequently molecular weights based on the gasification of

substances in a barometric vacuum. After having prepared the sulphoxides, Zaitsev discovered the sulphones, a new group of thiocompounds, in 1867, while Nikolai Nikolaevich Zinin (1812–80) prepared lepidene (tetraphenylfurfurane) in the course of his researches on benzoin.

In 1870, Albert Ladenburg, who together with Charles Friedel (1832–99) had already developed new syntheses of compounds with direct Si-C bonds, prepared silane, the parent compound of a family similar to the aliphatic hydrocarbons, as well as silico-ethane, alcohols, and ethers and attempted to extend his studies to tin, preparing $Sn_2(C_2H_5)_6$.

The Solvay Soda industry. Treatise-writing

Of extreme practical importance was the development of the ammonia process for producing soda, perfected by Ernest Solvay (1838–1922) in 1861. In 1866, Solvay established the first industrial soda plant in Belgium, definitively supplanting the outdated and expensive Leblanc process.

In 1860, Marcellin Berthelot published his *Chimie organique fondée sur la synthèse,* a work illustrating the state of chemical synthesis at the time, which has become a classic. Karl Remigius Fresenius (1818–97) founded the *Zeitschrift für analytische Chemie* in 1862. The first article contained a description of Kirchhoff's and Bunsen's spectral analyses of alkali and alkaliterrous metals. The first university course in organic chemistry, *Vvedenie k polnomu izucheniyu organicheskoy khimy* (Introduction to the Complete Study of Organic Chemistry) by Alexander Butlerov, came out in 1864–66.

LIFE SCIENCES

Claude Bernard's experimental medicine. The beginnings of endocrinology

This was the decade of Claude Bernard's *Introduction à l'étude de la médecine expérimentale,* the microbiological discoveries of Louis Pasteur (1822–95), and the researches of Gregor Mendel (1822–84) on the transmission of hereditary characteristics.

Bernard's work—really a "discourse on the method of the nineteenth century," as Henri Bergson later defined it—was the manifesto of a great school that embraced researchers from many countries: Élie de Cyon (1842–1912), Willy Kühne (1837–1900), Angelo Mosso (1846–1910), Luigi Vella, Peter Ludwig Panum (1820–85), John Dalton (1766–1844), Austin Flint (1812–86) and Silas Weir Mitchell (1829–1914). Physics, chemistry, and biology were strongly connected within the field of experimental medicine. Life was "creation," but its building-blocks were inorganic. On a methodological level, induction was completed by deduction, which provided hypothesis. The philosophical problem of the "primary cause" of natural events arose once again in the context of research approached from this point of view. But the physiologists' research in turn projected the "feeling of the complexity of things." In France, skillful researchers also came out of the laboratory of physiology of the Sorbonne: Paul Bert (1833–66), who studied the effects of barometric pressure on living organisms; Albert Dastre (1844–1917) and Jean Pierre Morat (1846–1920), who researched the problems of vasomotor innervation; L.C.A. Malassez (1842–1909), a haematologist;

and Arsène d'Arsonval (1851–1940), an electrophysiologist and specialist in studies of animal calorimetry.

The major figures outside Bernard's school were Édouard Brown-Séquard (1817–94), Alfred Vulpian (1826–87), Eugène Gley (1857–1930), and Jules Marey (1830–1904).

With his researches on the adrenal glands (1856), Brown-Séquard inaugurated the study of internal secretions. He also described the syndrome of transverse hemisection of the spinal cord. In 1868, together with Vulpian, he founded the Archives de physiologie normale et pathologique. Gley dealt with the physiology of the thyroid and the parathyroid, while Marey—who in 1864 created his own laboratory of physiology in Paris and in 1867 succeeded Flourens at the College de France—invented several clever devices for measuring cardiocirculatory dynamics (the sphygmograph, the cardiograph, the myograph, and the intracardiac catheter).

German physiology. Progress in physiological chemistry

German physiology registered new and fundamental contributions from Müller's school. Von Helmholtz's *Die Lehre von den Tonempfindungen als physiologische Grundlage für die Theorie der Musik* came out in 1863. It was reprinted a number of times over the following years, joined by other minor writings on physiological acoustics by the same author. A memoir of 1867 on the mechanics of the series of small bones in the *cavum tympani* stands out among these.

Ernst von Brücke (1819–92), director of the Institute of Physiology of Vienna, published works on the problems involved in the relations between physiology, esthetics, and linguistics; and, in 1873–74, he published a large manual on the discipline. Karl Ludwig also moved to Vienna in 1855. The two volumes of his *Lehrbuch der Physiologie* came out between 1852 and 1856. The declared intent of the author was to derive in a necessary manner corporeal functions from the elementary conditions inherent in the organism. Ludwig proposed an explicit and rigorous physicism in antithesis to vitalism. In 1865, Ludwig left Vienna for Leipzig, where he succeeded Ernst Heinric Weber and in 1869 inaugurated a new laboratory that became a model in Germany and abroad. His school achieved world-wide dimensions and prestige. In 1867, he invented the *stromuhr*, a device for measuring blood flow to the organs.

The physiology that was oriented towards the physico-chemical premises of vital functions found itself faced with a physiological chemistry armed with new instruments and a quantitative methodology. The major exponent of the new current was (Ernst) Felix Hoppe-Seyler (1825–1895), a professor of applied chemistry at Tübingen. In 1865, he applied spectral analysis to hemoglobin, but the range of his investigations extended to other substances as well (proteins, bile salts, chlorophyll) and even took on the problem of localizing oxidation processes. Another important figure of the German school of physiology was Edward W. Pflüger (1829–1910), a student of Du Bois-Reymond. Called to the chair in Bonn in 1860—one year after the publication of his *Untersuchungen über die Physiologie des Electrotonus*—he began a long series of experiments on the "teleological mechanisms" of the organism—in particular the problems of respiration and gasses in the blood—and reached the conclusion that oxidation took place in the tissues. In 1875,

Pflüger published a work on physiological combustion in living organisms that aroused considerable interest.

Physiological chemistry became "histo-chemistry" with Johann Friedrich Miescher (1844–95), a student of Hoppe-Seyler and a professor at Basel from 1871. Miescher initiated the study of nucleoproteins. Meanwhile, the foundations of Liebig's conception that muscular work was produced at the expense of the azotate substances in the muscle, while all the other substances contributed to the production of heat, began to totter. In 1860, Theodor Ludwig Bischoff and Carl von Voit (1831–1908), published *Die Gesetze der Ernährung des Fleischfressers durch neue Untersuchungen festgestellt*, an extreme attempt to justify Liebig's ideas by means of complicated hypotheses. In 1861, Ludwig Traube (1818–76) demonstrated that part of the energy of muscular activity came directly from food. In 1866, Adolphe Fick (1829–1901) and Johannes Wislicenus (1835–1902) reiterated that the proteic catabolism deducible from the excretion of urea did not provide enough energy for the work of the muscles. In a work of 1870, Liebig again attempted to sustain the idea that a "metamorphosis" of the azotate substances was the source of muscular energy; new experiments done by Voit did not contradict him.

Neurophysiology

Because of the bonds between physiology and medicine, the results of research in this field interested both pathology and chemistry. When in 1861 Paul Broca (1824–80) announced the discovery of a language center at the base of the third frontal convolution of the left hemisphere of the brain, on the one hand it was clear that Flourens's theory of cerebral unity had to be reexamined, while on the other some scientists proposed reformulating the clinical problem of aphasia (Armand Trousseau, 1864; John Huntington Jackson (1835–1911), 1868; Carl Wernicke, 1874). Jackson, of the National Hospital of London, described partial epilepsies caused by lesions of the encephalitic cortex. The field of neurology began to emerge within pathology; the first chair was established at the Saltpêtière—a hospice in Paris, later an asylum—for Jean Martin Charcot in 1882. Meanwhile, considerable progress was made in diagnostics with the introduction of laryngoscopy (Ludwig Türck, Johann Nepomuk Czermak, 1857), oesophagoscopy (Adolf Kussmaul, 1862), and ophthalmometry (L.E. Javal (1839–1907)).

Pasteur and the question of spontaneous generation

In December, 1862, Louis Pasteur, who regarded Claude Bernard with interest and sympathy, entered the Académie des Sciences of Paris, leaving behind the prestige and the polemics connected with his 1860 researches confuting the theory of "plant and animal proto-organisms born in artificial air and in oxygen gas" (Felix Pouchet, 1800–72). Supposed spontaneous generation was instead the development of germs present in the air. If these were destroyed by heat, no more micro-organisms would form. Pasteur was a skilled chemist—at Strasbourg he demonstrated that *Penicillium glaucum* developed at the expense of dextrogyrate tartaric acid—but was convinced that life possessed particular features, present on levels that had previously escaped observation. In 1865, the French government commissioned him to study an epidemic of pebrine, a silkworm disease that was raging through the farms of southern France. Pasteur discovered that the disease was caused by microscopic bodies and suggested effective preventive measures. Meanwhile, he was becoming more and more convinced of the need to ''prepare the way for serious research on the origin of diseases.'' Just as important were his studies on fermentation. His observations of the phenomenon, in fact, led to the discovery of the bacilli in lactic acid and the bacteria that cause butyric acid fermentation. A French physician, Casimir Joseph Davaine (1812–82), described the presence of rod-shaped bodies in the blood of animals suffering from anthrax (1865) and recognized their pathogenetic importance. In 1860, Ferdinand Cohn (1828–98) turned his attention to research on bacteria, and, contrary to Ehrenberg and Theodor Billroth, identified the various single bacteria as species and became the first to classify them according to their morphological characteristics. He recognized that the formation of spores, which he showed to be resistant to heat, was a general property of bacteria. He confuted the theory of spontaneous generation and recognized bacteria as agents of disease.

Haeckel's biogenetic law

Ernst H. Haeckel applied the criterion of phylogeny to the reformulation of comparative anatomy and embryology, publishing such successful synthesizing and popularizing work as the *Generelle Morphologie der Organismen* (1866) and the *Natürliche Schöpfungsgeschichte* (1868).

In the *Morphologie,* he stated what he called the ''fundamental biogenetic law,'' according to which ''ontogeny is the short and rapid recapitulation of phylogeny''; he distinguished between palingenetic, or ancestral, and cenogenetic, or successive, characters; he affirmed that ontogenetic recapitulation did not hold for single organs, but rather for the organism as a whole. Evolution began with a *Moner,* a precellular stage of life, from which it developed all the way up to the higher vertebrates.

Darwinism

Haeckel's theories, which sometimes lacked objective elements, stirred up lively criticism and impassioned support.

In 1870, Jean Louis Armand de Quatrefages de Bréau (1810–92), a professor of anthropology in Paris, attempted to reevaluate Lamarckism against Darwinism in his *Darwin et ses précurseurs français.* In Italy, Darwin's ideas were echoed by Filippo De Filippi (1814–67) in a sensational conference on *L'uomo e la scimmia (Man and the monkey)* in 1864.

Mendel and the inheritance of characters

In the convent of Brno, in Moravia, the abbot Gregor Mendel completed his researches, begun in 1858, on hybrids of several strains of *Pisum sativum*; in 1865, he published his results in a memoir entitled *Versuche über Pflanzenhybriden.* Starting with several pure strains characterized by the constancy of certain given characters, he crossed varieties with well-defined differences, obtaining a first, uniform generation that reproduced the characters of one of the two parents, while the other character, apparently masked in the previous generation, reappeared in one-fourth of the plants of the second generation. Mendel determined the percentages of the two types in successive generations

and concluded that the gametes of the hybrids each contained only one of the characters of the ancestral pair. Of the two opposite characters of the ancestors, one was called ''dominant'' and the other ''recessive.'' Heredity, therefore was assigned to a localized material and manifested itself in a discontinuous manner in hybrids. Mendel's researches remained unknown until the beginning of the following century. Charles Darwin continued to maintain that the characters of hybrids came from a mixture of the characters of their ancestors. As to the vector material of heredity, Darwin was a supporter of ''panspermia''; every tissue in the organism supposedly generated particles that he called ''gemmules,'' which were then sent to the reproductive organs. Francis Galton (1822–1911) attempted in vain to verify this hypothesis, and in a letter to Darwin dated December, 1875, he supported views similar to those of Mendel, the abbot of Brno, which he arrived at not by experimental means, but rather by statistical deduction from a theoretical model.

EARTH SCIENCES

Nappes

While William Nicol continued his studies on the Caledonian nappes of Scotland, similar formation from the Hercynian Age were discovered in the Franco-Belgian coal basin, where Devonian strata were on top of the more recent Carboniferous series. This discovery later had a great influence on the tectonic interpretation of the Alps and other mountain ranges.

Another observation that opened the way to studies in Alpine tectonics was made in 1862 by Angelo Sismonda (1807–78), who suggested that the ''modification gneiss'' (lime schist) of the Piedmontese Alps, formerly considered extremely ancient Precambrian, could instead have been formed during the Hurassic Age. Although it was disregarded at the time, Sismonda's intuition later proved to be of extreme importance in solving the structural problems connected with the formation of the Alps. In 1869, Heinrich Gerlach (1822–71) judged the structure formed by the gneiss of Antigorio to be a recumbent fold. This observation also revealed its importance during studies made at the beginning of the following century.

Glacio-isostasy

In 1865, Thomas Francis Jamieson offered some fundamental ideas on the alternating sinking and rising of regions covered by large icecaps, which then melted, relating these movements to the load represented by the masses of ice according to isostatic laws.

Paleontology and stratigraphy

Evolutionary ideas began to be applied to paleontology on a large scale. In 1862, Hermann von Meyer described the Archaeopteryx, the first known bird, which had many reptilian traits. The studies of Albert Gaudry (1827–1908) on the mammal fauna of Pikermi in Greece (*Animaux fossiles et géologie de l'Attique*) appeared in 1862–67. Edward Drinker Cope (1840–97) introduced the name Stegocephalia for fossil amphibians in 1869. In paleobotany, important work was done by Oswald Heer (1809–83), from 1864, and by Philipp Wilhelm Schimper (1808–80), who proposed the Paleocene period for the intermediate continental formations between the Cretaceous and Econeous marine strata of Western Europe. The ninth edition of Charles Lyell's *Principles of Geology* came out in 1866; it accepted and applied Charles Darwin's evolutionary ideas, which the author had already extended to humans in 1863 (*The Geological Evidences of the Antiquity of Man with Remarks on Theories of the Origin of Species by Variation*).

A treatise on the Tertiary of the Veneto region, written by Eduard Suess (1831–1914), appeared in 1868.

The affirmation of optical methods in petrography

The study of thin sections of rocks became increasingly more common during this period. The major exponent of this method was Ferdinand Zirkel (1838–1912), a German geologist who started a petrographic school in Vienna in 1863, setting the example for numerous other countries.

Systematic geological surveys

In order to promote geological surveys in their respective countries, Jean-Baptiste Élie de Beaumont established the Service Géologique de France (1868) and Quintin Sella (1827–84) founded the Comitato Geologico d'Italia in 1867 (it began operating in 1869).

Jules Marcou's geographical map of the world, a major work of compilation, appeared in 1862 (followed by a second revised edition in 1875).

Attempts at establishing an absolute geological chronology

On the basis of astronomic and thermodynamic considerations, William Thomson, Lord Kelvin (1824–1907) attempted to calculate the age of the sun and the age of the earth by estimating the amount of time needed for them to cool. His results appeared in two works of 1862: *On the Age of the Sun's Heat* and *On the Secular Cooling of the Earth*. For the age of our planet, he arrived at an incredibly low figure compared to the values considered possible by those who based their calculations on the evolution of organisms. His conclusions were later disproved by radiometric studies carried out on naturally radioactive elements.

The chemical composition of sea water

Another very important contribution was made in 1865 by the Danish chemist Johan Georg Forchhammer (1794–1865) in the area of the chemistry of sea water. Forchhammer's work *On the Composition of Sea Water in the Different Parts of the Ocean* summarized and concluded investigations and research that had been going on for almost twenty years.

Forchammer analyzed a great number of samples of sea water from various parts of the oceans, indicating their various compositions quantitatively with respect to their principal components. He used the most modern methods known, some of which he himself perfected or introduced. By comparing the weight of the sum of the components to the weight of chlorine, Forchammer discovered that the ratio between the two values was constant, or nearly so, regardless of where the water came from and was equal to approximately 1.81. Among other things, this fundamental observation clarified the origin of marine sal-

inity and the extremely long permanency of at least some of its components.

Forchammer's and later Dittmar's contribution finally made it possible to formulate a definition of salinity (by Knudsen).

Although chemical determination of salinity has now been replaced by the electrical method, it is important to remember the fundamental contribution made by these particular researches to the development and progress of oceanography.

Progress in oceanographic instruments

Several interesting instruments were devised during this decade. In 1869, Charles Wheatstone (1802–75) and William Siemens invented the electrical resistance thermometer and suggested using it for oceanographic measurements. As a matter of fact, this is the type of device used in oceanography today; the electrical measurement of temperature makes continuous *in situ* measurements possible, favoring the digital acquisition and successive elaboration of data to such an extent that it has totally supplanted the use of reversing mercury thermometers, except in special cases. At the time, however, the instruments were not good enough for taking readings at sea; the extremely sensitive galvanometers installed in Wheatstone bridges were also too sensitive to the rocking of the ships, so that measurements were rendered unusable or impossible. Wheatstone and Siemens's suggestion remained in suspension for decades. Miller and Casella also proposed their conventional thermometer in 1869. A considerable improvement over the Six thermometer, their instrument was one of the major protagonists of the *Challenger* expedition.

Bathymetric explorations, by this time very refined, made researches into detail possible and began to yield information on the real shape of the ocean floor. In 1863, James Dwight Dana, working with weighted sounding lines, discovered the underwater Hudson canyon, possibly the first piece of physical evidence in the field of underwater geology.

Biological research and oceanographic expeditions

Extremely important among the oceanographic explorations of this period were the English expeditions (essentially for biological purposes) of the *Lightning* (1864) and the *Porcupine* (1869), led by two of the most important figures in oceanography of this second half of the ninteenth century Charles Wyville Thomson and William Benjamin Carpenter (1813–85). Not least among the various phases of research was the investigation of the curious case of the Bathybius. During the earlier voyage of the *Cyclops* (1857), numerous biological specimens had been collected and preserved in alcohol. While examining some of these, the biologist Huxley noted a gelatinous substance that, after prolonged studies, was defined as an organism and baptized "Bathybius haeckeli." A wide variety of speculations were made on the origin of this organism, including that of spontaneous generation, i.e. a passage from abiotic to biotic that it was postlated could occur in the depths of the ocean. During the expedition of the *Lightning*, sea water was treated with alcohol in the same way and the Bathybius again appeared. During the expedition of the *Challenger*, the chemist Buchanan, after various analyses, showed that the Bathybius was nothing more than a colloidal suspension caused by the action of alcohol on some of the mineral components ($CaSO_4$) of salinity.

But it was not only the Bathybius that inspired the two English expeditions of 1868–69; there was a great need to make detailed studies of biology, marine populations, and the corresponding physico-chemical conditions of the sea. For example, it still had to be ascertained whether or not azoic conditions existed below a certain depth, and new problems that arose with the developement of modern oceanography had to be solved.

John Gwyn Jeffreys also collaborated in the expeditions of 1868–69 together with Wyville Thompson and Carpenter. The ocean floor was dredged in search of benthonic organisms; it was again confirmed that deep-water temperatures could fall below 4°C; and the influence of polar water on deepwater temperatures was observed. It was noticed that the influence of atmospheric conditions and variations affected only the surface layers of the ocean. Carpenter also studied the circulation of the Mediterranean Sea and quite accurately described its mutual exchanges with the Atlantic Ocean. In addition, he pointed out the importance of the distribution of water density in circulation, significantly anticipating the basic concepts of cynamic calculations (developed around thirty years later). The need to use a better ship in order to continue the research and extend it to all the oceans of the world led to the planning of the *Challenger* expedition.

A weather map

Robert Fitzroy (1805–65), who in 1859 had been the first person in the world to issue daily meteorogical forecasts, began to draw a daily weather map in 1861. In 1863, he published *The Weather Book: A Manual of Practical Meteorology*, which can be considered a summary of his findings in the field of meteorology. The volume was extremely useful to navigation during that period, giving various practical rules for making rudimentary weather forcasts in order to avoid storms and hurricanes of the kind Fitzroy had encountered many times on board the *Beagle*, which he had commanded for a number of years.

APPLIED SCIENCES

Prime movers

In most large factories of the epoch, mechanical energy was provided by steam engines. Incorporating the inventions proposed by Fredrick E. Sickels (1819–95) and George H. Corliss (1817–88) in the 1840s, horizontal engines were improved through the use of valve and regulator mechanisms that increased their power and speed. In 1863, the Porter-Allen steam engine was built in Philadelphia; it produced 168 horsepower at 350 revolutions per minute. Boilers were also improved: the most commonly used model was the one designed by the Americans G.H. Babcock and Stephen Wilcox, which had straight, vertical water tubes with natural circulation.

The mechanization of small industry, however, required the use of generators that produced less power than was convenient for an economical use of steam, a situation that led to research on other types of engines. At the beginning of the decade, a solution seemed within reach: Étienne Lenoir (1822–1900) built an engine that operated on an explosive mixture of gas and air and was very similar to the horizontal double-acting steam engine. Two years later, in 1862, Alphonse Beau de Rochas (1815–91) obtained a patent for a four-stroke engine, an idea that was taken up five years later by Nikolaus Otto (1832–91),

a German engineer who designed a vertical engine with atmospheric exhaust. Although the idea came from de Rochas, the four-stroke cycle was called the "Otto cycle" from then on.

The real solution, however, at least for certain applications, was provided by electrotechnology, which underwent significant developments during this very period. These developments were the subject of a memoir presented to the Royal Society by Henry Wilde (1833–1919) in 1866 on "A new and powerful generator of Dynamic Electricity." The phenomenon of residual magnetism, however, was not yet fully understood; and Wilde spoke of new and paradoxical phenomena, supposedly deriving from Michael Faraday's discovery of electromagnetic induction. The Phenomenon was clearly perceived by Varley, who at the end of that same year filed a patent for a self-exciting electromagnetic generator. However, he postponed the description of his patent until the following July, after a memoir by E. Werner von Siemens (1816–92) on the conversion of mechanical energy into electrical energy without the use of permanent magnets had already appeared. These self-exciting generators were later known as dynamo electric generators, a term proposed in 1867 by Charles Brooke (1804–79), while permanent magnet generators continued to be called magneto-electrical generators.

Various other inventors, including Anyos Jedlik (1800–95), Moses G. Farmer (1820–93), and Antoine Pacinotti (1841–1912), also discovered the principle of self-excitation during these same years. In 1860, for didactical purposes, Pacinotti presented an engine with new features, including a ring armature. The first dynamo of practical dimensions able to produce a continuous current was invented by Zénobe Théaphile Gramme (1826–1901) in 1869. It had a multiple-ring armature after the principle used in Pacinotti's experimental engine.

Machines

The increased demand for steam engines and other machines stimulated the invention of methods that would make it possible to produce large numbers of precision parts more economically. The inventive process was particularly lively in the United States, where the Civil War had decimated the already scarce labor supply. The first real universal milling machine able to mill various types of springs and gear teeth was built by the Brown and Sharpe Company in 1862, followed a few years later by the first universal grinding machine and, in 1865, by an automatic lathe for mass-producing screws. The American Civil War also gave a good boost to the mechanization of agriculture. The American farmer already had access to a completely automatic harvester and a grain sower; the harvester combine and the sower combine, two machines that became widely diffused only in the United States, appeared slightly later.

Raw materials

The International Exhibition, where all the major countries presented their best products, opened in Paris on April 1, 1867. This time Great Britain, which had played the lion's part in previous exhibitions, had to settle for just a few prizes, while the great majority went to the continental countries, where industrial development was beginning to bear its first fruits. England maintained its supremacy in those sectors where empirical inventiveness still ruled. This was essentially true in the iron-working industry, thanks, as we have already seen, to Bessemer's work on using hot air to eliminate the carbon from pig iron, and the work of Fredrick Siemens and Edward A. Cowper on heat regeneration.

However, British supremacy was coming to an end in this sector as well. In 1861, the German (Charles) William Siemens (1823–83) obtained a patent for the use of gas in place of the solid fuel employed in Frederick Siemens's open hearth furnace. In 1864, when Pierre Martin (1824–1915) noticed that the addition of iron ore caused the elimination of carbon from first-casting pig iron in Siemens's furnace, he got the idea of using scrap iron in his furnace at Angoulême. In 1866, a contract was drawn up between Martin and Siemens providing the basis for the successive expansion of the Martin-Siemens process.

Thanks to the development of metallography (especially after the introduction of the metallographic microscope in 1864), the properties of steel became better understood and were substantially improved with the use of suitable binders and heat treatments. In 1865, Julius Baur obtained a patent for manufacturing chromium steel, while Robert Forester Mushet (1811–91) prepared tungsten-magnesium alloy steels, which had a high carbon content and were hardened by air cooling.

Despite the improvements in fusion techniques and the quality of steels, blast furnaces still retained their principal feature, the forehearth, which allowed access to the hearth in order to remove slag. In 1867, the German engineer Fritz W. Lürmann (1834–1919) closed the hearth, devising a water-cooled canal for extracting slag. In this way, better heat retention was achieved, and blast-furnaces could be made higher; the average height arose from 15 m to 22 m. It was American technicians who developed Lürmann's invention on an industrial scale.

In 1868, Dmitri Konstantinovich Chernov discovered the critical temperatures at which steel underwent phases of transformation, i.e. when the structural properties of the metal changed substantially as a result of heating and cooling in the solid state. Placed on a thermometric scale, these temperatures formed a series on points (Chernov points) denoting the polymorphic transformations of iron.

In short, the foundations of the new iron-working processes were consolidated and developed on a technical level during this decade, although the industry actually developed during the following decade.

As a matter of fact, world steel production still only slightly exceeded half a million tons, most of which was produced in Great Britain (in 1867, the United States contributed only 22,000 tons). Steel found an immediate use in the construction of boilers, where thinner plates could now be used; ship-building, on the other hand, had to await the end of the century for the advent of steel.

Progress in lamination did not keep up with other inventions in the field or with the increased demand for laminates. For the most part, old unused patents were exhumed. (For example, a 1798 patent for a continuous cogging mill was reobtained in 1862 by George Bedson (1820–84)).

Transport

The advent of the availability of steel coincided with a huge increase in world railroad construction, especially in Russia, Germamy, and France. This development only began to slow down in the last two countries ten years after the end of the Franco-Prussian War. Great Britain remained the major supplier of tracks for the world, though only a small percentage of them were made of steel; the rest were still made of puddled iron. In ship-building, soldered iron was only replaced at the end of the century. However, the first step in the direction of the change-over was made with the *Banshee* and the *Clytemnestra*, two steel steamers launched in England in 1862–63.

Locomotives were improved even further. The most important advance came in boiler-filling, an operation that was originally done by a piston pump and required the continuous movement of the machine, which was placed on a special track inside the stations. Injectors designed by Henri Giffard (1825–82) and tested around the middle of the century began to be used during this period, along with other boiler-filling devices. The four-wheeled bogie, a model still in use today, with an extended wheel-base and lateral motion controlled by inclined planes, was patented by William Adams (1829–1904) in 1860. In order to eliminate stops for refilling, John Ramsbottom devised a water tank in the tender for refilling the boiler (1860).

The development of the steam locomotive also stimulated interest in steam-operated road vehicles, an interest that had lagged for two decades. Around 1860, Thomas Rickett introduced his two-ton car with gear transmission. It was followed by other models, designed by Yarrow, Pattison, and H.P. Holt. These vehicles reached speeds of thirteen to thirty kilometers per hour.

These were the last years of the great sailing ships. From this decade this type of propulsion declined continuously, though it still comprised fifty percent of tonnage in 1886, and only fell to seven percent in 1914. Motor propulsion had affirmed itself, although there was still some disagreement on whether to use paddle wheels or a propeller (or propellers). In the end, propellers won out. Even the English Cunard Company, the most conservative of them all, had a 2,500 ton propeller-driven steamship, the *China*, built in 1862. Construction in wood was also abandoned; mixed construction was adopted at first, passing slowly to a construction completely in iron. The size of the largest steamships remained at around 3,000–3,500 tons maximum, after the finanical failure of the *Great Eastern* (19,000 tons), launched in 1858 (which did, however, serve to lay the first transatlantic telegraph cable).

One of the most important technological achievements in the area of transport during this decade was the opening of the Suez Canal in 1869. By shortening sailing time, over the course of a few years it made a decisive contribution to the development of international trade.

It should also be noted that the opening of the canal was accompanied by extremely interesting geological observations. Among the famous figures who attended the inauguration was the Viennese geologist Edward Suess (1831–1914). In 1864, a commission led by Suess had solved the problem of supplying water to the city of Vienna, eliminating the constant source of epidemics that lay in the use of water from the Danube.

Industrial chemistry

The industry of synthetic dyestuffs, which arose in England and France, soon spread to Germany, where it encountered more favorable enconomic conditions. Huge chemical industries, built specifically for producing dyestuffs, sprung up in Germany: the Badische Anilin und Soda Fabrik in Ludwigshafen (1861), the Bayer factory in Leverkusen (1863), and the Meister, Lucius & Bruning company, later renamed Farbwerke Hoechst, at Hoechst (1863). The Lepetit, Dollfuss, and Gensser (1868), later known as Ledoga and finally Lepetit, on the other hand, originally manufactured tanning products. The Duffield, Parke & Co., which later became Parke & Davis, was founded in Detroit in 1866.

The major new dyestuffs produced during this period are listed in the section on CHEMICAL SCIENCES 1860–1870. Even the production of soda, extremely important to the textile industry, was revolutionized by the new process patented by Ernsest Solvay (1838–1922) in 1861. In the course of just a few years, the Solvay process completely replaced the Leblanc process: Solvay production went from 30,000 tons in 1874 to 1,650,000 tons in 1900, compared to a fall in Leblanc production from 495,000 to 10,000 tons.

During these same years, the processes devised by Walter Weldon (1832–85) and Henry Deacon (1822–76) for producing chlorine from hydrochloric acid began to be used (in 1866 and 1867, respectively).

The first artificial plastic material, celluloid, was prepared by John Wesley Hyatt (1837–1920) in 1869.

Building techniques

A considerable increase in the use of iron in construction had taken place during the last decade; it culminated with a patent filed by the Frenchman Joseph Monier (1823–1906) for reinforced concrete. The first iron semi-beam cantilever bridge was built at Hassfurt, in Bavaria, in 1867 by H. Gerber (1832–92). That same year, James B. Eads (1820–87) began building the St. Louis bridge over the Mississippi, one of the first two-story iron arch bridges for both car and train traffic.

Military technology

In 1867, Alfred Nobel discovered a way to make fossil meal absorb nitroglycerine, producing a mixture that he called dynamite (just as powerful as nitroglycerine, but not as dangerous to use). The discovery of dynamite soon transformed the mining industry, opening up completely new and greater possibilities.

It was during this period that most guns went from front-loading to breech-loading, with the use of oblong cartridges. Steel began to be used in artillery construction. These innovations were put to trial during the American Civil War, a war that marked the end of non-armored warships. From this time on, the use of steel, armor, and large guns without recoil progessively increased. There were no succcessors, however, to the gigantic Columbiads, a type of mortar used during the American Civil War. One of these weapons, built by Thomas Jackson Rodman, had a caliber of 507 millimeters weighed fifty-two tons, and used a fifty kilogram powder charge to shoot a 490 kilogram sphere a distance of 4,000 meters. Much more important was the invention of the underwater torpedo (1867), a device propelled by its own motor that was used to arm light-weight ships and submarines, when they were perfected, threatening the safety of huge warships. The ten-barrel Gatling machine gun also deserves mention; ten barrels were arranged in a parallel bundle turned by hand, and the gun was capable of a firing cadence of 100 rounds per minute. However, within twenty years multiple-barrel weapons were replaced by single-barrel, rapid-fire guns.

1870–1880

THOUGHT AND SCIENTIFIC LIFE

Basic transformations in educational organization

The decade opened with the political unification of Germany by Prussia. Germany soon achieved supremacy in continental Europe not only on a technical and economic level, but culturally as well (in the course of two decades the University of Berlin became the model for European universities), in governmental organization, and in the battle for progressive laicization. Even in its first social initiatives, Bismarckian Germany (involved in the *Kulturkampf* from 1872 to 1878) was a mandatory, though often controversial, reference point for more advanced European socio-political theoreticians. The increased prestige of Germany corresponded to the diminished influence of France: After the fall of the Second Empire and the repression of the Paris Commune by the government of Louis Adolphe Thiers, the Third Republic was faced with grave political problems and substantial economical difficulties. For this entire decade and beyond, a feeling of *revanche* toward Germany developed among many French intellectuals, often posing an obstacle to scientific communication between the two countries. Nevertheless, the policy of alliances, skillfully directed by Otto von Bismarck, ushered in a period of peace that lasted for nearly fifty years and, in the long run, favored cultural exchanges and the diffusion of scientific information among European countries, in certain cases making collaboration among researchers of different nationalities possible. In the Euro-American sphere, the phase of economic expansion that began in the middle of the century and reached its height in 1872–73 was soon followed by a severe recession. The international volume of trade contracted, while huge monopolies and oligopolies based on a protectionist ideology arose. Social tensions, which had been alleviated by substantial improvements in the European market brought about by the importation of meat and grain from the United States, were fully felt as a result of the crisis and found a more conscious direction within socialist parties, which by this time had become the standard-bearers of a mature theorization of the class struggle. European industry found itself faced with competition from companies that were bigger than anything that had ever before existed and at the same time had to deal with the social consequences of rising unemployment. It was in this context that scientific laboratories were set up in the heart of large industries, with research directly financed by the industrialists. Perhaps the prototype of the specialized research laboratory was the Menlo Park, New Jersey laboratory of Thomas Alva Edison (1847–1931) (with more than 100 researchers). Electricity set the example of a field of experimentation that, in its successive phases—from electrical telegraphy (as early as 1837), to Edison's discoveries and up to the first great developments in applied electrochemistry and electro-metallurgy—led to applications that soon changed the structure of society. This was also true for other areas of the applied sciences and technology (for example, the growing importance of techniques for working steel) and was soon reflected in the organization of education and cultural exchange. It was not a coincidence that Werner von Siemens (1816–92), the co-founder, together with J.G. Halske, of the first German industry of applied electrotechnics, also planned and financed the Berlin physico-technical institute. During the next decade, from 1887 on, under the direction of von Helmholtz, this school operated simultaneously in the vanguard of scientific research and in precision technology. In a more general vein, one of the most characteristic features of research after 1870 was the opening of new schools of engineering at the university level in Great Britain and Germany, where public opinion had become extremely sensitized to their need.

New academies and universities

The world of academies and universities was also undergoing profound transformations. In Rome, which by this time was the capital of Italy (1870), the Pontifica Academica dei Nuovi Lincei withdrew to the Vatican where it became the Pontifica Academia delle Scienze "Nuovi Lincei," while the Reale Accademia dei Lincei, arranged in two classes (one for the physical, mathematical, and natural sciences and the other for the moral, historical, and philosophical sciences), was founded in 1875.

In 1872, France also created an Association Française pour l'Avancement des Sciences. The Argentine Academy of the Sciences was inaugurated in Cordova in 1873, while the Imperial Academy of Japan in Tokyo and the Rumanian Academy of Bucharest were both founded in 1870. A Hungarian university was established at Cluj and a German university at Chernovtsy in 1878, both of which passed to Rumania in 1919. The University of Zagreb was established in 1879; that same year, in Switzerland, the University of Geneva was born out of the center for Calvinist studies.

Positivism and anti-positivism in France

Within the profound renewal of European and American scientific life just described, where tendencies toward specialization, an interest in the applicability of scientific theories to industrial production, and the need to popularize the great research programs arose, often in conflict with each other, the basic cultural point of reference was positivism in all its various shades. In many countries it was the standpoint from which the results obtained by science and technology were viewed— the form of philosophy to which the researcher, the popularizer, and the student all looked. It was not a hegemony without contrasts, however: the scientists often turned their backs on positivism when they developed programs that did not agree with the official science, and intellectual élites—scholars and artists in particular—often opposed positivism and returned to issues that it had tended to relegate to the margins of cultural life. In addition, especially in Catholic countries, the exponents of traditional religion displayed a hostility towards positivism, combatting it in the area of education.

French culture was especially varied. Hippolyte Taine's *Sur l'intelligence*, written in 1870, was extremely popular, and soon came to be considered one of the key texts of positivism.

In 1875, Taine began publication of a large five-volume history entitled *Origines de la France contemporaine* (1875–93), a work dedicated to the birth and development of the French Revolution

up to 1808. The author's portrait of the period is bitterly negative, such as would have pleased the bourgeoisie of the Third Republic, still fearful of the Commune and eager to emphasize the complete break between new Republican France and the institutions that arose during the Revolution.

The most effective barrier against the spread of positivism was spiritualism. This current, which had never completely died out after Biran and Cousin, had lost most of its earlier influence; now, however, it displayed new vitality, especially as a result of the work of certain authors, who soon achieved great fame: F. Ravisson-Mollien (1813–1900), for several years a professor at the Sorbonne, J. Lachelier (1823–99), a professor at the École Normale Supérieure, and the Swiss scholar Charles Secrétan (1815–95), a professor at the University of Lausanne but tied to French philosophical circles. The scientist Antoine Cournot (1801–77) was also strongly influenced by spiritualist philosophy, as can be seen in many of his works, for example in *Matérialisme, vitalisme, rationalisme; études sur l'emploi des données de la science en philosophie* (1875).

More complex was the position of Charles Bernard Renouvier (1815–1903) and Étienne Boutroux (1845–1921), who attempted on the one hand to apply positivism itself, and on the other to go beyond it in the name of the scientific spirit. The first tendency resulted in a sort of Kantianism; the second in a "contingentism" that opened the way to metaphysics and religion. Boutroux's most important work was his thesis for graduation from the Sorbonne (Ravaisson was his mentor), which bore the significant title *De la contingence des lois de la nature* (1874). It became an extremely popular work, referred to more or less directly by all those who wished to oppose determinism in order to preserve freedom.

The refinement of German positivism

In Germany, positivism underwent a process of refinement that accentuated its nature as an experimental philosophy. Two names deserve particular mention: those of Ernst Laas (1837–85) and Richard Avenarius (1843–96). In 1879, Laas began publishing a three volume work entitled *Idealismus und Positivismus (1879–84)*, which proposed to reduce the entire development of philosophy to a contest between these two currents.

Avenarius sent his famous work *Philosophie als Denken der Welt gemäss dem Prinzip des kleinsten Kraftmasses. Proloegomena zu einer Kritik der reinen Erfahrung* to press in 1877. The empirocriticism outlined in it was later developed and explored in detail in Avenarius's major work, *Kritik der reinen Erfahrung*, published in two volumes in 1888–90.

Both Lotze's spiritualism, which became more authoritative, and Kantianism, defended by a new school founded at Marburg, which later played a leading role in the history of German philosophical thought, rose up against positivism. The founder of the Marburg school was Hermann Cohen (1848–1918), who in 1871 published a *Kants Theorie der reinen Erfahrung*, and in 1877 a *Kants Begründung der Ethik*.

Marx and Engels

Karl Marx had written his *Grundrisse der Kritik der politischen Ökonomie* in 1857–58; the first volume of *Das Kapital*, the work to which Marx, actively involved in political struggle and the debate on socialism, dedicated a great part of his later life, came out in 1867 (the second and third volumes were published after Marx's death by Engels in 1885 and 1895, respectively). The work represented not only a point of reference for the workers' movement, but a fundamental step in the advancement of economic science and the elaboration of scientific socialism.

In 1873, Friedrich Engels began writing the *Naturdialektik*, a work that remained unfinished and has only recently been published in its incomplete form. In 1877–78 he published a work against the philosophy of Karl Eugen Dühring (1833–1921) that became fairly popular among the socialist ranks. Known as the *Anti-Dühring*, it was intended as a harsh critique of both dogmatism and agnosticism and aimed at the creation of a new form of materialism that would be more aware, on a philosophical level, than eighteenth–century mechanistic materialism or the "vulgar materialism" in vogue in Germany at the time. Engels developed this concept even further in his *Ludwig Feuerbach und der Ausgang der klassischen deutschen Philosophie* of 1886, which contains, among other things, a favorable evaluation of Joseph Dietzgen's work on the "nature of human brain work," which appeared in 1869. With these works, Engels established a fairly compact *corpus* (despite the fragmentariness of the text) of critical reflection on the achievements of the science of his times, reviewed in the light of a materialism that attempted to avoid the apriorism of which the scientists were accusing the Hegelian philosophy of nature. Familiar with contemporary scientific research (cell theory, thermodynamics, etc.), Engels insisted on the need for a theorization that could comprehend the significance of these programs, contending the field with those "philosophies of the scientists" that were moving from a worn-out positivism toward a revival of spiritualistic themes.

The rebirth of Hegelianism in England

Charles Darwin's *The Descent of Man*, which dealt with a very delicate subject (suffice it to recall the debate that raged during the last decade after Thomas Henry Huxley's *Man's Place in Nature* was published in 1863), was published in 1871 and immediately became the focus of harsh attacks, especially from the representatives of traditional religion. On the other hand, Herbert Spencer's moderate evolutionism, which even left room for a transcendent reality with its doctrine of the "unknowable," became especially popular among people of average learning and represented a compromise between the two views.

University circles became progressively more dominated by Hegelianism. The *Introduction to Hume's Treatise of Human Nature*, written by Thomas Hill Green (1836–82), who opposed Humian philosophy with an idealistic notion of reality, came out in 1874–75. The *Ethical Studies* of Francis Herbert Bradley (1846–1924), who soon became the most authoritative representative of English Hegelianism, was published in 1876.

In the United States, a subtle thinker, Charles Sanders Peirce (1839–1914), founded the pragmatist movement, which exerted a widespread influence even in Europe during the following years. The first formulation of Peirce's program can be found in two fundamental works published in *Popular Science Monthly* in 1877–78: "The Fixation of Belief" and "How to make Our Ideas Clear."

Positivism in Italy

The spread of positivism in Italy was facilitated by the support given to it by the democratic faction of the bourgeoisie, which

wanted finally to free itself from the domination of clerical culture. Rosmini and Gioberti were accused of having been papists, and of having assigned "philosophy the job of waiting on theology." Ardigò's thought was exalted as best responding to "the needs of present-day society." Ardigò published several works that openly adhered to positivist ideas: *La psicologia come scienza positiva* (1870), *La formazione naturale del fatto del sistema solare* (1877), and *La morale dei positivisti* (1878). The positivist atmosphere penetrated into a number of universities, especially in medicine. Cesare Lombroso (1836–1909) published various works that made him famous throughout Europe, such as his *Genio e follia* (1864) and *L'uomo delinquente* (1876). Several scholars of Hegelian formation, such as the physician Salvatore Tommasi (1813–88), also took a position of frank openness toward scientific philosophy.

MATHEMATICAL SCIENCES

The first systemizations of group theory

Group theory and topology, following the significant researches mentioned earlier, now began to take on a systematic form with notable developments. In 1870, Camille Jordan (1838–1922) investigated group theory, studying finite groups of movements and substitutions in his *Traité des substitutions et des équations algébriques*. He also developed Évariste Galois's clever anticipations, which up to then had, in many respects, remained obscure and not fully appreciated. The rigorous concept of transformation groups was then established in 1873 by Marius Sophus Lie (1842–99), who demonstrated its importance and varied applications in geometry. (He had already established the principle of transformation, with which he was able to reconduct the geometry of systems of spheres back to that of systems of straight lines, in 1870.) For this entire decade, he continued his researches on those particular groups, which were later named after him, in isolation.

Felix Klein's "Erlangen program"

The most celebrated application of these concepts and methodologists was the "Erlangen program" (1872), in which Felix Klein (1849–1925) classified the various geometries, based on the properties of figures that remained invariant with respect to certain groups of transformations of space itself. This program was of extreme historical and cultural importance because of the generality and abstractness of the unifying criteria proposed for the classification of geometrical theories, which no longer contained any reference to their more or less presumed objects. These criteria were based on the cardinal concept of transformation groups, in the sense that, according to Klein, every geometry can be identified on the basis of the transformation group that is characteristic of it. Thus traditional elementary geometry was identified as the study of the properties of figures invariant with respect to the congruence group, and similar geometry was defined as the study of invariants of the group of similitudes (broader than the first); like geometry and projective geometry were also characterized in the same way. Not only that, a given geometry could also, in a certain sense, be included in another of a more general nature. For example, the group that characterized elementary geometry could be considered a subgroup of projective transformations that left a certain figure, called an absolute, invaried. The right choice of the absolute could lead to a Euclidean geometry, or to one of the two non-Euclidean geometries. It is obvious that a great deal of light was shed on the age-old question of the possibility of different geometries and their mutual admissibility by this type of perspective. We can also see that, in the order of ideas put forth here, topology also became better defined. It could now be conceived of as the study of the geometrical properties that remained invariant with respect to the group of continuous transformations to which the figures could be subjected (later called the homeomorphism group). This discipline, however, found the real beginnings of its modern direction in a memoir by Enrico Betti of 1871 on spaces of any number of dimensions, a topic that Henri Poincaré returned to thirty years later. Betti had already achieved important results in the theory of elliptic functions and had made contributions to the mathematical theory of elasticity several years earlier.

Definitions of real numbers and the foundation of real analysis

Meanwhile, in their search for increasingly more rigorous formulations that did not resort to geometrical and physical intuitions, analytic mathematicians began to realize that the typical object of analysis was real numbers. Attempts began to be made to construct them, at first starting from rational numbers, then finally from natural numbers (such extensions of numerical range had particularly been studied during the previous decade, mainly in relation to the need to provide a rigorous justification for the theory of complex and hypercomplex numbers, such as quaternions). The most exhaustive systemization of these efforts was given in a work by Hermann Hankel (1839–73) entitled *Theorie der complexen Zahlensysteme* (1867). It included the well-known principle of permanency of the formal properties of the laws of calculus. Thus it happened that in the same year, 1872, various definitions of real numbers were proposed: Julius Wilhelm Dedekind (1831–1916) in his *Stetigkeit und irrationale Zahlen*, provided his classical definition of real numbers as sections of the rational field; Georg Cantor (1845–1910), in a memoir on the extension of a theorem that he had established on the univocacy of the representation of a function of a real variable by means of a trigonometric series, presented his definition of real numbers as limits of convergent successions of rational numbers (a definition that was also given simultaneously by Charles Méray (1835–1911)); and E. Kossak, in one of his works, expounded Karl Weirstrass's theory of real numbers, which were defined by means of opportune series of rational powers. This very possibility of defining real numbers starting from natural numbers explains the idea of presenting analysis as the pure development of the theory of numbers, a program that was especially embraced by Weierstrass's school and by Leopold Kronecker (1823–91). It should be mentioned that, within the context of this common aspiration, profound differences divided scholars on how it should be carried out. Weierstrass, Cantor, Dedekind, and others more or less explicitly used the concept of any function or set to reduce real numbers to natural numbers, while Kronecker and researchers inspired by him actually proposed avoiding resorting to any such concept, and, in fact, even attempted to rid mathematics of these concepts altogether, proceding in a rigorous way to rebuild mathematics starting from the natural numbers, without using any more complex

logical tools than those needed to build arithmetic. The deeper the meaning of this contrast became apparent in the twentieth century when similar conflicting ideas were advanced within the debate on the foundations of mathematics.

All of European mathematics participated to some extent in this work of critically revising the foundations of analysis. One of the most important works presenting the results of this effort in the *Fondamenti per la teoria delle funzioni di variabili reali*, published by Ulisse Dini (1845–1918) in 1878.

This work of critical refinement did not, however, keep remarkable advancements of a purely technical nature from being made in analysis. In 1876, for example, Charles Hermite (1822–1901), who had already distinguished himself as an exceptional scholar of elliptic functions by finding the solution to the general quintic, demonstrated the transcendency of the number *e* (the base of natural logarithms) by applying his own ingenious researches on continuous algebraic functions and the theory of analytical functions of a complex variable. This led to the demonstration, in 1822, of the transcendency of π by Ferdinand Lindemann (1852–1939), bringing to an end the age-old problem of whether or not the circle could be squared with a ruler and compass. During this period, Hermite also began publishing important memoirs on elliptic functions and their applications. Various classes of special functions, such as spherical functions, the subject of an important work of 1878 by Franz Ernst Neumann (1798–1895), *Beiträge zu Theorie der Kugelfunktionen*, were also studied in depth (even in relation to physical problems).

The following year, Charles Emile Picard (1856–1941) demonstrated a theory, later named after him, on the behavior of a uniform analytical function around an isolated singular point. During these same years, Picard also made important studies of linear differential equations with periodic coefficients.

In addition, we must not fail to mention that the last three decades of the century were dominated by the scientific studies of Weierstrass, which were totally dedicated to analysis. Weierstrass's work, however, was not limited to the critical task of reconstruction mentioned above, but was also characterized by a large number of specific discoveries, often of exceptional value, such as in the theory of the functions of complex variables. Weierstrass's influence was also widely felt in Europe, as numerous mathematicians sought to spend a part of their early education with him in Berlin. Contributions to the theory of the functions of real variables founded by Weierstrauss were made by Georg Cantor, Heinrich Eduard Heine (1821–81), and Paul Du Bois-Reymond (1831–89) in Germany, Charles Hermite, Camille Jordan, and Jean Gaston Darboux in France, and Ulisse Dini, Guilio Ascoli (1843–96), Cesare Arzelà (1847–1912), and Giuseppe Peano (1853–1932) in Italy.

The Cantorian theory of sets

One of the things that makes this decade particularly important in the history of mathematics is the fact that at this time Georg Cantor created the theory of sets. Prepared by a number of preliminary memoirs (including the important memoir of 1873 in which Cantor defined the concepts of power of a set and numerable set), the theory was systematically expounded in a series of six memoirs that appeared in the *Mathematische Annalen* between 1879 and 1884 under the title "Über unendliche lineare Punktmannigfaltigkeiten." Though genetically linked to fairly contemporary researches in analysis, this series of writings immediately revealed an originality and audacity of new perspectives, such as the admission into mathematics of actual infinity (i.e. the acceptability of considering a totality of infinite elements as given). Despite the authoritative consent it received (from Weierstrauss, for example), these very innovations created a general climate of incomprehensiveness and hostility toward the new theory (mainly led by Kronecker), keeping it from being accepted for quite a long time.

Frege's work in logic

The same year in which Cantor began his memorable series of works (1879), (Friedrich Ludwig) Gottlob Frege (1848–1925), another great scientist who had also long remained unknown and not understood published his *Begriffsschrift*, a fundamental work on mathematical logic that, with unprecedented rigor, proposed the idea of perfect formal deduction, ensured by the use of an unexceptionable symbolic tool. In this way mathematical logic, which for several decades had been considered merely a branch of mathematics, was assigned the responsibility of explicating and controlling the rigor with which mathematics itself was constructed and, in fact, of providing the proper instruments for this sort of rigorous construction. The creation of a detailed symbolism (and, in general, of a number of technical tools that modern mathematical logic owes to him) was not, therefore, an end in itself for Frege. In fact, he had in mind a grandiose program in which logic would provide not only the proper tool for building mathematical theories, but the very foundation of mathematics itself, in the sense that mathematics had to be obtainable starting from pure logic. In this way, the reductionistic program that, as we have seen, had gone so far as to reduce analysis to the theory of natural numbers, now received further impetus from the proposal to reduce the very concept of natural numbers to purely logical notions. First arithmetic, then all of mathematics, could thus be conceived of as branches of logic. Frege provided an initial outline of this logistic program in his *Die Grundlagen der Arithmetik* (1884). He then developed it extensively in the two volumes of his *Die Grundgesetze der Arithmetik,* which came out between 1893 and 1903.

ASTRONOMY

The Society of Italian Spectroscopists

The enormous growth of astrophysical researches and the remarkable activity of the sun in 1870, which was closely watched by a great many of the astronomers of the new generation, gave Angelo Secchi and Pietro Tacchini (1838–1905) the idea of founding a society for coordinating the researches into solar phenomena and celestial physics of observatories all over the world and publishing their results. Thus the society of Italian Spectroscopists (so named because the spectroscope was the instrument most used in these researches), the first institution of its kind in the world, was born in 1871. In 1920, it was restructured and became the present-day Società Astronomica Italiana.

The canals of Mars

A series of events soon drew the attention of astronomers and the general public to Mars and a particular chapter of planetary astronomy. Following the observations of Wilhelm Beer (1797–1850) and Johann Heinrich von Mädler (1794–1874), who together drew the first map of Mars, other observations were made, especially during oppositions, and other maps were drawn by Kaiser, Proctor, and Flammarion. In 1877, during an important opposition Giovanni Virginio Schaparelli made a large number of observations with the twenty-two-centimeter refractor of the observatory of Brera, and for the first time drew an accurate map. It was no longer based on sight drawings, as past maps had been, but on a special triangulation, obtained by connecting sixty-two fundamental points chosen on the most visible and stable spots. The famous "canals," thin dark lines of sorts that crossed the lighter areas and connected the darker ones, were discovered during these observations. In 1879, Schaparelli drew a new, even more detailed map, and in 1881–82 discovered the doubling (gemination) of many of the thin lines. The question of the canals, the source of many debates and speculations that were only resolved negatively by space probe operations almost a century later increased the public's interest in astronomy even more, strengthening a theory that had been dear to human imagination ever since antiquity: the possibility that other worlds, populated by intelligent beings, existed in the heavens. Schiaparelli, whose methodology served as an example, also provided the nomenclature that became universally adopted.

Another important discovery was made by Asaph Hall during the opposition of 1877. While observing the red planet on the nights of 11 and 12 August through the sixty-six-centimeter refractor of the Naval Observatory in Washington, Hall discovered two satellites, later named Phobos and Deimos.

Schmidt's map of the moon

A new map of the moon, larger and more detailed than any of the previous ones, was published in 1878. It was the product of thousands of observations made by Johann Friedrich Schmidt (1825–84) over a period of more than forty years, using various instruments. Schmidt's map was almost two meters in diameter and showed almost 33,000 craters. The work long remained unsurpassed, and from this time on, attempts began to be made to compile lunar atlases using photographic methods.

New observatories

This period saw the birth of two important new observatories, the Observatory of Potsdam in 1876–79, and the Lick Observatory, built on Mt. Hamilton at the initiative of the philanthropist James Lick. It was at this second observatory that the largest refractor in the world was built in 1888 (surpassed by the Yerkes refractor in 1896); its benefactor is buried under its pillar. The Lick Observatory soon became one of the most productive in the world—espceially in stellar astronomy—a role that it maintained throughout the entire next century. This was the first of the great Californian observatories.

A project for modernizing the Moscow Observatory was begun in 1873. Under the new direction of the astronomer-physicist Fedor Aleksandrovich Bredikhin (1831–1904), photographic and spectro-heliographic observations, as well as researches on the physical nature of the sun, comets, and meteorite showers, were conducted there. Bredikhin founded the Russian school of astrophysics, which over the years counted among its numbers such prominent figures as Vitold Karlovich Tserasky, and Sergey Konstantinovich Kostinski.

Bredikhin and the study of comets

Bredikhin began his research on comets, in 1861 and continued it until his death. On the basis of available observations, he elaborated the mechanical theory of the shape of comets, the foundations of which had been laid by Leonhard Euler and Friedrich Wilhelm Bessel. In 1877–79, to complete this theory Bredikhin published the physical theory of types of comet tails and the theory of the passage of comets and meteorite showers, which he based on the works of Giovanni Virginio Schiaparelli.

Silver bromide plates

In 1879, a new area in observation was opened up by a non-astronomical technical innovation: the introduction of dry photographic plates that used silver bromide. The higher speed of the new material and the fact that a relatively long period of time could elapse between its preparation and its use led to the rapid diffusion of the use of photographic plates as a means of observation. Henry Draper immediately photographed around fifty spectra of bright stars (his previous attempt at photographing Vega, the brightest star in the boreal skies, using colloid plates in 1872, had also been successful). In 1880, Draper photographed the nebula of Orion, the first nebula to ever be photographed; a year later he succeeded in photographing its spectrum as well. Photographs of comets and their spectra, much better than those obtained from 1858 on using wet plates, were also made in 1881. Finally in 1882 Draper began taking photographs of stellar spectra with a prism lens, already used by Lorenzo Respighi and Angelo Secchi for visual observations made in Rome in 1869. The application of this method of photography proved extremely valuable in the compilation of the first general catalogue of stellar spectrum types, published in 1890 by the Harvard Observatory. The catalogue was based on material gathered by Draper, who died in 1882, and is called the *Henry Draper Catalogue*.

PHYSICAL SCIENCES: MECHANICS

The principle of cause

During this decade, studies in mechanics proceded along the lines already set forth during the 1860s. This led, on one hand, to a progressive clarification of the fundamental notions and logical internal correlations of the theory of motion, and, on the other, to an intensification of the debate on the role of mechanics in the physical sciences as a whole. This second aspect clearly reflected the influence of the eclectic philosophical ideas with which many physicists attempted to find an area of reconciliation between science and religion. The intricate debate that took place both in England and in continental Europe on the relations between the principles of the conservation of mechanical energy and the continuity of nature was a product of this tendency. In a series of conferences held between 1868 and 1874, John Tyndall (1820–93) stated that the principle of conservation of energy implied an extremely general causal connection that took place everywhere in nature, and he attempted to demonstrate that, as a consequence, life may have originated from inanimate matter.

In any case, the principle of continuity to which Tyndall referred lent itself, as Baden Powell had already suggested in 1860, to being interpreted as limiting the field of scientific investigation and validating theological investigations into the origin of the universe. Tyndall's ideas were openly disputed by Balfour Stewart (1828–87) and Peter Guthrie Tait (1831–1901), authors of *The Unseen Universe: or Physical Speculations on a Future State*, a harsh antimaterialist polemic that was published in 1875. Knowledge about mechanical energy showed that the observable universe was moving toward an inevitable heat death. On the basis of the principle of continuity and the principle of conservation of energy, Stewart and Tait, in order to preserve the theory of immortality, proposed imagining that another, unobservable universe existed that was bound to the first by energy exchanges. In this exchange of mechanical energy, the law of conservation, which even the work of God obeyed out of rigorous respect for the casual correlation among events, was followed. Alexander Bain expressed fairly similar ideas with respect to the connection between the principles of uniformity and causality in his two-volume *Logic* of 1870. To the extent that this connection was seen as essentially founded on the principle of mechanics, it exerted considerable influence as a critical tool that was used against the deductions of statistical mechanics. These deductions, in fact, were attacked increasingly more often as lacking in rigor and were counterposed by the claim that the dynamic method was absolutely valid (see PHYSICAL SCIENCES: THERMODYNAMICS and STATISTICAL MECHANICS, 1870–1880, 1880–1890, and 1890–1900).

On the fundamental notions of mechanics

The illusiveness of the attempts to reduce all natural phenomena to mechanics was in Ernst Mach's mind when he wrote his *Die Geschichte und die Wurzel des Satzes von der Erhaltung der Arbeit*, a historico-critical work on mechanics in relation to the history and roots of the principle on the conservation of work, which was published in Prague in 1872. Mach again confronted these same problems in his great work of 1883, when his ideas were more fully matured (see PHYSICAL SCIENCES: MECHANICS, 1880–1890). Interesting considerations on space and time as basic categories of the theory of motion can be found in James Clerk Maxwell's *Matter and Motion*, published in 1876. In it Maxwell observed that these categories were not absolute, and wrote: "But as there is nothing to distinguish one portion of time from another except the different events which occur in them, so there is nothing to distinguish one part of space from another except its relation to the place of material bodies. We cannot describe the time of an event except by reference to some other event, or the place of a body except by reference to some other event, or the place of a body except by reference to some other body. All our knowledge, both of time and place, is essentially relative."

Kirchoff's lessons on mathematics, physics, and mechanics (*Vorlesungen über Mathematik, Physik und Mechanik*), which contained several basic indications for a logical reorganization of mechanics, appeared in 1874 and 1876. The most characteristic features of Kirchoff's studies were a restructuring of mechanics on the basis of a need for completeness and simplicity and a critique of any attempt to assign the science of motion the task of identifying the primary causes of motion itself. Mechanics, in short, was a purely descriptive science and should be based exclusively on primitive notions—space, time, and matter—without presuming to ''explain'' forces. Mach's and Heinrich Rudolf Hertz's analyses both connected up in different ways, with Kirchoff's program.

PHYSICAL SCIENCES: THERMODYNAMICS AND STATISTICAL MECHANICS

The interpretation of the second principle and Maxwell's "demon"

Numerous memoirs explicitly dealing with the various interpretations of the second principle were written in 1871. That same year, James Clerk Maxwell published his *Theory of Heat*. Unlike many English physicists, Maxwell did not believe that the new ideas introduced by Rudolf Clausius constituted an obstacle to the progress of thermodynamics. However, he did put forth a very particular interpretation of the second principle, which was based on the operative possibility of a creature that could distinguish the motion of individual molecules and therefore create the physical situations that validated the second principle. This creature, which other physicists immediately baptized ''Maxwell's Demon,'' made choices that were not limited, as they were with human beings, by the fact that interaction with matter necessarily takes place at a macroscopic level. ''For we have seen,'' wrote Maxwell, ''that the molecules, in a vessel full of air at uniform temperature are moving with velocities by no means uniform, though the mean velocity of any great number of them, arbitrarily selected, is almost exactly uniform. Now let us suppose that such a vessel is divided into two portions A and B, by a division in which there is a small hole, and that a being who can see the individual molecules, opens and closes this hole, so as to allow only the swifter molecules to pass from A to B, and only the slower ones to pass from B to A. He will thus, without expenditure of work, raise the temperature of B and lower that of A, in contradiction to the second law of thermodynamics.'' The basic question, according to Maxwell, was that experimental observations never regarded the perception of the motions of single molecules: it followed that, as he himself emphasized, ''we are compelled to adopt what I have described as the statistical methods of calculation and to abandon the strict dynamical method, in which we follow every motion by the calculus.'' An interesting debate concerning the determination obtainable using statistical methods immediately arose around this renunciation.

The rationalization of the kinetic theory

Rudolf Clausius and Ludwig Boltzmann also sent several

important communications on the second principle to press in 1871, distinguishing their position from the positions taken by Maxwell, and even more explicitly, by Tait, Kelvin and the other scientists who opposed mechanico-probabilistic models. Returning to several ideas developed by Boltzmann, Clausius analyzed the correlations between the disgregation function and the irregularity of molecular motion, developing his virial theorem. Boltzmann published three memoirs on the kinetic theory of gasses in which he upheld the need to abandon the official form of the equations that appear in the theory of heat and proceded towards a radical mathematical revision of the essential notions of molecular models. Boltzmann's rationalization of the models involved a formal redefinition of the terms ''state,'' and ''entropy.'' With these works, Boltzmann laid the foundations for the development of an explicitly probabilistic program that soon led him to formulate his first statements of the famous H theorem. Toward the end of 1871, Clausius's and Boltzmann's analyses were interpreted by C. Szily as mere attempts to reduce thermodynamics to mechanics. Szily suggested completing these researches by making a detailed examination of the relations between the second principle and Hamilton's principle. He concluded that the second proposition of the theory of heat was reduced to a universal principle of dynamics. It is interesting to note that in a memoir entitled *On the Connections Between the Second Proposition of the Mechanical Theory of Heat and Hamilton's Principle* (1871), Clausius answered Szily, criticizing his lack of knowledge of the ''special considerations'' needed to analyze the different relations existing between mechanics and thermodynamics.

In 1872, an important debate arose between Clausius and Tait. According to Tait, the notions introduced by Clausius, such as disgregation, were nothing more than obstacles to the further development of thermodynamics. Clausius replied that, on the contrary, the new notions had ''a universal mechanical meaning,'' a conclusion also drawn in Boltzmann's studies. Nevertheless, the debate continued over the years, becoming even more intense and involving Boltzmann as well. It gradually made the mistrust that had been stirred up by the notion of entropy and its meaning in probabilistic terms more explicit.

Probabilistic developments in Boltzmann's program

A fundamental monograph by Ludwig Boltzmann on new studies in the thermic equilibrium of gas molecules (*Weitere Studien über das Wärmegleichgewicht unter Gasmolekülen*) appeared in 1872. In it, Boltzmann defended the rigor used in applying the calculus of probabilities to physics, maintaining that physics consisted in the necessary consequences of determined premises and was thus totally free of indeterminateness. Thus the calculus of probabilities, according to Boltzmann, made it possible to ''take possession of an exact theory'' that was more complete than the researches of James Clerk Maxwell and other scholars of thermodynamics. First, Boltzmann demonstrated that, if the distribution $f(x,t)$ of molecular energy x at time t was Maxwellian, then that distribution would remain unchanged in time, since it obeyed the relation:

$$\frac{\partial f(x,t)}{\partial t} = 0.$$

Secondly, a quantity E, defined as follows, could be introduced:

$$E = \int_0^\infty f(x,t) \left\{ \log \left[\frac{f(x,t)}{\sqrt{x}} - 1 \right] \right\} dx.$$

This quantity was characterized by the fact that it never increased with variations in the conditions of the gas. E tended toward a minimum value, with dE/dt 0 for a Maxwellian distribution of f(x,t). Boltzmann concluded that this was a rigorous proof that, no matter what the distribution of the living forces at the beginning point, over a very long period of time this distribution would always and necessarily tend toward the one found by Maxwell. For this reason, according to Boltzmann, there always existed a given quantity that, after atomic motion had taken place, could not increase; the problem was to identify the relations between this quantity and entropy.

In discussing the properties of E, Boltzmann used a model that defined kinetic molecular energy as discreet energy, according to the hypothesis that molecules are not able to accept a continuous series of living forces, but only those living forces that are multiples of a given quantity. Boltzmann used this radical mathematical expedient to analyze a model for gasses where the molecules had energies with values that were whole multiples of an elementary quantity ε—(ε, 2ε, . . .,$p\varepsilon$)—, and redefined the quantity E as the following summation:

$$E = u_1 \log u_1 + \sqrt{2} u_2 \log u_2 + \ldots + \sqrt{p} u_p \log u_p,$$

where the u_i's are functions of the numbers $N_{e\lambda}^{ik}$ of collisions per unit volume in very small intervals of time between molecules with discrete energies $i\varepsilon$, $k\varepsilon$, $l\varepsilon$, and $\lambda\varepsilon$.

The quantity E thus redefined "must constantly decrease," and its properties coincide with those of the extreme case in which ε is "infinitely small" and p is "infinitely large." A complete analogy was thus reached between models with continuous energy and models with discrete energy.

In 1877, Boltzmann again took up these revolutionary mathematical rationalizations, generalizing them even further, in a monograph on the relation between the second principle of the mechanical theory of heat and the calculus of probabilities (über die Beziehung zwischen dem zweiten Haupsatz der mechanischen Wärmetheorie und der Wahrscheinlichkeitsrechnung respektive den Sätzen das Wärmegleichgewicht). In this monograph, to which Max Planck referred in an article written in 1900 on the constant h, Boltzmann correlated entropy to a probabilistic interpretation based on "complexions" and particularly emphasized the connection that existed between the application of probabilistic inferences to physical problems and the finite concept of mathematics, to which he adhered.

Reversibility and irreversibility

However, these developments in Boltzmann's program coincided with a progressive isolation surrounding his ideas, which were being attacked from all sides, for a number of reasons.

As early as 1874, in Nature magazine, the influential Kelvin published his "Kinetic Theory of the Dissipation of Energy," in which he discussed the situation of the logical relations between mechanics and the second principle of thermodynamics. As Kelvin wrote: "In abstract dynamics an instantaneous reversal of the motion of every moving particle of a system causes the system to move backwards, each particle of it along its old path, and at the same speed as before when again in the same position—that is to say, in mathematical language, any solution remains a solution when t is changed into −t. In physical dynamics, this simple and perfect reversibility fails on account of forces depending on friction of solids." This gave rise to the problem of the irreconcilability of purely mechanical discussions of gasses, which implied reversibility, with statistical discussions, which claimed to provide an explanation of irreversible phenomena.

Joseph Willard Gibbs (1839–1903) and Samuel Hawksley Burbury (1831–1911) both intervened on the problem of irreversibility, the first with a work entitled On the Equilibrium of Heterogenous Substances (1876), and the second with a communication entitled Equilibrium of Temperature in a Vertical Column of Gas (1875). Joseph Loschmidt (1821–1895) took advantage of the questions raised by the reversal of molecular velocities to present a critique of the generalizations developed by Boltzmann and Maxwell regarding the equipartition of energy and of Maxwell's distributon function. Between 1876 and 1877, Loschmidt, who in 1855 had determined the number of molecules in one cubic centimeter of a gas at zero degrees temperature and one atmosphere pressure, published four memoirs under the same title (Über den Zustand des Wärmegleichgewichtes eines System von Köpern mit Rücksicht auf die Schwerkraft) in which he discussed the need to reduce considerably the very possibility of applying the calculus of probabilities to the kinetic theory of gasses. According to Loschmidt, Boltzmann's general theory was, in fact, debatable, since in his opinion Boltzmann used hypotheses that could not be verified except by acritically using probabilistic inferences. Loschmidt stated that, given a conservative mechanical system, each of its transformations in which the quantity E decreased from E_1 to E_n correspond to another transformation in which E increased from E_n to E_1 with the instantaneous reversal of all molecular velocities.

The equation of the state of real gasses

During these years, studies on gasses had as their point of reference the model of a perfect gas, the states of which were characterized by Clapeyron's equation $PV = RT$. However, experimental researches made it increasingly evident that equations of state had to be found to describe the behavior of real gasses, and that the differences between this behavior and what was predicted by Clapeyron's equation had to be taken into account. The investigations carried out during this period by Johann Diderik van der Waals (1837–1923) were extremely important in this context. In 1873, van der Waals formulated the equation that still bears his name, which introduced both the co-volume b and the cohesion constant a of the gas being considered:

$$\left(P + \frac{a}{V^2} \right) (V - 6) = RT.$$

Meanwhile, the researches begun by Thomas Andrews (1813–85) on the critical temperature above which it is impossible to liquify gasses were continued. They culminated in 1877 in the production of liquid oxygen by Louis Paul Cailletet (1832–1913) and the investigations of Raoul Pictet (1846–1929).

Statistical mechanics according to Maxwell

In 1879, Maxwell reelaborated Boltzmann's 1868 analysis on the distribution of molecular energy, which had been the object of great perplexity for years. With a memoir entitled *On Boltzmann's Theorem on the Average Distribution of Energy in a System of Material Points*, Maxwell observed that Boltzmann's treatment, together with the successive research described by Henry William Watson in his treatise *The Kinetic Theory of Gasses* (1876), were inadequate and gave the impression that the generality of the solution was limited. The rigorously statistical approach suggested by Maxwell for overcoming this situation focused on the notion of "phase." Maxwell expounded it as follows: "I have found it convenient, instead of considering one system of material particles, to consider a large number of systems similar to each other in all respects except in the initial circumstances of the motion, which are supposed to vary from system to system, the total energy being the same in all. In the statistical investigation of the motion, we confine our attention to the *number* of these systems which at a given time are in a phase such that the variables which define it lie within given limits."

This memoir by Maxwell marked the beginning of a new trend in the probabilistic interpretation of thermodynamics, and provided the premises for the programs of complete mathematization that were first synthesized in the work of J.W. Gibbs and the studies conducted by Boltzmann.

PHYSICAL SCIENCES: ELECTROMAGNETIC THEORIES

Maxwell's Treatise

In 1873, Maxwell gave an ample and detailed exposition of the results reached in his memoirs on electromagnetic theory with his famous *Treatise on Electricity and Magnetism*. This treatise responded to a number of needs, including the lack of texts in which students and engineers could find a detailed discussion of the basic theoretical notions for understanding electromagnetic phenomena and provided an opportunity for making ideas on action by contact accessible to a vast scientific public. As to the first point, Maxwell specifically stated that he wanted to present an analysis that would be understandable to readers who were not used to consulting specialized journals. With respect to the second, he maintained that, although the mathematical results reached by continental physicists with the theory of action at a distance were of indubitable value, it was extremely important, from a philosophical point of view, to compare the two methods in order to make the most of the mathematical consequences of Faraday's notion. Despite repeated affirmations that the equations of field were essentialy derived from experimental evidence, Maxwell did not avoid the task of attempting to "bring electromagnetic phenomena within the province of dynamics" in the pages of his treatise. For this purpose, he used the formal apparatus of Lagrangean dynamics, without, however, claiming to explain mechanically the electromagnetic theory of light. In fact, as he himself wrote, the value of the Lagrangean formalism actually depended on the fact that "we have dynamic

methods of investigation which do not require a knowledge of the system." Maxwell interpreted the relation between Lagrangean dynamics and the concept of field as a function of the following two factors: first, if it was true that the equations of field were known, it was also true that the nature of the connections of the parts of this system were not. Secondly, Lagrangean dynamics depended solely on calculus, and as such made it possible to arrive at alternative formal schemes to those proposed by Weber and the other supporters of action at a distance. On the basis of the idea that models imitate the real connections of the field without copying them, Maxwell believed that the problem of finding nesessary mechanism for establishing a given connection among the movements of the various parts of a system always admitted an infinity of solutions. The difficulty of *Treatise*, with respect to both this connection between mechanics and electromagnetism and the use of vectorial notations, was the major cause for the belated diffusion of Maxwell's ideas.

Helmholtz's electrodynamics

Helmholtz also referred back to Lagrange's scheme in order to elaborate an electrodynamic theory based on the principle of the conservation of energy. However, in his analytical approach, Helmholtz did not follow the indications provided by the physics of action at a distance; instead, he attempted to reduce the so-called continental theories to particular cases of general structure. In effect, Helmholtz's work also proposed to include Maxwell's equations as an extreme case. Helmholtz's program, which exerted a profound influence on Hertz, went so far as to criticize the theories of Weber, Riemann, and Clausius and accept the "fundamental concepts" of Faraday. Helmholtz made a decisive contribution to the diffusion of Maxwell's electromagnetic theory within German science with his memoirs on electrodynamics (two of which were *Über die Theorie der Elektrodynamik*, 1870, and *Die electrodynamischen Kräften bewegten Leitern*, 1874). In 1879, following a suggestion from Helmholtz, the Academy of Berlin announced a competition to discover "a relation between electromagnetic forces an dielectric polarization in insulators." Hertz was one of those who worked on this project (see PHYSICAL SCIENCES: ELECTROMAGNETIC THEORIES, 1880–1890).

The researches of Lorentz and Fitzgerald

In his early studies, Hendrik Antoon Lorentz (1853–1928) examined Maxwell's memoirs on electromagnetism and Augustin Jean Fresnel's researches on the wave theory of light, guided by the perspectives elaborated by Helmholtz. In his doctoral thesis, presented in 1875, Lorentz discussed the phenomena of reflection and refraction and suggested an explanation that was connected to Maxwell's approach. The importance of Lorentz's researches, which dealt with a problem that was full of difficulties for the electromagnetic theory of light, was pointed out by Maxwell himself, in the context of an evaluation of the independent studies of George Francis Fitzgerald (1851–1901).

Fitzgerald's research (1878) led to a communication entitled *On the Electromagnetic Theory of the Reflection of Light*, published in 1880. In it, Fitzgerald returned to Maxwell's *Treatise* and linked the fundamental elements of the electromagnetic theory to MacCullagh's ether (see PHYSICAL SCIENCES: ELECTROMAGNE-

TISM AND OPTICS, 1830–1840), expounding his own ideas in terms of both quaternions and the more common Cartesian notation. Apart from a deduction of the laws of reflection and refraction, Fitzgerald's results included a generalization of MacCullagh's ether, which Joseph Larmor referred to around the end of the century (see PHYSICAL SCIENCES: ELECTROMAGNETIC THEORIES, 1890–1900).

Ether and field

The rapid development of electromagnetic theories made the problem of the relations between the notion of field and the need to explain electromagnetic phenomena on the basis of a treatment of the ether increasingly more urgent. Fitzgerald himself, in the studies mentioned above, was guided by the conviction that the theory of the ether was the indispensable basis for an effective understanding of electromagnetism and optics, a conviction that was quite widely shared by other scientists of the times. The ether, as Maxwell wrote in 1876, was the largest, the most uniform, and probably the most constant object known at that time. It was a sort of reality hidden behind phenomena, and an explanation of these phenomena depended on the development of a theory that could identify major characteristics of the etherial medium. As we shall see much more clearly in the next decade, Maxwell's equations were judged too abstract because this correlation with the hypothetical dynamic structure of the universal medium in which it was believed that electromagnetic waves were propagated had not been found. Maxwell's subtle distinctions of models that imitated nature without copying it, and his deferral in indicating hidden masses as a future source for a more profound understanding of electromagnetic phenomena, were not sufficient, in this decade, to ensure the affirmation of the general ideas expounded in the *Treatise* of 1873. Even English physicists alluded skeptically to Maxwell's equations, referring to the "so-called" electromagnetic theory of light.

Improvements in experimental research

A process of industrialization of experimental researches in the area of electromagnetism took place in England during this decade. A typical product of this tendency was the foundation of the great Cavendish laboratory in Cambridge, the direction of which was taken on by Maxwell, after William Thomson Kelvin had refused it. This laboratory, which was later directed by Lord Rayleigh (1842–1919) and Joseph John Thomson (1856–1940), provided a model of organized research for all of Europe, and toward the end of the century its equipment made it possible to determine the relation between charge and mass in electrons.

John Kerr (1824–1907) and Edwin Herbert Hall (1855–1938) also obtained important experimental results that seemed to provide support for Maxwell's theory. In 1875, Kerr rendered observable certain interactions between electricity and light that Michael Faraday had already studied in the course of his magneto-optical investigations (without, however, obtaining positive results). In 1879, Hall determined the principal characteristics of the action of magnetic fields on currents, discovering the phenomenon that is still known as "the Hall effect." Hall published his results, which were correlated to researches carried out by Henry Augustus Rowland (1848–1901) on the effects of a rotating electrified disk on a magnetic needle, in a memoir entitled *On a New Action of the Magnet on Electric Currents*.

A new field of investigation in optics was opened up with the discovery of anomalous dispersion by C. Christiansen (1843–1917) and August Kundt (1839–94). With respect to the development of electrochemistry, the work of Rudolf-Hermann Kohlrausch (1809–1858), which contributed to the studies on migration of ions already initiated by Johann Wilhelm Hittorf (1824–1914), deserves particular mention.

PHYSICAL SCIENCES: RADIATION AND MATTER

Hypotheses on the nature of cathode radiation

Great progress was made in the physics of cathode radiations early in the decade, thanks especially to the researches of Eugen Goldstein (1850–1930) and Cromwell Varley (1828–83). In 1871, the characteristics of cathode radiation already observed by Julius Plücker and Johann Wilhelm Hittorf were joined by Goldstein's discoveries. Goldstein experimentally proved that the rays in question had the following properties: a) they were always emitted perpendicularly to the surface of the cathode; b) their behavior was independent of the nature of the material from which the cathode was built; and c) the new rays could produce chemical reactions somewhat similar to photochemical reactions. A wave interpretation seemed to explain the data gathered by Goldstein, though such an explanation involved overcoming unforeseen difficulties that became even greater over the following years.

That same year, 1871, Varley proposed a completely different interpretation. According to him, the phenomena of fluorescence produced by cathode radiation in the glass of vacuum tubes was caused by the collision of particle "agents." These particles had to be considered carriers of an electrical charge (negative), since Plücker and Hittorf had shown that the trajectory of cathode rays was deflected in the presence of magnetic fields. However, Varley's corpuscular model, like Goldstein's, was not without its problems: if the rays were really made up of bundles of charged particles, they should be deflected not only by magnetic fields, but by electrical fields as well. Numerous experiments were carried out on this second type of deflection, the observability of which depended on how much of a vacuum was obtained in the discharge tube. Scientists such as Heinrich Rudolf Hertz and Joseph John Thomson attempted to turn these experiments into crucial proofs for or against the wave or corpuscular models (see PHYSICAL SCIENCES: RADIATION AND THE STRUCTURE OF MATTER, 1880–1890, 1890–1900).

The "fourth state" of matter

During the second half of the decade, William Crookes embraced Varley's ideas. An extremely skillful experimenter, Crookes was attracted by the unresolved problems of cathode radiation, studying them with the technical knowledge he had acquired during earlier spectroscopic investigations. He focused his attention in particular on observations of the mechanical action of the cathode radiation inside high vacuum tubes; in 1875, he made public his first results obtain with the radiometer in a memoir entitled *On Attraction and Repulsion Resulting from Radiation*. Crookes determined two more new characteristics of the rays: they would heat a thin plate on which they were focused, and they produced mechanical effects on delicate instruments such as the radiometer. Crookes maintained that these new phenomena were the manifestations of the properties of matter under high vacuums. He believed that phenomena caused by high values of the mean free path would prevail under these conditions, revealing a new state of matter itself. The phenomena in these high vacuum tubes, he felt revealed to science a new world—a world in which matter existed in a fourth state, where the corpuscular theory of light is valid, a world mankind could never enter and in the face of which the scientist must be satisfied with observing and experimenting from the outside. Crookes outlined his results in a memoir of 1876 entitled *On the Illuminations of Lines of Molecular Pressure, and the Trajectory of Molecules.*

Spectra and radiation

The new cathode phenomena, together with the accumulation of data from spectroscopic researches, posed fairly serious theoretical problems to physics. In 1878, an analysis expounded by George Francis Fitzgerald (1851–1901) in a memoir entitled *On the Mechanical Theory of Crookes' Force*, pointed out the difficulties encountered in attempts to interpret the pressure exerted by cathode rays in terms of statistical mechanics. These problems were effectively rendered complex by the need to refer to the pressure that could be predicted by means of electromagnetic considerations. Eilhard Weidemann also applied kinetic theories to the analysis of spectra and the luminosity observable in gasses with electrical discharges travelling through them. Wiedemann's considerations focused in particular on the possibility of attributing the cause of spectra to oscillations of the atoms that made up molecules. On the other hand, the complex question of the interactions between the ether and matter was also very much alive during these years. In 1876, Boltzmann attempted to overcome the paradox that arose from the possibility that molecular energies were dispersed within the ether, and in 1879, in a memoir entitled *The Existence of the Luminiferous Ether*, Ernest H. Cook hinted at a radical solution involving the formulation of a molecular theory.

Another extremely important result was obtained in 1879 by Josef Stefan (1835–93), who suggested a semi-empirical correlation between the heat radiated by a body and the fourth power of its absolute temperature, (*Über die Beziehung zwischen der Wärmestrahlung und der Temperatur*). A few years later, Boltzmann deduced Stefan's law, basing his calculations on the kinetic theory and the electromagnetic theory of light (see PHYSICAL SCIENCES: RADIATION AND MATTER, 1880–1890). Stefan's law constituted a fundamental point of reference for all later investigations of black bodies.

CHEMICAL SCIENCES

New classes of dyestuffs and the theory of chromophores

During this period, the production of synthetic dyestuffs was in a phase of continuous expansion that stimulated the preparation of numerous new intermediates in organic chemistry. Azoic dyestuffs were widely developed, since they were economical, easy to use, and relatively colorfast. Among these were Ponceaux and the oranges, obtained by Zacharie Roussin (1827–94) and Otto Nikolaus Witt (1853–1915) in 1875–78, and the tetrazoics,

such as Biebrich scarlet, which was prepared by Rudolf Nietzki (1847–1917) in 1879, and the famous butter yellow (Witt, 1876). In 1870, Adolf von Baeyer and Adolf Emmerling succeeded in synthesizing indigo from isatin. That same year, Friedrich August Kekulé and Coloman Hidegh discovered the copulation reaction of diazonium salts with phenols, which later became the basis for the preparation of azoic dyestuffs. Baeyer also discovered phenolphtalein and fluorescein, the first representatives of the phthalein class, in 1871. Phenolphthalein and methyl orange (Johann Peter Griess, 1875) also became widely used as indicators of acidity. Heinrich Caro and Witt prepared chrysoidine, the first azoic dyestuff of any practical importance obtained by means of the copulation reaction, in 1875–76. Other dyestuffs discovered during this period were methyl blue (Caro, 1876), an important thiazinic dyestuff, and copper green, a basic dyestuff of the triphenylmethane series, discovered by Otto Philipp Fischer (1852–1932) in 1877. In 1876, Otto Fischer and Emil Fischer identified the composition of fuschine. Charles Lauth prepared lauth violet, the first thiazinic dyestuff. It was during these same years that Witt formulated his theory on the chromophore and auxochrome groups.

The proliferation of organic preparations

Among the new organic compounds of this period were phenanthrene (Rudolph Fittig and Eugen Ostermeyer and Karl Graebe, 1872), anthraquinone (Kekulé and Antoine Paul Franchimont (1844–1919) in 1872) and butyrolactone (Aleksandr Mikhailovich Zaitsev, 1873). New synthetic compounds were aldol (Charles Adolphe Wurtz, 1848–97), and nitroparaffin (Victor Meyer, 1872), obtained from the action of silver nitrite on the aliphatic halogen derivatives. In 1875, Meyer treated the aliphatic halogen derivatives with nitrous acid, obtaining two new classes of azotate compounds, the nitrolic acids and the pseudonitrols. DDT was synthesized by Othmar Zeidler in 1874, and phenylhydrazine by Emil Fischer in 1875 (Fischer used this reagent to obtain osazones, though it is more generally used to synthesize, for identification purposes, derivatives of the carbonyl group). Saccharin was prepared by Constantin Fahlberg (1850–1910) and Ira Remsen (1846–1927) in 1879. Known compounds were also synthesized, including glycerin (C. Friedel and R.D. Silva in 1872), vanillin and other phenol aldehydes (Carl Ludwig Reimer (1856–1921) and Ferdinand Tiemann (1848–1899) in 1876) and various homologs of benzene (Friedel and James Mason Crafts (1839–1917) in 1877–78). This last synthesis was carried out in the presence of anhydrous aluminum chloride, following a widely-used method. A. Michael (1853–1942) synthesized the first glucosides in 1879. From the dry distillation of rubber, G. Bouchardat (1842–1918) obtained, among other things, isoprene (already obtained by C.H. Williams; by treating the isoprene with hydrochloric acid, Bouchardat obtained a polymer that behaved like natural rubber. In 1873, Fittig identified the structure of benzoquinone. In 1876, Alexander P. Borodin (1833–87) discovered the aldol condensation reaction and the successive dehydration to crotonaldehyde.

Aromaticity and isomerism in the benzene derivatives

Once the equivalence of the six carbon atoms in benzene was recognized (Kekulé and Albert Ladenberg), new structural formulas for the aromatic compounds were proposed (James Dewar and Wilhelm Körner succeeded in establishing the formula for pyridine), while Kekulé spoke of oscillation between double and simple bonds (1872). A fundamental problem that still remained unsolved in aromatic chemistry was that of the relative positions of the substitutions in the benzine ring. It was known that three forms of isomers, named *ortho, meta,* and *para,* existed for the di-substitution derivatives, but it was impossible to decide which was the 1.2, which the 1.3, and which the 1.4 isomer. Körner solved the problem in a rather simple and elegant way on the basis of the number of tri-substitution derivatives that could be obtained from each of the di-substitution products by the introduction of a third substitution, since the 1.2 isomer would be one to generate two tri-substitution derivatives, the 1.3 would give three and the 1.4 one only. Körner's theorem was put forth in 1867 in Körner's *Faits pour servir à la determination du lieu chimique dans la série aromatique;* it appeared in its complete form in 1874 in the classical memoir "Studi nell'isomeria delle cosidette sostanze aromatiche a sei atomi di carbonia," published in the *Gazzetta Chimica Italiana.* The *Gazzetta* first came out in 1871 under the direction of Stanislao Canizzaro, who published a number of articles about sanotin in its pages between 1873 and 1896.

Optical and geometrical isomerism

The passage from two-dimensional formulas on a plane to three-dimensional formulas in space was already beginning to take place: in 1874, Joseph Achille Le Bel (1847–1930) and Jacobus Henricus Van't Hoff (1852–1911) published the theory of optical activity in terms of the tetrahedral distribution of carbon valences, correlating optical stereoisomerism with the presence of asymmetrical carbon atoms. That same year, Van't Hoff predicted the existence of the geometrical cis-trans isomerism for ethylene derivatives; in 1887, he ascribed the isomerism displayed by maleic acids and fumaric acids to the same phenomenon.

Confirmation of the validity of the periodic table

Dimitri Ivanovich Mendeleev and his periodic system gained even more prestige at this point from the discovery of two elements that filled the respective voids and had the properties that the Russian chemist had predicted: gallium, discovered by P.E. de Poisbaudran (1884–1906) in 1875 and scandium discovered by L. F. Nilson (1840–99) in 1879. Other lanthanides— ytterbium, samarium, holmium, and thulium—were discovered in 1878–79.

Developments in physical chemistry

Physical chemistry acquired the law of the independent migration of ions, theorized by Friedrich Kohlrausch (1840–1910) in 1875, the equation of Johannes van der Waals (1837–1923) for real gasses (1873), the phase rule, discovered by Josiah Willard Gibbs (1839–1903) in 1876, and the laws of osmotic pressure, stated by Wilhelm Pfeffer (1845–1920) in 1877. François Marie Raoult (1830–1901) published the first results of his studies on solutions in 1878, bringing to light the proportionality of the relative lowering of vapor pressure, cryoscopic depression, and ebullioscopic elevation.

In 1877, N.A. Menschutkin (1842–1907) did research on chemical kinetics, attempting to evaluate the influence of the isomerism of alcohols and acids on the speed of esterification. He established that the reactivity of *m*-substitution aromatic and heterocyclic compounds was greater than that of *o*-substitution compounds, and discovered the importance of the solvent on variations in the speed of a given reaction.

LIFE SCIENCES

The nucleus and cell division

This decade was characterized by developments in cytology, progress in microbiology, and the debate on the evolution of living species. In 1861, Carl Gegenbaur (1826–1903) and Albert von Kölliker (1817–1905) likened the egg to a cell that gave origin to the organism by way of successive cleavages, following a hypothesis outlined by Theodor Schwann (1810–82) in his *Recherches.* Kölliker also carried out investigative experiments on the cellular nature of spermatozoa. But the decisive contribution of embryology to cytology came in 1875: Oscar Hertwig (1849–1922), in Berlin, observed the fusion of the egg and sperm nucleii in a sea urchin egg after fecundation. The phenomenon was named "karyogamy." In 1879, Hermann Fol (1845–92) succeeded in observing the penetration of sperm under the microscope, again in a sea urchin egg. Eduard Strasburger (1844–1912), a professor of botany in Bonn, and Walther Flemming (1845–1912), a professor of anatomy in Prague and Kiel, began studying the nucleus and the cell reproduction process. Two works, one by Strasburger, entitled *Über Zellbildung und Zellteilung* (1875), the other by Flemming, *Zellsubstanz, Kern und Zellteilung* (1882), created the foundations of cytology as a distinct branch of histology. It was found that the division of the nucleus preceded cell division. "Karyokinesis" (Flemming, 1879) was formed by distinct bodies which arranged themselves in "spindle" fashion during the karyokinetic process.

Bacteriology, the study of fermentation and antisepsis

Microscopic life had by this time been recognized as having profound value on biology; its importance was further confirmed by research in bacteriology. In March, 1878, in a note on the influence of Louis Pasteur's studies on the advancement of surgery, presented to the Academy of Sciences in Paris, Charles Sédillot (1804–83) introduced the neologism "microbe." The term was accepted by Pasteur and became universal. Pasteur's researches on yeast, begun just after France was defeated in the war against Prussia, enabled him to describe various types of leavenings and to devise a procedure for manufacturing and preserving them, which he patented in June, 1871. In essence the process consisted in fermentation in the absence of any contact with air. The Carlsberg brewery in Copenhagen adopted Pasteur's system.

In the field of medicine, Pasteur's influence affected antisepsis, which by now was widely applied. A forerunner in this area was the English surgeon Joseph Lister (1827–1912), who acknowledged his debt to Pasteur's ideas. He was followed by the French scientists Just Lucas-Championnière (1843–1913) and Stéphane Tarnier (1828–97), and by the German Theodor Billroth (1829–94).

Anthrax and vaccination

During the eighth decade of this century, Pasteur began study-ing the pathological and epidemic problem of anthrax, confirm-ing Casimir Joseph Davaine's observation that there was a micro-organism in the blood of animals affected by the disease. He identified two forms of the microorganism—a bacteria and a spore—and succeeded in obtaining weakened strains of it. In the course of a memorable experiment, carried out at Pouilly-le-Fort between May and June, 1881, Pasteur demonstrated the effectiveness of vaccination in preventing disease. Robert Koch (1843–1910), a student of the histologist Henle, was also active during this period. In 1876, in a classical work on the etiology of anthrax based on the growth cycle of *Bacillus anthracis*, Koch established the concept of explaining the pathological process by means of the pathogenous micro-organism.

Koch followed these studies with important researches on the infection of wounds (1878). Meanwhile, he also introduced into bacteriology the technique of staining bacteria and the use of gel as a culture medium.

The debate on Darwinism

The debate over the ideas of Charles Darwin—who published *The Descent of Man*, in which he maintained that the human species originated from inferior, monkey-like forms, in 1871, and a year later published *The Expression of the Emotions in Man and Animals*—raged hotly during this period. In England and the United States, Darwin's ideas were opposed by the anatomist Richard Owen, a naturalist at the British Museum, and by Louis Agassiz, a Swiss-born professor of zoology at Harvard. The major supporters of Darwinism were Thomas Henry Huxley, a zool-ogist and later professor of biology at the Royal College of Science, and the botanist Asa Gray. The geologist Charles Lyell, whose ''actualistic'' ideas had influenced the young Darwin and who was already hostile toward Lamarckism, was also converted to the theory of evolution.

The scientific debate and ideological disagreement about Dar-winism was even more heated in Germany. One of the first Darwinists was Fritz Müller (1821–97), the author of a work entitled *Für Darwin* that hinted at a correspondence between embryonic development and evolutionary history. A harsh critic of Darwin, on the other hand, was the botanist A. Wiegand (1821–66), whose work *Der Darwinismus* (1874–77) emphasized the element of chance involved in Darwin's phylogeny. But the predominant champion of the ''theory of descendency,'' as Ger-man authors called the theory of evolution, was Ernst Heinrich Haeckel, a student of Virchow and Kölliker and a professor of zoology at Jena. His *Generelle Morphologie der Organismen* (1866) and his *Natürliche Schöpfungsgeschichte* (1868) were followed by his *Anthropogenie oder Entwicklungsgeschichte des Menschen* (1874). Darwin's views on the history of life rose to the level of a philosophical system: evolutionism became the antithesis of tra-ditional metaphysics. Organisms had supposedly evolved from a primitive *Moner*—created by spontaneous generation—into human beings. According to Haeckel, the development of the individual, ontogeny, briefly recapitulated the development of the species, or phylogeny (Haeckel's fundamental biogenetic law).

Haeckel's genealogies, however, were strongly contested by embryologists and zoologists. French scientific circles gave very little indication to being disposed to accept ''transformism.'' In essence, Haeckel's affirmations, which were often carried to the extreme, ended up contributing, though indirectly, to the development of the critical attitude of modern biologists and helped get rid of the excessively mechanistic notions that ruled almost unchallenged throughout the entire nineteenth century.

Comparative anatomy, paleontology, and evolution

Descriptive and comparative morphology were given new life with the advent of the theory of evolution. The decade opened with the second edition (1870) of Carl Gegenbaur's *Grundzüge der vergleichenden Anatomy*. The relationships among the various organisms were expressed in the homologies among their respec-tive parts: comparative anatomy had to discover ''the clues to the genetic connections in animal organization.'' Gegenbaur thus reevaluated Etienne Geoffroy Saint-Hilaire's geometrical point of view in relation to Cuvier's functionalist views. Other authors, on the other hand, polemicized against evolutionism, and against Darwin's ideas in particular. The second volume of Karl Ernst von Baer's discourses (*Reden gehalten in wissenschaft-lichen Versammlugen und kleinere Aufsätze vermischten Inhalts*) was published in 1876, with an ample memoir on Darwin's theories. Von Baer reiterated the major objection to Darwinism: it did not explain the correlations among the various parts of living organisms.

Paleontology also felt the influence of evolutionary theory: the intermediate forms between species, genera, and classes were sought, starting from the premise that all organisms derived from a single structure through successive differentiations. Albert Gaudry (1827–1902) was inspired by such a conviction: according to him, Linnaeus's hypothesis on the fixity of species could be interpreted in two ways, both of which were absurd: the spontaneous organization of inert matter or the development of germs that had been in a latent state. In 1878, after publication of his *Animaux fossiles et géologie de l'Attique* (1862), Gaudry began his *Les enchaînements du monde animal dans les temps géologiques*, a vast work that he did not finish until 1890. More limited, but still outstanding, were the contributions made to paleontology by the Swiss scientist C. Ludwig Rutimeyer (1825–95), who conducted researches on the tooth structures of fossil horses, and by Vladimir O. Kovalevski (1842–83), who in 1873 pub-lished an important memoir on the natural classification of fossil equids (*Monographie der Gattung Antrococotherium Cuv. un Versuch einer naturlichen Klassifikation der fossilen Hufttiere*). In the United States, Othniel Marsh (1831–99) and Eduard D. Cope (1840–97) carried out excavations in the Far West territories, bringing to light a large number of unknown forms, mammals of the Mesozoic period.

Publication of the *Handbuch der Paläontologie*, by Karl von Zittel (1839–1904), an evolutionist who nevertheless openly acknowledged the gaps in the presumed history of life, was begun in 1876.

Zoogeography

In 1876, Alfred Russel Wallace completed his large work on zoogeography, *The Geographical Distribution of Animals*. Wallace's work confirmed the existence of this new discipline, which was born in 1853 with the publication of another work on the geographical distribution of animals written by the Austrian Ludwig Schmarda (1819–1908). Thus a new perspective was introduced in the study of the animal kingdom, and the concept of "zoological region" entered into use.

Phytogeography

Tendencies opposed to evolution appeared in phytogeography, a discipline that had been initiated by Alexander von Humboldt and had subsequently transcended the boundaries of systematics to take on problems in general biology. The transformist bias of the *Versuch einer Entwicklungsgesschichte der Pflanzenwelt insbesondere der Florngebiete, seit der Tertiärperiode* (1879–82), by Heinrich Gustav Adolf Engler (1844–1930), a professor of botany at Breslau, was countered by the approach of August Grisbach (1814–79) of Göttingen, who attributed plant varieties to environmental factors in his *Die Vegetation der Erde* (1878).

Wundt: the birth of experimental psychology

The *Grundzüge der physiologischen Psychlolgie*, by Wilhelm Wundt (1832–1920), appeared in 1873–74: the six editions published during the author's lifetime stand as evidence of the popularity of this work, which marked the birth of experimental psychology from physiology. Initially active at Heidelberg during von Helmholtz's years there, Wundt was called in 1875 to a chair of philosophy at the University of Leipzig, where four years later he founded the first institute of psychology. From 1881, Wundt's school also published a periodical, the *Philosophische Studien*. In Wundt's program, psychology became an "experimental science," free of the metaphysical presuppositions that were present even in Herbert and Fechner. The psychological experiment was seen as immediate, and the data of psychology considered intuitive. Almost all the major European and American psychologists—G. Stanley Hall, James McKeen Cattel, Edward Scripture, Frank Angell, Edward Bradford Titchener, Ludwig Lange, Oswald Külpe, Emil Kraepelin, Alfred Lehmann, and Ernst Meumann—passed through Wundt's laboratory. Meanwhile, another current, of a descriptive tendency, was forming within the study of psychology; its major representative during this period was Karl Eward Hering (1834–1918). Following publication of his *Die Lehre vom binokularen Sehen* in 1868, he was called to succeed Purkyně in the chair of physiology at the University of Prague (1870). The publication of his *Zur Lehre vom Lictsinne* between 1872 and 1878 led to an extremely heated debate with Hermann Ludwig von Helmholtz.

Cell kinetics

During these years, Rudolf Heidenhain (1834–97), a physiologist at Breslau, and Élie Mechnikov (1845–1916), a professor at Odessa and later at Paris, developed the concept of cellular kinetics in the epithelia of secretionary organs and in polynucleate white blood cells. Meanwhile, Cohnheim described the process of diapedesis, and Mechnikov discovered phagocytosis.

EARTH SCIENCES

Geological organizations

The International Geological Congress, which held its first session in Paris, was founded in 1878. The purpose of this institution was to promote discussion and comparison of results of geological research from various countries. The congress was to meet every three, then every four years, and the commissions working within it had the task of defining and settling any difficulties that arose as a result of the independence and diversity of studies. Besides scientific communications, field excursions were soon organized to explore the region around the congress's headquarters. Except for the years of the two world wars, the congress has met periodically ever since (it is now affiliated with the I.U.G.S., the International Union of Geological Sciences), offering a particularly favorable occasion for exchanging points of view and promoting studies in the field of earth sciences.

The Geological Survey of the United States of America was established in 1879 under the direction of Clarence King. It was preceded, in 1873, by a similar organization for Prussia and Thuringia. Within a short time, all the major countries had their own institutions for cartography and regional geological illustration.

In 1873, the State Geological Committee of Italy began a geological survey of Sicily, with the goal of preparing a 1:100,000 scale geological map of the national territory.

Studies on orogenetic phenomena and the birth of the geosynclinal theory

In 1873, James Dwight Dana introduced the term geosyncline to indicate the great underwater depressions where thick layers of sediment accumulated and from which folded mountain ranges, the major ranges of the earth's history originated. Two years later, in his *Die Entstehung der Alpen*, Eduard Suess maintained that horizontal movements had prevailed in the genesis of the great Alpine system and other similar mountain ranges and demonstrated the asymmetry of the structure of the Alps, where the folds were inverted toward the north and the west, indicating tangential rather than vertical movement. Von Buch's upheaval theory was definitively abandoned at this point. In 1878, Albert Heim (1849–1937) reiterated the idea of horizontal movement, spoke of recumbent folds, and demonstrated the passive role of magnetic masses in orogenesis.

Regional and stratigraphic geology

The geological studies conducted by Angelo Sismonda (1807–78) for the drilling of the Frejus tunnel were published in 1871 in a work entitled *Sur le tunnel de Fréjus*. Sismonda had already published other regional geological studies on the Piedmont, Liguria, and Savoy regions between 1862 and 1866.

Other important studies in this field were made by Dana, who introduced the Archean period (correcting Murchison's Azoic, 1876) and by Ferdinand von Richthofen (1833–1905) and Edmund Mojsisovics (1839–1907), who studied the Triassic cliffs of the southern Alps. Von Richthofen also attributed the loess of China to aeolian origins. Alexander Karpinski (1847–1936) contributed to knowledge on the Permocarboniferous, and Charles Lory dated the micaceous alceschists of the Savoy and Piedmont regions to the Triassic.

Studies in paleontology

Albert Gaudry worked on his great treatise on vertebrate paleontology, *Les enchaînements du monde animal dans les temps géologiques,* from 1878 to 1890, while Karl von Zittel (1839–1904) began publication of his fundamental treatise on paleontology in 1872.

The most important work of this period, however, was written by Charles Darwin and dealt with human paleontology: *The Descent of Man,* which also started the great debate on the evolution of the human species, was published in 1871.

Studies on volcanism

The second edition of George Poulett-Scrope's general work on volcanoes, which contained descriptions of all the known active volcanoes, came out in 1872. This same author had previously conducted studies on the ancient volcanoes of central France. In 1877, Eduard Reyer studied the differences between volcanoes and plutons, with particular attention to laccoliths, which he unfortunately interpreted in a rather unconvincing way.

Researches on glacialism

Archibald Geikie (1835–1924) indicated the existence of glaciers in Scotland during the Cambrian Age. James Croll (1875) and James Geikie (1839–1915) studied the relations between climate and glacialism in the earth's history (1877). In 1879, Albright Penck (1858–1945) began studying the Quaternary glaciations of the Alps. These studies later led him to differentiate the well-known climatic events of the last geological age.

Studies in petrography

In 1877, Karl Harry Rosenbusch (1836–1914) subdivided eruptive rocks on the basis of their feldspar content, introducing the method that is still in use today in petrography. Ferdinand Fouqué (1828–1904) and Auguste Michel-Lévy (1844–1911) presented the systematics of French eruptive rocks, based on studies under a microscope, in a long treatise of 1879.

Planetary geology

Hermann Ludwig von Helmholtz's studies on the solar system (1871), together with various studies on the planets, such as the researches conducted by Giovanni Virginio Schiaparelli (1835–1910) on Mars (1878), stimulated the formulation of new hypotheses on the possible primitive history of the planet earth. Heinrich Otto Làng proposed interesting ideas on the subject (1873), while Robert Mallet took into consideration the internal heat of the earth. In 1872, Johann Benedict Listing (1808–82) proposed the term ''geoid'' for the figure of the earth.

The voyage of the Challenger

The voyage of the *Challenger* was the greatest oceanographic expedition ever undertaken, opening the way to modern oceanographic research. In the period between 1872 and 1876, it covered all the oceans, for a total distance of almost 69,000 miles. 492 depth soundings, 263 temperature stations, and 133 dredgings of the bottom were made during the voyage, and approximately 600 cases of samples were taken home. The material gathered consisted of an enormous number of marine plants and animals, birds, and samples of water, coastal rock

and sediment from the bottom. Far away, practically unexplored islands of the Pacific were reached, and extremely interesting geographical and anthropological studies were made. The findings of the voyage were collected into fifty volumes, forty-five of which dealt with biological topics, while two contained various descriptions, two were on physico-chemical topics, and one described the samples taken from the ocean bottom. This documentation constituted the largest mass of oceanographic data gathered during that period. The themes of the research were to study the conditions of the ocean depths, including the nature of the bottom, to study the chemical composition of sea water, and above all to collect the largest possible amount of information on the life of marine plants and animals and the largest possible number of actual organisms.
animals and the largest possible number of actual organisms.

Promoter and leader of the expedition was Charles Wyville Thomson, while the inspiration for this particular expedition may have been the need to verify once and for all whether the deep waters of the ocean constituted an azoic environment or not. Ideas on the subject were very confused at that time: on the one hand was Forbes's assertion that no living creatures existed below three hundred leagues; on the other were authoritative observations, such as those of James Ross, who had found living creatures at a depth of even 1,000 leagues. While visiting The Norwegian biologist Michael Sars, Wyville Thomson had been able to observe a collection of rare animals found at a depth of around 900 meters in the vicinity of the Lofoten islands, and he became convinced of the need to do away with any and all doubts on the question. The voyage was sponsored by the Royal Society of London, and, in addition to Wyville Thomson, participating researchers were Henry Nottidge Moseley, Rudolph von Willemöes-Suhm (who died during the expedition), and John Murray, who succeeded Thomson in arranging the materials collected. The expedition's chemist was John Young Buchanan. Until 1874, the ship was commanded by Captain George Nares, who was succeeded by Frank Tourle Thomson. The *Challenger* was a three-masted English navy corvette with auxillary steam engines. She was built of wood, and was sixty-nine meters long, with a displacement of 2,306 tons. The ship was modified to accommodate research laboratories and storerooms: for one thing, all the cannon but two were removed. The ship had a crew of twenty-three plus six researchers.

The most delicate measurements were those of temperature, which were mostly done with Miller-Casella Six-type maximum-minimum thermometers (the expedition was also equipped with electric thermometers), since a number of instruments were often dropped along the same vertical to obtain thermometric profiles. The expedition received its first Negretti and Zambra reversing thermometers in 1874. Corrections had to be made later to compensate for the effects of pressure on the temperature readings. Water samples were collected in bottles designed by Buchanan, and soundings were made using Baillie instruments. Special St. Andrew's crosses were used for making current measurements.

One of the most important results of the research was that the water samples collected later enabled William Dittmar (1833–92) to establish definitively the law of the constant composition of sea water (already ascertained by Forchhammer). It was observed that the temperature remained constant below 200 meters, regardless of seasonal changes, and that abyssal temperatures varied from zone to zone as a result of upcasts in the ocean floor that hindered circulation. It was also found that the abyssal temperature could drop well below four degrees Celsius as had also been hypothesized (D'Urville, following de Saussure's measurements). The presence of the mid-Atlantic dorsal ridge was detected, and its location was fairly accurately given. Samples of the ocean bottom revealed the extraordinary extension of red clay deposits, especially in the more depressed areas, and nodules of manganese were discovered. Hundreds of species of oceanic flora and fauna were also described.

The expedition closed the old epoch of oceanographic research and opened the modern one, which is today in a phase of maximum expansion. Even though the leader and participants in the expediton were British, the study of the material collected was opened to scholars of all countries, thanks to Wyville Thomson's far-sighted wishes, providing one of the first examples of the type of international collaboration among specialists of various disciplines that is essential to modern oceanography.

Other advances in oceanography

Other voyages also took place during this decade. In 1877, Alexander Agassiz (1835–1910) continued a previous expedition in the Gulf of Mexico (1869), re-exploring it and the Caribbean Sea. Agassiz's expeditions, like those of Carpenter and Wyville Thomson, in addition to making important contributions to general oceanography and physics, essentially marked the beginning of hydrobiological explorations and direct studies of marine flora and fauna in their own environment. (In the past, except for coastal material, biological samples had been gathered by fishing boats and whalers.)

An important exploration of the Pacific, with studies of the bottom for the purpose of laying underwater cables, was made in 1874 by the American ship *Tuscarora*. Other scientific voyages were made by the *Silva* (1869–70) and the *Vega* in 1879. In the *Tuscarora*'s explorations, a new and extremely practical steel sounding line was used, replacing Lord Kelvin's recently proposed ingenious violin-string device. The use of strong, less cumbersome suspensions in place of the old hemp ropes made it possible to obtain quicker and more accurate measurements. Simple innovations in instrumentation such as these led to rapid advancements in knowledge about the morphology of the oceans.

Nergretti and Zambra's new reversing thermometers appeared in 1874–75 (the definitive version came out in 1887) for use on the various voyages. A number of them were sent to Hong Kong in 1874 to be delivered to the *Challenger*, which was docked there at the time during a break in the expedition.

With respect to theoretical work, James Croll's investigations on ocean currents in relation to the thermal conditions of the planet (1870) deserve mention. Croll also embarked upon a violent dispute with Carpenter over the problem of what caused ocean currents. The study of tides also progressed along new lines, after a special English committee, presided over by Lord Kelvin in 1867, promoted the study of harmonic methods for analyzing observed tides, in order to devise accurate systems of forecasting the phenomenon. Lord Kelvin himself designed a tide-forecasting instrument in 1872 (one of the first mechanical automatic calculators, a forerunner of the modern electronic calculator). The instrument was built by Robert and Lege. In 1879, Lord Kelvin and his brother invented a mechanical tide

analyzer that provided a quick harmonic analysis of the recordings. Another tide forecaster, capable of composing twelve components, was built by William Ferrel in America in 1880.

Systematic measurements with magetometers

The first systematic measurements with magetometers to detect iron ore deposits were begun almost simultaneously in Sweden (1876) and the United States (1873), by R. Thalén. In 1882, Carl Barus used the measurement of differences in electrical potential to search for underground ores.

The International Meteorological Organization

The time was now ripe for establishing a permanent international meteorological organization, as had already been proposed by Heinrich Wilhelm Dove (1803–79) in 1863 and by the Frenchmen Émilien Renou (1815–1902) and Edmé Hippolyte Marié-Davy (1820–93) in 1868. A preparatory meeting—known as the Second International Meteorological Conference—was held in Leipzig in 1872; it agreed upon several standardizations and convened the First International Meteorological Congress for the following year in Vienna.

While the Leipzig gathering was an unofficial meeting of scientists, the Vienna Congress was intergovernmental. More progress toward unification was made in Vienna: a Permanent Committee was elected, and it was proposed that an international meteorological fund be set up. On the last topic Wladimir Köppen (1846–1940), who for at least six decades remained one of the most influential figures in the meteorological world, presented a very detailed proposal. The Permanent Committee met four times: in Vienna immediately after the Congress, in Utrecht in 1874, in London in 1876, and again in Utrecht in 1878. Each session contributed to the further standardization of meteorology; the first also organized a Maritime Meteorological Conference (which was held in London in August, 1874 and reviewed and completed the work of the Brussels conference); and the fourth convened the Second International Meteorological Congress for 1879 in Rome, to which it decided to submit an outline of the "Statute of the International Meteorological Organization." Thus the abbreviation I.M.O. was born, though its use spread very slowly. The Congress of Rome decided to entrust the direction of the I.M.O. to an International Meteorological Committee. Following an unofficial congress held in Paris in 1889 with 174 participants (as compared to thirty-two in Vienna and forty in Rome), a new structure was established in 1891: the Conference of Directors of the various meteorological services would act as the assembly and elect the executive organ, an International Meteorological Committee. These bodies governed the I.M.O. until 1950. Most of the work was done by technical commissions (which increased in number over the years, from one in 1891 to twelve in 1929) and the secretariat, with the aid of a small secretarial staff.

In meteorological theory, particular mention should be made of an important work published in 1876, *Études sur les mouvements de l'atmosphère*, by the Norwegians Cato Maximillian Guldberg (1836–1902) and Henrik Mohn (1835–1916).

Instruments that were remarkable in all respects, but were especially important for their ability to translate angular motion into time intervals, were presented to the Academy of Brussels in 1875 and to the Congress of Electricity in Paris in 1881 by the Belgian F. van Rysselberghe. The Dutchman Eduard Heinrich Baumhauer independently proposed similar instruments in 1874.

Baumhauer's mechanism acquired notable historic importance when H. Olland, an instrument builder in Utrecht, adopted it for the many meteorographs he built between 1875 and 1895. Olland's instruments proved to be extremely durable and became quite famous. Along with them, what became universally known as the Olland system (though it would be more appropriate to call it the Baumhauer system) also acquired great fame and was used to build the first radiosondes, which included the Italian Siap-Ducati model.

APPLIED SCIENCES

Raw materials

Thanks to the Martin-Siemens process, scrap iron could now be used for producing steel. However, the new method had one drawback in common with the Bessemer process—only rich ores could be treated. The silica bricks used to line the convertors would remove carbon, silicon, and manganese from the ore, but not phosphorus, which made the steel brittle. The big step that made it possible to exploit less rich ferrous ores was made in 1875 by Sidney Gilchrist Thomas (1850–1885), who succeeded in dephosphorizing the metal bath with a basic lining and the addition of lime. Thomas described and offered his invention to the British Iron and Steel Institute, which rejected it, probably because England possessed important deposits of phosphorous-free ores, but also because of the blindness of its technicians. However, Thomas's invention, which could also be applied to the Siemens open-hearth furnace, was immediately put to use by the German steel industry, which began to exploit the great phosphorus-contaminated deposits of the Lorraine that it had just snatched away from France. This resulted in a shift in the geographical center of heavy industry towards Germany, which was soon producing more steel than England, reaching a quota of twice as much by the end of the century. (In 1900, 8 million tons were produced in Germany, as compared to 4.9 million tons in England.)

Prime movers

Meanwhile, an important energy compensation was made in favor of countries that were rich in water resources but poor in fossil fuel deposits when advances in the technology of electricity made it possible to run an electric generator on water power. A generator built by Zénobe Théophile Gramme eliminated the major defect of the generators that had been used up to then: i.e., the creation of eddy currents, which resulted in losses in yield and the production of excessive heat. Gramme's engine, however, ran without overheating, and by 1874 it had already been installed on a number of French and American ships. Gramme's commercial success, which was above all linked to his use of the ring armature, especially worried the competitive German industrialists, the brothers Karl Siemens and Werner Siemens, who used the double-T armature for their generators. The researches begun by the Siemens brothers, directed by the designer F. von

Hefner-Alteneck (1845–1904), led to the invention of the drum armature, which turned the "war" of the generators in their favor, despite Emil Bürgin's improvements on Gramme's machine.

At first, generators of electrical energy were only used for illumination. Arc Lamps, which were very reliable, were first installed in Gramme's own works, then in several French railway stations. However, they proved to be too expensive, since no more than one lamp could be operated by each dynamo.

An alternative was found when Joseph Swan (1828–1914) made use of the experiments of William Crookes (1832–1919) on electricity in vacuums to further research on the incandescent lamp. Swan presented his carbon filament lamp in 1878, but did not patent it, so he was beaten to the gun by Thomas Alva Edison, who took out a patent for an incandescent lamp with a platinum filament in 1878. Swan patented an improved carbon lamp in 1880.

Even though dynamos and alternators were now available, it was assumed that much less energy could be obtained from an electromechanical system than what was put into it. This was thought to be true because of the numerous losses that occurred in the various transformations. However, a public demonstration of the possibility of using electrical energy for powering machine tools was given at the Vienna Exhibition of 1873. Gramme and Fontaine displayed a reversible combination of two Gramme machines, on acting as a generator and the other as an engine (with continuous current). As one visitor to Gramme's factory a year later noted, the lathes, the machine tools, and the other engines were powered by one of his small electrical engines. This was still an isolated instance, however.

In addition to the electrical engine, various other factors concurred to undermine the predominance of the steam engine. The very nature of the dynamo posed serious problems to the builders of steam engines. Horizontal engines that operated at a low number of revolutions, which were then multiplied by drive belts, were being used to run dynamos. The problems involved were partly overcome in 1871 when Peter Brotherhood (1838–1902) patented a new type of steam engine adapted to the purpose. His vertical three-cylinder engine produced from 1.25 to 55 horsepower and turned with an angular velocity of between 500 and 1,000 revolutions per minute. The engine was directly connected to a dynamo that was specially designed to operate at the same speed.

Socio-economic factors combined with technical ones to supplant stationary steam-operated engines. As Franz Reuleaux (1829–1905) pointed out in 1875, only large capital could afford to buy and use the powerful steam engines, around which the rest of the installation, which also required capital but was not inseparable from it, was arranged. Energy, he felt, had to be rendered independent of capital. This was also the plan of Nikolaus Otto and his associates Langen and Deutz, who wanted to build a dam against the inundating, arbitrary power of capital, to strengthen small industry for the harsh struggles of economic life, and to return the criteria of production to within limits that would ensure the tranquil and continued development of civilization. It was in this context that research on the internal-combustion engine was born: the disadvantages of Otto's first engine (excessive weight, cumbersomeness, and a high noise level) were overcome in 1877 with the four-stroke engine. At first, these motors were dependent on the gas distribution network, but their application to vehicles was already being predicted.

Otto's engine was the pride of the German pavilion at the Philadelphia Exhibition of 1876, while European visitors were impressed by the American pavilion. F. Reuleux, one of the keenest observers, was surprised not only by the beauty and variety of American steam engines (with the Corliss 1,500 horsepower engine towering above them all), but by the fact that their parts were mass-produced automatically and could therefore be interchanged.

In the area of machine tools, American superiority by now appeared unquestionable. The need to work steel for its many applications led to a high demand for tools, which were built from steel alloys with the introduction of high-speed steel tools by Robert Forester Mushet (1811–91) at the beginning of the decade, the features of the new materials could be applied to machines, increasing strength and power and eliminating vibrations.

In France, chromium steel began to be produced in 1877; it was mainly used to build armor and shells for projectiles.

Food-producing techniques: canning and experiments with freezing meat

The rise in industrial production in Europe and the United States made it possible to exchange industrial products for food products and raw materials. The importation of foodstuffs into Europe was particularly great from Australia, New Zealand, and Argentina, as well as from the United States, where over a million and a half hectares of land were turned to cultivation between 1860 and 1900, while livestock farms developed very quickly, especially during this decade. By 1870, the American canned-meat industry, centered in Chicago, was able to supply the European market. From this point of view, advances made in the techniques for preserving foods were extremely important. Australia had been one of the first countries to export canned meat. The first experiments in freezing meat were carried out in Melbourne around 1873 by James Harrison, and the following year, Thomas Sutcliffe Mort founded the New South Whales Fresh Food and Ice Company in Sidney, where compression machines with ammonia as the refrigerating fluid were used.

Trans-oceanic shipments of frozen meat from Australia and then from the United States and Argentina to England, began as early as 1877. These shipments were not always successful, mainly because navigation was so slow.

Civil engineering

Progress in communications was characterized by both the extension of the railway network (great transcontinental lines) and particularly bold feats of engineering. The Mt. Cenis tunnel, the first of the great tunnels through the Alps, was completed in 1871. It was built using mechanical drilling machines designed by Germaine Sommeiller (1815–71). In 1872, work was begun on the fifteen kilometers-long St. Gotthard tunnel. McKean type mechanical drilling machines were used in building it. The tunnel, completed in 1880, connected the two stations of Göschen and Airolo and was opened to railway traffic on January 1, 1882.

The railroads also opened passages underneath the great rivers the first large two-track tunnel built below water-level passed

under the Severn and was begun in 1873. In 1874, work was begun on the tunnel under the Hudson using compressed air caissons, which had previously been used for building dams.

The same year, bridges began to be built with steel rather than iron sections.

The construction of more and more daring building projects (tunnels, high buildings, etc.) required the use of increasingly complex and more accurate mathematical calculations. As a result, researches in applied mechanics, and especially in the theory of elasticity, were developed, and their results were immediately brought to the attention of the engineers. Of notable interest were the *Elemente der graphischen Statik* (1871), by Johann Bauschinger (1834–93), the *Theoretische Maschinenlehre* (1871–86), by Franz Grashof (1826–93), and one of the most typical treatises of the epoch, the *Théorie de l'équilibre des systémes élastiques et ses applications* (1879), by Alberto Castigliano (1847–84).

The telephone and the phonograph

The great advances in communications culminated in the invention of the telephone. The original invention was the work of Antonio Meucci (1804–89), an Italian immigrant to the United States, who filed a patent in 1871; however, due to the inventor's scarce financial resources, the patent remained unused. Five years later, in 1876, a very similar patent taken out by Alexander Graham Bell (1847–1922) was immediately exploited on a commercial level. According to the Supreme Court of the United States, called in to settle the question of priority, Bell's invention was not based on fundamental researches into the reproduction of sound, but on a pure and simple appropriation of Meucci's patent.

In 1878, the first telephone company was established in Great Britain; a year later, a switchboard was installed that enabled customers to converse telephonically, though for the moment the distance between the two speakers remained limited.

In 1877, while working on the telephone microphone, Thomas Alva Edison invented the phonograph, a device that was able to record and reproduce sounds.

Basic and refined chemistry

During this decade, chemists perfected the materials and techniques needed to develop the art of photography. In 1873, John Wesley Hyatt (1837–1920) patented celluloid, which he had obtained a number of years earlier, for use in producing solid objects, for example as a substitute for ivory in the manufacture of billiard balls. The possibility of using it as a support for the sensitive material in photography did not occur to him, however. In 1871, Richard Leach Maddox (1816–1902) discovered gelatin emulsion, a substance that was twenty times more sensitive than wet colloid. Within a decade, wet colloid plates had disappeared, replaced by a new process that can be said to have initiated instantaneous photography. Thus the art of photography was democratized, becoming accessible to almost everyone. For a discussion of the use of these new photographic techniques in astronomical observations, see the section ASTRONOMY 1870–1880.

As to discoveries in the field of dyestuffs, which were extremely important during this period, see CHEMICAL SCIENCES 1870–1880.

1880–1890

THOUGHT AND SCIENTIFIC LIFE

International scientific collaboration. New academies and universities

The lasting peace, during which the great powers were mainly involved in new and ambitious colonial conquests, favored a climate of international scientific collaboration. This was manifested in the numerous congresses held by the various disciplines toward the end of the century: There were congresses on botany (Genoa, 1892) geography (Antwerp, 1871; Paris 1875; Venice, 1881; Paris, 1889; Berne, 1899), geology (Paris, 1878; Bologna, 1881; Berlin, 1885; London, 1888; Washington, 1891; Zurich, 1894; St. Petersburg, 1897), zoology (Paris, 1889; Moscow, 1892; Leiden, 1895; London, 1898) and electricity (Paris, 1881).

In our discussion of the previous decade we already mentioned the cultural superiority and prestige of the German universities, which served as models for a number of continental European countries (for example, many Italian scientists referred to the German institutions and a productive exchange took place between researchers from the two countries). The Flemish Academy in Gand and the Royal Serbian Academy in Belgrade were both founded in 1866. Universities were inaugurated in Sofia in Bulgaria and in Göteborg in Sweden. The Australian Society for the Advancement of Science, modelled after the analogous English and American societies, was founded in Sidney in 1899.

The vicissitudes of positivism in Italy and Europe

Positivism flourished in Italy, exerting an ever-increasing influence on various areas of the culture. In pedagogy, for example, its effects were felt not only in theoretical elaboration, but in the practical reforms advanced by this discipline as well. The country found itself faced with the extremely serious problem of illiteracy, and in order to wipe out this social scourge, the bourgeoisie took elementary education out of the hands of the church and entrusted it to the state. Laicism, and consequently positivism, became the cornerstone of the new elementray school system. More or less orthodox positivists, such as Aristide Gabelli (1830–91) and Saverio de Dominicis (1846–1930) became the foremost pedagogists of the epoch.

Outside Italy, on the other hand, the spread of positivism began to encounter increasingly greater obstacles, often as a result of dogmatic form in which many of its followers presented it and their faith in science and progress, which was seen as a direct consequence of scientific discoveries and their technical applications. Despite its rapid decline, positivism brought to light the need to give a scientific structure to a number of disiplines that had traditionally been excluded from the realm of "real science" and to place the problem of scientific knowledge at the center of philosophical reflection in European culture.

The first of these two needs resulted in efforts to give a scientific structure to psychology and sociology, even on the part of thinkers who were only indirectly connected to positivism. With regard to psychology, special mention should be made of Wilhelm Wundt (1832–1920), who in 1879 founded the famous laboratory of psychology at the University of Leipzig. Sociology,

on the other hand, was given a huge boost by Émile Durkheim (1859–1917), who adamantly maintained that societies were autonomous entities that should be investigated by direct observation. In 1896, Durkeim founded and directed the *Année sociologique*, an important journal that for over a decade served to coordinate all research conducted on the subject.

Epistemological reflection: the work of Mach

There were two basic approaches to the problem of integrating science into philosophical thought. Both were of notable interest: one, provided by the scientists themselves, was direct, in that a new order of researches (which furnished the starting point for modern epistemology) was promoted to establish the precise meaning and value of scientific propositions and the methods used to demonstrate them: the other, which came from the philosophers, consisted in basing the entire theory of knowledge on a critical examination of scientific knowledge.

Ernst Mach (1838–1916) held an extremmly important position among the "scientist-epistemologists," making a decisive contribution to the critique of mechanics with his *Die Mechanik in ihrer Entwicklung historisch-kritisch dargestellt* (1883). Mach also wrote two other works in this same historical-critical vein, one on the principles of thermology, *Die Prinzipen der Wärmelehre historisch-kritisch entwichelt* (1896), and the other on the principles of physical optics, *Die Prinzipien der physikalischen Optik historisch und erkenntnis-psychologisch entwickelt* (published posthumously in 1921). Both of these, especially the first, had a considerable influence on the critical debate regarding the accomplishments of eighteenth-century physics. Meanwhile, he also published *Die Analyse der Empfindungen und das Verhältnis des Physischen Zum Psychischen*, an analysis of feelings that summarized his prevous studies on the problems of perception. This work also achieved considerable success.

The connection between the anti-mechanistic polemic that Mach developed in all these works and the anti-metaphysical battle fought by Comte in his time cannot be denied. But in Mach, the anti-metaphysical need was a methodological need within science. In fact, Mach did not consider metaphysics a system of beliefs in and of itself, historically surpassed in all aspects by science; rather it had insidiously hidden itself inside the most fully accepted scientific theories (even Galilean-Newtonian mechanics). It was to this theme that the Viennese neoempirical philosophers returned in the 1920s, giving rise to the Ernst Mach Verein.

Although investigations were being made into the fundamentals of mathematics that provided a counterpart, in the formal sciences, to the investigations into the fundamentals of physics, these only indirectly fell into the general atmosphere of positivism. They took a decisive turn during this period with the work of the great logician Friedrich Ludwig Frege, who was mentioned in connection with the previous decade. Two of his significant works were the *Begriffsschrift* of 1879 and *Die Grundlagen der Arithmetik* of 1884.

The theory of knowledge

Not all the philosophers of this period acknowledged the centrality of the problem of science were positivists; for example, one of the most authoritative neo-Hegelians, Francis Herbert Bradley (1846–1924), admitted the importance of reflection on science. One of his fundamental works, *The Prin-*

ciples of Logic (1883), attests to this interest. But it was mainly in Germany that the problem of (mathematical-physical) science assumed particular importance in a school that was opposed to positivism: the neo-Kantian school of Marburg, founded by Hermann Cohen (1842–1918). Cohen's epoch-making work on the fundamentals of infinitesimal analysis, *Das Prinzip der infinitesimal methode und seine Geschichte*, published in 1883, stands as evidence of his interest in science. A similar commitment to philosophical reflection on mathematics and physics can also be seen in Cohen's collaborator and colleague, Paul Natorp (1854–1924). Natorp's *Descartes' Erkenntnislehre* (1882) and *Einleitung in die Psychologie nach Kritischer Methode* (1888) both deserve mention. Cohen's and Natorp's approach passed unchanged to their disciple Ernst Cassirer (1874–1945).

Other currents

Meanwhile, another neo-Kantian school with very different interests from those of the Marburg school was established by Wilhelm Windelband (1848–1915), who studied problems of values. Windelband's first important writings, *Geschichte der neueren Philosophie* (1878–80) and *Geschichte der Philosophie* (1889–92), were of an explicitly historical nature.

In antithesis to all the currents mentioned up to now (from positivism to English neo-Hegelianism and German neo-Kantianism) was the irrational Friedrich Nietzsche (1844–1900), which was strongly influenced by the thought of Schopenhauer. Nietzsche's philosophy went through three remarkably different phases, all of which, however, shared the particular themes of revolt against reason, exaltation of the survival instinct, and struggle against the old dogmas of metaphysics, science, and ethics. Nietzsche's thought was characterized by an implacable denunciation of the crisis that had befallen modern society, despite its apparent successes. Of the author's numerous writings, we shall only mention *Die fröhliche Wissenschaft* (1882), *Also sprach Zarathustra*, a philosophical polemic written in 1883–85 but only published in 1891, and *Jenseits von Gut und Bö* (1885).

MATHEMATICAL SCIENCES

The apogee of arithmetical research

The widespread interest in arithmetic, which many mathematicians pursued out of a logico-critical need, was cultivated by others on a strictly technical level, in continuation of the great tradition established by the researches of Gauss, Dirichlet and Kummer. This was especially true of Leopold Kronecker, Krummer's successor in his Berlin chair, who had achieved great renown in 1858 when simultaneously with Charles Hermite, he discovered the solution to the general quintic. In 1882, Kronecker published his *Grundzüge einer arithmetischen Theorie der algebraischen Grössen*, introducing, among other things, the concepts of field of rationality and field of integrity, which he used to define and extend the concept of irreducibility of algbraic equations.

Prominent technical contributions to arithmetic were also made by Hermann Minkowski (1864–1909). In a memoir of 1883, he used the theory of quadratic forms with integral coefficients to advance the theory of so-called diophantine approximations along lines successfully developed by Hermite as well. He also inaugurated a new area in higher arithmetic with his so-called *Geometrie der Zahlen*, published in 1896.

Meanwhile other scholars such as Julius Wilhelm Dedekind and Giuseppe Peano (1858–1932) combined the two approaches and, in addition to studying and developing these technical questions, worked on building a logically rigorous foundation for arithmetic, a project that, as we have seen, interested many mathematicians of this period (particular attention was given to justifying the principle of induction, which was becoming increasingly more appreciated as a characteristic element of the structure of natural numbers). In 1888, Dedekind published his *Was sind und was sollen die Zahlen*, in which the concept of natural numbers was based on ordinal numbers and arithmetic was brought back to a few elementary principles. Peano had already distinguished himself in 1886 by solving the problem of the integrability of a certain first degree differential equation using several conceptual definitions, such as upper and lower limit, that he himself introduced. In 1888, Peano published his *Calcolo geometrico*, which gave rise to modern vector calculus, and in 1889 he presented the famous axiomatization of arithmetic that still bears his name. In it, natural numbers were implicitly characterized by five axioms (which in substance coincided with those proposed by Dedekind), including the axiom of induction. Starting from this formalization of arithmetic, Peano attempted to proceed to a gradual reconstruction of the various branches of mathematics; the results of his efforts began to appear in 1895 with the publication of the first edition of the *Formulario matematico*, in which techniques of axiomatization and symbolic logic were used to carry out the reconstructions. In this way, the program of arithmetizing mathematics reached its crowning point via the explicit edification of various mathematical theories, beginning with arithmetic.

Meanwhile, important developments were also taking place in analysis. In 1882, Felix Klein published a pamphlet on Riemann's theory of algebraic functions, in which he presented the results of important researches on the Riemannian current's theory of functions. In 1885, Vito Volterra (1860–1940) published his first works on the theory of integral equations; in 1887, he published the memoir that marked the birth of functional analysis.

The systemization of differential geometry

In the field of geometry, it should be mentioned that a work by Moritz Pasch (1843–1930), *Vorlesungen über neuere Geometrie* (1882), inaugurated the axiomatic and formal reconstruction of elementary geometry that especially characterized the following decade, but which Peano already developed to a considerable degree as early as 1889 in his *I principi di geometria*. Meanwhile, a systematic understanding of all the properties of curves and surfaces, not as whole units, but around one of their points— a problem that had already been studied by Georg Friedrich Bernhard Riemann, Eugenio Beltrami, Elwin Bruno Christoffel and R.O. Lipschitz—was reached. The study of these properties was given the name differential geometry, and in 1886 Luigi Bianchi (1856–1928) published his *Lezioni di geometria differenziale*, an important treatise that reorganized and revived all the material on the subject.

The possibility of extending the principal theories of elementary and analytical geometry to hyperspaces, even without the aid of coordinates, was rigorously established by Giuseppe Veronese (1854–1917) in a memoir entitled *Behandlung der projectivischen Verhältnisse der Räume von verschiedenen Dimensionen durch das Prinzip des Projicirens und Schneidens* (1882), which marked the beginning of an important line of research. In Stockholm, in 1882, Gosta Mittag-Leffler (1846–1927), a well-known analyst, founded the *Acta mathematica,* in which Georg Cantor's second series of articles on set theory was published.

ASTRONOMY

Astronomy via telegram

The increasing frequency with which unexpected phenomena were being observed in the skies (apparitions of comets and novae, discoveries of planetoids, etc.) called for the institution of a data center where astronomical information could be gathered and dispatched via telegraph, so that phenomena that were taking place could be observed before they ended. An international center of this type was founded in Kiel in 1881.

Langley's bolometer

New methods also began to be used to investigate the sun; they were directed at quantitative and qualitative studies of the sun both when it was active and at rest. In 1878, Samuel Pierpont Langley (1834–1906), a physicist, astronomer, writer, and pioneer of aviation, had devised an instrument that made it possible to measure the energy at every point of the solar spectrum, even in the infrared zone that was invisible to the eye. Several years later, Langley used his instrument, which he called a bolometer, to measure the energy radiated by the sun at various wavelengths, thus founding spectrophotometry. With a photographic recording system, he also obtained a map of the infrared spectrum up to 5.5μm. Given that the energy emitted could be deduced by measuring the energy received when the distance from the source and any intermediate absorption were known, Langley extended his researches to problems of absorption by the earth's atmosphere and determining the solar constant.

Rowland's gratings

The localization of the lines on the solar spectrum, which Anders Jonas Ångström (1814–72) first worked on in 1868, measuring the positions of 150 fundamental lines and about a thousand others on solar spectra obtained with rudimentary gratings, was resumed by Henry Augustus Rowland (1848–1901) in Baltimore around 1881. After having invented a procedure for manufacturing screws with a uniform thread, he succeeded in producing diffraction gratings, scoring approximately six hundred lines per millimeter on flat (later concave) surfaces. Using these gratings, he was able to obtain normal solar spectra (that is, where dispersion did not vary with wavelength), which he recorded on photographic plates. By 1888, he was able to draw a map of the solar spectrum between 6,850 and 2,967 Å; altogether, it was thirteen meters long and included more than twenty thousand lines. Revisions of the map and extensions to other zones of the spectrum were made in 1928 and 1957.

Between 1888 and 1901, Nils Duner, director of the Observatory of Uppsala, conducted researches aimed at determining the sun's rotation by means of the spectroscope.

The Carte du Ciel

The great advantages offered by the new method of dry-plate photography opened the way to undertakings that were impossible using direct observation. Around 1880, in their work at the Paris Observatory on the great map of the ecliptic with stars of up to 13^m (begun by Charcornac in view of the researches on asteroids), the brothers Paul Henry (1848–1905) and Prosper Henry (1849–1903) reached the zone where the zodiac crosses the Milky Way. Realizing the impossibility of visually recording all the stars of that zone, they successfully experimented with photographic plates. Impressed by the results, Admiral Ernest Barthélémy Mouchez (1821–92), director of the observatory, commissioned the Henrys to build a special photographic equatorial telescope with an aperture of thirty-four centimeters. Using this instrument, the Henry brothers took exceptional photographs of the moon, the planets, and the star fields. On November 16, 1885, while photographing the Pleiades, they not only recorded 1,421 stars in the course of a few hours, but discovered a nebula that was invisible to the eye. These results filled Mouchez with enthusiasm, and in 1887 he convened in Paris the first international congress for compiling a photographic map and catalogue of the skies. The work was organized and distributed among eighteen observatories all over the world, where photographs were to be taken in different zones but using absolutely identical instruments, photographic materials, and methods. Unfortunately, this first great collective effort demonstrated not only the great spirit of collaboration but the limitations of group work, since all the data was collected but has not as yet been completely reduced for the catalogue.

A supernova in Andromeda

In 1884, a supernova was observed in the nucleus of the Andromeda nebula. While the spectra of other nebulae revealed their gaseous natures, the galactic or extra-galactic nature of Andromeda had not yet been determined. Due to the inadequacy of the means available at the time, this problem remained unsolved.

Cosmology

On the basis of the Kant-Laplace hypothesis on the evolution of the stars from galactic nebulosity, and of advances made in the physical sciences (laws of conservation of energy and the first and second principles of thermodynamics), cosmological theories on the contraction and heating of the sun were being developed (Hermann von Helmholtz, 1854). In 1879–81, George Darwin studied the role of tidal friction in the evolution of the earth-moon system. The problems of planetary cosmology were further developed in the theory of the equilibrium and stability of solids of rotation, elaborated by Jules Henri Poincaré (1885) and, in a more exhaustive form, by Aleksandr Mikhailovich Lyapunov (1896).

PHYSICAL SCIENCES: MECHANICS

The problem of the ether and Michelson's first experiment

The questions raised by the increased interest in the fundamentals of mechanics continually brought the attention of the scientists back to the problem of the existence of privileged reference systems. This problem especially involved the various concepts of the ether, which many saw as an absolute reference system with characteristics that made it the focus of both mechanics and the electromagnetic theory of light. During these years, the possibility of rendering the earth's motion observable by means of the ether was linked to the possibility of experimentally determining the results of measurements connected to the ratio between the velocity v of the planet and the velocity c of light. But the numerical value of this ratio was very small, since $v/c \approx 10^{-4}$. On the other hand, the electromagnetic theory of light led to the conclusion that, by means of special optical experiments, so-called "second-order effects" could be predicted, and therefore depended on $(v/c)^2$. Even James Clerk Maxwell, in an entry on the ether written for the *Encyclopedia Britannica*, had maintained the impossibility of measuring these effects, due to insurmountable technical difficulties; in 1880, *Nature* magazine published a brief posthumous piece in which Maxwell reiterated this impossibility. Albert Abraham Michelson (1852–1931), a United States expert in experimental optics, dedicated his attention to the problem. After moving to von Helmholtz's laboratories in Berlin in 1880, Michelson perfected a technique based on the phenomenon of interference and set out to measure "the motion of the earth in relation to the ether" by rendering second-order effects observable. By the spring of 1881, following a series of interferometric measurements discussed with von Helmholtz and carried out at Potsdam, Michelson presented a memoir that was more problematic than ever. His results did not confirm the predicted effect, that is, a displacement of interference fringes. And since the anticipated effect depended on assumptions concerning the ether, Michelson concluded that since the interpretation of his results showed that there was no displacement of the interference fringes, his measurements therefore demonstrated the incorrectness of the consequence of the hypothesis of a stationary ether. Despite the interest it stirred up among many authoritative scientists, including Hendrik Antoon Lorentz, William Thomas Kelvin, Lord Rayleigh, and Josiah Willard Gibbs, the outcome of the Potsdam experiment did not appear conclusive, in that Fresnel's hypothesis on the ether was amply confirmed by other types of experimental data. Michelson himself was aware of the situation, and after Kelvin's lessons in Baltimore (see PHYSICAL SCIENCES, ELECTROMAGNETIC THEORIES, 1880–1890) in 1884, which he attended together with Edward Williams Morley (1838–1923), he set out not so much to repeat the Potsdam experiment as to check out the results obtained by Armand Hippolyte Fizeau (1819–96) in 1851, which seemed to support the validity of Fresnel's hypothesis on the ether.

Mechanics according to Mach

While investigations into the ether and second-order effects involved the establishment of new correlations between mechan-

ics and electromagnetic theories, Ernst Mach's analyses on the so-called absolute concepts of mechanics led to new correlations between this science and thermodynamics. In his fundamental work on the development of mechanics presented from a historico-critical point of view, *Die Mechanik in ihrer Entwicklung historisch-kritisch dargestellt*, published in 1883, Mach maintained that the idea that mechanics is the foundation of all other sectors of physics was a prejudice. According to him, the unification of the physical sciences had to be based in thermodynamics, seen as a general science of energy transformations. This theme, which as we know influenced not only strictly scientific, but philosophical culture as well, found according to Mach, a rational basis in a historico-critical analysis of the formation of scientific concepts and involved a concise critique of mechanism. Mach upheld the need to redefine the notions of mass and force in such a way as to reduce them to mathematical expressions, thus depriving them of their traditional metaphysical content and linking them instead to evaluations of an empirical nature. Since making these links involved measuring spatial and temporal parameters, Mach's critique was also directed toward the concepts of space and time. These concepts could no longer be judged as absolutes, since the history of physics pointed out their empirical content and demonstrated what Mach, for example, called "the conceptual monstrosity of absolute space." Rather than speak of absolute space and time, according to Mach, one should speak of physical space and physical time: "Space and time are well-ordered systems of sets of sensations. The quantities stated in mechanical equations are simply ordinal symbols, representing those members of these sets that are to be mentally isolated and emphasized. The equations express the form of interdependence of these ordinal symbols." For this reason, the theory of motion rested not on absolute categories, but on altogether relative notations, since there was no difference between relative and absolute that human beings could grasp with their senses.

The reconstruction of mechanics as proposed by Mach had to develop along different lines from those outlined in Kirchhoff's program (see PHYSICAL SCIENCES: MECHANICS, 1870–1880) and led to a context where the problem was to establish correlations among phenomena that could be evaluated in terms of economy. Mach insisted on the conventional nature of physical laws and the empirical origin of their relative validity.

In keeping with his own program, Mach disagreed with the supporters of the statistical approach, to the extent that their theories depended on metaphysical hypotheses concerning the existence of atoms and molecules. In this sense, although he distinguished his positions from those of Wilhelm Ostwald and the more fervent energists, Mach was objectively an ally of those physicists and philosophers who were conducting a heated battle against atomistic hypotheses and the research programs supported by Ludwig Boltzmann (1844–1906) and other scientists who adhered to materialistic notions.

The law of inertia

A further contribution to clarifying the concept of space was provided by Ludwig Lange's *Über die wissenschaftliche Fassung des Galileischen Beharrungsgesetzes*, a work on the scientific formulation of Galileo's law of inertia that was published in 1885.

The difficulties connected with this concept could be resolved, according to Lange, through a rigorous definition of reference systems for which the law of inertia was valid. These inertial systems would provide a better understanding of the physical content of the law of inertia, thus avoiding having to resort to what Lange described as the "spiritual" nature of absolute space.

The variety of themes studied in relation to mechanics continued to grow toward the end of the decade, thanks to the rapid developments that took place in statistical mechanics, the electromagnetic theories of light and researches into the structure of matter and the ether. These developments have to be considered in order to understand why, during the last decade of the nineteenth century, reflections on mechanics were dominated by Heinrich Rudolf Hertz's elaborations of them.

PHYSICAL SCIENCES: THERMODYNAMICS AND STATISTICAL MECHANICS

The deduction of Stefan's law

In 1879, Josef Stefan (1835–93) found the correlation between the heat radiated by bodies and the fourth power of the temperature (see PHYSICAL SCIENCES: RADIATION AND MATTER, 1870–1880). Stefan's analysis was above all based on a critique of the law established by Dulong and Petit at the beginning of the century; however, it did not seem to find sufficient authoritative confirmation, and by the beginning of the eighties, only a very few physicists, among whom were J. Violle and Ludwig Valentin Lorenz, made reference to it. It was Boltzmann who provided proof of the logical necessity of Stefan's law, demonstrating that it was a consequence of the second principle of thermodynamics and Maxwell's electromagnetic theory. Drawing in part on several ideas advanced by the Italian physicist Adolfo Bartoli, in 1884 Boltzmann published a brief essay in which, instead of examining the phenomenon studied by Stefan (i.e. how bodies cooled), he studied the behavior of a certain amount of energy within a given space that was maintained at a constant temperature. Thus on the basis of physical concepts that were very different from those involved in Stefan's reflections, Boltzmann arrived at the same result, opening the way to further researches on what now appeared to be a fundamental law of nature. As a matter of fact, a comparison of Stefan's hypotheses and Boltzmann's mathematical deductions shows that the same law is valid in relation to both the cooling of a solid and the density of energy of heat radiation in space. This new path appeared in the rapidly developing theoretical framework of studies related to the distribution of energy in black bodies. Still another decade had to pass before Stefan's law was experimentally confirmed.

Heat conduction in rarefied gasses

During the first half of this decade, William Crookes (1832–1919) and James Thomson Bottomley made futher contributions to knowledge about the ductability of heat in rarefied gasses. Pursuing his program of research on what he by this time defined as "the fourth state of matter," which was closely connected to studies on cathode rays, Crookes, in 1881, published experimental data on the interdependence of heat radiation and the degree of vacuum. He observed that the cooling time of a glass bulb surrounded by an environment with controllable pressure increased suddenly at a pressure of approximately fifteen millionths of an atmosphere. He connected this occurrence to a sudden variation in the mean free path: in his opinion, heat was surrendered by conduction at low pressures. Crookes continued his researches up to two millionths of an atmosphere and found that they supported a corpuscular interpretation of the nature of cathode rays (see PHYSICAL SCIENCES: RADIATION AND MATTER, 1880–1890).

Planck's first theoretical program

In his first studies on thermodynamics, Max Planck (1858–1947) attempted to elaborate a fundamental macroscopic theory based on thermodynamics and the electromagnetic theory and therefore to find a very general theoretical framework that would rationally accommodate all statements on irreversibility. In doing this, Planck, for a number of years, remained critical of Boltzmann-type programs. In a memoir of 1882, he maintained that the second law of thermodynamics "is incompatible with the hypothesis of finite atoms," and in a work of 1887 on the principle of entropy (Über das Princip der Verhaltung der Entropie) rejected all assumptions on molecular motion in order to focus his attention on elaborating methods for calculating the values of Clausius's function. Although Planck joined forces with Wilhelm Ostwald (1853–1932) in criticizing Ludwig Boltzmann at the congress of German scientists held at Halle in 1891, it should be pointed out that his position was not at all the same as that of the energeticists. In fact, in his opinion, it was not possible to derive the second law of thermodynamics from mere energetic considerations. In the course of a few years, this divergency became the source of harsh attacks against the physicists who supported energetics (see PHYSICAL AND STATISTICAL MECHANICS, 1890–1900).

Irreversibility and the calculus of probabilities

In his demonstration of the theorem that was later known as the "H theorem," Boltzmann summarized the physical meaning of the second law of thermodynamics in his two memoirs of 1872 and 1877 as follows: In the majority of cases the initial state would be extremely improbable; the system goes from this state to increasingly more probable states, i.e. the state of thermal equilibrium. From this point of view, the entropy S of the system was proportional to the probability W of the macroscopic states and, therefore, to the number of molecular combinations. This proportionality (which Planck later gave the form $S = k \ln W$) implied that the notion of irreversibility should be interpreted as the tendency of physical systems to pass from less probable to more probable states and that it could be elaborated through deductions based on a probability concept of the behavior of the molecules. Nevertheless, in reviewing James Clerk Maxwell's monograph of 1879 on the mean distribution of energy in a system of material points (see PHYSICAL SCIENCES: THERMODYNAMICS AND STATISTICAL MECHANICS, 1870–1880), Boltzmann emphasized that the probabilistic approach could not be reduced to a single general method, but rather differed significantly from one research program to the next. There existed a difference in method between Maxwell and Boltzmann, noted the Viennese physicist in his review of 1881, in that Boltzmann measured the probability of a state on the basis of the amount of time during which the system is at an average in that state, while Maxwell took into consideration a huge number of analogously structured systems in all their possible initial conditions; the probability in question then being given the ratio between the number of systems that were in that condition and the total number of systems. These differences were related to different interpretations of the physical significance of both the second law of thermodynamics and the notions that were used to attempt to define molecular behavior. It can thus be seen how difficult the historical process was that led to a gradual clarification of the various aspects of the propositions regarding the irregularity of molecular motion, and how these aspects were connected to the possible interpretations of probability itself. During this period, the debate on these problems took the form of the ratio that could be found between probabilistic and dynamic representations of microscopic states, and this particular form led to the rise of paradoxical situations. We have already mentioned Kelvin's and Loschmidt's thoughts on the subject. During this particular debate, other scientists also tackled the problem of the possible correlations between dynamic reversibility and thermodynamic irreversibility. In a memoir of 1887, George Johnstone Stoney reproposed the same problem, suggesting that the second law of thermodynamics should not in any way be considered "a true law of dynamics" and that time could not be given a value as something existing apart from physical events. However, it should be pointed out that most physicists tended to debate Boltzmann's ideas, not so much confronting them in all their mathematical complexity as reducing them to more or less intuitive considerations. Thus various situations of controversy arose where, instead of dealing with more or less rigorous definitions of the category of probability, only general opinions, such as the inevitability of the thermic death of the universe, were confronted (often in heated tones). However, even these debates were important to a clarification of the statistical approach to mechanics and formed several of the premises on which the great debate of the final decade of the century was based; it was out of that debate that Planck's original ideas on quantum theory emerged.

The Tait-Boltzmann-Burbury controversy

Between 1886 and 1887, with a memoir entitled *On the Foundation of the Kinetic Theory of Gases*, Tait gave rise to a debate that, in addition to bringing out various concepts of thermodynamics, also made evident the existence of different ideas on the role that mathematics should play in the formulation of concepts and scientific laws. Tait's attack against Maxwell's and Boltzmann's ideas was more explicit and polemical than ever, even though it was partially veiled by an attempt to improve Maxwell's demonstration of the law of distribution of molecular velocities. One of Tait's arguments was directed against the rigorous formalism of Boltzmann's memoirs; Tait accused Boltzmann of a mathematism that constituted "a game of symbols" rather than "a reasoning conducted by successive steps." When Burbury became involved in the polemic, he limited his reaction to a dry accusation of "inconsistency" with respect to Tait's reasoning, but Boltzmann's reaction was more violent. In the first place, Tait's attempt to demonstrate the validity of certain inferences related to Maxwell's function was nothing more than

a *circulus vitiosus*, in Boltzmann's opinion, since it boiled down to an attempt to demonstrate what was already contained in the premises of the treatment. In the second place, again according to Boltzmann, a sort of unjustifiable mistrust of more complex mathematical formulas that, commented the Viennese physicist, were only unclear to those who could not understand them, lay at the root of Tait's errors. The polemic continued until 1888, and it is interesting to note that, while on the one hand Burbury attacked Tait on the grounds that his arguments had no other basis than the "weight of authority," on the other Tait was forced to admit that he had intervened upon Kelvin's request, although he never renounced his idea that Boltzmann used mathematics "in substitution of thought," and not for the purpose of "aiding thought" itself.

Irreversibility and periodicity

The controversy about the kinetic theory of gasses had a variety of roots. The theory and its advancement were opposed not only by those physicists who upheld the canons of energetism, but by researchers who, although they were convinced supporters of the existence of molecules and the possibility of understanding matter scientifically, did not admit a need in the physical sciences for probabilistic interpretations such as those that played a predominant role in the programs of Boltzmann and the other statisticians. At the congress of the mathematical and physical section of the British Association held in Montreal in 1884, Kelvin read a long paper on the subject entitled *Steps Toward a Kinetic Theory of Matter*. Although he admitted that the theory elaborated by Clausius, Maxwell, and Boltzmann was a step in the direction of understanding "a definite and perfectly intelligible order of events within Nature," Kelvin maintained that the model in which the theory was based was a source of irresolvable contradictions. In his opinion, if molecules were elastic solids that continuously collided with each other, all the translational energy must in the end be transformed into vibrational energy at increasingly higher frequencies. These unresolved difficulties, therefore, ought to elicit new investigations on the possibility of building an alternative to the model that was commonly accepted in kinetic theory, an alternative that would carefully avoid attributing elasticity to the elements that made up matter, turning instead to the physics of vortices.

In the years that followed, Kelvin carried out assiduous investigations in that direction, working with several complex theorems on the stability of stationary and periodical motion and on the maximums and minimums of energy in whirling motion. Toward the end of the decade, he also attempted to relate his results to the conclusions reached by Jules Henri Poincaré (1854–1912) in his 1880 memoir on the three-body problem (*Sur le problème des trois corps et les équations de la dynamique*). As we shall see, Kelvin abandoned some of his objections against the kinetic theory of gasses during the following decade, only to raise others. In any case, the physico-mathematical work that he and other scientists developed during the second half of this decade constituted a major point of reference for the debate on statistical mechanics that began with Culverwell's objection in 1890 and lasted until Planck formulated his quantum hypothesis in 1900. It also helped to improve the formulation of all the unresolved problems that arose when scientists were forced to examine the historical relations between researches on mechanics, thermodynamics, and electromagnetic theory in the context of the reelaboration of the physical sciences. For a discussion of chemical thermodynamics, see CHEMICAL SCIENCES, 1880–1890.

PHYSICAL SCIENCES: ELECTROMAGNETIC THEORIES

The observability of electromagnetic waves and their propagation in space

The association between field and electromagnetic waves was made in 1880 in one of the first treatises on electricity and magnetism that accepted James Clerk Maxwell's ideas. James Edward Gordon's *A Physical Treatise on Electricity and Magnetism* described the "propagation through space" of electromagnetic induction as though no reference was made to electromagnetic radiation emitted by sources of an electrical nature. The entire problem, as it manifested itself among Maxwell's first pupils and most of his followers from the time of his great memoir of 1864 to Hertz's famous experiments—carried out during the second half of this decade—was conditioned by various theoretical considerations, mainly concerning the nature of electricity. It is a well-known fact that Maxwell did not believe in looking for new models on the nature of electricity and electrical charge, and instead turned his attention to the relations between ether and matter, which in his opinion, were at the base of all electromagnetic phenomena. The failure of some of his students' first attempts to measure the velocity of propagation of electromagnetic events in the laboratories of the Cavendish Foundation after 1874 can mainly be attributed to this basic feature of his research program. On the other hand, George Francis Fitzgerald reiterated in 1879 that it was impossible to produce wave perturbations propagating into space from any system of mobile or fixed conductors charged in any way. These ideas, defended in a work significantly entitled *On the Possibility of Originating Wave Disturbances in the Ether by Means of Electrical Forces*, were later rejected by the same author in 1882 and 1883. In a communication entitled *On a Method of Producing Electromagnetic Disturbances of Comparatively Short Wavelengths*, presented to the 1883 congress of the British Association, Fitzgerald affirmed that, on the basis of Maxwell's equations, it could be predicted that sinusoidal currents radiate, and that the radiation could be rendered observable in the laboratory through experiments on the oscillations that accompany condensor discharge. Fitzgerald's ideas, together with similar considerations advanced by Oliver Lodge and Hendrik Antoon Lorentz, represent the first attempts at analyzing the observability of electromagnetic waves and their propagation in space.

Uncertainties on the nature of electrical charge

As Helmholtz wrote in a communication of 1881, dedicated to the development of Faraday's notions on electricity, considerable difficulties arose as soon as one attempted to clarify the meaning of "quantity of electricity" as used in Maxwell's theory: it did not seem possible, he observed, to explain its meaning without the use of mathematical formulas. Just a few years later, at the 1885 congress of the British Association, even the physicist Joseph John Thomson, in his *Report on Electrical Theories,*

acknowledged that it was rather difficult to see what was meant in Maxwell's theory by the expression "quantity of electricity."

In relation to all this, it is interesting to note that the scientists who were closest to the electromagnetic theory of light during these years tended to reject all attempts at providing a better explanation of the nature of electrical charge independent of the relations between ether and matter. In a communication of 1885 entitled *Electromagnetic Induction*, Oliver Heaviside (1850–1925) voiced "an enormous repugnance" for any approach that depended essentially on the notion of electricial fluid. And, as Lodge reiterated in the pages of his *Modern Views on Electricity,* published in 1889, the fundamental issue was not the problem of the nature of electricity, but a need to understand the nature of the ether.

On the other hand, various theoretical developments favoring a discreet interpretation of electricity began to appear. The speech given by George Johnstone Stoney (1826–1911) at the Belfast congress of 1874, in which he maintained that the fundamental units of matter should include a single and defined quantity of electricity that is independent of the bodies on which it acts and whose charge could also be calculated, was published in 1881. Important contributions were also provided by the development in Lorentz's program, which as early as 1878 maintained that the electricity contained in molecules was associated with corpuscular microstructures.

Developments in the physics of the ether

The problem of the interactions between matter and the ether in view of elaborating a dynamic theory of electromagnetic and optical phenomena was the focus for the famous lessons that Kelvin taught at Johns Hopkins University in Baltimore in 1884. These lessons, conducted in the form of a seminar for a very restricted group of specialists, represented a phase of reelaborating and relaunching programs aimed at generalizing the mathematical physics of the ether to include the various interactions of atoms in matter. In short, these lessons were a sort of critical response to the growing interest in James Clerk Maxwell's theory. As a matter of fact, the Baltimore lessons intended to contrapose the dynamics and physics of elastic solids—a dynamics and physics that, according to Kelvin, constituted "the only sustainable foundation for the wave theory of light in the present state of our understanding," to the "so-called electromagnetic theory of light." These seminars laid the premises for the particular developments that took place in the physics of the ether toward the end of the decade, in connection with the so-called Green and Kelvin ether. The mathematization of this ether utilized ideas related to the theory of gyroscopes that made it possible to formulate mechanical explanations for various optical phenomena. Complex models of the gyroscopic ether were also used to facilitate studies of it, but, as Kelvin himself observed more than once, these models were to be considered analogies or intuitive instruments and not reproductions of the real ether. At the same time, the Baltimore seminars pointed out the need to design experiments that could render the motion of the earth observable with respect to the ether. Albert Abraham Michelson (1852–1931) and Edward Williams Morley (1838–1923) both participated in these closed seminars (see PHYSICAL SCIENCES: MECHANICS, 1880–1890).

Heaviside's program

Accurate mathematical studies were made in Maxwellian physics toward the middle of this decade. The work of Oliver Heaviside deserves particular mention in this context. In a memoir of 1885 entitled *On the Electromagnetic Wave-surface*, Heaviside sought the correlation between Maxwell's equations and Fresnel's wave surfaces, suggesting the use of "higher methods" from a mathematical point of view. In place of the usual Cartesian notations, he used vectorial calculus, which enabled him not only to abbreviate the complex formalisms of Maxwell's theory, but above all to "clarify the meaning and mutal connections of the more complicated formulas." The simplifications that Heaviside introduced into Maxwell's formalisms were meant to lead to the renunciation of what he called "obscure and almost metaphysical reasoning." He reiterated similar ideas in 1888 in a memoir entitled *On Electromagnetic Waves,* which upheld the validity of "abstract electromagnetic theory" in relation to wave theory. On that occasion, Heaviside did not neglect to emphasize the notable contrast between Maxwell's ideas and those supported by Kelvin and other scientists who were attempting to formulate non-electromagnetic wave theories.

The Michelson and Morley experiments

The discourse launched by Michelson with his Potsdam observations of 1881 (see PHYSICAL SCIENCES: MECHANICS, 1880–1890) acquired new vigor after the Baltimore seminars and from the ideas proposed by Kelvin and Lord Rayleigh with particular reference to the problems connected to Fresnel's drag coefficient. The interest in Michelson's interferometric method rightly derived from the fact that it considered second order effects in v/c, and in any case was to attack delicate optical measurements with a much higher degree of accuracy than had been achieved in Armand Hippolyte Fizeau's observations of 1851.

In 1886, in Cleveland, Michelson and Morley completed a series of experiments aimed at testing the results obtained by Fizeau in his time. Writing to Kelvin and Joseph Willard Gibbs, they stated that Fizeau's work had been fully confirmed. Lorentz commented on the publication of Michelson and Morley's results and, in an article of 1887 on the influence of the earth's motion on light phenomena (*De l'influence du mouvement de la Terre sur les phénomènes lumineux),* attempted to connect the new observations regarding Fresnel's coefficient with the results of Michelson's 1881 experiment within a coherent theoretical framework. Lorentz's approach was characterized by a partial acceptance of Stokes's theory, to which he suggested adding a correction based on the consequences of accepting Fresnel's ideas. Instead of assuming that the ether had the velocity originally indicated by Stokes, Lorentz held that the earth partially interacted with the ether in such a way as to satisfy the value of Fresnel's coefficient in transparent media. From this point of view, the results of 1881 could be reinterpreted, since Michelson's calculations of that time contained an error that Lorentz pointed out and corrected. It was then a question of repeating the Potsdam experiment more accurately, as Rayleigh suggested, and in April, 1887, Michelson and Morley began making new measurements using a refined interferometric device. In August of that same year, Michelson reported to Rayleigh on their progress, writing that

the experiments on the relative motion of the earth and the ether had been completed, and the result was decisively negative. On the basis of the predictions, the fringes of interference with respect to zero should have been 0.40 of a fringe: but the maximum displacement was 0.02, and the average displacement was much less than 0.01, and did not occur in the proper zone. Thus the physical sciences were faced with profound questions that simultaneously invested mechanics, electromagnetic theories, and the dynamic theories of light. These problems finally found solution in the relativity theory of Albert Einstein (1879–1955).

Hertz's researches

As we have already seen, Helmholtz's teachings and researches on electrical and magnetic phenomena contributed to the spread of Maxwellian concepts on the Continent (see PHYSICAL SCIENCES: ELECTROMAGNETIC THEORIES, 1870–1880). And it is following a line of research that in many ways referred back to Helmholtz's ideas that we arrive at Heinrich Rudolf Hertz's first work on the observability of electromagnetic wave propagation in space (1887). To the physical sciences of the time, Hertz's publications, which appeared between 1887 and 1890, offered proof of the validity of the electromagnetic theory of light.

As Hertz himself pointed out in a memoir of 1888 on rays of electrical force (*Über strahlende elektrische Kraft*), he had been able to demonstrate that the action of an electrical oscillation expands in space like a wave. In his systematic study of these wave phenomena, Hertz meticulously analyzed rectilinear propagation, polarization, and the reflection and refraction of rays produced by oscillators and collected a large body of experimental data. Nevertheless, Hertz's intervention cannot be reduced to a mere proof of the validity of Maxwell's theory. Starting from the empirical data he collected after 1887 and an accurate knowledge of theories on electromagnetic phenomena, Hertz developed his own theoretical elaboration, which he expounded in two fundamental memoirs of 1890, dealing respectively with the electrodynamic equations for bodies at rest (*Über die Grundleichungen der Elektrodynamik für ruhende Körper*) and for bodies in motion (*Über die Grundleichungen der Elektrodynamik für bewegte Körper*). Hertz proposed to develop a clarification of Maxwell's ideas through a new mathematization of the electromagnetic theory. As we shall see in our discussion of electromagnetic theories during the last decade of the nineteenth century, Hertz presented a general outline of this program in his introduction to a collection of writings on electromagnetic waves that was published in 1892.

PHYSICAL SCIENCES: RADIATION AND MATTER

Models for cathode radiation

The corpuscular model proposed by Varley and defended by William Crookes during the previous decade was based on a hypothesis, particularly dear to Crookes, according to which cathode rays were a sort of flood of negatively charged molecules that were repelled by the cathode itself. In this sense, the "fourth state of matter" referred to by Crookes was the extremely rarefied state of matter inside cathode tubes, where molecular collisions were extremely rare. The hypothetical charge-carrying molecules had to move at very high speeds and, as Tait observed in 1880, an examination of the spectral lines of gasses inside discharge tubes should make it possible to measure a displacement caused by the Doppler effect. That this phenomenon could not be observed placed doubts on the validity of Crookes's model, already under attack by many continental physicists who favored an electromagnetic wave interpretation of cathode radiation. The credibility of the wave model found ample confirmation during this decade, thanks to the experiments of Eugen Goldstein and Hertz, while other experiments by Crookes and Arthur Schuster supported the possibility of consistently modifying Crookes's model. This debate on models led to the clarification of a number of problems related to the structure of cathode radiation and played a positive role in the development of research programs on electrons during the following decade.

The wave interpretation

Goldstein's idea that cathode rays were electromagnetic rays rested on the following experimental arguments: first, as measurements related to the Doppler effect showed, Crookes's hypothetical molecules had to travel at a much slower speed than was required by the corpuscular model; and second, as calculations based on the mean free path of these hypothetical particles showed, they should be deflected. However, no such effect was observable, since cathode rays travelled in a straight line. These considerations, published in 1880, were joined in 1883 by Hertz's results. In particular, using various arrangements of the cathode and the anode, Hertz demonstrated that the electric current between the two electrodes did not necessarily follow the direction of the cathode rays: thus the hypothesis that the rays were carriers of electrical charge could not be sustained. On the other hand, contary to all predictions based on Crookes's model, Hertz himself was not able to observe any deflection of cathode rays by electrostatic fields and therefore deduced that the corpuscular model ought to be abandoned.

The corpuscular interpretation

The experiments and ideas developed by A. Schuster beginning in 1884 seemed to suggest the possibility of maintaining a corpuscular model by making several changes in Crookes's original model. According to Schuster, the non-observability of special Doppler effects could be explained by presuming that cathode rays were not made up of molecules that became negatively charged when they came into contact with the cathode, but were rather the result of a dissociation of the gas molecules into positive and negative ions. The light emitted from the discharge tube therefore came from molecules that were hit by the rays and not from the particles that actually made up the rays themselves. In addition, Schuster demonstrated that the deflection of cathode rays in the presence of magnetic fields corresponded to predictions based on the corpuscular model. During the years that followed, Schuster conducted researches aimed at finding values for the ratio between charge and mass for the cathode charge carriers. Crookes's investigations particularly focused on the area of spectroscopy and led to attempts at generalizing his concepts, even to the point of proposing a sort of evolutionary theory of matter. One of the most interesting questions that Crookes tried to answer concerned the surprising complexity in the basic structures of matter revealed

by the new spectroscopic techniques. According to Crookes, researches on cathode rays and spectra should lead to reflections on the possibility of defining truly "elementary" objects. All elements seemed to be further divisible, so in an 1887 conference of the Royal Institution, Crookes asked where the true final element was. The more scientists advanced, he said, the more it moved away like a mirage. The very idea of an element in the sense of something absolutely primary and absolute seemed to become increasingly clear and meaningful. The situation in which the physical sciences found themselves thus posed radically new problems, which were not reducible to absolute explanations nor solvable by means of crucial experiments.

CHEMICAL SCIENCES

Properties of solutions: osmotic pressure and the solubility of electrolytes

Pursuing his studies, François Marie Raoult discovered the laws of cryoscopic depression (1882), ebullioscopic elevation, and the relative depression of vapor pressure (1886–88). From these he derived a method for measuring molecular weight, which was one of the first used by Raffaello Nasini (1854–1931) and Emanuele Paternò in 1886, and which was rendered accurate and practical by an apparatus devised by Ernst Beckmann (1853–1923) in 1888. In 1885, Jacobus Hendricus Van't Hoff demonstrated that these properties of solution were related to osmotic pressure, which was parallel to gasseous pressure. Hugo De Vries (1848–1935) proposed plant cells and H. J. Hamburger (1859–1924) red blood cells for measuring osmotic pressure (1884); osmometers with semi-permeable artificial membranes were described by Gustav Tammann (1861–1938) in 1884, following earlier descriptions proposed by Moritz Traube (1826–94) in 1867 and Wilhelm Pfeffer (1845–1920) in 1877.

Since inorganic substances deviated from Raoult's and Van't Hoff's laws, the latter introduced the empirical correction coefficient i. His theory of dilute solutions was published in 1886, and then in 1887 in Volume I of the *Zeitschrift für physikalische Chemie*, the same work in which Svante August Arrhenius (1859–1927) published an article on electrolytic dissociation that explained why electrolytes in solution behaved differently from non-electrolytes, on the basis of the degree of dissociation. Wilhêlm Ostwald (1853–1932), who in 1885 had already observed that the degree of dissociation increased with dilution, stated in 1887 the law of dilution, which he expressed as an equation obtained by applying the law of masses. This same law was used by Henry Edward Armstrong (1848–1937), in 1884 and by Arrhenius in 1890 to explain the hydrolysis of certain salts and to found the solubility product theory.

The researches of Walter H. Nernst (1864–1941), who stated the theory of electromotive force (e.m.f.) produced by voltaic cells and in 1889 provided the formula for electrode potentials, were fundamental to electrochemistry.

Thermochemistry

Thermodynamics also owes much to Van't Hoff, whose *Études de dynamique chemique* (1884) laid the foundations for chemical kinetics and chemical thermodynamics. The relation between equilibrium constant and temperature, a particular case of the principle of mobile equilibrium stated by Henry Le Châtelier (1850–1936) in 1884, was mainly studied by Arrhenius.

Equilibrium diagrams

Equilibrium diagrams began to be used: Van't Hoff gave the diagram for sulphur in 1884, while Hendrik Willem Bakhuis Roozebroom (1854–1907) furnished ones for various mixtures (beginning in 1888). In 1899, Roozebroom also provided a useful classification for eutectic mixtures (particular forms of equilibrium between two or more substances in the solid state that was important for metal alloys, hydrated salts, organic isomers, etc.

Fluorine and other new elements

The discovery of germanium Clemens Winkler (1838–1904), 1886) followed that of gallium to confirm the validity of the periodic system, while the discovery of other elements in the rare earths group (gadolinium, neodymium, praseodymium, dysprosium) did not help solve the problem of the lanthanides. In 1886, Henri Moissan (1852–1907) isolated fluorine, which no one up to then had ever been able to separate from its compounds, known for more than two centuries.

Emil Fischer and the chemistry of carbohydrates

Organic chemistry made important progress in the chemistry of sugars, especially after Emil Fischer introduced the use of phenylhydrazine. The masterly researches of Fischer, Bernhard Tollens (1841–1918), and H. Kiliani (1855–1945), who in 1885 developed the Kiliani reaction for lengthening the carbon chain in sugars by making hydrocyanic acid to react with the carbonyl group of the corresponding -onic acid, made it possible to establish the aldehyde (or ketone) and hemiacetalic formulas of the principal sugars.

In 1887, Karl Friedrich Von Auwers (1863–1939) and Victor Meyer turned their attention to the study of stereochemistry. In 1888, they introduced the term "stereochemistry," which replaced Van't Hoff's "chemistry in space." Auwers and Meyer were unable to explain the stereochemistry of the benzyl dioximes, which Meyer had prepared in 1882, by means of Johann Wislicensus's concept of geometric isomerism. In order to justify the large number of isomers, they thought that isomerism might be established between compounds that contained neither double $=C$ bonds nor asymmetrical C atoms. However, they attributed isomerism to a limited rotation around a simple C-C bond, until Arthur Hantzsch (1857–1935) and Alfred Werner (1866–1919) extended the concept of asymmetry, established for the carbon atom by Joseph Achille Le Bel and Van't Hoff, to the nitrogen atom.

Ehrlich and chemotherapy

The prodigious development of dyestuffs had a decisive influence first on histology, then on therapy. During this period, histologists noticed that some dyestuffs had an elective affinity for certain tissues and cell formations. With clever intuition, Paul Ehrlich (1854–1915) created chemotherapy. But already the accumulation of certain byproducts forced the industry to search for ways of using them; for example, the tons of *p*-nitrophenol (a byproduct in the production of *o*-nitrophenol, an intermediate for dyestuffs) stored in the warehouses of the Bayer company finally found a use in the synthesis of phenacetin (Oscar Hinsberg, 1883); simultaneously, the same company created acetanilide or antifebrin.

Fundamental syntheses in organic chemistry

Among the new organic products were thiophen (Victor Meyer, in 1882–83), dulcin (Berlinerblau, 1883), phenthiazine (Heinrich August Bernthsen (1885–1931)), sphingosine and other substances obtained from the brain (Johann Ludwig Thudichum, 1880–84), and adenine (Albrecht Kossel, 1885). In 1882, Albert Ladenburg synthesized tropine, which together with tropic acid formed atropine. In 1886, Ladenburg succeeded in synthesizing an alkaloid, conine.

Among the achievements of organic chemistry were the synthesis of quinoline (Zdenko Hans Skraup, 1880); the extension of the pyrrole ring to a pyridine ring by means of a reaction discovered in 1881 by Ciamician; the identification of the structure of phthaliacetic acid (1884) by Siegmund Gabriel (1851–1924), who in 1886 synthesized and determined the structure of isoquinoline, discovered a year earlier by Sebastian Hoogewerff and Willem Anne Van Dorp; Claisen's various syntheses (1881–87); the synthesis of benzonitrile by Hermann von Fehling (1812–85), in 1884 (the first element in a series that von Fehling named nitriles); Sandmeyer's reaction for obtaining variously substituted benzene nuclei from the diazonium salts; the transposition of the ketoximes by Ernst Beckmann (1853–1923) in 1886–87; and Gabriel's synthesis of primary amines without secondary amines from the reaction of phthalimide with alkyl halides followed by acid or basic hydrolysis. In 1889, Gabriel extended this synthesis to the preparation of amino acids. In 1887, Eugen Bamberger (1857–1932) demonstrated that the 1.8 and 4.5 disubstitution products of naphthaline presented specific properties similar to those of the *ortho* position and named the two positions *peri*. In 1882, Egor E. Wagner (1849–1903) discovered a method for determining the number and position of double bonds in complex unsaturated compounds (Wagner's oxidation) and used it to identify the structure of limonene, pinene, and other terpenes.

Two years later, Aleksei Evgrafovich Favorski (1860–1945) discovered the isomerization of acetylenic hydrocarbons, which he believed was caused by allenic compounds and vinyl ethers. While studying the effects of nitrous acid on the ethyl ester of glycine, Theodor Curtius (1857–1928) discovered the diazoacetic ester (1883), a highly reactive compound and the first representative of the aliphatic diazocompounds. Again among the organic derivatives of nitrogen, Wilhelm Losen (1838–1906) obtained the hydroxamic acids.

In 1887, Charles Friedel and James Mason Crafts (1839–1917) began researches that led to the discovery of the alkylation and acylation reactions of aromatic substrata with alkyl and acyl halides respectively, in the presence of Lewis acids as catalysts.

New analytical methods

An important contribution to analytical chemistry was made by Johan Gustav Kjeldahl (1849–1900), who devised the well-known method for detecting combined nitrogen. In 1881, August Wilhelm von Hofmann discovered the method of exhaustive methylation followed by the elimination of trimethylamine for determining the structure of cyclic azotate systems, which he used to study the structure of piperidine. That same year, he discovered the degradation reaction of the acid amines (now universally knowm as Hofmann's degradation), which, in addition to providing a new method for synthesizing the primary amines, also made it possible to reduce the number of carbon atoms in the fatty acid series.

The existence of optical antipodes in a single organism

With regard to natural substances, Arnaldo Piutti (1857–1928) isolated *d*-asparagine from vetch sprouts in 1884. Vetch sprouts also contain large amounts of left enantiomer; this was the first time two optical antipodes had ever been found in the same organism. Piutti later synthesized *d*-asparagine, thus providing its structure.

Studies on the origin of petroleum. Soda

In the area of applied chemistry Carl Oswald Engler (1842–1925) began his studies on petroleum. Engler can be considered the creator of the science of petroleum; he was the first to hypothesize that petrols were of organic origin, an idea supported by his experiments on the formation of hydrocarbon mixtures from animal fats. He devised methods and apparatus for determining the chemico-physical properties of petrols that are still in use today. The Solvay soda-producing process was introduced in the United States in 1884; before then, soda had been made by treating cryolite with lime.

Fundamental treatises

During this period, publication was begun on two monumental works that, with their successive updatings, became indispensable tools for generations of chemists: the *Handbuch der organischen Chemie,* by Friedrich Beilstein (1880), and the *Physikalisch-chemische Tabellen,* by Hans Landolt (1831–1910) and Richard Leopold Börnstein (1852–1913), published in 1883. The first came out of a reorganization of the author's personal notes, with updatings on new substances that had been discovered. In 1881, Dmitri Petrovich Konovalov published his classic memoir on the vapor pressure of liquids, contributing to the understanding of liquid-vapor equilibria.

The *Traité de chimie biologique* (1885) by Charles Adolphe Wurtz and the *Lehrbuch der allgemeinen Chemie* (1885–87) by Wilhelm Ostwald also deserve particular mention.

LIFE SCIENCES

Chromosomes as the carriers of hereditary units

Developments in cytology and bacteriology were again at the forefront in biological research during this ninth decade of the century. In 1883, Walther Fleming (1845–1905) became the first to observe the splitting in half of chromatin filaments, which in 1888 Wilhelm Waldeyer (1836–1921) named chromosomes. Wilhelm Roux (1850–1924) advanced the hypothesis that these phenomena were responsible for the preservation of a qualitative patrimony. His observations were confirmed for

plants by Léon Guignard (1852–1928), and Eduard Strasburger. In 1883, Guignard demonstrated that plant chromosomes, like animal chromosomes, divided longitudinally, while Strasburger had thought that they divided transversally. Guignard also demonstrated the constancy of the number of chromosomes and the existence of chromatine reduction in plants, and (independently of Sergei Navashin (1857–1930)) discovered the double fecundation of angiospermae. The first chromosome counts were made by Moritz Nussbaum and Eduard Van Beneden, while Carl Rabl maintained (1885) that each chromosome possessed individuality and continuity.

Hertwig's fundamental affirmation of 1875 that "fecundation consists of the fusion of sexually differentiated cell nucleii" led to increasingly more important findings. Van Beneden (1846–1910) observed that the nucleus of a fertilized *Ascaris megalocephala* egg had twice as many chromosomes as the male and female gametes, and identified the "reductive" phase of gametogenesis ("meiosis"). Theodor Boveri (1862–1915) determined the function of the centrosome in the karyokinesis of the fertilized egg and demonstrated that the two parent nuclei contributed equally in the ontogenetic process.

At this point, cytology was able to make a substantial contribution to researches on heredity: an essay by Roux on the significance of the figures of nuclear division (1883), which hypothesized that the chromosomes were carriers of hereditary units, established an explicit connection between two fields that up to then had remained autonomous.

Experimental embryology

Experimental embryology was founded by Wilhelm Roux, who as early as 1883 affirmed that the cells that originated from the first divisions of the fertilized egg had different evolutionary potentials. This new branch of biological research returned even more resolutely to the problems of heredity after Roux's experiments of 1888 on frog embryos, which established the concept of "mosaic" development. (In fact, when one of the two blastomers of the first division of the zygote was killed by heat, a hemiembryo was produced.)

Biometry: natural heredity

Meanwhile, researches into the phenomena of heredity progressed considerably from the time of Darwin's *On the Origin of Species* through the work of Francis Galton (1822–1911) and August Weismann (1834–1914).

A naturalist and explorer—and Darwin's second cousin—Galton introduced a quantitative criterion into the analysis of processes that appeared to be exclusively qualitative in nature, founding biometry together with Karl Pearson (1857–1936). Galton's *Natural Inheritance* came out in 1889, providing a statistical representation that, since Gregor Mendel's contribution had gone unnoticed, had to await a more precise knowledge of the facts in order to be confirmed.

Neo-Darwinism and the theory of mutation

Moving from experimentation to theory-building, fundamental results and hypotheses were obtained by August Weismann (1834–1914), who denied the inheritance of acquired characters. He maintained that germ cells became differentiated at a very early stage in ontogeny, and this was how they propagated themselves from generation to generation. He postulated an "idioplasm"—a term coined by the botanist Karl Nägeli (1817–91)—that carried the "determinators" of characters, and hypothesized increasingly complex structures from "determinators" to "ids" and "idants." In 1885, Weismann traced the idioplasm to nuclear chromatine.

Neo-Darwinism was born as Darwin's theory of descendence was rid of its Lamarckian residues, the inheritance of acquired characters in particular. Following Weismann, variability was attributed to the mixture of the two hereditary sets in the fertilized ovicell (amphimixis). But during this time, (1886), the botanist Hugo de Vries (1848–1945) began his observation of Oenothera lamarckiana, which led to the theory of mutation. This theory maintained the casualistic assumption of Darwinism: mutation only increased the variability of species. Theodor Eimer (1843–1908), on the other hand, upheld the concept of a directed evolution, formulating the principles of an "orthogenesis" in his Die Entstehung der Arten auf Grund von Vererben erworbener Eigenschaften (1888–1901). Similar ideas were advanced by the Italian biologist Daniele Rosa (1857–1944), with his theory of ologenesis. In 1883, Galton outlined the program of a new discipline, "eugenetics," which studied the possibilities of improving the patrimony of human heredity.

Philogenesis was introduced into botany by August W. Eichler (1839–87) and Heinrich Gustav Adolf Engler (1844–1930). Engler began publishing his monumental Die natürlichen Pflanzenfamilien in 1887.

Marine zoology stations and limnology

The opening of the Zoological Station of Naples in 1873 can be considered a sign of the zeal with which systematic and comparative studies were being approached during this period. The founder of the station, Anton Dohrn (1840–1909), a student of Haeckel and Gegenbaur, was a strong supporter of the theory of evolution, and the center of studies that he created in Naples reflected his conviction that the primitive forms of life should be sought among marine fauna. In 1887, Victor Hensen (1835–1934) created the term "plankton" to designate the microscopic flora and fauna that lived on the surface of oceans. A "plankton expedition" was organized by a group of German biologists in 1889 aboard the ship National. The ocean floor was explored after the discovery of living concretions on a telegraphic cable laid in very deep waters (1859) exploded the hypothesis that life could not exist below a depth of 600 meters. In 1881, Joseph Henri de Lacaze-Duthiers (1821–1901), who in 1872 had founded the seaside stations in marine zoology at Roscoff, opened the Arago Laboratory of Marine Biology at Banyuls-sur-mer. In 1885, Albert I of Monaco explored shallow waters, and with the material he collected created the principality's Oceanographic Museum. "Limnology"—a term coined in 1892 by François A. Forel (1841–1912), who explored Lake Leman—arose together with marine biology to study the flora and fauna of lakes. In 1887, Stephen Alfred Forbes (1844–1930) published a volume significantly entitled The Lake as a Microcosm.

Symbiosis, commensalism, ecology, and ethology

Morphology also acquired an interest in associations among living organisms—involving relations not only of function, but of form as well—and environmental problems. The most important studies on the symbiosis between ants and bacteria and ants and fungi were made during this decade and the next, while researches on "commensalism" continued. On the opposite end of the spectrum, but suggested by the same interests, were researches on the struggle for existence between certain species: an exotic mealybug, Icerva purchasi, which was devastating the orange groves of California, was eliminated by introducing another mealybug, Novius cardinalis, that fed on the Icerva (1886). The term "ecology" had been created in 1866 by Ernst Haeckel to indicate the science of animal behavior; "ecology" later came to be used to indicate the science of the relations between life and the environment, while another discipline, "ethology" (with the meaning given to the term by Isidore Geoffroy Saint-Hilaire in 1854) dealt with behavior. Karl Semper (1832–93) published a vast work on animal life from an ecological point of view (1879–81), while Moriz Wagner (1813–87) confirmed Charles Darwin's ideas in his Die Enstehung der Arten durch räumliche Sonderung (1889). In ethology, J. Henri Fabre (1823–1915) published the first volumes of his Souvenirs entologiques (1879–1907), a work written from an anti-evolutionary point of view.

Koch and the pathogenetic agent of tuberculosis

Substantial progress was made in bacteriology during this period. In 1882, Robert Koch discovered the microbic agent of tuberculosis, a bacillus that was named after him. A year later, during an epidemic at Alexandria in Egypt, he discovered the cholera vibrion. A school of adept students who perfected the new staining and culturing techniques developed by Koch flourished around him. Carl Joseph Eberth (1835–1926) described the typhoid bacillus (1880) and C. Friedlander described the pneumobacillus (1882). In 1883–84, Edwin Klebs (1834–1913) and Friedrich August Löffler (1852–1915) identified the diptheria bacillus. Theodor Escherich (1857–1911) indicated the presence of colibacillus in the intestines of newborn babies (1866); Karl Fränkel (1861–1915) identified pneumococcus (1886), and Anton Weichselbaum (1845–1920) meningicoccus (1887). In 1888, August Gärtner (1848–1934) identified salmonella enteritidis and in 1892 Richard Pfeiffer (1858–1945) isolated Haemophilus influenzae. Victor Babes (1854–1926), considered the founder of the Rumanian school of microbiology and pathological anatomy and one of the precursors of immunotherapy, discovered the genus Babesia, and in 1885, in collaboration with Andre Victor Cornil, published Les bactéries, the first complete treatise of bacteriology. Babes discovered more than fifty pathogenetic germs, described the Babes-Ernst bodies and the Babes-Negri bodies in nerve tissues, and pointed out the presence of the metachromatic Babes-Ernst granulations in the cytoplasm of some bacteria. Babes also became a precursor of sierotherapy in 1889, when he demonstrated the possibility of performing vaccinations with the liquids and cells of immunized animals. In 1883, Hans Christian Gram (1853–1938) discovered a new method for staining bacteria.

Louis Pasteur and a cure for rabies

On July 6, 1885 in Paris, Louis Pasteur tried out a vaccine made of a weak virus on a boy who had been bitten by a rabid dog. This experiment, together with others carried out over the following months, were crowned with success. The Pasteur Institute for the study and cure of rabies was established in November, 1888.

Bizzozero and tissue classification

These were years of great experimental and theoretical affirmation for the Italian school of pathology. In several memoirs published between 1883 and 1884, Giulio Bizzozero (1846–1901), a professor of general pathology, demonstrated that platelets were a constant element of blood. In 1894, Bizzozero proposed a classification that divided tissues into groups with labile, stabile, and perennial elements. One of his pupils, Camillo Golgi (1843–1926), devised a histo-chemical method for studying nerve tissue under the microscope. Based on chromium-silver impregnation, this method made new morphological information available.

New trends in medicine

Remarkably important events took place in the field of medicine. Jean Martin Charcot's *Leç sur les maladies du système nerveux*, a work documenting neurology, a branch of medicine that by this time had acquired its own autonomy and included a good part of psychiatry and psychology, came out in 1880. Augustus Désiré Waller (1856–1922) recorded the electrical activity of the human heart (previous experiments had always been done on an open heart) and laid the premises of electrocardiography (1887). Following the work of Douglas Argyll Robertson, Wilhelm Heinrich Erb, Carl Friedrich Westphal, and Pierre Marie (1853–1940), "reflexology," which was systematized by Joseph Babinski (1857–1932), was introduced into semiotics.

EARTH SCIENCES

The inside of the Earth and isostasy

In 1881, Philipp von Jolly (1809–84) determined the average mass of the earth. Four years later, comparing the composition of meteorites to the possible composition of the inside of the earth, Eduard Suess advanced the well-known hypothesis of concentric shells of Sial (originally Sal), Sima, and Nife in order to explain the earth's average density in relation to observations made on its outermost layers. In 1889, Clarence E. Dutton (1841–1912) proposed the name isostasy for the equilibrium between the lighter surface parts and the denser, underlying, inner parts of our planet. The breakdown of that equilibrium for whatever reason, such as erosion, sedimentation, or the accumulation of glacial masses, causes compensation movements, which are one of the major sources of tectonic dislocations. Thus the possibility of great vertical displacements also occurring inside the earth was proposed as an alternative to horizontal movements.

In 1887, Hugh Robert Mill (1861–1950) detected the continental shelf, the site of sedimentation and alternating oceanic and continental events.

The earth's magnetism

Important studies on the earth's magnetism, its hypothetical causes and the distribution of the earth's magnetic field were conducted in Germany by Edmund Naumann (1887) and Georg von Neumayer (1891).

Tectonic concepts

Between 1883 and 1888, Eduard Suess published his treatise *Das Antlitz der Erde*, a summary and interpretation of all the current geological knowledge on the earth's entire crust in the light of the Viennese author's new tectonic concepts. It reiterated the importance of horizontal displacement in the formation of folded mountain ranges. Among the many annotations is one concerning eustatic movements, a term Suess used to indicate the alternating variations in sea level during glaciations and interglacial periods.

With regard to nappes, Marcel Bertrand (1847–1907) offered a new structural interpretation of the Alps: in 1884, he compared the tectonics of the Glarus zone in Switzerland with the tectonics of the Franco-Belgian coal basin, for which he indicated the existence of important nappe phenomena. Eduard Reyer attempted to recreate experimentally several tectonic processes that he believed took place (1885), while Luigi Bombicci Porta (1833–1903) recalled attention to the importance of gravity in tectonics (1887).

Crystal lattices and diagenesis

In 1880, Leonhard Sohncke (1842–97) indicated that the particles situated in the nodes of crystal lattices could be atoms. A short while later, from 1885 onward, Evgraf Federov (1853–1919) and Arthur Schoenflis scientifically examined the various possible crystal lattices. Petrographic studies, both under the microscope and by chemical analysis, were systematically developed by Karl Harry Rosenbusch in 1888 and by Auguste Michel-Lévy and François Antoine Lacroix (1863–1948) in 1889. Gustav Tschermak (1836–1927) studied meteorites in 1883. Orazio Silvestri published his studies on the chemical geology of the eruptions of Etna from 1866 to 1881. F. Guido Sandberger developed the theory of lateral secretions from wall rock to explain the origin of veins of metal. Returning to an 1840 work by Dana, Dutton published an ample monograph on the volcanoes of the Hawaiian islands. Suess proposed the well-known hypothesis of the fusion and digestion of preexisting rocks to explain the genesis of large granite plutons, which he called batholiths. E. Reyer continued to make a distinction between plutonites and volcanites.

Karl W. von Gümbel (1823–98) used the term diagenesis to indicate the lithification process that takes place in sediments (*Diagenese*, 1888; the term was defined in more detail by Johannes Walther in 1894).

Paleontological discoveries

Between 1884 and 1890, Charles D. Walcott (1850–1927) described his researches on the Cambrian trilobite fauna of Canada and the United States. In 1884, Oswald Heer completed his paleontological study of Arctic flora, which he had begun in 1869.

In 1882, Gustav Steinmann (1856–1929) made geological explorations of the coastal regions of Peru and the Tierra del Fuego.

The stratigraphic scale

At the Second International Geological Congress, held in Bologna in 1881 under the scientific patronage of Giovanni Capellini, Miksa (Max von) Hantken (1821–93) described the dimorphism of nummolites; a universal stratigraphic scale was established and the colors for indicating each of the different periods on geological maps were codified. Hantken also published his major work on the *Clavulina Szabóli* foraminifera of the Paleogene system and determined their presence in the Euganean Hills near Padua as well as in the Maritime Alps around Nice (1884). He also discovered the first nummulite in America (1886). The Italian Geological Society was also founded during the Bologna Congress. The Russian Geological Committee was established in 1884.

The eruption of Krakatoa

In August, 1883, the eruption of the insular volcano of Krakatoa, between Java and Sumatra, caused a huge tidal wave that killed more than 36,000 people. Minor tidal waves were propagated in every ocean (some were even registered on the tide gauge of London). The fine dust sent up by the explosion made several trips around the world in the form of reddish clouds before it fell to the earth.

The morphology of the ocean floor

Many of the studies of this period were directed towards elaborating the data gathered during the *Challenger* expedition. After the premature death of Charles Wyville Thomson in 1882, John Murray took over as coordinator of the studies. The morphology of the floor of the Atlantic Ocean was by this time being studied in detail; Murray and Alphonse F. Renard proposed the first classification for bottom sediments (1884). In 1881, while reviewing information on depth distribution in the ocean, Murray pointed out that the greatest depths occurred very rarely, appearing only marginally as trenches. All the ancient legends on incommensurable depths were definitively disproved, and the morphology of ocean floors began to be clearly outlined. It was also evident by now that continental masses were sharply separated from the oceans by a ''slope'' located at the end of a varying offshore extension of the ''continential shelf,'' where the water was fairly shallow (less than 200 meters deep). It was while studying the African slope during an expedition of the *Buccaneer* in 1887 that John Young Buchanan and Anthony Thomson discovered the second underwater canyon known at the time, the Congo canyon.

Oceanographic explorations at the end of the century

A number of important voyages were made during this decade. Possibly the largest was that of the Russian ship *Vitjaz* (1886–89), under the leadership of Stepan Osipovich Makarov (1849–1904). It went all the way around the world, but focused above all on the Pacific Ocean, where soundings and investigations in physical and biological oceanography were made. Another group of oceanographic explorations was made by the *Albatross* (1888–97). Investigations in marine biology were made during voyages of the *Travailleur* (1880) and the *Talisman* (1883).

In 1885, Albert I of Monaco inaugurated a series of oceanographic campaigns into the Mediterranean and the North Atlantic. By 1910, he had made twenty-eight voyages with the *Hirondelle I* and *II* and the *Princess Alice I* and *II,* during which hydrological and biological material was gathered.

Hydrodynamics and marine chemistry

Notable contributions were made to hydrodynamic theory during this period. In 1889, Hermann von Helmholtz advanced the hypothesis that undulations were formed when fluids in motion came into contact with one another: surface waves were formed when air and water met, and internal waves were formed within masses of water when waters with different temperatures or salinities came into contact. The first clarifications of the turbulent nature of some fluid motions (most ocean-water motions) were provided, both theoretically and experimentally, by Osborne Reynolds (1842–1912), who introduced the concept of the ''Reynolds number'' (1883), which specifies the conditions of turbulence. In 1885, Henrik Mohn (1835–1916) gave the first formula for indirectly determining currents on the basis of the angle of isobaric surfaces (anticipating the contributions of Vilhelm Bjerknes (1862–1951), Bjorn Helland-Hansen and others to dynamic oceanography). These originated from air currents (winds), and the angle was that of the atmospheric pressure isobars at ground level.

Another extremely important achievement of this decade was the work done by William Dittmar (1833–94) on the composition of sea water and salinity. Dittmar, who had been an assistant to the great chemist Robert Bunsen, later moved to England, where he became interested in analyzing the water samples collected by the *Challenger*. These had been only partially examined by J.Y. Buchanan. There were not a huge number of samples (only seventy-seven, with seventeen taken from surface waters and thirty-four from depths of more than 2,000 meters), but they came from a wide range of places. Dittmar was thus able to refine Forchhammer's earlier discoveries and make a definitive statement on the constancy of the ratios among the major components of salinity. The average ratio between all the other components and chlorine was established at 1.0859, so that it could be said that water with a chlorine content of 19,149 had a salinity of 34,526. Dittmar's analyses, carried out with increasingly more refined methods, involved the determination of Cl, Br, SO_4, CO_3, Mg, Ca, Na, K, and Sr. Dittmar also studied the alkalinity of sea water and of dissolved gasses. His major work was published in 1884.

The establishment of meteorology and climatology

William Ferrel's psychrometric tables for calculating the humidity of the air from temperatures read on wet and dry bulb thermometers were first published in 1886.

In 1887, Ricard Assmann (1845–1918) invented the instrument for measuring humidity in the air that is known as the psychrometer. Atmospheric circulation was also studied by elaborating the data gathered by the *Challenger* (Alexander Buchan, 1889). The distribution of temperature over the globe and the vicissitudes of heat exchange were studied by Wladimir Köppen (1846–1940), who in 1884 presented his famous classification of climates, which he continued to refine until 1931. It is still the most commonly used classification, although there is now need for a better one. The distribution of heat over the earth's surface was also studied (1883–84) by Charles Alfred Angot

(1848–1924) and Samuel Langley, who investigated its correlation with solar radiation. The distribution of winds and atmospheric pressure was analyzed by Julius Hann (1839–1921), who in 1883 published the *Handbuch der Klimatologie*, and in 1887 the *Atlas der Meteorologie*, a part of the great *Physikalischer Atlas* of Hermann Berghaus (1797–1884).

It would be fair to say that, following the sporadic contributions made in the past, meteorology began to assume its true form as a physical science during this decade, especially at the end of the nineteenth century. In fact, this particular decade was characterized by the most intense theoretical development that had ever taken place in the history of meteorology. Helmholtz, who in the past had already published various memoirs of extreme importance to dynamic meteorology (*Über Integrale der hydrodynamischen Gleichungen, welche den Wirbelbewegungen entsprechen*, 1858; *Über diskontinuirlich Flüssigkeitsbewegungen*, 1868; *Über ein Theorem, geometrisch ähnliche Bewegungen flüssiger Körper betreffend nebst Anwendung auf das Problem Luftballons zu lenken*, 1873), now wrote two dealing explicitly with the motion of the atmosphere (*Über atmosphärische Bewegungen*, 1888–89). These studies gave rise to modern meteorology: the main trend in this field, from the work of Vilhelm Bjerknes (1862–1951) to that of Carl-Gustaf Rossby (1898–1957), and even to the worldwide program of researches on the atmosphere of the 1970s, has essentially consisted in applying the concepts contained in them to physical reality. Though not all scholars had the fortune of influencing future developments quite so directly, a number of other outstanding scientists participated during this decade in building what was then called mathematical meteorology: the Germans Wilhelm von Bezold (1837–1907), Hertz (a student of Helmholtz, and Bjerknes's teacher), and Anton Oberbeck (1846–1900), the Galician-Austrian Max Margules (1856–1920), and Lord Rayleigh. Particularly interesting is the case of the Italian Luigi de Marchi (1857–1936), a professor of physical geography, though he had the mentality and training of a physicist. Two of his memoirs, published in the annals of the Central Bureau of Meteorology (*Richerche sulla teoria matematica dei venti*, 1884, and *Sulla costanza della rotazione totale in un sistema di venti*, 1886), supported the hypothesis that "total rotation" (today we would say "absolute vorticity") was constant and illustrated its significance in a number of particular cases. De Marchi's theory—criticized by Margules and Oberbeck in the *Meteorologische Zeitschrift*—was again taken up (in all likelihood, independently) at a later date by Rossby (around 1940) and provided the basis for the development of meteorology for a period of twenty years.

Leon Teisserenc de Bort (1855–1913) made what was perhaps the last substantial advance in isobaric synoptic meteorology, studying the fundamental aspects of atmospheric circulation and introducing the concept and term "centers of action." These were the huge, semi-permanent anti-cyclones and low pressure areas that determined the general features of circulation in the lower atmosphere, and therefore the formation and movement of perturbations, for weeks and even for months. His concepts, outlined in his *Étude sur les causes qui déterminent la circulation de l'atmosphère* and *Sur les grands centres d'action de l'atmosphère* (1880–81) ruled synoptic meteorology for decades, and even today dominate the thinking of the practical meteorologist. The Scottish meteorologist John Aitken (1839–1919) experimented with condensation in atmospheric air and in filtered air, concluding

that water vapor in the atmosphere always condenses around a solid core ("On Dust, Fog and Clouds," published in the *Transactions of the Royal Society of Edinburgh*, XXX, 1881), and built the first condensation core counter ("On the Number of Dust Particles in the Atmosphere," *Transactions of the Royal Society of Edinburgh*, XXXV, 1888).

APPLIED SCIENCES

Prime movers

The final step in the evolution of the internal combustion engine was made during this decade. The initial preference for gas was abandoned, and in 1883 Gottlieb Daimler (1834–1900) and Wilhelm Maybach (1847–1929) built their first internal combustion (vertical single-cylinder) gasoline engine, which produced a velocity of 900 revolutions per minute. From 1883 on, inventions followed upon each other in quick succession. In 1885, Daimler built the first motorcycle and Karl Benz (1844–1929) created the first automobile, a lightweight, three-wheeled vehicle run by a horizontal, single-cylinder gasoline engine. The first four-wheeled vehicle with a Daimler engine came out the following year. In 1887, in the United States, George Baldwin Selden (1846–1922) developed a three-cylinder internal combustion engine that ran on liquid hydrocarbons. In 1889, John Boyd Dunlop (1840–1921) perfected the rubber tire, introducing the inner tube combined with an outer tire with tread, while Daimler patented a two-cylinder engine that was adopted by many continental engineers. Among these was Armand Peugeot (1849–1915), who as early as 1890 applied it to the first French-made vehicle.

The tumultuous development of the electrical industry proceeded. The electric engines developed up to this time had operated on direct current, but with the installation of the first lines carrying alternating current, the advantages of which we shall return to shortly, the problem arose of devising an engine that would operate on that system. The solution was found by Galileo Ferraris (1847–97) and Nikola Tesla (1856–1943). Ferraris discovered the principle of rotating magnetic fields on the basis of purely theoretical considerations, while Tesla arrived at the same result by imagining a rotation of magnetic poles. Ferraris made his discovery in 1885, but his memoir entitled *Rotazioni elettrodinamiche prodotte per mezzo di corrente alternata* only came out in March of 1888. Tesla's first patents, on the other hand, were taken out in October, 1887, though they were only made public the following May.

At the same time the new alternating current engine was being developed, the field of direct current generators was profoundly revolutionized by Charles A. Parsons (1854–1931). We have already mentioned the attempts made to adapt steam engines to run dynamos. In 1881, a new model, the Armington Sims engine, was proposed; one of these horizontal steam engines, which produced 175 horsepower and operated at a speed of 1,350 revolutions per minute, was used to run the dynamo at the Pearl Street Station. P.W. Willans built another high-speed single-acting steam engine, which was patented in 1884. However, despite these improvements, the reciprocating steam engine clearly had its limitations as a means for operating dynamos. It received competition from the steam turbine.

Despite the mechanical difficulties inherent in the higher speed of the steam turbine, its advantages were clear: higher yields and unit output and the elimination of the difficult process of exploiting steam power in the form of reciprocating motion, which then had to be converted into rotary motion. In 1884, following up on his earlier successes, Parsons built the first practical steam turbine, which utilized the reaction of the fluid coming out through an opening to push back vanes mounted on a disc. Since the maximum angular velocity of the dynamo was only about 1,300 revolutions per minute, Parsons also designed a new high-speed dynamo. These two inventions were patented in 1884. That same year, Parsons built the direct current turbogenerator, which turned at a speed of 18,000 revolutions per minute. Parsons installed his first turbine in an electrical station in 1888.

Electrical power plants

The first central lighting station was put into operation in 1881 at Appleton, Wisconsin; the following year, the famous Pearl Street Station was inaugurated in New York, introducing the practice of supplying electricity to the individual consumer. The first stations (which operated on low-voltage direct current) could only distribute electricity within a radius of a few hundred meters, and had to be installed at the center of the zone of distribution, creating an extremely restrictive situation. To be used economically, electricity had to be generated in places where fuel or water resources were accessible and then sent to the areas to be served. In order to do this, current had to be carried at high voltages; however, to increase the voltage after generating the current and then lower it upon arrival, so that it would be safe to distribute, alternating current had to be adopted, since transformers could be used with it. In order to use direct current, mechanically coupled engines and electrical generators had to be used to transform quantities of current and voltage.

Sebastian Ziani Ferranti (1864–1930), who in 1884 patented a single-phase alternator with a continuous insulated copper tape winding, played an important role in the initial stage of the "war of the systems." Ferranti was assigned the task of designing a large power station at Deptford that would supply electricity to London. In 1889, when insulating capacity was no more than 2,500 volts, Ferranti installed an alternator that generated alternating current at 10,000 volts. The use of electricity changed the face of the cities, not to mention revolutionizing railway transport. In fact, Thomas Alva Edison presented his first electric locomotive in 1880, while the first electric trolleys appeared at the end of the decade. Public lighting became widely diffused, thanks to Edison's work, and above all to Joseph Swan. After his patent of 1880, Swan perfected his lamp, and in 1883 used a plastic substance, obtained by dissolving nitrocellulose in acetic acid, then reducing it to thin fibers, for his filaments. This process transformed and greatly improved the production of lamps.

Raw materials

World production of nickel and aluminum remained almost negligible until 1880 (in 1875, no more than 500 tons of the

first and ten tons of the second were produced). The ore deposits of Sudbury, rich in nickel, were discovered in 1883, the same year that Marbrau obtained nickel steel, a strong and impact-resistant alloy. However, the Sudbury deposits were not immediately exploited, so that the first nickel steel for industrial use was only produced in France in 1888. The process for extracting nickel was also improved at that time, thanks to the researches of James Riley and Ludwig Mond (1839–1909). In 1890, Mond discovered nickel tetracarbonyl, a volatile compound of nickel that decomposed at a low temperature leaving metallic nickel.

The production of aluminum had been held up until now by the high cost of potassium and sodium, used as reducers. Hamilton Young Castner (1858–99) finally succeeded in overcoming the difficulties involved in producing sodium, and began to utilize his invention when Charles Martin Hall (1863–1914) patented a process for producing aluminum by electrolyzing the oxide in molten cryolite (1886). Castner turned to other uses for sodium, producing sodium peroxide, used for bleaching wool, and sodium cyanide, which he sold as "130% potassium cyanide," a product very much in demand among the owners of gold mines.

Despite the invention proposed by Fritz W. Lürmann (1834–1919), blast-furnace hearths were not immediately enlarged, and the unitary production of steel did not increase fast enough to meet the rising demand. The first enlargement was made in 1880 in furnace B of the Edgar Thomson Works in Pittsburg, where in a hearth 3.35 meters in diameter and 24.4 meters high, weekly production rose from 800–900 tons to 1,200 tons. That same year, William Siemens obtained steel in an electric furnace, which Paul Louis Héroult began perfecting in 1887. Two types of electric furnaces finally emerged: Héroult's direct-arc furnace, where the arc was produced between large carbon retort electrodes and the metal bath, and one devised by E. Stassano (1859–1922), where the arc was produced between carbon feeds placed above the bath, which was heated by radiation. The rapid spread of the use of steel during this period also led to new research methods in iron working and metallurgy.

As for special steels, in 1882 Robert Hadfield (1859–1940) experimented with steel with a high manganese content, which was non-magnetic and had a high resistance to wear.

Chemical industries

We have already mentioned the flourishing of the dye-making industry and the chemical industry in general in Germany during the decade that went from 1860 to 1870. The phenomenon became even more widespread between 1880 and 1890: industries primarily concerned with the production of dyestuffs and pharmaceutical substances were established in Switzerland (CIBA in 1885 and Sandoz in 1886, both in Basel), while in Italy a true chemical industry arose in 1888 with the foundation of the Montecatini company, which was favored by the new customs law of 1887 (see CHEMICAL SCIENCES, 1880–1890).

Construction

Steel began to be used in all construction projects. The Brooklyn Bridge, the first bridge in the world to use steel suspension cables, was opened to traffic in 1883. Construction of the bridge was initiated and completed by the Roeblings—John Augustus (1806–69) and Washington Augustus (1837–1926). That same year, John Fowler (1817–98) and Benjamin Baker (1840–1907) began building the famous cantilever bridge over the Firth of Forth in Scotland. It was built completely of steel, with two main spans of 520 meters each, and weighed approximately 50,000 tons. Steel was also used in one of the boldest constructions of the period, the tower built by Gustave Eiffel (1832–1923) for the Paris Exhibition of 1889.

The first time steel was used for the entire superstructure of a factory was at the end of the decade in the United States. Buildings with more than five floors had already been built there, beginning in 1882, especially after Elisha Graves Otis (1811–61) invented a safety device for elevators. The Home Insurance Building of Chicago was the first important building to use steel in its construction. The use of brick supporting walls for such a tall structure would have required very thick walls on the bottom floors and very small windows. But the use of steel made it possible to reduce the encumbrance of the bottom floors and provide sufficient light as well.

The engineering needs of these new constructions led to increasingly more detailed studies on materials and their behavior under stress. Otto Mohr (1835–1918) investigated the problem, making contributions to graphic statistics and to the study, published in 1882, of the stresses at any one point in a continuous material system, no matter how it is stressed.

In 1885, Gustav A. Wayss (1851–1917) purchased the right to use the inventions of Joseph Monier (1823–1906) in Germany, and developed a complete system for building with reinforced concrete, using it in bridge-building as well. Mathias Koenen (1849–1924) elaborated the theory for this type of construction in his *Das System Monier* (1887).

Food processing techniques

As we have already mentioned, the first experiments with freezing meat dated back to the decade 1870–1880. Although they were not completely successful at first, they served to indicate one of the most important paths to be followed in solving the problem of food preservation. Freezing techniques were considerably improved over the next decade; in Europe, the importation of meat (from New Zealand, Argentina, and North America) increased rapidly, with well-known economic consequences for both the exporting and the importing countries. The technique of preserving food by refrigeration was extended to other products: in 1884, the first long-distance shipment of a large quantity of refrigerated apples was made from Tasmania to London. There was a great need for technical devices that could automatically keep temperatures at the desired level (which obviously differed from one product to another). The problem was only satisfactorily solved during the early decades of the twentieth century.

Military technology

Smokeless powder was invented, and all armies began to use repeating rifles. In 1889, the American engineer Hiram Maxim (1840–1916) built the first real automatic machine gun. He was followed a short time later by his compatriot John M. Browning (1855–1926).

1890–1900

THOUGHT AND SCIENTIFIC LIFE

The age of imperialism

If the Holy Alliance had assured almost thirty years of peace in Europe during the early decades of the nineteenth century (1816–46) after the fall of the great Napoleonic ventures, the so-called Concert of Europe achieved analogous results on a world-wide scale during the final decades of the century. At the same time, social and economic progress revolutionized the map of the world. This great political fact was the result of the emergence of a specific entity, *haute finance*, which constituted the perfect link between the political and the economical organization of international life: balances and armaments, foreign trade and supplies of raw materials, national independence and sovereignty were now the functions of money and credit. The repercussions of the London money market were followed by businessmen all over the world; governments discussed plans for the future in the light of the state of major world markets; the political equilibrium guaranteed the economic system the peace it needed.

But if the age of Otto von Bismarck (1860–90) marked the high point of the European concert, it should be noted that by the end of the 1860s, Germany's use of the gold standard opened an era of protectionism and economic expansion. In the 1890s, Germany reinforced its position by forming an alliance with Austria-Hungary and Italy; Bismarck lost control of policy in the Reich, and Great Britain became the principal guarantor of peace during this period.

Colonial expansion took a decisively imperialistic turn under the prevailing influence of economic protectionism: in Africa, the European powers pushed inland; in Asia, colonial expansion reached the Far East, where strategic coastal areas of China were subjected to penetration by Great Britain, Germany, and France, all eager for trade advantages. Even czarist Russia, which sought a solution to its economic and social contradictions in the march eastward, and Japan, which was rapidly becoming modernized through efforts promoted by the imperial authorities, were pressing at the borders of the "Celestial Empire." While rivalry between Russia and Japan increased (leading to open conflict during the early years of the twentieth century), the United States was also looking toward the Far East, where some political circles anticipated in the "adventure beyond the Pacific" a direct continuation of the conquest of the west that had played such an important role in the revival that followed the war of succession. But the United States was extremely active on the American continent as well: while the Spanish American War wiped out the last of the Spanish empire in the Americas, the long arm of America began to weigh more and more on Mexico and the small countries of Central America.

New aspects of culture

In this changed context, the liberal and democratic ideal developed during the first half of the century had by this time been abandoned in the face of tighter bourgeois control over state and society. Even the policy of containment promoted by popular movements and exhorted by the growth of productive structures and the incessant extension of the market, allowed room for nationalism.

Pan-Germanic doctrines became popular in the Germany of Wilhelm II; in Austria-Hungary, intellectual currents that reflected the conflict among the various opposing nationalities within the empire rose up against the aging, but still very prestigious, centralized bureaucracy. In republican France, although attempts at restoring the monarchy had been thwarted, chauvinist and conservative sectors were growing in strength (as attested to by a growing spirit of *revanche* against Germany and the Dreyfus affair with its resulting wave of racism at the end of the century). English propaganda exalted the "imperial and colonizing mission," and in Russia, the "conservative utopia" of the pro-Slavs had become quite widely diffused. In the United States growing numbers of the population were abandoning isolationism to support the Yankees' role as leader for the two Americas.

At the same time, there was an abrupt turning away from those currents that had predicted the progressive and optimistic evolution of knowledge and society. A sense of rebellion against the most adamantly asserted themes of both positivist and idealist traditions were pervasive among many intellectuals, especially scholars and artists. In philosophy, this rebellion was perhaps most strongly expressed by Friedrich Nietzsche, whose vitalistic irrationalism became increasingly more widely diffused, and not only among German intellectuals. During the early decades of the twentieth century, numerous Nietzschian themes, often extremely simplified, such as the famous myth of the superman and the exaltation of the will to power, were adopted by politicians, artists, and scholars.

But during the 1890s, that sense of the end that dominated the intellectual climate of the years immediately preceding the first world war was still a minority view. We must remember that faith in reason and progress had not yet declined. It was, in fact, very much present, though in a number of different forms, in vast intellectual circles where the emphasis was on a need to step up scientific and technological efforts within the framework of tighter international collaboration. It is in this context that one of the most significant events of scientific life of this period—the establishment of the Nobel prizes—took place. Sweden, which had risen to the vanguard of social progress with the introduction of universal suffrage and its advanced labor legislation, approved the statutes of the Nobel foundation in 1898. Five prizes were set up: the first for "the most important discovery or invention in the field of physics"; the second, in the field of chemistry; the third in medicine; the fourth for the author of a literary work "of idealistic inspiration"; and the fifth for a person distinguished for "maintaining and promoting peace." The Nobel prizes were first awarded in 1901 and have, since then, contributed to identifying the most significant directions in scientific research.

All through the 1890s, *haute finance* remained at its peak and peace seemed more secure than ever: although British and French interests conflicted in Africa, and British and Russian interests competed in Asia, the powers interested in a world balance were still numerous. Scientific and cultural collaboration, the possibility of investing large sums of money in research, and unified multinational efforts in huge scientific and technical endeavors seemed at the time a decisive element in keeping peace among nations. Only in the early years of the twentieth century did an inversion of these cooperative tendencies manifest itself openly: especially after 1904, the distinct contraposition of the various imperial forces, grouped into two hostile coalitions (the powers of the Triple Entente on one side and the Triple Alliance on the other), the breakdown of the balance of power, and the corresponding dissolution of the economic structure appreciably hindered scientific collaboration among the various nations, increasing the tendency for technology to be controlled by military interests and allowing the more extreme nationalistic tendencies to prevail. The new generations, which were by now far removed from the democratic experiences of the nineteenth century, often saw the irrational thinkers of that era, such as Schopenhauer and Nietzsche, as prophets of the "refusal" of faith in progress and the "religion of science." But the late nineteenth-century intellectual scene was not limited to these manifestations of a ruthless critique of traditional values: the germs of fruitful rethinkings and important searchings that, often in opposition to those who simply denounced an ethical and political crisis of values, led to new concepts of society, the state, and the role of science and technology were also present. Here we shall only briefly mention the complex themes of socialist thought and, more in general, of the entire theoretical reflection of the working-class movement. These currents were extremely sensitive to the great debates and cultural themes of the epoch (the return to Kant, as well as phenomenalism, Machian empiricism, the critique of positivism, etc.), and were no less attentive to the great transformations taking place in society, which culminated dramatically in a world war on the one hand and in the first great revolutionary events of the twentieth century (Russia, Mexico, Ireland, China) on the other.

The "crisis of the sciences" and developments in epistemological reflection

Specifically, in the area of epistemological reflection, it should be observed that the idea of a "general defeat of principles," to use Jules Henri Poincaré's expression, became increasingly more widespread, to the extent that more than one historian in later generations considered it fitting to speak of a "crisis of the sciences." It is important to note, however, that during these same years, new roads were being opened in mathematics, physics, and biology. In fact, the scientists who were most aware of the fundamental changes taking place in their fields of research were often those who best understood that the appearance of new problems, far from representing failure, sowed the seeds of further theoretical development. Thus there were those who not only did not admit the defeat mentioned by Poincaré, but even perceived new lines of research in the decline of particular theories. Many cultural figures, on the other hand, even some exponents of scientific research and teaching, viewed the old scientific philosophies as inadequate, both for understanding the new discoveries of science and for dealing with the ethical, political, and esthetic problems raised by the new generations. Thus new philosophical currents arose in almost every country, denying any connection whatsoever with the concepts that had dominated up to then, especially positivism.

In the face of several spectacular achievements of physics and chemistry (soon to be joined by those of biology), that form of "mechanistic" materialism that had reached its peak in the work of Pierre Simon de Laplace at the beginning of the century and, in the course of the nineteenth century, had represented "a dream, an ambitious plan" (Pierre Duhem) for the extreme variety of scientific schools that ventured into new fields of research and were often in open conflict with each other, progressively lost its prestige. The need to change points of reference was by now being felt very strongly, thanks in part to the diffusion of Mach's critique of mechanics and of radical methodologies that polemicized against the "metaphysics of matter." Many of these tendencies, which often developed in a unilateral way, ended up making a *tout court* condemnation of materialism, frequently progressing from a denouncement of the limits of mechanistic materialism to idealistic interpretations of scientific results, providing arguments for spiritualist and even irrationalist currents. However, both in scientific circles and in other sectors of cultural life there were resolute stances being taken against these interpretations and in defense of the rationality and objectivity of scientific knowledge. Important in this context were both the polemic conducted by Ludwig Boltzmann (1844–1906) against phenomenological physics and the phenomenalist tendencies referred to by both Max Planck (1858–1947), the creator of the quantum theory of radiation, in his polemic against the Machian concept of science (1910), and Lenin, in his *Materialism and Empirocriticism*, where he criticized both Russia and western European "Machism."

Philosophy in France. Conventionalism. Bergson

Possibly the most important French critique of the concept of traditional science was provided during this period by the critico-methodological studies of Jules Henri Poincaré (1854–1912), a teacher of physics, mathematics, and celestial mechanics at the University of Paris. His epistemological contributions constituted the nucleus of several works, published in the twentieth century—*La science et l'hypothèse* (1902) and *La valeur de la science* (1905) among others—that later influenced numerous epistemologists, including some of the exponents of the logical positivism of the 1920s and thirties. These thinkers drew from Poincaré not only the idea of criticizing science from within science and a positive attitude toward scientific method, but several specific themes of his "conventionalism."

Nor were other critiques of the mechanistic concept lacking in French culture. For example, Étienne Boutroux (1845–1921), in his *De l'idée de loi naturelle dans la science et dans la philosophie contemporaines* (1895), upheld the idea of a "contingency" of natural laws, based on moral reasons which were much different from Poincaré's. As long as the dominion of science was not very extensive, "one could admit that there was room for freedom outside it," but in the epoch in which science became more extended, it was necessary, according to Boutroux, to reclaim autonomy for certain areas that science might incorporate, such as the psychic aspects of the feeling of human freedom. This led to a critique of traditional scientific determinism.

An important and innovative contribution to philosophical thought was also made by Henri Bergson (1859–1941). While Boutroux's followers often insisted on contraposing positivism and spiritualism, and even science and spiritualism, Bergson (who during this decade published several works, such as *Matière et mémoire* (1896) and *Le rire* (1899)) directed his attention against positivism, though trying to open the method of the positive

sciences to considerations of "conscience." Like Mach, Bergson espoused an "economic" concept of science. But unlike Mach he abandoned the physico-mathematical model for explaining science, in that he found its applications in the area of psychology misleading. This stance was accentuated in his later works; for example in his *Évolution créatrice* (1907), he again took up the evolutionary themes that had excited him so much in his youth when he read the works of Spencer. As the title of the work suggests, Bergson's position was by then far removed from Spencerian evolutionism. Bergson's ideas became widely diffused through reviews, polemics, and debates.

To give an idea of the wide range of positions that coexisted in French culture at the time, we shall only mention the nominalist and instrumentalist ideas of Édouard Leroy (1870–1954) (the term "instrumentalist," as used here, should not be confused with the instrumentalism of John Dewey; in this case it refers to the notion that scientific theories were mere "instruments" for "capturing" reality). In the wake of Bergson, Leroy maintained that science could not teach people anything about truth, but could only serve as a rule for action. In direct opposition to Leroy, Poincaré reiterated the need to discriminate in some way between rules that succeed, at least in general, and rules that lead to nothing, and tried to show that the conventions of science were not arbitrary, that science works continuously in front of our eyes; and that would not be possible if it did not make us understand something about reality. These polemics, in any case, favored the diffusion of conventionalism in France. Poincaré's conventionalism was soon joined by the much more radical conventionalism of the great historian of science Pierre Duhem (1861–1916). Duhem brought Novalis's motto "theories are nets: those who throw them, catch fish" back into vogue in the context of a global, sometimes extremely subtle critique of the "mechanist concept of the world."

Philosophy in Italy

In Italy, the process of moving beyond positivism was somewhat slower, so that many scholars remained faithful to the current, making an effort to spread the teachings of the most famous Italian and foreign positivists (Spencer in particular). Ardigò published several rather important works, such as *Il vero* (1891), *La ragione* (1894), and *L'unita della coscienza* (1898). Other thinkers, however, worked in completely different areas of research. For example, there was Antonio Labriola (1843–1904), a professor of philosophy at the University of Rome, who, after having approached Herbertism and Hegelianism in his youth, now embraced Marxism, attempting to differentiate it clearly from positivist positions. Two of his writings, *In memoria del manifesto dei comunisti* (1896) and *Del materialismo storico* (1897), are particularly significant in this context. Through Labriola's mediation, Benedetto Croce (1866–1952) also approached Marxism around 1895, dividing his philosophical interests between researches in esthetics and researches on historical materialism. The first fruits of these studies, *Tesi fondamentale di un'estetica come scienza dell'espressione e linguistica generale*, and *Materialismo storico ed economia politica*, came out in 1900.

The first works of Giovanni Gentile (1875–1944), *La filosofia di Marx* (1899) and *Rosmini e Gioberti* (1898), appeared at almost the same time. Both philosophers shared a dislike of positivism and an almost global condemnation of the culture that had given rise to that movement. Both Croce and Gentile, in strong opposition to the positivists, insisted on a drastic separation of science and philosophy, relegating science to the margins of culture.

Philosophical research in Germany

Evidence of the diversity of the cultural scene in Germany at the end of the century is provided by the fact that the period of Nietzsche's irrationalism and Wilhelm Dilthey's historicism was also one of interesting developments in positivism and materialism. This was the decade of *Die Welträtsel* (1899), a work on "the enigmas of the world" by Ernst Heinrich Haeckel (1834–1919), which sold more than four hundred thousand copies. Haeckel's doctrines had a huge influence on vast sectors of European society up to the time of the First World War. "Naturalist monism" left no room for faith or transcendency, and did not admit those "enigmas" reached by the agnosticism of the likes of Du Bois-Reymond.

The most widely diffused philosophy in Germany at this time was the doctrine of Kant, which tended to become the philosophy *par excellence* of university teaching. While the neo-Kantism of Marburg grew stronger, the neo-Kantism initiated by Windelband also became more influential. One of Windelband's students, Heinrich Rickert (1863–1936) even came to place an emphasis on the study of values, making a clear distinction between philosophy and scientific problems. Rickert ended up theorizing a complete separation between the two fields of investigation in his *Kulturwissenschaft und Naturwissenschaft* (1899).

Meanwhile, Edmund Husserl (1859–1938) began his philosophical career with the publication of his *Philosophie der Arithmetik* (1891). In this work the author still dealt with problems closely tied to those of the prevalent scientific philosophies; however, the lines of Husserl's later thought, where the central theme was a resolute affirmation of the superiority of immediate knowledge over mediated, categorical knowledge—the only kind that could be reached through conceptual (scientific) knowledge—are already visible in it.

Philosophy in the United States and England

In the United States, pragmatism, which in the work of Charles Sanders Pierce had had a markedly epistemological character (i.e. aimed at clarifying certain scientific concepts), underwent a radical change with the work of William James (1842–1910). James, who in 1890 had published an important work entitled *The Principles of Psychology*, published his *The Will to Believe* in 1897; in it, pragmatism became an exaltation of belief, which was considered an essential feeling for all human activity, including the elaboration of cognitive areas.

Meanwhile, pragmatism became diffused in England as well, where the modifications brought to it by James were accentuated even further. This took place mainly as a result of the work of Ferdinand Schiller (1864–1937), who in the first decade of the 20th century published several important works aimed at developing pragmatism in a "humanistic" direction, assigning human endeavor an active, leading role in the very organization of the universe. These works are collected in a single volume entitled *Humanism: Philosophical Essays,* which was published in 1907.

Idealism took on a tone that became further and further removed from the markedly rationalistic position of Bradley's neoHegelianism. Though initially influenced by Bradley, the American Josiah Royce (1855–1916) broke with him on enough points that his philosophy no longer qualified as Hegelian, but rather as an absolute voluntarism. In this context, Royce's important works were *The Conception of God* (1895) and *The World and the Individual* (1900–1902).

The Englishman John McTaggart (1866–1925) author of a number of studies on Hegel, such as *Studies in the Hegelian Dialectic* (1896), *A Commentary on Hegel's Logic* (1910) and others, came to the surprising conclusion that dialectics was not the constituent element of Hegelian procedure, but only represented the incompleteness of the finite, so that the absolute (perfectly complete in itself) was totally extraneous to it.

Schiller's and Royce's theological conclusions quickly gained the approval of numerous scholars. In fact, they fully answered the new needs of a bourgeoisie that had lost all faith in rationality (in particular in the rationality of science and technology) and aspired to escape the historical reality in which it found itself by adopting modern forms of religosity.

MATHEMATICAL SCIENCES

The beginnings of Hilbert's work

The figure of David Hilbert (1862–1943) began to emerge during this period. In 1890, Hilbert published a memoir on the theory of algebraic forms, which, together with his work on complete systems of invariants, crowned his fundamental researches on the theory of invariants. Then, in 1897, he published his *Zahlbericht*, which marked a decisive step forward in the theory of algebraic fields, which he succeeded in generalizing even further in a memoir of the following year. Finally, in 1899, he published the first edition of his famous *Grundlagen der Geometrie*, which proposed a purely formal axiomatization of elementary geometry, with the presentation of numerous critical questions on the compatibility and mutual independence of the axioms. This was the first detailed and complete example of a theory organized in accordance with the modern concept of axiomatic method, where the axiomatization of a mathematical theory no longer consisted of a search for several primitive propositions, guaranteed by their particular obviousness, from which all the other propositions that held true for that theory could be derived by deduction. Following the "crisis of evidence" brought on by the creation of non-Euclidian geometries and the perspectives on the plurality of possible geometries that emerged from Felix Klein's Erlangen program, it no longer seemed to make sense to speak of a true geometry; rather it seemed that every geometry could consider itself a body of propositions derived from axioms that were simply stated at the beginning and were such that they implicitly defined the primitive concepts that appeared in them. This line of thought was, as we have seen, consistent with the one being pursued in other areas, by Giuseppe Peano and his school with regard to all of mathematics, which tended to be presented in all its complexity as a purely hypothetical-deductive understanding (this characteristic expression was first used by one of Peano's students, Mario Pieri (1860–1913) in a work of 1899 significantly entitled *Della geo-*

metria elementare come sistema ipotetico-deduttivo). The *Grundlagen der Geometrie* thus foreshadowed Hilbert's productive interest in the fundamentals of mathematics, which he mainly developed around 1920 and which made him the leader of the formalist current and one of the major mathematical logicians of our time. This interest in the fundamentals of geometry, moreover, was very much alive in Italy too, even outside Peano's school: for example, in 1891 Giuseppe Veronese published his *Fondamenti di geometria a più dimensioni*, in which, among other things, he constructed an example of a non-Archimedean geometry, rendering Archimedes's postulate independent of the rest of elementary geometry (Hilbert's *Grundlagen*, published over a number of years, provided various examples of "non-geometries," in concomitance with the construction of proofs of the independence of various axioms).

Logic and set theory

All the above-mentioned work of critical revision was made possible in part by the fact that mathematical logic received constant attention from scholars: the three volumes of lessons on the algebra of logic (*Abriss der Algebra der Logik*), published by Friedrich Wilhelm Karl Ernst Schroder (1841–1902) in 1890–95, constituted an almost complete summary and systematization of the vast researches in the algebra of logic conducted over the preceding decades. But the area from which the algebra of logic ended up drawing most of the inspiration for its development was that of set theory, due to the many critical problems that arose in it. This theory was reformulated and further developed by its creator, Georg Cantor, between 1893 and 1897, in a series of articles entitled *Beiträge zur Begründung der transfiniten Mengenlehre*. These articles systematically dealt with abstract set theory and, in particular, built the great system of both ordinal and cardinal transfinite numbers, opening new and even more daring perspectives in mathematics than those contained in Cantor's articles of 1879–84 (these were limited to sets of points, even though they introduced various ideas that were later taken up, sometimes from different angles, by set topology). In 1897, Cesare Burali-Forti (1861–1931) discovered the first antinomy in set theory (that of the largest ordinal number), preluding the crisis of fundamentals that exploded after Bertrand Russell (1872–1970) discovered the antinomy of classes in 1902.

For the time being, Russell made himself known through an important essay on the fundamentals of geometry (1897), though his work really began to unfold at the beginning of the twentieth century (especially after the stimulus he received from his first meeting with Peano and his students at the World Congress of Mathematics in Paris in 1900), with his *Principles of Mathematics* (1903) and *Principia mathematica* (1910–13). These works were written in collaboration with Alfred North Whitehead (1861–1947) and in many respects remain the bibles of mathematical logic.

Preludes to twentieth-century research

The seeds of many of the theories and researches of contemporary mathematics were sown during these last few years of the nineteenth century. For example, operational calculus was invented in 1893 by Oliver Heaviside (1850–1924), foreshadowing modern distribution theory. Then there were the topological studies of Camille Jordan, who rigorously defined the concept of curve (1892), while Peano constructed a curve that satisfied the criteria of Jordan's rigorous definition and could fill a square surface.

These scholars were also responsible for the first statement of the concept of measuring a set, which was modified and broadened in 1894 by Félix Édouard Borel (1871–1956) to provide the foundation on which Henri Leon Lebesgue (1875–1941) built his theory of integration.

Federigo Enriques (1871–1949) and (Jules) Henri Poincaré (1854–1912), who both left lasting marks in the area of philosophical reflection on mathematics, made important technical studies during this period. In 1892, Enriques elaborated the general theory of the invariants of algebraic surfaces, while Poincaré, who was already famous for his extremely valuable researches in almost all sectors of mathematics, celestial mechanics, and mathematical physics, from 1895 onward extended his topological studies to multi-dimensional geometries and introduced the concept of homeomorphism for biunivocal and continuous correspondences. In 1899, Curbastro Gregorio Ricci (1853–1925), together with Tullio Levi-Civita (1873–1941), published an important memoir expounding the principles of absolute differential calculus, later called tensorial calculus, the fundamental mathematical tool used by Albert Einstein to state his theory of general relativity.

During the last decade of the century, the abstract theory of groups underwent a particular surge of development, synthesizing and generalizing the many studies made and results obtained in this field by mathematicians such as Jordan, Klein, Lie, Cayley, Dedekind, Kronecker, and Picard (who had all studied groups within relatively specialized contexts, i.e. the theory of equations, number theory, and transformation groups), and moving on to study groups in an autonomous and abstract way. Ludwig Otto Holder (1859–1937) and Ferdinand Georg Frobenius (1849–1917) were the major representatives of this line of research, which mainly flourished during the early decades of the twentieth century.

ASTRONOMY

The planetoids

The end of 1891 brought a decisive turn in research on asteroids. Maxmilian Wolf (1836–1932) of Heidelberg introduced a photographic method that made it possible to detect asteroids by the marks they left on photographic plates as they moved among the stars during exposure time. The first planetoid discovered using this technique was Brucia, so named by its discoverer in honor of Miss Bruce, a patron of astronomy. It was also the first planetoid discovered by Wolf, who established a record in the field by finding 230 of them. On 13 August 1898, the astrophile Witt, of the Urania Society of Berlin, photographically discovered Eros, a planetoid that became very important, since it came much closer to earth than any other and was therefore chosen for use in determining the solar parallax. In order accurately to determine the distance of the planetoid, a number of international campaigns were organized during oppositions up to 1949.

Hale and Deslandres's spectroheliograph

In 1891–92, George Ellery Hale (1868–1938) in the United States and Henri Alexandre Deslandres (1853–1948) in France independently reached a revolutionary technical milestone in the observation of the sun with the invention of the spectroheliograph. This new instrument could photograph the image of the sun in correspondence with a single spectral line, for example the H_α hydrogen line, observing only and all the configurations that emitted or absorbed at that wavelength.

This made it possible to photograph the protruberances at any time, at the limb or even within the solar disc, where they appeared as dark filaments.

Variable stars

Important discoveries were also made in the area of variable stars. Father Johann Georg Hagen, who in 1906 became director of the Specola Vaticana, reopened during this period by Pope Leo XIII, prepared the first general atlas of variable stars (which remained the only one of its kind), the *Atlas stellarum variabilium*, the first series of which came out in 1899. In 1894, Aristarch Apolonovich Belopolsky (1854–1934) discovered that the radial velocity of δ Cepheus varies with the same period at its magnitude but displays a specular curve with respect to the curve of light. After much discussion and controversy, this discovery made it possible to determine the pulsating nature of the cepheids. Almost simultaneously, Bailey photographically discovered cepheid variables with short periods, later called cluster variables, in several globular clusters. Further discoveries and researches on these variables made them extremely valuable, in the course of the following century, in determining the distances of the farthest star clusters and, as a result, the structure and dimensions of the galaxy, as well as the position of the sun.

The lunar atlas

After having taken the initiative for the photographic map of the heavens, on which astronomers were already diligently at work, the Paris Observatory, this time alone, undertook the project of compiling a huge photographic atlas of the moon, in seventy-one large tables. Begun in 1894 by Loewy and Puiseux, it took almost twenty years of work, but it remained unsurpassed for more than half a century.

Astrometry

Astrometry also reached great heights during this period. Measurements of the meridian circle made by Karl F. Küstner (1856–1936) around 1900 gave the positions of the stars with an error of only 0.27 minutes. For an idea of the progress made in this area over the years, we can remember that Hipparchus's error was 240 minutes, Tycho Brahe's, 25 minutes; James Bradley's, 2 minutes; and Friedrich Willhelm Bessel's, 0.7 minutes.

Periodicals

The quality and quantity of the publications of the epoch bear witness to the explosion in astronomical research that took place at the end of this century. In fact, two of the most authoritative modern periodicals, the *Astronomical Journal* and

the *Astrophysical Journal*, both American, were founded during this period. Worldwide production in the field, on the other hand, became so frequent and prolific that it could not easily be followed without a guide. Jean Charles Houzeau de Lehaie (1820–1888) and Albert Lancaster had just finished a three-volume *Bibliographie générale de l'astronomie* in 1889, listing all the astronomical works published from the introduction of printing to 1880. By 1899, a volume was needed every year to report on all the astronomical literature of the world. This periodical, originally called *Astronomischer Jahresbericht*, is still being published today, having become *Astronomy and Astrophysics Abstracts* in 1969.

PHYSICAL SCIENCES: MECHANICS

The gyrostatic ether

We have already seen that the problem of theoretically determining the relations between the ether and matter constituted one of the central themes in the physical sciences during the decade 1880–1890. The development of the themes introduced by Lord Kelvin in his Baltimore lessons (see PHYSICAL SCIENCES: ELECTROMAGNETIC THEORIES, 1880–1890) involved the elaboration of a "dynamic" theory of light and reached a particularly important point in two communications published by Kelvin in 1890. In one of these, on the constitution of an "ideal substance called ether for brevity," Kelvin did not deny that "the triple alliance" of ether, electricity, and ponderable matter was the result of scientists' lack of knowledge rather than a reality of nature.

Despite this awareness of the difficulties surrounding the interaction between the ether and matter, Kelvin reelaborated the gyrostatic model to accommodate the transmission of transversal waves in a medium with an inner potential of zero (*On a Gyrostatic Adynamic Constitution for "Ether"*). However, it should be mentioned that, at the end of the decade, Kelvin attempted a different approach, based on the difficulties brought up by Michelson and Morley's measurements.

Hertz's Mechanics

In 1894, Heinrich Rudolf Hertz (1857–1894) published his work on the principles of mechanics (*Die Principien der Mechanik in neuen Zusammenhange dargestellt*). As Hermann Ludwig von Helmholtz (1821–1894) observed in his preface, Hertz's work was closely tied to the developments that had taken place in electrodynamics. In this sense, even though there were connections between mechanics according to Hertz and mechanics according to Kirchhoff, the 1894 work was different from the 1874 work (see PHYSICAL SCIENCES: MECHANICS, 1870–1880). Both attempts to reorganize the science of motion shared the concept that mechanical theory was not interested in knowing what existed beyond the observable. Mechanics should thus be viewed as a hypothetical deductive system that operated, according to Hertz, on "symbols of external objects." This operation on symbols led to purely logical consequences that, in Hertz's opinion, should be considered "the necessary consequences in nature of the objects represented." In the reorganization of mechanics, it was therefore essential to emphasize the inner consistency of the theory and the formulability of rules of correspondence between some symbols and what we call external objects. The goal was to arrive at a "well-refined, complete, and law-abiding image of the universe." Using the formalism of Lagrangean mathematical physics, Hertz proposed reducing forces and energies to consequences of mass and motion. But in order to do this, it was impossible not to hypothesize on what was hidden in the universe. As Hertz wrote, hidden masses and hidden motions did not differ in principle from visible masses and motions. In his *Mechanics* of 1894, Hertz elaborated the correlations between observable actions and those caused by hidden masses, using the Lagrangean scheme. The theme of the ether was still at the basis of Hertz's program. In fact, according to Hertz, the hypothesis that an understanding of natural phenomena implied a resystemization of mechanics and electromagnetism necessarily involved a clarification of the essential characteristics of matter itself, i.e. an understanding of gravity and inertia. But, as Hertz had already stated in 1889, this clarification also implied an understanding of the properties of the ether. However, it should be noted that Hertz's critique of the concept of force, and the resulting relegation of the forces of mechanics to "dormant companions that leave the scene when real facts have to be represented," was not weakened by the introduction of hidden masses. In fact, Hertz pointed out that we assume that it is possible to unite the masses of the visible universe with other masses that obey the same law. In short, hidden masses were not intended as idealizations, they had the same character that Hertz perceived in forces and energies.

The debate on matter

The richness of the arguments directed against the very foundations of mechanics included a series of heated debates on the cognitive power of the physical sciences, in that the basic role played by the theory of motion in cognitive activity was being questioned. In 1894, in a communication entitled *Quelques réflexions au sujet de la physique expérimentale*, Pierre Duhem maintained that scientific knowledge had no other basis than the understandings of common sense. In 1895, Wilhelm Ostwald (1853–1932) published a work on the defeat of contemporary atomism (*La déroute de l'atomisme contemporaine*), where, as in other works, he affirmed that the defeat of mechanism implied the defeat of scientific materialism. The science of energeticists tended to become a philosophical concept of the world, and Ostwald himself, in his critique of materialism, launched upon extra-scientific digressions that led him into the great fold of neo-spiritualism, to the obvious delight of the more superficial metaphysicists (who wished to gain strength from the presumed "crisis of science"). Louis Marcel Brillouin's harsh reply to Ostwald, published in 1895 with the title *Pour la matière*, deserves mention. That same year, in a congress held at Lubeck, the ideas of Ostwald and the energeticists emerged victorious in a clash where the great loser seemed to be Ludwig Boltzmann's program in mathematical physics. In 1896, Max Planck (1858–1947) published an attack against the energeticist conception (*Gegen die neuere Energetik*), in which he wrote that energetics, because of the inaccuracy of its concepts, was not essentially

in a condition to contribute a new result that can be subjected to experimental control. This, felt Planck, was the harshest critique of all; because a theory that, in order to maintain its own existence, had to avoid real problems, no longer had its roots within the reign of natural science, but in the area of metaphysics, where empirical tools were no longer useful. Helm presented a historical reconstruction of the "purification" of the concept of energy of its mechanist implications in 1898 in his *Die Energetik*, which also made an attempt to reconstruct all of mechanics in energetic terms. An opposing view was presented in Boltzmann's publications, especially in *Über die Principien der Mechanik* (1900).

The clouds of the nineteenth century

The decade closed with a famous address entitled *Nineteenth Century Clouds over the Dynamical Theory of Heat and Light*, read by William Thomson Kelvin before the Royal Institute in April of 1900. The two major problems existing within the physical sciences, according to Kelvin, were connected with the difficulties of Michelson and Morley's experiment and with the developments in statistical mechanics that were based on the principle of equipartition of energy (see PHYSICAL SCIENCES: THERMODYNAMICS AND STATISTICAL MECHANICS, 1890–1900). The first problem concerned the constitution of the ether, and, in Kelvin's opinion, could not be resolved with hypotheses on the contraction of matter.

Even if, as Kelvin himself admitted, the suggestions put forth by Hendrik Antoon Lorentz and George Francis Fitzgerald seemed to provide a possibility of escaping the conclusion that presented itself as a result of Michelson and Morley's measurements, it was still true that the cloud that had fallen over the theory of dynamics had to be considered very thick. In a communication presented to the Scientific Congress held in August of 1900 in connection with the great international exhibition in Paris, Kelvin elaborated the problem of the motion of ponderable bodies in an infinite space occupied by an elastic solid. As he himself stated, it was "a pure problem of abstract mathematical dynamics, without any indications for a physical application." In the conclusion of the work, he could not avoid repeating the ideas already advanced in his April allocution. However, what radically differentiated Kelvin's awareness of the difficulties that existed in mechanics from the arguments that were rapidly spreading in many scientific circles was that, in these very real difficulties, Kelvin saw not the signs of a crisis, but rather the need to develop theories that were more complete and general.

A hundred years after critical and historical analysis had made the first clarifications of mechanics possible, the basic problems of mechanics appeared complex. They were inevitably correlated to problems and difficulties that had arisen in other sectors of the physical sciences as well. They were not isolated problems, but rather seemed to force scholars and philosophers to deal with changes that had taken place in all the scientific disciplines. These changes had been brought about in part by both the sudden increase in information on the structure of matter, and the intertwining of various theoretical schemes, which were no longer reducible to a single mechanistic structure and were strongly characterized by a progressive mathematization.

PHYSICAL SCIENCES: THERMODYNAMICS AND STATISTICAL MECHANICS

Kelvin's crucial thought experiments

Between the summer of 1891 and the spring of 1892, Lord Kelvin wrote several communications that returned to the polemics on irreversibility, generalizing them and above all questioning the principle of equipartition of energy. In the first work, entitled *On Some Test Cases for the Maxwell–Boltzmann Doctrine regarding Distribution of Energy*, Kelvin examined some elementary cases of motion and demonstrated that, in general, they did not provide confirmation of the principle of equipartition. On the contrary, even in the case of systems with kinetic energies that could be expressed as sums of squares, Kelvin maintained that under certain express conditions, they irremediably violated the general Boltzmann-Maxwell concept. Although he admitted that the critiques concerning the tendency of all translational energy to be converted into vibrational energy were erroneous (see PHYSICAL SCIENCES: THERMODYNAMICS AND STATISTICAL MECHANICS, 1880–1890), Kelvin reiterated that the fundamental principles if the kinetic theory of gasses were also in error. He elaborated counterexamples of a purely physico-mathematical character. As he emphasized in another work of the same year, *On Periodic Motion of a Finite Conservative System*, these counterexamples were based on a generalization of the theory that studied all physical systems in which the potential energy was quadratic function of the coordinates and the kinetic energy was a quadratic function of the velocities. These were thus generalized problems of dynamics and therefore, as Kelvin pointed out, admitted a rigorous discussion, as Jules Henri Poincaré's independent mathematical researches demonstrated. In a brief note on an ideal and crucial experiment (*On a Decisive Test-Case Disproving the Maxwell–Boltzmann Doctrine Regarding Distribution of Kinetic Energy*, 1892), which was based on a thought experiment using three particles only, Kelvin attempted definitively to confute the principles of statistical mechanics. In his opinion, "the assumption that the temperature of a solid or a liquid is equal to its mean kinetic energy per atom" was completely erroneous, was not at all derived from theorems, and did not in any way constitute a fundamental proposition of thermodynamics.

Culverwell's objections

Along with Kelvin's critiques, the analysis made by Edward P. Culverwell, based on the affirmation that "there is nothing in the nature of dynamical *equations* in virtue of which their solutions tend to a permanent average distribution of energy" deserves mention. This affirmation, which was related to the debates of the previous decade, and appeared in his 1890 *Note on Boltzmann's Kinetic Theory of Gases and on Sir W. Thomson's Address to Section A, British Association: 1884*, led Culverwell to maintain further in the same paper that "it is impossible to prove in general that a set of particles will tend to the Boltzmann configuration..." If such a tendency were real, according to Culverwell, it had to depend on a physical principle that was extraneous to a purely dynamic context. Culverwell asserted "that permanence is a consequence of the interaction of molecules and aether." "But," he notes, "the difficulties are only avoided; all that is done is to reduce the number of properties

explained by the kinetic theory, and to shift the temperature difficulties to the aether, so that in some shape or other they will crop up again."

Rayleigh's analysis

In 1892, Lord Rayleigh (1842–1919) published a memoir in defense of Maxwell's and Boltzmann's concepts entitled *Remarks on Maxwell's Investigations Regarding Boltzmann's Theorem*. In it, he pointed out that the critiques made against Maxwell and Boltzmann concerned a basic question that was present "more in the premises than in the conclusions" of the kinetic theory of gasses. The hypotheses of equiprobability that operated in these premises could not be considered always valid. Thus, Rayleigh believed, it was necessary to discern up to what point cases that were "too exceptional" could be disregarded, and to check out what one intended to affirm through propositions of a probabilistic nature. The problem, therefore, was not that of the particular form in which molecular kinetic energy was expressed, but rather consisted in the need to clarify the problems connected with the ergodic systems referred to by Boltzmann in 1887. This analysis was also important in the light of Rayleigh's later work, done toward the end of the decade, to which we shall return.

The great debate in Nature magazine

In 1894 and 1895, *Nature* magazine hosted a long debate on the possible interpretations of Boltzmann's H theorem. The tone of the papers presented by the journal was openly polemical: for example, in 1894 Culverwell maintained that it was almost impossible that the theorem in question had any significance whatsoever and stated that, in his opinion, it was very difficult to understand how any demonstration could exist for it. Returning to several arguments that he had already advanced in 1890, Burbury, in a note of 1894 entitled *Boltzmann's Minimum Function*, proposed attributing the variations of Boltzmann's function to "external causes," so that the stationary states could be evaluated in terms of the equation:

$$\frac{dH}{dt} + \frac{\partial H}{\partial t} = 0 \, ,$$

where dH/dt stood for the variations produced by collisions, while $\partial H/\partial t$ stood for those that could be ascribed to external causes. Simultaneously, however, in a note entitled *The Kinetic Theory of Gases* (1894), Culverwell pointed out another problem: radical differences existed between the provisions of the kinetic theory and experimental data on spectra. Although they were "logical ," the answers provided by Boltzmann could not, in Culverwell's opinion, be considered satisfactory. Boltzmann's position, in fact, seemed to boil down to that of someone who states a theorem in a purely mathematical context, then asks the physicists to see up to what point it is applicable to gasses. The debate was enriched when Boltzmann firmly declared that the H theorem (and therefore the statistical interpretation of the concept of entropy) was not "a dynamic theorem but a theorem concerning probabilities." With two papers published in April and July of 1895, Boltzmann defended the statistical interpretation, criticizing Burbury's attempts to resort to external causes. Nevertheless, toward the end of 1895, Boltzmann's position seemed increasingly more isolated.

Zermelo's objection

In 1896, Ernst Zermelo (1871–1953) published a work on one of the principles of dynamics and the mechanical theory of heat (*Über einen Satz der Dynamik und der mechanischen Wärmetheorie*). Basing his ideas on Poincaré's researches into the conditions of stability of stationary motions, Zermelo maintained that Boltzmann's reflections on the H theorem were contradictory and affirmed that the systems to which the theorem would be applicable were subject to almost periodic motion.

Boltzmann replied immediately to this fundamental objection (*Entgegnung auf die wärmetheoretischen Betrachtungen des Hrn. E. Zermelo, 1869*), accusing Zermelo of not having understood the meaning of the propositions that he intended to criticize. In the first place, it was not possible to understand the H theorem outside a statistical context; in the second place, the accusation of periodicity did not constitute an objection, since the H theorem did not exclude a return of the system to the initial state, but only prescribed a very small degree of probability for such a phenomenon. Boltzmann returned to these same themes and generalized them in his two-volume theory of gasses (*Vorlesungen über Gastheorie*), published between 1896 and 1898.

Irreversibility according to Planck

Max Planck's interest in irreversible phenomena reached a level of particular intensity in 1894. At the height of the debate on Boltzmann's ideas, Planck proposed to investigate radiation phenomena, since they seemed to offer a fertile terrain for theoretical researches based on an electromagnetic concept of nature. In this way, as Planck repeated in the introduction to his 1897 text on thermodynamics (*Vorlesungen über Thermodynamik*), it might be possible to found the second law of thermodynamics on the electromagnetic theory and not on statistical approaches to mechanics. In the second half of the decade, Planck's attitude towards energy became explicit critical; however, even though in the 1897 work he recognized that the kinetic theory "penetrated more deeply into the nature of the processes examined," Planck still maintained that the difficulties presented by the theory in the mechanical interpretation of the two principles were "insurmountable." This is why, according to Planck, it was necessary to introduce the argument that irreversibility should be analyzed in reference to oscillators capable of emitting and absorbing electromagnetic radiation. Planck's attempt immediately encountered the objections of Boltzmann. In fact, that same year (1897), Boltzmann maintained that irreversibility could not be a sort of consequence of the interaction between oscillators and electromagnetic waves, since that interaction was described by equations that implied reversibility.

It was in the context of this discussion with Boltzmann that Planck's memorable series of communications on radiation and irreversability (*Über irreversible Strahlungsvorgänge*), published between 1897 and 1900 and culminating in a proposal concerning the constant h, originated and matured. While at the beginning of Planck's new program we find the suggestion, mentioned earlier, of correlating irreversibility to the electromagnetic theory, Boltzmann's critiques gradually forced this program to include inferences based on a hypothesis of disorder. By 1899, Planck found himself supporting positions not far from those of Boltzmann. According to Planck, the concept of entropy should be considered theoretically privileged with respect to the concept of temperature; and as he himself pointed out in a lesson held in Munich in 1899, the mathematical analysis of oscillators not only required the electromagnetic theory, but was also in need of a hypothesis on "natural radiation" where the energy of the radiation itself was irregularly distributed. And, as Planck noted, the hypothesis on natural radiation was exactly the same as Boltzmann's hypothesis on molecular disorder.

Wien's theoretical researches

The problem of the distribution of energy in the black body spectrum became increasingly important following Boltzmann's deduction of Stefan's law (see PHYSICAL SCIENCES: THERMODYNAMICS AND STATISTICAL MECHANICS, 1880–1890). As early as 1889, Lord Rayleigh had attempted to develop a probabilistic wave model for radiation: in a communication entitled *On the Character of the Complete Radiation at a Given Temperature*, Rayleigh brought up the need for determining the form of Kirchoff's F (v) function, and suggested building a suitable model that would "verify whether or not it was possible to define a type of impulse that in irregular sequence could represent the complete radiation for every temperature." Wilhelm Wien (1864–1928), on the other hand, followed a completely different line of study. His memoir of 1894 on the temperature and entropy of radiation (*Temperatur und Entropie der Strahlung*) provided the basis for the so-called displacement law, according to which maximum radiated energy occurs at a wavelength that is inversely proportional to the absolute temperature of the black body. Two years later, Wien published a work on the energy problems of black body emission (*Über die Energievertheilung in Emissionsspektrum eines schwarzen Körpers*), in which he pointed out that, due to total ignorance of the causes of radiation, a particular hypothesis had to be introduced if an evaluation of the distribution of energy in that radiation were to be reached. According to Wien, when a certain correlation was set up between intensity and the wavelength of the vibrations emitted by each molecule, the Stefan-Boltzmann Law could be used to determine the relation linking intensity, wavelength, and temperature. This form was first confirmed by experimental data gathered by Friedrich Paschen. Meanwhile, Planck elaborated a different deduction of Wien's law of radiation in his memoir *Über irreversible Strahlungsvorgänge*.

Rayleigh's law

In 1900, following a different theoretical procedure from Wien's and Planck's, Lord Rayleigh demonstrated a law of radiation that had a very distinct form with respect to the one just mentioned. On the basis of considerations that had as their point of reference the principle of equipartition of energy (to which he dedicated another interesting memoir of 1900 entitled *The Law of Partition of Kinetic Energy*), Rayleigh obtained the relation:

$$u = \frac{8\pi v^2}{c^3} k T ,$$

which he presented in his *Remarks upon the Law of Complete Radiation*. This differed profoundly from the formula derived earlier by Wien and Planck:

$$u = \frac{8\pi v^2}{c^3} U(v, T),$$

where:

$$U(v, T) = a\, e^{-bv/T}.$$

The more profound significance of this divergence, which was above all a consequence of the different premises used respectively by Rayleigh and Planck, was not perceived at once. Instead, more attention was given to the numerous experimental studies conducted by various researchers, especially Paschen, Otto Lummer, and Ernst Pringsheim, between 1897 and 1900. These studies were concerned with verifying the consequences deduced from the Stefan-Boltzmann and Wien-Planck laws: as Planck observed in a paper of 1900 on the reelaboration of Wien's law (*Über eine Verbesserung der Wienschen Spektralgleichung*), these experiments proved that the law of distribution of energy given by Wien was not generally valid, and that at most, this law had the character of an extreme case whose simple form was due to a restriction on short wavelengths and law temperatures. Thus the blackbody problem had to be reformulated (see PHYSICAL SCIENCES: RADIATION AND MATTER, 1890–1900).

PHYSICAL SCIENCES: ELECTROMAGNETIC THEORIES

General lines of Hertz's program

A collection of Hertz's experimental and theoretical writings appeared in 1892 under the title *Untersuchungen über die Ausbreitung der elektrischen Kraft*, with a preface by Helmholtz. The following year, an English translation of the work, entitled *Electric Waves, Being Researches on the Propagation of Electric Action with Finite Velocity through Space*, was published with an introductory note by Kelvin. Hertz's long and detailed introduction to this collection was of extraordinary importance to the further development of electromagnetic theories and their interpretations. In the first place, Hertz claimed that he had provided experimental proof of the validity of the fundamental hypotheses of what he called "the Faraday-Maxwell theory," hypotheses that, as he himself observed, "contradicted the usual views and did not rest upon the evidence of decisive experiments." In the second place, this experimental confirmation alone was not sufficient, in Hertz's opinion, to answer the question, "What is Maxwell's theory?" According to Hertz, there were at least three different ways of expounding this theory. Besides Maxwell's original treatment, there was Helmholtz's, to which Hertz claimed to have adhered in planning his own experiments: "But, unfortunately, in the special limiting case of Helmholtz's theory which leads to Maxwell's equations, and to which the experiments pointed, the physical basis of Helmholtz's theory disappears, as indeed it always does, as soon as action-at-a-distance is disregarded." Then there was the problem of a third treatment, which Hertz developed in his two theoretical memoirs of 1890 (see PHYSICAL SCIENCES: ELECTROMAGNETIC THEORIES, 1880–1890). Now, these three different expositions "have substantially the same inner significance," and it was in reference to this "common significance" that Hertz answered the question posed above, affirming that "Maxwell's theory is Maxwell's system of equations." In Hertz's opinion, the theory had to be reorganized in order to simplify it as much as possible, thus pursuing a program already outlined in his *Mechanics*. In his first theoretical memoir of 1890, placed at the end of the collection, Hertz first of all outlined "the fundamental ideas and the formulae by which they are connected." Naturally, these formulae required "explanations," but, wrote Hertz, "these explanations are not to be regarded as proofs of the formulae. The statements," he observed, "will rather be given as facts derived from experience; and experience must be regarded as their proof. It is true, meanwhile, that each separate formula cannot be specially tested by experience, but only the system as a whole. But practically the same holds good for the system of equations of ordinary dynamics."

Electron theory according to Lorentz

From his very first works, which were directly focused on Maxwell's theory of electromagnetic field, Lorentz insisted on the fact that the electromagnetic theory of light was one of the means by which human understanding could be extended to the world of molecules. Lorentz provided a complex elaboration of this idea in a memoir of 1892 entitled *La théorie électromagnétique de Maxwell et son application aux corps mouvants*, in which his exposition of electron theory made reference to hypothetical charged particles present in large numbers in ponderable bodies. In Lorentz's approach, the problem of the ether existed in a particular form, to the extent that the ether was defined as a very particular substance, always at rest and with physical properties that were ruled by the laws of electromagnetic field. In Lorentz's theory of 1892, the electrons had the role of intermediaries between ponderable matter and the ether. The theory was thus based on five fundamental equations, the first four being the Maxwell-Hertz equations, which expressed the state of the ether, while the fifth was an equation devised to define the forces acting on the charged particles. As we can see, Lorentz's theory implied a corpuscular conception with respect to matter, and a continuous conception for the ether. A second memoir of 1892, specifically dedicated to the problems raised by Michelson and Morley's experiment, was written in the same context. According to Lorentz, the experiment necessarily implied a critical reexamination of both Stokes's and Fresnel's hypotheses on the ether. In order to reconcile Fresnel's hypothesis with Michelson and Morley's result, Lorentz introduced the famous assumption on contraction, i.e. the hypothesis that the line that connects two points of a solid body, if initially parallel to the direction of the earth's movement, does not stay the same length when it is successively rotated by ninety degrees. In 1895, all these themes were brought together in a fundamental memoir by Lorentz entitled *Versuch einer Theorie der elektrischen und optischen Erscheinungen in bewegten Körpern*, which in 1899 was followed by another treatise on the simplified theory of electrical and optical phenomena in bodies in motion. Here Lorentz developed his own concepts, elaborating the so-called "theorem of corresponding states" and introducing "local time."

The 1899 work was particularly important, in that in it Lorentz explicitly posed the problem of variable transformations, i.e. those transformations that are known even today as "Lorentz transformations," and which played an extraordinarily important role in studies on relativity.

The problem of the ether in Larmor's researches

The studies of Joseph Larmor (1857–1942) during the period 1893 to 1900 started out from the idea that it should be possible to interpret electromagnetic events correctly on the basis of a mechanical theory of radiation. Larmor also raised the problem of electrons, and, as we shall soon see, by the end of the decade he had reached a sophisticated characterization of the ether that depended on the need to take into account the carriers of electricity. Larmor's program, in its initial stages, was closely linked to George Francis Fitzgerald's treatments of the ether and was developed over the years by returning to Hamilton's principle and aiming at a global attempt to reduce the mechanics on material bodies to the properties of an essentially electromagnetic ether. This was still a reductionistic endeavor, but it should be noted that it was based, on the one hand, on the possibility of characterizing the ether by means of Hamilton's principle and, on the other, on the possibility of defining the electrons as like the singularity of that ether. One cannot, therefore, speak of ingenuous mechanism, but rather of a program that developed around a general electromagnetic conception, even though it was not able to provide an image of the ether except in terms of "something that is prior to matter." In 1897, Larmor generalized Lorentz's ideas regarding the transformations for the equations of the electron theory, introducing second order terms in v/c and explicitly linking these transformations to Fitzgerald's and Lorentz's hypotheses on contraction. Larmor outlined the entire significance of these researches in 1900 in his *Aether and Matter*, in which he maintained that Hamilton's principle made it possible to state every dynamic problem as a variational problem, in order to help reduce to dynamics the physical theories for which the intimate dynamic mechanisms are more or less hidden from direct observation. From this point of view, which implied the possibility of deriving the equations of electromagnetic field from Hamilton's scheme, the electron was, in Larmor's opinion, a singularity, in the sense that an electron e appeared in his analysis as a singular point in the ether analogous in reality to what is defined as a simple pole in the two-dimensional representation used in the theory of a complex variable function. Larmor himself was well aware of the abstractness and incompleteness of his theory. It is interesting to note how he approached the problem: The goal of theoretical physics, he felt, was not to provide a complete and exhaustive conquest of the *modus operandi* of natural phenomena. Such a goal would be hopeless and unreachable, if only because the mental apparatus with which the research would be conducted was itself, in one of its aspects, part of that scheme of nature that it was endeavoring to unveil.

The electromagnetic conception of nature

The affirmation of Maxwell's theory and the richness of its developments led to a series of elaborations that were of extreme interest to European culture. After 1894, many scientists leaned toward a general concept of the physical world in which all of reality consisted in mobile electrical particles within an electromagnetic ether. This electromagnetic conception of nature developed above all within the context of the possibility of reducing mass to an electromagnetic mass. Thus for example Emil Wiechert, in two memoirs of 1896 and 1898, respectively entitled *Über die Grundlagen der Elektrodynamik* and *Hypothesen für eine Theorie de elektrischen und magnetischen Erscheinungen*, upheld the validity of an electron theory similar to the one elaborated by Lorentz and based on the assumption that the electromagnetic ether was the only reality. The basic tendency of this program— and others similar to it—was to eliminate all mechanical concepts through a critique of the notion of mass. Pierre Duhem also took a stand against these same concepts in 1892, in the third volume of his *Leçons sur l'électricité et le magnetism*, but his critique came from an entirely different point of view that was not electromagnetic. While Duhem's goal was ultimately to develop a treatment based on thermodynamics, studies such as Wiechert's were aimed at replacing the notions of mechanics with those of electromagnetism. Lorentz's contributions to the spread of an electromagnetic conception of nature should not be undervalued. For example, in 1900, in a work entitled *Considérations sur la pesanteur*, Lorentz hinted at the possibility of connecting the explanation of gravitational phenomena to an essentially electromagnetic theory. That same year, Wilhelm Wien (1864–1928) referred even more explicitly to the same concept.

Poincaré's ideas

Poincare's critical contribution to the development of electromagnetic theories was elaborated in his *Électricité et optique*, 1890 and 1901. His ideas were expounded in a particularly enlightening way, especially in the second edition of the work. He observed that none of the principal formulations of the electromagnetic theory was able to satisfy the following three needs: to provide an explanation of Fizeau's experiments, to be consistent with the principle of conservation of electricity and magnetism, and to be consistent with the Newtonian principle of the equality of action and reaction. According to Poincaré, Lorentz's electron theory encountered serious difficulties with respect to the third need, even though Lorentz's theory was the best one available. Poincaré offered a possibility of surmounting these difficulties in a memoir of 1900 entitled *La théorie de Lorentz et le principe de reaction*. In this memoir, Poincaré established a connection between the themes of *Électricité et optique* and a significant part of Lorentz's program, and laid the premises for the statement of the principles of relativity and constancy of the speed of light that were the nucleus of the relation he presented at the Congress of Arts and Sciences in St. Louis in 1904. All of Poincaré's ideas matured within the context of the "physics of principles," and could not in any way lessen the importance of Einstein's early works, which came out of considerations that were fairly different from Poincaré's and were permeated with a profound awareness that a generalization of the principle of relativity to electromagnetism implied a global reexamination of classical mechanics and its concepts.

PHYSICAL SCIENCES: RADIATION AND MATTER

Charged particles

During this decade, the debate on the nature of cathode rays became interwoven with the reflections on the corpuscular nature of electricity on which various theoreticians based their programs for reelaborating Maxwell's theory. In this way, the criteria for choosing between the wave concept and the corpuscular concept were enriched by new and more articulated arguments. In opposition to the supposedly crucial experiments of Hertz and others, which tended to confirm wave theories and disprove corpuscular ones, were others, which provided new evidence in support of the corpuscular concept. Altogether, during the final decade of the century, the problem of the possibility that atoms were made up of charged particles was posed, and the assumption that atoms, long thought of as the "elements" of matter, were simple and indivisible fell. In his doctoral dissertation, Jean Baptiste Perrin (1870–1942) designed an experiment aimed at overturning Hertz's interpretation of the relations between electrical charge and cathode radiation (1895). Perrin's device rendered observable a series of effects that corroborated the assumption that cathode rays were made up of particles and that these particles were carriers of a negative electrical charge. The explanatory force of this assumption was illustrated by Joseph John Thomson, who used the data published by Phillipp Lenard (1862–1947) on the ability of cathode rays to travel determined distances in air and considerations on mean free path to arrive at the conclusion that cathode particles had to be much smaller than air molecules. In Thomson's opinion, the diameter of the particles was 10^5 times smaller than the diameter of the molecules.

Röntgen's radiation

One of the major results that emerged from the investigations on cathode radiation came from the observations conducted by Wilhelm Konrad Röntgen (1845–1923) on a new kind of radiation.

In December of 1895 Röntgen published a preliminary communication entitled *Über eine neue Art von Strahlen*, in which he declared that he had observed an active agent provoked by the discharge in a cathode tube and capable of producing fluorescence. In addition, wrote Röntgen, "all bodies are transparent with respect to this agent." The "agent" was given the name "X-rays." According to Röntgen, these particular "phenomena of the ether" were completely different from cathode phenomena, even though they were produced by cathode rays. In particular, Röntgen suggested that they could be longitudinal vibrations in the ether.

Becquerel's radiation

Just a few months after the discovery of Röntgen's rays, Antoine Henry Becquerel (1852–1908) published a memoir on the rays emitted by phosphorescence (*Sur les radiations émises par phosphorescence*, 1896), in which he reported the initial results of his investigations on uranium salts. Becquerel's point of departure was the earlier work of J. Henry and Niewenglowski on the radiation emitted by phosphorescent salts and the behavior of those rays with respect to opaque bodies. Becquerel's observations indicated that salts containing uranium were able to emit a penetrating radiation without being excited by any outside force and that radiation was not dependent upon the physical state of the salts but was a consequence of the presence of the element uranium in them. As Becquerel pointed out, the new radiation was emitted continuously and produced effects somewhat similar to those of X-rays.

The researches of the Cavendish Laboratory

The new discoveries in the area of radiation were immediately submitted to testing and verification in the well-equipped Cavendish Laboratory of Cambridge, which was directed by Joseph John Thomson. Thomson's interest in the behavior of ions in gasses and the theoretical development of Maxwell's concepts, as well as his adherence to corpuscular positions, found in these radiations an extremely important subject for study. With the collaboration of Ernest Rutherford (1871–1937), Thomson analyzed the ionizing power of the new rays and drew valuable conclusions concerning the possibility of determining the relation between charge and mass for the "atoms of electricity" (electrons) that make up cathode rays.

From these researches, Rutherford derived the first clues on the constitution of Becquerel's rays. In a communication of 1899, he expounded results that could be interpreted by assuming that the rays were formed by one not very penetrating, easily absorbed radiation and one more penetrating radiation. Thomson, on the other hand, immediately began a series of scrupulous observations and measurements of cathode ray deflection in the presence of electrical and magnetic fields. Making use of inferences on the ionizing power of the rays, he was able to refute the objections concerning the non-observability of deflection in the presence of an electrical field, advanced by Hertz in his time, and in his famous monograph of 1897, *On Cathode Rays*, he fully supported the validity of the corpuscular model. According to Thomson, it was impossible to escape the conclusion that cathode rays were formed by negative charges of electricity transported by particles of matter. These material carriers of charge did not depend on the gas that was present in the cathode tube, and their mean free path did not depend on the density of the medium they travelled through. They were, therefore, "primordial atoms," the velocity and charge/mass ratio of which could both be calculated. As Thomson wrote, this was a question of a new state of matter—a state in which the subdivision of matter was carried much further than what occurs in the common gaseous state, a state in which all matter was of a single type and was the substance of which all the chemical elements were made. Consequently, there existed the possibility of constructing models of the inner structure of atoms starting from electrons and of developing a new and revolutionary explanation for the regularities that appeared in the table of the elements.

Wilson's chamber

The values that Thomson obtained for the ratio between the charge and mass of electrons were not sufficient to determine the value of the charge carried by each particle. The solution to this problem required the creation of special experiments, which became possible as early as 1897, thanks to a fundamental method elaborated by one of Thomson's collaborators, Charles Thomson Wilson (1869–1959). Wilson's method was based on the fact that an ion can act as the nucleus for condensation in an environment supersaturated with moisture. The trails of charged particles could thus be rendered observable in such an environment. The perfection of this experimental technique enabled Thomson to evaluate electron charge (1899). Wilson's technique was later developed even further.

The Zeeman effect

Further information on the structure of matter was obtained in 1896 by Pieter Zeeman (1865–1943) from researches aimed at verifying the effect of a magnetic field on the wavelength of light emitted by incandescent bodies and observed through spectroscopic devices. Using sodium vapor as a source, Zeeman was able to observe a widening of the D line; a short time later, Oliver Lodge demonstrated that this effect was caused by a splitting of the observed line into three lines. An explanation of the Zeeman effect was supplied by Lorentz, with the aid of a model based on the hypothesis that the emission of light depended on charged particles rotating within atomic structures: the magnetic field in this case perturbed the orbits, and these perturbations led to a displacement in the wavelengths of the rays emitted. This explanation implied a certain value for the charge/mass ratio of the hypothetical orbiting particles, and in 1897 Larmor pointed out the coincidence between the theoretical and experimental values determined by Joseph John Thomson.

Photoelectric particles

Perfecting several observations made by Hertz in 1887 in the context of a corpuscular interpretation, Thomson attempted to evaluate the charge/mass ratio for the rays emitted by certain metals under ultraviolet light. In 1899, using an experimental set-up designed to deflect the trajectory of these photoelectric rays magnetically, Thomson was able to determine the above ratio, again arriving at a value that was almost the same as that already established for the electrons in cathode rays. The same held true for the rays emitted by incandescent metals.

The discovery of polonium and radium

Following Becquerel's investigations on the radiation emitted by uranium salts, the researches of Marie Curie (1867–1934) and Pierre Curie (1859–1906) led to the discovery of two other radioactive elements, polonium and radium.

The problem of radiation rapidly became a widespread issue, raising very serious theoretical questions, especially regarding the possible mechanisms that caused radiation. We have already mentioned the discovery of α and β rays: now, in 1899, various researchers showed that β radiation was deflected by a magnetic field and had the same charge/mass ratio as the electron. Finally, in 1899, Paul-Ulrich Villard discovered that the radiation emitted by radioactive bodies contained a third, highly penetrative component, γ rays.

Canal rays

Again in the area of cathode-ray research, Eugen Goldstein had in 1886 observed a weak radiation that manifested itself in the presence of perforated cathodes. This radiation, which was given the name "Kanalstrahlen," (canal rays) was accurately investigated by Wilhelm Wien in 1898. In a classical memoir entitled *Untersuchungen über die elektrische Entladung in verdünnten Gasen* (1898), Wien outlined the results he had obtained using the method of electrostatic and magnetostatic deflection and con-

cluded that canal rays were carriers of a positive charge, that they travelled at a speed of approximately 3.6×10^7 centimeters per second, and that their relative mass/charge ratio was 3.2×10^{-3}.

Planck's constant

By the end of the decade, the situation was extremely complex. Profound transformations were taking place at both theoretical and experimental levels, in that the objective world was revealing itself as much more richly articulated than had been thinkable at the beginning of the century. Interpretations of experimental data, which were closely linked to the contradictory developments of theoretical systems, became increasingly more difficult; and in theory, powerful mathematizations were attempted, which were much more refined than those (already extremely advanced) that had characterized the programs of the mathematical physicists of the first half of the nineteenth century. It was in this situation that, on 14 December 1900, Max Planck (1858–1947) presented his theory of the law of energy distribution in the normal spectrum (*Zur Theorie des Gesetzes der Energieverteilung in Normalspektrum*). According to Planck, a natural constant h—which he had already studied in May of the same year—had to be introduced into the theory of radiation as a companion to the constant k of the kinetic theory of gasses (see PHYSICAL SCIENCES: THERMODYNAMICS AND STATISTICAL MECHANICS, 1890–1900). Developing a model similar to the one discussed by Boltzmann in 1872 and 1877, Planck hypothesized that the energy E of a system of oscillators was composed of a well-defined number of equal parts, that is, of "energy elements." Each energy element was given by the relation:

$$\varepsilon = h\nu.$$

Linking this special asumption to a discussion of Boltzmann's complexes Planck obtained a spectral formula of the type:

$$u_\nu \, d\nu = \frac{8\pi\nu^3}{c^3} - \frac{d\nu}{e^{h\nu/kT} - 1}$$

This formula corresponded to the one that Planck had derived just a few weeks earlier from semi-empirical considerations (see PHYSICAL SCIENCES: THERMODYNAMICS AND STATISTICAL MECHANICS, 1890–1900), and thus received a considerable degree of confirmation from the experimental data gathered between 1897 and 1900 on black body radiation. The new formula, however, was not only in a sense more general, as Planck wrote, but, as was shown in the years that followed, because of its extraordinary implications constituted an innovative tool for a better understanding of microphenomena.

CHEMICAL SCIENCES

The first international rules for chemical nomenclature

During the early years of the decade, the need to establish some order in chemical nomenclature became pressing. In September, 1892, a congress held in Geneva, attended by thirty-four chemists from nine different European countries, set up a system of nomenclature for the aliphatic compounds, and suggested another be established for the aromatic and heterocyclic compounds. More precise rules were established from 1922 on by an international committee of editors and directors of chemical journals.

The discovery of radioactivity

The most salient event of this period, and fundamental to both physics and chemistry, was the discovery of radioactivity, which Antoine Henri Becquerel (1852–1908) observed in a salt of uranium in 1896 and Marie Curie (1867–1934) discovered in 1898 in thorium and, above all, in another element that she suspected of being a component of pitchblende (April, 1898). The new element was isolated by Marie and Pierre Curie in June, 1898, and named polonium. It was joined by radium in December, 1898. A short time later, André Louis Debierne (1874–1949), an assistant to the Curies, isolated actinium. The chemical, physical, and biological properties of the radioactive elements were studied, especially the nature of the rays they emitted, which were distinguished as α, β and γ rays by Ernest Rutherford (1871–1937), Paul-Ulrich Villard (1860–1934) and others.

The zero group: the noble gasses

At almost the same time as the radioactive metals, another group of elements, the noble gasses, which displayed almost no radioactivity, were discovered. The first of these, and the most plentiful in the air, was argon, found by William Ramsay (1852–1916) and Lord Rayleigh (1842–1919) in 1894. In 1895, helium was discovered in cleveite by Ramsay and, independently, by Nils Abraham Langelet in the laboratories of Per Theodor Cleve (1840–1905). William Crookes (1832–1919) discovered that its spectrum was identical to that of the gas that Joseph Norman Lockyer (1836–1920) had hypothesized existed in the sun, on the basis of a new line observed in the spectrum of the solar chromosphere by the astronomer Pierre Jules Janssen (1824–1907) in 1868. The same spectral line had also been found by Luigi Palmieri (1807–96) in 1881 in the gasses emitted by Vesuvius.

The discovery of these two elements led to the postulation of a new group in the periodic system, which was baptized the zero group and was soon completed with neon, krypton, xenon (Ramsey and Morris William Travers), and finally radon, (Friedrich Ernst Dorn in 1900).

In 1892, the chemist and spectrographer Paul Emil de Boisbaudran discovered unidentified lines in the spectrum of a sample of samarium-gadolinium. In 1902 Eugène Demarcay (1852–1903) attributed these lines to europium.

The discovery of hydrazoic and nitrohydroxylamic acids

Two new acids of nitrogen were discovered: hydrazoic acid, which Theodor Curtius obtained in 1890 by hydrolyzing the azides of the acids, and nitrohydroxylamic acid, which Angelo Angeli (1854–1931) obtained in 1896 by the nitration of hydrozylamine with ethyl nitrate in an alkaline environment. The decomposition of this acid to nitroxyl formed the basis of the

Angeli-Rimini reaction (after the chemist Enrico Rimini (1874–1917)) for distinguishing aldehydes from ketones (1904).

Alfred Werner's theory of complex compounds

Valency theory became the object of profound speculation in an attempt to explain, among other things, the formation of complex compounds. In 1891–93, Alfred Werner presented a theory of "primary" and "secondary" or "coordination" valencies. Johannes Thiele (1865–1918) proposed the theory of "partial" valencies (1899) to explain certain reactions of organic chemistry.

Various types of electrodes

Physical chemistry again benefitted from the work of Wilhelm Ostwald, who among other things proposed a theory of indicators, modified in 1908–14 by Arthur Rudolf Hantzsh in 1894. Ostwald also proposed the normal calomel electrode in 1893, and in 1900 Walther Hermann Nernst proposed the normal hydrogen electrode as the electrode of reference.

From 1890, Paul Sabatier (1854–1941), together with Jean Baptiste Senderens (1856–90), worked on problems of heterogeneous catalytic reactions with finely divided metals in the gaseous state.

Connections between chemistry, physics, and biology

In 1893, Ostwald proposed a classification from the point of view of physical chemistry of enzymes as catalyzers. Other points of contact between biological and physical chemistry emerged from the researches of Harold Picton and Samuel Ernest Lindon, who discovered the electrical charge of colloidal particles, and of William Hardy (1864–1934), who introduced the concept of isoelectric points in colloids. Biological chemistry achieved new and admirable goals: The first crystallization of a protein, albumin, was obtained by Franz Hofmeister (1850–1922) in 1890; that same year, Richard Neumeister (1854–1905) discovered another important amino-acid, tryptophan, which was isolated by Gowland Hopkins (1861–1947) and Sidney William Cole in 1902; in 1895, Eugen Baumann (1846–96) isolated a substance containing iodine from the thyroid, and in 1897 Adolf Oswald demonstrated that it came from the partial hydrolysis of the thyroid protein, thyreoglobulin; and in 1897, the long controversy over the nature of enzymes was resolved by Eduard Buchner, who demonstrated that by killing yeast cells by trituration (instead of by heat), a liquid that was still capable of activating alcoholic fermentation could be obtained.

The chemistry of sugars

By this time the chemistry of sugars was reaching completion: In 1891, Emil Fischer and Oscar Piloty (1866–1915) discovered epimerization and derived the sixteen possible stereoisomers of the aldohexoses; in 1893, Alfred Wohl found a method for transforming the hexoses into pentoses, while Otto Ruff (1871–1930) proposed another for shortening the carbon chain in 1898; Charles Joseph Tanret (1847–1917) isolated two stereoisomerous forms of glucose with different optical activities and explained mutarotation. At this point, with the field of sugars exhausted, Fischer turned to the more difficult area of proteins, and in 1899 began synthesizing polypeptides.

The terpenes

Another group of natural substances in which truly remarkable results were obtained during this period was that of the terpenes: in 1892, Baeyer synthesized the first terpene, dihydrocymeme; in 1897, Wilhelm Euler and, independently, Vladimir Niko-laivich Ipat'ev (1867–1952), synthesized isoprene, the basic nucleus of the terpenes of rubber and other natural substances; in 1893, Conrad Julius Bredt (1855–1937) gave the correct structural formula for camphor, after approximately thirty inac-curate ones had been proposed. Structural formulas were also found for some alkaloids, such as nicotine (Adolf Pinner 1842–1909 in 1891) and atropine (Richard Martin Willstätter 1872–1942 in 1897): both were confirmed by synthesis over the next few years. By this time, organic chemistry was advanced enough to discover the structure of and synthesize molecules of a certain complexity. Willstätter and Stanislaus von Kostanecki (1860–1910 were responsible for the structural formulas and syntheses of various anthocyanins, the flavone-derived coloring matters in flowers (1895 on). Siegmund Gabriel (1851–1924) prepared phthalazine in 1891 and, together with James Colman, pyrim-idine in 1899. In 1900 he succeeded in obtaining the latter by a simple synthesis from barbituric acid.

Discovery of the ketoenolic tautomerism in organic chemistry

In 1890, Sergei Nikolaivich Reformatski (1860–1934) dis-covered the synthesis reaction of β-ketone oxyacids, α-brom-oesters, and metallic zinc that bears his name. Hans von Pechmann (1850–1902) discovered diazomethane, the first dia-zoparaffin and a versatile reagent and aid to synthesis in organic chemistry, in 1894. That same year, Theodor Curtius described a new method for synthesizing the primary amines and degrading fatty acids based on the degradation of the azides. Organic chemistry was furthered by the work of Ludwig Gattermann (1860–1920), who devised methods for preparing halogen deriv-atives and aromatic nitriles from diazonium salts in the presence of copper dust, and for synthesizing aromatic aldehydes from hydrocarbons, carbonic oxide, and hydrochloric acid in the presence of aluminum chloride and aromatic oxyaldehydes from phenols, hydrochloric acid, and hydrocyanic acid (1898). In 1891, Ludwig Claisen obtained the first isoxazolone, and in his classical memoir of that same year, Über Isoxazole, founded the chemistry of this heterocyclic compound. In 1892, Alexander Crum Brown proposed the elementary rule for the order in which substitution groups enter into the benzene ring. In 1896, Ludwig Knorr isolated the ketone and enole forms of the diace-tylsuccinic ester and found that they generated a tautomeric equilibrium.

In 1893, Richard Anschütz (1852–1937) concluded studies begun in 1885 on the action of phosphorus oxychloride on salicylic acid. As a result of these studies, he began the researches on salicylic and dithiosalicylic acids that led to the isolation of a crystalline additive compound—salicylide-chloroform alde-hyde—which when heated gave a very pure chloroform (Anschutz chloroform), used in anesthesia.

The transferral of the study of the atom from chemistry to physics

Studies on the mechanism of organic reactions, such as those on radioactivity and the electrical charge of colloids, led to the conviction that positive and negative electrical charges existed within atoms. The discovery that the electron was a component of the atom, however, was more an achievement of physics than of chemistry.

LIFE SCIENCES

August Weismann and the germ plasm theory

This was the decade in which researches conducted in various fields—the inheritance of somatic characters, fecundation, and cell structure—converged to prepare the way for the birth of genetics. The work of Oscar Hertwig, Edouard van Beneden, and Otto Bütschli (1848–1920) during the eighth decade of the century had demonstrated how the male gamete penetrated into the ovicell; Hertwig and Eduard Adolf Strasburger had then succeeded in showing that, in both animals and plants, fecun-dation consisted in the fusion of the nuclei of the two gametes. In the following decade, Walther Flemming, E. van Beneden, and Theodor Bovéri had contributed to formulating the law of constancy of the number of chromosomes in fecundation and determining that each of the gametes contributed the same number of chromosomes. Neo-Darwinism had also arisen during the ninth decade of the century, with August Weismann's germ plasm theory. In 1885, Weismann had stated that only the nuclear substance could be the carrier of hereditary tendencies. Evolution, according to Weismann, depended upon the indi-vidual varieties that arose from the union of the two gametes, but also on "hereditary modifications" of the germ plasm. Weis-mann's theory, which exerted a profound influence on genetics, was fully elaborated between 1892 and 1895 in his Das Keim-plasma. Eine Theorie der Vererbung (1892), Aufsätze über Vererbung und verwandte biologische Fragen (1892), and Neue Gedanken zur Vererbungsfrage. Eine Antwort an H. Spencer (1895).

The significance of meiosis and the rediscovery of Mendel's law

In 1890, following van Beneden's researches on spermato-genesis, which made it possible to identify the reductive phase—or "meiosis"—of the number of chromosomes in the formation of the cells of a germ line, Oscar Hertwig explained the process as the need to maintain the constancy of the idioplasm, which was represented by the nuclear chromatin. The rediscovery of Mendel's law was close at hand: W. Haacke (1855–1921), a student of Haeckel, attributed the same Mendelian statistical distribution to hybrids, localizing several hereditary factors in the cytoplasm (1895). In 1899, William Bateson (1861–1926)—who in 1907 became the first to use the word genetics to indicate the science of hereditary phenomena—in turn reaf-firmed the two theoretical postulates of Gregor Mendel's pre-cursory research: that each single character had to be followed distinctly and that the results obtained had to be elaborated statistically. In 1900, three researchers—the Dutchman Hugo de Vries, the Austrian Erich Tschermak von Seysenegg, and the

German Carl Erich Corren (1864–1933)—all working independently on plant hybrids, rediscovered Mendel's works. Corren made them the subject of an illustrative memoir, while the others cited them.

The perseverance of chromosomes

In 1890, Theodor Boveri (1862–1915) developed the theory of chromosome individuality and victoriously defended it against those authors who maintained that chromosomes dissolved after mitosis. In 1891, Hermann Henking reported his observations of a mass of chromatin that seemed to behave in a particular way in the organizational lineage of several insects. Edmund Beecher Wilson (1856–1939), one of the authors who at the beginning of the next century supported the "chromosome theory of heredity," interpreted this as the discovery of the chromosomes carrying the sexual determinants (heterochromosomes). In 1894, Strasburger demonstrated that meiosis also took place in the gametogenesis of plants.

Malaria and its transmission

French and English tropical medicine and the Italian school of biology determined the life cycle of the malaria plasmodii and the agent that transmitted the disease to birds and humans. Charles Louis Alphonse Laveran (1845–1922)—a great parasitologist of the period—was the first to observe the presence of periglobular bodies in the bloodstream of victims of the disease. Ettore Marchiafava (1847–1935), Angelo Celli (1857–1914), and Camillo Golgi described several stages in the development of the parasite in the human bloodstream (1885–90). Golgi, in particular, distinguished various parasites related to tertian, quartan, and quotidian malaria. Patrick Manson (1844–1922), who had noted that the mosquito was the intermediate host of filaria, hypothesized that this same diptera transmitted bird malaria. In 1897, Ronald Ross (1857–1932) discovered cysts, which he recognized as evolutionary phases of the malaria parasite, in the intestine of a *Culex* mosquito. In 1898, he succeeded in transmitting the disease to birds through this genus of mosquito. Zoological researches carried out by Giovanni Battista Grassi (1854–1925) demonstrated that the agent that transmitted malaria to humans was the genus *Anopheles*. The bird and human parasites were studied and placed among the sporozoa, partly on the suggestion of Heinrich Hermann Robert Koch. In 1899, Grassi, Giuseppe Bastianelli, and Amico Bignami described the sexual cycle of the parasites.

Histology and the nervous system

During this same period, Golgi's name was associated with histological studies on the nervous system, which were made possible with the use of the histo-chemical technique of silver impregnation. In 1898, the Italian scholar described a reticular body—later called the Goldi body—in ganglion cells; this body was later found in all other tissue cells and came to be considered a basic element of cell structure. Meanwhile, the controversy on the neuron theory began in 1891 when Wilhelm von Waldeyer-Hartz (1836–1921) used the term neuron to designate the anatomical and functional unit of the nerve cell. Santiago Ramón y Cajal (1852–1934) supported the neuron theory, while Golgi contested it on the basis of the morphological patterns of a diffused nervous system as revealed by the silver reaction (one of the special staining techniques developed for study of the nervous system).

Sigmund Freud and psychopathology

While the foundations of traditional neurology were strengthened by these developments, psychopathology (which recognized the peculiarity of mental processes and their role in somatic manifestations of a sensory and motor nature) branched off from it with the work of Sigmund Freud (1856–1939).

In 1893, in a memoir on ideas for a comparative study of organic and hysterical motor paralyses, Freud confirmed the existence of paralytic situations of psychic origin and described how they differed from organic paralyses. His studies on hysteria, published in 1895 in collaboration with Josef Breuer (1842–1925), opened up an even wider perspective on the transformation of psychic disturbances into somatic symptoms.

Hemoculture, serodiagnosis, and serotherapy

Hemoculture, serodiagnosis, and serotherapy were the great innovations in the field of bacteriology. Charles Talamon (1850–1929), Isidore Straus, and Johannas Petruschky perfected the technique of aseptically drawing and biologically culturing blood from individuals with infections (1893–94). In the following years, hemoculture found a wide variety of applications in typhoid and Malta fever. Serodiagnosis was born out of new insights into the processes of immunity. In researches conducted at Messina and Odessa (1882–83), Elie Metchnikoff had already identified white blood cells as the active agents (phagocytes) in the organism's defense against microbes. In 1892, while working at the Pasteur Institute, he made comparative studies on inflammation, and in 1901, in disagreement with the humoralists, he reaffirmed the primary importance of phagocytosis in a treatise on immunity in the process of infection. Paul Ehrlich, a disciple of Koch, demonstrated that the reaction between toxins and specific anti-toxins obeyed the same laws as chemical reactions. Cellularists and humoralists both agreed that substances capable of causing agglutination and lysis existed in the blood in places where germs had passed. The serodiagnosis of typhoid, proposed by Fernand Widal (1862–1929), one of the major representatives of French medicine of the epoch, in 1896, was based on this fact. In 1897, Almroth Edward Wright (1861–1947) devised a similar diagnostic procedure for brucellosis. The first scientist to think of passively transferring immunity may have been Charles Robert Richet (1850–1935), the discoverer of anaphylaxis: this illustrates the relations that were being established between the physiology and pathology of immunity phenomena.

In 1890, Emil von Behring (1854–1917) and Shibasaburo Kitasato (1852–1931) planned to transfer passively the protective power of serum from horses treated with injections of diphtheria toxin to humans. Pierre Paul Emile Roux, Louis Martin (1864–1934), and August Chaillou reported the process to a German congress held at Budapest in 1894; thus an antidote was discovered for one of the most serious and widespread diseases of the time. Between 1890 and 1893, Behring, Roux, and Louis Vaillard devised the procedure for anti-tetanus serotherapy, which proved extraordinarily effective. Effective serotherapeutic measures were also discovered during these years against plague (Alexandre Yersin, 1894), anthrax (A. Scalvo and E. Machoux, 1895) and cholera (Roux, Metchnikoff and A. Taurelli-Salimbeni, 1896). Vaccinations against typhoid and paratyphoid diseases (1896), plague (1895), and cholera (1888) were also developed.

The theory of evolution and paleontology

Progress was made in other sectors of biology as well. The theory of evolution offered a new perspective to paleontology. Knowledge about the long history of life was broadened by Othniel Charles Marsh (1831–99), who discovered the intermediate forms between reptiles and birds, and by Edward Drinker Cope (1840–1914) and Louis Dollo (1857–1911), who described the giant reptiles of Kansas and Belgium. At the basis of these researches on the history of life was a comparative morphology that had also become open to the principle of phylogeny. The major representative of this current was Carl Gegenbaur, a professor at Jena and Heidelberg, who in 1899 published a large work entitled *Vergleichende Anatomie der Wirbeltiere mit Bercksichtgung der Wirbellosen*. Paleoanthropologists began to investigate the origins of *Homo sapiens* through fossil remains. In 1892, in Java, the Dutch physician Eugéne Dubois (1858–1940) discovered the remains of a form with human and monkey-like characteristics, which he called *Pithecanthropus erectus*.

Experimental embryology, 'regulative' and 'mosaic' development

A classical experiment in embryology was performed in 1891 by Hans Driesch (1867–1941), a student of Ernst Heinrich Haeckel and Wilhelm Roux, at the Zoological Station of Naples, on the fertilized egg of a sea urchin at the two-blastomere stage. By separating the two blastomeres, Driesch obtained two complete embryos. From this experiment, he formulated the concept of "regulative" development, as opposed to Roux's "mosaic" development, and drew his inspiration for a reaffirmation of vitalism.

Radiology and diagnostic techniques

The field of medicine was teeming with events, some of which took place on the borderline between medicine and experimental physiopathology. Scipione Riva-Rocci (1863–1937) perfected the method for measuring blood pressure and introduced the sphygmomanometer into clinical practice, while Heinrich Irenäus Quincke (1842–1922) developed a technique for drawing cephalorachidian fluid by means of lumbar puncture (1891). In November, 1895, Wilhelm Conrad Röntgen identified a new type of radiation that was invisible to the eye but could pass through opaque bodies and expose a photographic plate (see PHYSICAL SCIENCES: RADIATION AND MATTER).

He immediately utilized these X-rays—so-called because of their unknown nature—for photographing segments of the human skeleton. Before the end of the century, "radiography" was applied in diagnosing lesions of the skull, the heart, and the lungs, and in renal pathology. Antoine Béclère (1856–1939) inaugurated a course in radiology in Paris in 1897. That same year, opaque substances were first used to radiograph the digestive system.

Endocrinology and medical pathology

The premises of endocrinology were also developed during this period. In 1891, from experiments on organ grafting, Édouard Brown-Séquard formulated the idea of a chemical connection among the various organs and systems. Meanwhile, medical pathology identified specific syndromes, such as goiter cretinism and pachydermic cachexia, or myxoedema, which could be traced to thyroid disfunctions, osteitis fibrosa/fibrocystic disease, discovered by F. von Recklinghausen (1833–1910) and attributed to dysfunctions of the parathyroid (1904), acute superarenal insufficiency, pancreatic diabetes (Joseph von Mering and Oscar Minkowski, 1886–92), and acromegaly (Pierre Marie, 1886; Augusto Tamburini, 1894), which was beginning to be understood to be connected to the endocrine glands. In 1895, Thomas Richard Frazer (1841–1920) discovered the cardiotonic effect of strophantus, and in 1896 R. Marie (1868–1952) described the symptomology and course of myocardial infarction.

EARTH SCIENCES

Interpreting the tectonic structure of the Alps

The fundamental premises that in the first decade of the following century led to a clarification of the tectonic structure of the Alps and, by analogy, of other mountain ranges created by folds and nappes, were laid during this final decade of the nineteenth century.

Between 1890 and 1893, W. Hans Schardt (1858–1931) interpreted the *klippen* of the Lake of the Four Cantons in Switzerland as the remains of a great recumbent fold, or the edges of a nappe. Maurice Lugeon (1870–1953) came out in support of Schardt's views in 1896. In 1898, in Val Maira in the Piedmont, Seconde Franchi (1859–1932) discovered Jurassic fossils in lime schists, incontestible proof that the formation containing them was Mesozoic and not, as had previously been thought, extremely ancient, Prepaleozoic.

Clarence Dutton outlined his isostatic theories in more detail.

Karst phenomena

Two works published in 1894, *Höhlenkunde* by Paul Krause and *Les abîmes* by Eduard Alfred Martel, foreshadowed researches on karst phenomena, to which extremely imposing manifestations, such as the formation of grottoes in calcareous and dolomite rocks, were linked.

Opinions on the age of the earth

Lord Kelvin reiterated his figures for the age of the earth, deduced from the probable cooling processes that took place and, in fact, reduced his estimate to no more than twenty-four million years (1899), which was in agreement with the calculations of Clarence Rivers King (1842–1901). Thomas Chrowder Chamberlin, champion of the plantesimal hypothesis, opposed these views, and although he admitted that Lord Kelvin was right in assuming that geological time was not unlimited, he believed it to be quite a bit longer than what the English scholar proposed. Chamberlin based his arguments on radioactive phenomena, which had just recently been discovered (1896) by the French physicist Henri Becquerel (see PHYSICAL SCIENCES: RADIATION AND THE STRUCTURE OF MATTER 1890–1900).

The cosmological ideas involved in these problems were nourished by the studies of Emil Cohen (1894) and others, such as Grove Karl Gilbert (1843–1918), who investigated the composition of meteorites and their role in the creation of lunar craters.

Studies in stratigraphy and paleontology

The third edition of the *Traité de géologie* by Albert August de Lapparent (1839–1908), which contained systematic and comparative geological schemes for all the periods in various countries, was published in 1893. That same year, Louis Dollo (1857–1931) formulated the law of irreversibility of evolutionary phenomena that bears his name. Karl Alfred von Zittel also completed his treatise on paleontology in 1893. In 1892, Charles Barrois (1851–1939) and Lucien Cayeux (1864–1944) discovered Precambrian fossils in Brittany. In 1895, Eugène Dubois described the *Pithecanthropus*, which had been discovered on Java three years earlier. In 1896, Othniel Charles Marsh published *The Dinosaures of North America*.

A fundamental method for dating recent geological events in years was proposed by Gerald Jacob De Geer (1858–1943) in 1896 and developed considerably by him over the years that followed. The varve method made it possible to calculate the age of deposits in peri-glacial lakes, which were very extensive in the Northern European Quaternary, by observing the number of seasonal bands of varying widths, grains, and colors that had been formed in the summer, when the ice melted, and in the winter when very little debris was deposited. Wilhelm Ramsay also began his studies on the Scandinavian Quaternary during this period.

In 1893, Carlo Fabrizio Parona (1855–1939) published *Sugli scisti silicei a Radiolarie di Cesana presso il Monginevro*. In 1899, von Zittel compiled an excellent history of geology and paleontology (*Geschichte der Geologie und Paläontologie*) up to his time.

The mineralogical goniometer

Beginning in 1890, Evgraf Steparovich Federov and Victor Goldschmidt (1853–1933) built crystallographic goniometers with which notable advances were made in the study of crystalline forms.

Nansen's explorations and the Austro-Hungarian expedition

The most complete exploration of the Arctic Ocean was begun in 1893 by Fridtjof Nansen (1861–1930) with the ship *Fram*. Nansen made substantial contributions to knowledge about that ocean and to hydrological, meterological, and climatic knowledge about the polar zones in general. Among other things, he discovered the phenomenon of ice drift, providing the basis for the later fundamental theoretical studies of Vagn Walfrid Ekman (1874–1954) on drift currents.

During the same period, the Austro-Hungarians made several important oceanographic explorations in the Eastern Mediter-

ranean and the Red Sea (1889–98) with the ship *Pola*, following previous campaigns that were mainly limited to the Adriatic. These explorations were important in that they clarified the hydrological conditions of a region that had not been considered by any of the great oceanographic campaigns and had therefore remained virtually unknown. In addition, the chemist Konrad Natterer (1821–1901) had the chance during these voyages to refine the methods of thalassographic chemistry, in the wake of the fundamental work of Forchhammer, Usiglio, and Dittmar, and the equally accurate as well as innovative studies of Vierthaler, which unfortunately had been limited to the waters of the Adriatic.

Oceanography at the end of the century. Bjerknes's theorem. The birth of aerology

In 1895, Alexander Buchan conducted theoretical and experimental studies on currents, in an attempt to trace the development of ocean currents following the investigations of the *Challenger* expedition. Gustaf Ekman and Otto Petterson made current measurements in the Baltic, and Petterson elaborated a theory on water from melted ice as the source of motion for currents.

John Young Buchanan carried out chemical analyses of manganese nodules from 1890 to 1891, revealing their composition. The first work by V. Bjerknes (1862–1951) on circulation, which had a decisive influence on the evolution of meteorology and oceanography during the following decades, came out in 1898.

In 1892, the Frenchmen Gustave Hermite and George Besançon began systematically launching small unequipped aerostatic balloons of varying sizes; soon thereafter, they began fitting them out with instruments for recording the pressure and temperature extremes encountered. Similar experiments were also carried out in Germany, Russia, and Sweden beginning in 1893 and 1894. These initiatives attracted the attention of meteorologists, and as early as 1896 the Second Conference of Directors set up a Commission of Scientific Aerostations (the term aerostation was a Gallicism, now no longer in use, that meant "aerostatic navigation"), and Leon Teisseranc de Bort, with his own means and on his own property, founded the Observatoire de Météorologie Dynamique at Trappes for the purpose of studying atmospheric conditions at relatively high altitudes. His excellent theoretical background and remarkable technical skills soon brought him success. In 1899, to his great surprise, he discovered that between eight and eleven kilometers the temperature ceased to fall as altitude increased. He immediately ran more trials, in order to compare them with other findings and discuss them critically with his foreign colleagues. In 1902, at the Académie des Sciences of Paris, he officially announced the existence of what for the time being he called superior inversion. In 1908, he outlined the scheme of the atmosphere that is still in use today, introducing the concepts and terms troposphere and stratosphere. The *International Cloud Atlas*, a work by Hugo Hildebrand-Hildebrandsson (1838–1925), with the collaboration of Albert Riggenbach-Burchhardt (1854–1921) and Teisseranc de Bort, who paid for the cost of printing, also came out in 1896. By the second edition (1907), which was edited by Hildebrand-Hildebrandsson and Teisseranc de Bort, the problem of cloud classification could be considered solved.

APPLIED SCIENCES

Prime movers

This was a particularly fortunate time for the development of the applied sciences. In 1889, Charles Parsons lost his patents on the steam turbine, a fact that encouraged others—such as Carl Gustav de Laval in Sweden, C.E.A. Rateau (1836–1930) in France and C.G. Curtis (1860–1953) in the United States—to make improvements on it.

Turbines were first put to use on ships. In 1894, de Laval suggested installing a small fifteen-horsepower turbine on a sloop, while Parsons built an experimental ship. the *Turbinia*, with a displacement of forty-four tons on which a radial-flow turbine was installed to run the drive shaft. At the Spithead naval show in 1897, the *Turbinia* reached the exceptional speed of 34.5 knots. At this point, the turbine attracted the attention of the military (for use on large armored vessels), the ship-building industry (for use on transatlantic cruisers) and builders of electrical power plants.

In 1893, while the success of the internal combustion engine was spreading, Rudolf Diesel (1858–1913) published his *Theorie und Konstruktion eines rationellen Wärmemotors*. Diesel proposed a thermodynamic cycle in which the comburent gas—air—was compressed to such a high pressure that it became hot enough to cause the inflammable gas—petroleum—which was injected into the combustion chamber, to ignite. The following year, the rights of Diesel's engine were acquired by A. Bush, who founded the Diesel Motor Company of America.

Meanwhile, internal combustion engines were even further perfected. In 1893, Wilhelm Maybech patented a float carburetor that was used in Daimler engines. In 1896, D.A. Bosch (1861–1942) built his low-tension magneto, which was used with the "spark-plug" for ignition. Gottlieb Daimler's "beehive" radiator appeared that same year.

A very light-weight, high-speed gasoline engine, which operated on the Otto cycle and was somewhat similar to Daimler's first vertical engine, was built by the Compte de Dion (1856–1946) and George B. Bouton (1847–1938) in 1895.

Transport

This was the last decade of the absolute supremacy of the steam locomotive. Railway signal and automation systems were perfected. The improvements made in the internal combustion engine, which by now ran smoothly, made it possible to install them in motor vehicles, which began to be produced commercially.

In 1894, Karl Benz (1844–1929) built his first four-wheeled automobile with a four-stroke engine; the following year, the Cannstatt works began producing Daimler vehicles (the Daimler-Cannstatt vehicles later took on the name Mercedes). Herbert Austin (1866–1941) designed his first three-wheeled motor vehicle with a horizontal opposed-cylinder engine in 1895, while Henry Ford (1863–1947) presented his first model, which had a two-cylinder, four-stroke water-cooled engine, belt transmission, bar steering, and spoked wheels, in 1896. Tires were still solid, but modern pneumatic tires with inner tubes began to be manufactured that same year (1896) following Dunlop's model. In France, after having previously used Daimler V-two engines in their motor vehicles, the Peugeot brothers adopted a horizontal engine of their own design in 1891.

In 1899, the Renault brothers patented their first vehicle, which had a three-speed gearbox, a universal drive shaft, and a differential gear in the drive axle, an innovation that brought about the rapid decline of belt transmissions.

Steam-powered road vehicles, however, did not for this reason cease to be built: in the United States, the Stanley brothers succeeded in building powerful but light-weight vehicles (365 kilograms) with boilers heated by a petroleum burner.

Battery-powered electrical vehicles were also built, but, like steam vehicles, they could not withstand the competition of the internal combustion engine. In Italy, A. Faccioli (1849–1920) designed and built the first FIAT automobile in 1899. Electric traction on tracks began to be developed in cities, where the use of electric trolleys powered by overhead lines spread quickly. After numerous trials, the power for these vehicles was standardized at around 600 volts. The first electric locomotive was used in Baltimore in 1895.

In the area of transport by water, the *Campania* and the *Lucania*, both equipped with 30,000 horsepower reciprocating engines and two propellers, represented the ultimate in ship-building. By this time, the future belonged to the steam turbine.

The first practical attempts at "heavier than air" flight also had a glimmer of success. In 1895, in Germany, Otto Lilienthal (1848–96) conducted experiments with gliders, while the Wright brothers, Orville (1871–1948) and Wilbur (1867–1912), who had understood the nature of the forces that caused an airplane to stay up in the air, introduced the vertical rudder and movable wingtips to control yaw and especially rolling.

It was a natural step for the internal combustion engine to be applied to dirigibles in 1898. Thanks to the efforts of Ferdinand von Zeppelin (1838–1917), who flew his first dirigible in July, 1900, and the Brazilian Alberto Santos-Dumont (1873–1932), dirigibles achieved effective flight autonomy for the first time.

Power plants

With electricity, humanity acquired a form of energy that had an extraordinary capacity for capillary distribution. During this decade, the generation of electrical energy by reciprocal engines coupled with dynamos, which produced direct current with a maximum power of 7,500 kilowatts, reached its limits and had to surrender to alternating current. In 1891, the Brown Boveri company of Switzerland showed that it was possible to transport 225 kilowatts over a distance of 179 kilometers with a thirty kilovolts alternating current. In 1892, in Russia, N.N. Benardos (1842–1905) designed one of the first plans for a high-power alternating current hydroelectric plant to be built on the Neva. In St. Petersburg, V.N. Chikolev (1845–98) directed the project of illuminating the Litcjnij bridge as well as the construction of a three-phase alternating current hydroelectric plant for the Ohtensk factory. M.O. Dolivo-Dobrowolsky, who developed the principle of rotary magnetic field, discovered by G. Ferraris, theoretically demonstrated the possibility of obtaining that type of field with a three-phase current system. In 1890, he built the first three-phase transformer. A demonstration of his discoveries on a large scale was given at the World Electrotechnical Exhibition of 1891 held in Frankfurt am Main, where an electrical transmitting installation was built from the Lauffen Falls to Frankfurt, for a total distance of 170 kilometers at an average power of 15.2 kilovolts. This demonstration led to the adoption of three-phase current.

Models demonstrating the use of water power to produce electricity in various ways were displayed at the Chicago Exhibition in 1893. That same year, the Westinghouse Company was commissioned to build the alternators and the auxiliary system for the great Niagara Falls hydroelectric station. Meanwhile, the British Westinghouse Electrical and Manufacturing Company began building Parsons turbines for coupling with alternators. By the end of the decade—and the century—its Trafford Park factory was considered one of the greatest feats of mechanical engineering of the era. Despite the support of William Thomson Kelvin and Thomas Alva Edison, the tenacity of Sebastian Ziani Ferranti (1864–1930), Nikola Tesla, and George Westinghouse (1846–1914) won out; by this time alternating current power plants had become more powerful than direct current systems.

Germany was the giant among the various national electrical industries during this decade, having by this time far surpassed England.

Raw materials

This was still the epoch of coal and steel, as supremacy in the production of steel passed from the United Kingdom to the United States in 1890. The German steel industry also underwent enormous growth. Steel production was stimulated by the advent of electricity: Henri F. Moissan perfected the electrical furnace.

Charles Martin Hall's process for electrically producing aluminum became economically competitive after the improvements made on it by Paul Louis Héroult (1863–1914). At Foyers in Scotland, the British Aluminium Company built an electrical power station for producing aluminum, which could not have been perfected at a more opportune time. At the same time, the Leblanc process for producing soda began its decline. Especially diffused in Great Britain, this process was forced to give way to the Solvay method.

Still in the field of metallurgy, in 1894 Goldsmith introduced the alluminothermic process, which made it possible to prepare metals such as chrome, molybdenum, and vanadium in the pure state, free of carbon. It was also applied in the welding of iron and its alloys. In 1895, Carl von Linde succeeded in liquefying air on an industrial scale.

Researches into new synthetic materials, influenced by industrial and demographic growth, proceeded, especially in Germany and England. While conducting experiments to find a suitable filament for electric bulb lamps, two chemists from Manchester, Charles Frederick Cross (1855–1935) and Edward Bevan (1856–1921), discovered that wood pulp dissolved in caustic soda expanded. When treated with carbon disulphide it was converted into a new compound that, dissolved in water or diluted in a solution of caustic soda, produced a viscous, yellow solution, which they named viscose. Patented in 1892, viscose was used the following year by Stearn to produce filaments for lamps. But its future was in the textile field, where it gave rise to the first artificial fiber, rayon.

Communications

The profound changes that were taking place in the transport system were in a certain sense facilitated and complemented by improvements in the telephone system. The first automatic installation was set up in 1892 in La Porte, Indiana.

The invention of the wireless telegraph was one of incalculable importance, in 1895, using radiations of longer wavelength than those discovered by Hertz, Guglielmo Marconi (1874–1937) succeeded in sending signals over a distance of 1,600 meters. The following year, he moved to England, where he succeeded in sending signals over a distance of 15,000 meters. In 1897, Marconi founded his Wireless Telegraph Company, and Popov discovered that electromagnetic waves were reflected by metallic objects. After four years of uninterrupted experiments, Marconi finally succeeded in receiving and transmitting signals across the Atlantic, between Ireland and Newfoundland.

Nor should we forget the importance to improving communications of the introduction into printing of a new device for composition, the linotype machine. The linotype had been invented during the previous decade by the German-American Ottmar Mergenthaler (1854–99), but it was only after 1890, (when improvements were made on it) that it came into general use, revolutionizing the typographic industry.

Construction

During the final years of the century, the skyscraper was adopted by the building industry. In 1892, the twenty-one-story Masonic Building was erected in Chicago. In the course of a decade, the methods and techniques of steel-skeleton construction, not much different from those in use today, were perfected. In this way, the Chicago school of architecture developed an esthetic type of design, no longer based on the heavy shapes of traditional structures in brick. It was during this epoch that New York began to become the skyscraper capital of America and the world.

At the same time, the use of reinforced concrete was spreading in both Europe and the United States. Simultaneously, the first light-weight elliptical vaults were built, initially on an empirical basis, while a theoretical understanding of how they worked was developed.

Other technologies

Demographic growth also increased industry's interest in the problem of food preservation. Together with refrigeration, where considerable progress had been made over the previous two decades, dehydration techniques, the use of chemical additives, and canning methods were also perfected.

The use of chemical additives led to numerous debates over the possible dangers to the health of the consumer. Various laws were therefore passed to control their use. A qualitative advance was made in canning techniques in 1895 with the systematic application of recent findings made in bacteriology.

At the end of 1895, the Lumière brothers of France introduced an invention that they called the "cinematograph," a method for reproducing motion that revolutionized spectator arts in the twentieth century.

In the area of weaponry, all armies adopted the use of repeating rifles, while the machine gun, a truly automatic weapon, was introduced on a large scale as a weapon for the infantry. During this same decade, so-called "smokeless powder," an explosive with a nitrocellulose base, came into general use as gunpowder.

In 1897, Stepan Osipovich Makarov published his *Discussion of Questions in Naval Tactics*, a fundamental work on naval warfare conducted using a fleet of steamships. In 1903 he followed this treatise with another essay on the development of submarine fleets and their role in future wars.

The first icebreaker, the *Ermak*, designed by Makarov, was also launched in 1897. The *Ermak* made two voyages out of Kronstadt in 1899 for the scientific exploration of the Arctic.

Meanwhile, after many earlier attempts, submarine design was brought to its modern conception (with electric battery propulsion during immersion) around 1888 by G.A. Zédé (1825–91). Submarines now appeared on the scene, armed with torpedoes, threatening the immunity of surface vessels.

On the Threshold of the Twentieth Century

What was the social impact of this tremendous series of technological innovations, centered in the last quarter of the nineteenth century and continuing with uninterrupted intensity until the outbreak of the first world war? To have an idea, we can compare the developments that took place in mass technology in two historically significant epochs: 1870, the year that marked the end of the political stabilization of nineteenth-century Europe, and 1914, the beginning of World War I.

The length of the world railway network quadrupled between 1870 and 1914, reaching more than one million kilometers, just short of its length today. And although Western Europe and the United States (as opposed to Russia) had already completed a large part of their railway construction before 1890, the major technical improvements (more powerful locomotives, more comfortable cars, more frequent service) came at a later date. By 1914, trains were traveling at an average speed that was only slightly less than today's. One could go from Lisbon to Vladivostok, or cross North or South America, by train. The railroads became the axis of European colonization in Africa and Asia. And, as we have already seen, the first electric railways appeared around the turn of the century.

The mercantile fleet grew from around twelve million tons, three-quarters of which was wind-driven, to fifty million tons, only seven percent of which was sailing ships. But transport capacity was more than proportional to tonnage, since average speed had been increased by fifty percent. In a frenetic race for constantly larger size, the mastodonic transatlantic cruisers were created and had their golden age during this period.

In roadway transportation, the contest, as we have seen, was won by the internal combustion engine. By 1900, more than ten thousand motor vehicles were produced throughout the world, even though their diffusion was often hindered by existing legislation. In England, for example, it was illegal until 1896 to use motorized vehicles on roadways unless there were a driver and a substitute aboard, while a third person on foot had to run at least twenty meters ahead of the vehicle carrying a flag to signal its approach. In any case, the speed of the vehicle was limited to not more than six kilometers per hour. A law passed in 1903 eliminated the runner, and the speed limit was raised to eighteen kilometers per hour. Only in 1906, with the introduction of the driver's license, was the speed limit raised to thirty-two kilometers per hour. But no law, however far-sighted, could keep up with the progress being made in this field, as evidenced by the growing number of automobile competitions in which speed records were established. Thus while the Paris-Rouen race of 1894 was won at an average speed of eighteen kilometers per hour, in 1906 the first twenty-four hour Le Mans race was won by a Renault at 100 kilometers per hour. In 1914, over half a million automobiles were in circulation in Western Europe, while American production, which barely reached 4,000 vehicles in 1900, had risen to 550,000 by 1914. The spread of individual means of motorized transportation (including motorcycles, which had fully developed by 1914) was facilitated by improved road surfaces and the use of tires with inner tubes; these had begun to be used on bicycles around 1890. Bicycles were now produced by the millions, having proved their popularity as a cheap and practical means of transportation from the time of their appearance.

The electrical industry, which at the beginning of the twentieth century had just been born, grew with breathtaking speed: by 1907, the German electrical industry counted 110,000 employees, only slightly less than the 150,000 of the chemical industry, which had a much longer history. The development of the electrical industry was extremely lively from the very beginning, virtually quadrupling every twenty years. Unlike railway construction, no end could be foreseen to the demand for electrical energy. From nothing in 1880, the total electrical power installed in the world amounted to around 20×10^6 kilowatts by 1914; during that year, between 50 and 60×10^9 kilowatts were produced. In many countries, such as Italy and Japan, the development of the electrical industry replaced the development of the railway system (by now brought to a halt) as the major element in the process of national development.

The no-less-surprising progress made in the transmission of speech and thought led to the installation of 200,000 telephone sets in the United States in 1890, a figure that had risen to ten million by 1914.

We shall not dwell here on the technical applications of vacuums, which led to the invention of X-ray tubes and, within a few years, to the manufacture of diodes and triodes, extremely essential elements in the perfection of not only the wireless telegraph, but the telephone and the radiophone as well.

Similar technological advances were made in the area of armaments. By the end of the century, all field forces were armed with individual repeating weapons, which with very few modifications continued to be used until the Second World War. A forerunner of graver developments was the invention of the machine gun, an automatic weapon that even in its earliest models was able to fire more than 600 rounds per minute. Advances in metallurgy made it possible to improve rapid-fire artillery, while firing accuracy was immensely increased by using helicoidally rifled barrels and controlled amounts of explosives.

All the technological innovations of the last quarter of the nineteenth century—the steam turbine, the use of fuel oil, the perfection of optical instruments, radiotelegraphy—found a simultaneous application in the designing of huge single-caliber tanks, the first of which was built in England in 1904. By 1900, the concept of the submarine, which soon found widespread applications, had also been consolidated. The perfection of the light-weight internal combustion engine was also responsible for the development of aviation, especially heavier-than-air flight with airplanes.

The nineteenth-century economy was characterized by what was historically a rather unusual phenomenon. Between 1815 and 1900, an almost continuous process of deflation was taking place. The explanation for this anomaly seems to lie in increased productivity, which facilitated economic expansion and drastically reduced production costs. The first effects of this process were felt in the price of handwork, which could finally be produced industrially, rather than by artisans.

But the drop in prices also extended to agriculture, and the cost of food, when the revolution in land and sea transport made otherwise remote lands accessible for cultivation. The decline in prices over the course of the nineteenth century (while salaries remained low but stable) was thus an indication of the spread of industrialization. In the last quarter of the nineteenth century, this process gained new vigor and took on a somewhat different aspect with the advent of a second phase (in which we are still living), commonly called the "second industrialization."

This second industrialization consisted in a union between science and technology, which up to this time had existed separately, divided by a vacuum. The first industrial revolution (which had found its major inspiration in mechanical industry, the aspects of which were more accessible to concrete experience and common sense) had not had strong ties with science: firearms and steam engines were invented long before the laws of thermodynamics were known. The invention of electrical engines or radiotransmission, on the other hand, could never, as we have seen, have occurred without prior knowledge of the laws of electromagnetism.

Thus the last quarter of the nineteenth century saw this union of science and technology, which led to a new expansion in the material progress of humanity. The technological explosion drew its vital force from the fundamentals of science, and what had before been done by instinct and intuition could now be planned, using more solid criteria, and therefore perfected. This led to new improvements and the application of known scientific principles to newly conceived products. Everyone benefitted, to different degrees, from the second industrial revolution. The case of England was typical: By around 1870, the impetus implicit in the innovations of the first revolution and their development during it had just about died out, and the English economy was already showing obvious signs of decline, when around 1890 the recession abated and expansion slowly resumed. The first industrial revolution had begun later in Germany and the United States, and its final positive effects merged smoothly into the much more important second phase. Thus national income continued to rise, without the temporary stagnation that had tormented Great Britain. In France, the potential transition crisis was diluted by the slowness of the industrialization process there as a result of many complex factors, not the least of which

was the conservatism innate in French temperament. In any case, the economic decline of the end of the century was overcome in France, as well, by general growth, which began to make headway in 1900 and continued until the outbreak of the war.

The first and second industrial revolutions were superimposed in the birth of industry in Italy, Russia, and Austria-Hungary. However, these countries were only able to acquire some of the benefits of modern technology during that period, since they were only partially able to overcome backwardness on a national scale. Since industrial production provided only a small fraction of their national income, even accelerated expansion did not have a very great influence. However, within the industrial sector, they developed more rapidly than the more economically advanced countries, including Germany. As for Japan, even though industrial development was late in coming, it was enthusiastically promoted by the Japanese people.

In conclusion, two phenomena on a world-wide scale were taking place during the last quarter of the nineteenth century: the passage from an industrial monarchy (Great Britain) to an oligarchy (made up of new industrialized countries, with their economic satellites); and the process of adaptation required by the advent of the second industrial revolution.

Civil society absorbed the fruits of the second industrial revolution quite well. Administrative structures were able to keep up and adjust gradually to its political and social consequences, i.e. to the affirmation of a popular democratic system. But the dominant oligarchies, which were the expression of a still fairly restricted suffrage, did not understand all the aspects of the significance of this revolution. They did their best on a day-to-day level, but did not realize that the second industrial revolution was something essentially different from the first; much more revolutionary in nature, it tended to create and increase what is today called "technological separation"; in short, it tended to set up a hierarchy of power that could only be measured by a previously unknown standard. The colonial policy of the last quarter of the nineteenth century and the few years that followed, during which at least thirty million square kilometers of colonial territory fell under European control, is indicative of the fact that square kilometers were still considered units of power. These annexations marked the peak of European expansion, which had begun in the eleventh century on the plains of Elba, on the Castillian plateau, and in the Mediterranean. But they illustrated well the lack of understanding of an emerging reality: the territories beyond the seas, except for a few points of strategic interest, were of altogether secondary economic importance, given the poverty and backwardness of those continents. By the eve of the first world war, the industrial states had become one another's first and best customers, as well as their own best customers. As the importance of overseas markets declined, the weight of internal markets increased. This was where the richest consumers were, and their numbers and individual wealth were increasing rapidly even in backward areas. Each country found in its own market consumers who could be placed in a position to acquire products and services beyond the level of mere subsistence. In short, the foundations of a consumer society were laid during the final years of the nineteenth century, and by 1914 it was an accomplished fact.

Among the social services offered to the masses by the advent of scientific technology were compulsory education (at least at an elementary level) and an abundance of informational services. In the more advanced countries, daily newspapers were transformed from papers for the élite into papers for the masses. Three factors made this possible: a broadening of the circle of "developed" peoples, a higher degree of literacy among the lower classes, and technical progress in the typographical industry, which made large low-cost printings possible. Proof of this transformation is given by the number of newspapers sold daily, which went from seven million in 1870 (for the 350 million inhabitants of the "developed" world), to seventy million in 1914 (for the 650 million inhabitants of the same area). The introduction of new machines made it possible to reduce costs and selling prices drastically. From a record 25,000 copies of the daily *La Presse* sold in 1840, by 1880 the *Daily Telegraph* had already reached 300,000 copies and the *Petit Journal* 65,000. The prophetic milestone of a million copies per day was reached by the *Daily Mail* during the Boer War.

The fortune of news agencies was also established during this same epoch. The European agencies (Reuter Havas, Wolff, Stefani) had been founded around the middle of the nineteenth century, but it was only in 1893 that the Associated Press in the United States was reorganized on a much larger scale.

As we have seen, the first large leisure-time industry, the cinema, was born at the turn of the century. Its enormous explosion can be documented by just a few figures: On the eve of the war, in non-colonial countries, there was one movie house for every 10,000–20,000 people and yearly attendance was more than twice the population.

Through the utilization of electromagnetic waves and thanks to clever ideas elaborated at the end of the nineteenth century, other mass audiovisual media—radio and television—were already in an embryonic stage; however, they were only developed at a much later date.

In short, the achievements and promises of technology seemed to justify the positivistic optimism that permeated society at the dawn of the twentieth century, even though the latent dangers of conflict undermined its unconscious sense of security. The explanation is simple, since until a few generations earlier, production and productivity had been so low that they could only support existence on a subsistence level. Of course populations were generally ruled over by a very small oligarchy that lived in pomp and luxury, but this oligarchy was so small that the distribution of all its wealth over the entire population would not have improved conditions perceptibly. But now, for the first time, people realized that the had a "margin" of well-being. A poll taken in England at the beginning of the 1900s showed that sixteen percent of the population lived below the "poverty level" (i.e. had a lower-than-subsistence-level income). This meant that the other eighty-four percent lived above it, while a century earlier, the reverse would have been true.

The diffusion of wealth also reduced mortality and increased the average human lifespan, so that there was a rise in demographic growth; this was mainly concentrated in the more advanced countries, and was equal to one percent per annum averaged over the entire world (that is, world population doubled in seventy years). In order to support this demographic expansion, other sciences and technologies were needed to guarantee sufficient food production—since the land could no longer supply enough unless it was mechanically and intensively cultivated and stimulated with increasingly larger quantities of chemical fertilizers—and a wider availability of luxury items, which humanity no longer wanted to do without.

On the other hand, even the more wealthy and advanced countries (Great Britain and the United States) were still, in the end, poor. Poverty was so evident all around that abolishing it could not be considered anything but a goal worthy of society. The foundations of an economy directed towards increasing material wealth thus went unquestioned. An industrial development that appropriated the discoveries born out of technology was considered a fundamental means for emerging from the straits of a peasant society.

Thus it was natural that only the positive aspects of an industrial society were emphasized, while the negative ones, which in any case were not yet very visible, were relegated to the background. Malthus's theory, which had warned of the dangers and limits of uncontrolled development, seemed to have been definitively defeated, and humanity prepared itself, at the dawn of the twentieth century, to live in the "belle époque" of the final years of a century of peace.